# Nonlinear Phenomena, Chaos, Control and Applications to Engineering and Science and Experimental Aspects of Complex Systems

# Nonlinear Phenomena, Chaos, Control and Applications to Engineering and Science and Experimental Aspects of Complex Systems

Topic Editors

**José Manoel Balthazar**
**Angelo Marcelo Tusset**
**Átila Madureira Bueno**
**Diego Colón**
**Marcus Varanis**

MDPI

Basel • Beijing • Wuhan • Barcelona • Belgrade • Novi Sad • Cluj • Manchester

*Topic Editors*

José Manoel Balthazar
Department of Mechanical Engineering
São Paulo State University
Bauru
Brazil

Angelo Marcelo Tusset
Department of Mathematics
Federal University of Technology—Paraná (UTFPR)
Ponta Grossa
Brazil

Átila Madureira Bueno
Telecom and Control Department
University of São Paulo (USP)
Sorocaba
Brazil

Diego Colón
Telecom and Control Department
University of São Paulo
São Paulo
Brazil

Marcus Varanis
Physics Institute
Federal University of Mato Grosso do Sul (UFMS)
Campo Grande
Brazil

*Editorial Office*
MDPI AG
Grosspeteranlage 5
4052 Basel, Switzerland

This is a reprint of the Topic, published open access by the journals *Applied Sciences* (ISSN 2076-3417), *Axioms* (ISSN 2075-1680), *Entropy* (ISSN 1099-4300), *Mathematics* (ISSN 2227-7390), *Symmetry* (ISSN 2073-8994), *Dynamics* (ISSN 2673-8716) and *Machines* (ISSN 2075-1702), freely accessible at: https://www.mdpi.com/topics/nonlinear_chaos_engineering.

For citation purposes, cite each article independently as indicated on the article page online and as indicated below:

Lastname, A.A.; Lastname, B.B. Article Title. *Journal Name* **Year**, *Volume Number*, Page Range.

**ISBN 978-3-7258-5005-1 (Hbk)**
**ISBN 978-3-7258-5006-8 (PDF)**
**https://doi.org/10.3390/books978-3-7258-5006-8**

# Contents

# Preface

Nonlinear dynamics and chaos theory provide an understanding of the complex behaviours exhibited by engineering and science systems across various scales, from macro- to MEMS scales. We also highlighted that sensitivity to initial conditions is characteristic of chaotic motions. It is known that chaos theory is a profound branch of mathematics that delves into the intricate dynamics of deterministic systems exhibiting sensitive dependence on initial conditions. This Reprint focuses on the recent developments, findings and progress in theories, principles, methodology, computational techniques in nonlinear systems, and mathematics with engineering and scientific applications. The main contribution of this Reprint, which contains 34 (thirty-four) chapters, is to provide an interdisciplinary forum for this area, relating to the nonlinear dynamics of complex systems encountered in engineering and sciences, including phenomena, modelling and complexity, presenting new ideas, principles and techniques across a broad interdisciplinary front.

**José Manoel Balthazar, Angelo Marcelo Tusset, Átila Madureira Bueno, Diego Colón, and Marcus Varanis**
*Topic Editors*

mathematics

MDPI

itorial

# Jonlinear Phenomena, Chaos, Control and Applications to ngineering and Science and Experimental Aspects of Complex ystems: 1st Edition

sé Manoel Balthazar [1,2,*], Angelo Marcelo Tusset [3], Átila Madureira Bueno [4], Diego Colón [4] and Marcus Varanis [5]

1 Faculty of Mechanical Engineering, São Paulo State University, Bauru 17033-360, SP, Brazil
2 Department of Electronics, Federal University of Technology—Paraná (UTFPR), Ponta Grossa 84017-220, PR, Brazil
3 Department of Mathematics, Federal University of Technology—Paraná (UTFPR), Ponta Grossa 84016-210, PR, Brazil; a.m.tusset@gmail.com
4 Automations and Control Lab (LAC), Department of Telecommunications and Control Engineering (PTC), Polytechnic School of University of São Paulo (EP USP), São Paulo 05508-010, SP, Brazil; atila@lac.usp.br (Á.M.B.); diego@lac.usp.br (D.C.)
5 Physics Institute, Federal University of Mato Grosso do Sul (UFMS), Campo Grande 79070-900, MS, Brazil; marcusvaranis@gmail.com
* Correspondence: jmbaltha@gmail.com

## 1. Summary

In this Topic, 34 (thirty-four) papers were selected and published on the following subthemes: Modeling and Nonlinear Systems (Contributions 1–5); Control, Synchronization, and Optimization (Contributions 6–13); Chaos and Hyperchaos (Contributions 3 and 14–20); Complex Systems (Contributions 21–25); and Applications to Engineering and Sciences (Contributions 22 and 26–32).

The distribution of papers among MDPI journals is as follows: *Symmetry* (5 papers), *Entropy* (5 papers), *Machines* (3 papers), *Dynamics* (2 papers), *Applied Sciences* (11 papers), *Mathematics* (7 papers), and *Axioms* (1 paper).

The authors came from the following countries: China, Norway, Hong Kong, Japan, Romania, Turkey, Russia, Saudi Arabia, Italy, the United States, and Thailand.

This Editorial is organized as follows: the main contributions to this Topic are described in Section 2, and future work in development is presented in Section 3.

## 2. Contributed Papers

In this section, we will summarize the 34 (thirty-four) contributions to this Topic, considering the following: the formulation or evolution of overarching research goals and aims, the development or design of methodology, the creation of models, and the application of mathematical, computational, experimental, or other formal techniques to analyze or synthesize the study data.

In Contribution 1, Yuexin Wang et al. dealt with the ability of each oscillator in the coupled Stuart–Landau oscillator system, with Erdős–Rényi random graph network topology, to affect the dynamic survivability. Following this analysis, the maintenance of symmetry phenomenon synchronization was considered as the key mission to be completed. Inspired by the study of dynamic survivability, a new centrality index called dynamic survivability centrality was constructed, which is determined by the degree of the oscillator and the size of the system. The proposed index compensates for the limitation of traditional indices that solely consider network structure, thereby enabling a more accurate assessment of node influence on the normal operation of dynamic systems.

In Contribution 6, Zhifu Jia addressed dynamic game problems involving uncertainty, delays, and jumps using uncertain delay differential equations with jumps. A linear quadratic differential game was studied in this context, transforming the problem into a finite-dimensional form using the $Z(x,y)$ transform. The existence of a Nash equilibrium was shown to depend on solutions to cross-coupled Riccati equations. A saddle point equilibrium strategy was also derived and applied to a carbon emission reduction game.

An industrial robot experiment was designed to test the vibration suppression method proposed by Liang Liang et al. in Contribution 7. It can reduce the maximum acceleration amplitude of residual vibration by 62% and the vibration duration by 71%. Compared with the input shaping method, the method proposed can also reduce the terminal residual vibration more effectively and ensure the consistency of running time and trajectory.

In Contribution 26, Yong Zhang et al. presented a study on rubbing-induced vibration characteristics considering the flexibility of coated casings and blades. The obtained results showed that the coating can diminish the severity level of rubbing and the material and thickness of the coating changed the nodal diameter vibrations of the casings (NDVCs) induced by rubbing. This study provided valuable guidance for the optimization and design of blade–casing systems.

A special vibration damper was proposed by Chun-Nam Wong et al. in Contribution 27 for Young's modulus and hardness identification through a scratching process on extruded aluminum. The paper presented the design and working principle of the scratching damper based on a scratching device, and a non-contact electromagnetic shaker was used to generate the shaking force for the test sample vibration.

In Contribution 28, Parindra Kusriantoko et al. studied different CFD-codes for fluidized beds. The authors highlighted the significant influence of restitution coefficients on bed expansion due to their effect on particle–particle collisions, with a value of 0.9 deemed optimal for balancing simulation accuracy and computational efficiency. Conversely, the secularity coefficient, impacting particle–wall interactions, exhibited a more subtle effect on bed dynamics. This finding emphasized the critical role of carefully choosing these coefficients to effectively simulate the nuanced behaviors of fluidized beds.

In Contribution 21, Jiehui Liu et al. studied PMT fluorescence signal denoising processing. The focus was on the impact of noise interference on weak electrical signals and wavelet transform was used to denoise the signals. At the same time, considering that the photoelectric system was susceptible to temperature changes, a multiparameter fitting model was constructed, and a BP neural network was used to further process the signal, separating the real data from the original data. Finally, a high-precision and wide-range trace level sulfur dioxide concentration detection device and algorithm were obtained with success.

The attached flow method for solving reaction–diffusion processes described by a generalized Dodd–Bullough–Mikhailov equation was studied by Carmen Ionescu and Iulian Petrisor in Contribution 29. The method involves reducing the equation's order and explicitly introducing the flow variable through decomposition. This approach avoids complex integral solutions and enables accurate analytical results. It is applied to the generalized Dodd–Bullough–Mikhailov equation with nonlinearities up to the fifth order, with solutions analyzed based on parameter values.

In Contribution 22, Akio Tsuneda presented an analysis of the auto-correlation functions of chaotic binary sequences generated from a one-dimensional chaotic map using two binary functions applied alternately. Two similar methods were described, deriving the theoretical auto-correlation functions of the resulting binary sequences, expressed in terms of the auto- and cross-correlation functions of the sequences generated by each binary function individually. Numerical experiments were also conducted to validate the theoretical results.

In Contribution 33, Houxiang Wang and Jiaqing Chen proposed an improved switching linear dynamical system model that captures complex dynamics using higher-order autoregressive processes and flexible state transitions via stick-breaking logistic regression. Efficient inference is enabled through Pólya-gamma augmentation. The model outperforms existing methods on synthetic and fMRI data.

A new chaos-based image encryption algorithm that combines discrete Fourier transform and Joseph traversal to encrypt images in both frequency and spatial domains was proposed by Mingxu Wang, Xianping Fu, Xiaopeng Yan, and Lin Teng in Contribution 23. It uses a logistic map for chaos generation, variable-step Joseph traversal for scrambling, and block diffusion. The method enhances security and shows strong resistance to common attacks.

In Contribution 14, Erman Ozpolat andArif Gulten, explored the synchronization and implementation of a novel hyperchaotic system using an adaptive observer. Hyperchaotic systems, known for possessing a greater number of positive Lyapunov exponents compared to chaotic systems, present unique challenges and opportunities in control and synchronization. In this study, the authors introduced a novel hyperchaotic system, thoroughly examining its dynamic properties and conducting a comprehensive phase space analysis. The proposed hyperchaotic system underwent validation through circuit simulation to confirm its behavior. By introducing an adaptive observer synchronization technique, the authors successfully synchronized the dynamics of the novel hyperchaotic system with an identical counterpart. Importantly, Contribution 14 extends the application of this synchronization method to the domain of secure communication, showcasing its practical usage.

In Contribution 34, Maxim Shishlenin, Andrey Kozelkov, and Nikita Novikov showed the fundamental possibility of constructing a computational digital twin of the acoustic tomograph within the framework of a unified physics–mathematical model, based on the Navier–Stokes equations. The authors suggested that the size of the modeling area is quite small, sound waves are waves of "small" disturbance, and given that a person consists of more than 60% water, human organs can be modeled using a liquid model, considering their density. The results showed that the developed technology can be used to probe the human body in medical acoustic tomographs and determine its acoustic parameters to detect neoplasms.

In Contribution 8, Zihao Lu, Na Wang, and Shigui Dong considered nonlinear discrete systems with dual unknown inputs and proposed two new improved squared-root cubature Kalman filtering (ISRCKF) algorithms to estimate system states and dual unknown inputs. Improved square-root cubature Kalman filtering 1 (ISRCKF1) introduces an innovation that first obtains the unknown input estimated from the measurement equation, then updates the innovation to derive the unknown input estimates from the state equation, and uses the already obtained estimates of the dual unknown inputs to corrected the one-step estimate of the state; finally, the minimum variance unbiased estimate of the state is obtained. For many years, chaotic maps have been widely used in the design of various algorithms in cryptographic systems.

In Contribution 15, Lijun Zhang, Caochuan Ma, Yuxiang Zhao, and Wenbo Zhao introduced a compound chaotic system using quantum random walks controlled by a hyperchaotic map and proposed a novel method for dynamically generating S-Boxes. The system exhibits high randomness and complexity. The generated S-Boxes meet key cryptographic criteria and demonstrate excellent performance, making the method suitable for secure information systems. In addition, the proposed dynamic S-Box construction technique can be used to generate cryptographically strong substitution-boxes in practical information security systems.

An investigation into the dynamic states and relative intensity noise of laser diodes subjected to double optical feedback was conducted by Salah Abdulrhmann, Abu Mohamed Alhasan, and A. Y. Madkhli in Contribution 30. They employed modified and improved time-delay rate equations to account for double external optical feedback. The dynamic states and noise of lasers were investigated using bifurcation diagrams of the output photon number, its temporal variations, and the intensity noise of the laser. This analysis considered feedback strengths due to the double external cavity and their spacing from the front facet of the laser with and without phase due to feedback. The results revealed that considering phase causes significant variations in laser intensity and a phase shift in the temporal variations in the laser output. This results in relative intensity noise suppression and a frequency shift in the intensity noise spectrum. These findings represented new contributions to our understanding of the reliance of lasing frequency shift on the phase due to feedback, regardless of whether feedback originates from a single or double external cavity.

Xiaonan Yao, Gang Zhang, Song Yang, and Yuehua Chen used the sliding-mode control-based adaptive anti-disturbance strategy with success in Contribution 9, and the simulation results demonstrated that the control strategy proposed in this paper shows resistance to system uncertainties and unknown external disturbances. Furthermore, it reduces the model accuracy requirements for the controller and proves to be suitable for accurately controlling dissolved oxygen in circulating water systems.

In Contribution 16, Li-Jun Ouyang, Bing-Qing Xie, and Bin Ding analytically explored approximate lag and anticipating synchronization between two unidirectionally coupled hyperchaotic Chen systems without time-delay coupling. By treating lag synchronization as a form of generalized synchronization, the authors derived conditions independent of time delay. Using Taylor expansions, they showed how to recreate past or predict future signals, offering a simple method to counteract the effects of signal transmission delays in hyperchaotic systems.

In Contribution 31, Ruoshi Yang et al. considered a mechanical model of an elastic bar embedded in an elastic and viscoelastic medium, fixed at the base and free at the top, which was established based on the Winkler foundation assumption. The deflection function of a bar subjected to both axial force and locally distributed horizontal load was derived using the Rayleigh–Ritz method. Utilizing the elastic–viscoelastic correspondence principle, the viscoelastic medium surrounding the bar was modeled as an elastic medium in which the ground reaction coefficient varies within phase space formulation. This study provides a robust theoretical foundation for soft soil foundation engineering projects and fills a significant gap in the literature by offering a comprehensive framework for understanding displacement in elastic bars within viscoelastic media. Drawing upon the derivation of the deformation function for elastic rods within a viscoelastic medium, the findings of this research hold significant applicability across a range of domains. These include, but are not limited to, the expansion of roadways in regions characterized by coastal soft soil, as well as the successful monitoring of deformation and lifespan in bridge pile foundations.

Xiaochun Zhang, Jianhan Hu, and Shuyang Chen presented a mechanical model for an elastic bar embedded in a viscoelastic medium, reflecting the behavior of structures in coastal soft soil regions of China in Contribution 32. Using the Rayleigh–Ritz method and the Winkler foundation assumption, they derived the bar's deflection under axial and horizontal loads and applied the elastic–viscoelastic correspondence principle to model soil behavior, providing a solid theoretical foundation for infrastructure projects on soft soil, such as road expansion and bridge foundation monitoring.

In Contribution 10, Simone Fiori, Luca Ciaramicoli, and Giovanni Berluti proposed a novel control strategy for a DC-DC electrical converter realized as a switched circuit. The present endeavor was based on early work by Leonard and Krishnaprasad where a prototypical DC-DC converter was modeled as a state space dynamical system and controlled by an open-loop strategy based on Lie group theory. In the present work, the authors introduced a closed-loop control strategy based on maximum convergence rate, studied some features of the novel method, and illustrated its behavior by numerical simulations. A numerical comparison of the two methods complemented the paper and showed how the proposed feedback control method outperforms the static one in terms of convergence rate and resiliency against parameter mismatch.

In Contribution 2, Dan Gu, Yufeng Xie, Shulin Liu, Yuan Wei, and Jiayi Shen investigated subtle faults in high-speed motorized spindles. A dynamical model considering unbalanced mass and spindle inclination faults was developed. The natural frequencies and mode shapes were analyzed using the whole transfer matrix method (WTMM). The results showed that spindle deflection varies with fault type and speed, with maximum deflection occurring at the third natural frequency. This paper also showed that different faults produce distinct mode shapes even at the same speed, which is helpful information in fault identification and diagnosis.

In Contribution 3, Zun Li, Binqiang Xue, and Youyuan Chen dealt with the proposal of an event-triggered state estimation method for parameter-uncertain systems with a binary encoding transmission scheme. The stability analysis showed that the state estimation error of the proposed method was bound. Finally, the practical value of the method proposed in this paper was confirmed through numerical simulations.

Bin Zhen and Yu Zhang proposed a new approach to realize generalized function projective synchronization (GFPS) between two different chaotic systems with uncertain parameters in Contribution 17. The GFPS condition was derived by converting the differential equations describing the synchronization error systems into a series of Volterra integral equations.

The hourglass buoy was compared by Houssein Yassin to a spherical buoy to illustrate differences in their response to regular waves and energy extraction when using the same control laws in Contribution 4. A spherical buoy diameter of 7.5 m was required to obtain the same power output as a 5 m tall hourglass buoy. A power-force-amplitude (PFA) metric was introduced to compare energy extraction performance and power take-off requirements. The hourglass buoy's PFA was 13% larger than the spherical buoy, implying that it could produce similar power but with less control effort.

A dynamic model of the working mechanism of a compound pendulum jaw crusher with clearance was established by the Lagrange multiplier method (LMD) based on the L-N contact force model and modified Coulomb friction force model in Contribution 18. The results obtained by Shenpeng Wang, Yan Cui, and Chune Wang showed that, under the premise of meeting the requirements of crusher production capacity, increasing the driving speed and reducing the clearance size can weaken the shock and chaos phenomenon during the operation of the mechanism, improve the dynamic performance, and enhance the operation stability, and the service life of foundation was prolonged with the stability of vibration force. Furthermore, the optimal combination of driving speed and clearance size was determined, which provides a theoretical basis for the optimization of the vibration force of the jaw crusher.

An investigation of the formation problem of an array of large scale mobile sensor networks was carried out by Xueming Qian and Baotong Cui in Contribution 11. A new framework for the dynamic of mobile sensors as a continuum described by the parabolic system with boundary disturbance was proposed. The communication topology of agents is a chain graph and fixed. Leader feedback laws which were designed in a manner to the boundary control of large-scale mobile sensor networks allow the mobile sensors to achieve the formation steadily. By referring to the Lyapunov functional method and employing a boundary control approach, a new protocol was established to deal with a formation problem for the large-scale mobile sensor networks.

In Contribution 24, Hongrui Wang et al. proposed a distributed double-layer model predictive control algorithm based on dual decomposition for multivariate constrained systems to reduce the computational complexity of process control, giving the optimal solution formulation of the optimization problem consisting of this objective function and the constraints. The algorithm proposed in this paper achieved control goals while significantly reducing the computational complexity and has research significance for promoting the industrial implementation of double-layer model predictive control with success.

A three-dimensional (3D) autonomous chaotic system was analyzed by Meng Liu, Zhaoyan Wu, and Xinchu Fu in Contribution 19. In the system, each equation contains a quadratic cross product. Additionally, an equivalent electronic circuit for the proposed chaotic system was designed according to Kirchhoff's Law, and a corresponding response electronic circuit was also designed for identifying the unknown parameters or monitoring the changes in the system parameters. Moreover, numerical simulations were presented and complemented the theoretical results.

In Contribution 20, Qian Wei, Peng Su, Lin Zhou, and Wentao Shiin proposed a feedback optimization strategy of residual compensation to correct the trajectory prediction values and improve prediction accuracy. For the initial value sensitivity problem of the PSO model identification algorithm, a new initial population strategy was proposed, which improved the effectiveness of initial parameters on model identification. Through simulation experiment analysis, it was verified that the proposed OTP-SSCT algorithm achieves better prediction accuracy and faster response time.

In Contribution 12, Dharma Raj Khatiwad dealt with the synchronization of an ensemble of oscillators, that is, a phenomenon that is present in systems of different fields, ranging from social to physical and biological systems. This phenomenon was often described mathematically by the Kuramoto model, which assumes oscillators of fixed natural frequencies connected by an equal and uniform coupling strength, with an analytical solution only possible for an infinite number of oscillators. However, most real-life synchronization systems consist of a finite number of oscillators and are often perturbed by external fields. The results obtained showed that the identical trends observed in the two cases give an important indication that the synchrony persists even under the influence of external factors, something very plausible in the context of real-life synchronization events. The occasional boosting of coupling strength is also enough to keep the assembly of oscillators in a synchronized state persistently.

In Contribution 5, Kornphong Chonsiripong et al. studied the N-species competitive coexistence model with direct effect on habitat destruction, to analyze the behaviors of abundant and extinct species in the system caused by habitat loss. In this paper, the proportional abundances of 50 species were numerically portrayed in each case. The authors found that the survival of a species is guaranteed if its corresponding eigenvalue is positive. However, the fact that a species has a negative corresponding eigenvalue does not guarantee its extinction, as this also depends on the initial number of that species.

The relationship between the Chinese market and the US market is of great interest for researchers and investors. In Contribution 25, Chun-Xiao Nie and Jing Xiao used transfer entropy and local random permutation (LRPLRP) surrogates to detect the information flow dynamics between the two markets. In this case, dynamic analysis based on weekly data showed that the information flow evolves and includes three significant periods between

2004 and 2021. They also used daily data to analyze the dynamics of information flow in detail over the three periods and found that changes in the intensity of information flow were accompanied by major events affecting the markets, such as the 2008 financial crisis and the COVID-19 pandemic period. In particular, the impact of the S&P500 index on different industry indices in the Chinese market was analyzed, and it was found that the dynamics of information flow exhibited multiple patterns. This study revealed the complex information flow between the two markets from the perspective of nonlinear dynamics, thereby helping to analyze the impact of major events and providing quantitative analysis tools for investment practice.

Finally, in Contribution 13, Pu Yang, Yu Ding, Ziwei Shen, and Kejia Feng studied the consensus fault-tolerant control problem of a class of second-order leader–follower multi-agent systems with unknown disturbance and actuator faults, and proposed an integral non-singular terminal sliding mode control algorithm based on a finite-time observer. First, a finite-time disturbance observer was designed based on a combination of high-order sliding mode and dual-layer adaptive rules to realize fast estimation and compensation of disturbance and faults. Then, a sliding surface with additional integral links was designed based on the conventional sliding surface, and an integral non-singular terminal sliding mode controller was proposed to realize the robust consensus in finite time and accurately diminish the chattering phenomena. Finally, a numerical example and simulation verified effectiveness.

The contributions are listed below:

1. Wang, Y.; Sun, Z.; Ye, S.; Zhao, T.; Zhang, X.; Xu, W. Dynamic Survivability Centrality in Nonlinear Oscillator Systems. *Symmetry* **2024**, *16*, 1661. https://doi.org/10.3390/sym16121661.
2. Gu, D.; Xie, Y.; Liu, S.; Wei, Y.; Shen, J. Dynamic Modeling of Motorized Spindle System with Unbalanced Mass and Spindle Inclination. *Appl. Sci.* **2023**, *13*, 10053. https://doi.org/10.3390/app131810053.
3. Li, Z.; Xue, B.; Chen, Y. Event-Triggered State Estimation for Uncertain Systems with Binary Encoding Transmission Scheme. *Mathematics* **2023**, *11*, 3679. https://doi.org/10.3390/math11173679.
4. Yassin, H.; Gonzalez, T.D.; Parker, G.; Wilson, D. Effect of the Dynamic Froude–Krylov Force on Energy Extraction from a Point Absorber Wave Energy Converter with an Hourglass-Shaped Buoy. *Appl. Sci.* **2023**, *13*, 4316. https://doi.org/10.3390/app13074316.
5. Chonsiripong, K.; Duangdai, E.; Pongvuthithum, R.; Likasiri, C. Habitat Destruction and Restoration in Relation to Extinction and Survival of Species in Competitive Communities. *Appl. Sci.* **2022**, *12*, 8693. https://doi.org/10.3390/app12178693.
6. Jia, Z. Equilibrium Control in Uncertain Linear Quadratic Differential Games with V-Jumps and State Delays: A Case Study on Carbon Emission Reduction. *Entropy* **2024**, *26*, 943. https://doi.org/10.3390/e26110943.
7. Liang, L.; Wu, C.; Liu, S. A Terminal Residual Vibration Suppression Method of a Robot Based on Joint Trajectory Optimization. *Machines* **2024**, *12*, 537. https://doi.org/10.3390/machines12080537.
8. Lu, Z.; Wang, N.; Dong, S. Improved Square-Root Cubature Kalman Filtering Algorithm for Nonlinear Systems with Dual Unknown Inputs. *Mathematics* **2024**, *12*, 99. https://doi.org/10.3390/math12010099.
9. Yao, X.; Zhang, G.; Yang, S.; Chen, Y. Adaptive Anti-Disturbance Control of Dissolved Oxygen in Circulating Water Culture Systems. *Symmetry* **2023**, *15*, 2015. https://doi.org/10.3390/sym15112015.

10. Fiori, S.; Ciaramicoli, L.; Berluti, G. Maximum Convergence Rate Control of a Switched Electrical Network with Symmetries. *Symmetry* **2023**, *15*, 1861. https://doi.org/10.3390/sym15101861.
11. Qian, X.; Cui, B. A PDE Model Approach to Formation Control of Large-Scale Mobile Sensor Networks with Boundary Uncertainties. *Axioms* **2023**, *12*, 9. https://doi.org/10.3390/axioms12010009.
12. Khatiwad, D.R. Numerical Solution of Finite Kuramoto Model with Time-Dependent Coupling Strength: Addressing Synchronization Events of Nature. *Mathematics* **2022**, *10*, 3633. https://doi.org/10.3390/math10193633.
13. Yang, P.; Ding, Y.; Shen, Z.; Feng, K. Integral Non-Singular Terminal Sliding Mode Consensus Control for Multi-Agent Systems with Disturbance and Actuator Faults Based on Finite-Time Observer. *Entropy* **2022**, *24*, 1068. https://doi.org/10.3390/e24081068.
14. Ozpolat, E.; Gulten, A. Synchronization and Application of a Novel Hyperchaotic System Based on Adaptive Observers. *Appl. Sci.* **2024**, *14*, 1311. https://doi.org/10.3390/app14031311.
15. Zhang, L.; Ma, C.; Zhao, Y.; Zhao, W. A Novel Dynamic S-Box Generation Scheme Based on Quantum Random Walks Controlled by a Hyper-Chaotic Map. *Mathematics* **2024**, *12*, 84. https://doi.org/10.3390/math12010084.
16. Ouyang, L.-J.; Xie, B.-Q.; Ding, B. Analytical Studies on Approximate Lag and Anticipating Synchronization in Two Unidirectionally Coupled Hyperchaotic Chen Systems without Time Delay. *Appl. Sci.* **2023**, *13*, 11949. https://doi.org/10.3390/app132111949.
17. Zhen, B.; Zhang, Y. Generalized Function Projective Synchronization of Two Different Chaotic Systems with Uncertain Parameters. *Appl. Sci.* **2023**, *13*, 8135. https://doi.org/10.3390/app13148135.
18. Wang, S.; Cui, Y.; Wang, C. Dynamics Analysis and Chaos Identification of Compound Pendulum Jaw Crusher with Joint Clearance. *Appl. Sci.* **2023**, *13*, 238. https://doi.org/10.3390/app13010238.
19. Liu, M.; Wu, Z.; Fu, X. Dynamical Analysis of a One- and Two-Scroll Chaotic System. *Mathematics* **2022**, *10*, 4682. https://doi.org/10.3390/math10244682.
20. Wei, Q.; Su, P.; Zhou, L.; Shi, W. Online Tracking of Maneuvering Target Trajectory Based on Chaotic Time Series Prediction. *Entropy* **2022**, *24*, 1668. https://doi.org/10.3390/e24111668.
21. Liu, J.; Zhang, Y.; Li, J.; Zhao, Y.; Guo, J.; Yang, L.; Zhao, H. PMT fluorescence signal denoising processing based on wavelet transform and BP neural network. *Appl. Sci.* **2024**, *14*, 4866. https://doi.org/10.3390/app14114866.
22. Tsuneda, A. Auto-Correlation Functions of Chaotic Binary Sequences Obtained by Alternating Two Binary Functions Bayesian Inference of Recurrent Switching Linear Dynamical Systems with Higher-Order Dependence. *Dynamics* **2024**, *4*, 272–286. https://doi.org/10.3390/dynamics4020016.
23. Wang, M.; Fu, X.; Teng, X.Y.A. A new chaos-based image encryption algorithm based on discrete fourier transform and improved Joseph traversal. *Mathematics* **2024**, *12*, 638. https://doi.org/10.3390/math12050638.
24. Wang, H.; Zhang, P.; Yang, Z.; Zou, T. Decomposition–Coordination of Double-Layer MPC for Constrained Systems. *Entropy* **2023**, *25*, 17. https://doi.org/10.3390/e25010017.
25. Nie, C.-X.; Xiao, J. Dynamics of Information Flow between the Chinese A-Share Market and the U.S. Stock Market: From the 2008 Crisis to the COVID-19 Pandemic Period. *Entropy* **2022**, *24*, 1102. https://doi.org/10.3390/e24081102.

26. Zhang, Y.; Yang, S.; Tai, X.; Ma, H.; Guan, H.; Mu, Q.; Qu, L.; Ding, X. Study on Rubbing-Induced Vibration Characteristics Considering the Flexibility of Coated Casings and Blades. *Machines* **2024**, *12*, 481. https://doi.org/10.3390/machines12070481.
27. Wong, C.-N.; Vyas, A.; Wong, W.-O.; Sun, R. Young's Modulus and Hardness Identification of Extruded Aluminum by Scratching Damper. *Machines* **2024**, *12*, 413. https://doi.org/10.3390/machines12060413.
28. Kusriantoko, P.; Daun, P.F.; Einarsrud, K.E. A Comparative Study of Different CFD-codes for Fluidized Beds. *Dynamics* **2024**, *4*, 475–498. https://doi.org/10.3390/dynamics4020025.
29. Ionescu, C.; Petrisor, I. Attached Flows for Reaction—Diffusion Processes Described by a Generalized Dodd—Bullough—Mikhailov Equation. *Symmetry* **2024**, *16*, 531. https://doi.org/10.3390/sym16050531.
30. Abdulrhmann, S.; Alhasan, A.M.; Madkhli, A.Y. Dynamics of Semiconductor Laser Coupled with Two External Cavities. *Appl. Sci.* **2023**, *13*, 12827. https://doi.org/10.3390/app132312827.
31. Yang, R.; Sun, W.; Le, M.; Zhang, H. The Chinese Aviation Network: An Empirical Temporal Analysis on Its Structural Properties and Robustness. *Appl. Sci.* **2023**, *13*, 11627. https://doi.org/10.3390/app132111627.
32. Zhang, X.; Hu, J.; Chen, S. Study on Stability of Elastic Compression Bending Bar in Viscoelastic Medium. *Appl. Sci.* **2023**, *13*, 11111. https://doi.org/10.3390/app131911111.
33. Wang, H.; Chen, J. Bayesian Inference of Recurrent Switching Linear Dynamical Systems with Higher-Order Dependence. *Symmetry* **2024**, *16*, 474. https://doi.org/10.3390/sym16040474.
34. Shishlenin, M.; Kozelkov, A.; Novikov, N. Nonlinear Medical Ultrasound Tomography: 3D Modeling of Sound Wave Propagation in Human Tissues. *Mathematics* **2024**, *12*, 212. https://doi.org/10.3390/math12020212.


## 3. Future Work in Development

A second edition of the Topic "Nonlinear Phenomena, Chaos, Control and Applications to Engineering and Science and Experimental Aspects of Complex Systems" is in development. The collection will also provide a great opportunity for disseminating recent advances in analytical and numerical techniques, discussing novel phenomena and behaviors regarding several aspects of nonlinear dynamics and control, as was the case with the first edition.

**Acknowledgments:** The authors would like to express their sincere gratitude to all authors whose valuable works were published in this Topic. Thanks are due to all the authors, co-authors, and reviewers for all their time, collaboration, and enthusiasm in supporting this Topic. The authors are also grateful for the assistance provided by Jenny Lin, the Topic Specialist, throughout the process. Finally, the authors acknowledge the Brazilian funding agency CNPq (Conselho Nacional de Desenvolvimento Científico e Tecnológico).

**Conflicts of Interest:** The authors declare no conflicts of interest.

symmetry

MDPI

Article

# Dynamic Survivability Centrality in Nonlinear Oscillator Systems

**Yuexin Wang [1], Zhongkui Sun [2], Sijun Ye [1], Tao Zhao [1,*], Xinshuai Zhang [1] and Wei Xu [2]**

[1] Xi'an Modern Control Technology Research Institute, Xi'an 710065, China; 15991711809@163.com (Y.W.); 13324573779@163.com (S.Y.); 13819106281@163.com (X.Z.)

[2] School of Mathematics and Statistics, Northwestern Polytechnical University, Xi'an 710129, China; dynsun@126.com (Z.S.); weixunpu@nwpu.edu.cn (W.X.)

* Correspondence: 17319954944@163.com

**Abstract:** In light of the fact that existing centrality indexes disregard the influence of dynamic characteristics and lack generalizability due to standard diversification, this study investigates dynamic survivability centrality, which enables quantification of oscillators' capacity to impact the dynamic survivability of nonlinear oscillator systems. Taking an Erdős–Rényi random graph system consisting of Stuart–Landau oscillators as an illustrative example, the typical symmetry synchronization is considered as the key mission to be accomplished in light of the study and the dynamic survivability centrality value is found to be dependent on both the system size and connection density. Starting with a small scale system, the correctness of the theoretical results and the superiority in comparison to traditional indexes are verified. Further, we present the quantitative results by means of error analysis, distribution comparison of various indexes and relationship with system structure exploration, and give the position of the key oscillator. The results demonstrate a negligible error between the theoretical and numerical outcomes, and highlighting that the distribution of dynamic survivability centrality closely resembles the distribution of system state changes. The conclusions serve as evidence for the accuracy and validity of the proposed index. The findings provide an effective approach to protect systems to improve dynamic survivability.

**Keywords:** dynamic survivability centrality; key position; system collapse; topology structure

## 1. Introduction

The advancement of network science has played a pivotal role in enhancing our comprehension of the physical world [1,2], with diverse topological structures revealing an abundance of phenomena that have garnered significant attention [3,4]. However, the occurrence of element damage leads to substantial losses, prompting scholars to conduct extensive research on this matter [5]. Currently, one of the most prominent areas of investigation pertains to the security aspects associated with dynamic systems [6,7].

Dynamic survivability, an important concept for dynamic network systems, describes a system's ability to complete its key mission when attacks, failures or accidents occur [8–12]. A system with low dynamic survivability fails to mitigate substantial losses resulting from its collapse, so this topic has garnered increasing attention. Researchers have explored it under different dynamic behaviors, including single-layer and two-layer network topologies [8,13–15].

The findings of these studies suggest a strong correlation between dynamic survivability and the oscillators (nodes), i.e., the failure of oscillators leads to a reduction in dynamic survivability. Nevertheless, it is evidently impossible to protect all oscillators to enhance the dynamic survivability of dynamic system. As widely acknowledged, the functionality of various system mechanisms, including cascade failure, diffusion propagation, and synchronization, will be affected by the action of a small number of nodes [16–21]. It follows naturally that the same is true of dynamic survivability. Identifying these nodes with significant effects both theoretical significance and practical value. In fact, this is what is known today as centrality, which reflects the nodes occupying key positions in the entire network system, and the definition of centrality is well understood. In real life, centrality is always associated with excellent leadership, popularity or reputation [22,23]. Moreover, the greater the centrality attributed to individuals, the higher their access to convenience and influence within social networks [24,25].

Up to now, researchers have introduced various centrality indexes of a node in the network system to quantify the importance of the nodes in different situations, among which degree centrality [26,27], closeness centrality [28,29], betweenness centrality [30], eccentricity centrality [31], eigenvector centrality [32–34], and so on are widely used. It is a pity that these indexes have various standards and are put forward according to specific application scenarios, so they cannot be generalized. In addition, the current studies on node centrality are mainly based on the structural information of the systems, but neglect the dynamic properties of the systems. In view of this, Zhang et al. [35] proposed the concept of resilience centrality to quantify the ability of a single node to affect the resilience of the corresponding system. Theoretically, the resilience centrality was derived from the dynamic equation of the corresponding complex system, and its effectiveness was verified by comparison with other traditional indexes. However, all those centrality indexes cannot capture the contribution of an oscillator in dynamic survivability of the dynamic system, prompting us to propose a new centrality index.

Motivated by these facts, our objective in this paper is to offer a concept of dynamic survivability centrality to identify the ability of an oscillator to affect dynamic survivability of the dynamic system, thereby addressing the limitations inherent in conventional centrality indices.The definition of dynamic survivability centrality is first reported in Section 2, and the expression is derived from the coupled Stuart–Landau oscillator systems when the key mission to complete is synchronization. In Section 3, the dynamic survivability centrality analysis and the comparison with classic indexes to test the performance of our index are illustrated. The results demonstrate that the proposed index is more effective in assessing the significance of elements within dynamic systems, while also providing insights into the factors influencing this index. And Section 4 summarizes the conclusions. The diagram block for methodology is depicted in Figure 1.

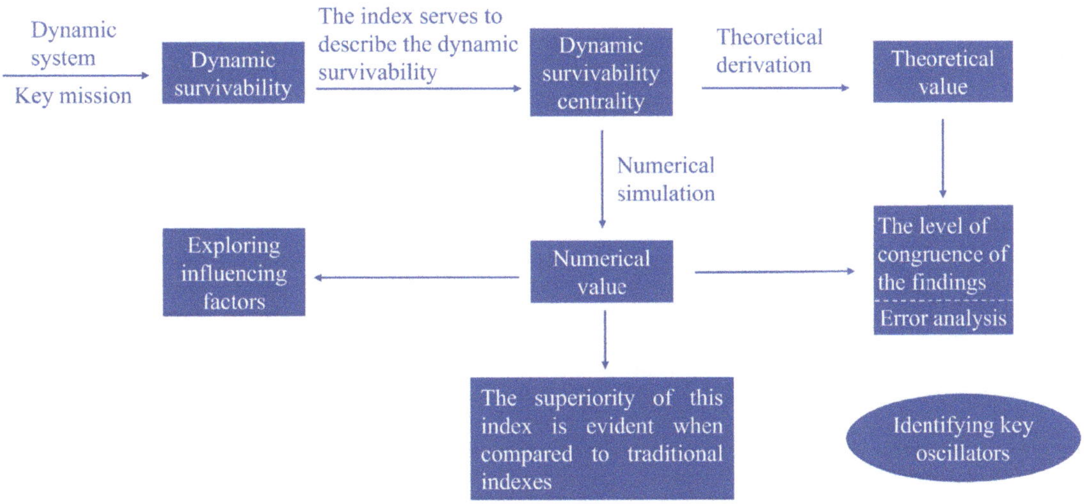


**Figure 1.** The diagram block for methodology.

## 2. Dynamic Survivability Centrality of Dynamic Systems

### 2.1. Dynamic Survivability of Dynamic Systems

The definition of dynamic survivability for nonlinear oscillator system is introduced by Sun et al. in [8]: The dynamic survivability is the capability that a dynamic system can complete its key mission in the event of attack, failure, or accident. Additionally, the mission completion probability $S$ is considered by authors to quantify the dynamic survivability, which characterizes the probability that a dynamic system can complete key mission under attack as follows:

$$S = P(C_c > C) = e^{-UC}, \quad (1)$$

where $C_c$ represents critical attack cost at which system loses its capacity to complete the key mission, which can serve as a criterion for distinguishing between favorable and adverse dynamic survivability. $C$ is a broad definition called attack cost describing any attack, failure or accident. The mission function $U$ is obtained from the key mission selected to completed, which is generally defined as the ratio of the initial value of the indicator that reflects the state of the system to complete the key mission to the value of the indicator at the time of the attack. Considering a nonlinear oscillator system composed of $N$ Stuart–Landau oscillators whose topology structure is determined by the Erdős–Rényi random graph (hereafter abbreviated as ER), which plays an important role and is widely used to study the dynamic behavior,

$$\dot{z}_j = (\alpha_j + i\Omega - |z_j|^2)z_j + \varepsilon \sum_{k=1}^{N} A_{jk}H(z_k - z_j), j = 1, 2, \cdots, N. \quad (2)$$

Therein, the complex state variable of $j$th oscillator $z_j = x_j + iy_j$. Setting bifurcation parameter $\alpha_j = 1$ and natural frequency $\omega = 3$. $\epsilon$ denotes the coupling strength and $H$ indicates the internal coupling function between the state variables of each oscillator, which is selected as the unit matrix. The adjacency matrix $A = (A_{jk})$ describes interaction between oscillators, where $A_{jk} = A_{kj} = 1 (j \neq k)$ if oscillator $j$ and oscillator $k$ have a connection and $A_{jk} = A_{kj} = 0$ $(j \neq k)$ if there is not a link from oscillator $j$ to oscillator $k$. Additionally, one can obtain the degree $k_j$ of oscillator $j$ according to $k_j = \sum_{k=1}^{N} A_{jk}$.

Furthermore, let $B_{jj} = -k_j$, $B_{jk} = B_{kj} = 1 (j \neq k)$ if there is a link from the $j$th and the $k$th oscillators, and $B_{jk} = B_{kj} = 0 (j \neq k)$ otherwise. Then, Equation (2) can be mapped the original dynamic system into an equivalent model as

$$\dot{z}_j = (\alpha_j + i\Omega - |z_j|^2)z_j + \varepsilon \sum_{k=1}^{N} B_{jk}Hz_k, j = 1, 2, \cdots, N, \quad (3)$$

where $B = (B_{jk})$ is called the coupling matrix, which is a negative Laplace matrix actually.

As demonstrated by dynamic survivability, it is necessary to determine the key mission and attack cost subsequently. Since the fact that the collective dynamical behaviors of the dynamic system changes when it is disturbed echoes the loss caused by the change in the damaged performance of the system in real world, the key mission for the dynamic system is to maintain the state of its collective dynamical behaviors. Admittedly, dynamic systems perform a wealth of collective dynamical behaviors, such as synchronization [3,4,36–38], explosive phenomena [39–42], and aging phase transition [43], etc., while the meaning of survivability only requires the most crucial one. The phenomenon of synchronization, being a prototypical manifestation of symmetry in physics, has been unequivocally established as profoundly significant, which is the first of the collective behavior of the oscillator system to receive attention and receive in-depth research. Based on the above truth, in this paper, we choose synchronization as the key mission to complete, and attack the system by removing the oscillators.

As the classical method to analyze synchronization, the master stability function (MSF) method [4,44] gives the condition $\varepsilon \geq |l/\lambda_2|$ needed to be satisfied, where $l$ is the maximum Lyapunov index of an isolated oscillator and $\lambda_2$ is the second-largest eigenvalue of the corresponding coupling matrix of Equation (3). When oscillators in the system are attacked, the $\lambda_2$ decreases and becomes zero. In other words, $\lambda_2$ is the parameter that can directly reflect the key mission completion state of the coupled nonlinear oscillator system. Then, we write the mission completion probability $S$ as

$$S = P(C_c > C) = e^{-UC} = e^{-\frac{\lambda_2(B(0))}{\lambda_2(B(C))}C}, \tag{4}$$

where $\lambda_2(B(0))$ is the initial second-largest eigenvalue of the coupling matrix $B$ when the system is not attacked and $\lambda_2(B(C))$ is the second-largest eigenvalue of the coupling matrix $B$ when the proportion of removing oscillators is $C$. The decrease of $S$ from 1 to 0, resulting by an increase in $C$, implies that no matter how the network topology of the system is changed due to the cause of removing oscillators, the impact on the dynamic survivability of the system can be expressed by the corresponding change in the second-largest eigenvalue of the coupling matrix $B$, so that the impact of arbitrary attack on the system can be described quantitatively.

### 2.2. Dynamic Survivability Centrality of Dynamic Systems

As previously mentioned, the majority of real-world systems evolve accompanied with time [45], while traditional centrality indexes solely consider static network topology changes. The capacity of individual nodes in a dynamic system to influence dynamic phenomena is thus inadequately reflected. Consequently, this section presents the definition and measurement of dynamic survivability centrality.

The dynamic survivability centrality reflects the important position of oscillators in the whole system by exploring the change extent of dynamic survivability after removing different oscillators. In this regard, we have already stated that the second-largest eigenvalue change in the coupling matrix after the removal of oscillators can be used to fully capture and quantify the change in dynamic survivability of the system. In summary, we define the change in the second-largest eigenvalue of the coupling matrix $\lambda_2(B)$ after the oscillator is removed as dynamic survivability centrality to quantify the importance of the oscillator in the system and provide solutions for protecting the dynamic survivability of the system. Specifically, the dynamic survivability centrality of oscillator $i$ is

$$SC(i) = \Delta\lambda_2(B) = \lambda_2^0(B) - \lambda_2^i(B), \tag{5}$$

where $\lambda_2^0(B)$ is the second-largest eigenvalue of coupling matrix of the original system, and $\lambda_2^i(B)$ corresponds to the second-largest eigenvalue of coupling matrix of the system after oscillator $i$ is removed. Certainly, when considering other dynamic phenomena, system parameters, and network topology, it suffices to modify the second-largest eigenvalue of the coupling matrix to account for the corresponding factor influencing changes in dynamic survivability, which is a general framework.

Next, the index (5) will be normalized in order to compare the importance of oscillators in different coupled nonlinear oscillator systems with different $\lambda_2(B)$. The normalized dynamic survivability centrality is denoted as $SC'$,

$$SC'(i) = \frac{\lambda_2^0(B) - \lambda_2^i(B)}{\lambda_2^0(B)}. \tag{6}$$

Taking into account the homogeneity of the ER system with total number of oscillators $N$, $d$ signifies the connectivity density, and its expression is $d \equiv \langle k \rangle/(N-1) \sim \langle k \rangle/N$

for large $N$, where $\langle k \rangle$ is the average degree of oscillators. As per the information cited in [46–48], the eigenvalue spectrum converges to the following semicircle rate:

$$\rho(\lambda(A)) = \begin{cases} \frac{\sqrt{4Nd(1-d)-\lambda(A)^2}}{2\pi Nd(1-d)} \text{if} |\lambda(A)| < 2\sqrt{Nd(1-d)} \\ 0 \quad \text{otherwise} \end{cases}. \tag{7}$$

This indicates that the second-largest eigenvalue of the adjacency matrix corresponding to the coupled nonlinear oscillator system is approximated as follows:

$$\lambda_2^0(A) \approx 2\sqrt{Nd^0(1-d^0)}. \tag{8}$$

Introducing the condition $\lambda_2^0(B) = \lambda_2^0(A) - \langle k \rangle^0$, where $\langle k \rangle^0$ represents the average degree of oscillators of the original system, then an approximation of the second-largest eigenvalue of the coupling matrix can be obtained,

$$\lambda_2^0(B) \approx 2\sqrt{Nd^0(1-d^0)} - d^0 N. \tag{9}$$

Finally, the dynamic survivability centrality index and the normalized dynamic oscillator centrality index are, respectively, written as follows:

$$\begin{aligned} SC(i) &= \Delta\lambda_2(B) = \lambda_2^0(B) - \lambda_2^i(B) \\ &= \left(2\sqrt{Nd^0(1-d^0)} - d^0 N\right) - \left(2\sqrt{(N-1)d^i(1-d^i)} - d^i(N-1)\right), \end{aligned} \tag{10}$$

and

$$\begin{aligned} SC'(i) &= \frac{\lambda_2^0(B) - \lambda_2^i(B)}{\lambda_2^0(B)} \\ &= \frac{\left(2\sqrt{Nd^0(1-d^0)} - d^0 N\right) - \left(2\sqrt{(N-1)d^i(1-d^i)} - d^i(N-1)\right)}{\left(2\sqrt{Nd^0(1-d^0)} - d^0 N\right)}. \end{aligned} \tag{11}$$

The significance of the oscillator can be ascertained by computing the dynamic survivability centrality index subsequent to its removal. The greater the value of this index indicates a greater impact on the system's dynamic survivability due to the attacked oscillator, thereby highlighting its importance.

## 3. Analysis and Results

### 3.1. *Dynamic Survivability Centrality Analysis of ER System with $N = 10$*

In order to illustrate the correctness and necessity of the constructing index, the dynamic survivability centrality analysis of an ER system with ten oscillators is firstly carried out. As shown in Figure 2a, the mission completion probability is $S = 1$ when the oscillator is not attacked, which depicts the probability of the system completing the key mission of synchronization is 1, i.e., the system is highly survivable. Then, each oscillator is removed separately. It can be seen that the value of the normalized dynamic survivability centrality index $SC'(i)$ is different when different oscillators are removed. The larger the $SC'(i)$ is, the smaller the corresponding mission completion probability $S$ is, which means that the dynamic survivability of the coupled nonlinear oscillator system is changed more. From the perspective of system dynamic survivability, the corresponding oscillator needs more attention in the system. The black solid line represents the result of theoretical derivation, which is in good agreement with the numerical result, verifying its correctness.

A schematic diagram of the ten-oscillator ER system is illustrated in Figure 2b according to the calculated importance of each oscillator of Figure 2a. Each dot represents an oscillator, and the larger the shape of the dot, the more important the oscillator. The diagram provides a more intuitive representation of the relative importance ranking among each oscillator, i.e., the oscillator 9 holds a pivotal position within this system due to its removal resulting in the most rapid decline in survivability, while the sixth oscillator exhibits the least impact on the dynamic survivability of the system. The rationale can be found in [8], where it has been demonstrated that removal of edges with minimal degree exhibit the most significant impact on dynamic survivability. As depicted in Figure 2b, the elimination of oscillator 6 corresponds to removing the edge with the minimum degree, while removing oscillator 9 conversely, which supports our conclusion regarding oscillator's importance.

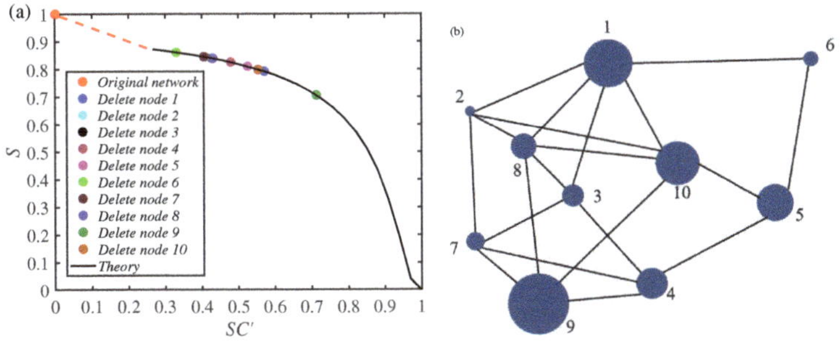


**Figure 2.** Dynamic survivability centrality of nonlinear oscillator system with ER structures ($N = 10$, $d = 0.4$), which display the excellent agreement of theoretical and numerical results. (**a**) The normalized dynamic survivability centrality index $SC'(i)$ and the corresponding mission completion probability S when each oscillator is removed. (**b**) Schematic of the system.

For comparison, three classic centrality indexes: degree centrality, closeness centrality, and betweenness centrality are used to describe the influence ability of each oscillator in our dynamic survivability scenario. Figure 3 displays the corresponding value calculated by each index definition when each oscillator is removed, respectively. As shown in Figure 3b, the most important oscillators are 1, 8, and 10, and the least important oscillator is oscillator 2. The closeness centrality in Figure 3c determines that oscillators 1, 8, and 10 are the most important and oscillator 2 is the least important, which is the same as the conclusion of degree centrality. And in Figure 3d, oscillator 1 is the most critical and oscillator 2 is the least important. These conclusions are different from what is observed in Figure 3a, and oscillators with the same centrality value can be generated without being able to further capture which oscillator has the greatest influence, indicating that dynamic survivability centrality can more accurately describe the oscillator importance ranking. In summary, the dynamic survivability centrality proposed in this paper accurately quantifies the ability of oscillators to affect dynamic survivability in coupled nonlinear oscillator systems, and provides better performance than other existing centrality indexes, which addresses the limitation of conventional centrality indices that solely consider static network topology.

In fact, although the demonstration is given with a primary focus on synchronization, for other key missions, it suffices to substitute the second-largest eigenvalue of the coupling matrix with other influence factors. In essence, this implies that the dynamic survivability centrality can be extended to encompass any complex system, thereby ensuring its universality.

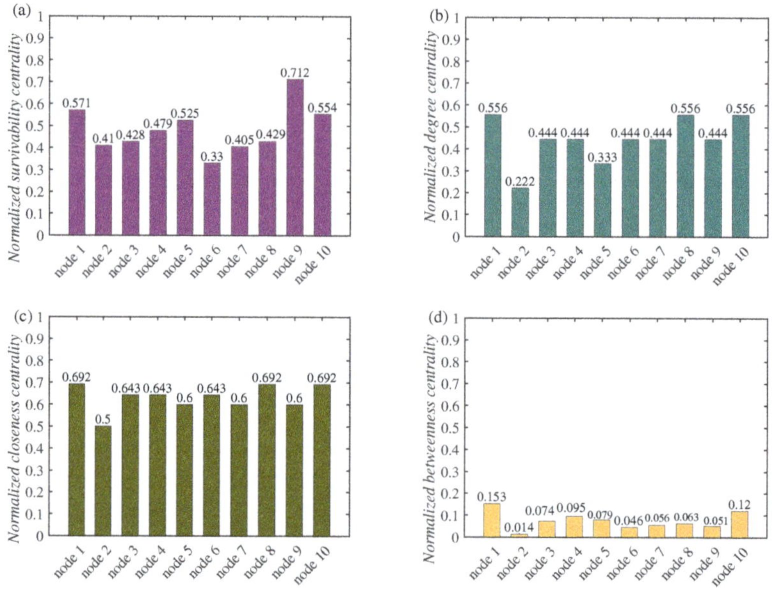


**Figure 3.** Comparison of dynamic survivability centrality with other centrality indexes ($N = 10$, $d = 0.4$), which prove the superiority of dynamic survivability centrality. (**a**) Dynamic survivability centrality. (**b**) Degree centrality. (**c**) Closeness centrality. (**d**) Betweennesss centrality.

### 3.2. Dynamic Survivability Centrality Analysis of ER System with $N = 100$

To validate the generalizability of our research and provide a credible framework to maintain the dynamic survivability of system, the dynamic survivability centrality of nonlinear oscillator system coupled with ER network, whose total number of oscillators $N$ is 100, will be discussed, and more phenomena will be explored in this section.

The results presented in Figure 4 demonstrate the normalized dynamic survivability centrality index $SC'$ obtained by removing each oscillator under increasing connection density $d = 0.08$, $d = 0.1$ and $d = 0.4$. It is observed that when the connection density is relatively small, the red and blue lines calculated according to Equation (6) change obviously, which means the importance of each oscillator is relatively different. And the value of dynamic survivability centrality will become closer accompanied by the increase in the connection density, which is shown in green curve. That elucidates that the reduction in connection density of the system diminishes the significant disparities among oscillators. The reason behind this phenomenon is that as the connection density of the ER network increases, it approaches a globally coupled network, resulting in a more uniform distribution of oscillator degrees and closer proximity of oscillators within the system.

Subsequently, to visually demonstrate the crucial significance of each oscillator within the system, we plotted the importance ranking of oscillators by color in Figure 5 when the connection density is $d = 0.08$ according to the normalized dynamic survivability centrality in Figure 4. The colors in Figure 5a change gradually from dark blue, to blue, green, brown, yellow, then light yellow, indicating that oscillators play an increasingly important role. The system diagram in Figure 5b shows that the more important oscillator is, the darker the color is.

In Section 2, based on the change in the second-largest eigenvalue of the coupling matrix, the importance of oscillators in the dynamic survivability of the coupled nonlinear oscillator system is quantified, i.e., Equations (5) and (6). The approximate forms (10) and (11) are obtained by simplified derivation of the elements in the equation, which are used to calculate the dynamic survivability centrality theoretically, and also show the dependence on the topology of the system. Since the approximation process will inevitably bring errors, we consider the error value obtained by the difference between (6) and (11). And the error distribution of dynamic survivability centrality derived from the theory and the results obtained from the real numerical simulation is analyzed in Figure 6. The results show that the error values of all oscillators are less than 0.03, and most error values are even less than 0.015, showing the reliability of the results.

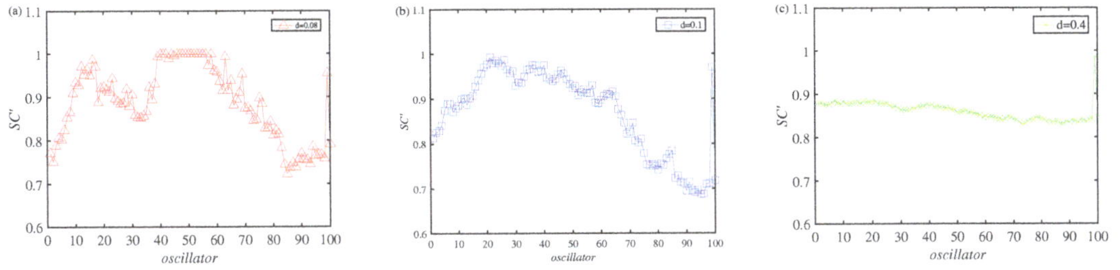


**Figure 4.** The dynamic survivability centrality of the coupled nonlinear oscillator system with ER structures under different connection densities ($N = 100$), which reflects the influence of connection density. (**a**) $d = 0.08$. (**b**) $d = 0.1$. (**c**) $d = 0.4$

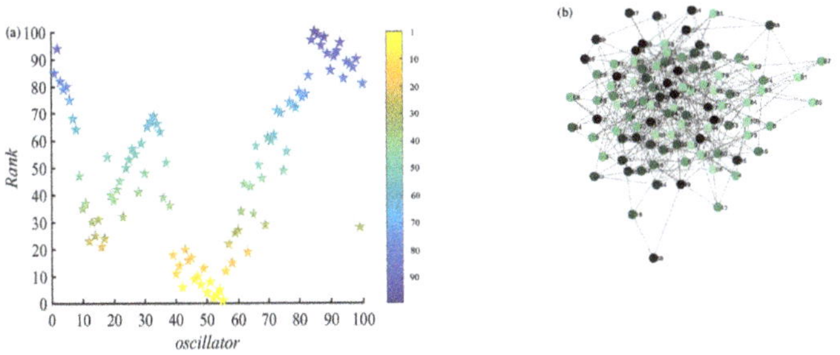


**Figure 5.** Visualizing the importance ranking of oscillators in the ER system with $N = 100$ and $d = 0.08$ according to the dynamic survivability centrality. (**a**) Order of importance of each oscillator. (**b**) Schematic of the system.

Using the variance of second-largest eigenvalue of the coupling matrix $\lambda_2$, we test the performance of dynamic survivability centrality. However, the synchronization performance of the dynamic system can be directly reflected by judging whether the following formula is equal to 0 from the system state point of view,

$$Z = \frac{\sum_{j=1}^{N} \left( \|x_j - x_1\|_2 + \|y_j - y_1\|_2 \right)}{N}. \tag{12}$$

Condition (12) equals 0, which implies that the state of all oscillators is equal to the first oscillator, i.e., the system is in a synchronous state. Therefore, the correlation between the change in $Z$ when each oscillator is removed and its centrality value can be compared, thereby reaffirming the validity of the proposed index in quantifying an oscillator's influence on dynamic survivability. The results shown in Figure 7 compare the distribution of the change in $Z$ when each oscillator is removed with the distributions of several types of centrality indexes. It is observed that the dynamic survivability centrality distribution is most similar to that of $\Delta Z$, demonstrating the accuracy of the proposed dynamic survivability centrality in describing the critical significance of system oscillators. Numerical integrations in this work were performed for random initial conditions by means of the fourth order Runge–Kutta method with time step 0.1.

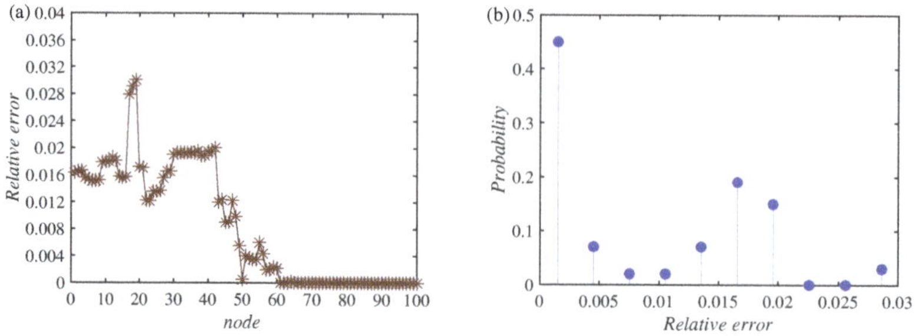


**Figure 6.** Error analysis of dynamic survivability centrality ($N = 100$, $d = 0.08$), embodying the correctness of theoretical derivation. (**a**) Relative error value corresponding to each oscillator. (**b**) Distribution of errors.

At the end of this section, the impact mechanism of system topology parameters on the dynamic survivability centrality is investigated. We visualize Equation (11) by contour plots in Figure 8. An increase in connection density $d^i$ calculated after attacking each oscillator leads to a smaller value of normalized dynamic survivability centrality $SC'$ with fixed initial connection density $d^0$. It tells us that connection density $d^i$ has a weakening effect on dynamic survivability centrality. The reason is that the larger the $d^i$ is, the closer it approaches $d^0$, indicating a minimal change in system connection density upon removal of oscillator $i$. This implies that the effect on the dynamic survivability is weak, so the oscillator is of little importance, thereby causing a small value of $SC'$. Additionally, the increase in system size will promote the dynamic survivability centrality, albeit the magnitude of this beneficial impact is relatively small exhibited by flat growth curve. While the impact of initial connection density $d^0$ on dynamic survivability centrality depends on the value of $d^i$. Specifically, $d^0$ enhances the dynamic survivability centrality when $d^i$ is high, while $SC'$ remains unchanged when $d^i$ is low.

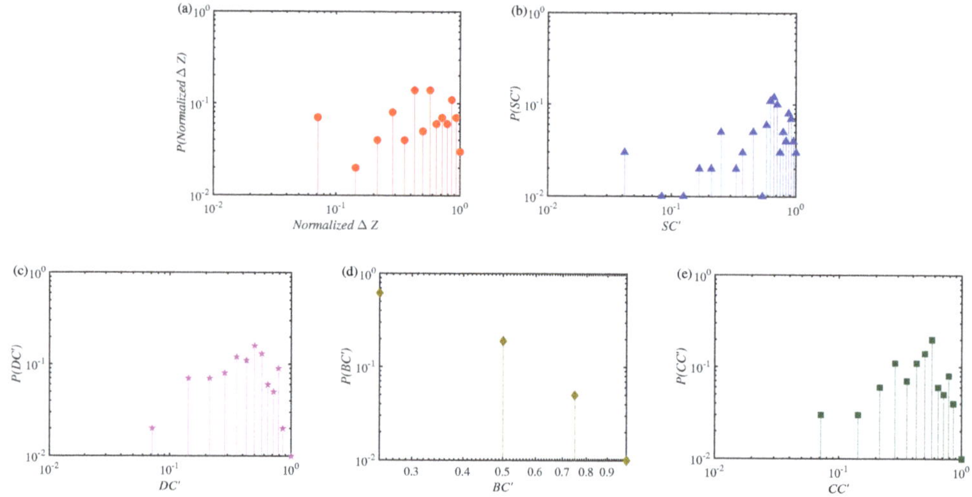


**Figure 7.** The distribution diagrams of state of system and centrality indexes ($N = 100$, $d = 0.1$), which prove the superiority of dynamic survivability centrality. (**a**) Normalized distribution of $\Delta Z$. (**b**) Normalized distribution of dynamic survivability centrality. (**c**) Normalized distribution of degree centrality. (**d**) Normalized distribution of closeness centrality. (**e**) Normalized distribution of betweenness centrality.

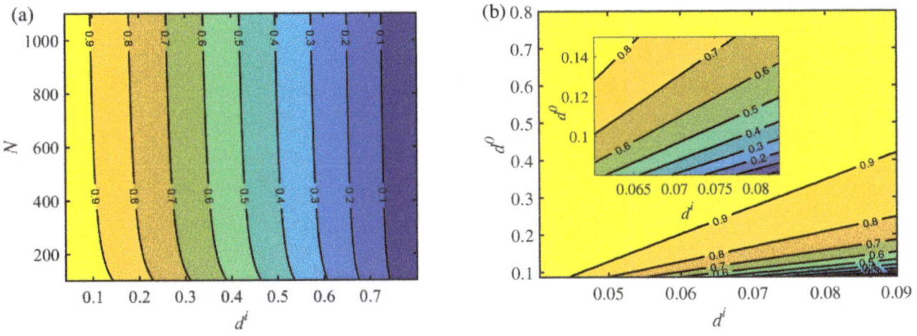

**Figure 8.** The influence of system topology parameters on the dynamic survivability centrality. (**a**) $\left(d^i, N\right)$ parameter plane contour plot ($d^0 = 0.8$). (**b**) $\left(d^i, d^0\right)$ parameter plane contour plot ($N = 100$).

## 4. Conclusions

In this paper, the ability of each oscillator in the coupled Stuart–Landau oscillator system with Erdős–Rényi random graph network topology to affect the dynamic survivability is analyzed, and the maintenance of symmetry phenomenon synchronization is considered as the key mission to be completed. Inspired by the study of dynamic survivability, a new centrality index called dynamic survivability centrality is constructed, which is determined by the degree of oscillator and the size of system. The proposed index compensates for the limitation of traditional indices that solely consider network structure, thereby enabling a more accurate assessment of node influence on the normal operation of dynamic systems.

In the case of a ten-oscillator ER system, the impact of each oscillator on the centrality of dynamic survivability is demonstrated. It is found that the theoretical results agree well with the numerical results, and the performance of the proposed dynamic survivability centrality is better than other classical centrality indexes, which proves the effectiveness of our method. Further study on the system with $N = 100$ shows that the value of dynamic survivability centrality will become closer when the connection density continues to increase. And the results of error distribution indicate that both the numerical value and theoretical value have errors below 0.03, thereby verifying the reliability and correctness of the constructed index. Moreover, the distribution of dynamic survivability centrality is the most similar with the distribution of change in system state caused by removing each oscillator in contrast to other centrality indexes. This finding further proves that the dynamic survivability centrality index has a good performance in measuring the influence of oscillators on dynamic survivability research.

Finally, our results indicate that the size of system and initial degree of oscillator have positive effects on dynamic survivability centrality, while the degree of oscillator after removing the oscillator has opposite effects.

These studies could provide a theoretical basis for protecting the key oscillators of the system to design a survivable system to complete missions, which assumes paramount significance in preserving the normal dynamic behaviors for uninterrupted service provision even under adverse conditions, thus constituting an indispensable determinant within the realm of complex network system security. Subsequent research will investigate different dynamic phenomena as key missions, exploring diverse network topologies and providing targeted protection without the need to identify key oscillators (nodes).

**Author Contributions:** Conceptualization, Y.W.; methodology, Y.W. and Z.S.; validation, Y.W., S.Y. and T.Z.; formal analysis, Y.W. and X.Z.; resources, Z.S. and T.Z.; data curation, Y.W.; writing—original draft preparation, Y.W.; writing—review and editing, Y.W., Z.S. and W.X.; visualization, Y.W.; supervision, Z.S., S.Y. and T.Z.; project administration, Z.S.; funding acquisition, Z.S. All authors have read and agreed to the published version of the manuscript.

**Funding:** This research was funded by the National Natural Science Foundation of China (Grant Nos. 12472032, 12272295).

**Data Availability Statement:** The original contributions presented in this study are included in the article. Further inquiries can be directed to the corresponding authors.

**Conflicts of Interest:** The authors declare no conflicts of interest.

## References


1. Leiner, B.M.; Cerf, V.G.; Clark, D.D.; Kahn, R.E.; Kleinrock, L.; Lynch, D.C.; Postel, J.; Roberts, L.G.; Wolff, S. A brief history of th Internet. *Comput. Commun. Rev.* **2009**, *39*, 22–31. [CrossRef]
2. Sun, K. Complex networks theory: A new method of research in power grid. In Proceedings of the 2005 IEEE/PES Transmissio & Distribution Conference & Exposition: Asia and Pacific, Dalian, China, 18 August 2005; pp. 1–6.
3. Wang, X.F.; Chen, G.R. Synchronization in scale-free dynamical networks: Robustness and fragility. *IEEE Trans. Circuits Sys* **2002**, *49*, 54–62. [CrossRef]
4. Wang, X.F.; Chen, G.R. Synchronization in small-world dynamical networks. *Int. J. Bifurcat. Chaos* **2002**, *12*, 187–192. [CrossRe
5. Mikaberidze, G.; Chowdhury, S.N.; Hastings, A.; D'Souza, R.M. Consensus formation among mobile agents in networks c heterogeneous interaction venues. *Chaos Solitons Fractals* **2024**, *178*, 114298. [CrossRef]
6. Albert, R.; Jeong, H.; Barabási, A. Error and attack tolerance of complex networks. *Nature* **2000**, *406*, 378–382. [CrossRe [PubMed]
7. Gao, J.X.; Barzel, B.; Barabási, A. Universal resilience patterns in complex networks. *Nature* **2016**, *530*, 307–312. [CrossRe [PubMed]
8. Sun, Z.K.; Wang, Y.X. Dynamic survivability in oscillator systems. *Phys. Scr.* **2023**, *98*, 095208. [CrossRef]
9. Ellison, R.J.; Linger, R.C.; Longstaff, T.; Mead, N.R. Survivable network system analysis: A case study. *IEEE Softw.* **1999**, *16*, 70–7 [CrossRef]
10. Holme, P.; Kim, B.J.; Yoon, C.N.; Han, S.K. Attack vulnerability of complex networks. *Phys. Rev. E* **2002**, *65*, 056109. [CrossRe [PubMed]
11. Yin, Y.; Liu, Q.; Zhang, C.Y.; Zhou, J. Survivability analysis of weighted-edge attacks on complex networks with incomplet information. *Phys. A* **2019**, *531*, 120957. [CrossRef]
12. Ingale, M.; Shekatkar, S.M. Resource dependency and survivability in complex networks. *Phys. Rev. E* **2020**, *102*, 06230 [CrossRef]
13. Wang, Y.X.; Sun, Z.K.; Liu, S.T.; Zhou, Y.N.; Xu, W. Dynamic Survivability in Nonlinear Oscillation Systems with Attractive Repulsive Interaction. *Int. J. Bifurc. Chaos* **2023**, *33*, 2350049. [CrossRef]
14. Wang, Y.X.; Sun, Z.K.; Zhang, H.Q.; Liu, S.T.; Xu, W. Dynamic survivability of two-layer networks with different topologies. *Eu Phys. J. Plus* **2024**, *139*, 94. [CrossRef]
15. Wang, Y.X.; Sun, Z.K.; Zhang, H.Q.; Zhou, Y.N.; Liu, S.T.; Xu, W. Dynamic survivability of two-layer networks: The role o interlayer coupling. *Chaos Solitons Fractals* **2024**, *180*, 114571. [CrossRef]
16. Motter, A.E. Cascade control and defense in complex networks. *Phys. Rev. Lett.* **2004**, *93*, 098701. [CrossRef] [PubMed]
17. Zhou, T.; Wang, B.H. Catastrophes in scale-free networks. *Chin. Phys. Lett.* **2005**, *22*, 1072.
18. Pastor, S.R.; Vespignani, A. Immunization of complex networks. *Phys. Rev. E* **2002**, *65*, 036104. [CrossRef] [PubMed]
19. Zhao, M.; Zhou, T.; Wang, B.H.; Wang, W.X. Enhanced synchronizability by structural perturbations. *Phys. Rev. E* **2005**, *72*, 05710 [CrossRef]
20. Zemanová, L.; Zhou, C.S.; Kurths, J. Structural and functional clusters of complex brain networks. *Phys. D* **2006**, *224*, 202–21 [CrossRef]
21. Zamora, L.; Zhou, C.S.; Kurths, J. Cortical hubs form a module for multisensory integration on top of the hierarchy of cortica networks. *Front. Neuroinform.* **2010**, *4*, 613.
22. Knoke, D.; Yang, S. *Social Network Analysis*; Sage: Thousand Oaks, CA, USA, 2008; Number 154.
23. David, K.; BDaniel J, B. Intraorganizational networks. In *Advances in Social Network Analysis*; Sage Publishing: Thousand Oak CA, USA, 1994; pp. 207–229.
24. Sparrowe, R.T.; Liden, R.C.; Wayne, S.J.; Kraimer, M.L. Social networks and the performance of individuals and groups. *Acac Manag. J.* **2001**, *44*, 316–325. [CrossRef]
25. Hochberg, Y.V.; Ljungqvist, A.; Lu, Y. Whom you know matters: Venture capital networks and investment performance. *J. Finan* **2007**, *62*, 251–301. [CrossRef]
26. Bolland, J.M. Sorting out centrality: An analysis of the performance of four centrality models in real and simulated network *Soc. Netw.* **1988**, *10*, 233–253. [CrossRef]
27. Lü, L.Y.; Chen, D.B.; Ren, X.L.; Zhang, Q.M.; Zhang, Y.C.; Zhou, T. Vital nodes identification in complex networks. *Phys. Rep* **2016**, *650*, 1–63. [CrossRef]
28. Beauchamp, M.A. An improved index of centrality. *Behav. Sci.* **1965**, *10*, 161–163. [CrossRef] [PubMed]
29. Sabidussi, G. The centrality index of a graph. *Psychometrika* **1966**, *31*, 581–603. [CrossRef]
30. Freeman, L.C. A set of measures of centrality based on betweenness. *Sociometry* **1977**, *40* , 35–41. [CrossRef]

Hage, P.; Harary, F. Eccentricity and centrality in networks. *Soc. Netw.* **1995**, *17*, 57–63. [CrossRef]
Bonacich, P. Power and centrality: A family of measures. *Am. J. Sociol.* **1987**, *92*, 1170–1182. [CrossRef]
Bonacich, P.; Lloyd, P. Eigenvector-like measures of centrality for asymmetric relations. *Soc. Netw.* **2001**, *23*, 191–201. [CrossRef]
Martin, T.; Zhang, X.; Newman, M.E.J. Localization and centrality in networks. *Phys. Rev. E* **2014**, *90*, 052808. [CrossRef] [PubMed]
Zhang, Y.T.; Shao, C.Q.; He, S.B.; Gao, J.X. Resilience centrality in complex networks. *Phys. Rev. E* **2020**, *101*, 022304. [CrossRef]
Alzabut, J.; Dhineshbabu, R.; Selvam, A.G.M.; Gómez-Aguilar, J.F.; Khan, H. Existence, uniqueness and synchronization of a fractional tumor growth model in discrete time with numerical results. *Results Phys.* **2023**, *54*, 107030. [CrossRef]
Nag Chowdhury, S.; Rakshit, S.; Hens, C.; Ghosh, D. Interlayer antisynchronization in degree-biased duplex networks. *Phys. Rev. E* **2023**, *107*, 034313. [CrossRef]
Pikovsky, A.; Rosenblum, M.; Kurths, J. *Synchronization;* Camb. Univ. Press: Cambridge, UK, 2001; Volume 12.
Vlasov, V.; Zou, Y.; Pereira, T. Explosive synchronization is discontinuous. *Phys. Rev. E* **2015**, *92*, 012904. [CrossRef] [PubMed]
Liu, S.T.; Sun, Z.K.; Zhao, N.N.; Xu, W. Explosive transition in coupled oscillators through mixed attractive-repulsive interactions. *Int. J. Bifurcat. Chaos* **2022**, *32*, 2250018. [CrossRef]
Liu, S.T.; Sun, Z.K.; Yan, L.Y.; Zhao, N.N.; Xu, W. Explosive behaviors on coupled fractional-order system. *Nonlinear Dyn.* **2022**, *110*, 2741–2751. [CrossRef]
Boccaletti, S.; Almendral, J.A.; Guan, S.; Leyva, I.; Liu, Z.; Sendiña-Nadal, I.; Wang, Z.; Zou, Y. Explosive transitions in complex networks' structure and dynamics: Percolation and synchronization. *Phys. Rep.* **2016**, *660*, 1–94. [CrossRef]
Liu, Y.Y.; Sun, Z.K.; Yang, X.L.; Xu, W. Dynamical robustness and firing modes in multilayer memristive neural networks of nonidentical neurons. *Appl. Math. Comput.* **2021**, *409*, 126384. [CrossRef]
Pecora, L.M.; Carroll, T.L. Master stability functions for synchronized coupled systems. *Phys. Rev. Lett.* **1998**, *80*, 2109. [CrossRef]
Dhineshbabu, R.; Alzabut, J.; Selvam, A.G.M.; Etemad, S.; Rezapour, S. Modeling and Qualitative Dynamics of the Effects of Internal and External Storage device in a Discrete Fractional Computer Virus. *Qual. Theor. Dyn. Syst.* **2024**, *23*, 182. [CrossRef]
Wigner, E.P. On the distribution of the roots of certain symmetric matrices. *Ann. Math.* **1958**, *67*, 325–327. [CrossRef]
Wigner, E.P. Characteristic vectors of bordered matrices with infinite dimensions II. In *The Collected Works of Eugene Paul Wigner: Part A: The Scientific Papers;* Springer: Berlin/Heidelberg, Germany, 1993; pp. 541–545.
Wigner, E.P. Characteristic vectors of bordered matrices with infinite dimensions i. In *The Collected Works of Eugene Paul Wigner: Part A: The Scientific Papers;* Springer: Berlin/Heidelberg, Germany, 1993; pp. 524–540.

entropy

MDPI

*Article*

# Equilibrium Control in Uncertain Linear Quadratic Differential Games with V-Jumps and State Delays: A Case Study on Carbon Emission Reduction

**Zhifu Jia**

School of Mathematics and Sciences, Suqian University, Suqian 223800, China; jzflzbx@nuaa.edu.cn

**Abstract:** Uncertainty, time delays, and jumps often coexist in dynamic game problems due to the complexity of the environment. To address such issues, we can utilize uncertain delay differential equations with jumps to depict the dynamic changes in differential game problems that involve uncertain noise, delays, and jumps. In this paper, we first examine a linear quadratic differential game optimistic value problem within an uncertain environment characterized by jumps and delays. By applying the $Z(x,y)$ transform, we convert the infinite-dimensional problem into a finite-dimensional one. We then demonstrate that the condition for the existence of a Nash equilibrium strategy is equivalent to the existence of solutions to two cross-coupled matrix Riccati equations. Furthermore, we explore the saddle point equilibrium strategy of the linear quadratic differential game optimistic value model and derive the corresponding saddle point equilibrium solution. Finally, we apply our results to solve a carbon emission reduction game problem.

**Keywords:** differential game optimistic value problems; Nash equilibrium; saddle point equilibrium

## 1. Introduction

The uncertainty of experts' experience, particularly in relation to belief degrees [1] as expressed by experts, is a significant factor when dealing with phenomena that are indeterminate. This uncertainty can have profound implications on decision making and modeling processes, especially when the phenomena in question do not have clear, definitive outcomes. In addressing such indeterminate scenarios, incorporating a measure of this uncertainty becomes essential for more accurate analysis and predictions. To effectively capture the dynamics of phenomena that are uncertain in nature, particularly with respect to varying belief degrees, Liu [2] introduced a novel mathematical framework known as the uncertain differential equation, which is grounded in what is now called the Liu process. This process provided a structured way to model the uncertainty inherent in these phenomena. Subsequent research [3] expanded upon Liu's initial proposal, exploring and deepening the understanding of the properties and behaviors of the Liu process, and allowing for more robust applications by uncertain differential systems driven by it. Uncertain differential equations (UDEs) and stochastic differential equations (SDEs) both deal with differential systems that exhibit some degree of randomness or uncertainty. However, they differ in their mathematical formulation and interpretation. SDEs utilize probability theory and Brownian motion to model randomness, effectively capturing the intrinsic variability and unpredictability inherent in certain phenomena. In contrast, UDEs utilize uncertainty theory, which is based on a non-additivity measure called belief degree, to describe systems with inherent indeterminacy that cannot be fully characterized by probabilistic methods. This distinction makes UDEs particularly suitable for situations where traditional probabilistic models may be inadequate or inappropriate.

In later studies, uncertain delay differential equations emerged as a significant area of interest. These equations, which incorporate time delays into uncertain systems, offered a more nuanced approach to modeling scenarios where the effects of a current state are delayed and manifest over time. This complexity attracted attention for its realism in capturing delayed responses in uncertain environments. In uncertain delay differential equations, "delay" signifies that the rate of change of the dependent variable at any time depends not only on its current value but also on a prior state. The delay represents processes where present dynamics are influenced by past states, a crucial feature for modeling temporal dependencies in control systems. Research on uncertain delay differential equations concentrated on four main aspects: solution existence, solution uniqueness, system stability under uncertainty and delays, and potential applications. Understanding these characteristics was vital to establishing whether solutions were reliable, how they evolved, and how systems could be controlled to achieve desired outcomes in the presence of uncertainty and delays. Addressing these phenomena, Barbacioru [4] utilized uncertainty functional differential equations to model volatility in financial markets, particularly for dynamic systems influenced by past states. Ge et al. [5] established existence and uniqueness theorems for uncertain delay differential equations under specific conditions, forming a theoretical foundation that ensured meaningful and interpretable solutions. Building on prior work, Wang et al. [6,7] introduced concepts such as stability in measure, mean, and p-th moment to describe the stability of uncertain delay differential equations, while Jia et al. [8] focused on stability in distribution. In this context, stability indicates whether the system remains predictable and well behaved over time, even with uncertainty. Their contributions included stability theorems that clarified conditions for system integrity despite fluctuating degrees of uncertainty. Chen, Zhu [9], and Jiang [10] investigated optimal control in uncertain systems with time delays, excluding sudden jumps. They employed expected value criteria, a common method where decisions are based on the weighted average of possible outcomes. Their research sought optimal control strategies to minimize risks and maximize desired outcomes, even in the face of uncertainty and delays.

It is widely recognized that uncertain processes can be subject to sudden, unpredictable changes brought on by emergencies or crises. These events can include economic downturns, pandemics, natural disasters like earthquakes, or geopolitical events such as wars. Such unexpected occurrences can drastically alter the behavior of a system, introducing a new layer of complexity and requiring models that can accommodate these sudden shifts. In response to the need for models that could account for abrupt changes in uncertain systems, Deng et al. [11] introduced the concept of $V$-jumps. These jumps represent significant, discrete changes in the state of the system that occur unpredictably. The $V$-jumps framework has since been widely adopted as an effective tool for describing the evolution of phenomena that experience sudden shifts in response to emergencies or other unforeseen events. Furthering the theoretical foundation of $V$-jumps, Deng et al. [12] provided mathematical proof of the existence and uniqueness of solutions to uncertain differential equations that incorporate $V$-jumps. This result was crucial because it ensured that the models used to describe systems with jumps could produce reliable and consistent outcomes, even when dealing with abrupt changes in system behavior. Following the development of uncertain differential equations with $V$-jumps, these mathematical models became widely adopted in the field of optimal control, particularly in situations where sudden, unpredictable changes, or jumps, occur within a system. The application of $V$-jumps allowed for more realistic and nuanced modeling of systems that are prone to abrupt shifts in state, making them especially useful in optimizing control strategies under uncertain conditions. One notable example of this application is found in the work of Deng [13–15], who explored uncertain optimal control models incorporating jumps. Deng also extended this investigation to linear quadratic models with jumps, all under the expected value evaluation criteria. These models were designed to minimize the cost or maximize the efficiency of control in systems where both linear dynamics and sudden jumps played significant roles, with the expected value approach providing a means of balancing the trade offs between different outcomes. Further extending the reach of uncertain jump models, Chen et al. [16] studied the optimal control of uncertain uncertain dynamic systems with jumps, applying their findings to practical scenarios such as advertising models. This work demonstrated the usefulness of such models in optimizing strategies for dynamic, unpredictable environments like the advertising market, where shifts in consumer behavior or market conditions can cause sudden changes in outcomes. In general terms, the expected value model is an effective measurement tool when assessing uncertain systems. By using the concept of a weighted average, this model provides a pragmatic way to account for various possible outcomes, assigning each one a belief degree and calculating the expected or average result. This approach works particularly well in situations where the range of possible outcomes is not extreme, and where a balanced, middle-ground prediction is appropriate for decision making. However, in cases where data polarization, meaning the outcomes are highly skewed or concentrated around extreme values, becomes a significant factor, or when other unique characteristics of uncertain variables must be considered, the optimistic value model offers a more suitable alternative. Unlike the expected value model, which seeks a middle ground, the optimistic value model focuses on the most favorable outcomes, making it more appropriate when decision makers prioritize the best case scenarios in uncertain environments. In contrast to the expected value approach, Deng and Chen [17,18], along with Deng et al. [19], conducted extensive research into optimistic value models within the context of uncertain optimal control with jumps. Their work aimed to explore how optimal control strategies could be devised when the focus shifts from an average, balanced outcome to a more optimistic outlook, prioritizing the potential for the best possible results despite the presence of uncertainties and sudden jumps in the system.

Building on the foundation of uncertain optimal control, Sun et al. [20] and Li et al. [21] expanded their research to address differential game problems under uncertainty, focusing specifically on the optimistic value criterion. Their study examined a counter-terror economic game model in an uncertain setting, though without the added complexity of jumps or delays. This work aimed to apply the optimistic value criterion in game theory, enabling strategic decision making in scenarios dominated by uncertainty. Jia et al. [22,23] subsequently broadened the scope by investigating uncertain differential games involving multiple factors and, notably, uncertain jumps in a multidimensional context. Their research offered significant insights into modeling and control of strategic interactions where multiple dimensions and abrupt changes introduce further unpredictability. In the realm of uncertain stochastic jump systems, Jia and Liu [24] explored a hybrid differential game system with V-n jumps, a form of sudden shift in the system. They introduced equilibrium strategies such as saddle point equilibria, indicating states where neither player can unilaterally gain an advantage. This theoretical advancement was applied to a practical case of an advertising duopoly game, where two firms compete in a market marked by stochastic elements and sudden shifts. Chen et al. [25] further advanced the field by analyzing optimal control and zero-sum games in systems characterized by multifactor uncertainty, where each factor could experience sudden jumps. Their research tackled these challenges, presenting strategies to optimize outcomes in highly unpredictable conditions. Besides their work on differential games, Jia et al. [26] contributed to the mathematical foundation by proving theorems on the existence and uniqueness of solutions for uncertain delay differential equations with jumps, ensuring that these models are mathematically robust under specified conditions. Real-world differential game problems are often characterized by combinations of uncertainty, time delays, and abrupt events. This complex setting requires adaptive models that account for uncertain, delayed, and sudden occurrences. To address these complex game problems, we will employ uncertain delay differential equations with jumps. These equations capture the dynamics of differential games influenced by Liu noises, time delays, and abrupt changes, all under the optimistic value criterion. This approach enables a focus on favorable outcomes in a highly uncertain environment, allowing for more strategic and optimistic decision making.

The contributions of this paper are: (1) to obtain the Nash equilibrium of an uncertain delay linear quadratic differential game optimistic value model with a jump; (2) to obtain the saddle equilibrium of an uncertain delay linear quadratic differential game optimistic value model with a jump. According to different optimization purposes, the expected value and optimistic value are different evaluation criterion. The results [9] of the expected model for uncertain control systems cannot be directly used for an optimistic value model. Compared with the models in Ref. [9], we extend the results to the uncertain delay linear quadratic differential game optimistic value model with a jump. Compared with the models in Ref. [20], we introduce jump factors and delays, and extend the model to the multidimensional case. In addition, we also consider Nash equilibrium solutions, and the optimistic value models in our paper are more complex and more practical due to the existences of jumps and delays; and (3) to apply the model to solve a carbon emission reduction game problem.

The structure of this paper is organized as follows. Section 2 will collect some preliminary results including the $\beta$-optimistic value, Liu process, V-jump process, and so on, which are essential for our analysis. In Section 3, we pose a linear quadratic differential game optimistic value problem under an uncertain jump and delay environment. By $Z(x,y)$ transform, the infinite dimension problem is transformed into a finite dimension; then, we obtain the existence condition of the Nash equilibrium strategy that is equivalent to the existence solution of two cross-coupled matrix Riccati equations. Section 4 is devoted to the study of the saddle point equilibrium problem of the uncertain delay linear quadratic differential game optimistic value model with a jump. The application is given to show the validity of the obtained results in Section 5.

## 2. Preliminaries

In this section, we review some preliminaries including the $\beta$-optimistic value, Liu process, V-jump process, and so on.

**Definition 1** ([4])**.** *Let $\xi$ be an uncertain variable, $\mathcal{M}$ be an uncertain measure, and the confidence level $\beta \in (0,1]$. Then, $\xi_{\sup}(\beta) = \sup\{r|\mathcal{M}\{\xi \geq r\} \geq \beta\}$ is called the $\beta$-optimistic value to $\xi$; and $\xi_{\inf}(\beta) = \inf\{r|\mathcal{M}\{\xi \leq r\} \geq \beta\}$ is called the $\beta$-pessimistic value to $\xi$.*

**Definition 2** ([4])**.** *An uncertain process $C_t$ is a Liu process with respect to time t if it satisfies the following:*

*(a)* $C_0 = 0$ *and almost all sample paths are Lipschitz continuous;*
*(b)* $C_t$ *is in independent increments and stationary;*
*(c)* *Every increment $C_{s+t} - C_s$ is a normal uncertain variable with expected value 0 and variance $t^2$, and its uncertainty distribution is*

$$\Phi(x) = \left(1 + \exp\left(\frac{-\pi x}{\sqrt{3}t}\right)\right)^{-1}, \quad x \in R.$$

**Remark 1.** *Liu [27] challenged the widely accepted presumption of Ito's calculus in 2013, and believed that Liu's calculus is a potential tool for the stock market. For example, Liu et al. [28] considered the real Alibaba stock prices from 1 January 2019 to 30 June 2020. Ye et al. [29] considered USD–CNY exchange rates from 1 October 2019 to 30 June 2021. The experiments showed that the residuals are far from frequency stability, and should not be regarded as random variables, but rather as uncertain variables. At the same time, the uncertain differential equations driven by the Liu process were very suitable for simulating these two real cases.*

The definition of $V$-jump uncertain process is as follow:

**Definition 3** ([11])**.** *A V-jump uncertain process with parameters $\vartheta_1$ and $\vartheta_2$ $(0 < \vartheta_1 < \vartheta_2 < 1)$ for $t \geq 0$ is defined if*

*(i)* $V_0 = 0$;
*(ii)* $V_t$ *has stationary and independent increments;*

(iii) *For any given time $t > 0$, every increment $V_{s+t} - V_s$ is a Z jump uncertain variable $\xi \sim Z(\vartheta_1, \vartheta_2, t)$ for $\forall s > 0$, whose uncertainty distribution is*

$$\Phi(x) = \begin{cases} 0 & if\ x < 0, \\ \dfrac{2\vartheta_1}{t}x & if\ 0 \le x < \dfrac{t}{2}, \\ \vartheta_2 + \dfrac{2(1-\vartheta_2)}{t}(x - \dfrac{t}{2}) & if\ \dfrac{t}{2} \le x < t, \\ 1 & if\ x \ge t. \end{cases}$$

**Definition 4** ([12]). *Suppose that $C_t$ is an uncertain Liu process with respect to time $t$, $V_t$ is an uncertain V-jump process with respect to time $t$, and $\rho, \varsigma$, and $\nu$ are some given functions about $X_t$ and $t$. Then,*

$$dX_t = \rho(t, X_t)dt + \varsigma(t, X_t)dC_t + \nu(t, X_t)dV_t$$

*is called an uncertain differential equation with a V-jump.*

Deng et al. [19] presented the following general multidimensional uncertain optimal control model with V-n jumps:

$$\begin{cases} J(t,\mathbf{x}) = \sup\limits_{\omega_s \in \Omega} \left\{ \int_t^T F(s, \mathbf{X}_s, \omega_s)ds + h(T, \mathbf{X}_T) \right\}_{\sup} (\beta) \\ \text{s. t.} \\ \quad \mathbf{dX}_s = \rho(s, \mathbf{X}_s, \omega_s)ds + \varsigma(s, \mathbf{X}_s, \omega_s)\mathbf{dC}_s + \nu(s, \mathbf{X}_s, \omega_s)\mathbf{dV}_s \\ \quad \mathbf{X}_t = \mathbf{x}, \end{cases} \tag{1}$$

where $\mathbf{X}_s = (X_{s1}, X_{s2}, \cdots, X_{sn})$ is an n-dimensional state vector with the initial condition $\mathbf{X}_t = x = (x_1, x_2, \cdots, x_n)$ at time $t$, $\omega_s$ is the decision vector of dimension $m$ (represents the control function $\omega(s, \mathbf{X}_s)$ of time $s$ and state $\mathbf{X}_s$) in a domain $\Omega$, $F : [0, T) \times R^n \times R^m \to R$ the objective function, and $h : [0, T) \times R^n \to R$ is the terminal reward function. In addition, $\rho : [0, T) \times R^n \times R^m \to R$ is a column-vector function about $s$, $\mathbf{X}_s$, and $\omega_s$, $\varsigma : [0, T) \times R^n \times R^m \to R^n \times R^l$ and $\nu : [0, T) \times R^n \times R^m \to R^n \times R^l$ are two matrix functions about $s$, $\mathbf{X}_s$, and $\omega_s$, and $\mathbf{C}_s = (C_{s1}, C_{s2}, \cdots, C_{sl})$, $\mathbf{V}_s = (V_{s1}, V_{s2}, \cdots, V_{sl})$, where $C_{s1}, C_{s2}, \cdots, C_{sl}$ are independent Liu processes, $V_{s1}, V_{s2}, \cdots, V_{sl}$ are independent V-n jumps processes, and $C_{si}$ and $V_{sj}$ for any $i, j = 1, 2, \cdots, l (i \ne j)$ are independent, and the final time $T > 0$ is fixed or free.

**Definition 5** ([19]). *Suppose $V_t$ is a V-jump process with parameters $\vartheta_1$ and $\vartheta_2$, and let $\eta = V_t = V_{t+\Delta t} - V_t$. Then, the $\beta$-optimistic value and $\beta$-pessimistic value are defined for any belief degree $\beta \in (0, 1)$*

$$\eta_{\sup}(\beta) = \begin{cases} (1 - \dfrac{\beta}{2(1-\vartheta_2)})\Delta t, & if\ 0 < \beta < 1 - \vartheta_2 \\ \dfrac{\Delta t}{2}, & if\ 1 - \vartheta_2 \le \beta < 1 - \vartheta_1 \\ \dfrac{1-\beta}{2\vartheta_1}\Delta t, & if\ 1 - \vartheta_1 \le \beta < 1, \end{cases}$$

*and*

$$\eta_{\inf}(\beta) = \begin{cases} \dfrac{\beta}{2\vartheta_1}\Delta t, & if\ 0 < \beta \le \vartheta_1 \\ \dfrac{\Delta t}{2}, & if\ \vartheta_1 < \beta \le \vartheta_2 \\ (1 - \dfrac{1-\beta}{2(1-\vartheta_2)})\Delta t, & if\ \vartheta_2 < \beta < 1, \end{cases}$$

*respectively.*

**Lemma 1 ([19]).** *If $J(t,\mathbf{x})$ exists and is twice differentiable on $[0,T]\times R^n$. Then, we have*

$$-J_t = \sup_{\omega_t}\left\{F + \nabla_{\mathbf{x}}J^\tau \rho + \frac{\sqrt{3}}{\pi}\ln\frac{1-\beta}{\beta}\|\nabla_{\mathbf{x}}J^\tau\varsigma\|_1 + k\|\nabla_{\mathbf{x}}J^\tau \nu\|\right\}, \tag{2}$$

*where $J_t(t,\mathbf{x})$ is the partial derivatives of the function $J(t,\mathbf{x})$ in $t$. $\nabla_{\mathbf{x}}J(t,\mathbf{x})$ is the gradient of $J(t,\mathbf{x})$ in $\mathbf{x}$, $\tau$ represents the transpose, and $\|\cdot\|_1$ and $\|\cdot\|$ are the 1-norm for vectors, that is, $\|\cdot\|_1 = \sum_{i=1}^{n}|\kappa_i|$ and $\|\cdot\| = \sum_{i=1}^{n}\kappa_i$ for $\kappa = (\kappa_1,\kappa_2,\cdots,\kappa_k)$, respectively. $\rho,\varsigma,\nu$ are defined the same as Equation (1), and $F,\rho,\varsigma,\nu,J_t,\nabla_{\mathbf{x}}J$ denote $F(t,\mathbf{x},\omega_t)$, $\rho(t,\mathbf{x},\omega_t)$, $\varsigma(t,\mathbf{x},\omega_t)$, $\nu(t,\mathbf{x},\omega_t)$, $J_t(t,\mathbf{x})$, $\nabla_{\mathbf{x}}J(t,\mathbf{x})$, respectively, and*

(1) *if $\nabla_{\mathbf{x}}J^\tau\nu \geq 0$ (all terms are nonnegative), then $k = 1 - \frac{\beta}{2(1-\vartheta_2)}(0 < \beta < 1-\vartheta_2)$ or $k = \frac{1}{2}(1-\vartheta_2 \leq \beta < 1-\vartheta_1)$ or $k = \frac{1-\beta}{2\vartheta_1}(1-\vartheta_1 \leq \beta < 1)$;*

(2) *if $\nabla_{\mathbf{x}}J^\tau\nu < 0$ (all terms are negative), then $k = \frac{\beta}{2\vartheta_1}(0 < \beta \leq \vartheta_1)$ or $k = \frac{1}{2}(\vartheta_1 < \beta \leq \vartheta_2)$ or $k = 1 - \frac{1-\beta}{2(1-\vartheta_2)}(\vartheta_2 < \beta < 1)$.*

**Lemma 2.** *Suppose $C_t$ is a Liu process and $V_t$ is a V-jump process, and let $\psi(t,c,v)$ be a continuously differentiable function. Then, $Y_t = \psi(t,C_t,V_t)$ is an uncertain process and*

$$dY_t = \frac{\partial\psi}{t}(t,C_t,V_t)dt + \frac{\partial\psi}{c}(t,C_t,V_t)dC_t + \frac{\partial\psi}{v}(t,C_t,V_t)dV_t.$$

Similar to the proof of Theorem 4 of Ref. [3], Lemma 2 can be easily obtained.

## 3. Main Results

Real-world differential game problems are often characterized by combinations of uncertainty, time delays, and abrupt events. This complex setting requires adaptive models that account for uncertain, delayed, and sudden occurrences. To address these complex game problems, we consider the following linear uncertain delay differential systems with V-n jumps. We separately outline the system's dynamics, constraints, and objectives for each player in the Nash equilibrium framework.

$$\begin{cases} dX_s = \{A_1(s)X_s + A_2(s)Y_s + A_3\varrho_s + B_1(s)\omega_1(s) + B_2(s)\omega_2(s)\}ds + \{A_4(s)X_s \\ \quad +A_5(s)Y_s\}dC_s + \{A_6(s)X_s + A_7(s)Y_s\}dV_s, \quad s\in[t,T] \\ Y_s = \int_{-d}^{0} e^{\lambda u}X_{s+u}du, \quad \varrho_s = X_{s-d}, \quad s\in[t,T] \\ X_s = \varphi_t(s-t), \quad t-d\leq s\leq t, \end{cases} \tag{3}$$

where the matrices $A_1(s), A_2(s), A_3, B_1(s), B_1(s), A_4(s), A_5(s), A_6(s)$, and $A_7(s)$ are appropriate size matrix functions, state $X_s(s\geq -d)\in R^n$ that started at time $-d<0$, $d$ represents the delay, and $\lambda$ is a constant. $C_t = (C_{t1},C_{t2},\cdots,C_{tl})^\tau$ where $C_{t1},C_{t2},\cdots,C_{tl}$ are independent Liu processes. $V_t = (V_{t1},V_{t2},\cdots,V_{tl})^\tau$ where $V_{t1},V_{t2},\cdots,V_{tl}$ are independent V-jump processes. The feasible strategies of the two players are: $\omega_1(s)\in\Omega_1$ and $\omega_2(s)\in\Omega_2$; $\Omega_1$ is the feasible strategy set of player 1 and $\Omega_2$ is the feasible strategy set of player 2. For a given confidence level $\beta\in(0,1)$, the two players seek their respective optimal feasible strategies to maximize the following cost functions under the constraint of the state Equation (3), namely

$$\begin{aligned} J_1(t,\varphi_t) = \sup_{\omega_1(s)\in\Omega_1}\Big\{&\int_t^T[(X_s + e^{\lambda d}A_3Y_s)^\tau D_1(s)(X_s + e^{\lambda d}A_3Y_s) + \omega_1^\tau(s)R_{11}\omega_1(s) \\ &+ \omega_2^\tau(s)R_{12}\omega_2(s)]ds + (X_T + e^{\lambda d}A_3Y_T)^\tau G_1(X_T + e^{\lambda d}A_3Y_T)\Big\}_{\sup}(\beta), \end{aligned} \tag{4}$$

$$J_2(t,\varphi_t) = \sup_{\omega_2(s)\in\Omega_2} \{\int_t^T [(X_s + e^{\lambda d}A_3Y_s)^\tau D_2(s)(X_s + e^{\lambda d}A_3Y_s) + \omega_1^\tau(s)R_{21}\omega_1(s)$$
$$+ \omega_2^\tau(s)R_{22}\omega_2(s)]ds + (X_T + e^{\lambda d}A_3Y_T)^\tau G_2(X_T + e^{\lambda d}A_3Y_T)\}_{\sup}(\beta), \quad (5)$$

where $D_1, R_{11}, R_{12}, G_1, D_2, R_{21}, R_{22}, G_2$ are appropriate size matrix functions, $D_1, D_2$ are the symmetric nonnegative definite matrix, and $R_{11}, R_{12}, R_{21}, R_{22}$ are the symmetric positive definite matrix. Actually, the cost functions change with the initial time $t$ and the initial state $\varphi_t$ change. Thus, we denote

$$\overline{J}_1(\omega_1(s),\omega_2(s)) = J_1(t,\varphi_t),$$

$$\overline{J}_2(\omega_1(s),\omega_2(s)) = J_2(t,\varphi_t),$$

where $\varphi_t \in C_R[-d,0]$ $(t=0, \varphi_0 \in C_R[-d,0])$ are the given initial functions and $C_R[-d,0]$ represents the space of all continuous functions on $[-d,0]$ taking values in $R$.

The so-called Nash equilibrium problem is to find the feasible control strategies $\omega_1^*(t)$ and $\omega_2^*(t)$ under the system state (3)–(5) constraints such that

$$\overline{J}_1(\omega_1^*(t),\omega_2^*(t)) \geq \overline{J}_1(\omega_1(t),\omega_2^*(t)), \forall \omega_1(t) \in \Omega_1, \quad (6)$$

$$\overline{J}_2(\omega_1^*(t),\omega_2^*(t)) \geq \overline{J}_2(\omega_1^*(t),\omega_2(t)), \forall \omega_2(t) \in \Omega_2. \quad (7)$$

We note that the value functions $J_1(t,\varphi_t)$ and $J_2(t,\varphi_t)$ in (4) and (5) were defined in an infinite-dimensional space, so it is very difficult to solve it directly. We will transform the problem into a finite-dimensional space by constructing a new state. For this purpose, we use the same method as Ref. [9]. For Questions (3)–(7), we denote that

$$\begin{cases} \rho_1(t,x,y,\omega_1,\omega_2) = A_1x + A_2y + B_1\omega_1 + B_2\omega_2, \\ \rho_2(x,y) = A_3, \\ F_i(t,x,y,\omega_1,\omega_2) = (x + e^{\lambda d}A_3y)^\tau D_i(t)(x + e^{\lambda d}A_3y) + \omega_1^\tau R_{i1}\omega_1 + \omega_2^\tau R_{i2}\omega_2, i = 1,2, \\ h_i(t,x,y,\omega_1,\omega_2) = (x_T + e^{\lambda d}A_3y_T)^\tau G_i(x_T + e^{\lambda d}A_3y_T), i = 1,2, \\ \varsigma(t,x,y,\omega_1,\omega_2) = A_4x + A_5y, \\ v(t,x,y,\omega_1,\omega_2) = A_6x + A_7y. \end{cases} \quad (8)$$

**Assumption 1.** *Given the feasible strategy pairs $\omega_1(t) \in \Omega_1$ and $\omega_2(t) \in \Omega_2$, if $(X_s, Y_s)$ follows $dX_s = \rho_1(s,X_s,Y_s,\omega_1,\omega_2)ds + \rho_2(X_s,Y_s)\varrho_s ds + \varsigma(s,X_s,Y_s,\omega_1,\omega_2)dC_s + v(s,X_s,Y_s,\omega_1,\omega_2)dV_s$, there exists an operator $Z(x,y): R^n \times R^k \to R^n$ such that*

$$e^{\lambda d}\Xi_x Z(x,y)\rho_2(x,y) - \Xi_y Z(x,y) = 0, \quad \forall (x,y) \in R^n \times R^k, \quad (9)$$

*where $\Xi_x Z(x,y)$ and $\Xi_y Z(x,y)$ present the Jacobi matrices of $Z(x,y)$ in $x$ and in $y$, respectively.*

*Let $L = R \times y(C_R[-d,0])$. For $\psi \in C_R[-d,0]$, we denote $x(\psi) = \psi(0), y(\psi) = \int_{-d}^0 e^{\lambda s}\mathcal{F}(\psi(s))ds$, and $\varrho(\psi) = \mathcal{F}(\psi(-d))$, where $\mathcal{F}: R^n \to R^k$ is a differentiable function. Then, the new state process $Z_t = Z(X_t,Y_t) \in Z(L)$.*

*Define*

*$\widetilde{\rho}: [0,+\infty) \times R^n \times R^k \times \Omega_1 \times \Omega_2 \to R^n$ by*

$$\widetilde{\rho}(t,x,y,\omega_1,\omega_2) = \Xi_x Z(x,y)\rho_1(t,x,y,\omega_1,\omega_2) + \Xi_y Z(x,y)(\mathcal{F}(x) - \lambda y),$$

*$\widetilde{\varsigma}: [0,+\infty) \times R^n \times R^k \times \Omega_1 \times \Omega_2 \to R^{n\times l}$ by*

$$\widetilde{\varsigma}(t,x,y,\omega_1,\omega_2) = \Xi_x Z(x,y)\varsigma(t,x,y,\omega_1,\omega_2),$$

*and* $\widetilde{v} : [0, +\infty) \times R^n \times R^k \times \Omega_1 \times \Omega_2 \to R^{n\times l}$ *by*

$$\widetilde{v}(t, x, y, \omega_1, \omega_2) = \Xi_x Z(x, y) v(t, x, y, \omega_1, \omega_2).$$

*If the functions* $\widetilde{\rho}$, $\widetilde{\varsigma}$, $\widetilde{v}$, $F_i$, *and* $h_i$ *by* $Z(x, y)$ *transform, then a finite-dimensional model can be obtained.*

**Assumption 2.** *There are functions*

$$\begin{aligned} &\overline{\rho} : [0, +\infty) \times R^n \times \Omega_1 \times \Omega_2 \to R^n, \quad \overline{\varsigma} : [0, +\infty) \times R^n \times \Omega_1 \times \Omega_2 \to R^{n\times l}, \\ &\overline{v} : [0, +\infty) \times R^n \times \Omega_1 \times \Omega_2 \to R^{n\times l}, \overline{F_i} : [0, +\infty) \times R^n \times \Omega_1 \times \Omega_2 \to R, \quad \overline{h_i} : R^n \to R, i = 1, 2. \end{aligned} \tag{10}$$

*such that for all* $t \in [0, T], \omega_1 \in \Omega_1, \omega_2 \in \Omega_2, (x, y) \in R^n \times R^k$, *and it holds that*

$$\begin{aligned} \overline{\rho}(t, Z(x, y), \omega_1, \omega_2) &= \widetilde{\rho}(t, x, y, \omega_1, \omega_2), \quad \overline{\varsigma}(t, Z(x, y), \omega_1, \omega_2) = \widetilde{\varsigma}(t, x, y, \omega_1, \omega_2), \\ \overline{v}(t, Z(x, y), \omega_1, \omega_2) &= \widetilde{v}(t, x, y, \omega_1, \omega_2), \\ \overline{F_i}(t, Z(x, y), \omega_1, \omega_2) &= F_i(t, x, y, \omega_1, \omega_2), \quad \overline{h_i}(Z(x, y)) = h_i(x, y). \end{aligned} \tag{11}$$

*Now, we can transform the infinite-dimensional differential Nash equilibrium problems* (6) *and* (7) *into finite-dimensional differential Nash equilibrium problems by* $Z$. *For* $\varphi_t \in C_R[-d, 0]$, *define* $z = Z(x(\varphi_t), y(\varphi_t)) \in Z(L)$. *Then, for* $t \in [0, T]$, *the Nash equilibrium problem can be transformed into the following* (12)–(14).

$$\begin{cases} \hat{J}_1(\omega_1(t), \omega_2(t)) = \overline{J}_1(t, z) = \sup\limits_{\omega_1(t)\in\Omega_1} \left\{ \int_t^T \overline{F}_1(s, z_s, \omega_1(s), \omega_2(s)) ds + \overline{h}_1(Z_T) \right\}_{\sup}(\beta), \\ \hat{J}_2(\omega_1(t), \omega_2(t)) = \overline{J}_2(t, z) = \sup\limits_{\omega_2(t)\in\Omega_2} \left\{ \int_t^T \overline{F}_2(s, z_s, \omega_1(s), \omega_2(s)) ds + \overline{h}_2(Z_T) \right\}_{\sup}(\beta), \end{cases} \tag{12}$$

$$\begin{cases} \hat{J}_1(\omega_1^*(t), \omega_2^*(t)) \geq \hat{J}_1(\omega_1(t), \omega_2^*(t)), \forall \omega_1(t) \in \Omega_1, \\ \hat{J}_2(\omega_1^*(t), \omega_2^*(t)) \geq \hat{J}_2(\omega_1^*(t), \omega_2(t)), \forall \omega_2(t) \in \Omega_2, \end{cases} \tag{13}$$

$$\begin{cases} dZ_s = \overline{\rho}(s, Z_s, \omega_1(s), \omega_2(s)) ds + \overline{\varsigma}(s, Z_s, \omega_1(s), \omega_2(s)) dC_s + \overline{v}(s, Z_s, \omega_1(s), \omega_2(s)) dV_s, \\ Z_t = z, \\ \omega_1(s) \in \Omega_1, \omega_2(s) \in \Omega_2, \quad s \in [t, T]. \end{cases} \tag{14}$$

**Theorem 1.** *If* $A_2(t) = e^{\lambda d}(\lambda I + A_1(t) + e^{\lambda d} A_3(t)) A_3(t)$, $A_5(t) = e^{\lambda d} A_4(t) A_3(t)$, $A_7(t) = e^{\lambda d} A_6(t) A_3(t)$ *hold in an uncertain delay linear quadratic differential game model with jumps* (3), *then the Nash equilibrium strategies are*

$$\begin{cases} \omega_1^*(t) = -R_{11}^{-1} B_1^\tau Q_1(t) z \\ \omega_2^*(t) = -R_{22}^{-1} B_2^\tau Q_2(t) z \end{cases} \tag{15}$$

*and the optimal value of* (3) *with initial value* $Z_0 = z_0$ *is*

$$J_1(0, \varphi_0) = (z_0^*)^\tau Q_1(0) z_0^*, J_2(0, \varphi_0) = (z_0^*)^\tau Q_2(0) z_0^*$$

*where* $z = x + e^{\lambda d} A_3 y$, *and* $Q_1$ *and* $Q_2$ *satisfy the following Riccati equation.*

$$
\begin{cases}
\dot{Q}_1(t) = -\{D_1 - Q_1^\tau B_1 R_{11}^{-1} B_1^\tau Q_1 + Q_2^\tau B_2 R_{22}^{-1} R_{12} R_{22}^{-1} B_2^\tau Q_2 + Q_1^\tau (A_1 + e^{\lambda d} A_3 + \frac{\sqrt{3}}{\pi} \ln \frac{1-\beta}{\beta} A_4 \\
\quad + kA_6) + (A_1 + e^{\lambda d} A_3 + \frac{\sqrt{3}}{\pi} \ln \frac{1-\beta}{\beta} A_4 + kA_6)^\tau Q_1 - 2Q_1^\tau B_2 R_{22}^{-1} B_2^\tau Q_2\},\ if\ (t,z) \in \Delta_1 \\
\dot{Q}_1(t) = -\{D_1 - Q_1^\tau B_1 R_{11}^{-1} B_1^\tau Q_1 + Q_2^\tau B_2 R_{22}^{-1} R_{12} R_{22}^{-1} B_2^\tau Q_2 + Q_1^\tau (A_1 + e^{\lambda d} A_3 - \frac{\sqrt{3}}{\pi} \ln \frac{1-\beta}{\beta} A_4 \\
\quad + kA_6) + (A_1 + e^{\lambda d} A_3 - \frac{\sqrt{3}}{\pi} \ln \frac{1-\beta}{\beta} A_4 + kA_6)^\tau Q_1 - 2Q_1^\tau B_2 R_{22}^{-1} B_2^\tau Q_2\},\ if\ (t,z) \in \Delta_2 \\
Q_1(T) = G_1 \\
\dot{Q}_2(t) = -\{D_2 - Q_2^\tau B_2 R_{22}^{-1} B_2^\tau Q_2 + Q_1^\tau B_1 R_{11}^{-1} R_{21} R_{11}^{-1} B_1^\tau Q_1 + Q_2^\tau (A_1 + e^{\lambda d} A_3 + \frac{\sqrt{3}}{\pi} \ln \frac{1-\beta}{\beta} A_4 \\
\quad + kA_6) + (A_1 + e^{\lambda d} A_3 + \frac{\sqrt{3}}{\pi} \ln \frac{1-\beta}{\beta} A_4 + kA_6)^\tau Q_2 - 2Q_2^\tau B_1 R_{11}^{-1} B_1^\tau Q_1\},\ if\ (t,z) \in \Delta_1 \\
\dot{Q}_2(t) = -\{D_2 - Q_2^\tau B_2 R_{22}^{-1} B_2^\tau Q_2 + Q_1^\tau B_1 R_{11}^{-1} R_{21} R_{11}^{-1} B_1^\tau Q_1 + Q_2^\tau (A_1 + e^{\lambda d} A_3 - \frac{\sqrt{3}}{\pi} \ln \frac{1-\beta}{\beta} A_4 \\
\quad + kA_6) + (A_1 + e^{\lambda d} A_3 - \frac{\sqrt{3}}{\pi} \ln \frac{1-\beta}{\beta} A_4 + kA_6)^\tau Q_2 - 2Q_2^\tau B_1 R_{11}^{-1} B_1^\tau Q_1\},\ if\ (t,z) \in \Delta_2 \\
Q_2(T) = G_2
\end{cases}
\tag{16}
$$

*where*

(1) *if* $A_6(t) \geq 0$, *then* $k = 1 - \frac{\beta}{2(1-\vartheta_2)} (0 < \beta < 1 - \vartheta_2)$ *or* $k = \frac{1}{2}(1 - \vartheta_2 \leq \beta < 1 - \vartheta_1)$ *or* $k = \frac{1-\beta}{2\vartheta_1}(1 - \vartheta_1 \leq \beta < 1)$;

(2) *if* $A_6(t) < 0$, *then* $k = \frac{\beta}{2\vartheta_1}(0 < \beta \leq \vartheta_1)$ *or* $k = \frac{1}{2}(\vartheta_1 < \beta \leq \vartheta_2)$ *or* $k = 1 - \frac{1-\beta}{2(1-\vartheta_2)}(\vartheta_2 < \beta < 1)$.

$$
\begin{aligned}
\Delta_1 = &\{(t,z) \mid z^\tau Q_i(t) A_4(t) z \geq 0, (t,z) \in [0,T] \times [a,b]^n\}, \\
\Delta_2 = &\{(t,z) \mid z^\tau Q_i(t) A_4(t) z < 0, (t,z) \in [0,T] \times [a,b]^n\}.
\end{aligned}
$$

See Appendix A for the proof process.

## 4. A Linear Quadratic Game Model Under Uncertain Jump and Delay Environment

In two-person uncertain Nash differential games, when the sum of the two players' cost function is zero, i.e., $J_1 = -J_2$, the game degenerates to a two-person zero-sum uncertain differential game problem. In this section, we consider the saddle point problem of a linear quadratic two-person zero-sum game model under an uncertain jump and delay environment. The uncertain jump and delay dynamic system is as follows:

$$
\begin{cases}
dX_s = \{A_1(s)X_s + A_2(s)Y_s + A_3\varrho_s + B_1(s)\omega_1(s) + B_2(s)\omega_2(s)\}ds + \{A_4(s)X_s \\
\quad + A_5(s)Y_s\}dC_s + \{A_6(s)X_s + A_7(s)Y_s\}dV_s, \quad s \in [t,T] \\
Y_s = \int_{-d}^{0} e^{\lambda u} X_{s+u} du, \quad \varrho_s = X_{s-d}, \quad s \in [t,T] \\
X_s = \varphi_t(s-t), \quad t - d \leq s \leq t.
\end{cases}
\tag{17}
$$

In the above equation, $\omega_1 \in \Omega_1[t,T], \omega_2 \in \Omega_2[t,T]$ are the control functions or decision functions. The admissible control sets are defined by $\Omega_i[t,T] = \{\omega_i : [t,T] \to \Omega_i | \omega_i$, which is a measurable function, where $\omega_1 \subset R^p$ and $\omega_2 \subset R^q$ are nonempty closed sets. $X_s, A_1(s), A_2(s), A_3, B_1(s), B_1(s), A_4(s), A_5(s), A_6(s), A_7(s), d, \lambda, \varphi_t, C_t = (C_{t1}, C_{t2}, \cdots, C_{tl})^\tau$, and $V_t = (V_{t1}, V_{t2}, \cdots, V_{tl})^\tau$ are defined the same as in Model (3). For any $0 < t < T$ and some given confidence level $\beta \in (0,1)$, we choose the objective function as follows:

$$
J(t,\varphi_t) = J(\omega_1,\omega_2) = \left\{ \int_t^T F(s, X_s, Y_s, \omega_1, \omega_2) ds + h(X_T, T) \right\}_{\sup} (\beta), \tag{18}
$$

where $F : [0,T] \times R^n \times R^k \times R^p \times R^q \to R$ is the continuous function of state and control, and $h : [0,T] \times R^n \to R$ is the continuous function of terminal reward. If there exists a pair $(\omega_1{}^*, \omega_2{}^*)$ satisfying

$$J(\omega_1{}^*, \omega_2) \leq J(\omega_1{}^*, \omega_2{}^*) \leq J(\omega_1, \omega_2{}^*),$$

then the pair $(\omega_1{}^*, \omega_2{}^*)$ are called the saddle point of the zero-sum uncertain differential game.

We take $J(\omega_1, \omega_2)$ as follows.

$$J(\omega_1, \omega_2) = \left\{ \frac{1}{2} \int_0^T ((X_s + e^{\lambda d} A_3 Y_s)^\tau P(s)(X_s + e^{\lambda d} A_3 Y_s) + \omega_2^\tau G(s)\omega_2 - \omega_1^\tau H(s)\omega_1) ds + X_T^\tau \Psi X_T \right\}_{\sup} (\beta), \quad (19)$$

where $P(s), G(s), H(s), \Psi$ are appropriate size matrix functions, and $P(s)$ is the symmetric nonnegative definite matrix, and $G(s), H(s)$ are the symmetric positive definite matrix.

**Theorem 2.** *For the saddle equilibrium problem (17), if $A_2(t) = e^{\lambda d}(\lambda I + A_1(t) + e^{\lambda d} A_3(t))A_3(t)$, $A_5(t) = e^{\lambda d} A_4(t) A_3(t)$, $A_7(t) = e^{\lambda d} A_6(t) A_3(t)$ hold and $Q(t)$ is the solution of the following Riccati differential Equation (20), then the control strategy pair $(\omega_1^*, \omega_2^*)$ of (21) are the equilibrium solutions of problem (17), and the equilibrium value is*

$$J(0, x_0) = \frac{1}{2} z_0^\tau Q(0) z_0.$$

$$\begin{cases} \dot{Q}(t) + (A_1(t) + e^{\lambda d} A_3)^\tau Q(t) + Q(t)(A_1(t) + e^{\lambda d} A_3) + P(t) + \frac{\sqrt{3}}{\pi} \ln \frac{1-\beta}{\beta} Q(t) A_4(t) \\ + \frac{\sqrt{3}}{\pi} \ln \frac{1-\beta}{\beta} A_4^\tau(t) Q(t) + k Q(t) A_6(t) + k A_6^\tau(t) Q(t) \\ + Q(t) B_1(t) H^{-1}(t) B_1^\tau(t) Q(t) - Q(t) B_2(t) G^{-1}(t) B_2^\tau(t) Q(t) = 0, \quad \text{if } (t,z) \in \Delta_1 \\ \dot{Q}(t) + (A_1(t) + e^{\lambda d} A_3)^\tau Q(t) + Q(t)(A_1(t) + e^{\lambda d} A_3) + P(t) - \frac{\sqrt{3}}{\pi} \ln \frac{1-\beta}{\beta} Q(t) A_4(t) \\ - \frac{\sqrt{3}}{\pi} \ln \frac{1-\beta}{\beta} A_4^\tau(t) Q(t) + k Q(t) A_6(t) + k A_6^\tau(t) Q(t) \\ + Q(t) B_1(t) H^{-1}(t) B_1^\tau(t) Q(t) - Q(t) B_2(t) G^{-1}(t) B_2^\tau(t) Q(t) = 0, \quad \text{if } (t,z) \in \Delta_2 \\ Q(T) = 2\Psi, \end{cases} \quad (20)$$

*and*

$$\omega_1^* = H^{-1}(t) B_1^\tau(t) Q(t) z, \omega_2^* = -G^{-1}(t) B_2^\tau(t) Q(t) z, \quad (21)$$

*where $z = x + e^{\lambda d} A_3 y$,*

(1) *if $A_6(t) \geq 0$, then $k = 1 - \frac{\beta}{2(1-\vartheta_2)} (0 < \beta < 1 - \vartheta_2)$ or $k = \frac{1}{2}(1 - \vartheta_2 \leq \beta < 1 - \vartheta_1)$ or $k = \frac{1-\beta}{2\vartheta_1}(1 - \vartheta_1 \leq \beta < 1)$;*

(2) *if $A_6(t) < 0$, then $k = \frac{\beta}{2\vartheta_1}(0 < \beta \leq \vartheta_1)$ or $k = \frac{1}{2}(\vartheta_1 < \beta \leq \vartheta_2)$ or $k = 1 - \frac{1-\beta}{2(1-\vartheta_2)} (\vartheta_2 < \beta < 1)$.*

$$\Delta_1 = \{(t,z) \mid z^\tau Q(t) A_4(t) z \geq 0, (t,z) \in [0,T] \times [a,b]^n\},$$
$$\Delta_2 = \{(t,z) \mid z^\tau Q(t) A_4(t) z < 0, (t,z) \in [0,T] \times [a,b]^n\}.$$

**Proof.** Since the saddle point equilibrium of the problem (17) is a special case of Nash equilibrium of the problem (3), similar to the proof of Theorem 1, we end here. □

**Remark 2.** *Comparing with the model in Ref. [20], we introduce jump factors and delays, and extend the model to the multidimensional case. In order to obtain $Q(t)$, we need to judge the sign of $z^\tau Q(t) A_4(t) z$, which undoubtedly increases the difficulty of solving the problem. Further, when we know the positive or negative of $Q(t) A_4(t)$, then this problem is easier to solve. Therefore,* *we consider a necessary and sufficient condition for the existence of the saddle point of a special one-dimensional model that can be used in later application.*

$$\begin{cases} \text{Find } (\omega_1^*, \omega_2^*) \text{ such that} \\ J(\omega_1, \omega_2^*) \leq J(\omega_1^*, \omega_2^*) \leq J(\omega_1^*, \omega_2) \\ \text{subject to:} \\ \qquad dX_s = \{A_1(s)X_s + A_2(s)Y_s + A_3\varrho_s + B_1(s)\omega_1(s) + B_2(s)\omega_2(s)\}ds + \{A_4(s)X_s \\ \qquad\qquad +A_5(s)Y_s\}dC_s + \{A_6(s)X_s + A_7(s)Y_s\}dV_s, \quad s \in [0, T] \\ Y_s = \int_{-d}^{0} e^{\lambda u} X_{s+u} du, \quad \varrho_s = X_{s-d}, \quad s \in [0, T] \\ X_s = \varphi_0(s), \quad -d \leq s \leq 0, \end{cases} \tag{22}$$

where

$$J(0, \varphi_0) = J(\omega_1, \omega_2) = \left\{ \frac{1}{2} \int_0^T (P(s)(X_s + e^{\lambda d} A_3 Y_s)^2 + G(s)\omega_2^2 - H(s)\omega_1^2) ds + \frac{1}{2} \Psi_f (X_T + e^{\lambda d} A_3 Y_T)^2 \right\}_{\sup} (\beta). \tag{23}$$

In the above model, $P(s) \geq 0$, $G(s) > 0$, $H(s) > 0$, $A_1(s)$, $A_2(s)$, $B_1(s)$, $B_2(s)$, $A_4(s) > 0$, $A_5(s)$, $A_6(s) \geq 0$, $A_7(s)$ are continuous bounded functions, and $A_3$ and $\Psi_f$ are the given constants. $J(0, \varphi_0)$ represents the equilibrium reward obtainable in $[0, T]$.

We note that the value functions $J(0, \varphi_0)$ in (23) were defined in an infinite-dimensional space, so we will use the idea of Section 3 to transform the problem into a finite-dimensional space by constructing a new state $Z$. For this purpose, we adopt the above Assumption 1 again. For Questions (22) and (23), we denote that

$$\begin{cases} \rho_1(t, x, y, \omega_1, \omega_2) = A_1 x + A_2 y + B_1 \omega_1 + B_2 \omega_2, \\ \rho_2(x, y) = A_3, \\ F(t, x, y, \omega_1, \omega_2) = P(x + e^{\lambda d} A_3 y)^2 + G\omega_2^2 - H\omega_1^2, \\ h(t, x, y, \omega_1, \omega_2) = \frac{1}{2} \Psi_f (x + e^{\lambda d} A_3 y)^2, \\ \varsigma(t, x, y, \omega_1, \omega_2) = A_4 x + A_5 y, \\ v(t, x, y, \omega_1, \omega_2) = A_6 x + A_7 y. \end{cases} \tag{24}$$

Define $\widetilde{\rho} : [0, +\infty) \times R \times R \times \Omega_1 \times \Omega_2 \to R$ by

$$\widetilde{\rho}(t, x, y, \omega_1, \omega_2) = Z_x(x, y)\rho_1(t, x, y, \omega_1, \omega_2) + Z_y(x, y)(\mathcal{F}(x) - \lambda y),$$

$\widetilde{\varsigma} : [0, +\infty) \times R \times R \times \Omega_1 \times \Omega_2 \to R$ by

$$\widetilde{\varsigma}(t, x, y, \omega_1, \omega_2) = Z_x(x, y)\varsigma(t, x, y, \omega_1, \omega_2),$$

and $\widetilde{v} : [0, +\infty) \times R \times R \times \Omega_1 \times \Omega_2 \to R$ by

$$\widetilde{v}(t, x, y, \omega_1, \omega_2) = Z_x(x, y)v(t, x, y, \omega_1, \omega_2).$$

If the functions $\widetilde{\rho}$, $\widetilde{\varsigma}$, and $\widetilde{v}$,$F$ as well as $h$ depend on $(x, y)$ through $Z(x, y)$ only, then the problem could be reduced to a finite-dimensional problem.

**Assumption 3.** *There are functions*

$$\overline{\rho} : [0, +\infty) \times R \times \Omega_1 \times \Omega_2 \to R, \quad \overline{\varsigma} : [0, +\infty) \times R \times \Omega_1 \times \Omega_2 \to R,$$
$$\overline{v} : [0, +\infty) \times R \times \Omega_1 \times \Omega_2 \to R, \overline{F} : [0, +\infty) \times R \times \Omega_1 \times \Omega_2 \to R, \quad \overline{h} : R \to R. \tag{25}$$

*such that for all $t \in [0,T], \omega_1 \in \Omega_1, \omega_2 \in \Omega_2, (x,y) \in R \times R$, it holds that*

$$\begin{aligned}
&\overline{\rho}(t,Z(x,y),\omega_1,\omega_2) = \widetilde{\rho}(t,x,y,\omega_1,\omega_2), \quad \overline{\varsigma}(t,Z(x,y),\omega_1,\omega_2) = \widetilde{\varsigma}(t,x,y,\omega_1,\omega_2),\\
&\overline{v}(t,Z(x,y),\omega_1,\omega_2) = \widetilde{v}(t,x,y,\omega_1,\omega_2),\\
&\overline{F}(t,Z(x,y),\omega_1,\omega_2) = F(t,x,y,\omega_1,\omega_2), \quad \overline{h}(Z(x,y)) = h(x,y).
\end{aligned} \tag{26}$$

*Now, we can transform the infinite-dimensional saddle point equilibrium problems into finite-dimensional saddle point equilibrium problems by Z. For $\varphi_t \in C_R[-d,0]$, define $z = Z(x(\varphi_t), y(\varphi_t)) \in Z(L)$. Then, for $t \in [0,T]$, the saddle equilibrium problem of the one-dimensional case of the objective function* (18) *and model* (22) *is equivalent to solving* (27)–(29).

$$\hat{J}(\omega_1(t),\omega_2(t)) = \overline{J}(t,z) = \left\{ \int_t^T \overline{F}(s,z_s,\omega_1(s),\omega_2(s))ds + \overline{h}(Z_T) \right\}_{\sup}(\beta) \tag{27}$$

$$\hat{J}(\omega_1{}^*,\omega_2) \le \hat{J}(\omega_1{}^*,\omega_2{}^*) \le \hat{J}(\omega_1,\omega_2{}^*), \tag{28}$$

$$\begin{cases} dZ_s = \overline{\rho}(s,Z_s,\omega_1(s),\omega_2(s))ds + \overline{\varsigma}(s,Z_s,\omega_1(s),\omega_2(s))dC_s + \overline{v}(s,Z_s,\omega_1(s),\omega_2(s))dV_s \\ Z_t = z, \\ \omega_1(s) \in \Omega_1, \omega_2(s) \in \Omega_2, \quad s \in [t,T] \end{cases} \tag{29}$$

**Theorem 3.** *If $A_2(t) = e^{\lambda d}(\lambda + A_1(t) + e^{\lambda d}A_3)A_3$, $A_5(t) = e^{\lambda d}A_4(t)A_3$, and $A_7(t) = e^{\lambda d}A_6(t)A_3$ hold in Model* (22)*, then a necessary and sufficient condition for the equilibrium solutions to Problem* (22) *is*

$$\omega_1^* = \frac{Q(t)B_1(t)}{H(t)}z, \omega_2^* = -\frac{Q(t)B_2(t)}{G(t)}z$$

*where $z = x + e^{\lambda d}A_3 y$, and $Q(t)$ satisfies:*

$$\begin{cases} \frac{dQ(t)}{dt} + 2Q(t)(A_1(t) + e^{\lambda d}A_3) + P + \frac{Q^2(t)B_1^2(t)}{H(t)} - \frac{Q^2(t)B_2^2(t)}{G(t)} \\ \quad + \frac{2\sqrt{3}}{\pi}\ln\frac{1-\beta}{\beta}Q(t)A_4(t) + 2kA_6(t)Q(t) = 0, \\ Q(T) = \Psi_f, \end{cases}$$

*where $k = 1 - \frac{\beta}{2(1-\vartheta_2)} (0 < \beta < 1-\vartheta_2)$ or $k = \frac{1}{2}(1-\vartheta_2 \le \beta < 1-\vartheta_1)$ or $k = \frac{1-\beta}{2\vartheta_1}(1-\vartheta_1 \le \beta < 1)$;*

*The equilibrium value is*

$$J(0,x_0) = \frac{1}{2}Q(0)(x_0 + e^{\lambda d}A_3\int_{-d}^{0} e^{\lambda u}X_u du)^2.$$

See Appendix A for the proof process.

## 5. Carbon Emission Reduction Problem

Delays and jumps are ubiquitous in economic and management issues, such as in consumption and carbon emission reduction. We will analyze a carbon emission reduction economic game model under an uncertain jump and delay environment. The problem is crucial for addressing global climate change and is widely studied in environmental economics, policy design, and sustainability sciences. A carbon emission reduction game problem involves modeling the strategic interactions between different players, such as countries, companies, or industries, who aim to reduce carbon emissions while balancing their economic interests. The problem typically focuses on how these players can optimize their strategies to achieve emission reduction targets, considering costs and regulatory policies. In the following carbon emission reduction model, $X_t$ represents the carbon-emitting resources stock of carbon-emitting enterprises, which not only represents the assets owned by carbon-emitting enterprises, but also includes a comprehensive quantified value of electricity, coal, oil, and natural gas. It is a widely recognized fact that the provision of essential energy resources such as electricity, coal, oil, and natural gas can experience delays. These delays may stem from a variety of causes including logistical challenges, infrastructure limitations, regulatory hurdles, and unpredictable events such as natural disasters or geopolitical conflicts. Thus, the assets are not only related to the time $t$, but also affected by $Y_t = \int_{-d}^{0} e^{\lambda s} X_{t+s} ds$, $\varrho_t = X_{t-d}$, $t \geq 0$ at time $t-d$, and $\varphi_t(s) = X_{t+s}$, $s \in [-d, 0]$ indicates the assets of the time period $[-d, 0]$, $\lambda$ are a constant. The growth rate of $X_t$ is a linear function of $X_t$, $Y_t$, carbon emissions, and government input measures. Moreover, despite the fact that governments and enterprises from various countries are actively taking measures to reduce carbon emissions, address global climate change, and promote sustainable development, the reality is often cruel. Certain groups or individuals, motivated by varied interests such as political gain, economic advantage, or other benefits, secretly endorse practices that increase carbon emissions. The influence of this part of "funding behind" cannot be ignored and therefore must be taken into account in analytical frameworks. However, their behavior is secretive and therefore not easily detected, and we cannot obtain sufficiently detailed data on their behavior. In this context, the expertise of criminal psychology professionals or specialists in carbon emission mitigation, particularly those focused on carbon policy, becomes invaluable. Their knowledge and experience are essential for evaluating the impact of this "funding behind" on the resources of carbon-emitting enterprises. In dealing with expert opinions, we adopt the uncertainty theory as a powerful tool, and consider that the carbon-emitting event itself is an emergency, so we describe the influence of "funding behind" as an uncertain jump process. In addition, considering the harmfulness and destructiveness of carbon-emitting enterprises, the average evaluation criterion of expected value is inappropriate for modeling the carbon emission reduction economic because it often ignores the interests of the majority of people. For the carbon emission reduction economic model, we believe that the optimistic value model is a relatively more appropriate choice because a certain confidence level (here, the confidence level is the success rate of the decision) can be selected, which also takes into account the interests of the majority. Thus, the model in a limited time horizon is as follows:

$$\begin{cases} \text{Find } (\omega_1^*, \omega_2^*) \text{ such that} \\ J(\omega_1, \omega_2^*) \leq J(\omega_1^*, \omega_2^*) \leq J(\omega_1^*, \omega_2) \\ \text{subject to:} \\ \quad dX_s = (\epsilon X_s + e^{\lambda d}(e^{\lambda d} + \lambda + \epsilon) Y_s + \varrho_s + \omega_1 - \omega_2) ds + \varsigma (X_s + e^{\lambda d} Y_s) dC_s \\ \qquad\quad + \nu (X_s + e^{\lambda d} Y_s) dV_s, \quad s \in [0, T] \\ \quad Y_s = \int_{-d}^{0} e^{\lambda u} X_{s+u} du, \quad \varrho_s = X_{s-d}, \quad s \in [0, T] \\ \quad X_s = \varphi_0(s), \quad -d \leq s \leq 0, \end{cases}$$

where

$$J(\omega_1, \omega_2) = \left\{ \frac{1}{2} \lambda_1 \int_0^T ((X_s + e^{\lambda d} Y_s)^2 + \omega_2^2 - \omega_1^2) ds + \frac{1}{2} \lambda_2 (X_T + e^{\lambda d} Y_T)^2 \right\}_{\sup} (\beta).$$

where $\omega_1$ is the frequency of corporate carbon emissions and $\omega_2$ is government input measures. The performance index represents the economic losses caused by carbon emissions, $\lambda_1, \lambda_2 > 0$ and $\omega_1, \omega_2 \geq 0$, and $\epsilon$, $\varsigma$, and $\nu$ are the given positive constants. Let $Z(x, y) = x + e^{\lambda d} y$; then, it is obvious that $e^{\lambda d} Z_x(x, y) - Z_y(x, y) = 0$ holds, and that Assumption 1 holds. It is easy to see that Assumption 3 holds. Using Theorem 3, we can write the Riccati equation:

$$\begin{cases} \dot{Q} + 2(\epsilon + e^{\lambda d}) Q + \lambda_1 + \frac{2\sqrt{3}}{\pi} \ln \frac{1-\beta}{\beta} \varsigma Q + 2k\nu P = 0, \\ Q(t) = \lambda_2, \end{cases}$$

and we obtain

$$Q(t) = -\frac{\lambda_1}{2(\epsilon + e^{\lambda d}) + \frac{2\sqrt{3}}{\pi}\ln\frac{1-\beta}{\beta}\varsigma + 2kv} + (\lambda_2 + \frac{\lambda_1}{2(\epsilon + e^{\lambda d}) + \frac{2\sqrt{3}}{\pi}\ln\frac{1-\beta}{\beta}\varsigma + 2kv})$$
$$\exp(2(\epsilon + e^{\lambda d}) + \frac{2\sqrt{3}}{\pi}\ln\frac{1-\beta}{\beta}\varsigma + 2kv)(T-t),$$

where owing to $\overline{J}_z(vz) = v \geq 0$, then $k = 1 - \frac{\beta}{2(1-\vartheta_2)}(0 < \beta < 1 - \vartheta_2)$ or $k = \frac{1}{2}(1 - \vartheta_2 \leq \beta < 1 - \vartheta_1)$ or $k = \frac{1-\beta}{2\vartheta_1}(1 - \vartheta_1 \leq \beta < 1)$.

The equilibrium strategy is obtained as follows:

$$\omega_1^* = \omega_2^* = [-\frac{1}{2(\epsilon + e^{\lambda d}) + \frac{2\sqrt{3}}{\pi}\ln\frac{1-\beta}{\beta}\varsigma + 2kv} + (\frac{\lambda_2}{\lambda_1} + \frac{\lambda_1}{2(\epsilon + e^{\lambda d}) + \frac{2\sqrt{3}}{\pi}\ln\frac{1-\beta}{\beta}\varsigma + 2kv})$$
$$\exp((2(\epsilon + e^{\lambda d}) + \frac{2\sqrt{3}}{\pi}\ln\frac{1-\beta}{\beta}\varsigma + 2kv)(T-t))]z.$$

The equilibrium values are given as follows:

$$J(t,x) = \frac{1}{2}[-\frac{\lambda_1}{2(\epsilon + e^{\lambda d}) + \frac{2\sqrt{3}}{\pi}\ln\frac{1-\beta}{\beta}\varsigma + 2kv} + (\lambda_2 + \frac{\lambda_1}{2(\epsilon + e^{\lambda d}) + \frac{2\sqrt{3}}{\pi}\ln\frac{1-\beta}{\beta}\varsigma + 2kv})\cdot$$
$$\exp((2(\epsilon + e^{\lambda d}) + \frac{2\sqrt{3}}{\pi}\ln\frac{1-\beta}{\beta}\varsigma + 2kv)(T-t))]z^2,$$

and

$$J(0,x_0) = \frac{1}{2}[-\frac{\lambda_1}{2(\epsilon + e^{\lambda d}) + \frac{2\sqrt{3}}{\pi}\ln\frac{1-\beta}{\beta}\varsigma + 2kv} + (\lambda_2 + \frac{\lambda_1}{2(\epsilon + e^{\lambda d}) + \frac{2\sqrt{3}}{\pi}\ln\frac{1-\beta}{\beta}\varsigma + 2kv})\cdot$$
$$\exp((2(\epsilon + e^{\lambda d}) + \frac{2\sqrt{3}}{\pi}\ln\frac{1-\beta}{\beta}\varsigma + 2kv)T)]z_0^2,$$

where $z = x_t + e^{\lambda d}y_t$, and $z_0 = x_0 + e^{\lambda d}y_0$. In order to explain the problem clearly, we take the following parameters:

$$x_0 = 10, T = 3, \lambda = -20, d = 0.2, \epsilon = 0.05, \varsigma = 1, v = 0.3, \lambda_1 = \lambda_2 = 1.$$

We let $\varphi_0(s) = \cos\pi s$, then $y_0 = \int_{-0.2}^{0} e^{-20u}\cos\pi u du = \frac{\pi\sin(0.2\pi)+20\cos(0.2\pi)-20e^{-4}}{e^{-4}(\pi^2+400)}$ and $z_0 = 10 + \frac{\pi\sin(0.2\pi)+20\cos(0.2\pi)-20e^{-4}}{\pi^2+400}$. We take $\vartheta_1 = 0.3$, $\vartheta_2 = 0.6$, and then we obtain the equilibrium value of different belief degree $\beta$. Meanwhile, we give the dynamic change curve of the equilibrium strategy $(\omega_1^*, \omega_2^*)$ and equilibrium value $J(t,x)$ of different belief degree $\beta$. According to Ref. [20], the equilibrium value $J(0,x_0)$ without state delays and V-jumps process is

$$J(0,x_0) = \frac{1}{2}[-\frac{\lambda_1}{2\epsilon + \frac{2\sqrt{3}}{\pi}\ln\frac{1-\beta}{\beta}\varsigma} + (\lambda_2 + \frac{\lambda_1}{2\epsilon + \frac{2\sqrt{3}}{\pi}\ln\frac{1-\beta}{\beta}\varsigma})\cdot$$
$$\exp((2\epsilon + \frac{2\sqrt{3}}{\pi}\ln\frac{1-\beta}{\beta}\varsigma)T)]x_0^2,$$

By taking the same parameters, we give the equilibrium value of various belief degrees $\beta$ without state delays and the V-jumps process as the comparison.

To study the influence of parametric errors on the carbon emission reduction model, let us take the parameter $\beta$ as an example. Based on the equilibrium value and error curves analysis, we firstly performed a sensitivity analysis; we perturbed the parameter

$\beta$ by 10% to examine the effect of this change on the equilibrium values both with and without delays. The results indicate that the equilibrium values are highly sensitive to changes in $\beta$, especially in the lower $\beta$ range, where the variation is more pronounced.The analysis shows that in the "no delay" case, the equilibrium values exhibit more dramatic fluctuations in response to changes in $\beta$ compared to the "with delays" case. The sensitivity curves demonstrate that the model reacts differently to parameter perturbations depending on whether state delays are included or not.

Then, we performed an error propagation analysis to assess the impact of parameter uncertainty (e.g., $\beta \pm 5\%$) on the equilibrium values. The results display the error bounds (upper and lower limits) for both the "no delay" and "with delays" cases.The "no delay" scenario shows larger error bounds compared to the "with delays" case, indicating greater sensitivity to parameter errors. In contrast, the "with delays" model exhibits more robustness to parameter errors, with smaller variations in equilibrium values.

Figure 1a illustrates the smooth equilibrium values for both the "no delay" and "with delays" scenarios as functions of $\beta$. As $\beta$ increases, the equilibrium values decrease in both cases, but the decline is sharper in the "no delay" case. The graph also presents the comparison between the original equilibrium values and those after a 10% perturbation in $\beta$. The perturbed curves deviate more noticeably from the original in the "no delay" case, indicating higher sensitivity to parameter changes. Figure 1b displays the equilibrium values with a ±5% parameter error. The error bounds are plotted as smooth curves above and below the original equilibrium values. In the "no delay" scenario, the error bounds are wider, reflecting a higher sensitivity to parameter uncertainty. In contrast, the "with delays" scenario has narrower error bounds, indicating more stability and resistance to parameter error.

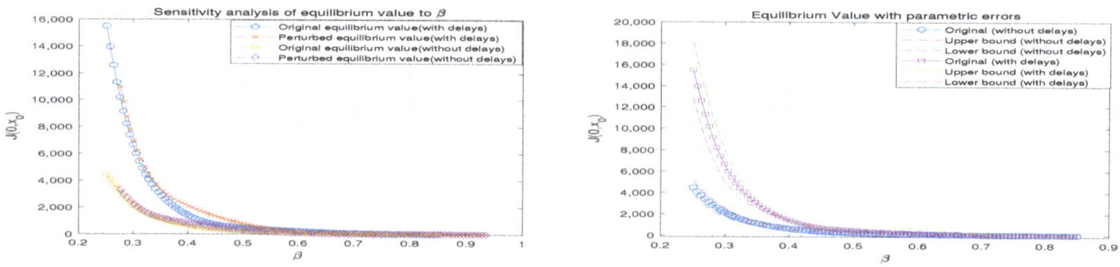


(a) (b)

**Figure 1.** The equilibrium value. (**a**) Sensitivity analysis of equilibrium value. (**b**) Equilibrium value with parametric errors.

Through sensitivity and error propagation analysis, it is evident that the "no delay" model is more sensitive to parameter changes and errors compared to the "with delays" model. This suggests that the "with delays" model demonstrates better robustness in the face of parameter uncertainties. The influence of $\beta$ on the equilibrium values is particularly significant in the lower range. The error analysis reveals that parameter errors can have a substantial impact on model predictions, especially in the "no delay" case.

Next, we present Figures 2–8, which intuitively illustrate the equilibrium strategy values and the equilibrium values.

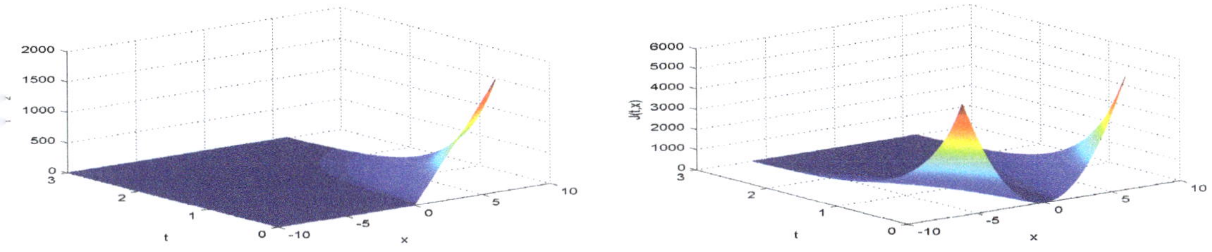


(a) (b)

**Figure 2.** The equilibrium strategy value (**a**) and the equilibrium value (**b**) when $\beta = 0.25$.

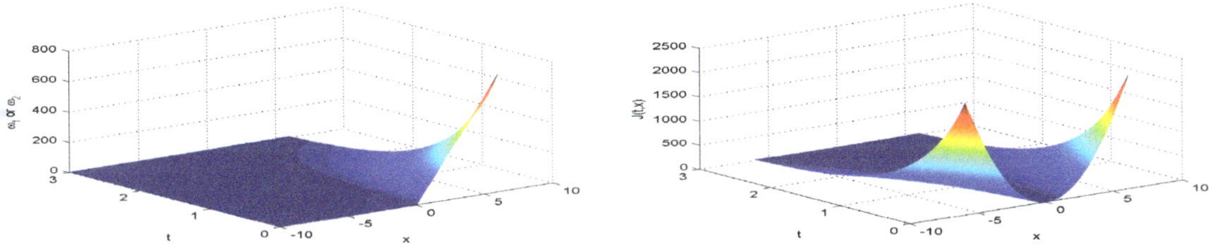


(a) (b)

**Figure 3.** The equilibrium strategy value (**a**) and the equilibrium value (**b**) when $\beta = 0.30$.

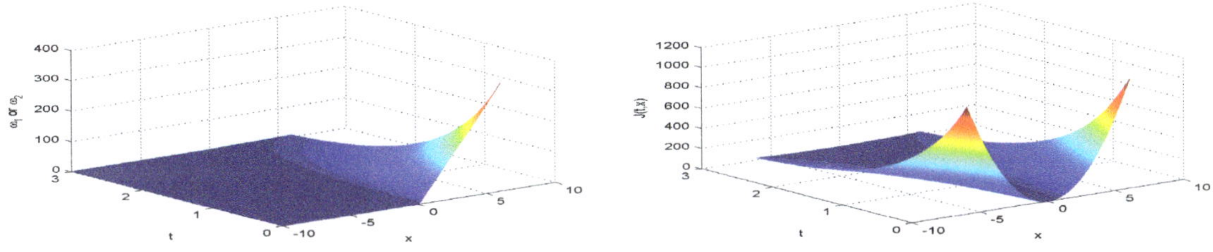


(a) (b)

**Figure 4.** The equilibrium strategy value (**a**) and the equilibrium value (**b**) when $\beta = 0.35$.

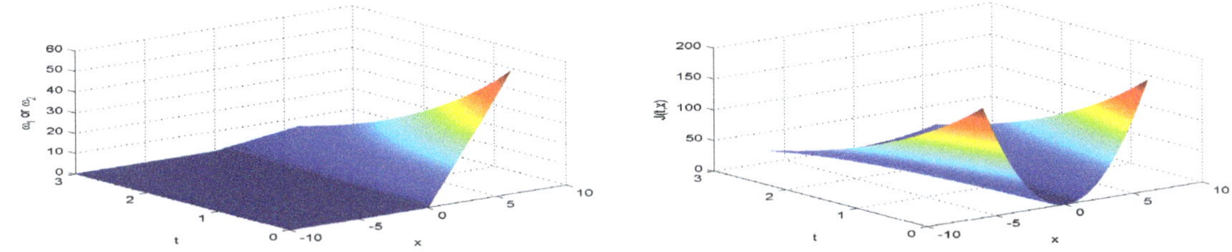


(a) (b)

**Figure 5.** The equilibrium strategy value (**a**) and the equilibrium value (**b**) when $\beta = 0.50$.

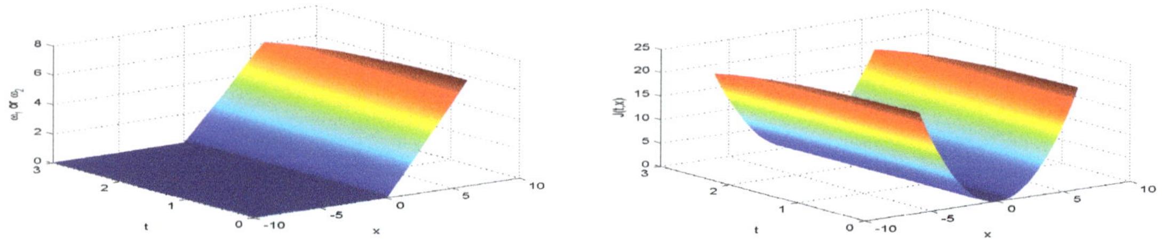


(**a**) (**b**)

**Figure 6.** The equilibrium strategy value (**a**) and the equilibrium value (**b**) when $\beta = 0.75$.

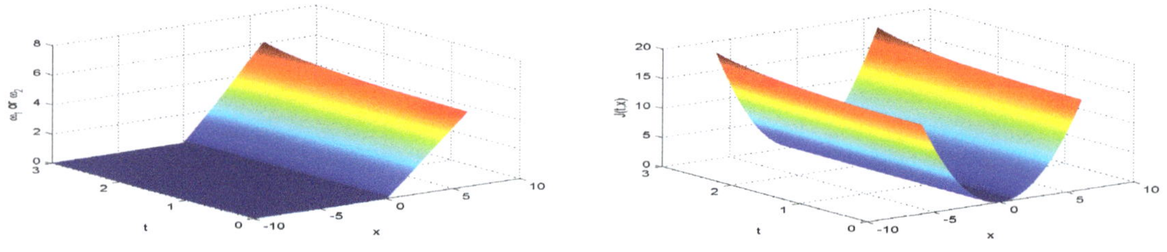


(**a**) (**b**)

**Figure 7.** The equilibrium strategy value (**a**) and the equilibrium value (**b**) when $\beta = 0.80$.

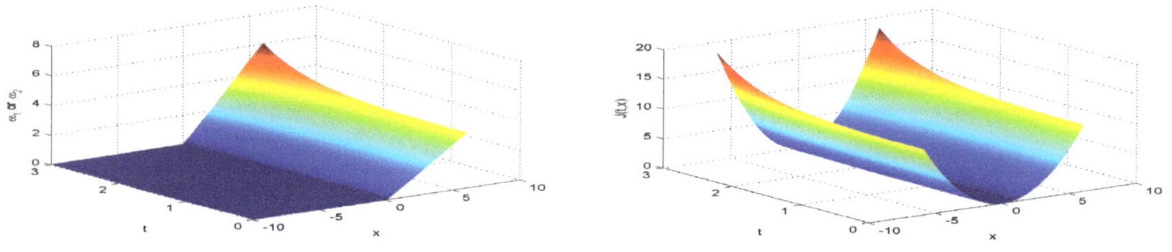


(**a**) (**b**)

**Figure 8.** The equilibrium strategy value (**a**) and the equilibrium value (**b**) when $\beta = 0.85$.

Table 1 shows that the belief degree $\beta$ represents the confidence level in reducing economic losses caused by carbon emissions. For instance, if carbon-emitting enterprises require an initial start-up fund of 10 million, they should prepare $10 + \frac{\pi \sin(0.2\pi)+20\cos(0.2\pi)-20e^{-4}}{\pi^2+400}$ million dollars considering delays. From Table 1, we observe that the equilibrium value $J(0, x_0) = 15,523.0$ when $\beta = 0.25$, resulting in government losses of 15,523.0 million. When $\beta = 0.30$, the equilibrium value is 6493.70, and the government losses are 6493.7 million. At $\beta = 0.35$, the equilibrium value is 2978.5, with corresponding government losses of 2978.5 million. When $\beta = 0.5$, the equilibrium value drops to 499.4, and the losses are 499.4 million. At $\beta = 0.75$, the equilibrium value further decreases to 60.2, with losses of 60.2 million. For $\beta = 0.80$, the equilibrium value is 42.5, and the losses are 42.5 million. Finally, at $\beta = 0.85$, the equilibrium value is 31.2, resulting in the smallest government losses of 31.2 million. This analysis reveals that both the government and carbon-emitting enterprises incur significant costs to achieve their objectives. It is also evident that due to the influence of jumps and delays, economic losses from terrorism are lower when $\beta > 0.5$ and higher when $\beta < 0.5$. Among the given belief degrees, $\beta = 0.85$ results in the least economic loss, and generally, higher $\beta$ values correlate with lower economic losses $J(0, x_0)$.

Additionally, the presence of jumps and delays increases the equilibrium value compared to scenarios without these factors when $\beta \in (0,1)$. This means that economic losses rise due to jumps and delays, aligning with the real-world dynamics and the objective laws of counter-terrorism economic issues. From Figures 2–8, we can intuitively observe the dynamic variation curves of the equilibrium strategy $(\omega_1^*, \omega_2^*)$ and the equilibrium value $J(t,x)$ at different confidence levels $\beta$. For governments, incorrect belief degrees may result in uneven resource allocation. For enterprises, mistakes in belief degrees can harm brand image and even lead to bankruptcy. Thus, improving the decision-making belief degree is crucial. Both governments and carbon-emitting enterprises must enhance the scientific and rational aspects of their decision-making processes, thoroughly considering various factors, conducting comprehensive research, and avoiding hasty decisions based on subjective assumptions. Establishing sound decision-making mechanisms and responsibility systems is also essential to ensure transparency, accountability, and support for sustainable economic and social development. In future development, governments and carbon-emitting enterprises should prioritize enhancing decision-making belief degrees by developing policies or strategies that both minimize emissions and maintain competitiveness and economic viability. Incorporating detailed economic analysis and being open to policy adjustments based on those insights will ensure that decisions are not only environmentally sound but also economically sustainable.

**Table 1.** The equilibrium value of various belief degree $\beta$.

| $\beta$ | $k$ | $J(0,x_0)$ Without State Delays and V-Jumps | Equilibrium Value $J(0,x_0)$ |
|---|---|---|---|
| 0.25 | 0.6875 | 4467.0 | 15,523.0 |
| 0.30 | 0.6250 | 2140.9 | 6493.7 |
| 0.35 | 0.5625 | 1127.7 | 2978.5 |
| 0.50 | 0.5000 | 242.4 | 499.4 |
| 0.75 | 0.4167 | 45.2 | 60.2 |
| 0.80 | 0.3333 | 35.2 | 42.5 |
| 0.85 | 0.2500 | 27.7 | 31.2 |

## 6. Conclusions

This study addressed the differential game optimistic value model for uncertain delay systems with jumps. By transforming the model into a finite-dimensional framework, the research provided valuable insights into the Nash equilibrium solution specifically for uncertain delay differential game optimistic value control models that incorporate jumps. Furthermore, the study identified the saddle equilibrium solution within a linear quadratic differential game optimistic value model under conditions characterized by uncertainty, delays and sudden jumps. To illustrate these concepts in a practical context, an example of a carbon emission reduction game was presented, highlighting the implications of the theoretical findings in real-world scenarios. This example not only demonstrated the applicability of the developed models but also underscored the critical role of strategic decision making in addressing pressing environmental challenges. Looking ahead, future research could delve deeper into both theoretical and practical dimensions. On the theoretical side, exploring more complex game structures involving singular systems, switched systems, or other non-linear systems could enhance our understanding of strategic interactions in uncertain environments. Investigating alternative criteria for decision making beyond the optimistic value might also yield new insights into equilibrium concepts and stability conditions. In terms of practical applications, expanding the scope of case studies to include diverse fields such as renewable energy adoption, transportation emissions, public policy, and financial markets could demonstrate the versatility of the proposed models. Additionally, incorporating real data and advanced computational techniques could facilitate dynamic modeling, allowing for more responsive and adaptive strategies in uncertain environments.

**Funding:** The authors are supported by the Suqian Sci & Tech Program (Grant No. K202332) and the Startup Foundation for Newly Recruited Employees of Suqian University (Grant No. 2024XRC004).

**Institutional Review Board Statement:** Not applicable

**Data Availability Statement:** Data are contained within the article.

**Conflicts of Interest:** The author declares no conflicts of interest.

## Appendix A

**Lemma A1.** *Let $Y(t,x,y):[0,+\infty)\times R^n\times R^k\to R^n$ be a continuously differentiable function, and given the feasible strategy pairs $\omega_1(t)\in\Omega_1$ and $\omega_2(t)\in\Omega_2$, here, $(X_t,Y_t)$ follows $dX_t=\rho_1(t,X_t,Y_t,\omega_1,\omega_2)dt+\rho_2(X_t,Y_t)\varrho_t dt+\varsigma(t,X_t,Y_t,\omega_1,\omega_2)dC_t+\nu(t,X_t,Y_t,\omega_1,\omega_2)dV_t$. Then, the uncertain process $Y(t,X_t,Y_t)$ satisfies*

$$\begin{aligned}dY(t,X_t,Y_t)=&\{Y_t(t,X_t,Y_t)+\Xi_x Y(t,X_t,Y_t)(\rho_1(t,X_t,Y_t,\omega_1,\omega_2)+\rho_2(X_t,Y_t)\varrho_t)\}dt\\&+\Xi_x Y(t,X_t,Y_t)\varsigma(t,X_t,Y_t,\omega_1,\omega_2)dC_t+\Xi_x Y(t,X_t,Y_t)\nu(t,X_t,Y_t,\omega_1,\omega_2)dV_t\\&+\Xi_y Y(t,X_t,Y_t)(\mathcal{F}(X_t)-e^{-\lambda d}\varrho_t-\lambda Y_t)dt.\end{aligned}\tag{A1}$$

**Proof.** Given the feasible strategy pairs $(\omega_1,\omega_2)$, and denoting $\widetilde{M}_t$ by $\widetilde{M}_t=\int_0^t\mathcal{F}(X_s)ds$, then, the process $Y_t$ has the representation

$$\begin{aligned}Y_t=&\int_{-d}^{0}e^{\lambda s}\mathcal{F}(X_{t+s})ds=\int_{-d}^{0}e^{\lambda s}d\widetilde{M}_{t+s}=e^{\lambda s}\widetilde{M}_{t+s}\mid_{-d}^{0}-\int_{-d}^{0}\widetilde{M}_{t+s}de^{\lambda s}\\=&\widetilde{M}_t-e^{-\lambda d}\widetilde{M}_{t-d}-\int_{-d}^{0}\lambda e^{\lambda s}\widetilde{M}_{t+s}ds\\=&\int_0^t\mathcal{F}(X_s)ds-e^{-\lambda d}\int_0^{t-d}\mathcal{F}(X_s)ds-\lambda\int_{-d}^{0}e^{\lambda s}\int_0^{t+s}\mathcal{F}(X_r)drds.\end{aligned}$$

Thus, $dY_t=(\mathcal{F}(X_t)-e^{-\lambda d}\mathcal{F}(X_{t-d})-\lambda Y_t)dt=(\mathcal{F}(X_t)-e^{-\lambda d}\varrho_t-\lambda Y_t)dt$. Applying Lemma 2 to $Y(t,X_t,Y_t)$, Equation (A1) follows.

Thus, the dynamics for $Z_t=Z(X_t,Y_t)$ by using (9) and (A1) can be expressed as follows.

$$\begin{aligned}dZ_t&=dZ(X_t,Y_t)\\&=\Xi_x Z(X_t,Y_t)(\rho_1(t,X_t,Y_t,\omega_1,\omega_2)+\rho_2(X_t,Y_t)\varrho_t)dt\\&\quad+\Xi_x Z(X_t,Y_t)\varsigma(t,X_t,Y_t,\omega_1,\omega_2)dC_t+\Xi_x Z(X_t,Y_t)\nu(t,X_t,Y_t,\omega_1,\omega_2)dV_t\\&\quad+\Xi_y Z(X_t,Y_t)(\mathcal{F}(X_t)-e^{-\lambda d}\varrho_t-\lambda Y_t)dt\\&=\Xi_x Z(X_t,Y_t)\rho_1(t,X_t,Y_t,\omega_1,\omega_2)dt+\Xi_y Z(X_t,Y_t)(\mathcal{F}(X_t)-\lambda Y_t)dt\\&\quad+\Xi_x Z(X_t,Y_t)\varsigma(t,X_t,Y_t,\omega_1,\omega_2)dC_t+\Xi_x Z(X_t,Y_t)\nu(t,X_t,Y_t,\omega_1,\omega_2)dV_t.\end{aligned}\tag{A2}$$

□

**Lemma A2.** *Suppose that Assumptions 1 and 2 hold and $\overline{J}_t(t,z)$ is twice differentiable on $[0,T]\times R^n$. Then, we have*

$$\begin{cases}-\overline{J}_{1t}(t,z)=\sup\limits_{\omega_1(t)\in\Omega_1}\{\overline{F}_1(t,z,\omega_1(t),\omega_2(t))+\nabla_z\overline{J}_1(t,z)^\tau\overline{\rho}(t,z,\omega_1(t),\omega_2(t))\\\qquad+\frac{\sqrt{3}}{\pi}\ln\frac{1-\beta}{\beta}\|\nabla_z\overline{J}_2(t,z)^\tau\overline{\varsigma}(t,z,\omega_1(t),\omega_2(t))\|_1+k\|\nabla_z\overline{J}_1(t,z)^\tau\overline{\nu}(t,z,\omega_1(t),\omega_2(t))\|\}\\\overline{J}_1(T,Z_T)=\overline{h}_1(Z_T),\end{cases}\tag{A3}$$

$$\begin{cases}-\overline{J}_{2t}(t,z)=\sup\limits_{\omega_2(t)\in\Omega_2}\{\overline{F}_2(t,z,\omega_1(t),\omega_2(t))+\nabla_z\overline{J}_2(t,z)^\tau\overline{\rho}(t,z,\omega_1(t),\omega_2(t))\\\qquad+\frac{\sqrt{3}}{\pi}\ln\frac{1-\beta}{\beta}\|\nabla_z\overline{J}_2(t,z)^\tau\overline{\varsigma}(t,z,\omega_1(t),\omega_2(t))\|_1+k\|\nabla_z\overline{J}_2(t,z)^\tau\overline{\nu}(t,z,\omega_1(t),\omega_2(t))\|\}\\\overline{J}_2(T,Z_T)=\overline{h}_2(Z_T),\end{cases}\tag{A4}$$

*and* $\overline{J}_i(t,z) = J_i(t,\varphi_t)$*, where* $\overline{J}_{it}(t,z)$ *is the partial derivative of the function* $\overline{J}_i(t,z)$ *in* $t$*, and* $\nabla_z\overline{J}_i(t,z)$ *is the gradient of* $\overline{J}_i(t,z)$ *in* $z$*.*

(1) *If* $\nabla_z\overline{J}_i^{\tau}\overline{v} \geq 0$*, then* $k = 1 - \frac{\beta}{2(1-\vartheta_2)}(0 < \beta < 1-\vartheta_2)$ *or* $k = \frac{1}{2}(1-\vartheta_2 \leq \beta < 1-\vartheta_1)$ *or* $k = \frac{1-\beta}{2\vartheta_1}(1-\vartheta_1 \leq \beta < 1)$*;*

(2) *if* $\nabla_z\overline{J}_i^{\tau}\overline{v} < 0$*, then* $k = \frac{\beta}{2\vartheta_1}(0 < \beta \leq \vartheta_1)$ *or* $k = \frac{1}{2}(\vartheta_1 < \beta \leq \vartheta_2)$ *or* $k = 1 - \frac{1-\beta}{2(1-\vartheta_2)}(\vartheta_2 < \beta < 1)$*.*

**Proof.** By Lemma 1, Equations (A3) and (A4) hold. Furthermore, if $\forall\omega_1(t) \in \Omega_1, \omega_2(t) \in \Omega_2$, it holds that

$$\begin{aligned}\overline{J}(t,z) &\geq \left\{\int_t^T \overline{F}(s,Z_s,\omega_1(s),\omega_2(s))ds + \overline{h}(Z_T)\right\}_{\sup}(\beta)\\ &= \left\{\int_t^T F(s,X_s,Y_s,\omega_1(s),\omega_2(s))ds + h(X_T,Y_T)\right\}_{\sup}(\beta).\end{aligned}$$

Thus, $\overline{J}(t,z) \geq \sup_{\omega_i\in\Omega_i}\left\{\int_t^T F(s,X_s,Y_s,\omega_1(s),\omega_2(s))ds + h(X_T,Y_T)\right\}_{\sup}(\beta) = J(t,\varphi_t)$, $i = 1,2$. In the same way, we obtain that $J(t,\varphi_t) \geq \overline{J}(t,z)$. □

**Proof of Theorem 1.** Let $Z(x,y) = x + e^{\lambda d}A_3y$; then, Assumption 1 holds. Further, we have

$$\begin{cases}\overline{\rho}(t,x,y,\omega_1,\omega_2) = \Xi_x Z(x,y)\rho_1(t,x,y,\omega_1,\omega_2) + \Xi_y Z(x,y)(x-\lambda y)\\ \quad = [A_1 + e^{\lambda d}A_3][Z(x,y) - e^{\lambda d}A_3y] + (A_2 - \lambda e^{\lambda d})y + B_1\omega_1 + B_2\omega_2\\ \quad = [A_1 + e^{\lambda d}A_3][x + e^{\lambda d}A_3y] - [A_1e^{\lambda d}A_3 - A_2 + e^{\lambda d}A_3(e^{\lambda d}A_3 + \lambda)]y\\ \quad\quad + B_1\omega_1 + B_2\omega_2\\ \overline{F}_i(t,z,\omega_1,\omega_2) = z^{\tau}D_iz + \omega_1^{\tau}R_{i1}\omega_1 + \omega_2^{\tau}R_{i2}\omega_2, i = 1,2,\\ \overline{h}_i(z) = z_T^{\tau}G_iz_T,\\ \overline{\varsigma}(t,x,y,\omega_1,\omega_2) = \Xi_x z(x,y)\varsigma(t,x,y,\omega_1,\omega_2) = A_4z + (A_5 - A_4e^{\lambda d}A_3)y\\ \overline{v}(t,x,y,\omega_1,\omega_2) = \Xi_x z(x,y)v(t,x,y,\omega_1,\omega_2) = A_6z + (A_7 - A_6e^{\lambda d}A_3)y.\end{cases} \quad (A5)$$

Therefore, Assumption 2 holds if and only if:

$$A_2(t) = e^{\lambda d}(\lambda I + A_1(t) + e^{\lambda d}A_3(t))A_3(t)$$

$$A_5(t) = e^{\lambda d}A_4(t)A_3(t)$$

$$A_7(t) = e^{\lambda d}A_6(t)A_3(t)$$

Then, the following finite-dimensional Nash equilibrium problem can be obtained by the transformation of $Z(x,y) = x + e^{\lambda d}A_3y$.

$$(\overline{NP})\left[\begin{array}{l}\overline{J}_i(t,z) = \sup_{\omega_i\in\Omega_i}\left\{\int_t^T [z^{\tau}D_iz + \omega_1^{\tau}R_{i1}\omega_1 + \omega_2^{\tau}R_{i2}\omega_2]ds + z_T^{\tau}G_iz_T\right\}_{\sup}(\beta)\\ \text{subject to}\\ \quad dZ_s = \left\{(A_1(s) + e^{\lambda d}A_3)Z_s + B_1\omega_1 + B_2\omega_2\right\}ds\\ \quad\quad + A_4(s)zdC_s + A_6(s)zdV_s, \quad s \in [t,T]\\ \quad Z_t = z,\\ \quad \omega_1(s) \in \Omega_1, \omega_2(s) \in \Omega_2, \quad s \in [t,T].\end{array}\right. \quad (A6)$$

It follows from the optimality condition of Lemma A2 that:

$$\begin{aligned}-\overline{J}_{it}(t,z) = &\sup\{\overline{F}_i(t,z,\omega_1(t),\omega_2(t)) + \nabla_z\overline{J}_i(t,z)^{\tau}\overline{\rho}(t,z,\omega_1(t),\omega_2(t))\}\\ &+ \nabla_z\overline{J}_i(t,z)^{\tau}k\overline{v}(t,z,\omega_1(t),\omega_2(t))\},\end{aligned}$$

that is,

$$
\begin{aligned}
-\bar{J}_{1t}(t,z) = &\sup_{\omega_1\in\Omega_1} \{z^\tau D_1 z + \omega_1^\tau R_{11}\omega_1 + \omega_2^\tau R_{12}\omega_2 + \nabla_z \bar{J}_1(t,z)^\tau[(A_1(s)+e^{\lambda d}A_3)z + B_1\omega_1 + B_2\omega_2] \\
&+ \nabla_z \bar{J}_1(t,z)^\tau k A_6(s)z\} \\
= &\sup_{\omega_1\in\Omega_1} g_1(z,\omega_1,\omega_2),
\end{aligned} \tag{A7}
$$

$$
-\bar{J}_{2t}(t,z) = \sup_{\omega_2\in\Omega_2} g_2(z,\omega_1,\omega_2), \tag{A8}
$$

where

$$
\begin{aligned}
g_1(z,\omega_1,\omega_2) =& z^\tau D_1 z + \omega_1^\tau R_{11}\omega_1 + \omega_2^\tau R_{12}\omega_2 + \nabla_z \bar{J}_1(t,z)^\tau[(A_1(s)+e^{\lambda d}A_3)z + B_1\omega_1 + B_2\omega_2] \\
&+ \frac{\sqrt{3}}{\pi}\ln\frac{1-\beta}{\beta}|\nabla_z \bar{J}_1(t,z)^\tau A_4(s)z| + \nabla_z \bar{J}_1(t,z)^\tau k A_6(s)z, \\
g_1(z,\omega_1,\omega_2) =& z^\tau D_2 z + \omega_1^\tau R_{21}\omega_1 + \omega_2^\tau R_{22}\omega_2 + \nabla_z \bar{J}_2(t,z)^\tau[(A_1(s)+e^{\lambda d}A_3)z + B_1\omega_1 + B_2\omega_2] \\
&+ \frac{\sqrt{3}}{\pi}\ln\frac{1-\beta}{\beta}|\nabla_z \bar{J}_2(t,z)^\tau A_4(s)z| + \nabla_z \bar{J}_2(t,z)^\tau k A_6(s)z,
\end{aligned}
$$

where $k$ depends on $A_6(t)$. Owing to $\nabla_z J^\tau A_6(t)z = A_6(t)$,

(1) if $A_6(t)\geq 0$, then $k = 1-\frac{\beta}{2(1-\vartheta_2)}(0<\beta<1-\vartheta_2)$ or $k=\frac{1}{2}(1-\vartheta_2\leq\beta<1-\vartheta_1)$ or $k=\frac{1-\beta}{2\vartheta_1}(1-\vartheta_1\leq\beta<1)$;

(2) if $A_6(t)<0$, then $k=\frac{\beta}{2\vartheta_1}(0<\beta\leq\vartheta_1)$ or $k=\frac{1}{2}(\vartheta_1<\beta\leq\vartheta_2)$ or $k=1-\frac{1-\beta}{2(1-\vartheta_2)}(\vartheta_2<\beta<1)$. Let

$$
\begin{cases}
\frac{\partial g_1(z,\omega_1,\omega_2)}{\partial\omega_1} = 2R_{11}\omega_1 + B_1^\tau\nabla_z\bar{J}_1 = 0 \\
\frac{\partial g_2(z,\omega_1,\omega_2)}{\partial\omega_2} = 2R_{22}\omega_2 + B_2^\tau\nabla_z\bar{J}_2 = 0,
\end{cases} \tag{A9}
$$

we can obtain that

$$
\begin{cases}
\omega_1^*(t) = -\frac{1}{2}R_{11}^{-1}B_1^\tau\nabla_z\bar{J}_1 \\
\omega_2^*(t) = -\frac{1}{2}R_{22}^{-1}B_2^\tau\nabla_z\bar{J}_2.
\end{cases} \tag{A10}
$$

Owing to $\bar{J}_i(T,z_T) = z_T^\tau G_i z_T, i=1,2$, we can guess

$$
\bar{J}_i(t,z) = z^\tau Q_i(t)z,
$$

Then,

$$
\nabla_z\bar{J}_i(t,z) = 2Q_i(t)z.
$$

Thus, (A10) can be written as:

$$
\begin{cases}
\omega_1^*(t) = -R_{11}^{-1}B_1^\tau Q_1(t)z \\
\omega_2^*(t) = -R_{22}^{-1}B_2^\tau Q_2(t)z.
\end{cases} \tag{A11}
$$

On the other hand, there is:

$$
\bar{J}_{it} = z^\tau\dot{Q}_i(t)z. \tag{A12}
$$

We denote

$$\Delta_1 = \{(t,z) \mid z^\tau Q_i(t) A_4(t) z \geq 0, (t,z) \in [0,T] \times [a,b]^n\},$$
$$\Delta_2 = \{(t,z) \mid z^\tau Q_i(t) A_4(t) z < 0, (t,z) \in [0,T] \times [a,b]^n\}.$$

If $(t,z) \in \Delta_1$, we substitute (A11) and (A12) into (A7) and (A8), and it holds that

$$\begin{aligned}-z^\tau \dot{Q}_1(t) z = & z^\tau \{D_1 + Q_1^\tau B_1 R_{11}^{-1} B_1^\tau Q_1 + Q_2^\tau B_2 R_{22}^{-1} R_{12} R_{22}^{-1} B_2^\tau Q_2 \\ & + 2Q_1^\tau (A_1 + e^{\lambda d} A_3 + \frac{\sqrt{3}}{\pi} \ln \frac{1-\beta}{\beta} A_4 + kA_6) - 2Q_1^\tau B_1 R_{11}^{-1} B_1^\tau Q_1 - 2Q_1^\tau B_2 R_{22}^{-1} B_2^\tau Q_2\} z,\end{aligned}$$

$$\begin{aligned}-z^\tau \dot{Q}_2(t) z = & z^\tau \{D_2 + Q_2^\tau B_2 R_{22}^{-1} B_2^\tau Q_2 + Q_1^\tau B_1 R_{11}^{-1} R_{21} R_{11}^{-1} B_1^\tau Q_1 \\ & + 2Q_2^\tau (A_1 + e^{\lambda d} A_3 + \frac{\sqrt{3}}{\pi} \ln \frac{1-\beta}{\beta} A_4 + kA_6) - 2Q_2^\tau B_1 R_{11}^{-1} B_1^\tau Q_1 - 2Q_2^\tau B_2 R_{22}^{-1} B_2^\tau Q_2\} z.\end{aligned}$$

Considering that $\bar{J}_i(T, z_T) = z_T^\tau G_i z_T$, we obtain

$$\begin{cases} \dot{Q}_1(t) = -\{D_1 - Q_1^\tau B_1 R_{11}^{-1} B_1^\tau Q_1 + Q_2^\tau B_2 R_{22}^{-1} R_{12} R_{22}^{-1} B_2^\tau Q_2 + Q_1^\tau (A_1 + e^{\lambda d} A_3 + \frac{\sqrt{3}}{\pi} \ln \frac{1-\beta}{\beta} A_4 \\ \quad + kA_6) + (A_1 + e^{\lambda d} A_3 + \frac{\sqrt{3}}{\pi} \ln \frac{1-\beta}{\beta} A_4 + kA_6)^\tau Q_1 - 2Q_1^\tau B_2 R_{22}^{-1} B_2^\tau Q_2\}, \\ Q_1(T) = G_1, \\ \dot{Q}_2(t) = -\{D_2 - Q_2^\tau B_2 R_{22}^{-1} B_2^\tau Q_2 + Q_1^\tau B_1 R_{11}^{-1} R_{21} R_{11}^{-1} B_1^\tau Q_1 + Q_2^\tau (A_1 + e^{\lambda d} A_3 + \frac{\sqrt{3}}{\pi} \ln \frac{1-\beta}{\beta} A_4 \\ \quad + kA_6) + (A_1 + e^{\lambda d} A_3 + \frac{\sqrt{3}}{\pi} \ln \frac{1-\beta}{\beta} A_4 + kA_6)^\tau Q_2 - 2Q_2^\tau B_1 R_{11}^{-1} B_1^\tau Q_1\}, \\ Q_2(T) = G_2. \end{cases}$$

If $(t,z) \in \Delta_2$, we substitute (A11) and (A12) into (A7) and (A8), and it holds that

$$\begin{aligned}-z^\tau \dot{Q}_1(t) z = & z^\tau \{D_1 + Q_1^\tau B_1 R_{11}^{-1} B_1^\tau Q_1 + Q_2^\tau B_2 R_{22}^{-1} R_{12} R_{22}^{-1} B_2^\tau Q_2 \\ & + 2Q_1^\tau (A_1 + e^{\lambda d} A_3 - \frac{\sqrt{3}}{\pi} \ln \frac{1-\beta}{\beta} A_4 + kA_6) - 2Q_1^\tau B_1 R_{11}^{-1} B_1^\tau Q_1 - 2Q_1^\tau B_2 R_{22}^{-1} B_2^\tau Q_2\} z,\end{aligned}$$

$$\begin{aligned}-z^\tau \dot{Q}_2(t) z = & z^\tau \{D_2 + Q_2^\tau B_2 R_{22}^{-1} B_2^\tau Q_2 + Q_1^\tau B_1 R_{11}^{-1} R_{21} R_{11}^{-1} B_1^\tau Q_1 \\ & + 2Q_2^\tau (A_1 + e^{\lambda d} A_3 - \frac{\sqrt{3}}{\pi} \ln \frac{1-\beta}{\beta} A_4 + kA_6) - 2Q_2^\tau B_1 R_{11}^{-1} B_1^\tau Q_1 - 2Q_2^\tau B_2 R_{22}^{-1} B_2^\tau Q_2\} z.\end{aligned}$$

Considering that $\bar{J}_i(T, z_T) = z_T^\tau G_i z_T$, we obtain

$$\begin{cases} \dot{Q}_1(t) = -\{D_1 - Q_1^\tau B_1 R_{11}^{-1} B_1^\tau Q_1 + Q_2^\tau B_2 R_{22}^{-1} R_{12} R_{22}^{-1} B_2^\tau Q_2 + Q_1^\tau (A_1 + e^{\lambda d} A_3 - \frac{\sqrt{3}}{\pi} \ln \frac{1-\beta}{\beta} A_4 \\ \quad + kA_6) + (A_1 + e^{\lambda d} A_3 - \frac{\sqrt{3}}{\pi} \ln \frac{1-\beta}{\beta} A_4 + kA_6)^\tau Q_1 - 2Q_1^\tau B_2 R_{22}^{-1} B_2^\tau Q_2\}, \\ Q_1(T) = G_1, \\ \dot{Q}_2(t) = -\{D_2 - Q_2^\tau B_2 R_{22}^{-1} B_2^\tau Q_2 + Q_1^\tau B_1 R_{11}^{-1} R_{21} R_{11}^{-1} B_1^\tau Q_1 + Q_2^\tau (A_1 + e^{\lambda d} A_3 - \frac{\sqrt{3}}{\pi} \ln \frac{1-\beta}{\beta} A_4 \\ \quad + kA_6) + (A_1 + e^{\lambda d} A_3 - \frac{\sqrt{3}}{\pi} \ln \frac{1-\beta}{\beta} A_4 + kA_6)^\tau Q_2 - 2Q_2^\tau B_1 R_{11}^{-1} B_1^\tau Q_1\}, \\ Q_2(T) = G_2. \end{cases}$$

The proof is completed. □

For problem (27)–(29), we first give the equilibrium equation of the optimistic value model as follows:

**Lemma A3.** *Suppose that Assumptions 1 and 3 hold and $\overline{J}(t,z)$ is twice differentiable on $[0,T]\times R$. Then, the equilibrium equation holds:*

$$-\overline{J}_t(t,z) = \max_{\omega_1\in\Omega_1}\min_{\omega_2\in\Omega_2}\{\overline{J}_z(t,z)^\tau\overline{\rho}(t,z,\omega_1,\omega_2)+\overline{F}(t,z,\omega_1,\omega_2)$$
$$+\frac{\sqrt{3}}{\pi}\ln\frac{1-\beta}{\beta}|\overline{J}_z(t,z)^\tau\overline{\varsigma}(t,z,\omega_1,\omega_2)|+k\overline{J}_z(t,z)^\tau\overline{v}(t,z,\omega_1,\omega_2)\} \quad \text{(A13)}$$

$$= \min_{\omega_2\in\Omega_2}\max_{\omega_1\in\Omega_1}\{\overline{J}_z(t,z)^\tau\overline{\rho}(t,z,\omega_1,\omega_2)+\overline{F}(t,z,\omega_1,\omega_2)$$
$$+\frac{\sqrt{3}}{\pi}\ln\frac{1-\beta}{\beta}|\overline{J}_z(t,z)^\tau\overline{\varsigma}(t,z,\omega_1,\omega_2)|+k\overline{J}_z(t,z)^\tau\overline{v}(t,z,\omega_1,\omega_2)\}. \quad \text{(A14)}$$

**Proof.** Assume that $(\omega_1^*,\omega_2^*)$ are the strategy pairs for (27)–(29). In addition, $\omega_1^*$ and $\omega_2^*$ satisfy (P1) and (P2):

$$\text{(P1)}\begin{cases}\overline{J}(t,z)\equiv\max\limits_{\omega_1\in\Omega_1}\hat{J}(\omega_1,\omega_2^*)\\ \text{subject to:}\\ dz_s=\overline{\rho}(s,z_s,\omega_1,\omega_2^*)ds+\overline{\varsigma}(s,z_s,\omega_1,\omega_2^*)dC_s+\overline{v}(s,z_s,\omega_1,\omega_2^*)dV_s \quad \text{and} \quad z_t=z,\end{cases}$$

and

$$\text{(P2)}\begin{cases}\overline{J}(t,z)\equiv\min\limits_{\omega_2\in\Omega_2}\hat{J}(\omega_1^*,\omega_2)\\ \text{subject to:}\\ dz_s=\overline{\rho}(s,z_s,\omega_1^*,\omega_2)ds+\overline{\varsigma}(s,z_s,\omega_1^*,\omega_2)dC_s+\overline{v}(s,z_s,\omega_1^*,\omega_2)dV_s \quad \text{and} \quad z_t=z.\end{cases}$$

Applying Lemma 1 to (P1) and (P2), it holds that

$$-\overline{J}_t(t,z) = \max_{\omega_1\in\Omega_1}\{\overline{J}_z(t,z)^\tau\overline{\rho}(t,z,\omega_1,\omega_2^*)+\overline{F}(t,z,\omega_1,\omega_2^*)$$
$$+\frac{\sqrt{3}}{\pi}\ln\frac{1-\beta}{\beta}|\overline{J}_z(t,z)^\tau\overline{\varsigma}(t,z,\omega_1,\omega_2^*)|+k\overline{J}_z(t,x)^\tau\overline{v}(t,z,\omega_1,\omega_2^*)\} \quad \text{(A15)}$$

$$= \min_{\omega_2\in\Omega_2}\{\overline{J}_z(t,z)^\tau\overline{\rho}(t,z,\omega_1^*,\omega_2)+\overline{F}(t,z,\omega_1^*,\omega_2)$$
$$+\frac{\sqrt{3}}{\pi}\ln\frac{1-\beta}{\beta}|\overline{J}_z(t,z)^\tau\overline{\varsigma}(t,z,\omega_1^*,\omega_2)|+k\overline{J}_z(t,z)^\tau\overline{v}(t,z,\omega_1^*,\omega_2)\}. \quad \text{(A16)}$$

By (A15), it holds that

$$-\overline{J}_t(t,z) \ge \max_{\omega_1\in\Omega_1}\min_{\omega_2\in\Omega_2}\{\overline{J}_z(t,z)^\tau\overline{\rho}(t,z,\omega_1,\omega_2)+\overline{F}(t,z,\omega_1,\omega_2)$$
$$+\frac{\sqrt{3}}{\pi}\ln\frac{1-\beta}{\beta}|\overline{J}_z(t,z)^\tau\overline{\varsigma}(t,z,\omega_1,\omega_2)|+k\overline{J}_z(t,z)^\tau\overline{v}(t,z,\omega_1,\omega_2)\}. \quad \text{(A17)}$$

Similarly, from (A16), it holds that

$$-\overline{J}_t(t,z) \le \min_{\omega_2\in\Omega_2}\max_{\omega_1\in\Omega_1}\{\overline{J}_z(t,z)^\tau\overline{\rho}(t,z,\omega_1,\omega_2)+\overline{F}(t,z,\omega_1,\omega_2)$$
$$+\frac{\sqrt{3}}{\pi}\ln\frac{1-\beta}{\beta}|\overline{J}_z(t,z)^\tau\overline{\varsigma}(t,z,\omega_1,\omega_2)|+k\overline{J}_z(t,z)^\tau\overline{v}(t,z,\omega_1,\omega_2)\}. \quad \text{(A18)}$$

Let

$$L(\omega_1,\omega_2) =\bar{J}_z(t,z)^\tau\widetilde{p}(t,z,\omega_1,\omega_2)+\widetilde{F}(t,z,\omega_1,\omega_2) + \frac{\sqrt{3}}{\pi}\ln\frac{1-\beta}{\beta}|\bar{J}_z(t,z)^\tau\widetilde{\varsigma}(t,z,\omega_1,\omega_2)| + k\bar{J}_z(t,z)^\tau\widetilde{v}(t,z,\omega_1,\omega_2).$$

We note that

$$\max_{\omega_1\in\Omega_1}\min_{\omega_2\in\Omega_2} L(\omega_1,\omega_2) \geq \min_{\omega_2\in\Omega_2} L(\omega_1,\omega_2), \quad \forall\omega_1. \tag{A19}$$

By (A19), we obtain

$$\max_{\omega_1\in\Omega_1}\min_{\omega_2\in\Omega_2} L(\omega_1,\omega_2) \geq \min_{\omega_2\in\Omega_2}\max_{\omega_1\in\Omega_1} L(\omega_1,\omega_2).$$

Together with (A17) and (A18), the lemma is completed. □

**Proof of Theorem 3.** First, we verify the necessity. Let $Z(x,y) = x+e^{\lambda d}A_3y$, then Assumption 1 holds. Further, we have

$$\begin{cases} \widetilde{p}(t,x,y,\omega_1,\omega_2) = Z_x(x,y)p_1(t,x,y,\omega_1,\omega_2)+Z_y(x,y)(x-\lambda y) \\ \qquad = [A_1(t)+e^{\lambda d}A_3][Z(x,y)-e^{\lambda d}A_3y]+(A_2-\lambda e^{\lambda d})y+B_1\omega_1+B_2\omega_2 \\ \qquad = [A_1(t)+e^{\lambda d}A_3][x+e^{\lambda d}A_3y]-[A_1e^{\lambda d}A_3-A_2+e^{\lambda d}A_3(e^{\lambda d}A_3+\lambda)]y \\ \qquad\quad +B_1\omega_1+B_2\omega_2 \\ \widetilde{F}(t,z,\omega_1,\omega_2) = Pz^2+G\omega_2^2-H\omega_1^2, \\ \widetilde{h}(z) = \frac{1}{2}\Psi_f z^2, \\ \widetilde{\varsigma}(t,x,y,\omega_1,\omega_2) = z_x(x,y)\varsigma(t,x,y,\omega_1,\omega_2) = A_4z+(A_5-A_4e^{\lambda d}A_3)y \\ \widetilde{v}(t,x,y,\omega_1,\omega_2) = z_x(x,y)v(t,x,y,\omega_1,\omega_2) = A_6z+(A_7-A_6e^{\lambda d}A_3)y. \end{cases} \tag{A20}$$

Therefore, Assumption 3 holds if and only if $A_2(t) = e^{\lambda d}(\lambda+A_1(t)+e^{\lambda d}A_3)A_3$ and $A_5(t) = e^{\lambda d}A_4(t)A_3$, $A_7(t) = e^{\lambda d}A_6(t)A_3$ hold. Then, the following finite-dimensional saddle equilibrium problem $(\overline{NP})$ can be obtained by the transformation of $Z(x,y) = x+e^{\lambda d}A_3y$.

$$(\overline{NP})\left[\begin{array}{l} \bar{J}(\omega_1,\omega_2) = \left\{\frac{1}{2}\int_0^T(P(s)Z^2+G(s)\omega_2^2-H(s)\omega_1^2)\mathrm{d}s+\frac{1}{2}\Psi_f Z_T^2\right\}_{\sup}(\beta) \\ \text{subject to} \\ \mathrm{d}Z_s = \Big\{(A_1(s)+e^{\lambda d}A_3)Z_s+B_1(s)\omega_1+B_2(s)\omega_2\Big\}\mathrm{d}s \\ \qquad +A_4(s)z\mathrm{d}C_s+A_6(s)z\mathrm{d}V_s, \quad s\in[t,T] \\ Z_t = z, \\ \omega_1(s)\in\Omega_1, \omega_2(s)\in\Omega_2, \quad s\in[t,T]. \end{array}\right. \tag{A21}$$

Since $\bar{J}(T,Z_T) = \frac{1}{2}\Psi_f Z_T^2$ is quadratic about $Z_T$. We guess

$$\bar{J}(t,z) = \frac{1}{2}Q(t)z^2$$

and

$$Q(t) = \Psi_f.$$

Applying the equilibrium Equations (A13) and (A14) of Lemma A3, we have

$$-\frac{1}{2}Q^{'}(t)z^2 = \max_{\omega_1}\min_{\omega_2}\{Q(t)z[(A_1(t)+e^{\lambda d}A_3)z + B_1(t)\omega_1 + B_2(t)\omega_2] + \frac{1}{2}P(t)z^2 + \frac{1}{2}G(t)u_2^2 - \frac{1}{2}H(t)\omega_1^2 + \frac{\sqrt{3}}{\pi}\ln\frac{1-\beta}{\beta}|Q(t)z\cdot A_4(t)z| + kQ(t)z\cdot A_6(t)z\}.$$

By the property of optimistic value and the assumptions $P(s)\geq 0$, $A_4(s)>0$,$G(s)>0$, $H(s)>0$, it is easy to see $Q(t)\geq 0$. Thus,

$$\begin{aligned}-\frac{1}{2}Q^{'}(t)z^2 =& Q(t)(A_1(t)+e^{\lambda d}A_3)z^2 + \frac{1}{2}P(t)z^2 + \frac{\sqrt{3}}{\pi}\ln\frac{1-\beta}{\beta}Q(t)A_4(t)z^2 + kQ(t)A_6(t)z^2\\ &+\max_{\omega_1}[Q(t)z\cdot B_1(t)\omega_1 - \frac{1}{2}H(t)\omega_1^2] + \min_{\omega_2}[Q(t)z\cdot B_2(t)\omega_2 + \frac{1}{2}G(t)\omega_2^2]\\ =&\chi_1 + \max_{\omega_1}\chi_2 + \min_{\omega_2}\chi_3.\end{aligned} \quad (A22)$$

Since $\chi_2$ is strictly concave about $\omega_1$, $\chi_3$ is strictly convex about $\omega_2$. Thus, the equilibrium controls $\omega_1$ and $\omega_2$ satisfy

$$\frac{\partial\chi_2}{\partial\omega_1} = Q(t)z\cdot B_1(t) - H(t)\omega_1 = 0,$$

$$\frac{\partial\chi_3}{\partial\omega_2} = Q(t)z\cdot B_2(t) + G(t)\omega_2 = 0.$$

Thus, we obtain the strategy pairs

$$\omega_1^* = \frac{Q(t)B_1(t)}{H(t)}z, \omega_2^* = -\frac{Q(t)B_2(t)}{G(t)}z. \quad (A23)$$

Substituting (A23) into (A22), we obtain

$$-\frac{1}{2}Q^{'}(t)z^2 = -\frac{1}{2}z^2(-2Q(t)(A_1(t)+e^{\lambda d}A_3) - P(t) - \frac{2\sqrt{3}}{\pi}\ln\frac{1-\beta}{\beta}QA_4(t) - 2kQ(t)A_6(t) - \frac{Q^2(t)B_1^2(t)}{H(t)} + \frac{Q^2(t)B_2^2(t)}{G(t)}).$$

Thus,it holds that

$$\begin{cases}\frac{dQ(t)}{dt} + 2Q(t)(A_1(t)+e^{\lambda d}A_3) + P + \frac{Q^2(t)B_1^2(t)}{H(t)} - \frac{Q^2(t)B_2^2(t)}{G(t)}\\ \quad + \frac{2\sqrt{3}}{\pi}\ln\frac{1-\beta}{\beta}Q(t)A_4(t) + 2kA_6(t)Q(t) = 0,\\ Q(T) = \Psi_f,\end{cases} \quad (A24)$$

Owing to $\bar{J}_z A_6(t)z = A_6(t)$, and $A_6(t)\geq 0$, then $k = 1 - \frac{\beta}{2(1-\vartheta_2)}(0<\beta<1-\vartheta_2)$ or $k=\frac{1}{2}(1-\vartheta_2\leq\beta<1-\vartheta_1)$ or $k=\frac{1-\beta}{2\vartheta_1}(1-\vartheta_1\leq\beta<1)$.

Next, we verify the sufficient condition. We let $Q(t)$ satisfy Equation (A24). Then, it holds that

$$\begin{aligned}
d(QZ_s^2) =& Z_s^2\dot{Q}ds + 2QZ_s dZ_s \\
=& Z_s^2\dot{Q}ds + 2QZ_s((A_1(s) + e^{\lambda d}A_3)Z_s + B_1\omega_1 + B_2\omega_2)ds + 2QA_4Z_s^2dC_s + 2QA_6Z_s^2dV_s \\
=& Z_s^2(\dot{Q} + 2Q(A_1(s) + e^{\lambda d}A_3))ds + 2QZ_s(B_1\omega_1 + B_2\omega_2)ds + 2QA_4Z_s^2dC_s + 2QA_6Z_s^2dV_s \\
=& Z_s^2(-P - \frac{Q^2B_1^2}{H} + \frac{Q^2B_2^2}{G} - \frac{2\sqrt{3}}{\pi}\ln\frac{1-\beta}{\beta}Q|A_4| - 2kA_6Q)ds \\
&+ 2QZ_s(B_1\omega_1 + B_2\omega_2)ds + 2QA_4Z_s^2dC_s + 2QA_6Z_s^2dV_s.
\end{aligned}$$

From 0 to $T$, integrating the above equation leads to

$$\begin{aligned}
\frac{1}{2}\Psi_f Z_T^2 =& \frac{1}{2}Q(0)z_0^2 + \frac{1}{2}\int_0^T Z_s^2(-P - \frac{Q^2B_1^2}{H} + \frac{Q^2B_2^2}{G} - \frac{2\sqrt{3}}{\pi}\ln\frac{1-\beta}{\beta}Q|A_4| - 2kA_6Q)ds \\
&+ \int_0^T QZ_s(B_1\omega_1 + B_2\omega_2)ds + \int_0^T QA_4Z_s^2dC_s + \int_0^T QA_6Z_s^2dV_s
\end{aligned}$$

Substituting it into $\bar{J}(\omega_1, \omega_2)$ of (A21), we obtain

$$\begin{aligned}
\bar{J}(\omega_1,\omega_2) =& [\frac{1}{2}\int_0^T (PZ_s^2 + G\omega_2^2 - H\omega_1^2)ds + \frac{1}{2}Q(0)z_0^2 \\
&+ \frac{1}{2}\int_0^T Z_s^2(-P - \frac{Q^2B_1^2}{H} + \frac{Q^2B_2^2}{G} - \frac{2\sqrt{3}}{\pi}\ln\frac{1-\beta}{\beta}QA_4 - 2kA_6Q)ds \\
&+ \int_0^T QZ_s(B_1\omega_1 + B_2\omega_2)ds + \int_0^T QA_4Z_s^2dC_s + \int_0^T QA_6Z_s^2dV_s]_{\sup}(\beta) \\
=& [\frac{1}{2}\int_0^T [G(\omega_2 + \frac{QB_2Z_s}{G})^2 - H(\omega_1 - \frac{QB_1Z_s}{H})^2]ds + \frac{1}{2}Q(0)z_0^2 + \int_0^T QA_4Z_s^2dC_s \\
&+ \int_0^T QA_6Z_s^2dV_s - \frac{\sqrt{3}}{\pi}\ln\frac{1-\beta}{\beta}\int_0^T QA_4Z_s^2ds - k\int_0^T QA_6Z_s^2ds]_{\sup}(\beta).
\end{aligned}$$

Obviously, the equilibrium solutions are $(\omega_1^*, \omega_2^*)$ of (A23). Then, the theorem is completed. □

## eferences


Liu, B. *Uncertainty Theory: A Branch of Mathematics for Modeling Human Uncertainty*; Springer: Berlin, Germany, 2010.
Liu, B. Fuzzy process, hybrid process and uncertain process. *J. Uncertain Syst.* **2008**, *2*, 3–16.
Liu, B. Some research problems in uncertainty theory. *J. Uncertain Syst.* **2009**, *3*, 3–10.
Barbacioru, I. Uncertainty functional differential equations for finance. *Surv. Math. Its Appl.* **2010**, *5*, 275–284.
Ge, X.; Zhu, Y. Existence and uniqueness theorem for uncertain delay differential equations. *J. Comput. Inf. Syst.* **2012**, *8*, 41–83.
Wang, X.; Ning, Y. Stability of uncertain delay differential equations. *J. Intell. Fuzzy Syst.* **2017**, *32*, 2655–2664. [CrossRef]
Wang, X.; Ning, Y. A new stability analysis of uncertain delay differential equations. *Math. Probl. Eng.* **2019**, *2019*, 1257386. [CrossRef]
Jia, L.; Sheng, Y. Stability in distribution for uncertain delay differential equation. *Appl. Math. Comput.* **2019**, *343*, 49–56. [CrossRef]
Chen, R.; Zhu, Y. An optimal control model for uncertain systems with time-delay. *J. Oper. Res. Soc. Jpn.* **2013**, *54*, 243–256. [CrossRef]
). Jiang, Y.; Yan, Y.; Zhu, Y. Optimal control problem for uncertain linear systems with multiple input delays. *J. Uncertain Syst.* **2016**, *4*, 5. [CrossRef]
1. Deng, L.; Zhu, Y. Uncertain optimal control with jump. *ICIC Exp. Lett.* **2012**, *3*, 419–424.
2. Deng, L.; Zhu, Y. Existence and uniqueness theorem of solution for uncertain differential equations with jump. *ICIC Exp. Lett.* **2012**, *6*, 2693–2698.
3. Deng, L. Multidimensional uncertain optimal control of linear quadratic models with jump. *J. Comput. Inf. Syst.* **2012**, *8*, 7441–7448.
4. Deng, L.; Zhu, Y. An uncertain optimal control model with n jumps and application. *Comput. Sci. Inf. Syst.* **2012**, *9*, 1453–1468. [CrossRef]

15. Deng, L.; Zhu, Y. Uncertain optimal control of linear quadratic models with jump. *Math. Comput. Model.* **2013**, *57*, 2432–244 [CrossRef]
16. Chen, X.; Zhu, Y.; Sheng, L. Optimal control for uncertain stochastic dynamic systems with jump and application to an advertisin model. *Appl. Math. Comput.* **2021**, *407*, 126337. [CrossRef]
17. Deng, L.; Zhu, Y. Optimistic value model of uncertain linear quadratic optimal control with jump. *J. Adv. Comput. Intell. Intell. In* **2016**, *20*, 189–196. [CrossRef]
18. Deng, L.; Chen, Y. Optimal control of uncertain systems with jump under optimistic value criterion. *Eur. J. Control* **2017**, *38*, 7–1 [CrossRef]
19. Deng, L.; You, Z.; Chen, Y. Optimistic value model of multidimensional uncertain optimal control with jump. *Eur. J. Control* **201** *39*, 1–7. [CrossRef]
20. Sun, Y.; Ding, C.; Zhu, Y. Saddle point equilibrium under uncertain environment. *Math. Methods. Appl. Sci.* **2018**, *41*, 2063–207 [CrossRef]
21. Li, B.; Zhang, R.; Jin, T.; Shu, Y. Parametric approximate optimal control of uncertain differential game with application to counte terror. *Chaos Solitons Fractals* **2021**, *146*, 110940. [CrossRef]
22. Jia, Z.; Liu, X. Optimal control of multifactor uncertain system with jumps. *Int. J. Control* **2023**, *96*, 1272–1287. [CrossRef]
23. Jia, Z.; Liu, X. Optimistic value models of saddle point equilibrium control problems for uncertain jump systems. *Optimizatio* **2024**, 1–35. [CrossRef]
24. Jia, Z.; Liu, X. Uncertain stochastic hybrid differential game system with V-n jumps:Saddle point equilibrium, strategies an application to advertising duopoly game. *Chaos Solitons Fractals* **2023**, *171*, 113490. [CrossRef]
25. Chen, X.; Tian, C.; Jin, T. Optimal control and zero-sum game subject to multifactor uncertain random systems with jum *Optimization* **2023**, 1–42. [CrossRef]
26. Jia, Z.; Liu, X.; Li, C. Existence, uniqueness, and stability of uncertain delay differential equations with V-jump. *Adv. Differ. Eq* **2020**, *2020*, 440. [CrossRef]
27. Liu, B. Toward uncertain finance theory. *J. Uncertain. Anal. Appl.* **2013**, *1*, 1–15. [CrossRef]
28. Liu, Y.; Liu, B. Residual analysis and parameter estimation of uncertain differential equations. *Fuzzy Optim. Decis. Mak.* **2022**, *2* 513–530. [CrossRef]
29. Ye, T.; Liu, B.; Li, C. Uncertain hypothesis test for uncertain differential equations. *Fuzzy Optim. Decis. Mak.* **2023**, *22*, 195–21 [CrossRef]

machines

MDPI

'ticle

# . Terminal Residual Vibration Suppression Method of a Robot ;ased on Joint Trajectory Optimization

ang Liang [1,2,*], Chengdong Wu [3] and Shichang Liu [1,2]

[1] College of Information Science and Engineering, Northeastern University, Shenyang 110819, China; liushichang@siasun.com
[2] SIASUN Robot & Automation Co., Ltd., Shenyang 110168, China
[3] Faculty of Robot Science and Engineering, Northeastern University, Shenyang 110169, China; wuchengdong@mail.neu.edu.cn
* Correspondence: liangliang@siasun.com

**Abstract:** Vibration problems have become one of the most important factors affecting robot performance. To this end, a terminal residual vibration suppression method based on joint trajectory optimization is proposed to improve the accuracy and stability of robot motion. Firstly, based on the characteristics of the friction nonlinearity due to joint coupling and physical feasibility of dynamic parameters, a semidefinite programming method is used to identify dynamic parameters with actual physical meaning, thereby obtaining an accurate dynamic model. Then, based on the result of the residual vibration time domain analysis, a joint trajectory optimization model with the goal of minimizing joint tracking error is established. The Chebyshev collocation method is used to discretize the optimization model. The dynamic model is used as the optimization constraint, and barycentric interpolation is used to obtain the optimized joint motion trajectory. Finally, industrial robot experiments prove that the vibration suppression method proposed in this article can reduce the maximum acceleration amplitude of residual vibration by 62% and the vibration duration by 71%. Compared with the input shaping method, the method proposed in this paper can reduce the terminal residual vibration more effectively and ensure the consistency of running time and trajectory.

**Keywords:** industrial robot; parameter identification; vibration suppression; optimal trajectory; Chebyshev collocation

## 1. Introduction

With the development of science and technology and the rise in industrial automation, industrial robots perform tasks through efficient and precise movements, improving productivity, reducing costs, and creating a safer working environment for humans [1–3]. The flexible characteristics of industrial robot joint transmission parts like the RV reducers and gears reduce the overall stiffness of the robot, making it have a more complex vibration mode and easier to vibrate during and after the movement [4,5]. Such vibration problems have become one of the most important factors affecting the performance and stability of robots. Robot vibration suppression technology reduces or eliminates robot vibration through advanced control strategies and design methods, thereby improving motion accuracy and stability.

With an accurate system model, motion profiles or feedforward control can be generated based on optimal trajectory planning to minimize overshoot and residual vibration. Based on the chaotic particle swarm optimization algorithm, the trajectory interpolation parameters can be optimized with the minimum vibrational deformation of the flexible end as the objective function [6]. In reference [7], the dynamic equation of the flexible curve connecting rod was derived by using the absolute nodal coordinate formula, and the optimal trajectory method was used to generate the trajectory with the least vibration of the flexible curve connecting rod. In reference [8], the trajectory planning problem of suppressing the residual vibration of the robot was transformed into the optimal control performance problem, and the Fourier series was used to fit the discrete optimal trajectory. The authors of [9] proposed a torque feedforward damping strategy based on the chaotic regression tree dynamics model to improve the motion accuracy of the robot in the process of low-speed motion and suppress the vibration of the robot. In reference [10], coprime factorization-based feedforward control and H∞ control are cooperatively applied to the quasi-full-closed-loop control structure, forming a fundamental control strategy with improved reference tracking and residual vibration suppression capability. Reference [11] adopted a fully compensated feedforward torque control method based on the dynamic model, and compared with other feedforward methods, the robot has the fastest response speed and the lowest joint trajectory tracking error and jitter. However, the above methods have high requirements for the model and parameter accuracy of the system, otherwise the interference caused by the uncertain disturbance inside the robot will affect the algorithm control effect.

If an accurate model is not available but the states associated with elastic vibrations can be measured or observed, singular perturbations [12,13] and nonlinear feedback control [14,15] can be implemented to accommodate unmodeled dynamics and perturbations. Reference [16] proposed a novel high-performance multilayer neurocontroller which can simultaneously reject matched and mismatched disturbances like vibrations. Iterative learning control is another option to solve these problems, and optimal feedforward control can be implemented through learning when robots perform the same tasks repeatedly [17]. Compared with the above methods, the input shaping method can effectively minimize overshoot and residual vibration, and it does not require complex dynamic models, there is no need for the direct measurement of elastic vibrations for online feedback, and no need for repetitive tasks. However, one disadvantage of the use of input shaping in industrial robots is the time delay, which can slow down the overall task for applications with fast stationary to stationary motion such as spot welding, handling, etc. References [18,19] proposed an improved zero-delay input shaping so that no delay was introduced. However, this kind of shaping algorithm depends on the vibration frequency parameter of the robot body, and the frequency change in different postures of the robot will affect the shaping effect. In addition, input shaping may change the original motion trajectory, and for applications with high trajectory accuracy requirements such as arc welding and gluing, changing the motion trajectory will reduce the quality of the work.

In order to solve the above problems, this paper proposes a terminal residual vibration suppression method based on joint trajectory optimization. Firstly, a piecewise nonlinear friction model is designed according to the coupling characteristics of the robot joints, and the semi-definite programming is used to identify the dynamic parameters and friction parameters with physical significance synchronously, so as to obtain the accurate dynamic characteristics of the robot. Then, a method to transform Cartesian space trajectory vibration suppression into joint space trajectory optimization is proposed, and the optimization of the initial state and objective function is designed, which can ensure the consistency of trajectory optimization on the basis of terminal vibration suppression.

## 2. Identification of Dynamic Parameters Based on Physical Feasibility Constraints

Due to the complexity of the robot components and assembly errors, it is difficult to know the robot dynamics accurately. However, it was found that the dynamic characteristics of the robot are a linear function of its inertial parameters [20], so the problem of obtaining the inertial parameters of the robot can be transformed into a system of linear equations. First, the robot dynamics equation is rewritten into a linear parametric form, as shown in (1)

$$\tau = Y(q, \dot{q}, \ddot{q})\beta \tag{1}$$

where $\tau$ is the vector of joint torque, $\beta \in \mathbb{R}^{N_b}$ is the reorganization parameter vector which is called minimum inertia parameter, and $Y(q, \dot{q}, \ddot{q}) \in \mathbb{R}^{n \times N_b}$ is the full-rank observation matrix obtained after reorganization. Afterwards, by controlling robot by joint excitation trajectory and collecting the torque value of each joint in the process to obtain a group of test data, Formula (2) is obtained according to (1)

$$\Gamma = W\beta + \rho \tag{2}$$

where $W = \left[Y(q(t_1), \dot{q}(t_1), \ddot{q}(t_1))_{n\times N_b}, \ldots, Y(q(t_M), \dot{q}(t_M), \ddot{q}(t_M))_{n\times N_b}\right]^T$ is the observation matrix, $\rho$ is the observation error due to various reasons, and $\Gamma = \left[\tau(t_1)^T \quad \tau(t_2)^T \quad \ldots \quad \tau(t_M)^T\right]^T$ is the joint torque value corresponding to the test data. Finally, according to the least squares method, the minimum inertia parameter estimation vector $\hat{\beta}$ can be obtained.

Constructing an accurate friction model is an important part of improving the accuracy of the dynamic model, and it is difficult for a single friction model to fully fit the diverse joint features. For coupling joints of an industrial robot, the friction between two joints interacting with each other produces characteristics that differ from the standard friction model. Therefore, the friction model $f_j$ of the coupling joint $j$ is designed as Equation (3)

$$f_j = \begin{cases} Fc_{j-1} \cdot \mathrm{sign}(\dot{q}_j) + Fv_{j-1} \cdot \dot{q}_j, \dot{q}_s \leq \left|\dot{q}_j\right| < \dot{q}_l \\ Fc_{j-2} \cdot \mathrm{sign}(\dot{q}_j) + Fv_{j-2} \cdot \dot{q}_j, \left|\dot{q}_j\right| \geq \dot{q}_l \end{cases} \tag{3}$$

where $\dot{q}_s$ and $\dot{q}_l$ are pre-identified parameters for the friction model. $Fc_{j-m}$ and $Fv_{j-n}$ are the parameters to be identified with minimum inertial parameters synchronously.

The minimum inertial parameters calculated by the least squares method can only identify linear combinations of parameters and cannot assert the physical feasibility directly. Therefore, it is necessary to add feasibility constraints to the calculation of dynamic parameter identification to obtain the standard parameter form that satisfies the definition of physics. For the connecting rod k, the parameter $\delta_k$ to be identified is defined as

$$\delta_k = \left[L_{k,xx}, L_{k,xy}, L_{k,xz}, L_{k,yy}, L_{k,yz}, L_{k,zz}, l_{k,x}, l_{k,y}, l_{k,z}, m_k\right] \tag{4}$$

For a six-joint industrial robot, the mass of connecting rod $k$ must be positive, and the inertial tensor is symmetrical and positive definite. The set of available physical feasibility constraints for the robot is expressed as

$$\Omega = \left\{\delta \in \mathbb{R}^n : m_k > 0, L_k - \frac{1}{m_k} S(l_k)^T S(l_k) \succ 0 \mid k = 1, \ldots, 6\right\} \tag{5}$$

where $L_k$ is the inertial tensor of the connecting rod $k$ with respect to the coordinate system of the connecting rod, $l_k$ is the first moment of the vector of inertia, and $S(\cdot)$ is a partial symmetric matrix operator. Combined with the parameters of the friction model $f_{p,i}$ of each joint, the linear matrix inequality of the physical feasibility dynamic parameters is obtained as

$$D(\delta) = diag(D(\delta_1), \ldots, D(\delta_6), f_{p,1}, \ldots, f_{p,6}) \succ 0 \tag{6}$$

Since there is a one-to-one mapping relationship $G$ between the minimum inertia parameters $\beta$, the independent terms $\delta_d \in \delta$ in the parameters to be identified, and the linkage dynamics parameters [21], the physical feasibility constraints can be transformed into the constraints of the minimum inertia parameters and independent parameters, as shown in (7)

$$D_\beta(\beta, \delta_d) = D(\delta) = D\left(G^{-1}(\beta, \delta_d)\right) \tag{7}$$

$u$ is defined as the distance of the actual and identified values of the minimum inertial parameters as shown in (8)

$$u = \|\Gamma - W\beta\|^2 = (\Gamma - W\beta)^T(\Gamma - W\beta) \tag{8}$$

Convert (8) into a linear matrix inequality form as (9)

$$U_\omega(u,\beta) = \begin{bmatrix} u & (\Gamma - W\beta)^T \\ \Gamma - W\beta & 1 \end{bmatrix} \geq 0 \tag{9}$$

In summary, the problem of solving the dynamic parameters based on the constraints of physical feasibility can be expressed as the following form of SDP:

$$\min_{(u,\beta,\delta_d)} u \tag{10}$$

$$s.t. \quad F(u,\beta,\delta_d) \geq 0 \tag{11}$$

where the linear matrix inequality constraints are defined as (12)

$$F(u,\beta,\delta_d) = \begin{bmatrix} U_\omega(u,\beta) & \mathbf{0} \\ \mathbf{0} & D_\beta(u,\beta) \end{bmatrix} \tag{12}$$

For the optimization problem described in (10), (11), and (12), the optimal solution $u^*$ is the sum of the identification errors under constraints, and the standard parameters $\delta^*$ of robot dynamics satisfying the physical feasibility constraints can be calculated from the mapping of the $\delta_d^*$ and $\beta^*$.

## 3. Residual Vibration Suppression Based on the Optimal Trajectory Principle

Unlike the conventional position motion of the robot, the residual vibration is characterized by a small oscillation with the position at the end of the trajectory as the equilibrium point. Bringing the joint vibration response into the forward kinematics model of the robot, it can be seen that the wrist position of the robot during residual vibration is [8]

$$\begin{cases} p_{xe} = \sum\limits_{p=1}^{N} E_p \cos(\omega_p t - \phi_p) \\ p_{ye} = \sum\limits_{p=1}^{N} F_p \cos(\omega_p t - \phi_p) \\ p_{ze} = \sum\limits_{p=1}^{N} G_p \cos(\omega_p t - \phi_p) \end{cases} \tag{13}$$

where $E_p$, $F_p$, and $G_p$ are the amplitude of item $p$. The position error of the robot $P_e(t)$ in cartesian space is a function of the joint position error and velocity error at the end time. Therefore, the problem of reducing residual vibration can be transformed into a problem of reducing the position and velocity error at the end time.

On the basis of the initial value of the trajectory, the optimal trajectory problem can be obtained by setting up objective functions and constraints to meet the requirements [22]. The optimal trajectory algorithm designed in this paper includes three steps: initial trajectory discretization, establishment and solution of objective function and constraints, and barycentric interpolation.

### 3.1. Initial Trajectory Discretization

According to the robot kinematic model, due to the nonlinear mapping of the terminal pose and the joint angle, the acceleration discontinuity or abrupt change will still occur in the joint space, so the optimization calculation will be carried out in the joint space. For cartesian space trajectory $P(t)$, $0 \leq t \leq t_f$, the angle of each joint in each control cycle is $q_{p,i}(t)$, on which Chebyshev collocation [22] is performed as shown in (14)

$$q_{p,i}(T_k) = q_{p,i}\left(t_f[\cos(k\pi/N) + 1]/2\right), k = 0, \ldots, N \tag{14}$$

where $q_{p,i}(T_0)$ and $q_{p,i}(T_N)$ are the joint values corresponding to the initial and ending positions of the trajectory respectively.

One of the advantages of using the polynomial coordination method is that the differential matrix D can be used to derive the variables on the nodes quickly and accurately, so the differential of the position and torque of each joint are shown in (15)

$$\begin{cases} dq_{p,i}(T_k)/dT = 2D \cdot q_{p,i}(T_k)/t_f \\ d\tau_{p,i}(T_k)/dT = 2D \cdot \tau_{p,i}(T_k)/t_f \end{cases} \tag{15}$$

### 3.2. Objective Functions and Constraints

Define the position and velocity of each joint as the optimal problem state variables, and the joint torque as the control variable, as shown in (16)

$$\begin{cases} x(T_k) = [q(T_k), \dot{q}(T_k)]^T \\ u(T_k) = \tau(T_k) \end{cases} \tag{16}$$

Under the premise that the tracking performance of the control system is guaranteed, the problem of minimizing the position and speed error of the whole process and the start and stop points can be further simplified to the minimum acceleration change rate during motion, so the discretized optimization objective function is obtained as follows

$$J = \sum_{k=1}^{N} \dot{u}(T_k)^T W_R \dot{u}(T_k) \tag{17}$$

where $W_R$ is the weighted diagonal matrix. Theoretically, the greater the inertia, the greater the joint weight is.

The dynamic parameters identified by (10)–(12) can establish more accurate dynamic constraints for each joint, as shown in (18)

$$u(T_k) = \mathrm{DYN}(x(T_k), \delta^*) \tag{18}$$

In addition, the position and velocity of the start and end moments of each joint are used as equation constraints, and the position limit, velocity limit, and acceleration limit of each joint are used as inequality constraints at nodes to ensure that the robot starts and stops running. After the above discretization and design of the optimization objective function, the problem of the minimum error of position and velocity at the end point has been transformed into a typical nonlinear programming (NLP) problem with constraints. The algorithm based on the gradient descent principle can be used to solve the optimal control quantity $u^*(T_k)$ and state quantity $x^*(T_k)$ at the node.

### 3.3. Barycentric Interpolation

Since the calculated optimal state quantity is the value at each node after discretization, it is necessary to select an interpolation algorithm to convert it into the joint position value of each control cycle. The node-based polynomial fitting is achieved by barycentric interpolation [23], which can be expressed as a linear combination of Lagrange polynomials, and the optimized angles of each joint are obtained as shown in (19)

$$\hat{q}(t) = \sum_{j=0}^{N} q^*(T_j) l_j(t) \tag{19}$$

where $l_j(t)$ is the $j$-th Lagrange polynomial, defined as (20)

$$l_j(t) = \prod_{k=0}^{N} (t - T_k) \frac{v_j}{t - T_j} \tag{20}$$

where $v_j$ is the imputation weight corresponding to node $j$. For node-based polynomial interpolation, the error of the optimized trajectory and the original trajectory at the node is 0, and the error between the nodes can be obtained by (21)

$$\varepsilon(t) = \mathrm{P}(t) - \mathrm{KINE}(\hat{q}(t)) \tag{21}$$

Based on the above calculations, the motion trajectory with the same number of interpolation points as the pre-optimization trajectory is obtained again, which not only ensures that the optimized start–stop position is the same as the original trajectory but also ensures the consistency between the optimized running trajectory and the original trajectory. The joints of the robot are controlled to complete the movement according to the optimized trajectory, so as to reduce the amplitude and duration of residual vibration.

## 4. Experimental Verification and Analysis of Results

The industrial robot T12B is selected for experimental analysis, and the inertial sensor is selected for vibration testing. The parameters are shown in Table 1. The experiment is divided into two steps, the dynamic parameters are identified first, and then the residual vibration suppression experiment is carried out based on the identification results.

**Table 1.** Experimental platform parameters.

| Equipment | Parameter Items | Parameter Value |
|---|---|---|
| Robot | Type | SIASUN T12B |
| | Degree of freedom | 6 |
| | Standard load | 14 KG |
| | Scope of work | 1465 mm |
| Inertial sensor | Type | XSENS MTI-100-2A8G4 |

### *4.1. Identification of Dynamic Parameters*

The excitation trajectory based on the Fourier series function [24] is selected to control the robot motion, the joint position and torque information are collected, and the dynamic parameters are calculated according to the identification algorithm. Then, the verification trajectory different from the excitation trajectory is selected, the actual torque of each joint during the movement is recorded, and the theoretical torque value is calculated according to the identification parameters and the difference between the two torques is compared to verify the correctness and accuracy of the identification results.

The minimum inertial parameters without considering the constraints of physical feasibility are used, and the dynamic parameter identification results are calculated by the least squares method. The actual torque of each joint and the theoretical torque based on the identified parameters are plotted as shown in Figure 1. It can be seen from the figure that the theoretical torque and the actual torque have a large spike protrusion when the velocity direction changes, and there is always a certain range of error in the constant velocity section.

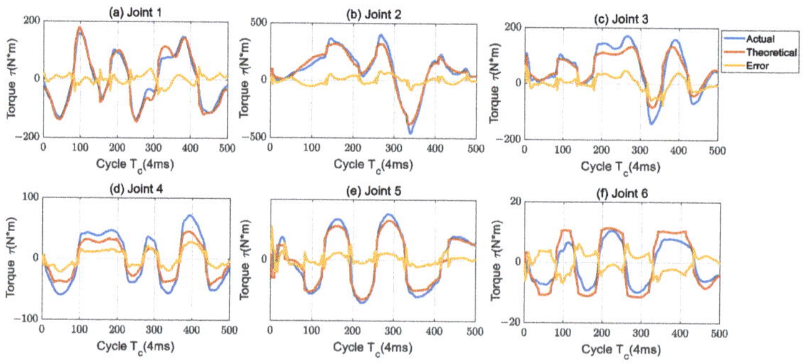


**Figure 1.** Theoretical and actual curve of joint torque based on identified parameters.

According to the coupling characteristics of the T12B robot and the unified identification method of joint friction, the friction model such as Equation (3) is selected, and the friction model is included in the identification parameter set. The same excitation trajectory is selected, and the test data are substituted into Equations (10)–(12) to calculate the kinetic parameters that meet the constraints of physical feasibility. The recalculated kinetic parameters are shown in Table 2, and the identified friction parameters are shown in Table 3.

**Table 2.** Identification results of the dynamic parameters based on physical feasibility constraints.

| Joint | $L_{xx}$ | $L_{xy}$ | $L_{xz}$ | $L_{yy}$ | $L_{yz}$ | $L_{zz}$ | $l_x$ | $l_y$ | $l_z$ | $m$ | $J_m$ |
|---|---|---|---|---|---|---|---|---|---|---|---|
| 1 | 44.876 | 0 | 0 | 44.876 | 0 | 1.476 | 0 | 0 | 0 | 19.965 | $8.8 \times 10^{-7}$ |
| 2 | 3.102 | −0.843 | −0.696 | 10.001 | −1.549 | 10.089 | 8.828 | 0.517 | 0.371 | 9.261 | 4.245 |
| 3 | 4.140 | 1.608 | 0.873 | 3.305 | 0.215 | 4.170 | 3.899 | 6.325 | 0.811 | 16.031 | 0.757 |
| 4 | 1.724 | −0.126 | −0.393 | 1.758 | −0.313 | 0.257 | 1.505 | 0.018 | −1.768 | 4.489 | 1.505 |
| 5 | 0.480 | 0.015 | −0.099 | 0.489 | 0.074 | 0.032 | 0.007 | −0.005 | 0.035 | 0.002 | 0.156 |
| 6 | 0.005 | −0.007 | 0.006 | 0.019 | 0.002 | 0.020 | 0.129 | 0.052 | −0.049 | 0.965 | 4.422 |

**Table 3.** Identification results of the friction parameters based on physical feasibility constraints.

| Joint | $f_{v1}$ | $f_{c1}$ | $f_{v2}$ | $f_{c2}$ |
|---|---|---|---|---|
| 1 | 123.382 | 26.457 | 51.442 | 70.107 |
| 2 | 41.718 | 32.308 | 0 | 0 |
| 3 | 35.108 | 32.500 | 0 | 0 |
| 4 | 65.267 | 8.969 | 24.356 | 33.791 |
| 5 | 15.213 | 3.546 | 2.692 | 10.961 |
| 6 | 9.032 | 2.277 | 1.713 | 7.442 |

The parameters of Table 2 are brought into the dynamic formula, and the friction of the joint is calculated according to the friction parameters in Table 3. The actual output torque of each joint and the theoretical torque based on the identified parameters are plotted as shown in Figure 2. Compared with Figure 1, it can be seen that the peak protrusion of the theoretical and actual torque is significantly reduced when the velocity direction changes, and the accuracy in the uniform velocity section is significantly improved, especially the torque errors of joints 3, 4, and 6 are improved significantly.

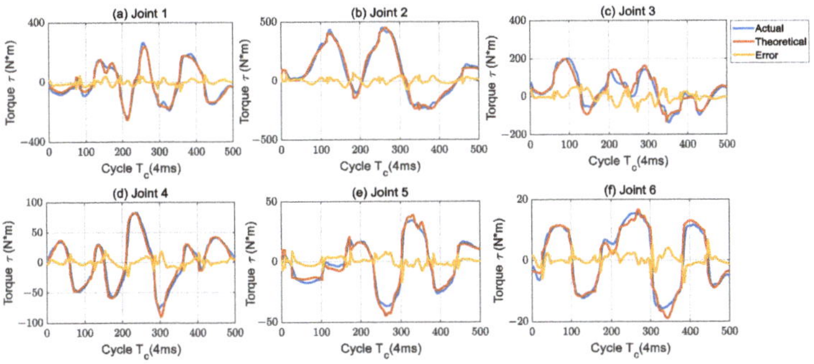


**Figure 2.** Theoretical and actual curve of joint torque based on the identification algorithm in this paper.

The root mean square error (RMS) of the theoretical torque and the actual torque of the identification model are shown in Table 4. According to the RMS results, the theoretical value of the torque in the whole process is closer to the actual torque value identified by the constraint-based inertia parameters. Compared with the identification results of the unconstrained and frictionless model, the results of the 4, 5, and 6 joints proposed in this paper are improved by about 80% compared with the original algorithm, which shows that the algorithm has a significant effect on coupling joints and friction-sensitive joints.

**Table 4.** RMS of dynamic parameter identification results.

| Parameter Type | Index | Joint 1 | Joint 2 | Joint 3 | Joint 4 | Joint 5 |
|---|---|---|---|---|---|---|
| P1 [1] | RMS (R1) | 14.650 | 22.683 | 13.826 | 6.547 | 2.680 |
| P2 [2] | RMS (R2) | 29.388 | 25.806 | 16.875 | 37.783 | 14.737 |
| P1 to P2 | (R2−R1)/R2 | 50.150% | 12.102% | 18.068% | 82.672% | 81.814% |

[1] Parameters with feasibility constraint; [2] Parameters without constraint.

## *4.2. Verification of the Performance of Vibration Suppression*

As shown in Figure 3, the robot moves in a straight line from position P1 to position P2 with a linear speed of 200 mm/s. P1 and P2 are (590, 0, 294) and (1317, 0, 294), respectively, under the base coordinate, and the unit is mm. An inertial sensor is installed at the flange to measure vibration during motion.

(**a**) (**b**)

**Figure 3.** Trajectory of the residual vibration experiment. (**a**) Robot experiment path. (**b**) Inertial sensor.

The experimental trajectory is the linear motion in cartesian space with no displacement in the *Y*-axis and *Z*-axis directions. The velocity in the *X*-axis direction is an S-shaped curve, which ensures a smooth velocity in cartesian space. From the kinematic model, it can be seen that the *X*-axis direction movement is mainly generated by the linkage of joints 2, 3, and 5, as shown in Figure 4.

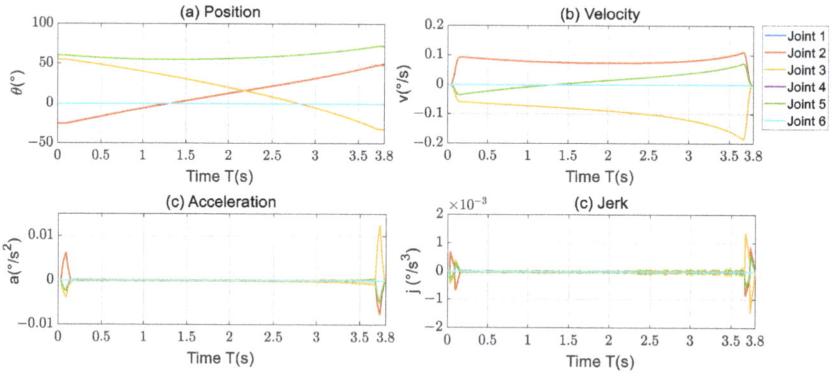


**Figure 4.** Joint motion profiles.

However, due to the nonlinearity of the kinematic inverse solution, the above-mentioned joints produce a large velocity step at the start–stop point. In particular, in the deceleration stage of joint 3 at the stop point, the joint acceleration produces a peak value of 0.012, which is about 4 times the peak acceleration value of the initial stage. The sudden change in acceleration of the deceleration section leads to a large tracking error, which causes residual vibration at the end of the robot.

The acceleration curve in the Z direction of the terminal is obtained by the inertial sensor, as shown in Figure 5. As can be seen from the figure, the vibration with an amplitude of less than 1 in the Z direction at the beginning of the motion quickly converges to less than 0.5 with the motion. In the stop deceleration phase, the end amplitude is about three times that of the starting segment due to the abrupt change in acceleration, and a significant damping oscillation is generated.

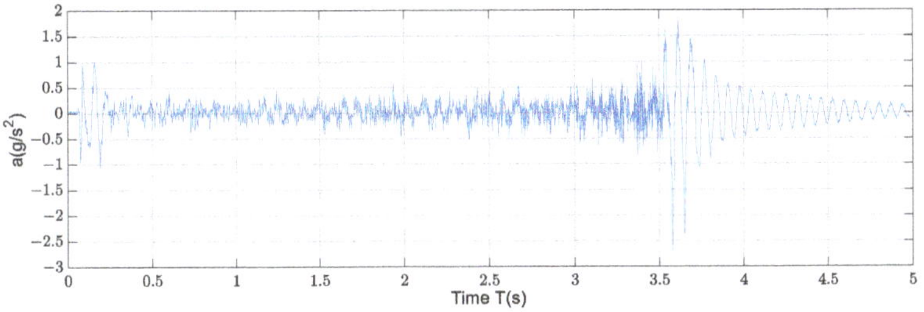


**Figure 5.** Acceleration curve in the Z direction of the robot flange in the motion process.

According to the optimization model established in this paper, the kinetic parameters identified in Table 1 are brought into the dynamic constraints, the number of collocation nodes is selected as N = 12, the threshold of the number of iterations is 10, the solution time is 0.2 s, and the motion curve of the optimized joint is obtained as shown in Figure 6.

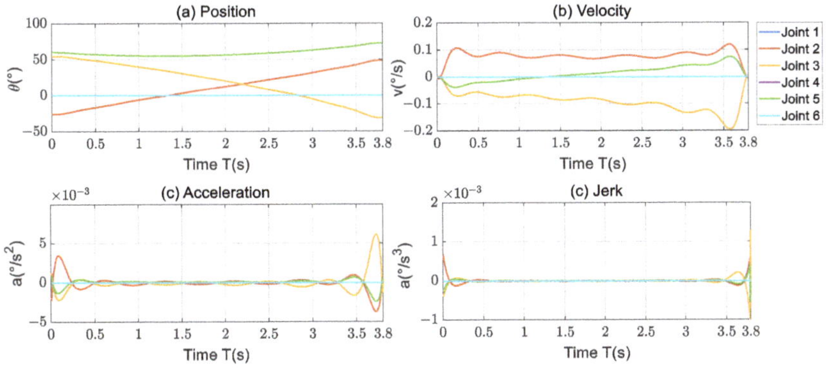


**Figure 6.** Joint motion profiles after optimization.

After optimization, the start and stop points of the position curve of each joint are the same as those of the original curve, and the running time is the same. From Equation (21), it can be seen that there is a maximum error of 0.5 mm between the trajectory and the original trajectory during the process, which ensures the consistency of the running trajectory. The acceleration amplitude of joint 2 and joint 3 in the stop phase is about 0.5 times that of the original trajectory, and the jerk amplitude is basically the same as that of the original trajectory.

After optimization, the vibration at the end of the robot during operation was re-measured, there was no visible vibration at the end of the entire operation stage, and the vibration curve of the stop section is shown in Figure 7. At the same time, the joint input shaping method is used to obtain the movement trajectory, the input shaping frequency is 8 Hz, the damping ratio is 0.4, and the delay of about 0.4 s is generated compared with the original trajectory, as shown in Figure 7.

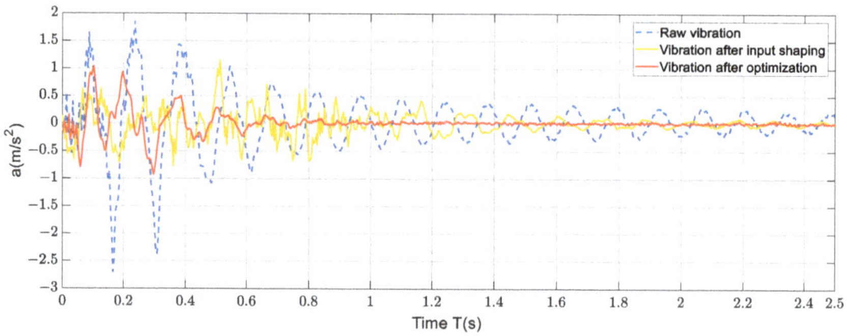


**Figure 7.** Residual vibration suppression effect curve in the Z direction of the flange.

As can be seen from the figure, the maximum amplitude of the original vibration acceleration curve is 2.71, and the amplitude converges to less than 0.3 after 1.58 s. After optimization, the maximum amplitude of the curve is 1.04, the amplitude converges to less than 0.3 after 0.46 s, the vibration stops completely at 0.7 s, the maximum amplitude decreases by 61.62%, and the vibration duration decreases by 70.89%. The input shaping vibration curve has a significant delay compared with the original curve with the maximum amplitude of 1.15 at 0.51 s. The maximum amplitude converges to within 0.3 after 0.69 s, but there is still a small vibration after that. The amplitude converges to within 0.1 at 1.64 s. From the above analysis, it can be seen that compared with the input shaping method, the proposed method can reduce the residual vibration more effectively.

## 5. Conclusions

In this paper, a terminal vibration suppression method based on joint trajectory optimization is proposed. This method is based on the identification results of dynamic parameters with physical feasibility constraints and takes the joint position and velocity tracking difference as the optimization goals. The trajectory obtained by optimization reduces the terminal vibration acceleration amplitude and convergence time and ensures the consistency between the running trajectory and the original trajectory. The experimental results show that compared with the unconstrained least squares method, the torque error is significantly reduced at the joint velocity reversal, and the RMS of the coupling joint torque is reduced by 82.67% at most. Based on the identification results as dynamic constraint parameters, the maximum acceleration amplitude can be reduced by 61.62% and the vibration duration by 70.89%.

**Author Contributions:** Methodology, L.L. and C.W.; Software, L.L. and S.L.; Validation, L.L. and C.W.; Writing—review and editing, L.L. and S.L. All authors have read and agreed to the published version of the manuscript.

**Funding:** This research was funded by the National Natural Science Foundation of China (U20A20197).

**Data Availability Statement:** Data are contained within the article. The vibration suppression test is openly available in Github at https://github.com/ErenYeager0/Terminal-Residual-Vibration-Suppression (accessed on 27 July 2024).

**Conflicts of Interest:** Authors Liang Liang and Shichang Liu were employed by the company SIASUN Robot & Automation Co., Ltd. The remaining authors declare that the research was conducted in the absence of any commercial or financial relationships that could be construed as a potential conflict of interest.

## eferences


Bluemel, S.; Bastick, S.; Staehr, R. Robot based remote laser cutting of three-dimensional automotive composite parts with thicknesses up to 5mm. *Procedia CIRP* **2018**, *74*, 417–420. [CrossRef]

Tran, C.C.; Lin, C.Y. An intelligent path planning of welding robot based on multisensor interaction. *IEEE Sens. J.* **2023**, *23*, 8591–8604. [CrossRef]

Liu, Y.; Zi, B.; Wang, Z.; You, W.; Zheng, L. Research Progress and Trend of Key Technology of Intelligent Spraying Robot. *J. Mech. Eng.* **2022**, *58*, 53–74.

Shabana, A.A.; Bai, Z. Actuation and motion control of flexible robots: Small deformation problem. *J. Mech. Robot.* **2022**, *14*, 011002. [CrossRef]

Yan, Y.; Guo, Y.; Liu, X. Tooth root crack detection of planet gear in industrial robot RV reducer. *Meas. Control* **2023**, *56*, 1720–1731. [CrossRef]

Long, T.; Li, E.; Yang, G.; Yang, L.; Fan, J.; Liang, Z. Trajectory planning of vibration suppression for hybrid structure flexible manipulator based on PSO non-uniform spline interpolation. *J. Intell. Fuzzy Syst.* **2018**, *33*, 978–988.

Zhao, L.; Wang, H.; Chen, W. Optimal trajectory planning for manipulators with flexible curved links. In *Intelligent Autonomous Systems 14, Proceedings of the 14th International Conference IAS-14, Shanghai, China, 3–7 July 2016*; Springer: Cham, Switzerland, 2017; pp. 1013–1025.

Chen, J. Dynamic Performance-Oriented Industrial Robot Control Technology Research. Ph.D. Thesis, Harbin Institute of Technology, Harbin, China, 2015.

Jia, B.; Chen, L.; Zhang, L.; Fu, Y.; Zhang, Q.; Pan, H. Vibration suppression of welding robot based on chaos-regression tree dynamic model. *Nonlinear Dyn.* **2024**, *112*, 4393–4407. [CrossRef]

). Trung, T.V.; Iwasaki, M. Fast and precise positioning with coupling torque compensation for a flexible lightweight two-link manipulator with elastic joints. *IEEE ASME Trans. Mechatron.* **2022**, *28*, 1025–1036. [CrossRef]

. Zhang, T.; Hong, L.; Qiang, K. Real-time feedforward torque control of an industrial robot based on the dynamics model. *Univ. Politeh. Buchar. Sci. Bull. D Mech. Eng.* **2020**, *82*, 3–18.

2. Zheng, K.; Zhang, Q.; Zeng, S. Trajectory control and vibration suppression of rigid-flexible parallel robot based on singular perturbation method. *Asian J. Control* **2022**, *24*, 3006–3021. [CrossRef]

3. Khan, R.F.A.; Rsetam, K.; Cao, Z.; Man, Z. Singular perturbation-based adaptive integral sliding mode control for flexible joint robots. *IEEE Trans. Ind. Electron.* **2022**, *70*, 10516–10525. [CrossRef]

4. Rigatos, G.; Abbaszadeh, M. Nonlinear optimal control for multi-DOF robotic manipulators with flexible joints. *Optim. Control Appl. Methods* **2021**, *42*, 1708–1733. [CrossRef]

5. Alam, W.; Mehmood, A.; Ali, K.; Javaid, U.; Alharbi, S.; Iqbal, J. Nonlinear control of a flexible joint robotic manipulator with experimental validation. *Stroj. Vestn. J. Mech. Eng.* **2018**, *64*, 47–55.

5. Yang, G.; Yao, J. Multilayer neurocontrol of high-order uncertain nonlinear systems with active disturbance rejection. *Int. J. Robust Nonlinear Control.* **2024**, *34*, 2972–2987. [CrossRef]

7. Wang, Z.; Zhou, R.; Hu, C.; Zhu, Y. Online iterative learning compensation method based on model prediction for trajectory tracking control systems. *IEEE Trans. Ind. Inform.* **2021**, *18*, 415–425. [CrossRef]

3. Zhao, Y.; Chen, W.; Tang, T. Zero time delay input shaping for smooth settling of industrial robots. In Proceedings of the IEEE International Conference on Automation Science and Engineering, Fort Worth, TX, USA, 21–25 August 2016; pp. 620–625.

9. Zhao, Y.; Tomizuka, M. Modified zero time delay input shaping for industrial robot with flexibility. In Proceedings of the Dynamic Systems and Control Conference, Tysons, VA, USA, 11–13 October 2017; ISBN 978-0-7918-5829-5, Paper No: DSCC2017-5219, V003T22A003.

0. Han, Y.; Wu, J.; Liu, C.; Xiong, Z. An iterative approach for accurate dynamic model identification of industrial robots. *IEEE Trans. Robot.* **2020**, *36*, 1577–1594. [CrossRef]

1. Sousa, C.D.; Cortesao, R. Physical feasibility of robot base inertial parameter identification: A linear matrix inequality approach. *Int. J. Rob. Res.* **2014**, *33*, 931–944. [CrossRef]

2. Kelly, M. An introduction to trajectory optimization: How to do your own direct collocation. *SIAM Rev.* **2017**, *59*, 849–904. [CrossRef]

3. Berrut, J.P.; Trefethen, L.N. Barycentric lagrange interpolation. *SIAM Rev.* **2004**, *46*, 501–517. [CrossRef]

4. Swevers, J.; Verdonck, W.; De, S.J. Dynamic model identification for industrial robots. *IEEE Control Syst. Mag.* **2007**, *27*, 58–71.

machines

MDPI

*Article*

# Study on Rubbing-Induced Vibration Characteristics Considering the Flexibility of Coated Casings and Blades

**Yong Zhang [1,2], Shuhua Yang [1], Xingyu Tai [1], Hui Ma [3,4,*], Hong Guan [3,*], Qinqin Mu [5], Lin Qu [6] and Xiangfu Ding**

1 Shengu Group Co., Ltd., Shenyang 110869, China; zhangyong@shengu.com.cn (Y.Z.); yangshuhua@shengu.com.cn (S.Y.); taixingyu@shengu.com.cn (X.T.)
2 School of Energy and Power Engineering, Xi'an Jiaotong University, Xi'an 710049, China
3 School of Mechanical Engineering and Automation, Northeastern University, Shenyang 110819, China; 2300362@stu.neu.edu.cn
4 Key Laboratory of Vibration and Control of Aero-Propulsion Systems Ministry of Education of China, Northeastern University, Shenyang 110819, China
5 Aircraft Strength Institute of China, Xi'an 710065, China; muqinqin198465@163.com
6 Sunny Optical Technology (Group) Co. Ltd., Ningbo 315400, China; lindsyqu@163.com
* Correspondence: huima@me.neu.edu.cn (H.M.); 2310094@stu.neu.edu.cn (H.G.)

**Abstract:** Rubbing between a blade and its coated casing is one of the main failures in aero-engine systems. This paper aims to study the effects of coated casings on rubbing-induced dynamic responses considering the flexibility of the coated casing and the flexibility of the blade. Firstly, an actual compressor blade is established by the shell element and verified by the experiment and ANSYS 19.2 software. Subsequently, a new dynamic model for the coated casing is proposed based on the laminated shell element, and the proposed dynamic model for the coated casing is verified by comparing the natural characteristics calculated by ANSYS software. Moreover, a comprehensive analysis is conducted to analyze the influences of the casing model, coating parameters, and casing parameters on vibration characteristics. Finally, the results show that the coating can diminish the severity level of rubbing. Notably, the material and thickness of the coating can change the nodal diameter vibrations of the casings (NDVCs) induced by rubbing. This study provides valuable guidance for the optimization and design of blade–casing systems.

**Keywords:** blade-coated casing system; rubbing; vibration characteristic; finite element method; flexible casing

## 1. Introduction

To enhance engine efficiency by changing airtightness, the radial clearance between blades and casings has been gradually reduced [1–3]. However, this reduction has aroused the issue of blade–casing rubbing faults, further resulting in economic losses and catastrophic accidents. Technological advancements have led to the application of coatings on the casing to enhance wear resistance and reduce vibration induced by rubbing. Consequently, attention has gradually shifted from blade–casing rubbing to the blade-coated casing, and the latter has become a main concern in aero-engine operation. Rubbing may have adverse effects on blades and casings, as depicted in Figure 1. Consequently, a comprehensive analysis of blade-coated casing rubbing is important to ensure the secure operation of aero-engines.

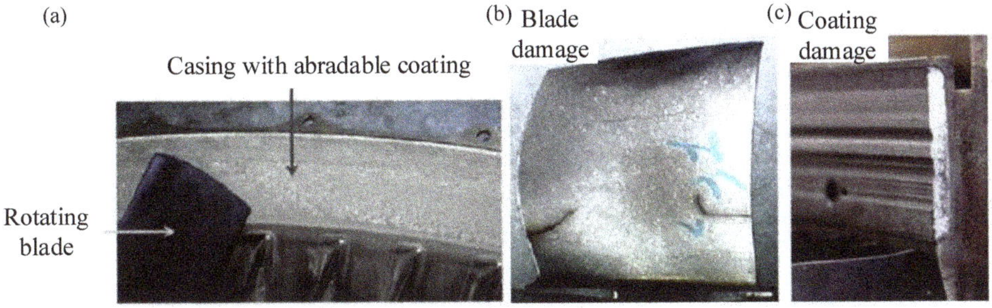


**Figure 1.** Schematic of a blade–casing system and the hazards induced by rubbing: (**a**) blade-coated casing system, (**b**) blade damage in Ref. [4]; (**c**) coating damage in Ref. [5].

The dynamic modeling of blades has been a prominent focus in recent decades, with a variety of approaches using the finite element (FE) method, such as the cantilever beam model [6–16], the cantilever plate model [17–22], the shell model [23–31], and the solid element model [32–36]. It is a challenge to accurately capture the true morphological characteristics of compressor blades when beam element models are employed. Although the solid element model can compensate for this shortcoming, the use of solid elements often results in lower computational efficiency. Hence, the adoption of shell element modeling emerges as a compromise. Based on the shell element, various researchers have made significant contributions to the establishment of dynamic models for rotating blades. For instance, Sinha and Zylka [23] derived the control differential equation for rotating pre-twisted airfoil blades, transforming it into matrix eigenvalue form using the Rayleigh–Ritz method. Sun et al. [24] proposed a modeling method for pre-twisted rotating blades based on shell theory and the Hamilton principle, and they studied the vibration characteristics under different stagger angles and rotational speeds. Kee and Shin [26] developed a modeling approach using composite blades with a mixed element, combining an eight-node solid element with a shell element. Yangui et al. [27] used a three-node triangular shell element with six degrees of freedom to conduct the dynamic modeling of blades. Rafiee et al. [29] used the Shell99 element to establish an FE model of fan blades and focused on the study of the aeroelastic characteristics of blades. Kee and Kim [30] established an FE model for pre-twisted rotating composite blades using a degenerated shell element with nine nodes. They studied the effects of composite material lamination and fiber orientation parameters on the dynamic characteristics of blades. Additionally, Xiong et al. [31] utilized shell elements to establish a dynamic model for actual compressor blades and analyzed crack-induced characteristics. The aforementioned literature underlines the maturity of employing shell elements in blade modeling. Thus, based on our previous study (Ref. [31]), shell elements are utilized to establish compressor blades to investigate the rubbing fault of compressor-blade-coated casings in this paper.

There is a great deal of interest in the modeling of casings for blade–casing systems, with the lumped-mass model and FE model being approaches of interest. To simplify calculations, casings are built by the lumped-mass method. For instance, Sinha [37] simplified the casing as a lumped-mass point and investigated the transient dynamic response considering rubbing. Ma et al. [38] developed an FE model of a rotor–rigid casing and analyzed the vibration response induced by rubbing, and the casing was considered as the lump-mass point. While such simplified models serve certain analytical purposes, the actual casing, which is a thin shell structure, undergoes local deformation due to rubbing. Therefore, the accurate modeling of casings needs to consider casing flexibility. Zeng et al. [39] used Timoshenko beam elements to establish an FE model of casings and studied rubbing-induced dynamic responses. Lesaffre et al. [40] studied the stability of a rotating beam-flexible casing considering rubbing based on the Routh–Hurwitz criterion. Guo et al. [41] introduced an elastic-supported casing model according to the Sanders shell theory and mainly investigated the rubbing-induced vibration response of casings. Zeng et al. [42] also proposed a modeling approach of a rotor-disk-flexible casing considering elastic support using ANSYS/LS-DYNA software. In particular, Batailly and Legrand [43] analyzed the vibration response caused by the unilateral contact between blades and flexible casings based on the contact algorithm. In the context of coated casings, which are prevalent in modern aero-engine designs, more attention is paid to composite casings [44–51]. To enhance calculation accuracy, the FE method is widely employed to establish composite cylindrical shells (i.e., coated casings). For example, Zhang et al. [49,50] used a four-node shell element to conduct the dynamic modeling of a hard-coated cylindrical shell and investigated nonlinear vibration characteristics considering the base excitation. Additionally, Khan et al. [51] also studied the dynamic characteristics of laminated cylindrical shells via the FE method. Despite these advancements, the literature indicates a need for further exploration into utilizing more accurate shell elements in building dynamic models of coated casings. The development of coated casing models using the finite shell element method is still a field worthy of further research to improve accuracy and reliability.

In addition, scholars have also turned their attention to the study of the rubbing dynamic response in blade–casing systems. For example, Ma et al. [52,53] considered the bending deformation of blades and established an improved normal rubbing force model. Zhou et al. [54] investigated rubbing-induced vibration characteristics and found that harmonic frequencies near the natural frequency are more susceptible to the effect of rubbing. Hou et al. [55] analyzed frequency modulation considering rubbing in a rotor system and proposed an indicator to evaluate the severity of rubbing. Wu et al. [56] established a dynamic model with large rotation in a rotor-blade system and found the natural frequency in the spectrum. Shao et al. [57] also investigated the vibration characteristics caused by looseness and rubbing. Li et al. [58] proposed a rubbing model considering the local deformation of the casing during the rubbing process and studied the vibration characteristics of casings. Yang et al. [59–61] proposed a novel hysteresis force model to characterize contact force and studied the bending–torsional coupling vibration of the blade–casing system under non-uniform initial clearance conditions. Zeng et al. [62] primarily analyzed the rubbing-induced dynamic responses of blades. However, casing vibration has not received substantial attention in Refs. [63–70]. Although the vibration of an uncoated casing was considered in our previous paper [41], coatings can alter the natural characteristics of a casing and affect the dynamic response. Notably, coating wear has garnered significant attention, including in relation to the wear process [71–73], blade–coating interactions [74–76], wear models [77,78], and so on. While current research on blade–casing rubbing predominantly analyzes the vibration characteristics of blades, that of casing is often overlooked. Although the rubbing-induced dynamic response of blade–casing systems has gained widespread attention, it is essential to shift focus towards the vibration characteristics of coated casings and the effect of coatings on rubbing, with regard to coating material, coating thickness, etc.

The primary motivation of this paper is to study the rubbing-induced dynamic responses in a blade-coated casing system. The following contributions are listed in this paper on the basis of previous works: (1) A modeling method of the coated casing is developed based on laminated shell elements. Subsequently, a dynamic model of the system with rubbing is established. (2) The influence of the casing model (lumped-mass model, cylindrical shell model, and laminated casing model) and casing parameters (casing support stiffness and casing length) on the dynamic response of the system is systematically analyzed. (3) Notably, this study delves into the analysis of the effects of coatings on dynamic responses considering rubbing, including coating thickness and coating materials. In summary, this article primarily focuses on the response of coated casings and explores the influence of coatings on rubbing-induced dynamic responses, providing valuable insights into the field.

The layout of this article is organized as follows: The establishment of the model is described in Section 2, including blade dynamic modeling, casing dynamic modeling, rubbing force modeling, and system dynamic modeling. Section 3 mainly includes numerical simulation analyses, such as a comparison of casing models and investigations into the influence of coating parameters and casing parameters. Finally, Section 4 displays some conclusions.

## 2. Dynamic Modeling of the Blade-Coated Casing System with Rubbing

The dynamic model proposed is displayed in Figure 2. Its main components include blade dynamic modeling, casing dynamic modeling, and so on. Throughout the modeling process, several assumptions are considered as follows:

(1) Similar to Ref. [59], coating wear and temperature effects are not considered when rubbing occurs.
(2) The origin of the coordinate system is located at the center of mass of the rigid disk.

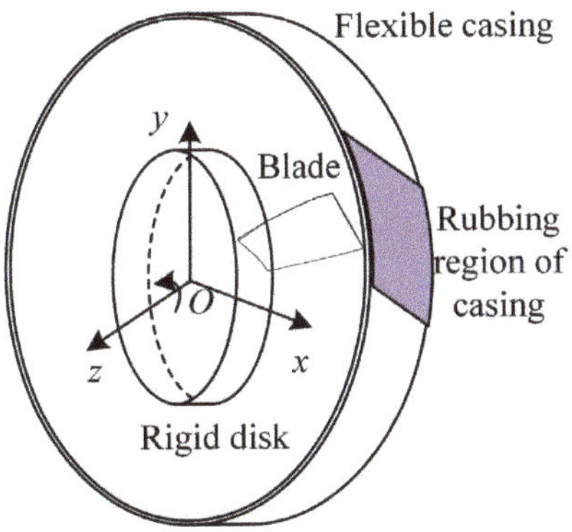


**Figure 2.** Schematic of the flexible blade–casing system.

### 2.1. Dynamic Model of the Blade

The compressor blade is shown in Figure 3 and the blade is established by the FE method. To accurately describe the blade shape, a three-dimensional shell element with eight nodes is used to establish the FE model of the blade. This FE model has a total of 400 (20 × 20) elements. *Oxyz* and *O′x′y′z′* are the global and local coordinate systems, respectively. $R_d$ represents the radius of the rigid disk.

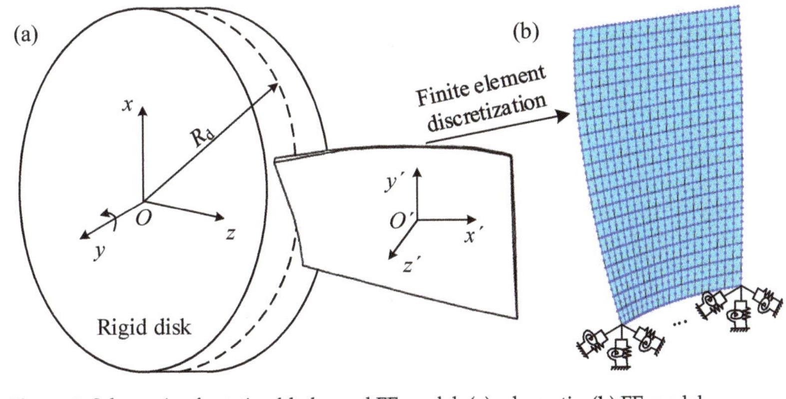


**Figure 3.** Schematic of rotating blades and FE model: (**a**) schematic; (**b**) FE model.

On the basis of Ref. [31], the dynamic model of the blade is built. Detailed parameters can be found in Ref. [31]. A spring element is applied at each node of the blade root to simulate the elastic support, as depicted in Figure 3. It is essential to note that the shell element possesses only five degrees of freedom. Consequently, the last torsional stiffness of the spring element is filled with zero N·m/rad. Consequently, the differential equation of motion of the blade with elastic support is written as

$$\boldsymbol{M}\ddot{\boldsymbol{x}} + (\boldsymbol{G} + \boldsymbol{C})\dot{\boldsymbol{x}} + (\boldsymbol{K}_{\mathrm{e}} + \boldsymbol{K}_{\mathrm{ce}} + \boldsymbol{K}_{\mathrm{co}} - \boldsymbol{K}_{\mathrm{s}})\boldsymbol{x} = \boldsymbol{0} \quad (1)$$

where $\boldsymbol{M}$, $\boldsymbol{C}$, $\boldsymbol{G}$, $\boldsymbol{K}_{\mathrm{e}}$, $\boldsymbol{K}_{\mathrm{ce}}$, $\boldsymbol{K}_{\mathrm{co}}$, and $\boldsymbol{K}_{\mathrm{s}}$ represent the mass, damping, Coriolis force, structure stiffness, centrifugal stiffening, elastic support stiffness, and spin softening matrixes of the blade. Detailed modeling processes and descriptions are displayed in Appendix A. In this paper, the stiffness of springs to simulate the elastic support is calculated and the results are found to be $k_x = 4.001 \times 10^8$ N/m, $k_y = 4.61 \times 10^7$ N/m, $k_z = 1.997 \times 10^8$ N/m, $k_{\theta x} = 1.437 \times 10^{11}$ N·m/rad, and $k_{\theta y} = 2.118 \times 10^{10}$ N·m/rad during the analysis process in Section 3.

The proposed model of the compressor blade is verified through comparisons of natural frequency, which is determined through the frequency-sweep experiment, as shown in Figure 4. Additionally, ANSYS software is also employed to calculate the natural frequency. Throughout the experiment, the blade is securely fixed in the electromagnetic excitation shake by the fixture and an accelerometer is placed in the middle of the blade to measure the vibration signal. The vibration signal is collected by the DH5956 data acquisition instrument and the mobile workstation is used to process the collected data. By applying loads with different frequencies, the response amplitude is amplified if the resonance phenomenon occurs. The frequency approximately corresponds to the natural frequency of the blade.

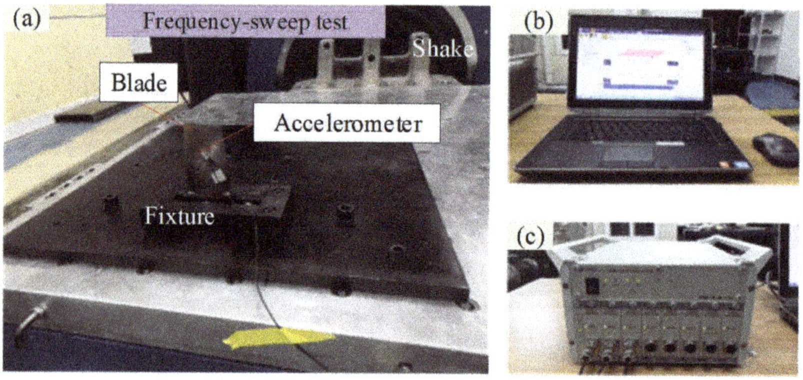


**Figure 4.** Test rig for measuring natural frequencies [31]: (**a**) test bench; (**b**) mobile workstation; (**c**) DH5956 data-collecting and -processing system.

The comparative results of the first three natural frequencies (NFs) of the blade are listed in Table 1. All frequency errors are less than 5%. This verifies the proposed dynamic model of the blade. Sources of error are analyzed as follows: (1) Material property errors: Physical properties of materials, such as elastic modulus, density, and Poisson's ratio, may have measurement errors or variability. (2) Geometric dimension errors: The geometric dimensions used in the model (such as thickness, length, width, etc.) may deviate from the actual dimensions. (3) Boundary condition errors: The boundary conditions applied in the simulation (for example, constraints) may differ from the actual conditions. (4) Manufacturing errors: Tolerances in the manufacturing process can cause the actual dimensions of structural components to differ from the design dimensions. (5) Numerical errors: Approximation and rounding errors are induced in numerical calculations.

**Table 1.** The first three NFs of the blade with fixed support.

| Orders | Experiment [31] | ANSYS Model | Proposed Model | Errors between the Experiment and Proposed Model (%) | Errors between ANSYS Model and Proposed Model (%) |
|---|---|---|---|---|---|
| $f_{\mathrm{n1b}}$ (Hz) | 657.5 | 680.1 | 679.0 | 3.27 | −0.15 |
| $f_{\mathrm{n2b}}$ (Hz) | 2008.0 | 1993.8 | 1976.1 | −1.60 | −0.89 |
| $f_{\mathrm{n3b}}$ (Hz) | 2409.0 | 2531.4 | 2500.0 | 3.77 | −1.22 |

It is difficult to guarantee the same elastic modulus and density in a simulation and an experiment. Thus, these two typical parameter errors are analyzed, as shown in Figure 5. As the density increases, the first natural frequencies of the blade and error linearly decrease. However, with the increase in elasticity, the first natural frequencies of the blade and error increase linearly.

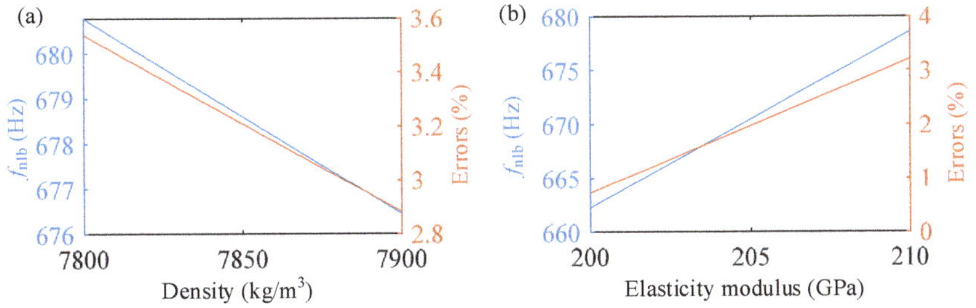


**Figure 5.** Study of the influence of parametric errors: (**a**) the influence of density, (**b**) the influence of elasticity modulus.

## 2.2. Dynamic Model for the Coated Casing

Although the lumped-mass modelling of casings can improve computational efficiency, it fails to consider the local deformation of the casing and potentially leads to errors in calculating the radial clearance at the blade tips. Thus, the laminated shell element is introduced to establish the FE model for flexible coated casings. By defining the distinct material parameters of each casing layer, the stiffness and mass matrices of laminated shell elements are calculated.

Within the laminated shell element, material parameters change in the thickness direction. This leads to different densities, strain matrices, and elastic matrices in different layers. Consequently, when calculating the mass and stiffness matrices of the laminated shell element, integration by parts is employed for precision. To simplify the computation process, a new local coordinate system $\xi\eta\zeta'$ is established in each layer of the shell, as depicted in Figure 6. Numbers 1–8 represent 8 nodes within a shell element. The mass matrix and stiffness matrix of each material layer are computed using Gaussian integration in the new local coordinate system $\xi\eta\zeta'$, and the corresponding position of the $p$-th layer is expressed as

$$\zeta_{p-1} = -1 + 2 \times \frac{h_{p-1}}{h_{\mathrm{t}}},\ \zeta_p = -1 + 2 \times \frac{h_p}{h_{\mathrm{t}}}, \tag{2}$$

where $\zeta_{p-1}$ and $\zeta_p$ correspond to $-1$ and 1 in the local coordinate system $\xi\eta\zeta'$, and $h_{\mathrm{t}}$ represents the total thickness of the laminated shell. $h_{p-1}$ and $h_p$ represent the upper and lower boundaries of the $p$th layer.

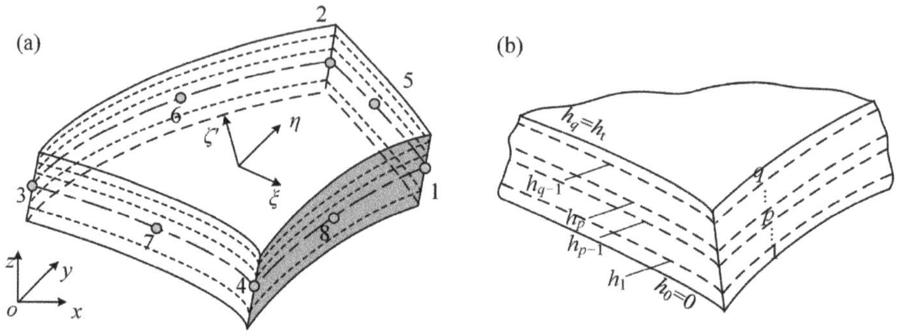


**Figure 6.** Schematic of laminated shell elements: (**a**) shell element; (**b**) lamination diagram.

The relationship between coordinate system $\xi\eta\zeta'$ and coordinate system $\xi\eta\zeta$ is denoted as

$$\zeta = -1 + \frac{(\zeta' + 1)(\zeta_p - \zeta_{p-1})}{2},\ d\zeta = \frac{\zeta_p - \zeta_{p-1}}{2} d\zeta'. \quad (3)$$

The mass matrix of the laminated shell element is written as

$$M^e = \int_V \rho N^T N dV = \sum_{p=1}^{q} \int_{-1}^{1} \int_{-1}^{1} \int_{-1}^{1} \rho_p N_p{}^T N_p |J| \left|J'_p\right| d\xi d\eta d\zeta' \quad (4)$$

where $\zeta_{p-1}$ and $\zeta_p$ are the material density of the $p$-th layer and $q$ is the total number of layers. $N_p$ denotes the shape function matrix along the thickness direction $\zeta'$. $\left|J'_p\right|$ is written as $\left|J'_p\right| = \frac{\zeta_p - \zeta_{p-1}}{2}$.

The stiffness matrix of the laminated shell element can be written as

$$K^e = \int_V B'^T D B' dV = \sum_{p=1}^{q} \int_{-1}^{1} \int_{-1}^{1} \int_{-1}^{1} B'_p{}^T D_p B'_p |J| \left|J'_p\right| d\xi d\eta d\zeta', \quad (5)$$

where $B'_p$ and $D_p$ denote the strain and elastic matrices in the $p$-th layer of elements.

By assembling the stiffness and mass matrices of the element, those of the coated casing can be obtained. The equation of motion of the coated casing can be written as

$$M_c \ddot{u}_c + C_c \dot{u}_c + K_c u_c = 0, \quad (6)$$

where $M_c$, $C_c$, and $K_c$ denote the mass, damping, and stiffness matrices of the coated casing. The expression for the damping matrix is denoted as

$$C_c = \alpha_c M_c + \beta_c K_c, \quad (7)$$

where $\alpha_c = \frac{4\pi f_{n1c} f_{n2c} (f_{n1c}\xi_2 - f_{n2c}\xi_1)}{(f_{n1c}^2 - f_{n2c}^2)}$ and $\beta_c = \frac{f_{n2c}\xi_2 - f_{n1c}\xi_1}{\pi(f_{n2c}^2 - f_{n1c}^2)}$. $f_{n1c}$ (Hz) and $f_{n2c}$ (Hz) represent the first two NFs of the casings. $\xi_1$ and $\xi_2$ represent the damping ratios of the first and second modes, respectively.

The proposed laminated shell model for the coated casing is validated by comparing the NFs with those obtained from the ANSYS software. The coating material is Metco 313 NS, and the material parameters are listed in Table 2. The first four NFs are presented in Table 3 and the mode shapes are presented in Figure 7. A comparison with the ANSYS results indicates that the errors in the first four NFs are all within 0.2%, and the mode shapes exhibit good consistency. This verifies the proposed dynamic model for the coated casing.

**Table 2.** Parameters of the coated casing.

| Parameters | Value | Parameters | Value | Parameters | Value |
|---|---|---|---|---|---|
| Thickness of coating $h_a$ (mm) | 1 | Thickness of casing $h_c$ (mm) | 2 | Length of casing $L_c$ (m) | 0.1 |
| Density of coating $\rho_a$ (kg/m$^3$) | 2000 | Density of casing $\rho_c$ (kg/m$^3$) | 4370 | Inner diameter of casing $R_c$ (m) | 0.30682 |
| Elastic modulus of coating $E_a$ (GPa) | 3.11 | Elastic modulus of casing $E_c$ (GPa) | 125 | Poisson ratio $v$ | 0.3 |

**Table 3.** The first four NFs of the coated casing with fixed support.

| Orders | $f_{n1c}$ (Hz) | $f_{n2c}$ (Hz) | $f_{n3c}$ (Hz) | $f_{n4c}$ (Hz) |
|---|---|---|---|---|
| Proposed model | 1929.1 | 1930.3 | 1950.7 | 1954.9 |
| ANSYS model | 1929.8 | 1930.8 | 1951.4 | 1955.4 |
| Errors (%) | −0.036 | −0.026 | −0.036 | −0.026 |

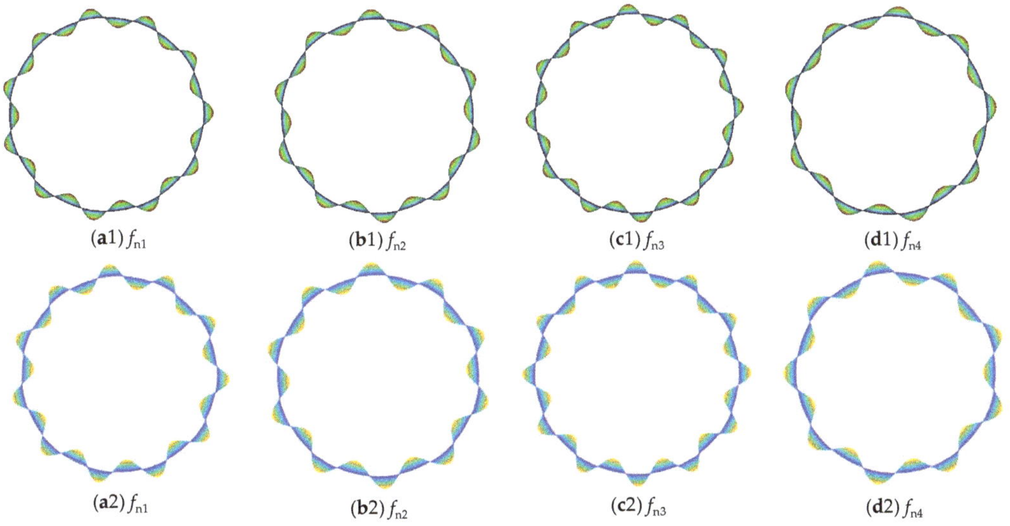


**Figure 7.** Mode shapes of the first four modes of the coated casing (the first row comprises the mode shapes from the ANSYS software, and the second row comprises the mode shapes from the proposed model).

Three-dimensional linear spring elements are used at each node of both ends of the casing to simulate its elastic support. Spring elements are incorporated into the cylindrical coordinate system. This requires transformation to the cartesian coordinate system during stiffness matrix calculations. The stiffness values in each direction in the cylindrical coordinate system are $k_r = 5 \times 10^7$ N/m, $k_t = 4 \times 10^7$ N/m, $k_a = 2 \times 10^8$ N/m, $k_{\theta r} = 1 \times 10^6$ N·m/rad, and $k_{\theta t} = 1 \times 10^6$ N·m/rad. The FE model for the coated casing with elastic support is shown in Figure 8. The first six NFs of the casing with elastic support are shown in Table 4, and those of the proposed model show good consistency with the results from ANSYS. Thus, the proposed model for coated casings is verified considering elastic support. The error in the natural frequency of the casing is very small, so the source of this error may be numerical, including approximation and rounding errors.

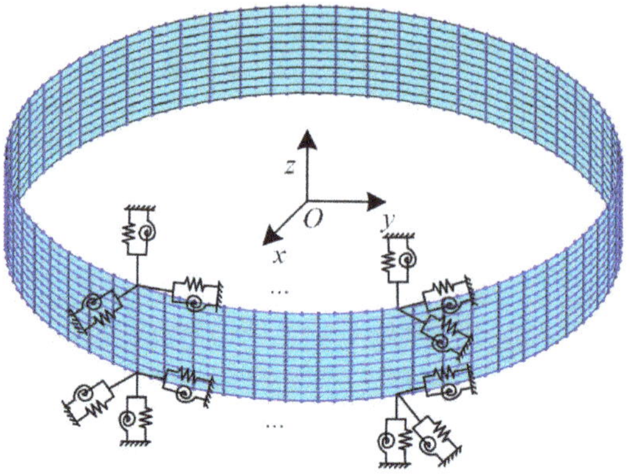


**Figure 8.** FE model of the coated casing with elastic support.

**Table 4.** The first six NFs of the coated casing with elastic support.

| Orders | $f_{n1c}$ (Hz) | $f_{n2c}$ (Hz) | $f_{n3c}$ (Hz) | $f_{n4c}$ (Hz) | $f_{n5c}$ (Hz) | $f_{n6c}$ (Hz) |
|---|---|---|---|---|---|---|
| Proposed model | 823.54 | 826.63 | 827.93 | 839.19 | 842.85 | 858.30 |
| ANSYS model | 823.41 | 826.47 | 827.83 | 839.12 | 842.65 | 858.29 |
| Errors (%) | 0.0157 | 0.0193 | 0.0120 | 0.0080 | 0.0230 | 0.0012 |

The proposed casing model consists of 3584 elements and 11,264 nodes (i.e., 67,584 degrees of freedom), which causes a low calculation speed. To enhance computational efficiency, the Craig–Bampton method is used to establish the reduced dynamic model. The reduced dynamic model for the casing is depicted as

$$\hat{M}_c\ddot{q}_c + \hat{C}_c\dot{q}_c + \hat{K}_c q_c = \hat{F}_c \tag{8}$$

The reduced casing has 320 degrees of freedom. Table 5 displays the first five NFs of the reduced dynamic model. The maximum error of the natural frequency is 0.307%. This meets the required accuracy standards and verifies the reduced dynamic model.

**Table 5.** Comparison of NFs between the reduced model and unreduced model.

| Orders | $f_{n1c}$ (Hz) | $f_{n2c}$ (Hz) | $f_{n3c}$ (Hz) | $f_{n4c}$ (Hz) | $f_{n5c}$ (Hz) |
|---|---|---|---|---|---|
| Unreduced model | 823.54 | 826.63 | 827.93 | 839.19 | 842.85 |
| Reduced model | 824.23 | 827.32 | 830.21 | 840.39 | 845.44 |
| Errors (%) | 0.084 | 0.084 | 0.275 | 0.140 | 0.307 |

### 2.3. *Dynamic Model of the Blade-Coated Casing System*

Considering static misalignment $e_c$, the schematic of a rotating blade-flexible coated casing is shown in Figure 9, considering rubbing. $O_c$ and $O$ represent the centers of the casing and disk in a stationary state, respectively. Mesh refinement is implemented at locations susceptible to rubbing, as depicted in Figure 9a. In order to correspond to the node of the blade, the casing in the rubbing region is divided into 20 elements in the length direction. The non-rubbing regions on both sides of the casing are divided into two elements. Thus, the casing has a total of 24 elements in the length direction in Figure 9a. In Figure 9b, the casing position and blade tip orbit are displayed by the blue and pink lines when the rubbing fault occurs. By analyzing the gap $c^i_{\mathrm{rub}}$ between the blade tip and the casing, whether the rubbing fault occurs is judged and the normal rubbing force is calculated. The calculation process of $c^i_{\mathrm{rub}}$ is shown in Appendix B, and the expression of $c^i_{\mathrm{rub}}$ is shown in Equation (A29). Additionally, Figure 9c mainly shows the local contact between the blade and the casing.

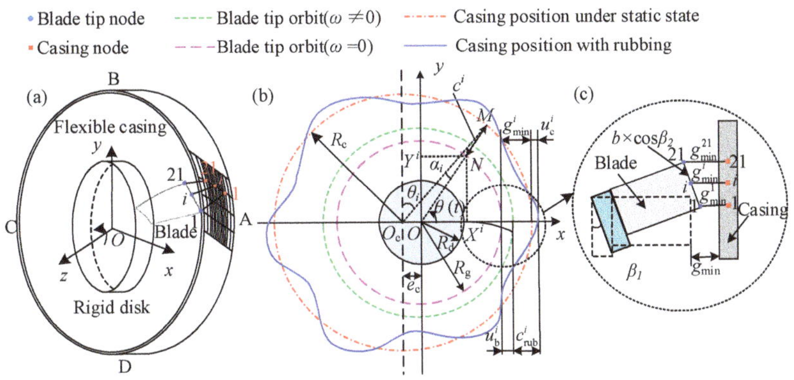


**Figure 9.** Schematic of rubbing model [79]: (**a**) schematic of blade–casing rubbing; (**b**) blade–casing rubbing model; (**c**) partial enlarged drawing.

The rubbing model is shown in Figure 10. Figure 10a displays the blade-coated casing system. This rubbing model simultaneously considers the contact deformation between the blades and the casing, the structural deformation of each component, and the energy dissipation phenomenon caused by contact damping. The rubbing stiffness of the hysteresis contact-force model includes the structural stiffnesses $k_c$ and $k_b$, as well as the local contact stiffness $k_h^*$. In other words, the rubbing stiffness is equal to the series stiffness of these three stiffness values, as shown in Figure 10b.

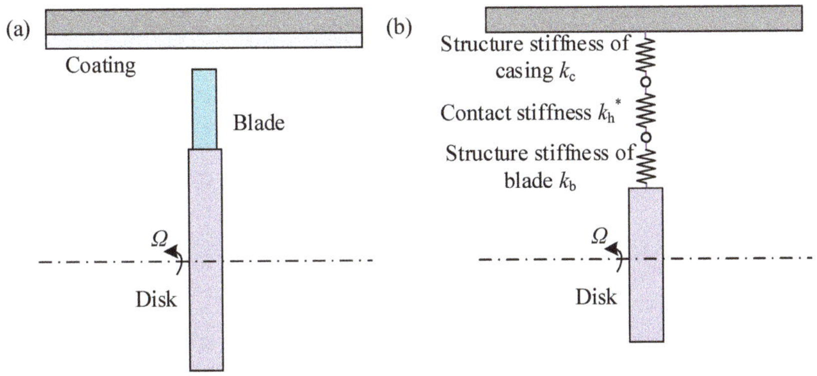


**Figure 10.** Schematic of the hysteretic contact-force models: (**a**) blade-coated casing system; (**b**) equivalent contact diagram.

The detailed derivation of the rubbing force is shown in Appendix B.

During the operation of an aircraft engine, the blade is subjected to aerodynamic forces, and the overall aerodynamic load vector $Q_p$ is written as

$$Q_p = \sum_{i=1}^{n} Q_i^{p,ext}, \quad (9)$$

where $Q_i^{\mathrm{p,ext}}$ is the aerodynamic load of an element. The detailed expression can be found in Ref. [79].

Taking node pair 1 in Figure 9 as an example, the friction force is

$$F_{\mathrm{t}}^{1} = \mu F_{\mathrm{n}}^{1}, \tag{10}$$

where $\mu$ represents the frictional coefficient.

The rubbing forces $F_{\mathrm{n}}^{i}$ and $F_{\mathrm{t}}^{i}$ ($F_{\mathrm{t}}^{i} = \mu F_{\mathrm{n}}^{i}$) are projected in the $x$ and $y$ directions. $\mu$ represents the friction coefficient. The resultant forces $F_x^i$ and $F_y^i$ are written as

$$\begin{aligned} F_x^i &= F_{\mathrm{n}}^i \cos(\omega t + \phi_{\mathrm{b}}) - F_{\mathrm{t}}^i \sin(\omega t + \phi_{\mathrm{b}}) \\ F_y^i &= F_{\mathrm{n}}^i \sin(\omega t + \phi_{\mathrm{b}}) + F_{\mathrm{t}}^i \cos(\omega t + \phi_{\mathrm{b}}) \end{aligned}, \tag{11}$$

where $\omega$ represents the rotating speed and $\phi_{\mathrm{b}}$ represents the blade phase.

The expressions of $\boldsymbol{F}_{\mathrm{rb}}$ and $\boldsymbol{F}_{\mathrm{rc}}$ are

$$\begin{aligned} \boldsymbol{F}_{\mathrm{rb}} &= \begin{bmatrix} \cdots & F_x^1 & F_y^1 & 0 & 0 & 0 & \cdots & F_x^{21} & F_y^{21} & 0 & 0 & 0 & 0 & \cdots \end{bmatrix} \\ \boldsymbol{F}_{\mathrm{rc}} &= -\boldsymbol{F}_{\mathrm{rb}} \end{aligned}. \tag{12}$$

Thus, considering the rubbing forces, the dynamic model of the blade-coated casing is written as

$$\begin{bmatrix} \boldsymbol{K}_{\mathrm{e}} + \boldsymbol{K}_{\mathrm{ce}} + \boldsymbol{K}_{\mathrm{co}} - \boldsymbol{K}_{\mathrm{s}} & \boldsymbol{0} \\ \boldsymbol{0} & \hat{\boldsymbol{K}}_{\mathrm{c}} \end{bmatrix} \begin{bmatrix} \boldsymbol{x} \\ \boldsymbol{q}_{\mathrm{c}} \end{bmatrix} + \begin{bmatrix} \boldsymbol{G} + \boldsymbol{C} & \boldsymbol{0} \\ \boldsymbol{0} & \hat{\boldsymbol{C}}_{\mathrm{c}} \end{bmatrix} \begin{bmatrix} \dot{\boldsymbol{x}} \\ \dot{\boldsymbol{q}}_{\mathrm{c}} \end{bmatrix} + \begin{bmatrix} \boldsymbol{M} & \boldsymbol{0} \\ \boldsymbol{0} & \hat{\boldsymbol{M}}_{\mathrm{c}} \end{bmatrix} \begin{bmatrix} \ddot{\boldsymbol{x}} \\ \ddot{\boldsymbol{q}}_{\mathrm{c}} \end{bmatrix} = \begin{bmatrix} \boldsymbol{F}_{\mathrm{c}} \\ \boldsymbol{0} \end{bmatrix} + \begin{bmatrix} \boldsymbol{Q}_{\mathrm{p}} \\ \boldsymbol{0} \end{bmatrix} + \begin{bmatrix} \boldsymbol{F}_{\mathrm{rb}} \\ \boldsymbol{F}_{\mathrm{rc}} \end{bmatrix} \tag{13}$$

where $\boldsymbol{F}_{\mathrm{rb}}$, $\boldsymbol{F}_{\mathrm{rc}}$, and $\boldsymbol{F}_{\mathrm{c}}$ represent the rubbing force vectors of the blade and casing and the centrifugal force vector of the blade. The rubbing force is nonlinear, so once the system has a rubbing fault, the system will show obvious nonlinear characteristics.

## 3. Numerical Analysis

The casing is established using a laminated shell element with elastic support, while the blade is built as detailed in Section 2.2. The influence of the parameters related to the coating casing on the dynamic response of the system with rubbing is analyzed in this section.

### 3.1. Influence of the Casing Model

Guo et al. [41] analyzed the effects of casing models on rubbing, encompassing the lump-mass casing model and the cylindrical shell casing model. Inspired by Ref. [41], this subsection further compares the vibration responses of different casing models based on the hysteresis contact-force model, including the lump-mass casing model, cylindrical shell casing model, and laminated casing model. This comparison is conducted under the same geometric structure parameters and rubbing parameters. The simulation parameters of the blade are detailed in Table 6, and those of the casing are provided in Table 7. The dynamic responses are calculated by the Newmark-$\beta$ method, and are detailed in Appendix C.

**Table 6.** Simulation parameters of the blade.

| Parameters | Value | Parameters | Value |
|---|---|---|---|
| Frictional coefficient $\mu$ | 0.3 | Rotating speed $\omega$ (r/min) | 10,000 |
| Static parallel misalignment $e_{\mathrm{c}}$ (mm) | 1.6 | Minimum clearance $g_{\min}$ (mm) | 0.1 |
| Misalignment angle $\beta_1$ (°) | 0.15 | Blade tip staggered angle $\beta_2$ (°) | 54.68 |
| Radius of the blade tip $R_{\mathrm{d}}$ (mm) | 216.52 | Inner diameter of casing $R_{\mathrm{c}}$ (mm) | 306.82 |

**Table 7.** Simulation parameters of the casing.

| Models | Parameters |
|---|---|
| Lumped-blade mass casing model | $m_c = 1.69$ kg, $k_c = 1.34 \times 10^6$ N/m |
| Cylindrical-blade shell casing model | $L_c$=0.1 m, $h_c = 2$ mm, $E_c = 125$ GPa, $\rho_c = 4370$ kg/m$^3$ |
| Laminated-blade casing model | $L_c = 0.1$ m, $h_c = 2$ mm, $h_a = 1$ mm, $E_a = 3.11$ Gpa, $E_c = 125$ GPa, $\rho_a = 2000$ kg/m$^3$, $\rho_c = 4370$ kg/m$^3$ |

Figure 11 displays the time-domain waveform (TDW) of the rubbing forces and vibration response at the 21st node of the blade tips under three casing models. Notably, the different casing models exhibit a minimal difference in bending displacement and radial displacement. The frequency components remain consistent among these three models. Amplitude amplification is observed at frequency $5f_r$ because frequency $5f_r$ is close to the first natural frequency of the blade. Comparatively, the cylindrical shell casing model demonstrates a slight reduction in the maximum normal rubbing force, radial displacement, and bending displacement in contrast to the lumped-mass casing model. This phenomenon can be attributed to the localized deformation of the cylindrical shell casing model under the action of the normal rubbing force. This deformation increases the gap between the blade and the casing in subsequent load steps, ultimately reducing blade penetration and the severity level of rubbing.

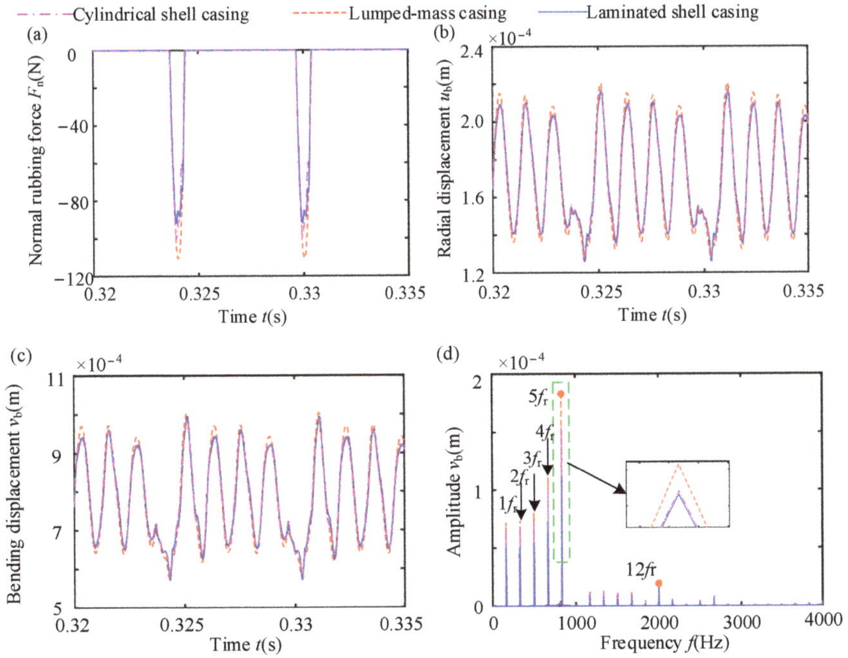


**Figure 11.** NRFs and vibration responses of blades for various casing models: (**a**) NRF; (**b**) TDW of radial displacement; (**c**) TDW of bending displacement; (**d**) spectrum.

The natural characteristics and radial equivalent stiffness of the casing are directly influenced by changing the material parameters. The first NFs of the cylindrical shell casing and laminated shell casing are 906.12 Hz and 823.55 Hz, respectively. The equivalent radial stiffnesses are $1.396 \times 10^6$ N/m and $1.336 \times 10^6$ N/m, respectively. The equivalent radial stiffness of the laminated shell casing is greater than that of the cylindrical shell casing, but its maximum normal rubbing force, radial displacement, and bending displacement are slightly reduced compared to the cylindrical shell casing model (see Figure 11a), indicating that having a coating on the surface of the casing can reduce the severity level of blade–casing rubbing.

The vibration responses of the blade at the 11th and 1st nodes of the blade tip are shown in Figure 12. When comparing Figures 11 and 12, a notable difference in the radial and bending displacements can be found among the blade tips. Specifically, the radial and bending displacements of the blade tip nodes exhibit a gradual decrease with the change in the position at the blade tip. This is because the thickness of the blade is non-uniform, and the elongation of each node at the blade tip is different under the action of the centrifugal force. Near the 21st node, contact between the blade tip and the casing appears, causing a reduction in the radial and bending displacements. Of course, an obvious conclusion is that changes in the casing model will not cause changes in frequency components, only in amplitude.

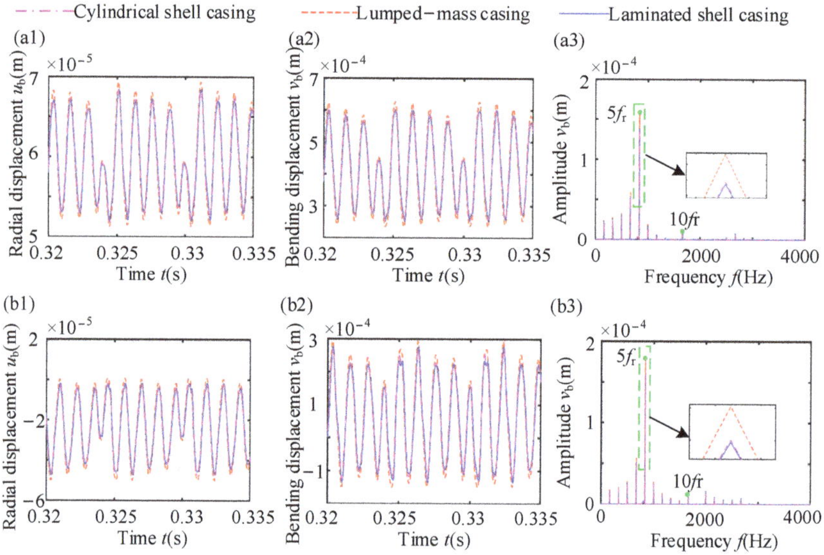


**Figure 12.** Vibration responses of different nodes at blade tips: (**a1**,**b1**) radial displacement; (**a2**,**b2**) bending displacement; (**c1**,**c2**) spectrum. The first row shows the response at the 11th node of the blade tip. The first row shows the response at the 1st node of the blade tip.

The blade passes the coated casing at four positions: points A (0), B (π/2), C (π), and D (3π/2), the deformation diagram of the casing is shown in Figure 13. The definition of the points can be found in Figure 9a. At position A, blade–casing rubbing can obviously be found. When rubbing occurs, the deformation of laminated shell models at point A is greater than that of the cylindrical shell models, and the deformation of the two models at that point is significantly greater than that at the other points. However, for the lumped-mass casing, the overall casing is translated to the right due to rubbing forces. Additionally, obvious NDVCs can be observed in the cylindrical shell model and the laminated shell casing model. However, the induced NDVCs are also different among these two flexible casing models.

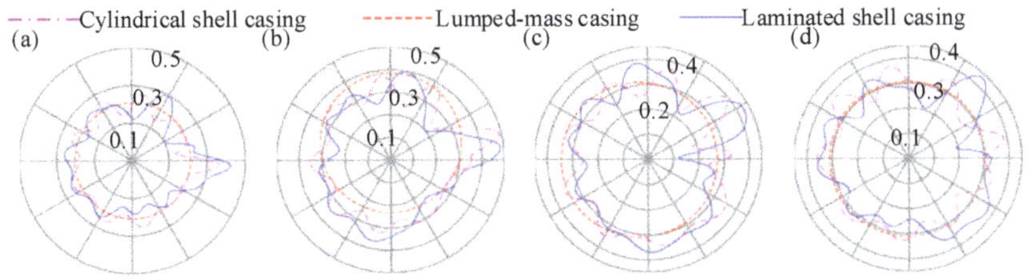


**Figure 13.** Deformation diagrams of the casing at different moments (unit: m): (**a**) $t$ = 0.324 s, corresponding to point A; (**b**) $t$ = 0.3255 s, corresponding to point B; (**c**) $t$=0.3247 s, corresponding to point C; (**d**) $t$ = 0.3285 s, corresponding to point D.

The time-domain waveforms and the spectra of radial vibration at points A, B, C, and D are shown in Figures 14–16. In these figures, the radial displacement of the rigid casing is much smaller than that of the cylindrical shell casing and laminated shell casing. The difference in radial displacement between the cylindrical shell casing and the laminated shell casing at the rubbing point is not significant, but at the non-rubbing position, the laminated shell casing has a larger vibration displacement compared to the cylindrical shell casing. Additionally, there is a significant difference in the frequency components between the two flexible casing models. The cylindrical shell casing exhibits amplitude amplification at frequency $6f_r$, while the laminated shell casing exhibits amplitude amplification at frequency $5f_r$, and exhibits dense frequency characteristics at rubbing point A. At non-rubbing points (points B, C, and D), the cylindrical shell casing and the laminated shell casing also exhibit amplitude amplification at $6f_r$ and $5f_r$, respectively. The main reason for the above differences is the difference in natural characteristics between the cylindrical shell casing and the laminated shell casing.

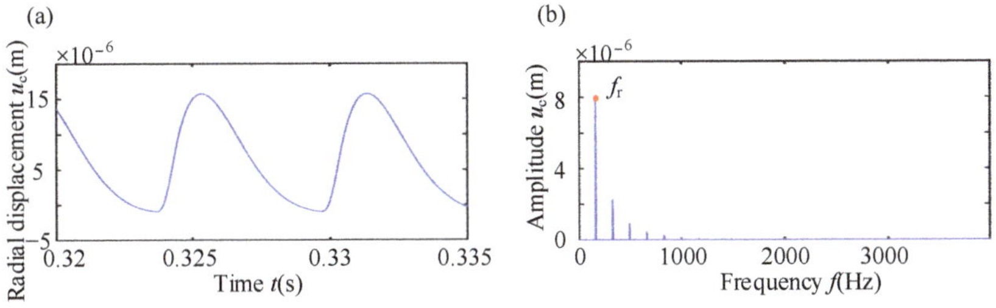


**Figure 14.** Vibration response of rigid casing: (**a**) time-domain waveform; (**b**) spectrum.

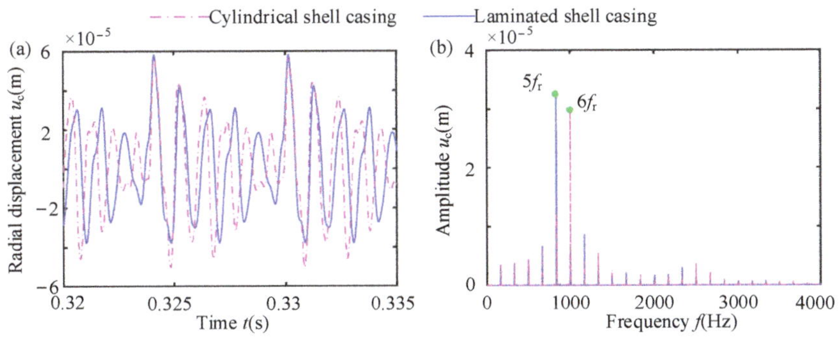


**Figure 15.** Vibration response of rubbing point A with the flexible casing: (**a**) time-domain waveform; (**b**) spectrum.

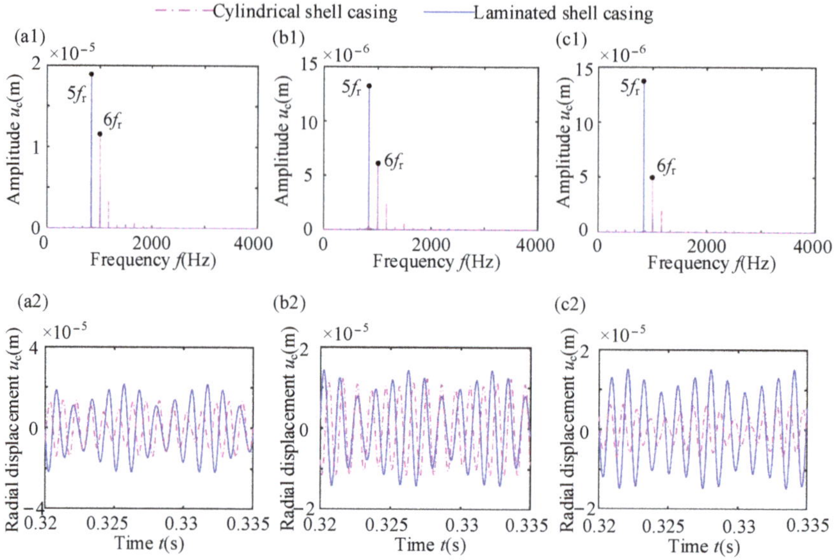


**Figure 16.** Vibration responses of rubbing points B, C, and D with flexible casing: (**a1**,**a2**) point B; (**b1**,**b2**) point C; (**c1**,**c2**) point D.

### 3.2. Influence of Coating Thickness

Changes in coating thickness directly affect the natural characteristics and radial stiffness of a casing. Therefore, the influences of coating thickness are investigated.

The NRF and dynamic response at the 21st node are shown in Figure 17. The equivalent radial stiffnesses of casings with coating thicknesses of 0.5 mm, 1 mm, and 2 mm are $1.357 \times 10^6$ N/m, $1.396 \times 10^6$ N/m, and $1.54 \times 10^6$ N/m, respectively. Correspondingly, the first NFs are 860.68 Hz, 823.55 Hz, and 768.99 Hz, respectively. Notably, the equivalent radial stiffness increases with the increase in coating thickness, which causes rubbing force to increase. However, in Figure 18, the rubbing force at a 0.5 mm thickness is greater than that at a 1 mm thickness. The main reason is that the radial displacement of the 1 mm coating thickness at the rubbing point is greater than that of the 0.5 mm coating thickness during the initial rubbing time period from 0.3238 s to 0.324 s, resulting in a smaller penetration between the blade and casing, and ultimately a smaller rubbing force. The coating thickness has a slight effect on the response amplitude of the blade.

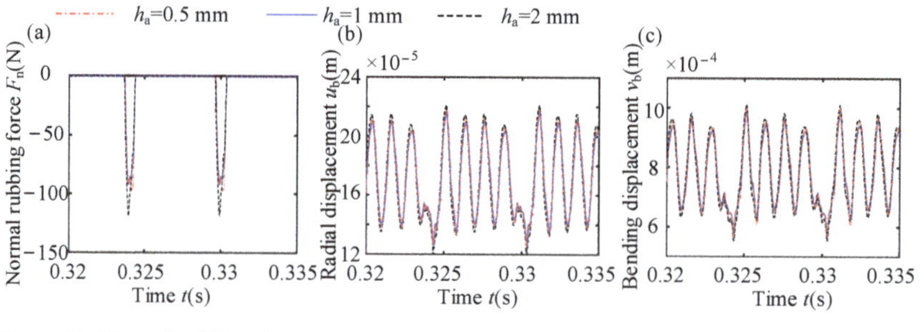


**Figure 17.** Normal rubbing forces and vibration responses at various coating thicknesses: (**a**) NRF; (**b**) TDW of radial displacement; (**c**) TDW of bending displacement.

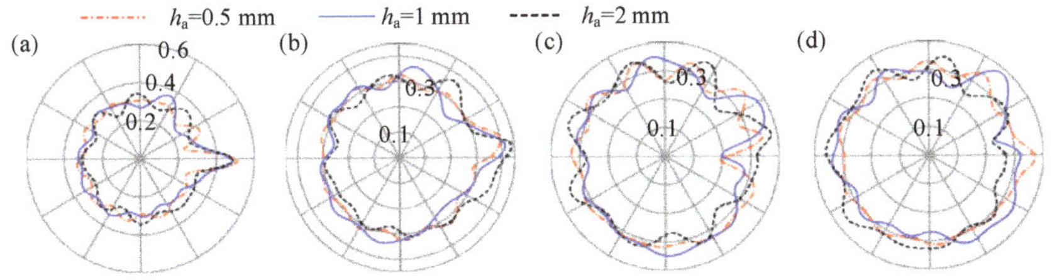


**Figure 18.** Deformation diagrams of the casing at different moments (unit: m): (**a**) $t = 0.324$ s; (**b**) $t = 0.3255$ s; (**c**) $t = 0.3247$ s; (**d**) $t = 0.3285$ s.

From Figures 18 and 19, it is evident that the NDVCs induced by rubbing under different coating thicknesses are different. At rubbing point A, the radial displacement and frequency components of the casing under the three coating thicknesses show minimal differences, and an amplitude amplification phenomenon is observed at frequency $5f_r$. Conversely, at non-rubbing point B, there is a notable difference in the radial displacement and frequency components of the casing. Specifically, when the coating thicknesses are 1 mm and 2 mm, the amplitude amplification phenomenon occurs at frequency $5f_r$. In contrast, when the coating thickness is 0.5 mm, the amplitude amplification phenomenon appears at frequency $6f_r$ in Figure 19d.

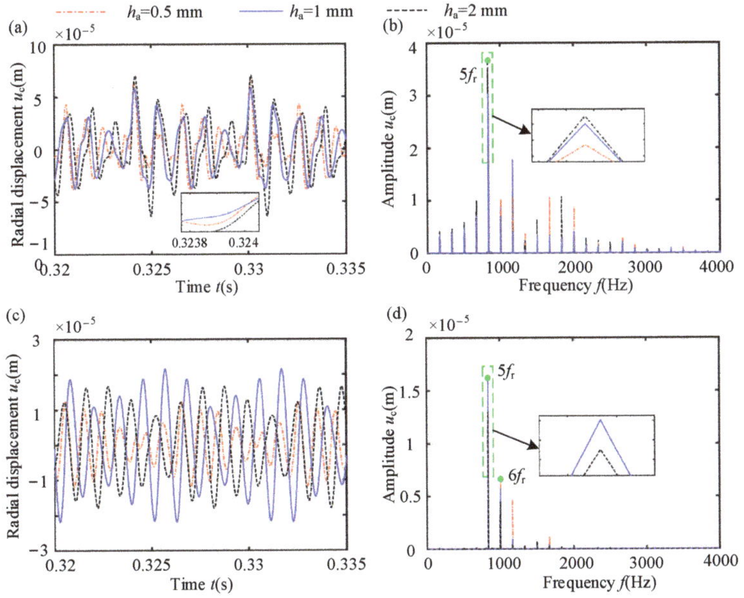


**Figure 19.** Dynamic responses of casings at various coating thicknesses: (**a**) TDW of displacement at point A; (**b**) displacement spectrum at point A; (**c**) TDW of displacement at point B; (**d**) displacement spectrum at point B.

### 3.3. Influence of Coating Material

A comparative analysis of NRF and vibration response under three coating materials (Metco 308 NS, Metco 601 NS, and Metco 313 NS) is conducted. The corresponding material parameters are detailed in Table 8. Variations in coating materials lead to alterations in the NF and equivalent radial stiffness of the casing, and the results are shown in Table 8.

**Table 8.** Material parameters, NF, and equivalent radial stiffness of coatings.

| Materials | Ingredient | Density $\rho_a$ (kg/m$^3$) | Elasticity Modulus $E_a$ (GPa) | Natural Frequency $f$ (Hz) | Equivalent Radial stiffness (N/m) |
|---|---|---|---|---|---|
| Metco 308NS | Ni-graphite | 4900 | 3.95 | 841.32 | $1.376 \times 10^6$ |
| Metco 601NS | Al-Si polymer | 1500 | 2.08 | 731.76 | $1.412 \times 10^6$ |
| Metco 313NS | Al-Si-graphite | 2000 | 3.11 | 823.55 | $1.396 \times 10^6$ |

Figures 20–22 display the TDW of NRF, the vibration response of the blade, casing deformation, and the vibration response of the casing at points A and B under different coating materials. Observations from these figures reveal that, when the coating material is Metco 308 NS, the NRF is the largest, and the NDVCs induced by rubbing are more pronounced. Conversely, the normal rubbing force, vibration response of the blade, and vibration response of the casing at points A and B are roughly the same when the coating materials are Metco 601 NS and Metco 313 NS. However, it is noteworthy that Metco 601 NS is primarily suitable for low-pressure compressors at low-temperature conditions. Therefore, for the purposes of this simulation and analysis, Metco 313 NS is chosen as the coating material to study in this paper. The vibration response induced by rubbing can serve as a crucial criterion for the selection of coating materials in the design process of blade-coated casing systems.

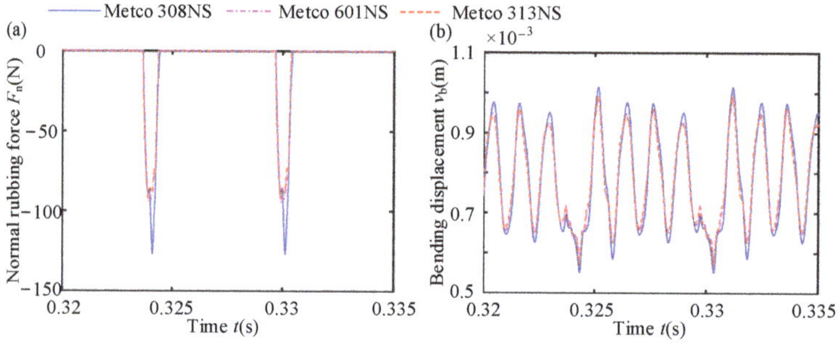


**Figure 20.** Normal rubbing forces and dynamic responses of blades with various coating materials: (**a**) normal rubbing force; (**b**) bending displacement of the 21st node of the blade tip.

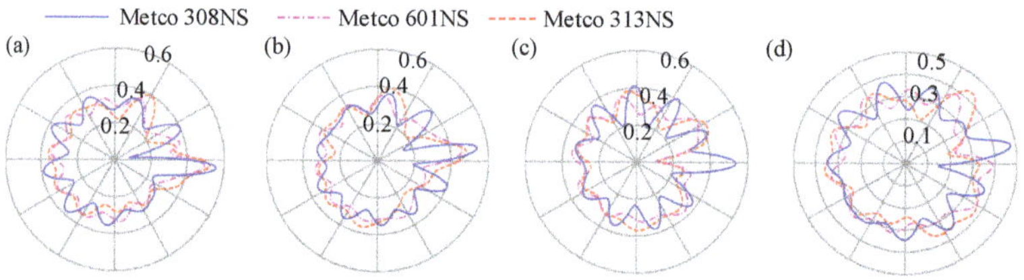


**Figure 21.** Deformation diagrams of the casing at different moments: (**a**) $t = 0.324$ s, (**b**) $t = 0.3255$ s, (**c**) $t = 0.3247$ s, (**d**) $t = 0.3285$ s.

Figures 21 and 22 reveal variations in radial displacements, frequency components, and induced NDVCs under different coating materials. At rubbing point A, an amplitude amplification phenomenon is consistently observed at frequency $5f_r$ for all three materials. However, at non-rubbing point B, the amplitude amplification phenomenon occurs at frequency $5f_r$ for both Metco 308 NS and Metco 313 NS materials, while for the Metco 601 NS material casing, the amplitude amplification phenomenon appears at frequency $6f_r$.

These differences highlight the sensitivity of the vibration response of casings to the choice of coating materials, emphasizing the importance of this factor in material selection for good performance when rubbing occurs in a blade-coated casing system.

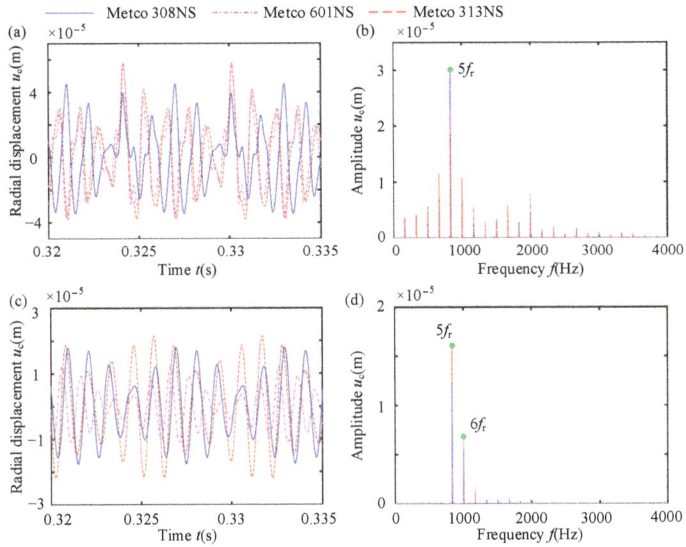


**Figure 22.** Dynamic responses of casings with various coating materials: (**a**) TDW of displacement at point A; (**b**) displacement spectrum at point A; (**c**) TDW of displacement at point B; (**d**) displacement spectrum at point B.

*3.4. Influence of the Support Stiffness of the Coated Casing*

The elastic support stiffness of casings can influence natural characteristics and the equivalent radial stiffness of the casing. Therefore, this section discusses the effects of the radial, tangential, and axial elastic support stiffnesses of casings on the equivalent radial stiffness of the casing and the vibration response of the system. Simulation parameters are listed in Table 9.

**Table 9.** Simulation parameters.

| Varying Parameters | Ranges | Constant Parameters |
|---|---|---|
| Radial support stiffness $k_r$ | $[1 \times 10^6, 1 \times 10^9]$ N/m | $k_t = 1 \times 10^8$ N/m, $k_a = 1 \times 10^8$ N/m, $L_c = 0.1$ m, $E_c = 125$ GPa, $\rho_c = 4370$ kg/m$^3$, $h_c = 2$ mm, $E_a = 3.11$ GPa, $\rho_a = 2000$ kg/m$^3$, $h_a = 1$ mm |
| Tangential support stiffness $k_t$ | $[1 \times 10^6, 1 \times 10^9]$ N/m | $k_r =1 \times 10^8$ N/m, $k_a = 1 \times 10^8$ N/m, $L_c = 0.1$ m, $E_c = 125$ GPa, $\rho_c = 4370$ kg/m$^3$, $h_c = 2$ mm, $E_a = 3.11$ GPa, $\rho_a = 2000$ kg/m$^3$, $h_a = 1$ mm, |
| Axial support stiffness $k_a$ | $[1 \times 10^6, 1 \times 10^9]$ N/m | $k_r =1 \times 10^8$ N/m, $k_t = 1 \times 10^8$ N/m, $L_c = 0.1$ m, $E_c = 125$ GPa, $\rho_c = 4370$ kg/m$^3$, $h_c = 2$ mm, $E_a = 3.11$ GPa, $\rho_a = 2000$ kg/m$^3$, $h_a = 1$ mm, |

Figures 23–25 show the equivalent radial stiffnesses and rubbing forces of the casings under different elastic support stiffnesses. The vibration response of the 21st node of the blade tip and the vibration response at point A of the casing are analyzed. Some conclusions are listed as follows:

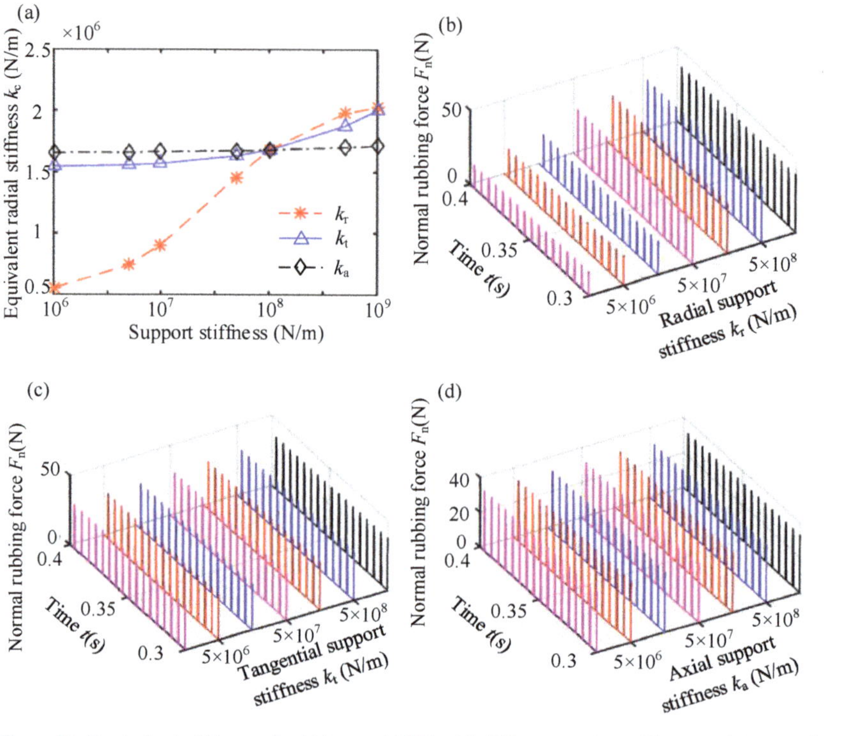


**Figure 23.** Equivalent stiffness of rubbing and NRF with different spring stiffnesses: (**a**) equivalent radial stiffness of the coated casing; (**b**) influence of radial support stiffness; (**c**) influence of tangential support stiffness; (**d**) influence of axial support stiffness.

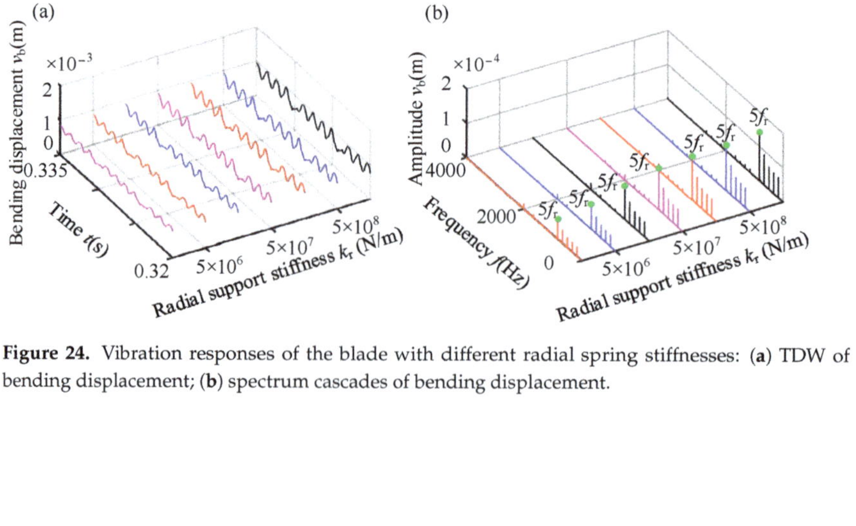


**Figure 24.** Vibration responses of the blade with different radial spring stiffnesses: (**a**) TDW of bending displacement; (**b**) spectrum cascades of bending displacement.

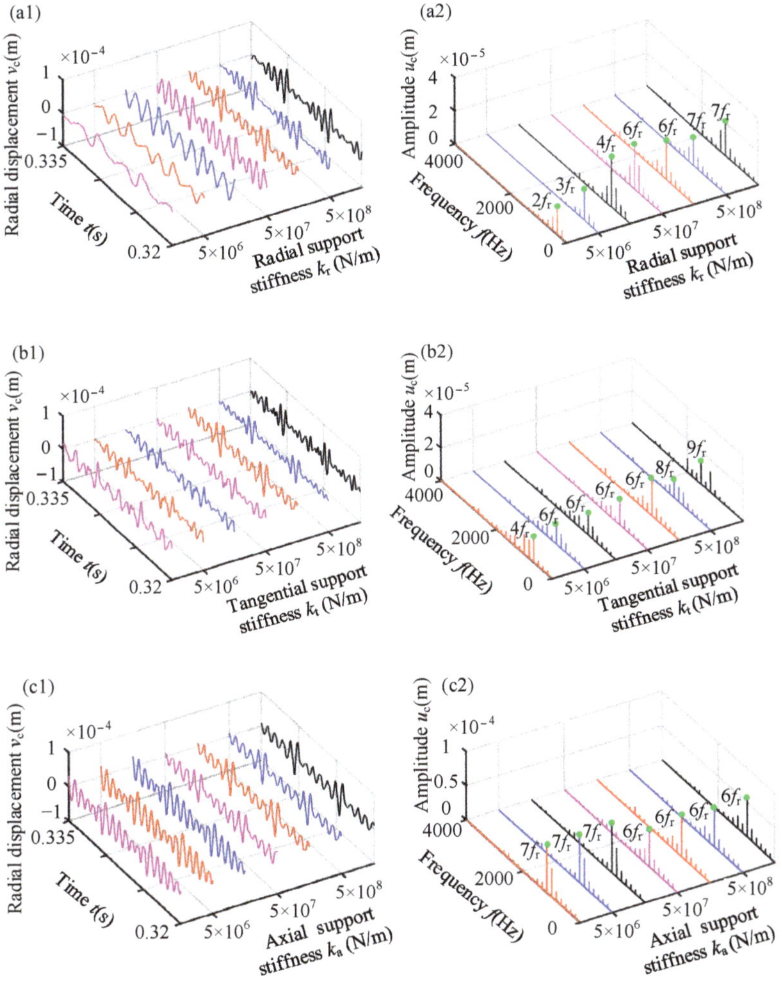


**Figure 25.** Dynamic responses of point A with various spring stiffnesses: (**a1**,**a2**) influence of radial support stiffness; (**b1**,**b2**) influence of tangential support stiffness; (**c1**,**c2**) influence of axial support stiffness. The **left** side of the figure shows the TDW of the displacement and the **right** shows the spectrum.

(1) Equivalent radial stiffness exhibits an increase with the rise in elastic support stiffness. Specifically, the equivalent radial stiffness and maximum rubbing force are primarily influenced by the radial and tangential elastic support stiffnesses, and the axial elastic support stiffness has a minimal effect on the equivalent radial stiffness and maximum normal rubbing force. The equivalent radial stiffness undergoes significant variations when the radial elastic support stiffness ranges from $1 \times 10^6$ N/m to $1 \times 10^8$ N/m and the tangential elastic support stiffness ranges from $1 \times 10^8$ N/m to $1 \times 10^9$ N/m. Overall, the radial elastic support stiffness has the most obvious influence on the equivalent radial stiffness of the coated casing.

(2) As the radial elastic stiffness of the casing increases, there is an obvious increase in the amplitudes of rubbing force and vibration displacement, while the frequency component remains unchanged. The support stiffness in three directions exerts a notable influence on the radial vibration response at point A of the casing. Specifically, with an increase in

radial and tangential stiffness, the frequency corresponding to the amplitude amplification phenomenon also increases. This is attributed to the difference in the natural frequency of the casing induced by changes in the support stiffness of the coated casing.

### 3.5. Influence of the Length of the Coated Casing

The length of the casing directly affects the natural frequency and equivalent radial stiffness of the coated casing. Therefore, this section analyzes the effect of casing length on vibration response. The vibration response of the 21st node of the blade tip and the vibration response at point A of the casing are shown in Figure 26. The first natural frequency, equivalent radial stiffness, and maximum normal rubbing force of the 21st node under different casing lengths are listed in Table 10.

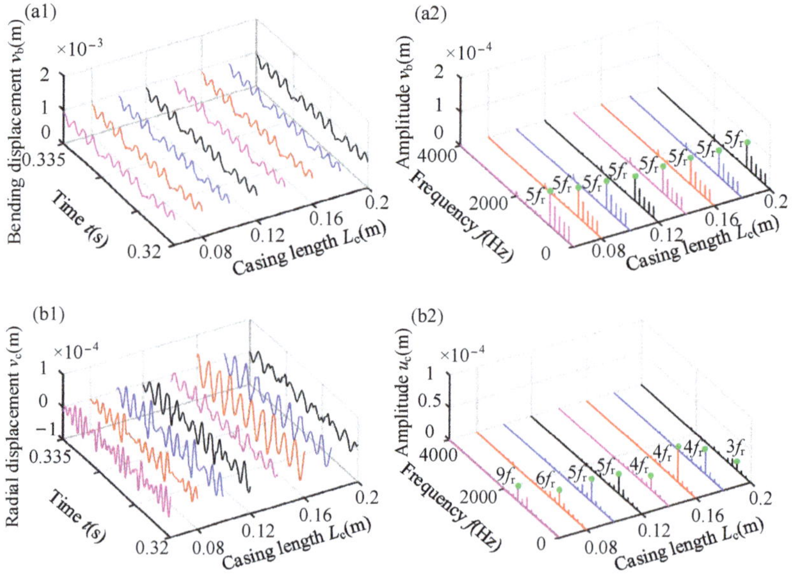


**Figure 26.** Vibration responses of blade–casing rubbing with different casing lengths: (**a1**,**a2**) 21st node of the blade tip; (**b1**,**b2**) point A of the casing.

**Table 10.** The first NF and equivalent radial stiffness of the casing at different lengths.

| Casing Length $L_c$(m) | 0.06 | 0.08 | 0.1 | 0.12 | 0.14 | 0.16 | 0.18 | 0.2 |
|---|---|---|---|---|---|---|---|---|
| Equivalent radial stiffness $k_c$ ($10^6$ N/m) | 1.543 | 1.558 | 1.396 | 1.308 | 1.270 | 1.248 | 1.228 | 1.208 |
| 1st natural frequency $f_{n1c}$ (Hz) | 1146.20 | 939.93 | 860.64 | 844.81 | 684.28 | 633.77 | 591.70 | 555.67 |
| Maximum normal rubbing force $F_n$ (N) | 33.27 | 31.36 | 33.21 | 31.09 | 28.22 | 29.16 | 28.63 | 28.87 |

As shown in Figure 26, a change in the length of the casing only slightly affects the amplitude of the bending vibration of the blade and does not affect the frequency component. It can be also observed that the length of the casing has a significant influence on the radial displacement at point A of the casing. As the length of the casing increases, the amplitude amplification gradually decreases from frequency $9f_r$ to frequency $3f_r$, which is caused by the difference in the natural frequency of the casing.

From Table 10, as the length of the casing increases, both the first natural frequency and equivalent radial stiffness of the casing decrease. However, the maximum normal rubbing force does not decrease with the reduction in the radial stiffness of the casing. The primary reason lies in the varying deformation of the casing with different lengths, leading to different penetrations between the blade and casing. Consequently, the rubbing force does not decrease proportionally with the decline in the radial stiffness of the casing. As depicted in Figure 26, changes in the length of the casing only slightly affect the amplitude of the bending vibration of the blade and have no obvious effect on the frequency components. Simultaneously, the length of the casing significantly influences the radial displacement at point A of the casing. With the increase in casing length, the amplitude amplification gradually decreases from frequency $9f_r$ to frequency $3f_r$.

## 4. Conclusions

To investigate blade-coated casing rubbing, an FE model of a laminated shell casing is established using eight-node shell elements. The accuracy of the model is validated by comparing its natural characteristics with the results obtained from ANSYS software. Additionally, an actual compressor blade with varying sections and a twisted shape is also established using the shell element. Subsequently, based on hysteretic contact force, a dynamic model with rubbing is established to investigate rubbing-induced vibration characteristics. Some key conclusions can be summarized as follows:

(1) The vibration responses of the blade–casing system with rubbing are analyzed among three casing models: the lumped-mass model, the cylindrical shell model, and the laminated shell model. The study reveals that the vibration response amplitude of the laminated-blade shell casing with rubbing is the smallest. This indicates that the coating can reduce the severity level of rubbing. Additionally, the frequency corresponding to the amplitude amplification phenomenon of the cylindrical shell model and the laminated shell model at the rubbing and non-rubbing points are different due to variations in natural characteristics.

(2) Furthermore, the material and thickness of the coating can change the NDVCs induced by rubbing. The frequency corresponding to amplitude amplification is also influenced by the casing support stiffness and casing length, as these parameters change the natural characteristics of the casing.

The primary contribution of this article lies in the establishment of a dynamic model for an actual blade-coated casing, taking into account rubbing faults. The study emphasizes understanding the effects of coating and casing on dynamic responses in the presence of rubbing, with a specific focus on analyzing casing vibration. It is important to note that this article does not address coating wear, which stands out as a potential focus for future work. The study of these dynamics considering wear would provide a more comprehensive understanding of dynamic behavior under operational conditions.

**Author Contributions:** Conceptualization, Y.Z. and S.Y.; methodology, L.Q.; software, H.G.; validation, L.Q.; investigation, X.T.; resources, Y.Z.; data curation, X.D.; writing—original draft preparation, H.G.; writing—review and editing, H.M. and Q.M.; visualization, L.Q.; supervision, H.M.; project administration, Y.Z.; funding acquisition, H.M. All authors have read and agreed to the published version of the manuscript.

**Funding:** This work was financially supported by the National Science and Technology Major Project (J2022-IV-0005-0022) and Aero Science Foundation of China (20230015050001).

**Data Availability Statement:** Data will be made available on reasonable request.

**Conflicts of Interest:** Authors Y.Z., S.Y. and X.T. were employed by Shengu Group Co., Ltd. Author L.Q. was employed by Sunny Optical Technology (Group) Co. Ltd. The remaining authors declare that the research was conducted in the absence of any commercial or financial relationships that could be construed as a potential conflict of interest.

## Abbreviation

All symbols are listed as follows:

**Nomenclature**

| | |
|---|---|
| $b$ | Blade width |
| $\boldsymbol{B}'_p$ | Strain matrix in the local coordinate system $\xi\eta\zeta'$ of the shell |
| $c^i$ | Initial clearance between the blades and the casing |
| $c^i_{\text{rub}}$ | Dynamic clearance between blade tip node $i$ and casing |
| $c^{i,j}_{\text{rub}}$, $c^{i,j-1}_{\text{rub}}$ | Clearance between blade tip node $i$ and casing at the $j$th and $j$-1th load steps |
| $c_{\text{e}}$ | Collision recovery coefficient |
| $\boldsymbol{C}$ | Damping matrix of the blade |
| $\boldsymbol{C}_{\text{c}}$ | Damping matrix of the coated casing |
| $\hat{\boldsymbol{C}}_{\text{c}}$ | Damping matrix of the reduced coated casing |
| $\boldsymbol{D}_p$ | Elastic matrix in the local coordinate system $\xi\eta\zeta'$ of the shell |
| $e_{\text{c}}$ | Misalignment distance |
| $E_{\text{a}}$ | Elastic modulus of the coating |
| $E_{\text{b}}$ | Elastic modulus of the blade |
| $E_{\text{c}}$ | Elastic modulus of the casing |
| $f_{\text{n1c}}, f_{\text{n2c}}$ | First and second NFs of the casing |
| $F_{\text{e1}}$ | Aerodynamic amplitude |
| $F^i_{\text{n}}$ | Normal rubbing force of the $i$th node at the blade tip |
| $\hat{\boldsymbol{F}}_{\text{c}}$ | External force vector of the reduced coated casing |
| $\boldsymbol{F}_{\text{rb}}$, $\boldsymbol{F}_{\text{rc}}$ | Rubbing force vector of the blade and casing |
| $g^1_{\text{min}}$ | Minimum clearance between the first node of the blade tip and the casing |
| $g_{\text{min}}$ | Minimum clearance between the blade tip and the casing at $\beta_1 = 0°$ |
| $\boldsymbol{G}$ | Coriolis force matrix of the blade |
| $h_{\text{t}}$ | Total thickness of the laminated shell |
| $h_{p\text{-}1}$, $h_p$ | Upper and lower boundaries of the $p$-th layer |
| $h_{\text{a}}$ | Thickness of the coating |
| $h_{\text{c}}$ | Thickness of the casing |
| $k_{\text{h}}$ | Hertz contact stiffness |
| $k_{\text{h}}^*$ | Contact stiffness between the blade and casing |
| $k_{\text{b}}$, $k_{\text{c}}$ | Structure stiffness of the blade and casing |
| $k_{\text{s}}$ | Structural stiffness of the blade in series with the casing |
| $k_x, k_y, k_z, k_{\theta x}, k_{\theta y}, k_{\theta z}$ | Root constraint stiffness of the blade |
| $k_{\text{r}}, k_{\text{t}}, k_{\text{a}}, k_{\theta\text{r}}, k_{\theta\text{t}}$ | Root constraint stiffness of the casing |
| $k_{\text{h(eq)}}$, $k_{\text{l(eq)}}$ | Equivalent rubbing stiffness |
| $\boldsymbol{K}_{\text{e}}, \boldsymbol{K}_{\text{ce}}, \boldsymbol{K}_{\text{co}}, \boldsymbol{K}_{\text{s}}$ | Structure stiffness, centrifugal stiffening, elastic support stiffness, and spin softening matrixes of the blade |
| $\boldsymbol{K}^{\text{e}}$ | Stiffness matrix of the shell element |
| $\boldsymbol{K}_{\text{c}}$ | Stiffness matrix of the coated casing |
| $\hat{\boldsymbol{K}}_{\text{c}}$ | Stiffness matrix of the reduced coated casing |
| $L_{\text{c}}$ | Length of casing |
| $m_{\text{c}}$ | Mass of the casing |
| $\boldsymbol{M}$ | Mass matrix of the blade |
| $\boldsymbol{M}^{\text{e}}$ | Mass matrix of the shell element |
| $\boldsymbol{M}_{\text{c}}$ | Mass matrix of the coated casing |
| $\hat{\boldsymbol{M}}_{\text{c}}$ | Mass matrix of the reduced coated casing |
| $q$ | Total number of floors |
| $\boldsymbol{q}_{\text{c}}, \dot{\boldsymbol{q}}_{\text{c}}, \ddot{\boldsymbol{q}}_{\text{c}}$ | Displacement, velocity, and acceleration vectors of the reduced casing |
| $Q_{\text{P}}$ | Aerodynamic load |
| $R_{\text{d}}$ | Radius of the disk |
| $R_{\text{c}}$ | Inner diameter of casing |
| $R_{\text{g}}$ | Orbit radius of the blade tip in a stationary state |
| $Z^i$, $X^i$ | Initial coordinate of the blade tip in the $x$ and $z$ directions |
| $u^i_{\text{c}}$, $u^i_{\text{b}}$ | Radial displacement of the $i$-th node of the casing and the $i$-th node of the blade tip |
| $\boldsymbol{u}, \dot{\boldsymbol{u}}, \ddot{\boldsymbol{u}}$ | Displacement, velocity, and acceleration vectors of the casing |
| $\boldsymbol{x}, \dot{\boldsymbol{x}}, \ddot{\boldsymbol{x}}$ | Displacement, velocity, and acceleration vectors of the blade |

**Greek symbols**

| | |
|---|---|
| $\beta_1$ | Misalignment angle |
| $\beta_2$ | Stagger angle of the blade tip |
| $\dot{\delta}^{i-}$, $\dot{\delta}^{i}$ | Relative initial collision velocity and the relative collision velocity at any moment |
| $\delta_h^i$ | Local deformation of the coating |
| $\delta^i$ | Overall penetration deformation between the blade and the casing |
| $\xi_1$, $\xi_2$ | Modal damping ratio |
| $\zeta_{p-1}$, $\zeta_p$ | Corresponding to $-1$ and 1 in the local coordinate system $\xi\eta\zeta'$ |
| $\omega$ | Rotating speed of the rotating blade |
| $\rho_p$ | Material density of the $p$-th layer |
| $\rho_a$ | Density of coating |
| $\rho_c$ | Density of casing |
| $v$ | Poisson's ratio |
| $v_b$, $v_a$ | Poisson's ratio of the blade and coating |

**Abbreviation**

| | |
|---|---|
| FE | Finite element |
| NDVCs | Nodal diameter vibrations of the casings |
| NRF | Normal rubbing force |
| NFs | Natural frequencies |
| TDW | Time-domain waveform |

## Appendix A

In this paper, a detailed description of the Mindlin–Reissner shell is provided. Figure A1 shows a representative shell element (Mindlin–Reissner shell). This shell element is characterized by two surfaces above and below, with a wraparound surface bounded by a straight line in the direction of shell thickness. Notably, the nodes of the element are located on the mid-surface, with a total of eight nodes comprising the four corner points of the mid-surface and the mid-points of the four edges. The coordinate system of the element and the positions of the nodes are shown in Figure A1. Numbers 1–8 represent 8 nodes within a shell element.

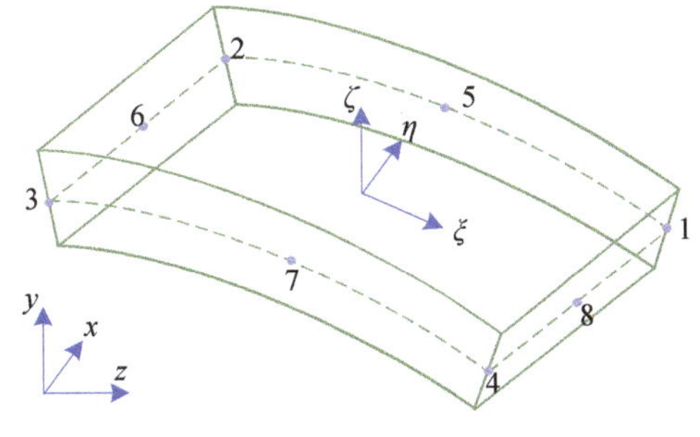


**Figure A1.** Schematic of the degenerated shell element is displayed [80]. The element is degraded from a 3D solid element and has some of the properties of a solid element, enclosed by two surfaces above and below and a wraparound surface bushed by a straight line in the direction of shell thickness. It has eight nodes. $\xi$ and $\eta$ are the surface coordinates of the shell, and $\zeta$ is the linear coordinate in the thickness direction.

In Figure A1, the nodes of the element are numbered, and the coordinates of node $i$ ($i$ = 1, 2, ..., 8) under the global coordinate system are ($x_i$, $y_i$, $z_i$); then, the coordinate values

$\xi$ and $\eta$ at any point on the element surface can be expressed by interpolation in the global coordinates $(x, y, z)$. The specific expressions are as follows:

$$x = \sum_{i=1}^{8} N_i x_i,\ y = \sum_{i=1}^{8} N_i y_i,\ z = \sum_{i=1}^{8} N_i z_i, \tag{A1}$$

where $N_i$ is the shape function of the element, and the specific expression is

$$\begin{cases} N_i = \frac{1}{4}(1+\xi\xi_i)(1+\eta\eta_i)(\xi\xi_i+\eta\eta_i-1)\ , & i = 1,2,3,4 \quad \text{(Corner Point)} \\ N_i = \frac{\xi_i^2}{2}(1+\xi\xi_i)(1-\eta^2)+\frac{\eta_i^2}{2}(1+\eta\eta_i)(1-\xi^2)\ , & i = 5,6,7,8 \ \text{(Middle Point)} \end{cases}. \tag{A2}$$

Within the element, displacements of points can be expressed according to node displacements and geometric relationships in the element of the middle surface, as follows:

$$\begin{aligned} \begin{bmatrix} u \\ v \\ w \end{bmatrix} &= \sum_{i=1}^{8} N_i(\xi,\eta) \begin{bmatrix} u_i \\ v_i \\ w_i \end{bmatrix} + \sum_{i=1}^{8} N_i(\xi,\eta)\ \zeta\frac{t_i}{2} \begin{bmatrix} 0 & n_{3i} & -m_{3i} \\ -n_{3i} & 0 & l_{3i} \\ m_{3i} & -l_{3i} & 0 \end{bmatrix} \begin{bmatrix} \theta_{xi} \\ \theta_{yi} \\ \theta_{zi} \end{bmatrix} \\ &= \begin{bmatrix} \boldsymbol{N}_1 & \boldsymbol{N}_2 & \ldots & \boldsymbol{N}_8 \end{bmatrix} \begin{bmatrix} \boldsymbol{a}_1 \\ \boldsymbol{a}_2 \\ \ldots \\ \boldsymbol{a}_8 \end{bmatrix} \end{aligned}, \tag{A3}$$

where $\boldsymbol{a}_i$ and $\boldsymbol{N}_i$ can be represented as

$$\boldsymbol{a}_i = [u_i\ v_i\ w_i\ \theta_{xi}\ \theta_{yi}\ \theta_{zi}]^{\mathrm{T}},\ (i = 1, 2, \ldots, 8), \tag{A4}$$

$$\boldsymbol{N}_i = \begin{bmatrix} N_i & 0 & 0 & 0 & N_i\zeta\frac{t_i}{2}n_{3i} & -N_i\zeta\frac{t_i}{2}m_{3i} \\ 0 & N_i & 0 & -N_i\zeta\frac{t_i}{2}n_{3i} & 0 & N_i\zeta\frac{t_i}{2}l_{3i} \\ 0 & 0 & N_i & N_i\zeta\frac{t_i}{2}m_{3i} & -N_i\zeta\frac{t_i}{2}l_{3i} & 0 \end{bmatrix}. \tag{A5}$$

Matrix $\boldsymbol{B}$ can be written as

$$\boldsymbol{B} = \boldsymbol{G}\bar{\boldsymbol{J}} \begin{bmatrix} \boldsymbol{M}_{\xi 1} & \boldsymbol{M}_{\xi 2} & \cdots & \boldsymbol{M}_{\xi 8} \\ \boldsymbol{M}_{\eta 1} & \boldsymbol{M}_{\eta 2} & \cdots & \boldsymbol{M}_{\eta 8} \\ \boldsymbol{M}_{\zeta 1} & \boldsymbol{M}_{\zeta 2} & \cdots & \boldsymbol{M}_{\zeta 8} \end{bmatrix}, \tag{A6}$$

where matrix $\boldsymbol{G}$ is a constant matrix, and can be represented as

$$\boldsymbol{G} = \begin{bmatrix} 1 & 0 & 0 & 0 & 0 & 0 & 0 & 0 & 0 \\ 0 & 0 & 0 & 0 & 1 & 0 & 0 & 0 & 0 \\ 0 & 0 & 0 & 0 & 0 & 0 & 0 & 0 & 1 \\ 0 & 1 & 0 & 1 & 0 & 0 & 0 & 0 & 0 \\ 0 & 0 & 0 & 0 & 0 & 1 & 0 & 1 & 0 \\ 0 & 0 & 1 & 0 & 0 & 0 & 1 & 0 & 0 \end{bmatrix}. \tag{A7}$$

The Jacobi matrix $\boldsymbol{J}$ is

$$\boldsymbol{J} = \begin{bmatrix} \frac{\partial x}{\partial \xi} & \frac{\partial y}{\partial \xi} & \frac{\partial z}{\partial \xi} \\ \frac{\partial x}{\partial \eta} & \frac{\partial y}{\partial \eta} & \frac{\partial z}{\partial \eta} \\ \frac{\partial x}{\partial \zeta} & \frac{\partial y}{\partial \zeta} & \frac{\partial z}{\partial \zeta} \end{bmatrix}. \tag{A8}$$

Thus, the matrix $\bar{\boldsymbol{J}}$ can be written as

$$\bar{\boldsymbol{J}} = \boldsymbol{J} \otimes \boldsymbol{E}, \tag{A9}$$

where $\boldsymbol{E}$ is an identity matrix of order 3 × 3 and the symbol "⊗" denotes the Kronecker product between two matrices.

Additionally, matrices $\boldsymbol{M}_{\xi i}$, $\boldsymbol{M}_{\eta i}$, and $\boldsymbol{M}_{\zeta i}$can be written as

$$\boldsymbol{M}_{\xi i} = \frac{\partial N_i}{\partial \xi}\begin{bmatrix} 1 & 0 & 0 & 0 & \zeta\frac{t_i}{2}n_{3i} & -\zeta\frac{t_i}{2}m_{3i} \\ 0 & 1 & 0 & -\zeta\frac{t_i}{2}n_{3i} & 0 & \zeta\frac{t_i}{2}l_{3i} \\ 0 & 0 & 1 & \zeta\frac{t_i}{2}m_{3i} & -\zeta\frac{t_i}{2}l_{3i} & 0 \end{bmatrix}, \tag{A10}$$

$$\boldsymbol{M}_{\eta i} = \frac{\partial N_i}{\partial \eta}\begin{bmatrix} 1 & 0 & 0 & 0 & \zeta\frac{t_i}{2}n_{3i} & -\zeta\frac{t_i}{2}m_{3i} \\ 0 & 1 & 0 & -\zeta\frac{t_i}{2}n_{3i} & 0 & \zeta\frac{t_i}{2}l_{3i} \\ 0 & 0 & 1 & \zeta\frac{t_i}{2}m_{3i} & -\zeta\frac{t_i}{2}l_{3i} & 0 \end{bmatrix}, \tag{A11}$$

$$\boldsymbol{M}_{\zeta i} = N_i\begin{bmatrix} 0 & 0 & 0 & 0 & \frac{t_i}{2}n_{3i} & -\frac{t_i}{2}m_{3i} \\ 0 & 0 & 0 & -\frac{t_i}{2}n_{3i} & 0 & \frac{t_i}{2}l_{3i} \\ 0 & 0 & 0 & \frac{t_i}{2}m_{3i} & -\frac{t_i}{2}l_{3i} & 0 \end{bmatrix}, \tag{A12}$$

The relationship between strain and displacement in the global coordinate system can be written as

$$\begin{aligned} \boldsymbol{\varepsilon} &= \begin{bmatrix} \varepsilon_x\ \varepsilon_y\ \varepsilon_z\ \gamma_{xy}\ \gamma_{yz}\ \gamma_{zx} \end{bmatrix}^{\mathrm{T}} \\ &= \begin{bmatrix} \frac{\partial u}{\partial x}\ \frac{\partial v}{\partial y}\ \frac{\partial w}{\partial z}\ \frac{\partial u}{\partial y} + \frac{\partial v}{\partial x}\ \frac{\partial v}{\partial z} + \frac{\partial w}{\partial y}\ \frac{\partial u}{\partial z} + \frac{\partial w}{\partial x} \end{bmatrix} \\ &= \boldsymbol{B}\begin{bmatrix} \boldsymbol{a}_1 & \boldsymbol{a}_2 & \ldots & \boldsymbol{a}_8 \end{bmatrix}^{\mathrm{T}} \end{aligned} \tag{A13}$$

where $\boldsymbol{a}_i (i = 1, 2, \ldots 8)$ represents the displacement vector of node $i$ within a shell element. $\boldsymbol{\varepsilon}$ represents the strain matrix. $\varepsilon_x$, $\varepsilon_y$, and $\varepsilon_z$ denote the principal strain values. $\gamma_{xy}$, $\gamma_{yz}$, and $\gamma_{zx}$ represent the shear strain values.

In the local coordinate system, the elasticity matrix $\boldsymbol{D}'$ of the material is written as

$$D' = \frac{E}{1-\mu^2}\begin{bmatrix} 1 & \mu & 0 & 0 & 0 & 0 \\ \mu & 1 & 0 & 0 & 0 & 0 \\ 0 & 0 & 0 & 0 & 0 & 0 \\ 0 & 0 & 0 & \frac{1-\mu}{2} & 0 & 0 \\ 0 & 0 & 0 & 0 & \frac{1-\mu}{2k} & 0 \\ 0 & 0 & 0 & 0 & 0 & \frac{1-\mu}{2k} \end{bmatrix}, \tag{A14}$$

where $E$ and $\mu$ represent the elastic modulus and Poisson's ratio. $k$ ($k = 1.2$) is a constant value.

Thus, in the global system, the elasticity matrix $\boldsymbol{D}$ of the material is denoted as

$$\boldsymbol{D} = \boldsymbol{T}^{\mathrm{T}}\boldsymbol{D}'\boldsymbol{T}, \tag{A15}$$

where the specific expression for matrix $\boldsymbol{T}$ is given in Ref. [81].

The position coordinate of any point on the rotation body (i.e., the blade in this paper) in the global coordinate system is denoted as $\boldsymbol{r}_0$, the corresponding elastic deformation is denoted as $\boldsymbol{r}_A$, and the position of any point relative to the fixed coordinate system is written as

$$\boldsymbol{r} = \boldsymbol{r}_0 + \boldsymbol{r}_A = \begin{bmatrix} x_0 \\ y_0 \\ z_0 \end{bmatrix} + \begin{bmatrix} u \\ v \\ w \end{bmatrix}, \tag{A16}$$

Thus, the velocity at this point is written as

$$\boldsymbol{v} = \boldsymbol{\Omega} \times \boldsymbol{r} + \dot{\boldsymbol{r}}. \tag{A17}$$

The absolute velocity at this point is

$$\boldsymbol{v}_A = \dot{\boldsymbol{r}}_A + \boldsymbol{\Omega} \times \boldsymbol{r}, \tag{A18}$$

where

$$\Omega = \begin{bmatrix} \Omega_x \\ \Omega_y \\ \Omega_z \end{bmatrix}, \tag{A19}$$

$$\Omega \times r = \underline{\Omega} r = \begin{bmatrix} 0 & -\Omega_z & \Omega_y \\ \Omega_z & 0 & -\Omega_x \\ -\Omega_y & \Omega_x & 0 \end{bmatrix} r, \tag{A20}$$

where $\Omega_x$, $\Omega_y$, and $\Omega_z$ are the rotational speeds of the element around the $x$, $y$, and $z$ axes, respectively.

The kinetic energy of the element is

$$\begin{aligned} T_m^e &= \frac{1}{2}\rho \iiint_V (v_A)^{\mathrm{T}} \cdot (v_A) \mathrm{d}V \\ &= \frac{1}{2}(\dot{a}^e)^{\mathrm{T}} \int_V \rho N^{\mathrm{T}} N \mathrm{d}V \, \dot{a}^e + \frac{1}{2}\int_V \rho r_0{}^{\mathrm{T}} \underline{\Omega}^{\mathrm{T}} \underline{\Omega} \, r_0 \mathrm{d}V + (a^e)^{\mathrm{T}} \int_V \rho N^{\mathrm{T}} \underline{\Omega}^{\mathrm{T}} \underline{\Omega} \, r_0 \mathrm{d}V \\ &+ \frac{1}{2}(a^e)^{\mathrm{T}} \int_V \rho N^{\mathrm{T}} \underline{\Omega}^{\mathrm{T}} \underline{\Omega} \, N \mathrm{d}V a^e + (\dot{a}^e)^{\mathrm{T}} \int_V \rho N^{\mathrm{T}} \underline{\Omega} \, r_0 \mathrm{d}V + (\dot{a}^e)^{\mathrm{T}} \int_V \rho N^{\mathrm{T}} \underline{\Omega} N \mathrm{d}V \, a^e \end{aligned} . \tag{A21}$$

The tensile and bending strain energy of the element is

$$\begin{aligned} U_\varepsilon^e &= \frac{1}{2}\int_V (\sigma_x \varepsilon_x + \sigma_y \varepsilon_y + \sigma_z \varepsilon_z + \tau_{xy}\gamma_{xy} + \tau_{yz}\gamma_{yz} + \tau_{zx}\gamma_{zx}) \mathrm{d}V \\ &= \frac{1}{2}(a^e)^{\mathrm{T}} \int_V (B_L)^{\mathrm{T}} D B_L \mathrm{d}V a^e + \frac{1}{2}(a^e)^{\mathrm{T}} \int_V S^{\mathrm{T}} \bar{\sigma} S \mathrm{d}V a^e \end{aligned}, \tag{A22}$$

where

$$\bar{\sigma} = \begin{pmatrix} \sigma_{xx}E & \tau_{xy}E & \tau_{xz}E \\ \tau_{yx}E & \sigma_{yy}E & \tau_{yz}E \\ \tau_{zx}E & \tau_{zy}E & \sigma_{zz}E \end{pmatrix}. \tag{A23}$$

By substituting the kinetic energy (Equation (A21)) and strain energy (Equation (A22)) of the element into the Lagrange equation, we can obtain

$$\frac{\mathrm{d}}{\mathrm{d}t}\left(\frac{\partial T_m^e}{\partial \dot{a}}\right) - \frac{\partial T_m^e}{\partial a} + \frac{\partial V_m^e}{\partial a} = F^e, \tag{A24}$$

Therefore, the differential equation of the motion of the element is

$$\begin{aligned} &\int_V \rho N^{\mathrm{T}} N \mathrm{d}V \, \ddot{a}^e - \int_V \rho N^{\mathrm{T}} \underline{\Omega}^{\mathrm{T}} \underline{\Omega} \, r_0 \mathrm{d}V - \int_V \rho N^{\mathrm{T}} \underline{\Omega}^{\mathrm{T}} \underline{\Omega} \, N \mathrm{d}V a^e + 2\int_V \rho N^{\mathrm{T}} \underline{\Omega} N \mathrm{d}V \dot{a}^e \\ &+ \int_V (B_L)^{\mathrm{T}} D B_L \mathrm{d}V a^e + \int_V S^{\mathrm{T}} \bar{\sigma} S \mathrm{d}V a^e = F^e \end{aligned} . \tag{A25}$$

Through assembling the stiffness matrix, mass matrix, and so on, the differential equation of the motion of the blade is established. When the external force is 0 N, the differential equation of the motion of the blade can be written as

$$M\ddot{x} + (G + C)\dot{x} + (K_{\mathrm{e}} + K_{\mathrm{ce}} + K_{\mathrm{co}} - K_{\mathrm{s}})x = 0. \tag{A26}$$

where $M = \sum_{el} \int_V \rho N^{\mathrm{T}} N \mathrm{d}V$ is the mass matrix of the blade, $G = 2\sum_{el} \int_V \rho N^{\mathrm{T}} \underline{\Omega} N \mathrm{d}V$ is the Coriolis force matrix of the blade, $K_{\mathrm{e}} = \sum_{el} \int_V (B_L)^{\mathrm{T}} D B_L \mathrm{d}V$ is the stiffness matrix of the blade, $K_\sigma = \sum_{el} \int_V S^{\mathrm{T}} \bar{\sigma} S \mathrm{d}V$ is the centrifugal stiffening matrix, and $K_{\mathrm{s}} = \sum_{el} \int_V \rho N^{\mathrm{T}} \underline{\Omega}^{\mathrm{T}} \underline{\Omega} \, N \mathrm{d}V$ is the spin-softening matrix. $C$ represents the Rayleigh damping matrix, and its expression can be found in Equation (7).

## Appendix B

Considering the influence of angular misalignment, the minimum clearance between the $i$th node of the blade tip and the casing is written as [28]

$$g_{\min}^{i} = g_{\min}^{1} + (21 - i) \times b \times \cos\beta_2/20 \times \sin\beta_1, \tag{A27}$$

where $b$ denotes the blade width, $\beta_1$ represents the misalignment angle, and $\beta_2$ represents the stagger angle of the blade tips. $g_{\min}^{1}$ represents minimum clearance at the first node pair. The schematic is shown in Figure 9b.

Based on the geometric relationship $ON + c^i = OM$, the initial clearance is [28]

$$c^i = R_c \sin(\pi/2 - \theta_i(t))/\sin\theta(t) - R_g, \tag{A28}$$

where $R_g$ is the static orbit radius of the blade tips. $\theta(t) = \omega t + \varphi_{i0}$, $\varphi_{i0} = \arctan(Z^i/X^i)$ and $\omega$ represent the rotational speeds of the blades. $Z^i$ and $X^i$ are the initial coordinates of the blade tip in the $x$ and $z$ directions. $R_c$ is the inner radius of the casing and $\theta_i(t) = \pi/2 - \theta(t) + \alpha_i(t)$, where $\alpha_i(t) = \arcsin((R_c - g_{\min}^{i} - R_g)\sin\theta(t)/R_c)$.

The dynamic clearance $c_{\mathrm{rub}}^{i}$ between the $i$th node pair is expressed as

$$c_{\mathrm{rub}}^{i} = c^i + u_c^i - u_b^i, \tag{A29}$$

where $u_c^i$ and $u_b^i$ are the radial displacement of the $i$th node of the casing and the blade tip, respectively.

Cao et al. [59–61] established a hysteresis contact-force model to simulate the contact between the blade and the coated casing. The rubbing stiffness is expressed as

$$k = \frac{k_b k_c k_h^*}{k_b k_c + k_b k_h^* + k_c k_h^*}, \tag{A30}$$

where $k_b$ and $k_c$ denote the structure stiffnesses of the blade and the casing, and $k_h^*$ ($k_h^* = k_h \delta_h^{1/2}$) denotes the contact stiffness. $\delta_h$ is the local deformation of the coating. $k_h$ is the Hertz contact stiffness between the blade and the casing, and it can be written as

$$k_h = \frac{4}{3\left(\frac{1-v_b^2}{E_b} + \frac{1-v_a^2}{E_a}\right)} \left(\frac{R_g R_c}{R_g + R_c}\right)^{\frac{1}{2}}, \tag{A31}$$

where $v_b$ and $v_a$ denote the Poisson's ratio of the blade and the coating, $E_b$ and $E_a$ are the elastic moduli of the blade and the coating, and $R_g$ and $R_c$ are the rotation radius of the blade tip node and the inner diameter of the casing.

The local deformation $\delta_h^i$ of coatings is denoted as [59]

$$\delta_h^i = \left(\frac{4^{1/3} f(\delta^i)}{6k_h} + \frac{k_s^2 4^{2/3}}{6k_h f(\delta^i)} - \frac{k_s}{3k_h}\right)^2, \tag{A32}$$

where $k_s$ ($k_s = k_b k_a/(k_b + k_a)$) is the structural stiffness of the blade in series with the casing. $f(\delta^i)$ is the penetration function, and its expression is

$$f\left(\delta^i\right) = \left(k_s\left(27\delta^i k_h^2 - 2k_s^2 + 3\sqrt{3}\sqrt{\delta(27\delta^i k_h^2 - 4k_s^2)}k_h\right)\right)^{1/3}. \tag{A33}$$

To calculate the damping force, Cao et al. [61] derived a normal hysteresis force model using the laws of energy conservation and momentum conservation. The normal rubbing force considering viscous damping is represented as

$$f_{\mathrm{n}}^{i} = k\delta^{i}\left(1+\frac{3(1-c_{\mathrm{e}}^{2})\dot{\delta}^{i}}{4\dot{\delta}^{i-}}\right) = \frac{k_{\mathrm{b}}k_{\mathrm{c}}k_{\mathrm{h}}^{*}}{k_{\mathrm{b}}k_{\mathrm{c}}+k_{\mathrm{b}}k_{\mathrm{h}}^{*}+k_{\mathrm{c}}k_{\mathrm{h}}^{*}}\delta^{i}\left(1+\frac{3(1-c_{\mathrm{e}}^{2})\dot{\delta}^{i}}{4\dot{\delta}^{i-}}\right), \tag{A34}$$

where $\dot{\delta}^{i-}$ is the relative initial collision velocity and $\dot{\delta}^{i}$ is the relative collision velocity at any moment. It is solved by the difference method and $\dot{\delta}^{i} = \left(c_{\mathrm{rub}}^{i,j} - c_{\mathrm{rub}}^{i,j-1}\right)/dt$, where $c_{\mathrm{rub}}^{i,j}$ and $c_{\mathrm{rub}}^{i,j-1}$ are the clearance between blade tip node $i$ and the casing at the $j$th and $j-1$th load steps. $c_{\mathrm{e}}$ is the collision recovery coefficient, and $c_{\mathrm{e}} = 0.8$ in this paper.

When the local deformation of the coated casing meets $\delta_{\mathrm{h}}^{i} \geq 4k_{\mathrm{s}}^{2}/\left(27k_{\mathrm{h}}^{2}\right)$, a hysteresis contact-force model is used to simulate the rubbing force. The overall penetration deformation meets

$$\delta^{i} = \delta_{\mathrm{c}}^{i}+\delta_{\mathrm{h}}^{i} = \frac{k_{\mathrm{h}}^{*}\delta_{\mathrm{h}}^{i}}{k_{\mathrm{s}}}+\delta_{\mathrm{h}}^{i} \geq \left(1+\frac{2}{\sqrt{27}}\right)\frac{4k_{\mathrm{s}}^{2}}{27k_{\mathrm{h}}^{2}}. \tag{A35}$$

When the local contact stiffness of the coating is much smaller than the stiffness of the blade and casing, the rubbing stiffness is mainly determined by the local contact stiffness. The overall penetration deformation of the blade and casing is approximately equal to the local deformation of the coating. Thus, a modified Hertz contact model can be used to simulate the rubbing process between the blade and the coated casing. The overall penetration deformation meets $\delta \approx \delta_{\mathrm{h}}^{i} < \alpha k_{\mathrm{s}}^{2}/27k_{\mathrm{h}}^{2}$ ($\alpha$ = 1%). The equivalent rubbing stiffness of this model is written as [59]

$$k_{\mathrm{h(eq)}} = \left(\frac{2\sqrt{27}}{\sqrt{\alpha}}-\frac{54}{\alpha}\ln\left(\sqrt{27}+\sqrt{\alpha}\right)+\frac{54}{\alpha}\ln\sqrt{27}\right)k_{\mathrm{h}}^{*}. \tag{A36}$$

When the local deformation of coatings meets $\frac{\alpha k_{\mathrm{s}}^{2}}{27k_{\mathrm{h}}^{2}} < \delta_{\mathrm{h}}^{i} < \frac{4k_{\mathrm{s}}^{2}}{27k_{\mathrm{h}}^{2}}$, the modified linear model is used, and the equivalent rubbing stiffness is [59]

$$k_{\mathrm{l(eq)}} = \frac{2\left(4^{3/2}-\alpha^{3/2}\right)}{3\sqrt{27}(4-\alpha)+2\left(4^{3/2}-\alpha^{3/2}\right)}k_{\mathrm{s}}. \tag{A37}$$

Therefore, the normal rubbing force is

$$f_{\mathrm{n}}^{i} = \begin{cases} k_{\mathrm{h(eq)}}\delta^{i}\left(1+\frac{3(1-c_{\mathrm{e}}^{2})\dot{\delta}^{i}}{4\dot{\delta}^{i-}}\right) & \delta^{i} < \frac{\alpha k_{\mathrm{s}}^{2}}{27k_{\mathrm{h}}^{2}} \\ k_{\mathrm{l(eq)}}\delta^{i}\left(1+\frac{3(1-c_{\mathrm{e}}^{2})\dot{\delta}^{i}}{4\dot{\delta}^{i-}}\right) & \frac{\alpha k_{\mathrm{s}}^{2}}{27k_{\mathrm{h}}^{2}} \leq \delta^{i} < \left(1+\frac{2}{\sqrt{27}}\right)\frac{4k_{\mathrm{s}}^{2}}{27k_{\mathrm{h}}^{2}} \\ \frac{k_{\mathrm{b}}k_{\mathrm{c}}k_{\mathrm{h}}^{*}}{k_{\mathrm{b}}k_{\mathrm{c}}+k_{\mathrm{b}}k_{\mathrm{h}}^{*}+k_{\mathrm{c}}k_{\mathrm{h}}^{*}}\delta^{i}\left(1+\frac{3(1-c_{\mathrm{e}}^{2})\dot{\delta}^{i}}{4\dot{\delta}^{i-}}\right) & \delta^{i} \geq \left(1+\frac{2}{\sqrt{27}}\right)\frac{4k_{\mathrm{s}}^{2}}{27k_{\mathrm{h}}^{2}} \end{cases}. \tag{A38}$$

The thickness of airfoil blades varies. The radial displacement and penetration depth of each node are different under the action of the centrifugal forces. Therefore, it is necessary to modify the model. The normal rubbing force of the $i$th node is modified as

$$F_{\mathrm{n}}^{i} = f_{\mathrm{n}}^{i}\delta^{i}/\delta, \tag{A39}$$

where $\delta$ ($\delta = \sum_{i=1}^{21}\delta^{i}$) is the sum of total penetration at each node of the blade tip.

## Appendix C

The dynamic responses are calculated based on the Newmark-$\beta$ method, and the coefficients utilized in the Newmark-$\beta$ method are listed as

$$c_0 = 1/(\beta_0 \Delta t^2) \tag{A40}$$

$$c_1 = \gamma_0/(\beta_0 \Delta t) \tag{A41}$$

$$c_2 = 1/(\beta_0 \Delta t) \tag{A42}$$

$$c_3 = 1/\beta_0^2 - 1 \tag{A43}$$

$$c_4 = \gamma_0/\beta_0 - 1 \tag{A44}$$

$$c_5 = \frac{1}{2}\Delta t(\gamma_0/\beta_0 - 2) \tag{A45}$$

$$c_6 = \Delta t(1 - \gamma_0) \tag{A46}$$

$$c_7 = \Delta t \gamma_0 \tag{A47}$$

where $\Delta t$ is the time step of the numerical integration. For the average acceleration method, $\gamma_0$ = ½ and $\beta_0 = 1/4$.

The effective stiffness matrix of the conventional Newmark-$\beta$ method is written as

$$\boldsymbol{K}_{\text{eff}}(t) = c_0 \boldsymbol{M}_{\text{o}} + c_1(\boldsymbol{C}_{\text{o}} + \boldsymbol{G}_{\text{o}}) + \boldsymbol{K}_{\text{o}} \tag{A48}$$

where $\boldsymbol{M}_{\text{o}}$, $\boldsymbol{C}_{\text{o}}$, $\boldsymbol{G}_{\text{o}}$, and $\boldsymbol{K}_{\text{o}}$ represent the total mass, damping, gyroscope, and stiffness matrices of the system.

The effective load vector is written as

$$\boldsymbol{F}_{\text{eff}}(t) = \boldsymbol{F}_{\text{c}} + \boldsymbol{M}_{\text{o}}(c_0\boldsymbol{X}_{(i-1)} + c_2\dot{\boldsymbol{X}}_{(i-1)} + c_3\ddot{\boldsymbol{X}}_{(i-1)}) + (\boldsymbol{C}_{\text{o}} + \boldsymbol{G}_{\text{o}})(c_1\boldsymbol{X}_{(i-1)} + c_4\dot{\boldsymbol{X}}_{(i-1)} + c_5\ddot{\boldsymbol{X}}_{(i-1)}) \tag{A49}$$

where $\boldsymbol{F}_{\text{e}}$ represents the total force vector.

The iterative scheme of the Newmark-$\beta$ method is

$$\text{Displacement}: \boldsymbol{X}_{(i)} = \boldsymbol{K}_{\text{eff}}^{-1}\boldsymbol{F}_{\text{eff}} \tag{A50}$$

$$\text{Acceleration}: \ddot{\boldsymbol{X}}_{(i)} = c_0(\boldsymbol{X}_{(i)} - \boldsymbol{X}_{(i-1)}) - c_2\dot{\boldsymbol{X}}_{(i-1)} - c_3\ddot{\boldsymbol{X}}_{(i-1)} \tag{A51}$$

$$\text{Velocity}: \dot{\boldsymbol{X}}_{(i)} = \boldsymbol{X}_{(i-1)} + c_6\ddot{\boldsymbol{X}}_{(i-1)} + c_7\ddot{\boldsymbol{X}}_{(i)} \tag{A52}$$

## References


Wang, N.F.; Liu, C.; Jiang, D.X. Prediction of transient vibration response of dual-rotor-blade-casing system with blade off. *Proc. Inst. Mech. Eng. Part G J. Aerosp. Eng.* **2019**, *8*, 351–364. [CrossRef]

Yu, M.Y.; Feng, Z.G.; Huang, J.J.; Zhu, L.L. A new characteristic analysis method for aero-engine rotor–stator rubbing. *Proc. Inst. Mech. Eng. Part G J. Aerosp. Eng.* **2017**, *232*, 1120–1133. [CrossRef]

Yang, Y.; Tang, J.; Chen, G.; Yang, Y.R.; Cao, D.Q. Rub-impact investigation of a single-rotor system considering coating effect and coating hardness. *J. Vib. Eng. Technol.* **2021**, *9*, 491–505. [CrossRef]

Millecamps, A.; Batailly, A.; Legrand, M.; Garcin, F. Snecma's viewpoint on the numerical and experimental simulation of blade-tip/casing unilateral contacts. In Proceedings of the ASME Turbo Expo 2015: Turbine Technical Conference and Exposition, Montreal, QC, Canada, 15–19 June 2015.

Ma, H.; Yin, F.L.; Guo, Y.Z.; Tai, X.Y.; Wen, B.C. A review on dynamic characteristics of blade–casing rubbing. *Nonlinear Dyn.* **2016**, *84*, 437–472. [CrossRef]

Hamdi, H.; Farah, K. Beam finite element model of a vibrate wind blade in large elastic deformation. *Wind. Struct.* **2018**, *26*, 25–34. [CrossRef]

Arumugam, A.B.; Rajamohan, V.; Pandey, A.; Sudhagar, P. Finite element vibration analysis of rotating laminated composite beam with varying cross-section using HSDT. *Int. J. Interact. Des. Manuf.* **2017**, *11*, 703–712. [CrossRef]

Stoykov, S.; Ribeiro, P. Vibration analysis of rotating 3D beams by the p-version finite element method. *Finite Elem. Anal. Des.* **2013**, *65*, 76–88. [CrossRef]

9. Gunda, J.B.; Ganguli, R. New rational interpolation functions for finite element analysis of rotating beams. *Int. J. Mech. Sci.* **2007**, *50*, 578–588. [CrossRef]
10. Carassale, L.; Maurici, M.; Traversone, L. Reduced-Order Modeling of Compressor Blades by 1D Finite Elements. In Proceeding of the ASME Turbo Expo: Turbine Technical Conference and Exposition, Montréal, QC, Canada, 15–19 June 2015. [CrossRef]
11. Lee, S.L. Equivalent beam modeling scheme based on three-dimensional structure with blade element concept. *Adv. Compos. Mater.* **2020**, *29*, 191–203. [CrossRef]
12. Sakar, G.; Sabuncu, M. Dynamic stability of a rotating asymmetric cross-section blade subjected to an axial periodic force. *Int. J. Mech. Sci.* **2003**, *45*, 1467–1482. [CrossRef]
13. Guan, H.; Ni, K.X.; Ma, H.; Xiong, Q.; Wang, W.W.; Wang, H.J. Dynamic modeling and verification of rotating compressor blade with crack based on beam element. *Appl. Math. Model.* **2024**, *133*, 367–393. [CrossRef]
14. Sabuncu, M.; Evran, K. The dynamic stability of a rotating pre-twisted asymmetric cross-section blade subjected to lateral parametric excitation. *Finite Elem. Anal. Des.* **2006**, *42*, 1113–1122. [CrossRef]
15. Sabuncu, M.; Evran, K. Dynamic stability of a rotating pre-twisted asymmetric cross-section Timoshenko beam subjected to an axial periodic force. *Int. J. Mech. Sci.* **2006**, *48*, 579–590. [CrossRef]
16. Oh, Y.; Yoo, H.H. Vibration analysis of a rotating pre-twisted blade considering the coupling effects of stretching, bending, and torsion. *J. Sound Vib.* **2018**, *431*, 20–39. [CrossRef]
17. Kou, H.J.; Du, J.J.; Liang, M.X.; Zhu, L.; Zeng, L.; Zhu, Z.D.; Zhang, F. Nonlinear characteristics of contact-induced vibrations o the rotating variable thickness plate under large deformations. *Eur. J. Mech./A Solids* **2019**, *77*, 103801. [CrossRef]
18. Yao, M.H.; Ma, L.; Zhang, W. Nonlinear dynamics of the high-speed rotating plate. *Int. J. Aerosp. Eng.* **2018**, *2018*, 5610915 [CrossRef]
19. Sun, J.; Kari, L.; Arteaga, I.L. A dynamic rotating blade model at an arbitrary stagger angle based on classical plate theory and the Hamilton's principle. *J. Sound Vib.* **2013**, *332*, 1355–1371. [CrossRef]
20. Rostami, H.; Ranji, A.R.; Bakhtiari-Nejad, F. Vibration characteristics of rotating orthotropic cantilever plates using analytica approaches: A comprehensive parametric study. *Arch. Appl. Mech.* **2018**, *88*, 481–502. [CrossRef]
21. Luo, Z.Z.; Zhu, Y.P.; Liu, H.P.; Wang, D.Y. Dynamic similitude design method of the distorted model on variable thickness cantilever plates. *Appl. Sci.* **2016**, *6*, 228. [CrossRef]
22. Chen, Y.Z.; Zhang, D.G.; Li, L. Dynamics analysis of a rotating plate with a setting angle by using the absolute nodal coordinate formulation. *Eur. J. Mech./A Solids* **2019**, *74*, 257–271. [CrossRef]
23. Sinha, S.K.; Zylka, R.P. Vibration analysis of composite airfoil blade using orthotropic thin shell bending theory. *Int. J. Mech. Sci.* **2017**, *121*, 90–105. [CrossRef]
24. Sun, J.; Arteaga, I.L.; Kari, L. General shell model for a rotating pretwisted blade. *J. Sound Vib.* **2013**, *332*, 5804–5820. [CrossRef]
25. Júnior, C.J.F.; Cardozo, A.C.P.; Júnior, V.M.; Neto, A.G. Modeling wind turbine blades by geometrically-exact beam and shell elements: A comparative approach. *Eng. Struct.* **2019**, *180*, 357–378. [CrossRef]
26. Kee, Y.J.; Shin, S.J. Structural dynamic modeling for rotating blades using three dimensional finite elements. *J. Mech. Sci. Technol.* **2015**, *29*, 1607–1618. [CrossRef]
27. Yangui, M.; Bouaziz, S.; Taktak, M.; Haddar, M.; El-Sabbagh, A. Nonlinear analysis of twisted wind turbine blade. *J. Mech.* **2016**, *34*, 269–278. [CrossRef]
28. Sun, Q.; Ma, H.; Zhu, Y.P.; Han, Q.K.; Wen, B.C. Comparison of rubbing induced vibration responses using varying-thicknessp twisted shell and solid-element blade models. *Mech. Syst. Signal Process.* **2018**, *108*, 1–20. [CrossRef]
29. Rafiee, R.; Tahani, M.; Moradi, M. Simulation of aeroelastic behavior in a composite wind turbine blade. *J. Wind. Eng. Ind. Aerodyn.* **2016**, *151*, 60–69. [CrossRef]
30. Kee, Y.J.; Kim, J.H. Vibration characteristics of initially twisted rotating shell type composite blades. *Compos. Struct.* **2003**, *64*, 151–159. [CrossRef]
31. Xiong, Q.; Guan, H.; Ma, H.; Wu, Z.Y.; Guo, X.M.; Wang, W.W. Dynamic characteristic analysis of rotating blade with breathing crack. *Mech. Syst. Signal Process.* **2023**, *196*, 110325. [CrossRef]
32. Zhao, Q.; Liu, Z.L.; He, Y.H.; Yao, H.L.; Wen, B.C. Dynamic stress prediction method for rubbing blades. *J. Vibroengineering* **2016**, *18*, 1–12. Available online: https://www.extrica.com/article/16087 (accessed on 7 April 2024).
33. Guan, H.; Xiong, Q.; Ma, H.; Wang, W.W.; Ni, K.X.; Wu, Z.Y.; Yin, X.M.; Zhao, S.T.; Zhang, X.X. Comparison of nonlinear vibration responses induced by edge crack and surface crack of compressor blades. *Mech. Syst. Signal Process.* **2024**, *216*, 111465. [CrossRef]
34. Wang, W.W.; Ma, H.; Zhao, C.G.; Wu, Z.Y.; Wang, H.J. Dynamic contact characteristics of a rotating twisted variable-section blade with breathing crack. *J. Cent. South Univ.* **2024**, *31*, 858–877. [CrossRef]
35. Ma, H.; Wang, D.; Tai, X.Y.; Wen, B.C. Vibration response analysis of blade-disk dovetail structure under blade tip rubbing condition. *J. Vib. Control* **2017**, *23*, 252–271. [CrossRef]
36. Jones, S.; Legrand, M. Forced vibrations of a turbine blade undergoing regularized unilateral contact conditions through the wavelet balance method. *Int. J. Numer. Methods Eng.* **2015**, *101*, 351–374. [CrossRef]
37. Sinha, S.K. Combined torsional-bending-axial dynamics of a twisted rotating cantilever Timoshenko beam with contact-impact loads at the free end. *J. Appl. Mech.* **2007**, *74*, 505–522. [CrossRef]
38. Ma, X.X.; Ma, H.; Zeng, J.; Piao, Y.H. Rubbing-induced vibration response analysis of dual-rotor-casing system. *Trans. Nanjing Univ. Aeronaut. Astronaut.* **2018**, *35*, 101–108. [CrossRef]

Zeng, J.; Ma, H.; Yu, K.; Guo, X.M.; Wen, B.C. Rubbing response comparisons between single blade and flexible ring using different rubbing force models. *Int. J. Mech. Sci.* **2019**, *164*, 105164. [CrossRef]

Lesaffre, N.; Sinou, J.J.; Thouverez, F. Stability analysis of rotating beams rubbing on an elastic circular structure. *J. Sound Vib.* **2006**, *299*, 1005–1032. [CrossRef]

Guo, X.M.; Zeng, J.; Ma, H.; Zhao, C.G.; Yu, X. A dynamic model for simulating rubbing between blade and flexible casing. *J. Sound Vib.* **2020**, *466*, 115036. [CrossRef]

Zeng, J.; Ma, H.; Ma, X.X.; Wu, Z.Y.; Qin, Z.Y. Blade-loss-caused rubbing dynamic characteristics of rotor-bladed disk-casing system. *Trans. Nanjing Univ. Aeronaut. Astronaut.* **2018**, *35*, 116–125. [CrossRef]

Batailly, A.; Legrand, M. Unilateral contact induced blade/casing vibratory interactions in impellers: Analysis for flexible casings with friction and abradable coating. *J. Sound Vib.* **2015**, *348*, 344–364. [CrossRef]

Zhong, R.; Tang, J.Y.; Wang, A.L.; Shuai, C.J.; Wang, Q.S. An exact solution for free vibration of cross-ply laminated composite cylindrical shells with elastic restraint ends. *Comput. Math. Appl.* **2019**, *77*, 641–661. [CrossRef]

Sun, W.W.; Zhu, M.W.; Wang, Z. Free vibration analysis of a hard-coating cantilever cylindrical shell with elastic constraints. *Aerosp. Sci. Technol.* **2017**, *63*, 232–244. [CrossRef]

Li, C.F.; Zhong, B.F.; Shen, Z.C.; Zhang, J.R. Modeling and nonlinear vibration characteristics analysis of symmetrically 3-layer composite thin circular cylindrical shells with arbitrary boundary conditions. *Thin-Walled Struct.* **2019**, *142*, 311–321. [CrossRef]

Dey, T.; Ramachandra, L.S. Non-linear vibration analysis of laminated composite circular cylindrical shells. *Compos. Struct.* **2017**, *163*, 89–100. [CrossRef]

Song, X.Y.; Han, Q.K.; Zhai, J.Y. Vibration analyses of symmetrically laminated composite cylindrical shells with arbitrary boundaries conditions via Rayleigh-Ritz method. *Compos. Struct.* **2015**, *134*, 820–830. [CrossRef]

Zhang, Y.; Sun, W.; Yang, J.; Han, Q.K. Nonlinear vibration analysis of a hard-coating cylindrical shell with elastic constraints by finite element method. *Nonlinear Dyn.* **2017**, *90*, 2879–2891. [CrossRef]

Zhang, Y.; Sun, W.; Yang, J. A new finite element formulation for nonlinear vibration analysis of the hard-coating cylindrical shell. *Coatings* **2017**, *7*, 70. [CrossRef]

Khan, K.; Patel, B.P.; Nath, Y. Free and forced vibration characteristics of bimodular composite laminated circular cylindrical shells. *Compos. Struct.* **2015**, *126*, 386–397. [CrossRef]

Ma, H.; Tai, X.Y.; Han, Q.K.; Wu, Z.Y.; Wang, D.; Wen, B.C. A revised model for rubbing between rotating blade and elastic casing. *J. Sound Vib.* **2015**, *337*, 301–320. [CrossRef]

Ma, H.; Yin, F.L.; Tai, X.Y.; Wang, D.; Wen, B.C. Vibration response analysis caused by rubbing between rotating blade and casing. *J. Mech. Sci. Technol.* **2016**, *30*, 1983–1995. [CrossRef]

Zhou, T.; Jia, Y.F.; Zou, L.M.; Jiang, Z.N.; Wang, W.M.; Hu, M.H. Vibration characteristics of blade-casing rubbing fault considering rotor–stator coupling. *Mech. Syst. Signal Process.* **2024**, *218*, 111589. [CrossRef]

Hou, Y.H.; Cao, S.Q.; Kang, Y.H. Study on the frequency modulation phenomenon in the rotor system with blade-casing rub-impact fault. *Int. J. Non-Linear Mech.* **2024**, *159*, 104626. [CrossRef]

Wu, Z.Y.; Zhao, L.C.; Yan, H.; Yan, G.; Chen, A.; Zhang, W.M. Multi-blade rubbing characteristics of the shaft-disk-blade-casing system with large rotation. *Appl. Math. Mech.* **2024**, *45*, 111–136. [CrossRef]

Shao, J.; Wu, J.G.; Yang, K.; Zhang, Y. Dynamic characteristic analysis of a twin-spool rotor–casing system with looseness and intershaft rubbing coupling faults. *J. Mech. Sci. Technol.* **2024**, *38*, 101–120. [CrossRef]

Li, B.Q.; Ma, H.; Zeng, J.; Guo, X.M.; Wen, B.C. Rotating blade-casing rubbing simulation considering casing flexibility. *Int. J. Mech. Sci.* **2018**, *148*, 118–134. [CrossRef]

Yang, Y.; Cao, D.Q.; Wang, D.Y. Investigation of dynamic characteristics of a rotor system with surface coatings. *Mech. Syst. Signal Process.* **2017**, *84*, 469–484. [CrossRef]

Yang, Y.; Ouyang, H.J.; Wu, X.L.; Jin, Y.L.; Yang, Y.R.; Cao, D.Q. Bending-torsional coupled vibration of a rotor-bearing-system due to blade-casing rub in presence of non-uniform initial gap. *Mech. Mach. Theory* **2019**, *140*, 170–193. [CrossRef]

Cao, D.Q.; Yang, Y.; Chen, H.T.; Wang, D.Y.; Jiang, G.Y.; Li, C.G.; Wei, J.; Zhao, K. A novel contact force model for the impact analysis of structures with coating and its experimental verification. *Mech. Syst. Signal Process.* **2016**, *70*, 1056–1072. [CrossRef]

Zeng, J.; Zhao, C.G.; Ma, H.; Yu, K.; Wen, B.C. Rubbing dynamic characteristics of the blisk-casing system with elastic supports. *Aerosp. Sci. Technol.* **2019**, *95*, 105481. [CrossRef]

Kang, Y.H.; Cao, S.Q.; Gao, T.; You, Z.Z. Development and validation of a rotating blade-casing rubbing model by considering the blade deformation and abradable coating. *J. Sound Vib.* **2023**, *563*, 117853. [CrossRef]

Kang, Y.H.; Cao, S.Q.; Hou, Y.H.; Chen, N.; Li, B. Dynamics research on the rubbing process and rubbing forms of rotor–blade–casing systems. *Int. J. Non-Linear Mech.* **2022**, *147*, 104242. [CrossRef]

Jin, M.; Wang, A.L.; Wang, Q.S.; Wang, L.K. Rub-impact dynamic analysis of the central tie rod rotor-blade-casing coupling system with the Hirth couplings connection. *J. Vib. Eng. Technol.* **2024**, *12*, 1479–1503. [CrossRef]

Tang, T.; Wang, Y.; Chen, Z.Y.; Wang, S.; Zhang, M.Q. Study on rubbing characteristics of blade-casing model considering transverse cracks. *J. Sound Vib.* **2023**, *567*, 117928. [CrossRef]

Wang, N.F.; Liu, C.; Jiang, D.X.; Kamran, B. Casing vibration response prediction of dual-rotor-blade-casing system with blade-casing rubbing. *Mech. Syst. Signal Process.* **2019**, *118*, 61–77. [CrossRef]

68. Wang, N.F.; Liu, C.; Jiang, D.X. Experimental analysis of dual-rotor-support-casing system with blade-casing rubbing. *Eng. Fai Anal.* **2021**, *123*, 105306. [CrossRef]
69. Zhang, J.H.; Lu, X.; Lin, J.W.; Ma, L.; Dai, H.W. Dynamic characteristics analysis of blade-casing rubbing faults with abradabl coatings. *Proc. Inst. Mech. Eng. Part C J. Mech. Eng. Sci.* **2021**, *235*, 975–987. [CrossRef]
70. Jin, Y.L.; Liu, Z.W.; Yang, Y.; Li, F.S.; Chen, Y.S. Nonlinear vibrations of a dual-rotor-bearing-coupling misalignment system wit blade-casing rubbing. *J. Sound Vib.* **2021**, *497*, 115948. [CrossRef]
71. Salvat, N.; Batailly, A.; Legrand, M. Modeling of abradable coating removal in aircraft engines through delay differential equation *J. Eng. Gas Turbines Power* **2013**, *135*, 102102. [CrossRef]
72. Kammer, A.S.; Olgac, N. Blade/Casing rub interaction in turbomachinery: Structural parameters' influence on stability. *J. Propuls Power* **2016**, *32*, 928–938. [CrossRef]
73. Berthoul, B.; Batailly, A.; Stainier, L.; Legrand, M.; Cartraud, P. Phenomenological modeling of abradable wear in turbomachine *Mech. Syst. Signal Process.* **2018**, *98*, 770–785. [CrossRef]
74. Legrand, M.; Batailly, A.; Pierre, C. Numerical investigation of abradable coating removal in aircraft engines through plasti constitutive law. *J. Comput. Nonlinear Dyn.* **2012**, *7*, 011010. [CrossRef]
75. Batailly, A.; Legrand, M.; Millecamps, A.; Garcin, F. Numerical-experimental comparison in the simulation of rotor/stato interaction through blade-tip/abradable coating contact. *J. Eng. Gas Turbines Power* **2012**, *134*, 082504. [CrossRef]
76. Batailly, A.; Cuny, M.; Legrand, M.; Philippon, S. Numerical-experimental confrontation in the simulation of tool/abradabl material interaction. *J. Eng. Gas Turbines Power* **2013**, *135*, 062102. [CrossRef]
77. Almeida, P.; Gibert, C.; Thouverez, F.; Leblanc, X.; Ousty, L.P. Numerical analysis of bladed disk–casing contact with friction anc wear. *J. Eng. Gas Turbines Power* **2016**, *138*, 122802. [CrossRef]
78. Almeida, P.; Gibert, C.; Thouverez, F. On Some Physical Phenomena Involved in Blade-Casing Contact. In Proceeding of the Eurodyn-International Conference on Structural Dynamics, Porto, Portugal, 30 June—2 July 2014; Available online https://hal.science/hal-02121521/ (accessed on 7 April 2024).
79. Ma, H.; Guan, H.; Qu, L.; Yang, T.R.; Zeng, Y.; Chen, Y.Y.; Zhu, Z.M.; Wang, H.J. Blade-coating-casing rubbing induced vibratior responses and wear characteristics. *Tribol. Int.* **2024**, *194*, 109571. [CrossRef]
80. Guan, H.; Ma, H.; Qu, X.C.; Wu, Z.; Zeng, J.; Xiong, Q.; Wang, H. Dynamic stress analysis of cracked rectangular blade: Simulatior and experiment. *Int. J. Mech. Sci.* **2024**, *267*, 109015. [CrossRef]
81. Zhao, T.Y.; Li, K.; Ma, H. Study on dynamic characteristics of a rotating cylindrical shell with uncertain parameters. *Anal. Math Phys.* **2022**, *12*, 97. [CrossRef]

dynamics

MDPI

Article

# A Comparative Study of Different CFD Codes for Fluidized Beds

Parindra Kusriantoko *, Per Fredrik Daun and Kristian Etienne Einarsrud

Department of Materials Science and Engineering, NTNU Norwegian University of Science and Technology, 7491 Trondheim, Norway; fredrik.daun@hydro.com (P.F.D.); kristian.e.einarsrud@ntnu.no (K.E.E.)
* Correspondence: parindra.kusriantoko@ntnu.no

**Abstract:** Fluidized beds are pivotal in the process industry and chemical engineering, with Computational Fluid Dynamics (CFD) playing a crucial role in their design and optimization. Challenges in CFD modeling stem from the scarcity or inconsistency of experimental data for validation, along with the uncertainties introduced by numerous parameters and assumptions across different CFD codes. This study navigates these complexities by comparing simulation results from the open-source MFIX and OpenFOAM, and the commercial ANSYS FLUENT, against experimental data. Utilizing a Eulerian–Eulerian framework and the kinetic theory of granular flow (KTGF), the investigation focuses on solid-phase properties through the classical drag laws of Gidaspow and Syamlal–O'Brien across varied parameters. Findings indicate that ANSYS Fluent, MFiX, and OpenFOAM can achieve reasonable agreement with experimental benchmarks, each showcasing distinct strengths and weaknesses. The study also emphasizes that both the Syamlal–O'Brien and Gidaspow drag models exhibit reasonable agreement with experimental benchmarks across the examined CFD codes, suggesting a moderated sensitivity to the choice of drag model. Moreover, analyses were also carried out for 2D and 3D simulations, revealing that the dimensional approach impacts the predictive accuracy to a certain extent, with both models adapting well to the complexities of each simulation environment. The study highlights the significant influence of restitution coefficients on bed expansion due to their effect on particle–particle collisions, with a value of 0.9 deemed optimal for balancing simulation accuracy and computational efficiency. Conversely, the specularity coefficient, impacting particle–wall interactions, exhibits a more subtle effect on bed dynamics. This finding emphasizes the critical role of carefully choosing these coefficients to effectively simulate the nuanced behaviors of fluidized beds.

**Keywords:** fluidized beds; CFD; multiphase flow; fluidization; fluent; OpenFOAM; MFiX

## 1. Introduction

Fluidized beds are one of the most widely used chemical reactors in many industry areas. They have been used mostly for gas–solid reactions as they provide some advantages including their easiness of handling and transport of solids due to fluid-like behavior, intensive mixing of solids leads to a consistent temperature distribution and favorable conditions for heat and mass transfer [1]. Despite their advantages, they also have some drawbacks which heavily rely upon the complexity of hydrodynamics behavior and the difficulty in scaling up the reactor [1,2]. Historically, empirical methodologies derived from laboratory-scale investigations of fluidized bed reactors have been instrumental in calculating key operational parameters, such as minimum fluidization velocity, bubble dynamics, particle attrition rates, and the axial distribution of solid mass fractions [3].

Numerical simulations offer a shortcut to circumvent the extensive process involved in designing and executing experiments, enabling rapid assessment of relevant fields at the industrial scale [4]. One of the numerical methods widely used is so-called Computational Fluid Dynamics (CFD). Several CFD models exist for simulating gas–solid flows, including the two-fluid model, discrete-particle model, hybrid model, and direct numerical simulations. The two-fluid method, particularly its most utilized form, the Eulerian–Eulerian

(E-E) approach, considers both phases as interpenetrating continua. Closure relations based on the kinetic theory of granular flows (KTGF) help describe the solid phase with gas-like properties, considering the particles' motion influenced by interactions within the gas phase and with other particles or walls [5]. In the discrete-particle approach, the fluid phase is treated as a continuum, while the solid phase is considered discrete, tracking individual particles within the computational domain. This method, notably utilized in the Eulerian–Lagrangian (E-L) framework, is recognized for its unique particle collision detection, which can be either stochastic or deterministic, facilitating detailed analysis of particle dynamics [6]. In the hybrid methods, the two-fluid and discrete-particle are integrated to leverage both models' benefits. This sub-domain method simplifies the computational domain based on an equilibrium index, optimizing simulation efficiency and accuracy [7]. In Direct Numerical Simulation (DNS), it solves directly the Navier–Stokes equation aiming to capture the full range of scales and dynamics by using high-resolution grids and accurate numerical schemes. However, DNS is also limited by its high computational cost and low Reynolds numbers [8].

Over the past 20 years, numerous studies on gas–solid CFD models have been conducted using commercial CFD software , e.g., ANSYS FLUENT [9], ANSYS CFX [10], OpenFOAM [11], MFiX [12], BARRACUDA Virtual Reactor [13], CFDEM coupling [14], COMSOL Multiphysics [15], STAR-CCM [16], and EDEM [17]. Given the fluidized bed's complex hydrodynamics, researchers focus on examining various parameters that influence gas–solid flow dynamics. Essential factors to consider in numerical models include superficial gas velocity, drag, specularity and restitution coefficients, and frictional viscosity, as these significantly affect the behavior of fluidized beds [18].

Taghipour et al. [19] compared hydrodynamics in a fluidized bed through experiments and FLUENT 6.0 simulations. The study used a pseudo-2D experimental setup and a 2D Eulerian–Eulerian model incorporating Gidaspow [5], Syamlal–O'Brien [20], and Wen–Yu drag models. Results showed the bed reaching steady state in 3 s, with consistent bed expansion and bubble formation. While the drag models closely matched experimental observations, discrepancies in voidage and bed expansion highlighted the need for further CFD model validation.

Cornelissen et al. [21] investigated the impact of mesh size, time step, and convergence criteria on the accuracy of a liquid–solid fluidized bed model in FLUENT 6.1.22, benchmarking their findings against experimental data. They determined that a convergence criterion of $10^{-4}$ led to faster, convergent results compared to the standard $10^{-3}$. Additionally, they noted that Courant numbers within the range of 0.03 to 0.3 produced reliable outcomes regardless of mesh size and time step.

Li et al. [22] explored how solid phase boundary conditions affect simulations by adjusting specularity and particle–wall restitution coefficients in both 2D and 3D Eulerian–Eulerian models using MFiX software, comparing results to a pseudo-2D experiment. They discovered that the restitution coefficient's impact was minimal compared to the specularity coefficient, with specularity coefficients under 0.05 leading to notable deviations in solid flow near walls and altering circulation patterns. While both 2D and 3D simulations captured the general flow behavior, 3D simulations aligned more closely with experimental data, emphasizing the significance of wall effects in pseudo-2D experiments for CFD model validation.

Herzog et al. [23] conducted a comparative analysis using MFIX, OpenFOAM versions 1.6/2.0, and FLUENT 6.3 against experimental findings by Taghipour, focusing on a 2D Eulerian–Eulerian using Gidaspow and Syamlal O'Brien drag models. The study revealed that FLUENT and MFIX closely matched experimental data for both drag models, whereas OpenFOAM's predictions were not satisfactory, suggesting OpenFOAM requires further refinement and validation for fluidization prediction.

Shi et al. [18] carried out research to align 2D and 3D models with Taghipour et al. [19]'s experimental findings, utilizing Eulerian–Eulerian frameworks in FLUENT 16.2. Their investigation spanned dimensions, flow regimes, boundary conditions, mesh granularity, and drag models. The optimal grid size emerged as 18 times the particle diameter for the closest experimental alignment, with finer meshes diminishing accuracy. Minor variances were observed within specific specularity and restitution coefficient ranges, deeming them suitable. Turbulence's impact was minimal within the bed but influenced gas flow above, recommending laminar models for non-reactive, isothermal beds and RANS $k - \epsilon$ models otherwise. Syamlal–O'Brien drag models, particularly with Johnson–Jackson boundary conditions and turbulent RANS models, showed superior experimental congruence. They noted 3D models matched experimental data more closely than 2D, cautioning that certain 2D model parameter combinations might mislead due to unrealistic assumptions. Consequently, they advocated for 3D validations, reserving 2D analyses for pre-validated model-sensitivity studies.

Reyes-Urrutia et al. [24] compared OpenFOAM and MFIX for simulating bubbling fluidized bed systems under varying temperature and heat transfer conditions. Both codes accurately predicted minimum fluidization velocity and wall-to-bed heat transfer coefficients. Despite structural similarities, differences in models and meshing techniques led to variations in predictions. MFiX showed slightly more accurate results for heat transfer and fluidization patterns but had higher computational costs and less versatility for meshing curved geometries compared to OpenFOAM.

Mofakham and Rasteh [25] aimed to enhance the accuracy of simulating gas–solid fluidized bed hydrodynamics using Fluent and MFIX. By applying a 2D Eulerian model with KTGF and two drag models, they predicted pressure drop, voidage, and solid-phase velocities. Results showed Fluent reduced error by 3.17% for time-averaged voidage (Gidaspow) and 17.35% for pressure drop (Syamlal-O'Brien) compared to previous works. MFiX excelled in predicting time-averaged voidage with AARD% values below 9%.

In the current study, the fluidized bed columns from Taghipour et al. [19] were analyzed in both 2D and 3D simulations using three different CFD software codes: ANSYS Fluent 2023 R3, MFiX 23.1.1, and OpenFOAM 10. Simulations in MFiX and OpenFOAM were realized as part of the MSc thesis of Daun [26]. The current work extends the analysis presented by [26], and compares results further with new simulations realized in ANSYS Fluent. This investigation included the examination of two drag models, Syamlal–O'Brien and Gidaspow, across varied superficial gas velocities. Additionally, the study explored the effects of varying restitution and specularity coefficients. Simulations employed a Eulerian–Eulerian framework underpinned by the kinetic theory of granular flow (KTGF) aiming to highlight and address the subtleties inherent in inter-software comparisons.

## 2. Problem Description

The observation of several parameters as outlined previously was done by replicating the experimental case from Taghipour et al. [19] as it enables modeling validation by comparing to the experimental results therein. The model's geometry consists of a rectangular column, measuring 0.28 m in width, 1 m in height, and 0.025 m in depth. The lower section of the column, extending up to 0.4 m, is densely packed with granular particles with mean particle size of 275 µm at a packing density of 0.6, illustrated in Figure 1. According to Taghipour et al. [19] the minimum fluidization velocity ($U_{mf}$) is 0.065 m/s, whereas the system exhibited an overall pressure drop of 4400 Pa, a bed expansion ratio of 1.1, and a voidage (gas-phase volume fraction) of 0.5.

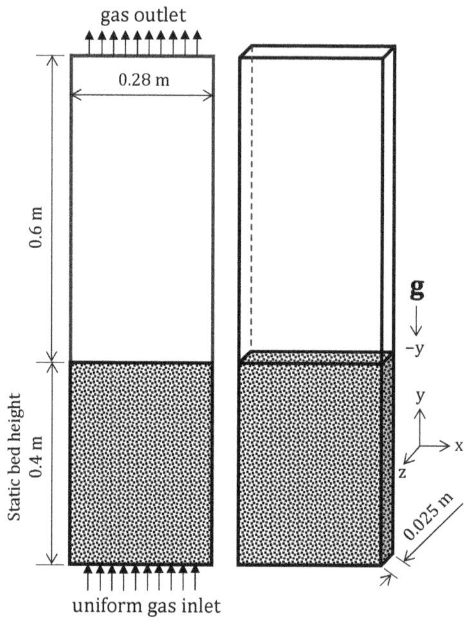


**Figure 1.** The geometric setup used for simulations, adapted from Taghipour et al. [19].

## 3. Computational Model

In this work, the simulations were carried out in the framework of Eulerian–Eulerian (E-E) by utilizing the kinetic theory of granular flow (KTGF). Three different CFD codes, namely, ANSYS Fluent version 2023 R3, MFiX 23.1.1 and OpenFOAM 10, were evaluated along with the implementation of Syamlal–O'Brien and Gidaspow drag model by varying superficial gas velocities up to 0.03, 0.1, 0.38, 0.46, and 0.51 m/s, equivalent to 0.5, 1.5, 6, 7, and 8 times the minimum fluidization velocity ($U_{mf}$). The values of gas velocity were also adapted from Taghipour et al. [19]. The complete selected simulation model parameters are summarized in Table 1. The governing equation and simulation setting of ANSYS Fluent, MFiX, and OpenFOAM for developing the fluidized bed simulation model are explained below.

**Table 1.** Parameters of simulation cases performed in this study, inspired by Taghipour et al. [19].

| Description | Value | Remark |
|---|---|---|
| Bed width | 0.28 m | Fixed value |
| Bed height | 1 m | Fixed value |
| Static bed height | 0.4 m | Fixed value |
| Grid interval spacing | 0.005 m | Specified |
| Particle density | 2500 kg/m$^3$ | Glass beads |
| Gas density | 1.225 kg/m$^3$ | Air |
| Gas kinematic viscosity | $1.485 \times 10^{-5}$ m$^2$/s | Fixed value |
| Mean particle diameter | 275 μm | Uniformly distributed |
| Initial solids packing | 0.6 | Fixed value |
| Maximum solids packing | 0.63 | Fixed value |
| Superficial gas velocity | 0.03–0.51 m/s | Parametized |
| Restitution coefficient | 0.9; 0.95; 0.99 | Parametized |
| Specularity coefficient | 0.05; 0.1; 0.2 | Parametized |
| Min. fluidization velocity | 0.065 m/s | Experiment |
| Viscous model | Laminar | Specified |

**Table 1.** *Cont.*

| Description | Value | Remark |
|---|---|---|
| Drag model | Syamlal–O'Brien; Gidaspow | Parametized [5,20] |
| Granular bulk viscosity | Lun et al. | [27,28] |
| Kinetic viscosity | Syamlal–O'Brien; Gidaspow | Parametized |
| Granular temperature | Algebraic; PDE; PDE | ANSYS Fluent; OpenFOAM; MFiX |
| Frictional viscosity | Schaeffer | [27,29] |
| Frictional pressure | Based-ktgf | [27] |
| Radial distribution | Lun et al. | [27,28] |
| Inlet boundary condition | Velocity | Superficial gas velocity |
| Outlet boundary condition | Outflow | Fully developed flow |
| Time steps | 0.001 s | Specified |
| Time averaged range | 3–12 s | Specified |
| Transient formulation | First order | Specified |
| Solution algorithm | SIMPLE; PIMPLE; MFiX algorithm | ANSYS Fluent; OpenFOAM; MFiX |

### 3.1. Governing Equation

The governing equations used in this simulation of fluidized bed for air and glass bead cases can be summarized as follows:

#### 3.1.1. Mass Balance Equations

$$\frac{\partial}{\partial t}(\alpha_g \rho_g) + \nabla \cdot (\alpha_g \rho_g \vec{u}_g) = 0 \tag{1}$$

$$\frac{\partial}{\partial t}(\alpha_s \rho_s) + \nabla \cdot (\alpha_s \rho_s \vec{u}_s) = 0 \tag{2}$$

where $\alpha_g$, $\alpha_s$, $\vec{u}_g$, $\vec{u}_s$, $\rho_g$, and $\rho_s$ represent the volume fraction, velocity and density of gas phase and solid phase, respectively. The total volume fraction of gas and solid phase is equal to unity:

$$\alpha_g + \alpha_s = 1 \tag{3}$$

#### 3.1.2. Momentum Balance Equations

$$\frac{\partial}{\partial t}(\alpha_g \rho_g \vec{u}_g) + \nabla \cdot (\alpha_g \rho_g \vec{u}_g \vec{u}_g) = -\alpha_g \nabla p + \nabla \cdot \overline{\overline{\tau}}_g + \alpha_g \rho_g \vec{g} + K_{gs}(\vec{u}_g - \vec{u}_s) \tag{4}$$

$$\frac{\partial}{\partial t}(\alpha_s \rho_s \vec{u}_s) + \nabla \cdot (\alpha_s \rho_s \vec{u}_s \vec{u}_s) = -\alpha_s \nabla p - \nabla p_s + \nabla \cdot \overline{\overline{\tau}}_s + \alpha_s \rho_s \vec{g} + K_{gs}(\vec{u}_g - \vec{u}_s) \tag{5}$$

where $K_{gs}$ represents the momentum exchange coefficient or interphase drag force coefficient which will be given by following Syamlal–O'Brien and Gidaspow drag models. The terms $\overline{\overline{\tau}}_g$ and $\overline{\overline{\tau}}_s$ represent the deviatoric stress tensors for the gas and solid phases, respectively. Pressure gradients are accounted for by $\nabla p$ in the gas phase and $\nabla p_s$ in the solid phase. The gravitational force is represented by $\vec{g}$. The velocity of gas phase and solid phase are given by $\vec{u}_g$ and $\vec{u}_s$.

#### 3.1.3. Syamlal–O'Brien Drag Model [20]

$$K_{gs} = \frac{3}{4}\frac{\alpha_s \alpha_g \rho_g}{v_{r,s}^2 d_s} C_D\left(\frac{Re_s}{u_{r,s}}\right)|\vec{u}_s - \vec{u}_g| \tag{6}$$

where

$$C_D = \left(0.63 + \frac{4.8}{\sqrt{Re_s/u_{r,s}}}\right)^2 \tag{7}$$

and

$$u_{r,s} = 0.5\left(A - 0.06Re_s + \sqrt{(0.06Re_s)^2 + 0.12Re_s(2B - A) + A^2}\right) \quad (8)$$

where

$$Re_s = \frac{\rho_g d_s |\vec{u}_s - \vec{u}_g|}{\mu_g} \quad (9)$$

with

A = $\alpha_g^{4.14}$, B = $0.8\alpha_g^{1.28}$ for $\alpha_g \leqslant 0.85$, or A = $\alpha_g^{4.14}$, B = $\alpha_g^{2.65}$ for $\alpha_g > 0.85$

3.1.4. Gidaspow Drag Model [5]

For $\alpha_g > 0.8$

$$K_{gs} = \frac{3}{4}C_D \frac{\alpha_s \alpha_g \rho_g |\vec{u}_s - \vec{u}_g|}{d_s} \alpha_g^{-2.65} \quad (10)$$

for $\alpha_g \leqslant 0.8$

$$K_{gs} = 150\frac{\alpha_s^2 \mu_g}{\alpha_g d_s^2} + 1.75\frac{\alpha_s \rho_g |\vec{u}_s - \vec{u}_g|}{d_s} \quad (11)$$

where

$$C_D = \frac{24}{\alpha_g Re_s}\left(1 + 0.15(\alpha_g Re_s)^{0.687}\right) \quad (12)$$

3.1.5. Solid Phase Fluctuating Energy Equations [30]

$$\frac{3}{2}\left[\frac{\partial}{\partial t}(\rho_s \alpha_s \Theta_s) + \nabla \cdot (\rho_s \alpha_s \vec{u}_s \Theta_s)\right] = \left(-p_s \bar{\bar{I}} + \bar{\bar{\tau}}_s\right) : \nabla \vec{u}_s + \nabla \cdot (k_{\Theta_s} \nabla \Theta_s) - \gamma_{\Theta_s} + \varphi_{gs} \quad (13)$$

In order to attain closure of the solid stress terms in Equations (4), (5) and (13), KTGF was used to derive the constitutive equations provided in in Table A1. KTGF provides values for solid-phase shear and bulk viscosities, as well as solid pressure. This theory evolves from Ogawa et al. [31]. It assumes particles are smooth, with inelastic and instantaneous binary collisions, as detailed by Gidaspow [5]. Drag models from Syamlal and O'Brien [20] and Gidaspow [5] were used in this study to calculate the momentum exchange coefficients as shown in Equations (6)–(12). Conversely, the Gidaspow drag model merges the Wen–Yu drag model with the Ergun equation that is pertinent for dense beds, linking drag to pressure drop across porous media.

### 3.2. Simulation Setup

The OpenFOAM simulations adopted the fluidized bed tutorial case on multiphase EulerFoam for fluidized Bed, whereas MFiX simulations used the 2D TFM fluid bed preset, and ANSYS Fluent based its setup on a 2D fluidized bed case from the training materials [32]. The computational domains for 2D and 3D models were discretized into 11,200 and 56,000 rectangular cells, respectively, with a grid width of $\Delta x = 0.005$ m. Boundary conditions for all codes were set to default. Inlet velocities were varied, 0.03, 0.1, 0.38, 0.46, and 0.51 m/s. For the gas phase at the wall, a no-slip condition for the gas phase was applied at the walls. All three CFD codes facilitate the application of the Johnson and Jackson partial slip boundary condition for particles, enabling nuanced simulation of particle–wall interactions [33]. This formulation, articulated by Johnson and Jackson [33], prescribes boundary conditions for velocity and granular temperature in the solid phase, satisfying the specified requirements as outlined in Equations (14) and (15). The default specularity coefficient of 0.2 and restitution coefficient of 0.9 for all codes.

$$\mu_s \frac{\partial \vec{u}_s}{\partial x} = \frac{\pi K \rho_s \alpha_s g_{0,ss} \sqrt{\Theta_s}}{2\sqrt{3}\alpha_s^{\max}} \vec{u}_s \quad (14)$$

$$\kappa_s \frac{\partial K}{\partial x} = -\frac{\pi K \vec{u}_s^2 \rho_s \alpha_s g_{0,ss} \sqrt{\Theta_s}}{2\sqrt{3}\alpha_s^{\max}} - \frac{\pi \sqrt{3} \rho_s \alpha_s g_{0,ss} \left(1 - e_{ss}^2\right) \sqrt{\Theta_s}}{4\alpha_s^{\max}} \Theta_s \tag{15}$$

where $K$ and $e_{ss}$ represent the specularity and restitution coefficients respectively, their values are constrained between 0 and 1. A specularity coefficient ($K$) of 0 denotes perfectly specular collisions, whereas a value of 1 indicates perfectly diffuse collisions. Similarly, a restitution coefficient ($e_{ss}$) of 0 corresponds to perfectly inelastic collisions, and a value of 1 represents perfectly elastic collisions. Motivated by the clear graphical analysis provided by [18], which elucidates the influence of these coefficients on fluidized bed flow behavior, this study set forth to investigate the effects of several restitution (0.9, 0.95, and 0.99) and specularity (0.05, 0.1, and 0.2) coefficient values to evaluate their impact on simulation results. In total, 50 simulation cases were executed, employing specific parameters i.e., CFD codes, superficial gas velocities, drag models, restitution coefficient, specularity coefficient and effect of 2D and 3D geometry as delineated in Table 2, to thoroughly investigate their effects.

**Table 2.** Parameters of simulation cases performed in this study.

| Case | Software | μ, m/s | Drag Force | $e_{ss}$ | $K$ | Geomery | $H/H_0$ | $\Delta P$, Pa |
|---|---|---|---|---|---|---|---|---|
| 1 | Fluent | 0.03 | Syamlal–O'Brien | 0.9 | 0.2 | 2D | 0.975 | 5800 |
| 2 | Fluent | 0.1 | Syamlal–O'Brien | 0.9 | 0.2 | 2D | 1.162 | 5647 |
| 3 | Fluent | 0.2 | Syamlal–O'Brien | 0.9 | 0.2 | 2D | 1.388 | 5581 |
| 4 | Fluent | 0.38 | Syamlal–O'Brien | 0.9 | 0.2 | 2D | 1.700 | 5657 |
| 5 | Fluent | 0.46 | Syamlal–O'Brien | 0.9 | 0.2 | 2D | 1.875 | 5697 |
| 6 | Fluent | 0.51 | Syamlal–O'Brien | 0.9 | 0.2 | 2D | 1.963 | 5628 |
| 7 | Fluent | 0.38 | Syamlal–O'Brien | 0.95 | 0.2 | 2D | 1.762 | 5613 |
| 8 | Fluent | 0.38 | Syamlal–O'Brien | 0.99 | 0.2 | 2D | 1.813 | 5577 |
| 9 | Fluent | 0.38 | Syamlal–O'Brien | 0.9 | 0.05 | 2D | 1.700 | 5633 |
| 10 | Fluent | 0.38 | Syamlal–O'Brien | 0.9 | 0.1 | 2D | 1.700 | 5514 |
| 11 | Fluent | 0.38 | Syamlal–O'Brien | 0.9 | 0.2 | 3D | 1.513 | 5551 |
| 12 | Fluent | 0.03 | Gidaspow | 0.9 | 0.2 | 2D | 0.975 | 5776 |
| 13 | Fluent | 0.1 | Gidaspow | 0.9 | 0.2 | 2D | 1.087 | 5731 |
| 14 | Fluent | 0.2 | Gidaspow | 0.9 | 0.2 | 2D | 1.275 | 5710 |
| 15 | Fluent | 0.38 | Gidaspow | 0.9 | 0.2 | 2D | 1.563 | 5697 |
| 16 | Fluent | 0.46 | Gidaspow | 0.9 | 0.2 | 2D | 1.663 | 5663 |
| 17 | Fluent | 0.51 | Gidaspow | 0.9 | 0.2 | 2D | 1.762 | 5644 |
| 18 | Fluent | 0.38 | Gidaspow | 0.95 | 0.2 | 2D | 1.550 | 5608 |
| 19 | Fluent | 0.38 | Gidaspow | 0.99 | 0.2 | 2D | 1.625 | 5661 |
| 20 | Fluent | 0.38 | Gidaspow | 0.9 | 0.05 | 2D | 1.525 | 5554 |
| 21 | Fluent | 0.38 | Gidaspow | 0.9 | 0.1 | 2D | 1.539 | 5604 |
| 22 | Fluent | 0.38 | Gidaspow | 0.9 | 0.2 | 3D | 1.625 | 5622 |
| 23 | OpenFOAM | 0.03 | Syamlal–O'Brien | 0.9 | 0.2 | 2D | 1.000 | 1450 |
| 24 | OpenFOAM | 0.1 | Syamlal–O'Brien | 0.9 | 0.2 | 2D | 1.020 | 4710 |
| 25 | OpenFOAM | 0.2 | Syamlal–O'Brien | 0.9 | 0.2 | 2D | 1.260 | 5480 |
| 26 | OpenFOAM | 0.38 | Syamlal–O'Brien | 0.9 | 0.2 | 2D | 1.630 | 5300 |
| 27 | OpenFOAM | 0.46 | Syamlal–O'Brien | 0.9 | 0.2 | 2D | 1.800 | 5240 |
| 28 | OpenFOAM | 0.51 | Syamlal–O'Brien | 0.9 | 0.2 | 2D | 1.860 | 5180 |
| 29 | OpenFOAM | 0.38 | Syamlal–O'Brien | 0.9 | 0.2 | 3D | 1.520 | 5190 |
| 30 | OpenFOAM | 0.03 | Gidaspow | 0.9 | 0.2 | 2D | 1.000 | 2700 |
| 31 | OpenFOAM | 0.1 | Gidaspow | 0.9 | 0.2 | 2D | 1.160 | 5520 |
| 32 | OpenFOAM | 0.2 | Gidaspow | 0.9 | 0.2 | 2D | 1.350 | 5220 |
| 33 | OpenFOAM | 0.38 | Gidaspow | 0.9 | 0.2 | 2D | 1.720 | 5080 |
| 34 | OpenFOAM | 0.46 | Gidaspow | 0.9 | 0.2 | 2D | 1.890 | 4850 |

**Table 2.** *Cont.*

| Case | Software | μ, m/s | Drag Force | $e_{ss}$ | $K$ | Geomery | $H/H_0$ | $\Delta P$, Pa |
|---|---|---|---|---|---|---|---|---|
| 35 | OpenFOAM | 0.51 | Gidaspow | 0.9 | 0.2 | 2D | 1.980 | 5190 |
| 36 | OpenFOAM | 0.38 | Gidaspow | 0.9 | 0.2 | 3D | 1.520 | 4910 |
| 37 | MFiX | 0.03 | Syamlal–O'Brien | 0.9 | 0.2 | 2D | 0.951 | 1620 |
| 38 | MFiX | 0.1 | Syamlal–O'Brien | 0.9 | 0.2 | 2D | 1.040 | 5070 |
| 39 | MFiX | 0.2 | Syamlal–O'Brien | 0.9 | 0.2 | 2D | 1.200 | 5790 |
| 40 | MFiX | 0.38 | Syamlal–O'Brien | 0.9 | 0.2 | 2D | 1.520 | 5620 |
| 41 | MFiX | 0.46 | Syamlal–O'Brien | 0.9 | 0.2 | 2D | 1.700 | 5590 |
| 42 | MFiX | 0.51 | Syamlal–O'Brien | 0.9 | 0.2 | 2D | 1.790 | 5580 |
| 43 | MFiX | 0.38 | Syamlal–O'Brien | 0.9 | 0.2 | 3D | 1.460 | 5860 |
| 44 | MFiX | 0.03 | Gidaspow | 0.9 | 0.2 | 2D | 0.947 | 3130 |
| 45 | MFiX | 0.1 | Gidaspow | 0.9 | 0.2 | 2D | 1.080 | 5840 |
| 46 | MFiX | 0.2 | Gidaspow | 0.9 | 0.2 | 2D | 1.310 | 5730 |
| 47 | MFiX | 0.38 | Gidaspow | 0.9 | 0.2 | 2D | 1.640 | 5610 |
| 48 | MFiX | 0.46 | Gidaspow | 0.9 | 0.2 | 2D | 1.730 | 5580 |
| 49 | MFiX | 0.51 | Gidaspow | 0.9 | 0.2 | 2D | 1.910 | 5580 |
| 50 | MFiX | 0.38 | Gidaspow | 0.9 | 0.2 | 3D | 1.460 | 5990 |

Based on the sensitivity analysis conducted by Taghipour et al. [19] regarding time step and spatial discretization, minimal differences were observed between time steps of 0.0005 s and 0.001 s, and between second and first-order discretizations. Consequently, this study adopted a time step of 0.001 s. In the discretization settings, ANSYS Fluent employed a first-order upwind and PRESTO for spatial discretization, alongside a first-order implicit method for temporal discretization. MFiX utilized a high order upwind scheme with a downwind factor for spatial discretization, incorporating the Superbee flux limiter. This approach was complemented by an implicit Euler method for temporal aspects. OpenFOAM's strategy involved Gaussian discretization, applying flux limiters like "vanLeer" or "limitedLinear" based on the variable, highlighting a targeted approach to discretization.

The numerical settings included the use of the AMG solver coupled with the SIMPLE algorithm in ANSYS Fluent to address the software's fluid flow challenges. MFiX and OpenFOAM utilized the PBiCGStab solver. MFiX employed a specific algorithm designed for its simulations [34], whereas OpenFOAM used the PIMPLE algorithm, indicating distinct computational strategies adopted by each software to navigate the complexities of multiphase flow simulations. Regarding convergence criteria, ANSYS Fluent was set to default settings for all variables at a tolerance of 0.001. OpenFOAM adopted strict tolerances, with $1 \times 10^{-9}$ for volume fraction (*alpha.**), $1 \times 10^{-8}$ for pressure head (*p_rgh*) and velocity (*U.**), and for temperature-related fields (*Theta.**), with turbulence kinetic energy and its dissipation rate (*k* | *epsilon.**) at $1 \times 10^{-5}$ . MFiX set its default continuity tolerance at 0.001, with tolerances for energy, species, granular energy, and scalar variables at 0.0001. Under-relaxation factors were kept at default settings for ANSYS Fluent, OpenFOAM, and MFiX, reflecting a standard approach to ensuring numerical stability within each software framework.

The simulation duration was set to 12 s, with time-averaging from 3 to 12 s, as suggested by Herzog et al. [23] and Taghipour et al. [19], to capture a stable and representative state of the system being analyzed.

## 4. Results

The results, derived from various simulations, are delineated in terms of bed expansion ratio ($H/H_0$), pressure drop, solid volume fraction contour plots, solid velocity at $y = 0.2$ m, and voidage at $y = 0.2$ m. These results were obtained using three distinct CFD codes and evaluated across two drag models. The bed expansion ratio and pressure drop data encompass all examined superficial velocities, 0.03, 0.1, 0.38, 0.46, and 0.51 m/s, corresponding to 0.5, 1.5, 6, 7, and 8 times $U_{mf}$, respectively. In contrast, the solid volume fraction contour plots specifically showcase three selected superficial velocities, 0.2, 0.38, and 0.46 m/s, at times of 0.4 s, 1 s, and 3 s. Additionally, the analyses of solid velocity and voidage at $y = 0.2$ m encapsulate the effects of variations across both drag models and the three CFD codes at 0.38 m/s of superficial gas velocity. The simulation outcomes, including bed expansion ratio, pressure drop, and voidage, were evaluated against the experimental results from Taghipour et al. [19], thereby assessing the simulations' accuracy by using the formula for standard deviation as elucidated in Shi et al. [18].

### 4.1. Bed Expansion Ratio and Pressure Drop

The expansion of the bed was assessed by charting the time-averaged solid fraction at the bed's center vertically. A criterion was established where a solid fraction value below 1% indicated the extent of bed expansion. This procedure was applied across various superficial gas velocities to observe how bed expansion varied with changes in inlet velocity. The summary of the results of bed expansion and pressure drop is given in Table 1. Figure 2 presents the bed expansions determined using the Syamlal–O'Brien drag model within ANSYS Fluent, MFiX, and OpenFOAM. From this figure, it is evident that all CFD codes forecast minimal bed expansion at lower velocities, with a linear increase in expansion noted for velocities exceeding 0.38 m/s. In 3D scenarios, bed expansion appears reduced compared to analogous 2D situations. Moreover, the figure illustrates that ANSYS Fluent tends to forecast greater bed expansion compared to MFiX and OpenFOAM. Figure 3 illustrates the bed expansions observed when employing the Gidaspow drag model in MFiX and OpenFOAM. From this figure, it is apparent that three software packages register bed expansion at all velocities above the minimum fluidization velocity. The expansion appears to increase linearly in relation to the superficial gas velocity. The 3D simulations show a notably lesser degree of bed expansion compared to their 2D counterparts. A comparison with Figure 2 reveals that the expansions resulting from the Gidaspow model from OpenFOAM and MFiX exceed those observed with the Syamlal–O'Brien model. Surprisingly, ANSYS Fluent forecasts lower bed expansions than OpenFOAM and MFiX when utilizing the Gidaspow model. Additionally, a contraction in the bed at the lowest gas velocity is noted in ANSYS Fluent and MFiX with both drag models, a phenomenon not observed in OpenFOAM simulations.

The pressure drop across the bed was assessed at the outlet of the fluidized bed reactor for each tested superficial gas velocity. Figures 4 and 5 illustrate a pattern where the pressure drop initially spikes to a peak value before gradually diminishing as the superficial gas velocity increases. It is observed that with the Gidaspow model, the peak pressure drop occurs at lower superficial gas velocities compared to the Syamlal–O'Brien model. Additionally, both figures indicate that MFiX and ANSYS Fluent estimates a greater pressure drop across all tested superficial gas velocities than OpenFOAM, in both 2D and 3D simulations. A comparison between 2D and 3D results shows that ANSYS Fluent and OpenFOAM reports a smaller pressure drop in 3D compared to 2D, whereas MFiX exhibits an increase in pressure drop from 2D to 3D simulations. It is also worth mentioning, specifically for ANSYS Fluent, that the pressure drop at superficial velocities lower than the fluidization velocity tends to predict a very high pressure drop. This observation aligns with the findings from Taghipour et al. [19], El Ajouri et al. [35], suggesting that the primary cause could be attributed to the solids not being fully fluidized at these lower velocities, thereby being predominantly influenced by interparticle frictional forces. Such forces are not adequately predicted by the multifluid model used for simulating gas–solid phases in ANSYS Fluent, which may not incorporate these frictional interactions effectively.

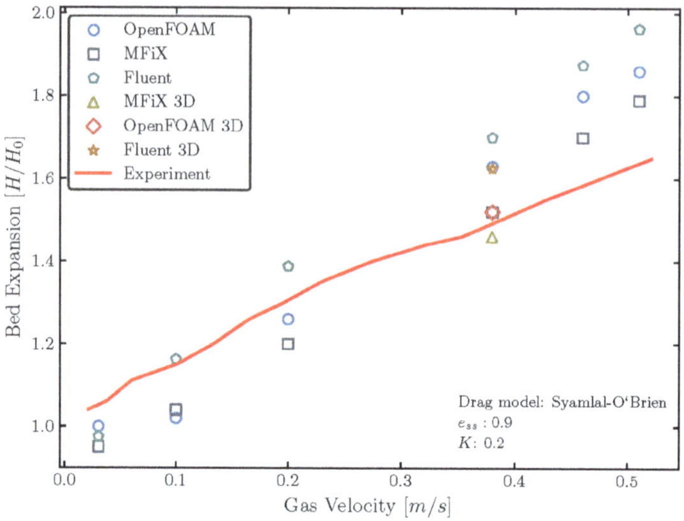


**Figure 2.** Bed expansion for different superficial gas velocities with the Syamlal–O'Brien drag model in ANSYS Fluent, MFiX, and OpenFOAM. Experiment data were taken from [19], with permission from Elsevier.

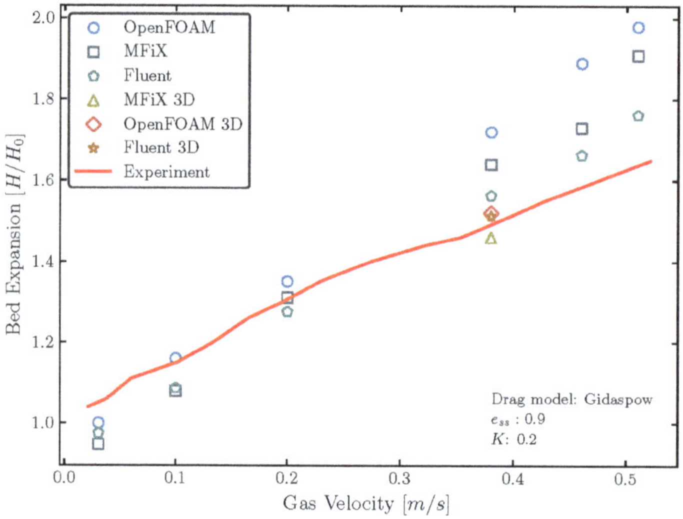


**Figure 3.** Bed expansion for different superficial gas velocities with the Gidaspow drag model in ANSYS Fluent, MFiX, and OpenFOAM. Experiment data were taken from [19], with permission from Elsevier.

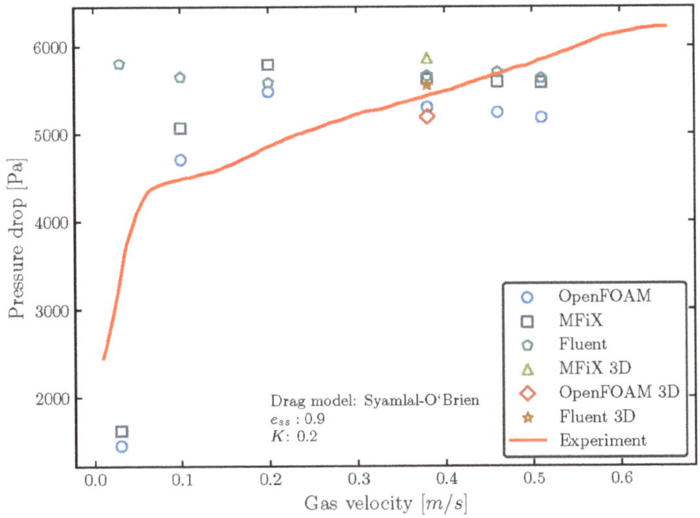


**Figure 4.** Pressure drop for different superficial gas velocities with the Syamlal–O'Brien drag model in ANSYS Fluent, MFiX, and OpenFOAM. Experiment data were taken from [19], with permission from Elsevier.

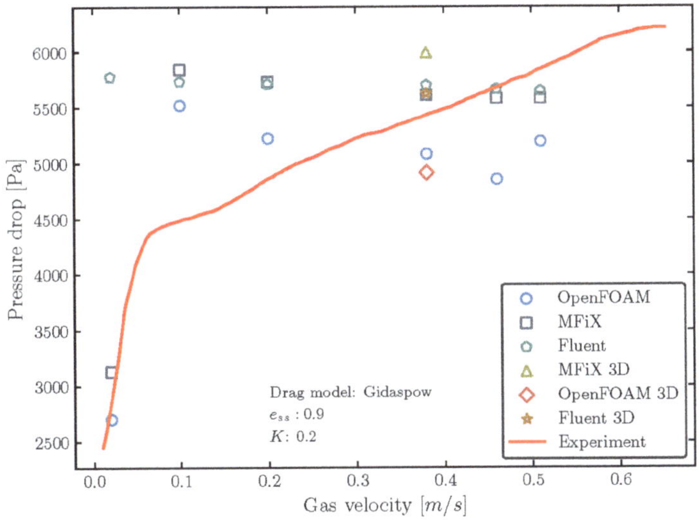


**Figure 5.** Pressure drop for different superficial gas velocities with the Gidaspow drag model in ANSYS Fluent, MFiX, and OpenFOAM. Experiment data were taken from [19], with permission from Elsevier.

### *4.2. Solid Velocity and Voidage*

Figure 6 displays the time-averaged solid particle velocity in the y-direction at a height of 0.2 m, with a superficial gas velocity of 0.38 m/s, utilizing both Syamlal–O'Brien drag model in Fluent, MFiX, and OpenFOAM. The observed trend across all data lines reveals a downward velocity near the walls transitioning to a parabolic upward velocity near the bed's center. Notably, Fluent predicts significantly higher velocities than MFiX, particularly near the walls and the reactor's center. Furthermore, for 3D simulations, velocities were averaged across the z-direction cells to align with the 2D results, offering a comprehensive view of the domain. While MFiX's 3D outcomes closely resemble its 2D counterparts, OpenFOAM and Fluent markedly diverge, maintaining a relatively constant velocity across the bed as opposed to the parabolic profile observed in 2D simulations. In contrast to the Syamlal–O'Brien drag model, Figure 7 illustrates the time-averaged solid particle velocity using the Gidaspow drag model, processed similarly to Figure 6. The outcomes indicate that Fluent and MFiX yield nearly identical predictions, with velocities marginally higher at the bed's center but displaying a somewhat flattened profile across both 3D and 2D simulations. Conversely, OpenFOAM's results diverge slightly, showing higher velocities near the walls and at the center of the bed in both 3D and 2D scenarios. Notably, the 3D results in OpenFOAM exhibit a slightly flatter velocity profile compared to their 2D counterparts, suggesting a distinct behavior in spatial velocity distribution when using the Gidaspow model.

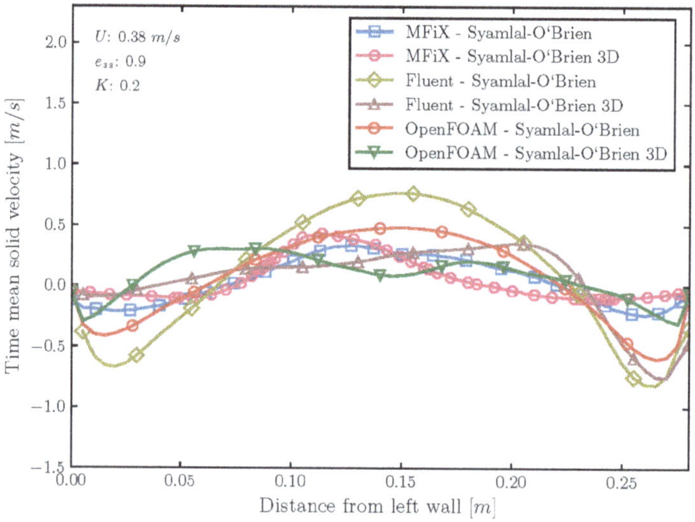


**Figure 6.** Time-averaged solid velocities recorded between 3 and 12 s for a superficial gas velocity of 0.38 m/s with Syamlal–O'Brien drag model. These measurements were conducted at a height of y = 0.2 m.

Further results, as shown in Figure 8, compare the time-averaged voidage for 2D and 3D simulations at the same bed height and gas velocity, employing the Syamlal–O'Brien drag model. This figure indicates that Fluent displays the highest voidage in 3D simulations compared to others, presenting an asymmetrical appearance relative to the rest. In OpenFOAM, the 3D simulations show a maximum voidage nearly identical to their 2D counterparts. Conversely, MFiX's 3D simulations exhibit a zone of low voidage extending from the wall, which sharply increases to a peak significantly greater than that observed in 2D simulations at the center of the bed. On the other hand, Figure 9 showcases the time-averaged voidage for 2D and 3D simulations at the same bed height and gas velocity, utilizing the Gidaspow drag model. From this figure, it is evident that MFiX's 3D simulations replicate the behavior observed in the Syamlal–O'Brien case, with voidage sharply increasing at the center of the bed in 3D simulations. For the Fluent simulations, both 2D and 3D present almost identical patterns, though the 3D simulations exhibit slightly higher voidage at the left wall. In OpenFOAM, the 3D results are slightly higher than 2D, maintaining a nearly similar pattern.

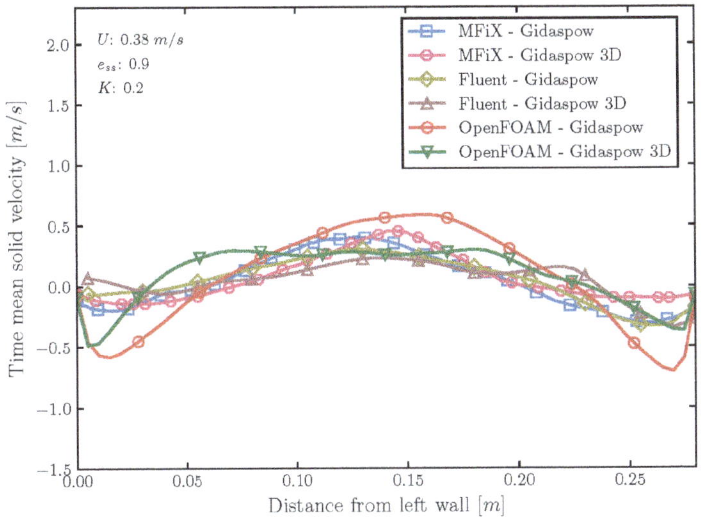


**Figure 7.** Time-averaged solid velocities recorded between 3 and 12 s for a superficial gas velocity of 0.38 m/s with Gidaspow drag model. These measurements were conducted at a height of y = 0.2 m.

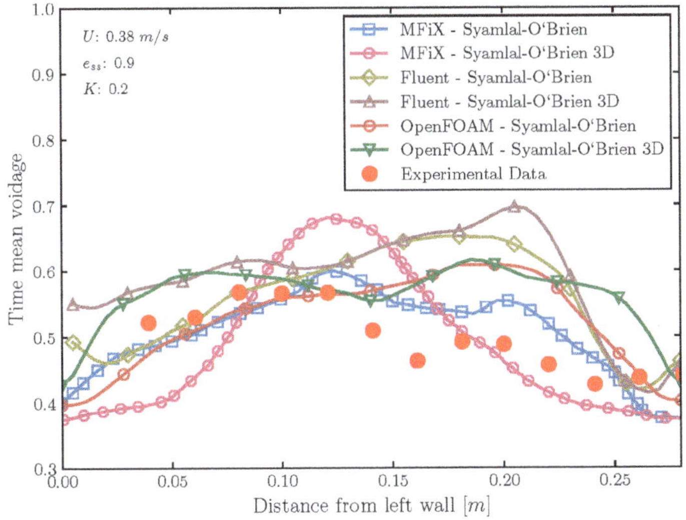


**Figure 8.** Time-averaged voidage recorded between 3 and 12 s for a superficial gas velocity of 0.38 m/s with Syamlal–O'Brien drag model. These measurements were conducted at a height of y = 0.2 m. Experiment data were taken from [19], with permission from Elsevier.

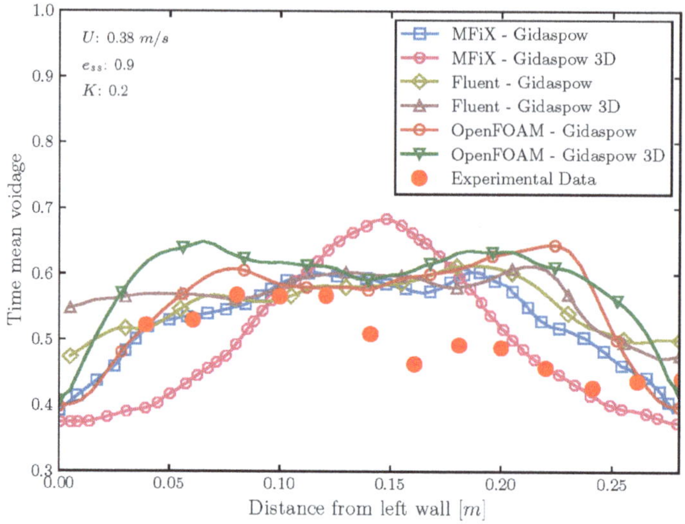


**Figure 9.** Time-averaged voidage recorded between 3 and 12 s for a superficial gas velocity of 0.38 m/s with Gidaspow drag model. These measurements were conducted at a height of y = 0.2 m. Experiment data were taken from [19], with permission from Elsevier.

### *4.3. Solid Volume Fraction*

The evolution of the fluidized beds and time-averaged solid volume fractions with three chosen superficial gas velocities equal to 0.2, 0.38, and 0.46 m/s as the results from three different CFD codes with Syamlal–O'Brien drag model as well as Gidaspow drag model are depicted in Figures 10 and 11. Initially, the bed height rose with the formation of bubbles until it stabilized at a constant steady-state height. The solid flow patterns inside the bed are relatively symmetrical for all cases before t < 3 s. The chaotic and transient bubble formation process at 3 s indicates that the statistical steady-state condition was achieved. Therefore, it confirmed the previous findings that a simulation duration of 3 s is sufficient to attain statistical steady-state behavior across all drag function models [19,23].

At the beginning of simulation, Fluent and MFiX display comparable patterns, with OpenFOAM showing slight deviations. As the flow progresses, the patterns of flow and bubble formation for both Syamlal–O'Brien and Gidaspow models remain similar in Fluent and MFiX but differ in OpenFOAM, highlighting that the numerical results of MFiX and Fluent are quite aligned, as also confirmed by Herzog et al. [23]. A closer examination reveals good agreement in the time-averaged solid volume fraction among all codes. All cases show a lower voidage observed near the wall and at the base of the bed, whereas an increased voidage is noted in the bed's central area. This pattern of solid volume fraction distribution indicates a higher formation of bubbles in the central region as opposed to the bottom. Specifically, in the Syamlal–O'Brien model, Fluent exhibits a higher voidage and greater bed expansion compared to MFiX and OpenFOAM. Conversely, in the Gidaspow model, OpenFOAM demonstrates larger bed expansion and more voidage. The case studies with three superficial velocities further reveal that MFiX tends to display a more distinct phase border than OpenFOAM and Fluent, where phase interfaces blend more smoothly. This distinction appears to lead to the formation of more, but smaller, bubbles in OpenFOAM, whereas MFiX features fewer but significantly larger bubbles, highlighting the intricate dynamics of fluidized beds as captured by different simulation tools and drag models.

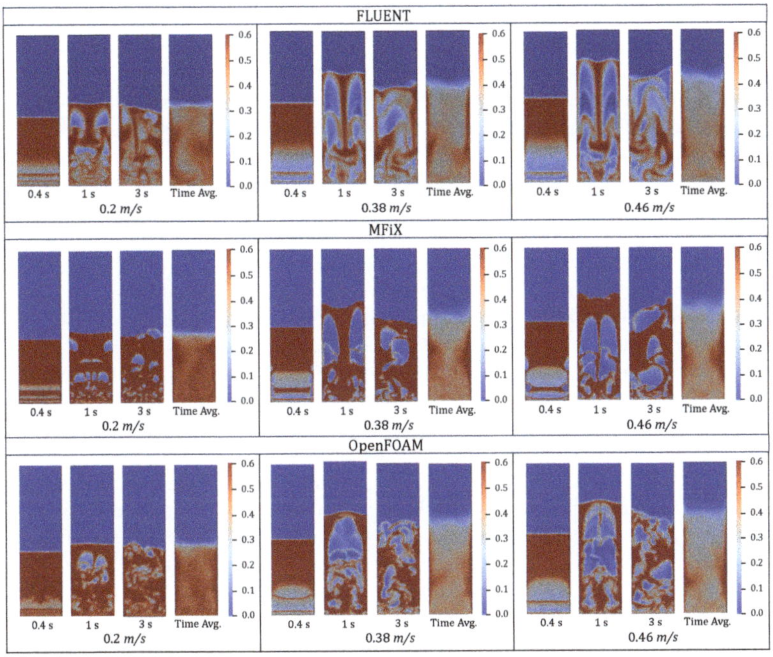


**Figure 10.** Particle distribution throughout the simulation with the Syamlal–O'Brien drag model with various superficial gas velocities in ANSYS Fluent, MFiX, and OpenFOAM.

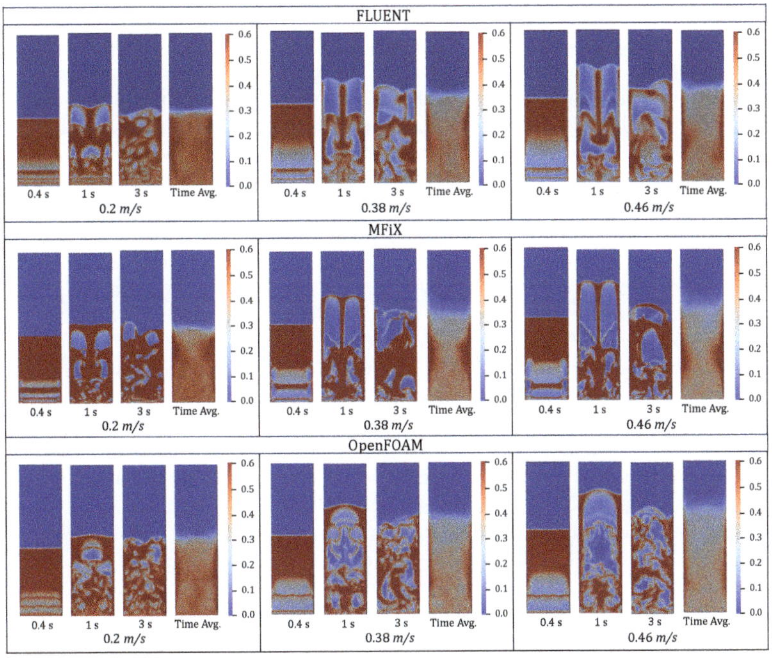


**Figure 11.** Particle distribution throughout the simulation with the Gidaspow drag model with various superficial gas velocities in ANSYS Fluent, MFiX, and OpenFOAM.

## 5. Discussion

Fluidized beds continue to be indispensable in the industrial sector, a fact underscored by the extensive volume of research utilizing Computational Fluid Dynamics (CFD) codes for their study, as elaborately discussed by Alobaid et al. [4]. This significant focus on fluidized beds through CFD simulations brings to the forefront the necessity of determining the most suitable CFD software for accurately capturing the complex hydrodynamics within these systems. Among the available CFD tools, ANSYS Fluent and MFiX have been recognized for their maturity and reliability in modeling fluidized bed dynamics. In contrast, OpenFOAM, despite its potential, has faced skepticism regarding its accuracy and applicability for fluidized bed simulations, as highlighted by Herzog et al. [23] (2012). In this context, our study undertook on a comparative evaluation of these three CFD codes by simulating a standard fluidized bed configuration and comparing the outcomes with the experimental findings reported by Taghipour et al. [19]. In addition to that, the examination of each software's numerical predictions of voidage fraction by employing the standard deviation formula as a measure of accuracy was also presented.

### 5.1. Comparative Study of Different Codes

Figures 2 and 3 showed a comparison between simulation results and experimental data for Syamlal–O'Brien and Gidaspow drag model, respectively. It is crucial to note that a challenge with the experimental data lies in the absence of a detailed description of how bed expansion measurements were conducted, limiting the direct comparability to the current simulation findings. Despite this, a general consistency in trends between simulated and experimental results is observable. For superficial velocities below 0.38 m/s using the Syamlal–O'Brien drag model, in the case of using Syamlal–O'Brien drag model, ANSYS Fluent tends to overestimate bed expansion, whereas MFiX and ANSYS Fluent show a slight underestimation. However, starting from a superficial velocity of 0.38 m/s, all three software items overestimate bed expansion. It is noteworthy that results from 3D simulations presented lower bed expansion compared to their 2D counterparts, underscoring the significance of incorporating friction along the front and back walls, regardless of the drag model applied. While MFiX and OpenFOAM align closely with experimental outcomes, ANSYS Fluent consistently overestimates bed expansion, highlighting discrepancies across different CFD tools in predicting fluidized bed dynamics. This can also be clearly observed from the solid phase distribution pattern depicted in Figure 10, which shows that ANSYS Fluent, using the Syamlal–O'Brien drag model, exhibits higher bed expansion and significantly lower particle occupancy, or higher voidage, appearing much more fluidized compared to other software utilizing the same drag model. Conversely, with the Gidaspow drag model, the tendency of simulation results to overestimate bed expansion at and above 0.38 m/s persists. However, in this scenario, ANSYS Fluent typically underpredicts bed expansion, followed closely by MFiX, while OpenFOAM tends to slightly overpredict. Nonetheless, starting from a superficial velocity of 0.38 m/s, all three software solutions overestimate the bed expansion. Despite these tendencies, it is observed that findings from the 3D simulations align well with the experimental data.

The overprediction of bed expansion in all CFD codes, particularly at higher gas velocities, is due to the failure to account for the effects of non-uniform structures on sub-grid drag. The impact of mesoscale heterogeneous structures on drag coefficient prediction is significant under coarse-grid conditions. As the superficial gas velocity increases, mesoscale structures such as bubbles and particle clusters gradually form in the bed. Therefore, at high gas velocities, the bed expansion becomes increasingly overestimated compared to experimental values [36,37]. A grid sensitivity analysis was carried out using ANSYS Fluent with the Syamlal–O'Brien drag model at a gas velocity of 0.38 m/s, as presented in Figure 12 to analyze its effect on bed height expansion. The results clearly show a decrease in bed height as the number of cells increases. The change from coarse (0.01 m, 2800 cells) to medium (0.005 m, 11,200 cells) is significant, around 5.9%. However, the change from medium to fine (0.0025 m, 44,800 cells) is less pronounced, around 1.5%, suggesting that a default grid size of 0.005 m is reasonable, as suggested by Shi et al. [18], which is 18 times the particle size.

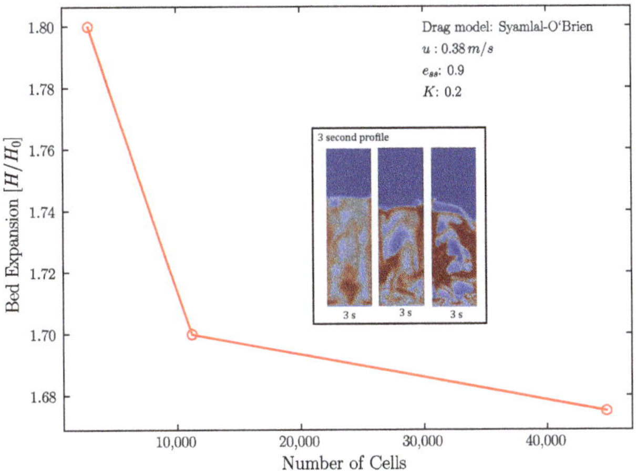


**Figure 12.** Effect of number of cells to the bed expansion ratio at a superficial gas velocity of 0.38 m/s using the Syamlal–O'Brien drag model in ANSYS Fluent (Time-averaged from 3 to 10.5 s).

The comparison of simulation outcomes with experimental data on pressure drop, employing the Syamlal–O'Brien and Gidaspow models, is illustrated in Figures 4 and 5, respectively. Experimentally, the pressure drop is observed to sharply increase until reaching the minimum fluidization velocity, thereafter ascending more gradually with the rise in superficial gas velocity. In contrast, numerical simulations diverge from this pattern, showing a sharp increase in pressure drop around the minimum fluidization velocity, followed by a gradual decrease as superficial gas velocity increases.

Notably, previous observations from OpenFOAM, as reported by Herzog et al. [23], indicated a significant overestimation of pressure drop. However, current results demonstrate that OpenFOAM now mirrors the trend seen in MFiX, with a pressure drop approximately 500 Pa lower. The introduction of a third dimension reveals differing impacts on pressure drop predictions between MFiX and OpenFOAM across both drag models. MFiX predicts a higher, and OpenFOAM a lower, pressure drop than their 2D simulations, deviating further from the experimental data. Conversely, with ANSYS Fluent, a distinct discrepancy is noted at a superficial velocity of 0.03 m/s across both drag models as previously explained. Across the board, ANSYS Fluent tends to overestimate pressure drops initially, then adjusts to underestimations. It is evident that no simulation perfectly aligns with the experimental data, with all predicted void fractions being overly large. Despite this, the outcomes could still be deemed in reasonable concordance with experimental observations. However, ANSYS Fluent gives a better fit to the experimental data at superficial velocities of 0.38 m/s and above. Furthermore, the introduction of the third dimension in ANSYS Fluent even results in a much better fit to the experimental result compared to MFiX and OpenFOAM. Notably, while the Syamlal–O'Brien and Gidaspow models yield broadly similar results, the Syamlal–O'Brien model slightly edges closer to the experimental data than Gidaspow in terms of pressure drop. It also aligns with the report from Aboudaoud et al. [38] and Liu and Hinrichsen [39] which implied that the Syamlal–O'Brien model is more suitable for pressure drop prediction compared to Gidaspow.

### *5.2. Simulation Error*

To assess the accuracy of the numerical predictions of voidage fraction against experimental data from Taghipour et al. [19], the standard deviation profiles for both 2D and 3D simulations were calculated and are presented in Figure 13. The formula for the standard deviation is given as follows:

$$\sigma_{est}\% = \sqrt{\frac{\sum\left(Y_{simulation} - Y_{experiment}\right)^2}{N}} \times 100 \quad (16)$$

where $\sigma_{est}$ represents the estimated standard error, and $Y_{experiment}$ and $Y_{simulation}$ are the experimental and predicted points of the voidage fraction along the x-coordinate at y = 0.2 m, respectively. The standard deviation serves as a reliable measure of error compared to other metrics since this method is derived from Linear Least Squares Regression commonly used in curve fitting. The standard deviation for 3D and 2D models falls within the range of 5.06–13.57%. The findings suggest that, by validating against the experimental voidage data, neither 3D nor 2D simulations inherently guarantee closer agreement with experimental data. Interestingly, certain 2D simulations exhibit better alignment with the experimental data than their 3D counterparts, indicating that with appropriate parameter selection, 2D simulations can also be a viable option, as supported by Shi et al. [18].

By examining both the qualitative aspects presented in Figures 8 and 9 and the quantitative data outlined in Figure 13, one can discern the impact of varying software and drag models on simulation outcomes. The results demonstrate that MFiX generally yields lower simulation discrepancies, followed by ANSYS Fluent and OpenFOAM. Notably, in the 3D context, MFiX and ANSYS Fluent exhibit similar levels of simulation error. Regarding the choice of drag model, the Syamlal–O'Brien model tends to produce lower simulation errors when used with MFiX and OpenFOAM, whereas the Gidaspow model appears to minimize simulation errors more effectively when applied in ANSYS Fluent. This result is consistent with the findings reported by Mofakham and Rasteh [25]. Consequently, while OpenFOAM holds potential for various applications, its performance in fluidized bed simulations appears to lag slightly behind that of MFiX and ANSYS Fluent, highlighting the importance of selecting the appropriate software and drag model to achieve the best simulation accuracy.

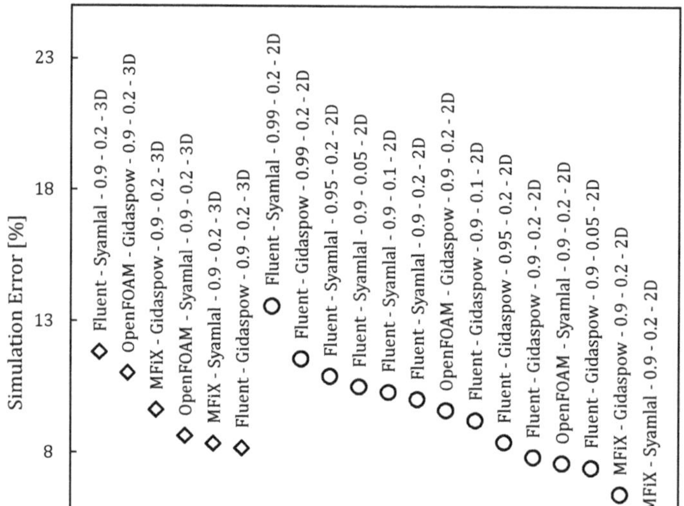


**Figure 13.** Simulation Error based on experimental data of Taghipour et al. [19] at y = 0.2 and $u$ = 0.38 m/s, with permission from Elsevier.

This observation is consistent with the most significant differences in flow evolution between OpenFOAM and ANSYS Fluent and MFiX, as illustrated in Figures 10 and 11. A previously discussed critical distinction lies in the phase interfaces within the simulations. OpenFOAM's tendency to generate numerous and diffuse bubbles suggests that gas flow actively propels particles upwards across almost the entire domain. This phenomenon is visible in time-averaged plots from OpenFOAM, where a narrow region of particle accumulation along the wall is the only area where particles can descend. In contrast, MFiX and ANSYS Fluent simulations depict a different behavior, where smaller bubbles at the bed's bottom coalesce into relatively larger and fewer distinct bubbles that ascend through the bed's center.

### 5.3. *The Effect of Restitution and Specularity Coefficient*

The restitution coefficient is indicative of the collision behavior between particles in the fluidized bed, while the specularity coefficient represents the nature of collisions between particles and the walls. In this study, restitution coefficient values of 0.9, 0.95, and 0.99 were explored. According to the results presented in Table 1, a higher restitution coefficient leads to increased bed expansion in the Syamlal–O'Brien drag model, although it does not significantly affect the pressure drop. This trend is consistent with the Gidaspow drag model, where increasing the restitution coefficient also increases bed expansion, albeit with a slight decrease (0.013) between 0.9 and 0.95, followed by an increase at 0.99.

Analyzing the simulation results based on voidage measurement at $y = 0.2$ m at a superficial velocity of 0.38 m/s, it was observed that with both drag models, as the restitution coefficient increases (Syamlal–O'Brien: 10.03, 10.90, 13.57%; Gidaspow: 7.84, 8.40, 11.56%), the simulation error also increases; this aligns with the result reported by Shi et al. [18]. This suggests that a restitution coefficient of 0.9 most accurately represents the conditions.

Conversely, regarding the impact of the specularity coefficient, which was varied between 0.05, 0.1, and 0.2, the results in the Syamlal–O'Brien drag model indicate that an increase in the specularity coefficient does not significantly alter bed expansion. However, in the Gidaspow drag model, a slight reduction in bed expansion was observed. The simulation error analysis within the Syamlal–O'Brien model correlates with bed expansion outcomes, showing minimal change across specularity coefficients (Syamlal–O'Brien: 10.51, 10.31, 10.03), suggesting that the influence of specularity coefficient variations on voidage predictions can be neglected, as also noted by Shi et al. [18]. Nonetheless, employing the Gidaspow model (Gidaspow: 7.45, 9.24, 7.84) reveals slight differences in simulation error values, with a notable increase at 0.1, indicating a sensitivity to specularity coefficient variations within the Gidaspow drag model.

## 6. Conclusions

In the current study, the gas–solid flow within 2D and 3D fluidized bed columns, replicated from Taghipour et al. [19], was simulated using the Eulerian–Eulerian framework coupled with the Kinetic Theory of Granular Flow (KTGF). This investigation aimed to assess the capabilities of three prominent CFD software packages—ANSYS Fluent, MFiX, and OpenFOAM—in modeling fluidized bed dynamics. The study specifically focused on the application of the Syamlal–O'Brien and Gidaspow drag models across varying superficial velocities, restitution coefficients, and specularity coefficients to provide a comprehensive evaluation of these tools. The key findings of this research are summarized as follows:

1. The assessment of ANSYS Fluent, MFiX, and OpenFOAM showed they could achieve reasonable agreement with experimental data across both drag models, offering insights into fluidized bed hydrodynamics. This comparison underscored specific strengths and weaknesses, revealing MFiX and ANSYS Fluent as more consistent and accurate in simulating fluidized bed dynamics, thereby emerging as preferable choices. OpenFOAM has shown maturity in predicting fluidized bed phenomena,

yet it necessitates further validation and exploration to fully grasp the impact of various numerical settings and boundary conditions on its outcomes. This approach underscores the importance of fine-tuning and accurately assessing OpenFOAM's capabilities for enhanced reliability in fluidized bed simulations.

2. The analysis of the Syamlal–O'Brien and Gidaspow drag models revealed significant differences in their predictions, particularly in terms of bed expansion and pressure drop. The Syamlal–O'Brien model, when used with MFiX and OpenFOAM, typically resulted in lower simulation errors compared to the Gidaspow model. However, ANSYS Fluent showed a preference for the Gidaspow model, indicating that the choice of drag model can significantly influence simulation outcomes and should be carefully considered based on the CFD software being used.
3. The transition from 2D to 3D simulations introduced notable variations in bed expansion and pressure drop predictions towards better agreement with the experimental data. While 3D simulations offer a more detailed representation of fluidized bed dynamics, they do not always guarantee improved alignment with experimental data, particularly when validating against experimental voidage data. In some cases, 2D simulations provided better fits, suggesting that the dimensionality of the simulation should be chosen based on specific study requirements and the expected flow behavior.
4. The study showed that while higher restitution coefficients led to increased bed expansion, indicating their significant impact on particle–particle collisions, variations in the specularity coefficient, which affects particle–wall interactions, had a subtler influence on bed dynamics, particularly noted in the Gidaspow drag model. A restitution coefficient of 0.9 was identified as optimal, balancing accuracy and computational efficiency, whereas the specularity coefficient's effect, though present, was less pronounced within the examined range. This emphasizes the necessity of carefully selecting these coefficients to accurately model the complex behaviors in fluidized beds.

**Author Contributions:** Conceptualization, K.E.E., P.F.D. and P.K.; methodology, K.E.E., P.F.D. and P.K.; software, P.F.D. and P.K.; validation, P.F.D. and P.K.; formal analysis, P.F.D. and P.K.; investigation, P.F.D. and P.K.; resources, K.E.E.; data curation, P.F.D. and P.K.; writing—original draft preparation, P.K.; writing—review and editing, K.E.E., P.F.D. and P.K.; visualization, P.F.D. and P.K.; supervision, K.E.E.; project administration, K.E.E.; funding acquisition, K.E.E. All authors have read and agreed to the published version of the manuscript.

**Funding:** Financial support of the Norwegian Research Council, grant 337007, is acknowledged.

**Data Availability Statement:** The data can be made available upon request from the corresponding author.

**Conflicts of Interest:** The authors declare no conflicts of interest.

## Abbreviations

The following abbreviations are used in this manuscript:

| | |
|---|---|
| $\alpha_g$ | Volume fraction of the gas phase |
| $\alpha_s$ | Volume fraction of the solid phase |
| $\alpha_s^{max}$ | The maximum packing fraction of the solid phase |
| $C_D$ | Drag coefficient |
| $d_s$ | Diameter of the solid particles |
| $D_{s,ij}$ | The rate of deformation tensor components for the solidphase |
| $\eta$ | A factor incorporating the restitution coefficient |
| $e_{ss}$ | The restitution coefficient for solid-solid collisions |
| $\gamma_{\Theta_s}$ | Dissipation rate of fluctuating energy in the solidphase |
| $\gamma_{\Theta m}$ | Collision dissipation of energy |
| $g_{0,ss}$ | The radial distribution function at contact for solid-solidinteractions |
| $\vec{g}$ | Gravitational acceleration vector |

| | |
|---|---|
| $I_{2D}$ | The second invariant of the deformation tensor |
| $K$ | The specularity coefficient |
| $K_{gs}$ | Momentum exchange coefficient or interphase drag forcecoefficient |
| $k_{\Theta_s}$ | Thermal conductivity of the solid phase for thefluctuating energy |
| $\kappa_s$ | The thermal conductivity of the solid phase for granulartemperature |
| $\lambda_s$ | Granular bulk viscosity |
| $\mu_g$ | Dynamic viscosity of the gas phase |
| $\mu_s$ | The viscosity coefficient of the solid phase |
| $\mu_{s,col}$ | Solid collision viscosity |
| $\mu_{s,fr}$ | Solid frictional viscosity |
| $\mu_{s,kin}$ | Solid kinetic viscosity |
| $\nabla$ | Nabla operator, representing vector differential operations |
| $p$ | Pressure in the gas phase |
| $P_s$ | Solid phase pressure |
| $\vec{u}_g$ | Velocity vector of the gas phase |
| $\vec{u}_s$ | Velocity vector of the solid phase |
| $\overline{\tau}_g$ | Deviatoric stress tensor of the gas phase |
| $\overline{\tau}s$ | Solid phase shear stress tensor |
| $\overline{\overline{I}}$ | The identity tensor for the gas phase |
| $\overline{\overline{\tau}}g$ | Gas phase shear stress tensor |
| $\overline{\overline{\tau}}s$ | Deviatoric stress tensor of the solidphase |
| $\rho_g$ | Density of the gas phase |
| $\rho_s$ | Density of the solid phase |
| $\varnothing$ | The angle of internal friction |
| $t$ | Time |
| $\Theta_s$ | Fluctuating energy of the solid phase |
| $vr, s$ | Relative velocity between phases |
| $\varphi gs$ | Energy exchange rate between gas and solid phases |
| $\varphi gs$ | Transfer of kinetic energy between gas and solid phases |
| $x$ | The spatial coordinate in the direction of the gradient |
| $(\nabla\vec{u}_g)^T$ | The transpose of the velocitygradient tensor of the gas phase |
| $Re_s$ | Reynolds number for the solid phase, characterizing the flowregime |

## Appendix A

The constitutive equations derived from KGTF are provided in the table below.

**Table A1.** Constitutive equations used in this study.

Gas phase shear stress tensor

$$\overline{\overline{\tau}}_g = \alpha_g \mu_g \left[ \nabla\vec{u}_g + (\nabla\vec{u}_g)^T - \frac{2}{3}(\nabla \cdot \vec{u}_g)\overline{\overline{I}} \right] \quad \text{(A1)}$$

Solid phase shear stress tensor

$$\overline{\overline{\tau}}_s = \alpha_s \mu_s \left[ \nabla\vec{u}_s + (\nabla\vec{u}_s)^T + \left(\frac{2}{3} - \frac{\lambda_s}{\mu_s}\right)(\nabla \cdot \vec{u}_s)\overline{\overline{I}} \right] \quad \text{(A2)}$$

Granular bulk viscosity

$$\lambda_s = \frac{4}{3}\alpha_s \rho_s d_s g_{0,ss}(1 + e_{ss})\left(\frac{\Theta_s}{\pi}\right)^{1/2} \quad \text{(A3)}$$

Solid shear viscosity

$$\mu_s = \mu_{s,col} + \mu_{s,kin} + \mu_{s,fr} \quad \text{(A4)}$$

**Table A1.** *Cont.*

Solid collision viscosity [34]

$$\mu_{s,col} = \frac{4}{3}\alpha_s^2\rho_s d_s g_{0,ss}(1+e_{ss})\left(\frac{\Theta_s}{\pi}\right)^{1/2} \tag{A5}$$

Solid kinetic viscosity
Syamlal–O'Brien [34]

$$\mu_{s,kin} = \frac{d_s\rho_s\sqrt{\Theta_s\pi}}{6(3-e_{ss})}\left[1+\frac{2}{5}(1+e_{ss})(3e_{ss}-1)\alpha_s g_{0,ss}\right] \tag{A6}$$

Gidaspow [40]

$$\mu_{s,kin} = \frac{10\rho_s d_s\sqrt{\Theta_s\pi}}{96\alpha_s(1+e_{ss})g_{0,ss}}\left[1+\frac{4}{5}g_{0,ss}\alpha_s(1+e_{ss})\right]^2 \tag{A7}$$

Solid frictional viscosity [29]

$$\mu_s^{fr} = \frac{P_s\sin\varnothing}{2\sqrt{I_{2D}}} \tag{A8}$$

$$I_{2D} = \frac{1}{6}\left[(D_{s,11}-D_{s,22})^2+(D_{s,22}-D_{s,33})^2+(D_{s,33}-D_{s,11})^2\right] + D_{s,12}^2+D_{s,23}^2+D_{s,31}^2 \tag{A9}$$

$$D_{s,ij} = \frac{1}{2}\left(\frac{\partial\vec{u}_{s,i}}{\partial x_j}+\frac{\partial\vec{u}_{s,j}}{\partial x_i}\right) \tag{A10}$$

Solid phase pressure [28]

$$p_s = \alpha_s\rho_s\Theta_s + 2\rho_s(1+e_{ss})\alpha_s^2 g_{0,ss}\Theta_s \tag{A11}$$

Radial distribution function [28,41]

$$g_{0,ss} = \frac{1}{1-\alpha_s}+\frac{3}{2}\frac{\alpha_s}{(1-\alpha_s)^2}+\frac{1}{2}\frac{\alpha_s^2}{(1-\alpha_s)^3} \tag{A12}$$

Diffussion coefficient of granular temperature
Syamlal–O'Brien [34]

$$k_{\Theta_s} = \frac{15d_s\rho_s\alpha_s\sqrt{\Theta_s\pi}}{4(41-33\eta)} \left[1+\frac{12}{5}\eta^2(4\eta-3)\alpha_s g_{0,ss}+\frac{16}{15\pi}(41-33\eta)\eta\alpha_s g_{0,s}\right] \tag{A13}$$

where

$$\eta = \frac{1}{2}(1+e_{ss}) \tag{A14}$$

Gidaspow [40]

$$k_{\Theta_s} = \frac{150\rho_s d_s\sqrt{(\Theta\pi)}}{384(1+e_{ss})g_{0,ss}}\left[1+\frac{6}{5}\alpha_s g_{0,s}(1+e_s)\right]^2 + 2\rho_s\alpha_s^2 d_s(1+e_{ss})g_{0,ss}\sqrt{\frac{\Theta_s}{\pi}} \tag{A15}$$

Collision dissipation of energy [28]

$$\gamma_{\Theta m} = \frac{12(1-e_{ss}^2)g_{0,ss}}{d_s\sqrt{\pi}}\rho_s\alpha_s^2\Theta_s^{3/2} \tag{A16}$$

Transfer of kinetic energy [40]

$$\varphi_{gs} = -3K_{gs}\Theta_s \tag{A17}$$

## References


1. Werther, J. Fluidized Bed Reactors. In *Ullmann's Encyclopedia of Industrial Chemistry*; John Wiley & Sons, Ltd.: Hoboken, NJ, USA, 2007.
2. Witt, P.M.; Hickman, D.A. Fluidized bed reactor scale-up: Reaction kinetics required. *AIChE J.* **2022**, *68*, e17803. [CrossRef]

13. Kunii, D.; Levenspiel, O. *Fluidization Engineering*; Butterworth-Heinemann: Oxford, UK, 1991.
14. Alobaid, F.; Almohammed, N.; Massoudi Farid, M.; May, J.; Rößger, P.; Richter, A.; Epple, B. Progress in CFD Simulations of Fluidized Beds for Chemical and Energy Process Engineering. *Prog. Energy Combust. Sci.* **2022**, *91*, 100930.
15. Gidaspow, D. *Multiphase Flow and Fluidization: Continuum and Kinetic Theory Descriptions*; Academic Press: Cambridge, MA, USA, 1994.
16. Stroh, A.; Alobaid, F.; Hasenzahl, M.T.; Hilz, J.; Ströhle, J.; Epple, B. Comparison of three different CFD methods for dense fluidized beds and validation by a cold flow experiment. *Particuology* **2016**, *29*, 34–47. [CrossRef]
17. Grüner, C. *Kopplung des Einzelpartikel-und des Zwei-Kontinua-Verfahrens für die Simulation von Gas-Feststoff-Strömungen*; Shaker Verlag: Aachen, Germany, 2004.
18. Moin, P.; Mahesh, K. Direct Numerical Simulation: A Tool in Turbulence Research. *Annu. Rev. Fluid Mech.* **1998**, *30*, 539–578. [CrossRef]
19. *ANSYS FLUENT 2023 R3*; Commercial Software. ANSYS Inc.: Canonsburg, PA, USA, 2023.
20. *ANSYS CFX*; Commercial Software. ANSYS Inc.: Canonsburg, PA, USA, 2023.
21. *OpenFOAM 10*; General Public License. The OpenFOAM Foundation: Newark, DE, USA, 2023.
22. *MFiX 23.1.1*; General Public License. National Energy Technology Laboratory: Morgantown, WV, USA, 2023.
23. *BARRACUDA Virtual Reactor*; Commercial Software. CPFD Software: Houston, TX, USA, 2023.
24. *LIGGGHTS CFDEM Coupling*; General Public License. CFDEM Project: Linz, Austria, 2023.
25. *COMSOL Multiphysics*; Commercial Software. COMSOL Inc.: Burlington, MA, USA, 2023.
26. *STAR-CCM+*; Commercial Software. CD-adapco: Melville, NY, USA , 2023.
27. *EDEM*; Commercial Software . Altair Engineering: Singapore, 2023.
28. Shi, H.; Komrakova, A.; Nikrityuk, P. Fluidized beds modeling: Validation of 2D and 3D simulations against experiments. *Powder Technol.* **2019**, *343*, 479–494. [CrossRef]
29. Taghipour, F.; Ellis, N.; Wong, C. Experimental and computational study of gas–solid fluidized bed hydrodynamics. *Chem. Eng. Sci.* **2005**, *60*, 6857–6867. [CrossRef]
30. Syamlal, M.; O'Brien, T.J. Simulation of granular layer inversion in liquid fluidized beds. *Int. J. Multiph. Flow* **1988**, *14*, 473–481. [CrossRef]
31. Cornelissen, J.T.; Taghipour, F.; Escudié, R.; Ellis, N.; Grace, J.R. CFD modelling of a liquid–solid fluidized bed. *Chem. Eng. Sci.* **2007**, *62*, 6334–6348. [CrossRef]
32. Li, T.; Grace, J.; Bi, X. Study of wall boundary condition in numerical simulations of bubbling fluidized beds. *Powder Technol.* **2010**, *203*, 447–457. [CrossRef]
33. Herzog, N.; Schreiber, M.; Egbers, C.; Krautz, H.J. A comparative study of different CFD-codes for numerical simulation of gas–solid fluidized bed hydrodynamics. *Comput. Chem. Eng.* **2012**, *39*, 41–46. [CrossRef]
34. Reyes-Urrutia, A.; Venier, C.; Mariani, N.J.; Nigro, N.; Rodriguez, R.; Mazza, G. A CFD Comparative Study of Bubbling Fluidized Bed Behavior with Thermal Effects Using the Open-Source Platforms MFiX and OpenFOAM. *Fluids* **2022**, *7*, 1. [CrossRef]
35. Mofakham, A.S.; Rasteh, M. CFD simulation of gas–solid fluidized bed hydrodynamics; prediction accuracy study. *Int. J. Chem. React. Eng.* **2023**, *21*, 391–407. [CrossRef]
36. Daun, P.F. A Comparative Study of Different CFD-Codes for Fluidized Beds. Master's Thesis, Norwegian University of Science and Technology, Materials Science and Engineering, Trondheim, Norway, 2023.
37. ANSYS, Inc. *Ansys Fluent Theory Guide*; ANSYS, Inc.: Canonsburg, PA, USA, 2021.
38. Lun, C.K.K.; Savage, S.B.; Jeffrey, D.J.; Chepurniy, N. Kinetic theories for granular flow: Inelastic particles in Couette flow and slightly inelastic particles in a general flowfield. *J. Fluid Mech.* **1984**, *140*, 223–256. [CrossRef]
39. Schaeffer, D.G. Instability in the evolution equations describing incompressible granular flow. *J. Differ. Equ.* **1987**, *66*, 19–50. [CrossRef]
40. Ding, J.; Gidaspow, D. A bubbling fluidization model using kinetic theory of granular flow. *AIChE J.* **1990**, *36*, 523–538. [CrossRef]
41. Ogawa, S.; Umemura, A.; Oshima, N. On the equations of fully fluidized granular materials. *Z. Angew. Math. Phys. ZAMP* **1980**, *31*, 483–493. [CrossRef]
42. ANSYS, Inc. *Workshop: Modeling Uniform Fluidization in a 2D Fluidized Bed*; ANSYS, Inc.: Canonsburg, PA, USA, 2023.
43. Johnson, P.C.; Jackson, R. Frictional–collisional constitutive relations for granular materials, with application to plane shearing. *J. Fluid Mech.* **1987**, *176*, 67–93. [CrossRef]
44. Syamlal, M.; Rogers, W.; O'Brien, T. *MFIX Documentation Theory Guide*; Technical Report; National Energy Technology Laboratory, U.S. Department of Energy: Morgantown, WV, USA, 1993.
45. El Ajouri, O.; Lahlaouti, M.L.; Kharbouch, B. CFD-simulation of gas–solid flow in bubbling fluidized bed reactor. *E3S Web Conf.* **2022**, *336*, 00059. [CrossRef]
46. Zhu, L.T.; Ouyang, B.; Lei, H.; Luo, Z.H. Conventional and data-driven modeling of filtered drag, heat transfer, and reaction rate in gas–particle flows. *AIChE J.* **2021**, *67*, e17299. [CrossRef]
47. Zhu, L.T.; Liu, Y.X.; Tang, J.X.; Luo, Z.H. A material-property-dependent sub-grid drag model for coarse-grained simulation of 3D large-scale CFB risers. *Chem. Eng. Sci.* **2019**, *204*, 228–245. [CrossRef]
48. Aboudaoud, S.; Touzani, S.; Abderafi, S.; Cheddadi, A. CFD Simulation of Air-Glass Beads Fluidized Bed Hydrodynamics. *J. Appl. Fluid Mech.* **2023**, *16*, 1778–1791.

39. Liu, Y.; Hinrichsen, O. CFD modeling of bubbling fluidized beds using OpenFOAM®: Model validation and comparison of TVI differencing schemes. *Comput. Chem. Eng.* **2014**, *69*, 75–88. [CrossRef]
40. Gidaspow, D.; Bezburuah, R.; Ding, J. Hydrodynamics of Circulating Fluidized Beds: Kinetic Theory Approach. In Proceeding of the 7th International Conference on Fluidization, Gold Coast, Australia, 3–8 May 1992.
41. Carnahan, N.F.; Starling, K.E. Equation of State for Nonattracting Rigid Spheres. *J. Chem. Phys.* **1969**, *51*, 635–636. [CrossRef]

applied sciences

MDPI

Article

# PMT Fluorescence Signal Denoising Processing Based on Wavelet Transform and BP Neural Network

Jiehui Liu [1,2], Yunhan Zhang [1,2,3], Jianshen Li [1,2,*], Yadong Zhao [3], Jinxi Guo [1,2], Lijie Yang [1,2] and Haichao Zhao [4]

1 School of Mechanical and Equipment Engineering, Hebei University of Engineering, Handan 056038, China; jiehuiliu@hebeu.edu.cn (J.L.); 13363295401@163.com (Y.Z.); guojinxi@hebeu.edu.cn (J.G.); yanglijie255@163.com (L.Y.)
2 Key Laboratory of Intelligent Industrial Equipment Technology of Hebei Province, Hebei University of Engineering, Handan 056038, China
3 School of Mechanical Engineering, Anyang Institute of Technology, Anyang 455000, China; zhaoyd@ayit.edu.cn
4 Handan Environmental Protection Research Institute, Handan 056038, China; 15833082296@163.com
* Correspondence: lijianshen@hebeu.edu.cn; Tel.: +86-18231098913

**Featured Application: This study is aimed at the detection of sulfur dioxide concentration in the real environment. Due to the vast territory of China and the large temperature difference span between the north and the south, the current sulfur dioxide detection devices find it difficult to meet the requirements of portability, low-cost delivery, and temperature adaptability. A reliable real-time monitoring device and detection algorithm for sulfur dioxide concentration using PMT are proposed.**

**Abstract:** Air is the environmental foundation for human life and production, and its composition changes are closely related to human activities. Sulfur dioxide ($SO_2$) is one of the main atmospheric pollutants, mainly derived from the combustion of fossil fuels. But $SO_2$ is a trace gas in the atmosphere, and its concentration may be less than one part per billion (ppb). This paper is based on the principle of photoluminescence and uses a photomultiplier tube (PMT) as a photoelectric converter to develop a device for real-time detection of $SO_2$ concentration in the atmosphere. This paper focuses on the impact of noise interference on weak electrical signals and uses wavelet transform to denoise the signals. At the same time, considering that the photoelectric system is susceptible to temperature changes, a multi parameter fitting model is constructed, and a BP neural network is used to further process the signal, separating the real data from the original data. Finally, a high-precision and wide-range trace level sulfur dioxide concentration detection device and algorithm were obtained.

**Keywords:** BP neural network; data fusion; temperature drift; weak current signal detection; wavelet transform

## 1. Introduction

With the rapid development of modern industry, the harmful gases produced in chemical industry production have attracted people's attention. Among them, sulfur dioxide is the main component that causes sulfuric acid rain. Usually, based on different detection principles and implementation methods, gas detection techniques can be divided into two categories: optical detection methods and non-optical detection methods. The development and application of non-optical detection methods occurred earlier; they directly detect and quantitatively analyze the measured gas by detecting the physical and chemical reactions between substances. There exist electrochemical methods, chemical fluorescence methods, gas chromatography, and mass spectrometry methods. However, the above-mentioned detection methods often fail to meet the requirements of real-time monitoring of air pollution and harsh environmental conditions in some application scenarios due to poor sensitivity, selectivity, and linearity, as well as long measurement times and the need for offline detection.

In recent years, optical-based trace gas detection methods have made significant progress. According to different detection principles, absorption spectroscopy is mainly divided into direct absorption spectroscopy technology and indirect absorption spectroscopy technology. There are various types of direct absorption spectroscopy techniques, and there are many mature ones that have been studied and applied, such as tunable diode laser absorption spectroscopy (TDLAS) [1,2], cavity ring down absorption spectroscopy (CRDS) [3,4], and cavity enhancement technology [5,6]. Typical indirect absorption spectroscopy techniques include fluorescence spectroscopy [7], photothermal spectroscopy (PTS) [8–10], and photoacoustic spectroscopy (PAS).

The ultraviolet fluorescence method based on the principle of photoluminescence has become the mainstream solution due to its advantages of high accuracy, high reliability, and no other pollution. Regarding UV fluorescence detection technology for sulfur dioxide gas concentration, Hideo Okabe analyzed and studied the relationship between excitation wavelength and fluorescence intensity in the spectral range of 200 nm–230 nm. The experiment showed that under appropriate excitation wavelength, the fluorescence intensity detected by the detector can be used to measure the content of sulfur dioxide in the gas. This experiment also became the basic experiment for sulfur dioxide ultraviolet fluorescence detection, and it was later verified through experimental methods that 213.8 nm is the optimal excitation wavelength for detecting sulfur dioxide fluorescence [11]. Afterwards, Bradshaw first used an Nd: YAG laser to induce fluorescence in sulfur dioxide [12]. G. Somesfalean et al. achieved a detection sensitivity of 20 ppm for $SO_2$ gas using a UV diode laser [13]. Yutaka Matsumi used a broadband optical parametric oscillator to generate double frequency light in a laser-induced fluorophore, achieving a detection limit of 5 ppt for $SO_2$ in the instrument [14]. However, due to the extremely low content of sulfur dioxide gas in the air, the excited fluorescence signal was very weak. Meanwhile, the process of photoelectric signal detection is inevitably affected by temperature drift [15,16] and electronic noise, which poses high requirements for the acquisition circuit and subsequent data processing. Therefore, it is of great significance to develop new gas detection technologies and methods with high sensitivity, good selectivity, high detection efficiency, and real-time online monitoring. This paper achieves the effective capture and amplification of weak fluorescence signals by combining appropriate optical paths and acquisition circuits. At the same time, wavelet transform [17] is used to denoise the PMT signal, and a BP neural network [18] is used to process the temperature drift of PMT. After verification, it is found the detection device [19] has a high accuracy and strong anti-interference ability.

## 2. Detection System Section

### 2.1. Optical Path Section

Firstly, based on the principle of photoluminescence, sulfur dioxide molecules are irradiated by ultraviolet light with a wavelength of 200 nm–220 nm to produce excited sulfur dioxide molecules, which emit fluorescence with a wavelength of 240 nm–420 nm during their return to the ground state. Within a certain concentration range, the fluorescence intensity is directly proportional to the concentration of sulfur dioxide in the sample air. The optical path and assembly diagram of the detection equipment are shown in Figure 1.

The excitation light source selected is the L12745-01 pulse xenon lamp [20] from Hamamatsu Photonics Co., Ltd. in Iwata City, Japan. It has a rated power of 20 W and can provide spectra in the 185 nm–2500 nm wavelength range. The duration of a single excitation signal is about 5 microseconds.

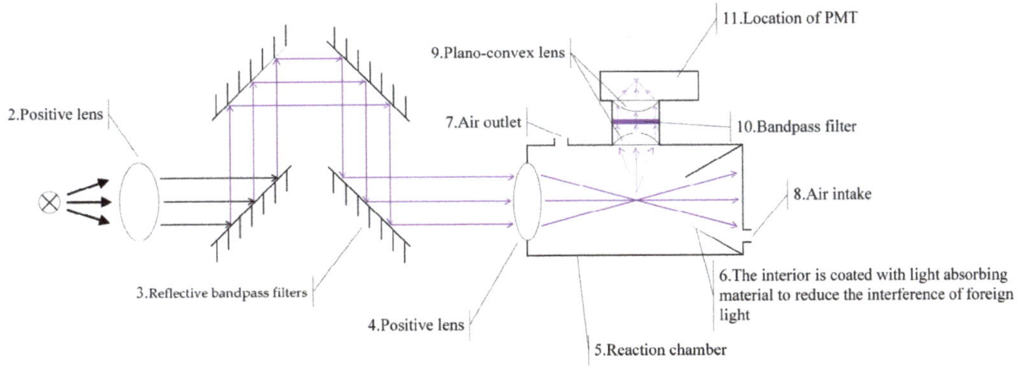


**Figure 1.** The optical path and assembly diagram of the detection device.

In order to obtain the required wavelength range of ultraviolet light for the experiment and reduce the interference of impurity light, four reflective bandpass filters are placed on the excitation light path, with a wavelength range of 209 nm–219 nm [21], the incident angle is 45°. Reflective filters have a stronger retention effect on the required spectral intensity than transmissive filters, which can increase the proportion of spectral intensity required and reduce interference from impurity light. Afterwards, through the combination of multiple positive lens, the light entering the reaction chamber is concentrated directly below the PMT, which can obtain the strongest fluorescence signal. In addition, a conical structure needs to be designed at the rear end of the chamber, with the inner side sprayed with absorbing material. The purpose of doing so is to absorb the light entering the chamber as much as possible, avoid diffuse reflection in the chamber, and be received by PMT again, which will reduce the proportion of fluorescence signal.

The photoelectric converter selected is the R11558 photomultiplier tube from Hamamatsu Photonics Co., Ltd. in Japan, equipped with a magnetic shielding shell, with a response band of 300 nm–650 nm. The PMT power supply module adopts the original C12597-01 high-voltage base. Due to the introduction of power supply, it will bring more electromagnetic crosstalk, which will add impurities to the output of PMT. Therefore, resistor programming has been chosen for high-voltage base power supply. The high-voltage base provides a reference power line that can provide gain voltage to the PMT through resistive voltage division, thereby avoiding the introduction of more electromagnetic crosstalk. A bandpass filter with a wavelength range of 280 nm–400 nm is placed directly below the PMT, which can effectively detect the fluorescence emitted by sulfur dioxide molecules during their return to the ground state and reduce the interference of some impurity light. Of course, it is also necessary to place a combination of plano-convex lenses to integrate the optical path, ensuring that the generated fluorescence signal is focused on the effective detection part of PMT, in order to obtain the strongest fluorescence signal.

### *2.2. Circuit Section*

The acquisition circuit is composed of the STM32 platform and the FPGA platform [22,23]. The STM32 part selects STM32F407ZGT6 (STMicroelectronics, Plan-les-Ouates, Switzerland) as the main control chip, responsible for communicating with the FPGA platform and upper computer. The FPGA platform chooses EP4CE6E22C8N (Intel, Santa Clara, CA, USA) as the main control chip, which can drive high-performance ADCs with its bandwidth and slew rate. It is also responsible for joint communication with the STM32 main control chip. LTC2387-18FA (Analog Devices, Wilmington, MA, USA) is chosen as the analog-to-digital converter, which is a high-performance ADC with 18-bit depth and 10 M speed.

Due to the fact that the original signal output by PMT is a reverse signal, the signal first enters an inverted operational amplifier circuit and, then, outputs a signal with adjustable size and positive direction. Then, the subtractor circuit is entered. This process is shown in Figure 2, where the red part represents the fluorescence signal of sulfur dioxide. After passing through this part of the circuit, the proportion of sulfur dioxide fluorescence signal significantly increases, which helps to improve the sensitivity and resolution of the detection device.

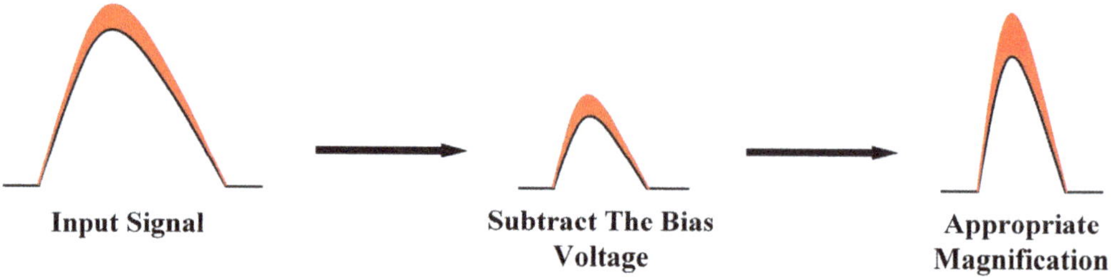


**Figure 2.** Subtraction circuit diagram.

Then, the signal enters the differential amplification circuit, which divides the signal into two paths with different DC biases, opposite directions, and no phase deviation. The difference between the two signals is used as an analog signal to enter the ADC for analog-to-digital conversion. The two signals are transmitted by two signal lines with extremely short physical distances, and it can be considered that they are subjected to the same external interference, which can effectively reduce the impact of external interference and common mode noise on the electrical signal [24]. In addition, a four-layer circuit board is used in the experiment, with the power supply and grounding parts located in separate board layers. Compared to common double-layer circuit boards, it has better electrical performance. The physical image of the collection circuit is shown in Figure 3.

**Figure 3.** Collect physical circuit diagram.

## 3. Wavelet Transform

Due to sulfur dioxide being a trace gas in the atmosphere, the number of fluorescent photons generated by excitation is very small, making it susceptible to interference from electronic noise. Therefore, noise reduction processing is the key to improving the accuracy of equipment detection.

Wavelet transform [25,26] is a powerful signal processing tool and a mathematical method widely used in various fields such as signal processing and seismic exploration. It decomposes and analyzes signals at different scales and positions using wavelet functions.

Before reaching the optimal decomposition level of the signal, the relationship between signal-to-noise ratio and decomposition level is expressed as an approximate positive correlation. After reaching the optimal decomposition level, the signal-to-noise ratio of the signal will not increase again but will instead decrease. Applying threshold denoising techniques to each sub band [27] can effectively remove noise and preserve the true signal.

The MATLAB multi-resolution analysis is an important tool for wavelet transform analysis. In order to distinguish effective signals from noisy signals, a series of detail signals with different resolutions can be used for hierarchical representation. The process of wavelet decomposition is shown in Figure 4.

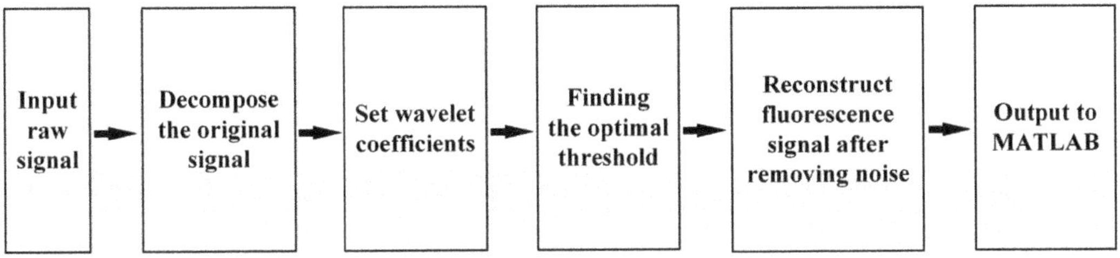


**Figure 4.** Wavelet denoising process.

In this fluorescence detection system, the effective signal is usually a low-frequency signal, while electronic noise is usually a high-frequency signal. Therefore, the original signal can be decomposed a certain number of times through wavelet transform to isolate the effective signal from noise, thereby removing the interference of high-frequency electronic noise and improving the signal-to-noise ratio. The wavelet transform denoising used in this article can be divided into the following three steps:

(1) Original signal decomposition. Firstly, select multiple original signals, perform N-layer wavelet decomposition on the signals, identify common trends in the signals, and modify and compare the parameters.

(2) Noise signal removal. Adjust and optimize the high-frequency noise threshold coefficients for each layer from layer 1 to layer N, and select the optimal noise reduction coefficient to reduce the impact of electronic noise.

(3) Effective signal reconstruction. Perform wavelet reconstruction of fluorescence signals based on the low-frequency coefficients of the Nth layer of wavelet decomposition and the high-frequency coefficients of the 1st to Nth layers after quantization processing.

The most crucial step in these three steps is the determination of the threshold, which aims to remove electronic noise as much as possible while ensuring that the effective signal is not distorted. Due to the discontinuity of the hard threshold function at the selected threshold, it can cause system oscillations and cause distortion. A soft threshold will cause some singularities to disappear [28,29], resulting in the loss of key details of the signal, so this paper proposes an improved threshold wavelet decomposition algorithm for this experimental environment.

The collection of fluorescent signals faces noise problems, mainly due to the following reasons:

(1) The signal itself is weak. Due to the concentration of sulfur dioxide in this experiment being at the ppb level, the excited fluorescence signal is very weak. If not properly processed, such a signal is easily masked by other noise.

(2) Device noise. Electronic components usually introduce a certain degree of noise, such as operational amplifiers, filters, voltage stabilizing chips, etc. These noises may come

from the internal circuits of the device or from the interaction between the device and the external electromagnetic environment.

(3) Environmental noise. There may be various sources of interference in the experimental environment, such as power crosstalk, electromagnetic radiation, thermal noise, etc. These interference sources will introduce additional noise, affecting the accuracy of the signal and the analysis of the data.

In the experiment, due to the inability to achieve complete darkness, the photomultiplier tube will still output small electrical signals when not exposed to pulsed xenon lamps. After passing through the operational amplifier circuit, it will output static noise as shown in Figure 5. After calculation, the standard deviation of static noise is 4404.932, and the average absolute deviation is 3533.375. This will have a significant impact on the accuracy of subsequent experiments.

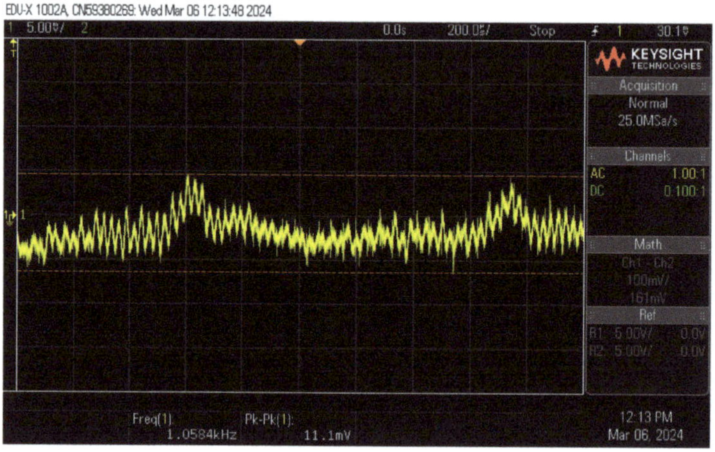


**Figure 5.** Static noise of photomultiplier tubes.

Six wavelet algorithms were selected for denoising, and the results are shown in Table 1 below. Finally, the sym8 wavelet was selected for 5-layer decomposition.

**Table 1.** Performance comparison of wavelet algorithms.

| Wavelet Algorithm | Decomposition Level | Evaluation Index | Hard Threshold | Soft Threshold | Improvement Threshold |
|---|---|---|---|---|---|
| db10 | 4 | SNR | 37.68 | 39.58 | 40.29 |
| | | Mean standard deviation | 3451 | 3102 | 2993 |
| sym8 | 5 | SNR | 36.65 | 42.68 | 44.26 |
| | | Mean standard deviation | 2864 | 2685 | 2697 |
| coif5 | 4 | SNR | 38.61 | 39.31 | 39.98 |
| | | Mean standard deviation | 3504 | 3220 | 3302 |
| bior5.5 | 5 | SNR | 39.41 | 40.24 | 41.11 |
| | | Mean standard deviation | 2845 | 2811 | 2774 |
| ribo6.8 | 5 | SNR | 34.43 | 36.68 | 37.40 |
| | | Mean standard deviation | 3102 | 2992 | 2864 |
| fk22 | 5 | SNR | 38.86 | 40.24 | 40.82 |
| | | Mean standard deviation | 3305 | 3105 | 3102 |

After repeated experiments and multiple adjustments to the threshold, the thresholds for the first to fifth layers were determined to be 2233.843, 3237.675, 5129.732, 8748.381, and 9686.040, respectively. Figure 6 shows the process of wavelet denoising, and Figure 7 shows the comparison before and after denoising.

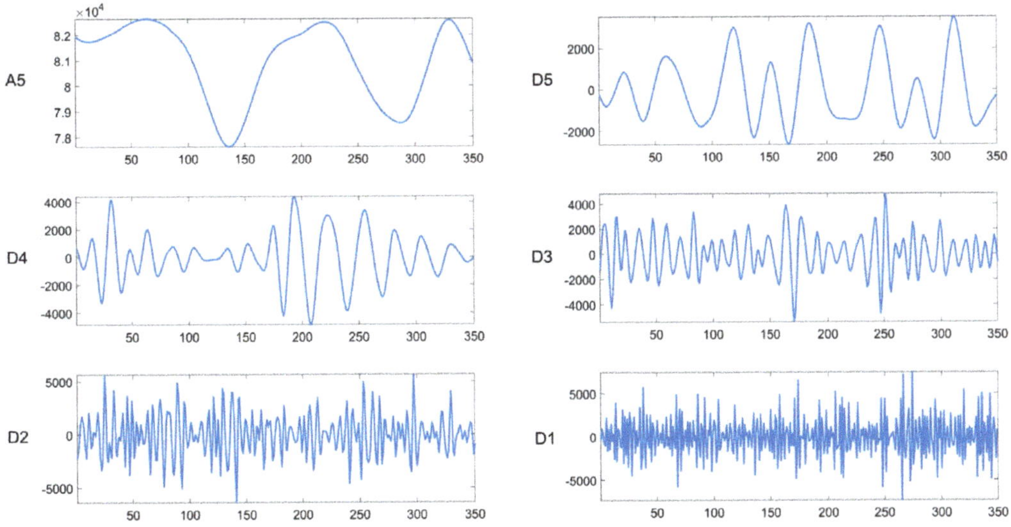


**Figure 6.** Static noise of photomultiplier tubes. Horizontal data points with an interval of 0.1 microseconds. (**D1–D5,A5**) represent the waveform of the signal in that frequency domain, and the higher the number, the higher the frequency interval. Therefore, from top to bottom and from left to right, the waveform in each frequency domain is shown as the frequency interval gradually decreases.

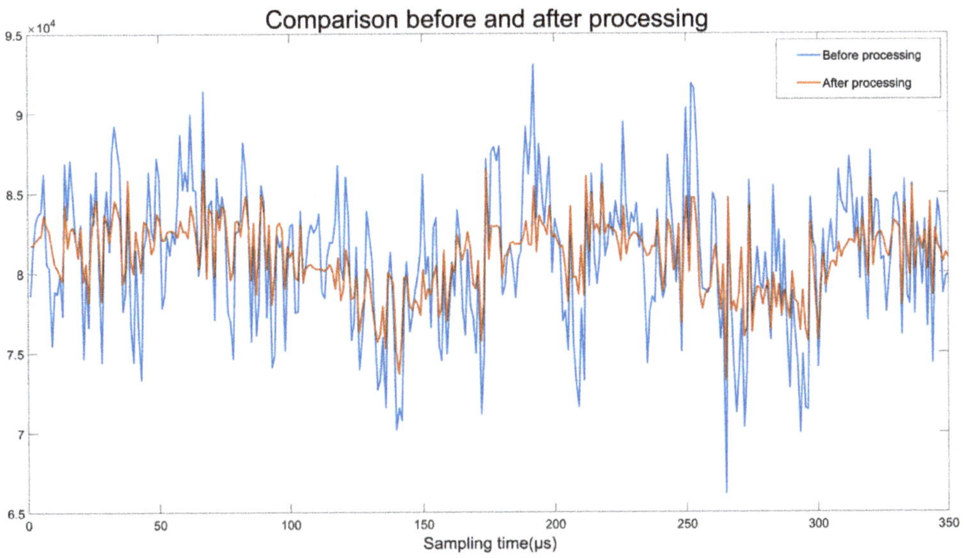


**Figure 7.** Comparison of static noise before and after wavelet denoising. Horizontal data points with an interval of 0.1 microseconds.

The denoising effect and signal fidelity can be determined by performing fast Fourier transform (FFT) analysis on the processed signal. The FFT image after denoising is shown in Figure 8. The results indicate that noise reduction processing effectively reduces noise interference without causing serious signal distortion.

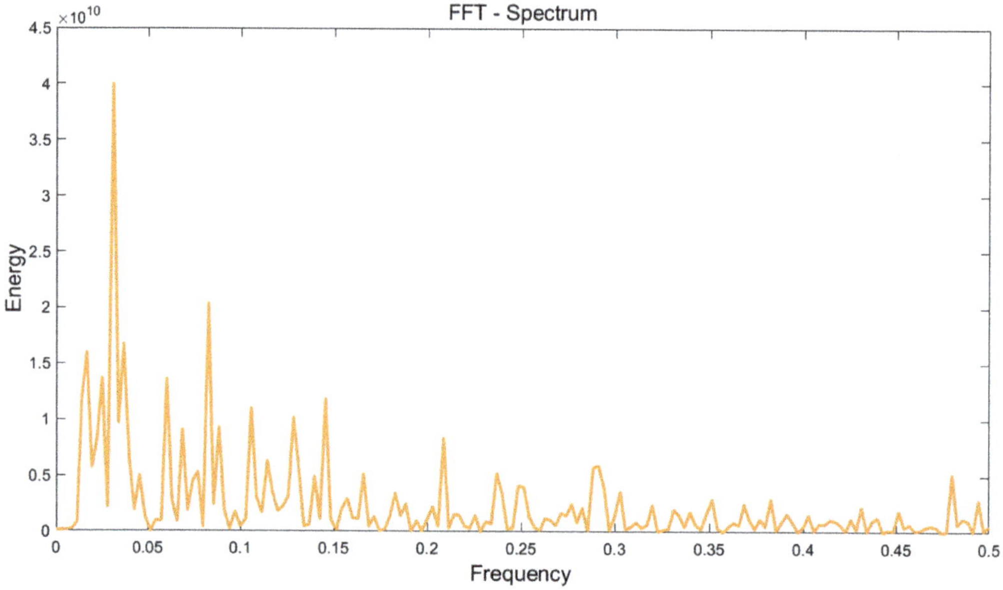


**Figure 8.** Fourier spectrum of signal after wavelet denoising. The horizontal axis unit is MHz.

The advantages of wavelet transform in time-frequency localization and frequency band segmentation can effectively distinguish between effective signals and noisy signals. Its excellent zoom performance enables the method based on multi-resolution analysis and wavelet transform to have fine scale division ability, which can effectively improve the detection accuracy [30] of this experiment. Embedding the above MATLAB code in Quartus and Keil can achieve real-time data processing without the need to import it into a PC for analysis, thus simplifying the experimental process.

## 4. Temperature Drift of PMT

### *4.1. Data Collection*

Firstly, verify the feasibility of the detection device. Use the square wave signal generated by flipping the pin level of the STM32 main control chip (STMicroelectronics) to drive the excitation light source. Set the working frequency of the pulse xenon lamp to 10 Hz, the number of data acquisition points to 600, and the gas flow rate to be measured to 6 L/min. Compared with ordinary electronic devices such as resistors and capacitors, photomultiplier tubes are more susceptible to the influence of environmental temperature, mainly manifested as nonlinear changes in the gain rate with temperature. Record the peak value of the fluorescence waveform as PMT Output Value, the static reference value when not irradiated as PMT Base Value, and the difference between the two as valid data, denoted as AD Value [31], as shown in Figure 9. This can, to some extent, reduce the impact of temperature changes on the detection results.

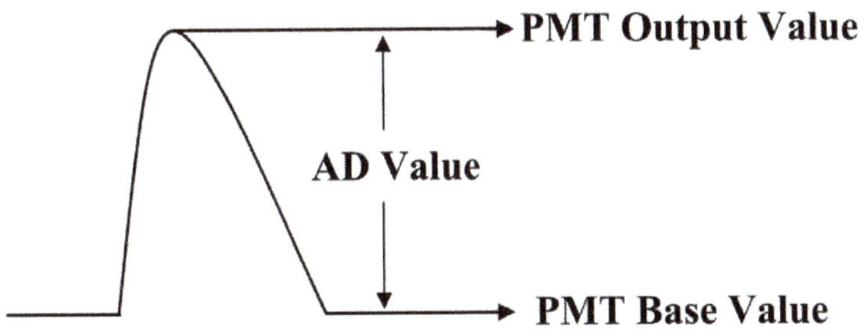


**Figure 9.** Data acquisition process.

The temperature sensor is selected as MLX90614 (Melexis, Shenzhen, China) with a resolution of 0.02 °C. Use it to monitor the temperature of PMT. Inject pure nitrogen gas into the reaction chamber, and after a period of experimentation, obtain the temperature data as shown in Figure 10.

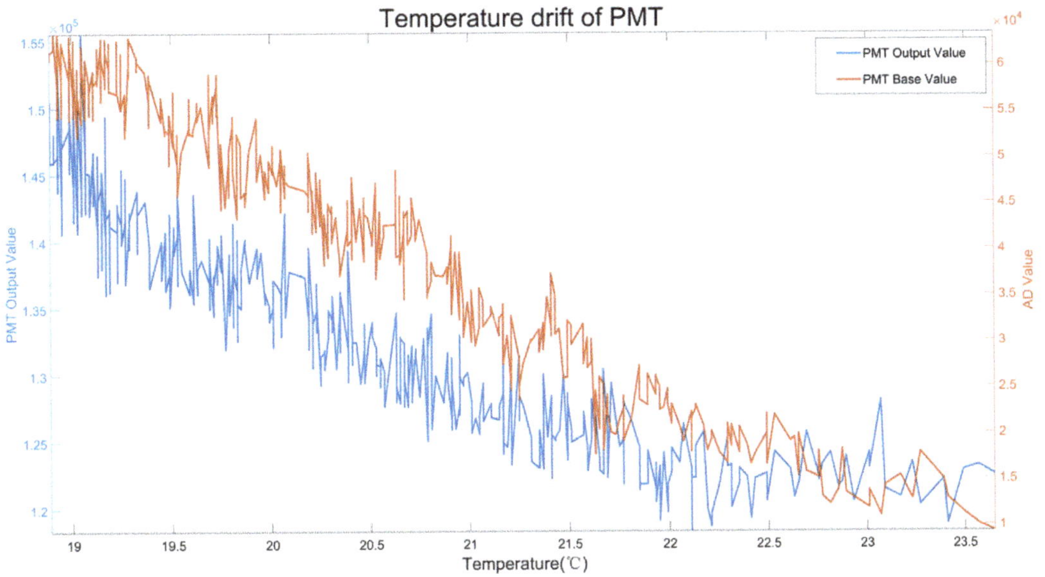


**Figure 10.** Temperature drift of PMT.

It can be observed that as the ambient temperature gradually increased during the detection process, the gain rate of PMT changed [32–34], manifested as a decrease in both the optical PMT Output Value and PMT Base Value, and their change rates were different. In order to mitigate the impact of temperature changes, different weights were assigned to the PMT Output Value and PMT Base Value by calculating their respective proportions of change. Finally, the following weighted comparison Figure 11 was obtained. Due to the fact that the gain rate of PMT increased as the temperature decreased, the AD Value should also increase as the temperature decreases, which is consistent with the test results. Before processing, the standard deviation of AD Value was 5843.183, and the average absolute error was 5030.428.

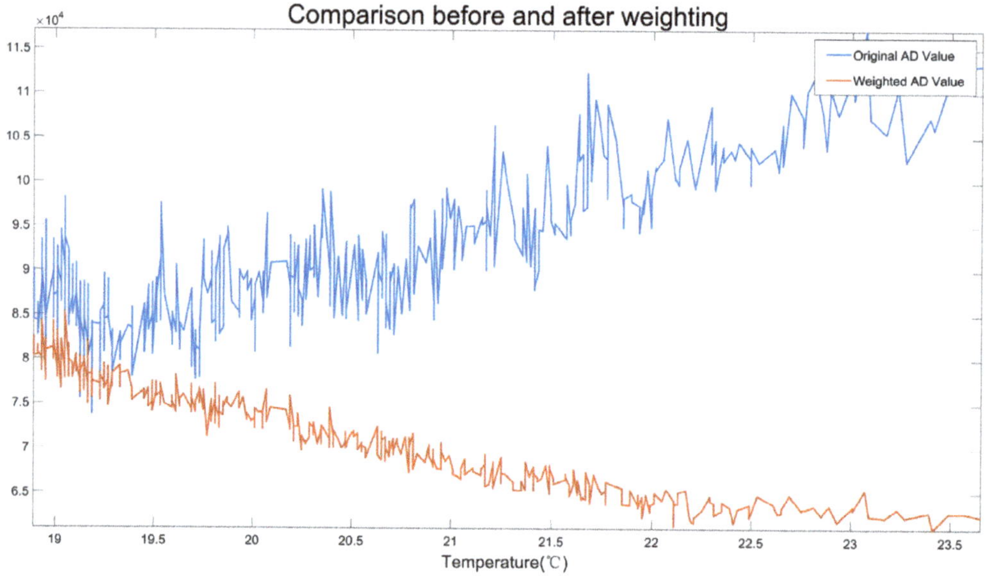


**Figure 11.** Comparison of AD values before and after weighting.

*4.2. Data Processing*

In the selection of algorithms, the commonly used Empirical Mode Decomposition (EMD) algorithm is first considered. However, the EMD algorithm still has many shortcomings in practical calculation and application, including mode mixing, endpoint effects, and screening iteration stopping standards. Among them, when there are intermittent phenomena caused by abnormal events (such as intermittent signals, pulse interference, and noise) in the signal, the decomposition results of EMD will exhibit mode aliasing. The occurrence of mode aliasing not only leads to false time-frequency distribution but also causes IMF to lose its physical meaning.

The widely used Kalman Filter (KF) [35] is an earlier time-domain denoising algorithm, mainly applied in fields such as communication, navigation, sensor networks, etc., but it can only be applied to linear systems. The later proposed Extended Kalman Filter (EKF) has theoretical limitations that cannot be overcome by themselves, such as low first-order linearization approximation accuracy, the need to calculate Jacobian matrices, and the requirement for continuous differentiability of nonlinear functions. Secondly, EKF is sensitive to the selection of initial estimation and process noise. When there are complex situations such as model mismatch, measurement interference, measurement loss, measurement delay, or state mutation in the system, EKF filtering accuracy is poor, and numerical stability is poor.

Although Particle Filters (PFs) are also suitable for nonlinear and non-Gaussian systems, their computational cost is high and increases with the increase in state dimensions. At the same time, a large number of particles are required to calculate the posterior distribution, and in high-dimensional state spaces, PF are more susceptible to problems such as insufficient sample size and particle degradation, leading to unstable estimation results. The limitation of its high computational complexity is also not suitable for long-term operation on platforms with limited computing and storage performance, such as embedded systems.

The characteristic of an ELMAN neural network [36] is that the output of the hidden layer is delayed and stored by the receiving layer, which makes it sensitive to historical state data. The addition of internal feedback network enhances the network's ability to process dynamic information, thereby achieving the goal of dynamic modeling. In addition,

ELMAN neural networks can approximate any nonlinear mapping with arbitrary accuracy, without considering the specific form of external noise on the system. If the input and output data of the system are correct, the system can be modeled. However, due to the fact that the learning process of ELMAN neural networks is based on gradient descent, this may result in the network only being able to find local optimal solutions and unable to find global optimal solutions.

BP neural networks [37,38] are suitable for solving problems that traditional algorithms or computational methods find difficult to solve. A BP neural network is a mature nonlinear mapping method for solving real-world problems. In addition, it has the ability to classify any complex pattern and can solve XOR problems that simple sensors cannot solve. A BP neural network does not require mathematical relationships to determine the mapping relationship between input and output. It only imports input and output data into the training set, trains itself, finds exercises between the input and output, establishes a model, and finally imports input values into the model to obtain the latest expected output values.

The transmission process of the BP neural network consists of two parts, namely forward transmission and backward transmission. In the forward transmission process, data are processed layer by layer from the entry layer to the hidden layer and, then, transferred to the output layer. The condition of each level of neuron directly affects the condition of the next level of neuron. Assuming that the expected output value cannot be obtained at the output layer, it enters the opposite direction of transmission, regresses the bias information along the original interface channel, and adjusts the weights of each neuron to minimize the bias information.

This paper utilizes the principle of the BP neural network algorithm and uses MATLAB R2021b to train the neural network for AD value and temperature. The comparison results of the algorithms are shown in Table 2. In this experiment, the quantify conjugate gradient algorithm is selected as the fitting algorithm [39].

**Table 2.** Comparison of training algorithms.

| Training Algorithm | Layers | MSE | R |
|---|---|---|---|
| Levenberg–Marquardt | 12 | $1.4808 \times 10^6$ | 0.9775 |
| Bayesian regularization | 14 | $1.4032 \times 10^6$ | 0.9767 |
| Quantify conjugate gradient | 4 | $1.1334 \times 10^6$ | 0.9821 |

A total of 350 sets of data were randomly selected from the 500 sets of data collected above for training. The training/testing/validation ratio was 7:1.5:1.5, and a BP neural network was used to train the data. The hidden layer was set to 4, and 41 repeated training sessions were conducted. The training situation is shown in Figures 12 and 13. The best performance was obtained in the 35th round of training.

The training, validation, and testing of the neural network are shown in Figure 14.

The comparison after neural network training is shown in Figure 15. After calculation, the standard deviation of static noise is 5686.011, and the average absolute deviation is 4850.8.

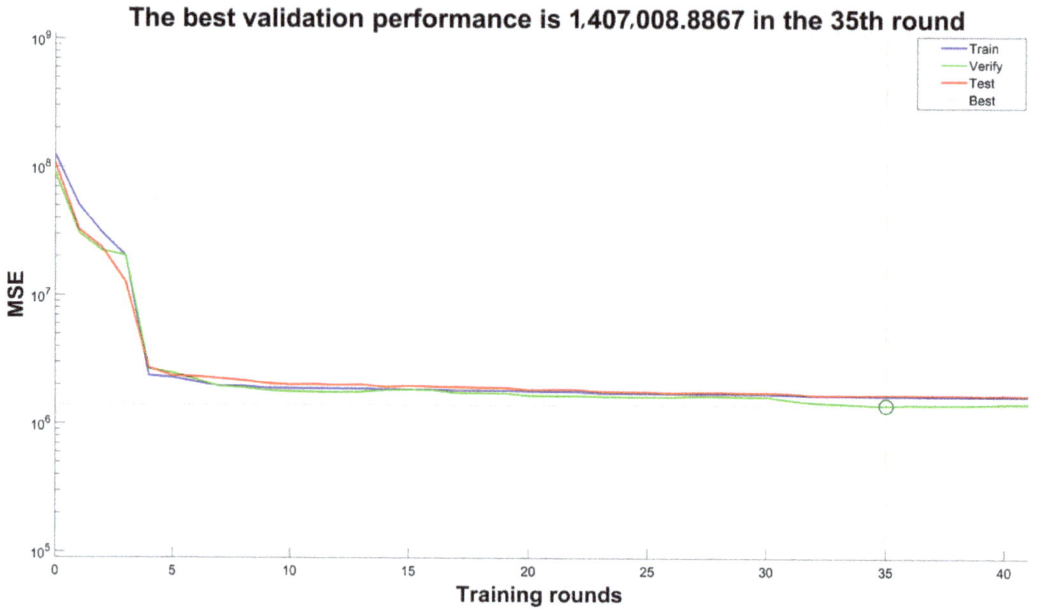


**Figure 12.** Training status of neural networks.

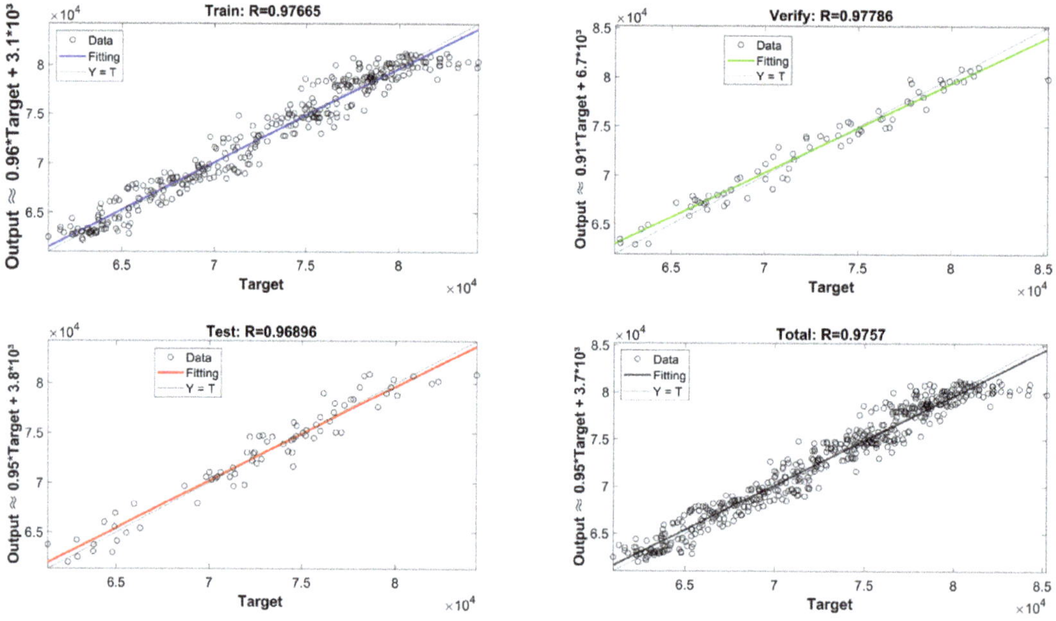


**Figure 13.** Neural network regression situation.

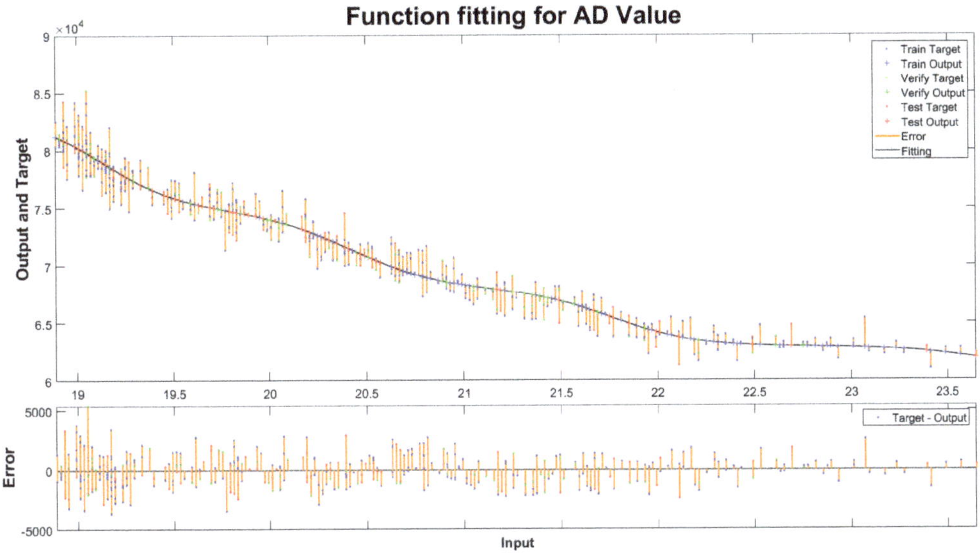


**Figure 14.** Neural network fitting situation.

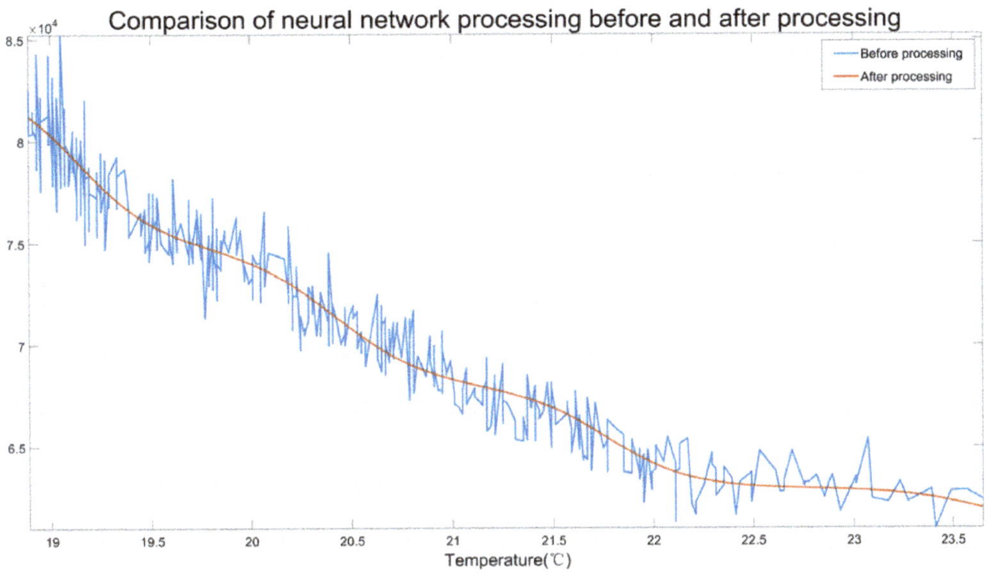


**Figure 15.** Comparison of neural network processing before and after processing.

## 5. Experimental Verification

According to the National Environmental Protection Standard HJ 1044-2019 [40] of the People's Republic of China, when the concentrations of methane, nitric oxide, and hydrogen sulfide in the sample air are 2155 ppb, 123 ppb, and 6939 ppb, respectively, the determination results of sulfur dioxide will be affected by 3 ppb, 3 ppb, and no more than 1 ppb. When aromatic hydrocarbons are present in the sample air, it can affect the measurement results and can be removed through a hydrocarbon remover. Under normal circumstances, the interference of the aforementioned gases in the ambient air can be ignored. In terms of gas configuration, the TR1BKD dynamic calibration instrument from Zhongke Tianrong (Beijing) Technology Co., Ltd. (Beijing, China) was used, which can configure standard gases mixed with nitrogen and sulfur dioxide with a resolution of 0.01 ppb. Different concentrations of sulfur dioxide gas were gradually introduced, and data were obtained. In Figure 16, two fitted data represent the detection accuracy before and after processing, respectively. This indicates that the detection device proposed in this article has good linear detection accuracy.

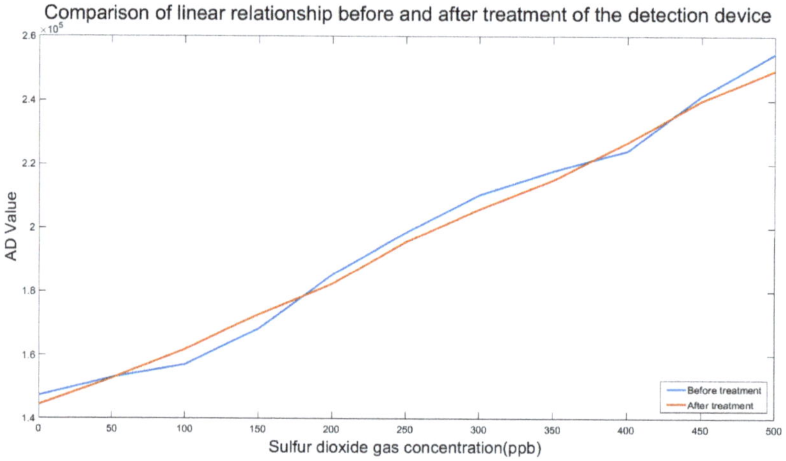


**Figure 16.** Fitting curves for different sulfur dioxide concentrations.

It is also necessary to consider the working conditions in the actual environment. Select five sets of standard gases with different concentrations; feed them into the detector for multiple sampling, and observe their stability. The results are shown in Figure 17. The coefficient of variation is a commonly used statistical indicator in statistics. It is a relative statistical measure of data dispersion, mainly used to compare the dispersion of sample data, using the ratio of standard deviation to mean for comparison. After calculation, the dispersion coefficients of 0 ppb, 100 ppb, 200 ppb, 300 ppb, 400 ppb, and 500 ppb after treatment are 0.2656%, 0.2468%, 0.2895%, 0.3334%, 0.3715%, and 0.3156%, respectively, proving the good reliability of the detector.

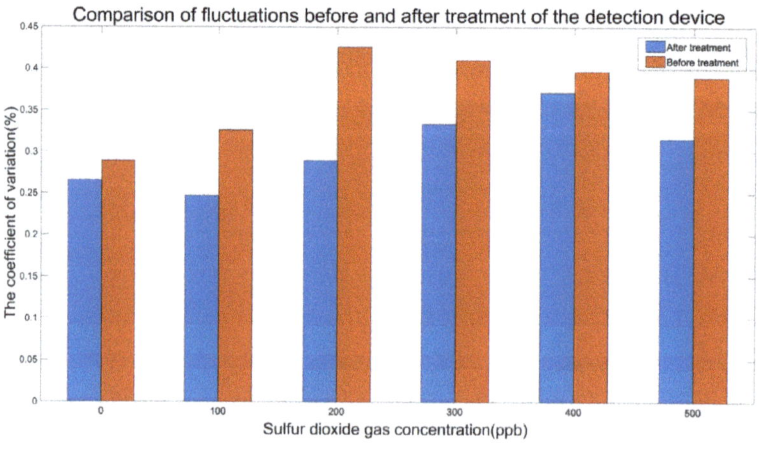


**Figure 17.** Stability test of the same sulfur dioxide concentration.

## 6. Conclusions

This paper develops a sulfur dioxide concentration detection device based on PMT. Using wavelet transform to denoise PMT signals and using a BP neural network algorithm to denoise the temperature drift phenomenon of PMT, the signal-to-noise ratio is improved. After verification, the detection results are close to the actual values, with good resolution and reliability. In the future, it may be possible to apply Unscented Kalman Filters (UKFs) for multi-sensor data fusion to further improve the performance of detection device.

**Author Contributions:** Conceptualization, H.Z. and Y.Z. (Yadong Zhao); methodology, J.L. (Jianshen Li); software, L.Y.; validation, J.G., H.Z. and Y.Z. (Yadong Zhao); formal analysis, Y.Z. (Yunhan Zhang); investigation, J.L. (Jianshen Li); resources, J.L. (Jiehui Liu); data curation, J.L. (Jiehui Liu); writing—original draft preparation, Y.Z. (Yunhan Zhang); writing—review and editing, L.Y.; visualization, J.G.; supervision, H.Z.; project administration, J.L. (Jiehui Liu); funding acquisition, J.L. (Jianshen Li). All authors have read and agreed to the published version of the manuscript.

**Funding:** This research received no external funding.

**Institutional Review Board Statement:** Not applicable.

**Informed Consent Statement:** Not applicable.

**Data Availability Statement:** The original contributions presented in the study are included in the article, further inquiries can be directed to the corresponding author.

**Acknowledgments:** Thanks to China Energy Saving Tianrong Technology Co., Ltd. for providing technical guidance and experimental equipment support for this study.

**Conflicts of Interest:** The authors declare no conflicts of interest.

## eferences


Dong, L.; Tittel, F.K.; Li, C.; Sanchez, N.P.; Wu, H.; Zheng, C.; Yu, Y.; Sampaolo, A.; Griffin, R.J. Compact TDLAS based sensor design using interband cascade lasers for mid-IR trace gas sensing. *Opt. Express* **2016**, *24*, A528–A535. [CrossRef] [PubMed]

Liu, C.; Xu, L.; Chen, J.; Cao, Z.; Lin, Y.; Cai, W. Development of a fan-beam TDLAS-based tomographic sensor for rapid imaging of temperature and gas concentration. *Opt. Express* **2015**, *23*, 22494–22511. [CrossRef] [PubMed]

Romanini, D.; Kachanov, A.; Sadeghi, N.; Stoeckel, F. CW cavity ring down spectroscopy. *Chem. Phys. Lett.* **1997**, *264*, 316–322. [CrossRef]

Wheeler, M.D.; Newman, S.M.; Orr-Ewing, A.J.; Ashfold, M.N. Cavity ring-down spectroscopy. *J. Chem. Soc. Faraday Trans.* **1998**, *94*, 337–351. [CrossRef]

Engeln, R.; Berden, G.; Peeters, R.; Meijer, G. Cavity enhanced absorption and cavity enhanced magnetic rotation spectroscopy. *Rev. Sci. Instrum.* **1998**, *69*, 3763–3769. [CrossRef]

Gupta, M.; Jiao, H.; O'keefe, A. Cavity-enhanced spectroscopy in optical fibers. *Opt. Lett.* **2002**, *27*, 1878–1880. [CrossRef] [PubMed]

Rodgers, M.O.; Asai, K.; Davis, D.D. Photofragmentation-laser induced fluorescence: A new method for detecting atmospheric trace gases. *Appl. Opt.* **1980**, *19*, 3597–3605. [CrossRef] [PubMed]

Waclawek, J.P.; Moser, H.; Lendl, B. Balanced-detection interferometric cavity-assisted photothermal spectroscopy employing an all-fiber-coupled probe laser configuration. *Opt. Express* **2021**, *29*, 7794–7808. [CrossRef] [PubMed]

Zhao, P.; Zhao, Y.; Bao, H.; Ho, H.L.; Jin, W.; Fan, S.; Gao, S.; Wang, Y.; Wang, P. Mode-phase-difference photothermal spectroscopy for gas detection with an anti-resonant hollow-core optical fiber. *Nat. Commun.* **2020**, *11*, 847. [CrossRef]

). Pinto, D.; Moser, H.; Waclawek, J.P.; Russo, S.D.; Patimisco, P.; Spagnolo, V.; Lendl, B. Parts-per-billion detection of carbon monoxide: A comparison between uartz-enhanced photoacoustic and photothermal spectroscopy. *Photoacoustics* **2021**, *22*, 100244. [CrossRef]

. Okabe, H.; Paul, L.S.; Joseph, J.B. Ambient and Source $SO_2$ Detector Based on a Fluorescence Method. *J. Air Waste Mang. Assoc.* **1973**, *23*, 514–516.

2. Bradshaw, J.D.; Rodgers, M.O.; Davis, D.D. Single-photon laser-induced fluorescence detection of NO and $SO_2$ for atomospheric conditions of composition and pressure: Errata. *Appl. Opt.* **1982**, *21*, 2493–2500. [CrossRef] [PubMed]

3. Somesfalean, G.; Zhang, Z.G.; Sjöholm, M.; Svanberg, S. All-diode-laser ultraviolet absorption spectroscopy for sulfur dioxide detection. *Appl. Phys. B* **2005**, *80*, 1021–1025. [CrossRef]

14. Matsumi, Y.; Shigemori, H.; Takahashi, K. Laser-induced fluorescence instrument for measuring atmospheric $SO_2$. *Atmos. Enviror* **2005**, *39*, 3177–3185. [CrossRef]
15. Dorosz, P.; Baszczyk, M.; Glab, S.; Kucewicz, W.; Mik, L.; Sapor, M. Silicon photomultiplier's gain stabilization by bias correctioı for compensation of the temperature fluctuations. *Nucl. Instrum. Methods Phys. Res. Sect. A Accel. Spectrometers Detect. Assoc Equip.* **2013**, *718*, 202–204. [CrossRef]
16. Kuznetsov, E. Temperature-compensated silicon photomultiplier. *Nucl. Instrum. Methods Phys. Res. Sect. A Accel. Spectrometer Detect. Assoc. Equip.* **2018**, *912*, 226–230. [CrossRef]
17. Dong, X.; Li, G.; Jia, Y.; Li, B.; He, K. Non-iterative denoising algorithm for mechanical vibration signal using spectral grapl wavelet transform and detrended fluctuation analysis. *Mech. Syst. Signal Process.* **2021**, *149*, 107202. [CrossRef]
18. Dong, Y.; Fu, Z.; Peng, Y.; Zheng, Y.; Yan, H.; Li, X. Precision fertilization method of field crops based on the Wavelet-BP neura network in China. *J. Clean. Prod.* **2020**, *246*, 118735. [CrossRef]
19. Ahmad, S.; Blin, S.; Callier, S.; Cizel, J.B.; Conforti, S.; de La Taille, C.; Dulucq, F.; Fleury, J.; Martin-Chassard, G.; Raux, L.; et a OMEGA SiPM readout ASICs. *Nucl. Instrum. Methods Phys. Res. Sect. A Accel. Spectrometers Detect. Assoc. Equip.* **2021**, *986*, 16462 [CrossRef]
20. Paredes, B.L.; Araújo, H.M.; Froborg, F.; Marangou, N.; Olcina, I.; Sumner, T.J.; Taylor, R.; Tomás, A.; Vacheret, A. Response o photomultiplier tubes to xenon scintillation light. *Astropart. Phys.* **2018**, *102*, 56–66. [CrossRef]
21. Wang, L.; Zhang, Y.; Zhou, X.; Qin, F.; Zhang, Z. Optical sulfur dioxide sensor based on broadband absorption spectroscopy iı the wavelength range of 198–222 nm. *Sens. Actuators B Chem.* **2017**, *241*, 146–150. [CrossRef]
22. Liu, D.; Ma, M.; Zhang, Y. Research of pulsed Xenon lamp power supply based on STM32. *J. Phys. Conf. Ser.* **2021**, *1754*, 01201€ [CrossRef]
23. Holík, M.; Ahmadov, F.; Ahmadov, G.; Akbarov, R.; Berikov, D.; Mora, Y.; Nuruyev, S.; Pridal, P.; Sadigov, A.; Sadygov, Z.; et a Miniaturized read-out interface "Spectrig MAPD" dedicated for silicon photomultipliers. *Nucl. Instrum. Methods Phys. Res. Sec A Accel. Spectrometers Detect. Assoc. Equip.* **2020**, *978*, 164440. [CrossRef]
24. Serrano-Finetti, E.; Casas, O.; Pallàs Areny, R. Common mode electronic noise in differential circuits. *Measurement* **2019**, *14C* 207–214. [CrossRef]
25. Lan, T.; Zeng, Z.; Han, L.; Zeng, J. Seismic Data Denoising Based on Wavelet Transform and the Residual Neural Network. *App Sci.* **2023**, *13*, 655. [CrossRef]
26. Golestani, A.; Kolbadi SM, S.; Heshmati, A.A. Localization and de-noising seismic signals on SASW measurement by wavele transform. *J. Appl. Geophys.* **2013**, *98*, 124–133. [CrossRef]
27. Kabir, M.A.; Shahnaz, C. Denoising of ECG signals based on noise reduction algorithms in EMD and wavelet domains. *Biomea Signal Process. Control* **2012**, *7*, 481–489. [CrossRef]
28. Xu, X.; Luo, M.; Tan, Z.; Pei, R. Echo signal extraction method of laser radar based on improved singular value decompositior and wavelet threshold denoising. *Infrared Phys. Technol.* **2018**, *92*, 327–335. [CrossRef]
29. Patil, R. Noise Reduction using Wavelet Transform and Singular Vector Decomposition. *Procedia Comput. Sci.* **2015**, *54*, 849–85: [CrossRef]
30. Liu, Z.; Feng, X.; Dong, C.; Jiao, M. Study on Denoising Method of Photoionization Detector Based on Wavelet Packet Transforr *Chemosensors* **2023**, *11*, 146. [CrossRef]
31. Hu, M.; Zhang, S.; Dong, W.; Xu, F.; Liu, H. Adaptive denoising algorithm using peak statistics-based thresholding and nove adaptive complementary ensemble empirical mode decomposition. *Inf. Sci.* **2021**, *563*, 269–289. [CrossRef]
32. Dorosz, P.; Baszczyk, M.; Kucewicz, W.; Mik, L. Real-time measurement system with automatic gain detection and autocalibratior for silicon photomultipliers. *Nucl. Instrum. Methods Phys. Res. Sect. A Accel. Spectrometers Detect. Assoc. Equip.* **2019**, *936*, 574–57 [CrossRef]
33. Ginzburg, D.; Kopeika, N.; Paran, J.; Cohen-Zada, I.; Ghelman, M.; Pushkarsky, V.; Marcus, E.; Manor, A.; Mazor, T.; Kadmon, Y et al. Optimizing the design of a silicon photomultiplier-based radiation detector. *Nucl. Instrum. Methods Phys. Res. Sect. A Acce Spectrometers Detect. Assoc. Equip.* **2011**, *652*, 474–478. [CrossRef]
34. Peng, B.; Gao, C.; Zhou, Y.; Guo, Y. Temperature-compensated ppb-level sulfur dioxide detection system based on fourie transform ultraviolet differential optical absorption spectrum method. *Sens. Actuators B Chem.* **2020**, *312*, 127988. [CrossRef]
35. Safari, S.; Shabani, F.; Simon, D. Multirate multisensor data fusion for linear systems using Kalman filters and a neural network *Aerosp. Sci. Technol.* **2014**, *39*, 465–471. [CrossRef]
36. Chong, S.; Rui, S.; Jie, L.; Xiaoming, Z.; Jun, T.; Yunbo, S.; Jun, L.; Huiliang, C. Temperature drift modeling of MEMS gyroscop based on genetic-Elman neural network. *Mech. Syst. Signal Process.* **2016**, *72–73*, 897–905. [CrossRef]
37. Li, J.; Yao, X.; Wang, X.; Yu, Q.; Zhang, Y. Multiscale local features learning based on BP neural network for rolling bearing intelligent fault diagnosis. *Measurement* **2020**, *153*, 107419. [CrossRef]
38. Ma, W.; Li, Q.; Li, J.; Ding, L.; Yu, Q. A method for weighing broiler chickens using improved amplitude-limiting filterin algorithm and BP neural networks. *Inf. Process. Agric.* **2021**, *8*, 299–309. [CrossRef]

. Chen, C.; Chen, Q.; Li, G.; He, M.; Dong, J.; Yan, H.; Wang, Z.; Duan, Z. A novel multi-source data fusion method based on Bayesian inference for accurate estimation of chlorophyll-a concentration over eutrophic lakes. *Environ. Model. Softw.* **2021**, *141*, 105057. [CrossRef]
. *HJ 1044-2019*; Ambient Air—Automatic Determination of Sulfur Dioxide—Ultraviolet Fluorescence Method. Springer Nature: Cham, Switzerland, 2019.

machines

MDPI

Article

# Young's Modulus and Hardness Identification of Extruded Aluminum by Scratching Damper

Chun-Nam Wong [1], Anand Vyas [2], Wai-On Wong [1,*] and Ruqi Sun [3]

[1] Department of Mechanical Engineering, The Hong Kong Polytechnic University, Hong Kong, China; chun-nam.wong@connect.polyu.hk
[2] Division of Science, Engineering and Health Studies, CPCE, The Hong Kong Polytechnic University, Hong Kong, China; anand.vyas@cpce-polyu.edu.hk
[3] School of Automobile, Chang'an University, Xi'an 710064, China; sunruqi@chd.edu.cn
* Correspondence: wai.on.wong@connect.polyu.hk

**Abstract:** A special vibration damper is proposed for Young's modulus and hardness identification through a scratching process on extruded aluminum. This paper presents the design and working principle of a scratching damper based on a scratching device. A non-contact electromagnetic shaker is used to generate the shaking force for test sample vibration. The required forces on the scratched material during the scratching process are generated by an adjustable compression spring. The proposed damper is designed and tested on an extruded aluminum 3004 sample for the determination of its Young's modulus and hardness, and validation is performed using the standard test instruments. The physical dimensions of the scratching tracks are measured using a microscope and utilized to compute the scratching energy factor. Load curves are obtained at different divisions of the scratching process. The loop energy during the scratching process of the tested object is measured and used for the determination of sample material properties. Furthermore, the energy conservation law, scratch energy release rate of semi-conical scratch head, and loop energy release rate are established to determine the Young's modulus and hardness of the sample. Their estimation accuracies are evaluated. The proposed method has several advantages over the traditional methods, including low cost, directness, and high repeatability. The results suggest this to be used as an alternative to the standard modulus and hardness tester.

**Keywords:** scratching damper; energy release rate; scratch energy conservation; extruded aluminum 3004; Brinell hardness

## 1. Introduction

In order to acquire the material property via a scratch test, different experimental processes and analytical methods have been implemented. Kamplade and Biermann [1] utilized scratch tests to analyze the grain performance while grinding thermoplastics. Corundum grains were used in two different scratch test set-ups for fundamental analysis of the material removal process. The scratching process was evaluated according to the relative material removal volumes of the scratch grooves. In addition to that, the forces occurring during the process and the required specific scratch energy were analyzed. Scratch tests using a Rockwell-C diamond cone indenter were reported by Varga et al. [2], where the test load was induced with a slip-stick mechanism. Line scars were produced on the specimen after scratching. Scratch topographies and average profiles for increasing the scratch load were studied, and their cross-sections were analyzed. Moreover, studies on numerical simulations of the scratch process have also been reported using smooth particle hydrodynamics [3].

The energy required to create a unit fracture surface was computed by Akono and Ulm [4] using an energetic contour-independent J-integral. The energy release rate was linked to stress intensity factors in mode I and II via the Griffith–Irwin equation [5]. Their equation on mode I failure coincided with our plain strain analysis. Finite element meshing was performed on crack tip. Using the fracture criterion, the forces and tool geometry were linked to the plane strain fracture toughness [6]. Cross-sectional investigations were carried out and analyzed. Moreover, it is reported [7] that the elastic modulus and hardness of silicon carbide particle-reinforced aluminum matrix composites, SiCp/Al, were determined through the indentation test by loading onto an Al matrix and SiC particles generating three material phases during the deformation process. More complicated effects were reflected in the applied load scratch depth (P-h) curves. A relative accurate result is obtained from a single-phase material in our test. In ultrasonic vibration-assisted machining tests, precision or ultra-precision machining was combined with micro-scale (1–15 μm) and high-frequency (16–40 kHz) tool or workpiece vibration. In this paper, the authors use the test dimension at the millimeter scale up to several mm and lower frequency around 10 Hz.

A scratch test on glass ceramics [8] has shown that crack propagation increases linearly with the material removal volume. Wakeel and Hubler [9] introduced heterogeneity into the micro-scratch test fracture. A similar formulation for the J-integral is utilized in our scratch energy derivation. Akono and Ulm [10] established the fracture scaling relations for scratch tests of axisymmetric shape of a scratch indenter in which the equivalent force is composed of the scratch force and indentation force. Kamplade and Biermann [1] examined the material removal of unreinforced, thermoplastic polymers by scratch tests to analyze the grain engagement. A helical scratch groove was produced using a single grain. Also, the indentation force and specific scratch energy were measured at a specific wheel speed. In this damper test, multiple averaged tracks were formed instead of a more irregular single groove. Smooth finishing on the sample surface was achieved. Measurements of scratch dimensions, such as scratch depth, contact length, and removal volume, were enhanced.

To determine the Young's modulus of aluminum, ultimate stress, Young's modulus, and the ultimate strains of normal and high strength were considered for aluminum 6063-T5 and 6061-T6 at elevated temperatures [11]. Their ratios were related systematically to the temperature ratio using the Ramberg and Osgood equation [12]. Janeczek and Fydrych [13] compared the effect of the shape of a tool on the joint to obtain the values of the friction stir welding parameters. A tensile strength test of aluminum alloy 3004 was conducted. The influence of the tool pin shape on the contact intensity was studied. The hardness of the forged precipitation-hardened aluminum alloy EN AW-2618A for different aging times and temperatures was obtained by Rockenhäuser et al. [14]. The Brinell hardness number was computed by DIN EN 6506-1 [15].

In this investigation, Young's modulus and the hardness of aluminum 3004 are estimated using scratching theory. In a traditional scratch test, a static load is applied to obtain the fracture surface when the test is conducted one time in a certain direction and is not repeatable. There might be unstable indenter movement along the path, depending on the material–indenter interaction. In our case, cyclic vibration load was used instead, generating a repeatable load-path loop. As a result, a stable smooth fracture surface was created. This generated refined sub-layer scratching within the test material, facilitating in-depth property analysis. Moreover, in the computation of the scratch damping ratio, the load profile at scratch path, test speed, and scratching force at indenter tip could be accurately measured by vibration sensors. Furthermore, the indenter was changeable from a round-head shape for damping motion to a pin-head shape for significant scratching motion.

## 2. Design and Working Principle of Scratching Damper

An innovative damper was modified from the existing EMSD-RCFD DVA developed by Sun, Wong, and Cheng [16]. A divisional scratching test was conducted through scratch damping, where the required load could be adjusted by dial knot divisions. The test-rig schematic of the scratching damper is illustrated in Figure 1a. A non-contact electromagnetic exciter was mounted on top of the mounting frame that provided the excitation force without introducing any additional stiffness to the dynamic system, as shown in Figure 1b. The exciter coil was fixed on the mounting frame, while permanent magnets were fixed on the magnet stringer rob, which can be regarded as part of the system mass. The mass, stiffness, and natural frequency of the damper system were $M = 6.87$ kg, $K = 20.961$ N/mm, and $ft = 9$ Hz respectively.

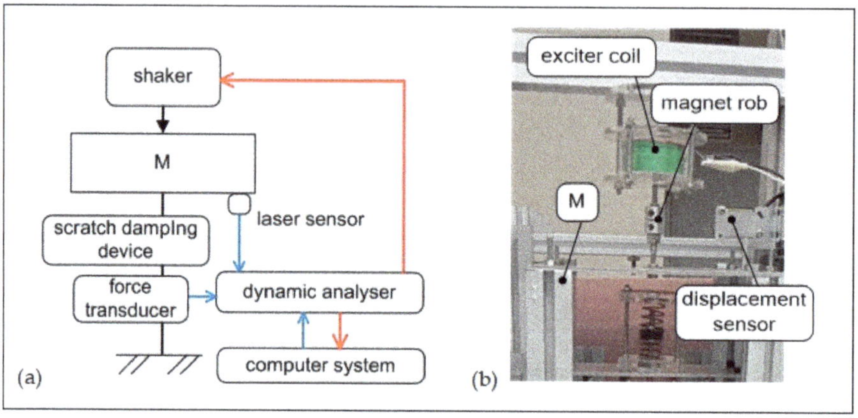


**Figure 1.** (**a**) Equipment schematic of scratching damper; (**b**) electromagnetic exciter and its mounting to system mass. The blue and red arrows show the input and output signals to the dynamic analyser, respectively.

The linear guide was composed of four rotation bearings on each side face, and the smooth glassy surface for bearing motion allowed for a reduction in the parasitic damping of the damper system. To measure the damper force, a force sensor was fastened to the damper rob, which was grounded to a steel table (Figure 2a). A laser sensor was used to measure the absolute displacement of the system's mass. A nylon tube, with a pre-compressed spring inside, was fixed onto the tuning mechanism of the scratching device to provide a rough linear guide for the mass vibration. The data acquisition and signal generation were conducted with the B&K PULSE 7767 system (Figure 1a). The driving signal to the exciter was amplified by the B&K 2712 power amplifier before being sent to the non-contact exciter. The device was installed in the middle part of the system.

Under the principle of scratch energy conservation, divisional scratching energy was computed by the energy equation within the same interval test:

$$U^{ji} = \wp_j^i + \Psi^{ji} \quad (1)$$

where $U^{ji}$ is $j$ divisional total energy of scratching motion at the $i$th interval, $\Psi^{ji}$ is $j$ divisional energy of the damping motion, $\wp_j^i = F_v^{ji} \cdot v_j^i$ is $j$ divisional scratching energy, $F_v^{ji}$ is $j$ divisional scratching energy factor, and $v_j^i$ is $j$ divisional material removal volume. Applying the energetic contour-independent $J$-integral [4,9], the scratch energy release rate, which is the energy required to create a unit scratching surface, is

$$\wp = \frac{1}{p}\int_S \left(\Psi n_x - \tau_x \frac{\partial U_x}{\partial x}\right) dS \quad (2)$$

where $\Psi = \frac{\kappa}{2E}\sigma_{xx}^2$ is the free energy density with $\kappa = 1 - \nu^2$, $\nu$ is Poisson's ratio, $\frac{\partial U_x}{\partial x}$ is the displacement gradient, $p$ is the scratch head perimeter, and $\tau_x = \sigma_{xx} n_x$ is the surface traction. $ds$ is the differential line element and $dS$ is the surface element, related as

$$ds = \sqrt{1 + \cot\theta^2},\ dS = r d\phi ds \quad (3)$$

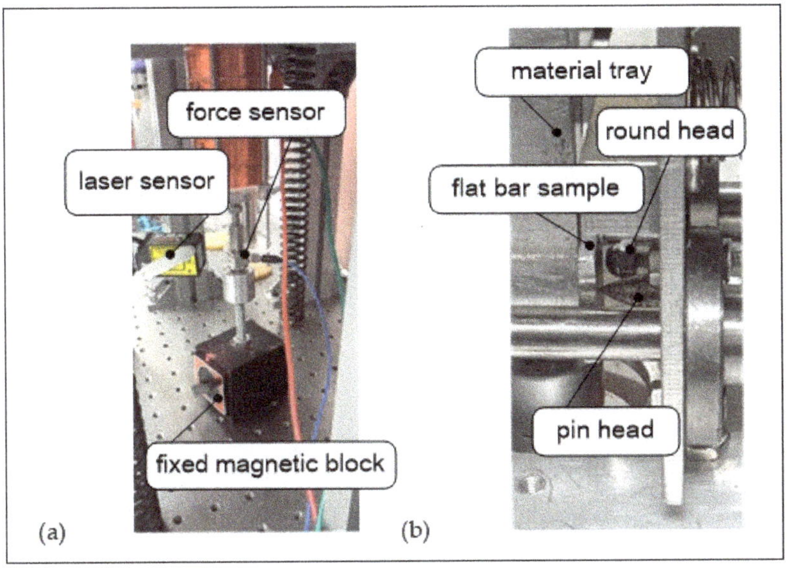


**Figure 2.** (**a**) Force transducer assembly of damper rob and magnetic block of scratching damper; (**b**) assembly components of scratching device in scratching damper.

For tangential track generation from the scratch head, the Airy stress function was established as [10]

$$\Gamma(x,z) = -ax\left(z^3 - \frac{3}{4}zd^2\right) + bz^2 \tag{4}$$

where $a$ and $b$ are constants. Its symmetric stress tensors were computed by

$$\sigma_{zz} = \frac{\partial^2\Gamma}{\partial z^2} = -6axz + 2b \tag{5}$$

$$\sigma_{xz} = \frac{\partial^2\Gamma}{\partial x \partial z} = a\left(3z^2 - \frac{3}{4}d^2\right) \tag{6}$$

$$\sigma_{zz} = \frac{\partial^2\Gamma}{\partial x^2} = 0 \tag{7}$$

The displacement gradient at the scratch head–material interface was determined by the potential function:

$$\Theta = -\frac{a}{2}\left[-\frac{x^4 + z^4}{2} + 3(xz)^2\right] + 2bxz \tag{8}$$

which satisfies $\Theta_{,zz} = 0$, $\Theta_{,xz} = \sigma_{xx}$. Afterwards, the displacement gradient is calculated by:

$$\frac{\partial U_x}{\partial x} = \frac{1}{E}\left[-(1+v)\frac{\partial^2\Gamma}{\partial x^2} + \kappa\frac{\partial^2\Theta}{\partial z \partial x}\right] = -6axz + b = \frac{\kappa}{E}\sigma_{xx} \tag{9}$$

Substituting this to Equation (2), one can obtain

$$\wp = \frac{1}{p}\int_S -\frac{\kappa}{2E}\sigma_{xx}^2 n_x dS = \frac{\kappa}{2pE}\int_A \sigma_{xx}^2 dA' \tag{10}$$

where $E$ is Young's modulus and $p$ and $A$ are the scratch head perimeter and projected contact area, respectively. On the other hand, the scratch force is given by:

$$F_t = \int_S (\sigma_{xx}n_x + \sigma_{xz}n_z)dS = \int_A \sigma_{xx}dA',\ F_n = -\int_S \sigma_{xz}n_x dS \tag{11}$$

Substituting Equations (5)–(7) into Equation (11), $a$ and $b$ are determined to be

$$a = -\frac{2\cot\theta F_n}{d^4},\ b = \frac{\cot\theta F_t}{2d^2} \tag{12}$$

Assuming that the stress field over the scratch head is constant, its linear stress average is related to the quadratic stress average on the projected contact area as the integral in Equation (11):

$$\int_A \sigma_{xx}^2 dA' \simeq \left(\int_A \sigma_{xx}dA'\right)^2 / \int_A dA' = \frac{F_t^2}{A} \tag{13}$$

Hence, we have the scratch energy release rate, created as:

$$\wp = \frac{1-\nu^2}{E}\frac{F_t^2}{2pA} \tag{14}$$

The semi-conical scratch head shape function, defined by Figure 3, can be generated by:

$$z = r\cot\theta \tag{15}$$

in which $r$ is the radius of the head, $z \subset [0, d]$ is the height, and $\theta$ is the conical angle. Firstly, $p$ projected onto the scratch direction is generated from Equation (2):

$$p = \int ds = \int_0^{d/\cot\theta} \sqrt{1+\cot\theta^2}dr = \frac{d}{\cot\theta}\sqrt{1+\cot\theta^2} = \frac{d}{\cos\theta} \tag{16}$$

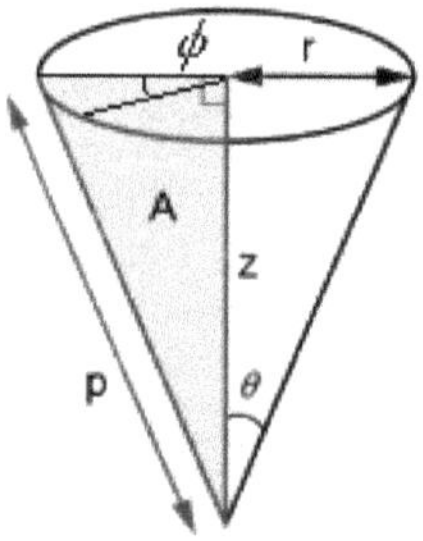


**Figure 3.** Dimensions of semi-conical scratch pin head.

Secondly, the scratch surface projected in the scratch direction becomes:

$$A = \int n_x dS = \int \hat{n}\cdot e_x r\sqrt{1+\cot\theta^2}d\phi dr \tag{17}$$

where $\hat{n} = \frac{\cot\theta\cos\phi}{\sqrt{1+\cot\theta^2}}e_r + \frac{1}{\sqrt{1+\cot\theta^2}}e_z$. For the semi-conical head, the cylindrical angle is $\phi \subset [0, \pi/2]$. Therefore, $A$ can be computed using the cylindrical co-ordinate as:

$$A = \int_0^{d/\cot\theta}\int_0^{\pi/2} \cot\theta(\cos\phi r)d\phi dr = \frac{\cot\theta}{2}\left[\sin\phi|_0^{\pi/2}\right]\left[r^2\Big|_0^{d/\cot\theta}\right] = \frac{d^2\tan\theta}{2} \tag{18}$$

Substituting Equations (16) and (18) into Equation (14), one obtains the unique $\wp$ for semi-conical pin head:

$$\wp = \frac{1-\nu^2}{E}\frac{F_t^2}{d^3 \sin\theta} \tag{19}$$

## 3. Results and Discussion

### *3.1. Young's Modulus of Extruded Aluminum Flat Bar*

Upon the scratch test of the extruded aluminum 3004, a flat bar sample, of the size 15(L) × 10(W) × 4(T) in mm, as shown in Figure 2b, was prepared. Special machining was not required. And the preparation was simple, consisting of cutting and fitting. Using the scratching damper, the following tests in two to four dial knot divisions of the compression spring were conducted to investigate its material properties (such as Young's modulus and hardness), as shown in Figure 4.

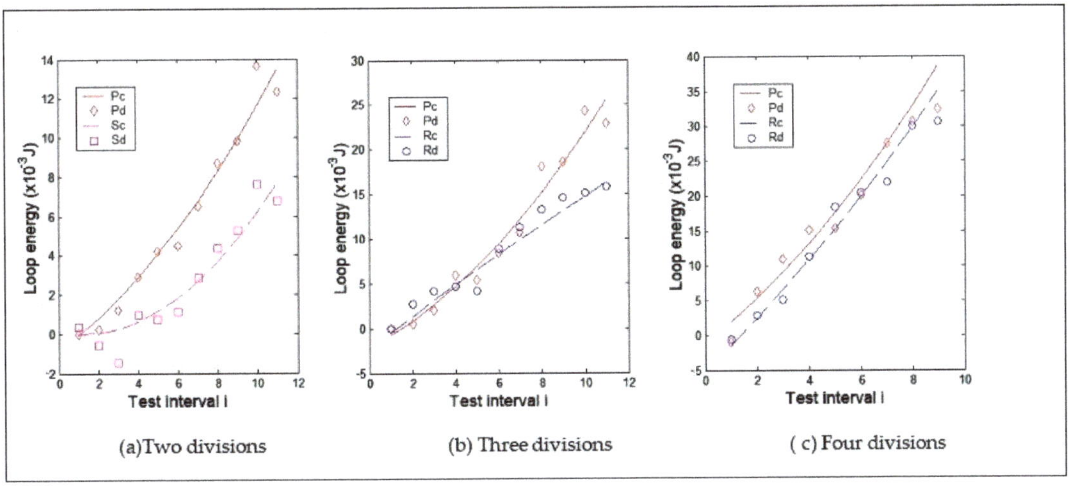


**Figure 4.** Divisional loop energy curves of extruded aluminum (— Pc ◇ Pd curve fit and test data of $U^j_{pin}$, — Rc ○ Rd curve fit and test data of $\Psi^j_{round}$, — Sc □ Sd curve fit and test data of $\wp_j$).

These load curves indicate that the loop energy increased consecutively with the interval load [5–8]. The loop energy is the area integration of the $F_t$- displacement loop measured by the force transducer and laser sensor in Figure 2a. Figure 4a complies with Akono and Ulm [10] in that the scratching energy release rate increased with the test interval in a quadratic curve [1,11]. Meanwhile, the damping energy increased with the test interval in a linear curve, as shown in Figure 4a,b. The summation of loop energy per interval with the conical pinhead is the divisional total energy, expressed as:

$$U^j_{pin} = \sum_{i=1}^{N} U^{ji}_{pin} \cdot j \cdot f_t \cdot l_t/2 \tag{20}$$

where $U^{ji}_{pin}$ is the single-loop energy with the pin head, $i$ is the number of the test interval, $j$ is the dial division of the test, $f_t$ = 9 Hz is the fixed-sine test frequency, and $l_t$ = 48 s is the test period. It is interesting to observe that the shape of the loop is highly repeatable within the same test period, with nearly 500 cycles. Meanwhile, the summation of the loop damping energy per interval with the round head is expressed as:

$$\Psi^{j}_{round} = \sum_{i=1}^{N} \Psi^{ji}_{round} \cdot j \cdot f_t \cdot l_t/2 \tag{21}$$

where $\Psi^{ji}_{round}$ is the single-loop energy with the round head. According to Equation (1), the scratching energy per interval with *j* divisions is

$$\wp_j = U^{j}_{pin} - \Psi^{j}_{round} \tag{22}$$

The divisional scratching energy release rate is

$$\wp_j = \sum_{i=1}^{i=N} \frac{\wp^{i}_{j}}{N \cdot j} \tag{23}$$

Meanwhile, according to Equation (14), the divisional Young's modulus stands as:

$$E_j = \frac{1 - v^2}{\wp_j} \times \frac{Feq^2_j}{2p_j A_j} \tag{24}$$

with notations in *j* divisions. The results from two to four divisions are given in Table 1. In general, their values decreased slightly with the scratch performance. Two divisions were the best at 29.5%. Thus, through the two-divisional test, the *i*th interval $E^{i}_{2}$ was generated. We observed that it increased to a maximum, then dropped steadily to a stable value. Hence, from the *N*th data, the ultimate Young's modulus was obtained as $E^{N}_{2} = 54.6$ GPa.

**Table 1.** Extruded aluminum material property comparison from scratching test.

| Data\Dial Division | 2 | 3 | 4 |
|---|---|---|---|
| Normal force $F^{j}_{n}$ (N) | 0.16 | 0.24 | 0.32 |
| Total energy $U^{j}$ (J) | 35.2 | 56.5 | 78.8 |
| Damping motion energy $\Psi^{j}$ (J) | 24.8 | 47.8 | 70.3 |
| Scratch energy $\wp_j$ (J) | 10.4 | 8.69 | 8.51 |
| Scratch performance $\wp_j/U^{j}$ (%) | 29.5 | 15.4 | 10.8 |
| Young's modulus $E_j$ (GPa) | 54.6 | 56.2 | 56.8 |
| Hardness $H^{B}_{j}$ | 73.4 | 76.1 | 77.5 |
| Scratch depth $D_j$ (mm) | 0.193 | 0.176 | 0.180 |
| Scratch length $L^{j}$ (mm) | 3.38 | 3.10 | 2.92 |
| Scratch volume $V^{j}$ (mm$^3$) | 0.183 | 0.160 | 0.141 |
| Scratch energy factor $F^{j}_{v}$ (GJ/m$^3$) | 56.8 | 54.5 | 60.2 |

A tensile test is utilized to validate the Young's modulus of extruded aluminum. Five samples of nominal size—57.4(L) × 19.1(W) × 2.76(T) in mm—were cut perpendicular to their extruded directions. The tensile testing machine was set to pull typical samples until a fracture was exhibited (Figure 5a). A stress–elongation curve was generated by the machine as shown in Figure 5b.

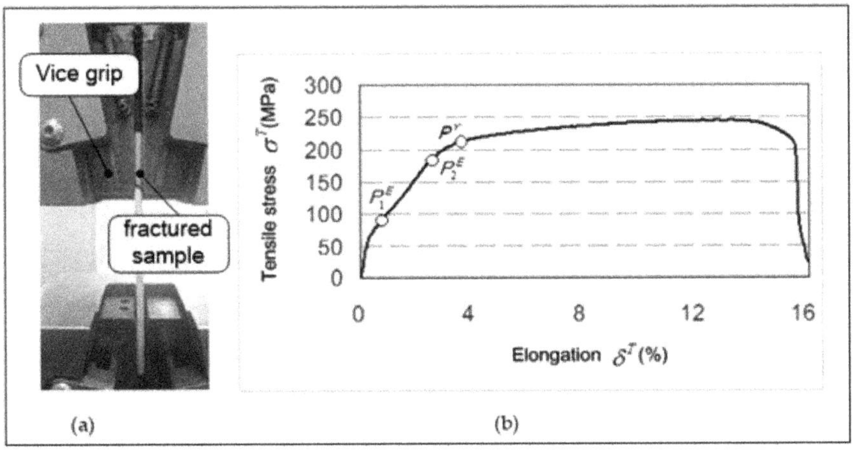


**Figure 5.** (**a**) Fracture of aluminum sample during tensile test; (**b**) stress–elongation curve generated by tensile testing machine ($P_1^E(92.9, 0.78)$, $P_2^E(179, 2.42)$, $P^Y(210, 3.5)$).

According to the analysis of Su and Young [11], Young's modulus of extruded aluminum varies linearly in initial stage of this curve. Changes in tensile stress $\Delta\sigma$ and elongation $\Delta\delta$ at separate positions $P_1^E = (92.9, 0.78)$, $P_2^E = (179, 2.42)$ of the elastic region were computed. The slope of the curve was used to calculate the Young's modulus of this sample as:

$$E^T = \frac{\Delta\sigma}{\Delta\delta} = \frac{\sigma_2^T - \sigma_1^T}{\delta_2^T - \delta_1^T} \times \frac{L}{WT} \tag{25}$$

The generated value was $E^T = 57.2$ GPa, +4.55% larger than $E_2^N$. As reported by Janeczek and Fydrych [13], a positive deviation arose at the excessive fracture, and material removal occurred at a high scratch performance using the pin head tool. This tended to underestimate $E_2^i$ at a slightly lower tensile strength. $\Delta\delta$ was 1.64%, which is close to the 3% of Tamadon [17].

For extruded aluminum, its normalized dependence of the actual Young's modulus ratio $E_r = E_a/E_o$ on the actual density ratio $\rho_r = \rho_a/\rho_o$ for the flat bar sample [18] can be formed from the following exponential equation [12,19,20]:

$$E_r = w\rho_r^u \tag{26}$$

where $w = 0.974$ is fitting constant, $u = 1.862$ is fitting exponent as given by Kovacik and Simancik [18], and $E_o = 68$ GPa for pure aluminum 3004 [13]. Using a digital weight test, $\rho_a = 2475$ kg/m$^3$. From the wrought aluminum alloy data of WA-3004 [19], $\rho_o = 2720$ kg/m$^3$. Thus, the extruded density $\rho_r = 2475/2720 = 0.91$ was found to be 0.91. Substituting $\rho_r$ and $E_o$ into this equation, $E_a = 55.6$ GPa was determined for the extruded sample. This only deviated from the measured $E^T$ by $-2.9$%. In the tool pin shape comparison [13,21–23], Young's modulus of the conical pin was 70.8 GPa, 24% higher than the 43.6 GPa of the thread pin. This trend reveals that the conical pin tended to provide more intense contact [13,24,25]. The semi-conical pin fell within the range providing medium-intensity contact. Janeczek and Fydrych [13] reported the tensile stress of aluminum alloy 3004 to be in the range of 175–198 MPa, with elongation of 10–16%. These are close to our test results, within a 10% deviation [26,27].

*3.2. Hardness Estimation and Brinell Test*

The hardness of the extruded aluminum was computed as follows. The divisional contact stiffness is given as

$$\frac{dP_j^i}{dH_j^i} = \frac{\beta}{\sqrt{\pi}} \times \frac{Feq_j^{i\,2}}{p_j^i \wp_j^i \sqrt{A_j^i}} \tag{27}$$

where $\beta$ is the scratch head correction factor. From this equation, $P_j$ is computed by its integration. Zheng et al. [7] showed that the test load increased with the square of the scratch depth. This complies with our data showing that $P_j$ varied quadratically with $A_j$. Initially, it gradually increased, and then increased more rapidly afterwards. Meanwhile, divisional tensile stress was computed as:

$$H_j^i = \frac{P_j^i}{A_j^i} \tag{28}$$

Their tensile stresses were quadratic curve-fitted to the Brinell hardness number as illustrated by Ali et al [28]. Using four divisional tests, $H_4^i$ increased to a stable peak at $H_4^N$. According to definition of the $P_j - A_j$ curve, hardness was evaluated at the peak load. Thus, tensile stress was calculated using its $N$th interval as $H_4^N = P_4^N / A_4^N = 215$ MPa. This value was within the 40–270 MPa range of the aluminum 3000 series [29]. As a common metal in the industrial market, a wrought aluminum flat bar is prepared by the extrusion plastic-forming process [30,31]. Its yield strength was raised by the strain rate and temperature during this extrusion process [30,32,33]. As recorded by the Aluminum Association, the yield strength increased from 50 MPa to 350 MPa as the temperature decreased from 450 °C to 25 °C. Despite the decrease in Young's modulus due to the decrease in extruded density, the hardness was relatively higher throughout this process. Thus, the tensile stress at 0.2% yield point $P^Y$ of Figure 5b, $\sigma^Y = 210$ MPa, was relatively high at the 74th percentile [27]. Meanwhile, the hardness was 75 using the Brinell tester, as in Equation (30). From the analysis of Ali [28], the hardness–tensile stress quadratic curve was modified as:

$$H_j^N = 0.14 H_j^{B2} - 0.97 H_j^B + 206 \tag{29}$$

in which unit of $H_j^N$ is in MPa. Substituting $H_4^N$ into this equation, its Brinell hardness number was estimated as $H_4^B = 77.5$. Using load data from two and three divisions, their hardnesses are listed in Table 1. As stated by Tamadon [17], the hardness of AA-1100 varied from the conical pin of 140 MPa to the thread pin of 200 MPa at the yield point, which was 8% lower than $H_4^N$.

For existing standard tests, hardness can be measured through the Brinell hardness tester. Considering the tester in Figure 6a, an indenting load of 100 kgf was exerted on the surface of the extruded Al flat bar through a hardened steel ball 1/16 inch in diameter.

Multiple tests were carried out using the same three size samples, each with four indentations. The Brinell hardness number was used to compare the actual hardnesses. The diameter of the resulting permanent impression in a typical measured sample was 594.12 µm, as shown in Figure 6b. According to DIN EN 6506-1 [15], the Brinell hardness number was computed as

$$B_h = 2P_b / \left( \pi D_s \left( D_s - \sqrt{D_s^2 - d_I^2} \right) \right) = 75.0 \tag{30}$$

where $P_b = 100$ kgf is the load on the indenting tool, $D_s = 1.59$ mm is the diameter of the steel ball, and $d_I = 1.19$ mm is the measured diameter at the rim of indentation. This value fell into the range of 50–140, as measured by Rockenhäuser [14] and Ali [28]. Compared with $H_4^5$, the percentage difference was −3.37%. Negative deviation arose at the estimation of higher $P_j$ under higher $F_n^j$. Due to the lower scratch performance, $A_j$ became relatively lower. Thus, $B_h$ tended to be slightly overestimated.

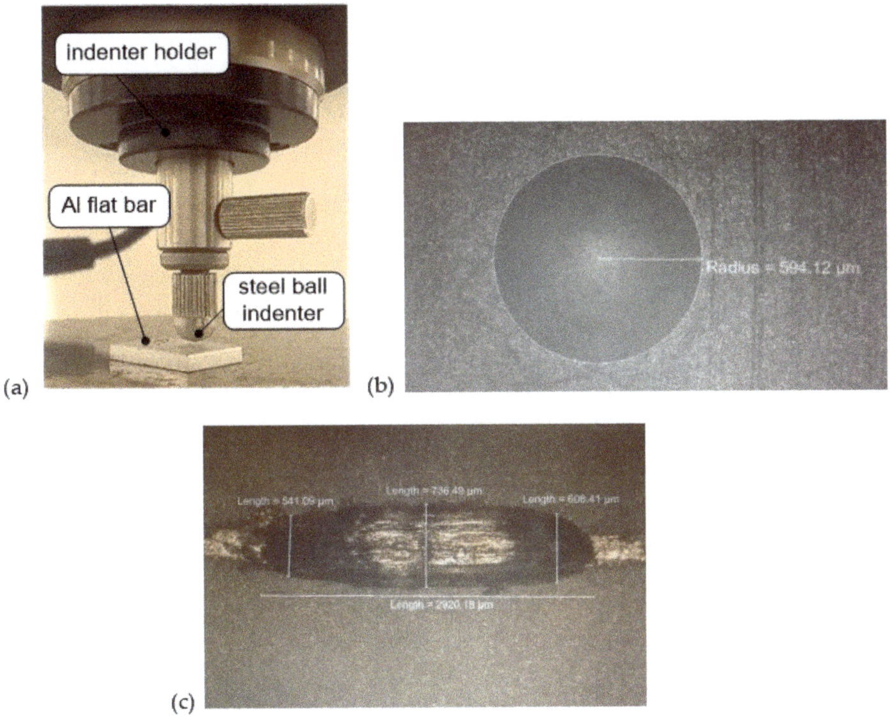


(a) (b) (c)

**Figure 6.** (**a**) Hardness tester using steel ball indenter; (**b**) microscopic measurement of steel ball indentation; (**c**) scratching track of extruded aluminum 3004.

### 3.3. Dimension Inspection and Performance Evaluation

In order to check the physical depth $d = D_j$ of the scratching track, a laser sensor was used to measure the depth of each track on the divisional samples. Moreover, a microscope is used to determine the physical dimensions $L^j$, $V^j$ of the scratching track (Figure 6c), which were smooth and measurable. The measured data are given in Table 1 and were used to compute the material removal volume $v$. The scratching energy factor at $j$ division, $F_v$, was computed by Equation (1), as listed in Table 1. It was stable around 56.8 GJ/m$^3$. From Table 1, it can be observed that $U^j$ increased linearly with $j$. Simultaneously, $\Psi^j$ increased with $j$ at a higher linear rate. As a result, $\wp_j$ decreased with $j$, but was saturated at the four divisions test. The overall scratching performance $\wp_j/U^j$, decreased with $j$. This is in agreement with the slightly increasing trend seen in $E_j$ and $H_j$ estimations. On the other hand, $F_v^j$ fluctuated around a constant value and was independent of the performance. Hence, it was independent of $i$ or $j$ and was in agreement with the energy conservation principle in Equation (1).

## 4. Conclusions

In this paper, a scratching damper is developed based on a scratching device. A cyclic vibration load is applied by the damper, creating a highly repeatable load-path loop. A smooth fracture surface was obtained, which was measurable at the microscopic scale. A round head indenter was used to provide a pure damping motion, while a pin head indenter was used to produce significant scratching motion. The scratching energy release rate was generated using free energy density and the displacement gradient of the semi-conical pin head. For comparison, the Young's modulus, $E_2^N$, was slightly underestimated as $E^T$ due to a fracture and material removal upon scratching. Using a weight test, the extruded $\rho_r$ was correctly calculated by the normalized dependence equation. For the hardness number, $B_h$ was slightly less than the $H_4^N$ caused by the lower scratching performance. $F_v^j$ was found to be stable, agreeing with the energy conservation principle. The results discussed here further the study of innovative methods and the development of devices and approaches to determine material properties.

**Author Contributions:** Conceptualization, C.-N.W. and W.-O.W.; methodology, C.-N.W. and A.V.; software, C.-N.W.; validation, C.-N.W., W.-O.W. and A.V.; formal analysis, C.-N.W.; investigation, C.-N.W.; resources, W.-O.W. and R.S.; data curation, C.-N.W. and A.V.; writing—original draft preparation, C.-N.W.; writing—review and editing, A.V. and W.-O.W.; visualization, C.-N.W. and A.V.; supervision, W.-O.W.; project administration, W.-O.W. All authors have read and agreed to the published version of the manuscript.

**Funding:** This research received no external funding.

**Data Availability Statement:** Research data may be shared upon request.

**Acknowledgments:** The authors would like to thank The Hong Kong Polytechnic University for the laboratory support of the experiments in this research.

**Conflicts of Interest:** The authors declare no conflicts of interest.

## References


1. Kamplade, K.; Biermann, D. Examination of the Material Removal of unreinforced, thermoplastic Polymers by Scratch Tests *Prod. Eng.* **2019**, *13*, 713–719. [CrossRef]
2. Varga, M.; Leroch, S.; Rojacz, H.; Ripoll, M.R. Study of wear mechanisms at high temperature scratch testing. *Wear* **2017**, *388–389* 112–118. [CrossRef]
3. Li, H.; Zhang, X.; Yang, X. Numerical study of liquid sloshing using smoothed particle hydrodynamics with adaptive spatia resolution. *Eng. Anal. Bound. Elem.* **2024**, *159*, 272–287. [CrossRef]
4. Akono, A.; Ulm, F. An improved technique for characterizing the fracture toughness crack growth. *Wear* **2014**, *313*, 117–124 [CrossRef]
5. Torabi, A.R.; Shahbazian, B. Notch tip plastic zone determination by extending Irwin's model. *Theor. Appl. Fract. Mech.* **2020**, *108* 102643. [CrossRef]
6. Souza, J.M.; Yoshimura, H.N.; Peres, F.M.; Schön, C.G. Effect of sample pre-cracking method and notch geometry in plane strai fracture toughness tests as applied to a PMMA resin. *Polym. Test.* **2012**, *31*, 834–840. [CrossRef]
7. Zheng, W.; Wang, Y.; Zhou, M.; Wang, Q.; Ling, L. Material deformation and removal mechanism of SiCp/Al composites in ultrasonic vibration assisted scratch test. *Ceram. Int.* **2018**, *44*, 15133–15144. [CrossRef]
8. Qiu, Z.; Liu, C.; Wang, H.; Yang, X.; Fang, F.; Tang, J. Crack propagation and the material removal mechanism of glass–ceramics by the scratch test. *Mechnical Behav. Biomed. Mater.* **2016**, *64*, 75–85. [CrossRef]
9. Al Wakeel, S.; Hubler, M.H. Introducing heterogeneity into the micro-scratch test fracture toughness relation for brittle particle composites. *Exp. Mech.* **2018**, *58*, 1237–1247. [CrossRef]
10. Akono, A.; Ulm, F. Fracture scaling relations for scratch tests of axisymmetric shape. *Mech. Phys. Solids* **2012**, *60*, 379–390 [CrossRef]
11. Su, M.; Young, B. Material properties of normal and high strength aluminium alloys at elevated temperatures. *Thin-Walled Struc* **2019**, *137*, 463–471. [CrossRef]
12. Ramberg, W.; Osgood, W.R. Description of Stress–Strain Curves by Three Parameters. NACA Tech. Note 902, 1943. Available online: https://ntrs.nasa.gov/citations/19930081614 (accessed on 10 June 2024).
13. Janeczek, A.; Fydrych, D. The Influence of Tool Shape and Process Parameters on the Mechanical Properties of AW-3004 Aluminium Alloy Friction Stir Welded Joints. *Materials* **2021**, *14*, 3244. [CrossRef] [PubMed]
14. Rockenhäuser, C.; Hartrott, P.; Skrotzki, B. Brinell-Hardness data (HBW 2.5/62.5) of aluminum alloy EN AW-2618A after differen aging times and temperatures. *Data Brief* **2023**, *46*, 108830. [CrossRef] [PubMed]
15. *DIN EN ISO 6506-1:2015-02*; Metallic Materials—Brinell Hardness Test—Part 1: Test Method. Beuth Verlag GmbH: Berlin Germany, 2015.
16. Sun, R.; Wong, W.; Cheng, L. Tunable electromagnetic shunt damper with opposing magnets configuration. *Smart Mater. Struct* **2020**, *29*, 115034. [CrossRef]

. Tamadon, A.; Baghestani, A.; Bajgholi, M.E. Influence of WC-based pin tool profile on microstructure and mechanical properties of AA1100 FSW welds. *Technologies* **2020**, *8*, 34. [CrossRef]
. Kovacik, J.; Simancik, F. Aluminium foam—Modulus of elasticity and electrical conductivity according to percolation theory. *Scr. Mater.* **1998**, *39*, 239–246. [CrossRef]
. Isichenko, M. Percolation Statistical Topography and Transport in Random Media. *Rev. Mod. Phys.* **1992**, *64*, 961. [CrossRef]
. Kováčik, J. The tensile behaviour of porous metals made by GASAR process. *Acta Mater.* **1998**, *46*, 5413–5422. [CrossRef]
. Su, H.; Xue, L.; Wu, C. Optimizing the tool pin with three flats in friction stir welding of aluminum alloy. *Int. J. Adv. Manuf. Technol.* **2020**, *108*, 721–733. [CrossRef]
. Eyvazian, A.; Hamouda, A.M.; Aghajani, H. Study on the effects of tool tile angle, offset and plunge depth on friction stir welding of poly(methyl methacrylate)T-joint. *Proc. Inst. Mech. Eng. Part B J. Eng. Manuf.* **2019**, *234*, 773–787. [CrossRef]
. Hemmouche, L.; Chicot, D.; Amrouche, A.; Iost, A.; Belouchrani, M.A.; Decoopman, X.; Louis, G.; Puchi-Cabrera, E.S. An analysis of the elastic properties of a porous aluminium oxide film by means of indentation techniques. *Mater. Sci. Eng. A* **2013**, *585*, 155–164. [CrossRef]
. Sharmaa, N.; Siddiqueea, A.N.; Khana, Z.A. Material stirring during FSW of Al–Cu: Effect of pin profile. *Mater. Manuf. Process.* **2018**, *33*, 786–794. [CrossRef]
. Shah, P.H.; Badheka, V.J. Friction stir welding of aluminium alloys: An overview of experimental findings—Process, variables, development and applications. *Proc. Inst. Mech. Eng. Part L J. Mater. Des. Appl.* **2017**, *233*, 1191–1226. [CrossRef]
. Aluminum Association (AA). *Aluminum Design Manual (ADM)*; Aluminum Association (AA): Washington, DC, USA, 2015.
. European Committee for Standardization, Eurocode 9: Design of aluminum structures—Part 1-2: Structural fire design, BS EN 1999-1-2, Brussels, 2007. Available online: https://www.phd.eng.br/wp-content/uploads/2014/11/en.1999.1.2.2007.pdf (accessed on 29 April 2024).
. Ali, M.M.; Omran, A.N.M.; Mohamed, M.A. Prediction the correlations between hardness and tensile properties of aluminium-silicon alloys produced by various modifiers and grain refineries using regression analysis and an artificial neural network model. *Eng. Sci. Technol.* **2021**, *24*, 105–111.
. *International Alloy Designations and Chemical Composition Limits for Wrought Aluminum and Wrought Aluminum Alloys*; The Aluminum Association: Washington, DC, USA, 2018; ISSN 2377-6692.
. Li, Q.; Zhang, X.; Wang, L.; Qiao, J. The Effect of Extrusion and Heat Treatment on the Microstructure and Tensile Properties of 2024 Aluminum Alloy. *Materials* **2022**, *15*, 7566. [CrossRef] [PubMed]
. Ma, Y.; Liu, C.; Miao, K. Effects of cooling rate and cryogenic temperature on the mechanical properties and deformation characteristics of an Al-Mg-Si-Fe-Cr alloy. *J. Alloys Compd.* **2023**, *947*, 169559. [CrossRef]
. Wang, Y.; Xiong, L.; Feng, D. Investigation of the penetration performance of the radial forging process for Wrought Aluminium Alloy. *Materials* **2024**, *17*, 2065. [CrossRef]
. Wang, Y.G.Y.; Zhao, S. Experimental investigation and optimization of the semisolid multicavity squeeze casting process for wrought aluminum alloy scroll. *Materials* **2020**, *13*, 5278. [CrossRef]

dynamics

MDPI

Article

# Auto-Correlation Functions of Chaotic Binary Sequences Obtained by Alternating Two Binary Functions

**Akio Tsuneda**

Division of Informatics and Energy, Faculty of Advanced Science and Technology, Kumamoto University, Kumaomto 860-8555, Japan; tsuneda@cs.kumamoto-u.ac.jp

**Abstract:** This paper discusses the auto-correlation functions of chaotic binary sequences obtained by a one-dimensional chaotic map and two binary functions. The two binary functions are alternately used to obtain a binary sequence from a chaotic real-valued sequence. We consider two similar methods and give the theoretical auto-correlation functions of the new binary sequences, which are expressed by the auto-/cross-correlation functions of the two chaotic binary sequences generated by a single binary function. Furthermore, some numerical experiments are performed to confirm the validity of the theoretical auto-correlation functions.

**Keywords:** chaotic binary sequence; one-dimensional chaotic map; auto-correlation function; cross-correlation function; binary function

## 1. Introduction

One-dimensional (1-D) nonlinear maps can generate chaotic real-valued sequences [1,2] and they can be used as random numbers due to their simplicity, despite their complex behavior [3–6]. Chaotic real-valued sequences can be transformed into discrete-valued (e.g., binary) sequences using discrete-valued functions, and they have been used in some applications (e.g., CDMA communications as spreading codes) [7]. In 1988, chaotic binary sequences based on a 1-D chaotic map (tent map) were proposed and proven to be independent and identically distributed *(i.i.d.)* [8]. In 1997, the statistical properties of chaotic binary sequences based on 1-D chaotic maps were deeply discussed and a sufficient condition to generate *i.i.d.* binary sequences was given for a class of chaotic maps and binary functions [9].

As described above, ideal random numbers are often assumed to be *i.i.d.* since the most prominent application of chaos-based random numbers is in the area of security, such as cryptography [3–6]. On the other hand, correlated chaotic sequences have also been discussed, and they are useful in some applications other than cryptography. The auto-correlation functions of chaotic real-valued sequences generated by the skew tent map were theoretically derived in [10], where the auto-correlation functions were exponentially decreasing. Chaotic discrete-valued sequences with exponential auto-correlation functions were discussed in [11–13], and some of them can be applied to CDMA communications as spreading codes. In Monte Carlo integration, the convergence rate can be drastically improved by using chaotic sequences with proper auto-correlations, which is called *super-efficient chaotic Monte-Carlo simulation* [14]. Thus, in applications of chaos-based random numbers, the controllability of the statistical properties is quite important. It should be noted that the desired (or optimal) statistical properties of the random numbers are different depending on the application.

Still, there have been many studies on chaos-based random number generation in both theoretical and experimental contexts (e.g., [15–17]). In order to adapt chaos-based random numbers for many applications, we have been attempting to realize chaotic binary (or discrete-valued) sequences with various auto-correlation properties [18]. To generate a chaotic binary sequence, a chaotic map and a binary function are normally used. Of course, there are many combinations of a chaotic map and a binary (or discrete-valued) function, which implies that various statistical properties can be realized. In this paper, however, we propose a different approach to generating chaotic binary sequences using 1-D chaotic maps. We use one chaotic map and two binary functions to generate a chaotic binary sequence, where the two binary functions are alternately used. The theoretical average and auto-correlation function of the new chaotic binary sequence are given, which is followed by numerical experiments using some chaotic maps and binary functions.

The rest of this paper is organized as follows. In Section 2, a brief review of conventional chaotic binary sequences based on 1-D chaotic map is given, where the theoretical average and auto-correlation function are defined. In Section 3, we propose two methods to generate chaotic binary sequences, where a chaotic map and two binary functions are used to generate a chaotic binary sequence. The two binary functions are alternately used. We derive the theoretical auto-correlation functions of the chaotic binary sequences generated by the proposed methods. Numerical results obtained using some chaotic maps and binary functions are shown in Section 4. Finally, Section 5 concludes this paper.

## 2. Chaotic Binary Sequences Based on a One-Dimensional Map and a Binary Function

A one-dimensional nonlinear difference equation defined by

$$x_{n+1} = \tau(x_n), \quad x_n \in I = [0,1], \quad n = 0,1,2,\cdots \tag{1}$$

can generate a *chaotic* real-valued sequence $\{x_n\}_{n=0}^{\infty}$ for a chaotic map $\tau(x)$ [1,2]. Moreover, we can obtain a binary sequence $\{B(x_n)\}_{n=0}^{\infty}$ using a binary function $B(x)(\in \{0,1\})$ from a real-valued sequence $\{x_n\}_{n=0}^{\infty}$. Then, the theoretical auto-correlation function of the binary sequence $\{B(x_n)\}_{n=0}^{\infty}$ is defined by

$$A(\ell; B) = \int_I (B(x) - E[B])(B(\tau^{\ell}(x)) - E[B])f^*(x)dx, \tag{2}$$

under the assumption that $\tau(x)$ has an invariant density function $f^*(x)$, where $\tau^{\ell}(x)$ is the $\ell$-th iterate of the map $\tau$ starting from an initial value $x = x_0$, and $E[B]$ denotes the average of the binary sequence $\{B(x_n)\}_{n=0}^{\infty}$ defined by

$$E[B] = \int_I B(x)f^*(x)dx. \tag{3}$$

We also define the normalized auto-correlation function by

$$\widetilde{A}(\ell; B) = \frac{A(\ell; B)}{A(0; B)}, \tag{4}$$

where $A(0; B)$ is the variance of $\{B(x_n)\}_{n=0}^{\infty}$. On the other hand, the auto-correlation function of $\{B(x_n)\}_{n=0}^{\infty}$ in time-average form is defined by

$$\widehat{A}_N(\ell; B(x_n)) = \frac{1}{N}\sum_{n=0}^{N-1}(B(x_n) - E[B])(B(x_{n+\ell}) - E[B]), \tag{5}$$

which can be used in numerical calculations of auto-correlation functions. According to the Birkhoff individual ergodic theorem [1,2], we have

$$\lim_{N\to\infty} \widehat{A}_N(\ell; B(x_n)) = A(\ell; B) \quad \text{for almost all } x_0. \tag{6}$$

Note that the time average of $\{B(x_n)\}_{n=0}^{\infty}$ is defined by

$$\widehat{E}_N[B(x_n)] = \frac{1}{N}\sum_{n=0}^{N-1} B(x_n), \tag{7}$$

and we also have

$$\lim_{N\to\infty} \widehat{E}_N[B(x_n)] = E[B] \quad \text{for almost all } x_0. \tag{8}$$

Next, for two chaotic binary sequences $\{B_1(x_n)\}_{n=0}^{\infty}$ and $\{B_2(x_n)\}_{n=0}^{\infty}$ generated from a common real-valued sequence $\{x_n\}_{n=0}^{\infty}$, their cross-correlation function is defined by

$$C(\ell; B_1, B_2) = \int_I (B_1(x) - E[B_1])(B_2(\tau^{\ell}(x)) - E[B_2])f^*(x)dx. \tag{9}$$

Note that $C(\ell; B_1, B_2) = B(\ell; B_2, B_1)$ does not always hold. The normalized cross-correlation function is defined by

$$\widetilde{C}(\ell; B_1, B_2) = \frac{C(\ell; B_1, B_2)}{\sqrt{A(0; B_1)}\sqrt{A(0; B_2)}}. \tag{10}$$

The cross-correlation function between $\{B_1(x_n)\}_{n=0}^{\infty}$ and $\{B_2(x_n)\}_{n=0}^{\infty}$ in time-average form is defined by

$$\widehat{C}_N(\ell; B_1(x_n), B_2(x_n)) = \frac{1}{N}\sum_{n=0}^{N-1} (B_1(x_n) - E[B_1])(B_2(x_{n+\ell}) - E[B_2]). \tag{11}$$

Similar to (6), we have

$$\lim_{N\to\infty} \widehat{C}_N(\ell; B_1(x_n), B_2(x_n)) = C(\ell; B_1, B_2) \quad \text{for almost all } x_0. \tag{12}$$

## 3. Chaotic Binary Sequences Obtained by Alternating Two Binary Functions

We use two binary functions $B_1(x)$ and $B_2(x)$ to generate a new binary sequence from a chaotic real-valued sequence $\{x_n\}_{n=0}^{\infty}$. We propose the following two methods and discuss the auto-correlation functions of the generated binary sequences.

### 3.1. Method 1

Using $B_1(x)$ and $B_2(x)$ alternately, we generate a new binary sequence $\{D_n^{(1)}\}_{n=0}^{\infty}$ as

$$B_1(x_0), B_2(x_1), B_1(x_2), B_2(x_3), \cdots, \tag{13}$$

that is,

$$D_n^{(1)} = \begin{cases} B_1(x_n) & (n = 0, 2, 4, \cdots), \\ B_2(x_n) & (n = 1, 3, 5, \cdots). \end{cases} \tag{14}$$

Here, we assume $E[B_1] = E[B_2]$, which gives

$$E[D_n^{(1)}] = E[B_1] = E[B_2]. \tag{15}$$

Next, consider the auto-correlation function of the binary sequence $\{D_n^{(1)}\}_{n=0}^{\infty}$. First, the auto-correlation function in time-average form is expressed by

$$\widehat{A}_N(\ell; D_n^{(1)}) = \begin{cases} \frac{1}{N}(B_1(x_0)B_1(x_\ell) + B_2(x_1)B_2(x_{1+\ell}) + B_1(x_2)B_1(x_{2+\ell}) + B_2(x_3)B_2(x_{3+\ell}) + \cdots) \\ \qquad (\ell = 0, 2, 4, \cdots), \\ \frac{1}{N}(B_1(x_0)B_2(x_\ell) + B_2(x_1)B_1(x_{1+\ell}) + B_1(x_2)B_2(x_{2+\ell}) + B_2(x_3)B_1(x_{3+\ell}) + \cdots) \\ \qquad (\ell = 1, 3, 5, \cdots). \end{cases} \tag{16}$$

Thus, the theoretical auto-correlation function of $\{D_n^{(1)}\}_{n=0}^{\infty}$ is given by

$$A(\ell; D_n^{(1)}) = \begin{cases} \frac{1}{2}(A(\ell; B_1) + A(\ell; B_2)) & (\ell = 0, 2, 4, \cdots), \\ \frac{1}{2}(C(\ell; B_1, B_2) + C(\ell; B_2, B_1)) & (\ell = 1, 3, 5, \cdots). \end{cases} \tag{17}$$

Note that $A(\ell; D_n^{(1)})$ is invariant under exchanging $B_1(x)$ and $B_2(x)$.

*3.2. Method 2*

Similar to Method 1, we generate a new binary sequence $\{D_n^{(2)}\}_{n=0}^{\infty}$ using $B_1(x)$ and $B_2(x)$ alternately as

$$B_1(x_0), B_2(x_0), B_1(x_1), B_2(x_1), \cdots. \tag{18}$$

That is, for each real value $x_n$, two binary values $B_1(x_n)$, $B_2(x_n)$ are generated and $D_n^{(2)}$ is expressed by

$$D_n^{(2)} = \begin{cases} B_1(x_{\frac{n}{2}}) & (n = 0, 2, 4, \cdots), \\ B_2(x_{\frac{n-1}{2}}) & (n = 1, 3, 5, \cdots). \end{cases} \tag{19}$$

We also assume $E[B_1] = E[B_2]$, which gives

$$E[D_n^{(2)}] = E[B_1] = E[B_2]. \tag{20}$$

The auto-correlation function of $\{D_n^{(2)}\}_{n=0}^{\infty}$ in time-average form is expressed by

$$\widehat{A}_N(\ell; D_n^{(2)}) = \begin{cases} \frac{1}{N}(B_1(x_0)B_1(x_{\frac{\ell}{2}}) + B_2(x_0)B_2(x_{\frac{\ell}{2}}) + B_1(x_1)B_1(x_{1+\frac{\ell}{2}}) + B_2(x_1)B_2(x_{1+\frac{\ell}{2}}) + \cdots) \\ \qquad (\ell = 0, 2, 4, \cdots), \\ \frac{1}{N}(B_1(x_0)B_2(x_{\frac{\ell-1}{2}}) + B_2(x_0)B_1(x_{\frac{\ell+1}{2}}) + B_1(x_1)B_2(x_{1+\frac{\ell-1}{2}}) + B_2(x_1)B_1(x_{1+\frac{\ell+1}{2}}) + \cdots) \\ \qquad (\ell = 1, 3, 5, \cdots). \end{cases} \tag{21}$$

Thus, the theoretical auto-correlation function of $\{D_n^{(2)}\}_{n=0}^{\infty}$ is given by

$$A(\ell; D_n^{(2)}) = \begin{cases} \frac{1}{2}(A(\frac{\ell}{2}; B_1) + A(\frac{\ell}{2}; B_2)) & (\ell = 0, 2, 4, \cdots), \\ \frac{1}{2}(C(\frac{\ell-1}{2}; B_1, B_2) + C(\frac{\ell+1}{2}; B_2, B_1)) & (\ell = 1, 3, 5, \cdots). \end{cases} \tag{22}$$

Note that $A(\ell; D_n^{(2)})$ is not invariant under the exchange of $B_1(x)$ and $B_2(x)$.

## 4. Numerical Experiments

We perform numerical experiments on Method 1 and Method 2 using three chaotic maps ($I = [0, 1], f^*(x) = 1$) and some binary functions. Note that the chaotic maps and the binary functions are chosen as examples to obtain some interesting (or unique) auto-/cross-correlation functions.

### 4.1. Bernoulli Map and Binary Functions

The Bernoulli map with $I = [0,1]$ is defined by [1,2]

$$\tau_B(x) = \begin{cases} 2x & (0 \leq x < \frac{1}{2}), \\ 2x - 1 & (\frac{1}{2} \leq x \leq 1). \end{cases} \tag{23}$$

For this map, we use the following two binary functions:

$$b_1(x) = \Theta_{\frac{1}{8}}(x) - \Theta_{\frac{1}{4}}(x) + \Theta_{\frac{1}{2}}(x) - \Theta_{\frac{3}{4}}(x) + \Theta_{\frac{7}{8}}(x), \tag{24}$$

$$b_2(x) = \Theta_{\frac{1}{4}}(x) - \Theta_{\frac{3}{8}}(x) + \Theta_{\frac{1}{2}}(x) - \Theta_{\frac{5}{8}}(x) + \Theta_{\frac{3}{4}}(x), \tag{25}$$

where $\Theta_t(x)$ is a *threshold function* with a threshold $t$ defined by

$$\Theta_t(x) = \begin{cases} 0 & (x < t), \\ 1 & (x \geq t). \end{cases} \tag{26}$$

The Bernoulli map $\tau_B(x)$ and the binary functions $b_1(x)$, $b_2(x)$ are illustrated in Figure 1. Since the Bernoulli map has $f^*(x) = 1$ (uniform density), we have

$$E[b_1] = E[b_2] = \frac{1}{2}, \tag{27}$$

that is, the binary sequences $\{b_1(\tau_B^n(x))\}_{n=0}^{\infty}$ and $\{b_2(\tau_B^n(x))\}_{n=0}^{\infty}$ are balanced sequences. The theoretical correlation functions of the binary sequences can be derived by referring to [9,13] (see Appendix A for details). The normalized theoretical auto-/cross-correlation functions of the two binary sequences are summarized in Table 1 and illustrated in Figure 2. In Figure 2, numerical auto-/cross-correlation functions calculated by (5) and (11) with $N = 10^6$ are also shown.

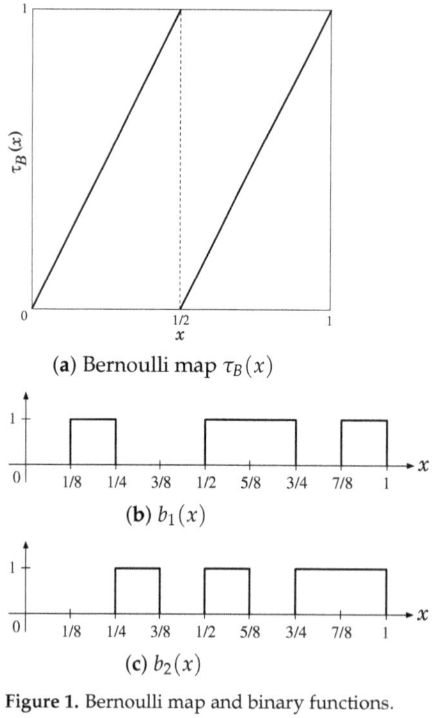


**Figure 1.** Bernoulli map and binary functions.

**Table 1.** Normalized theoretical auto-/cross-correlation functions of $\{b_1(\tau_B^n(x))\}_{n=0}^{\infty}$ and $\{b_2(\tau_B^n(x))\}_{n=0}^{\infty}$.

| | $\ell = 0$ | $\ell = 1$ | $\ell = 2$ | $\ell \geq 3$ |
|---|---|---|---|---|
| $\widetilde{A}(\ell; b_1)$ | 1 | $-\frac{1}{2}$ | $\frac{1}{4}$ | 0 |
| $\widetilde{A}(\ell; b_2)$ | 1 | 0 | $-\frac{1}{4}$ | 0 |
| $\widetilde{C}(\ell; b_1, b_2)$ | 0 | 0 | $\frac{1}{4}$ | 0 |
| $\widetilde{C}(\ell; b_2, b_1)$ | 0 | $\frac{1}{2}$ | $-\frac{1}{4}$ | 0 |

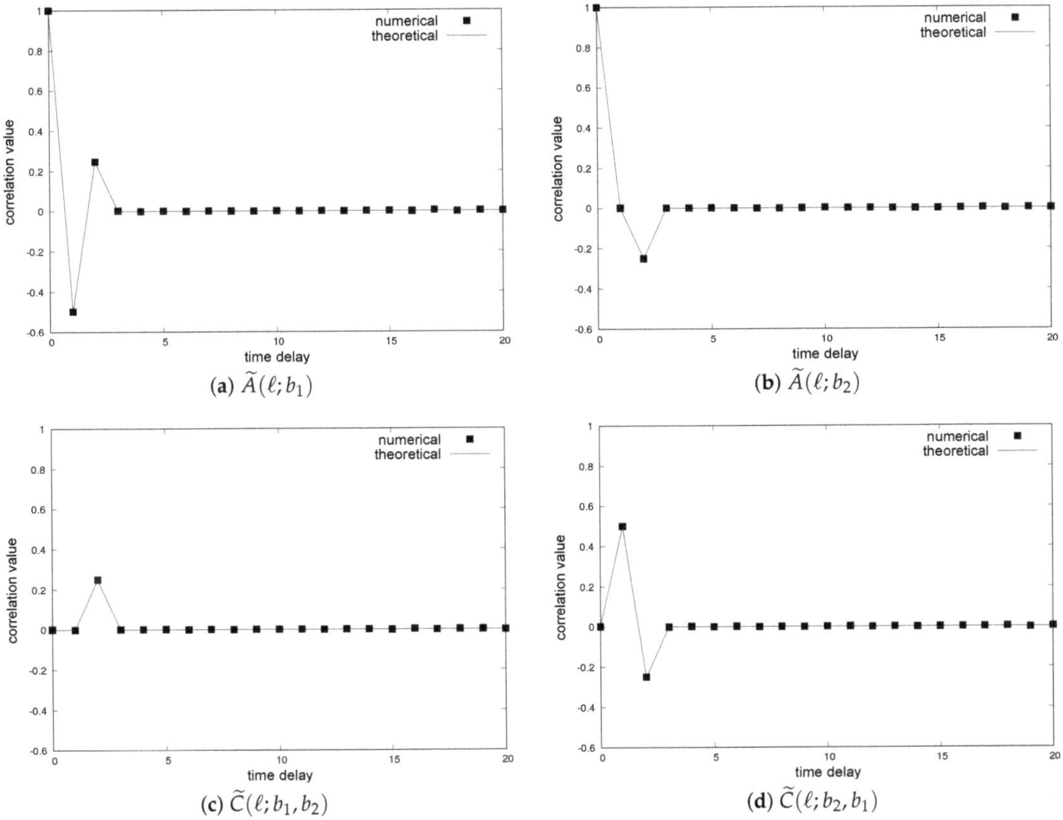


(a) $\widetilde{A}(\ell; b_1)$ (b) $\widetilde{A}(\ell; b_2)$ (c) $\widetilde{C}(\ell; b_1, b_2)$ (d) $\widetilde{C}(\ell; b_2, b_1)$

**Figure 2.** Normalized auto-/cross-correlation functions of $\{b_1(\tau_B^n(x))\}_{n=0}^{\infty}$ and $\{b_2(\tau_B^n(x))\}_{n=0}^{\infty}$.

### 4.2. Piecewise Linear Map with Three Sections and Binary Functions

Define a fully stretching piecewise linear (PL) map with $I = [0, 1]$ by [13]

$$\tau_{PL}(x) = \begin{cases} -4x + 1 & (0 \leq x < \frac{1}{4}), \\ \frac{12}{5}x - \frac{3}{5} & (\frac{1}{4} \leq x < \frac{2}{3}), \\ -3x + 3 & (\frac{2}{3} \leq x \leq 1). \end{cases} \tag{28}$$

For this map, we use the following two binary functions:

$$b_3(x) = \Theta_{\frac{3}{4}}(x), \tag{29}$$

$$b_4(x) = 1 - \Theta_{\frac{1}{4}}(x). \tag{30}$$

The PL map $\tau_{PL}(x)$ and the binary functions $b_3(x)$, $b_4(x)$ are illustrated in Figure 3. Since the PL map $\tau_{PL}(x)$ also has $f^*(x) = 1$, we have

$$E[b_3] = E[b_4] = \frac{1}{4}, \tag{31}$$

that is, the binary sequences $\{b_3(\tau_{PL}^n(x))\}_{n=0}^{\infty}$ and $\{b_4(\tau_{PL}^n(x))\}_{n=0}^{\infty}$ are unbalanced sequences. The theoretical correlation functions of the binary sequences can be derived by referring to [9,13] (see Appendix A for details). The normalized theoretical auto-/cross-correlation functions of the two binary sequences are summarized in Table 2 and illustrated in Figure 4. In Figure 4, numerical auto-/cross-correlation functions calculated by (5) and (11) with $N = 10^6$ are also shown.

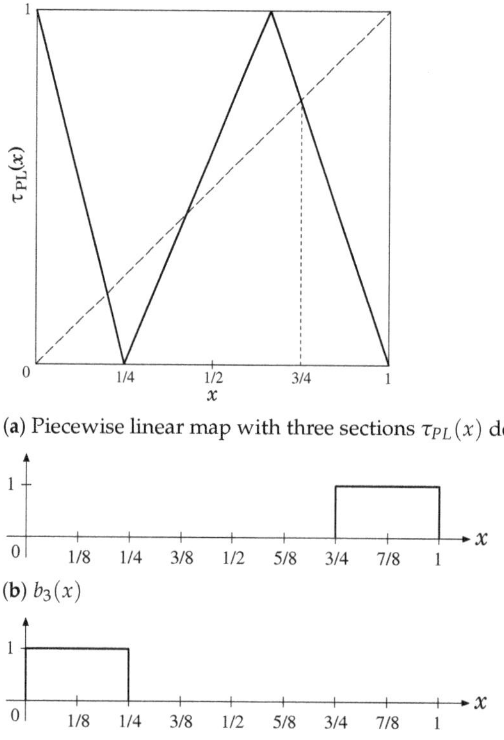


(**a**) Piecewise linear map with three sections $\tau_{PL}(x)$ defined by (28)

(**b**) $b_3(x)$

(**c**) $b_4(x)$

**Figure 3.** Piecewise linear map with three sections and binary functions.

**Table 2.** Normalized theoretical auto-/cross-correlation functions of $\{b_3(\tau_{PL}^n(x))\}_{n=0}^{\infty}$ and $\{b_4(\tau_{PL}^n(x))\}_{n=0}^{\infty}$.

<table>
<tr><th></th><th>$\ell = 0$</th><th>$\ell = 1$</th><th>$\ell = 2$</th><th>$\ell \geq 3$</th></tr>
<tr><td>$\widetilde{A}(\ell; b_3)$</td><td colspan="4">$(-3)^{-\ell}$</td></tr>
<tr><td>$\widetilde{A}(\ell; b_4)$</td><td>1</td><td>0</td><td>0</td><td>0</td></tr>
<tr><td>$\widetilde{C}(\ell; b_3, b_4)$</td><td colspan="4">$(-3)^{-\ell-1}$</td></tr>
<tr><td>$\widetilde{C}(\ell; b_4, b_3)$</td><td>$-\frac{1}{3}$</td><td>0</td><td>0</td><td>0</td></tr>
</table>

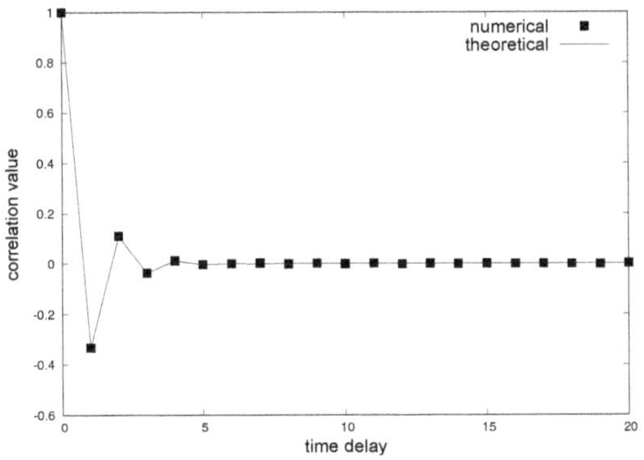


(a) $\widetilde{A}(\ell; b_3)$

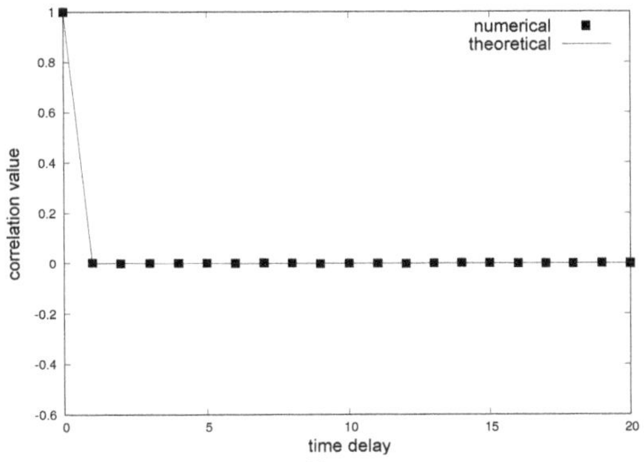


(b) $\widetilde{A}(\ell; b_4)$

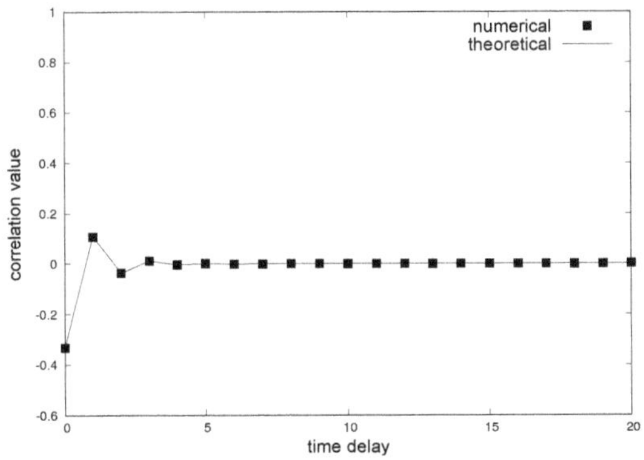


(c) $\widetilde{C}(\ell; b_3, b_4)$

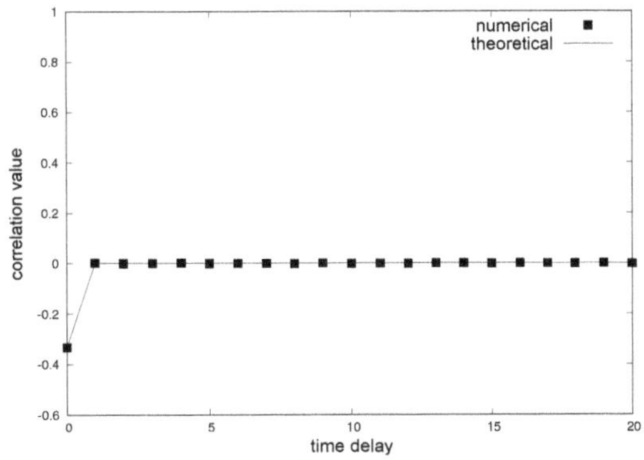


(d) $\widetilde{C}(\ell; b_4, b_3)$

**Figure 4.** Auto-/cross-correlation functions of $\{b_3(\tau_{PL}^n(x))\}_{n=0}^{\infty}$ and $\{b_4(\tau_{PL}^n(x))\}_{n=0}^{\infty}$.

### *4.3. Tent Map and Binary Functions*

Define the tent map with $I = [0, 1]$ by [1,2,10]

$$\tau_T(x) = \begin{cases} 2x & (0 \le x < \frac{1}{2}), \\ 2(1-x) & (\frac{1}{2} \le x \le 1). \end{cases} \tag{32}$$

For this map, we use the following two binary functions:

$$b_5(x) = \Theta_{\frac{1}{2}}(x), \tag{33}$$

$$b_6(x, d) = \Theta_d(x) - \Theta_{\frac{1}{2}}(x) + \Theta_{1-d}(x) \ (0 < d \le \frac{1}{2}). \tag{34}$$

The tent map $\tau_T(x)$ and the binary functions $b_5(x)$, $b_6(x)$ are illustrated in Figure 5. Since the tent map $\tau_T(x)$ also has $f^*(x) = 1$, we have

$$E[b_5] = E[b_6] = \frac{1}{2}, \tag{35}$$

that is, the binary sequences $\{b_5(\tau_T^n(x))\}_{n=0}^{\infty}$ and $\{b_6(\tau_T^n(x), d)\}_{n=0}^{\infty}$ are balanced sequences. The theoretical correlation functions of the binary sequences can be derived by referring to [9,13] (see Appendix A for details). Actually, $\{b_5(\tau_T^n(x))\}_{n=0}^{\infty}$ and $\{b_6(\tau_T^n(x), d)\}_{n=0}^{\infty}$ are *i.i.d.* and uncorrelated to each other for $\ell \ge 1$. Note that the cross-correlation function for $\ell = 0$, $C(0; b_5, b_6)$, can be controlled by $d$ (parameter of the binary function $b_6(x, d)$).

The normalized theoretical auto-/cross-correlation functions of the two binary sequences are summarized in Table 3 and illustrated in Figure 6. In Figure 6, numerical auto-/cross-correlation functions calculated by (5) and (11) with $N = 10^6$ are also shown.

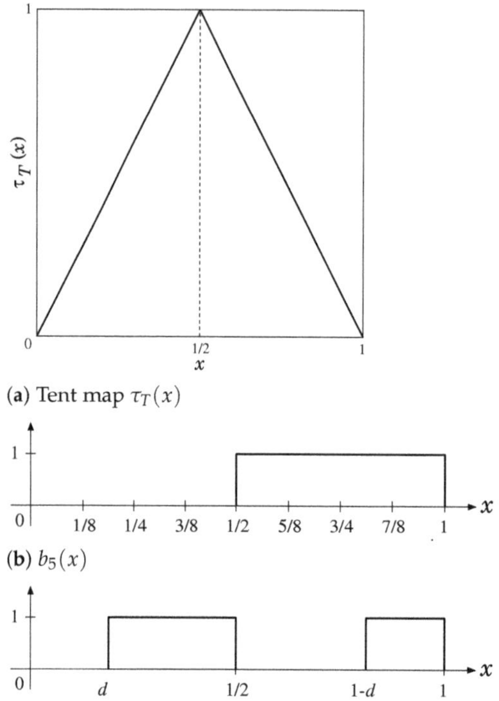


(**a**) Tent map $\tau_T(x)$

(**b**) $b_5(x)$

(**c**) $b_6(x, d)$ $(0 < d \le \frac{1}{2})$

**Figure 5.** Tent map and binary functions.

**Table 3.** Normalized theoretical auto-/cross-correlation functions of $\{b_5(\tau_T^n(x))\}_{n=0}^{\infty}$ and $\{b_6(\tau_T^n(x), d)\}_{n=0}^{\infty}$.

| | $\ell = 0$ | $\ell \geq 1$ |
|---|---|---|
| $\widetilde{A}(\ell; b_5)$ | 1 | 0 |
| $\widetilde{A}(\ell; b_6)$ | 1 | 0 |
| $\widetilde{C}(\ell; b_5, b_6)$ | $4d - 1$ | 0 |

*4.4. Auto-Correlation Functions of New Binary Sequences by Method 1 and Method 2*

First, Figure 7 shows the normalized auto-correlation functions of new binary sequences $\{D_n^{(1)}\}_{n=0}^{\infty}$ in Method 1, where the theoretical auto-correlation functions are calculated by (17) and the numerical ones are calculated by (5), where $N = 10^6$. We can find that the auto-correlation functions of $\{D_n^{(1)}\}_{n=0}^{\infty}$ given in Figure 7a,b are different from those of the original binary sequences, but Figure 7c,d show the uncorrelated property, which is the same as in the original two binary sequences. We also confirm that the theoretical and numerical ones are in good agreement.

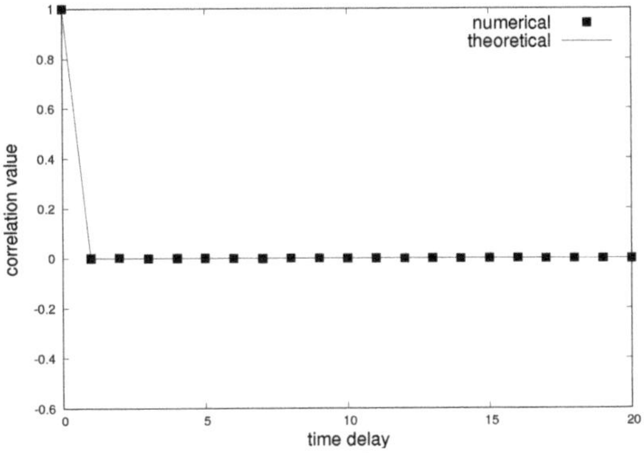


**(a)** $\widetilde{A}(\ell; b_5), \widetilde{A}(\ell; b_6)$

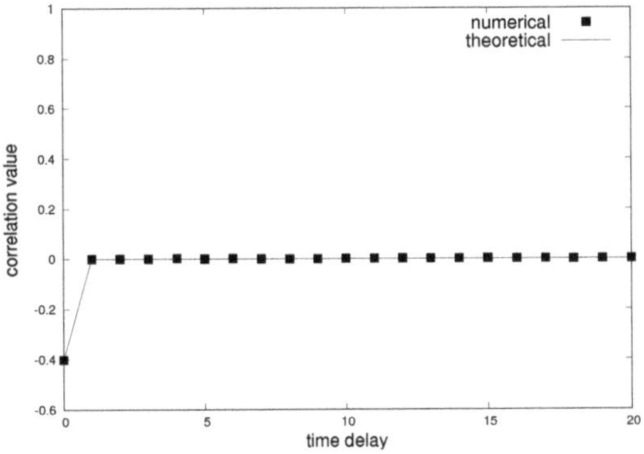


**(b)** $\widetilde{C}(\ell; b_5, b_6)$ $(d = 0.15)$

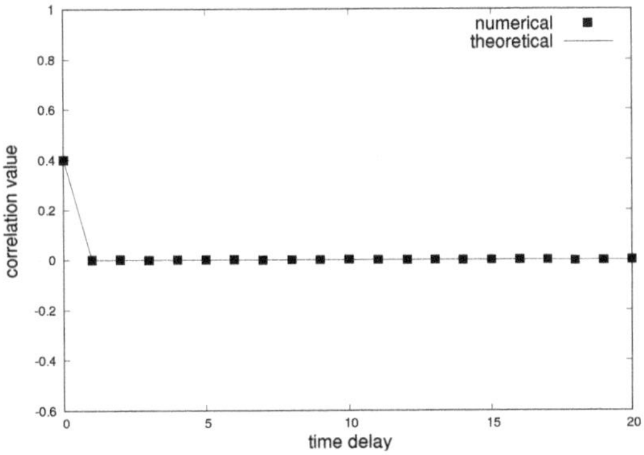


**(c)** $\widetilde{C}(\ell; b_5, b_6)$ $(d = 0.35)$

**Figure 6.** Auto-/cross-correlation functions of $\{b_5(\tau_T^n(x))\}_{n=0}^{\infty}$ and $\{b_6(\tau_T^n(x), d)\}_{n=0}^{\infty}$ $(d = 0.15, 0.35)$.

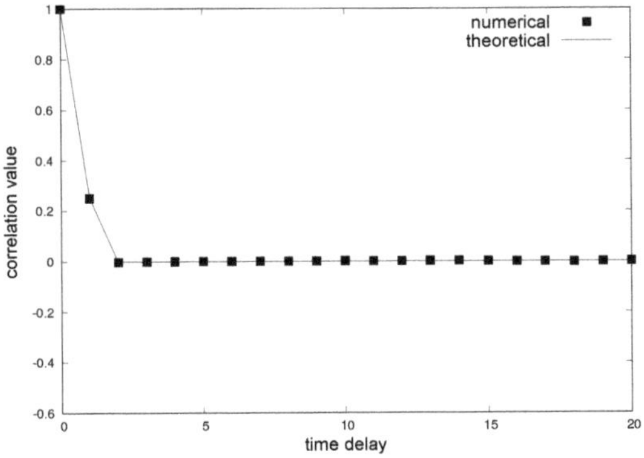


**(a)** $\tau_B(x)$, $B_1(x) = b_1(x)$, $B_2(x) = b_2(x)$

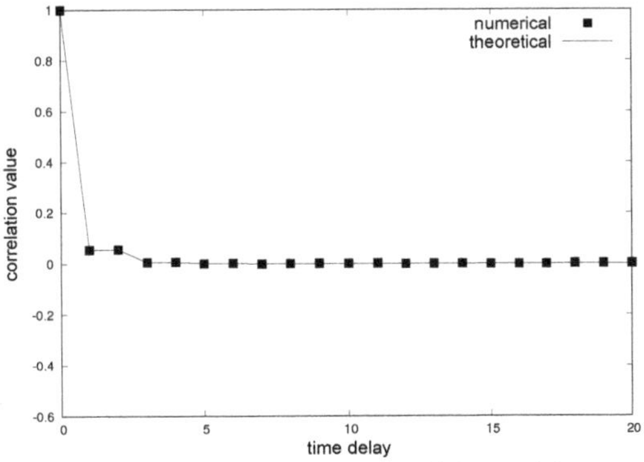


**(b)** $\tau_{PL}(x)$, $B_1(x) = b_3(x)$, $B_2(x) = b_4(x)$

**Figure 7.** *Cont.*

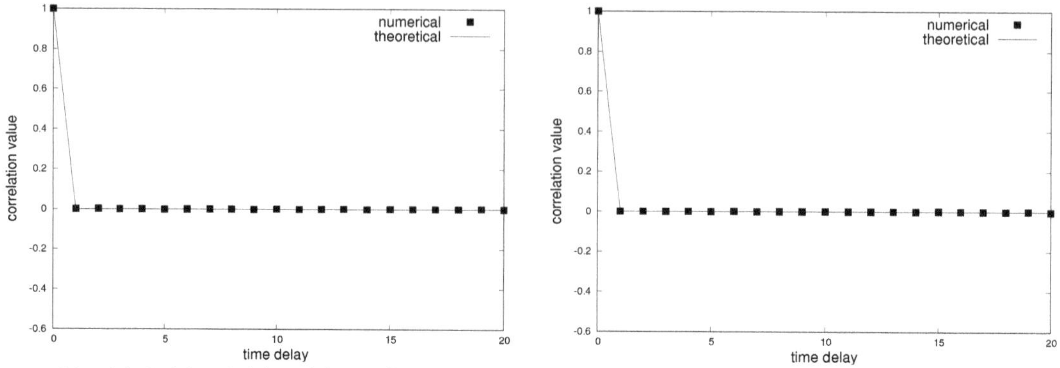


(c) $\tau_T(x)$, $B_1(x) = b_5(x)$, $B_2(x) = b_6(x, 0.15)$ (d) $\tau_T(x)$, $B_1(x) = b_5(x)$, $B_2(x) = b_6(x, 0.35)$

**Figure 7.** Normalized auto-correlation functions of new binary sequences $\{D_n^{(1)}\}_{n=0}^{\infty}$ in Method 1.

Next, Figure 8 shows the normalized auto-correlation functions of new binary sequences $\{D_n^{(2)}\}_{n=0}^{\infty}$ in Method 2, where the theoretical auto-correlation functions are calculated by (22) and the numerical ones are calculated by (5), where $N = 10^6$. We can find that the auto-correlation functions of $\{D_n^{(2)}\}_{n=0}^{\infty}$ are different from those of the original binary sequences, and the theoretical and numerical ones are in good agreement. Moreover, it is confirmed that the auto-correlation functions change if $B_1(x)$ and $B_2(x)$ are exchanged. It should be noted that the auto-correlation value at $\ell = 1$ in (e) and (f) (Figure 8) can be controlled by the parameter $d$.

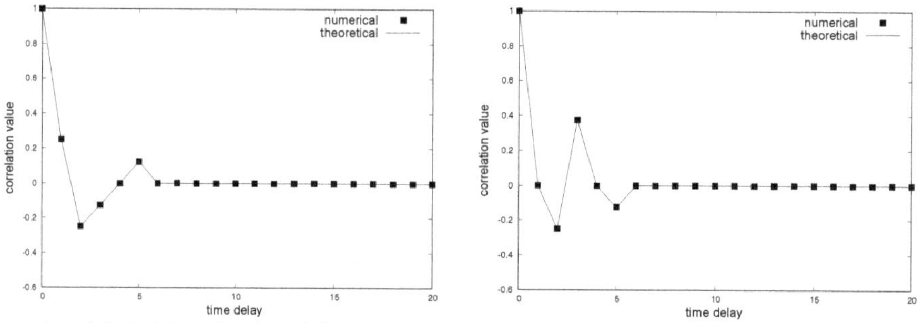


(a) $\tau_B(x)$, $B_1(x) = b_1(x)$, $B_2(x) = b_2(x)$ (b) $\tau_B(x)$, $B_1(x) = b_2(x)$, $B_2(x) = b_1(x)$

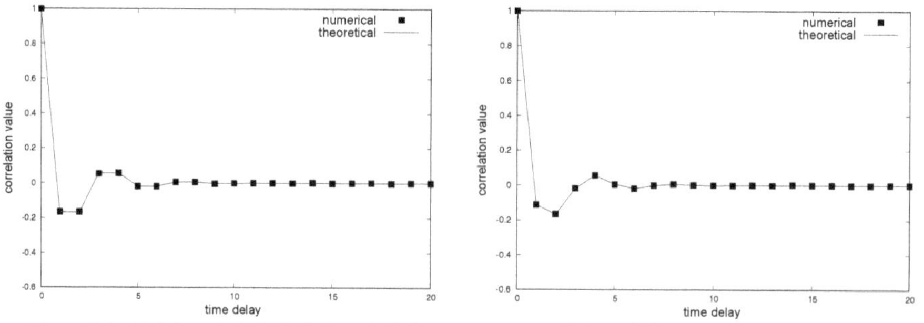


(c) $\tau_{PL}(x)$, $B_1(x) = b_1(x)$, $B_2(x) = b_2(x)$ (d) $\tau_{PL}(x)$, $B_1(x) = b_2(x)$, $B_2(x) = b_1(x)$

**Figure 8.** *Cont.*

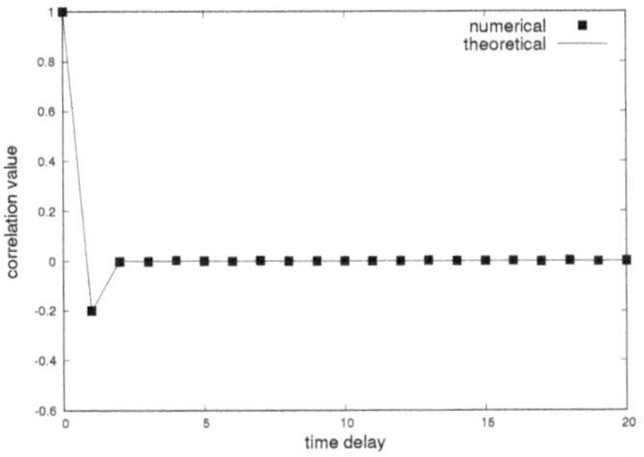

(e) $\tau_T(x)$, $B_1(x) = b_5(x)$, $B_2(x) = b_6(x, 0.15)$

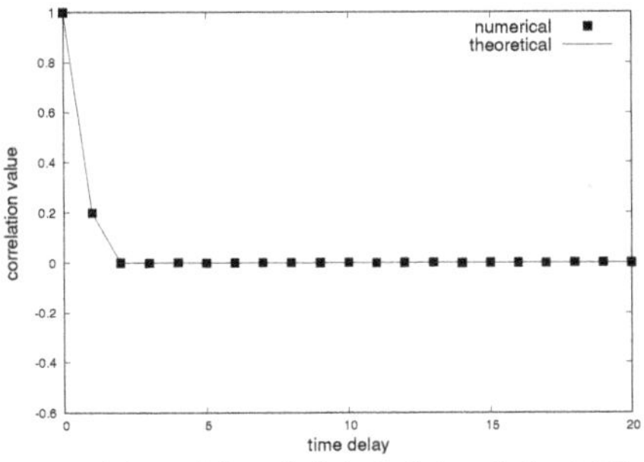

(f) $\tau_T(x)$, $B_1(x) = b_5(x)$, $B_2(x) = b_6(x, 0.35)$

**Figure 8.** Normalized auto-correlation functions of new binary sequences $\{D_n^{(2)}\}_{n=0}^{\infty}$ in Method 2.

## 5. Conclusions

The auto-correlation functions of chaotic binary sequences obtained by alternating two binary functions are discussed. Their theoretical auto-correlation function is given and verified by numerical experiments. The proposed methods give more flexibility in designing chaotic sequences with various auto-correlation properties (for example, they can be applied to the method in [18]). The number of binary functions can be extended to three or larger numbers, which will be discussed in future work.

**Funding:** This work was supported by JSPS KAKENHI, Grant Number JP19K12158.

**Data Availability Statement:** Data are contained within the article.

**Conflicts of Interest:** The author declare no conflict of interest.

## Appendix A. Derivation of Auto-/Cross-Correlation Functions of Chaotic Binary Sequences Used in Numerical Experiments

First, we define the Perron–Frobenius (PF) operator $P_\tau$ of the map $\tau$ with an interval $I = [0, 1]$ by

$$P_\tau G(x) = \frac{d}{dx} \int_{\tau^{-1}([0,x])} G(y)dy \tag{A1}$$

which can be rewritten as

$$P_\tau G(x) = \sum_i |g_i'(x)| G(g_i(x)), \tag{A2}$$

where $g_i(x)$ is the $i$-th preimage of the map $\tau(\cdot)$ [1,2]. The PF operator is very useful in analyzing correlation functions because it has the following important property [1,2]:

$$\int_I G(x) P_\tau\{H(x)\}dx = \int_I G(\tau(x)) H(x)dx. \tag{A3}$$

Using (A3), the cross-correlation function defined by (9) can be rewritten as

$$C(\ell; B_1, B_2) = \int_I P_\tau^\ell\{(B_1(x) - E[B_1])f^*(x)\}(B_2(x) - E[B_2])dx. \tag{A4}$$

Putting $B_1(x) = B_2(x) = B(x)$ in (A4), we have the auto-correlation function. Thus, it is important to calculate $P_\tau^\ell\{(B(x) - E[B])f^*(x)\}$ $(\ell = 1, 2, \cdots)$ for the analysis of the auto-/cross-correlation functions.

*Appendix A.1. Correlation Functions of Chaotic Binary Sequences Generated by Bernoulli Map*

For Bernoulli map $\tau_B(x)$ defined by (23) and the threshold function $\Theta_t(x)$ defined by (26), we have [9]

$$P_\tau\{\Theta_t(x) - E[\Theta_t]\} = \frac{1}{2}(\Theta_{\tau(t)}(x) - E[\Theta_{\tau(t)}]), \tag{A5}$$

where $f^*(x) = 1$ is taken into account. For binary functions $b_1(x)$, $b_2(x)$ defined by (24), (25), we have [9]

$$\begin{cases} P_\tau\{b_1(x) - E[b_1]\} = P_\tau\{\Theta_{\frac{1}{8}}(x) - E[\Theta_{\frac{1}{8}}]\} - P_\tau\{\Theta_{\frac{1}{4}}(x) - E[\Theta_{\frac{1}{4}}]\} + P_\tau\{\Theta_{\frac{1}{2}}(x) - E[\Theta_{\frac{1}{2}}]\} \\ \qquad -P_\tau\{\Theta_{\frac{3}{4}}(x) - E[\Theta_{\frac{3}{4}}]\} + P_\tau\{\Theta_{\frac{7}{8}}(x) - E[\Theta_{\frac{7}{8}}]\}, \\ \qquad = \frac{1}{2}\{(\Theta_{\frac{1}{4}}(x) - E[\Theta_{\frac{1}{4}}]) - 2(\Theta_{\frac{1}{2}}(x) - E[\Theta_{\frac{1}{2}}]) + (\Theta_{\frac{3}{4}}(x) - E[\Theta_{\frac{3}{4}}])\}, \\ P_\tau^2\{b_1(x) - E[b_1]\} = \frac{1}{2}(\Theta_{\frac{1}{2}}(x) - E[\Theta_{\frac{1}{2}}]), \\ P_\tau^\ell\{b_1(x) - E[b_1]\} = 0 \quad (\ell \geq 3), \end{cases} \tag{A6}$$

$$\begin{cases} P_\tau\{b_2(x) - E[b_2]\} = P_\tau\{\Theta_{\frac{1}{4}}(x) - E[\Theta_{\frac{1}{4}}]\} - P_\tau\{\Theta_{\frac{3}{8}}(x) - E[\Theta_{\frac{3}{8}}]\} + P_\tau\{\Theta_{\frac{1}{2}}(x) - E[\Theta_{\frac{1}{2}}]\} \\ \qquad -P_\tau\{\Theta_{\frac{5}{8}}(x) - E[\Theta_{\frac{5}{8}}]\} + P_\tau\{\Theta_{\frac{3}{4}}(x) - E[\Theta_{\frac{3}{4}}]\}, \\ \qquad = \frac{1}{2}\{2(\Theta_{\frac{1}{2}}(x) - E[\Theta_{\frac{1}{2}}]) - (\Theta_{\frac{3}{4}}(x) - E[\Theta_{\frac{3}{4}}]) - (\Theta_{\frac{1}{4}}(x) - E[\Theta_{\frac{1}{4}}])\}, \\ P_\tau^2\{b_2(x) - E[b_2]\} = -\frac{1}{2}(\Theta_{\frac{1}{2}}(x) - E[\Theta_{\frac{1}{2}}]), \\ P_\tau^\ell\{b_2(x) - E[b_2]\} = 0 \quad (\ell \geq 3). \end{cases} \tag{A7}$$

Using (A6) and (A7), we can obtain

$$A(\ell; b_1) = \int_0^1 P_\tau^\ell\{(b_1(x) - E[b_1])\}(b_1(x) - E[b_1])dx = \begin{cases} \frac{1}{4} & (\ell = 0) \\ -\frac{1}{8} & (\ell = 1) \\ \frac{1}{16} & (\ell = 2) \\ 0 & (\ell \geq 3) \end{cases}, \tag{A8}$$

$$A(\ell; b_2) = \int_0^1 P_\tau^\ell\{(b_2(x) - E[b_2])\}(b_2(x) - E[b_2])dx = \begin{cases} \frac{1}{4} & (\ell = 0) \\ 0 & (\ell = 1) \\ -\frac{1}{16} & (\ell = 2) \\ 0 & (\ell \geq 3) \end{cases}, \tag{A9}$$

$$C(\ell; b_1, b_2) = \int_0^1 P_\tau^\ell\{(b_1(x) - E[b_1])\}(b_2(x) - E[b_2])dx = \begin{cases} 0 & (\ell = 0) \\ 0 & (\ell = 1) \\ \frac{1}{16} & (\ell = 2) \\ 0 & (\ell \geq 3) \end{cases}, \tag{A10}$$

$$C(\ell; b_2, b_1) = \int_0^1 P_\tau^\ell\{(b_2(x) - E[b_2])\}(b_1(x) - E[b_1])dx = \begin{cases} 0 & (\ell = 0) \\ \frac{1}{8} & (\ell = 1) \\ -\frac{1}{16} & (\ell = 2) \\ 0 & (\ell \geq 3) \end{cases}, \tag{A11}$$

which give the normalized auto-/cross-correlation functions given in Table 1.

*Appendix A.2. Correlation Functions of Chaotic Binary Sequences Generated by PL Map Defined by (28)*

For the PL map $\tau_{PL}(x)$ defined by (28) and the threshold function $\Theta_t(x)$ defined by (26), we have [13]

$$P_\tau\{\Theta_{\frac{3}{4}}(x) - E[\Theta_{\frac{3}{4}}]\} = -\frac{1}{3}(\Theta_{\frac{3}{4}}(x) - E[\Theta_{\frac{3}{4}}]), \tag{A12}$$

$$P_\tau\{\Theta_{\frac{1}{4}}(x) - E[\Theta_{\frac{1}{4}}]\} = 0, \tag{A13}$$

where $f^*(x) = 1$, $\tau_{PL}(\frac{3}{4}) = \frac{3}{4}$ (i.e., $x = \frac{3}{4}$ is a fixed point on the map), and $\tau_{PL}(\frac{1}{4}) = 0$ are taken into account. For binary functions $b_3(x)$, $b_4(x)$ defined by (29), (30), we immediately obtain, from (A12) and (A13),

$$P_\tau^\ell\{b_3(x) - E[b_3]\} = P_\tau^\ell\{\Theta_{\frac{3}{4}}(x) - E[\Theta_{\frac{3}{4}}]\} = \frac{1}{(-3)^\ell}(\Theta_{\frac{3}{4}}(x) - E[\Theta_{\frac{3}{4}}]), \tag{A14}$$

$$P_\tau^\ell\{b_4(x) - E[b_4]\} = P_\tau^\ell\{1 - \Theta_{\frac{1}{4}}(x) - (1 - E[\Theta_{\frac{1}{4}}])\} = 0 \quad (\ell \geq 1). \tag{A15}$$

Using (A14) and (A15), we can obtain

$$A(\ell; b_3) = \int_0^1 P_\tau^\ell\{(b_3(x) - E[b_3])\}(b_3(x) - E[b_3])dx = \frac{3}{16}(-3)^{-\ell}, \tag{A16}$$

$$A(\ell; b_4) = \int_0^1 P_\tau^\ell\{(b_4(x) - E[b_4])\}(b_4(x) - E[b_4])dx = \begin{cases} \frac{3}{16} & (\ell = 0) \\ 0 & (\ell \geq 1) \end{cases}, \tag{A17}$$

$$C(\ell; b_3, b_4) = \int_0^1 P_\tau^\ell\{(b_3(x) - E[b_3])\}(b_4(x) - E[b_4])dx = -\frac{1}{16}(-3)^{-\ell}, \tag{A18}$$

$$C(\ell; b_4, b_3) = \int_0^1 P_\tau^\ell\{(b_4(x) - E[b_3])\}(b_3(x) - E[b_3])dx = \begin{cases} -\frac{1}{16} & (\ell = 0) \\ 0 & (\ell \geq 1) \end{cases}, \tag{A19}$$

which give the normalized auto-/cross-correlation functions given in Table 2.

*Appendix A.3. Correlation Functions of Chaotic Binary Sequences Generated by Tent Map*

For the tent map $\tau_T(x)$ defined by (32) and the threshold function $\Theta_t(x)$ defined by (26), we have [9]

$$P_\tau\{\Theta_t(x) - E[\Theta_t]\} = \begin{cases} \frac{1}{2}(\Theta_{\tau(t)}(x) - E[\Theta_{\tau(t)}]) & (0 < t < \frac{1}{2}) \\ -\frac{1}{2}(\Theta_{\tau(t)}(x) - E[\Theta_{\tau(t)}]) & (\frac{1}{2} \leq t < 1) \end{cases}, \tag{A20}$$

where $f^*(x) = 1$ is taken into account. For binary functions $b_5(x)$, $b_6(x, d)$ defined by (33), (34), we have

$$P_\tau\{b_5(x) - E[b_5]\} = 0, \tag{A21}$$

$$P_\tau\{b_6(x, d) - E[b_6]\} = 0, \tag{A22}$$

where $\tau_T(d) = \tau_T(1 - d)$ is used. That is, $b_5(x)$ and $b_6(x, d)$ for the tent map satisfy the sufficient condition for the generation of *i.i.d.* binary sequences [9]. Using (A21) and (A22), the auto-correlation functions are given by

$$A(\ell; b_5) = \begin{cases} \frac{1}{4} & (\ell = 0) \\ 0 & (\ell \geq 1) \end{cases}, \tag{A23}$$

$$A(\ell; b_6) = \begin{cases} \frac{1}{4} & (\ell = 0) \\ 0 & (\ell \geq 1) \end{cases}, \quad (0 < d \leq \frac{1}{2}). \tag{A24}$$

From (A21) and (A22), it is obvious that

$$C(\ell; b_5, b_6) = C(\ell; b_6, b_5) = 0 \quad (\ell \geq 1). \tag{A25}$$

Additionally, we have

$$C(0; b_5, b_6) = C(0; b_6, b_5) = d - \frac{1}{4}. \tag{A26}$$

Thus, the normalized auto-/cross-correlation functions given in Table 3 have been derived.

## References


1. Lasota, A.; Mackey, M.C. *Chaos, Fractals, and Noise*; Springer: New York, NY, USA, 1994.
2. Boyarsky, A.; Góra, P. *Laws of Chaos*; Birkhäuser Boston: Boston, MA, USA, 1997.
3. Gerosa, A.; Bernardini, R.; Pietri, S. A fully integrated chaotic system for the generation of truly random numbers. *IEEE Trans. Circuits Syst. I* **2001**, *49*, 993–1000. [CrossRef]
4. Stojanovski, T.; Kocarev, L. Chaos-based random number generators—Part I: Analysis. *IEEE Trans. Circuits Syst. I* **2001**, *48*, 281–288. [CrossRef]
5. Cicek, I.; Pusane, A.E.; Dundar, G. A novel design method for discrete time chaos based true random number generators. *Integr. VLSI J.* **2014**, *47*, 38–47. [CrossRef]
6. Li, C.; Feng, B.; Li, S.; Kurths, J.; Chen, G. Dynamic analysis of digital chaotic maps via state-mapping networks. *IEEE Trans. Circuits Syst. I* **2019**, *66*, 2322–2335. [CrossRef]
7. Kennedy, M.P.; Rovatti, R.; Setti, G. (Eds.) *Chaotic Electronics in Telecommunications*; CRC: Boca Raton, FL, USA, 2000.
8. Liu, Z.; Tang, J.; Yu, J. An application of chaos: Generating binary pseudo-random sequences. In Proceedings of the 1988 IEEE International Symposium on Circuits and Systems, Espoo, Finland, 7–9 June 1988; pp. 1–3.
9. Kohda, T.; Tsuneda, A. Statistics of chaotic binary sequences. *IEEE Trans. Inf. Theory* **1997**, *43*, 104–112. [CrossRef]
10. Sakai, H.; Tokumaru, H. Autocorrelations of a certain chaos. *IEEE Trans. Acoust. Speech Signal Process* **1980**, *28*, 588–590. [CrossRef]
11. Rovatti, R.; Mazzini, G. Interference in DS-CDMA systems with exponentially vanishing autocorrelations: Chaos-based spreading is optimal. *Electron. Lett.* **1998**, *34*, 1911–1913. [CrossRef]
12. Mazzini, G.; Rovatti, R.; Setti, G. Interference minimization by autocorrelation shaping in asynchronous DS-CDMA systems: Chaos-based spreading is nearly optimal. *Electron. Lett.* **1999**, *35*, 1054–1055. [CrossRef]
13. Tsuneda, A. Design of binary sequences with tunable exponential autocorrelations and run statistics based on one-dimensional chaotic maps. *IEEE Trans. Circuits Syst. I* **2005**, *52*, 454–462. [CrossRef]
14. Yang, C.A.; Yao, K.; Umeno, K.; Biglieri, E. Using deterministic chaos for superefficient Monte-Carlo simulations. *IEEE Circuits Syst. Mag.* **2013**, *13*, 26–35. [CrossRef]
15. Souza, C.E.C.; Chaves, D.P.B.; Pimentel, C. One-dimensional pseudo-chaotic sequences based on the discrete Arnold's cat map over $\mathbb{Z}_{3^m}$. *IEEE Trans. Circuits Syst. II* **2021**, *68*, 491–495.
16. Paul, P.S.; Sadia, M.; Hossain, M.R.; Muldrey, B.; Hasan, M.S. Cascading CMOS-based chaotic maps for improved performance and its application in efficient RNG design. *IEEE Access* **2022**, *10*, 33758–33770. [CrossRef]
17. Tang, J.; Zhang, Z.; Chen, P.; Huang, Z.; Huang, T. A simple chaotic model with complex chaotic behaviors and its hardware implementation. *IEEE Trans. Circuits Syst. I* **2023**, *70*, 3676–3688. [CrossRef]
18. Tsuneda, A. Various Auto-correlation functions of $m$-bit random numbers generated from chaotic binary sequences. *Entropy* **2021**, *23*, 1295. [CrossRef]

symmetry

MDPI

*rticle*

# Bayesian Inference of Recurrent Switching Linear Dynamical Systems with Higher-Order Dependence

ouxiang Wang [1] and Jiaqing Chen [1,2,*]

1 School of Science, Wuhan University of Technology, Wuhan 430071, China; 273212@whut.edu.cn
2 Hubei Longzhong Laboratory, Xiangyang 441100, China
* Correspondence: 9523@whut.edu.cn

**Abstract:** Many complicated dynamical events may be broken down into simpler pieces and efficiently described by a system that shifts among a variety of conditionally dynamical modes. Building on switching linear dynamical systems, we develop a new model that extends the switching linear dynamical systems for better discovering these dynamical modes. In the proposed model, the linear dynamics of latent variables can be described by a higher-order vector autoregressive process, which makes it feasible to evaluate the higher-order dependency relationships in the dynamics. In addition, the transition of switching states is determined by a stick-breaking logistic regression, overcoming the limitation of a restricted geometric state duration and recovering the symmetric dependency between the switching states and the latent variables from asymmetric relationships. Furthermore, logistic regression evidence potentials can appear as conditionally Gaussian potentials by utilizing the Pólya-gamma augmentation strategy. Filtering and smoothing algorithms and Bayesian inference for parameter learning in the proposed model are presented. The utility and versatility of the proposed model are demonstrated on synthetic data and public functional magnetic resonance imaging data. Our model improves the current methods for learning the switching linear dynamical modes, which will facilitate the identification and assessment of the dynamics of complex systems.

**Keywords:** dynamic systems; state space methods; Kalman filter; Bayesian inference

## 1. Introduction

Complex systems frequently exhibit multiple levels of abstraction in their descriptions [1]. For example, a computer program can be characterized by the collection of functions it calls, the sequence in which it carries out statements or the assembly instructions it sends to the CPU. This assertion is true for a multitude of natural systems. Brain activity can be classified based on either broad psychological states or the activation of individual ion channels. The necessary amount of specificity may differ based on the particular task being performed [2]. Through the identification of these behavioral units and their interdependence, we can obtain a deeper understanding of the intricate mechanisms that give rise to complex natural occurrences. Furthermore, modern machine learning offers powerful tools to help model the dynamics of complex systems. The toolbox has recently been improved to incorporate more versatile elements, such as Gaussian processes [3] and neural networks [4], into probabilistic time series models.

Time series analysis encompasses several methodologies and models, such as stationary process models, spectral models, state space models, and non-linear models [5]. Among them, the state space methods are regarded as more versatile and adept at addressing a broader range of problems compared to the other models. Hidden Markov models (HMM) and switching linear dynamical systems (SLDS) are two well-known state space models. In numerous real-world time series situations, the present condition of a dynamic system is connected to its condition in prior time intervals. Both HMM and SLDS can address such problems and have been widely used in several problem fields. Examples include human motion [6,7], computer vision [8], speech recognition [9], econometrics [10], machine learning [11] and neuroscience [12,13].

HMM is a discrete Markov process with doubly embedded stochastic models, namely, (i) an unobservable hidden process characterized by a Markov chain and (ii) an observation process determined by hidden states. HMM-based models have garnered significant interest and demonstrated their utility in the signal processing community for accurately simulating the intricate temporal progression of signals. More precisely, HMM have a lengthy track record in signal processing, particularly in the field of speech processing where they have achieved notable success [14]. The SLDS allows for the modeling of nonlinear time series data by dividing it into subsequences controlled by linear dynamics. Indeed, the generative model of SLDS is slightly different from HMM, where there is an additional set of latent variables between the switching states and the observations. In addition, a SLDS differs from a HMM by choosing from a collection of linear Gaussian dynamics that evolve continuously, instead of a standard Gaussian mixture density as in HMM. The SLDS can be viewed as an expansion of the HMM, where each HMM state, or mode, is linked to a linear dynamical process. One further benefit of SLDS is its ability to handle time series data with high dimensionality. From a generative model perspective, it can be claimed that high-dimensional time series can be cheaply represented by a dynamical process specified on a low-dimensional manifold. This representation benefits from the model structure of SLDS. Nevertheless, employing HMM directly on time series with high dimensionality is likely to result in overfitting. [15]. Therefore, SLDS offers the potential for increased descriptive capability [1].

Although SLDS offers advanced predictive models, the dynamics from the data are only stated under certain conditions, which poses some shortcomings of the generative model and may limit its application. First, the continuously evolving linear Gaussian dynamics of latent variables is essentially equivalent to a first-order switching vector autoregressive (VAR) process, i.e., VAR(1) process, which may not be suitable for the case when there are higher-order dependency relationships in the time series. Higher-order autoregressive interactions are more common in time series data [16,17]. Furthermore, determining an appropriate autoregressive order must trade off the model likelihood and complexity, constituting a non-trivial task in the applications [18]. Indeed, a restricted VAR(1) process involved in the linear dynamics of latent variables limits the ability of SLDS to recover the higher-order dependency relationships in dynamical phenomena.

Second, the switching state transition in SLDS follows the Markov assumption, which means that the duration time (i.e., time interval spent in a certain state) is restricted to a geometric distribution [19]. Thus, more weight is given to shorter consecutive time periods within a certain state, which means states switch frequently. This may not be appropriate when the system spends a long time in one state [20]. In addition, the state transitions are based only on the previous state and are unrelated to the latent variables. If a discrete switch happens when the system enters a particular area of the state space, the SLDS cannot recognize this relationship. While hidden semi-Markov model (HSMM) based on the semi-Markov assumption is an extension of the HMM that alleviates this issue, it should be noted that state transitions in HSMM still have no associations with the previous observations [20].

To address the mentioned limitations, an extension of SLDS, the recurrent switching linear dynamical systems with higher-order dependence (HO-rSLDS), is presented. The HO-rSLDS provides a new method that improves the modeling of dynamics from complex systems, by enabling a higher-order dependence in the linear Gaussian dynamics of latent variables and allowing the switching state transition probabilities to hinge on the preceding values of the latent variables. Specifically, to address the first limitation, we enhance the generative process of latent variables by directly making the current latent variable depend on the state values at the previous several time steps, rather than only the previous one time step. Therefore, the proposed model is feasible for capturing the higher-order autoregressive relationships in the data. To address the second limitation, we utilize the stick-breaking logistic regression to determine the switching state transition [21]. In addition, we leverage a Pólya-gamma augmentation approach [22] to improve block inference updating. This enhancement transforms specific logistic regression evidence potentials into conditionally Gaussian potentials, facilitating efficient Bayesian inference procedures. In this way, the transition of switching states is directly associated with the previous switching state and latent variable. Moreover, since the switching states and the latent variables are mutually interdependent in HO-rSLDS, the symmetric dependency among them is recovered, enhancing the generative process of both switching states and latent variables.

Similar to SLDS, the exact inference in an HO-rSLDS is intractable, which impedes efficient model estimation and parameter learning [8]. In this study, the HO-rSLDS is learned through variational inference, which is an approximate scheme to maximize the evidence lower bound, i.e., minimize the Kullback–Leibler divergence between a restricted family of functions of the model parameters and the actual posterior distribution [23]. Moreover, we propose message-passing algorithms about switching states and latent variables in HO-rSLDS to evaluate sufficient statistics for further parameter updates.

The primary contributions of the novel approach suggested in this study are outlined below:

1. HO-rSLDS assumes the linear dynamics of latent variables can be described by a higher-order VAR process, which makes it feasible to dig out and evaluate the long-term dependency relationships from dynamical phenomena;
2. Stick-breaking logistic regression is applied to determine the switching state transition in HO-rSLDS. By this means, the transition probabilities are time-varying and can be adjusted according to the previous latent variable, which overcomes the limitation of restricted geometric state duration time by Markov assumption and recovers the symmetric dependency between the switching states and the latent variable;
3. The Pólya-gamma augmentation strategy permits efficient Bayesian inference algorithms. In addition, we propose message-passing algorithms, including a forward Kalman filter and backward Kalman smoother for HO-rSLDS, which facilitates the parameter update in variational inference.

The rest of this study is organized as follows. In Section 2.3, we discuss the related models. In Section 3, we present the model learning algorithms in detail. In Section 4, we demonstrate the performance of the proposed HO-rSLDS on simulated data. Our method is applied to public functional magnetic resonance imaging (fMRI) data as well, and the results are described in Section 5. Finally, we wrap up our conclusions in Section 6.

## 2. Methods

### 2.1. HMM and HSMM

HMMs are mathematical models that analyze dynamic systems by using random processes with hidden states. A HMM is constructed using a Markov chain and is composed of two layers: a layer representing hidden states and a layer representing observations. The observed sequences are allocated to their respective hidden states according to the observation probability distribution. In the context of HMM, the observations are solely influenced by the current state and are not influenced by any preceding states [20]. The entire state sequence can be presented as $z_{1:T} = \{z_t\}_{t=1}^{T}$, which can be viewed as a sequence over a finite set $\mathcal{Z}$ with cardinality $|\mathcal{Z}|$. The state transition probability from state $i$ to $j$ is defined as $\pi_{ij} = p(z_{t+1} = j|z_t = i)$. The distribution of observations $y_t \in \mathcal{R}^n$ given a specific state is denoted by $p(y_t|z_t, \theta_j)$, where $\theta_j$ denotes the emission parameters of state $j$. Above all, the HMM can be described as:

$$z_t|z_{t-1} \sim \pi_{z_{t-1}}, \tag{1}$$

$$y_t|z_t \sim F(\theta_{z_t}), \tag{2}$$

where $F(\cdot)$ denotes emission distribution.

Compared to HMM, the HSMM formalism improves upon the standard HMM by incorporating a random state length time, chosen from a state-specific distribution, into the generative process. The state remains constant until the period elapses, at which juncture there is a Markov transition to a fresh state. By using this approach, the distribution of state durations is not limited to a geometric form [24,25]. The random variable $D_t$ represents the length of time that a state is entered at time $t$. The associated probability mass function is denoted as $p(D_t|z_t = j)$ and the HSMM can be described as

$$\nu_s|\nu_{s-1} \sim \pi_{\nu_{s-1}}, \tag{3}$$

$$D_s \sim g(\omega_s), \tag{4}$$

$$z_{t_s^1:t_s^{D_s+1}} = \nu_s, \tag{5}$$

$$y_{t_s^1:t_s^{D_s+1}}|z_{t_s^1:t_s^{D_s+1}} \sim F(\theta_{\nu_s}, D_s), \tag{6}$$

where $\nu_s$ indexes the state shared by state segment $s$. A graphical illustration of HMM and HSMM is shown in Figure 1. HMM or HSMM typically uses a Gaussian mixture as the emission distribution, which fails to capture potential dependencies in observed time series. An effective approach to tackle this problem is to employ the VAR emission model, which describes the behavior of time series through linear historical interactions among the observed time series. In this way, the emission model in HMM or HSMM can be specifically expressed as:

$$y_t|z_t = k \sim \mathcal{N}(\sum_{l=1}^{p} W_{k,l}y_{t-l}, \Sigma_k), \tag{7}$$

where $p$ denotes the maximum lag order, $W_{k,l}$ are $n \times n$ dimensional matrices representing the $k$-th state autoregressive coefficient matrices for lag $l$ and the covariance matrix is $\Sigma_k$. Subsequently, we refer to the H(S)MM with VAR(p) emission model as the AR(p)-H(S)MM.

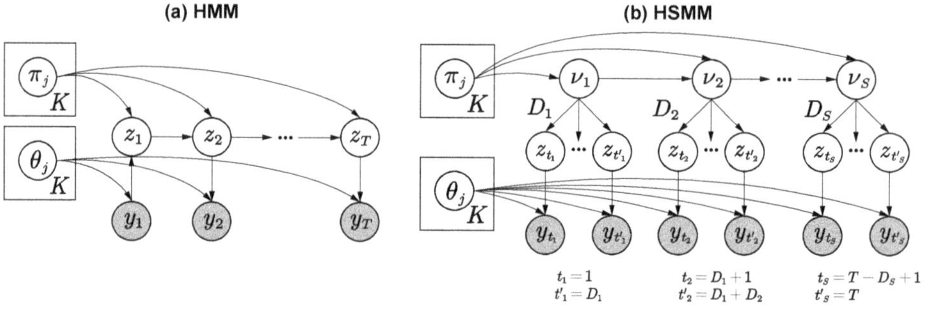


**Figure 1.** Graphical model of HMM (**a**) and HSMM (**b**). An HMM typically consists of two layers: a hidden state layer based on a Markov chain and an observation layer dependent on the current state. An HSMM enhances the generative process of the standard HMM with a random state duration time, which is drawn from some state-specific distribution when the state is entered. $K$ denotes the cardinality of the hidden state space, which can be a finite or infinite value.

### 2.2. SLDS

A SLDS Markov process differs from a hidden Markov model (HMM) by choosing from a collection of linear Gaussian dynamics that evolve continuously, instead of a standard Gaussian mixture density as in HMM. Therefore, SLDS offers the potential for increased descriptive capability.

Similar to HMM, at each time $t = 1, 2, \ldots, T$ there is a discrete switching state $z_t \in \{1, 2, \ldots, K\}$ that evolves according to Markov dynamics:

$$z_{t+1} | z_t, \{\pi_k\}_{k=1}^K \sim \pi_{z_t}, \tag{8}$$

$\{\pi_k\}_{k=1}^K$ constitutes the Markov transition matrix and $\pi_k \in [0,1]^K$ is its $k$-th row. The $z_1$ is assumed to be sampled from an initial distribution $\pi_0$ parameterized by prior parameter $\alpha_0$, i.e., $z_1|\alpha_0 \sim \pi_0(\alpha_0)$. Furthermore, a continuous latent variable $x_t \in \mathbb{R}^m$ follows conditionally linear dynamics depend on the state values with at the previous time step, and the switching state $z_t$ determines the linear dynamical system at time $t$ by:

$$x_{t+1} = A_{z_{t+1}}^{(0)} x_t + v_t, \quad v_t \overset{\text{i.i.d}}{\sim} \mathcal{N}(0, Q_{z_{t+1}}^{(0)}), \tag{9}$$

where $A_{z_{t+1}}^{(0)} \in \mathbb{R}^{m \times m}$ denotes the coefficient matrix and $Q_{z_{t+1}}^{(0)} \in \mathbb{R}^{m \times m}$ is covariance matrix. At each time $t$, a linear Gaussian observation $y_t \in \mathbb{R}^n$ is produced from the associated latent continuous variable:

$$y_t = C^{(0)} x_t + w_t, \quad w_t \overset{\text{i.i.d}}{\sim} \mathcal{N}(0, S^{(0)}), \tag{10}$$

where $C^{(0)} \in \mathbb{R}^{n \times m}$, $S^{(0)} \in \mathbb{R}^{n \times n}$. Usually, $n$ is much larger than $m$, such that the dependence structure of high-dimensional observation data can be evaluated based on the low-dimensional latent variable space. A graphical illustration of SLDS is given in Figure 2.

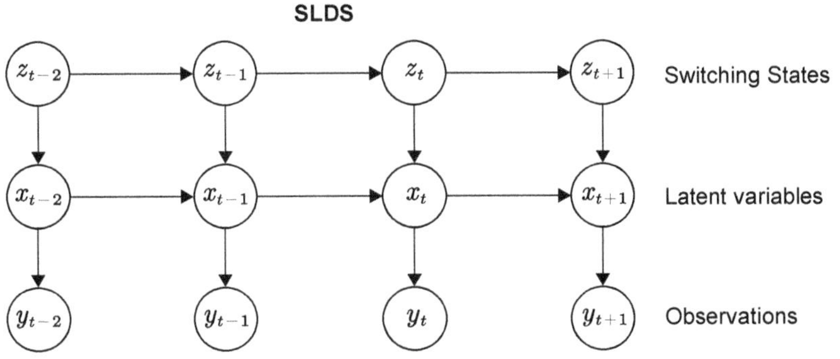


**Figure 2.** The graphical model of the SLDS, which consists of switching states ($z_t$), latent variables ($x_t$) and observations ($y_t$). Note that we assume it does not contain a link from $z_t$ to $y_t$.

In classical form, however, there are some shortcomings in the generative model of SLDS, which may limit its application. First, the continuously-evolving linear Gaussian dynamics of $x_t$ is essentially equivalent to a first-order switching vector autoregressive (VAR) process, i.e., the VAR(1) process, which may not be suitable for the case when there are higher-order dependency relationships in $x_t$ or $y_t$. Substitute Equation (9) into Equation (10) and assume ${C^{(0)}}^\top C^{(0)} = I$ and $w_t$ approximately zero, one can immediately obtain the conclusion that SLDS describes the dependency relationships by a VAR(1) process since $y_t = C^{(0)} A_{z_t}^{(0)} {C^{(0)}}^\top y_{t-1} + C^{(0)} v_t$. Second, the state transition of $z_t$ follows the Markov assumption, which means that the duration time is restricted to a geometric distribution [19]. Thus, more weight is given to shorter consecutive time periods within a certain state, which means states switch frequently. This may not be appropriate when the system spends a long time in one state [20]. In addition, the state $z_{t+1}$ transitions based only on the previous state $z_t$, and $z_{t+1}|z_t$ is unrelated to the latent variable $x_t$. If a discrete switch happens when the system enters a particular area of the state space, the SLDS will not be able to recognize this relationship. To address these issues, the proposed

HO-rSLDS enhances the generative process of $x_t$ and the switching state transition, which is demonstrated in the following subsection.

### 2.3. *Recurrent Switching Linear Dynamic System with Higher-Order Dependence*

In this section, we describe the proposed HO-rSLDS comprising two main components that are switching linear dynamic systems with higher-order dependence (HO-SLDS, without recurrence) and the stick-breaking logistic regression as well as the Pólya-gamma augmentation for HO-SLDS. Each component is demonstrated below.

#### 2.3.1. HO-SLDS

The switching state transition in HO-SLDS is the same as that in SLDS. However, the continuous latent variable $x_t \in \mathcal{R}^m$ follows conditionally linear dynamics depending on the state values at the previous $p$ time steps, rather than only the previous one time step, and the switching state $z_t$ determines the linear dynamical system at time $t$ by:

$$x_{t+1} = A_{z_{t+1}} \bar{x}_t + v_t, \quad v_t \overset{\text{i.i.d}}{\sim} \mathcal{N}(0, Q_{z_{t+1}}), \tag{11}$$

where $\bar{x}_t = [x_{t-1}^\top, \dots, x_{t-p}^\top]^\top \in \mathcal{R}^{pm}$, $A_{z_{t+1}} \in \mathcal{R}^{m \times pm}$ and $Q_{z_{t+1}} \in \mathcal{R}^{m \times m}$. Indeed, the evolution of the latent variable dynamics in $x_t$ can be considered as a VAR$(p)$ process. The initial $\bar{x}_0$ given the initial discrete state $z_0$ is supposed to be normally distributed with mean $\mu_0$ and covariance $\Sigma_0$, i.e., $\bar{x}_0 \sim \mathcal{N}(\mu_0, \Sigma_0)$. Similarly, at each time $t$, a linear Gaussian observation $y_t \in \mathcal{R}^n$ is produced from the associated latent continuous variable:

$$y_t = C_{z_t} x_t + w_t, \quad w_t \overset{\text{i.i.d}}{\sim} \mathcal{N}(0, S_{z_t}), \tag{12}$$

where $C_{z_t} \in \mathcal{R}^{n \times m}$, $S_{z_t} \in \mathcal{R}^{n \times n}$. Note that here we assume $C_{z_t}$ and $S_{z_t}$ are related to $z_t$. Usually, $n$ is much larger than $m$, such that the dependence structure of high-dimensional observation data can be evaluated based on the low-dimensional latent variable space. The system parameters consist of the discrete Markov transition matrix and the collection of linear dynamical system matrices, denoted as:

$$\theta = \pi_0 \cup \{\mu_0, \Sigma_0, \pi_k, A_k, Q_k, C_k, S_k\}_{k=1}^K. \tag{13}$$

#### 2.3.2. Stick-Breaking Logistic Regression and Pólya-Gamma Augmentation for HO-SLDS

Another component included in HO-rSLDS is a stick-breaking logistic regression. We utilize a Pólya-gamma augmentation technique to improve block inference updates [21]. This approach allows certain logistic regression evidence potentials to transform to conditionally Gaussian potentials in an augmented distribution, which facilitates the Bayesian inference algorithms.

Consider a logistic regression model with regressors $x \in \mathcal{R}^m$ that maps to a categorical distribution over the switching state $z \in \{1, 2, \dots, K\}$, denoted as:

$$z | x \sim \pi_{\text{SB}}(\nu), \quad \nu = Rx + r, \tag{14}$$

where $R \in \mathcal{R}^{K-1 \times M}$ represents a weight matrix and $r \in \mathcal{R}^{K-1}$ denotes a bias vector. We use a stick-breaking link function $\pi_{\text{SB}} : \mathcal{R}^{K-1} \to [0,1]^K$, which transforms a real vector into a normalized probability vector based on the stick-breaking process:

$$\pi_{\text{SB}}(\nu) = \left(\pi_{\text{SB}}(\nu)^{(1)}, \dots, \pi_{\text{SB}}(\nu)^{(K)}\right), \tag{15}$$

$$\pi_{\text{SB}}^{(k)} = \sigma(\nu_k) \prod_{j<K} \sigma(-\nu_j), \tag{16}$$

for $k = 1, \cdots, K-1$ and $\pi_{\text{SB}}^{(K)} = \prod_{k=1}^{K-1} \sigma(-\nu_k)$ where $\nu_k$ represents the $k$th component of $\nu$ and $\sigma(\nu) = (1 + e^{-\nu})^{-1}$ is the logistic function. The probability mass function $p(z|x)$ is given by

$$p(z|x) = \prod_{k=1}^{K-1} \sigma(\nu_k)^{\mathbb{I}(z=k)} \sigma(-\nu_k)^{\mathbb{I}(z>k)}, \tag{17}$$

where $\mathbb{I}(\cdot)$ is the indicator function.

The posterior density $p(x|z)$ is non-Gaussian and does not allow straightforward Bayesian inference for this regression model due to the likelihood $p(z|x)$ not aligning with a Gaussian prior density $p(x)$. To solve this problem, we introduce Pólya-gamma auxiliary variables $\omega = \{\omega_k\}_{k=1}^{K}$ so that the conditional density $p(x|z, \omega)$ becomes Gaussian [22]. Specifically, the conditional density of $\omega_k$ is distributed according to a Pólya-Gamma distribution. A random variable X has a Pólya-Gamma distribution with parameters $b > 0$ and $c \in \mathcal{R}$, denoted as $X \sim \text{PG}(b, c)$, if

$$X \overset{D}{=} \frac{1}{2\pi^2} \sum_{k=1}^{\infty} \frac{g_k}{(k - 1/2)^2 + c^2/(4\pi^2)}, \tag{18}$$

where the $g_k \sim \text{Gamma}(b, 1)$ are independent Gamma random variables, and where $\overset{D}{=}$ indicates equality in distribution. In particular, by choosing $\omega_k | x, z \sim \text{PG}(\mathbb{I}(z > k), \nu_k)$, we have the following:

$$x|z, \omega \sim \mathcal{N}(\Omega^{-1}\kappa, \Omega^{-1}), \tag{19}$$

where the mean vector $\Omega^{-1}\kappa$ and covariance matrix $\Omega^{-1}$ are calculated based on the following:

$$\Omega = \text{diag}(\omega), \tag{20}$$

$$\kappa_k = \mathbb{I}(z = k) - \frac{1}{2}\mathbb{I}(z > k). \tag{21}$$

Thus allowing efficient block updates while maintaining Gaussian posterior distribution $p(x|z)$ in Bayesian inference.

The HO-rSLDS splits the latent space into $K$ sections, with each section following its own linear dynamics by including the recurrence (Equation (17)) in the transition density of $z_t$, which can be described as follows:

$$z_{t+1} | z_t, x_t, R_{z_t}, r_{z_t} \sim \pi_{\text{SB}}(\nu_t), \tag{22}$$

$$\nu_t = R_{z_t} x_t + r_{z_t}. \tag{23}$$

Moreover, we place Gaussian priors on $R_{z_t}$ and $r_{z_t}$. In this way, the probability mass function $p(z_{t+1}|x_t)$ can be expressed as:

$$\begin{aligned} p(z_{t+1}|x_t) &= \prod_{k=1}^{K-1} \sigma(\nu_k)^{\mathbb{I}(z_{t+1}=k)} \sigma(-\nu_k)^{\mathbb{I}(z_{t+1}>k)} \\ &= \prod_{k=1}^{K-1} \frac{(e^{\nu_{t,k}})^{\mathbb{I}(z_{t+1}=k)}}{(1 + e^{\nu_{t,k}})^{\mathbb{I}(z_{t+1}>k)}}. \end{aligned} \tag{24}$$

The Pólya-gamma augmentation focuses on exactly the above densities, utilizing the following integral identity:

$$\frac{(e^{\nu})^a}{(1 + e^{\nu})^b} = 2^{-b} e^{\kappa \nu} \int_0^{\infty} e^{-\omega \nu^2} p_{\text{PG}}(\omega | b, 0) d\omega, \tag{25}$$

where $\kappa = a - b/2$ and $p_{\text{PG}}(\omega|b,0)$ refers to the density of the Pólya-gamma distribution, $\text{PG}(b,0)$, which is independent of $\nu$ [22]. Combing Equations (24) and (25), $p(z_{t+1}|x_t)$ can be expressed as the marginalization of a component in the probability distribution $p(z_{t+1}|x_t,\omega_t)$, where $\omega_t \in \mathbb{R}^{K-1}$ represents a vector of auxiliary variables. Dependent on $\nu_t$, there is

$$p(z_{t+1}, x_t, \omega_t) \propto \prod_{k=1}^{K-1} \exp\left\{-\frac{1}{2}\omega_{t,k}\nu_{t,k}^2 + \kappa_{t+1,k}\nu_{t,k}\right\}, \tag{26}$$

where $\kappa_{t+1,k} = \mathbb{I}(z_{t+1} = k) - \frac{1}{2}\mathbb{I}(z_{t+1} > k)$. Hence,

$$p(z_{t+1}, x_t, \omega_t) \propto \mathcal{N}(\nu_t|\Omega_t^{-1}\kappa_{t+1}, \Omega_t^{-1}), \tag{27}$$

where $\Omega_t = \text{diag}(\omega_t)$ and $\kappa_{t+1} = (\kappa_{t+1,1}, \ldots, \kappa_{t+1,K-1})$. After augmentation, the conditional density on $x_t$ becomes virtually Gaussian, simplifying the evaluation of the integrals needed for message passing during the Bayesian inference. Finally, we summarize the proposed HO-rSLDS as

$$\begin{aligned}
&z_1|\theta \sim \pi_0(\alpha_0),\\
&\bar{x}_0|\theta \sim \mathcal{N}(\bar{x}_0|\mu_0, \Sigma_0),\\
&z_{t+1}|z_t, x_t, \theta \sim \pi_{\text{SB}}(\nu_t), \quad \nu_t = R_{z_t}x_t + r_{z_t},\\
&x_{t+1}|\bar{x}_t, z_{t+1}, \theta \sim \mathcal{N}(x_{t+1}|A_{z_{t+1}}\bar{x}_t, Q_{z_{t+1}}),\\
&y_t|x_t, z_t, \theta \sim \mathcal{N}(y_t|C_{z_t}x_t, S_{z_t}),\\
&\theta = \{\alpha_0, \mu_0, \Sigma_0\} \cup \{A_k, Q_k, C_k, S_k, R_k, r_k\}_{k=1}^{K}.
\end{aligned} \tag{28}$$

To learn the HO-rSLDS utilizing Bayesian inference, we assign conjugate priors to each component of the model parameters $\theta$, given by

$$\begin{aligned}
&\pi_0|\alpha_0 \sim \text{Dir}(\alpha_0),\\
&(A_k, Q_k)|\eta \sim \text{MNIW}(\eta),\\
&(C_k, S_k)|\lambda \sim \text{MNIW}(\lambda),\\
&R_k|\zeta \sim \prod_{i=1}^{K-1}\mathcal{N}(R_{k,i}|\zeta_i),\\
&r_k|\rho \sim \mathcal{N}(r_k|\rho),
\end{aligned} \tag{29}$$

where Dir and MNIW denote Dirichelet and matrix normal inverse Wishart distribution, respectively, $R_{k,i}$ denotes the $i$th row of $R_k$. A graphical illustration of HO-rSLDS is given in Figure 3.

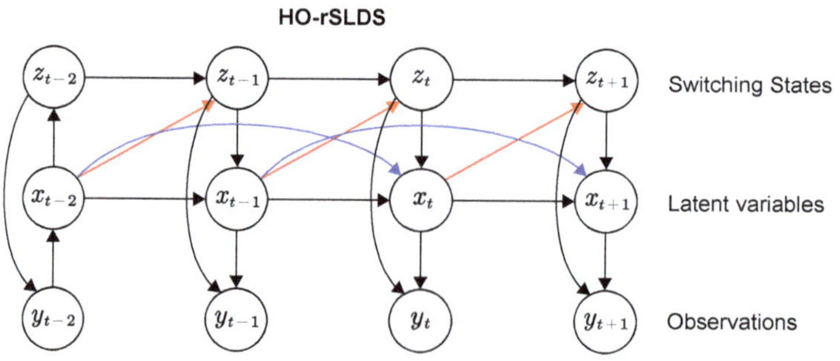


**Figure 3.** The graphical model of the HO-rSLDS, where there is higher-order dependence involved in $x_t$ (blue) and the recurrent dependence from $x_{t-1}$ to $z_t$ (red).

## 3. Variational Inference of HO-rSLDS

Now we describe the model learning inference of HO-rSLDS, which is fulfilled by variational inference. In variational inference, a parameter is inferred by maximizing the evidence's lower bound, hence minimizing the Kullback–Leibler divergence as well as the variational free energy. Specifically, we denote $\mathcal{Y} := y_{1:T}$, $\Theta := \{z_{1:T}, (\bar{x}_0, x_{1:T}), \omega_{1:T}, \theta\}$, then the following decomposition holds:

$$\ln p(\mathcal{Y}) = \mathcal{L}(q(\Theta)) + \mathrm{KL}(q(\Theta)||p(\Theta|\mathcal{Y})), \tag{30}$$

where

$$\mathcal{L}(q(\Theta)) = \int q(\Theta) \log \frac{p(\mathcal{Y}, \Theta)}{q(\Theta)} d\Theta, \tag{31}$$

$$\mathrm{KL}(q(\Theta)||p(\Theta|\mathcal{Y})) = -\int q(\Theta) \log \frac{p(\Theta|\mathcal{Y})}{q(\Theta)} d\Theta. \tag{32}$$

The exact evaluation of $p(z_{1:T}, x_{0:T}, \omega_{1:T}|y_{1:T})$ where $x_{0:T} := \bar{x}_0 \cup x_{1:T}$ in HO-rSLDS is intractable. In this study, we leverage an approximate variational inference method to assess the approximate posterior $q(z_{1:T}, x_{0:T}, \omega_{1:T}|y_{1:T})$ in the proposed HO-rSLDS. In order to obtain a tractable inference, a structured mean field approximation is adopted on the generative model, given by

$$q(z_{1:T}, x_{0:T}, \omega_{1:T}, \theta|y_{1:T}) = q(z_{1:T})q(x_{0:T})q(\omega_{1:T})q(\theta). \tag{33}$$

Moreover, $q(\omega_{1:T})$ further factorizes as:

$$q(\omega_{1:T}) = \prod_{t=1}^{T}\prod_{k=1}^{K-1} q(\omega_{t,k}). \tag{34}$$

Given the factorized forms in Equations (33) and (34), we can obtain the optimal solution of the factorized variational posteriors by maximizing the evidence lower bound in Equation (31), i.e., $q(z_{1:T}), q(x_{0:T}), q(\omega_{1:T})$ and $q(\theta)$, given by

$$\ln q(z_{1:T}) \propto \mathbb{E}_{-q(z_{1:T})} \ln p(z_{1:T}, x_{0:T}, \omega_{1:T}, \theta, y_{1:T}), \tag{35}$$

$$\ln q(x_{0:T}) \propto \mathbb{E}_{-q(x_{0:T})} \ln p(z_{1:T}, x_{0:T}, \omega_{1:T}, \theta, y_{1:T}), \tag{36}$$

$$\ln q(\omega_{1:T}) \propto \mathbb{E}_{-q(\omega_{1:T})} \ln p(z_{1:T}, x_{0:T}, \omega_{1:T}, \theta, y_{1:T}), \tag{37}$$

$$\ln q(\theta) \propto \mathbb{E}_{-q(\theta)} \ln p(z_{1:T}, x_{0:T}, \omega_{1:T}, \theta, y_{1:T}), \tag{38}$$

where $\mathbb{E}_{-q(z_{1:T})}(\cdot) := \mathbb{E}_{q(x_{0:T})q(\omega_{1:T})q(\theta)}(\cdot)$. Hence the joint log-likelihood is

$$\ln p(z_{1:T}, x_{0:T}, \omega_{1:T}, \theta, y_{1:T}) = \ln p(z_{1:T}, x_{0:T}, \omega_{1:T}, y_{1:T}|\theta) + \ln p(\theta). \tag{39}$$

The conditional log-likelihood $\ln p(z_{1:T}, x_{0:T}, \omega_{1:T}, y_{1:T}|\theta)$ can be written as

$$\begin{aligned}
&\ln p(z_{1:T}, x_{0:T}, \omega_{1:T}, y_{1:T}|\theta) = \\
&\ln p(z_1|\theta) + \ln p(\bar{x}_0|\theta) + \sum_{t=1}^{T}[\ln p(z_t|x_{t-1}, z_{t-1}, \omega_{t-1}, \theta)\mathbb{I}(t > 1) + \ln(x_t|\bar{x}_{t-1}, z_t, \theta)] \\
&+ \sum_{t=1}^{T} p(y_t|z_t, x_t, \theta)
\end{aligned} \tag{40}$$

$$\begin{aligned}
&\propto \ln p(z_1|\theta) - \frac{1}{2}(\bar{x}_0 - \mu_0)^\top(\Sigma_0)^{-1}(\bar{x}_0 - \mu_0) \\
&- \frac{1}{2}\sum_{t=1}^{T}\Big[(\nu_{t-1} - \Omega_{t-1}^{-1}\kappa_t)^\top\Omega_{t-1}(\nu_{t-1} - \Omega_{t-1}^{-1}\kappa_t)\mathbb{I}(t > 1) \\
&\qquad + (x_t - A_{z_t}\bar{x}_{t-1})^\top Q_{z_t}^{-1}(x_t - A_{z_t}\bar{x}_{t-1})\Big] \\
&- \frac{1}{2}\sum_{t=1}^{T}(y_t - C_{z_t}x_t)^\top S_{z_t}^{-1}(y_t - C_{z_t}x_t).
\end{aligned} \tag{41}$$

The exact optimal solutions for $q(z_{1:T})$ and $q(x_{0:T})$ will not be explicitly evaluated; rather, we use message passing algorithms to obtain the sufficient statistics with respect to these factors. In the following subsections, we present the involved message passing algorithms and the detailed derivations of update formula for each factor.

### 3.1. *Update for* $q(x_{0:T})$

Since $\mathbb{E}_{-q(x_{0:T})} \ln p(\theta)$ is not a function of $x_{0:T}$, Equation (36) can be written as

$$
\begin{aligned}
\ln q(x_{0:T}) &\propto \mathbb{E}_{-q(x_{0:T})} \ln p(z_{1:T}, x_{0:T}, \omega_{1:T}, y_{1:T}|\theta), \\
&\propto \mathbb{E}_{-q(x_{0:T})}\left[-\frac{1}{2}(\bar{x}_0 - \mu_0)^\top (\Sigma_0)^{-1}(\bar{x}_0 - \mu^0)\right] \\
&+ \mathbb{E}_{-q(x_{0:T})}\left[-\frac{1}{2}\sum_{t=2}^{T}(\nu_{t-1} - \Omega_{t-1}^{-1}\kappa_t)^\top \Omega_{t-1}(\nu_{t-1} - \Omega_{t-1}^{-1}\kappa_t)\right] \\
&+ \mathbb{E}_{-q(x_{0:T})}\left[-\frac{1}{2}\sum_{t=1}^{T}(x_t - A_{z_t}\bar{x}_{t-1})^\top Q_{z_t}^{-1}(x_t - A_{z_t}\bar{x}_{t-1})\right] \\
&+ \mathbb{E}_{-q(x_{0:T})}\left[-\frac{1}{2}\sum_{t=1}^{T}(y_t - C_{z_t}x_t)^\top S_{z_t}^{-1}(y_t - C_{z_t}x_t)\right].
\end{aligned} \tag{42}
$$

From Equation (42) it can be observed that the structure of the variational posterior $\ln q(x_{0:T})$ resembles the joint log-likelihood function of a time-varying linear dynamical system with higher-order dependence (HO-LDS). Thus, the $\ln q(x_{0:T})$ can be re-expressed as

$$
\begin{aligned}
\ln q(x_{0:T}) \propto &-\frac{1}{2}(\bar{x}_0 - \hat{\mu}_0)^\top (\hat{\Sigma}_0)^{-1}(\bar{x}_0 - \hat{\mu}_0) \\
&-\frac{1}{2}\sum_{t=1}^{T}(x_t - \hat{A}_t\bar{x}_{t-1})^\top \hat{Q}_t^{-1}(x_t - \hat{A}_t\bar{x}_{t-1}) \\
&-\frac{1}{2}\sum_{t=1}^{T}(y_t - \hat{C}_t x_t)^\top \hat{S}_t^{-1}(y_t - \hat{C}_t x_t),
\end{aligned} \tag{43}
$$

and the set of variational parameters involved in Equation (43) are defined as

$$
\hat{\theta}_{x_{0:T}} := \left\{\hat{\mu}_0, \hat{\Sigma}_0, \{\hat{A}_t\}_{t=1}^{T}, \{\hat{Q}_t\}_{t=1}^{T}, \{\hat{C}_t\}_{t=1}^{T}, \{\hat{S}_t\}_{t=1}^{T}\right\}. \tag{44}
$$

Each component in the $\hat{\theta}_{x_{0:T}}$ can be precisely determined by matching the coefficients of each term in Equations (42) and (43). Algorithm 1 outlines an efficient scheme for this purpose.

---

**Algorithm 1:** Obtaining $\hat{\theta}_{x_{0:T}}$

---

**for** $t = T : 1$ **do**
- $\hat{S}_t^{-1} \leftarrow \mathbb{E}_{-q(x_{0:T})}[S_{z_t}^{-1}]$;
- $\hat{C}_t \leftarrow \hat{S}_t \mathbb{E}_{-q(x_{0:T})}[S_{z_t}^{-1} C_{z_t}]$;
- $\hat{Q}_t^{-1} \leftarrow$
$$
\begin{cases}
\mathbb{E}_{-q(x_{0:T})}\left[Q_{z_t}^{-1} + C_{z_t}^\top S_{z_t}^{-1} C_{z_t}\right] - \hat{C}_T^\top \hat{S}_T^{-1}\hat{C}_T, & t = T \\
\mathbb{E}_{-q(x_{0:T})}\left[Q_{z_t}^{-1} + \sum_{j=1}^{p} A_{z_{t+j}}^\top(j) Q_{z_{t+j}}^{-1} A_{z_{t+j}}(j) + C_{z_t}^\top S_{z_t}^{-1} C_{z_t} + R_{z_t}^\top \Omega_t R_{z_t}\right] & ; \\
\quad - \sum_{j=1}^{p} \hat{A}_{t+j}^\top(j)\hat{Q}_{t+j}^{-1}\hat{A}_{t+j}(j) - \hat{C}_t^\top \hat{S}_t^{-1}\hat{C}_t, & t < T
\end{cases}
$$
- $\hat{A}_t \leftarrow \hat{Q}_t \mathbb{E}_{-q(x_{0:T})}[Q_{z_t}^{-1} A_{z_t}]$;

**end**

$\hat{\Sigma}_0^{-1} \leftarrow \mathbb{E}_{-q(x_{0:T})}[\Sigma^0 + A_{z_1}^\top Q_{z_1} A_{z_1}] - \hat{A}_1^\top \hat{Q}_1 \hat{A}_1$;

$\hat{\mu}_0 \leftarrow \hat{\Sigma}_0 \mathbb{E}_{-q(x_{0:T})}[(\Sigma^0)^{-1}\mu^0]$;

**Output:** $\hat{\theta}_{x_{0:T}}$

---

**Remark 1.** *$A_{z_t}\bar{x}_{t-1} = \sum_{j=1}^{p} A_{z_t}(j)x_{t-j}$ where $A_{z_t}(j)$ is defined as the submatrix of $A_{z_t}$ associated with $x_{t-j}$ in $A_{z_t}\bar{x}_{t-1}$.*

Thus, we perform message passing on a time-varing HO-LDS parameterized by $\hat{\theta}_{x_{0:T}}$ to obtain the posterior distribution $q(x_{0:T})$ and sufficient statistics used in the update for the other factors. To be concrete, the message passing comprises the generalized Kalman filter and Kalman smoother for the time-varing HO-LDS, which are demonstrated by Theorems 1 and 2. To prove the theorems, we need the following lemmas [26]:

**Lemma 1.** *If random variables X and Y follow the Gaussian probability distributions:*

$$X \sim \mathcal{N}(m, P), \tag{45}$$
$$Y|X \sim \mathcal{N}(HX + u, R), \tag{46}$$

*then the joint distribution of X and Y, as well as the marginal distribution of Y, are provided by*

$$\begin{aligned} \begin{pmatrix} X \\ Y \end{pmatrix} &\sim \mathcal{N}\left(\begin{pmatrix} m \\ Hm + u \end{pmatrix}, \begin{pmatrix} P & PH^\top \\ HP & HPH^\top + R \end{pmatrix}\right), \\ Y &\sim \mathcal{N}\left(Hm + u, HPH^\top + R\right). \end{aligned} \tag{47}$$

**Lemma 2.** *If the random variables X and Y follow the joint Gaussian probability distribution*

$$\begin{pmatrix} X \\ Y \end{pmatrix} \sim \mathcal{N}\left(\begin{pmatrix} a \\ b \end{pmatrix}, \begin{pmatrix} A & C \\ C^\top & B \end{pmatrix}\right), \tag{48}$$

*then the marginal and conditional distributions of X and Y are as follows:*

$$\begin{aligned} X &\sim \mathcal{N}(a, A), \\ Y &\sim \mathcal{N}(b, B), \\ X|Y &\sim \mathcal{N}\left(a + CB^{-1}(Y - b), A - CB^{-1}C^\top\right), \\ Y|X &\sim \mathcal{N}\left(b + C^\top A^{-1}(X - a), B - C^\top A^{-1}C\right). \end{aligned} \tag{49}$$

Now we demonstrate Theorems 1 and 2 in turn, which provides the Kalman filter and smoother equations about $q(x_{0:T})$ in the time-varying HO-LDS.

**Theorem 1.** *The Kalman filter equations for the time-varing HO-LDS (Equation (43)) parameterized by $\hat{\theta}_{x_{0:T}}$ can be evaluated in closed form, which results in Gaussian distributions:*

$$p(x_t|y_{1:t-1}) = \mathcal{N}(x_t|\mu_t^{(p)}, \Sigma_t^{(p)}), \tag{50}$$
$$p(x_t|y_{1:t}) = \mathcal{N}(x_t|\mu_t^{(u)}, \Sigma_t^{(u)}), \tag{51}$$
$$p(\bar{x}_t|y_{1:t}) = \mathcal{N}(\bar{x}_t|\mu_t^{(m)}, \Sigma_t^{(m)}). \tag{52}$$

*The distribution parameters can be calculated using the following Kalman filter prediction, update and merging steps detailed in the following proof. The recursion begins with the prior mean $\mu_0^m$ and covariance $\Sigma_0^m$ (i.e., $\hat{\mu}_0$ and $\hat{\Sigma}_0$).*

**Proof.** The Gaussian filter distributions (Equations (50)–(52)) can be obtained by the following steps:

1. Prediction step. By Lemma 1, the joint distribution of $x_t$ and $\bar{x}_{t-1}$ given $y_{1:t-1}$ is

$$\begin{aligned} p(x_t, \bar{x}_{t-1} | y_{1:t-1}) &= p(x_t | \bar{x}_{t-1}) p(\bar{x}_{t-1} | y_{1:t-1}) \\ &= \mathcal{N}(x_t | \hat{A}_t \bar{x}_{t-1}, \hat{Q}_t) \mathcal{N}(\bar{x}_{t-1} | \mu_{t-1}^{(m)}, \Sigma_{t-1}^{(m)}) \\ &= \mathcal{N}\left( \begin{pmatrix} \bar{x}_{t-1} \\ x_t \end{pmatrix} \middle| \mu^{(1)}, \Sigma^{(1)} \right), \end{aligned} \tag{53}$$

where

$$\mu^{(1)} = \begin{pmatrix} \mu_{t-1}^{(m)} \\ \hat{A}_t \mu_{t-1}^{(m)} \end{pmatrix}, \quad \Sigma^{(1)} = \begin{pmatrix} \Sigma_{t-1}^{(m)} & \Sigma_{t-1}^{(m)} \hat{A}_t^\top \\ \hat{A}_t \Sigma_{t-1}^{(m)} & \hat{A}_t \Sigma_{t-1}^{(m)} \hat{A}_t^\top + \hat{Q}_t \end{pmatrix}, \tag{54}$$

and by Lemma 2 the marginal distribution of $x_t$ is given by

$$p(x_t | y_{1:t-1}) = \mathcal{N}(x_t | \mu_t^{(p)}, \Sigma_t^{(p)}), \tag{55}$$

where

$$\mu_t^{(p)} = \hat{A}_t \mu_{t-1}^{(m)}, \quad \Sigma_t^{(p)} = \hat{A}_t \Sigma_{t-1}^{(m)} \hat{A}_t^\top + \hat{Q}_t. \tag{56}$$

2. Update step. By Lemma 1, the joint distribution of $y_t$ and $x_t$ given $y_{1:t-1}$ is

$$\begin{aligned} p(x_t, y_t | y_{1:t-1}) &= p(y_t | x_t) p(x_t | y_{1:t-1}) \\ &= \mathcal{N}(y_t | \hat{C}_t x_t, \hat{S}_t) \mathcal{N}(x_t | \mu_t^{(p)}, \Sigma_t^{(p)}) \\ &= \mathcal{N}\left( \begin{pmatrix} x_t \\ y_t \end{pmatrix} \middle| \mu^{(2)}, \Sigma^{(2)} \right), \end{aligned} \tag{57}$$

where

$$\mu^{(2)} = \begin{pmatrix} \mu_t^{(p)} \\ \hat{C}_t \mu_t^{(p)} \end{pmatrix}, \quad \Sigma^{(2)} = \begin{pmatrix} \Sigma_t^{(p)} & \Sigma_t^{(p)} \hat{C}_t^\top \\ \hat{C}_t \Sigma_t^{(p)} & \hat{C}_t \Sigma_t^{p} \hat{C}_t^\top + \hat{S}_t \end{pmatrix}. \tag{58}$$

By Lemma 2 the conditional distribution of $x_t$ given $y_{1:t}$ is

$$\begin{aligned} p(x_t | y_t, y_{1:t-1}) &= p(x_t | y_{1:t}) \\ &= \mathcal{N}(x_t | \mu^{(u)}, \Sigma^{(u)}), \end{aligned} \tag{59}$$

where

$$\mu^{(u)} = \mu_t^{(p)} + \Sigma_t^{(p)} \hat{C}_t^\top (\hat{C}_t \Sigma_t^{(p)} \hat{C}_t^\top + \hat{S}_t)^{-1} (y_t - \hat{C}_t \mu_t^{(p)}), \tag{60}$$

$$\Sigma^{(u)} = \Sigma_t^{(p)} - \Sigma_t^{(p)} \hat{C}_t^\top (\hat{C}_t \Sigma_t^{(p)} \hat{C}_t^\top + \bar{S}_t)^{-1} \hat{C}_t \Sigma_t^{(p)}. \tag{61}$$

3. Merging step. By Lemma 1, the joint distribution of $\bar{x}_{t-1}, x_t, y_t$ given $y_{1:t-1}$ is

$$\begin{aligned} p(\bar{x}_{t-1}, x_t, y_t | y_{1:t-1}) &= p(y_t | x_t) p(x_t, \bar{x}_{t-1} | y_{1:t-1}) \\ &= \mathcal{N}(\hat{C}_t x_t, \hat{S}_t) \mathcal{N}\left( \begin{pmatrix} \bar{x}_{t-1} \\ x_t \end{pmatrix} \middle| \mu^{(1)}, \Sigma^{(1)} \right), \\ &= \mathcal{N}\left( \begin{pmatrix} \bar{x}_{t-1} \\ x_t \\ y_t \end{pmatrix} \middle| \mu^{(3)}, \Sigma^{(3)} \right), \end{aligned} \tag{62}$$

where

$$\mu^{(3)} = \begin{pmatrix} \mu_1^{(3)} \\ \mu_2^{(3)} \\ \mu_3^{(3)} \end{pmatrix} = \begin{pmatrix} \mu_{t-1}^{(m)} \\ \hat{A}_t \mu_{t-1}^{(m)} \\ \hat{C}_t \hat{A}_t \mu_{t-1}^{(m)} \end{pmatrix}, \tag{63}$$

$$\Sigma^{(3)} = \begin{pmatrix} \Sigma_{11}^{(3)} & \Sigma_{12}^{(3)} & \Sigma_{13}^{(3)} \\ \Sigma_{21}^{(3)} & \Sigma_{21}^{(3)} & \Sigma_{23}^{(3)} \\ \Sigma_{31}^{(3)} & \Sigma_{32}^{(3)} & \Sigma_{33}^{(3)} \end{pmatrix},$$
$$= \begin{pmatrix} \Sigma_{t-1}^{(m)} & \Sigma_{t-1}^{(m)} \hat{A}_t^\top & \Sigma_{t-1}^{(m)} \hat{A}_t^\top \hat{C}_t^\top \\ \hat{A}_t \Sigma_{t-1}^{(m)} & \hat{A}_t \Sigma_{t-1}^{(m)} \hat{A}_t^\top + \hat{Q}_t & (\hat{A}_t \Sigma_{t-1}^{(m)} \hat{A}_t^\top + \hat{Q}_t) \hat{C}_t^\top \\ \hat{C}_t \hat{A}_t \sigma_{t-1}^{(m)} & \hat{C}_t (\hat{A}_t \Sigma_{t-1}^{(m)} \hat{A}^\top + \hat{Q}_t) & \hat{C}_t (\hat{A}_t \Sigma_{t-1}^{(m)} \hat{A}_t^\top + \hat{Q}_t) \hat{C}_t^\top + \hat{S}_t \end{pmatrix}. \tag{64}$$

By Lemma 2, the conditional distribution of $\bar{x}_{t-1}$ and $x_t$ given $y_{1:t}$ is

$$\begin{aligned} p(\bar{x}_{t-1}, x_t | y_t, y_{1:t-1}) &= p(\bar{x}_{t-1}, x_t | y_{1:t}) \\ &= \mathcal{N}(\bar{x}_{t-1}, x_t | \mu^{(4)}, \Sigma^{(4)}), \end{aligned} \tag{65}$$

where

$$\mu^{(4)} = \begin{pmatrix} \mu_1^{(3)} \\ \mu_2^{(3)} \end{pmatrix} + \begin{pmatrix} \Sigma_{13}^{(3)} \\ \Sigma_{23}^{(3)} \end{pmatrix} (\Sigma_{33}^{(3)})^{-1} (y_t - \mu_3^{(3)}), \tag{66}$$

$$\Sigma^{(4)} = \Sigma_{11}^{(3)} - \begin{pmatrix} \Sigma_{13}^{(3)} \\ \Sigma_{23}^{(3)} \end{pmatrix} (\Sigma_{33}^{(3)})^{-1} \begin{pmatrix} \Sigma_{13}^{(3)} & \Sigma_{23}^{(3)} \end{pmatrix}. \tag{67}$$

Finally, the marginal distribution of $\bar{x}_t$ is

$$\begin{aligned} p(\bar{x}_t | y_{1:t}) &= \mathcal{N}(\bar{x}_t | \mu_t^{(m)}, \Sigma_t^{(m)}) \\ &= \mathcal{N}(\bar{x}_t | \mu^{(4)}(\bar{x}_t), \Sigma^{(4)}(\bar{x}_t)), \end{aligned} \tag{68}$$

where $\mu^{(4)}(\bar{x}_t)$ and $\Sigma^{(4)}(\bar{x}_t)$ are marginal mean vector and covariance matrix of $\bar{x}_t$. □

The Kalman filter for the time-varying HO-LDS in Theorem 1 calculates the estimates based on measurements collected up to and including time step $t$. After obtaining the filtering posterior state distributions, Theorem 2 outlines the results for calculating the marginal posterior distributions for each time step based on all data up to the final time step $T$, which is described as the following:

**Theorem 2.** *The Kalman smoother equations for the time-varing HO-LDS (Equation (43)) parameterized by $\hat{\theta}_{x_{0:T}}$ can be evaluated in closed form, which results Gaussian distributions:*

$$p(x_t | y_{1:T}) = \mathcal{N}(x_t | \mu_t^{(s_1)}, \Sigma_t^{(s_1)}), \tag{69}$$

$$p(\bar{x}_t | y_{1:T}) = \mathcal{N}(\bar{x}_t | \mu_t^{(s_2)}, \Sigma_t^{(s_2)}), \tag{70}$$

$$p(\bar{x}_{t-1}, x_t | y_{1:T}) = \mathcal{N}((\bar{x}_{t-1}, x_t) | \mu_t^{(s_3)}, \Sigma_t^{(s_3)}). \tag{71}$$

*The smoothing distribution and filtering distribution of the last time step are identical, allowing for the recursive computation of the smoothing distribution for all time steps by beginning from the last step and moving backward.*

**Proof.** Firstly by Lemma 1, the joint distribution of $\bar{x}_{t-1}$ and $x_t$ given $y_{1:t-1}$ is

$$\begin{aligned} p(x_t, \bar{x}_{t-1}|y_{1:t-1}) &= p(x_t|\bar{x}_{t-1})p(\bar{x}_{t-1}|y_{1:t-1}) \\ &= \mathcal{N}(x_t|\hat{A}_t\bar{x}_{t-1}, \hat{Q}_t)\mathcal{N}(\bar{x}_{t-1}|\mu_{t-1}^{(m)}, \Sigma_{t-1}^{(m)}) \\ &= \mathcal{N}\left(\begin{pmatrix} \bar{x}_{t-1} \\ x_t \end{pmatrix}\middle|\mu^{(1)}, \Sigma^{(1)}\right), \end{aligned} \tag{72}$$

where

$$\mu^{(1)} = \begin{pmatrix} \mu_{t-1}^{(m)} \\ \hat{A}_t\mu_{t-1}^{(m)} \end{pmatrix}, \quad \Sigma^{(1)} = \begin{pmatrix} \Sigma_{t-1}^{(m)} & \Sigma_{t-1}^{(m)}\hat{A}_t^\top \\ \hat{A}_t\Sigma_{t-1}^{(m)} & \hat{A}_t\Sigma_{t-1}^{(m)}\hat{A}_t^\top + \hat{Q}_t \end{pmatrix}. \tag{73}$$

According to the Markov property of the states, we obtain

$$p(\bar{x}_{t-1}|x_t, y_{1:T}) = p(\bar{x}_{t-1}|x_t, y_{t-1}). \tag{74}$$

Therefore, we have the conditional distribution as

$$\begin{aligned} p(\bar{x}_{t-1}|x_t, y_{1:T}) &= p(\bar{x}_{t-1}|x_t, y_{t-1}), \\ &= \mathcal{N}(\bar{x}_{t-1}|\tilde{\mu}^{(2)}, \tilde{\Sigma}^{(2)}), \end{aligned} \tag{75}$$

where

$$\begin{aligned} G_t &= \Sigma_{t-1}^{(m)}\hat{A}_t^\top(\hat{A}_t\Sigma_{t-1}^{(m)}\hat{A}_t^\top + \hat{Q}_t)^{-1}, \\ \tilde{\mu}^{(2)} &= \mu_{t-1}^{(m)} + G_t(x_t - \hat{A}_t\mu_{t-1}^{(m)}), \\ \tilde{\Sigma}^{(2)} &= \Sigma_{t-1}^{(m)} - G_t\hat{A}_t\Sigma_{t-1}^{(m)}. \end{aligned} \tag{76}$$

Then, the joint distribution of $\bar{x}_{t-1}$ and $x_t$ given all the data is

$$\begin{aligned} p(\bar{x}_{t-1}, x_t|y_{1:T}) &= p(\bar{x}_{t-1}|x_t, y_{1:T})p(x_t|y_{1:T}) \\ &= \mathcal{N}(\bar{x}_{t-1}|\tilde{\mu}^{(2)}, \tilde{\Sigma}^{(2)})\mathcal{N}(x_t|\mu_t^{(s_1)}, \Sigma_t^{(s_1)}) \\ &= \mathcal{N}\left(\begin{pmatrix} \bar{x}_{t-1} \\ x_t \end{pmatrix}\middle|\mu_t^{(s_3)}, \Sigma_t^{(s_3)}\right), \end{aligned} \tag{77}$$

where

$$\mu_t^{(s_3)} = \begin{pmatrix} G_t\mu_t^{(s_1)} + \mu_{t-1}^{(m)} - G_t\hat{A}_t\mu_{t-1}^{(m)} \\ \mu_t^{(s_1)} \end{pmatrix}, \tag{78}$$

$$\Sigma_t^{s3} = \begin{pmatrix} G_t\Sigma_t^{(s_1)}G_t^\top + \tilde{\Sigma}^{(2)} & \Sigma_t^{(s_1)}G_t^\top \\ G_t\Sigma_t^{(s_1)} & \Sigma_t^{(s_1)} \end{pmatrix}. \tag{79}$$

Thus, the marginal distribution of $\bar{x}_{t-1}$ is given as

$$p(\bar{x}_{t-1}|y_{1:T}) = \mathcal{N}(\bar{x}_{t-1}|\mu_t^{(s_2)}, \Sigma_t^{(s_2)}), \tag{80}$$

where

$$\mu_t^{(s_2)} = G_t\mu_t^{(s_1)} + \mu_{t-1}^{(m)} - G_t\hat{A}_t\mu_{t-1}^{(m)}, \tag{81}$$

$$\Sigma_t^{(s_2)} = G_t\Sigma_t^{(s_1)}G_t^\top + \tilde{\Sigma}^{(2)}. \tag{82}$$

The $p(x_{t-1}|y_{1:T})$ will be directly obtained from $p(\bar{x}_{t-1}|y_{1:T})$. Therefore, we have derived the concrete expressions of the Gaussian smoother distributions (Equations (69)–(71)). □

Thus, we can obtain the posterior distributions of $q(x_{0:T}|y_{1:T})$ as well as the expectations involved in the update for the other factors, which are $\mathbb{E}_{q(x_{0:T})}(x_t), \mathbb{E}_{q(x_{0:T})}(x_t x_t^\top)$ and $\mathbb{E}_{q(x_{0:T})}(\bar{x}_t x_{t+1}^\top)$.

### 3.2. *Update for* $q(\omega_{1:T})$

$q(\omega_{1:T})$ can be updated using a closed-form expression by the Pólya-gamma augmentation. Equations (37) and (41) results in the following update for $q(\omega_{1:T}) = \prod_{t=1}^{T}\prod_{k=1}^{K-1} q(\omega_{t,k})$:

$$\begin{aligned}\ln q(\omega_{t,k}) &\propto \mathbb{E}_{-q(\omega_{1:T})} \ln p(z_{t+1}, \omega_t|x_t) \\ &\propto \mathbb{E}_{-q(\omega_{1:T})} \ln p(z_{t+1}|\omega_t, x_t) + \mathbb{E}_{-q(\omega_{1:T})} \ln \mathrm{PG}(\omega_{t,k}|\mathbb{I}(z_{t+1} > k), 0).\end{aligned} \tag{83}$$

Since $\mathbb{E}_{-q(\omega_{1:T})} \ln p(z_{t+1}|\omega_t, x_t) \propto -\frac{1}{2}\mathbb{E}_{-q(\omega_{1:T})}(\nu_{t,k}^2)\omega_{t,k}$, we have

$$q(\omega_{t,k}) = \mathrm{PG}(\omega_{t,k}|\mathbb{E}_{q(z_{1:T})}\mathbb{I}(z_{t+1} > k), \mathbb{E}_{-q(\omega_{1:T})}(\nu_{t,k}^2)). \tag{84}$$

Moreover, the expectation of the posterior Pólya-gamma distribution is available in closed form [22]:

$$\mathbb{E}(\omega_{t,k}) = \frac{\mathbb{E}_{q(z_{1:T})}\mathbb{I}(z_{t+1} > k)}{2\mathbb{E}_{q(z_{1:T})q(x_{1:T})q(\theta)}(\nu_{t,k}^2)} \tanh\left[\frac{1}{2}\mathbb{E}_{q(z_{1:T})q(x_{1:T})q(\theta)}(\nu_{t,k}^2)\right]. \tag{85}$$

### 3.3. *Update for* $q(z_{1:T})$

Since $\mathbb{E}_{-q(z_{1:T})} \ln p(\theta)$ is not a function of $z_{1:T}$, Equation (35) can be written as

$$\begin{aligned}\ln q(z_{1:T}) &\propto \mathbb{E}_{-q(z_{1:T})} \ln p(z_{1:T}, x_{0:T}, \omega_{1:T}, y_{1:T}|\theta) \\ &\propto \mathbb{E}_{-q(z_{1:T})}\left[\ln p(z_1|\theta) + \sum_{t=2}^{T} \ln p(z_t|x_{t-1}, z_{t-1}, \omega_{t-1}, \theta) + \sum_{t=1}^{T} \ln p(x_t|z_t, \theta)\right].\end{aligned} \tag{86}$$

From Equation (86) it is evident that the resulting factor $q(z_{1:T})$ has the form of an HMM parameterized by $q(x_{1:T}, \omega_{1:T}, \theta)$, and the expected sufficient statistics required for updating the other factors can be obtained by message passing algorithms. We define $\ln \tilde{\pi}_{z_1} := \mathbb{E}_{-q(z_{1:T})} \ln p(z_1|\theta)$, $\ln \tilde{a}_{z_t z_{t+1}} := \mathbb{E}_{-q(z_{1:T})} \ln p(z_{t+1}|z_t, x_t, \omega_t, \theta)$, $\ln \tilde{b}_{z_t}(x_t) := \mathbb{E}_{-q(z_{1:T})} \ln p(x_t|z_t, \theta)$ and the set of the HMM parameters is denoted as $\hat{\theta}_{z_{1:T}}$. Thus, we have

$$\ln q(z_{1:T}) \propto \ln \tilde{\pi}_{z_1} + \sum_{t=2}^{T} \ln \tilde{a}_{z_{t-1} z_t}(t-1) + \sum_{t=1}^{T} \ln \tilde{b}_{z_t}(x_t). \tag{87}$$

Note that the densities $p(z_1|\theta), p(z_{t+1}|z_t, x_t, \omega_t, \theta), p(x_t|z_t, \theta)$ belong to the exponential family, these expectations can be evaluated in closed form. The computation of the HMM parameters by Equation (86) is summarized in Algorithm 2.

We apply a standard forward-backward algorithm for an HMM where the forward recursion is:

$$\alpha_t(i) := p(z_t = i|x_{1:t}) \propto \tilde{b}_i(x_t) \sum_{j=1}^{K} \alpha_{t-1}(j), \tag{88}$$

where $\alpha_1(i) \propto \tilde{\pi}_i \tilde{b}_i(x_1)$ and the backward recursion is:

$$\beta_t(i) := p(x_{t+1:T}|z_t = i) \propto \sum_{j=1}^{K} \beta_t(j)\tilde{b}_j(x_t)\tilde{a}_{ij}(t). \tag{89}$$

**Algorithm 2:** Obtaining $\hat{\theta}_{z_{1:T}}$

**for** $k = 1 : K$ **do**

    $\tilde{\pi}_k \propto \exp\left\{\mathbb{E}_{q(z_1)}(\ln p(z_1 = k|\theta))\right\}$

**end**

**for** $i, j = 1 : K$ **do**

    **for** $t = 1 : T$ **do**

        $\tilde{a}_{ij}(t) \propto \exp\Big\{ -\frac{1}{2}\sum_{i=1}^{K} \mathbb{E}_{q(\omega_t)} \ln \omega_{t,i} - \frac{1}{2}\mathbb{E}_{q(\omega_t,x_t,\theta)}(x_t^\top R_i^\top \Omega_t R_i x_t) -$

        $\mathbb{E}_{q(\omega_t,x_t,\theta)}(x_t^\top R_i^\top \omega_t r_i - x_t^\top R_i^\top \kappa(z_{t+1} = j)) -$

        $\frac{1}{2}\mathbb{E}_{q(\omega_t,\theta)}\Big[(r_i - \Omega_t^{-1}\kappa(z_{t+1} = j))^\top \Omega^t (r_i - \Omega_t^{-1}\kappa(z_{t+1} = j))\Big]\Big\}$

    **end**

**end**

**for** $k = 1 : K$ **do**

    **for** $t = 1 : T$ **do**

        $\tilde{b}_k(x_t) \propto \exp\left\{\mathbb{E}_{q(x_{1:T},\theta)}\left[-\frac{1}{2}\ln|Q_k| - \frac{1}{2}(x_t - A_k\bar{x}_{t-1})^\top Q_k (x_t - A_k\bar{x}_{t-1})\right]\right\}$

    **end**

**end**

Then we can obtain the sufficient statistics used in the update for the other factors using the following equations:

$$\gamma_t(i) := p(z_t = i|x_{1:T}) \propto \alpha_t(i)\beta_t(i), \tag{90}$$

$$\xi_t(i,j) := p(z_t = i, z_{t+1} = j|x_{1:T}) \propto \alpha_t(i)\tilde{a}_{ij}(t)\tilde{b}_j(x_{t+1})\beta_{t+1}(j). \tag{91}$$

### 3.4. Update for $q(\theta)$

Given the sufficient statistics of $q(z_{1:T})$ and $q(x_{1:T})$ obtained by message passing, the conjugate posterior distribution of the model parameters can be updated in closed form by evaluating Equation (38). For each parameter, we present the optimized variational distribution in the following:

- Update $q(\pi_0)$. Simplifying Equation (38) and extracting the terms related to $\pi_0$, we have

$$\ln q(\pi_0) \propto \mathbb{E}_{q(-\theta)} \ln p(z_1|\theta) + \ln p(\pi_0), \tag{92}$$

which results in

$$q(\pi_0) \sim \mathrm{Dir}(\pi_0|\alpha'_0), \tag{93}$$

where $\alpha'_0 = \left[\alpha'_{0,1}, \ldots, \alpha'_{0,k}, \ldots, \alpha'_{0,K}\right] \in \mathbb{R}^K$ and

$$\alpha'_{0,k} = \alpha_{0,k} + \gamma_1(k), \tag{94}$$

$\alpha_{0,k}$ are prior parameters of $p(\pi_0)$.

- Update $(\mu_0, \Sigma_0)$. The posterior $q(\mu_0, \Sigma_0)$ can be obtained by the expectations of $q(x_{0:T})$, given by

$$\mu_0 \leftarrow \mathbb{E}_{q(x_{0:T})}(\bar{x}_0), \tag{95}$$

$$\Sigma_0 \leftarrow \mathbb{E}_{q(x_{0:T})}\left(\bar{x}_0\bar{x}_0^\top\right) - \mu_0(\mu_0)^\top. \tag{96}$$

- Update $q(A_k, Q_k)$. Simplifying Equation (38) and extracting the terms related to $(A_k, Q_k)$, we have

$$\begin{aligned}
&\ln q(A_k, Q_k) \propto \\
&\mathbb{E}_{-q(\theta)}\left[-\frac{\sum_{t=1}^{T}\gamma_t(k)}{2}\ln|Q_k| - \frac{1}{2}\sum_{t=1}^{T}\gamma_t(k)(x_t - A_k\bar{x}_{t-1})^\top Q_k^{-1}(x_t - A_k\bar{x}_{t-1})\right] \\
&+ \ln p(A_k, Q_k),
\end{aligned} \tag{97}$$

which results in

$$q(A_k, Q_k) \sim \mathrm{MNIW}(A_k, Q_k|\eta_k'), \tag{98}$$

where $\eta_k' := \left\{M_k^{(\eta')}, \Phi_k^{(\eta')}, \iota_k^{(\eta')}, \Delta_k^{(\eta')}\right\}$,

$$\begin{aligned}
q(A_k, Q_k) \propto |Q_k|^{-\frac{\iota_k^{(\eta')}+(p+1)m+1}{2}}\exp\Big\{ &-\frac{1}{2}\mathrm{Tr}(\Delta_k^{(\eta')}Q_k^{-1}) \\
&-\frac{1}{2}\mathrm{Tr}\left[(\Phi_k^{(\eta')})^{-1}(A_k - M_k^{(\eta')})^\top Q_k^{-1}(A_k - M_k^{(\eta')})\right]\Big\}.
\end{aligned} \tag{99}$$

The variational posterior parameters can be optimized as

$$\begin{aligned}
\Phi_k^{(\eta')} &= \left((\Phi_k^{(\eta)})^{-1} + \sum_{t=1}^{T}\mathbb{E}_{q(x_{0:T})}(\gamma_t(k)\bar{x}_{t-1}\bar{x}_{t-1}^\top)\right)^{-1}, \\
M_k^{(\eta')} &= \left(M_k^{(\eta)}(\Phi_k^{(\eta)})^{-1} + \sum_{t=1}^{T}\mathbb{E}_{q(x_{0:T})}(\gamma_t(k)x_t\bar{x}_{t-1}^\top)\right)\Phi_k^{(\eta')}, \\
\Delta_k^{(\eta')} &= \Delta_k^{(\eta)} + \sum_{t=1}^{T}\mathbb{E}_{q(x_{0:T})}(\gamma_t(k)x_t x_t^\top) + M_k^{(\eta)}(\Phi_k^{(\eta)})^{-1}(M_k^{(\eta)})^\top - M_k^{(\eta')}(\Phi_k^{(\eta')})^{-1}(M_k^{(\eta')})^\top, \\
\iota_k^{(\eta')} &= \iota_k^{(\eta)} + \sum_{t=1}^{T}\gamma_t(k),
\end{aligned} \tag{100}$$

where $M_k^{(\eta)}, \Phi_k^{(\eta)}, \iota_k^{(\eta)}, \Delta_k^{(\eta)}$ are prior parameters of $p(A_k, Q_k)$.

- Update $q(C_k, S_k)$. Simplifying Equation (38) and extracting the terms related to $(C_k, S_k)$, we have

$$\begin{aligned}
&\ln q(C_k, S_k) \propto \\
&\mathbb{E}_{-q(\theta)}\left[-\frac{\sum_{t=1}^{T}\gamma_t(k)}{2}\ln|S_k| - \frac{1}{2}\sum_{t=1}^{T}(y_t - C_k x_t)^\top S_k^{-1}(y_t - C_k x_t)\right] \\
&+ \ln p(C_k, S_k),
\end{aligned} \tag{101}$$

which results in

$$q(C_k, S_k) \sim \mathrm{MNIW}(C_k, S_k|\lambda_k'), \tag{102}$$

where $\lambda'_k := \left\{M_k^{(\lambda')}, \Phi_k^{(\lambda')}, \iota_k^{(\lambda')}, \Delta_k^{(\lambda')}\right\}$,

$$q(C_k, S_k) \propto |S_k|^{-\frac{\iota_k^{(\lambda')}+(p+1)m+1}{2}} \exp\left\{ -\frac{1}{2}\mathrm{Tr}(\Delta_k^{(\lambda')} S_k^{-1}) - \frac{1}{2}\mathrm{Tr}\left[(\Phi_k^{(\lambda')})^{-1}(C_k - M_k^{(\lambda')})^\top Q_k^{-1}(C_k - M_k^{(\lambda')})\right]\right\}. \tag{103}$$

The variational posterior parameters can be optimized as

$$\begin{aligned}
\Phi_k^{(\lambda')} &= \left((\Phi_k^{(\lambda)})^{-1} + \sum_{t=1}^{T} \mathbb{E}_{q(x_{0:T})}(\gamma_t(k) x_t x_t^\top)\right)^{-1}, \\
M_k^{(\lambda')} &= \left(M_k^{(\lambda)}(\Phi_k^{(\lambda)})^{-1} + \sum_{t=1}^{T} \mathbb{E}_{q(x_{0:T})}(\gamma_t(k) y_t x_t^\top)\right)\Phi_k^{(\lambda')}, \\
\Delta_k^{(\lambda')} &= \Delta_k^{(\lambda)} + \sum_{t=1}^{T} \gamma_t(k) y_t y_t^\top + M_k^{(\lambda)}(\Phi_k^{(\lambda)})^{-1}(M_k^{(\lambda)})^\top - M_k^{(\lambda')}(\Phi_k^{(\lambda')})^{-1}(M_k^{(\lambda')})^\top, \\
\iota_k^{(\lambda')} &= \iota_k^{(\lambda)} + \sum_{t=1}^{T} \gamma_t(k),
\end{aligned} \tag{104}$$

where $M_k^{(\lambda)}, \Phi_k^{(\lambda)}, \iota_k^{(\lambda)}, \Delta_k^{(\lambda)}$ are prior parameters of $p(C_k, S_k)$.

- Update $q(R_k)$. Simplifying Equation (38) and extracting the terms related to $R_k$, we have

$$\begin{aligned}
&\ln q(R_k) \propto \\
&\mathbb{E}_{-q(\theta)}\left[-\frac{1}{2}\sum_{t=1}^{T-1} \gamma_t(k)(R_k x_t + r_k - \Omega_t^{-1}\kappa_{t+1})^\top \Omega_t (R_k x_t + r_k - \Omega_t^{-1}\kappa_{t+1})\right] \\
&+ \sum_{i=1}^{K-1} \ln p(R_{k,i}),
\end{aligned} \tag{105}$$

which results in

$$q(R_k) \sim \prod_{i=1}^{K-1} \mathcal{N}(R_{k,i}|\zeta'_{k,i}), \tag{106}$$

where $\zeta'_{k,i} := \left\{\mu_{k,i}^{(\zeta')}, \Sigma_{k,i}^{(\zeta')}\right\}$, which can be optimized as

$$\Sigma_{k,i}^{(\zeta')} = \left[\left(\Sigma_{k,i}^{(\zeta)}\right)^{-1} + \sum_{t=1}^{T} \mathbb{E}_{q(\omega_{1:T}, x_{0:T})}\left(x_t \gamma_t(k) \Omega_{t,i} x_t^\top\right)\right]^{-1}, \tag{107}$$

$$\mu_{k,i}^{(\zeta')} = \Sigma_{k,i}^{(\zeta')}\left[\sum_{t=1}^{T-1} \mathbb{E}_{q(\omega_{1:T})q(x_{0:T})}(\gamma_t(k)\Omega_{t,i} H_{t,k,i} x_t) + \left(\Sigma_{k,i}^{(\zeta)}\right)^{-1}\mu_{k,i}^{(\zeta)}\right], \tag{108}$$

$$H_{t,k} = [H_{t,k,1}, \cdots, H_{t,k,i}, \cdots, H_{t,k,K-1}]^\top = \mathbb{E}_{q(r_k)q(\omega_{1:T})}\left(r_k - \Omega_t^{-1}\kappa_{t+1}\right), \tag{109}$$

and $\mu_{k,i}^{(\zeta)}, \Sigma_{k,i}^{(\zeta)}$ are prior parameters of $p(R_{k,i})$. To derive Equation (107), just note that $R_k^\top \Omega_t R_k = \sum_{i=1}^{K-1} R_{k,i}\Omega_{k,i} R_{k,i}^\top$ and use the matrix trace.

- Update $q(r_k)$. Simplifying Equation (38) and extracting the terms related to $r_k$, we have

$$\begin{aligned}
&\ln q(r_k) \propto \\
&\mathbb{E}_{-q(\theta)}\left[-\frac{1}{2}\sum_{t=1}^{T-1} \gamma_t(k)(R_k x_t + r_k - \Omega_t^{-1}\kappa_{t+1})^\top \Omega_t (R_k x_t + r_k - \Omega_t^{-1}\kappa_{t+1})\right] \\
&+ \ln p(r_k),
\end{aligned} \tag{110}$$

which results in

$$q(r_k) \sim \mathcal{N}(r_k|\rho'_k), \tag{111}$$

where $\rho'_k := \left\{\mu_k^{(\rho')}, \Sigma_k^{(\rho')}\right\}$, which can be optimized as

$$\Sigma_k^{(\rho')} = \left[\left(\Sigma_k^{(\rho)}\right)^{-1} + \sum_{t=1}^{T-1} \mathbb{E}_{q(\omega_{1:T})}(\gamma_t(k)\Omega_t)\right]^{-1}, \tag{112}$$

$$\mu_k^{(\rho')} = \Sigma_k^{(\rho')}\left[\left(\Sigma_k^{(\rho)}\right)^{-1}\mu_k^{(\rho)} + \sum_{t=1}^{T-1} \mathbb{E}_{q(R_k)q(x_{0:T})q(\omega_{1:T})}\left(\gamma_t(k)\Omega_t\left(R_k x_t - \Omega_t^{-1}\kappa_{t+1}\right)\right)\right], \tag{113}$$

where $\mu_k^{(\rho)}, \Sigma_k^{(\rho)}$ are prior parameters of $p(r_k)$.

### 3.5. Initialization

To obtain a reliable variational inference, we initialized the model parameters and latent states with reasonable values by using the following initialization procedure:

1. Probabilistic Principal Component Analysis (PPCA) [27] is conducted on the data, to initialize the continuous latent variables, $x_{0:T}$ and the parameters $C$;
2. Fit an AR(p)-HMM to initialize the switching states, $z_{1:T}$ and the parameters, $\{A_k, Q_k\}$. The autoregressive order, $p$, is determined by AIC, BIC, and HQ criterion;
3. To alleviate the possible and undesirable dependence on ordering that arises from the stick-breaking formulation during the inference, we adopt a strategy of greedily fitting a decision list [21] to identify the most suitable permutation of the switching states for the stick-breaking process. Specifically, we start by performing a greedy search on permutations by creating a decision list based on $(x_t, z_t), z_{t+1}$ pairs, given by

$$z_{t+1} = \begin{cases} o_1 & \text{if } p_1 \\ o_2 & \text{if } \neg p_1 \cap p_2 \\ \vdots & \\ o_K & \text{if } \cap_{i=1}^{K-1}(\neg p_i) \end{cases}, \tag{114}$$

where $(o_1, \ldots, o_K)$ is a permutation of $(1, \ldots, K)$, and $p_1, \ldots, p_k$ are predicates that rely on $(x_t, z_t)$ and provide a true or false. In our framework, these predicates are defined by logistic functions:

$$p_j = \sigma\left(R_{0,j}^{\top} x_t\right) > \varepsilon, \tag{115}$$

where $\varepsilon \in (0,1)$ should be predetermined. To determine $o_1$ and $r_1$, we used the maximum a posteriori estimate of the model for each of the $K$ potential states. For the $k$th logistic regression, the inputs are $x_{0:T}$ and the outputs are $y_t = \mathbb{I}(z_{t+1} = k)$. We chose the logistic regression model with the largest likelihood as the first output. Then we excluded time points where $z_{t+1} = o_1$ from the data and proceeded to $K-1$ logistic regressions to predict the subsequent output, $o_2$, and so on. After cycling through all $K$ results, we obtained the permutation of $z_{1:T}$. In addition, the $\{R_{0,j}\}_{j=1}^{K-1}$ served as an initialization of the recurrence weights ($R$).

## 4. Numerical Experiments

To investigate the performance of the proposed HO-rSLDS, we generated synthetic datasets through the following steps:

1. Benchmark model settings.We take AR-HSMM as a benchmark model to generate synthetic time series. Specifically, the dimensionality of the synthetic time series is $n = 5$, and the total length is $T = 1000$. The state number is $K = 3$, and the autoregressive order $p$ takes values $\{1, 2, 3\}$.
2. Generating $(A_k, Q_k)$. $(A_k, Q_k)$ corresponds to the emission parameters involved in the VAR process of latent variables $x_t$ (Equation (11)). The autoregressive coefficient matrices $A_k$ are generated with 50% sparsity (defined as the proportion of the zero elements), i.e., 50% of the elements in $A_k$ are 0. The non-zero elements of $A_k$ are generated to be positive or negative with equal probability, and their absolute value is sampled uniformly between 0.2 and 0.5. The covariance matrices $Q_k$ are generated as $Q_k = P_k^\top \mathrm{diag}(\sigma_1, \ldots, \sigma_d) P_k$, with $P_k$ being an orthogonal matrix and $\sigma_i$ assumed positive. The matrix $P_k$ is constructed by orthogonalizing a random matrix whose entries are simulated from a standard Normal, while each $\sigma_i$ is uniformly sampled in the interval $[1, 3]$. Additionally, a rejection step is done to check that the sampled $(A_k, Q_k)$ constituted a stable VAR process.
3. Generating switching state sequence $z_{1:T}$. To generate $z_{1:T}$, we set the state transition probabilities $\pi_{jk}$ to 0.5, for $j \neq k = 1, 2, 3$. Note that in an HSMM, the self-transition probability is 0. Furthermore, we simulate the state duration time in two cases. In the first case, the state duration time is sampled from a geometric distribution (the between-state transition probabilities are set to 0.1). In the second case, the state duration time is directly sampled from $\{1, \ldots, 10\}$ with specified sampling probabilities. Since the second case does not correspond to a parametric distribution, we denote these two cases as geometric and nonparametric, respectively. In addition, the initial state is uniformly sampled from $\{1, \ldots, 10\}$.
4. Generating $(C_k, S_k)$. $(C_k, S_k)$ corresponds to the emission parameters generating the Gaussian observations $y_t$ (Equation (12)). We generate $(C_k, S_k)$ in the same manner as $(A_k, Q_k)$.
5. Repeat the above procedure 100 times to obtain 100 synthetic data by setting different random seeds when generating $(A_k, Q_k, C_k, S_k)$.

Figure 4 displays all dimensions and the true state sequence of one synthetic dataset.

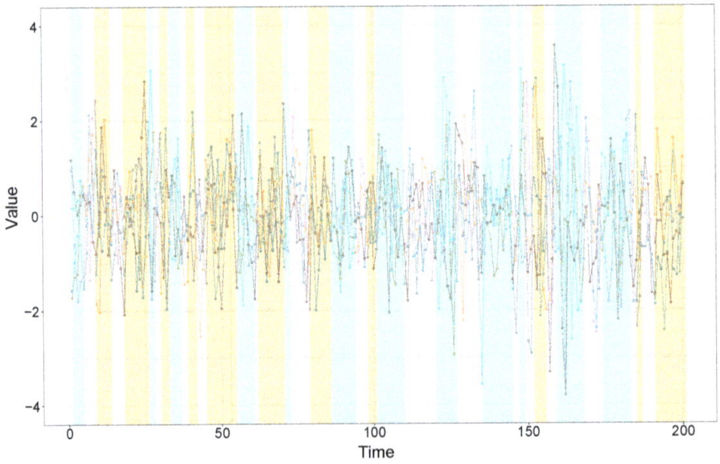


**Figure 4.** Simulated time series with dimension $d = 5$ and total time points $T = 1000$ (the last 200 time points are exhibited). The five dimensions are displayed in different colors. The color intervals indicate the state sequence.

We investigated the performance of HO-rSLDS by comparing with the competing models, while the estimation accuracy is evaluated by the normalized mean squared error (NMSE) between the true and estimated emission model parameters and the signal-to-noise ratio (SNR) of the inferred optimized switching state sequence ($\hat{z}_{1:T}$), given by

$$\text{NMSE} = \frac{1}{K}\sum_{k=1}^{K}\frac{||\hat{C}_k - C_k||_\text{F}}{||C_k||_\text{F}}, \tag{116}$$

$$\text{SNR} = \frac{\sum_{t=1}^{T}\mathbb{I}(z_t = \hat{z}_t)}{T}, \tag{117}$$

where $||\cdot||_\text{F}$ denotes Frobenius norm, $\hat{C}_k$ and $\hat{z}_{1:T}$ are estimated values. For AR-HMM, the NMSE is evaluated based on the true and estimated VAR emission parameters. To obtain $\hat{z}_{1:T}$, we set $\hat{z}_t = \arg\max_k \gamma_t(k)$. In addition, we avoid the optimization about discrete state number ($K$) and the dependence order ($p$) in HO-rSLDS by directly setting them to the true value.

The results of estimated NMSE and SNR for the three models are summarized in Tables 1 and 2, respectively. It is observed that HO-rSLDS outperforms AR-HMM and SLDS with lower average NMSE and higher average SNR for all the autoregressive orders and state duration distribution combinations. In addition, when $p = 1$, for both geometric and nonparametric state duration, SLDS outperforms AR-HMM slightly, which may be because the synthetic data are generated by a simulated extensive SLDS structure. However, when $p > 1$ and there is long-term dependence involved in the latent variables or observation data, it is just the opposite, since AR-HMM with higher $p$ can capture the temporal relationships in the observation data while SLDS cannot. Moreover, when the transition of the underlying discrete states is characterized by a nonparametric duration distribution, the estimation accuracy of AR-HMM and SLDS is reduced, while HO-rSLDS remains stable with good estimation accuracy. This is because state transitions in both AR-HMM and SLDS are characterized by a Markov assumption and restricted to geometric duration distribution. Thus, more weight is given to shorter consecutive time periods within a certain state, which means the regimes switch frequently, which may not be appropriate in cases without sufficient prior information about a geometric state duration. By comparison, HO-rSLDS exhibits robustness to the state duration distributions since it improves the state transition by leveraging the stick-breaking logistic regression. The current state $z_t$ no longer depends on only the preceding state $z_{t-1}$, but also the latent variable $x_{t-1}$, and, furthermore, the previous observations.

A more exhaustive comparison is illustrated in Figure 5 where HO-rSLDS outperforms the competing models with better estimation accuracy and statistical significance in all the cases ($t$-test, ** $p < 0.01$).

**Table 1.** NMSE derived from AR-HMM, SLDS, and HO-rSLDS based on the synthetic time series with different state duration distribution and autoregressive order in the generative model. All the results are the average of the 100 experiments (mean ± std).

| NMSE | | $p = 1$ | $p = 2$ | $p = 3$ |
|---|---|---|---|---|
| Geometric | AR-HMM | $0.25 \pm 0.06$ | $0.33 \pm 0.07$ | $0.37 \pm 0.09$ |
| | SLDS | $0.23 \pm 0.05$ | $0.39 \pm 0.09$ | $0.44 \pm 0.11$ |
| | HO-rSLDS | $0.16 \pm 0.04$ | $0.17 \pm 0.05$ | $0.20 \pm 0.06$ |
| Nonparametric | AR-HMM | $0.32 \pm 0.07$ | $0.41 \pm 0.09$ | $0.48 \pm 0.11$ |
| | SLDS | $0.31 \pm 0.09$ | $0.44 \pm 0.11$ | $0.49 \pm 0.13$ |
| | HO-rSLDS | $0.18 \pm 0.06$ | $0.20 \pm 0.05$ | $0.21 \pm 0.06$ |

**Table 2.** SNR derived from AR-HMM, SLDS, and HO-rSLDS based on the synthetic time series with different state duration distribution and autoregressive order in the generative model. All the results are the average of the 100 experiments (mean ± std).

| SNR | | $p = 1$ | $p = 2$ | $p = 3$ |
|---|---|---|---|---|
| Geometric | AR-HMM | 0.79 ± 0.08 | 0.76 ± 0.07 | 0.69 ± 0.08 |
| | SLDS | 0.85 ± 0.06 | 0.68 ± 0.09 | 0.64 ± 0.09 |
| | HO-rSLDS | 0.93 ± 0.04 | 0.92 ± 0.05 | 0.90 ± 0.05 |
| Nonparametric | AR-HMM | 0.73 ± 0.08 | 0.65 ± 0.07 | 0.59 ± 0.10 |
| | SLDS | 0.80 ± 0.06 | 0.60 ± 0.09 | 0.54 ± 0.11 |
| | HO-rSLDS | 0.91 ± 0.04 | 0.87 ± 0.05 | 0.86 ± 0.05 |

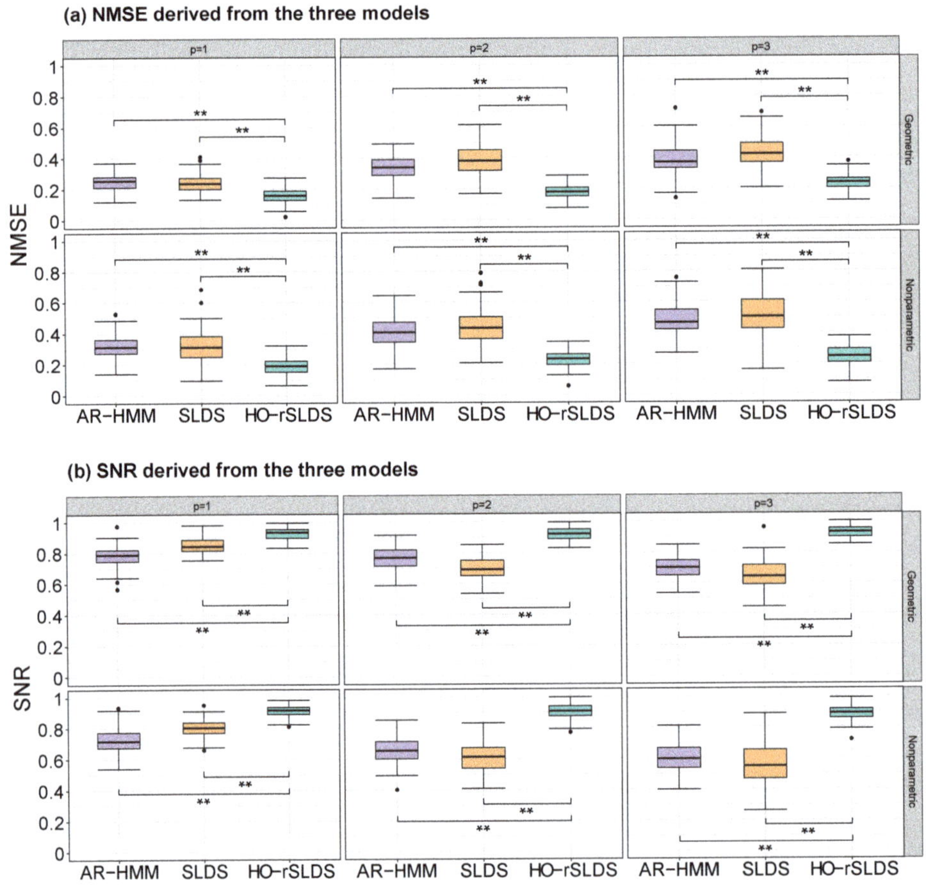


**Figure 5.** Box-plot of NMSE and SNR derived from AR-HMM, SLDS and HO-rSLDS based on the 100 numerical experiments. HO-rSLDS outperforms the competing models with significant lower NMSE and higher SNR on average for all the autoregressive order and state duration distribution combination. ($t$-test, ** $p < 0.01$).

## 5. Dynamic Functional Connectivity Analysis in fMRI Data

To show that HO-rSLDS can also work well with real-world data, we evaluate the performance of HO-rSLDS by comparing it with standard SLDS based on public functional magnetic resonance imaging (fMRI) data. The data are collected in the Human Connectome

Project (HCP) 1200 Parcellation Timeseries Netmats (HCP1200-TPN) dataset (Available at https://www.humanconnectome.org, (accessed on 6 March 2024)). Please see [28] for a detailed explanation of the entire acquisition protocol. In this study, we apply HO-rSLDS to data from 20 unrelated subjects with dimension $d = 15$ and times series length $T_0 = 4800$ for each subject included in the HCP1200-PTN release to perform dynamic functional connectivity (DFC) analysis, which can be used to analysis the dynamical temporal coherence among endogenous fluctuations in distributed brain regions [29,30]. In general, HO-rSLDS and SLDS assume that there are metastable states with characteristic connectivity pattern in the brain, while the connectivity pattern can be represented by the statistical correlations directly. Specifically, by Lemma 1, the functional connectivity pattern of state $k$ in HO-rSLDS and SLDS can be obtained as

$$\mathrm{DFC}_k = \frac{\sum_{t=1}^{T} \gamma_t(k) \left[ \mathbb{E}C_k \left( \mathbb{E}x_t x_t^\top - \mathbb{E}x_t \mathbb{E}x_t^\top \right) \mathbb{E}C_k^\top + \mathbb{E}S_k \right]}{\sum_{t=1}^{T} \gamma_t(k)}. \tag{118}$$

In addition, there are two unknown parameters: the state number $K$ and the dependence order $p$. We use the mixture minimum description length (MMDL) considering optimal code lengths for each state in a mixed model [31], given by

$$\mathrm{MMDL}(K) := -\ln \mathcal{L}(K) + \frac{1}{2} \ln(T) K(K-1) + \sum_{k=1}^{K} \frac{1}{2} \ln(T \bar{\gamma}_k) d^2, \tag{119}$$

where $\mathcal{L}(K)$ is the variational lower bound evaluated during the model inference and $\bar{\gamma}_k := \sum_{t=1}^{T} \gamma_t(k)$. The concept of MMDL may be elucidated by the minimal description length principle in [32]. The state number $K$ is determined by minimizing the MMDL on a specified list of $K$. In addition, we fit VAR model on the fMRI time series and the optimal dependence order $p$ will be determined by AIC, BIC and HQ criteria [33], given by

$$\mathrm{AIC} := \ln\big(\det(\hat{\Sigma})\big) + \frac{2pd^2}{T}, \tag{120}$$

$$\mathrm{BIC} := \ln\big(\det(\hat{\Sigma})\big) + \frac{\ln(T) pd^2}{T}, \tag{121}$$

$$\mathrm{HQ} := \ln(\det(\Sigma)) + \frac{2\ln(\ln T) pd^2}{T}, \tag{122}$$

where $\hat{\Sigma}$ is the estimated covariance matrix of noise in VAR model. To avoid the model order too large to permit feasible computation, we choose the lag order on the condition when AIC/BIC/HQ show no further substantial decreases at higher orders [34]. As illustrated in Figure 6, we set $K = 4$ according to the minimal MMDL and $p = 4$ since the downward trend of AIC, BIC and HQ tends to be flat with higher order.

In Figure 7, we present the DFC patterns derived from the proposed HO-rSLDS and the standard SLDS. The 15 brain regions producing the fMRI time series are indexed by numbers 1 to 15. The states are ordered by their fractional occupancy (defined as the percentage of time allocated to that specific state); it is evident that the 4 states from SLDS have a more average fractional occupancy distribution. Moreover, it is observed that DFC patterns derived from SLDS are more similar than those from HO-rSLDS, where there is only a slight difference in DFC patterns between State4 and the other three states from SLDS. We use cosine similarity to measure the difference of the DFC patterns characterized by each state, given by

$$\cos(F_1, F_2) = \frac{\mathrm{Tr}(F_1 F_2)}{||F_1||_\mathrm{F} ||F_2||_\mathrm{F}}, \tag{123}$$

where $F_1$ and $F_2$ are square matrices. By this means, the averaged cosine similarity for SLDS is 0.9189 while that for HO-rSLDS is 0.6731. Although the real DFC patterns of this fMRI data are unknown, compared to standard SLDS, HO-rSLDS can indeed dig out more information about the DFC patterns by the model superiority. For example, the State4 in HO-rSLDS, with the lowest fractional occupancy, can be considered as a strongly connected state where the absolute functional connectivity is significantly larger than that of the other three states. Moreover, compared with standard SLDS, there is significantly positive or negative functional connectivity in the four states derived from HO-rSLDS. In addition, there is similar functional connectivity between State1 and State2, which results in a higher cosine similarity than the averaged value for HO-rSLDS (0.7776). The similar functional connectivity is not exhibited obviously in State3 (such as the negative connectivity between region 8 and regions 11, 12, 13). However, these findings cannot be obtained by standard SLDS.

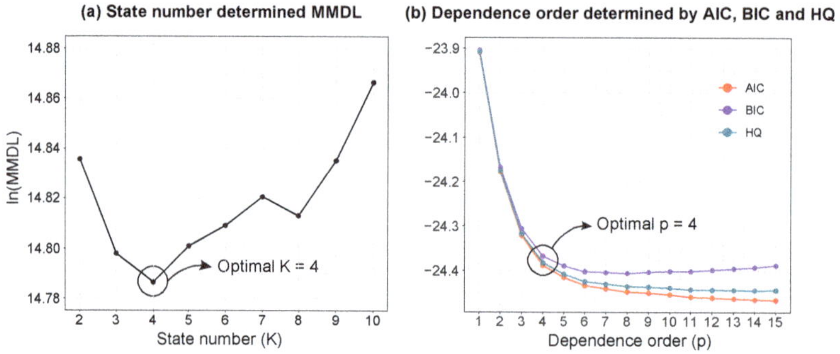


**Figure 6.** (**a**) State number ($K$) determined by MMDL. The best $K = 4$ is chosen with respect to the minimal MMDL. The y-axis has been logarithmically scaled to highlight the variation. (**b**) Dependence order $p$ is determined by AIC, BIC, and HQ criteria. The ideal $p$ is set to 4 since there is no further substantial decrease with higher order.

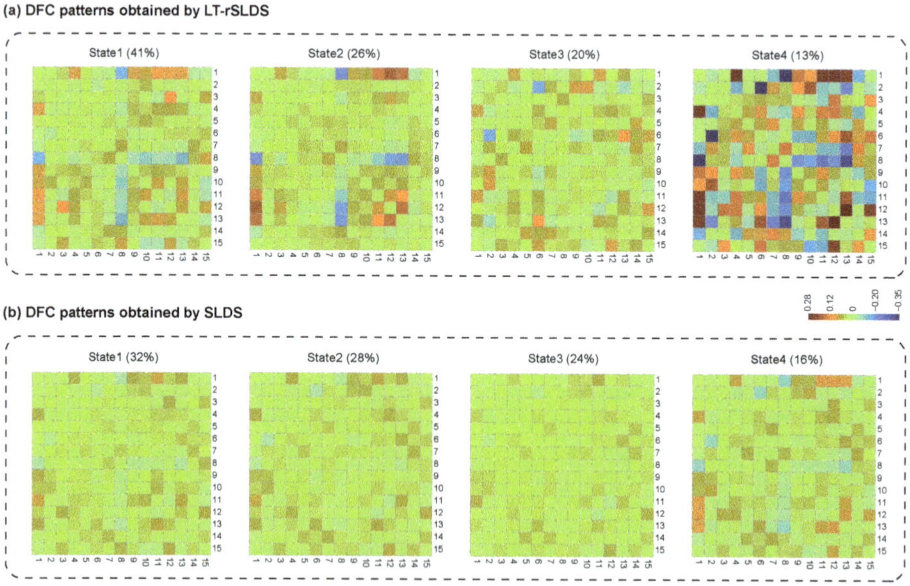


**Figure 7.** DFC patterns derived from HO-rSLDS (**a**) and standard SLDS (**b**). The 15 brain regions are indexed by the numbers 1 to 15, and the states are ordered by their fractional occupancy (number in the brackets). The color bar is shared by the two group heatmaps.

## 6. Conclusions

In this study, we have established a new method named HO-rSLDS as an extension of SLDS that improves the statistical modeling of dynamics from complex systems. First, HO-rSLDS addresses the problem of discovering a higher-order temporal dependence involved in the latent variables or observation data by allowing the current latent variable to be associated with the measurements of several previous time steps as in AR-HMM. Second, HO-rSLDS improves the switching state transition by utilizing stick-breaking logistic regression and Pólya-gamma augmentation, making the transition dependent on the latent variable and thus overcoming the limitations of the Markov assumption, leading to a geometric distribution of state durations. Moreover, the symmetric dependency between the switching states and the latent variable is recovered from the asymmetric relationships as in standard SLDS. We have presented the detailed inference algorithms of the HO-rSLDS for message passing and parameter learning. In numerical experiments, we concluded that HO-rSLDS outperforms the competing models with higher estimation accuracy and robustness. The utility and versatility of the developed HO-rSLDS have been demonstrated on real-world data as well. The application of fMRI data for DFC analyses indicates the superiority of the method in discovering abundant information about DFC patterns. A potential limitation of the proposed HO-rSLDS is that the state number and dependency order need to be predetermined in applications, while the hyperparameter optimization is always a non-trivial task. Future research will improve the proposed model by utilizing Bayesian nonparametric methods so that the complicated hyperparameter optimization can be efficiently avoided [35]. The proposed model improves the current methods for learning the switching linear dynamical modes, which will facilitate the identification and assessment of the dynamics of complex systems.

**Author Contributions:** Conceptualization, H.W. and J.C.; methodology, H.W.; software, H.W.; validation, H.W.; formal analysis, H.W. and J.C.; investigation, H.W.; resources, H.W. and J.C.; data curation, J.C.; writing—original draft preparation, H.W.; writing—review and editing, J.C.; visualization, H.W.; supervision, J.C.; project administration, J.C.; funding acquisition, J.C. All authors have read and agreed to the published version of the manuscript.

**Funding:** This research was funded by the National Natural Science Foundation (grant number 81671633), and the project was supported by the Open Fund of Hubei Longzhong Laboratory.

**Data Availability Statement:** The data that supported the findings of this study are included in this article.

**Conflicts of Interest:** The authors declare no conflict of interest.

## Abbreviations

The following abbreviations are used in this manuscript:

| | |
|---|---|
| SLDS | Switching linear dynamical systems |
| HO-SLDS | Switching linear dynamical systems with higher-order dependence |
| HO-rSLDS | Recurrent switching linear dynamical systems with higher-order dependence |
| HMM | Hidden Markov model |
| VAR | Vector autoregressive |
| AR-HMM | Autoregressive hidden Markov model |
| AR-HSMM | Autoregressive hidden semi-Markov model |
| NMSE | Normalized mean squared error |
| SNR | Signal-to-noise ratio |
| fMRI | Functional magnetic resonance imaging |
| DFC | Dynamic functional connectivity |

## References

1. Fox, E.; Sudderth, E.B.; Jordan, M.I.; Willsky, A.S. Bayesian nonparametric inference of switching dynamic linear models. *IEEE Trans. Signal Process.* **2011**, *59*, 1569–1585. [CrossRef]
2. Pandarinath, C.; O'Shea, D.J.; Collins, J.; Jozefowicz, R.; Stavisky, S.D.; Kao, J.C.; Trautmann, E.M.; Kaufman, M.T.; Ryu, S.I.; Hochberg, L.R.; et al. Inferring single-trial neural population dynamics using sequential auto-encoders. *Nat. Methods* **2018**, *15*, 805–815. [CrossRef] [PubMed]
3. Frigola, R.; Chen, Y.; Rasmussen, C.E. Variational Gaussian process state-space models. In Proceedings of the Annual Conference on Neural Information Processing Systems 2014, Montreal, QC, Canada, 8–13 December 2014; Volume 27.
4. Krishnan, R.; Shalit, U.; Sontag, D. Structured inference networks for nonlinear state space models. In Proceedings of the AAAI Conference on Artificial Intelligence, San Francisco, CA, USA, 4–9 February 2017; Volume 31.
5. Hamilton, J.D. *Time Series Analysis*; Princeton University Press: Princeton, NJ, USA, 2020.
6. de Souza Baptista, R.; Bó, A.P.; Hayashibe, M. Automatic human movement assessment with switching linear dynamic system: Motion segmentation and motor performance. *IEEE Trans. Neural Syst. Rehabil. Eng.* **2016**, *25*, 628–640. [CrossRef] [PubMed]
7. Liu, H.; Wang, L. Human motion prediction for human-robot collaboration. *J. Manuf. Syst.* **2017**, *44*, 287–294. [CrossRef]
8. Alameda-Pineda, X.; Drouard, V.; Horaud, R.P. Variational inference and learning of piecewise linear dynamical systems. *IEEE Trans. Neural Netw. Learn. Syst.* **2021**, *33*, 3753–3764. [CrossRef] [PubMed]
9. Ma, T.; Srinivasan, S.; Lazarou, G.; Picone, J. Continuous speech recognition using linear dynamic models. *Int. J. Speech Technol.* **2014**, *17*, 11–16. [CrossRef]
10. Pagan, A.R.; Pesaran, M.H. Econometric analysis of structural systems with permanent and transitory shocks. *J. Econ. Dyn. Control* **2008**, *32*, 3376–3395. [CrossRef]
11. Haluszczynski, A.; Räth, C. Controlling nonlinear dynamical systems into arbitrary states using machine learning. *Sci. Rep.* **2021**, *11*, 12991. [CrossRef] [PubMed]
12. Smith, J.F.; Pillai, A.; Chen, K.; Horwitz, B. Identification and validation of effective connectivity networks in functional magnetic resonance imaging using switching linear dynamic systems. *Neuroimage* **2010**, *52*, 1027–1040. [CrossRef]
13. Wang, E.T.; Vannucci, M.; Haneef, Z.; Moss, R.; Rao, V.R.; Chiang, S. A Bayesian switching linear dynamical system for estimating seizure chronotypes. *Proc. Natl. Acad. Sci. USA* **2022**, *119*, e2200822119. [CrossRef]
14. Rabiner, L.R. A tutorial on hidden Markov models and selected applications in speech recognition. *Proc. IEEE* **1989**, *77*, 257–286. [CrossRef]
15. Vidaurre, D. A new model for simultaneous dimensionality reduction and time-varying functional connectivity estimation. *PLoS Comput. Biol.* **2021**, *17*, e1008580. [CrossRef] [PubMed]
16. Hamada, R.; Kubo, T.; Ikeda, K.; Zhang, Z.; Shibata, T.; Bando, T.; Hitomi, K.; Egawa, M. Modeling and prediction of driving behaviors using a nonparametric Bayesian method with AR models. *IEEE Trans. Intell. Veh.* **2016**, *1*, 131–138. [CrossRef]
17. Houpt, J.W.; Frame, M.E.; Blaha, L.M. Unsupervised parsing of gaze data with a beta-process vector auto-regressive hidden Markov model. *Behav. Res. Methods* **2018**, *50*, 2074–2096. [CrossRef] [PubMed]
18. Vidaurre, D.; Quinn, A.J.; Baker, A.P.; Dupret, D.; Tejero-Cantero, A.; Woolrich, M.W. Spectrally resolved fast transient brain states in electrophysiological data. *Neuroimage* **2016**, *126*, 81–95. [CrossRef] [PubMed]
19. Glennie, R.; Adam, T.; Leos-Barajas, V.; Michelot, T.; Photopoulou, T.; McClintock, B.T. Hidden Markov models: Pitfalls and opportunities in ecology. *Methods Ecol. Evol.* **2023**, *14*, 43–56. [CrossRef]
20. Johnson, M.J.; Willsky, A.S. Bayesian Nonparametric Hidden Semi-Markov Models. *J. Mach. Learn. Res.* **2013**, *14*, 673–701.
21. Linderman, S.W.; Johnson, M.J.; Miller, A.C.; Adams, R.P.; Blei, D.M.; Paninski, L. Bayesian Learning and Inference in Recurrent Switching Linear Dynamical Systems. In Proceedings of the 20th International Conference on Artificial Intelligence and Statistics (AISTATS), Fort Lauderdale, FL, USA, 20–22 April 2017; Volume 54, pp. 914–922.
22. Polson, N.G.; Scott, J.G.; Windle, J. Bayesian inference for logistic models using Pólya–Gamma latent variables. *J. Am. Stat. Assoc.* **2013**, *108*, 1339–1349. [CrossRef]
23. Jacobs, W.R.; Baldacchino, T.; Dodd, T.; Anderson, S.R. Sparse Bayesian nonlinear system identification using variational inference. *IEEE Trans. Autom. Control* **2018**, *63*, 4172–4187. [CrossRef]
24. Yu, S.Z. Hidden semi-Markov models. *Artif. Intell.* **2010**, *174*, 215–243. [CrossRef]
25. Liu, Q.; Liu, W.; Dong, M.; Li, Z.; Zheng, Y. Residual useful life prognosis of equipment based on modified hidden semi-Markov model with a co-evolutional optimization method. *Comput. Ind. Eng.* **2023**, *182*, 109433. [CrossRef]
26. Särkkä, S.; Svensson, L. *Bayesian Filtering and Smoothing*; Cambridge University Press: Cambridge, UK, 2023; Volume 17.
27. Tipping, M.E.; Bishop, C.M. Probabilistic principal component analysis. *J. R. Stat. Soc. Ser. B Stat. Methodol.* **1999**, *61*, 611–622. [CrossRef]
28. Glasser, M.F.; Sotiropoulos, S.N.; Wilson, J.A.; Coalson, T.S.; Fischl, B.; Andersson, J.L.; Xu, J.; Jbabdi, S.; Webster, M.; Polimeni, J.R.; et al. The minimal preprocessing pipelines for the Human Connectome Project. *Neuroimage* **2013**, *80*, 105–124. [CrossRef] [PubMed]
29. Hutchison, R.M.; Womelsdorf, T.; Allen, E.A.; Bandettini, P.A.; Calhoun, V.D.; Corbetta, M.; Della Penna, S.; Duyn, J.H.; Glover, G.H.; Gonzalez-Castillo, J.; et al. Dynamic functional connectivity: Promise, issues, and interpretations. *Neuroimage* **2013**, *80*, 360–378. [CrossRef] [PubMed]

30. Motlaghian, S.; Vahidi, V.; Baker, B.; Belger, A.; Bustillo, J.; Faghiri, A.; Ford, J.; Iraji, A.; Lim, K.; Mathalon, D.; et al. A method for estimating and characterizing explicitly nonlinear dynamic functional network connectivity in resting-state fMRI data. *J. Neurosci. Methods* **2023**, *389*, 109794. [CrossRef]
31. Zhang, G.; Cai, B.; Zhang, A.; Stephen, J.M.; Wilson, T.W.; Calhoun, V.D.; Wang, Y.P. Estimating dynamic functional brain connectivity with a sparse hidden Markov model. *IEEE Trans. Med. Imaging* **2019**, *39*, 488–498. [CrossRef] [PubMed]
32. Barron, A.; Rissanen, J.; Yu, B. The minimum description length principle in coding and modeling. *IEEE Trans. Inf. Theory* **1998**, *44*, 2743–2760. [CrossRef]
33. Gredenhoff, M.; Karlsson, S. Lag-length selection in VAR-models using equal and unequal lag-length procedures. *Comput. Stat.* **1999**, *14*, 171–187. [CrossRef]
34. Brovelli, A.; Ding, M.; Ledberg, A.; Chen, Y.; Nakamura, R.; Bressler, S.L. Beta oscillations in a large-scale sensorimotor cortical network: Directional influences revealed by Granger causality. *Proc. Natl. Acad. Sci. USA* **2004**, *101*, 9849–9854. [CrossRef]
35. Teh, Y.W.; Jordan, M.I.; Beal, M.J.; Blei, D.M. Hierarchical Dirichlet Processes. *J. Am. Stat. Assoc.* **2006**, *101*, 1566–1581. [CrossRef]

symmetry

MDPI

Article

# Attached Flows for Reaction–Diffusion Processes Described by a Generalized Dodd–Bullough–Mikhailov Equation

**Carmen Ionescu** † **and Iulian Petrisor** *,†

Department of Physics, University of Craiova, 13 A.I. Cuza, 200585 Craiova, Romania; carmen.ionescu@edu.ucv.ro
* Correspondence: iulian.petrisor@edu.ucv.ro
† These authors contributed equally to this work.

**Abstract:** This paper uses the attached flow method for solving nonlinear second-order differential equations of the reaction–diffusion type. The key steps of the method consist of the following: (i) reducing the differentiability order by defining the first derivative of the variable as a new variable called the flow and (ii) a forced decomposition of the derivative-free term so that the flow appears explicitly in it. The resulting reduced equation is solved using specific balancing rules. Only step (i) would lead to an Abel-type equation with complicated integral solutions. Completed with (ii) and with the graduation procedure, the attached flow method used in the paper, without requiring such a great effort, allows for the obtaining of accurate analytical solutions. The method is applied here to a subclass of reaction–diffusion equations, the generalized Dodd–Bulough–Mikhailov equation, which includes a translation of the variable and nonlinearities up to order five. The equation is solved for each order of nonlinearity, and the solutions are discussed following the values of the parameters involved in the equation.

**Keywords:** attached flow method; reaction–diffusion equations; generalized Dodd–Bulough–Mikhailov equation

## 1. Introduction

In a previous paper [1], we introduced an improved approach to the rather classical method for solving nonlinear second-order differential equations, based on the use of the first derivative as a new variable and on the investigation of the resulting first-order equation in this new variable [2]. Despite the apparent simplification generated via the reduction of the differentiability order, the method is not easy to apply, as it leads to an Abel-type equation [3]. Precisely to overcome this situation, the approach we proposed, named the attached flow method, contains an additional step: a forced decomposition of the term in the equation that does not contain derivatives. To be very specific, we consider a rather wide class of second-order equations, the reaction–diffusion equation:

$$A(u)\,u'' + B(u)\,u'^2 + C(u)\,u' + E(u) = 0. \tag{1}$$

Equation (1) is the explicit form of the ordinary differential equation generated after applying the wave transformation $\xi = x - Vt$ to the general nonlinear diffusion–convection–reaction equation with functional coefficients. In the $2D$ space defined by the coordinates $\{x, t\}$, such an equation can be written as follows:

$$u_t = (A(u)u_x)_x + [C(u) - V]u_x + E(u). \tag{2}$$

It is a general equation describing nonlinear phenomena from various fields, such as nonlinear optics [4], plasma physics [5,6], or heat and fluids diffusion [7]. From mathematical point of view, there are many interesting approaches related to the stability and convergence analysis of this $2D$ equation, to its variational structure, or to a positivity-preserving and energy-stable operator splitting scheme [8,9].

As we are now interested in the traveling waves solutions, we will consider here exclusively the $1D$ Equation (1).

In [1], we imposed $A(u)$, $B(u)$, and $C(u)$ to be monomials, with $A(u) \neq 0$, while $E(u)$ had the freedom to be a polynomial with minimum and maximum degrees in $u$ denoted as $n(E)$, respectively $N(E)$. How the method works was illustrated for the following cases: (i) $A(u) \neq 0$, $B(u) = C(u) = 0$, $E(u) \neq 0$; (ii) $A(u) \neq 0$, $B(u) = 0$, $C(u) \neq 0$, and $E(u) \neq 0$; (iii) none of the coefficients are vanishing, but $E(u)$ becomes a monomial. The case when $C(u)$ is the only vanishing coefficient was only announced as requiring special precautions, with the mention that it would be addressed later. The approach of this case represents the main objective of the current paper.

The paper is structured as follows: after these introductory notes, the attached flow method is revised in the second section. The third section is devoted to the presentation of a specific case of (1) with $C(u) = 0$, a new version of the generalized Dodd–Bulough–Mikhailov (gDBM) equation that, to avoid the solving difficulties, includes a simple translation of the variable. How the attached flow method has to be applied to this model will be investigated in the fourth section of the paper, while the last section will present a synthesis of the reported results with some supplementary comments on the method. We do not claim to report new DBM solutions here; the real objective of the work is to see, using this new gDBM as a toy model, how the general algorithm of the attached flow method should be applied to equations in this class.

## 2. Basic Facts on the Attached Flow Method

Let us now summarize the key aspects presented in [1] for the attached flow method. Starting from a general nonlinear second-order differential equation of reaction–diffusion type [10], the method supposes two reductions: (i) the reduction in the differentiability order by defining the first derivative of the variable [11,12] as a new variable called the flow, and (ii) the reduction in the nonlinearity degree by decomposing the derivative-free term so that the flow appears explicitly in it [1]. The reduction in the nonlinearity avoids reaching an Abel-type equation, which is difficult to solve analytically [11]. The previous two reductions lead to a simpler equation that, in our approach, is easily solved using specific balancing rules. This supplementary balancing makes the attached flow method very effective in looking for analytic solutions that, in other previous approaches, appear in complicated integral forms and need full effort to be obtained.

Explicitly, the method is based on three main requirements:

- The variable $u(\xi)$ must supplementarily satisfy a flow equation of the following form:

$$u'(\xi) = f(u). \tag{3}$$

Relation (3) represents the classic requirement for reducing the differentiability order recommended in textbooks. It transforms (1) in a particular case of the second kind of Abel equation in the variable $f(u)$:

$$A(u)f(u)\frac{df}{du} + B(u)f(u)^2 + C(u)f(u) + E(u) = 0, \tag{4}$$

- The flow $f(u)$ must be connected with $E(u)$ via a forced decomposition of the following form:

$$E(u) = f(u) \cdot h(u). \tag{5}$$

This is the novelty proposed by us, and it aims to avoid the direct solving of Equation (4) that usually only has implicit solutions. The relation (5) transforms (1) into the following first-order differential equation:

$$A(u)\frac{df}{du} + B(u)\,f(u) + C(u) + h(u) = 0, \tag{6}$$

- The two functions from (5), $f(u)$ and $h(u)$, are considered polynomials of the following form:

$$f(u) = \sum_{i=n(f)}^{N(f)} \alpha_i u^i;\ h(u) = \sum_{j=n(h)}^{N(h)} \beta_j u^j. \tag{7}$$

The summation limits in (7) are established through a balancing procedure, imposing the requirement to compensate the different powers of $u$ in relations (5) and (6).

The steps in solving (1) now become the following: (i) to determine the flow $f(u)$ and (ii) to find the solution $u(\xi)$ by integrating (3). The first step is supposed to consider (7) in Equations (5) and (6) and to determine the possible form of $f(u)$ via an adequate balancing procedure.

Two important clarifications need to be made at this point:

- It is clear that the decomposition (5) is not unique. Various solutions can be generated, supposing $f(u)$ and $h(u)$ as polynomials with degrees running from unit to unit, between $\{n(f)$ and $N(f)\}$, respectively $\{n(h)$ and $N(h)\}$. However, a single decomposition proves to be compatible with solving a certain nonlinear model. It is established in the attached flow approach through a specific balancing process, another element of originality of the method.
- In general $\{\alpha_i, \beta_j\}$ are constant coefficients. In specific cases that will be mentioned below, we will be obliged to consider them as functions of the independent variable:

$$\alpha_i = \alpha_i(\xi); \beta_j = \beta_j(\xi).$$

In these cases, the flow Equation (3) will become the following:

$$u'(\xi) = f(\xi, u(\xi)) = \sum_{i=n(f)}^{N(f)} \alpha_i(\xi) u^i(\xi). \tag{8}$$

With (8), the reduced Equation (6) becomes the following:

$$A(u)\left[\frac{\partial f}{\partial \xi} + \frac{df}{du}f\right] + B(u)\,f(u) + C(u) + h(u) = 0. \tag{9}$$

With these remarks in mind, let us now return to the problem of determining the flow, $f(u)$. This is now equivalent to determining the coefficients $\alpha_i$, and $N(f)$, $n(f)$, the maximum, and, respectively, the minimum degrees in $u$ from (7) or (8). The relation (5) imposes, as necessary but not sufficient conditions, the following requirements:

$$N(f) + N(h) = N(E); n(f) + n(h) = n(E). \tag{10}$$

The requirement for mutual compensation of the terms with different powers in $u$ that appear in (6) leads to an additional set of algebraic relations that, in principle, suggests the admissible forms of the flow $f(u)$ or more precisely leads to the following conclusions on the maximal degrees in $u$ of the flow $f(u)$ [1]:

$$\begin{aligned}
N(f) &= N(C) - N(A) + 1 \text{ if } N(B) \le N(A) - 1 \le 2N(C) - N(E);\\
N(f) &= \frac{N(E) - N(A) + 1}{2} \text{ if } \left\{ \begin{array}{c} N(B) \le N(A) - 1 \\ 2N(C) - N(E) \le N(A) - 1 \end{array} \right. ;\\
N(f) &= N(C) - N(B) \text{ if } N(A) - 1 \le N(B) \le 2N(C) - N(E);\\
N(f) &= \frac{N(E) - N(B)}{2} \text{ if } \left\{ \begin{array}{c} N(A) - 1 \le N(B) \\ 2N(C) - N(E) \le N(B) \end{array} \right. ;
\end{aligned} \tag{11}$$

$$N(f) = N(E) - N(C) \text{ if } \begin{cases} N(A) - 1 \leq 2N(C) - N(E) \\ N(B) \geq N(E) - 2N(C) \end{cases}.$$

These degrees are not compulsory integers. How exactly the mechanism is functioning will be illustrated in the fourth section of the paper with a specific sub-equation of (1) with $C(u) = 0$, namely the gDBM equation.

## 3. The DBM Model as a Reaction–Diffusion Equation

As announced, the aim of this article is to show how the attached flow as a solving method can be applied on reaction–diffusion equations of the form (1) with $C(u) = 0$. To be very specific, we will illustrate this using the DBM model, important in many problems from nonlinear optics, hydrodynamics, or quantum field theory. Despite numerous studies on the DBM equation and the various techniques used for solving it, the model is surprisingly rich, and depending on the number of parameters each approach introduces in the study, new classes of solutions can be generated. An important note has to be made here: there are maybe dozens of recent papers claiming that new DBM solutions have been obtained, but in many cases, these apparently "new" solutions are nothing more than solutions expressed as slightly different mathematical expressions belonging to already known classes of solutions. This is one of the current errors pointed out in [13], and this is why, to avoid misunderstandings, we have already stated that we will not introduce new solutions in this paper, but will only illustrate how the attached flow method can be used in the case of the DBM equation. A possible classification of the DBM solutions is suggested in [14].

In the $2D$ space of the coordinates $\{x, t\}$, the DBM model is described by a function, $v(x,t)$, which satisfies the following equation:

$$w_{xt} + e^{w} + e^{-2w} = 0. \tag{12}$$

At the classical level, (12) is known as a nonlinear integrable model describing both classical and quantum propagation phenomena. In the quantum context, the equation appears as the Bullough–Dodd model for scalar fields [15], and it is equivalent to an affine Toda field with zero curvature representation in twisted affine Kac–Moody algebras [16]. This feature assures the existence of soliton-like solutions.

Equation (12) can be included, in turn, in a more general equation, the generalized Tzitzeica–Dodd–Bullough–Mikhailov (gTDBM) equation [17]:

$$w_{xt} + pe^{\sigma w} + qe^{\rho w} = 0. \tag{13}$$

From this form, we can recover not only the simple DBM Equation (12) but also other interesting models, such as, for example, the *sinh*- or *cosh*–Gordon equations [18] and the Liouville equation, corresponding to $q = 0$.

Due to the non-polynomial form of (13), special techniques have to be applied in order to determine its solutions. The classical approach consists of using the transformation $w(x,t) = \ln \phi(x,t)$ with the master Equation (12) taking the following form:

$$\phi_{xt}\phi - \phi_x\phi_t + p\phi^{\sigma+2} + q\phi^{\rho+2} = 0. \tag{14}$$

Equation (14) contains four parameters: the real parameters $p, q$, respectively, the integers $\sigma, \rho$. This number can be reduced using suitable choices. We will discuss here various reduced equations with fewer parameters. The starting point will be a slightly different form of the previous equation, usually called the generalized DBM equation [14]. It corresponds to the following choices: $\sigma + 2 = m$, $\rho = -2$, choices that lead to the following:

$$\phi_{xt}\phi - \phi_x\phi_t + p\phi^{m} + q = 0. \tag{15}$$

This will be the main equation considered in this paper, to which we will apply the attached flow. The method allows the traveling wave solutions to be found, and it supposes, as a first step to transform (15) into a nonlinear ordinary differential equation (NODE) by taking into consideration the wave transformation, $\xi = x - Vt$. With it and using the notations $\phi(x,t) \rightarrow u(\xi)$, $u' = du/d\xi$, $u'' = d^2u/d\xi^2$, Equation (15) can be written as follows:

$$-Vuu'' + Vu'^2 + pu^m + q = 0. \quad (16)$$

A first remark is that the DBM model (12) corresponds to (16) with $p = 1, q = 1$, and $m = 3$:

$$-Vuu'' + Vu'^2 + u^3 + 1 = 0. \quad (17)$$

A second remark is that (16) can be generalized by including a translation of the dependent variable:

$$u \rightarrow v \equiv u + k, \ k = const. \quad (18)$$

The main equation to be studied becomes the following:

$$-V(u+k)u'' + Vu'^2 + p(u+k)^m + q = 0, \quad (19)$$

where $m$ is the nonlinearity order, $p$ and $q$ are the constant coefficients of the free derivative term, $V$ is the wave velocity, and $k$ is the translation constant.

Equation (19) is what we will understand here by the generalized Dodd–Bullough–Mikhailov equation. In its version with $k = 0$ , it was intensively studied before; see, for example, [17]. This version includes what is called the DBM equation as the particular case, $m = 3$. The translation with $k \neq 0$ is what our paper proposes as a novelty, and as we will see, it is essential to finding the minimal degree n(f) for the flow (7).

By translating the variable, the supplementary term $-Vku''$ breaks the initial symmetry between the minimal degrees of $f(u)$ and $h(u)$, making possible $n(f) \neq n(h)$, a key fact in solving the problem with our method. The derivative-free term generated in the same process now contains a full spectrum of powers in $u$ and those terms, together with derivative ones, allow the solving of the reduced equation: compatible expressions for the two functions may be generated now, when the specific balancing of the attached flow approach is applied. We will discuss in the next section how this can be done and what the results are when nonlinearities given via $m = \{1,2,3,4,5\}$ are considered.

Let us mention that the DBM equation has been solved so far with various other methods. The simplest approaches tend to use a predefined form of solutions, as in [19] or [20], where the tanh method is applied. More generally, traveling wave solutions for DBM have been generated via different methods belonging to the auxiliary equation technique: the $\left(\frac{G'}{G}\right)$-expansion [21], the functional expansion [22], or the sub-equation technique, considering the elliptic Jacobi as an auxiliary equation [15]. Soliton and compacton-like solutions were generated in [23] with the exponential function method, or in [24]with the generalized Kudryashov method and an improved F-expansion method. Other investigations were based on the truncated Painlevé expansion [25], Darboux transformation [26], or integral bifurcation [17]. Analytic solutions for DBM were reported in [27] using Lax operators, an approach that, together with Hirota's method, provides important tools for investigating the complete integrability of nonlinear models [28,29]. An invariant group of solutions could also be obtained using the symmetry reduction method [30]. In [31] the DBM equation was approached based on the first integral method, the closest approach to the attached flow that we will use here.

## 4. Solving DBM with Attached Flows

Let us consider now how the gDBM Equation (19) could be solved directly using the attached flow method. As it has the form of the generalized reaction–diffusion Equation (1), we can transfer the results mentioned for this last equation in Section 2 from above. It is simple to see that the following applies:

$$A(u) = -V(u+k); B(u) = V; C(u) = 0; E(u) = p(u+k)^m + q. \quad (20)$$

So, $A(u)$ is a polynomial with $N(A) = 1, n(A) = 0$, and $B(u)$ is a monomial with $N(B) \equiv n(B) = 0$, while $E(u)$ is a polynomial with $N(E) = m$ and $n(E) = 0$. We have a case where $N(A) - 1 \leq N(B)$ and $2N(C) - N(E) \leq N(B)$ from (11), so the general theory gives the following:

$$N(f) = N(h) = \frac{N(E)}{2} = \frac{m}{2}. \tag{21}$$

The reduced Equation (6) becomes the following:

$$-V(u+k)\frac{df}{du} + V f(u) + h(u) = 0. \tag{22}$$

An important remark is that $N(f)$ and $N(h)$ are multiples of $1/2$, integers only when $m$ is even. The minimal degrees also satisfy the constraint (10) with $n(E) = 0$, so the following applies:

$$n(h) = -n(f). \tag{23}$$

As we consider $n(f) \geq 0$, we conclude that $n(h) \leq 0$. More precisely, $n(h) = 0$ if and only if $n(f) = 0$. The flow $f(u)$ to be attached to (19) will be given by (7), but its expansion essentially depends on the considered value of $m$. For convenience, we will choose the following notations:

$$f(u) = \alpha_{\frac{m}{2}} u^{\frac{m}{2}} + \alpha_{\frac{m-2}{2}} u^{\frac{m-2}{2}} + \ldots + \alpha_0; \text{ for } m = even, \tag{24}$$

$$f(u) = \alpha_{\frac{m}{2}} u^{\frac{m}{2}} + \alpha_{\frac{m-2}{2}} u^{\frac{m-2}{2}} + \ldots + \alpha_1 u^{\frac{1}{2}}; \text{ for } m = odd. \tag{25}$$

Similarly, we will consider $h(u)$, given through the following:

$$h(u) = \beta_{\frac{m}{2}} u^{\frac{m}{2}} + \beta_{\frac{m-2}{2}} u^{\frac{m-2}{2}} + \ldots + \beta_0; \text{ for } m = even, \tag{26}$$

$$h(u) = \beta_{\frac{m}{2}} u^{\frac{m}{2}} + \beta_{\frac{m-2}{2}} u^{\frac{m-2}{2}} + \ldots + \beta_{-1} u^{-\frac{1}{2}}; \text{ for } m = odd. \tag{27}$$

The previous relations have to make compatible Equations (5) and (6), equations that, for gDBM, have the following specific form:

$$p(u+k)^m + q = f(u)\, h(u), \tag{28}$$

$$-V(u+k)\frac{df}{du} + Vf(u) = h(u). \tag{29}$$

Using the suitable expressions of $f(u)$ and $h(u)$ from above in these last two equations and asking for the canceling of the coefficients of various powers of $u$, we get an algebraic system that allows us to determine the coefficients $\alpha_i$ and $\beta_j$. We will see that very important are the last coefficients, $\{\alpha_0, \beta_0\}$ for $m = even$ and, respectively, $\{\alpha_1, \beta_{-1}\}$ for $m = odd$, and the last equations lead to the following relations among them:

$$\beta_0 = -V\alpha_0; \beta_{-1} = Vk\frac{\alpha_1}{2}. \tag{30}$$

Conclusion: To make the large number of mathematical derivations clearer, we can synthesize the flowchart that is used in the attached flow method to solve the gDBM equation:

- We start from the gTDBM Equation (13) and transform it into (15) via an adequate change in the variable and choice of parameters.
- The PDE (15) is reduced to the ODE (16) using the wave variable.
- The translation (18) leads to Equation (19), which will be studied for various values of $m$.
- We apply the first specific ingredient of the attached flow approach: the decomposition of $E(u)$, as in (5). The gDBM Equation (19) takes the reduced form (22).

- We apply the second ingredient of the attached flow: a balancing procedure that leads to explicit forms of $f(u)$ and $h(u)$, which depend on $m$: (i) if $m$ is even, with (24) and (26); (ii) if $m$ is odd, with (25) and (27).
- We use these expressions to find the solutions of (19) for $m = 1, \ldots, 5$.

### 4.1. Solutions for $m = 1$

Let us consider first that $m = 1$, so the equation to be solved is as follows:

$$-V(u+k)u'' + Vu'^2 + pu + (pk+q) = 0. \tag{31}$$

In this case, using (25) and (27) in (28), respectively (29), will yield the following:

$$f(u) = \alpha_1 u^{\frac{1}{2}}, \quad h(u) = -\frac{V\alpha_1}{2}\left(u^{\frac{1}{2}} - ku^{-\frac{1}{2}}\right), \tag{32}$$

with the following compatibility condition:

$$k = -\frac{q}{2p}, \; p \neq 0. \tag{33}$$

In conclusion, the gDBM equation can be solved for $m = 1$ using the attached flow method. A translation with $k$, given by (33), is required. The equation accepts, in this case, a solution of the following form:

$$u(\xi) = -\frac{p}{4V}\xi^2 - \frac{q}{2p}. \tag{34}$$

We can see from (33) that $k = 0$ when $q = 0$. No translation is needed in this case. Otherwise, the values of $k$ are influenced by the values of $q$ and $p$. As $p \neq 0$ the specific form of the final parabolic solution (34) depends on the parametric errors induced via $q$ and $V$.

### 4.2. Solutions for $m = 2$

When we consider $m = 2$ in (19), the equation we have to solve will be as follows:

$$-V(u+k)\,u'' + Vu'^2 + p(u+k)^2 + q = 0. \tag{35}$$

As $N(E) = 2 = even, n(E) = 0$, we have the following: $N(f) = N(h) = 1, n(f) = n(h) = 0$. A direct check shows that trying to apply the aforementioned solving procedure considering $\alpha_i$ as constants leads us to incompatibilities. It is not possible in this case to determine $f(u)$ and $h(u)$ in the form (7) with $\alpha_i$ and $\beta_j$ as simple constants due to the fact that the first two terms, $-Vu\frac{df}{du}$ and $Vf$, from the reduced Equation (22) cancel each other and lead to $h = 0$, incompatible with the decomposition $fh = E \neq 0$. For the procedure to work, we must now consider a flow with an explicit dependence on the independent variable, $f = f(\xi, u(\xi))$, as in (8). We have to apply the method of the attached flow with variable coefficients for $f$ and $h$. We will consider the following:

$$f = \alpha_1(\xi)u(\xi) + \alpha_0(\xi), \; h = \beta_1(\xi)u(\xi) + \beta_0(\xi). \tag{36}$$

Using these expressions in the decomposition $E = fh$ and in the reduced Equation (22), we generate a system of equations that allows, in principle, for finding the functions $\{\alpha_i(\xi), \beta_i(\xi), i = 0,1\}$ and the compatible values of the parameters $k, V, p, q$. From the decomposition $E = fh$, we get relations allowing to express $\beta_0, \beta_1$ in terms of $\alpha_0, \alpha_1$:

$$\begin{aligned} fh &= \alpha_1\beta_1 u^2 + (\alpha_1\beta_0 + \alpha_0\beta_1)u + \alpha_0\beta_0, \\ fh &= E = p(u+k)^2 + q = pu^2 + 2pku + pk^2 + q. \end{aligned}$$

$$\begin{aligned} \alpha_1\beta_1 &= p,\ \alpha_1\beta_0+\alpha_0\beta_1 = 2pk,\ \alpha_0\beta_0 = pk^2+q, \\ \alpha_1\beta_1 &= p \to \beta_1 = \frac{p}{\alpha_1}; \alpha_0\beta_0 = pk^2+q \to \beta_0 = \frac{pk^2+q}{\alpha_0}, \\ \alpha_1\beta_0+\alpha_0\beta_1 &= 2pk \to \frac{\alpha_1}{\alpha_0}(pk^2+q)+\frac{\alpha_0}{\alpha_1}p = 2pk. \end{aligned} \tag{37}$$

Introducing these expressions in (35) and canceling the coefficients of various powers of $u$, we get the following compatibility conditions:

$$\begin{aligned} \alpha_1 &= \frac{p}{V}\xi + C, \\ \alpha_0 &= \frac{kp}{V}\xi + 2kC = k\alpha_1, \\ q &= 0. \end{aligned}$$

Conclusion: In the case of $m=2$, the generalized DBM equation takes the form (35). It has a nontrivial solution for $q=0$ and $k=0$, with the corresponding attached flow that solves the equation being as follows:

$$f(\xi,u) = \frac{p}{V}\xi u(\xi) + \frac{kp}{V}\xi = \frac{p}{V}\xi u.$$

By integrating the flow equation, we get the following solution for the gDBM equation:

$$u(\xi) = Ce^{\frac{p\xi^2}{2V}}, C = const.$$

Remark: Numerical computations show that nontrivial solutions could be also obtained for $q \neq 0$, but around $\xi = 0$ only. Equation (35) could also admit a nontrivial solution for $q \neq 0$ if $p$ and $q$ also become functions of the independent variable $\xi$. In this case, the problem of the parametric errors becomes important and has to be properly considered.

### 4.3. Solutions for $m=3$

In the case of $m = N(E) = 3 = odd$, using (10) and (11), we conclude that $N(f) = 3/2$ and $N(h) = 3/2$. For the minimal degrees, as $n(E)=0$, we get $n(f) = 1/2$ and $n(h) = -1/2$. We are in a situation where the degrees of $f(u)$ and $h(u)$ are no longer integers. We will consider the following:

$$f(u) = \alpha_3 u^{\frac{3}{2}} + \alpha_1 u^{\frac{1}{2}},\ h(u) = \beta_3 u^{\frac{3}{2}} + \beta_1 u^{\frac{1}{2}} + \beta_{-1} u^{-\frac{1}{2}}.$$

We impose the compatibility conditions (28) and (29), and we get the following algebraic system:

$$\begin{aligned} \alpha_3\beta_3 = p,\ \alpha_3\beta_{-1} + \alpha_1\beta_1 = 3pk^2, \\ \alpha_3\beta_1 + \alpha_1\beta_3 = 3pk,\ \alpha_1\beta_{-1} = pk^3 + q \end{aligned}$$

with the coefficients $\beta$ given via the following:

$$\beta_3 = \frac{V\alpha_3}{2}; \beta_1 = \frac{3Vk}{2}\alpha_3 - \frac{V\alpha_1}{2}; \beta_{-1} = \frac{Vk\alpha_1}{2}.$$

These relations are compatible with the following:

$$k = \frac{1}{2}\left(\frac{q}{p}\right)^{\frac{1}{3}}. \tag{38}$$

We conclude that applying the attached flow method to the generalized DBM Equation (19) in the case of $m=3$ asks for a translation, $u \to u+k$, with $k \neq 0$. Depending on the values of the other parameters, the following solutions can be generated:

(i) For $V > 0$ and $q \neq 0$, we get periodic solutions:

$$u_1(x,t) = \frac{1}{2}\left(\frac{q}{p}\right)^{\frac{1}{3}} + \frac{3}{2}\left(\frac{q}{p}\right)^{\frac{1}{3}} tg^2\left(\sqrt{\frac{3}{4V}}p^{\frac{1}{3}}q^{\frac{1}{6}}(x-Vt)\right),$$

$$u_2(x,t) = \left(\frac{q}{p}\right)^{\frac{1}{3}}\left[\frac{1}{2} + \frac{3}{2}\cot^2\left(-\sqrt{\frac{3}{4V}}p^{\frac{1}{3}}q^{\frac{1}{6}}(x-Vt)\right)\right];$$

(ii) For $V < 0$ and $q \neq 0$, the solutions are hyperbolic:

$$u_3(x,t) = \left(\frac{q}{p}\right)^{\frac{1}{3}}\left[\frac{1}{2} - \frac{3}{2}th^2\left(\sqrt{\frac{3}{4V}}p^{\frac{1}{3}}q^{\frac{1}{6}}(x-Vt)\right)\right],$$

$$u_4(x,t) = \left(\frac{q}{p}\right)^{\frac{1}{3}}\left[\frac{1}{2} + \frac{3}{2}cth^2\left(-\sqrt{\frac{3}{4V}}p^{\frac{1}{3}}q^{\frac{1}{6}}(x-Vt)\right)\right];$$

(iii) For $q = 0$, we get a rational solution:

$$u_5(x,t) = \frac{2V}{(x-Vt)^2}$$

Let us mention that all of these solutions were obtained and reported in [14] but using other numerical methods, not directly as here. The solutions are influenced by different choices of $q$ and $p$. Other interesting numerical experiments on the DBM equation and the influence of the parameters on the solutions are included in [32,33]. As a numerical example, if $p < 0$ and $q = 1$, the hyperbolic solution could become a harmonic one, and vice versa. So, the parameter values can strongly affect the types of solutions.

### 4.4. Solutions for $m = 4$

Let now us consider Equation (19) with $m = 4$. The general relation (24) leads us to the idea that the flow should have the following form:

$$f(u) = \alpha_2 u^2 + \alpha_1 u + \alpha_0. \tag{39}$$

The compatibility of (28) and (29) can be obtained if and only if $k = 0$. The coefficients $\alpha_i$ from (39) satisfy, in this case, the following constraints:

$$V\alpha_2^2 = p, \alpha_1 = 0,\ V\alpha_0^2 = q.$$

As an explicit expression of the flow, we can use the following:

$$f(u) = \pm\sqrt{\frac{p}{V}}u^2 \pm \sqrt{-\frac{q}{V}}. \tag{40}$$

We can generate various numerical experiments using the previous expression to integrate the flow Equation (40):

- For $q \neq 0$ and $-\frac{pq}{V^2} > 0$, the solutions are periodic:

$$u_1(\xi) = \left(-\frac{q}{p}\right)^{\frac{1}{4}} + \left(-\frac{q}{p}\right)^{\frac{1}{4}} tg\left(-\frac{pq}{V^2}\right)^{\frac{1}{4}}\xi,$$

$$u_2(\xi) = \left(-\frac{q}{p}\right)^{\frac{1}{4}} + \left(-\frac{q}{p}\right)^{\frac{1}{4}} \cot\left[-\left(-\frac{pq}{V^2}\right)^{\frac{1}{4}}\xi\right]; \tag{41}$$

- For $q \neq 0$ and $-\frac{pq}{V^2} < 0$, the solutions are hyperbolic:

$$u_3(\xi) = \left(-\frac{q}{p}\right)^{\frac{1}{4}} + \left(-\frac{q}{p}\right)^{\frac{1}{4}} th\left(-\frac{pq}{V^2}\right)^{\frac{1}{4}}\xi,$$

$$u_4(\xi) = \left(-\frac{q}{p}\right)^{\frac{1}{4}} + \left(-\frac{q}{p}\right)^{\frac{1}{4}} cth\left[-\left(-\frac{pq}{V^2}\right)^{\frac{1}{4}}\xi\right]; \tag{42}$$

- For $q = 0$, the solution is rational:

$$u_5 = -\left(\frac{V}{p}\right)^{\frac{1}{2}} \frac{1}{\xi} \tag{43}$$

*4.5. Solution for $m = 5$*

Let us now consider that $m = 5$ in (19). We have, in the case of (25), the flow given by the following:

$$f(u) = \alpha_5 u^{\frac{5}{2}} + \alpha_3 u^{\frac{3}{2}} + \alpha_1 u^{\frac{1}{2}}.$$

The complementary function $h(u)$ will be as follows:

$$h(u) = \beta_5 u^{\frac{5}{2}} + \beta_3 u^{\frac{3}{2}} + \beta_1 u^{\frac{1}{2}} + \beta_{-1} u^{-\frac{1}{2}}.$$

The compatibility of the previous expressions with (28) leads to the following algebraic system:

$$\begin{aligned} \alpha_5\beta_5 &= p, \\ \alpha_5\beta_3 + \alpha_3\beta_5 &= 5pk, \\ \alpha_5\beta_1 + \alpha_3\beta_3 + \alpha_1\beta_5 &= 10pk^2, \\ \alpha_5\beta_{-1} + \alpha_3\beta_1 + \alpha_1\beta_3 &= 10pk^3, \\ \alpha_3\beta_{-1} + \alpha_1\beta_1 + \alpha_2\beta_0 &= 5pk^4, \\ \alpha_1\beta_{-1} &= pk^5 + q. \end{aligned} \tag{44}$$

The compatibility with (29) provides other supplementary requirements:

$$\begin{aligned} \beta_5 &= \frac{3V\alpha_5}{2}, \\ \beta_3 &= \frac{5\alpha_5 Vk}{2} + \frac{V\alpha_3}{2}, \\ \beta_1 &= \frac{3\alpha_3 Vk}{2} - \frac{V\gamma\alpha_1}{2}, \\ \beta_{-1} &= \frac{\alpha_1}{2} Vk. \end{aligned} \tag{45}$$

Direct computations show that the only case of the compatibility of (44) and (45) is for $q = 0$. In this case, $k = 0$, and the solution has the following form:

$$u(\xi) = \left(\frac{3p}{2V}\right)^{1/3} \xi^{2/3}.$$

Remark: Let us note that, as in the case of $m = 5$, for any $m > 5$ the gDBM equation may admit analytic solutions for $q = 0$ only. This is somehow related to the fact that the compatibility conditions that must be imposed lead to algebraic equations of orders higher than 5 and there are no general numerical formulas that provide analytic solutions expressed in radicals. In fact, as Abel and Ruffini established long ago [34,35], such formulas exist up to $m = 4$, not for quintic equations. These last equations could have solutions in terms of elliptic functions [36] or as very complicated expressions of no practical interest.

## 5. Comments on the Method and Results

This paper has dealt with applying the method of the attached flow proposed in [1] to a particular case of the reaction–diffusion Equation (1), namely the case of $C(u) = 0$. The key aspects of the attached flow method applied to a general equation of the form (1) is the decomposition $E(u) = f(u)h(u)$, where $f(u)$ represents the flow satisfying the equation $u' = f(u)$. The advantage of this decomposition is that it is always feasible, it brings the equation to the reduced form (6), and it allows the flow to be determined as polynomials of the form (7). The extreme values of the degrees, $n(f)$ and $N(f)$, are obtained via other important aspects of the method: the introduction of gradation rules and the requirement that terms of the same power in $u$ compensate each other. The graduation rules seem to introduce restrictions, being applicable only to equations with polynomial coefficients, but most models of physical interest, defined in various space–time dimensions [37], belong to this category. More sophisticated graduations appear when multidifferential complexes are considered [38,39].

In the case of gDBM Equation (16), the balancing requirement leads to a system of algebraic equations that must be solved. The graduation rules show us that, for some values of $m$, compatibility problems could appear in $n(f)$, and as a novelty brought about via the paper, a translation, $u \to u + k$, was taken into account so that we studied Equation (19):

$$-V(u+k)u'' + Vu'^2 + p(u+k)^m + q = 0.$$

Moreover, only for $m = 2$, that is, the differentiability order of gDMB, the compatibility in the highest degree, $N(f)$, asks for a dependency of $\xi$ in the flow equation, which could be written in the general form as follows:

$$u'(\xi) = f(\xi\delta_{m2}, u(\xi)),$$

where $\delta$ represents the Kronecker symbol.

The gDBM Equation (19) was written with five parameters: two real parameters, $p$ and $q$, the wave velocity, $V$ and two others of special interest in our study: $m$—the degree of nonlinearity of the free term, and the translation parameter $k$. Let us summarize the main solvable cases and the parametric error influence we have reported here, based on various numerical experiments. If we consider $q = 0$, the equation can be solved directly for any value of $m$, and its general solution does not depend on $k$. For $V \neq 0$, $p \neq 0$, the solution behaves as follows:

$$u(\xi) \sim \begin{cases} \xi^{-\frac{2}{m-2}}, & m \neq 2 \\ e^{\frac{p}{2V}\xi^2}, & m = 2 \end{cases}.$$

As we can see, a special situation appears for $m = 2$, and it is due to the fact that gDBM is a second-order differential equation. Trying to apply the attached flow in this case, we needed to consider flows of the form (8), depending explicitly on the independent variable, $\xi$.

For $q \neq 0$, depending on $m$, we obtained solutions with or without translations. Practically, no translation is necessary for $m = 2, 4$, and a translation, $k \neq 0$, is required for $m = 1, 3$. For $m = 5$, our method allowed us to find solutions in the case of $q = 0$ only. For $q \neq 0$, the system of algebraic equations to be solved contains an equation of the degree of 5, whose roots cannot be expressed with radicals. It cannot be compatibilized with the other equations of the system. This impossibility of compatibilization is valid for any $m > 6$, so all these equations cannot be analytically solved, not because of the attached flow method but, rather, due to the Abel–Ruffini theorem.

**Author Contributions:** The first author of this paper (C.I.) contributed the conceptualization, while the corresponding author (I.P.) contributed the validation of the results. All authors have read and agreed to the final version of the manuscript.

**Funding:** This research received no external funding.

**Data Availability Statement:** No new data were created or analyzed in this study.

**Conflicts of Interest:** The authors declare no conflict of interest.

## References

1. Ionescu, C.; Constantinescu, R. Solving Nonlinear Second-Order Differential Equations through the Attached Flow Method. *Mathematics* **2022**, *10*, 2811. [CrossRef]

2. Zwillinger, D. *Handbook of Differential Equations*, 3rd ed.; Academic Press: Cambridge, MA, USA, 1997.
3. Panayotounakos, D.E.; Zarmpoutis, T.I. Construction of Exact Parametric or Closed Form Solutions of Some Unsolvable Classes of Nonlinear ODEs (Abel's Nonlinear ODEs of the First Kind and Relative Degenerate Equations). *Int. J. Math. Sci.* **2011**, 387429. [CrossRef]
4. Agrawal, G.P. *Nonlinear Fiber Optics*, 4th ed.; Academic Press: Cambridge, MA, USA, 2006.
5. Shalchi, A.; Negrea, M.; Petrisor, I. Stochastic field-line wandering in magnetic turbulence with shear: I. Quasi-linear theory. *Phys. Plasmas* **2016**, *23*, 072306. [CrossRef]
6. Negrea, M.; Petrisor, I.; Shalchi, A. Stochastic field-line wandering in magnetic turbulence with shear. II. Decorrelation trajectory method. *Phys. Plasmas* **2017**, *24*, 112303. [CrossRef]
7. Anderson, J.D. *Fundamentals of Aerodynamics*, 4th ed.; McGraw–Hill: New York, NY, USA, 2007.
8. Liu, C.; Wang, Y. On Lagrangian schemes for porous medium type generalized diffusion equations: A discrete energetic variational approach. *J. Comput. Phys.* **2020**, *417*, 109566. [CrossRef]
9. Liu, C.; Wang, C.; Wang, Y. A structure-preserving, operator splitting scheme for reaction-diffusion equations with detailed balance. *J. Comput. Phys.* **2021**, *436*, 110253. [CrossRef]
10. Cherniha, R.; Davydovych, V. *Nonlinear Reaction-Diffusion Systems*; Springer: Berlin/Heidelberg, Germany, 2017.
11. Polyanin, A.D.; Zaitsev, V.F. *Handbook of Exact Solutions for Ordinary Differential Equations*; Chapman & Hall/CRC Press: Boca Raton, FL, USA, 2003.
12. Wang, X.Y. Exact and explicit solitary wave solutions for the generalised Fisher equation. *Phys. Lett. A* **1988**, *131*, 277–279. [CrossRef]
13. Kudryashov, N.A. Seven common errors in finding exact solutions of nonlinear differential equations. *Commun. Nonlinear Sci. Numer. Simul.* **2009**, *14*, 3507–3529. [CrossRef]
14. Islam, N.; Khan, K.; Islam, M.H. Travelling wave solution of Dodd-Bullough-Mikhailov equation: A comparative study between Generalized Kudryashov and improved F-expansion methods. *J. Phys. Commun.* **2019**, *3*, 055004. [CrossRef]
15. Cimpoiasu, R.; Constantinescu, R.; Pauna, A.S. Solutions of the Bullough–Dodd Model of Scalar Field through Jacobi-Type Equations. *Symmetry* **2021**, *13*, 1529. [CrossRef]
16. Assis, P.E.G.; Ferreira, L.A. The Bullough-Dodd model coupled to matter fields. *Nucl. Phys. B* **2008**, *800*, 409–449. [CrossRef]
17. Rui, W. Exact traveling wave solutions for a nonlinear evolution equation of generalized Tzitzéica-Dodd-Bullough-Mikhailov type. *J. Appl. Math.* **2013**, 395628. [CrossRef]
18. Fan, X.; Yang, S.; Yin, J.; Tian, L. Traveling wave solutions to the (n+1)-dimensional sinh–cosh–Gordon equation. *Comput. Math. Appl.* **2011**, *61*, 699–707. [CrossRef]
19. Wazwaz, A.M. The tanh method: Solitons and periodic solutions for the Dodd–Bullough–Mikhailov and the Tzitzeica–Dodd–Bullough equations. *Chaos Solitons Fractals* **2005**, *25*, 55–63. [CrossRef]
20. Constantinescu, R. New solutions of Dodd-Bullough-Mikhailov equation by using an improved tanh-method. *Rom. Rep. Phys.* **2017**, 69, 112.
21. Borhanifar, A.; Moghanlu, A.Z. Application of the $(G'/G)$-expansion method for the Zhiber–Shabat equation and other related equations. *Math. Comput. Model.* **2011**, *54*, 2109–2116. [CrossRef]
22. Ionescu, C.; Constantinescu, R.; Stoicescu, M. Functional expansions for finding traveling wave solutions. *JAAC* **2020** , *10*, 569–583. [CrossRef] [PubMed]
23. He, J.H.; Wu, X.H. Exp-function method for nonlinear wave equations. *Chaos Solitons Fractals* **2006**, *30*, 700 –708. [CrossRef]
24. Kudryashov, N.A. One method for finding exact solutions of nonlinear differential equations. *Commun. Nonlinear Sci. Numer. Simul.* **2012**, *17*, 2248–2253. [CrossRef]
25. Meleshko, S.V.; Schief, W.K. A truncated Painlevé expansion associated with the Tzitzéica equation: Consistency and general solution. *Phys. Lett. A* **2002**, *299*, 349–352. [CrossRef]
26. Brezhnev, Y.V. Darboux transformation and some multi-phase solutions of the Dodd–Bullough–Tzitzéica equation. *Phys. Lett. A* **1996**, *211*, 94–100. [CrossRef]
27. Babalic, C.N.; Constantinescu, R.; Gerdjikov, V.S. On Tzitzeica equation and spectral properties of related Lax operators. *Balk. J. Geom. Its Appl.* **2014**, *19*, 11–22.
28. Babalic, C.N. Complete integrability and complex solitons for generalized Volterra system with branched dispersion. *Int. J. Mod. Phys. B* **2020**, *34*, 2050274. [CrossRef]
29. Babalic, C.N. Integrable discretization of coupled Ablowitz-Ladik equations with branched dispersion. *Rom. J. Phys.* **2018**, *63*, 114.
30. Cimpoiasu, R.; Rezazadeh, H.; Florian, D.A.; Ahmad, H.; Nonlaopon, K.; Altanji, M. Symmetry reductions and invariant-group solutions for a two-dimensional Kundu-Mukherjee-Naskar model. *Results Phys.* **2021**, *8*, 104583. [CrossRef]
31. Shen, G.; Sun, Y.; Xiong, Y. New Travelling-Wave Solutions for Dodd-Bullough Equation. *J. Appl. Math.* **2013**, 364718. [CrossRef]
32. Seadawy, A.R.; Lu, D.; Khater, M.M.A. Bifurcations of traveling wave solutions for Dodd–Bullough–Mikhailov equation and coupled Higgs equation and their applications. *Chin. J. Phys.* **2017**, *55*, 1310–1318. [CrossRef]
33. Wentao, T.S.H. Bifurcations of Travelling Wave Solutions For The Generalized Dodd-Bullough-Mikhailov Equation. *Appl. Math. A J. Chin. Univ. Ser. B* **2007**, *22*, 21–28.
34. Ruffini, P. *Teoria Generale Delle Equazioni, in cui si Dimostra Impossibile la Soluzione Algebraica Delle Equazioni Generali di Grado Superiore al Quarto*; Stamperia di S. Tommaso d'Aquino: Bologna, Italy, 1799.

35. Abel, N.H. Démonstration de l'impossibilité de la résolution algébrique des équations générales qui passent le quatrième degré. In *Oeuvres completes de Niels Hendrik Abel I*, 2nd ed.; Sylow, L., Lie, S., Eds.; Grondahl & Son: Oslo, Norway, 1881; pp. 66–87.
36. Hermite, C. Sur la résolution de l'équation du cinquième degré. *C. R. Hebd. Seances Acad. Sci.* **1858**, *46*, 508–515.
37. Cimpoiasu, R.; Cimpoiasu, V.; Constantinescu, R. Nonlinear dynamical systems in various space–time dimensions. *Rom. J. Phys.* **2010**, *55*, 25–35.
38. Constantinescu, R.; Ionescu, C. Gauge fixing procedure in the extended BRST theory: The example of the abelian 2–forms. *Ann. Phys.* **2006**, *15*, 169–176. [CrossRef]
39. Constantinescu, R.; Ionescu, C. Multidifferential complexes and their application to gauge theories. *Int. J. Mod. Phys. A* **2006**, 21, 6629–6640. [CrossRef]

applied sciences

MDPI

*Article*

# Synchronization and Application of a Novel Hyperchaotic System Based on Adaptive Observers

**Erman Ozpolat [1,*] and Arif Gulten [2]**

1 Department of Electrical-Electronics Engineering, Mus Alparslan University, Mus 49100, Turkey
2 Department of Electrical-Electronics Engineering, Firat University, Elazig 23119, Turkey
* Correspondence: e.ozpolat@alparslan.edu.tr; Tel.: +90-436-249-49-49

**Abstract:** This paper explores the synchronization and implementation of a novel hyperchaotic system using an adaptive observer. Hyperchaotic systems, known for possessing a greater number of positive Lyapunov exponents compared to chaotic systems, present unique challenges and opportunities in control and synchronization. In this study, we introduce a novel hyperchaotic system, thoroughly examining its dynamic properties and conducting a comprehensive phase space analysis. The proposed hyperchaotic system undergoes validation through circuit simulation to confirm its behavior. Introducing an adaptive observer synchronization technique, we successfully synchronize the dynamics of the novel hyperchaotic system with an identical counterpart. Importantly, we extend the application of this synchronization method to the domain of secure communication, showcasing its practical usage. Simulation outcomes validate the effectiveness of our methodology, demonstrating favorable results in the realm of adaptive observer-based synchronization. This research contributes significantly to the understanding and application of hyperchaotic systems, offering insights into both the theoretical aspects and practical implementation. Our findings suggest potential advancements in the field of chaotic systems, particularly in their applications within secure communication systems. By presenting motivations, methods, results, conclusions and the significance of our work in a more appealing manner, we aim to engage readers and highlight the innovative contributions of this study.

**Keywords:** hyperchaotic systems; adaptive observers; synchronization; secure communication

## 1. Introduction

The understanding of chaotic behavior was greatly advanced in the 1960s when Edward Lorenz developed a computer program to simulate airflow, which is when chaos theory first emerged [1]. Extensive study was conducted in the 1970s when Lorenz discovered that even minimal changes in the initial circumstances might have a significant influence on weather forecasts. According to this study, system states varied significantly based on the basic characteristic of chaotic systems, which is the smallest deviation between two beginning circumstances [2].

The study of hyperchaotic systems was further developed in 1979 when Rössler discovered a four-dimensional hyperchaotic system [3]. Multidimensional chaotic and hyperchaotic systems became more well-known in the following years, providing useful applications in science and engineering [4]. Recently, there are many studies to be found in the literature on the dynamic analysis of chaotic systems [5–7].

In contrast to chaotic systems, hyperchaotic systems, which have multiple positive Lyapunov exponents, are more unpredictable and resistant to control. This property has drawn interest from the encryption and communication industries [8]. Multiple attractors improve security by making it more difficult for an attacker to determine the precise state of the system, which makes hyperchaotic systems more resistant to disruptions [9]. Secure applications of hyperchaotic systems have been a topic of great interest recently [10,11].

Circuit simulations of hyperchaotic systems facilitate real-world exploration and analysis of their dynamics, validating theoretical models and providing insights into system behavior under different conditions [12,13]. The implementation of hyperchaotic systems in circuits offers practical applications in secure communications, random number generation and nonlinear dynamics research [14,15].

Synchronization has gained attention in the context of hyperchaotic systems, especially with the growing focus on secure communication applications [16]. In order to attain coherence, synchronization entails changing the dynamic behaviors of two systems, referred to as the driver and reaction [17]. In an effort to compel the response system to synchronize with the driving system, the notion of synchronization in chaotic systems was first presented in 1990 [18]. To this end, a number of approaches have been put forth, such as adaptive design techniques and sampled data feedback synchronization [19,20].

Observer-based hyperchaotic system synchronization involves designing an observer (state estimator) and control strategy to achieve synchronization between two or more hyperchaotic systems [21]. Observers play a crucial role in estimating unmeasurable states, overcoming challenges posed by limited sensors or measurements, especially in hyperchaotic systems. Adaptive observers address uncertainties and variations in system parameters, enhancing the robustness and effectiveness of synchronization strategies [22].

The synchronization of adaptive observer-based hyperchaotic systems holds particular importance in secure communication systems [23]. The complex dynamics of hyperchaotic systems offer potential for encryption. Decryption and synchronization of these systems are vital for establishing secure communication channels [24]. The inclusion of adaptive observers adds robustness to the synchronization process, addressing uncertainties and variations in real-world applications [25].

The synchronization and implementation of adaptive observer-based hyperchaotic systems present a promising approach for secure communication. This approach, encompassing dynamic encryption keys, adaptation to changing conditions and resistance to attacks, proves valuable for applications where secure and robust communication is imperative. Typical examples are military communications, financial transactions, or confidential data transfer [26].

Xiong et al. [27] proposed a new 4-dimensional hyperchaotic system. They synchronized the proposed system with a reduced number of amplifiers. This synchronization has been used in secure communication applications. Zhang et al. [15] proposed a 4-dimensional hyperchaotic system based on cosine function. The authors have demonstrated the dynamic analysis and synchronization of the new system. They also presented an FPGA-based implementation of their proposed hyperchaotic system. Iskakova et al. [28] proposed a new nonlinear 4-dimensional hyperchaotic system. They showed in their publication that the proposed system has hidden attractors. Then, they applied it on FPAA to show the accuracy of their proposed system. Yang et al. [29] proposed a new 4-dimensional hyperchaotic system with hidden attractors and existing attractors without an equilibrium point. They performed active control-based synchronization of the proposed system.

Nowadays, secure and effective information communication is increasing and gaining in importance. In this context, chaotic systems and in particular hyperchaotic systems have the potential to offer new and effective solutions for communication security. Recent studies have underscored the challenges in synchronizing hyperchaotic systems, particularly in practical applications involving parameter uncertainty. In systems with parameter uncertainty, achieving synchronization is notably challenging. To address this difficulty, the use of adaptive-based observers is considered a viable solution for designing such systems.

The innovations presented in this study and the contributions of this work to the literature are listed as follows:

- First, a new hyperchaotic system is proposed as an addition to the literature. Comprehensive dynamic analysis of the system is carried out. As a result of the dynamic analysis, it is observed that the proposed system exhibits hyperchaotic behavior.

- Adaptive observer-based synchronization of the new system is designed. While performing the synchronization, parameter uncertainty is also taken into account. As a result of the design, it is observed that the synchronization errors reach zero in a short time, while at the same time the parameter uncertainty is eliminated.
- The adaptive observer-based synchronization method is tested in a secure communication application where chaotic system synchronization is usually used. The transmitted signal is first masked by the transmitting hyperchaotic system and then transmitted to the receiving hyperchaotic system.
- The analysis shows that the transmitted signal is successfully extracted from the masked signal received by the receiving hyperchaotic system within a short period of time.

In Section 2, the mathematical model of the proposed hyperchaotic system is introduced and the necessary dynamic analysis methods are applied. In Section 3, the circuit simulation is performed. In Section 4, adaptive observer-based synchronization of the new hyperchaotic system is implemented and simulated. In Section 5, the realized synchronization is used in a secure communication application. Section 6 presents the conclusions.

## 2. Materials and Methods

In this paper, various analyses of the proposed 4-dimensional hyperchaotic system were executed. In order to prove that the new hyperchaotic system exhibits chaotic and hyperchaotic behavior, dynamic analysis studies were carried out. Some of these analyses were performed using the MATLAB platform. The hyperchaotic circuit implementation of the system was created on the Multisim platform. The synchronization of the proposed hyperchaotic system was designed using an adaptive observer. The novel 4-dimensional hyperchaotic system synchronization was tested in a secure communication application and its execution was successfully proved. The block diagram of the material and the method proposed in this paper are given in Figure 1.

### 2.1. The Proposed 4-Dimensional Hyperchaotic System

The mathematical model of the proposed hyperchaotic system is shown in (1):

$$\begin{aligned} \dot{x}_1 &= -12x_1 + 30x_2 \\ \dot{x}_2 &= bx_1 - x_2 - x_1x_3 \\ \dot{x}_3 &= ax_2 + x_1x_2 - x_3 + x_4 \\ \dot{x}_4 &= -x_1b + x_1x_3 - 2x_4 \end{aligned} \quad (1)$$

In the system (1), the variables $x_1$, $x_2$, $x_3$ and $x_4$ represent states, while $a$ and $b$ are positive parameters. The system (1) demonstrates chaotic behavior for specific values of tuned parameters and the initial conditions that are provided. Notably, the system is distinguished by its independence from the initial conditions. The examination of various initial conditions reveals that under specific parameter values, the system can manifest chaotic, hyperchaotic, periodic and quasi-periodic behaviors.

### 2.2. Dynamic Analysis of the Proposed Hyperchaotic System

The bifurcation diagram of the proposed new hyperchaotic system is shown in Figure 2. While plotting the bifurcation diagram, the parameter $b$ is taken as 45 and the variation in the parameter $a$ between 14 and 16 is plotted. For the bifurcation diagram analysis, the initial values of the system are taken as $x_1(0) = 1,\ x_2(0) = 1,\ x_3(0) = 1,\ x_4(0) = 1$.

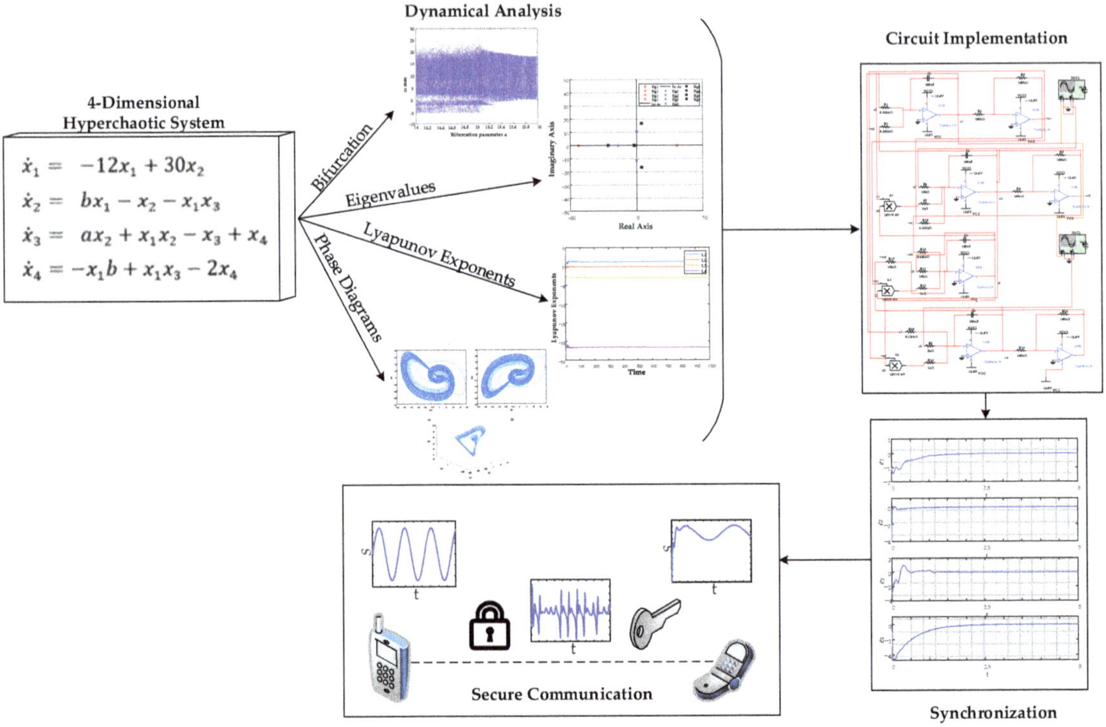


**Figure 1.** Illustration of the proposed methods.

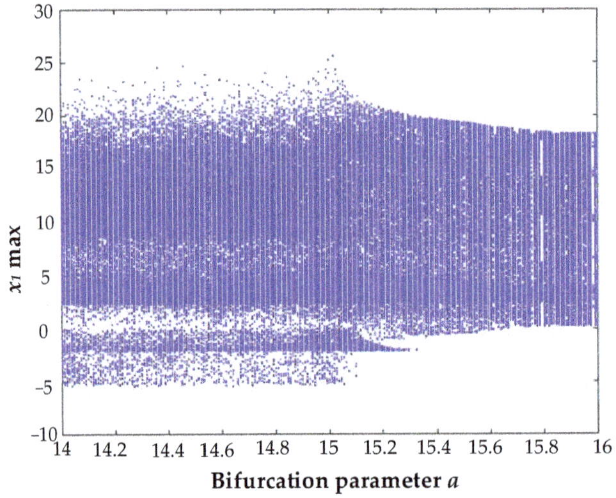


**Figure 2.** Bifurcation diagram based on *a* parameter.

As can be seen from the Figure 2, the value of the parameter *a* is taken as 15 for the proposed system.

Additionally, the 3D bifurcation diagram of the proposed hyperchaotic system is shown in Figure 3. While drawing this bifurcation diagram, in addition to parameter *a*, parameter $b$ was also assumed to be between 40 and 50. For the bifurcation diagram analysis, the initial values of the system are taken as $x_1(0) = 1,\ x_2(0) = 1,\ x_3(0) = 1,\ x_4(0) = 1$.

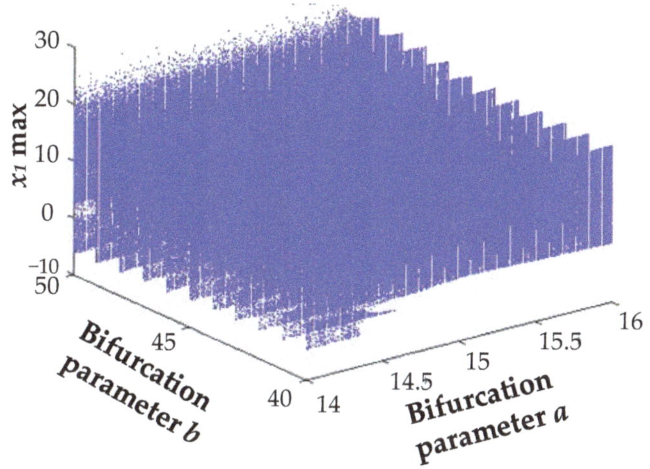


**Figure 3.** The 3D Bifurcation diagram based on $a$ and $b$ parameters.

Based on Figure 3, the $b$ parameter value of the proposed hyperchaotic system is taken as 45.

The proposed hyperchaotic system can be represented in vector notation as in (2):

$$\dot{x} = f(x) = \begin{bmatrix} f_1(x) \\ f_2(x) \\ f_3(x) \\ f_4(x) \end{bmatrix} \tag{2}$$

Based on this, the proposed system can be expressed as in (3):

$$\begin{aligned} f_1(x) &= -12x_1 + 30x_2 \\ f_2(x) &= bx_1 - x_2 - x_1x_3 \\ f_3(x) &= ax_2 + x_1x_2 - x_3 + x_4 \\ f_4(x) &= -x_1b + x_1x_3 - 2x_4 \end{aligned} \tag{3}$$

The divergence of the hyperchaotic system is investigated as (4) for the system [30]:

$$div\ f = \frac{\partial f_1(x)}{\partial x_1} + \frac{\partial f_2(x)}{\partial x_2} + \frac{\partial f_3(x)}{\partial x_3} + \frac{\partial f_4(x)}{\partial x_4} = -12 - 1 - 1 - 2 = -16 \tag{4}$$

For the divergence of the system, $div\ f < 0$ is required. As can be seen from (4), the system satisfies the condition.

To find the equilibrium points of the proposed hyperchaotic system, the system of equations must be equal to zero. This equality is shown in (5):

$$\begin{aligned} 0 &= -12x_1 + 30x_2 \\ 0 &= bx_1 - x_2 - x_1x_3 \\ 0 &= ax_2 + x_1x_2 - x_3 + x_4 \\ 0 &= -x_1b + x_1x_3 - 2x_4 \end{aligned} \tag{5}$$

When analyzing the bifurcation diagrams of the system, specific values of $a = 15$ and $b = 45$ were chosen to induce hyperchaotic behavior. Under these parameter settings, the system exhibits three equilibrium points, namely $E_1 = [0,0,0,0,0]$, $E_2 =$

$[5.55, 2.22, 44.6, -1.11]$ and $E_3 = [-20.05, -8.02, 44.6, 4.01]$. To ascertain the stability of the hyperchaotic system, it is imperative to determine the system's eigenvalues. For the specified values $a = 15$ and $b = 45$, the system's Jacobian matrix, essential for calculating the eigenvalues, is presented in (6).

$$J = \begin{bmatrix} -12 & 30 & 0 & 0 \\ b - x_3 & -1 & -x_1 & 0 \\ x_2 & x_1 + a & -1 & 1 \\ x_3 - b & 0 & x_1 & -2 \end{bmatrix} \quad (6)$$

The Jacobian matrix was solved, and the eigenvalues corresponding to the equilibrium points $E_1$, $E_2$ and $E_3$ were determined. Equation (7) was employed for the computation of these eigenvalues:

$$|J - \lambda I| = 0 \quad (7)$$

Table 1 provides the values of the equilibrium points and eigenvalues.

**Table 1.** Eigenvalues of the system and equilibrium points.

| $E_i$ | Equilibrium Points | Eigenvalues |
|---|---|---|
| $E_1$ | $[0, 0, 0, 0]$ | $(-2.0, -1.0, -43.65, 30.65)$ |
| $E_2$ | $[5.55, 2.22, 44.6, -1.11]$ | $(-0.0252 \pm j10.98, -2.03, -13.91)$ |
| $E_3$ | $[-20.05, -8.02, 44.6, 4.01]$ | $(3.68 \pm j16.71, -1.96, -21.40)$ |

The graph of the changes in the eigenvalues of the system is given in Figure 4.

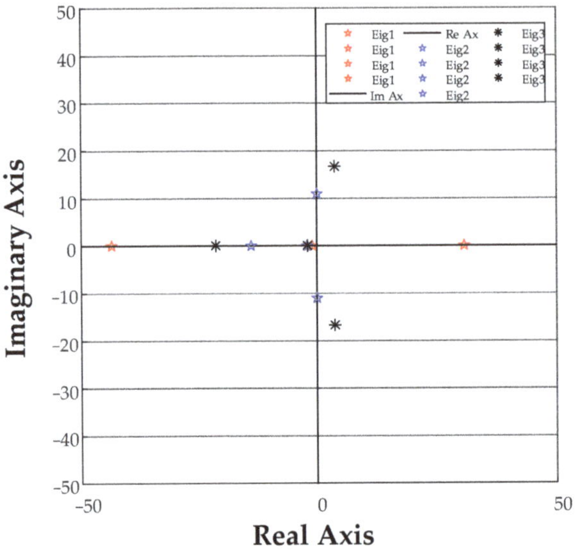


**Figure 4.** Eigenvalues ($\lambda$) of the hyperchaotic system.

Upon analyzing the eigenvalues ($\lambda$) of the system, it becomes evident that the system possesses positive eigenvalues, indicating its instability. Furthermore, it can be inferred that the hyperchaotic attractor of the new system is affiliated with the self-exciting family.

The Alan Wolf algorithm was employed for the computation of Lyapunov exponents [31]. For the first computation, throughout the calculation, the initial conditions were set as $x_1(0) = 1$, $x_2(0) = 1$, $x_3(0) = 1$, $x_4(0) = 1$, parameter $b$ was taken as 45 and parameter $a$ was assumed to vary between 14 and 16. The evolution of Lyapunov exponents according to the variation in $a$ parameter is shown in Figure 5.

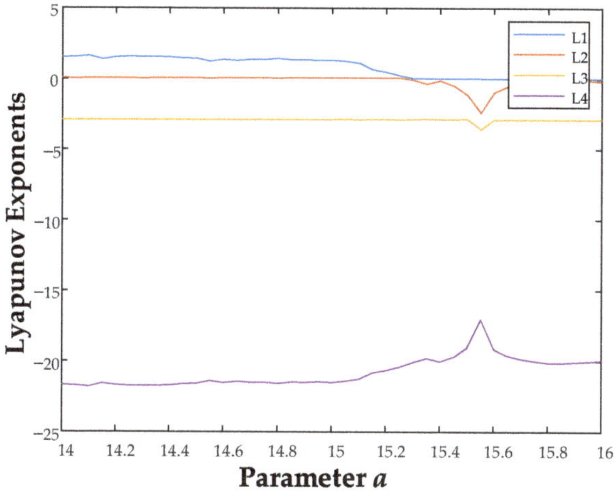


**Figure 5.** Lyapunov exponents according to the variation in $a$ parameter.

As can be seen from Figure 5, the system has two positive Lyapunov exponent values of $a$ parameter between 14 and 15.3. Therefore, taking the $a$ parameter value of the system as 15 and the $b$ parameter value as 45, the change in the Lyapunov exponents of the system with respect to time is plotted in Figure 6. The initial conditions were set as $x_1(0) = 1,\ x_2(0) = 1,\ x_3(0) = 1,\ x_4(0) = 1$.

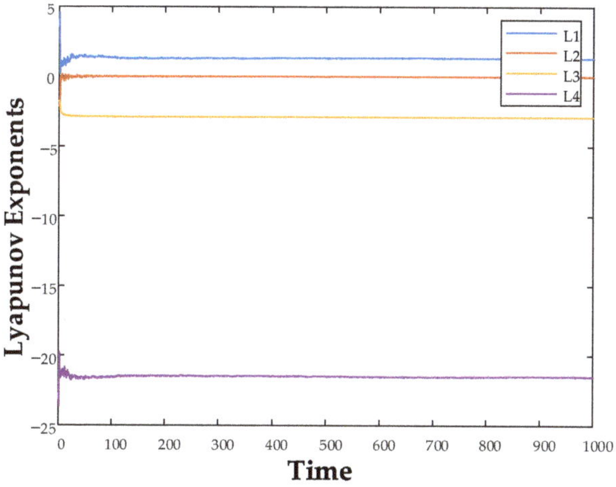


**Figure 6.** Diagram of Lyapunov exponents of the system.

The Lyapunov exponents are determined as $L_1 = 1.3264,\ L_2 = 0.0227,\ L_3 = -2.8806$ and $L_4 = -21.5094$. Two positive Lyapunov exponents are found when the calculated Lyapunov exponents of the recently suggested hyperchaotic system are examined. As a result, the behavior of the system is hyperchaotic.

Hyperchaotic systems are inevitably drawn to fractals. The degree of system predictability can be inferred from the fractal dimension of the attractor. The data would have been generated by a deterministic system if the fractal dimension is minimal. The data would have been produced by a random system if the fractal dimension is large. Because this structure is not like other geometrical formations, the researchers use fractal dimensions to explain it [32]. The proposed four-dimensional system exhibits a Lyapunov fractal dimension [33], which is computed using (8):

$$D_{KY} = j + \frac{\sum_{i=1}^{J} L_i}{|L_{j+1}|} = 3 + \frac{1.3264 + 0.0227 - 2.8806}{|-21.5094|} = 2.92 \quad (8)$$

This shows that the proposed system's Lyapunov dimension is fractional. The new system contains non-periodic orbits and close paths that diverge due to its fractal character. As a result, this nonlinear system is truly chaotic.

In engineering applications, chaotic phase behavior can pose challenges in terms of control and predictability. Systems exhibiting chaotic dynamics may be more difficult to control, as small changes in input or conditions can lead to unpredictable outcomes. Engineers need to account for and understand chaotic behavior in order to design robust and reliable systems. The phase space diagrams of the proposed hyperchaotic system in various dimensions are shown in Figure 7. For the phase diagram analysis, $a = 15, b = 45$ and the initial values of the system are taken as $x_1(0) = 1,\ x_2(0) = 1,\ x_3(0) = 1,\ x_4(0) = 1$.

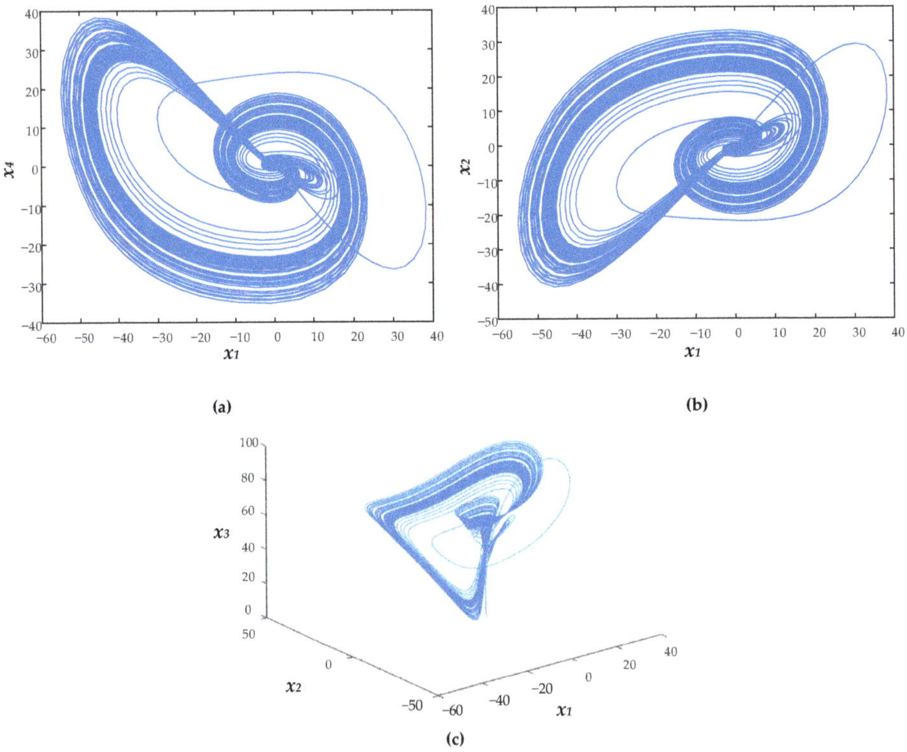


**Figure 7.** (**a**) $x_1$ $x_4$ phase diagram; (**b**) $x_1$ $x_2$ phase diagram; (**c**) $x_1$ $x_2$ $x_3$ phase diagram.

The system's phase diagrams reveal that it is a hyperchaotic system with dynamic oscillations and a variety of strange attractors. The phase diagrams for different $a$ and $b$ parameter values are shown in Figure 8. For the phase diagram analysis, $a = 5$ and $b = 6$ values were taken and initial values taken as $x_1(0) = 1,\ x_2(0) = 1,\ x_3(0) = 1,\ x_4(0) = 1$.

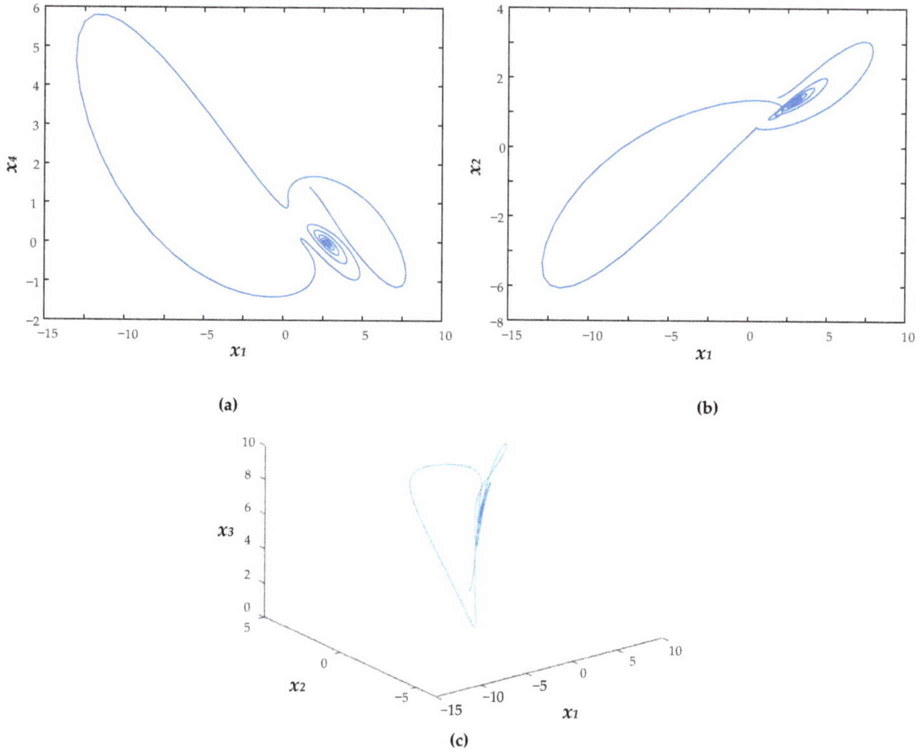


**Figure 8.** (**a**) $x_1$ $x_4$ phase diagram; (**b**) $x_1$ $x_2$ phase diagram; (**c**) $x_1$ $x_2$ $x_3$ phase diagram.

As can be seen from Figure 8, the chaotic properties of the system change for different parameter values.

## 3. Circuit Implementation of the Proposed Hyperchaotic System

Using Multisim 14.3 software, an analogous electronic circuit for the novel hyperchaotic system was created in order to evaluate the viability of the proposed system. The highly visual modeling environment offered by the electronic circuit simulation program Multisim is designed for hyperchaotic systems [34]. It makes the process of creating and fine-tuning circuit models easier and permits the monitoring of their behavior in real time. This real-time feedback forms the basis for simulating the implementation of hyperchaotic systems in embedded circuitry and helps to understand their dynamic features.

For simulating electronic circuits, a variety of software packages are available, including PSPICE 9.2, LTSPICE 17.1.15, Multisim 14.3 and Matlab Simulink 2021a. Multisim, the industry-standard circuit design and simulation program for analog, digital and power electronics in academic and research settings, was chosen for this thesis [35].

An operational amplifier (TL084ACN), a multiplier (IC AD633), linear resistors and capacitors driven by a ±15 V power source are all included in the developed circuit in this thesis. The operational amplifier adds, subtracts and integrates the circuits, and the multiplier, which combines the four state variables into a single cohesive unit, symbolizes the nonlinearity of the system. The system (1) presents the hyperchaotic system from which the electrical circuit equations are obtained, and (9) formulates it:

$$
\begin{aligned}
\dot{x}_1 &= \frac{1}{C_1}\left[\frac{1}{R_1}x_1 + \frac{1}{R_2}\left(\frac{R_9}{R_8}\right)(-x_2)\right] \\
\dot{x}_2 &= \frac{1}{C_2}\left[\frac{1}{R_{10}}\left(\frac{R_4}{R_3}\right)(-x_1) + \frac{1}{R_6}x_2 + \frac{1}{R_7}\frac{x_1x_3}{10}\right] \\
\dot{x}_3 &= \frac{1}{C_3}\left[\frac{1}{R_{14}}\left(\frac{R_9}{R_8}\right)(-x_2) + \frac{1}{R_{12}}\left(\frac{R_4}{R_3}\right)\frac{(-x_1)x_2}{10} + \frac{1}{R_{11}}x_3 + \frac{1}{R_{17}}\left(\frac{R_{16}}{R_{15}}\right)(-x_4)\right] \\
\dot{x}_4 &= \frac{1}{C_4}\left[\frac{1}{R_{20}}x_1 + \frac{1}{R_{13}}\left(\frac{R_4}{R_3}\right)\frac{(-x_1)x_3}{10} + \frac{1}{R_5}x_4\right]
\end{aligned}
\tag{9}
$$

After performing the necessary calculations, the values of the circuit elements were determined as follows: $R_1 = 0.833$ kΩ, $R_2 = 0.333$ kΩ, $R_3 = R_4 = R_8 = R_9 = R_{15} = R_{16} = 100$ kΩ, $R_5 = 5$ kΩ, $R_6 = R_{11} = R_{17} = 10$ kΩ, $R_7 = R_{12} = R_{13} = 1$ kΩ, $R_{10} = R_{20} = 0.222$ kΩ, $R_{14} = 0.666$ kΩ and $C_1 = C_2 = C_3 = C_4 = 100$ μF. The Multisim model of the circuit is illustrated in Figure 9. The circuit model encompasses seven operational amplifiers (TL084ACN), four capacitors, three multipliers (IC AD633), and eighteen resistors.

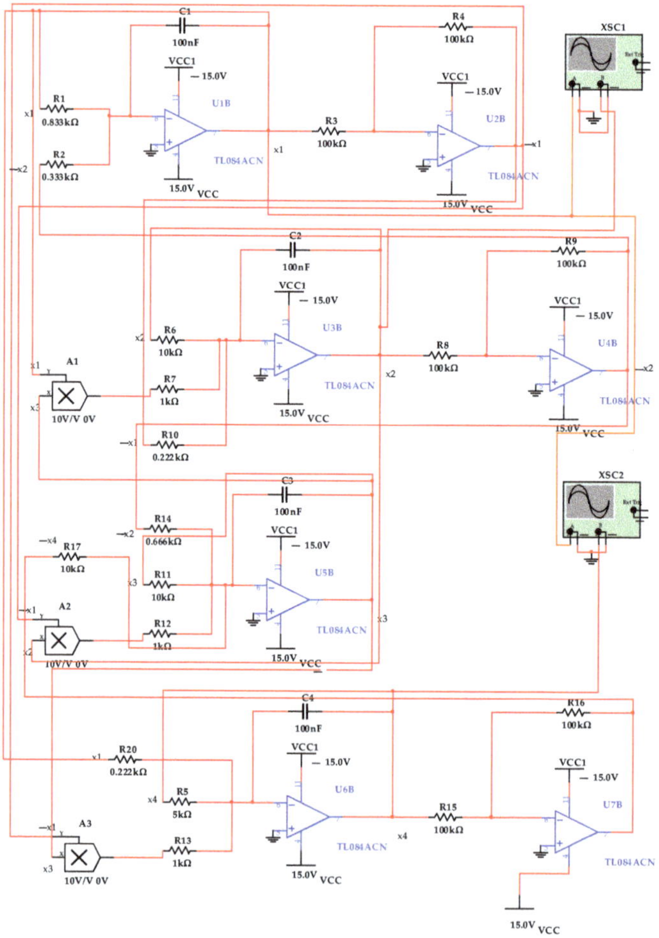


**Figure 9.** Electronic circuit diagram of the proposed hyperchaotic system.

When the circuit simulation is run, the oscilloscope images showing the phase diagrams of the states are as shown in Figure 10.

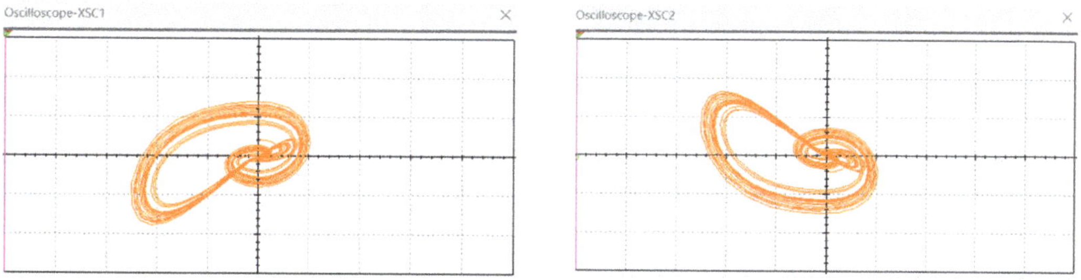


**(a)** **(b)**

**Figure 10.** (**a**) Multisim $x_1$ $x_2$ phase diagram of the hyperchaotic system; (**b**) Multisim $x_1$ $x_4$ phase diagram of the hyperchaotic system.

Similarity was observed between the graphs obtained from the numerical simulation program MATLAB 2021a and the electronic circuit simulation software. Multisim provides a strong confirmation for hyperchaotic attractors. Thoroughly comparing the output data from both software tools shows that there is a significant degree of consistency in the way attractors dynamically behave under different settings and how easily they alter. These results validate the presence of attractors in the hypothesized hyperchaotic system.

Finally, consistency between the electrical circuit simulation findings and numerical simulation results provides a solid theoretical foundation for the actual implementation of the novel hyperchaotic system, while also confirming the feasibility of the system. It provides a strong basis for the application of novel hyperchaotic systems in various domains.

## 4. Adaptive Observer Based Synchronization of the Proposed Hyperchaotic System

Hyperchaotic systems, which form a separate category within nonlinear systems, are typically characterized by a set of nonlinear differential equations. It is a prevalent practice to decompose a hyperchaotic dynamical system into two components: the linear component and the nonlinear component. In instances where the hyperchaotic dynamical system involves unknown parameters and is influenced by unidentified disturbances, it is often expressed in the form of (10) [36]:

$$\dot{x} = Ax + Bf(x) + Bg(x)\theta + B\eta(x,t) \tag{10}$$

Here, $x \in R^n$ represents the state vector; $\theta \in R^p$ is the vector of unknown parameters; $\eta \in R^r$ is the external disturbance vector; and $A \in R^{nxn}$ and $B \in R^{nxr}$ are known matrices. Additionally, $f(\cdot) \in R^r$ is a nonlinear function vector, and $g(\cdot) \in R^{rxp}$ is a matrix of functions. The matching condition is satisfied by the external disturbance vector, the parameter uncertainty vector and the nonlinear function vector.

It is noteworthy that many hyperchaotic systems can be reformulated into the same structure as (10). For instance, the Lorenz hyperchaotic system can be expressed in the form of (11):

$$\begin{bmatrix} \dot{x}_1 \\ \dot{x}_2 \\ \dot{x}_3 \\ \dot{x}_4 \end{bmatrix} = \begin{bmatrix} -a & a & 0 & 1 \\ 0 & -1 & 0 & 0 \\ 0 & 0 & 0 & 0 \\ 0 & 0 & 0 & r \end{bmatrix} \begin{bmatrix} x_1 \\ x_2 \\ x_3 \\ x_4 \end{bmatrix} + \begin{bmatrix} 0 & 0 & 0 \\ -1 & 0 & 0 \\ 0 & 1 & 0 \\ 0 & -1 & 0 \end{bmatrix} \begin{bmatrix} x_1x_3 \\ x_1x_2 \\ x_2x_3 \end{bmatrix} + \begin{bmatrix} 0 & 0 & 0 \\ -1 & 0 & 0 \\ 0 & 1 & 0 \\ 0 & -1 & 0 \end{bmatrix} \begin{bmatrix} -x_1 & 0 \\ 0 & -x_3 \\ 0 & 0 \end{bmatrix} \begin{bmatrix} c \\ b \end{bmatrix} \tag{11}$$

Here, $a$ is known but $c$ and $b$ are unknown.

We tackle the synchronization challenge of (10) employing the driver–response configuration. In this setup, if the system represented by (10) is designated as the drive system, a corresponding response system needs to be formulated to synchronize with the drive system using a designated driving signal. To achieve this, certain essential assumptions must be fulfilled.

Firstly, the Lipschitz condition is articulated. It is required that the nonlinear function vector $f(x)$ and the matrix $g(x)$ adhere to the Lipschitz condition:

$$\begin{aligned} ||f(x) - f(\hat{x})|| &\leq k_f||x - \hat{x}|| \\ ||g(x) - g(\hat{x})|| &\leq k_g||x - \hat{x}|| \end{aligned} \tag{12}$$

Here, the Lipschitz constants $k_f$ and $k_g$ are considered unknown.

Secondly, the uncertain parameter $\theta$ and the external disturbance $\eta$ are subject to norm bounds defined by two unknown positive constants, denoted as $\delta_\theta$ and $\delta_\eta$, respectively.

Thirdly, in pursuit of the objective of hyperchaotic synchronization, it is typically necessary to design artificial output information.

The initial step involves determining a matrix $C_1 \in \Re^{1xn}$ such that the matrix $(C_1, A)$ is observable, and there exists a matrix $L$ such that the Lyapunov equation $(A - LC_1)^T P + P(A - LC_1) = -Q$ has a positive definite matrix solution $P$ for a given positive definite matrix $Q$.

The second step is to specify $B^T P = C_2$. If there exists a matrix $H \in \Re^{rxp}$ such that $C_2 = HC_1$, then the output matrix of the drive system must be $C = C_1$; otherwise, it will be $C = (C_1^T\, C_2^T)^T$. In our study, we assume the absence of such an $H$ matrix. Therefore, we set our system output as $y = (y_1^T\, y_2^T)^T = (C_1^T\, C_2^T)^T x$.

To synchronize the system presented in (8), an adaptive observer-based response system is designed. This design is illustrated in (11):

$$\dot{\hat{x}} = A\hat{x} + Bf(\hat{x}) + Bg(\hat{x})\hat{\theta} + \frac{1}{2}\bar{k}B(y_2 - C_2\hat{x}) + L(y_1 - C_1\hat{x}) + \alpha(t) \tag{13}$$

Here, $\hat{\theta}$ and $\bar{k}$ symbolize the answers to two adaptation laws that are still to be found, and $\alpha(t)$ is a further control rule that will be defined later [37]. The error signals are $e = x - \hat{x}$, $\tilde{\theta} = \theta - \hat{\theta}$ and $\tilde{k} = k - \bar{k}$. Here, $k$ is a constant to be determined. Utilizing (10) and (13), the error dynamics are expressed as follows:

$$\dot{e} = (A - LC_1)e + B(f(x) - f(\hat{x})) + B(g(x)\theta - g(\hat{x})\hat{\theta}) - \frac{1}{2}\bar{k}B(y_2 - C_2\hat{x}) + B\eta - \alpha \tag{14}$$

The synchronization problem thus transforms into ensuring the stability of the error dynamics. The response system can achieve global synchronization with the drive system if it attains global stability at the origin.

We assume the validity of the assumptions for both the driver system and the response system. The response system can synchronize with the driver system under certain conditions.

Firstly, the adaptation laws for $\hat{\theta}$ and $\bar{k}$ are selected as given in (15) and (16):

$$\dot{\hat{\theta}} = l_\theta g(\hat{x})^T (y_2 - C_2\hat{x}) \tag{15}$$

$$\dot{\bar{k}} = l_k||y_2 - C_2\hat{x}||^2 \tag{16}$$

Here, the positive constants $l_\theta$ and $l_k$ are chosen appropriately.

The additional control law $\alpha(t)$ is chosen as (17):

$$\alpha(t) = r(t)B(y_2 - C_2\hat{x}) \tag{17}$$

The value of $r(t)$ used in (17) is shown in (18):

$$r(t) \geq \frac{\varepsilon_\eta}{2(y_2 - C_2\hat{x})^T} \tag{18}$$

Building upon Assumption 3 and the Kalman and Yakubovich Lemma, two positive definite matrices, denoted as $P = P^T$ and $Q = Q^T$ [38], are introduced. These matrices are considered valid for the algebraic equation.

$$(A - LC_1)^T P + P(A - LC_1) = -Q, \quad B^T P = C_2 \tag{19}$$

To prove the stability of the error dynamics given in (14), let us consider the Lyapunov function candidate given in (20):

$$V = e^T Pe + \frac{1}{2} l_k^{-1} \tilde{k}^2 + \frac{1}{l_\theta} \tilde{\theta}^2 \tag{20}$$

The derivative of (20) with respect to time becomes (21):

$$\begin{aligned} \dot{V} = \; & e^T \left[ (A - LC_1)^T P + P(A - LC_1) \right] e + l_k^{-1} \tilde{k}\dot{\tilde{k}} + \tfrac{2}{l_\theta} \tilde{\theta}^T \dot{\tilde{\theta}} + 2e^T PB(f(x) - f(\hat{x})) \\ & + 2e^T PB\eta - 2e^T P\alpha + 2e^T PB(g(x)\theta - g(\hat{x})\hat{\theta}) - \bar{k}e^T PB(y_2 - C_2\hat{x}) \end{aligned} \tag{21}$$

If we then rewrite Equation (21) taking into account the law of adaptation given in (15), we obtain (22):

$$\begin{aligned} \dot{V} = \; & -e^T Qe + l_k^{-1} \tilde{k}\dot{\tilde{k}} + 2e^T C_2^T (f(x) - f(\hat{x})) - \bar{k}e^T C_2^T C_2 e + 2e^T C_2^T \eta - 2e^T P\alpha \\ & + 2e^T C_2^T (g(x)\theta - g(\hat{x})\theta) \end{aligned} \tag{22}$$

Given the first assumption, we obtain the inequalities shown in (23):

$$\begin{aligned} & 2e^T C_2^T (f(x) - f(\hat{x})) \leq 2k_f \|C_2 e\| \cdot \|x - \hat{x}\| \leq \tfrac{k_f^2}{\varepsilon_f} \|C_2 e\|^2 + \varepsilon_f \|e\|^2 \\ & 2e^T C_2^T (g(x)\theta - g(\hat{x})\theta) \leq 2k_g \delta_\theta \|C_2 e\| \cdot \|e\| \leq \tfrac{k_g^2 \delta_\theta^2}{\varepsilon_g} \|C_2 e\|^2 + \varepsilon_g \|e\|^2 \\ & 2e^T C_2^T \eta \leq 2\delta_\eta \|C_2 e\| \leq \tfrac{\delta_\eta^2}{\varepsilon_\eta} \|C_2 e\|^2 + \varepsilon_\eta \end{aligned} \tag{23}$$

Here, $\varepsilon_f$, $\varepsilon_g$ and $\varepsilon_\eta$ are three appropriate positive constants. Equation (24) is then obtained:

$$\begin{aligned} \dot{V} \leq \; & e^T \left[ (A - LC_1)^T P + P(A - LC_1) + \varepsilon_f I_n + \varepsilon_g I_n \right] e + l_k^{-1} \tilde{k}\dot{\tilde{k}} \\ & + \left( \tfrac{k_f^2}{\varepsilon_f} + \tfrac{k_g^2 \delta_\theta^2}{\varepsilon_g} + \tfrac{\delta_\eta^2}{\varepsilon_\eta} \right) \|C_2 e\|^2 - \bar{k} \|C_2 e\|^2 + \varepsilon_\eta - 2e^T P\alpha \\ k = \; & \left( \tfrac{k_f^2}{\varepsilon_f} + \tfrac{k_g^2 \delta_\theta^2}{\varepsilon_g} + \tfrac{\delta_\eta^2}{\varepsilon_\eta} \right) \end{aligned} \tag{24}$$

If we modify (24) with the adaptation law given in (15) and the control law given in (17), we obtain (25):

$$\dot{V} \leq e^T \left[ (A - LC)^T P + P(A - LC) + \varepsilon_f I_n + \varepsilon_g I_n \right] e \tag{25}$$

From the inequality provided in (25), it can be deduced that the independent parameters $\varepsilon_f$ and $\varepsilon_g$ can be chosen to be sufficiently small, ensuring $\dot{V} < 0$ for all non-zero states. Consequently, the error dynamics presented in (14) can achieve asymptotic stability. In other words, the drive system given in (10) and the response system given in (13) can be synchronized.

In (24), not only the unknown Lipschitz constants $k_f$ and $k_g$, but also the unknown limits $\delta_\theta$ and $\delta_\eta$ can be encompassed within the constant $k$, which can be adaptively tuned by the adaptation law provided in (16). Moreover, if the constants $\varepsilon_f$ and $\varepsilon_g$ are chosen to be sufficiently small, the stability of the error dynamics in (14) can be guaranteed.

The initial concept of the control law can be traced back to the work by Guv and Poon [39]. Evidently, as the error $y - C\hat{x}$ approaches zero, the magnitude of $\alpha(t)$ will escalate indefinitely and render computational feasibility unattainable. In practical applications, we can substitute the control law $r(t)$ with (26):

$$\begin{array}{ll} r(t) \geq \frac{\varepsilon_\eta}{2\|y_2 - C_2\hat{x}\|^2} & \text{if } \|e\| \geq \varepsilon \\ r(t) = 0 & \text{if } \|e\| < \varepsilon \end{array} \tag{26}$$

Here, $\varepsilon$ is a sufficiently small positive constant [39]. Consequently, the state error will remain within a neighborhood of the origin.

Hua et al. [40] first proposed the idea of adaptation for unknown Lipschitz constants in order to manage chaotic systems. We expand on this idea in this thesis to synchronize a larger class of hyperchaotic systems that have external disturbances and parameter uncertainty. A control rule can be used to achieve synchronization if the hyperchaotic systems meet the rigorous positive real condition.

*Numerical Simulation*

In this section of the paper, the effectiveness of adaptive observer-based synchronization for the proposed hyperchaotic system will be demonstrated. The dynamics of the hyperchaotic system are depicted in (27):

$$\begin{aligned} \dot{x}_1 &= -12x_1 + 30x_2 \\ \dot{x}_2 &= bx_1 - x_2 - x_1x_3 \\ \dot{x}_3 &= ax_2 + x_1x_2 - x_3 + x_4 \\ \dot{x}_4 &= -x_1b + x_1x_3 - 2x_4 \end{aligned} \tag{27}$$

It is established that the system exhibits hyperchaotic behavior for $a = 15$ and $b = 45$. Let us consider that the parameter $b$ is unknown, and $a$ has a perturbation of $\Delta_a = 0.03\sin(2t)$. The relevant matrices $A$, $B$ and $C$ for the hyperchaotic system presented in system (27) are provided in (28)–(30):

$$A = \begin{bmatrix} -12 & 30 & 0 & 0 \\ 0 & -1 & 0 & 0 \\ 0 & a & -1 & 1 \\ 0 & 0 & 0 & -2 \end{bmatrix} \tag{28}$$

$$B = \begin{bmatrix} 0 & 0 \\ -1 & 0 \\ 0 & 1 \\ 1 & 0 \end{bmatrix} \tag{29}$$

$$C_1 = \begin{bmatrix} 1 & 1 & 1 & 0 \end{bmatrix} \tag{30}$$

Furthermore, we define $f(x) = \begin{bmatrix} x_1x_3 \\ x_1x_2 \end{bmatrix}$, $g(x) = \begin{bmatrix} -x_1 \\ 0 \end{bmatrix}$ and $\eta(x,t) = \begin{bmatrix} 0 \\ \Delta_a x_2 \end{bmatrix}$. It is evident that the pair $(C, A)$ is observable. Therefore, the gain matrix $L$ is selected as $L = \begin{bmatrix} -3.5107 & 0.1818 & 1.328 & -0.8 \end{bmatrix}^T$ such that the eigenvalues of the matrix $A - LC_1$ are $-6$, $-4$, $-2 + i$ and $-2 - i$. Moreover, the symmetric positive definite matrices provided in (31) and (32) satisfy (19):

$$P = \begin{bmatrix} 0.3475 & -0.3392 & 0.3903 & 0.0379 \\ -0.3392 & 120.7581 & 0.2553 & -3.0015 \\ 0.3903 & 0.2553 & 2.2728 & 0.2408 \\ 0.0379 & -3.0015 & 0.2408 & 4.6182 \end{bmatrix} \quad (31)$$

$$Q = \begin{bmatrix} 6.7527 & 4.4022 & 5.7821 & -3.9132 \\ 4.4022 & 305.9790 & -17.9037 & -18.0623 \\ 5.7821 & -17.9037 & 7.5493 & -5.6039 \\ 0.0379 & -18.0623 & -5.6039 & 17.9912 \end{bmatrix} \quad (32)$$

As mentioned before, the adaptive observer-based response system is designed as (33):

$$\begin{bmatrix} \dot{\hat{x}}_1 \\ \dot{\hat{x}}_2 \\ \dot{\hat{x}}_3 \\ \dot{\hat{x}}_4 \end{bmatrix} = \begin{bmatrix} -12 & 30 & 0 & 0 \\ 0 & -1 & 0 & 0 \\ 0 & a & -1 & 1 \\ 0 & 0 & 0 & -2 \end{bmatrix} \begin{bmatrix} \hat{x}_1 \\ \hat{x}_2 \\ \hat{x}_3 \\ \hat{x}_4 \end{bmatrix} + \begin{bmatrix} 0 & 0 \\ -1 & 0 \\ 0 & 1 \\ 1 & 0 \end{bmatrix} \begin{bmatrix} \hat{x}_1\hat{x}_3 \\ \hat{x}_1\hat{x}_2 \end{bmatrix} + \begin{bmatrix} 0 & 0 \\ -1 & 0 \\ 0 & 1 \\ 1 & 0 \end{bmatrix} \begin{bmatrix} -\hat{x}_1 \\ 0 \end{bmatrix} \left[\hat{b}\right] + \frac{1}{2}\bar{k} \begin{bmatrix} 0 & 0 \\ -1 & 0 \\ 0 & 1 \\ 1 & 0 \end{bmatrix} (y_2 - C_2\hat{x}) + L(y_1 - C_1\hat{x}) + r(t)B(y_2 - C_2\hat{x}) \quad (33)$$

The values of $r(t)$, $\varepsilon_\eta = 1$ and $\varepsilon = 0.01$ used here are related to the inequalities given in (26) and the laws of adaptation on $\hat{b}$ and $\bar{k}$ are (34) and (35):

$$\dot{\hat{b}} = l_b \begin{bmatrix} -\hat{x}_1 \\ 0 \end{bmatrix}^T (y_2 - C_2\hat{x}) \quad (34)$$

$$\dot{\bar{k}} = l_k \|y_2 - C_2\hat{x}\|^2 \quad (35)$$

Here, $l_b = 1$ and $l_k = 1$. The values for the free parameters $\varepsilon_f$ and $\varepsilon_g$, as given in (25), are set to 1. As mentioned earlier, $C_2 = \begin{bmatrix} 0.3771 & -123.7596 & -0.0145 & 7.6197 \\ 0.3903 & 0.2553 & 2.2728 & 0.2408 \end{bmatrix}$ based on the equation $B^T P = C_2$.

The initial conditions for the simulation are $x_1(0) = 1, x_2(0) = 1, x_3(0) = 1, x_4(0) = 1$ ve $\hat{x}_1(0) = 2, \hat{x}_2(0) = 5, \hat{x}_3(0) = 4, \hat{x}_4(0) = 2$ and $\hat{b}(0) = 10, \bar{k}(0) = 10$. The synchronization errors are depicted in Figure 11.

As can be seen in Figure 8, the synchronization error quickly goes to 0. Therefore, it can be said that the synchronization is successful. When other studies in the literature are examined, the synchronization speed and robustness have been very successful. The estimated $b$ value after synchronization is shown in Figure 12.

The $b$ parameter is equal to 45, consistent with the value in the drive system. This suggests that the response system can synchronize with the proposed hyperchaotic system, even in the presence of unknown Lipschitz constants on the function matrices and unknown bounds on uncertainties. As can be seen, the designed structure quickly resisted the parameter uncertainty and predicted the required $b$ value. In summary, the proposed adaptive observer-based response system effectively achieves synchronization with the hyperchaotic system.

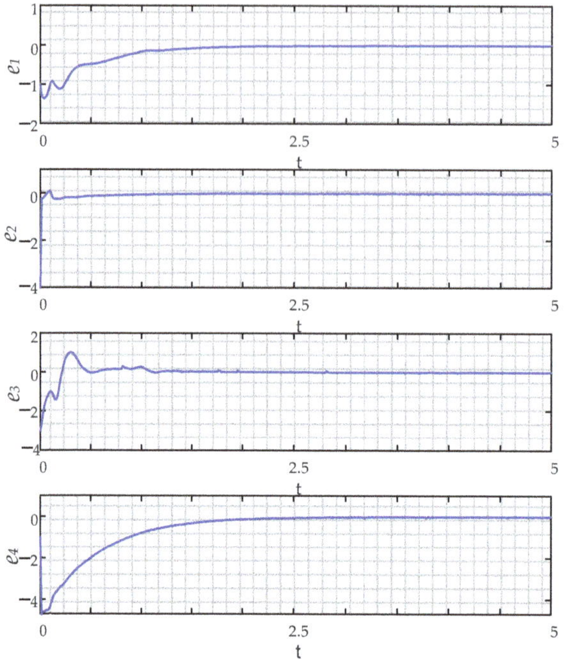


**Figure 11.** The synchronization errors.

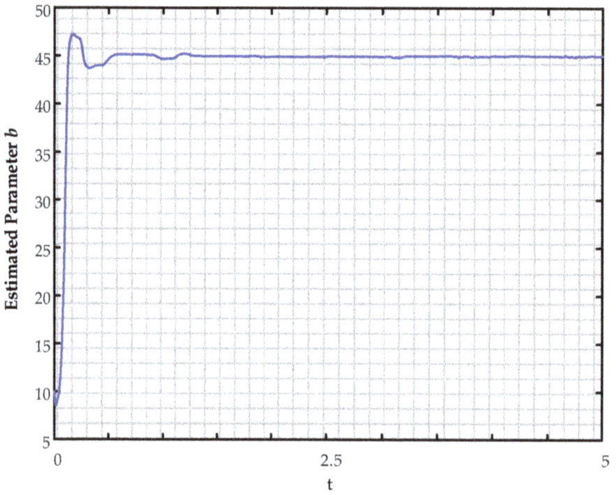


**Figure 12.** Estimation of unknown parameter $b$.

## 5. Application for Secure Communication

In secure communication applications, hyperchaotic synchronization is highly relevant. The simplicity of its implementation in secure communication systems serves as a crucial example of this. The suggested communication system is shown in Figure 13.

At the receiving end of the connection, there is a transmitter (driving) and a receiver (response).

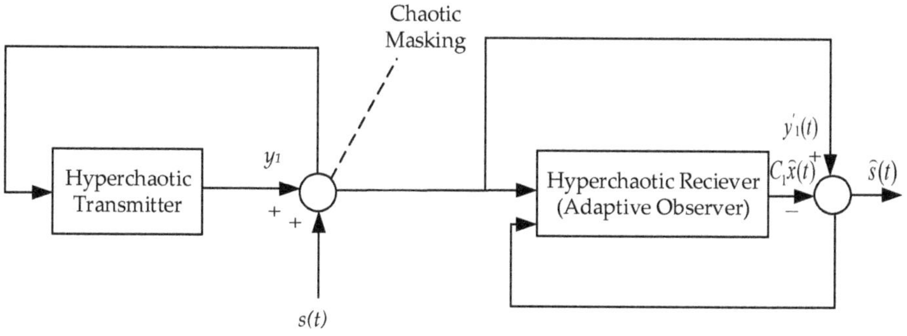


**Figure 13.** Secure communication system with chaotic masking.

The information signal and the chaotic transmitter's output combine to generate the sent signal. The information is deliberately included into the driving mechanism to create the transmitter. Therefore, the transmitter is the synchronized hyperchaotic system with the additional information signal. The system (36) expresses the used system:

$$\begin{aligned} \dot{x} &= Ax + Bf(x) + Bg(x)\theta + B\eta(x,t) \\ y_1' &= y_1 + s(t) \end{aligned} \tag{36}$$

Here, $s \in \mathfrak{R}$ represents the information signal and $y_1'(t)$ is one of the transmitted signals guiding the receiver in the hyperchaotic system. Another transmitted signal is formulated as $y_2' = y_2 = C_2x$. Following synchronization, the recovery of the information signal $\hat{s}(t)$ can be achieved as detailed in (37):

$$\hat{s}(t) = y_1'(t) - C_1\hat{x}(t) \tag{37}$$

We acknowledge that the error dynamics are governed by (14) during the synchronization design. Therefore, akin to its proof, it is concluded that the transmitter can be synchronized. Upon the completion of synchronization, the following function is attained (38):

$$\lim_{t\to\infty} \hat{s}(t) = \lim_{t\to\infty} (y_1'(t) - C_1\hat{x}(t)) = \lim_{t\to\infty} (C_1x(t) + s(t) - C_1\hat{x}(t)) = \lim_{t\to\infty} s(t) \tag{38}$$

This signifies the successful recovery of the information signal. For the simulation, the values employed during synchronization are utilized and $s(t) = 1.2\sin(2t)$ is considered as the information signal. The information signal is depicted and the two transmission signals are formulated as $y_1'(t) = C_1x(t) + s(t)$ and $y_2' = C_2x$. The information signal $s(t)$, the signal $y_1'$ masked (i.e., encrypted, with the transmitted information signal) and the $s(t)$ signal obtained after synchronization and the recovery error after secure communication are illustrated in Figure 14.

As can be seen in Figure 14c, the encrypted signal shown in Figure 14b is decrypted by the receiver in a very short time and the information signal is obtained. The simulation results prove that the secure communication design is successful.

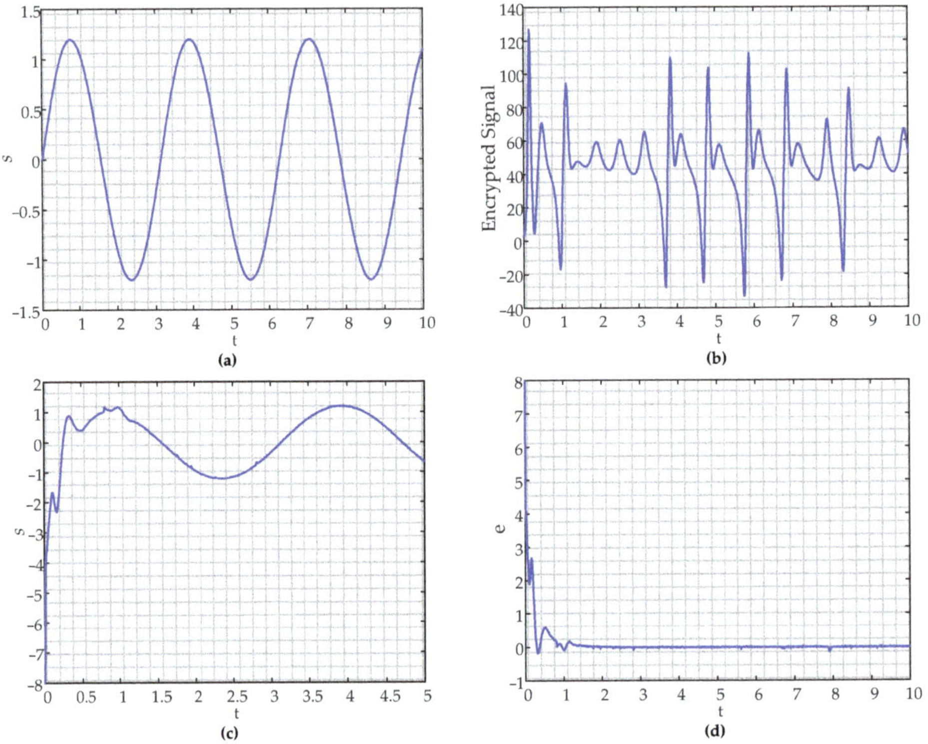


**Figure 14.** (**a**) Information signal; (**b**) transmitted encrypted signal; (**c**) the signal obtained after synchronization; (**d**) recovery error.

## 6. Conclusions

Our study addresses current challenges in chaotic systems, specifically focusing on the introduction and analysis of a novel hyperchaotic system. Through dynamic analysis, it is concluded that there is significant hyperchaotic behavior, paving the way for novel applications in secure communication. To validate the physical feasibility of our proposed system, an equivalent electronic circuit (see Figure 9) was created. The success of our adaptive observer-based synchronization under parameter uncertainty demonstrates a robust solution to real-world scenarios. This synchronization method, analyzed through Lyapunov stability theory, showcases its effectiveness by converging the synchronization error to 0 and accurately estimating unknown parameter values. The application of adaptive observer-based hyperchaotic system synchronization to secure communication, employing the chaotic masking method (see Figure 13), adds a layer of practical significance. Our system not only ensures the secure transmission of information but also successfully decodes the masked signal at the receiving end. Our study contributes to the understanding and application of hyperchaotic systems, with notable findings in both theoretical and practical domains. However, it is crucial to acknowledge limitations. While our methodology demonstrates success in simulation, future work should involve implementing the proposed system in hardware-based applications to bridge the gap between simulation and real-world scenarios. In conclusion, our research not only highlights the dynamic properties of the proposed hyperchaotic system but also showcases a synchronization method with promising applications in secure communication. By addressing current challenges, presenting innovative solutions and acknowledging the path forward, we contribute to the advancement of chaotic systems research and open avenues for future exploration and refinement.

**Author Contributions:** Conceptualization, E.O. and A.G.; methodology, E.O. and A.G.; validation, A.G.; formal analysis, E.O. and A.G.; investigation, E.O.; simulations, E.O.; writing—original draft preparation, E.O. and A.G.; writing—review and editing, E.O. and A.G.; visualization, E.O.; supervision, A.G. All authors have read and agreed to the published version of the manuscript.

**Funding:** This research received no external funding.

**Institutional Review Board Statement:** Not applicable.

**Informed Consent Statement:** Not applicable.

**Data Availability Statement:** The original contributions presented in the study are included in the article, further inquiries can be directed to the corresponding author.

**Conflicts of Interest:** The authors declare no conflicts of interest.

## References


1. Kolumban, G.; Kennedy, M.P.; Chua, L.O. The Role of Synchronization in Digital Communications Using Chaos. II. Chaotic Modulation and Chaotic Synchronization. *IEEE Trans. Circuits Syst. I Fundam. Theory Appl.* **1998**, *45*, 1129–1140. [CrossRef]
2. Lorenz, E.N. Deterministic Nonperiodic Flow. *J. Atmos. Sci.* **1963**, *20*, 130–141. [CrossRef]
3. Rossler, O.E. An Equation for Hyperchaos. *Phys. Lett. A* **1979**, *71*, 155–157. [CrossRef]
4. Emiroglu, S.; Akgül, A.; Adıyaman, Y.; Gümüş, T.E.; Uyaroglu, Y.; Yalçın, M.A. A New Hyperchaotic System from T Chaotic System: Dynamical Analysis, Circuit Implementation, Control and Synchronization. *Circuit World* **2022**, *48*, 265–277. [CrossRef]
5. Wei, Z.; Wang, F.; Li, H.; Zhang, W. Jacobi Stability Analysis and Impulsive Control of a 5D Self-Exciting Homopolar Disc Dynamo. *DCDS-B* **2022**, *27*, 5029. [CrossRef]
6. Wei, Z.; Zhang, W.; Moroz, I.; Kuznetsov, N.V. Codimension One and Two Bifurcations in Cattaneo-Christov Heat Flux Model. *Discret. Contin. Dyn. Syst. Ser. B* **2020**, *26*, 5305–5319. [CrossRef]
7. Wei, Z.; Li, Y.; Sang, B.; Liu, Y.; Zhang, W. Complex Dynamical Behaviors in a 3D Simple Chaotic Flow with 3D Stable or 3D Unstable Manifolds of a Single Equilibrium. *Int. J. Bifurc. Chaos* **2019**, *29*, 1950095. [CrossRef]
8. Xue, H.; Du, J.; Li, S.; Ma, W. Region of Interest Encryption for Color Images Based on a Hyperchaotic System with Three Positive Lyapunov Exponets. *Opt. Laser Technol.* **2018**, *106*, 506–516. [CrossRef]
9. Zelinka, I.; Senkerik, R. Chaotic Attractors of Discrete Dynamical Systems Used in the Core of Evolutionary Algorithms: State of Art and Perspectives. *J. Differ. Equ. Appl.* **2023**, *29*, 1202–1227. [CrossRef]
10. Feng, W.; Wang, Q.; Liu, H.; Ren, Y.; Zhang, J.; Zhang, S.; Qian, K.; Wen, H. Exploiting Newly Designed Fractional-Order 3D Lorenz Chaotic System and 2D Discrete Polynomial Hyper-Chaotic Map for High-Performance Multi-Image Encryption. *Fractal Fract.* **2023**, *7*, 887. [CrossRef]
11. Feng, W.; Zhao, X.; Zhang, J.; Qin, Z.; Zhang, J.; He, Y. Image Encryption Algorithm Based on Plane-Level Image Filtering and Discrete Logarithmic Transform. *Mathematics* **2022**, *10*, 2751. [CrossRef]
12. Matsumoto, T.; Chua, L.O.; Kobayashi, K. Hyper Chaos: Laboratory Experiment and Numerical Confirmation. *IEEE Trans. Circuits Syst.* **1986**, *33*, 1143–1147. [CrossRef]
13. Kopp, M. Hyperchaos, Adaptive Control, Synchronization, and Electronic Circuit Design of a Novel 6D Hyperchaotic Convective Dynamo System. *TechRxiv* **2022**. [CrossRef]
14. Yujun, N.; Xingyuan, W.; Mingjun, W.; Huaguang, Z. A New Hyperchaotic System and Its Circuit Implementation. *Commun. Nonlinear Sci. Numer. Simul.* **2010**, *15*, 3518–3524. [CrossRef]
15. Zhang, J.; Hou, J.; Xu, L.; Zhu, X.; Xie, Q. Dynamical Analysis, Circuit Implementation, and Simultaneous Application of a Novel Four-Dimensional Hyperchaotic System Based on Cosine Functions. *Microelectron. Eng.* **2023**, *271–272*, 111939. [CrossRef]
16. Aguilar-Bustos, A.Y.; Cruz-Hernández, C. Synchronization of Discrete-Time Hyperchaotic Systems: An Application in Communications. *Chaos Solitons Fractals* **2009**, *41*, 1301–1310. [CrossRef]
17. Wang, Y.; Xue, Y.; Wang, X.; Ceng, B.-l.; Qiao, Y.-f. Dynamic Behaviors in Two-Layer Coupled Oscillator System. *Chaos Solitons Fractals* **2021**, *144*, 110454. [CrossRef]
18. Pecora, L.M.; Carroll, T.L. Synchronization in Chaotic Systems. *Phys. Rev. Lett.* **1990**, *64*, 821. [CrossRef]
19. Chen, S.; Lü, J. Synchronization of an Uncertain Unified Chaotic System via Adaptive Control. *Chaos Solitons Fractals* **2002**, *14*, 643–647. [CrossRef]
20. Chen, S.; Lü, J. Parameters Identification and Synchronization of Chaotic Systems Based upon Adaptive Control. *Phys. Lett. A* **2002**, *299*, 353–358. [CrossRef]
21. Feki, M.; Robert, B. Observer-Based Chaotic Synchronization in the Presence of Unknown Inputs. *Chaos Solitons Fractals* **2003**, *15*, 831–840. [CrossRef]

22. Zhao, Y.; Zhang, W.; Su, H.; Yang, J. Observer-Based Synchronization of Chaotic Systems Satisfying Incremental Quadratic Constraints and Its Application in Secure Communication. *IEEE Trans. Syst. Man Cybern. Syst.* **2018**, *50*, 5221–5232. [CrossRef]
23. Liao, T.-L.; Tsai, S.-H. Adaptive Synchronization of Chaotic Systems and Its Application to Secure Communications. *Chaos Solitons Fractals* **2000**, *11*, 1387–1396. [CrossRef]
24. Shoreh, A.-H.; Kuznetsov, N.V.; Mokaev, T.N. New Adaptive Synchronization Algorithm for a General Class of Complex Hyperchaotic Systems with Unknown Parameters and Its Application to Secure Communication. *Phys. A Stat. Mech. Its Appl.* **2022**, *586*, 126466. [CrossRef]
25. Wu, X.-J.; Wang, H.; Lu, H.-T. Hyperchaotic Secure Communication via Generalized Function Projective Synchronization. *Nonlinear Anal. Real World Appl.* **2011**, *12*, 1288–1299. [CrossRef]
26. Hassan, M.F. A New Approach for Secure Communication Using Constrained Hyperchaotic Systems. *Appl. Math. Comput.* **2014**, *246*, 711–730. [CrossRef]
27. Xiong, L.; Liu, Z.; Zhang, X. Dynamical Analysis, Synchronization, Circuit Design, and Secure Communication of a Novel Hyperchaotic System. *Complexity* **2017**, *2017*, e4962739. [CrossRef]
28. Iskakova, K.; Alam, M.M.; Ahmad, S.; Saifullah, S.; Akgül, A.; Yılmaz, G. Dynamical Study of a Novel 4D Hyperchaotic System: An Integer and Fractional Order Analysis. *Math. Comput. Simul.* **2023**, *208*, 219–245. [CrossRef]
29. Yang, M.; Dong, C.; Sui, X. A New Four-Dimensional Hyperchaotic System with Hidden Attractors and Multistablity. *Phys. Scr.* **2023**, *98*, 125261. [CrossRef]
30. Shivamoggi, B.K. Chaos in Dissipative Systems. In *Nonlinear Dynamics and Chaotic Phenomena: An Introduction*; Fluid Mechanics and Its Applications; Springer: Dordrecht, The Netherlands, 2014; Volume 103, pp. 189–244. ISBN 978-94-007-7093-5.
31. Wolf, A.; Swift, J.B.; Swinney, H.L.; Vastano, J.A. Determining Lyapunov Exponents from a Time Series. *Phys. D Nonlinear Phenom.* **1985**, *16*, 285–317. [CrossRef]
32. Frederickson, P.; Kaplan, J.L.; Yorke, E.D.; Yorke, J.A. The Liapunov Dimension of Strange Attractors. *J. Differ. Equ.* **1983**, *49*, 185–207. [CrossRef]
33. Kuznetsov, N.V. The Lyapunov Dimension and Its Estimation via the Leonov Method. *Phys. Lett. A* **2016**, *380*, 2142–2149. [CrossRef]
34. Mobayen, S.; Vaidyanathan, S.; Sambas, A.; Kacar, S.; Çavuşoğlu, Ü. A Novel Chaotic System with Boomerang-Shaped Equilibrium, Its Circuit Implementation and Application to Sound Encryption. *Iran. J. Sci. Technol. Trans. Electr. Eng.* **2019**, *43*, 1–12. [CrossRef]
35. Fu, S.; Cheng, X.; Liu, J. Dynamics, Circuit Design, Feedback Control of a New Hyperchaotic System and Its Application in Audio Encryption. *Sci. Rep.* **2023**, *13*, 19385. [CrossRef] [PubMed]
36. Chen, M.; Zhou, D.; Shang, Y. Synchronizing a Class of Uncertain Chaotic Systems. *Phys. Lett. A* **2005**, *337*, 384–390. [CrossRef]
37. Zhu, F. Observer-Based Synchronization of Uncertain Chaotic System and Its Application to Secure Communications. *Chaos Solitons Fractals* **2009**, *40*, 2384–2391. [CrossRef]
38. Ioannou, P.A.; Sun, J. *Robust Adaptive Control*; PTR Prentice-Hall: Upper Saddle River, NJ, USA, 1996; Volume 1.
39. Gu, D.-W.; Poon, F.W. A Robust State Observer Scheme. *IEEE Trans. Autom. Control* **2001**, *46*, 1958–1963.
40. Hua, C.; Guan, X.; Li, X.; Shi, P. Adaptive Observer-Based Control for a Class of Chaotic Systems. *Chaos Solitons Fractals* **2004**, *22*, 103–110. [CrossRef]

*mathematics*

*Article*

# A New Chaos-Based Image Encryption Algorithm Based on Discrete Fourier Transform and Improved Joseph Traversal

**Mingxu Wang *, Xianping Fu *, Xiaopeng Yan and Lin Teng**

School of Information Science and Technology, Dalian Maritime University, Dalian 116026, China; yanxp@dlmu.edu.cn (X.Y.); tenglin@dlmu.edu.cn (L.T.)
* Correspondence: mxwang@dlmu.edu.cn (M.W.); fxp@dlmu.edu.cn (X.F.)

**Abstract:** To further enhance the security of image encryption, a new chaos-based image encryption algorithm (IEA) based on discrete Fourier transform and Joseph traversal is proposed to encrypt the plain image in both the frequency domain and space domain simultaneously. In the proposed IEA, the logistic map is used to generate the appropriate chaotic sequence, and the improved Joseph traversal is used to scramble the image in different starting positions and variable step sizes. Then, block diffusion is performed at the end. The main finding concerning the proposed IEA is that the combination of discrete Fourier transform and Joseph traversal can enhance the security of the image information, which has been validated by measuring the performance in resisting the common types of attacks.

**Keywords:** image encryption; Joseph traversal; Fourier transform; logistic map

**MSC:** 37N99

## 1. Introduction

With the fast development of computer information technology, the digital image as a kind of important data carrier is widely used and data security is receiving more attention, so the significance of data security research has become great. Image encryption technology has become an important part of information security, and it is used in many fields, such as engineering, industry and medical treatment. Therefore, to ensure the reliability and security of image data transmission and storage, the image encryption algorithm (IEA) has become a research hotspot [1–4]. The traditional IEAs, such as the data encryption standard [5] and RSA [6], have many defects when facing images with large data capacity and high redundancy, such as the relatively simple key, small key space and small parameters to determine the sequence, and low security performance of the IEAs. Therefore, there will be great disadvantages when it is applied to IEAs. Hence, several IEAs are proposed using different theories and technologies, such as deep learning [7], S-box [8], elliptic curve [9], Josephus traversal [10], DNA operation [11], compression sensing [12], quantum cellular Automata [13], fractional Fourier transform (FRFT) [14], and chaos theory [15–24].

Recently, some chaos-based IEAs have been proposed [25–35] due to the cryptographic properties of chaos theory, such as the initial conditions' sensitivity and unpredictability. Zhu et al. [25] constructed a new one-dimensional fractional chaotic map and applied it in an IEA using parallel DNA coding. The new chaotic map has a larger range of chaotic parameters and better chaotic characteristics, whereas the parallel DNA coding overcomes the shortcoming of common DNA coding-based IEAs. To enlarge the small parameter range and key space of the existing hyperchaotic maps, Zhong et al. [26] proposed a novel wide-range discrete hyperchaotic map (3D-SCC) based on the mathematical model of the Sine map, and they further designed a 3D-SCC-based IEA. Among them, the IEAs in the transform domain usually have better anti-interference performance and higher security [27–35]. Shao et al. [27] introduced a trinion discrete cosine transform that can process color images holistically, which can be computed by a combination of single-channel discrete cosine transform and then developing a robust color IEA jointing with a quantum logistic map and Josephus traversing. To strike a good balance between the security and efficiency of the IEA, Feng et al. [28] introduced an IEA based on plane-level image filtering and discrete logarithmic transformation. With the increase in the image transmission data, single IEAs are of low efficiency to meet the real-life applications. Hence, double-image and multi-image IEAs have designed [29–35]. Kaur et al. [29] applied a fractional Fourier transform for double image encryption and proposed a reality-preserving scheme to overcome the complex outcome of the transform, which becomes a limitation due to the requirement of double memory for storage and transmission besides computational complexity. By combining FRFT with discrete fractional angular transform, a double-image IEA was designed by Qiu et al. [30]. Shao et al. [31] introduced a double IEA mainly based on the symmetry of two-dimensional discrete Fourier transform and equal modulus decomposition. Tong et al. [32] designed a nonlinear multi-image IEA by combining the reality-preserving discrete fractional angular transform with the deoxyribonucleic acid sequence operations. Based on discrete Fourier transforms (DFTs) and logistic-exponent-sine map, Tang et al. [33] investigated an IEA for multiple color images. When encrypting multiple images, compression is a commonly used method to reduce the transmission burden, which may result in a loss of partial image data and reduce the quality of the decryption image. To address this issue, Wang et al. [34] proposed a double-image compression and encryption scheme based on a chaotic map and reality-preserving multi-order discrete fractional Fourier transform. A secure double-color IEA with a nonlinear operation and a plaintext-related joint permutation–diffusion mechanism was designed Xiao et al. [35] based on the quaternion multiple parameter discrete fractional angular transform.

Motivated by the above discussions, an IEA based on DFT and improved Joseph traversal (IEA-DIJT) is designed in this paper. IEA-DIJT can encrypt the plain image from the frequency domain and spatial domain with a high efficiency and low complexity. Through different analyses and experimental comparison, the proposed IEA-DIJT in this work has been validated to have good security and effectiveness.

The structure of this paper is organized as follows. Section 2 introduces some preliminary works. Section 3 provides the detailed IEA-DIJT. In Section 4, the simulation experiments and security analyses are performed. Finally, Section 5 summarizes this paper and discusses the future work.

## 2. Preliminary Works

### 2.1. Logistic Map

The chaotic map is sensitive to the initial value [10], which means that even if there is a tiny modification in the input parameters, the results obtained will vary greatly. This feature is very suitable for an IEA. The chaotic map used in the IEA-DIJT is the classical logistic map and its formula is:

$$x_{n+1} = \mu x_n(1 - x_n) \tag{1}$$

where $x_{n+1} \in (0, 1)$. $\mu$ is the control parameter. Figure 1 respectively depicts the bifurcation diagram and the Lyapunov exponent of Equation (1). As can be seen in Figure 1, Equation (1) performs chaotic behaviors and not periodic when $\mu \in (3.5699456,\ 4]$.

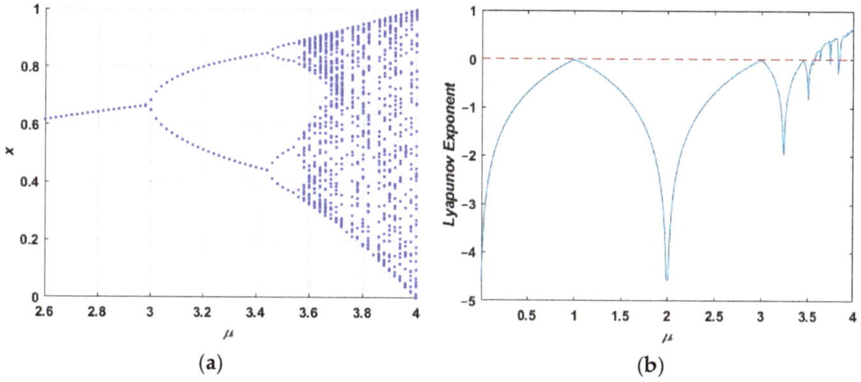


**Figure 1.** Performance evaluation of the logistic map: (**a**) bifurcation diagram and (**b**) Lyapunov exponent.

### 2.2. Fourier Transform

Fourier transform, as a classical signal-processing method, has excellent applicability in the field of IEAs. Encrypting images in the transform domain can ensure that the corresponding image information can still be decoded even if some information is lost. Moreover, IEAs based on Fourier transform have the characteristics of high efficiency and high security. Hence, this paper designs an IEA based on Fourier transform. This section details the DFT. For the finite length series $f(x)$, the one-dimensional DFT is:

$$F(u) = DFT[f(x)] = \sum_{x=0}^{N-1} f(x)W^{ux},\ u = 0, 1, \cdots, N-1 \tag{2}$$

and the corresponding inverse transformation is:

$$f(x) = IDFT[F(u)] = \frac{1}{N}\sum_{x=0}^{N-1} F(u)W^{-ux},\ x = 0, 1, \cdots, N-1 \tag{3}$$

where $W = e^{-j\frac{2\pi}{N}}$ is the transformation kernel.

The matrix model of the DFT is:

$$\begin{pmatrix} F(0) \\ F(1) \\ \vdots \\ F(N-1) \end{pmatrix} = \begin{pmatrix} W^0 & W^0 & W^0 & \cdots & W^0 \\ W^0 & W^{1\times1} & W^{2\times1} & \cdots & W^{(N-1)\times1} \\ \vdots & \vdots & \vdots & & \vdots\ \vdots \\ W^0 & W^{1\times(N-1)} & W^{2\times(N-1)} & \cdots & W^{(N-1)\times(N-1)} \end{pmatrix} \begin{pmatrix} f(0) \\ f(1) \\ \vdots \\ f(N-1) \end{pmatrix}, \tag{4}$$

and the corresponding inverse transformation is:

$$\begin{pmatrix} f(0) \\ f(1) \\ \vdots \\ f(N-1) \end{pmatrix} = \frac{1}{N} \begin{pmatrix} W^0 & W^0 & W^0 & \cdots & W^0 \\ W^0 & W^{-1\times1} & W^{-2\times1} & \cdots & W^{-(N-1)\times1} \\ \vdots & \vdots & \vdots & & \vdots\ \vdots \\ W^0 & W^{-1\times(N-1)} & W^{-2\times(N-1)} & \cdots & W^{-(N-1)\times(N-1)} \end{pmatrix} \begin{pmatrix} F(0) \\ F(1) \\ \vdots \\ F(N-1) \end{pmatrix} \tag{5}$$

### 2.3. Improved Joseph Traversal

This section describes the specific progress of the improved Joseph traversal. First, the Joseph traversal is a mathematical problem: encircling $n$ number in a ring, looping through them in order, and deleting the $k-1$th number, and iterating this process from the $k$th number until the last number is selected. The function of this process is $f(n, k)$. Assuming it starts from 0, it is obvious that the traditional Joseph traversal has a basic value, which is $f(1,k) = 0$. The first chosen number is $(k-1)\%n$, so the remaining array is 0, 1, 2, 3, ..., $k-3$, $k-2$, $k$, $k+1$. In other words, it needs to start counting from the $k$ again, but the array changes from 0, 1, 2, ..., $n-1$ to a new array $k$, $k+1$, $k+2$, ..., $n-1$, 0, 1, 2, ..., $k-2$, which is different from the original starting position, and $f(n,k)$ is no longer applicable to the new array. Hence, a new function $g(n-1,k)$ is established to represent the solution of the Joseph traversal for $n-1$ numbers starting from $k$. Since the new array is actually a subproblem, the solution of $g(n-1,k)$ and the solution of $f(n,k)$ have the same array number, $g(n-1,k) = f(n,k)$. However, the two functions are different and cannot be recursively expressed, so it needs to establish a mapping that converts $g(n-1,k)$ into the type of $f(n-1,k)$ by converting it between two arrays: $k \to 0$; $k+1 \to 1; \cdots; n-1 \to n-k-1; 0 \to n-k; 1 \to n-k+1; \cdots; k-2 \to n-2$. It can be observed that it can use a function $p(x) = (x+n-k)\%n$ to represent the mapping from left to right. Conversely, if a mapping from right to left is required, $q(x) = (x+k)\%n$ can be used. So, the Joseph traversal that operates after mapping is:

$$f(n-1,k) = p(g(n-1,k)) \tag{6}$$

$$g(n-1,k) = p^{-1}(f(n-1,k)) = q(f(n-1),k) = [f(n-1,k)+k]\%n \tag{7}$$

Hence, the recursive formula of the Joseph traversal has obtained:

$$f(n,k) = \begin{cases} 0 & n = 1 \\ [f(n-1,k)+k]\%n & n > 1 \end{cases} \tag{8}$$

Taking the traversal order of 8 numbers as an example, the traditional Joseph traversal is depicted in Figure 2.

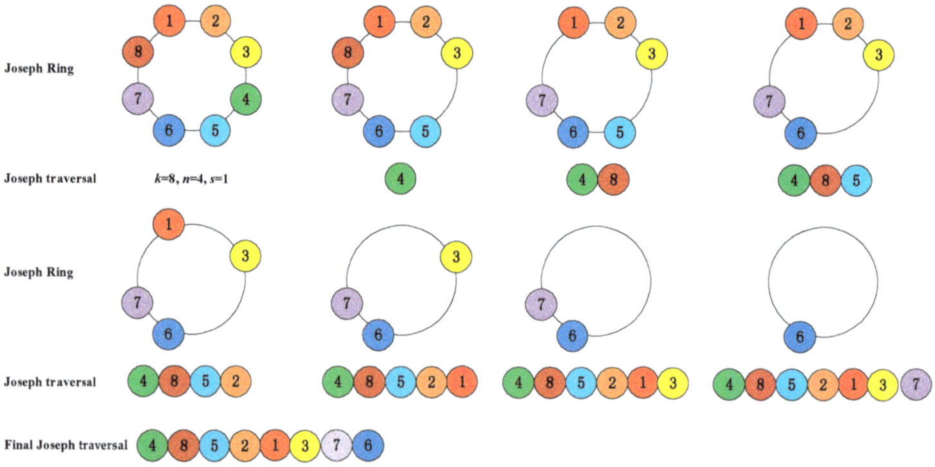


**Figure 2.** Joseph traversal.

For a good IEA, its encryption process should be flexible and sensitive. For the traditional Joseph mentioned above, it starts from a fixed starting point, traverses with a fixed step size, and finally, obtains a traversal sequence, which is the same for each encryption process. The chosen-plaintext and chosen-ciphertext attacks can easily crack this process, posing a threat to the IEA's security. Therefore, to obtain different and sensitive traversal sequences during each encryption process, it is totally important to improve the fixed starting point and step size of the traditional methods. Hence, this section proposes an improved Joseph traversal. The core idea of this method is that the initialization process of its starting position and traversal step size is highly adaptively associated with the plain image and key set. Assuming the length of the sequence $S$ to be traversed is $M \times N$, the specific traversal steps are as follows:

Step 1: Iterate a pseudo-random sequence of length $2 \times M \times N$ using Equation (1), and then standardize it via Equation (9):

$$Z(i) = floor[(z(i) \times 10^n)\text{mod}256] \tag{9}$$

Step 2: Divide $Z$ into two sub-sequences of length $M \times N$, named $Z_1$ and $Z_2$, respectively, and perform operations on the first element of $Z_1$ and $Z_2$ via:

$$\begin{cases} H_1 = (Z_1(1) + HS_1)\text{mod}M \\ H_2 = (Z_2(1) + HS_2)\text{mod}N \end{cases} \tag{10}$$

where $H_1$ and $H_2$ are the calculated starting position. $HS_1$ and $HS_2$ represent the hash values based on the plain image.

Step 3: Divide sequence $S$ into two sub-sequences based on the hash value $HS_3$, named $S_1$ and $S_2$, respectively. Perform the improved Joseph traversal on $S_1$ and $S_2$ based on the starting position $H_1$ and $H_2$. The variable step size and direction of traversal are determined based on the values and their parity. For sequence $S_1$, it is a traversal process with a variable step size determined by $Z_1$ and the parity determined by sequence $Z_2$. Conversely, $S_2$ is a traversal process with a variable step size determined by $Z_2$ and the parity determined by $Z_1$. The meaning of parity here is to traverse counterclockwise if the value is odd, and clockwise if not.

Step 4: Combine the two traversal subsequences into the final sequence.

The traversal results of the Joseph traversal are further expanded as a result of the above improvements. The original Joseph traversal is improved to a Joseph traversal model with the variable step sizes, directions, and starting positions.

### 2.4. Hash Function

By using hash functions, hash values with specific length and irreversibility can be obtained, which are highly in line with the requirements of image encryption algorithms and are often used to resist common attacks. However, in recent years, some IEAs using SHA-1 have shown weak security. By comparison, SHA-2-based IEAs still exhibit better security. As a hash function in SHA-2, SHA-512 performs tremendous collision resistance, making it difficult to observe two different input messages that produce the same hash value. Even if a slightly change to the input message occurs, the generated hash array is totally different. Moreover, due to the robust avalanche effect of SHA-512, it is widely utilized in the field of IEAs. Therefore, this paper uses the SHA-512 in the IEA-DIJT. It can produce a hexadecimal number with a length of 128, which is used as the main part of the key set for the IEA-DIJT.

## 3. The Specific Operations of the IEA-DIJT

This section details the whole operations of the IEA-DIJT in this work, which includes two main phases: key generation and encryption progress.

### 3.1. Key Generation

The key set of the IEA-DIJT is mainly the hash value keys, named $HV$, which is generated based on the plain image. Then, initialize the parameters of the IEA-DIJT using the obtained $HV$. The following content provides the specific initialization process:

Step 1: Generate the hash array $HV$ using the SHA-512 function. Read the plain image $P$ into the function to generate a hexadecimal number $HS$ with a length of 128, then divide $HS$ into 6 parts and convert into a decimal as follows:

$$HS = hash(PI, '\text{SHA} - 512') \tag{11}$$

$$HV = hex2dec(HS(i:i+21)), i = 1, 22, \cdots, 106 \tag{12}$$

Among Equations (11) and (12), the function $hash(PI, \text{'SHA} - 512\text{'})$ is used to calculate the SHA-512 hash value of the plain image $P$, whereas the function $hex2dec(\cdot)$ converts a hexadecimal string into the decimal format. $HS(i:i+21)$ are the values of $HS$ from the $i$-th to the $i+21$-th element.

Step 2: Select 6 hash values from $HV$ as the initial states $x_i(0)(i = 1, 2, 3)$ and control parameters $\mu_i(i = 1, 2, 3)$ of Equation (1) (please see Equation (13)), which are the secret keys. For $\mu_i(i = 1, 2, 3)$, this work limits its value to within $[3.59, 4]$ to ensure that the obtained sequence values have a good chaotic state.

$$\begin{cases} x_1(0) = Key_1 \\ x_2(0) = Key_2 \\ x_3(0) = Key_3 \\ \mu_1 = 3.59 + 0.11Key_4 \\ \mu_2 = 3.69 + 0.11Key_5 \\ \mu_3 = 3.79 + 0.11Key_6 \end{cases} . \tag{13}$$

Step 3: Iterate Equation (1) and transform the generated random sequence $A_i$ into a chaotic matrix of the same size as the plaintext matrix. Then, standardize the matrix $A_i$ via:

$$K_i = \text{mod}(floor(A_i \times 10^{13}), 255) \tag{14}$$

where $K_i$ is further reshaped into the key matrix $K$ via:

$$K = reshape(K_i, M, N) \tag{15}$$

where $M$ and $N$ are the height and width of the image matrix.

### 3.2. Description of the IEA-DIJT

Step 1. For the cipher-image hash operation, the SHA-512 function is used to calculate the standard hash sequence and select the corresponding hash values in the hash sequence to calculate the initial value and control parameters of Equation (1):

$$x = H(i). \tag{16}$$

$$\mu(i) = 3.89 + 0.11 \times H(i). \tag{17}$$

Step 2. Read the initial state and control parameter into Equation (1) to generate the chaotic sequence:

$$A(i) = \mu(i) \times x(i) \times (1 - x(i)). \tag{18}$$

Step 3. The discrete forward Fourier transform of the cipher image. By the Fourier transform, the gray distribution function $f(x, y)$ of the image is transformed into the frequency distribution function $F(u, v)$ of the image and obtains the size of the image frequency domain spectrum $(u, v)||F$ and phase spectrum $\Phi(u, v)$.

Step 4. According to the random sequence of Equation (1), the starting point and step size are calculated, and the amplitude diagram and phase diagram after the Fourier transform are respectively carried out for the forward and backward Joseph traversal.

Step 5. The inverse Fourier transform of the scrambled amplitude graph and phase graph is used to obtain a round scrambled result.

Step 6. The scrambled image $P$ is partitioned, and the image matrix is divided into matrix blocks of the same size, denoted as $F(I)$, where $I$ is the number of blocks.

Step 7. The chaotic sequence $A_1$ is used to sort and scramble the matrix, and the sequence $A_2$ is sorted to obtain the index sequence $C_1$. The matrix $F$ is transformed into a row of matrix $P_1$. The matrix $F$ is sorted and scrambled by the index, and the scrambling result $P_2$ is obtained, and $P_2$ is restored to the same size as the original matrix.

$$\begin{cases} C_1 = sort(A_1(i)) \\ P_1 = reshape(F,1,i) \\ P_2 = P_1(C_1(i)) \end{cases}. \quad (19)$$

Step 8. For the scrambled image, block diffusion is carried out. Firstly, the first block is diffused, and the diffusion key matrix is selected and generated from the chaotic sequence $A_3$.

Step 9. The first diffusion matrix is used as the key matrix and the rest of the parts are circularized.

$$\begin{cases} R(i) = P_3(i) \oplus T(i) \\ R(i+1) = R(i) \oplus P_3(i+1) \\ R(i-1) = R(i) \oplus P_3(i-1) \end{cases}. \quad (20)$$

This is the end of the proposed scheme. Figure 3 depicts the above-mentioned operations.

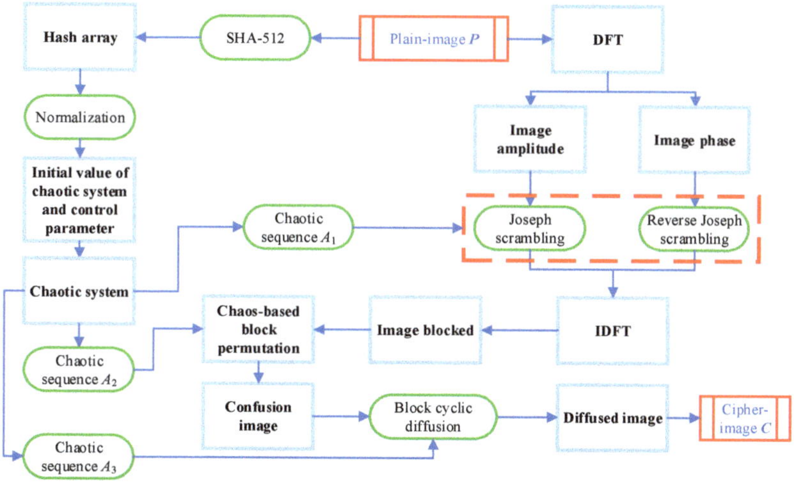


**Figure 3.** Flow chart of the IEA-DIJT.

## 4. Experimental Results and Security Analyses

To verify the effectiveness of the IEA-DIJT, this section performs an evaluation of the IEA-DIJT using the digital plain-images. The IEA-DIJT is implemented in the MATLAB R2023b and Windows 10 operating system. Different plain images generate different hash values $HV$. According to https://ccia.ugr.es/cvg/index2.php and https://sipi.usc.edu/database/database.php (accessed on 3 January 2024), the grayscale plain images were used for the simulation experiments and security analyses.

### 4.1. Simulation Results

First, Figure 4 plots the cipher images and decoded images of the plain images "Lena" (256 × 256), "Cameraman" (256 × 256), "Elaine" (512 × 512), and "Airport" (1024 × 1024).

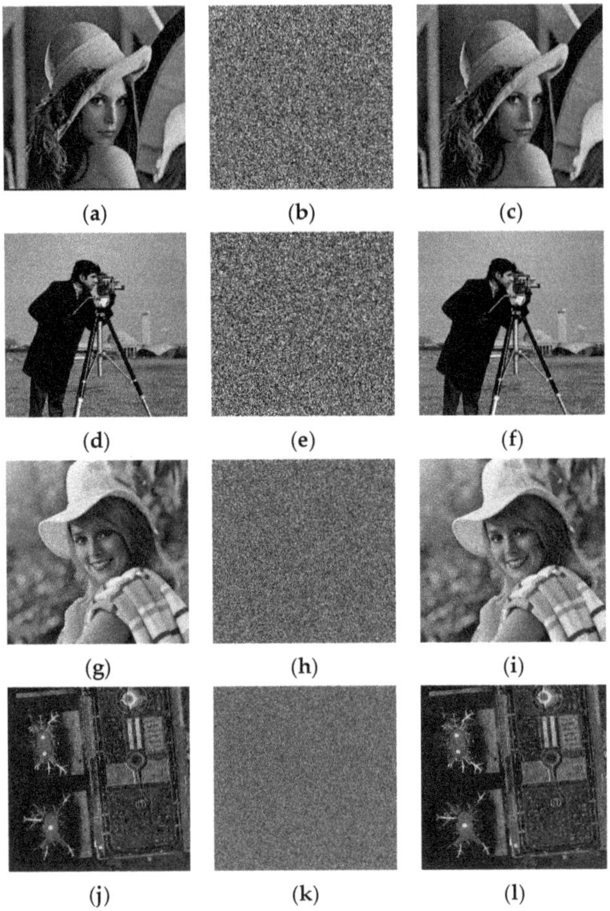


**Figure 4.** Experimental results: (**a**,**d**,**g**,**j**) are the plain images "Lena", "Cameraman", "Elaine", and "Airport", (**b**,**e**,**h**,**k**) are the corresponding cipher images of (**a**,**d**,**g**,**j**), and (**c**,**f**,**i**,**l**) are the corresponding decrypted images of (**b**,**e**,**h**,**k**).

### *4.2. Key Sensitivity*

If an IEA is highly sensitive to the key stream, it is a good IEA. On the one hand, due to the fact that hash functions are extremely sensitive to the input plain-image message, even a small change can result in a completely different hash value $HS$. This feature can effectively enhance the key security of the IEA-DIJT. On the other hand, this section further takes the key $Key_1$ as an example to test the key sensitivity of the IEA-DIJT using the plain image "Lena". The experimental results are shown in Figure 5. When the other keys remain unchanged, Figure 5a is the cipher image generated according to key $x_1(0)$. If the key $x_1(0)$ is changed from 0.8761 to 0.87610000000001, then the obtained cipher image is Figure 5b. Figure 5c is the difference between Figure 5a,b. Figure 5d is the decrypted image generated by decoding Figure 5b. With a minor modification of the secret key, the cipher image generated by the IEA-DIJT is extremely different from the cipher image using the original keys. Moreover, if a tiny change to a pixel is made, the cipher image is as shown in Figure 5e. Figure 5f is the difference between Figure 5a,e. It indicates that the IEA-DIJT is extremely sensitive to the key set.

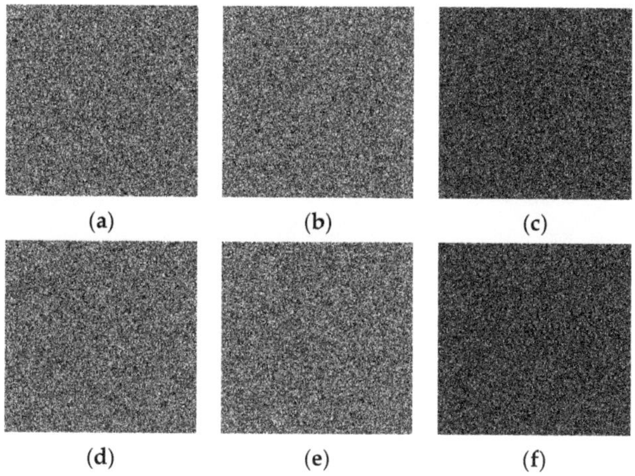

**Figure 5.** Key sensitivity. The cipher image of the plain image "Lena" using (**a**) $x_1(0)$ and (**b**) $x_1'(0)$, (**c**) the difference between (**a**,**b**,**d**) the decoded image using $x_1'(0)$, (**e**) the cipher image after a tiny change, and (**f**) the difference between (**a**,**e**).

### 4.3. Histogram

Digital images are composed of pixels, and generally speaking, unsatisfied images have very clear fluctuant properties in the pixel distribution. The pixel values of the cipher image must be roughly distributed to withstand statistical attacks. The histogram of an image provides an accurate and first-hand view of the pixel value spread, and it is often used to find out whether the cipher images obtained by one IEA are robust against statistical attacks. If the histogram of an image looks smooth, the pixel spread is evenly distributed and it is not easy to observe helpful message from the statistical analysis attacks. That is to say, the better the performance of an IEA, the smoother the histogram of the cipher image it obtains.

Figure 6 illustrates the histograms of the plain images "Lena", "Peppers", and "Cameraman" and their corresponding cipher images, respectively. Figure 6a,c,e depict the pixel distribution of the plaint images. For example, there are pixels with a huge number of elements in the image, while some pixels do not even show in the image. In contrast, the pixel distribution in the histogram of the cipher image shown in Figure 6b,d,f is presumably uniform and does not have any distribution features. Therefore, the attackers are unable to exploit statistical attacks to disrupt the IEA-DIJT.

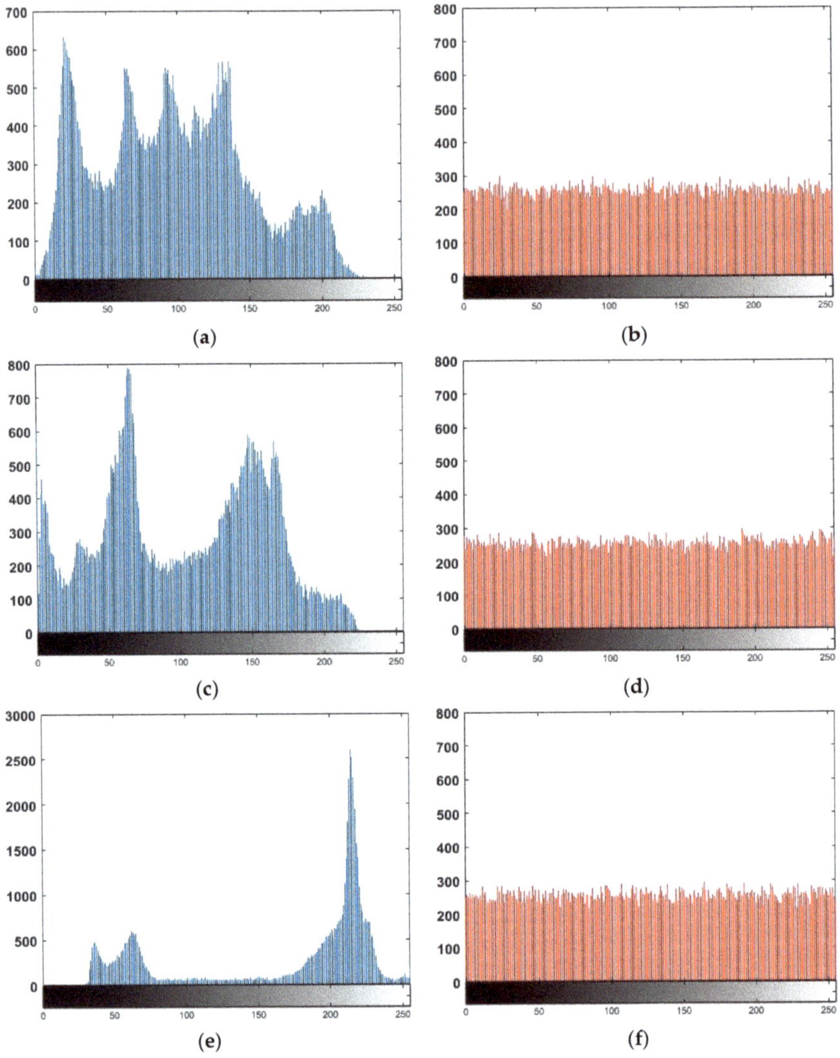

**Figure 6.** Histograms of the IEA-DIJT: (**a**,**c**,**e**) are the histograms of the plain images "Lena", "Peppers", and "Cameraman", respectively; and (**b**,**d**,**f**) are the histograms of the cipher images "Lena", "Peppers", and "Cameraman", respectively.

### 4.4. Chi-Square Test

To quantitatively measure the distribution situation of the histogram, this section further uses the chi-square ($\chi^2$) test as a metric. The description of the chi-square [36] is as follows:

$$\chi^2 = \sum_{i=0}^{255} \frac{(q_i - q)^2}{q}, \tag{21}$$

where $q_i$ means the number of pixels with a value of $i$ present in the image, and $q$ is calculated via Equation (22):

$$q = \frac{M \times N}{256}. \tag{22}$$

Table 1 tabulates the $\chi^2$ results of the cipher images, which demonstrate that the obtained $\chi^2$ result was below 293.2483 at a significance level of 5%. Therefore, the IEA-DIJT represents a superior behavior to resist statistical analysis attacks.

**Table 1.** The obtained $\chi^2$ results.

| Size | Image | $\chi^2$ Value | H (0 or 1) | Decision |
|---|---|---|---|---|
| 256 | Cameraman | 250.6172 | 0 | Accept |
| | Clock | 261.7734 | 0 | Accept |
| | House | 258.7813 | 0 | Accept |
| | Lena | 291.2969 | 0 | Accept |
| | Moonface | 287.2578 | 0 | Accept |
| | Peppers | 233.2734 | 0 | Accept |
| 512 | Baboon | 283.5293 | 0 | Accept |
| | Boat | 246.8066 | 0 | Accept |
| | Couple | 247.3340 | 0 | Accept |
| | Elaine | 263.0449 | 0 | Accept |
| | Lena | 262.1440 | 0 | Accept |
| | Peppers | 243.7363 | 0 | Accept |

### *4.5. Correlation*

The correlation coefficient between two neighboring pixels is used to measure the differences between the plain images and the cipher images. In many plain images, the correlation results between the pixels approach 1. An efficacious IEA should be proposed to break the correlation of the images. This section stochastically picks 10,000 pixels from the plain image "Lena" and the corresponding cipher image, and it estimates the correlations among the pixels in the horizontal (H), vertical (V), and diagonal (D) directions via [37]:

$$r_{xy} = \frac{\operatorname{cov}(x,y)}{\sqrt{D(x)}\sqrt{D(y)}} \tag{23}$$

where

$$\operatorname{cov}(x,y) = \frac{1}{N}\sum_{i=1}^{N}(x_i - E(x))(y_i - E(y)),\ D(x) = \frac{1}{N}\sum_{i=1}^{N}(x_i - E(x))^2,\ E(x) = \frac{1}{N}\sum_{i=1}^{N}x_i. \tag{24}$$

Figure 7 respectively plots the correlations of the plain images and the cipher images. Moreover, Tables 2 and 3 further tabulate the comparison results obtained by different IEAs. The correlation between two neighboring pixels of the plain mages is powerful. In contrast, it is quite weak in its ciphertext. It can be seen from the comparison results that the correlation results of the IEA-DIJT are generally lower compared with the existing IEAs [38–42]. Hence, the IEA-DIJT is effective enough to withstand statistical analysis attacks.

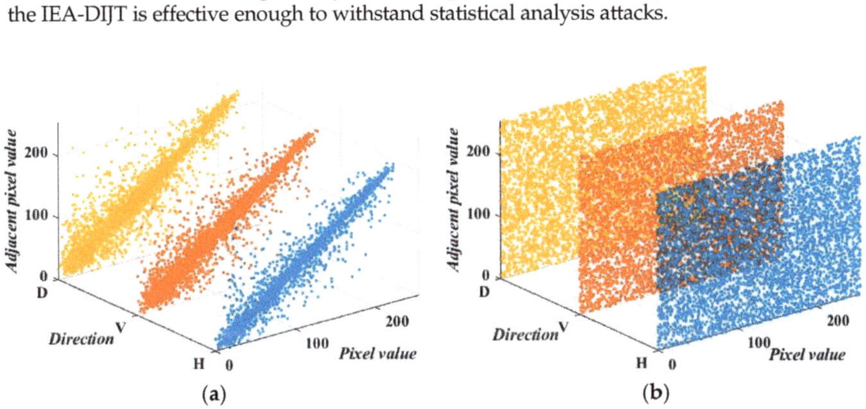


**Figure 7.** Correlation: the correlation of (**a**) Figure 4a and (**b**) Figure 4b in three directions.

**Table 2.** Correlation of neighboring pixels of the IEA-DIJT.

| Size | Image | Direction | IEA-DIJT | | Size | Image | Direction | IEA-DIJT | |
|---|---|---|---|---|---|---|---|---|---|
| | | | *P* | *C* | | | | *P* | *C* |
| 256 | Cameraman | H | 0.8950 | −0.0017 | 512 | Baboon | H | 0.7267 | 0.0010 |
| | | V | 0.8987 | 0.0022 | | | V | 0.7328 | −0.0019 |
| | | D | 0.9311 | 0.0011 | | | D | 0.6838 | 0.0011 |
| | Clock | H | 0.9340 | 0.000 | | Lena | H | 0.9719 | −0.0004 |
| | | V | 0.9431 | −0.0013 | | | V | 0.9850 | −0.0004 |
| | | D | 0.9469 | 0.0007 | | | D | 0.9593 | 0.0002 |
| | Lena | H | 0.9143 | 0.0013 | | Peppers | H | 0.9573 | 0.0008 |
| | | V | 0.9039 | 0.0009 | | | V | 0.9667 | 0.0018 |
| | | D | 0.9098 | 0.0007 | | | D | 0.9684 | 0.0007 |
| | Peppers | H | 0.9337 | −0.0031 | | | | | |
| | | V | 0.9464 | −0.0017 | | | | | |
| | | D | 0.9329 | −0.0028 | | | | | |

**Table 3.** Comparison results with other IEAs for the cipher image "Lena".

| Schemes | Directions | | |
|---|---|---|---|
| | **H** | **V** | **D** |
| IEA-DIJT | 0.0009 | 0.0007 | −0.0031 |
| Ref. [38] | −0.015595 | 0.028436 | −0.007102 |
| Ref. [39] | 0.016720 | 0.006996 | 0.011698 |
| Ref. [40] | −0.040583 | −0.027371 | −0.014449 |
| Ref. [41] | 0.004555 | −0.015623 | −0.005112 |
| Ref. [42] | 0.009998 | 0.001372 | −0.006567 |
| Ref. [43] | −0.001885 | −0.012793 | 0.007396 |

## 4.6. Information Entropy

### 4.6.1. Global Information Entropy

Information entropy is considered a measure of the degree of order in the motion state of an object. The higher the entropy, the higher the degree of disorder. Its calculation is Equation (25) [37]. Theoretically, the possibility of information leakage decreases gradually as the information entropy approaches 8.

$$H = \sum_{i=0}^{2^n-1} p(m_i) \log_2 \frac{1}{p(m_i)} \tag{25}$$

where $p(m_i)$ means the probability of $m_i$ happening. For image processing, entropy presents the amount of information contained in an image. When an image embodies $N$ grayscale values, if the grayscale values of each pixel are different, the entropy value is the maximum, $H = \log_2 N$, and the image has the biggest amount of message. It can be considered that each pixel is an independent object target with a maximum information content of $N$. Hence, the bigger the entropy $H$ of the image, the richer the grayscale of the pixels contained in the image, the more uniform the grayscale distribution, the more ground objects in the image, and the greater the message content of the image.

For a good IEA, the information entropies of the cipher images should be close to 8. Tables 4 and 5 list the computed results of the images and comparison results. As can be observed, all the entropy results [38–43] are close to 8. Moreover, the cipher images obtained by the IEA-DIJT are less likely to leak information and have better ability to resist statistical analysis attacks than other IEAs.

**Table 4.** Information entropy values of the IEA-DIJT.

| Size | Image | IEA-DIJT | | Size | Image | IEA-DIJT | |
|---|---|---|---|---|---|---|---|
| | | *P* | *C* | | | *P* | *C* |
| 256 | Cameraman | 6.9719 | 7.9973 | 512 | Baboon | 7.3585 | 7.9992 |
| | Clock | 6.7057 | 7.9972 | | Boat | 7.1914 | 7.9993 |
| | House | 6.4971 | 7.9971 | | Couple | 7.2010 | 7.9994 |
| | Lena | 7.5534 | 7.9968 | | Elaine | 7.5060 | 7.9993 |
| | Moonface | 6.7093 | 7.9968 | | Lena | 7.4474 | 7.9995 |
| | Peppers | 7.5819 | 7.9974 | | Peppers | 7.5925 | 7.9993 |

**Table 5.** Comparison results of the information entropy.

| Size | Image | IEA-DIJT | | Ref. [38] | Ref. [39] | Ref. [40] | Ref. [41] | Ref. [42] | Ref. [43] |
|---|---|---|---|---|---|---|---|---|---|
| | | *P* | *C* | | | | | | |
| 256 | Baboon | 7.5534 | 7.9973 | 7.9302 | 7.9973 | 7.9970 | 7.9973 | 7.9787 | 7.9967 |
| | Lena | 7.5819 | 7.9974 | 7.9144 | 7.9970 | 7.9969 | 7.9973 | 7.9950 | 7.9971 |
| 512 | Baboon | 7.3585 | 7.9992 | 7.9866 | 7.9985 | 7.9992 | 7.9993 | 7.9791 | 7.9993 |
| | Lena | 7.4474 | 7.9995 | 7.9729 | 7.9986 | 7.9992 | 7.9992 | 7.9788 | 7.9993 |

#### 4.6.2. Local Shannon Entropy

For better analysis, this section further conducts experimental comparisons of the local information entropy (LSE) via Equation (26):

$$H_{k,T_B}(S,L) = \sum_{i=1}^{k} \frac{H(S_{T_B},L)}{k}, \tag{26}$$

where $H(S_{T_B},L)$ denotes the Shannon entropy of non-overlapping image blocks $S_{T_B}$. It means that each of the $k = 30$ blocks contain $T_B = 1936$ pixels. The image passes the test if the LSE value is located in [7.901515698, 7.903422936].

The IEA-DIJT was evaluated on the plain images. The results, tabulated in Table 6 (Bold values indicate passing the test), demonstrated that the cipher images passed the LSE test with a passing rate of over 90%. This indicates that the IEA-DIJT provides a high degree of randomness and an even distribution of the cipher images.

**Table 6.** LSE values for different cipher images.

| Size | Image | LSE | Size | Image | LSE |
|---|---|---|---|---|---|
| 256 | Cameraman | **7.902814** | 512 | Baboon | **7.9032934** |
| | Clock | **7.9018914** | | Boat | **7.9020965** |
| | House | 7.9036226 | | Couple | **7.9013717** |
| | Lena | **7.9033111** | | Elaine | **7.9025729** |
| | Moonface | **7.9025067** | | Lena | **7.902469** |
| | Peppers | **7.9026073** | | Peppers | **7.9019517** |
| PASS/ALL | | | 11/12 | | |

### 4.7. *Differential Attack*

A differential attack can be used to find the vulnerabilities of an IEA. It can help the attackers to break an IEA by recovering the key set or plaintext message from the cipher image. Therefore, a strong IEA must be greatly sensitive to minor changes in the plaintext, which must spread throughout the entire encrypted image. The number of the pixel change rate ($NPCR$) and the unified average change intensity ($UACI$) are common metrics used to measure the ability of IEAs against those attacks. In this section, they are calculated via Equations (27)–(29) [44]:

$$NPCR = \frac{1}{W \times H} \sum_{i=1}^{W} \sum_{j=1}^{H} D(i,j) \times 100\%, \tag{27}$$

$$UACI = \frac{1}{W \times H} \left( \sum_{i=1}^{W} \sum_{j=1}^{H} \frac{|c_1(i,j) - c_2(i,j)|}{255} \right) \times 100\%, \tag{28}$$

$$D(i,j) = \begin{cases} 0, & c_1(i,j) = c_2(i,j) \\ 1, & c_1(i,j) \neq c_2(i,j) \end{cases}, \tag{29}$$

where $W \times H$ is the total number of pixels in an image. $c_1$ and $c_2$ are two encrypted images with a one-bit difference of their original image. The ideal $NPCR$ and $UACI$ scores are near 99.6093% and 33.4635%, respectively.

Moreover, the important values for determining whether the obtained $NPCR$ and $UACI$ have passed the test are in the light of their critical values. For $NPCR$, the critical value $N_\alpha^*$ for different image sizes is calculated via:

$$N_\alpha^* = \frac{L - \Phi^{-1}(\alpha)\sqrt{L/MN}}{L+1}, \tag{30}$$

where $\alpha$ represents the significance level, and $\Phi^{-1}(\alpha)$ denotes the inverse cumulative density function of normal distribution $N(0,1)$. If the obtained is $NPCR$ larger than $N_\alpha^*$, it is considered that the IEA has passed the test. For $UACI$, there are two strict critical values $U_\alpha^{*-}$ and $U_\alpha^{*+}$ for different image sizes. They are calculated via:

$$\begin{cases} U_\alpha^{*-} = \mu_u - \Phi^{-1}(\alpha/2) \times \sigma_u, \\ U_\alpha^{*+} = \mu_u + \Phi^{-1}(\alpha/2) \times \sigma_u, \end{cases} \tag{31}$$

where

$$\mu_u = \frac{L+2}{3L+3}, \tag{32}$$

$$\sigma_u^2 = \frac{(L+2)(L^2+2L+3)}{18MNL(L+1)^2}, \tag{33}$$

where $L$ denotes the gray level. $MN$ is the total number of pixels in an image.

Keeping the key set unchanged, this section calculates the $NPCR$ and $UACI$ of two encrypted images $c_1$ and $c_2$ (please see Table 7) and compares the obtained values with the existing IEAs [38–43] (please see Tables 8 and 9). As we can observe, most of the $NPCR$ and $UACI$ scores have passed the test, proving that the IEA-DIJT has a strong ability against the differential attacks.

**Table 7.** $NPCR$ and $UACI$ (%) scores of the IEA-DIJT.

| Size | Image | $NPCR$ | $UACI$ | Size | Image | $NPCR$ | $UACI$ |
|---|---|---|---|---|---|---|---|
| | | $N^*_{0.05} \geq 99.5693$ | $U^{*-}_{0.05}$=33.2824 $U^{*+}_{0.05}$=33.6447 | | | $N^*_{0.05} \geq 99.5893$ | $U^{*-}_{0.05}$=33.3730 $U^{*+}_{0.05}$=33.5541 |
| 256 | Cameraman | 99.6096 | 33.5167 | 512 | Baboon | 99.6131 | 33.5181 |
| | Clock | 99.6118 | 33.4908 | | Boat | 99.6265 | 33.5226 |
| | House | 99.6107 | 33.5008 | | Couple | 99.6128 | 33.5269 |
| | Lena | 99.6098 | 33.4754 | | Elaine | 99.6208 | 33.5054 |
| | Moonface | 99.6126 | 33.4708 | | Lena | 99.6085 | 33.5172 |
| | Peppers | 99.6170 | 33.5672 | | Peppers | 99.6143 | 33.5188 |

**Table 8.** Comparison results of the *NPCR* (%).

| Size | Image | IEA-DIJT | Ref. [38] | Ref. [39] | Ref. [40] | Ref. [41] | Ref. [42] | Ref. [43] |
|---|---|---|---|---|---|---|---|---|
| 256 | Baboon | 99.6096 | 99.1073 | 99.6216 | 99.5911 | 99.6231 | 99.5941 | 99.6307 |
| | Lena | 99.6098 | 98.9807 | 99.6346 | 99.6323 | 99.6154 | 99.6307 | 99.6506 |
| 512 | Baboon | 99.6131 | 99.4785 | 99.6268 | 99.6627 | 99.6299 | 99.5979 | 99.6147 |
| | Lena | 99.6085 | 99.5033 | 99.6436 | 99.6093 | 99.5845 | 99.6082 | 99.6059 |

**Table 9.** Comparison results of the *UACI* (%).

| Size | Image | IEA-DIJT | Ref. [38] | Ref. [39] | Ref. [40] | Ref. [41] | Ref. [42] | Ref. [43] |
|---|---|---|---|---|---|---|---|---|
| 256 | Baboon | 33.4804 | 33.7694 | 33.2025 | 33.5648 | 33.4941 | 33.1112 | 33.1598 |
| | Lena | 33.5172 | 33.0815 | 33.1984 | 33.2378 | 33.3377 | 33.0881 | 33.4858 |
| 512 | Baboon | 33.5181 | 33.7609 | 33.1173 | 33.4588 | 33.4316 | 33.0019 | 33.4921 |
| | Lena | 33.5172 | 33.2334 | 33.6028 | 33.4873 | 33.5262 | 33.0228 | 33.4209 |

### 4.8. Known-Plaintext and Chosen-Plaintext Attacks

The known-plaintext and chosen-plaintext attacks are commonly used security attacks to measure the performance of an IEA. A weak IEA will offer a chance to disclose the relation between the plaintexts and ciphertexts and even deduce the secret keys by choosing arbitrary plaintexts. To avoid this issue, part of key set of the IEA-DIJT is the hash values. It can generate a totally different cipher image when the IEA-DIJT is applied to the same plain image. As shown in Figure 8, the plain image "Lena" is encrypted twice with two different selected hash values and keeps the other keys unchanged. Figure 8a,b are the obtained cipher images, respectively. Figure 8c depicts the difference in each pixel between the two cipher images in Figure 8a,b. Figure 8d,e are the corresponding histograms. Obviously, Figure 8d,e are different. This means that the IEA-DIJT is able to resist the chosen plaintext attacks. Moreover, this section also tests special plain images, such as pure black or white images, with the size $256 \times 256$. Figure 9 depicts the obtained cipher images and their corresponding histograms. As can be observed, the IEA-DIJT can withstand the known-plaintext and chosen-plaintext attacks.

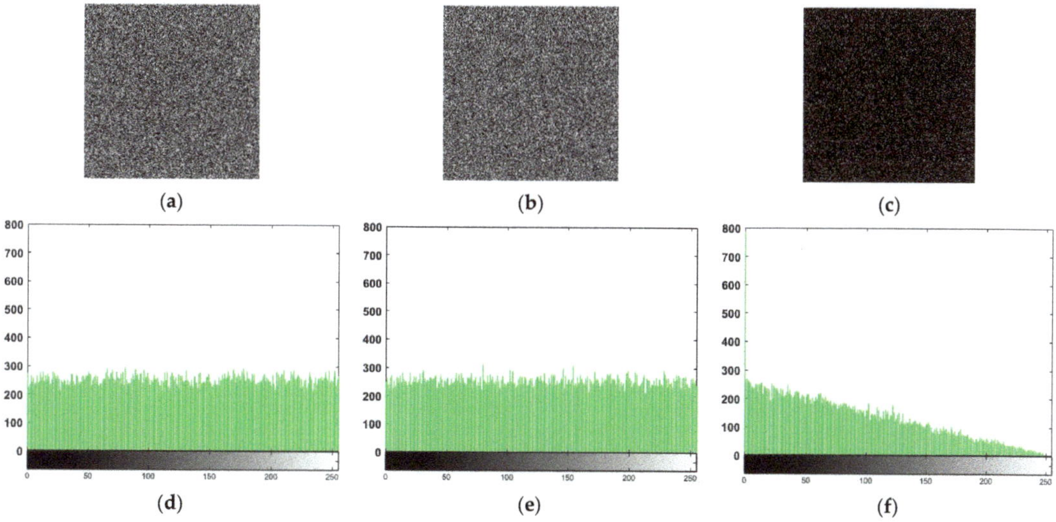


**Figure 8.** Known-plaintext and chosen-plaintext attacks: (**a**,**b**) are two cipher images with only a one-bit difference in the original image "Lena", and (**c**) the difference between (**a**,**b**,**d**–**f**) are the histograms of (**a**–**c**), respectively.

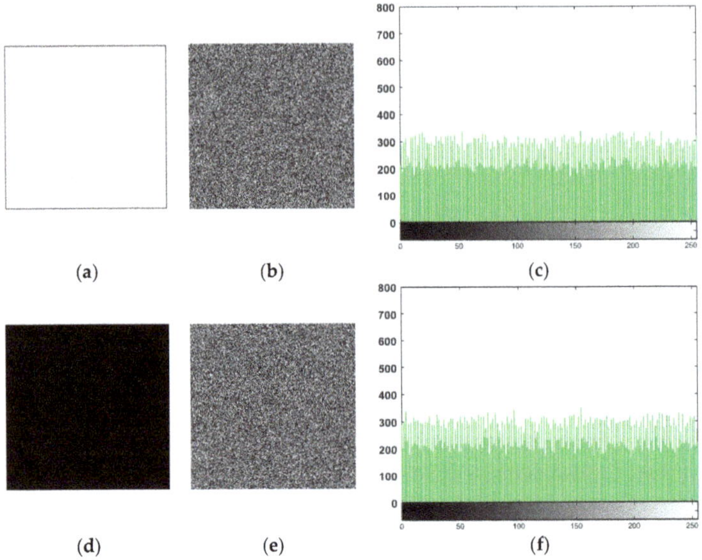


**Figure 9.** Experimental results of the pure white and black: (**a**) the pure white image, (**b**) the cipher image of (**a**,**c**) the histograms of (**b**,**d**) pure black image, and (**e**) the cipher-image of (**e**,**f**) the histograms of (**e**).

### *4.9. Robustness to Noise and Data Loss Attacks*

Robustness is a significant metric to evaluate the ability of anti-jamming for an IEA. This section tests it via noise and data loss attacks.

#### 4.9.1. Noise Attack

Noise is unavoidably added to the encrypted message in practical communication, which contains Gaussian noise and pepper and salt noise. An efficacious IEA must have the ability to resist these attacks. The salt–pepper noises of different intensities are added to the encrypted image. Figure 10 plots the test results. Figure 10a–c plot the cipher images with the intensity values of 0.01, 0.05, and 0.1, and the corresponding decoded images are depicted in Figure 10d–f. As we can observe, even if the noise intensity is 0.1, the original image can still be recognized. Hence, the IEA-DIJT has a good ability to resist noise attack.

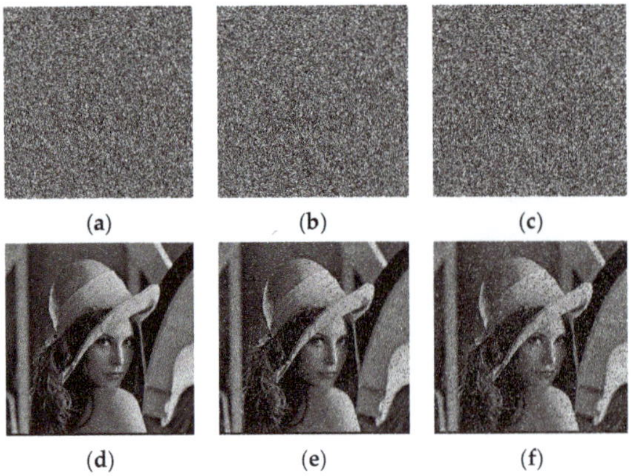


**Figure 10.** Noise attack. The cipher-image "Lena" with noise intensity (**a**) 0.01, (**b**) 0.05, and (**c**) 0.1, and the decoded images of (**a**–**f**).

#### 4.9.2. Data Loss Attack

As data can be lost during transmission, a data loss attack is necessary to assess how effectiveness the IEA in this paper is. The cipher images are cut off from a certain proportion of data to evaluate the robustness of the IEA, and the reconstruction is then carried out. Clearly, with the increasing proportion of messages, the quality of the decoded images decreases. Figure 11 plots the test results with the data loss proportions 6.08%, 25.32%, 40.20%, and 82.45%. As we can observe, the main information of the original image can be still recognized in the decoded images. Hence, the IEA-DIJT has a good ability to resist data loss attacks. Therefore, the IEA-DIJT is robust.

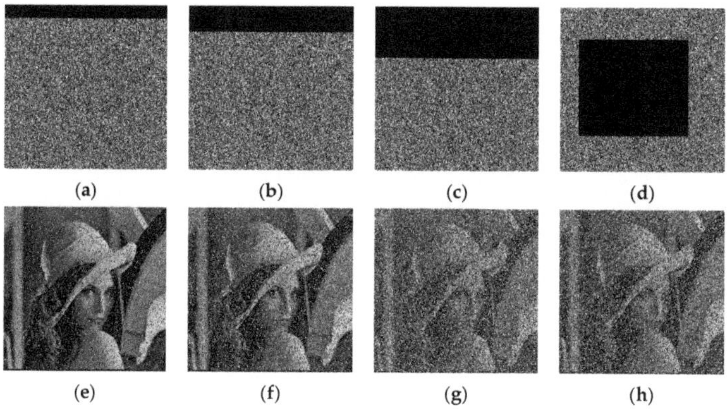

**Figure 11.** Data loss attack. The cipher-image "Lena" with data loss proportions (**a**) 6.08%, (**b**) 25.32%, (**c**) 40.20%, and (**d**) 82.45%, and the decoded image of (**a–h**).

### *4.10. Complexity and Time*

This section evaluates the time and complexity of the IEA-DIJT. Initially, the chaotic sequences are generated by iterating the logistic map. The time taken for this operation is $O(M \times N)$. In the confusion phase, it needs order $O\left(\frac{1}{4} \times M \times N\right)$ times. In the diffusion stage, the time complexity is $O(\frac{3}{4} \times M \times N)$, so the overall time complexity required to perform the IEA-DIJT is $O(M \times N)$. Moreover, Table 10 tabulates the encryption time of the IEA-DIJT, proving the acceptable speed of the IEA-DIJT.

**Table 10.** Encryption time (s) results.

| Size | Image | Time | Size | Image | Time |
|---|---|---|---|---|---|
| 256 | Cameraman | 1.005048 | 512 | Baboon | 5.208273 |
| | Clock | 0.974152 | | Boat | 5.439504 |
| | House | 0.98626 | | Couple | 5.362937 |
| | Lena | 1.006378 | | Elaine | 5.394718 |
| | Moonface | 0.976231 | | Lena | 5.522115 |
| | Peppers | 0.987105 | | Peppers | 5.487128 |

## 5. Conclusions

This work proposed a chaos-based IEA called the IEA-DIJT. Initially, the logistic map is used to generate chaotic sequences. Then, the plain image is decomposed into a phase map and an amplitude map using DFT, and different Joseph traversal operations are performed for scrambling. Moreover, the chaos-based diffusion is used to modify the pixel value to make the encryption effect deeper. Finally, an encrypted image is obtained. Through the simulation experiments and theoretical analyses, the security of the IEA-DIJT in resisting the common types of the attacks is verified. Under the premise of ensuring the security of the IEA, the efficiency of the IEA-DIJT is acceptable, which means there is room for improvement in the efficiency and security. For example, for the chaotic sequence, the IEA-DIJT uses the commonly used logistic map, which can further expand its chaotic region and obtain sequences with better chaotic characteristics. For the DFT-based confusion process, improving it to enhance the efficiency is another worthwhile research direction. Obviously, the improved Joseph traversal and its combination with DFT effectively enhances the security of the IEA-DIJT, increasing the application direction of the classic mathematical Joseph problem and expanding the research tools for IEAs, providing a direction for future research work.

**Author Contributions:** Methodology, M.W.; Validation, M.W.; Formal analysis, M.W., X.Y. and L.T.; Resources, M.W., X.F. and X.Y.; Writing—original draft, M.W.; Writing—review & editing, M.W. and X.F.; Visualization, X.Y.; Supervision, L.T.; Funding acquisition, X.F. and L.T. All authors have read and agreed to the published version of the manuscript.

**Funding:** This work was supported by the National Natural Science Foundation of China (Nos: 62176037 and 61701070), the Fundamental Research Funds for the Central Universities (Nos: 3132023252), and the China Postdoctoral Science Foundation (No: 2020M680933).

**Data Availability Statement:** No new data were created or analyzed in this study. Data sharing is not applicable to this article.

**Conflicts of Interest:** The authors declare that they have no conflicts of interest.

## References

1. Daoui, A.; Yamni, M.; Chelloug, S.A.; Wani, M.A.; El-Latif, A.A.A. Efficient Image Encryption Scheme Using Novel 1D Multiparametric Dynamical Tent Map and Parallel Computing. *Mathematics* **2023**, *11*, 1589. [CrossRef]
2. Wang, X.; Zhang, X.; Gao, M.; Tian, Y.; Wang, C.; Iu, H.H.-C. A Color Image Encryption Algorithm Based on Hash Table, Hilbert Curve and Hyper-Chaotic Synchronization. *Mathematics* **2023**, *11*, 567. [CrossRef]
3. Qiu, H.; Zhang, X.; Yue, H.; Liu, J. A Novel Eighth-Order Hyperchaotic System and Its Application in Image Encryption. *Mathematics* **2023**, *11*, 4099. [CrossRef]
4. Li, M.; Fang, X.; Ernest, A. A Color Image Encryption Method Based on Dynamic Selection Chaotic System and Singular Value Decomposition. *Mathematics* **2023**, *11*, 3274. [CrossRef]
5. Li, S.-Y.; Gai, Y.; Shih, K.-C.; Chen, C.-S. An Efficient Image Encryption Algorithm Based on Innovative DES Structure and Hyperchaotic Keys. *IEEE Trans. Circuits Syst. I Regul. Pap.* **2023**, *70*, 4103–4111. [CrossRef]
6. Sahin, M.E. Memristive chaotic system-based hybrid image encryption application with AES and RSA algorithms. *Phys. Scr.* **2023**, *98*, 075216. [CrossRef]
7. Ding, Y.; Wang, Z.; Qin, Z.; Zhou, E.; Zhu, G.; Qin, Z.; Choo, K.-K.R. Backdoor Attack on Deep Learning-Based Medical Image Encryption and Decryption Network. *IEEE Trans. Inf. Forensics Secur.* **2024**, *19*, 280–292. [CrossRef]
8. Ding, C.; Xue, R. Signal-sensing dynamic S-box image encryption with 2D Griewank–sin map. *Nonlinear Dyn.* **2023**, *111*, 22595–22620. [CrossRef]
9. Chen, W.-H.; Zhou, X.-F.; Li, M.-J.; Hu, M. Image encryption algorithm based on optical chaos and elliptic curve. *Eur. Phys. J. D* **2023**, *77*, 197. [CrossRef]
10. Wang, M.; Wang, X.; Wang, C.; Zhou, S.; Xia, Z.; Li, Q. Color image encryption based on 2D enhanced hyperchaotic logistic-sine map and two-way Josephus traversing. *Digit. Signal Process.* **2023**, *132*, 103818. [CrossRef]
11. Lone, P.N.; Mir, U.H.; Gaffar, A. Hyperchaotic image encryption using DNA coding and discrete cosine transform. *J. Frankl. Inst.* **2023**, *360*, 13318–13338. [CrossRef]
12. Wang, Y.; Chen, J.; Wang, J. Visually meaningful image encryption based on 2D compressive sensing and dynamic embedding. *J. Inf. Secur. Appl.* **2023**, *78*, 103613. [CrossRef]
13. Mohamed, N.A.E.-S.; El-Sayed, H.; Youssif, A. Mixed Multi-Chaos Quantum Image Encryption Scheme Based on Quantum Cellular Automata (QCA). *Fractal Fract.* **2023**, *7*, 734. [CrossRef]
14. Chen, H.; Liu, Z.; Zhu, L.; Tanougast, C.; Blondel, W. Asymmetric color cryptosystem using chaotic Ushiki map and equal modulus decomposition in fractional Fourier transform domains. *Opt. Lasers Eng.* **2019**, *112*, 7–15. [CrossRef]
15. Wang, M.; Wang, X.; Zhao, T.; Zhang, C.; Xia, Z.; Yao, N. Spatiotemporal chaos in improved cross coupled map lattice and its application in a bit-level image encryption scheme. *Inf. Sci.* **2020**, *544*, 1–24. [CrossRef]
16. Wang, X.-Y.; Li, Z.-M. A color image encryption algorithm based on Hopfield chaotic neural network. *Opt. Lasers Eng.* **2018**, *115*, 107–118. [CrossRef]
17. Gayathri, J.; Subashini, S. An efficient spatiotemporal chaotic image cipher with an improved scrambling algorithm driven by dynamic diffusion phase. *Inf. Sci.* **2019**, *489*, 227–254. [CrossRef]
18. Wang, X.; Gao, S. Image encryption algorithm for synchronously updating Boolean networks based on matrix semi-tensor product theory. *Inf. Sci.* **2020**, *507*, 16–36. [CrossRef]

19. Wang, X.; Feng, L.; Zhao, H. Fast image encryption algorithm based on parallel computing system. *Inf. Sci.* **2019**, *486*, 340–358. [CrossRef]
20. Jamal, S.S.; Hazzazi, M.M.; Khan, M.F.; Bassfar, Z.; Aljaedi, A.; Islam, Z.U. Region of interest-based medical image encryption technique based on chaotic S-boxes. *Expert Syst. Appl.* **2023**, *238*, 122030. [CrossRef]
21. Zhou, S.; Wang, X.; Wang, Z.; Zhang, C. A novel method based on the pseudo-orbits to calculate the largest Lyapunov exponent from chaotic equations. *Chaos Interdiscip. J. Nonlinear Sci.* **2019**, *29*, 033125. [CrossRef] [PubMed]
22. Teng, L.; Wang, X.; Xian, Y. Image encryption algorithm based on a 2D-CLSS hyperchaotic map using simultaneous permutation and diffusion. *Inf. Sci.* **2022**, *605*, 71–85. [CrossRef]
23. Wang, X.; Liu, P. A New Full Chaos Coupled Mapping Lattice and Its Application in Privacy Image Encryption. *IEEE Trans. Circuits Syst. I Regul. Pap.* **2021**, *69*, 1291–1301. [CrossRef]
24. Wang, M.; Wang, X.; Wang, C.; Xia, Z.; Zhao, H.; Gao, S.; Zhou, S.; Yao, N. Spatiotemporal chaos in cross coupled map lattice with dynamic coupling coefficient and its application in bit-level color image encryption. *Chaos Solitons Fractals* **2020**, *139*, 110028. [CrossRef]
25. Zhu, S.; Deng, X.; Zhang, W.; Zhu, C. Image Encryption Scheme Based on Newly Designed Chaotic Map and Parallel DNA Coding. *Mathematics* **2023**, *11*, 231. [CrossRef]
26. Zhong, H.; Li, G.; Xu, X.; Song, X. Image Encryption Algorithm Based on a Novel Wide-Range Discrete Hyperchaotic Map. *Mathematics* **2022**, *10*, 2583. [CrossRef]
27. Shao, Z.; Wang, X.; Tang, Y.; Shang, Y. Trinion discrete cosine transform with application to color image encryption. *Multimed. Tools Appl.* **2023**, *82*, 14633–14659. [CrossRef]
28. Feng, W.; Zhao, X.; Zhang, J.; Qin, Z.; Zhang, J.; He, Y. Image Encryption Algorithm Based on Plane-Level Image Filtering and Discrete Logarithmic Transform. *Mathematics* **2022**, *10*, 2751. [CrossRef]
29. Kaur, G.; Agarwal, R.; Patidar, V. Double Image Encryption Based on 2D Discrete Fractional Fourier Transform and Piecewise Nonlinear Chaotic Map. *Adv. Inform. Comput. Res.* **2019**, *955*, 519–530. [CrossRef]
30. Qiu, T.; Dai, W.-H.; Chen, S.-H.; Zhou, H.; Gong, L.-H. Double-image encryption algorithm based on discrete fractional angular transform and fractional Fourier transform. *Opt. Appl.* **2022**, *52*, 669–684. [CrossRef]
31. Shao, Z.; Tang, Y.; Liang, M.; Shang, Y.; Wang, F.; Wang, Y. Double image encryption based on symmetry of 2D-DFT and equal modulus decomposition. *Multimed. Tools Appl.* **2021**, *80*, 8973–8998. [CrossRef]
32. Tong, L.-J.; Zhou, N.-R.; Huang, Z.-J.; Xie, X.-W.; Liang, Y.-R. Nonlinear Multi-Image Encryption Scheme with the Reality-Preserving Discrete Fractional Angular Transform and DNA Sequences. *Secur. Commun. Netw.* **2021**, *2021*, 6650515. [CrossRef]
33. Tang, Y.; Shao, Z.; Zhao, X.; Shang, Y. Robust multiple color images encryption using discrete Fourier transforms and chaotic map. *Signal Process. Image Commun.* **2021**, *93*, 116168. [CrossRef]
34. Wang, M.-M.; Zhu, C.-N.; Li, L.; Xu, M.-T. Double-image compression and encryption scheme based on chaotic system and real-preserving multi-order discrete fractional Fourier transform. *J. Mod. Opt.* **2022**, *69*, 917–930. [CrossRef]
35. Xiao, M.; Tan, R.; Ye, H.; Gong, L.; Zhu, Z. Double-Color-Image Compression-Encryption Algorithm Based on Quaternion Multiple Parameter DFrAT and Feature Fusion with Preferable Restoration Quality. *Entropy* **2022**, *24*, 941. [CrossRef]
36. Wang, X.Y.; Yang, L.; Liu, R.; Kadir, A. A chaotic image encryption algorithm based on perceptron model. *Nonlinear Dyn.* **2010**, *62*, 615–621. [CrossRef]
37. Chai, X.; Chen, Y.; Broyde, L. A novel chaos-based image encryption algorithm using DNA sequence operations. *Opt. Lasers Eng.* **2017**, *88*, 197–213. [CrossRef]
38. Ye, G.; Huang, X. A secure image encryption algorithm based on chaotic maps and SHA-3. *Secur. Commun. Netw.* **2016**, *9*, 2015–2023. [CrossRef]
39. Zhang, W.; Wang, S.; Han, W.; Yu, H.; Zhu, Z. An Image Encryption Algorithm Based on Random Hamiltonian Path. *Entropy* **2020**, *22*, 73. [CrossRef]
40. Zheng, J.; Luo, Z.; Zeng, Q. An efficient image encryption algorithm based on multi chaotic system and random DAN coding. *Multimed. Tools Appl.* **2020**, *79*, 29901–29921. [CrossRef]
41. Erkan, U.; Toktas, A.; Toktas, F.; Alenezi, F. 2D e$\pi$-map for image encryption. *Inf. Sci.* **2022**, *589*, 770–789. [CrossRef]
42. Li, L.; Luo, Y.; Qiu, S.; Ouyang, X.; Cao, L.; Tang, S. Image encryption using chaotic map and cellular automata. *Multimed. Tools Appl.* **2022**, *81*, 40755–40773. [CrossRef]
43. Agarwal, V.; Kumar, D. Secure chaotic image encryption method using random graph traversal and three step diffusion. *Multimed. Tools Appl.* **2023**, 1–26. [CrossRef]
44. Xu, L.; Li, Z.; Li, J.; Hua, W. A novel bit-level image encryption algorithm based on chaotic maps. *Opt. Lasers Eng.* **2016**, *78*, 17–25. [CrossRef]

mathematics

MDPI

Article

# A Novel Dynamic S-Box Generation Scheme Based on Quantum Random Walks Controlled by a Hyper-Chaotic Map

Lijun Zhang [1], Caochuan Ma [2,3], Yuxiang Zhao [1] and Wenbo Zhao [1,*]

[1] School of Electronic Information and Electrical Engineering, Tianshui Normal University, Tianshui 741000, China; zhanglj_81@tsnu.edu.cn (L.Z.); zhaoyx@tsnu.edu.cn (Y.Z.)
[2] School of Mathematics and Statistics, Tianshui Normal University, Tianshui 741000, China; ccma@tsnu.edu.cn
[3] College of Mathematics and Physics, Wenzhou University, Wenzhou 325035, China
* Correspondence: zhaowenbo211211@163.com

**Abstract:** For many years, chaotic maps have been widely used in the design of various algorithms in cryptographic systems. In this paper, a new model (compound chaotic system) of quantum random walks controlled by a hyper-chaotic map is constructed and a novel scheme for constructing a dynamic S-Box based on the new model is proposed. Through aperiodic evaluation and statistical complexity measurement, it is shown that the compound chaotic system has features such as complex structure and stronger randomness than classical chaotic systems. Based on the chaotic sequence generated by the composite system, we design a dynamic S-Box generation mechanism. The mechanism can quickly generate high-security S-Boxes. Then, an example of randomly generating S-Boxes is given alongside an analytical evaluation of S-Box standard performance criteria such as bijection, boomerang uniformity, bit independence, nonlinearity, linear approximate probability, strict avalanche effect, differential uniformity, the and generalized majority logic criterion. The evaluation results confirm that the performance of the S-Box is excellent. Thus, the proposed dynamic S-Box construction technique can be used to generate cryptographically strong substitution-boxes in practical information security systems.

**Keywords:** compound chaotic system; quantum random walk; S-Box generation mechanism; standard performance criteria; information security

**MSC:** 94A60; 39A33; 68M25

## 1. Introduction

In recent years, with the rapid development of network communication technology, network security has become an increasingly important research field. Accordingly, information security technology, which is used for data transmission security, has attracted widespread attention. Scholars have proposed various secure communication mechanisms, including a large number of data encryption algorithms. The user data is converted to cipher text by encryption technology before data transmission. This means that the input data block (plaintext) is converted into insignificant output block (ciphertext) and the ciphertext is useless to an attacker. In symmetric cryptographic systems, block encryption algorithms are widely used as they are easy to implement and provide the required cipher strength [1]. In block cipher systems, as the only nonlinear component, the substitution box (abbreviated as S-Box) is an important nonlinear component in many block cipher systems such as the Data Encryption Standard (DES) and the Advanced Encryption Standard (AES). In these block cipher systems, that use of an S-Box provides a complex relationship between plaintext and ciphertext and plays a scrambling effect (chaotic effect). The security of the S-Box determines the security strength of the whole block cipher system. Thus, it is meaningful to study how to quickly generate high security S-Boxes and this has attracted more and more attention. Usually, a designer must evaluate the strength of respective S-Boxes in block cipher system. For S-Box security, there are some typical evaluation criteria, such as bijection, bit independence, nonlinearity, linear approximate probability, and strict avalanche effect. Some new attack technologies against S-Boxes have emerged with the development of new theories and technologies, such as the boomerang attack [2,3]. The authors of [4] summarize the many security attributes (including some new tests) of S-Boxes. A strong S-Box can resist cryptanalysis-based attacks if it can pass standard benchmarks tests. Thus, determining how to build a strong S-Box has received intensive attention.

However, the S-Box that is used in block cipher systems such as AES is a fixed component. That is, a block cipher has a static S-Box and employs the unchanged S-Box in each round. A static S-Box allows a third party to check the characteristics of the S-Box, which will lead to the attacker discovering its defects and eventually having the opportunity to cryptanalyze the ciphertext generated by the block cipher [5]. If high-quality S-Boxes (based on a user key) can be dynamically generated and applied, the encryption strength of the cryptosystem would be more able to resist various cryptanalytical attacks. A dynamically generated S-Box mechanism can be used to improve classic block cipher algorithms and design new block cipher algorithms. In recent years, researchers have explored elements of the concept of dynamic S-Box design, such as S-Box generation efficiency and randomness. In an actual cryptographic system, the S-Box is not only required to be safe but also the time and space efficiency of the algorithm for generating the S-Box must be high. Researchers have sought alternative S-Box design approaches. There are some approaches that are frequently applied to the construction of S-Boxes, such as the analytical approach [6], algebraic techniques [7], triangle groups [8], and chaotic maps [9–11]. Since chaos-based cryptography is considered highly secure, robust, and computationally powerful, and is complex enough to make cryptanalysis harder, chaos-based design emerges as the best method from the above approaches [11].

Chaotic dynamical systems have special nonlinear dynamic characteristics such as the sensitivity to initial conditions, unpredictability of long-term behavior, intrinsic randomness, and the ergodic nature of the processes. These characteristics of chaotic systems are equivalent to diffusion and confusion in traditional cryptographic systems [12]. Chaos cryptography is an important application field of chaos theory. For cryptosystems, there have many chaos-based algorithms proposed in recent years such as chaos-based image encryption, which was utilized in studies such as "Color image encryption using orthogonal Latin squares and a new 2D chaotic system" [13]; "Exploiting semi-tensor product compressed sensing and hybrid cloud for secure medical image transmission" [14]; and "A parallel image encryption algorithm using intrabitplane scrambling" [15]. For the development of modern cryptographic systems, chaotic maps have emerged as an alternative chaotic typology, and as part of this work, researchers have used chaos in the design of S-Boxes. An S-Box is essentially a random placement of $n \times n$ numbers from 0 to $n \times n - 1$. Chaotic systems with complex dynamic behaviors are able to generate high-quality pseudo random numbers. These pseudo random numbers can be used to design S-Box generation schemes. In the past few years, chaotic systems have been widely used in the design of S-Boxes due to the useful properties of chaotic maps, and analysis shows that an S-Box based on chaos has strong resistance to different attacks and higher space–time efficiency [16]. Scholars such as Ye and Zhimao [10], Masood et al. [11], Zheng and Zeng [17], Bin Faheem et al. [18], Zhu et al. [19], Lu et al. [20] have proposed many S-Box-based cryptographic systems that include effective S-Boxes that possess favorable characteristics such as bit independence, nonlinearity, etc.

However, problems such as unpredictability and short cycle may occur when implementing chaotic systems in digital devices (digital chaos) [21,22]. This will result in successful attacks on a cryptographic systems based on the original digital chaotic maps. Many cryptographic systems have serious security problems due to improper selection and application of chaotic maps [23]. Several researchers are working to rid chaotic maps of their weak points and further improve their chaotic behavior and randomness. Modified chaotic maps can be used to enhance the security characteristics of various encryption schemes. It is necessary to combat the degradation of chaotic dynamic performance and further optimize chaos. Lu et al. [20], proposed $8 \times 8$ S-Boxes based on a improved compound chaotic map (tent-logistic system) that has a large key space, better chaotic behavior, and chaos at a large scale. In another study [24], a novel method of constructing S-Boxes is proposed that can generate strong S-Boxes through the use of TSS maps and algebraic Mobius transformations.

Inspired by the above discussion, we aim to search for a compound chaotic map with complex dynamic behavior and develop a new S-Box generation algorithm that can improve the linear probability (LP) and other characteristics of S-Boxes and enhance the robustness of cryptographic systems to linear analysis attacks, differential cryptanalysis attacks and boomerang attacks. The innovations of this work include the following:

1. A new compound chaotic system based on a two-dimensional hyper-chaotic map and quantum random walks is proposed, which has a larger key space and better chaotic performance and is more suitable for practical applications;
2. A simple and efficient dynamic S-Box generation algorithm is proposed;
3. A comprehensive and detailed security analysis of the generated S-Box is made to evaluate it against cryptographic landscapes. The analytical results demonstrate that the S-Box can well meet multiple cryptographic criteria.

The rest of this paper is organized as follows. Section 2 proposes a new compound chaotic system based on a two-dimensional hyper-chaotic map and quantum random walks. Section 3 describes an efficient dynamic S-Box generation algorithm based on the compound chaotic system. In Section 4, we show safety analyses of the proposed S-Box generated by the algorithm and make a comparison with some recent S-Boxes from the literature.

## 2. A Compound Chaotic System Based on a Two-Dimensional Hyper-Chaotic Map and Quantum Random Walks

In this section, we select a known two-dimensional hyper-chaotic map whose output is used as the control source of quantum random walks on the ring graph. Thus, a compound chaotic system with a complex structure that is sensitive to initial conditions and can generate uniform sequences is obtained.

### 2.1. Two-Dimensional Hyper-Chaotic Map Controlling Quantum Random Walks

A hyper-chaotic system has at least two expanding directions due to a minimum of two positive LEs [25]. The expression of a two-dimensional hyper-chaotic map is as follows [26] (p. 49):

$$\begin{aligned} u_{n+1} &= k_{11} + k_{12}u_n + k_{13}u_n^2 + k_{14}v_n + k_{15}v_n^2 + k_{16}u_nv_n, \\ v_{n+1} &= k_{21} + k_{22}u_n + k_{23}u_n^2 + k_{24}v_n + k_{25}v_n^2 + k_{26}u_nv_n. \end{aligned} \tag{1}$$

We simplify Expression (1) to make most of the parameter coefficients zero, i.e., $k_{11} = k_{12} = k_{13} = k_{16} = k_{21} = k_{23} = k_{25} = k_{26} = 0$. Finally, we obtain a simple two-dimensional hyper-chaotic map:

$$\begin{aligned} u_{n+1} &= k_{14}v_n + k_{15}v_n^2, \\ v_{n+1} &= k_{22}u_n + k_{24}v_n, \end{aligned} \tag{2}$$

where $k_{14} = 1.55$, $k_{15} = -1.55$, $k_{24} = 0.1$ and $k_{22}$ is the control parameter. A fixed point of System (2) is $O = (0, 0)$, and the norm can be defined as follows:

$$\|(u, v)\| = \sqrt{u^2 + v^2}.$$

Let $h = (h_1, h_2)$, where

$$h_1(u, v) = -1.55v^2 + 1.55v, h_2(u, v) = k_{22}u + 0.1v.$$

$h$ is continuously differentiable on $R^2$, and the Jacobian matrix is

$$Dh(u, v) = \begin{pmatrix} 0 & -3.1v + 1.55 \\ k_{22} & 0.1 \end{pmatrix}.$$

For the fixed point $O = (0, 0)^T$, as $k_{22} < -1.003558730$ and $k_{22} > 1.003558730$, Map (2) may be chaotic if the absolute value of the eigenvalues of the matrix $Dh(O)$ is greater than 1. The randomness of the hyper-chaotic Map (2) will be investigated by trajectory diagram, Lyapunov exponent, and bifurcation diagrams with control parameters .

2.1.1. Trajectory Diagram

For a sequence generated by a dynamical system, in different stage from the initial state, values can be presented by using the trajectory diagram. If values occupy the entire phase space and occupy this space by choosing the correct initial conditions for a dynamical system, it can be concluded that the system appears chaotic if the correct initial conditions are chosen. The trajectory of Map (2) with the initial $(u_0, v_0) = (0.1, 0.1)$ is shown in Figure 1. In Figure 1, the effect of different $k_{22}$ control parameter values is assessed according to the values generated by the hyper-chaotic map. As shown in Figure 1, the occupancy of the phase space in the trajectory increases as $k_{22}$ grows increasingly negative.

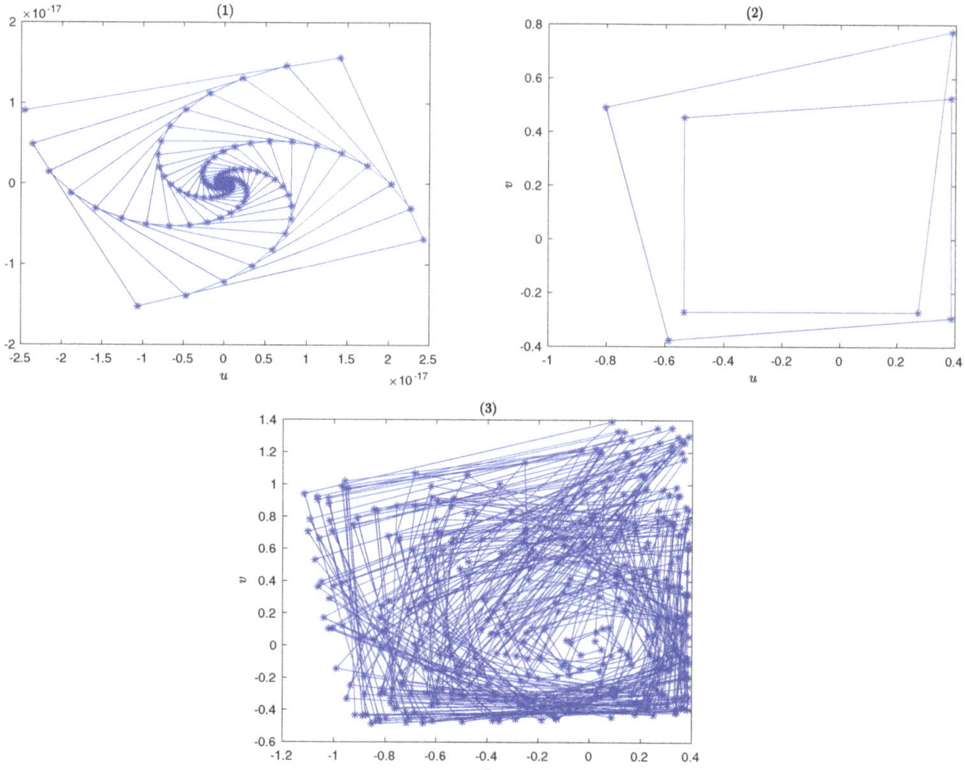


**Figure 1.** Trajectory diagram of map (2) with different values of the $k_{22}$ control parameter ($k_{22} = 0.1$ in subfigure (1), $k_{22} = -0.9$ in subfigure (2), and $k_{22} = -1.16$ in subfigure (3)).

2.1.2. Lyapunov Exponents

The Lyapunov exponents (LEs) are one of the most important criteria for checking the randomness of a map and it sensitivity to the initial conditions. If the motion is chaotic, the Lyapunov exponent will be positive. Let $K_{area} = [-1.165, \ -1.105] - [-1.137, \ -1.135] - [-1.121, \ -1.117] - [-1.091, \ -1.088]$. Figure 2(1–3) show the Lyapunov exponents of the hyper-chaotic map, in which sub-figures (2) is refined using the subdivision method by selecting sub intervals. From Figure 2(1,2), it can be seen that the Lyapunov exponent is positive as $k_{22} \in K_{area}$.

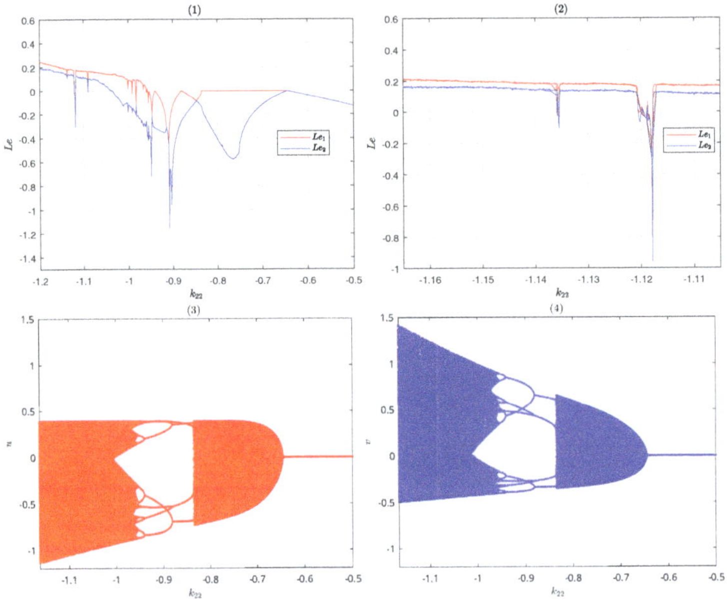


**Figure 2.** Lyapunov exponent and Bifurcation diagram of Sys (2) with different values of the $k_{22}$ control parameter (Lyapunov exponents are shown in subfigure (1,2) and Bifurcation diagrams are shown in subfigure (3,4)).

2.1.3. Bifurcation Diagram

As a visual tool, a bifurcation diagram displays the process of period-doubling through the change of nonlinear dynamical system parameters. A chaotic system can achieve the best chaotic state by adjustment of the control parameters. That is, filling the phase space can be considered as the optimal value of the control parameters. Bifurcation diagrams of the system with changing $k_{22}$ parameter are shown in Figure 2(3,4). It can be seen from Figure 2(3,4) that the filling space of the variable becomes greatest as $k_{22} \in K_{area}$.

*2.2. A Scheme for a Sequence Generated by Quantum Random Walks on a Cycle Graph*

$G$ is a $n$-cycle graph and the degree of every node is 2. The quantum random walks on the cycle graph $G$ contains two sub systems: Walker and Coin. Walker can be denoted as a position in an n-dimensional Hilbert space $H_p$ and the basis state is $\{|i\rangle, i \in \{0, 1, 2, \cdots, n\}\}$ that span $H_p$. Any position of Walker can be represented as $\sum_i k_i|i\rangle$ , and $\sum_i |k_i|^2 = 1$. Coin is a sub quantum system in two-dimensional dimensional Hilbert space $H_c$, and the canonical basis state is $\{|0\rangle, |1\rangle\}$. The state of Coin can be expressed as $a|0\rangle + b|1\rangle$, where $|a^2| + |b^2| = 1$. The joint state of Walker and Coin resides in $H_t = H_p \otimes H_c$, where $p$ and $c$ correspond to Walker and Coin, respectively.

The coin operator $\hat{C}$ has been extensively employed [27]:

$$\hat{C} = \cos\theta|0\rangle\langle 0| + \sin\theta|0\rangle\langle 1| + \sin\theta|1\rangle\langle 0| - \cos\theta|1\rangle\langle 1|.$$

The Walker shift operator has the following form:

$$\hat{S} = \sum_i (|i + forward(mod\ n)\rangle\langle i| \otimes |0\rangle\langle 0| + |i - back(mod\ n)\rangle\langle i| \otimes |1\rangle\langle 1|),$$

where *forward* means that Walker takes steps to the right when the accompanying coin state is $|0\rangle$, and *back* means steps to left when the coin state is $|1\rangle$. Each operator on the total Hilbert space can be expressed as

$$\hat{U} = \hat{S}(\hat{C} \otimes I). \quad (3)$$

The initial state of the total system is $|\varphi(0)\rangle$, after $t$ steps, the state is

$$|\varphi(t)\rangle = \hat{U}^t|\varphi(0)\rangle = \left(\hat{S}(\hat{C} \otimes I)\right)^t|\varphi(0)\rangle. \quad (4)$$

The probability of Walker at position $|v\rangle$ after $t$ steps is

$$P_t(v|\varphi(0)) = |\langle (v,1)|\varphi(t)\rangle|^2 + |\langle (v,0)|\varphi(t)\rangle|^2, \quad (5)$$

and the resulting probability distribution is as follows:

$$\mathbf{P}_t = [P_t(v_1|\varphi(0)),\ P_t(v_2|\varphi(0)), \cdots P_t(v_n|\varphi(0))]. \quad (6)$$

The limiting distribution of Walker at position $|v\rangle$ is

$$\lim_{T\to\infty} \bar{P}_T(v|\varphi(0)) = \lim_{T\to\infty} \frac{1}{T} \sum_{t=0}^{T-1} P_t(v|\varphi(0)). \quad (7)$$

According to Theorem 3.6 and Theorem 4.1 in the study [28], Equation (7) is close to a uniform distribution as $n$ is odd. It can be seen from Equation (5) that there is a nonlinear map between initial state $|\varphi(0)\rangle = (i_0, c_0)$ and the resulting probability distribution $\mathbf{P}_t$ that is highly sensitive to the initial conditions [29]. According to Equations (4) and (5), a uniformly distributed sequence can be generated. Following this, the steps of a sequence generation scheme using quantum random walks on a ring graph are as shown in Algorithm 1.

---

**Algorithm 1** Random number generation algorithm based on ring graph

---

Input: $(\theta,\ forward,\ back,\ n,\ i_0,\ c_0, r)$
Output: $AllOutputSeq$
(1) Init: $|\varphi(0)\rangle = (i_0,\ c_0)$,
    $\hat{C} = \sin\theta|0\rangle\langle 1| - \cos\theta|1\rangle\langle 1| + \cos\theta|0\rangle\langle 0| + \sin\theta|1\rangle\langle 0|$
    $\hat{S} = \sum_i (|i + forward(mod\ n)\rangle\langle i| \otimes |0\rangle\langle 0| + |i - back(mod\ n)\rangle\langle i| \otimes |1\rangle\langle 1|)$
    $\Phi = |\varphi(n)\rangle = \hat{U}^n|\varphi(0)\rangle = \left(\hat{S}(\hat{C} \otimes I)\right)^n|\varphi(0)\rangle$,
    $T_{max} = n,\ AllOutputSeq = \varnothing;$
(2) $for\ T = 1 : T_{max}$
  $\Phi = \hat{U}^r\Phi = \left(\hat{S}(\hat{C} \otimes I)\right)^r\Phi$,
  $for\ j = 1 : n$
   $p_j = |\langle (j,0)|\Phi\rangle|^2 + |\langle (j,1)|\Phi\rangle|^2$
  $endfor$
  $OutputSeq = [p_1, p_2, \cdots, p_n]$,
  $AllOutputSeq+ = OutputSeq$,
 $end\ for.$
(3) $return\ AllOutputSeq$

---

### 2.3. A New Compound Chaotic System Based on Quantum Random Walks Controlled by a Hyper-Chaotic Map

In this section, a new compound chaotic system will be constructed by using quantum random walks and a two-dimensional hyper-chaotic map. Let $Q(\cdot)$ be Algorithm 1, which is essentially a pseudo random number generator. However, $Q(\cdot)$ can only generate $n \times n$ random numbers at a time. To generate new random numbers, there are two options. One is to continue to perform step (2) of Algorithm 1, and the other is to modify the relevant parameters except $n$, $c_0$, and $i_0$ before executing step (2) of Algorithm 1. Obviously, the second method is more flexible. If a two-dimensional system is used to adjust the corresponding parameters, the key space can be enlarged on the basis of increasing the complexity. Thus, the output $(u, v)$ of hyper-chaotic map (2) is used to construct the parameter $\theta$ of $Q(\cdot)$. Let the normalization function be:

$$norm(x) = \frac{x - x_{min}}{x_{max} - x_{min}},$$

and the parameters $(\theta, forward)$ are constructed as follows:

$$\begin{aligned} \theta &= (1 - norm(v)) \times norm(u) + norm(v) \times \theta, \\ forward &= back = mod(floor(Z \times (norm(v) + norm(v))), \end{aligned} \tag{8}$$

where $floor(x)$ is the integral function and Z is an integer, which is generally larger to improve the randomness. The new compound chaotic system is based on quantum random walks controlled by a hyper-chaotic map. As the control parameter $k_{22} \in [-1.165, \ -1.105]$, the non-periodicity and statistical complexity of the compound chaotic system will be evaluated.

#### 2.3.1. Degree of Non-Periodicity

The Scale index, which is based on the continuous wavelet transform (CWT), is presented by Benítez et al. [30], and we will use the tool to study non-periodicity in sequences generated by the proposed compound chaotic system. For chaotic sequences, wavelets are more suitable for studying non-periodicity due to their non-stationary nature [31]. We can assume that sequence $f$ is a continuous function and $f \in L^2(\mathbb{R})$, which is defined over a finite time interval $I = [a, \ b]$. In Benítez et al. [30], the CWT of $f$ at scale $s$ and time $u$ is defined as

$$Wf(u,s) := \langle f, \psi_{u,s} \rangle = \int_{-\infty}^{+\infty} f(t)\psi^*_{u,s}(t)dt,$$

and from there it is possible to obtain the frequency component of $f$ corresponding to time location $u$ and scale $s$. The $f$ scalogram, which is the energy of the CWT of $f$ at scale $s$, is defined as

$$\mathcal{S}(s) = \|Wf(u,s)\| = \left( \int_{-\infty}^{+\infty} |Wf(u,s)|^2 du \right)^{\frac{1}{2}}.$$

Let $J(s) = [c(s), \ d(s)] \subseteq I$ and it is the maximal subinterval in $I$. The inner scalogram of $f$ at a scale $s$ can be defined by

$$\mathcal{S}^{inner}(s) = \|Wf(u,s)\|_{J(s)} = \left( \int_{c(s)}^{d(s)} |Wf(u,s)|^2 du \right)^{\frac{1}{2}},$$

where $I$ must be big enough and $b - a \gg sl$ [32]. In order to compare the values of the inner scalogram at different scales, the inner scalogram should be normalized:

$$\overline{\mathcal{S}}^{inner}(s) = \frac{\mathcal{S}^{inner}(s)}{(d(s) - c(s))^{\frac{1}{2}}}.$$

In the scale interval $[s_0, s_1] \subseteq I$, the Scale index of $f$ is given by

$$i_{scale} = \frac{\mathcal{S}^{inner}(s_{min})}{\mathcal{S}^{inner}(s_{max})},$$

where $s_{max} \in [s_0, s_1]$ is the maximal scale such that $\mathcal{S}^{inner}(s_{max}) \geq \mathcal{S}^{inner}(s)$ for all $s \in [s_0, s_1]$, and $s_{min} \in [s_{max}, 2s_1]$ is the smallest scale such that $\mathcal{S}^{inner}(s_{min}) \leq \mathcal{S}^{inner}(s)$ for all $s \in [s_{max}, 2s_1]$ [32]. For highly non-periodic sequences, the Scale index will be close to 1; while, it will close to 0 for periodic sequences [30].

We set $s_0 = 1$ and $s_1 = 20$, and choose a Haar wavelet as the mother wavelet function for calculating the Scale index. Figure 3 shows the Scale index of the compound chaotic system, the logistic map, and the Henon map. From a comparison of Figure 3a–c, we can obtain that the Scale index of the compound system is higher than those of the other two chaotic maps. Thus, we can conclude that the sequence generated by the compound system is more non-periodic.

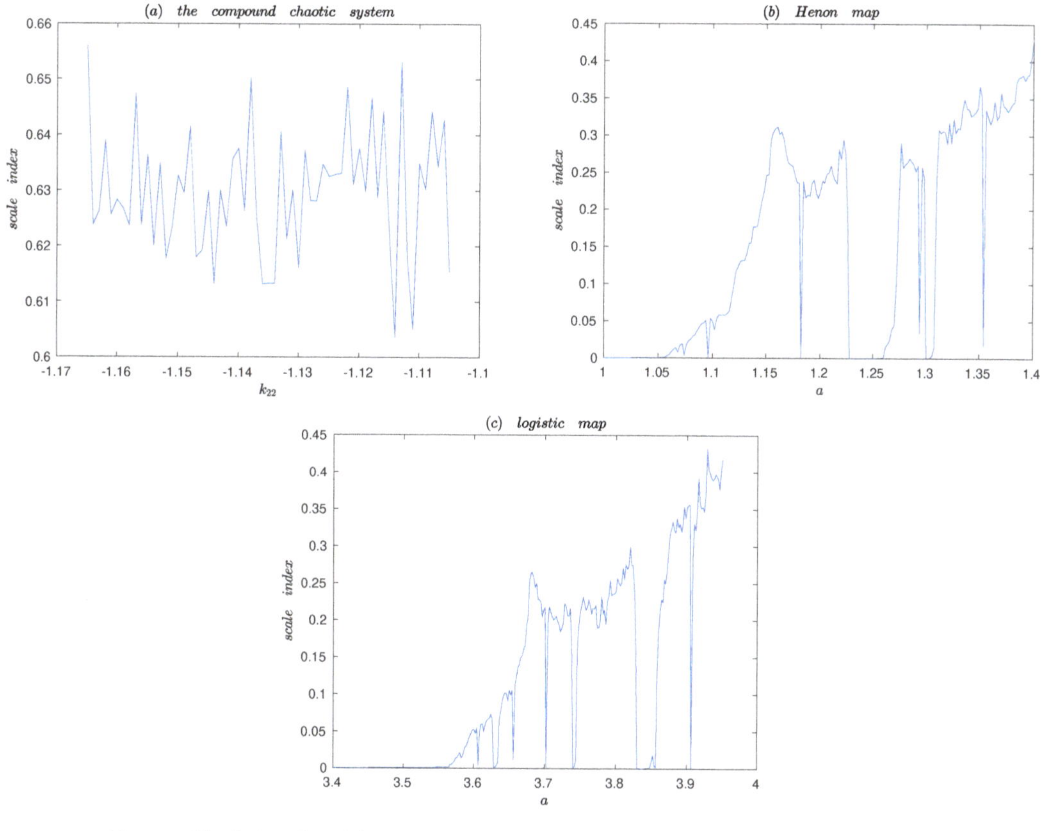


**Figure 3.** The Scale index of the Henon map, the logistic map, and the compound chaotic system.

### 2.3.2. Statistical Complexity Measures

The authors of [33], proposed statistical complexity measures (SCMs), which are used to quantify the degree of physical structure in a sequence. We use the the method of Larrondo et al. [34] to calculate the SCMs. The intensive SCM ($C_j[P]$) can be thought of as a quantity that represents the probability distribution $P$ that is associated with a sequence [35]. $C_j[P]$, introduced by Larrondo et al. [34], is defined as:

$$C_j[P] = H_S[P]Q_j \cdot [P, P_e], \tag{9}$$

where $Q_J$ is "disequilibrium" and $H_S[P]$ is the entropic measure (normalized entropy). $H_S[P]$ can be calculated by

$$H_S[P] = H[P]/H_{max}, \tag{10}$$

where $H$ is the Shannon entropy. The entropy of a byte sequence is given by the following:

$$H = \sum_{i=1}^{N} p(a_i) log_2 p^{-1}(a_i), \tag{11}$$

where $p(a_i)$ is the probability that $a_i$ occur in the sequence. For a byte sequence, the ideal value of entropy ($H_{max}$ ) is 8. $Q_J$ is an intensive quantity [35] and $Q_J$ is given by

$$Q_j[P, P_e] = Q_0 \cdot \{H[(P + P_e)/2] - H[P]/2 - H[P_e]/2\},$$

where $Q_0$ is a normalization coefficient, which can be calculated by

$$Q_0 = -2\left\{\left(\frac{N+1}{N}\right)\ln(N+1) - 2\ln(2N) + \ln N\right\}^{-1}.$$

If a PRNG can generate high-quality pseudo-random numbers, it can be expected that "no attractor" will be reconstructed. It is reasonable to obtain a homogeneity cloud of points with a strong performance that "fill" up a d-dimension space [34]. Consequently, the associated permutation probability distribution will be $P \simeq P_e$, so $H_S[P] \simeq 1$ and $C_J[P] \simeq 0$ and for periodic sequences will have $H_S[P] \simeq 0$ and $C_J[P] \simeq 0$ [34]. $H_S$ and $C_J$ as functions of the number of 8 bits are shown in Figure 4 based on the above calculation method. From Figure 4, it can be see that $C_j$ and $H_s$ tend to 0 and 1, respectively, when the number of words of the analyzed sequence increases. Thus, the randomness of the PRNG based on the proposed compound chaotic system is successfully verified by statistical complexity and the normalized Shannon entropy.

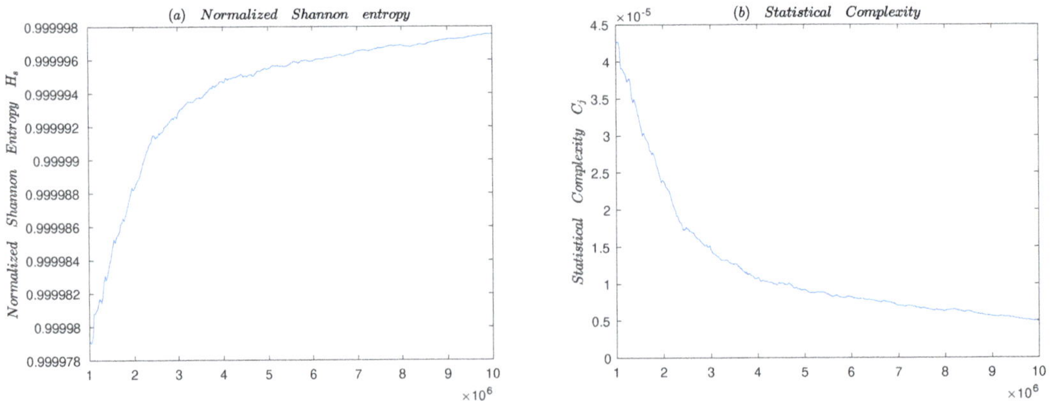


**Figure 4.** $H_s$ and $C_j$ for the proposed PRNG.

## 3. Dynamic S-Box Generation Algorithm Based on the Compound Chaotic System

### *3.1. Introduction of the S-Box*

In a symmetric block cipher system, the S-Box plays an important role as the only nonlinear component. An S-Box is essentially a nonlinear permutation function, which can be understood as an encrypted black box. Furthermore, it confuses the relationship between the ciphertext and plaintext. An S-Box can be abstracted as a vector Boolean function of $n$-bit input and $m$-bit output:

$$Sbox: \begin{array}{c} \mathbb{F}_2^n \rightarrow \mathbb{F}_2^m \\ \{0,1\}^n \rightarrow \{0,1\}^m \end{array}.$$

For most S-Boxes in cryptography, commonly the case of $n = m$ is used and an $n \times n$ S-Box is shown in Figure 5. We focused on the design algorithm for $8 \times 8$ S-Boxes, which are the most commonly used type in cryptography. An $8 \times 8$ S-Box is a set of integers $S_{set} = \{0, 1, 2, \cdots, 2^8 - 1\}$, which is unique for a matrix $Sb$: $Sb = \{Sb(i,j) \,|\, i = 1, 2, \ldots, 16;\ j = 1, 2, \ldots, 16\}$. The matrix $Sb$ can be obtained:

$$\begin{cases} i = floor(x/16),\ j = x \bmod 16 + 1, \\ y = Sbox(x) = Sb(i,j), \end{cases} \tag{12}$$

if the vector Boolean function corresponding to an $8 \times 8$ S-Box is $y = Sbox_{8\times 8}(x)$, where both x and y are elements in $S_{set}$. Conversely, it is known that an $8 \times 8$ S-Box corresponds to $Sbox_{8\times 8}^{-1}(y)$:

$$\begin{cases} y = Sb(i,j), \\ x = Sbox_{8\times 8}^{-1}(y) = 16(i-1) + j - 1. \end{cases} \tag{13}$$

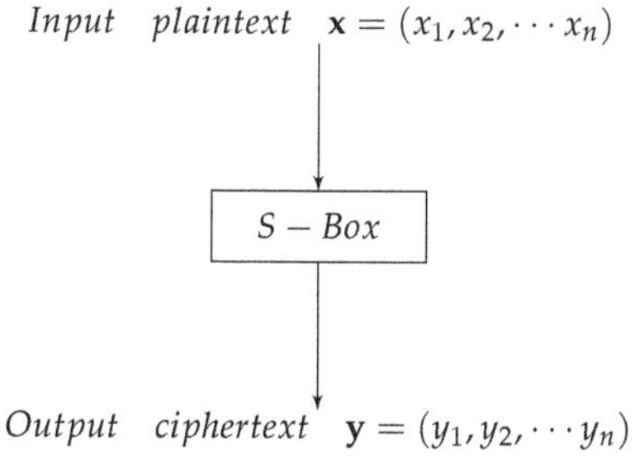


**Figure 5.** The function and basic principle of an $n \times n$ S-Box.

### 3.2. Pseudo-Random Number Generator (PRNG) Based on the Proposed Compound Chaotic System

In this section, a PRNG is designed based on the proposed compound chaotic system, followed by a NIST Statistical Test of the PRNG. The fixed-point algorithm [35] expressed by $P$-bit precision is adopted, and the specific steps of generating random number are as follows:

1. Initialization $(u_0, v_0, k_{22}, \theta,\ faward,\ back,\ n,\ i_0,\ c_0, r)$;
2. $Q_{Hy}^{1000}(u_0, v_0)$;
3. Let $Y_k = \sqrt{3}p_{3(k-1)} + \sqrt{5}\left(p_{3(k-1)+1}p_{3(k-1)+2}\right)$. Furthermore, the following formula is used to generate and output the random number $z_k$:

   $$z_n = \bmod\left(floor\left(\sqrt{2}\left(2^P - 1\right) \cdot 10^5 \cdot Y_k\right), 2^P\right),$$

   where $floor(\cdot)$ is an integral function.

#### NIST Statistical Test

We will use NIST SP800-22 [36] to estimate the randomness of sequences generated by the PRNG based on the compound system. In the NIST suite, there are 17 test items and every test item concentrates on one type of non-randomness that can exist in a sequence. There are two performance indicators, pass rate and $p$-value, to determine the random performance of the sequence. In our tests, the following settings were used: the number of sequences to be tested was 100, the sequence length was $10^6$ bits, and the significance level was $\alpha = 0.01$, which implies that if the sequence passed the test, it can be referred to as being random with a probability of 99%. For each statistical test item except for the random excursion test, the minimum pass rate is around 96 for sample capacity = 100 bit sequences. The results of NIST SP800–22 test for the PRNG are given in Table 1. The results of the sequence generated by proposed PRNG are all "Pass". Hence, the sequence generated by the PRNG be considered to have high randomness, and the PRNG can be used in information security systems.

**Table 1.** Results of the NIST SP800–22 for the PRNG.

| Test Name | *p*-Value | Pass Rate | Result |
|---|---|---|---|
| Frequency | 0.816537 | 99/100 | Pass |
| Block Frequency (m = 128) | 0.249284 | 98/100 | Pass |
| Cumulative Sums (Forward) | 0.289667 | 99/100 | Pass |
| Cumulative Sums (Reverse) | 0.319084 | 99/100 | Pass |
| Runs | 0.935716 | 98/100 | Pass |
| Longest Run of Ones | 0.419021 | 99/100 | Pass |
| Rank | 0.955835 | 100/100 | Pass |
| FFT | 0.202268 | 99/100 | Pass |
| Non-Overlapping Templates (m = 9, B = 000000001) | 0.964295 | 100/100 | Pass |
| Overlapping Templates (m = 9) | 0.595549 | 100/100 | Pass |
| Universal | 0.494392 | 99/100 | Pass |
| Approximate Entropy (m = 10) | 0.191687 | 98/100 | Pass |
| Random-Excursions (data1) | 0.848588 | 63/63 | Pass |
| Random-Excursions Variant Serial (data7) | 0.788728 | 63/63 | Pass |
| Serial Test 1 (m = 16) | 0.383827 | 99/100 | Pass |
| Serial Test 2 (m = 16) | 0.319084 | 100/100 | Pass |
| Linear complexity (M = 500) | 0.595549 | 100/100 | Pass |

### *3.3. Dynamic S-Box Generation Algorithm Based on the Proposed PRNG*

Generally, complex designs can generate S-Boxes with high cryptographic strength, but the time cost is very large while the efficiency is low. Here, we propose a simple and effective method of constructing strong $8 \times 8$ S-Boxes based on the proposed PRNG. The new method takes advantage of the excellent chaotic characteristics of the compound system.

For generating S-Boxes, the steps of algorithm based on the proposed PRNG are shown in Algorithm 2.

By the proposed method, the length of sequences generated by the PRNG is used to construct an S-Box that corresponds to a $16 \times 16$ matrix. The proposed new S-Boxes are generated by the above algorithm and the parameters are set as $\{u_0 = 0.1, v_0 = 0.1, k_{22} = -1.1, n = 15,\ r = 5,\ i_0 = 2,\ c_0 = 1,\ \theta = \frac{\pi}{r}, forward = back = 1\}$. The matrix $Sb$ corresponding to the first generated S-Box is shown in Table 2.

**Table 2.** The *Matrix* representing the proposed S-Box.

| *i*/*j* | 1 | 2 | 3 | 4 | 5 | 6 | 7 | 8 | 9 | 10 | 11 | 12 | 13 | 14 | 15 | 16 |
|---|---|---|---|---|---|---|---|---|---|---|---|---|---|---|---|---|
| 1 | 145 | 129 | 70 | 86 | 165 | 18 | 91 | 84 | 29 | 3 | 88 | 132 | 47 | 20 | 17 | 49 |
| 2 | 217 | 158 | 107 | 219 | 147 | 10 | 39 | 130 | 60 | 179 | 202 | 34 | 190 | 185 | 128 | 63 |
| 3 | 79 | 213 | 41 | 114 | 97 | 74 | 166 | 169 | 230 | 118 | 199 | 135 | 142 | 31 | 75 | 124 |
| 4 | 73 | 93 | 6 | 83 | 151 | 99 | 120 | 203 | 19 | 122 | 64 | 248 | 106 | 4 | 183 | 250 |
| 5 | 187 | 148 | 116 | 143 | 164 | 125 | 81 | 35 | 138 | 161 | 9 | 111 | 223 | 159 | 176 | 108 |
| 6 | 157 | 2 | 50 | 45 | 155 | 178 | 113 | 7 | 14 | 181 | 194 | 174 | 244 | 239 | 54 | 0 |
| 7 | 180 | 229 | 76 | 149 | 72 | 141 | 117 | 207 | 51 | 27 | 231 | 154 | 16 | 110 | 126 | 236 |
| 8 | 28 | 242 | 197 | 121 | 25 | 104 | 172 | 240 | 37 | 221 | 13 | 24 | 94 | 103 | 198 | 205 |
| 9 | 243 | 33 | 175 | 171 | 80 | 193 | 153 | 245 | 152 | 11 | 5 | 53 | 208 | 137 | 90 | 234 |
| 10 | 38 | 42 | 177 | 67 | 85 | 26 | 101 | 201 | 215 | 186 | 127 | 78 | 150 | 216 | 96 | 218 |
| 11 | 252 | 65 | 62 | 43 | 184 | 220 | 112 | 52 | 225 | 66 | 224 | 182 | 36 | 115 | 237 | 98 |
| 12 | 77 | 191 | 57 | 123 | 95 | 8 | 167 | 55 | 227 | 56 | 131 | 30 | 222 | 200 | 136 | 235 |
| 13 | 69 | 206 | 23 | 255 | 102 | 22 | 196 | 253 | 92 | 211 | 170 | 254 | 133 | 40 | 249 | 15 |
| 14 | 160 | 251 | 68 | 214 | 192 | 109 | 87 | 48 | 189 | 195 | 188 | 228 | 12 | 139 | 105 | 163 |
| 15 | 61 | 32 | 58 | 246 | 241 | 46 | 209 | 134 | 1 | 146 | 204 | 168 | 226 | 173 | 238 | 89 |
| 16 | 21 | 144 | 119 | 82 | 233 | 156 | 140 | 162 | 44 | 59 | 232 | 212 | 100 | 247 | 210 | 71 |

**Algorithm 2** The algorithm for generating S-Boxes

Input: $(N, r, i_0, c_0, \theta, farward, u_0, v_0, k_{22})$ and $m$ (number of S-Boxes required)
Output: $AllSboxes$
(1) $AllSboxes = \{\}$;
(2)
%%
Initialize the state of an S-Box:
%%
    $Sbox[i] = i,\ i = 0, 1, \cdots 2^8 - 1$;
(3) for $0 \leq i < 2^8$
        Obtain a number $j$ from the proposed PRNG;
        swap values of $Sbox\left[2^8 - 1 - i\right]$ and $Sbox[j]$;
    end for;
(4) $status = 1$;
%%
Check whether the S-Box meets the criteria of high diffusion and low differential uniformity [37].
%%
    for $0 \leq i < 2^8$
        if $\left|Sbox[i] - Sbox\left[\mathrm{mod}\left(i + 1, 2^8\right)\right]\right| \leq 2 \parallel \delta(Sbox) > 10$
%%
The operator $\delta()$ is used to obtain the differential uniformity value of an S-Box.
%%
            $status = 0$;
            break;
        end if
    end for;
    If $status == 1$
        $AllSboxes+ = Sbox$;
    else
        jump to (2);
(5) if $count(AllSboxes) < m$
        jump to (2);

## 4. Performance Tests of the Constructed S-Box

An S-Box is the only nonlinear component in a block encryption system (providing nonlinear mapping between plaintext and ciphertext), so it is very important to evaluate the robustness of encrypted data attacks by analyzing the nonlinear characteristics of the S-Box. We will evaluate the encryption strength of the proposed S-Box given in Table 2. We will use S-Box performance evaluation criteria such as strict avalanche criterion (SAC), bijection, nonlinearity, BIC, and Linear approximation probability (LAP).

### 4.1. Basic Cryptographic Evaluation Criteria

#### 4.1.1. Bijection

For a function or S-Box $Y = S(X): \ X \in \mathbb{F}_2^n \to Y \in \mathbb{F}_2^n$, it is bijective if it is a one-to-one map. From Table 2, it can be seen that a clearly different value of $Y$ corresponds to only one $X$. Conversely, one can obtain a distinct value $X$ by Equation (13) and Table 2. Hence, the obtained S-Box of Table 2 has bijective properties.

#### 4.1.2. Algebraic Degree

The algebraic degree of the S-Box can provide an upper limit for the algebraic degree of the whole cryptographic system [4].

**Definition 1.** *For a Boolean function $Y = S(X)$, the algebraic normal form of $Y$ is as the follows [4]:*

$$\mathrm{ANF}_S = \bigoplus_{u \in \mathbb{F}_2^n} \alpha_u \prod_{i=0}^{n-1} x_i^{u_i},$$

*where $\alpha_u \in \mathbb{F}_2$. The Algebraic degree of $S(X)$*

$$\mathrm{Deg}(S) \triangleq \max\{\mathrm{wt}(u) \mid u \in \mathbb{F}_2^n \ and \ \alpha_u \neq 0 \in \mathbb{F}_2^n \ in \ \mathrm{ANF}_S\}.$$

For an $8 \times 8$ S-Box, $S(X)$ is usually represented by a collection of 8 Boolean functions of 8 variables called the coordinates of $S$. Thus, $S(X)$ can be represented as

$$\mathrm{S}(X) = \mathrm{S}_1(X)\mathrm{S}_2(X)\ldots\mathrm{S}_8(X),$$

where $\mathrm{S}_i(X) \in \{0,1\}$. In practical calculation, $\mathrm{Deg}(S)$ is given by estimate the maximum among all degrees of the coordinate functions:

$$\mathrm{Degree}(S) = \max\{degree(S_i) | S(X) = (S_1(X), S_2(X), \cdots S_n(X))\}. \tag{14}$$

The degree of the coordinate functions of the listed S-Box is presented in Table 3 according to Equation (14). We can see that degree of every coordinate function is 7, the largest possible value.

**Table 3.** Degree of coordinate functions of the proposed S-Box.

| $Sbox_i$ | $S_1$ | $S_2$ | $S_3$ | $S_4$ | $S_5$ | $S_6$ | $S_7$ | $S_8$ |
|---|---|---|---|---|---|---|---|---|
| *degree* | 7 | 7 | 7 | 7 | 7 | 7 | 7 | 7 |

#### 4.1.3. Algebraic Complexity (Univariate Degree)

For a cryptosystem, the algebraic complexity of the S-Box can be used to evaluate the ability of the system to cope with interpolation attacks [4]. If the univariate degree of the S-Box is too low, it can lead to successful interpolation attacks [38].

It is clear from Bao et al. [4], [p. 17] that if $\mathbb{F}_{2^n}[X]/(f(X)) \cong \mathbb{F}_{2^n}$, $S(X)$ can be uniquely represented as a univariate polynomial in $\mathbb{F}_{2^n}[X]/(f(X))$ from:

$$S(X) = \sum_{i=0}^{2^n-1} A_i X^i.$$

$A_0 = S(0)$, $A_{2^n-1} = \sum_{x \in \mathbb{F}_{2^n}} S(x)$, and the other coefficien $A_i$ can be calculated by discrete Fourier transform of the values of $S$, that is:

$$A_i = \sum_{k=0}^{2^n-2} S\left(\alpha^k\right)\alpha^{-ki},\ 0 \le i \le 2^n - 2, \tag{15}$$

is a primitive element in $\mathbb{F}_{2^n}[X]/(f(X))$. The Algebraic Complexity AC of $S(X)$ is the number of non-zero coefficients in the linear polynomial representation of $S(X)$.

For an $8 \times 8$ S-Box, $f(X) = X^8 + X^4 + X^3 + X^2 + 1$ is an irreducible polynomial and $\beta$ is taken as a root of $f(X)$ in $\mathbb{F}_{2^8}$. $\beta$ is a primitive element and we identify $\mathbb{F}_2^8$ with $\mathbb{F}_{2^8} = \left\{0, 1, \beta, \beta^2, \cdots, \beta^{254}\right\}$ using the basis $\left\{1, \beta, \beta^2, \cdots, \beta^7\right\}$. Because $\mathbb{F}_{2^8}[X]/(f(X)) \cong \mathbb{F}_{2^8}(\beta) \cong \mathbb{F}_2 \oplus \mathbb{F}_2 \oplus \cdots \oplus \mathbb{F}_2$, we can obtain that the corresponding codes for $0, 1, X^1, \cdots, X^{254}$ shown in Table 4. Then, coefficient $A_i$ of $X^i$ in the univariate expression corresponding to the S-Box can be calculated by Equation (15), as shown in Table 5. From Table 5, We can see that the AC is 255, the largest possible value.

**Table 4.** Code of $X^t = X^{(m-1)\times 16+n}$.

| *m*/*n* | **1** | **2** | **3** | **4** | **5** | **6** | **7** | **8** | **9** | **10** | **11** | **12** | **13** | **14** | **15** | **16** |
|---|---|---|---|---|---|---|---|---|---|---|---|---|---|---|---|---|
| 1 | 0 | 1 | 2 | 4 | 8 | 16 | 32 | 64 | 128 | 29 | 58 | 116 | 232 | 205 | 135 | 19 |
| 2 | 38 | 76 | 152 | 45 | 90 | 180 | 117 | 234 | 201 | 143 | 3 | 6 | 12 | 24 | 48 | 96 |
| 3 | 192 | 157 | 39 | 78 | 156 | 37 | 74 | 148 | 53 | 106 | 212 | 181 | 119 | 238 | 193 | 159 |
| 4 | 35 | 70 | 140 | 5 | 10 | 20 | 40 | 80 | 160 | 93 | 186 | 105 | 210 | 185 | 111 | 222 |
| 5 | 161 | 95 | 190 | 97 | 194 | 153 | 47 | 94 | 188 | 101 | 202 | 137 | 15 | 30 | 60 | 120 |
| 6 | 240 | 253 | 231 | 211 | 187 | 107 | 214 | 177 | 127 | 254 | 225 | 223 | 163 | 91 | 182 | 113 |
| 7 | 226 | 217 | 175 | 67 | 134 | 17 | 34 | 68 | 136 | 13 | 26 | 52 | 104 | 208 | 189 | 103 |
| 8 | 206 | 129 | 31 | 62 | 124 | 248 | 237 | 199 | 147 | 59 | 118 | 236 | 197 | 151 | 51 | 102 |
| 9 | 204 | 133 | 23 | 46 | 92 | 184 | 109 | 218 | 169 | 79 | 158 | 33 | 66 | 132 | 21 | 42 |
| 10 | 84 | 168 | 77 | 154 | 41 | 82 | 164 | 85 | 170 | 73 | 146 | 57 | 114 | 228 | 213 | 183 |
| 11 | 115 | 230 | 209 | 191 | 99 | 198 | 145 | 63 | 126 | 252 | 229 | 215 | 179 | 123 | 246 | 241 |
| 12 | 255 | 227 | 219 | 171 | 75 | 150 | 49 | 98 | 196 | 149 | 55 | 110 | 220 | 165 | 87 | 174 |
| 13 | 65 | 130 | 25 | 50 | 100 | 200 | 141 | 7 | 14 | 28 | 56 | 112 | 224 | 221 | 167 | 83 |
| 14 | 166 | 81 | 162 | 89 | 178 | 121 | 242 | 249 | 239 | 195 | 155 | 43 | 86 | 172 | 69 | 138 |
| 15 | 9 | 18 | 36 | 72 | 144 | 61 | 122 | 244 | 245 | 247 | 243 | 251 | 235 | 203 | 139 | 11 |
| 16 | 22 | 44 | 88 | 176 | 125 | 250 | 233 | 207 | 131 | 27 | 54 | 108 | 216 | 173 | 71 | 142 |

**Table 5.** The coefficients $A_i = A_{(m-1)\times 16+n-1}$ (coefficient of univariate polynomial representation).

| *m*/*n* | **1** | **2** | **3** | **4** | **5** | **6** | **7** | **8** | **9** | **10** | **11** | **12** | **13** | **14** | **15** | **16** |
|---|---|---|---|---|---|---|---|---|---|---|---|---|---|---|---|---|
| 1 | 145 | 33 | 253 | 168 | 50 | 241 | 221 | 168 | 150 | 208 | 54 | 179 | 157 | 196 | 216 | 21 |
| 2 | 44 | 151 | 203 | 161 | 238 | 133 | 253 | 107 | 223 | 199 | 184 | 244 | 229 | 154 | 44 | 176 |
| 3 | 129 | 69 | 236 | 230 | 239 | 250 | 186 | 45 | 158 | 139 | 78 | 198 | 29 | 56 | 99 | 31 |
| 4 | 36 | 206 | 82 | 115 | 33 | 35 | 1 | 194 | 139 | 133 | 45 | 47 | 163 | 233 | 203 | 184 |
| 5 | 169 | 112 | 221 | 31 | 45 | 205 | 101 | 169 | 141 | 190 | 234 | 148 | 172 | 100 | 129 | 157 |
| 6 | 201 | 14 | 91 | 123 | 222 | 143 | 49 | 209 | 221 | 39 | 156 | 206 | 76 | 19 | 7 | 99 |
| 7 | 37 | 69 | 33 | 126 | 22 | 103 | 140 | 233 | 73 | 72 | 186 | 3 | 168 | 9 | 175 | 49 |
| 8 | 49 | 174 | 30 | 81 | 115 | 124 | 232 | 221 | 123 | 23 | 75 | 207 | 18 | 139 | 22 | 11 |

**Table 5.** *Cont.*

| m/n | 1 | 2 | 3 | 4 | 5 | 6 | 7 | 8 | 9 | 10 | 11 | 12 | 13 | 14 | 15 | 16 |
|---|---|---|---|---|---|---|---|---|---|---|---|---|---|---|---|---|
| 9 | 254 | 82 | 194 | 182 | 201 | 173 | 168 | 226 | 90 | 104 | 205 | 190 | 62 | 15 | 40 | 20 |
| 10 | 193 | 114 | 89 | 84 | 219 | 1 | 98 | 146 | 52 | 106 | 139 | 194 | 69 | 199 | 39 | 170 |
| 11 | 176 | 185 | 66 | 62 | 245 | 220 | 99 | 94 | 43 | 44 | 43 | 11 | 246 | 199 | 83 | 110 |
| 12 | 63 | 238 | 98 | 188 | 167 | 30 | 180 | 214 | 95 | 249 | 101 | 5 | 148 | 154 | 219 | 76 |
| 13 | 115 | 5 | 215 | 170 | 233 | 45 | 112 | 37 | 119 | 253 | 88 | 89 | 223 | 32 | 52 | 243 |
| 14 | 107 | 171 | 144 | 56 | 148 | 61 | 150 | 130 | 117 | 255 | 55 | 22 | 29 | 143 | 248 | 210 |
| 15 | 63 | 251 | 113 | 209 | 158 | 134 | 61 | 59 | 164 | 114 | 130 | 147 | 184 | 249 | 143 | 194 |
| 16 | 106 | 98 | 219 | 204 | 166 | 88 | 116 | 200 | 170 | 54 | 212 | 225 | 23 | 64 | 186 | 0 |

### 4.1.4. Nonlinearity

For an S-Box $S(X)$, in order to effectively resist linear cryptanalysis attacks, the relationship between outputs and inputs should be highly nonlinear. We will use the nonlinearity of every Boolean function $S_i$ to estimate the S-Box's nonlinear strength. Based on the Walsh transform of the Boolean function $S_i$, the evaluation criteria for nonlinearity is defined. The Walsh transform of a Boolean function $S_i$ is defined as [39]:

$$WS_{S_i}(w) = \sum_{x \in \mathbb{F}_2^n} (-1)^{w \cdot x \oplus S_i(X)},$$

where the $\cdot$ operator denotes the scalar product. For a Boolean function $S_i$, the nonlinearity calculation is based on the following equation [40]:

$$N_{S_i} = 2^{n-1} - 2^{-1} \max_{w \in F_2^n} WS_{S_i}(w).$$

The nonlinearity values of all constituent 8-bit Boolean functions of the proposed S-Box are provided in Table 6. From Table 6, we can see that the average nonlinearity was 106.5, the maximum nonlinearity was 112, and the minimum nonlinearity was 104.

**Table 6.** Nonlinearities of coordinate functions of the proposed S-Box.

| $S - Box/S_i$ | $S_1$ | $S_2$ | $S_3$ | $S_4$ | $S_5$ | $S_6$ | $S_7$ | $S_8$ | *Average* |
|---|---|---|---|---|---|---|---|---|---|
| *Nonlinearity* | 104 | 106 | 106 | 104 | 110 | 106 | 112 | 104 | 106.5 |

### 4.1.5. Strict Avalanche Criterion (SAC)

An imperative cryptographic feature for an S-Box is the Strict Avalanche Criterion (SAC), which was first introduced in Webster and Tavares [41]. If an S-Box satisfies the SAC, as a single bit is changed in the input, half the output bits should be modified. For an S-Box, the SAC property requires that all the values in the dependency matrix are close to the ideal value of 0.5. The SAC values provided by the dependency matrix of the proposed S-Box are listed in Table 7. From Table 7, it can be seen that the average value is equal to 0.5. Thus, the proposed S-Box has good SAC performance.

**Table 7.** *SAC_Dep* matrix of the proposed S-Box.

| | | | | | | | |
|---|---|---|---|---|---|---|---|
| 0.5746 | 0.5517 | 0.5188 | 0.4753 | 0.4772 | 0.5379 | 0.4979 | 0.5276 |
| 0.4719 | 0.4821 | 0.5756 | 0.4837 | 0.5895 | 0.5004 | 0.4463 | 0.4622 |
| 0.5407 | 0.4823 | 0.5816 | 0.5187 | 0.4307 | 0.4070 | 0.4942 | 0.4878 |
| 0.5474 | 0.4812 | 0.5162 | 0.4815 | 0.5114 | 0.5339 | 0.4757 | 0.4535 |
| 0.5065 | 0.4680 | 0.5365 | 0.5185 | 0.5197 | 0.4902 | 0.5321 | 0.4392 |
| 0.5367 | 0.5240 | 0.4952 | 0.5165 | 0.5976 | 0.5691 | 0.5347 | 0.5091 |
| 0.5654 | 0.5288 | 0.5458 | 0.5255 | 0.4708 | 0.4594 | 0.4648 | 0.5513 |
| 0.5229 | 0.5791 | 0.5664 | 0.5164 | 0.4262 | 0.6032 | 0.3924 | 0.4726 |

#### 4.1.6. Bit Independence Criterion (BIC)

Based on the reverse of plaintext, a set of avalanche vectors can be generated, and all avalanche variables should be paired and independent. The degree of independence $\rho$ can be measured by calculating the phase relation of two avalanche vectors $A$ and $B$:

$$\rho\{A,B\} = \frac{cov\{A,B\}}{\sigma\{A\}\sigma\{B\}},$$

where $\sigma^2\{A\} = E\{A^2\} - (E\{A\})^2$ and $cov\{A,B\} = E\{AB\} - E\{A\}E\{B\}$. If $\rho$ is 0, the two variables are independent of each other. Furthermore, the two variables are the "same" if $\rho$ is 1, while if it is equal to $-1$, the two variables are complementary.

A method of actually measuring BIC was proposed by C.Adams and S.Tavares in [41,42]: for the coordinate Boolean functions $S_j$ and $S_k (j \neq k)$ of a given S-Box, if $S_i \oplus S_k$ is highly nonlinear or $S_i \oplus S_k$ satisfies SAC as much as possible and the correlation coefficient of every output bit pair approaches 0 if only one input bit is inverted. By verifying whether $S_i \oplus S_k$ between any two output bits of the S-Box strictly satisfies the avalanche effect, or by calculating the nonlinearity of $S_i \oplus S_k$, whether the S-Box satisfies BIC can be tested. For 8 Boolean functions of the proposed S-Box, the nonlinearity and SAC values, respectively, are listed in Tables 8 and 9. We can see from Tables 8 and 9 that the SAC values and average nonlinearity for BIC are 0.50698 and 105.6, respectively. Thus, there is an extremely weak linear association among the output bits, and, thus, the obtained S-Box had good BIC performance.

**Table 8.** Bit independence criterion for SAC.

| | $S_1$ | $S_2$ | $S_3$ | $S_4$ | $S_5$ | $S_6$ | $S_7$ | $S_8$ |
|---|---|---|---|---|---|---|---|---|
| $S_1$ | - | 0.46875 | 0.54688 | 0.54688 | 0.51562 | 0.54688 | 0.56250 | 0.51562 |
| $S_2$ | 0.54688 | - | 0.48438 | 0.48438 | 0.46875 | 0.53125 | 0.53125 | 0.57812 |
| $S_3$ | 0.51562 | 0.57812 | - | 0.51562 | 0.54688 | 0.48438 | 0.54688 | 0.56250 |
| $S_4$ | 0.46875 | 0.48438 | 0.51562 | - | 0.51562 | 0.51562 | 0.53125 | 0.51562 |
| $S_5$ | 0.48438 | 0.59375 | 0.42188 | 0.50000 | - | 0.59375 | 0.46875 | 0.42188 |
| $S_6$ | 0.53125 | 0.50000 | 0.40625 | 0.53125 | 0.48438 | - | 0.45312 | 0.59375 |
| $S_7$ | 0.50000 | 0.43750 | 0.50000 | 0.46875 | 0.53125 | 0.53125 | - | 0.39062 |
| $S_8$ | 0.51562 | 0.46875 | 0.48438 | 0.45312 | 0.43750 | 0.51562 | 0.54688 | - |

**Table 9.** Bit independence criterion for nonlinearity.

| | $S_1$ | $S_2$ | $S_3$ | $S_4$ | $S_5$ | $S_6$ | $S_7$ | $S_8$ |
|---|---|---|---|---|---|---|---|---|
| $S_1$ | - | 108 | 108 | 108 | 106 | 104 | 112 | 106 |
| $S_2$ | 108 | - | 108 | 106 | 106 | 102 | 108 | 108 |
| $S_3$ | 108 | 108 | - | 102 | 102 | 108 | 100 | 104 |
| $S_4$ | 108 | 106 | 102 | - | 108 | 102 | 106 | 106 |
| $S_5$ | 106 | 106 | 102 | 108 | - | 106 | 102 | 104 |
| $S_6$ | 104 | 102 | 108 | 102 | 106 | - | 106 | 106 |
| $S_7$ | 112 | 108 | 100 | 106 | 102 | 106 | - | 106 |
| $S_8$ | 106 | 108 | 104 | 106 | 104 | 106 | 106 | - |

#### 4.1.7. Differential Probability

The XOR distribution of the input and output of an S-Box can be obtained from the differential probability [43] of the S-Box. By analyzing the differential probability, it is possible to obtain the original plaintext. Differential probability is computed by:

$$DP_S = \mathrm{Max}_{\Delta X \neq 0, \Delta Y}\left(\frac{\#\{X \in N \mid \Delta Y = S(X \oplus \Delta X) \oplus S(X)\}}{2^n}\right),$$

where $\Delta Y$ is the output difference and $\Delta X$ is the input difference. The lower the value of the differential probability, the stronger the ability of the S-Box to resist differential cryptanalysis. The maximum $DP$ value of the S-Box generated by this algorithm is only 0.0390625.

#### 4.1.8. Maximal Degree of the Product of $k$ Coordinates

For a cryptosystem, the degrees of the product of any $k$ coordinates of the S-Box can provide the degree of the superior upper bound. For an S-Box $S(X)$, to obtain the definition of maximal algebraic degree of the product of any $k$ coordinates, we go through generalized definitions of an S-Box.

**Definition 2.** *For an S-Box, the Walsh transform [39] of the corresponding $S(X)$ is defined as the following:*

$$\mathcal{W}_S(\alpha,\beta) = \sum_{x \in \mathbb{F}_2^n} (-1)^{\beta \cdot S(x) \oplus \alpha \cdot x}, \quad \alpha \in \mathbb{F}_2^n, \beta \in \mathbb{F}_2^n.$$

In matrix $\mathcal{W}_S$, the value of point $(\alpha,\beta)$ represent the Walsh coefficient at $(\alpha,\beta)$. The maximal algebraic degree of the product of any $k$ coordinates of $S(X)$ is given by Bao et al. [4]:

$$d_k(S) = \max_{K \subseteq \{1,\ldots,8\}, |K| \leq k} \deg\left(\prod_{i \in K} S_{e_i}\right),$$

where $S_{e_i}$ is called the ith coordinate of the S-Box and

$$S_{e_i}(x) = \frac{1}{2} - \frac{1}{2^{n+1}} \sum_{a \in \mathbb{F}_2^n} \mathcal{W}_S\left(a, 2^i\right)(-1)^{a \cdot x}.$$

The values of $d_k(S)$ are closely related to resistance against higher-order differential attacks [4]. This means that the higher their values, the stronger the S-Box's ability to resist high-order differential attacks . The $d_k(S)$ values of the proposed S-Box are shown in Table 10. We can see that all $d_k(S)$ values are 7, the largest possible value.

**Table 10.** Maximal degree of the product of $k$ coordinates of the proposed S-Box.

| $k$ | **1** | **2** | **3** | **4** | **5** | **6** | **7** |
|---|---|---|---|---|---|---|---|
| $d_k$ | 7 | 7 | 7 | 7 | 7 | 7 | 7 |

#### 4.1.9. Differential Uniformity

The differential uniformity of an S-Box is defined as [4]:

$$\delta_S = \max_{a \neq 0, b} \delta(a,b) = \max_{a \neq 0, b} \#\left\{X \in \mathbb{F}_2^8 : S(X \oplus a) \oplus S(X) = b\right\}, \tag{16}$$

where $a,\ b \in F_2^8$. Low differential uniformity is advantageous for a good S-Box. The smaller $\delta_S$ is, the better the S-Box approximates to being perfectly nonlinear [44]. $\delta_S$ of the proposed S-Box is 10 according to Equation (16).

#### 4.1.10. Linear Approximation Probability

The linear approximation probability (LAP) is mainly concerned with linear attacks on block encryption technology, and it is used to measure the ability of a cryptographic system to resist linear cryptanalysis. The linear approximation probability of an S-Box is the maximum unbalance value, which is presented as [43]:

$$\mathrm{LAP} = \max_{\alpha,\beta} \left| \frac{\#\{X \in \mathbb{GF}(2^n) \mid X \cdot \alpha = S(X) \cdot \beta\}}{2^n} - \frac{1}{2} \right|, \tag{17}$$

where $\alpha$ and $\beta$ are the input and output mask values, and $\cdot$ is the dot product operation on $\mathbb{F}^2$. The maximum value of LAP of the proposed S-Box is only 0.1328125.

### 4.2. Security Analysis

#### 4.2.1. Resistance to Algebraic Attacks

A cryptosystem can be regarded as an algebraic system and the fundamental principles of algebraic attacks can be traced back to Shannon's work [45], in which he said that the entire cryptosystem can be mathematically modeled and expressed as a multivariate algebraic equation. The resistance of cryptographic systems to algebraic attacks can be evaluated by calculating the nonlinearity, algebraic complexity, and algebraic degree.

It is clear from Tables 5 and 3 that AC and $\mathrm{Deg}(S)$ need to be high for the cryptosystem to resist algebraic attacks efficiently.

#### 4.2.2. Resistance to Differential Attack

A differential attack is an analysis of the distribution of output differences caused by an input with a fixed difference [4]. The security of an $8 \times 8$ S-Box against differential attack is quantified by computing its differential probability $DP(S)$ , maximal algebraic degree of the product of any $k$ coordinates $d_k(S)$, and differential uniformity $\delta_S$. The $DP(S)$ and $\delta_S$ of the proposed S-Box are small, while the $d_k(S)$ values are the largest possible. Thus, the S-Box has good resistance to differential attacks.

#### 4.2.3. Resistance to Linear Attack

The linear attack was proposed by Mitsuru Matsui [46]. A linear attack exploits the fact that a cryptosystem $C = E_K(P)$ can form a Boolean function that behaves as non-random (imbalanced) on the set of $P$ [46]. In a linear attack, the cyber attacker searches for the most biased linear relation between the input and output bits of the S-Box. Security against linear attack is quantified by computing LAP, nonlinearity, and AC. If LAP is small while the nonlinearity and AC are high, an S-Box can be highly resistant to linear attack. It can be clearly seen from the above nonlinearity of the proposed S-Box (in Table 6), AC value of 255, and LAP value of 0.1328125 that the NL and AC of the proposed S-Box and are sufficiently high, while the LAP is low. Thus, the S-Box has good resistance to linear attack.

#### 4.2.4. Resistance to Boomerang Attack

A differential-style attack called the boomerang attack was first described by Wagner [3]. The boomerang attack is considered an extension of classical differential attacks. In a classic boomerang attack, cryptosystem $E$ is regarded as a combination of two sub-cryptosystems $E_0$ and $E_1$ such that $E = E_1 \circ E_0$. There are two different inputs, $\alpha$ and $\gamma$; $\alpha$ is exported to $\beta$ by $E_0$ with probability $p$ and $\gamma$ is propagated to $\delta$ by $E_1$ with probability $q$. In the original study [3], the boomerang attack is based on the expectation probability:

$$\Pr\left[E^{-1}(E(x) \oplus b) \oplus E^{-1}(E(x \oplus a) \oplus b) = a\right] = p^2 q^2.$$

Since Wagner put forward this kind of attack, many improved boomerang attacks have been proposed [2]. In particular, the cryptosystem $E$ is decomposed into three modules: $E = E_1 \circ E_m \circ E_0$, and the $E_m$ corresponds to a simple change (usually an S-Box). The probability of generating a right quartet in each S-Box in the middle S-Box layer is given by the following equation [47]:

$$\frac{\#\{x \in \{0,1\}^n \mid S^{-1}(S(x) \oplus \nabla_o) \oplus S^{-1}(S(x \oplus \Delta_i) \oplus \nabla_o) = \Delta_i\}}{2^n},$$

where $(\nabla_o, \Delta_i)$ is a given differential pair. Cid et al. focused on the effect of S-Box construction on boomerang attacks, and constructed a pre-calculated table to evaluate this probability. This table is called the boomerang connectivity table (BCT), which is defined as follows.

For an S-Box, the boomerang connectivity table (BCT) of the corresponding $S(X)$ is defined by the following equation [47]:

$$\beta_S(a,b) \triangleq \#\left\{x \in \mathbb{F}_2^n \mid S^{-1}(S(x) \oplus b) \oplus S^{-1}(S(x \oplus a) \oplus b) = a\right\}. \tag{18}$$

The highest value in the BCT except for point $(0,0)$ is defined as boomerang uniformity [4]:

$$\mathcal{BU}(S) = \max(\beta_S(a,b)). \tag{19}$$

For an S-Box, the smaller the $\mathcal{BU}(S)$, the stronger the resistance to boomerang attack. The $\mathcal{BU}(S)$ value of the proposed S-Box is 20 according to Equations (18) and (19), which indicates that the proposed S-Box has strong resistance to boomerang attack.

### 4.3. *Performance Comparison with Different S-Boxes*

In this section, linear approximation probability (LAP), strict avalanche criterion (SAC), nonlinearity (NL), differential approximation probability (DP), and algebraic complexity (AC) are selected as the key cryptanalysis attributes of S-Boxes, and the cryptographic performance of the S-Box proposed in this paper is compared with some recently proposed S-Boxes. The key cryptanalysis results of the different S-Boxes are shown in Table 11. From the table, we can make the following observation: the S-Box generated by the proposed algorithm has good nonlinearity and DP value, and the nonlinear values of SAC, BIC-SAC, and BIC are close to the ideal values. Thus, the proposed algorithm to generated an S-Box can meet the needs of data encryption.

**Table 11.** Performance comparison of different S-Boxes.

| S-Box Method | Nonlinearity | | | SAC | BIC-NL | LAP | DP | AC |
|---|---|---|---|---|---|---|---|---|
| | Min | Max | Average | | | | | |
| Ref. [1] | 102 | 108 | 104.5 | 0.498 | 104.6 | 0.125 | 0.048 | 254 |
| Ref. [16] | 104 | 108 | 106.8 | 0.507 | 103.9 | 0.140 | 0.054 | 254 |
| Ref. [20] | 104 | 110 | 106 | 0.499 | 103.8 | 0.125 | 0.039 | 255 |
| Ref. [48] | 104 | 110 | 106.5 | 0.495 | 103.8 | 0.141 | 0.039 | 255 |
| Ref. [49] | 98 | 110 | 102 | 0.493 | 104.6 | 0.140 | 0.046 | 255 |
| Proposed S-Box | 104 | 112 | 106.5 | 0.506 | 105.6 | 0.132 | 0.039 | 255 |

## 5. Conclusions and Discussion

In a summary, we proposed a novel compound chaotic system based on quantum random walks controlled by a hyper-chaotic map. We have evaluated the new system by using some test tools such as Scale index and statistical complexity measures. The results show that the new system has complex dynamic behavior. We have proposed a new PRNG, which passes all the standard statistical tests in NIST SP800-22. Next, we built a new chaotic S-Box generation algorithm based on the PRNG. We have evaluated the performance of the S-Box so-generated by using various cryptographic criteria such as the bit independence criterion (BIC), bijection, strict avalanche criterion (SAC), nonlinearity, differential probability (DP), and linear probability (LP). Test results demonstrated that the S-Box can well meet multiple cryptographic criteria and the new chaotic S-Box generation algorithm is effective.

The results lead us to the following conclusions. By controlling or perturbing the parameters of two types of chaos based on the outputs of one chaotic map, a compound chaotic system can be constructed. Based on the compound chaotic system, designers can easily propose efficient S-Box generation mechanisms that can produce secure S-Boxes. Our proposed compound chaotic system has more complex dynamic behavior. Therefore, based on this dynamical system, it is easy to design modules for cryptosystems such as image encryption algorithms.

In future work, we will take advantage of the proposed compound chaotic system in applications such as image encryption schemes and video authentication.

**Author Contributions:** Conceptualization, L.Z. and W.Z.; formal analysis and investigation, L.Z. and W.Z.; discussion and suggestion, C.M. and Y.Z.; writing—original draft preparation, C.M. and W.Z.; writing—review and editing, C.M., W.Z. and L.Z. All authors have read and agreed to the published version of the manuscript.

**Funding:** This work is supported by the National Natural Science Foundation of China (No. 12161077); the Natural Science Foundation of Gansu Province (Nos. 23JRRE0737, 22JR11RE189, 22CX8GA075); and the Innovation Fund Project of Tianshui Teachers College (Nos. CXJ2021-04, CXJ2021-01, PTJ2022-01) and Tianshui Natural Science Foundation (No. 2020-FZJHK-9757).

**Data Availability Statement:** The data presented in this study are available on request from the corresponding author. The data are not publicly available due to the reason that data could have been generated by the algorithm provided in this paper.

**Conflicts of Interest:** The authors declare no conflicts of interest.

## Abbreviations

The following abbreviations are used in this manuscript:

| | |
|---|---|
| LAP | Linear approximation probability |
| SAC | Strict avalanche criterion |
| CWT | Continuous wavelet transform |
| DP | Differential approximation probability |
| SCM | Statistical complexity measure |

## References


1. Liu, L.; Zhang, Y.; Wang, X. A Novel Method for Constructing the S-Box Based on Spatiotemporal Chaotic Dynamics. *Appl. Sci.* **2018**, *8*, 2650. [CrossRef]
2. Boura, C.; Canteaut, A. On the Boomerang Uniformity of Cryptographic Sboxes. *IACR Trans. Symmetric Cryptol.* **2018**, *2018*, 290–310. [CrossRef]
3. Wagner, D. The Boomerang Attack. In *Fast Software Encryption*; Knudsen, L., Ed.; Springer: Berlin/Heidelberg, Germany, 1999; pp. 156–170.
4. Bao, Z.; Guo, J.; Ling, S.; Sasaki, Y. PEIGEN—A Platform for Evaluation, Implementation, and Generation of S-Boxes. *IACR Trans. Symmetric Cryptol.* **2019**, *2019*, 330–394. [CrossRef]
5. Katiyar, S.; Jeyanthi, N. Pure Dynamic S-Box Construction. *Int. J. Comput.* **2016**, *1*, 42–46.
6. Adams, C.; Tavares, S. Good S-Boxes Are Easy To Find. In *Advances in Cryptology—CRYPTO' 89 Proceedings*; Brassard, G., Ed.; Springer: New York, NY, USA, 1990; pp. 612–615.
7. Hussain, I.; Anees, A.; Al-Maadeed, T.A.; Mustafa, M.T. Construction of S-Box Based on Chaotic Map and Algebraic Structures. *Symmetry* **2019**, *11*, 351. [CrossRef]
8. Rafiq, A.; Khan, M. Construction of new S-Boxes based on triangle groups and its applications in copyright protection. *Multimed. Tools Appl.* **2019**, *78*, 15527–15544. [CrossRef]
9. Artuğer, F.; Özkaynak, F. SBOX-CGA: Substitution box generator based on chaos and genetic algorithm. *Neural Comput. Appl.* **2022**, *34*, 20203–20211. [CrossRef]
10. Ye, T.; Zhimao, L. Chaotic S-Box: Six-dimensional fractional Lorenz–Duffing chaotic system and O-shaped path scrambling. *Nonlinear Dyn.* **2018**, *94*, 2115–2126. [CrossRef]
11. Masood, F.; Boulila, W.; Alsaeedi, A.; Khan, J.S.; Ahmad, J.; Khan, M.A.; Rehman, S.U. A novel image encryption scheme based on Arnold cat map, Newton–Leipnik system and Logistic Gaussian map. *Multimed. Tools Appl.* **2022**, *81*, 30931–30959. [CrossRef]
12. Sambas, A.; Vaidyanathan, S.; Tlelo-Cuautle, E.; Abd-El-Atty, B.; El-Latif, A.A.A.; Guillén-Fernández, O.; Sukono; Hidayat, Y.; Gundara, G. A 3-D Multi-Stable System with a Peanut-Shaped Equilibrium Curve: Circuit Design, FPGA Realization, and an Application to Image Encryption. *IEEE Access* **2020**, *8*, 137116–137132. [CrossRef]
13. Hua, Z.; Zhu, Z.; Chen, Y.; Li, Y. Color image encryption using orthogonal Latin squares and a new 2D chaotic system. *Nonlinear Dyn.* **2021**, *104*, 4505–4522. [CrossRef]
14. Chai, X.; Fu, J.; Gan, Z.; Lu, Y.; Zhang, Y.; Han, D. Exploiting Semi-Tensor Product Compressed Sensing and Hybrid Cloud for Secure Medical Image Transmission. *IEEE Internet Things J.* **2023**, *10*, 7380–7392. [CrossRef]
15. Song, W.; Fu, C.; Zheng, Y.; Tie, M.; Liu, J.; Chen, J. A parallel image encryption algorithm using intra bitplane scrambling. *Math. Comput. Simul.* **2023**, *204*, 71–88. [CrossRef]

16. Zahid, A.H.; Arshad, M.J. An Innovative Design of Substitution-Boxes Using Cubic Polynomial Mapping. *Symmetry* **2019**, *11*, 437. [CrossRef]
17. Zheng, J.; Zeng, Q. An image encryption algorithm using a dynamic S-Box and chaotic maps. *Appl. Intell.* **2022**, *52*, 15703–15717. [CrossRef]
18. Bin Faheem, Z.; Ali, A.; Khan, M.A.; Ul-Haq, M.E.; Ahmad, W. Highly dispersive substitution box (S-Box) design using chaos. *ETRI J.* **2020**, *42*, 619–632. [CrossRef]
19. Zhu, H.; Tong, X.; Wang, Z.; Ma, J. A novel method of dynamic S-Box design based on combined chaotic map and fitness function. *Multimed. Tools Appl.* **2020**, *79*, 12329–12347. [CrossRef]
20. Lu, Q.; Zhu, C.; Wang, G. A Novel S-Box Design Algorithm Based on a New Compound Chaotic System. *Entropy* **2019**, *21*, 1004. [CrossRef]
21. Zhao, W.; Chang, Z.; Ma, C.; Shen, Z. A Pseudorandom Number Generator Based on the Chaotic Map and Quantum Random Walks. *Entropy* **2023**, *25*, 166. [CrossRef]
22. Sambas, A.; Vaidyanathan, S.; Zhang, S.; Abd El-Latif, A.A.; Mohamed, M.A.; Abd-El-Atty, B. Multistability Analysis and MultiSim Simulation of a 12-Term Double-Scroll Hyperchaos System with Three Nonlinear Terms, Bursting Oscillations and Its Cryptographic Applications. In *Cybersecurity: A New Approach Using Chaotic Systems*; Springer International Publishing: Cham, Switzerland, 2022; pp. 221–235. [CrossRef]
23. Ahmad, M.; Alam, M.Z.; Ansari, S.; Lambić, D.; AlSharari, H.D. Cryptanalysis of an image encryption algorithm based on PWLCM and inertial delayed neural network. *J. Intell. Fuzzy Syst.* **2018**, *34*, 1323–1332. [CrossRef]
24. Jamal, S.S.; Anees, A.; Ahmad, M.; Khan, M.F.; Hussain, I. Construction of Cryptographic S-Boxes Based on Mobius Transformation and Chaotic Tent-Sine System. *IEEE Access* **2019**, *7*, 173273–173285. [CrossRef]
25. Singh, J.P.; Roy, B. The nature of Lyapunov exponents is (+, +, −, −). Is it a hyperchaotic system? *Chaos Solitons Fractals* **2016**, *92*, 73–85. [CrossRef]
26. Guo, F.; Xu, L. *Applications of Chaos Theory to Cryptography*, 1st ed.; Beijing Institute of Technology Press: Beijing, China, 2015.
27. Venegas-Andraca, S.E. Quantum walks: A comprehensive review. *Quantum Inf. Process.* **2012**, *11*, 1015–1106. [CrossRef]
28. Aharonov, D.; Ambainis, A.; Kempe, J.; Vazirani, U. Quantum Walks on Graphs. In Proceedings of the Thirty-Third Annual ACM Symposium on Theory of Computing, Hersonissos, Greece, 6–8 July 2001; Association for Computing Machinery: New York, NY, USA, 2001; STOC '01, pp. 50–59. [CrossRef]
29. Yang, Y.G.; Zhao, Q.Q. Novel pseudo-random number generator based on quantum random walks. *Sci. Rep.* **2016**, *6*, 20362. [CrossRef] [PubMed]
30. Benítez, R.; Bolós, V.J.; Ramírez, M.E. A wavelet-based tool for studying non-periodicity. *Comput. Math. Appl.* **2010**, *60*, 634–641. [CrossRef]
31. Chandre, C.; Wiggins, S.; Uzer, T. Time–frequency analysis of chaotic systems. *Phys. D Nonlinear Phenom.* **2003**, *181*, 171–196. [CrossRef]
32. Bolós, V.J.; Benítez, R.; Ferrer, R. A New Wavelet Tool to Quantify Non-Periodicity of Non-Stationary Economic Time Series. *Mathematics* **2020**, *8*, 844. [CrossRef]
33. Martin, M.; Plastino, A.; Rosso, O. Statistical complexity and disequilibrium. *Phys. Lett. A* **2003**, *311*, 126–132. [CrossRef]
34. Larrondo, H.A.; González, C.M.; Martín, M.T.; Plastino, A.; Rosso, O.A. Intensive statistical complexity measure of pseudorandom number generators. *Phys. A Stat. Mech. Its Appl.* **2005**, *356*, 133–138. [CrossRef]
35. Akhshani, A.; Akhavan, A.; Mobaraki, A.; Lim, S.C.; Hassan, Z. Pseudo random number generator based on quantum chaotic map. *Commun. Nonlinear Sci. Numer. Simul.* **2014**, *19*, 101–111. [CrossRef]
36. Bassham, L.E.; Rukhin, A.L.; Soto, J.; Nechvatal, J.R.; Smid, M.E.; Barker, E.B.; Leigh, S.D.; Levenson, M.; Vangel, M.; Banks, D.L.; et al. SP 800-22 Rev. 1a. A Statistical Test Suite for Random and Pseudorandom Number Generators for Cryptographic Applications. Technical Report; NIST: Gaithersburg, MD, USA, 2010.
37. Biham, E.; Shamir, A. Differential Cryptanalysis of Feal and N-Hash. In *Advances in Cryptology—EUROCRYPT '91*; Davies, D.W., Ed.; Springer: Berlin/Heidelberg, Germany, 1991; pp. 1–16.
38. Jakobsen, T. Attacks on block ciphers of low algebraic degree. *J. Cryptol.* **2001**, *14*, 197–210. [CrossRef]
39. Carlet, C. Vectorial Boolean Functions for Cryptography. In *Boolean Models and Methods in Mathematics, Computer Science, and Engineering*; Encyclopedia of Mathematics and its Applications; Cambridge University Press: Cambridge, UK, 2010; pp. 398–470. [CrossRef]
40. Canteaut, A.; Carlet, C.; Charpin, P.; Fontaine, C. Propagation Characteristics and Correlation-Immunity of Highly Nonlinear Boolean Functions. In *Advances in Cryptology—EUROCRYPT 2000*; Preneel, B., Ed.; Springer: Berlin/Heidelberg, Germany, 2000; pp. 507–522.
41. Webster, A.F.; Tavares, S.E. On the Design of S-Boxes. In *Advances in Cryptology — CRYPTO '85 Proceedings*; Williams, H.C., Ed.; Springer: Berlin/Heidelberg, Germany, 1986; pp. 523–534.
42. Adams, C.; Tavares, S. The structured design of cryptographically good s-boxes. *J. Cryptol.* **1990**, *3*, 27–41. [CrossRef]
43. Hussain, I.; Shah, T.; Mahmood, H.; Gondal, M.A. Construction of S8 Liu J S-Boxes and their applications. *Comput. Math. Appl.* **2012**, *64*, 2450–2458. [CrossRef]
44. Nyberg, K.; Knudsen, L.R. Provable Security Against Differential Cryptanalysis. In *Advances in Cryptology—CRYPTO' 92*; Brickell, E.F., Ed.; Springer: Berlin/Heidelberg, Germany, 1993; pp. 566–574.

45. Shannon, C.E. Communication theory of secrecy systems. *Bell Syst. Tech. J.* **1949**, *28*, 656–715. [CrossRef]
46. Matsui, M. Linear Cryptanalysis Method for DES Cipher. In *Advances in Cryptology—EUROCRYPT '93*; Helleseth, T., Ed.; Springer: Berlin/Heidelberg, Germany, 1994; pp. 386–397.
47. Cid, C.; Huang, T.; Peyrin, T.; Sasaki, Y.; Song, L. Boomerang Connectivity Table: A New Cryptanalysis Tool. In *Advances in Cryptology—EUROCRYPT 2018*; Nielsen, J.B., Rijmen, V., Eds.; Springer International Publishing: Cham, Switzerland, 2018; pp. 683–714.
48. Wang, X.; Ünal, Ç.; Kaçar, S.; Akgul, A.; Pham, V.T.; Jafari, S.; Alsaadi, F.E.; Nguyen, X.Q. S-Box Based Image Encryption Application Using a Chaotic System without Equilibrium. *Appl. Sci.* **2019**, *9*, 781. [CrossRef]
49. Ali, T.S.; Ali, R. A novel color image encryption scheme based on a new dynamic compound chaotic map and S-Box. *Multimed. Tools Appl.* **2022**, *81*, 20585–20609. [CrossRef]

mathematics

MDPI

Article

# Nonlinear Medical Ultrasound Tomography: 3D Modeling of Sound Wave Propagation in Human Tissues

**Maxim Shishlenin [1,2,†], Andrey Kozelkov [3,4,5,†] and Nikita Novikov [1,2,*,†]**

1 Institute of Computational Mathematics and Mathematical Geophysics, 630090 Novosibirsk, Russia; m.a.shishlenin@mail.ru
2 Sobolev Institute of Mathematics, 630090 Novosibirsk, Russia
3 Russian Federal Nuclear Center, All-Russian Research Institute of Experimental Physics, 607188 Sarov, Russia; askozelkov@mail.ru
4 Nizhny Novgorod State Technical University n.a. R.E. Alekseev, 603155 Nizhny Novgorod, Russia
5 Sarov Institute of Physics and Technology—Branch of the National Research Nuclear University "MEPHI", 607186 Sarov, Russia
* Correspondence: novikov-1989@yandex.ru
† These authors contributed equally to this work.

**Abstract:** The article aimed to show the fundamental possibility of constructing a computational digital twin of the acoustic tomograph within the framework of a unified physics–mathematical model based on the Navier–Stokes equations. The authors suggested that the size of the modeling area is quite small, sound waves are waves of "small" disturbance, and given that a person consists of more than 60% water, human organs can be modeled using a liquid model, taking into account their density. During numerical experiments, we obtained the pressure registered in the receivers that are located on the side walls of the tomograph. The differences in pressure values are shown depending on the configuration of inclusions in the mannequin imitating internal organs. The results show that the developed technology can be used to probe the human body in medical acoustic tomographs and determine the acoustic parameters of the human body to detect neoplasms.

**Keywords:** nonlinear acoustics; acoustic tomography; ultrasound; Navier–Stokes equations

**MSC:** 76Q05; 65M08

## 1. Introducion

Currently, medical tomography is actively used to diagnose a pathologically altered area of the human organ at an early stage of the development of tumors and diseases. In order to increase its efficiency, new mathematical models, numerical solution methods and computer technologies in ultrasound tomography are being developed, which are based on the propagation of acoustic waves in a (liquid) medium. Therefore, the study of the generation, propagation and registration of acoustic waves in a liquid medium [1,2] is an important problem.

The development of methods and algorithms for ultrasound tomography for the diagnostics of breast cancer in the early stages has been actively studied recently [3–20]. Such problems are usually formulated in terms of coefficient inverse problems, when it is necessary to recover the parameters of the model (which in our case describes the propagation of ultrasound waves through the object) using measurement data [21–30].

At the moment, models based on systems of equations directly related to conservation laws, as well as models based on a second-order hyperbolic equation, are usually used to simulate ultrasound tomography processes. Models based on the second-order wave equation are usually easier to study, and therefore there are more ways to solve the forward problem efficiently, but there is no guarantee that the numerical solution is close to the physical one. The problem formulation based on conservation laws, which we consider in this article, requires more computational resources. The advantage is the close connection with the physics of wave propagation since the equations can be derived directly from conservation laws. Inverse problems for hyperbolic system were studied in [31].

Recently, there have been a lot of works on nonlinear acoustic tomography [32–39]. The development of an adequate mathematical model of medical acoustic tomography describing the propagation of waves in the human body, taking into account most physical effects, is an urgent problem.

In this paper, we use three-dimensional numerical modeling based on the solution of the Navier–Stokes equations to simulate the problem of acoustic wave propagation in human soft tissues [40]. The proposed approach allows to describe the process of forming, propagation and registration of the acoustic waves, taking into account non-stationary processes of turbulent mixing [41].

We were inspired by the work in which several aspects of ultrasound in medicine, like acoustic nonlinearity, finite amplitude distortion, shock wave formation, harmonic components, and non-linearly induced absorption were considered [42], as well as the saturation and the effect of these effects on ultrasound.

Let us consider some recent results, related to nonlinear acoustic tomography.

The method of measuring the acoustic properties of biological tissue elements at the microscopic level was developed on the basis of acoustic microscopy methods with mechanical scanning in the frequency range from 100 to 200 MHz in [43].

In [44], the authors investigated nonlinear effects, such as progressive distortion of the waveform, the generation of frequency harmonics and acoustic shocks, excessive energy release and acoustic saturation, during the propagation of ultrasound in liquids with relatively low acoustic attenuation. Similar effects are observed in soft tissues, although they are limited by absorption and scattering.

Sufficient conditions were given to recover both the sound speed of the medium being probed and the source [11] in thermoacoustic tomography and photoacoustic tomography, two coupled physics imaging modalitiesthat attempt to combine the high resolution of ultrasound and the high contrast capabilities of electromagnetic waves.

B. Kaltenbacher [45] investigated the issues of well-posedness and the effect of attenuation both for classical models of nonlinear acoustics and for some higher-order equations.

In [46], the authors proposed nonlinear acoustic tomography to obtain both the temperature and velocity fields from the time-of-flight measurements of sonic rays.

In [47], a complex magnetohydrodynamic model was investigated, which follows from the widely accepted dynamic theory of geomagnetism.

The theoretic basis of nonlinear acoustics was presented in [48].

The problem of the propagation of disturbances in a liquid medium from a harmonic oscillation source was considered [49]. Estimates of the required spatial and temporal resolutions were obtained to ensure acceptable accuracy of the solution.

In [50], the nonlinear acoustics was used to locate and visualize the defects of materials.

To test the developed numerical algorithms, it is very important to have phantom materials with acoustic parameters close to the parameters of the human body. In [51], the acoustic properties of seven phantom materials imitating tissues at different concentrations of their compounds and five phantom materials at room temperature were described and compared to determine the most suitable phantom material for contrast-enhanced ultrasound.

In [52], the 3D ultrasound computer tomography was used to increase the resolution in breast cancer imaging.

The inverse problem for a nonlinear wave equation with a damped and nonlinear term in medical acoustic sounding was investigated in [53].

It was established that 3D low-frequency ultrasound tomography is a safe, low-cost, high-resolution, whole-body medical imaging modality and gives high-resolution quantitatively accurate relevant images, which are potentially clinically useful [54,55].

The article discusses one of the most common numerical methods for modeling hydrodynamics: a finite-volume discretization method together with an iterative algorithm SIMPLE [40,56]. The final velocity of acoustic wave propagation is taken into account by using the equation of state for a liquid with a given compressibility coefficient. The numerical method is implemented on the LOGOS software ver. 5.3.22.837 [56]. The results of the numerical simulation of the propagation of sound waves in a digital acoustic tomograph, in the center of which is a digital mannequin of a person with internal inclusions (liver, adipose tissue), are presented in order to identify liver pathologies according to acoustic tomography. The mannequin is submerged in water. It is shown how the speed of sound affects the passage of acoustic waves inside the tomograph.

## 2. Governing Equations and the Numerical Method

Unsteady three-dimensional viscous fluid flows are described by a system of Navier–Stokes equations, which contain the continuity equation, the momentum conservation equation, and the energy conservation equation [40,57]. In a conservative form, in Cartesian coordinates, the system has the form:

$$\frac{\partial \rho}{\partial t} + \frac{\partial}{\partial x_j}(\rho u_i) = 0; \tag{1}$$

$$\frac{\partial \rho u_i}{\partial t} + \frac{\partial}{\partial x_j}(\rho u_i u_j) = -\frac{\partial p}{\partial x_i} + \frac{\partial}{\partial x_j}\tau_{ij}. \tag{2}$$

Here, $t > 0$ denotes the time, $u_i = (u_x, u_y, u_z) = (u, v, w)$ is the velocity, $\rho$ is the density, and $\tau_{ij}$ is the viscous stress tensor.

The final velocity of the acoustic wave propagation is taken into account by using the equation of state for a liquid with a given compressibility coefficient. The compression and expansion of the gas passes through an adiabatic process, where $k$ is the polytrope index and is defined as:

$$k = \frac{1}{\rho c^2}.$$

The system (1) and (2) is valid for both laminar and turbulent flows [40]. Their direct use for modeling flows is reduced to calculating the viscous stress tensor.

In addition, for the numerical solution, the system (1) and (2) must be supplemented by boundary conditions that depend, generally speaking, on the considered problem. In the case of incompressible flows, if the boundary of the calculation area is formed by solid walls, it is necessary to put all velocity components at the boundary equal to the corresponding components of the velocity of the solid surface, i.e., it is impossible for the liquid to slip along the "liquid–solid wall" boundary, or to move along the normal to it. On solid walls, the pressure gradient is zero:

$$\frac{\partial p}{\partial n} = 0.$$

The speed value is zero:

$$u = 0, \qquad v = 0, \qquad w = 0.$$

At the "liquid—liquid" boundary, the velocity and shear stresses must be continuous.

In the presence of an air boundary, zero static pressure is fixed on it, and the gradients of velocity and volume fractions are zero:

$$\frac{\partial u}{\partial n} = 0, \qquad \frac{\partial v}{\partial n} = 0, \qquad \frac{\partial w}{\partial n} = 0.$$

The non-reflective boundary conditions necessary for the free departure of waves from the computational domain are described in detail in [58].

Discretization of the system of Equations (1) and (2) is carried out by the finite volume method on an arbitrary unstructured grid, and for its numerical solution, a completely implicit method [59,60] based on the SIMPLE algorithm [40].

The modeling of flows with a free surface implies certain modifications of the SIMPLE algorithm. The description of the basic formulas of the modified SIMPLE algorithm, boundary conditions and implementation in the LOGOS software is described in detail [59,61].

We briefly present the discretization of the equations in the volume that is needed for further presentation. Let us consider the finite-volume scheme of the discretization of the equations used. The basic equation for solving the system (1) and (2) is the scalar quantity transfer equation:

$$\frac{\partial \rho \varphi}{\partial t} + \frac{\partial}{\partial x_j}(\rho \varphi u_j) = \frac{\partial}{\partial x_j}\tau_{ij} + Q. \tag{3}$$

The first term in (3) is a nonstationary term, the second is a convective term, and then a diffusion term. The equation may also contain sources and sinks represented by the last member of $Q$. The tensor $\tau_{ij}$ contains spatial derivatives of the desired value $\varphi$. We assume for simplicity that $\tau_j = \mu \frac{\partial \varphi}{\partial x_j}$. This significantly reduces calculations, but in no way reduces the generality of the methods described below.

We consider an arbitrary unstructured grid, the view of which is shown in Figure 1:

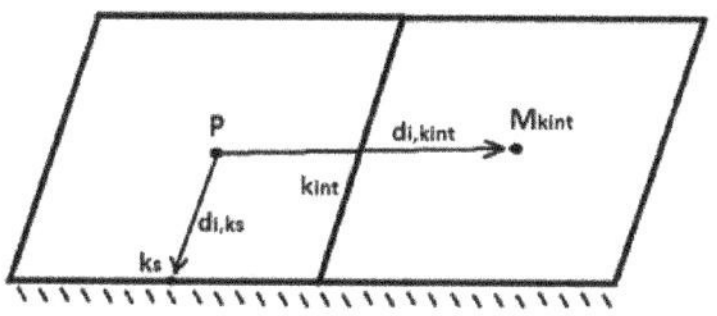


**Figure 1.** The calculated grid $k$ is the set of faces of the cell "P", which is composed of a set of inner faces of $k_{int}$ and a set of outer faces of $k_s$. The neighboring cell along the inner face of $k_{int}$ is denoted by $M_{k_{int}}$. The vector is the area of the face $k$—$S_{i,k}$, where $i = 0,1,2$ is the number of the vector component. The vector from the center of cell "P" to the center of cell "M" along the $k_{int}$ face is denoted as $d_{i,k_m} = r_{1,M} - r_{i,P}$, and the vector from the center of "P" to the center of the face is $d_{k_S} = r_{k_S} - r_P$, where $r_i$ is the radius vector.

Time discretization of Equation (3) according to the Adams–Beschfort scheme (the second-order scheme) is:

$$\frac{3\rho^{j+1}\varphi^{j+1} - 4\rho^j\varphi^j + \rho^{j-1}\varphi^{j-1}}{2\Delta t} + \left[\frac{\partial}{\partial x_j}(\rho\varphi u_j) - \frac{\partial}{\partial x_j}\tau_{ij} - Q\right]^{j+1} = 0.$$

To discretize Equation (3) in space, we integrate it in terms of the volume of the cell "P" and proceed to area integration for the convective and diffusion terms:

$$\int\limits_{V_P} \frac{3\rho^{j+1}\varphi^{j+1} - 4\rho^j\varphi^j + \rho^{j-1}\varphi^{j-1}}{2\Delta t} dV = \left(\frac{\rho^{j+1}\varphi^{j+1} - \rho^j\varphi^j}{\Delta t}\right)_P V_P,$$

$$\int\limits_{V_P} Q dV = Q_P V_P.$$

The discrete analog of the diffusion term is written in the following form:

$$\oint_{S_P} \mu \frac{\partial \varphi}{\partial x_j} dS_j \approx \sum_k \left( \mu \frac{\partial \varphi}{\partial x_j} \right)_k S_{j,K} = \sum_k \mu_k \left( \frac{\partial \varphi}{\partial n_k} \right)_k |S_K|.$$

Here, $n_k$ is the normal of the face $k$. In the right part, under the sign of the sum in the product, there is a derivative in the direction $\frac{\partial \varphi}{\partial n_k}$, which, in the case of an orthogonal calculation, can be defined as follows:

$$\frac{\partial \varphi}{\partial n_k} |S_K| = \frac{\varphi_M - \varphi_P}{|d_K|} |S_K|.$$

For approximation on a finite-volume grid, the convective term is written as:

$$\oint_{S_P} \rho \varphi u_j dS_j \approx \sum_k \rho_k \varphi_k u_{j,k} S_{j,k} = \sum_k \rho_k \varphi_k F_k.$$

Here, $F_k$ is the volumetric flow through the face $k$. The value on the face of $\varphi_k$ is determined by the applied sampling scheme of the convective term. There are a large number of sampling schemes that are applicable on arbitrary unstructured grids [62–65]. Among them, several schemes can be distinguished that have the "highest rating" of applicability in solving applied problems: counterflow scheme (Upwind Differences, UDs), counterflow scheme with linear interpolation (Linear Upwind Differences, LUDs), QUICK scheme, central difference scheme (Central Differences, CDs), schemes of the NVD family (Normalized Variable Diagram) and hybrid circuits (the above circuits mixed with a counterflow circuit to increase monotony).

Using this discretization, Equation (3) is replaced by a system of linear algebraic equations written for each calculation cell:

$$A_P \varphi_P + \sum_{k_{int}} A_{k_{int}} \varphi_{M_{k_{int}}} = R_P.$$

The resulting system of linear algebraic equations is solved using the AMG multigrid solver [56].

In an acoustic tomograph for the emission of acoustic waves, there are sources and a receiver that record the pressure. Sources and receivers are cylinders made of a material with acoustic parameters equal to water. The receiver records the entire pressure field inside the cylinder, and then the data of the inverse problem are averaged by volume.

To simulate an acoustic tomograph, the LOGOS software implements a technique that allows you to set the sources of acoustic waves, as well as process data in the receiver. The user can specify an arbitrary number of sources/receivers, specifying their location, size, the normal along which they are directed, and their status (that is, whether this "sensor" is a source or receiver). The source of acoustic waves is given as follows (Riker pulse):

$$I(t) = \left(1 - 2\left(\pi \nu_0 \left(t - \frac{1}{\nu_0}\right)\right)^2\right) e^{-\left(\pi \nu_0 \left(t - \frac{1}{\nu_0}\right)\right)^2}, \tag{4}$$

where $\nu_0$ is the frequency. The acoustic wave source $I(t)$ is included in the source $S_i$, which is added to the momentum Equation (2), and looks like this:

$$\vec{S} = A \cdot V \cdot \rho \cdot I(t) \cdot \vec{n}.$$

Here, $\rho$ is the cell density, $V$ is the cell volume, $A$ is the amplitude, and $\vec{n}$ is the normal in the direction of the signal. The pressure in each receiver is calculated by determining the cells that fall under the specified dimensions and location, and calculating the average pressure in the area of each receiver (volume integration).

## 3. Numerical Experiments

In this section, we consider a numerical simulation of the propagation of acoustic waves in the human body in order to study the possibility of determining liver pathologies with acoustic tomography.

Numerical calculations were performed for two cases: in the first case, two cubic inclusions are contained inside the mannequin symmetrically relative to the center—on the right, the inclusion parameters correspond to liver tissue, and on the left, adipose tissue; in the second case, two cubic inclusions are also located inside the mannequin, but the liver is surrounded by adipose tissue 0.5 cm thick. An acoustic tomograph is a cylinder filled with water. The parameters of the mannequin's body correspond to the muscle tissue (see Table 1). Liver pathologies include many diseases, including neoplasms. We considered fatty liver disease, in which there is an excessive accumulation of fat in the liver. To simplify the situation, we assumed that the liver is covered with an additional layer of adipose tissue.

**Table 1.** Acoustic parameters of the human body.

| | Speed of Sound, m/s | Density, kg/m$^3$ |
|---|---|---|
| Muscular tissue | 1550 | 994 |
| Adipose tissue | 1460 | 904 |
| Liver | 1570 | 1083 |

In Figure 2, we provide the scheme of the numerical experiment—the mannequin is located inside the tomograph (the cylinder):

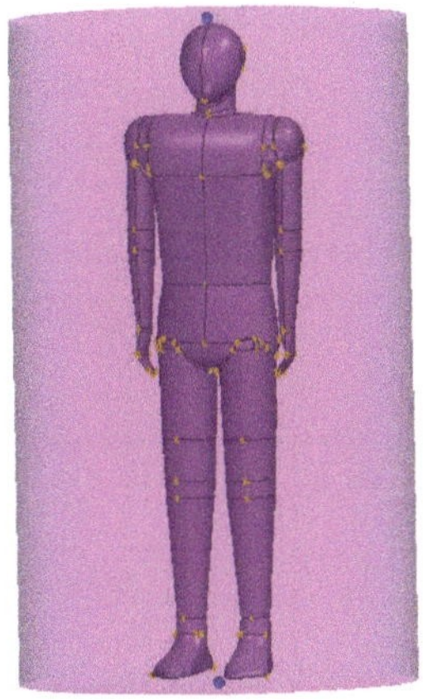

**Figure 2.** The digital mannequin of the human body inside the acoustic tomograph.

We use a block grid with a cell size of 0.004 m, consisting of 15 million cells, during the experiment. Figure 3 shows the resulting grid (in vertical and horizontal sections). At the initial moment of time, there are no disturbances, and the liquid in the tomograph is at rest.

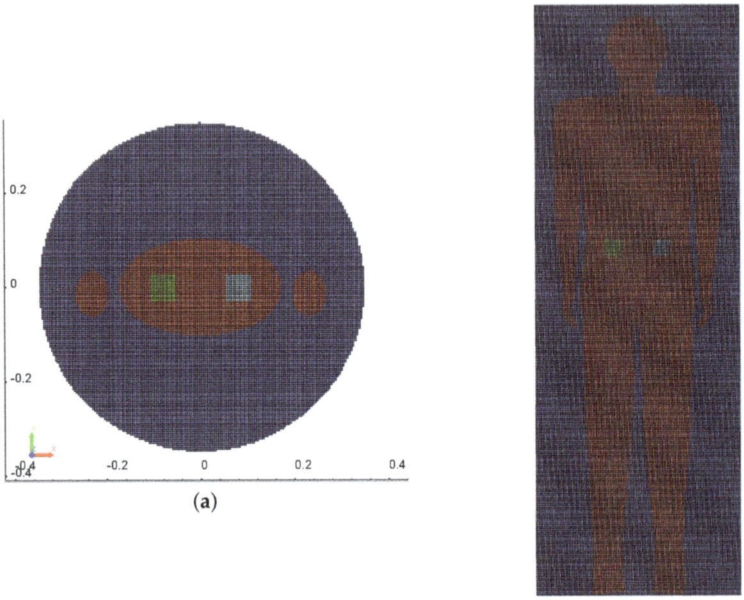


(a)

(b)

**Figure 3.** Constructed block grid: (**a**) calculated grid in the horizontal section, (**b**) calculated grid in the vertical section.

The digital mannequin parameters are a height of 1.95 m, shoulder width of 0.56 m, and thickness of 0.34 m. Parameters of the tomograph (cylinder) are a height of 2 m and diameter of 0.7 m. Table 1 shows the acoustic parameters of the human body.

The mean parameters of the human body, presented in Table 1, were taken in accordance with [43,51,66,67].

At the first stage, the calculation was carried out with two cubic inclusions, $0.05 \times 0.05 \times 0.05$ m, located symmetrically relative to the center inside the mannequin—liver (right) and adipose tissue (left). Eight transducers (sources and receivers) are located on the side of the tomograph at the level of the internal organs (Figure 4). The location of the source of the sounding wave in the calculations is marked with the number 3; all the others are receivers and record the pressure as a function of time. The source (receiver) is a cylinder with a diameter of 0.01 m and a height of 0.05 m made of a material with acoustic parameters equal to those of water. In the receiver, the pressure function captures the entire pressure field inside the cylinder and then averages by volume.

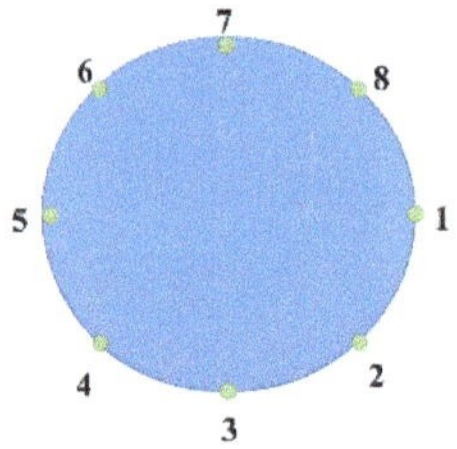


**Figure 4.** The location of sources and receivers.

During the computations, we used the Ricker wavelet (4) with the frequency $\nu_0$ = 20,000 Hz. In Figure 5, we demonstrate the pressure field at $t = 0.0004$ s in a section located at a height in the center of the internal organs.

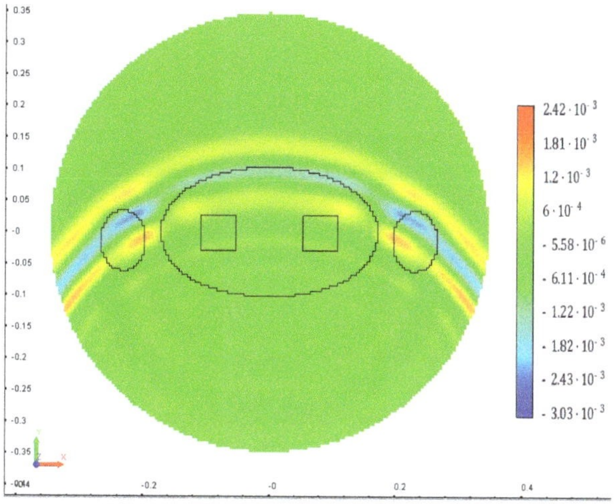


**Figure 5.** Experiment 1—the pressure field at $t = 0.0004$ s.

At the second stage, a calculation was carried out in which a cube with liver parameters is surrounded by a boundary layer with a thickness of 0.5 cm. Thus, there are two cubic inclusions in the mannequin, located symmetrically relative to the center inside the mannequin: adipose tissue (left) with a size of $0.05 \times 0.05 \times 0.05$ m, and liver (right) with a size of $0.06 \times 0.06 \times 0.06$ m. Figure 6 shows the pressure field at time 0.0004 s in a cross section.

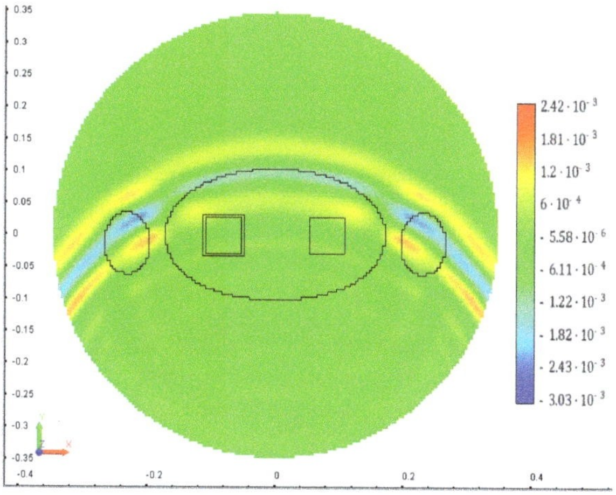


**Figure 6.** Experiment 2—the pressure field at $t = 0.0004$ s.

In Figure 7, one can see the pressure, registered in the receivers for both experiments.

Significant differences in pressure in the two calculations are observed in the 7th receiver. When the wave passes through the liver surrounded by adipose tissue, the pressure amplitude is increased compared to calculation 1, and the passage time of the acoustic waves is decreased due to the fact that the speed of sound in adipose tissue is higher than in water.

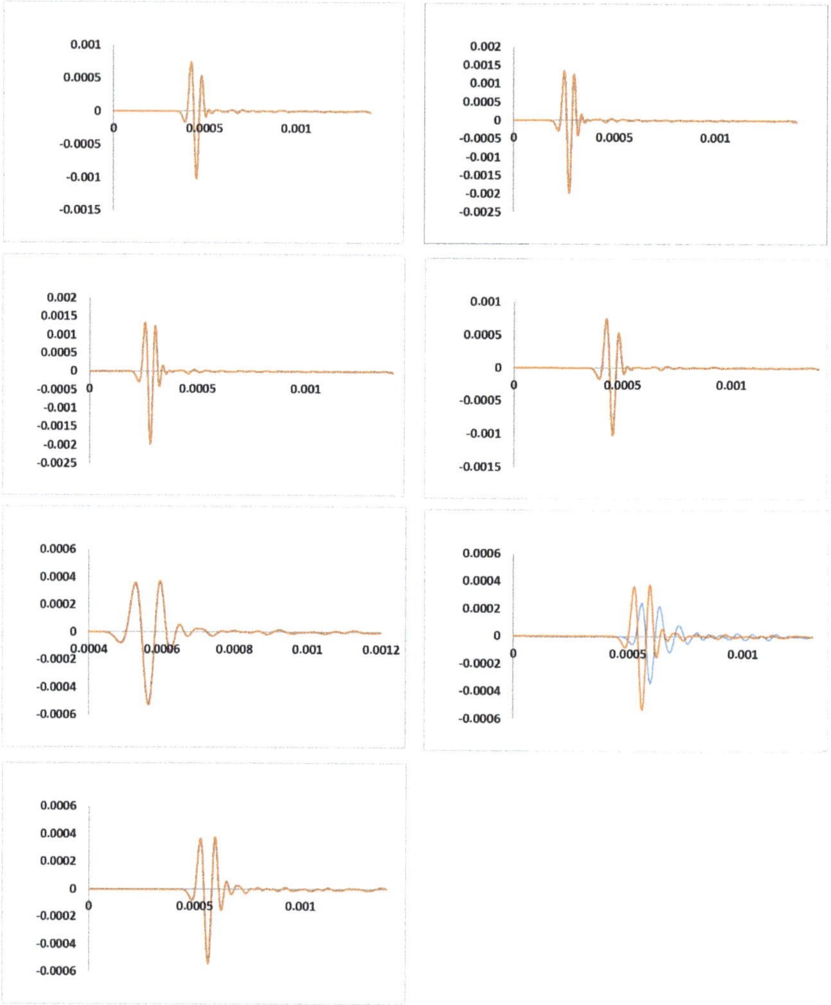


**Figure 7.** Comparison of two calculations (calculation 1—there is no layer of adipose tissue around the liver; calculation 2—the liver is surrounded by a layer of adipose tissue 0.01 m thick).

Minor differences are also observed in receiver 6 (Figure 8).

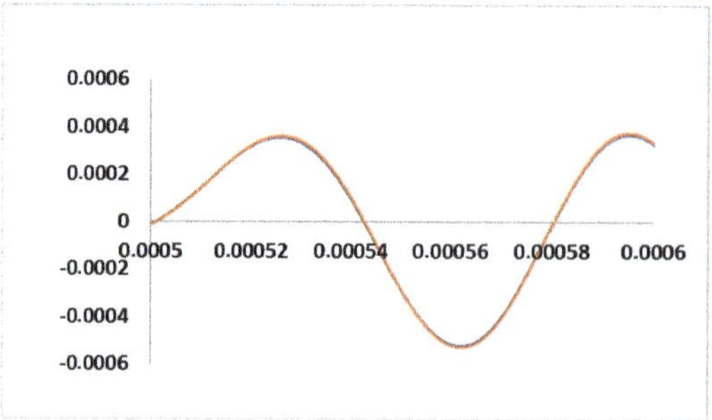


**Figure 8.** Comparison of pressure in receiver 6.

Figure 9 represents the quantitative difference of the registered pressure for the two experiments.

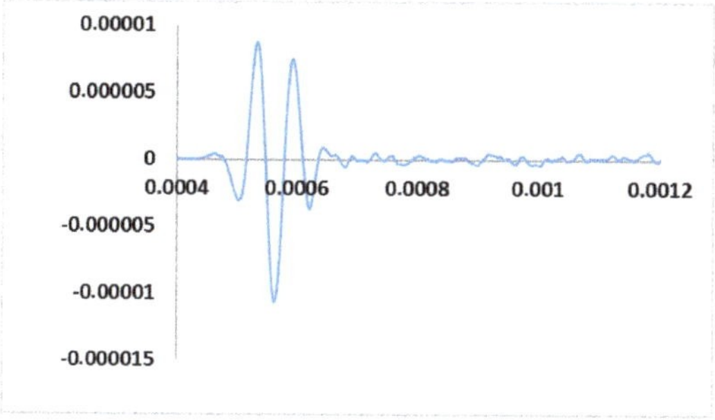


**Figure 9.** The registered pressure for the two experiments.

In order to test the efficiency of the technology, we carried out another numerical simulation, similar to calculation 2. The speed of sound in the adipose tissue surrounding the liver we set to 1000 m/s. Figure 10 shows graphs of pressure for three calculations in receivers 6 and 7, where the greatest difference is observed.

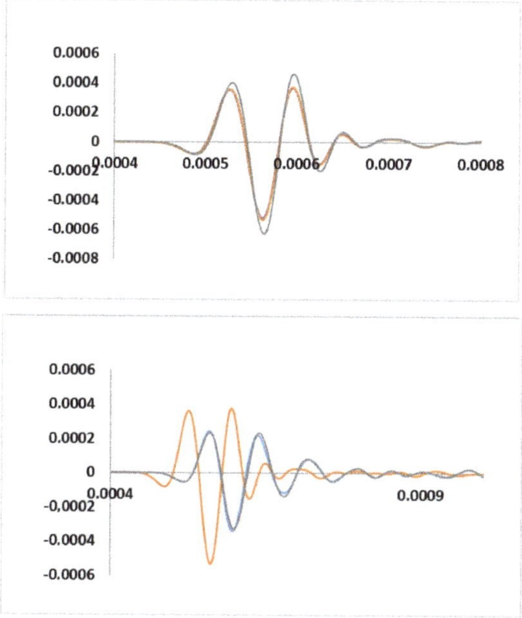


**Figure 10.** Comparison of two calculations (experiment 1—there is no layer of adipose tissue around the liver; experiment 2—the liver is surrounded by a layer of adipose tissue at a sound speed of 1460 m/s; experiment 3—the liver is surrounded by a layer of adipose tissue at a sound speed of 1000 m/s.

When a wave passes through the liver surrounded by adipose tissue at the speed of sound of 1000 m/s (experiment 3), the passage time of the acoustic waves to the receivers is increased compared to calculations 1 and 2. The pressure amplitude in receiver 6 is increased compared to experiments 1 and 2, and in receiver 7, it is lower compared to experiment 2.

## 4. Discussion

The article presents the results of the numerical simulation of acoustic wave propagation in the human body with nonlinear acoustic tomography based on the Navier–Stokes equations. This mathematical model takes into account most of the nonlinear effects of acoustic wave propagation, including shock waves. The computational technology presented in the article shows the fundamental possibility of using mathematical modeling methods based on the numerical solution of the complete hydrodynamic system of the Navier–Stokes equations to study the properties of acoustic waves during tomography of the human body. The introduction of such methods in the future will make it possible to understand almost all physical aspects of the functioning of such an ultrasound medical tomograph.

The article aimed to show the fundamental possibility of constructing a computational digital twin of the acoustic tomograph within the framework of a unified physics–mathematical model based on the Navier–Stokes equations. The authors proceeded from the fact that the size of the modeling area is quite small, sound waves are waves of "small" disturbance, and given that a human is more than 60% water, human organs can be modeled by a liquid model with their density. This is a kind of approximation in which, as in any model, there are assumptions, but it has the right to life and, in general, will allow us to determine the areas of change in the parameters of the sound wave to identify the required parameters. During numerical experiments, we obtained the pressure registered in the receivers that were located on the side walls of the tomograph. The differences in pressure values are shown depending on the configuration of inclusions in the mannequin imitating internal organs. The results show that the developed technology can be used to probe the human body in medical acoustic tomographs and determine the acoustic parameters of the human body to detect neoplasms. Numerical solution methods are currently being developed and implemented to solve inverse problems [68–72], as well as methods based on deep learning [73–75].

**Author Contributions:** Problem Formulation, M.S.; Methodology, A.K.; Software, A.K.; Validation, A.K.; Data curation, N.N.; Writing—original draft, A.K.; Writing—review & editing, M.S. and N.N.; Supervision, M.S.; Project administration, M.S.; Funding acquisition, M.S. All authors have read and agreed to the published version of the manuscript.

**Funding:** The work was supported by Russian Science Foundation, 19-11-00154 "Developing of new mathematical models of acoustic tomography in medicine. Numerical methods, HPC and software".

**Data Availability Statement:** Data are contained within the article.

**Conflicts of Interest:** The authors declare no conflicts of interest.

## References

1. Strutt, J.W. *The Theory of Sound*; Cambridge University Press: Cambridge, UK, 2011; Volume 1.
2. Strutt, J.W. *The Theory of Sound*; Cambridge University Press: Cambridge, UK, 2011; Volume 2.
3. Li, S.; Jackowski, M.; Dione, D.P.; Varslot, T.; Staib, L.H.; Mueller, K. Refraction corrected transmission ultrasound computed tomography for application in breast imaging. *Med. Phys.* **2010**, *37*, 2233–2246. [CrossRef] [PubMed]
4. Huthwaite, P.; Simonetti, F. High-resolution imaging without iteration: A fast and robust method for breast ultrasound tomography. *J. Acoust. Soc. Am.* **2011**, *130*, 1721–1734. [CrossRef] [PubMed]
5. Duric, N.; Littrup, P.; Li, C.; Roy, O.; Schmidt, S.; Janer, R.; Cheng, X.; Goll, J.; Rama, O.; Bey-Knight, L.; et al. Breast ultrasound tomography: Bridging the gap to clinical practice. *Proc. SPIE* **2012**, *8320*, 832000.
6. Jirik, R.; Peterlik, I.; Ruiter, N.; Fousek, J.; Dapp, R.; Zapf, M.; Jan, J. Sound-speed image reconstruction in sparse-aperture 3D ultrasound transmission tomography. *IEEE Trans. Ultrason. Ferroelectr. Freq. Control* **2012**, *59*, 254–264. [CrossRef] [PubMed]
7. Marmot, M.G.; Altman, D.G.; Cameron, D.A.; Dewar, J.A.; Thompson, S.G.; Wilcox, M. The benefits and harms of breast cancer screening: An independent review. *Br. J. Cancer* **2013**, *108*, 2205–2240. [CrossRef] [PubMed]
8. Birk, M.; Dapp, R.; Ruiter, N.V.; Becker, J. GPU-based iterative transmission reconstruction in 3D ultrasound computer tomography. *J. Parallel Distrib. Comput.* **2014**, *74*, 1730–1743. [CrossRef]
9. Burov, V.A.; Zotov, D.I.; Rumyantseva, O.D. Reconstruction of the sound velocity and absorption spatial distributions in soft biological tissue phantoms from experimental ultrasound tomography data. *Acoust. Phys.* **2015**, *61*, 231–248. [CrossRef]

10. Sandhu, G.Y.; Li, C.; Roy, O.; Schmidt, S.; Duric, N. Frequency domain ultrasound waveform tomography: Breast imaging using a ring transducer. *Phys. Med. Biol.* **2015**, *60*, 5381–5398. [CrossRef]
11. Liu, H.; Uhlmann, G. Determining both sound speed and internal source in thermo- and photo-acoustic tomography. *Inverse Probl.* **2015**, *31*, 105005. [CrossRef]
12. Huang, L.; Shin, J.; Chen, T.; Lin, Y.; Gao, K.; Intrator, M.; Hanson, K. Breast ultrasound tomography with two parallel transducer arrays. In *Medical Imaging 2016: Physics of Medical Imaging (International Society for Optics and Photonics)*; SPIE: Bellingham, WA, USA, 2016; Volume 9783, p. 97830C.
13. Matthews, T.P.; Wang, K.; Li, C.; Duric, N.; Anastasio, M.A. Regularized dual averaging image reconstruction for full-wave ultrasound computed tomography. *IEEE Trans. Ultrason. Ferroelectr. Freq. Control* **2017**, *64*, 811–825. [CrossRef]
14. Opieliński, K.J.; Pruchnicki, P.; Szymanowski, P.; Szepieniec, W.K.; Szweda, H.; Świś, E.; Jóźwik, M.; Tenderenda, M.; Bułkowski, M. Multimodal ultrasound computer-assisted tomography: An approach to the recognition of breast lesions. *Comput. Med. Imaging Graph.* **2018**, *65*, 102–114. [CrossRef]
15. Malik, B.; Terry, R.; Wiskin, J.; Lenox, M. Quantitative transmission ultrasound tomography: Imaging and performance characteristics. *Med. Phys.* **2018**, *45*, 3063–3075. [CrossRef] [PubMed]
16. Guo, R.; Lu, G.; Qin, B.; Fei, B. Ultrasound imaging technologies for breast cancer detection and management: A review. *Ultrasound Med. Biol.* **2018**, *44*, 37–70. [CrossRef] [PubMed]
17. Wiskin, J.; Malik, B.; Natesan, R.; Lenox, M. Quantitative assessment of breast density using transmission ultrasound tomography. *Med. Phys.* **2019**, *46*, 2610–2620. . [CrossRef] [PubMed]
18. Sood, R.; Rositch, A.F.; Shakoor, D.; Ambinder, E.; Pool, K.L.; Pollack, E.; Mollura, D.J.; Mullen, L.A.; Harvey, S.C. Ultrasound for breast cancer detection globally: A systematic review and meta-analysis. *J. Glob. Oncol.* **2019**, *5*, 1–17. [CrossRef] [PubMed]
19. Vourtsis, A. Three-dimensional automated breast ultrasound: Technical aspects and first results. *Diagn. Intervent. Radiol.* **2019**, *100*, 579–592. [CrossRef]
20. Park, C.K.S.; Xing, S.; Papernick, S.; Orlando, N.; Knull, E.; Toit, C.D.; Bax, J.S.; Gardi, L.; Barker, K.; Tessier, D.; et al. Spatially tracked whole-breast three-dimensional ultrasound system toward point-of-care breast cancer screening in high-risk women with dense breasts. *Med. Phys.* **2022**, *49*, 3944–3962. [CrossRef]
21. Wang, K.; Matthews, T.; Anis, F.; Li, C.; Duric, N.; Anastasio, M.A. Waveform inversion with source encoding for breast sound speed reconstruction in ultrasound computed tomography. *IEEE Trans. Ultrason. Ferroelectr. Freq. Control* **2015**, *62*, 475–493. [CrossRef]
22. Bernard, S.; Monteiller, V.; Komatitsch, D.; Lasaygues, P. Ultrasonic computed tomography based on full-waveform inversion for bone quantitative imaging. *Phys. Med. Biol.* **2017**, *62*, 7011–7035. [CrossRef]
23. Pérez-Liva, M.; Herraiz, J.L.; Udías, J.M.; Miller, E.; Cox, B.T.; Treeby, B.E. Time domain reconstruction of sound speed and attenuation in ultrasound computed tomography using full wave inversion. *J. Acoust. Soc. Am.* **2017**, *141*, 1595–1604. [CrossRef]
24. Filatova, V.; Danilin, A.; Nosikova, V.; Pestov, L. Supercomputer Simulations of the Medical Ultrasound Tomography Problem. *Commun. Comput. Inf. Sci.* **2019**, *1063*, 297–308.
25. Kabanikhin, S.I.; Klyuchinskiy, D.V.; Novikov, N.S.; Shishlenin, M.A. Numerics of acoustical 2D tomography based on the conservation laws. *J. Inverse Ill-Posed Probl.* **2020**, *28*, 287–297. [CrossRef]
26. Klyuchinskiy, D.; Novikov, N.; Shishlenin, M. Recovering density and speed of sound coefficients in the 2d hyperbolic system of acoustic equations of the first order by a finite number of observations. *Mathematics* **2021**, *9*, 199. [CrossRef]
27. Filatova, V.; Pestov, L.; Poddubskaya, A. Detection of velocity and attenuation inclusions in the medical ultrasound tomography. *J. Inverse Ill-Posed Probl.* **2021**, *29*, 459–466. [CrossRef]
28. Rumyantseva, O.D.; Shurup, A.S.; Zotov, D.I. Possibilities for separation of scalar and vector characteristics of acoustic scatterer in tomographic polychromatic regime. *J. Inverse Ill-Posed Probl.* **2021**, *29*, 407–420. [CrossRef]
29. Shurup, A.S. Numerical comparison of iterative and functional-analytical algorithms for inverse acoustic scattering. *Eurasian J. Math. Comput. Appl.* **2022**, *10*, 79–99. [CrossRef]
30. Lucka, F.; Pérez-Liva, M.; Treeby, B.E.; Cox, B.T. High resolution 3D ultrasonic breast imaging by time-domain full waveform inversion. *Inverse Probl.* **2022**, *38*, 025008. [CrossRef]
31. Romanov, V.G.; Kabanikhin S.I. *Inverse Problems for Maxwell's Equations*; VSP: Utrecht, The Netherlands, 1994.
32. Jafarzadeh, E.; Amini, M.H.; Sinclair, A.N. Spectral Shift Originating from Non-linear Ultrasonic Wave Propagation and Its Effect on Imaging Resolution. *Ultrasound Med. Biol.* **2021**, *47*, 1893–1903. [CrossRef]
33. Campo-Valera, M.; Villó-Pérez, I.; Fernández-Garrido, A.; Rodríguez-Rodríguez, I.; Asorey-Cacheda, R. Exploring the Parametric Effect in Nonlinear Acoustic Waves. *IEEE Access* **2023**, *11*, 97221–97238. [CrossRef]
34. Ichida, N.; Sato, T.; Linzer, M. Imaging the nonlinear ultrasonic parameter of a medium. *Ultrason. Imaging* **1983**, *5*, 295–299. [CrossRef]
35. Lewin, P.A.; Nowicki, A. 16—Nonlinear acoustics and its application to biomedical ultrasonics. In *Woodhead Publishing Series in Electronic and Optical Materials, Ultrasonic Transducers*; Nakamura, K., Ed.; Woodhead Publishing: Cambridge, UK, 2012; pp. 517–544. . [CrossRef]
36. Meliani, M.; Nikolić, V. Analysis of General Shape Optimization Problems in Nonlinear Acoustics. *Appl. Math. Optim.* **2022**, *86*, 39. [CrossRef]

37. Panfilova, A.; van Sloun, R.J.; Wijkstra, H.; Sapozhnikov, O.A.; Mischi, M. A review on B/A measurement methods with a clinical perspective. *J. Acoust. Soc. Am.* **2021**, *149*, 2200. [CrossRef]
38. Tabak, G.; Oelze, M.L.; Singer, A.C. Effects of acoustic nonlinearity on communication performance in soft tissues. *J. Acoust. Soc. Am.* **2022**, *152*, 3583. [CrossRef]
39. Kokumo, K.; Sharma, A.; Myers, K.; Ambinder, E.; Oluyemi, E.; Bell, M.A.L. Theoretical basis and experimental validation of harmonic coherence-based ultrasound imaging for breast mass diagnosis. In *Medical Imaging 2023: Physics of Medical Imaging*; SPIE: Bellingham, WA, USA, 2023; Volume 12463, pp. 95–104.
40. Ferziger, J.H.; Peric, M. *Computational Method for Fluid Dynamics*; Springer-Verlag: New York, NY, USA, 2002.
41. Mozer, D.; Kim, J.; Mansour, N.N. DNS of Turbulent Channel Flow. *J. Phys. Fluids* **1999**, *11*, 943–945.
42. Muir, T.G.; Carstensen, E.L. Prediction of nonlinear acoustic effects at biomedical frequencies and intensities. *Ultrasound Med. Biol.* **1980**, *6*, 345–357. [CrossRef] [PubMed]
43. Okawai, H.; Kobayashi, K.; Nitta, S. An approach to acoustic properties of biological tissues using acoustic micrographs of attenuation constant and sound speed. *J. Ultrasound Med.* **2001**, *20*, 891–907. [CrossRef] [PubMed]
44. Duck, F.A. Nonlinear acoustics in diagnostic ultrasound. *Ultrasound Med. Biol.* **2002**, *28*, 1–18. [CrossRef]
45. Kaltenbacher, B. Mathematics of nonlinear acoustics. *Evol. Equ. Control. Theory* **2015**, *4*, 447–491. [CrossRef]
46. Yu, T.; Cai, W. Simultaneous reconstruction of temperature and velocity fields using nonlinear acoustic tomography. *Appl. Phys. Lett.* **2019**, *115*, 104104. [CrossRef]
47. Deng, Y.; Li, J.; Liu, H. On Identifying Magnetized Anomalies Using Geomagnetic Monitoring Within a Magnetohydrodynamic model. *Arch. Ration. Mech. Anal.* **2020**, *235*, 691–721. [CrossRef]
48. Gan, W.S. *Nonlinear Acoustical Imaging*; Springer: Singapore, 2021.
49. Kozelkov, A.S.; Krutyakova, O.G.L.; Kurulin, V.V.; Strelets, D.Y.E.; Shishlenin, M.A. The accuracy of numerical simulation of the acoustic wave propagations in a liquid medium based on Navier-Stokes equations. *Sib. Electron. Math. Rep.* **2021**, *18*, 1238–1250. [CrossRef]
50. Maev, R.G.; Seviaryn, F. Applications of non-linear acoustics for quality control and material characterization. *J. Appl. Phys.* **2022**, *132*, 161101. [CrossRef]
51. Chen, P.; Pollet, A.M.; Panfilova, A.; Zhou, M.; Turco, S.; den Toonder, J.M.; Mischi, M. Acoustic characterization of tissue-mimicking materials for ultrasound perfusion imaging research. *Ultrasound Med. Biol.* **2022**, *48*, 124–142. [CrossRef]
52. Gemmeke, H.; Hopp, T.; Zapf, M.; Kaiser, C.; Ruiter, N.V. 3D ultrasound computer tomography: Hardware setup, reconstruction methods and first clinical results. *Nucl. Instrum. Methods Phys. Res. Sect. A Accel. Spectrometers Detect. Assoc. Equip.* **2017**, *873*, 59–65. [CrossRef]
53. Zhang, Y. Nonlinear acoustic imaging with damping. *arXiv* **2023**, arXiv.2302.14174.
54. Wiskin, J.; Malik, B.; Klock, J. Low frequency 3D transmission ultrasound tomography: Technical details and clinical implications. *Z. Med. Phys.* **2023**, *33*, 427–443. [CrossRef]
55. Wiskin, J.; Malik, B.; Ruoff, C.; Pirshafiey, N.; Lenox, M.; Klock, J. Whole-Body Imaging Using Low Frequency Transmission Ultrasound. *Acad. Radiol.* **2023**, *30*, 2674–2685. [CrossRef]
56. Lashkin, S.V.; Kozelkov, A.S.; Yalozo, A.V.; Gerasimov, V.Y.; Zelensky, D.K. Efficiency analysis of the parallel implementation of the simple algorithm on multiprocessor computers. *J. Appl. Mech. Tech. Phys.* **2017**, *58*, 1242–1259. [CrossRef]
57. Kozelkov, A.; Tyatyushkina, E.; Kurkin, A.; Kurulin, V.; Kurkina, O.; Kochetkova, O. Fluid flow simulation in a *T*-connection of square pipes using modern approaches to turbulence modeling. *Sib. Electron. Math. Rep.* **2023**, *20*, 25–46.
58. Kozelkov, A.; Kurkin, A.; Utkin, D.; Tyatyushkina, E.; Kurulin, V.; Strelets, D. Application of Non-Reflective Boundary Conditions in Three-Dimensional Numerical Simulations of Free-Surface Flow Problems. *Geosciences* **2022**, *12*, 427. [CrossRef]
59. Kozelkov, A.S.; Lashkin, S.V.; Efremov, V.R.; Volkov, K.N.; Tsibereva, Y.A.; Tarasova, N.V. An implicit algorithm of solving Navier–Stokes equations to simulate flows in anisotropic porous media. *Comput. Fluids* **2018**, *160*, 164–174. [CrossRef]
60. Chen, Z.J.; Przekwas, A.J. A coupled pressure-based computational method for incompressible/compressible flows. *J. Comput. Phys.* **2010**, *229*, 9150–9165. [CrossRef]
61. Moukalled, F.; Darwish, M. Pressure-Based Algorithms for Multi-Fluid Flow at All Speeds—Part I: Mass Conservation Formulation, Numerical Heat Transfer. *Part B Fundam.* **2004**, *45*, 495–522. [CrossRef]
62. Jasak, H. Error Analysis and Estimation for the Finite Volume Method with Applications to Fluid Flows. Ph.D. Thesis, Department of Mechanical Engineering, Imperial College of Science, London, UK, 1996.
63. Jasak, H.; Weller, H.G.; Gosman, A.D. High resolution NVD differencing scheme for arbitrarily unstructured meshes. *Int. J. Numer. Methods Fluids* **1999**, *31*, 431–449. [CrossRef]
64. Leonard, B.P. A stable and accurate convective modeling procedure based on quadratic upstream interpolation. *Comput. Methods Appl. Mech./Eng.* **1979**, *19*, 59–98. [CrossRef]
65. Gaskell, P.H. Curvature-compensated convective-transport—SMART, A new boundedness-preserving transport algorithm. *Int. J. Numer. Methods Fluids* **1988**, *8*, 617–641. [CrossRef]
66. Goss, S.A.; Johnston, R.L.; Dunn, F. Comprehensive compilation of empirical ultrasonic properties of mammalian tissues. *J. Acoust. Soc. Am.* **1978**, *64*, 423–457. [CrossRef]
67. Mast, T.D. Empirical relationships between acoustic parameters in human soft tissues. *Acoust. Res. Lett.* **2000**, *1*, 37–42. [CrossRef]

68. Klyuchinskiy, D.; Novikov, N.; Shishlenin, M. A Modification of gradient descent method for solving coefficient inverse problem for acoustics equations. *Computation* **2020**, *8*, 73. [CrossRef]
69. Shishlenin, M.A.; Savchenko, N.A.; Novikov, N.S.; Klyuchinskiy, D.V. On the reconstruction of the absorption coefficient for the 2D acoustic system. *Sib. Electron. Math. Rep.* **2023**, *20*, 1474–1489.
70. Klyuchinskiy, D.V.; Novikov, N.S.; Shishlenin, M.A. CPU-time and RAM memory optimization for solving dynamic inverse problems using gradient-based approach. *J. Comput. Phys.* **2021**, *439*, 110374. [CrossRef]
71. Jenaliyev, M.T.; Bektemesov, M.A.; Yergaliyev, M.G. On an inverse problem for a linearized system of Navier–Stokes equations with a final overdetermination condition. *J. Inverse Ill-Posed Probl.* **2023**, *31*, 611–624. [CrossRef]
72. Wang, Z.; Dulikravich, G. A numerical Method for Solving Cascade Inverse Problems Using Navier-Stokes Equations. 33rd Aerospace Sciences Meeting and Exhibit. 1995. Published Online: 22 August 2012. Available online: https://arc.aiaa.org/doi/abs/10.2514/6.1995-304 (accessed on 20 November 2023).
73. Arridge, S.; Maass, P.; Öktem, O.; Schönlieb, C. Solving inverse problems using data-driven models. *Acta Numer.* **2019**, *28*, 1–174. [CrossRef]
74. Basurto-Hurtado, J.A.; Cruz-Albarran, I.A.; Toledano-Ayala, M.; Ibarra-Manzano, M.A.; Morales-Hernandez, L.A.; Perez-Ramirez, C.A. Diagnostic Strategies for Breast Cancer Detection: From Image Generation to Classification Strategies Using Artificial Intelligence Algorithms. *Cancers* **2022**, *14*, 3442. [CrossRef]
75. Fan, Y.; Wang, H.; Gemmeke, H.; Hopp, T.; Hesser, J. Model-data-driven image reconstruction with neural networks for ultrasound computed tomography breast imaging. *Neurocomputing* **2022**, *467*, 10–21. [CrossRef]

symmetry

MDPI

Article

# Adaptive Anti-Disturbance Control of Dissolved Oxygen in Circulating Water Culture Systems

Xiaonan Yao, Gang Zhang *, Song Yang and Yuehua Chen

Faculty of Maritime and Transportation, Ningbo University, Ningbo 315211, China;
2111084045@nbu.edu.cn (X.Y.); 2011084039@nbu.edu.cn (S.Y.); chenyuehua@nbu.edu.cn (Y.C.)
* Correspondence: zhanggang@nbu.edu.cn

**Abstract:** In the three-dimensional culture model, the breeding basket of the culture area is symmetrical and it is important to control the dissolved oxygen in the symmetrical region to improve the culture efficiency. Practical engineering issues, such as the influence of flow rate, pH, water temperature, and biological oxygen consumption on the dissolved oxygen content in the circulating water culture system, must be considered along with the presence of modeling errors in the control model. The authors propose an adaptive anti-disturbance control strategy for dissolved oxygen that combines nonlinear disturbance observation with an adaptive sliding model control. Initially, a dynamic model for controlling dissolved oxygen in a recirculating water aquaculture system was developed. The model considers external disturbances like artificial oxygenation, abrupt changes in system flow, and variations in culture oxygen consumption. Secondly, to enhance the robustness and accuracy of controlling dissolved oxygen concentration, the paper introduces a nonlinear adaptive disturbance observer for real-time estimation and observation of external disturbances and system uncertainties. This is accompanied by a sliding-mode control-based adaptive anti-disturbance strategy. Lastly, the simulation results demonstrate that the control strategy proposed in this paper shows resistance to system uncertainties and unknown external disturbances. Furthermore, it reduces the model accuracy requirements for the controller and proves to be suitable for accurately controlling dissolved oxygen in circulating water systems.

**Keywords:** recirculating water farming; dissolved oxygen; dissolved oxygen control; nonlinear disturbance observer; slide model control

## 1. Introduction

In recent years, with the decline of marine resources, the fishing industry, which has been primarily based on marine fishing, has gradually shifted towards aquaculture. Traditional aquaculture practices are often extensive, resulting in low yields and high pollution, leading to continuous deterioration of the aquaculture environment. These drawbacks cannot meet the goals of sustainable development. In response to these challenges, the concept of "recirculating aquaculture systems" has emerged as a promising solution. Factory recirculating aquaculture is a novel farming model that offers advantages such as high productivity and water conservation [1]. In aquaculture, breeding baskets are symmetrically distributed; the quality of aquaculture water directly affects the survival of farmed organisms [2]. Therefore, real-time monitoring of water quality and regulation of the corresponding water treatment equipment to meet the requirements of aquaculture is a necessary prerequisite for ensuring the recyclability of water. Dissolved oxygen is a fundamental parameter in water quality and imbalances in its concentration can cause sickness or death in aquatic species [3,4]. As a result, accurate control of dissolved oxygen in recirculating water culture is critical. However, achieving such control is challenging due to uncertainties arising from factors like environmental temperature variations, biological activity, human behavior, and changes in system flow during water treatment processes [5].

Furthermore, there are clear nonlinear characteristics between dissolved oxygen content and aeration as well as other disturbances [6–8]. Moreover, because the influencing factors such as biomass and biological respiration rate are continually changing within a particular range, the dissolved oxygen control system has typical time-varying properties in the meantime. At the same time, to assure the quality of the flowing water, the oxygenation equipment generally operates in the continuous load mode for an extended length of time [9]. However, various water treatment loads occur at different periods during the breeding process and the fixed-load operating strategy has resulted in a significant energy burden. According to the literature [10], the intelligent control of the aeration equipment using frequency conversion technology is proposed. The oxygenator operates in a low-power mode when the dissolved oxygen concentration meets the requirements but the power increases when there is a significant deviation from the set value. However, the operating power is increased when the dissolved oxygen concentration deviates significantly from the set value. Compared to traditional control methods, the frequency conversion control method saves more than 42.3% of the power. As a result, variable load accurate regulation of dissolved oxygen concentration is an essential research subject to enhance water quality and save energy.

To achieve accurate regulation of dissolved oxygen, it is necessary to establish a mathematical model for dissolved oxygen. Numerous researchers are now working on dissolved oxygen modeling [11–13]. Huang et al. [14,15] employed a small indoor culture pond as a study object and the dissolved oxygen system was modeled as a transfer function while disregarding temperature and pH variations and the model parameters were obtained experimentally. Zhou et al. [16] created a dynamic model of dissolved oxygen in a recirculating water culture system, assuming constant system flow, culture mass, and biological oxygen consumption rate, and provided model parameters for dissolved oxygen without taking into account changes in internal and external factors via aeration experiments. Based on diffusion theory and material balance, Yin et al. [17] created a dynamic model of the daily change in dissolved oxygen at different depths in a crab pond. The model considers plant photosynthesis, biological respiration, mechanical oxygenation, and other variables that influence dissolved oxygen variations. The experimental findings suggest that the model's average absolute inaccuracy is around 6%. In the model research of dissolved oxygen in the recirculating water culture system, there are limitations to assuming constant parameters like circulating water flow, water temperature, pH, culture mass, and biological oxygen consumption, ignoring system uncertainties like disturbance factors and using a transfer function form as a mathematical model for dissolved oxygen in circulating water culture systems.

The traditional PID control [18,19] is more extensively employed in the dissolved oxygen control and has achieved certain successes but, since its control parameters tend to remain constant, the control performance in the dissolved oxygen system with uncertainty is low [20]. The literature [21–23] brought intelligent control approaches into the classic PID control, such as neural network techniques and fuzzy control methods. The adaptiveness of dissolved oxygen management is increased to some extent by using the neural network's nonlinear function approximation abilities and fuzzy inference rules based on expert knowledge to update PID parameters live. A fuzzy controller was devised for the problem of dissolved oxygen regulation in indoor shrimp growth systems in the literature [24] and the controller output was regulated based on the dissolved oxygen tracking error and the rate of change in the error. Circulating water culture systems are prone to nonlinear disturbances, parameter uncertainty, and other variables and fuzzy control lacks autonomous learning capability, requiring expert knowledge to construct inference rules [25]. A fuzzy neural network internal mode controller was proposed in the literature [26] to enhance dissolved oxygen management by online modification of filter settings using a fuzzy neural network. However, factors such as the network structure and initial value setting influence neural network performance had a significant impact on adaptive performance [27]. As a result, the problem of adaptive immunity regulation of dissolved oxygen system under internal and external disruptions and uncertainty environment requires more investigation.

This paper proposes a dissolved oxygen adaptive anti-disturbance control strategy for the problem of dissolved oxygen control in circulating water aquaculture systems with nonlinearity, time variation, and uncertainty which reduces water quality fluctuation while improving dissolved oxygen control accuracy. The following concerns were highlighted:

(1) The effects of temperature fluctuations, changes in circulating water flow, variations in biological respiration rates, modeling errors, and other factors were comprehensively considered to develop a mathematical model for dissolved oxygen in circulating water aquaculture systems that aligns with actual working conditions;
(2) To address the issue of uncertain disturbances in dissolved oxygen control of circulating water aquaculture systems, a nonlinear disturbance observer with a simple structure and real-time capability is introduced. This observer effectively eliminates the influence of disturbances and model uncertainty on dissolved oxygen. Building upon this, an adaptive sliding mode controller is designed to enhance dissolved oxygen control immunity and ensure water quality stability.

The structure of this paper is organized as follows. Section 1 presents a mathematical model for dissolved oxygen in the recirculating aquaculture system. Section 2 shows the problem of uncertain disturbances in dissolved oxygen control within the recirculating aquaculture system, which is addressed herein. This section introduces a nonlinear disturbance observer and designs an adaptive sliding mode controller to regulate the dissolved oxygen levels in the water. Section 3 conducts simulations to validate the performance of the controller.

## 2. Materials and Methods

### *2.1. Dissolved Oxygen Control Model*

#### 2.1.1. Recirculating Water Farming System

The main process structure of the system is shown in Figure 1, including culture ponds, sand filters, biofilters, and other water treatment systems. The water from the breeding pond undergoes purification through a microfilter which effectively removes fine suspended materials. The protein separator separates most of the organic impurities in the water, the sand filtering pond precipitates and filters the water impurities, and the breeding water then flows through the biofilter equipped with an aeration device and a biological filler to reduce the ammonia and nitrogen content of the breeding water; finally, the breeding water is pumped back into the breeding pond after sterilization and disinfection.

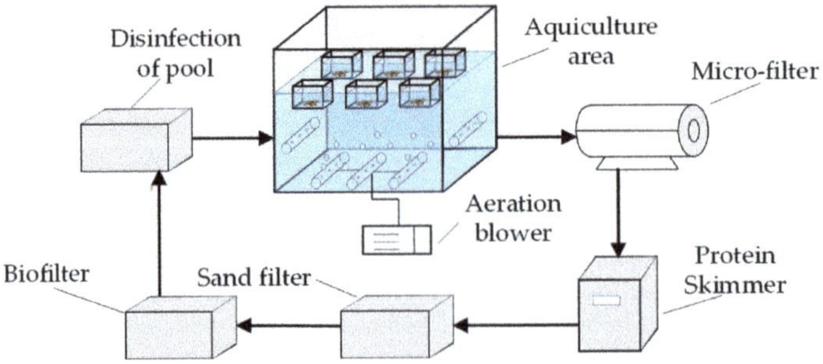


**Figure 1.** Diagram of circulating water system.

Temperature, pH, biotic and abiotic oxygen consumption, artificial oxygenation, and intake and outflow water flow all influence the concentration of dissolved oxygen in the culture pond in the circulating water culture area. It is difficult to precisely explain its dynamic features due to its usual nonlinear, coupled, and unpredictable qualities. As a result, before modeling dissolved oxygen in circulating water systems, acceptable assumptions must be established. The following assumptions are commonly used in current dissolved oxygen modeling research [14–16]:

**Hypothesis 1 (H1).** *Over a short period of time, there is a negligible increase in biological mass in the culture system.*

**Hypothesis 2 (H2).** *Changes in water temperature, pH, and biological respiration rate are negligible.*

**Hypothesis 3 (H3).** *In the culture region, the microporous aeration process is continuous and stable and the dissolved oxygen content in the water bodies is more equally distributed.*

However, when the culture system is in use, the biological quality, biological respiration rate, and water quality parameters are constantly changing. The model assumptions are decreased here in order to represent dissolved oxygen more accurately to the real circulating water culture conditions.

The hypothesis proposed in this study suggests that the microporous aeration process within the culture area maintains a continuous and stable supply of oxygen, resulting in a uniform distribution of dissolved oxygen content in the water.

Unlike the usual dissolved oxygen control modeling assumptions in aquaculture, in this paper, we consider changes in biomass, temperature, and biological respiration in modeling and treat the effects of these factors on dissolved oxygen as a composite perturbation of the dissolved oxygen model. According to the assumptions of this study, the entire breeding region is regarded as a system. The dynamics of dissolved oxygen in this system is a combination of dissolved oxygen in the water entering and leaving the culture area, artificial oxygenation of the culture area, oxygen consumption by respiratory metabolism of the culture, and dissolved oxygen influenced by other compound disturbances. As seen in Figure 2, this article first develops a notional model of dissolved oxygen including the three primary components of artificial oxygenation, circulating water inflow and output, and oxygen consumption by cultures in a recirculating water culture system. Based on this, the system composite perturbation is introduced to the model to boost its applicability.

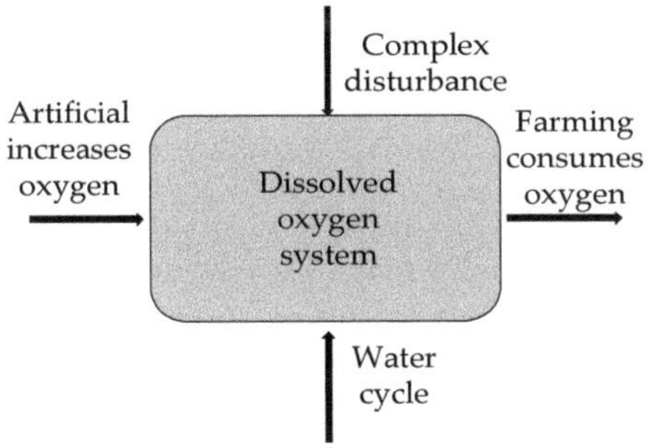


**Figure 2.** Dynamic of dissolved oxygen system.

2.1.2. Dissolved Oxygen Control Model with Uncertainty

(1) Circulating water in and out process

The aquaculture water is filtered and returned to the aquaculture area by sedimentation, biological filtration, sterilization, and other activities in the circulating water aquaculture system. Since there is no dissolved oxygen consumption throughout the disinfection process, the dissolved oxygen concentration of the biofilter effluent may be assumed to represent the dissolved oxygen concentration of the influent water in the breeding region. Based on the assumptions in this paper, the dissolved oxygen concentration value of the water outflowing from the breeding area can be used as the dissolved oxygen concentration in the breeding area, thus,

$$E_{io} = \frac{Q}{V}(C_{in} - y) \tag{1}$$

where $E_{io}$—dissolved oxygen change rate in a circulating water process (mg $\cdot L^{-1} \cdot h^{-1}$); $V$—volume of water in the breeding area ($m^3$); $Q$—circulating water system flow ($m^3 \cdot h^{-1}$); $C_{in}$—dissolved oxygen concentration in the influent water of the breeding area (mg $\cdot L^{-1}$); and $y$—dissolved oxygen concentration in the breeding area (mg $\cdot L^{-1}$).

To guarantee that the biochemical reaction of the biofilter runs normally, the dissolved oxygen content of the biofilter effluent should be more than 2.0 mg $\cdot L^{-1}$ [28]; $C_{in}$ is taken as 3.0 mg $\cdot L^{-1}$ in this article.

(2) Aeration process

Microporous aeration and oxygenation are one of the most common oxygenation technologies used in the recirculating water culture process. Microporous devices are evenly arranged at the bottom of the breeding area and the wall of the breeding pool. Aeration equipment, such as blowers, are linked to microporous air stones or microporous tubes to deliver oxygen straight into the breeding region for oxygen. Dissolved oxygen transfer processes for the aeration process [29,30].

$$A_m = K_{La} \cdot (C_{sat} - y) \tag{2}$$

where $A_m$—dissolved oxygen change rate during aeration (mg $\cdot L^{-1} \cdot h^{-1}$); $C_{sat}$—saturated dissolved oxygen concentration in the water bodies (mg $\cdot L^{-1}$); $K_{La}$—total oxygen transfer coefficient which is mostly related to water temperature and aeration, as described by the following equation [31]:

$$K_{La} = (c_1 + c_2 \cdot u(t)^{c_3}) \cdot \beta^{(T-20)} \tag{3}$$

In Equation (3), $c_1$—atmospheric reoxygenation factor, $c_2$—oxygen transfer rate of artificial aeration process, $u$—aeration volume ($m^3 \cdot h^{-1}$), and $\beta$—temperature correction factor. In recirculating water culture, $c_3$ is close to 1.0 [17] and the aeration of the artificial aeration component is roughly linear with the oxygen transfer coefficient [30–32]; hence, $c_3 = 1.0$ is used in this research.

(3) Oxygen consumption by respiration of cultures

The primary cause of the reduction in dissolved oxygen concentration in the culture region is the culture's respiratory metabolism. The respiration rate of cultures is influenced by a variety of internal and external factors. The relationship between the respiration rate, temperature, and body weight of farmed crabs is generally described by the following equation [33,34]:

$$R = \frac{A}{1 + e^{(2.351 - 0.138 \cdot T)}} \times M^{-0.5} \tag{4}$$

In Equation (4), $R$—oxygen consumption rate per unit mass of crab (mg $\cdot h^{-1} \cdot kg^{-1}$); $A$—upper limit of the biological respiration rate (mg $\cdot h^{-1}$); and $M$—biological quality (kg).

When the temperature and biological mass do not fluctuate much, the respiration rate per unit mass of culture may be regarded a constant using Equation (4). In this article, the culture's respiration rate is utilized as a constant to build a notional model of the system and the culture's oxygen consumption [35] is taken as

$$R_s = \frac{M \cdot R}{V} \tag{5}$$

where $R_s$—oxygen consumption rate of all cultures (mg $\cdot L^{-1} \cdot h^{-1}$) and $V$—volume of water in the breeding area ($m^3$).

In summary, the dynamic model of dissolved oxygen without considering other factors can be described as

$$\frac{dy}{dt} = E_{io} + A_m - R_s \tag{6}$$

Substituting into Equations (1)–(5) yields:

$$\frac{dy(t)}{dt} = \frac{Q}{V} \cdot (C_{in} - y(t)) + (c_1 + c_2 \cdot u(t)) \cdot \beta^{(T-20)}(C_{sat} - y(t)) - \frac{M \cdot R}{V} \tag{7}$$

The dissolved oxygen model (7) is defined as follows:

$$y(t) = x(t) \tag{8}$$

$$f_0(x(t)) = \frac{Q}{V} \cdot (C_{in} - x(t)) - \frac{M \cdot R}{V} + c_1 \cdot \beta^{(T-20)} \cdot (C_{sat} - x(t)) \tag{9}$$

$$g_0(x(t)) = c_2 \cdot \beta^{(T-20)} \cdot (C_{sat} - x(t)) \tag{10}$$

Then, the dissolved oxygen control state space model is:

$$\begin{cases} \dot{x}(t) = f_0(x(t)) + g_0(x(t))u(t) \\ y(t) = x(t) \end{cases} \tag{11}$$

The system flow rate $Q$, biological respiration rate $R$, and biological mass parameters $M$, and $c_1$ and $c_2$ in the dissolved oxygen equation obtained from the above analysis are nominal values. In the actual operation of the system, changes in water components, temperature, system flow, biological growth, and many other factors will lead to parameter uptake and external disturbances, introducing model parameter errors $\Delta Q$, $\Delta R$, $\Delta M$, $\Delta c_1$ $\Delta c_2$ and $\Delta C_{in}$. Substituting them into Equations (9) and (10) yields

$$\begin{aligned} f(x(t)) = {} & \tfrac{Q+\Delta Q}{V} \cdot ((C_{in} + \Delta C_{in}) - x(t)) - \tfrac{(M+\Delta M)\cdot(R+\Delta R)}{V} \\ & +(c_1 + \Delta c_1) \cdot \beta^{(T-20)} \cdot (C_{sat} - x(t)) \\ & = \tfrac{Q}{V} \cdot (C_{in} - x(t)) - \tfrac{M \cdot R}{V} + c_1 \cdot \beta^{(T-20)} \cdot (C_{sat} - x(t)) \\ & + \tfrac{\Delta Q}{V} \cdot (C_{in} - x(t) + \Delta C_{in}) + \tfrac{Q \cdot \Delta C_{in}}{V} - \tfrac{\Delta M \cdot R + M \cdot \Delta R + \Delta M \cdot \Delta R}{V} \\ & + \Delta c_1 \cdot \beta^{(T-20)} \cdot (C_{sat} - x(t)) \end{aligned} \tag{12}$$

$$\begin{aligned} g(x(t)) & = (c_2 + \Delta c_2) \cdot \beta^{(T-20)} \cdot (C_{sat} - x(t)) \\ & = c_2 \cdot \beta^{(T-20)} \cdot (C_{sat} - x(t)) + \Delta c_2 \cdot \beta^{(T-20)} \cdot (C_{sat} - x(t)) \end{aligned} \tag{13}$$

Organizing the above equation makes

$$\begin{aligned} f_n(x(t)) & = \tfrac{\Delta Q}{V} \cdot (C_{in} - x(t) + \Delta C_{in}) + \tfrac{Q \cdot \Delta C_{in}}{V} - \tfrac{\Delta M \cdot R + M \cdot \Delta R + \Delta M \cdot \Delta R}{V} \\ & + \Delta c_1 \cdot \beta^{(T-20)} \cdot (C_{sat} - x(t)) \end{aligned} \tag{14}$$

$$g_n(x(t)) = \Delta c_2 \cdot \beta^{(T-20)} \cdot (C_{sat} - x(t)) \tag{15}$$

Then, the original dissolved oxygen state space equation is

$$\begin{cases} \dot{x}(t) & = f(x(t)) + g(x(t))u(t) + \xi(t) \\ & = f_0(x(t)) + f_n(x(t)) + [g_0(x(t)) + g_n(x(t))] \cdot u(t) + \xi(t) \\ & = f_0(x(t)) + g_0(x(t)) \cdot u(t) + d \\ y(t) & = x(t) \end{cases} \tag{16}$$

where $d = f_n(x(t)) + g_n(x(t)) \cdot u(t) + \xi(t)$ and $f_n(x(t)) + g_n(x(t)) \cdot u(t)$ are internal perturbations due to parameter uptake and $\xi(t)$ are external perturbations due to time-varying and unmodeled parts of the system.

### 2.2. Dissolved Oxygen Controller Design

The dissolved oxygen adaptive immunity controller is proposed in this study for the dissolved oxygen state space Equation (16). The controller, as illustrated in Figure 3, is made up of a nonlinear disturbance observer and an adaptive sliding film controller. The nonlinear disturbance observer continuously estimates the real-time system disturbance and uncertainty while the adaptive law is designed to compensate for the disturbance before the convergence of the disturbance observation error. This approach enhances the speed and accuracy of the disturbance convergence. Finally, the sliding film controller is intended to improve the system's capacity to suppress disturbances and guarantee that the dissolved oxygen system maintains a consistent track of the set value. Therefore, the dissolved oxygen controller design objectives are summarized as follows:

① The nonlinear interference-based observer makes the observation error converge while the input gain design compensates for the control $u_d$.

② Based on ①, the adaptive law $\hat{f}$ and the sliding film control law $u_s$ are designed to make the system track the dissolved oxygen setting value asymptotically in a limited time.

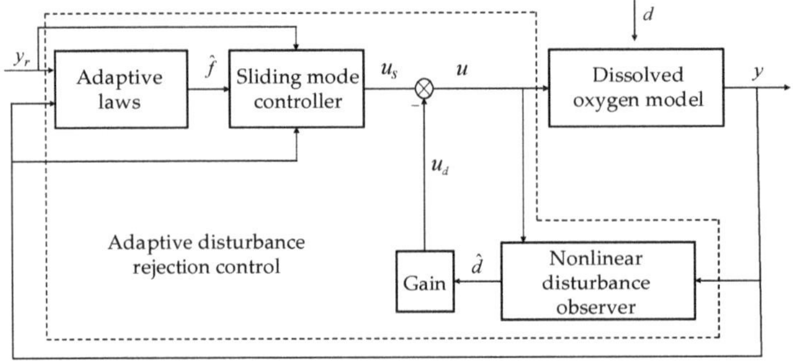


**Figure 3.** Dissolved oxygen control structure.

#### 2.2.1. Nonlinear Interference Observer Design

$d$ is a complicated system perturbation consisting of unmodeled dissolved oxygen system dynamics as well as internal and external disturbances; $d$ is bounded. The structural nonlinear interference observer [36] is designed.

$$\begin{cases} \hat{d} = z + p(x) \\ \dot{z} = -L(x) \cdot z - L(x) \cdot [f_0(x) + g_0(x) \cdot u + p(x)] \end{cases} \tag{17}$$

In Equation (17), $\hat{d}$—observer estimation of compound disturbances; $z$—auxiliary Variables; $p(x)$—the function to be designed; and $L(x)$—the nonlinear interference observer's gain as defined by Equation (18):

$$L(x) \cdot \dot{x} = \frac{dp(x)}{dt} \tag{18}$$

Since the prior information of the perturbation differential is unknown, it is assumed that the disturbance evolves slowly in relation to the observer's dynamic attributes, i.e., $\dot{d} \approx 0$. By defining the observation error $f$,

$$f = d - \hat{d} \tag{19}$$

From Equation (19), the dynamic characteristics of the observation error can be obtained as follows:

$$\begin{aligned} \dot{f} &= \dot{d} - \dot{\hat{d}} = -\dot{z} - L(x) \cdot \dot{x} \\ &= L(x) \cdot \hat{d} + L(x) \cdot (f_0(x) + g_0(x) \cdot u) - L(x) \cdot x \\ &= L(x) \cdot \hat{d} - L(x) \cdot d \\ &= -L(x) \cdot f \end{aligned} \tag{20}$$

Construct the Lyapunov function for Equation (17).

$$V_1 = \frac{1}{2} f^2 \tag{21}$$

Differentiating the two sides of (21) with respect to $t$:

$$\dot{V}_1 = f \cdot \dot{f} = -L(x) \cdot f^2 \tag{22}$$

Obviously, when the interference observer gains $L(x) > 0$, $\dot{V}_1 \le 0$, the observer estimation error converges exponentially, taking $p(x) = L(x) \cdot x$.

In this paper, we design the compensation control amount according to the model of dissolved oxygen system as

$$u_d = \frac{\hat{d}}{g_0(x)} \tag{23}$$

2.2.2. Adaptive Sliding Film Controller Design Based on Disturbance Estimation

Before convergence of the nonlinear disturbance observation error, the dissolved oxygen system can be expressed as $\dot{x} = f_0(x) + g_0(x) \cdot u_s + f$ where $f = d - \hat{d}$. This paper designs an adaptive sliding film controller for the existing dissolved oxygen system (16) based on a sliding film control to accelerate error convergence and improve system immunity. The control aim is to accomplish an accurate dissolved oxygen control in circulating water systems by reducing the error between system output and the fixed value of dissolved oxygen to zero. Let $x_d$ be the dissolved oxygen set value in the circulating water culture system and define the system control output error vector $e = x - x_d$. The control task is to create the control law so that the $\lim_{t \to \infty} e(t) = 0$ holds. Defining the slide surface:

$$s = e \tag{24}$$

Differentiating (24) yields the sliding mode dynamic equation as follows:

$$\dot{s} = \dot{x} - \dot{x}_d \tag{25}$$

Consider the synovial surface convergence law $\dot{s} = -\varepsilon \cdot \mathrm{sgn}(s) - k \cdot s$, $\varepsilon > 0$, and $k > 0$, where $-k \cdot s$ is the exponential convergence term which speeds up the convergence speed and $-\varepsilon \cdot \mathrm{sgn}(s)$ makes certain that the system status reaches the synovial surface in a limited time.

Define and differentiate the Lyapunov function $V_2 = \frac{1}{2} s^2$ as follows:

$$\dot{V}_2 = s \cdot \dot{s} = s \cdot (f_0(x) + g_0(x) \cdot u_s + f - \dot{x}_d) \tag{26}$$

In actual water cycle aquaculture systems, external disturbance signals contain multiple uncertainties. An adaptive compensator is designed using an adaptive strategy to compensate for the unobserved disturbance of the dissolved oxygen system and the disturbance signal before the convergence of the disturbance observation error to speed up the convergence of the disturbance observation error and improve the robustness of the dissolved oxygen control.

$\hat{f}$ is the estimated value of the perturbed signal $f$. Define the estimation error $\tilde{f} = f - \hat{f}$. Define the Lyapunov function:

$$V_3 = V_2 + \frac{1}{2\gamma}\tilde{f}^2 \tag{27}$$

where $\gamma$ is a normal number. Differentiate (27):

$$\begin{aligned} \dot{V}_3 &= \dot{V}_2 - \tfrac{1}{\gamma}\tilde{f} \cdot \dot{\hat{f}} \\ &= s \cdot (f_0(x) + g_0(x) \cdot u_s + f - \dot{x}_d) - \tfrac{1}{\gamma}\tilde{f} \cdot \dot{\hat{f}} \\ &= s \cdot (f_0(x) + g_0(x) \cdot u_s + \hat{f} + \tilde{f} - \dot{x}_d) - \tfrac{1}{\gamma}\tilde{f} \cdot \dot{\hat{f}} \\ &= s \cdot (f_0(x) + g_0(x) \cdot u_s + \hat{f} - \dot{x}_d) - \tfrac{1}{\gamma}\tilde{f} \cdot (\dot{\hat{f}} - \gamma \cdot s) \end{aligned} \tag{28}$$

Design a control law based on an Equation (28):

$$u_{eq} = -\frac{f_0(x) + \hat{f} - \dot{x}_d}{g_0(x)} \tag{29}$$

$$u_{vss} = -\frac{\varepsilon \cdot \mathrm{sgn}(s) + k \cdot s}{g_0(x)} \tag{30}$$

$$u_s = u_{eq} + u_{vss} \tag{31}$$

In Equation (31), $u_s$ is the sliding film control law, $u_{eq}$ is the equivalent control, and $u_{vss}$ is the switching control to compensate for system uncertainty and disturbances.

The adaptive law s

$$\dot{\hat{f}} = \gamma \cdot s \tag{32}$$

2.2.3. Stability Analysis

**Conclusion 1.** *When the sliding film convergence law parameter $\varepsilon > e^*$, the compensatory control law (23), the sliding film control law (31), and the adaptive law (32) are used such that the dissolved oxygen closed-loop system (16) asymptotically follows the dissolved oxygen setting for a short duration. The tracking time meets the following:*

$$t \leq \frac{\sqrt{2} \cdot \sqrt{V_2(0)}}{\varepsilon - e^*}$$

**Proof of Conclusion 1.** Design a Lyapunov function for the closed-loop system Equation (16).

$$V_4 = \frac{1}{2}s^2 + \frac{1}{2}f^2 + \frac{1}{2\gamma}\tilde{f}^2 \tag{33}$$

According to Equation (33),

$$\dot{V}_4 = s \cdot (f_0(x) + g_0(x) \cdot u_s + \hat{f} - \dot{x}_d) - \frac{1}{\gamma} \cdot \tilde{f} \cdot (\dot{\hat{f}} - \gamma \cdot s) + f \cdot \dot{f} \tag{34}$$

From (29)–(32), (34)

$$\dot{V}_4 = -\varepsilon \cdot |s| - k \cdot |s|^2 - L(x) \cdot f^2 \tag{35}$$

Clearly, there exists $k$,$\varepsilon$ and interference observer gain $L(x)$ make $\dot{V}_4(x) \leq 0$.

In addition, the system output tracks the set value along the sliding die surface for a finite time, for $\dot{V}_2$:

$$\dot{V}_2 = s \cdot \dot{s} = s \cdot (f_0(x) + g_0(x) \cdot u_s + f - \dot{x}_d) \tag{36}$$

It follows from (22) that the presence of the disturbance observer gain $L(x)$, which converges the system disturbance estimation error, with $\left|f-\tilde{f}\right| \le e^* = \sup\limits_{t>0}\left|f-\tilde{f}\right|$, holds. Substituting the control law (31) into Equation (36) yields the following:

$$\begin{aligned} \dot{V}_2 &= s \cdot \dot{s} \\ &= s \cdot (-\varepsilon \cdot \operatorname{sgn}(s) - k \cdot s + f - \tilde{f}) \\ &\le |s| \cdot (-\varepsilon + e^*) = -\sqrt{2}(\varepsilon - e^*) \cdot \sqrt{V_2} \end{aligned} \tag{37}$$

When $\varepsilon > e^*$,

$$\dot{V}_2 = -\sqrt{2}(\varepsilon - e^*) \cdot \sqrt{V_2} \le 0 \tag{38}$$

Integration of Equation (38),

$$\sqrt{V_2(t)} \le -\frac{\sqrt{2}}{2}(\varepsilon - e^*) \cdot t + \sqrt{V_2(0)} \tag{39}$$

From Equation (39),

$$t \le \frac{\sqrt{2} \cdot \sqrt{V_2(0)}}{\varepsilon - e^*}$$

In summary, through the design of the dissolved oxygen adaptive immunity controller, the system satisfies the Lyapunov stability theory condition and can track the dissolved oxygen setting asymptotically for a limited period of time. Proof complete. □

## 3. Results

### *Dissolved Oxygen Control Simulation*

To validate the efficiency of the dissolved oxygen adaptive anti-disturbance control approach presented in this study, simulation experiments with varying dissolved oxygen model parameters and uncertainty disturbances in the circulating water system are conducted. The effective volume of the breeding area of this circulating water system is 1.5 $m^3$. The total mass of farmed crab is about 3.25 kg. The dissolved oxygen concentration setting value is 6 $mg \cdot L^{-1}$. Based on data from relevant culture systems [16,37], the nominal model parameters of dissolved oxygen are shown in Table 1.

**Table 1.** Model parameters.

| Parameters | Numerical Value |
|---|---|
| Circulating water flow/($m^3 \cdot h^{-1}$) | 0.34 |
| Water volume of aquaculture area/$m^{-3}$ | 1.5 |
| Total quality of aquatic products/kg | 3.25 |
| Oxygen consumption rate/($mg \cdot h^{-1} \cdot kg^{-1}$) | 110.2 |
| Saturated dissolved oxygen/($mg \cdot L^{-1}$) | 8.89 |
| $c_1$ | 0.08046 |
| $c_2$ | 1.129 |
| $\beta$ | 1.014 |
| $T$ | 19.89 |
| $k_p$ | 1.1 |
| $k_i$ | 0.92 |
| $k_d$ | 0.71 |

To evaluate the stability and anti-interference performance of the controller, this paper compares and simulates the PID controller and the sliding mode controller designed in this study. The parameters of the sliding mode controller are typically determined based on the system's dynamic characteristics and control performance requirements. Considering that the controller's design objective in this paper is to achieve system stability and robustness while maintaining the desired response time and disturbance estimation speed, the initial values for the sliding mode controller's gain and the parameters of the nonlinear interference observer are set. Following several testing and changes, the controller parameters $k = 15$, $\varepsilon = 0.1$, $\gamma = 0.1$, and the gain of nonlinear disturbance observer $L(x) = 28$ are established with the goal of assuring system reaction speed and disturbance estimation speed. As indicated in Table 1, the PID controller parameters are first substantially modified using the PID Tuner toolbox in MATLAB and then fine-tuned in the order of first $k_p$, then $k_i$, and lastly $k_d$.

Changes in system parameters can be caused by variable flow rate management of circulating water [38] and weight increases in the aquaculture item during recirculating aquaculture. To test the durability of the control approach to changes in the dissolved oxygen model parameters, the circulating water flow rate was doubled at 180 min. The mass of the culture is doubled, all other simulation data are left alone, and the control simulation results are given in Figure 4. As shown in Figure 4, when the system parameters change, the controller overshoots by 0.3% and may be corrected to the set value within 10 min. The overrun of the PID control is 2%, the adjustment time is 180 min, and the adaptive capacity is low. In comparison, the controller in this study has minor overshoot, quick correction, and great self-adaptive capacity and the system's dissolved oxygen content can follow the set value stably under parameter uncertainty.

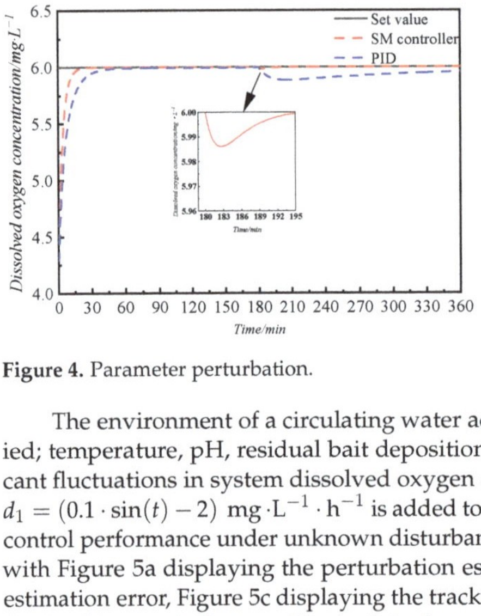


**Figure 4.** Parameter perturbation.

The environment of a circulating water aquaculture system is complicated and varied; temperature, pH, residual bait deposition, and other variables can all cause significant fluctuations in system dissolved oxygen content. A slow time-varying disturbance $d_1 = (0.1 \cdot \sin(t) - 2)$ mg$\cdot$L$^{-1} \cdot$ h$^{-1}$ is added to the system to further assess the controller's control performance under unknown disturbances; Figure 5 depicts the simulation results with Figure 5a displaying the perturbation estimation results, Figure 5b displaying the estimation error, Figure 5c displaying the tracking control effect, and Figure 5d displaying the tracking control error. Figure 5a,b shows that the estimation accuracy is great, the absolute estimation error is less than $\pm 0.002$, and the interference estimation converges rapidly. It is clear from the study of Figure 5c,d that the PID control of dissolved oxygen concentration struggles to maintain the set value while disturbances are present; at 360 min, the error between the dissolved oxygen concentration and set value is around $-0.04$ mg$\cdot$L$^{-1}$. The controller designed in this paper can track the set value in 20 min in the presence of disturbance and the tracking error is kept within $\pm 0.001$ mg$\cdot$L$^{-1}$, which has high control accuracy, and the dissolved oxygen concentration in the water body can maintain a stable state.

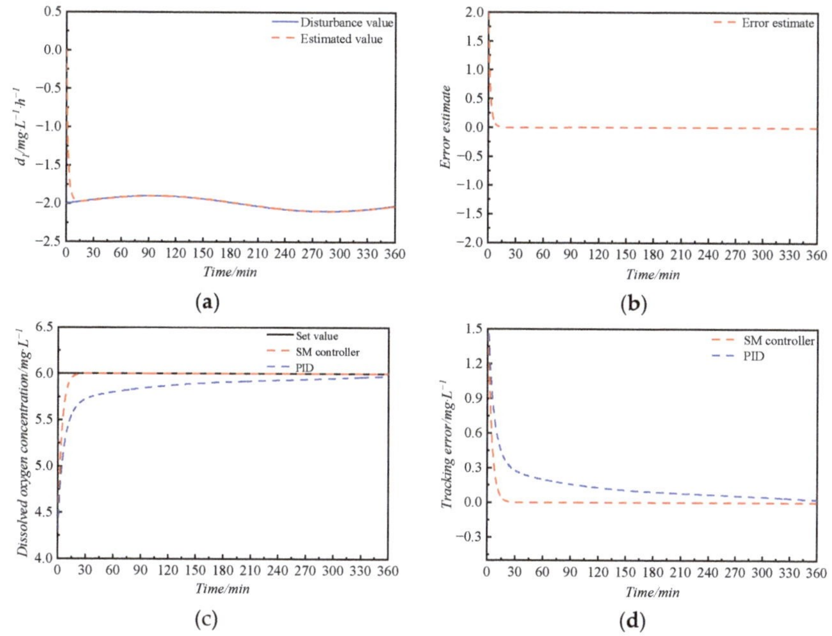


**Figure 5.** Control simulation under $d_1$ disturbance. (**a**) Disturbance estimation; (**b**) Disturbance estimation error; (**c**) Control under disturbance; (**d**) Control error under disturbance.

The amount of dissolved oxygen will rapidly decrease if there are too many microorganisms in the water body and they are actively consuming oxygen. In this regard, other simulation parameters are kept constant and a step disturbance $d_2$ is applied to the dissolved oxygen system. For $d_2$ at 0–120 min amplitude is $-0.2\ \text{mg}\cdot\text{L}^{-1}\cdot\text{h}^{-1}$, 120–240 min at amplitude is $-1.88\ \text{mg}\cdot\text{L}^{-1}\cdot\text{h}^{-1}$, and 240–360 min at amplitude is $-0.37\ \text{mg}\cdot\text{L}^{-1}\cdot\text{h}^{-1}$, the control simulation results under this perturbation are shown in Figure 6. Figure 6a shows the results of the perturbation estimation, Figure 6b shows the estimation error, Figure 6c shows the results of the tracking control, and Figure 6d shows the results of the tracking error. The analysis of Figure 6a,b demonstrates that the estimation accuracy is high, the nonlinear observer perturbation estimation absolute error is within $\pm 0.01$, and the disturbance estimation response is quick. It can be seen from Figure 6c,d that the water quality fluctuates significantly under PID control. At 135 min, the dissolved oxygen concentration decreased to roughly $5.75\ \text{mg}\cdot\text{L}^{-1}$ at which point the overshoot was 4.1%. After 120 min, the overshoot was reduced to be close to the set value but the tracking effect once more departed from the set value when the perturbation changed. Under the action of the controller designed in this paper, the tracking error of dissolved oxygen concentration is within $\pm 0.095\ \text{mg}\cdot\text{L}^{-1}$, and the tracking accuracy is high. The system's maximum overshoot is 0.28% at 135 min and the regulation time is 12 min. In the presence of perturbation steps, the set value is still steadily monitored. From the aforementioned simulation, it can be seen that the disturbance observer used in this paper also has a good step disturbance observation effect. The dissolved oxygen control based on the disturbance observer can successfully overcome system uncertainty and the controller has strong anti-disturbance ability.

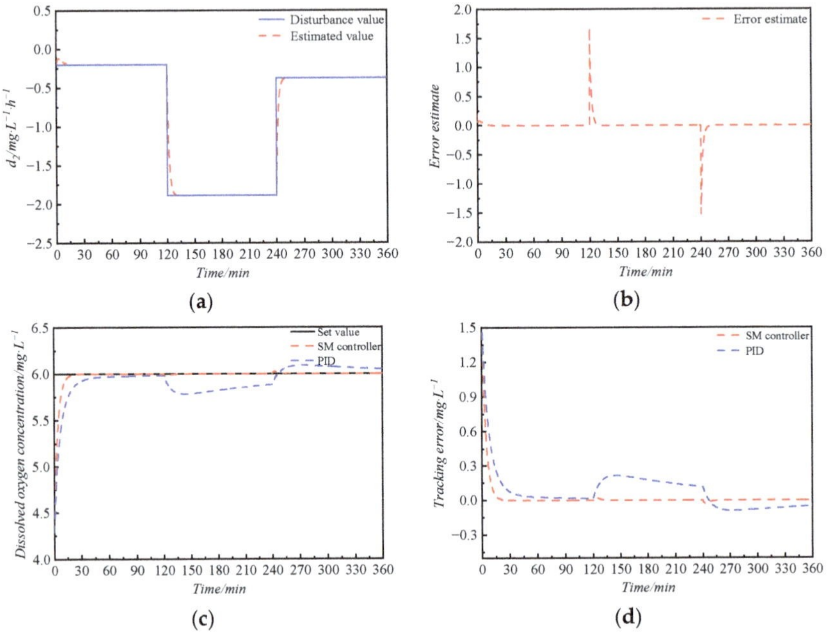


**Figure 6.** Control simulation under $d_2$ disturbance. (**a**) Disturbance estimation; (**b**) Disturbance estimation error; (**c**) Control under disturbance; (**d**) Control error under disturbance.

In aquaculture, the required level of dissolved oxygen concentration required by cultured organisms varies in different growth states [39]. The controller is simulated for this purpose using variable setpoint tracking. The dissolved oxygen is set to fluctuate in the range of 4.5 $\sim$ 6.5 mg $\cdot L^{-1}$. Figure 7 shows that when the dissolved oxygen set value changes, the controller described in this study has a quick reaction time, a small tracking error, and strong tracking capabilities for various set values.

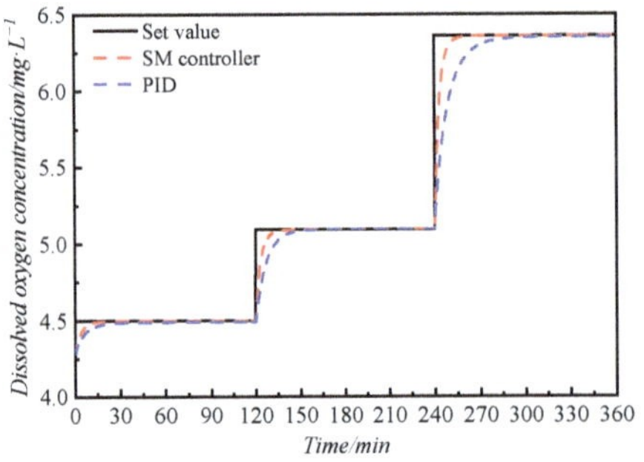


**Figure 7.** Variable set point control simulation.

In a complicated breeding environment, the controller can keep the concentration of dissolved oxygen at a steady level while reducing the need for the precision of the system model to accomplish exact control of dissolved oxygen. Two indicators, the integral of absolute value of error (IAE) and integral of squared error (ISE), were utilized to objectively monitor and evaluate the performance of the controller in order to manage the concentration of dissolved oxygen in recirculating water cultures.

$$IAE = \int_0^t |y - y_r| dt \tag{40}$$

$$ISE = \int_0^t (y - y_r)^2 dt \tag{41}$$

where $y$ is the dissolved oxygen concentration and $y_r$ is the set value. The controller performance index is calculated from Equations (40) and (41).

According to the performance indicators listed in Table 2, under the effect of $d_1$ disturbance, the IAE of this controller is 0.121, which is 85.76% lower than the IAE index of PID control, and the ISE index is 0.102, which is 66.88% lower. The IAE index of the PID controller is 0.727 and the ISE index is 0.243 under $d_2$ perturbation while the IAE index of the controller in this paper is 0.124, which is 82.9% lower, and the ISE index is 0.097, which is 60% lower. The controller in this research has lower IAE and ISE metrics under $d_3$ perturbation compared to the PID controller. In the dissolved oxygen variable setpoint tracking control, this controller has an ISE performance index of 0.049 and an IAE performance index of 0.112, both of which are 62.87% and 62.66% poorer than the PID control, respectively. In conclusion, the controller in this study can match the demands for aquaculture dissolved oxygen management and has good control precision and tracking capabilities in complicated aquaculture environments.

**Table 2.** Controller effect data comparison.

| | **This Strategy** | | | **PID** | | |
|---|---|---|---|---|---|---|
| | **Under $d_1$** | **Under $d_2$** | **Tracking Setpoint** | **Under $d_1$** | **Under $d_2$** | **Tracking Setpoint** |
| IAE | 0.121 | 0.124 | 0.112 | 0.85 | 0.727 | 0.3 |
| ISE | 0.102 | 0.097 | 0.049 | 0.308 | 0.243 | 0.132 |

## 4. Conclusions

(1) In this paper, we conducted an analysis of the influencing factors of dissolved oxygen from the characteristics of a circulating water aquaculture system, established a dynamic model of dissolved oxygen in a circulating water aquaculture system and improved the model's applicability by taking into account the influence of various disturbing factors such as temperature, pH, and circulating water flow. On the basis of this, a disturbance observer-based adaptive sliding film control method is proposed. Through disturbance observation and adaptive compensation, this approach controls the concentration of dissolved oxygen under unpredictable disturbances without relying on an exact mathematical model of the system. Theory and simulation are used to show that dissolved oxygen control is stable in the presence of unknown disturbances;

(2) When the system circulating water flow and biological quality parameters are modified, the controller overshoot in this study is 0.3% with a 10 min adjustment time but the PID control overshoot is 2% with a 180 min adjustment time. The controller has good anti-interference performance;

(3) The disturbance observer can estimate the unknown disturbance caused by the uncertainty of the circulating water culture system in real time while the adaptive slide film controller assures the stability of the closed-loop system and increases the immunity of dissolved oxygen concentration management. When applying a slow time-varying disturbance $d_1$ and a step disturbance $d_2$ to the system, this controller reduces the absolute error integral (IAE) index by 85.76% and 82.9% and the error squared integral (ISE) index by 66.88% and 60%, respectively, compared to the PID controller;

(4) This control technique efficiently overcomes the effects of model uncertainty, random interference, and other variables on dissolved oxygen precision control as well as the problem of water quality fluctuations caused by traditional controllers. The controller designed in this study can estimate compensating interference in real time, ensuring the stability of dissolved oxygen concentrations in circulating water culture systems and improving water quality.

In future work, we will delve into the following areas: (1) the implementation of the nonlinear disturbance observers and adaptive sliding mode control methods discussed in this article may introduce increased complexity to the control system which could potentially pose challenges during algorithm implementation. (2) To achieve optimal performance, the adaptive sliding mode control method necessitates the adjustment of controller parameters which may require expertise in order to attain desired outcomes. (3) This strategy heavily relies on real-time online estimation and observation of external disturbances which places stringent demands on computational power and real-time capabilities.

**Author Contributions:** Conceptualization, X.Y. and G.Z.; methodology, X.Y. and S.Y.; software, X.Y.; validation, X.Y., S.Y. and Y.C.; formal analysis, G.Z.; investigation, G.Z.; resources, S.Y.; data curation, S.Y.; writing—original draft preparation, X.Y.; writing—review and editing, X.Y.; visualization, S.Y.; supervision, G.Z.; project administration, G.Z.; funding acquisition, Y.C. All authors have read and agreed to the published version of the manuscript.

**Funding:** This research was funded by [Zhejiang Province Public Welfare Technology Application Research Project] grant number [No. LGN21C190007], and also by [Natural Science Foundation of Ningbo Municipality] grant number (No. 202003N4074).

**Data Availability Statement:** Not applicable.

**Conflicts of Interest:** The authors declare no conflict of interest.

## References


1. Ahmed, N.; Turchini, G.M. Recirculating aquaculture systems (RAS): Environmental solution and climate change adaptation. *J. Clean. Prod.* **2021**, *297*, 1–14. [CrossRef]
2. Wang, Z. "The 14th Five-Year Plan" Zhejiang maritime economic development thinking. *Zhejiang Econ.* **2019**, *24*, 46–47.
3. Fan, S.; Li, H.; Zhao, R. Effects of normoxic and hypoxic conditions on the immune response and gut microbiota of Bostrichthys sinensis. *Aquaculture* **2020**, *525*, 735336. [CrossRef]
4. Holan, A.B.; Good, C.; Powell, M.D. Health management in recirculating aquaculture systems (RAS). In *Aquaculture Health Management*; Academic Press: Cambridge, MA, USA, 2020; pp. 281–318.
5. Prüter, J.; Strauch, S.M.; Wenzel, L.C.; Klysubun, W.; Palm, H.W.; Leinweber, P. Organic Matter Composition and Phosphorus Speciation of Solid Waste from an African Catfish Recirculating Aquaculture System. *Agriculture* **2020**, *10*, 466. [CrossRef]
6. Li, C.; Li, Z.; Wu, J.; Zhu, L.; Yue, J. A Hybrid Model for Dissolved Oxygen Prediction in Aquaculture based on Multi-scale Features. *Inf. Process. Agric.* **2018**, *5*, 11–20. [CrossRef]
7. Zhu, C.; Liu, X.; Li, H. Optimization of prediction model of dissolved oxygen in industrial aquaculture. *Trans. Chin. Soc. Agric. Mach.* **2016**, *47*, 273–278.
8. Cheng, X.; Xie, J.; Yu, D.; Zeng, Y. Calculated analysis of oxygen transfer from air bubble-water interface and turbulent water surface in microporous aeration systems. *Trans. Chin. Soc. Agric. Eng.* **2013**, *29*, 190–199.
9. Xie, H.; Wang, Y.; Xu, G. Purification efficiency in an "in-pond raceway recirculating (IPRS)-shrimp-crab" tandem system aquaculture model. *J. Dalian Ocean. Univ.* **2021**, *36*, 118–126.
10. Zhao, S.; Liu, H.; Ding, W.; Tang, Y.; Li, Y. The development and experiment of aeration control system in pond. *J. Nanjing Agric. Univ.* **2015**, *38*, 682–688.
11. Han, J.; Wu, C. A study on dynamic model of dissolved oxygen in fish ponds. *Trans. CSAE* **1994**, *10*, 52–58.
12. Buonomo, B.; Falcucci, M.; Hull, V.; Rionero, S. A mathematical model for an integrated experimental aquaculture plant. *Ecol. Model.* **2005**, *183*, 11–28. [CrossRef]
13. Mandal, S.; Debnath, M.; Ray, S.; Ghosh, P.B.; Roy, M.; Ray, S. Dynamic modelling of dissolved oxygen in the creeks of Sagar island, Hooghly–Matla estuarine system, West Bengal, India. *Appl. Math. Model.* **2012**, *36*, 5952–5963. [CrossRef]
14. Huang, D.; Hu, J. Research on the dissolved oxygen fuzzy control system in the aquaculture. *J. Agric. Mech. Res.* **2013**, *34*, 58–61.
15. Hu, J.; Huang, D. The detection and control system of dissolves oxygen in the aquaculture based on Simulink. *J. Agric. Mech. Res.* **2009**, *31*, 47–49.
16. Zhou, X.; Wang, J.; Huang, L.; Li, D.; Duan, Q. Modelling and controlling dissolved oxygen in recirculating aquaculture systems based on mechanism analysis and an adaptive PID controller. *Comput. Electron. Agric.* **2022**, *192*, 106583. [CrossRef]

17. Yin, L.; Fu, L.; Wu, H.; Xia, Q.; Jiang, Y.; Tan, J.; Guo, Y. Modeling dissolved oxygen in a crab pond. *Ecol. Model.* **2021**, *440*, 109385. [CrossRef]
18. Wahab, N.A.; Katebi, R.; Balderud, J. Multivariable PID control design for activated sludge process with nitrification and denitrification. *Biochem. Eng. J.* **2009**, *45*, 239–248. [CrossRef]
19. Revollar, S.; Vilanova, R.; Francisco, M.; Vega, P. PI Dissolved Oxygen control in wastewater treatment plants for plantwide nitrogen removal efficiency. *IFAC-Pap.* **2018**, *51*, 450–455. [CrossRef]
20. Khan, O.; Madhuranthakam, C.M.; Douglas, P.; Lau, H.; Sun, J.; Farrell, P. Optimized PID controller for an industrial biological fermentation process. *J. Process Control.* **2018**, *71*, 75–89. [CrossRef]
21. Du, X.; Wang, J.; Jegatheesan, V.; Shi, G. Dissolved Oxygen Control in Activated Sludge Process Using a Neural Network-Based Adaptive PID Algorithm. *Appl. Sci.* **2018**, *8*, 261. [CrossRef]
22. Ning, N.; Ma, H. Dissolved oxygen concentration Fuzzy self-tuning PID control in sewage treatment aeration process. *J. Chang. Univ. Technol.* **2019**, *40*, 325–331.
23. Zhou, X.; Li, D.; Zhang, L.; Duan, Q. Application of an adaptive PID controller enhanced by a differential evolution algorithm for precise control of dissolved oxygen in recirculating aquaculture systems. *Biosyst. Eng.* **2021**, *208*, 186–198. [CrossRef]
24. Yuswantoro, D.; Natan, O.; Angga, A.N.; Gunawan, A.I.; Taufiqurrahman; Dewantara, B.S.B.; Kurniawan, M.A. Fuzzy logic-based control system for dissolved oxygen control on indoor shrimp cultivation. In Proceedings of the 2018 International Electronics Symposium on Engineering Technology and Applications (IES-ETA), Bali, Indonesia, 29–30 October 2018; pp. 37–42.
25. Zhang, X.X.; Zhao, L.R.; Li, J.J.; Cao, G.T.; Wang, B. Space-decomposition based 3D fuzzy control design for nonlinear spatially distributed systems with multiple control sources using multiple single-output SVR learning. *Appl. Soft Comput.* **2017**, *59*, 378–388. [CrossRef]
26. Ma, H.; Ning, N.; Chen, Z. Internal model control of dissolved oxygen in wastewater treatment based on improved fuzzy neural network. *Comput. Appl. Softw.* **2020**, *37*, 41–46.
27. Zhang, S.; Wang, H.; Chen, P.; Zhang, X.; Li, Q. Overview of the application of neural networks in the motion control of unmanned vehicles. *Chin. J. Eng.* **2022**, *44*, 235–243.
28. Liu, H.; Chen, J.; Ni, Q.; Xu, H. Design of a recirculating aquaculture system based on mass balance. *Trans. Chin. Soc. Agric. Eng.* **2009**, *25*, 161–166.
29. Haider, H.; Ali, W.; Haydar, S. Evaluation of various relationships of reaeration rate coefficient for modeling dissolved oxygen in a river with extreme flow variations in Pakistan. *Hydrol. Process.* **2013**, *27*, 3949–3963. [CrossRef]
30. Zhang, Y.; Nie, Y.; Chen, L. Adaptive Fuzzy Fault-Tolerant Control against Time-Varying Faults via a New Sliding Mode Observer Method. *Symmetry* **2021**, *13*, 1615. [CrossRef]
31. Chern, J.M.; Chou, S.R.; Shang, C.S. Effects of impurities on oxygen transfer rates in diffused aeration systems. *Water Res.* **2001**, *35*, 3041–3048. [CrossRef]
32. Cheng, X.; Zeng, Y.; Xie, J.; Gong, W. Impact of microporous aeration flow and aeration tube length on oxygen transfer performance in water. *Trans. Chin. Soc. Agric. Eng.* **2014**, *30*, 209–217.
33. Li, T.; Liu, X. Model-Free Non-Cascade Integral Sliding Mode Control of Permanent Magnet Synchronous Motor Drive with a Fast Reaching Law. *Symmetry* **2021**, *13*, 1680. [CrossRef]
34. Nejjari, F.; Dahhou, B.; Benhammou, A.; Roux, G. Non-linear multivariable adaptive control of an activated sludge wastewater treatment process. *Int. J. Adapt. Control. Signal Process.* **1999**, *13*, 347–365. [CrossRef]
35. Zou, E.; Du, N.; Wei, L. The Effects of Mass Temperature and Thermal Acclimation on The Respiration Rate of The Chinese Freshwater Crab *Eriocheir sinensis* (Crustacea: Decapod). *Zool. Res.* **1995**, *16*, 49–58.
36. Chen, W.H.; Ballance, D.J. A nonlinear disturbance observer for robotic manipulators. *Ind. Electron. IEEE Trans. Ind. Electron.* **2000**, *47*, 932–938. [CrossRef]
37. Wang, X.; Wang, F.; Lu, Y.; Dong, S. Effects of light intensity on the respiratory metabolism of swimming crab (*Portunus trituberculatus*). *J. Fish. China* **2014**, *38*, 237–243.
38. Shen, J.; Shi, M.; Ruan, Y.; Deng, Y. Effects of variable flow rates on water qualities in recirculating aquaculture system. *Trans. Chin. Soc. Agric. Mach.* **2016**, *47*, 309–314.
39. Abdel-Tawwab, M.; Hagras, A.E.; Elbaghdady, H.A.; Monier, M.N. Effects of dissolved oxygen and fish size on Nile tilapia, *Oreochromis niloticus* (L.): Growth performance, whole-body composition, and innate immunity. *Aquac. Int.* **2015**, *23*, 1261–1274. [CrossRef]

applied sciences

MDPI

Article

# Dynamics of Semiconductor Laser Coupled with Two External Cavities

**Salah Abdulrhmann [1,*], Abu Mohamed Alhasan [2,3,†] and A. Y. Madkhli [1]**

1 Department of Physics, College of Science, Jazan University, P.O. Box 114, Jazan 45142, Saudi Arabia; amadkhli@jazanu.edu.sa
2 Department of Physics, Faculty of Science, Assiut University, Assiut 71516, Egypt; am.alhasan.sq@gmail.com
3 Bağlar Mahallesi, 31500 Ryhanlı, Hatay, Turkey
* Correspondence: sabdulrhmann@jazanu.edu.sa
† Retired.

**Abstract:** An investigation into the dynamic states and relative intensity noise of laser diodes subjected to double optical feedback has been conducted. We employed modified and improved time-delay rate equations to account for double external optical feedback. The dynamic states and noise of lasers will be investigated using bifurcation diagrams of the output photon number, its temporal variations, and the intensity noise of the laser. This analysis considers feedback strengths due to the double external cavity and their spacing from the front facet of the laser with and without phase due to feedback. The results reveal that considering phase causes significant variations in laser intensity and a phase shift in the temporal variations of the laser output. This results in relative intensity noise suppression and a frequency shift in the intensity noise spectrum. These findings represent new contributions to our understanding of the reliance of lasing frequency shift on the phase due to feedback, regardless of whether feedback originates from a single or double external cavity. We investigated the optimal conditions corresponding to stable dynamic states of the laser with the lowest noise level. Additionally, we identified conditions that result in chaotic dynamics, where the spectrum does not convey information about the laser system. These insights have potential applications in chaotic and secure optical data encryption.

**Keywords:** semiconductor lasers; optical feedback; relative intensity noise; double external cavity; CW; pulsation; chaos; communication using chaos

## 1. Introduction

Laser diodes, known as LDs, offer numerous advantages when compared to other types of lasers [1]. These benefits include their compact size, cost-effectiveness, and the ability to control their output by adjusting the injected current. As a result, LDs play a crucial role as light sources in various aspects, such as fiber-optic communications and data processing.

However, a common challenge faced by many applications is experiencing external optical feedback (OFB) [2]. This feedback occurs when laser radiation reflects off an external reflector and re-enters the laser cavity. Understanding the impact of OFB is crucial, as it can significantly alter the laser's behavior, leading to various modes of operation [3].

One notable consequence of OFB is the induction of chaotic dynamics, characterized by irregularities in the laser's output and a loss of coherence [2]. This effect becomes particularly pronounced under conditions of moderate OFB and when the distance between the laser and reflector is extended, resulting in increased noise levels and reduced coherency.

Previous research mainly focused on LD behavior under a single OFB, which describes this phenomenon by introducing a term of time delay into the laser rate equation [4]. However, this model had limitations and was applicable only to situations with weak to moderate OFB. A theoretical model of time delay applicable over any strength of OFB, accounting for multiple round-trips and optical phases due to OFB, has been presented in Ref. [5], which was applied in investigating the dynamics and noise of pumping LDs used in fiber amplifier systems [5,6]. Moreover, the impacts of nonlinear gain, radiative and non-radiative lifetime, linewidth enhancement factor (LEF), and combined optical phase on the operation states and relative intensity noise (RIN) of 0.1550 μm InGaAsP/InP time-delayed LD have been studied [7–9].

To address more complex scenarios where LDs encounter external OFB from two reflectors, such as in fiber-optic communication systems, additional research is needed. This double OFB, typically a result of reflections on both facets of an optical fiber, has applications in encrypted systems of data, generating secured signals that do not reveal any data about the system itself [10]. Pal et al. investigated experimentally and theoretically the dynamics of the LD subjected to two filtered OFBs from two separate external cavities. They found that a second filtered OFB allows versatile manipulation of the LD frequency and provides an extra set of control parameters to get a stable laser [11].

In this paper, we extended a model in Ref. [5] based on the existing time-delay rate equation model from single OFB to encompass double OFB scenarios. This is completed using a traveling-wave approach to propagate the laser in the double cavities and their reflections between the laser front facet and the external reflectors [12]. After that, we propose a modified time delay term to describe double OFB and include it in the LD rate equation. This expansion represents a novel work in the investigations of OFB in LDs and their implementations in communication systems and other applications. LDs are, for instance, widely used in optical communication systems, including fiber-optic communication networks, where they serve as optical transmitters for data transmission. Another example is Optical Data Storage where LDs are used in devices such as CD and DVD players, as well as in Blu-ray technology, for reading and writing data on optical discs. The common issue for all these applications is the distortion of the signal as a result of OFB-induced signal distortion. OFB can, for instance, introduce timing errors and jitters in the laser output, which can lead to issues with data synchronization and the precise writing or reading of data on optical storage media. This can result in difficulty in accurately positioning the laser spot on the storage medium. Excessive OFB can cause data integrity problems. When writing data to optical storage media, any fluctuations in the laser power or position due to feedback can result in data being written incorrectly or with errors. Similarly, during data retrieval, the feedback-induced noise can lead to difficulty in accurately reading the stored data. In extreme cases, OFB-induced power fluctuations can lead to excessive heating of the media or optical components, potentially causing physical damage to the storage medium or laser diode.

In addition to the traveling wave approach, a numerical solution for the enhanced time-delay rate equation model by assigning arbitrary lengths to the external cavities will be introduced [13–15]. Using very fine time integration in combination with a long integration time interval of several microseconds will result in a stable solution. With the help of this solution, we will study the operational aspects, dynamic behavior, and noise characteristics of the semiconductor laser further. This exploration will involve categorizing these features based on the bifurcation diagram (BD) of the output power of the LD and the RIN power spectrum in the laser signal [16]. The dynamics of the laser under double OFB are characterized using various tools, including the time variations of the emitted photons and RIN. Our investigation will span various values of external OFB strengths originating from double external cavities, along with variations in their distance from the laser's facet.

We aim to identify the optimal conditions that result in stable laser operation with low noise for use in optical fiber communications. Additionally, we will identify the conditions that lead to chaotic dynamics, where the RIN spectrum does not reveal information about the laser system. This chaotic behavior holds potential significance in the realm of secure optical data encryption. Another application of our research is to better understand certain feedback processes and the resulting feedback noise, as these are the main limitations of Fiber-To-The-Home (FTTH) systems—for instance, to propose noise reduction solutions without the need for expensive optical isolators [17]. Using feedback control can involve adjusting the laser parameters or using them to minimize the impact of the feedback on system performance. In addition, optical attenuators can reduce the power of the feedback signal, which can help in reducing its impact on the laser, and the application of anti-reflective coatings to optical components can reduce the amount of reflection within the optical system, thereby minimizing the potential for feedback. The results of this work have a high impact on researching these topics further.

Our improved model of double OFB can be used with LDs operating in single and multi-longitudinal modes, including various types of LDs. We categorize the laser's behavior based on output power and RIN power spectrum, considering two distinct external cavity lengths and optical phase due to OFB [18].

We present an investigation of the dynamic states and RIN of LDs subject to double OFB. We used modified and improved time delay rate equations for the LDs to consider double external OFB [12]. The dynamic states and noise of the LDs will be determined using the output photon number BD, its temporal variations, and the RIN of the LD with OFB strengths from the double OFB, as well as their spacing from the laser facet and with and without optical phase due to OFB. The LD dynamics are studied within two regions: short and long double external cavities. The simulated results show that the single OFB LD changes its state from continuous wave (CW) to periodic oscillation (PO), period doubling (PD), coherence collapse, and chaotic operation. In a short double external cavity region, LD draws from each of the operations with the single OFB to a route-to-CW state. This route-to-CW state is a reflection of the identical route-from-chaos to the identical operation with a single OFB. In a long double OFB regime, increased double OFB strength takes the LD to a route-to-chaos. In comparison with a single OFB case, the double OFB external cavity produces smoother, more chaotic RIN power spectra, which have no specific relationship with the resonance frequencies of the external cavities. Considering optical phase due to OFB in single OFB and double OFB cavities causes significant intensity suppression and phase shift in the temporal photon output of the LD changes. This intensity suppression and phase shift induce RIN suppression and a frequency shift in the RIN spectrum. This dependence of the lasing frequency shift on the optical phase due to OFB is noticed for the first time in this article. RIN suppression is close to the solitary laser noise level when LD operates in the regimes of CW and PO. Double OFB and optical phase due to OFB lead the LD dynamic states to get better, i.e., the dynamic states change to a better state such as CW or PO.

This paper proceeds with an explanation of LD operation states and dynamics with double OFB, outlines numerical calculation procedures, presents simulations with short and long double OFB, and concludes with key findings.

## 2. Time-Delay Model of Two-Reflector OFB

A model of the rate equation, which includes multiple round-trips in the external cavity and optical phase due to OFB, has been established [5]. The model is extended to study the theoretically dynamic states of the LDs subject to double external OFB [12]. An improved rate equation of LD with double OFB from two external cavities for photon numbers, phase, and carrier numbers is formulated as follows:

$$\frac{dS}{dt} = \frac{1}{2}[G - G_{th}]\, S + \frac{a\xi}{V}N + F_S \tag{1}$$

$$\frac{d\theta}{dt} = \frac{1}{2}\alpha\frac{a\xi}{V}(N - \overline{N}) - \frac{c}{n_D L_D}\phi + F_\theta \tag{2}$$

$$\frac{dN}{dt} = \frac{I}{e} - \frac{N}{\tau_s} - \frac{a\xi}{V}(N - N_g)S + F_N \tag{3}$$

$$G = \frac{a\xi}{V}(N - N_g)S - BS, \tag{4}$$

where $G$ is the optical gain, $a$ is the slope of the optical gain with the changes in carrier number $N$, $\xi$ is the conferment factor, $V$ is the active region volume, $\overline{N}$ is the $N$ time-averaged value, $\alpha$ is the LEF, $I$ is the injection current, $\tau_s$ is the electron lifetime due to spontaneous emission, and $N_g$ is the carrier number at transparency. The universal constant $e$ defines the electron charge, and $B$ is the nonlinear gain coefficient, respectively [19–21]. The threshold gain is given using

$$G_{th} = G_{thD} - \frac{c}{n_D L_D} \ln|U|, \quad (5)$$

where $G_{thD}$ is the threshold gain [21]:

$$G_{thD} = \frac{c}{n_D}\left[\kappa + \frac{1}{2L_D} \ln \frac{1}{R_f R_b}\right], \quad (6)$$

where $k$, $L_D$, $R_f$, and $R_b$ are the internal loss, laser cavity length, and front and back facet reflectivity, respectively. The function $U$ expressing the double OFB in the rate equations is given using [12]:

$$\begin{aligned} U = \; & 1 - K_{ex1}\sqrt{\frac{S(t-\tau_1)}{S(t)}} e^{j[\theta(t-\tau_1)-\theta(t)-\omega\tau_1]} \\ & -K_{ex2}\sqrt{\frac{S(t-\tau_2)}{S(t)}} e^{j[\theta(t-\tau_2)-\theta(t)-j\omega\tau_2]} \end{aligned} \quad (7)$$

where $\omega$ is the angular frequency of the laser emission, $[\theta(t-\tau_1) - \theta(t)]$ and $[\theta(t-\tau_2) - \theta(t)]$ are the optical phase deviations due to chirping induced by delay in external cavity 1 and external cavity 2, respectively. $K_{ex1}$ and $K_{ex2}$ are the OFB strengths from the single external cavity and double external cavity, respectively, and are determined with the front facet reflectivity, first reflector reflectivity, and second reflector reflectivity, respectively [12]:

$$K_{ex1} = \left(1 - R_f\right)\frac{\sqrt{R_{ex1}}}{\sqrt{R_f}} \quad (8)$$

$$K_{ex2} = \left(1 - R_f\right)(1 - R_{ex1})\frac{\sqrt{R_{ex2}}}{\sqrt{R_f}} \quad (9)$$

Figure 1 illustrates a model scheme of semiconductor lasers subjected to double OFB. The laser emitted from the front travels a round-trip in the cavity between the front and first reflector, $R_{ex1}$. Then, a round-trip is made between the front facet and double reflector $R_{ex2}$, which then re-enters the laser cavity. That is, there are time delays $\tau_1$ and $\tau_2$ between the laser reflected by the front of the laser facet into the cavity of the laser and the laser entered back from single and double reflectors, respectively [12]. By supposing $L_{ex1}$ and $L_{ex2}$ to be the lengths of the single and double external cavities, respectively, and the external cavity refractive index $n_{ex}$, the feedback laser has time delays of

$$\tau_1 = \frac{2n_{ex1}L_{ex1}}{c}, \text{ and } \tau_2 = \frac{2n_{ex}L_{ex2}}{c} \quad (10)$$

The Langevin noise sources $F_S(t)$, $F_N(t)$, and $F_\theta(t)$, are Gaussian probability distributions with their correlation functions, $\langle F_x(t)\, F_y(t')\rangle = V_{xy}\,\delta(t-t')$, $x, y = S, N$ *or* $\theta$, and have zero mean values, $\langle F_S(t)\rangle = \langle F_N(t)\rangle = \langle F_\theta(t)\rangle = 0$. Where $\delta(t-t')$ is Dirac's $\delta$ function and $V_{xy}$ is the correlation variance [22–26]. Adding noise sources to the rate equations induces fluctuations on the photon and carrier numbers due to spontaneous emission.

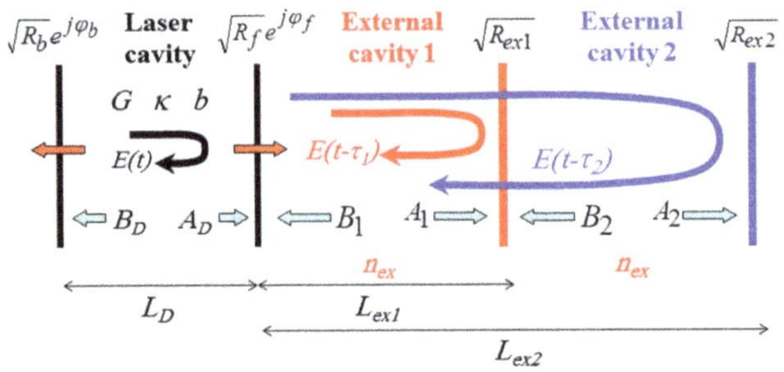


**Figure 1.** Scheme of laser diode under double optical feedback [12].

## 3. Numerical Calculations

The LD dynamics with double OFB are studied numerically by integrating the rate Equations (1)–(3) using the fourth-order Runge Kutta method [27]. AlGaAs LD emitting at a wavelength of 0.780 μm parameters was used in the calculations, where the numerical values of this LD are given in Ref. [12]. The corresponding threshold current $I_{thD}$ = 22.5 mA, and the threshold gain $G_{thD} = 2.82 \times 10^{11}\ s^{-1}$. In this article, a short external cavity has been used in the first external cavity $L_{ex1}$ = 3 cm, whereas two lengths are considered to be the second external cavity of $L_{ex2} = 3.5 \times 10^{-2}$ m (the short cavity) and $10^{-1}$ m (approaches to the long cavity). The refractive index of these external cavities is filled with air ($n_{ex}$ = 1.0). The resonance frequency separations of the external cavities are $f_{ex1} = 1.0/\tau_1 = 5$ GHz and $f_{ex2} = 1.0/\tau_2 = 4.286$ GHz. The states are investigated in terms of the strength of the OFB coefficients (intensity) $K_{ex1}$ and $K_{ex2}$. The integration time step was selected as 5 ps and was used over a period of 5 μs, for which the operation reached a steady state. The phase $\varphi$ of the OFB is given using

$$\phi = \tan^{-1}\frac{\mathrm{Im}U}{\mathrm{Re}U} + p\pi \quad (11)$$

where $p$ is an integer and is changed continuously for time evolution since the solution of the arc tangent is defined in the range of $-\pi/2.0$ to $+\pi/2.0$ in the simulation.

The values $\overline{S}$ and $\overline{N}$ of the photon and carrier number are specified from the steady-state solutions of the rate equation of the solitary laser [12]:

$$\overline{S} = \frac{I - I_{thD}}{eG_{thD}} \quad (12)$$

$$\overline{N} = \frac{N_{thD} + \frac{V}{a\xi}B\overline{S}}{\overline{S}+1} \quad (13)$$

where $N_{thD}$ is the threshold carrier number.

The spectra of the *RIN* are then calculated from the output photon number fluctuations $\delta S(t) = S(t) - \overline{S}$, obtained instantaneously using fast Fourier transform [28,29]:

$$RIN = \frac{1}{\overline{S}^2}\left\{\frac{1}{T}\int_0^T\left[\int_0^\infty \delta S(t)\delta S(t+\tau)e^{j\omega\tau}d\tau\right]dt\right\} = \frac{1}{\overline{S}^2}\left\{\frac{1}{T}\left|\int_0^T \delta S(\tau)e^{-j\omega\tau}d\tau\right|^2\right\} \quad (14)$$

where $\overline{S}$ is the dc-value of the photon number $S(t)$, which is timely averaged and also numerically calculated. The *RIN* simulations were carried out as the steady state reached to $t \approx 3$~5 μs.

## 4. Numerical Simulation Results and Discussions

We will start by showing the dynamic states of the laser induced with a single external cavity. The double OFB impacts on the LD dynamic states are then demonstrated by adding a second reflector to the external cavity. This is conducted by investigating the changes happening in each dynamical state produced with the single external cavity.

### *4.1. Laser Dynamic States under Single External Cavity*

LD dynamic states induced via the single external cavity at two values of optical phase $\psi_1 = 0.0$ (red color) and $\omega\tau_1$ (black color) are shown in Figure 2. The dynamic states are calculated with BD of the photon number $S(t)$, peak value, with the OFB strength $K_{ex1}$. Abdulrhmann et al. investigated the influence of including the noise sources in the rate equations on the amplitude of pulsation and found that it is larger when counting the noise sources in the rate equations [6]. Since including the noise sources in the rate equations has a very little variation on the dynamic states induced via OFB, especially in the transition border to another state [6]. Langevin noise sources are omitted from the rate Equations (1)–(3) in the calculations of the BDs to get a clear vision of the dynamic states of the laser. The LD continues oscillating with the CW state for sufficiently small values of OFB strength $K_{ex1}$, which is represented by no points in the figure. When $K_{ex1}$ goes beyond the HB point (pointed by the first arrow) of $K_{ex1} = 0.0129$ and 0.0117 in the case of optical phase $\psi_1 = 0.0$ and $\omega\tau_1$, respectively, the LD dynamic state is identified with PO output for relaxation oscillations undamped. This PO is of the one-period kind with frequency increases with an increase in $K_{ex1}$. The POs continue up to $K_{ex1} = 0.0227$ and 0.0231, with optical phases $\psi_1 = 0.0$ and $\omega\tau_1$, respectively. Beyond the PO regime, the bifurcation diagram shows two branches indicating the appearance of oscillations with period doubling (PD) (pointed by the second arrow). When $K_{ex1} > 0.027$ and 0.029 and optical phase $\psi_1 = 0.0$ and $\omega\tau_1$, respectively, the PD oscillation bifurcates more and shows many random points and tours, which represents a chaotic dynamic state (pointed by the third arrow). The shown PD route-to-chaos recognizes the LD with a short external cavity of $f_{ex}/f_r > 1.0$, as shown in Ref. [30]. By reducing the optical phase $\psi_1$ to zero the strength of the LD is decreased, and the dynamic states are shifted toward increasing the OFB intensity $K_{ex1}$ up to the PO region, which leads to more stability, as shown in Ref. [10]. However, after the PO region, the dynamical regimes such as PD and chaos start to appear at lower values of $K_{ex1}$. As shown in the figure, reducing the optical phase $\psi_1$ from $\omega\tau_1$ to 0.0 has no effect on the route-to-chaos, which keeps the PD route-to-chaos.

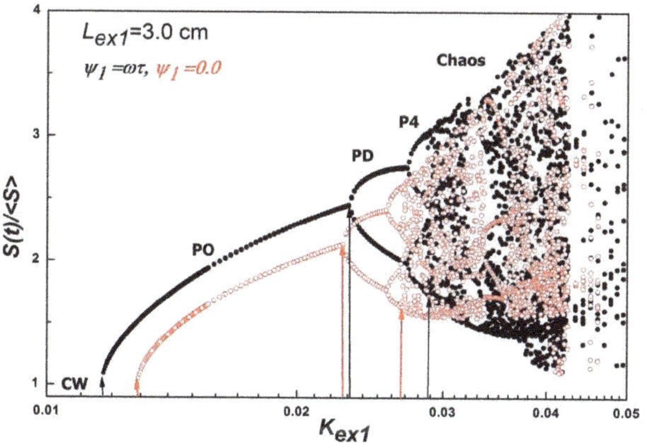


**Figure 2.** BD of the LD under single OFB at two values of optical phase $\psi_1 = 0.0$ (red color) and $\omega\tau_1$ (black color).

In the regime when OFB $K_{ex1}$ = 0.042–0.05, the laser starts to be attracted to a less chaotic or coherence collapse region, and this is due to entering the laser with a strong OFB, which induces stability in the dynamic states of the laser [5,6].

Examples of the different dynamic states of the single external cavity laser shown in Figure 2 are characterized in Figure 3, when optical phase $\psi_1 = 0.0$ and $\psi_1 = \omega\tau_1$; namely, the PO, PD, and chaos at OFB strength $K_{ex1}$ = 0.02, 0.025, and 0.04, respectively. Figure 3 plots the photon numbers $S(t)/S_{AV}$ time variations and the corresponding RIN power spectrum, which is calculated using the fast Fourier transform of $S(t)$ using Equation (14). The intensity of $S(t)/S_{av}$ is decreased, and the phase is shifted by decreasing the optical phase $\psi_1$ from $\omega\tau_1$ to zero in all laser states, as shown in Figure 3(a1–a3). The effect of varying optical phase $\psi_1$ due to OFB on the laser fluctuations has been noticed in Refs. [9,11,31] and confirmed experimentally in Ref [11,31]. The reduction in the intensity of $S(t)/S_{av}$ is shown in Figure 3b as a little reduction of RIN. As found in Figure 3(b1), RIN shows a sharp peak at PO frequency $f_{PO}$ = 3.221 GHz and its higher harmonic peaks, which represent uniform periodic oscillations with the relaxation frequency ~ $f_r$ = 3.22 GHz. The RIN power spectrum in the PD dynamic state is distinguished by two sharp peaks at PD frequency $f_{PD}$ = 3.2 GHz and its half harmonic, as shown in Figure 3(b2). An unequal time type of $S(t)$ variations represents the chaotic operation as displayed in Figure 3(b3). These inequalities in the output of the LD are created due to operating the LD unstably with two conditions: one of them identifies the LD with a frequency of $f_{Laser}$ = 3.4 GHz and the other to the compound cavity with a frequency of $f_{Comb}$ = 4.6 GHz [9].

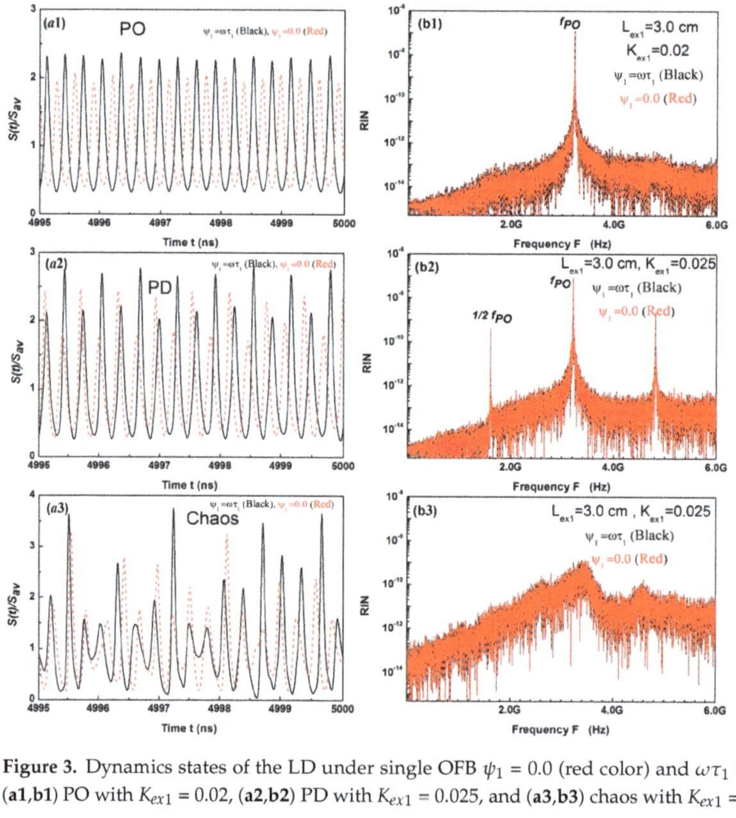


**Figure 3.** Dynamics states of the LD under single OFB $\psi_1$ = 0.0 (red color) and $\omega\tau_1$ (black color): (**a1**,**b1**) PO with $K_{ex1}$ = 0.02, (**a2**,**b2**) PD with $K_{ex1}$ = 0.025, and (**a3**,**b3**) chaos with $K_{ex1}$ = 0.04.

### 4.2. Laser Dynamic States under Double External Cavity

#### 4.2.1. Region of Short Double OFB ($L_{ex2}$ = 3.5 $\times$ $10^{-2}$ m)

To investigate the impact of the double external cavity on the dynamic states and RIN of the laser, bifurcation diagrams are used. We will start by assuming that the distance between the first and second reflectors is 5 $\times$ $10^{-3}$ m, making the cavity of the double OFB in the short cavity region ($L_{ex2}$ = 3.5 $\times$ $10^{-2}$ m). Dynamic states of lasers with short external cavities are highly influenced by the number of round-trips in short external cavities [32]. In this paper, we consider the number of round-trips in our simulations. Figures 4a, 5a and 6a plot the bifurcation diagrams and the corresponding average Low-frequency RIN ($f$ < 10 MHz) spectrum versus OFB intensity $K_{ex2}$ with optical phase $\psi_2 = \omega(\tau_1 + \tau_2)$ when the single OFB strength is set to dynamic states corresponding to PO, PD, and chaos states at single OFB strength $K_{ex1}$ = 0.02, 0.025, and 0.04, respectively.

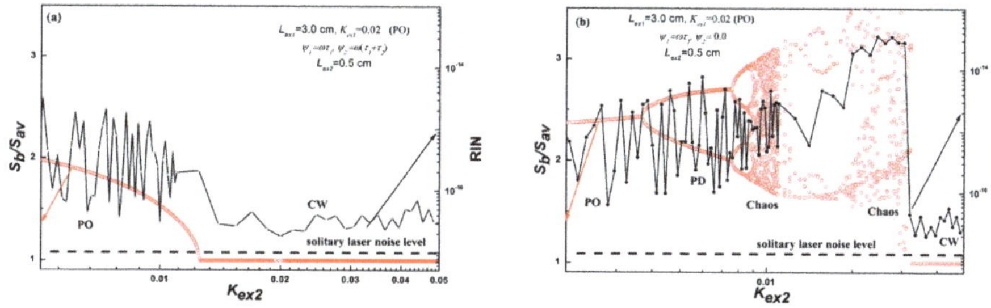


**Figure 4.** BDs of states of LD and average Low-frequency RIN ($f$ < 10 GHz) with double OFB at $K_{ex1}$ = 0.02, $L_{ex1}$ = 3 cm, $\psi_1 = \omega\tau_1$, and $L_{ex2}$ = 3.5 cm: (**a**) $\psi_2 = \omega(\tau_1 + \tau_2)$ and (**b**) $\psi_2$ = 0.0.

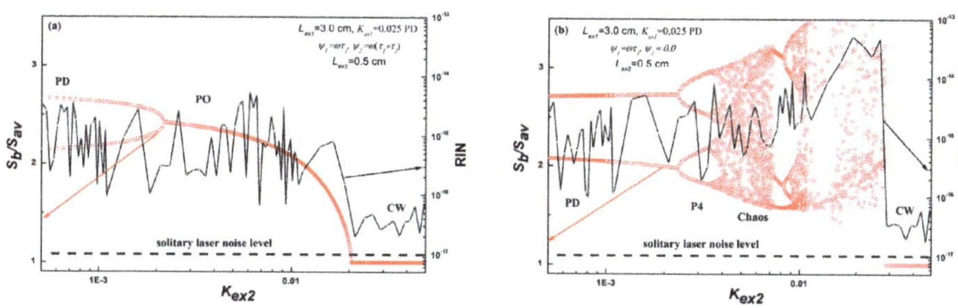


**Figure 5.** BDs of states of LD and average Low-frequency RIN ($f$ < 10 GHz) with double OFB at $K_{ex1}$ = 0.025, $L_{ex1}$ = 3 cm, $\psi_1 = \omega\tau_1$, and $L_{ex2}$ = 3.5 cm: (**a**) $\psi_2 = \omega(\tau_1 + \tau_2)$ and (**b**) $\psi_2$ = 0.0.

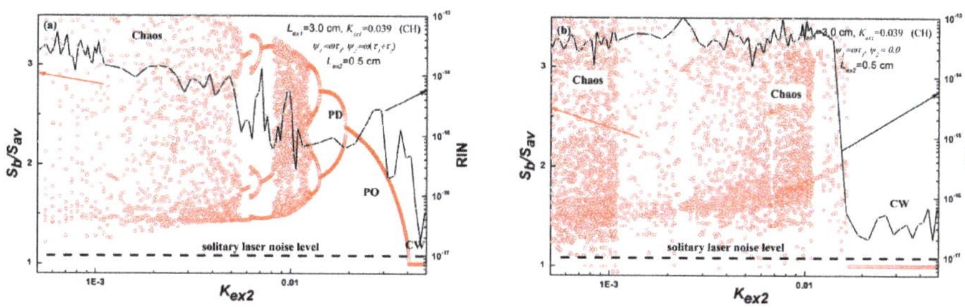


**Figure 6.** BDs of states of LD and average Low-frequency RIN ($f$ < 10 GHz) with double OFB at $K_{ex1}$ = 0.04, $L_{ex1}$ = 3 cm, $\psi_1 = \omega\tau_1$, and $L_{ex2}$ = 3.5 cm: (**a**) $\psi_2 = \omega(\tau_1 + \tau_2)$ and (**b**) $\psi_2$ = 0.0.

The shown figures are characterized by three features at each dynamic state: (1) the laser dynamic state in the weak OFB region begins with the dynamic state of a single OFB and (2) increases in the strength of the double OFB $K_{ex2}$, leading the LD to a route-to-CW (more stable dynamics). Finally, (3) this route-to-CW is a reflector to the identical route-from-CW to a pertinent dynamic at this level of single OFB strength $K_{ex1}$ in the bifurcation diagrams of Figure 2 under a single external cavity.

When the optical phase $\psi_2 = 0.0$, the LD dynamic states in the presence of double OFB are shown in Figures 4b, 5b and 6b. Figures show that reducing the optical phase $\psi_2$ to 0.0 moves the laser dynamic state to unstable states and changes the route-to-CW when $\psi_2 = \omega(\tau_1 + \tau_2)$ to the route-to-chaos state at $\psi_2 = 0.0$. The width of the CW dynamic is wider as $\psi_2 = \omega(\tau_1 + \tau_2)$ than that as $\psi_2 = 0.0$, as shown in Figure 4. Figure 5 shows the same scenario but in the PD state, where, by increasing the double OFB strength $K_{ex2}$, the dynamic states transfer from the PD state to tour and chaotic states, which is the route to the chaos state. Figure 6 shows the dynamic states of the double external cavity at ($K_{ex1} = 0.04$, chaos), the chaos state is directly the route to the CW dynamic state, and the CW state width is wider as $\psi_2 = 0.0$ than that as $\psi_2 = \omega(\tau_1 + \tau_2)$.

In the same Figures 4–6 and on the right *Y*-axis, the average RIN at Low-frequency ($f < 10$ MHz) versus the double cavity OFB intensity $K_{ex2}$ corresponds to the BDs shown. Figure 4a shows the Low-frequency RIN in the PO regime; as shown in the figure, it is almost in the range of $10^{-15}$ to $10^{-14}$ and decreases to approach a solitary laser noise level (dashed line) with increasing the double OFB strength $K_{ex2}$. As $\psi_2 = 0.0$, Figure 4b displays RIN enhancement when dynamic states transfer from PO, PD, and chaos regimes till reaching the CW state. The level of the RIN is reduced near the level of solitary laser noise. Figures 5b and 6a,b show the highest level of the RIN, which occurs almost in the chaos regime. We notice that by reducing the optical phase $\psi_2$ to 0.0, the RIN displays a lower level than that as of $\psi_2 = \omega(\tau_1 + \tau_2)$ in the CW, PO, and PD regimes.

As shown in the figures, when the optical phases $\psi_1 = \omega\tau_1$, and $\psi_2 = \omega(\tau_1 + \tau_2)$, the dynamic of the LD is more stable than that as $\psi_1 = \omega\tau_1$, and $\psi_2 = 0.0$, and the RIN suppresses to a level near the solitary laser noise, especially when the laser operates in the CW or PO regime. Figures show that the double external cavity leads to stabilizing the irregular states produced by the single external cavity and keeps the stable CW state, which reduces the RIN to lower levels near solitary laser noise level. That leads to improvements in the properties of the LD, especially in applications such as optical communication systems and others. The RIN corresponding to unstable operation states can be stabilized as rising double OFB strength $K_{ex2}$ and optical phase $\psi_2$, which helps in planning lasers at best dynamics and low noise level. Impacts of optical phase and double OFB strength $K_{ex2}$ on the RIN with adapting laser design to adjust values of $K_{ex2}$ and $\psi_2$ are presented here and have not been investigated before. It may be helpful in increasing/decreasing the stability/RIN of LDs, respectively.

The chaotic dynamic state under double OFB is important in many applications, such as secure communication systems. In this section, we have an interest in investigating the output photon number $S(t)/S_{av}$, time change, and the corresponding chaos dynamic RIN spectral characteristics under double OFB when $K_{ex1} = 0.04$ and $K_{ex2} = 0.005$ and at optical phase $\psi_1 = \omega\tau_1$, $\psi_2 = 0.0$, and $\omega(\tau_1 + \tau_2)$. Figure 7a plots the variations of $S(t)/S_{av}$, which are intensity-reduced and phase-shifted by increasing the optical phase $\psi_2$ from zero to $\omega(\tau_1 + \tau_2)$, which is similar to single OFB Figure 3(a3). The reduction of the intensity of $S(t)/S_{av}$ is translated into a decrease of RIN in Figure 7b. The RIN spectrum shows more enhanced peaks than the chaotic RIN with a single OFB shown in Figure 3(b3). The values of the peak frequency have no relationships and are random. The highest peak frequency is 3.4 GHz, which is equal to a single OFB peak frequency. Increasing the optical phase $\psi_2$ from zero to $\omega(\tau_1 + \tau_2)$ causes phase shift and suppression in the intensity of the RIN by about one order of magnitude.

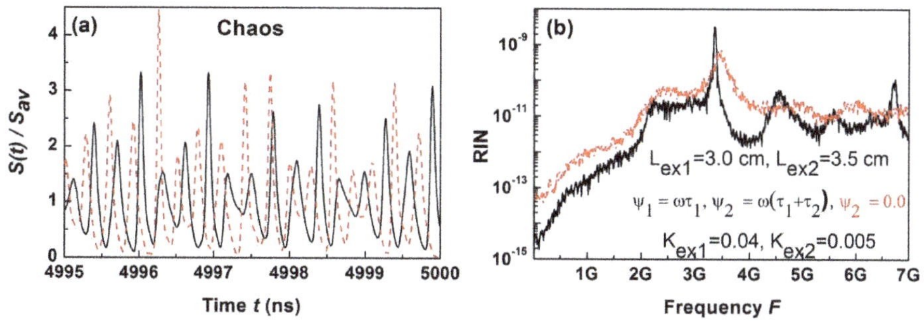


**Figure 7.** RIN of chaotic state with $K_{\rm ex1} = 4 \times 10^{-2}$ and $\psi_1 = \omega\tau_1$, at $K_{\rm ex2} = 5 \times 10^{-3}$ ($L_{\rm ex1} = 3 \times 10^{-2}$ m, $L_{\rm ex2} = 3.5 \times 10^{-2}$ m), $\psi_1 = \omega\tau_1$, $\psi_2 = 0.0$ (red color), and $\psi_2 = \omega(\tau_1 + \tau_2)$ (black color): (**a**) time variation and (**b**) RIN.

4.2.2. Region of Long Double OFB ($L_{ex2} = 10^{-1}$ m)

The impacts of double OFB on the dynamic states and the corresponding average Low-frequency RIN when f < 10 MHz of laser in a long external cavity will be investigated. We set the distance between the first and second reflectors at $7 \times 10^{-2}$ m, making the double OFB cavity approach the region of the long external cavity ($L_{ex2} = 10^{-1}$ m). Figures 8–10 plot the bifurcation diagrams and the corresponding average Low-frequency RIN ($f < 10$ MHz) when the single OFB induces the same dynamic states as Figures 4–6. The figures also indicate that the dynamics of the LD begin with the same dynamic of the LD with a single OFB in the region of weak OFB strength (at small $K_{ex2}$)—in both cases, when $\psi_2 = 0.0$ and $\omega(\tau_1 + \tau_2)$. However, the increase in the double OFB intensity of $K_{ex2}$ does not take the laser to a route-to-CW. It takes it to a route-to-chaos, as occurred in the case when $\psi_2 = 0.0$ in Figures 4b, 5b and 6b. The chaos dynamics range of $K_{ex2}$ is narrowest at $K_{ex1} = 2 \times 10^{-2}$ and $\psi_2 = 0.0$, which agree with PO subject to single OFB, and widest at $K_{ex1} = 4 \times 10^{-2}$, and $\psi_2 = 0.0$ that agree with chaos dynamic state under single OFB. On the contrary, Figure 10a shows similar behavior as shown in Figure 6a: the rise in the double OFB intensity $K_{ex2}$ takes the LD to a route-to-CW state when $\psi_2 = \omega(\tau_1 + \tau_2)$ and attracts the laser to a route-to-chaos state at $\psi_2 = 0.0$, which is cleared in Figure 10b. The figures show that these routes to the chaos state do not mirror the congruent path from CW to the relevant dynamic at single OFB strength $K_{ex1}$ in the bifurcation diagrams of Figure 2 under single OFB. The figures show that variations of PO frequency $f_{PO}$ with $K_{ex2}$ increase with increasing double OFB strength $K_{ex2}$ through each dynamic state (PO, PD, and chaos) when $\psi_2 = 0.0$ and $\omega(\tau_1 + \tau_2)$, as mentioned in Ref. [12].

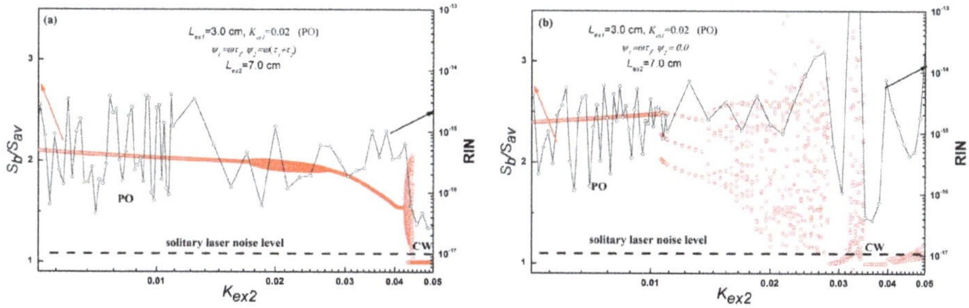


**Figure 8.** BDs of states of LD and average Low-frequency RIN ($f$ < 10 GHz) with double OFB at $K_{\rm ex1} = 0.02$, $L_{\rm ex1} = 3$ cm, $\psi_1 = \omega\tau_1$, and $L_{\rm ex2} = 10$ cm: (**a**) $\psi_2 = \omega(\tau_1 + \tau_2)$ and (**b**) $\psi_2 = 0.0$.

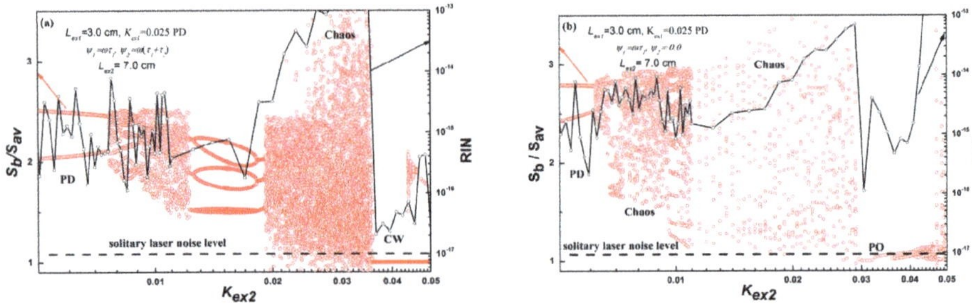


**Figure 9.** BDs of states of LD and average Low-frequency RIN ($f$ < 10 GHz) with double OFB at $K_{ex1}$ = 0.025, $L_{ex1}$ = 3 cm, $\psi_1 = \omega\tau_1$, and $L_{ex2}$ = 10 cm: (**a**) $\psi_2 = \omega(\tau_1 + \tau_2)$ and (**b**) $\psi_2$ = 0.0.

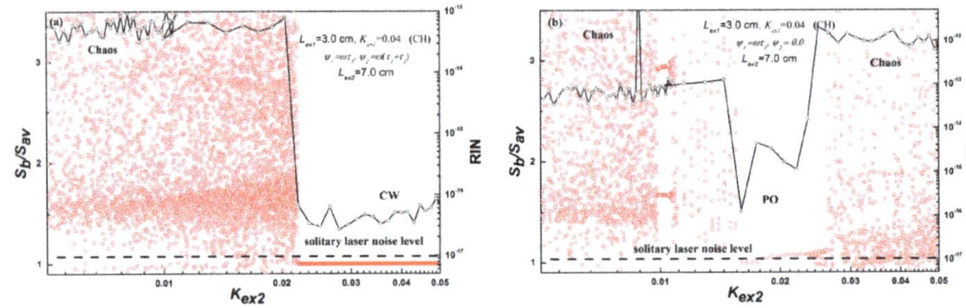


**Figure 10.** BDs of states of LD and average Low-frequency RIN ($f$ < 10 GHz) with double OFB at $K_{ex1}$ = 0.04, $L_{ex1}$ = 3 cm, $\psi_1 = \omega\tau_1$, and $L_{ex2}$ = 10 cm: (**a**) $\psi_2 = \omega(\tau_1 + \tau_2)$ and (**b**) $\psi_2$ = 0.0.

In Figures 8–10 and on the right $Y$-axis, the average Low-frequency RIN ($f$ < 10 MHz) variation with the double cavity OFB intensity $K_{ex2}$ corresponds to BDs shown in the same figures. The RIN, as $\psi_2$ = 0.0, enhances to higher noise levels when dynamic states transfer from PO and PD to the chaos regime and never approach the solitary laser noise level (dashed line). Figures show that when the optical phase $\psi_2$ = 0.0, the RIN displays a lower level than that as $\psi_2 = \omega(\tau_1 + \tau_2)$.

As shown in the figures, when the optical phases $\psi_1 = \omega\tau_1$, and $\psi_2 = \omega(\tau_1 + \tau_2)$, the dynamic states of the laser are more stable than when $\psi_1 = \omega\tau_1$, and $\psi_2$ = 0.0, and the RIN decreases to levels near quantum noise. Figures 4–6 (short external cavity regime) in comparison with Figures 8–10 (long external cavity regime) show that the double external cavity leads to setting the irregular states produced by the single external cavity and keeping the stable CW. This reduces the RIN to lower levels near the solitary laser noise level. This leads to improvements in the properties of the LD, especially in applications such as optical communication systems and others. The simulated results show that the RIN, which corresponds to unstable states of the LD, may be stabilized by operating the laser in a short external cavity regime and increasing the double OFB intensity $K_{ex2}$ and optical phase $\psi_2$. This overall helps in designing lasers with high dynamics and low noise levels. The impacts of the optical phase and double OFB $K_{ex2}$ on the RIN, with an adapting laser design to adjust the values of $K_{ex2}$ and $\psi_2$, are introduced for the first time. This can be useful in increasing/decreasing stability/RIN of lasers, respectively.

Figure 11 plots the temporal variation of $S(t)/S_{av}$ and the corresponding chaos RIN spectral characteristics under double OFB of the long external cavity when $K_{ex1}$ = 0.029 and $K_{ex2}$ = 0.018, and $K_{ex1}$ = 0.04 and $K_{ex2}$ = 0.015, at optical phase $\psi_1 = \omega\tau_1$, $\psi_2$ = 0.0 and $\omega(\tau_1 + \tau_2)$. Figure 11(a1,a2) show an intensity-suppression and phase shift in the fluctuations of $S(t)/S_{AV}$ on the increasing of the optical phase $\psi_2$ from zero to $\omega(\tau_1 + \tau_2)$, which is the same as the chaos state in single OFB and double OFB with short external cavities as shown in Figure 3(a3) and Figure 7a, respectively. The suppression and phase shift of the fluctuations of $S(t)/S_{AV}$ due to increasing optical phase $\psi_2$ from zero to $\omega(\tau_1 + \tau_2)$ are translated in Figure 11(b1,b2) into suppression and frequency shift of the RIN spectrum. By comparison, the chaotic RIN spectra in Figure 3(b3) and Figure 7b with the spectra in Figure 11(b1,b2) are smoother, with two weaker and broader peaks. Figure 11(b2) shows that the frequency of the higher peak is 3.6 GHz and 3.8 GHz, as $\psi_2 = 0.0$ and $\omega(\tau_1 + \tau_2)$, which is higher than the stated peak in the single OFB chaos RIN in Figure 3(b3) by 400 MHz. The lower peak is found at 6.5 GHz and 6.3 GHz, as $\psi_2 = 0.0$ and $\omega(\tau_1 + \tau_2)$, and has no relation to the primary peak frequency, as shown in Figure 11(b2). Moreover, these two peak frequencies are not related to the resonance frequencies of the external cavities, $f_{ex1} = 5$ GHz and $f_{ex2} = 4.286$ GHz, of the laser. That means no information related to the optical system shape can be obtained from the RIN spectra. This characteristic indicates that the RIN power spectrum of the chaotic dynamic state with a double OFB in the long external cavity is unequal to that describing the single OFB. Therefore, the laser with double OFB and a long external cavity may be utilized to obtain smooth chaos signals used in a secure communication system.

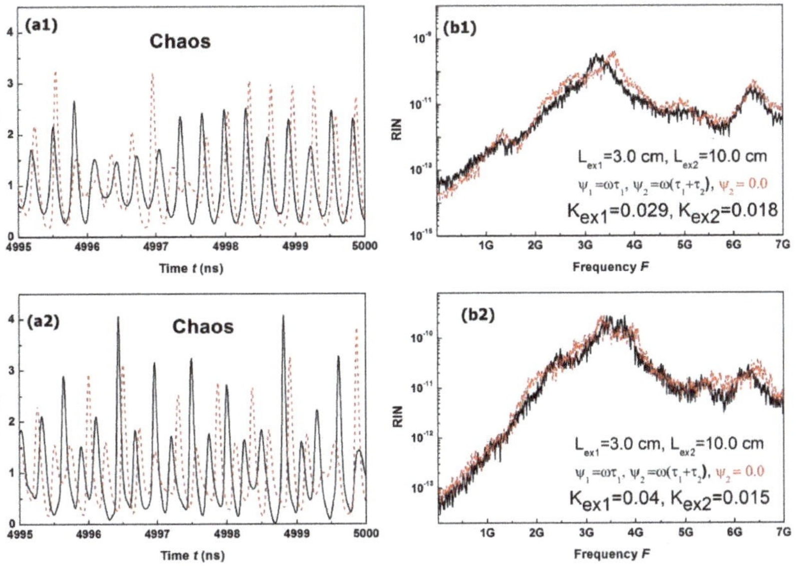


**Figure 11.** RIN of chaotic state with $L_{ex1} = 3 \times 10^{-2}$ m and $L_{ex2} = 10 \times 10^{-2}$ m, $\psi_1 = \omega\tau_1$, $\psi_2 = 0.0$ (red color), and $\psi_2 = \omega(\tau_1 + \tau_2)$ (black color): (**a1,b1**) time variations and RIN at $K_{ex1} = 0.029$ and $K_{ex2} = 0.018$; (**a2,b2**) time variations and RIN at $K_{ex1} = 0.04$ and $K_{ex2} = 0.015$.

## 5. Conclusions

We presented a numerical investigation of the dynamics, states, and RIN of LDs operating under a double OFB by providing our previous model of a single OFB external cavity. The dynamics of the LD were studied within two regions: the short double external cavity ($3.5 \times 10^{-2}$ m) and the long double cavity ($10^{-1}$ m). In the single OFB cavity, the first reflector has been set at $3 \times 10^{-2}$ m far from the LD front facet, which agrees with a PD route-to-chaos. The simulations show that the LD turns its state into POs, PD, and then a chaotic state. In the regime of a short double external cavity, the LD is taken from each of the states with a single OFB to a route-to-CW state. This route-to-CW is a reflection of the identical route-from-chaos to the identical state with a single OFB. In long double OFB, the rise in double OFB strength takes the LD to a route-to-chaos. In comparison with a single OFB case, the double OFB external cavity produces smoother chaotic RIN with two broad and weak peaks at frequencies = 3.6, 3.8, and 6.5, 6.3 GHz, as $\psi_2 = 0.0$ and $\omega(\tau_1 + \tau_2)$, respectively. These frequencies have no relation to the resonance frequencies of the external cavities. This kind of chaos can be used in a secure communication system. By considering the optical phase due to OFB in single OFB and double OFB cavities, it causes significant intensity suppression and phase shift in the LD output temporal changes. This intensity suppression and phase shift induced RIN suppression and a frequency shift in the RIN spectrum. These simulation results on the impacts of the optical phase due to double OFB on the lasing frequency shift in the RIN spectra are new contributions and have not been investigated before. The RIN is reduced near the solitary noise level as LD operates in the CW and PO regimes. Considering the double OFB and optical phase due to OFB, it causes a change in the dynamic states of the LD from a chaotic state to a more stable CW or PO. The dynamic states of LDs can get better by considering the effects of double OFB and optical phase due to OFB, which leads to a better drawing and a best-performance LD.

**Author Contributions:** Conceptualization, S.A.; methodology, S.A., A.M.A. and A.Y.M.; software, S.A.; validation, S.A.; formal analysis, S.A.; investigation, S.A., A.M.A. and A.Y.M.; resources, S.A., A.M.A. and A.Y.M.; data curation, S.A., A.M.A. and A.Y.M.; writing—original draft preparation, S.A.; writing—review and editing, S.A., A.M.A. and A.Y.M.; visualization, S.A., A.M.A. and A.Y.M.; supervision, S.A.; project administration, S.A. All authors have read and agreed to the published version of the manuscript.

**Funding:** This research was funded by Jazan University, Deanship of Scientific Research, Research Units Support Program, Support Number: RUP2-02, Jazan University, Jazan, Saudi Arabia.

**Data Availability Statement:** The data presented in this study are available in the article.

**Acknowledgments:** The authors extend their appreciation to Deanship of Scientific Research, Jazan University, for supporting this research work through the Research Units Support Program, Support Number: RUP2-02.

**Conflicts of Interest:** The authors declare no conflict of interest.

## References


1. Heeger, A.J. Light emission from semiconducting polymers: Light-emitting diodes, light-emitting electrochemical cells, lasers and white light for the future. *Solid State Commun.* **1998**, *107*, 673–679. [CrossRef]
2. Cardoza-Avendano, L.; Spirin, V.; López-Gutiérrez, R.M.; López-Mercado, C.A.; Cruz-Hernández, C. Experimental characterization of dfb and fp chaotic lasers with strong incoherent optical feedback. *Opt. Laser Technol.* **2011**, *43*, 949–955. [CrossRef]
3. Ahmed, M. Longitudinal mode competition in semiconductor lasers under optical feedback: Regime of short-external cavity. *Opt. Laser Technol.* **2009**, *41*, 53–63. [CrossRef]
4. Rusu, S.S.; Oloinic, T.; Tronciu, V.Z. Quantum dots lasers dynamics under the influence of double cavity external feedback. *Opt. Commun.* **2016**, *381*, 140–145. [CrossRef]
5. Abdulrhmann, S.; Ahmed, M.; Okamoto, T.; Ishimori, W.; Yamada, M. An improved analysis of semiconductor laser dynamics under strong optical feedback. *IEEE J. Sel. Top. Quantum Electron.* **2003**, *9*, 265–274. [CrossRef]
6. Abdulrhmann, S.; Ahmed, M.; Yamada, M. New model of analysis of semiconductor laser dynamics under strong optical feedback in fiber communication system. *Proc. SPIE* **2003**, *4986*, 490–501.
7. Abdulrhmann, S.; Hakami, J.; Alhasan, A.M.; Altowyan, A.S.; Afifi, G. Impact of α-factor, combined optical phase and gain saturation on the onset of chaos of short external cavity semiconductor lasers. *Opt.-Int. J. Light Electron Opt.* **2023**, *286*, 171030. [CrossRef]
8. Imran, S.; Yamada, M. Numerical Analysis of Suppression Effects on Optical Feedback Noise by Superposition of High Frequency Current in Semiconductor Lasers. *IEEE J. Quantum Electron.* **2013**, *49*, 196–204. [CrossRef]
9. Bakry, A. Modeling of Millimeter-Wave Modulation Characteristics of Semiconductor Lasers under Strong Optical Feedback. *Sci. World J.* **2014**, *2014*, 728458. [CrossRef]
10. de Mey, R.; Jolly, S.W.; Virte, M. Clarifying the impact of dual optical feedback on semiconductor lasers through analysis of the effective feedback phase. *arXiv* **2023**, arXiv:2306.04006.
11. Pal, V.; Suelzer, J.S.; Prasad, A.; Vemuri, G.; Ghosh, R. Semiconductor Laser Dynamics with Two Filtered Optical Feedbacks. *IEEE J. Quantum Electron.* **2013**, *49*, 340–349. [CrossRef]
12. Bakry, A.; Abdulrhmann, S.; Ahmed, M. Theoretical Modeling of the Dynamics of Semiconductor Laser Subject to Double-Reflector Optical Feedback. *J. Exp. Theor. Phys.* **2016**, *122*, 960–969. [CrossRef]
13. Tromborg, B.; Lassen, H.E.; Olesen, H. Traveling wave analysis of semiconductor lasers: Modulation responses, mode stability and quantum mechanical treatment of noise spectra. *IEEE J. Quantum Electron.* **1994**, *30*, 939–956. [CrossRef]

14. Schunk, N.; Petermann, K. Numerical analysis of the feedback regimes for a 3(i) an7single-mode semiconductor laser with external feedback. *IEEE J. Quantum Electron.* **1988**, *QE-24*, 1242–1247. [CrossRef]
15. Spiteri, R.J.; Wei, S. Fractional-step runge–kutta methods: Representation and linear stability analysis. *J. Comput. Phys.* **2023**, *476*, 111900. [CrossRef]
16. Chen, L.-H.; Zhang, Z.-X.; Cai, Z.-J.; Yang, Y.-T.; Liu, J.-H.; Luo, Z.-C. Period- doubling bifurcation to chaos in a pure-sextic soliton fiber laser. *Infrared Phys. Technol.* **2023**, *134*, 104912. [CrossRef]
17. Prat, J.; Balaguer, P.E.; Gené, J.M.; Díaz, O.; Figuerola, S. *Fiber-to-the-Home Technologies*; Springer: New York, NY, USA, 2004. [CrossRef]
18. Dang, L.; Zheng, B.; Cao, Y.; Zhang, C.; Huang, L.; Iroegbu, P.I.; Shi, L.; Lan, T.; Li, J.; Yin, G.; et al. Tunable ultra-narrow linewidth linear-cavity fiber lasers assisted by distributed external feedback. *Opt. Laser Technol.* **2023**, *166*, 109529. [CrossRef]
19. Ahmed, M.; Yamada, M. An infinite order perturbation approach to gain calculation in injection semiconductor lasers. *J. Appl. Phys.* **1998**, *84*, 3004. [CrossRef]
20. Yamada, M.; Suematsu, Y. Analysis of gain suppression in undoped injection lasers. *J. Appl. Phys.* **1981**, *52*, 2653. [CrossRef]
21. Nishimura, Y.; Kobayashi, K.; Ikegami, T.; Suematsu, Y. Hole-burning effect in semiconductor lasers. In Proceedings of the International Quantum Electronics Conference, Kyoto, Japan, 7–10 September 1970.
22. Agrawal, G.P. *Optical Fiber Communication Systems*; Chapter 6; Van Nostrand Reinhold: New York, NY, USA, 2003.
23. Petermann, K. *Laser Diode Modulation and Noise*; Chapter 7; Kluwer Academic: Dordrecht, The Netherlands, 1988; p. 152.
24. Yamamoto, Y. AM and FM quantum noise in semiconductor lasers—Part I: Theoretical analysis. *IEEE J. Quantum Electron.* **1983**, *19*, 34–46. [CrossRef]
25. Marcuse, D. Computer simulation of laser photon fluctuations: Theory of single-cavity laser. *IEEE J. Quantum Electron.* **1984**, *20*, 1139–1148. [CrossRef]
26. Yamada, M. Theoretical analysis of line broadening due to mode-competition and optical feedback in semiconductor injection lasers. *Trans. IEICE* **1988**, *E71*, 152–160.
27. Press, W.H.; Teukolsky, S.A.; Vetterling, W.T.; Flannery, B.P. *Numerical Recipes in Fortran: The Art of Scientific Computing*, 2nd ed.; Chapter 7; Cambridge University Press: New York, NY, USA, 1992.
28. Ahmed, M.; Yamada, M.; Saito, M. Numerical modeling of intensity and phase noise in semiconductor lasers. *IEEE J. Quantum Electron.* **2001**, *37*, 1600–1610. [CrossRef]
29. Mahmood, R.; Bin-Tarik, F.; Imran, S.M.S. Analysis of Intensity and Phase Noise of Solitary Semiconductor Lasers Operating in Single-Mode. *Int. J. Photonics Opt. Technol.* **2018**, *4*, 1–6.
30. Kao, Y.H.; Wang, N.M.; Chen, H.M. Mode description of routes to chaos in external-cavity coupled semiconductor lasers. *IEEE J. Quantum Electron.* **1994**, *30*, 1732–1739.
31. Nada, F.; Tarek, M.; Alaa, M. Impact of linewidth enhancement factor and gain suppression on chirp characteristics of high-speed laser diode and performance of 40 Gbps optical fiber links. *Appl. Phys. B* **2022**, *128*, 45.
32. Hao, C.; Zheng, W.; Liu, A. Dynamics and feedback sensitivity of vertical-cavity surface-emitting lasers with arbitrary optical feedback intensity. *J. Phys. D Appl. Phys.* **2023**, *56*, 215103. [CrossRef]

*applied sciences*

MDPI

*Article*

# Analytical Studies on Approximate Lag and Anticipating Synchronization in Two Unidirectionally Coupled Hyperchaotic Chen Systems without Time Delay

**Li-Jun Ouyang [1], Bing-Qing Xie [1] and Bin Ding [2,*]**

[1] School of Environment and Architecture, University of Shanghai for Science and Technology, Shanghai 200093, China; ouyang@usst.edu.cn (L.-J.O.); 18705276091@139.com (B.-Q.X.)

[2] Department of Civil Engineering, Wenzhou Polytechnic, Wenzhou 325035, China

* Correspondence: dingb0577@163.com

**Abstract:** In this paper, approximate lag synchronization (LS) and anticipating synchronization (AS) between two unidirectionally coupled hyperchaotic Chen systems without time-delay coupling are analytically investigated. Firstly, the synchronization condition for exact LS in two unidirectionally coupled hyperchaotic Chen systems with time delay in signal transmission is analytically obtained. Under such conditions, approximate LS and AS are discussed by replacing the true time-delay terms with their Taylor expansions up to the third order.Differently from other research studies, the condition for exact LS is derived by regarding LS as a special type of generalized synchronization (GS), which has nothing to do with the value of the time delay. It is convenient to individually change the value of the lag and anticipation time of approximate LS and AS without considering the synchronization condition. Our study shows the power of a new method for recreating the past signals or predicting the future signals of a hyperchaotic Chen system by using its current signals. The results provide a simple way to eliminate the negative effects of time delay in the signal transmission between two hyperchaotic systems.

**Keywords:** lag synchronization; anticipating synchronization; hyperchaotic Chen system; drive–response systems

## 1. Introduction

The synchronization of chaotic systems plays a tremendous role in nonlinear system analysis and has been applied in a large variety of physical, chemical, neural network and biological systems [1–3] since the pioneering work of Pecora and Carroll [4]. The synchronization of chaotic systems is a process wherein two chaotic systems adjust a certain characteristic of their motion to standard performance because of a coupling or a forcing. Many different synchronization regimes have been studied in the past several decades, for instance, complete (or identical) synchronization (CS) [5,6], phase synchronization (PS) [7], lag synchronization (LS) [8–12], generalized synchronization (GS) [13,14], anticipating synchronization (AS) [15–17] and so on. Among all types of synchronization, CS is the simplest one, where the states of all oscillators are identical. GS is considered to be the type of chaos synchronization most frequently occurring in natural systems and is characterized by the existence of a functional relation between the state variables of drive–response systems. Different types of synchronization, such as CS, LS and AS, can be viewed as special types of GS with different functional relations.

In technology and nature, the time delay is ubiquitous due to finite signal transmission times, switching speeds as well as memory effects. LS, typically appearing as a coincidence of the shifted-in-time states of two systems, has received extensive attention in secure communications [18], laser networks [9] and digital image encryption [19]. AS was firstly proposed by Voss [20,21], who considered two chaotic systems unidirectionally coupled in a drive–response configuration and showed that the current states of a response system can be synchronized with the future states of a drive system. Two different coupling schemes were suggested by Voss: One is for the case where there exists an internal delay in a drive system that defines the anticipation time of a synchronized state. The other is for the case where a drive system does not possess internal delays, while in a salve system, there exists time-delay self-feedback. The work by Voss has led to the proven possibility of a real-time predictor of chaotic dynamics for systems without delays. Moreover, the scheme allows one to change the anticipation time without altering the dynamics of a drive system. As a possible application, AS can be used to design control methods for avoiding unwanted pulses from a drive system. AS has been reported in many applications in semiconductor lasers [22], chaotic laser diodes and cryptography [23,24].

Both LS and AS are mostly observed in systems using a time-delay factor. However, approximate LS has been found in bidirectionally and unidirectionally coupled oscillators with parameter mismatch [25]. AS has been also observed in systems without time delay. Such a type of phenomenon was first reported in the classical Voss scheme, where the true time-delay term was replaced with its first-order approximation [26]. Nevertheless, AS is not exact, that is, the advanced trajectory of a drive system only approximately coincides with the current trajectory of a response system. In addition, the anticipation time, in this case, is small. Blakely et al. [25] considered a more general case of nonidentical systems without time delay in which the mean frequency of a free response system is greater than the mean frequency of a master system. They showed that AS may occur in the Voss coupling scheme with vanishing time delay. If the above condition is satisfied in neural systems, the chaotic spikes of a presynaptic neuron may be predicted by a postsynaptic neuron. The Voss scheme provides a simple memoryless method to achieve LS and AS between coupled oscillators. In previous research studies on approximate LS and AS between two unidirectionally coupled chaotic systems without time-delay coupling, the true time-delay terms are first replaced by using their first-order approximations, which results in parameter mismatch between the drive and response systems. However, the value of the lag or anticipation time in approximate LS and AS significantly affects the synchronization condition for the two chaotic systems with such parameter mismatches. To guarantee synchronization, the value has to be limited to a very finite range. Furthermore, extending the result derived for two chaotic systems to larger arrays must also take into account possible instabilities caused by additional spatial modes. To overcome the disadvantage, the nature of the evolution of approximate LS and AS needs to be further studied in detail.

In this paper, approximate LS and AS between two unidirectionally coupled hyperchaotic Chen systems [27] without time-delay coupling are investigated. A hyperchaotic system is classified as a chaotic system with more than one positive Lyapunov exponent or unstable direction, which generates high complexity. We first analyze the synchronization condition for exact LS in two unidirectionally coupled hyperchaotic Chen systems with time delay in the signal transmission from the drive systems. Differently from most research studies, LS is regarded as a special type of GS, which can be studied by using the auxiliary system approach [28,29]. We show that the synchronization condition for exact LS analytically obtained in this paper is effective for any value of time delay in the signal transmission between two hyperchaotic Chen systems. Under such conditions, approximate LS and AS are discussed by replacing the true time-delay terms with their Taylor expansions up to the third order. Our research provides an effective method for recreating the past signals or predicting the future signals of a hyperchaotic Chen system by using its current signals.

The rest of this paper is organized as follows: In Section 2, two unidirectionally coupled hyperchaotic Chen systems with time delay in signal transmission are introduced. In Section 3, the synchronization condition for exact LS in the drive–response system is derived by regarding LS as a special type of GS, in which the auxiliary system approach and Laplace transform, combined with the successive approximation approach [30] in the Volterra integral equation theory, are used. In Section 4, numerical simulations are presented to verify the effectiveness of the synchronization condition for exact LS obtained in this paper. Approximate LS and AS are discussed in Sections 5 and 6, respectively. The conclusions are finally drawn in Section 7.

## 2. System Description

The hyperchaotic Chen system [27] is described as follows:

$$\begin{aligned} \dot{u} &= a(v-u)+p, \\ \dot{v} &= du+cv-uw, \\ \dot{w} &= -bw+uv, \\ \dot{p} &= rp+vw, \end{aligned} \tag{1}$$

where $u, v, w$ and $p$ are state variables and $a, b, c, d$ and $r$ are real parameters. Assume that $(u_0, v_0, w_0, p_0)$ is one equilibrium point in system (1). For convenience, by moving the equilibrium to the origin, system (1) can be written as

$$\begin{aligned} \dot{x}_1 &= -ax_1+ax_2+x_4, \\ \dot{x}_2 &= \alpha x_1+cx_2-u_0x_3-x_1x_3, \\ \dot{x}_3 &= v_0x_1+u_0x_2-bx_3+x_1x_2, \\ \dot{x}_4 &= w_0x_2+v_0x_3+rx_4+x_2x_3, \end{aligned} \tag{2}$$

where $x_1 = u - u_0$, $x_2 = v - v_0$, $x_3 = w - w_0$, $x_4 = p - p_0$, and $\alpha = d - w_0$. System (2) is chosen as the drive system. The response system is given as:

$$\begin{aligned} \dot{y}_1 &= -ay_1+ay_2+y_4+k_1(x_1(t-\tau)-y_1), \\ \dot{y}_2 &= \alpha y_1+cy_2-u_0y_3-y_1y_3+k_2(x_2(t-\tau)-y_2), \\ \dot{y}_3 &= v_0y_1+u_0y_2-by_3+y_1y_2+k_3(x_3(t-\tau)-y_3), \\ \dot{y}_4 &= w_0y_2+v_0y_3+ry_4+y_2y_3+k_4(x_4(t-\tau)-y_4), \end{aligned} \tag{3}$$

where $y_{1,2,3,4}$ are state variables, $k_{1,2,3,4}$ are coupling strengths and $\tau$ is the time delay in signal transmission between systems (2) and (3). LS in systems (2) and (3) is reached if

$$||y_i - x_i(t-\tau)|| \to 0, \quad i = 1,2,3,4, \quad for \quad t \to \infty. \tag{4}$$

## 3. The Synchronization Condition for Exact LS between Systems (2) and (3)

Condition (4) is a functional relation between $y_i$ and $x_i$, $i = 1,2,3,4$. Then, LS between systems (2) and (3) can be regarded as a special type of GS between them, which can be investigated by using the auxiliary system approach [28]. The auxiliary system corresponding to systems (2) and (3) is given by

$$\begin{aligned} \dot{z}_1 &= -az_1+az_2+z_4+k_1(x_1(t-\tau)-z_1), \\ \dot{z}_2 &= \alpha z_1+cz_2-u_0z_3-z_1z_3+k_2(x_2(t-\tau)-z_2), \\ \dot{z}_3 &= v_0z_1+u_0z_2-bz_3+z_1z_2+k_3(x_3(t-\tau)-z_3), \\ \dot{z}_4 &= w_0z_2+v_0z_3+rz_4+z_2z_3+k_4(x_4(t-\tau)-z_4), \end{aligned} \tag{5}$$

where $z_{1,2,3,4}$ are state variables. The following condition is necessary to obtain condition (4):

$$||y_i - z_i|| \to 0, \quad i = 1,2,3,4, \quad for \quad t \to \infty. \tag{6}$$

By letting

$$e_i = \frac{y_i - z_i}{2}, \quad e_{i+4} = \frac{y_i + z_i}{2}, \qquad i = 1,2,3,4,$$

the synchronization error system for condition (6) can be given as

$$\begin{aligned}
\dot{e}_1 &= -(k_1 + a)e_1 + ae_2 + e_4, \\
\dot{e}_2 &= \alpha e_1 + (c - k_2)e_2 - u_0 e_3 - e_1 e_7 - e_3 e_5, \\
\dot{e}_3 &= v_0 e_1 + u_0 e_2 - (b + k_3)e_3 + e_1 e_6 + e_2 e_5, \\
\dot{e}_4 &= w_0 e_2 + v_0 e_3 + (r - k_4)e_4 + e_2 e_7 + e_3 e_6.
\end{aligned} \tag{7}$$

Condition (6) becomes

$$||e_i|| \to 0, \quad i = 1,2,3,4, \quad for \quad t \to \infty. \tag{8}$$

Consider the Laplace transform, defined as follows:

$$\begin{aligned}
E_i(s) = L[e_i] &= \int_0^{+\infty} e_i(t)e^{-st}dt, \\
e_i(t) = L^{-1}[E_i] &= \frac{1}{2\pi j}\int_{\sigma-j\infty}^{\sigma+j\infty} E_i(s)e^{st}ds, \quad i = 1,2,3,4.
\end{aligned} \tag{9}$$

By taking the Laplace transform on both sides of system (7), it yields

$$(sI_4 - M)\begin{bmatrix} E_1 \\ E_2 \\ E_3 \\ E_4 \end{bmatrix} = \begin{bmatrix} e_{10} \\ W_1 + e_{20} \\ W_2 + e_{30} \\ W_3 + e_{40} \end{bmatrix} \tag{10}$$

where $e_{0i}$, $i = 1,2,3,4$ are the given initial values for system (7), $I_4$ is the $4 \times 4$ identity real matrix,

$$M = \begin{bmatrix} -k_1 - a & a & 0 & 1 \\ \alpha & c - k_2 & -u_0 & 0 \\ v_0 & u_0 & -b - k_3 & 0 \\ 0 & w_0 & v_0 & r - k_4 \end{bmatrix} \tag{11}$$

and $W_i$, $i = 1,2,3$ are the Laplace transform of the nonlinear parts in the last three equations in system (7), respectively,

$$\begin{aligned}
W_1 &= -\int_0^{+\infty} [e_1 e_7 + e_3 e_5]e^{-st}dt, \\
W_2 &= \int_0^{+\infty} [e_1 e_6 + e_2 e_5]e^{-st}dt, \\
W_3 &= \int_0^{+\infty} [e_2 e_7 + e_3 e_6]e^{-st}dt.
\end{aligned}$$

Solving Equation (10) by using Cramer's rule, one has

$$\begin{aligned}
E_1 &= \frac{1}{D(s)}\begin{vmatrix} e_{10} & a & 0 & 1 \\ W_1 + e_{20} & c - k_2 & -u_0 & 0 \\ W_2 + e_{30} & u_0 & -b - k_3 & 0 \\ W_3 + e_{40} & w_0 & v_0 & r - k_4 \end{vmatrix}, \\
E_2 &= \frac{1}{D(s)}\begin{vmatrix} -k_1 - a & e_{10} & 0 & 1 \\ \alpha & W_1 + e_{20} & -u_0 & 0 \\ v_0 & W_2 + e_{30} & -b - k_3 & 0 \\ 0 & W_3 + e_{40} & v_0 & r - k_4 \end{vmatrix}, \\
E_3 &= \frac{1}{D(s)}\begin{vmatrix} -k_1 - a & a & e_{10} & 1 \\ \alpha & c - k_2 & W_1 + e_{20} & 0 \\ v_0 & u_0 & W_2 + e_{30} & 0 \\ 0 & w_0 & W_3 + e_{40} & r - k_4 \end{vmatrix},
\end{aligned}$$

$$E_4 = \frac{1}{D(s)} \begin{vmatrix} -k_1 - a & a & 0 & e_{10} \\ \alpha & c - k_2 & -u_0 & W_1 + e_{20} \\ v_0 & u_0 & -b - k_3 & W_2 + e_{30} \\ 0 & w_0 & v_0 & W_3 + e_{40} \end{vmatrix}, \tag{12}$$

where

$$D(s) = s^4 + \beta_1 s^3 + \beta_2 s^2 + \beta_3 s + \beta_4, \tag{13}$$

$$\begin{aligned}
\beta_1 &= a + b - c - k - r, \\
\beta_2 &= (b - \alpha - c - r)a - (c + k + r)b + (k + r)c + kr + {u_0}^2, \\
\beta_3 &= (a - k - r){u_0}^2 + a u_0 v_0 + ((\alpha + c)r - (\alpha + c + r)b)a + (ck + rc + kr)b \\
&\quad - krc - \alpha w_0 - {v_0}^2, \\
\beta_4 &= r(k - a){u_0}^2 + (w_0 - ar - \alpha)u_0 v_0 + c{v_0}^2 + (\alpha + c)abr - (krc + \alpha w_0)b.
\end{aligned}$$

$D(s)$ is just the characteristic polynomial of matrix (11). In fact, we do not need to know the actual formulas of the numerators of $E_i$, $i = 1,2,3,4$. They must have the following forms:

$$E_i = \frac{h_{i0}(s)}{D(s)} + \sum_{j=1}^{3} \frac{h_{ij}(s)}{D(s)} W_j, \quad i = 1,2,3,4, \tag{14}$$

in which $h_{ij}(s)$ ($i = 1,2,3,4$, $j = 0,1,2,3$) are polynomials of $s$ whose orders are less than four. Taking the inverse Laplace transform for $E_i$, $i = 1,2,3,4$ and considering the convolution theorem in the Laplace transform produces

$$\begin{aligned}
e_i = \gamma_{i0}(t) &- \int_0^t \gamma_{i1}(t-\tau)[e_1 e_7 + e_3 e_5]d\tau + \int_0^t \gamma_{i2}(t-\tau)[e_1 e_6 + e_2 e_5]d\tau \\
&+ \int_0^t \gamma_{i3}(t-\tau)[e_2 e_7 + e_3 e_6]d\tau, \quad i = 1,2,3,4,
\end{aligned} \tag{15}$$

where $\gamma_{ij}(t) = L^{-1}[\frac{h_{ij(s)}}{D(s)}]$, $i = 1,2,3,4$ and $j = 0,1,2,3$. If the denominator $D(s)$ in Equation (14) can be decomposed into $D(s) = Q_1(s)Q_2(s)$, and $Q_{1,2}$ have no common factors, then $h_{ij}(s)/D(s)$, $i \in \{1,2,3,4\}$, $j \in \{0,1,2,3\}$ can be decomposed into

$$\frac{h_{ij}(s)}{D(s)} = \frac{\bar{h}_{ij}(s)}{Q_1(s)} + \frac{\hat{h}_{ij}(s)}{Q_2(s)}, \quad i \in \{1,2,3,4\},\ j \in \{0,1,2,3\},$$

where $\bar{h}_{ij}(s)$ and $\hat{h}_{ij}(s)$ are polynomials of $s$. Moreover, the highest powers are less than those of $Q_1(s)$ and $Q_2(s)$. One can continue the process until only the following two classes of functions retain

$$\frac{A_1(s)}{(s-\eta_0)^m}, \quad \frac{A_2(s)}{(s^2 + \eta_1 s + \eta_2)^n},$$

in which $m \in \{1,2,3,4\}$, $n \in \{1,2\}$, $\eta_{0,1,2} \in R$, $\eta_1^2 - 4\eta_2 < 0$ and the highest powers of polynomials $A_{1,2}$ are less than $m$ and $n$, respectively. Consider the inverse Laplace transform in the following three cases:

- $\frac{A_1(s)}{(s-\eta_0)^m} = \frac{B_1}{s-\eta_0} + \frac{B_2}{(s-\eta_0)^2} + \cdots + \frac{B_m}{(s-\eta_0)^m}$,

  where $B_{(m-i)} = \frac{1}{i!}\frac{d^i}{ds^i}[\frac{A_1(s)}{(s-\eta_0)^m}(s-\eta_0)^m]\Big|_{s=\eta_0}$, $i = 1,2,\cdots,m-1$,

  $B_m = [\frac{A_1(s)}{(s-\eta_0)^m}(s-\eta_0)^m]\Big|_{s=\eta_0}$,

  $L^{-1}[\frac{A_1(s)}{(s-\eta_0)^m}] = (B_1 + B_2 t + B_3 t^2 + \cdots + B_m t^{m-1})e^{\eta_0 t}$

- $\frac{A_2(s)}{s^2+\eta_1 s+\eta_2} = \frac{A_2(s)}{(s-\omega_1-j\omega_2)(s-\omega_1+j\omega_2)} = \frac{B_1}{s-\omega_1-j\omega_2} + \frac{B_2}{s-\omega_1+j\omega_2}$,
  where $B_{1,2} = \frac{A_2(s)}{2s+\eta_1}\Big|_{s=\omega_1 \pm j\omega_2}$,
  $L^{-1}[\frac{A_2(s)}{s^2+\eta_1 s+\eta_2}] = B_1 e^{(\omega_1+j\omega_2)t} + B_2 e^{(\omega_1-j\omega_2)t}$
- $\frac{A_2(s)}{(s^2+\eta_1 s+\eta_2)^2} = \frac{B_1}{(s-\omega_1-j\omega_2)^2} + \frac{B_2}{s-\omega_1-j\omega_2} + \frac{B_3}{(s-\omega_1+j\omega_2)^2} + \frac{B_4}{s-\omega_1+j\omega_2}$,
  where $B_{1,3} = \frac{A_2(s)}{s^2+\eta_1 s+\eta_2}(s-\omega_1 \pm j\omega_2)^2\Big|_{s=\omega_1 \pm j\omega_2}$,
  $B_{2,4} = \frac{d}{ds}[\frac{A_2(s)}{s^2+\eta_1 s+\eta_2}(s-\omega_1 \pm j\omega_2)^2]\Big|_{s=\omega_1 \pm j\omega_2}$,
  $L^{-1}[\frac{A_2(s)}{(s^2+\eta_1 s+\eta_2)^2}] = (B_1 t + B_2)e^{(\omega_1+j\omega_2)t} + (B_3 t + B_4)e^{(\omega_1-j\omega_2)t}$

From Equation (15), the necessary conditions for $||e_i|| \to 0$, $i = 1,2,3,4$ are $\gamma_{ij}(t) \to 0$, $i \in \{1,2,3,4\}$ and $j \in \{0,1,2,3\}$. According to the above analysis for the inverse Laplace transform, all eigenvalues of matrix (11) should have negative real parts. Under such conditions, when $t \to 0$ system (15) becomes

$$\begin{aligned} e_i = & -\int_0^t \gamma_{i1}(t-\tau)[e_1e_7 + e_3e_5]d\tau + \int_0^t \gamma_{i2}(t-\tau)[e_1e_6 + e_2e_5]d\tau \\ & + \int_0^t \gamma_{i3}(t-\tau)[e_2e_7 + e_3e_6]d\tau, \quad i = 1,2,3,4. \end{aligned} \tag{16}$$

Actually, these are a set of Volterra integral equations in system (16) that can be solved based on the successive approximation method [30]. For the sake of clarity, we briefly introduce the successive approximation method below. Consider the integral equation of the following form

$$y(t) = \Phi(t) + \int_0^t g(t-\tau)h(\tau, y(\tau))d\tau, \tag{17}$$

where $g$ is an $n \times n$ matrix; $\Phi(t)$ and $h(t, y(t))$ are vectors with $n$ components. If the following conditions are satisfied, then Equation (17) has a unique continuous solution:

- $|y| < \infty$;
- $\Phi$ and $h$ are continuous for $0 < t < t_0$, in which $0 < t_0 < +\infty$;
- $|g| \in L[0, \epsilon]$ holds for any $0 < \epsilon < t_0$;
- For any $\xi > 0$, if $|y_1 - y_2| < \xi$ there must exist a constant $\epsilon > 0$ such that $|h(t, y_1) - h(t, y_2)| < \epsilon$.

Moreover, the following successive approximations

$$\omega_0(t) = 0, \quad \omega_{n+1}(t) = \Phi(t) + \int_0^t g(t-\tau)h(\tau, \omega_n(\tau))d\tau, \quad n = 0,1,2,\cdots \tag{18}$$

will uniformly converge to the unique continuous solution to Equation (17).

It is clear that the functions and parameters in Equation (16) satisfy the assumptions for Equation (17). Then, Equation (16) has the unique continuous solution $e_i = 0$, $i = 1,2,3,4$.

**Remark 1.** *The steps required to derive the synchronization condition for CS between systems (3) and (5) are listed below:*

- *System (7) is obtained with variable substitution;*
- *Through the Laplace transform and the convolution theorem, system (7) is converted into a set of Volterra integral equations (system(15)), which can be further simplified to Equation (16) based on the convergence requirement;*
- *The successive approximation method [30] is used to solve Equation (16). Then, the synchronization condition for CS between systems (3) and (5) is determined.*

**Remark 2.** *The condition that all roots of $D(s)$ defined in Equation (13) have negative real parts guarantees the occurrence of CS between systems (3) and (5). $D(s)$ is the characteristic polynomial of matrix (11). From the auxiliary system approach, under such conditions the functional relationship $y_i = x_i(t-\tau)$, $i = 1,2,3,4$ will be achieved between systems (2) and (3).*

**Remark 3.** *According to the Routh–Hurwitz criteria, the necessary and sufficient condition that all roots of $D(s)$ have negative real parts can be written as*

$$\beta_1 > 0, \quad \beta_3 > 0, \quad \beta_4 > 0, \quad and \quad \beta_1\beta_2\beta_3 > \beta_3^2 + \beta_1^2\beta_4. \tag{19}$$

*In particular, if $u_0 = v_0 = w_0 = p_0 = 0$, Equation (13) becomes*

$$D(s) = [s^2 + (k_1 + k_2 - c + a)s + (k_1 + a)(k_2 - c) - ad](s + k_3 + b)(s + k_4 - r). \tag{20}$$

*Condition (19) can be written as:*

$$k_1 + k_2 > a - c, \quad (k_1 + a)(k_2 - c) - ad > 0, \quad k_3 < -b, \quad and \quad k_4 > r. \tag{21}$$

**Remark 4.** *The time delay $\tau$ has nothing to do with conditions (19) or (21), which implies that the synchronization condition obtained in this paper is effective for any value of $\tau$. In the next section, we numerically verify the effectiveness of condition (21).*

## 4. Numerical Simulations

In this section, numerical simulations are carried out to verify the effectiveness of the criterion obtained in the previous section for the appearance of LS between systems (2) and (3). When $a = 35, b = 3, c = 12, d = 7$ and $0 \le r \le 0.085$, system (1) is chaotic; when $a = 35, b = 3, c = 12, d = 7$ and $0.085 < r \le 0.798$, system (1) is hyperchaotic; and when $a = 35, b = 3, c = 12, d = 7$ and $0.798 < r \le 0.90$, system (1) is periodic. In the following, we consider the hyperchaotic case. By calculation, when $a = 35, b = 3, c = 12, d = 7$ and $0.085 < r \le 0.798$, the hyperchaotic Chen system (1) only has one equilibrium $(0,0,0,0)$.

In the following simulations, the parameters in systems (2), (3) and (5) are taken as $a = 35$, $b = 3$, $c = 12$, $d = 7$ and $r = 0.5$. From condition (21), the functional relationship $y_i = x_i(t-\tau)$, $i = 1,2,3,4$ between systems (2) and (3) can be achieved if one chooses $k_1 = 0$, $k_2 = 19$, $k_3 = -2$ and $k_4 = 0.6$. The numerical results are presented in Figures 1–4 for different values of the time delay $\tau$, where the initial conditions are taken as $x_1(-\tau,0] = 0.1$; $x_2(-\tau,0] = 0.2$; $x_3(-\tau,0] = 0.3$; $x_4(-\tau,0] = 0.4$; $y_1(-\tau,0] = 0.2$; $y_2(-\tau,0] = 0.2$; $y_3(-\tau,0] = 0.3$; $y_4(-\tau,0] = 0.4$; $z_1(-\tau,0] = 0.3$; $z_2(-\tau,0] = 0.4$; $z_3(-\tau,0] = 0.3$; and $z_4(-\tau,0] = 0.4$. It is depicted that the synchronization errors $y_i - z_i$, $i = 1,2,3,4$ between systems (3) and (5) approach zero and the functional relation $y_i = x_i(t-\tau)$, $i = 1,2,3,4$ between systems (2) and (3) can be achieved for different values of the time delay: $\tau = 0, 0.2, 1$ and $2$, respectively. The time delay in the signal transmission results in the transition from CS to LS between two hyperchaotic Chen systems under condition (21). In fact, the synchronization condition obtained in this paper for exact LS between systems (2) and (3) is not involved with the value of $\tau$. We show in Figure 5 that even if the value of $\tau$ is extremely large, exact LS between systems (2) and (3) still can be achieved, where the initial conditions remain as those in Figures 1–4.

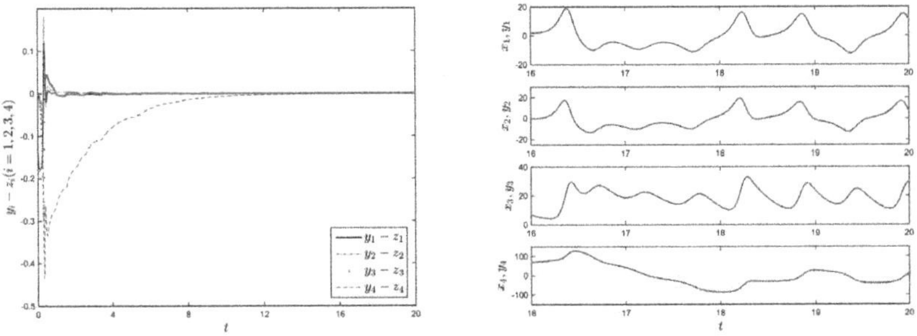


(a) Synchronization errors $y_i - z_i, i = 1,2,3,4$.

(b) $x_i$ (black line –), $y_i$ (blue dot ·), $i = 1,2,3,4$.

**Figure 1.** (**a**) Synchronization errors between systems (3) and (5); (**b**) the time series of $x_i$, $y_i$, $i = 1,2,3,4$ in systems (2) and (3) with $k_1 = 0, k_2 = 19, k_3 = -2, k_4 = 0.6$ and $\tau = 0$.

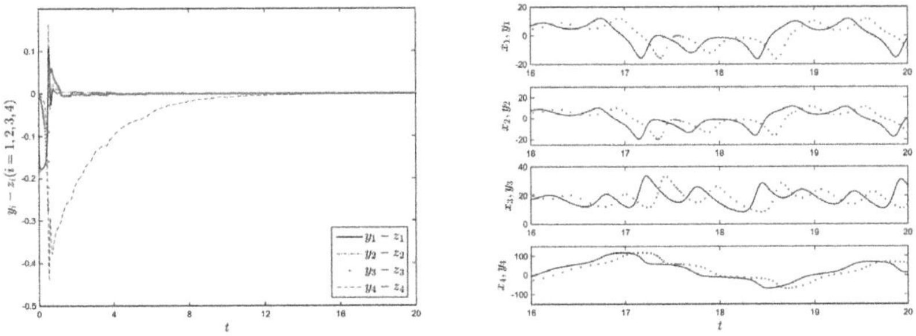


(a) Synchronization errors $y_i - z_i, i = 1,2,3,4$.

(b) $x_i$ (black line –), $y_i$ (blue dot ·), $i = 1,2,3,4$.

**Figure 2.** (**a**) Synchronization errors between systems (3) and (5); (**b**) the time series of $x_i$, $y_i$, $i = 1,2,3,4$, in systems (2) and (3) with $k_1 = 0, k_2 = 19, k_3 = -2, k_4 = 0.6$ and $\tau = 0.2$.

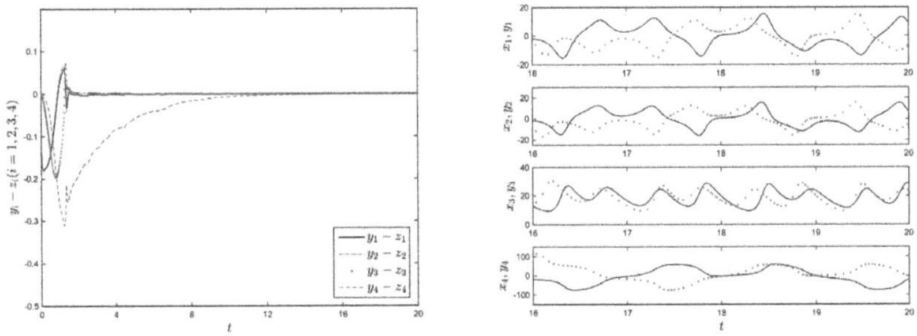


(a) Synchronization errors $y_i - z_i, i = 1,2,3,4$.

(b) $x_i$ (black line –), $y_i$ (blue dot ·), $i = 1,2,3,4$.

**Figure 3.** (**a**) Synchronizationerrors between systems (3) and (5); (**b**) The time series of $x_i$, $y_i$, $i = 1,2,3,4$, in systems (2) and (3) with $k_1 = 0, k_2 = 19, k_3 = -2, k_4 = 0.6, \tau = 1$.

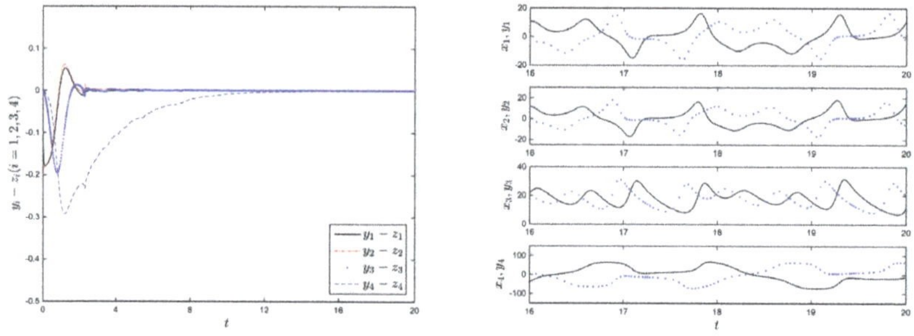


(a) Synchronization errors $y_i - z_i$, $i = 1, 2, 3, 4$.

(b) $x_i$ (black line –), $y_i$ (blue dot ·), $i = 1, 2, 3, 4$.

**Figure 4.** (**a**) Synchronization errors between systems (3) and (5); (**b**) the time series of $x_i$, $y_i$, $i = 1, 2, 3, 4$, in systems (2) and (3) with $k_1 = 0$, $k_2 = 19$, $k_3 = -2$, $k_4 = 0.6$ and $\tau = 2$.

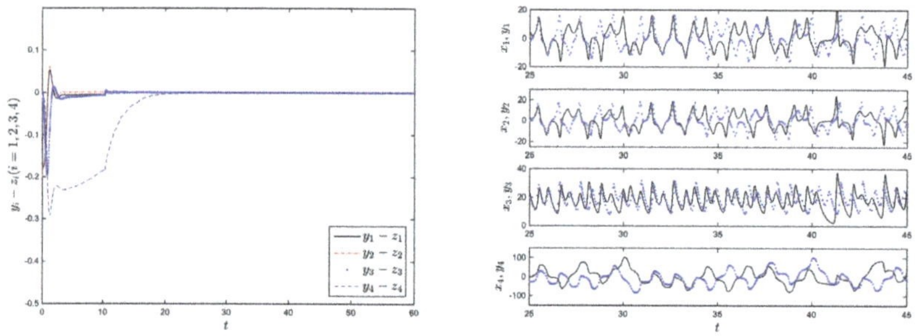


(a) Synchronization errors $y_i - z_i$, $i = 1, 2, 3, 4$.

(b) $x_i$ (black line –), $y_i$ (blue dot ·), $i = 1, 2, 3, 4$.

**Figure 5.** (**a**) Synchronization errors between systems (3) and (5); (**b**) the time series of $x_i$, $y_i$, $i = 1, 2, 3, 4$, in systems (2) and (3) with $k_1 = 0$, $k_2 = 19$, $k_3 = -2$, $k_4 = 0.6$ and $\tau = 10$.

## 5. The Approximation of LS between Two Hyperchaotic Chen Systems without Time-Delay Coupling

The synchronization condition for exact LS between two unidirectionally coupled hyperchaotic Chen systems with time-delay coupling serves as the foundation to realize approximate LS between two unidirectionally coupled hyperchaotic Chen systems without time-delay coupling. Synchronization condition (21) is valid for any value of the time delay. Under such conditions, we replace the time-delay terms in Equation (3) with their Taylor expansions to achieve approximate LS between Equations (2) and (3).

Near $\tau = 0$, $y_i = x_i(x - \tau)$, $i = 1, 2, 3, 4$, can be approximately expanded as

$$x_i(t-\tau) = x_i(t) - \frac{dx_i}{dt}\tau + \frac{1}{2!}\frac{d^2x_i}{dt^2}\tau^2 - \frac{1}{3!}\frac{d^3x_i}{dt^3}\tau^3 + O(\tau^3), \quad i = 1, 2, 3, 4, \tag{22}$$

where $O(\tau^3)$ represents the higher order terms of $\tau$. It is theoretically possible that Equation (22) holds for any value of $\tau$. To demonstrate the feasibility of approximate LS, we define the following function of $\tau$:

$$H_i(\tau) = x_i(t) - \frac{dx_i}{dt}\tau + \frac{1}{2!}\frac{d^2x_i}{dt^2}\tau^2 - \frac{1}{3!}\frac{d^3x_i}{dt^3}\tau^3, \quad i = 1, 2, 3, 4. \tag{23}$$

$H_i(\tau)$ is the truncation up to the third order of each $x_i(t - \tau)$ in Equation (22), $i = 1, 2, 3, 4$. From system (2), one has

$$
\begin{aligned}
\frac{dx_1}{dt} &= -ax_1 + ax_2 + x_4, \\
\frac{d^2x_1}{dt^2} &= \left(-ax_3 + a^2 + ad\right)x_1 + \left(-a^2 + ac + x_3\right)x_2 + (-a + r)x_4, \\
\frac{d^3x_1}{dt^3} &= -ax_2{x_1}^2 + \left({x_2}^2 - {x_3}^2 + \left(2\,a^2 + ab - ac + d\right)x_3 - \left(a^2 + ad\right)a + \left(-a^2 + ac\right)d\right)x_1 \\
&\quad + \left(\left(-a^2 - a - b + c + r\right)x_3 + \left(a^2 + ad\right)a + \left(-a^2 + ac\right)c\right)x_2 \\
&\quad - ax_4x_3 + \left(a^2 + ad + (-a + r)r\right)x_4,
\end{aligned}
\tag{24}
$$

$$
\begin{aligned}
\frac{dx_2}{dt} &= cx_2 + dx_1 - x_1x_3, \\
\frac{d^2x_2}{dt^2} &= c(cx_2 + dx_1 - x_1x_3) + (d - x_3)(-ax_1 + ax_2 + x_4) - x_1(-bx_3 + x_1x_2), \\
\frac{d^3x_2}{dt^3} &= c(c(cx_2 + dx_1 - x_1x_3) + (d - x_3)(-ax_1 + ax_2 + x_4) - x_1(-bx_3 + x_1x_2)) \\
&\quad + (d - x_3)(-a(-ax_1 + ax_2 + x_4) + a(cx_2 + dx_1 - x_1x_3) + rx_4 + x_3x_2) \\
&\quad - 2(-ax_1 + ax_2 + x_4)(-bx_3 + x_1x_2) - x_1(-b(-bx_3 + x_1x_2) \\
&\quad + (-ax_1 + ax_2 + x_4)x_2 + x_1(cx_2 + dx_1 - x_1x_3)),
\end{aligned}
\tag{25}
$$

$$
\begin{aligned}
\frac{dx_3}{dt} &= -bx_3 + x_1x_2, \\
\frac{d^2x_3}{dt^2} &= -b(-bx_3 + x_1x_2) + (-ax_1 + ax_2 + x_4)x_2 + x_1(cx_2 + dx_1 - x_1x_3), \\
\frac{d^3x_3}{dt^3} &= -b(-b(-bx_3 + x_1x_2) + (-ax_1 + ax_2 + x_4)x_2 + x_1(cx_2 + dx_1 - x_1x_3)) \\
&\quad + (-a(-ax_1 + ax_2 + x_4) + a(cx_2 + dx_1 - x_1x_3) + rx_4 + x_2x_3)x_2 \\
&\quad + 2(-ax_1 + ax_2 + x_4)(cx_2 + dx_1 - x_1x_3) + x_1(c(cx_2 + dx_1 - x_1x_3) \\
&\quad + d(-ax_1 + ax_2 + x_4) - (-ax_1 + ax_2 + x_4)x_3 - x_1(-bx_3 + x_1x_2)),
\end{aligned}
\tag{26}
$$

$$
\begin{aligned}
\frac{dx_4}{dt} &= rx_4 + x_2x_3, \\
\frac{d^2x_4}{dt^2} &= r(rx_4 + x_2x_3) + (cx_2 + dx_1 - x_1x_3)x_3 + x_2(-bx_3 + x_1x_2), \\
\frac{d^3x_4}{dt^3} &= (-4\,x_3 + 3\,d)x_2{x_1}^2 + ((r + 3\,c - a - b){x_2}^2 + (-r - c + a + 3\,b){x_3}^2 \\
&\quad + (-ad - 2\,db + cd + rd)x_3)x_1 + a{x_2}^3 + x_4{x_2}^2 - x_4{x_3}^2 + dx_4x_3 + r^3x_4 \\
&\quad + \left(-a{x_3}^2 + \left(r(-b + c + r) + ad + c^2 - 2\,cb + b^2\right)x_3\right)x_2,
\end{aligned}
\tag{27}
$$

Replacing the time-delay terms in Equation (3) with $x_i(t - \tau) = H_i(\tau)$, $i = 1,2,3,4,$ yields

$$
\begin{aligned}
\dot{y}_1 &= -ay_1 + ay_2 + y_4 + k_1(H_1(\tau) - y_1), \\
\dot{y}_2 &= \alpha y_1 + cy_2 - u_0y_3 - y_1y_3 + k_2(H_2(\tau) - y_2), \\
\dot{y}_3 &= v_0y_1 + u_0y_2 - by_3 + y_1y_2 + k_3(H_3(\tau) - y_3), \\
\dot{y}_4 &= w_0y_2 + v_0y_3 + ry_4 + y_2y_3 + k_4(H_4(\tau) - y_4).
\end{aligned}
\tag{28}
$$

Next, we investigate approximate LS between Equations (2) and (28) involving no time-delay terms. For convenience, we choose the following auxiliary system:

$$
\begin{aligned}
\dot{z}_1 &= -az_1 + az_2 + z_4 + k_1(H_1(\tau) - z_1), \\
\dot{z}_2 &= \alpha z_1 + cz_2 - u_0 z_3 - z_1 z_3 + k_2(H_2(\tau) - z_2), \\
\dot{z}_3 &= v_0 z_1 + u_0 z_2 - bz_3 + z_1 z_2 + k_3(H_3(\tau) - z_3), \\
\dot{z}_4 &= w_0 z_2 + v_0 z_3 + rz_4 + z_2 z_3 + k_4(H_4(\tau) - z_4).
\end{aligned}
\tag{29}
$$

By using the method presented in the previous section, GS between systems (2) and (28) with the functional relation $y_i = H_i(\tau)$, $i = 1,2,3,4$ can be achieved under condition (19). As long as $H_i$ approaches $x_i(t-\tau)$, $i = 1,2,3,4$, the GS between systems (2) and (28) means approximate LS between them. Numerical simulations are carried out to verify the conclusion, where the parameters remain the same as those in Figures 1–5, and the initial conditions are taken as $x_1(0) = 0.1$; $x_2(0) = 0.2$; $x_3(0) = 0.3$; $x_4(0) = 0.4$; $y_1(0) = 0.2$; $y_2(0) = 0.2$; $y_3(0) = 0.3$; $y_4(0) = 0.4$; $z_1(0) = 0.3$; $z_2(0) = 0.4$; $z_3(0) = 0.3$; and $y_4(0) = 0.4$. The numerical results are shown in Figures 6–9, in which the values of $\tau$ are chosen as $\tau = 0.02$, 0.05, 0.1 and 0.15, respectively. It is shown that for different values of $\tau$, GS always occurs between systems (2) and (28). If $\tau$ is very small (Figures 6b and 7b), LS can be approximately reached between systems (2) and (28). With the increase in $\tau$, approximate LS between systems (2) and (28) is destroyed (Figures 8b and 9b) because $H_i$ no longer approaches $x_i(t-\tau)$, $i = 1,2,3,4$. More higher order terms of $\tau$ need to be added into Equation (23) to keep approximate LS between systems (2) and (28). Figures 6–9 indicate that it is feasible to approximately reconstruct the past signals of a hyperchaotic Chen system by using its current signals.

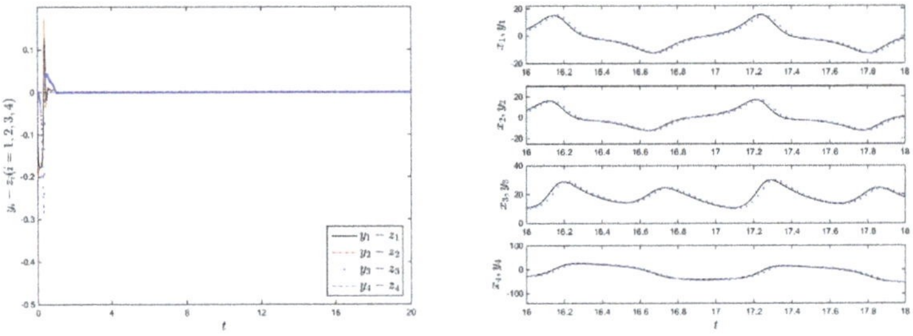


(a) Synchronization errors $y_i - z_i$, $i = 1,2,3,4$. (b) $x_i$ (black line –), $y_i$ (blue dot ·), $i = 1,2,3,4$.

**Figure 6.** (**a**) Synchronization errors between systems (28) and (29). (**b**) The time series of $x_i$ and $y_i$, $i = 1,2,3,4$ in systems (2) and (28) with $y_i = H_i(\tau)$; $i = 1,2,3,4$; $k_1 = 0$; $k_2 = 19$; $k_3 = -2$; $k_4 = 0.6$; and $\tau = 0.02$.

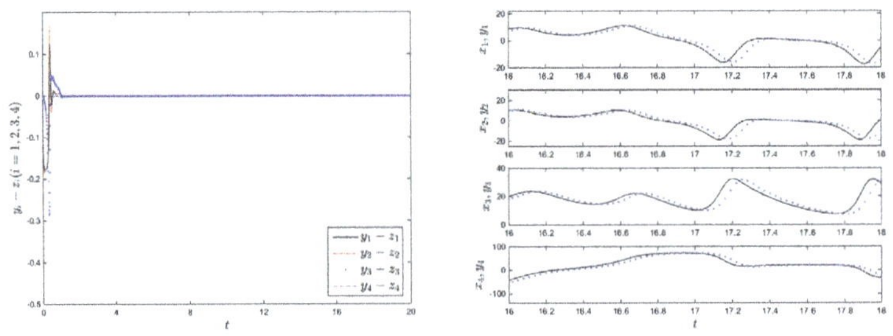


(a) Synchronization errors $y_i - z_i$, $i = 1,2,3,4$. (b) $x_i$ (black line –), $y_i$ (blue dot ·), $i = 1,2,3,4$.

**Figure 7.** (**a**) Synchronization errors between systems (28) and (29). (**b**) The time series of $x_i$ and $y_i$, $i = 1,2,3,4$ in systems (2) and (28) with $y_i = H_i(\tau)$; $i = 1,2,3,4$; $k_1 = 0$; $k_2 = 19$; $k_3 = -2$; $k_4 = 0.6$; and $\tau = 0.05$.

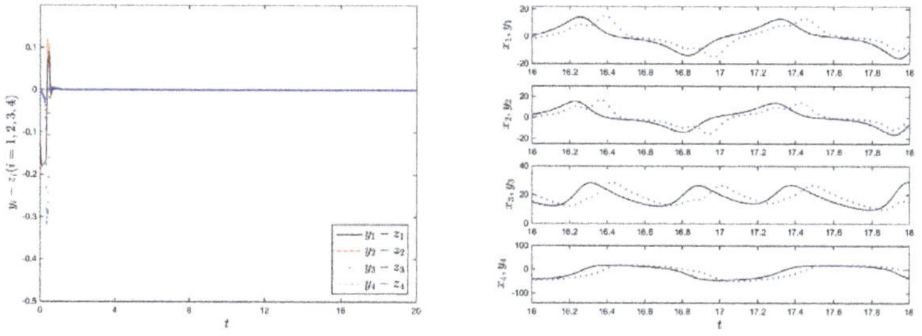

(a) Synchronization errors $y_i - z_i, i = 1,2,3,4$. (b) $x_i$ (black line −), $y_i$ (blue dot ·), $i = 1,2,3,4$.

**Figure 8.** (**a**) Synchronization errors between systems (28) and (29). (**b**) The time series of $x_i$ and $y_i$, $i = 1,2,3,4$ in systems (2) and (28) with $y_i = H_i(\tau)$; $i = 1,2,3,4$; $k_1 = 0$; $k_2 = 19$; $k_3 = -2$; $k_4 = 0.6$; and $\tau = 0.1$.

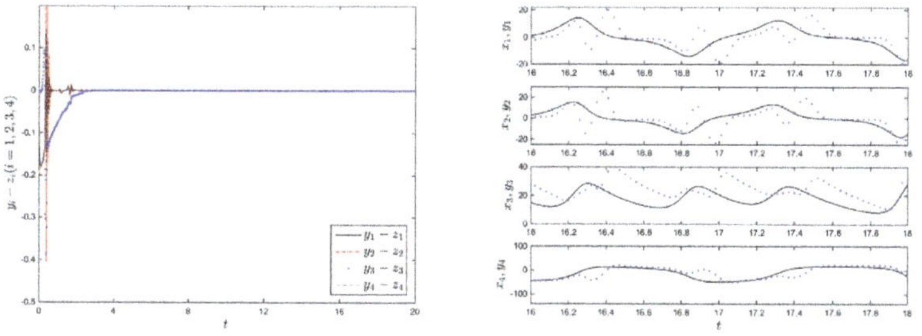

(a) Synchronization errors $y_i - z_i, i = 1,2,3,4$. (b) $x_i$ (black line −), $y_i$ (blue dot ·), $i = 1,2,3,4$.

**Figure 9.** (**a**) Synchronization errors between systems (28) and (29). (**b**) The time series of $x_i$ and $y_i$, $i = 1,2,3,4$ in systems (2) and (28) with $y_i = H_i(\tau)$; $i = 1,2,3,4$; $k_1 = 0$; $k_2 = 19$; $k_3 = -2$; $k_4 = 0.6$; and $\tau = 0.15$.

## 6. The Approximation of AS between Two Hyperchaotic Chen Systems without Time-Delay Coupling

Consider system (2) as the drive system; systems (28) and (29) are the response and auxiliary systems, respectively. Now, we consider $\tau < 0$ in the function $H_i(\tau), i = 1,2,3,4$ as performing the numerical simulations, where all other parameters and the initial conditions remain the same as those in the case of $\tau > 0$. The numerical results are depicted in Figures 10–13, in which the values of $\tau$ are chosen as $\tau = -0.02$, $-0.05$, $-0.1$ and $-0.15$, respectively. It is clear that GS still appears between systems (2) and (28) for different values of $\tau$ under condition (21). If $\tau$ approaches 0 (Figures 10b and 11b), AS can be approximately achieved between systems (2) and (28). With the decrease in $\tau$, the approximation begins to disappear (Figures 12b and 13b). To maintain approximate AS between systems (2) and (28), more higher order terms of $\tau$ need to be added into $H_i(\tau)$, $i = 1,2,3,4$ in Equation (23). Figures 10–13 imply that it is feasible to approximately predict the future signals of a hyperchaotic Chen system by using its current signals.

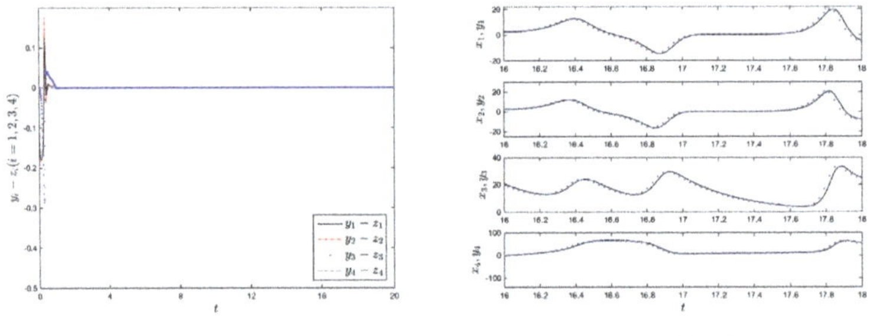

(a) Synchronization errors $y_i - z_i, i = 1,2,3,4$. (b) $x_i$ (black line –), $y_i$ (blue dot ·), $i = 1,2,3,4$.

**Figure 10.** (**a**) Synchronization errors between systems (28) and (29). (**b**) The time series of $x_i$ and $y_i$, $i = 1,2,3,4$ in systems (2) and (28) with $y_i = H_i(\tau)$; $i = 1,2,3,4$; $k_1 = 0$; $k_2 = 19$; $k_3 = -2$; $k_4 = 0.6$; and $\tau = -0.02$.

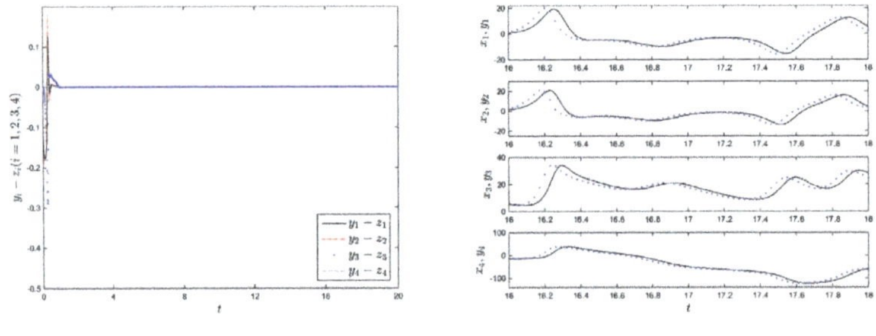

(a) Synchronization errors $y_i - z_i, i = 1,2,3,4$. (b) $x_i$ (black line –), $y_i$ (blue dot ·), $i = 1,2,3,4$.

**Figure 11.** (**a**) Synchronization errors between systems (28) and (29). (**b**) The time series of $x_i$ and $y_i$, $i = 1,2,3,4$ in systems (2) and (28) with $y_i = H_i(\tau)$; $i = 1,2,3,4$; $k_1 = 0$; $k_2 = 19$; $k_3 = -2$; $k_4 = 0.6$; and $\tau = -0.05$.

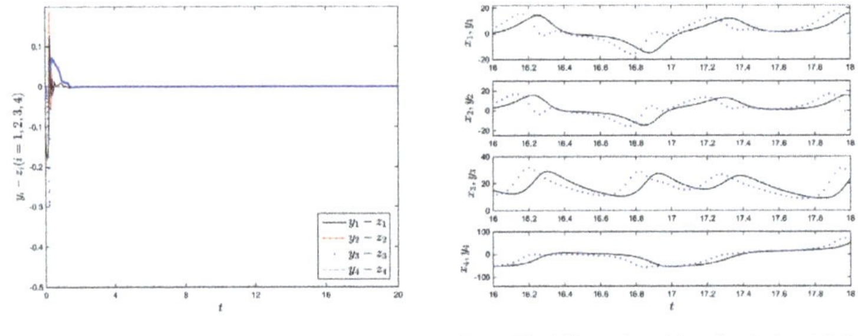

(a) Synchronization errors $y_i - z_i, i = 1,2,3,4$. (b) $x_i$ (black line –), $y_i$ (blue dot ·), $i = 1,2,3,4$.

**Figure 12.** (**a**) Synchronization errors between systems (28) and (29). (**b**) The time series of $x_i$ and $y_i$, $i = 1,2,3,4$ in systems (2) and (28) with $y_i = H_i(\tau)$; $i = 1,2,3,4$; $k_1 = 0$; $k_2 = 19$; $k_3 = -2$; $k_4 = 0.6$; and $\tau = -0.1$.

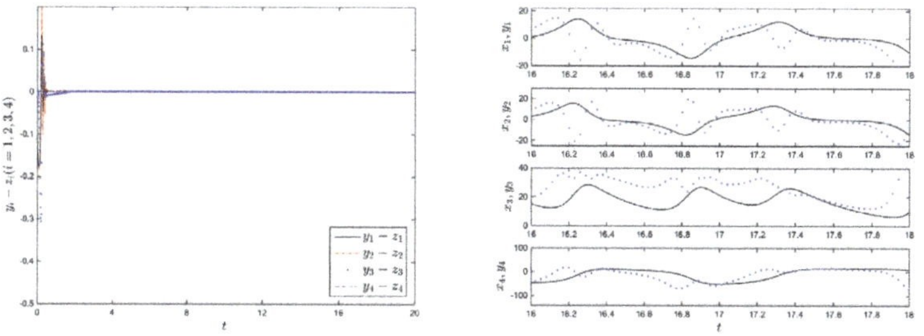


(a) Synchronization errors $y_i - z_i$, $i = 1, 2, 3, 4$.   (b) $x_i$ (black line –), $y_i$ (blue dot ·), $i = 1, 2, 3, 4$.

**Figure 13.** (**a**) Synchronization errors between systems (28) and (29). (**b**) The time series of $x_i$ and $y_i$, $i = 1, 2, 3, 4$ in systems (2) and (28) with $y_i = H_i(\tau)$; $i = 1, 2, 3, 4$; $k_1 = 0$; $k_2 = 19$; $k_3 = -2$; $k_4 = 0.6$; and $\tau = -0.15$.

## 7. Conclusions

In this paper, approximate LS and AS in two unidirectionally coupled hyperchaotic Chen systems without time-delay coupling are analytically investigated. Firstly, the synchronization condition for exact LS between two unidirectionally coupled hyperchaotic Chen systems with time delay in signal transmission is obtained. Differently from other studies, LS is viewed as a special type of GS, which can be analyzed by using the auxiliary system approach. Based on the Laplace transform and the successive approximation method, the synchronization condition for exact LS is analytically obtained, and its effectiveness is verified using numerical simulations. Our analysis shows that the condition for exact LS is valid for any value of the time delay in signal transmission. Under such conditions, approximate LS and AS in two unidirectionally coupled hyperchaotic Chen systems without time-delay coupling are discussed by replacing the true time-delay terms with their Taylor expansions up to the third order. For a small time delay, the approximate LS and AS are precise enough. Our research indicates the possibility of recreating the past signals or predicting the future signals of a hyperchaotic Chen system using its current signals.

In previous research studies, the true time-delay terms are first replaced by using their first-order approximations. Then, synchronization conditions for approximate LS and AS are discussed by considering two chaotic systems with parameter mismatch. The lag and anticipation time of approximate LS and AS have to be involved in the synchronization condition, which are difficult to change within a wide range. In our study, the synchronization condition for exact LS between the drive and response systems with time-delay coupling is firstly obtained, which has nothing to do with the time delay. That is, the condition for exact LS is valid for any value of the time delay. Then, the value of the lag and anticipation time in approximate LS and AS can be individually varied. It is theoretically possible that any value of the lag and anticipation time in approximate LS and AS can be reached as long as there are enough truncated terms in Equation (23). The calculation of terms, of course, is tedious with the increase in truncated terms. In addition, it will become difficult to physically realize such a controller in the real world [31,32]. This is the disadvantage of the proposed scheme achieving approximate LS and AS between two undirectionally coupled hyperchaotic Chen systems without time-delay coupling.

Our research provides a simple and effective way to achieve approximate LS and AS between two hyperchaotic systems without time-delay coupling, which can be applied to eliminate the negative effects of the time delay in signal transmission.

**Author Contributions:** Conceptualization, L.-J.O. and B.-Q.X.; methodology, B.D.; software, B.-Q.X.; validation, B.-Q.X.; formal analysis, L.-J.O.; investigation, B.D.; resources, B.D.; data curation, B.-Q.X.; writing—original draft preparation, L.-J.O.; writing—review and editing, B.D.; visualization, B.-Q.X.; supervision, B.D.; project administration, Bin Ding; funding acquisition, B.D. All authors have read and agreed to the published version of the manuscript.

**Funding:** The Zhejiang Provincial Natural Science Foundation of China (Nos:LY20E080017), the National Natural Science Foundation of China (Nos:51708349) and the Service Innovation Project of the Wenzhou Association for Science and Technology, China. (Grant No.KJFW53).

**Institutional Review Board Statement:** Not applicable.

**Informed Consent Statement:** Not applicable

**Data Availability Statement:** The study did not report any data.

**Acknowledgments:** The authors greatly appreciate the anonymous referees for their valuable suggestions and questions.

**Conflicts of Interest:** The authors declare no conflict of interest.

## References

1. Uchida, A.; Rogister, F.; García-Ojalvo, J.; Roy, R. Synchronization and communication with chaotic laser systems. *Prog. Opt.* **2005**, *48*, 203–341. [CrossRef]
2. Li, Y.N.; Chen, L.; Cai, Z.S.; Zhao, X. Experimental study of chaos synchronization in the Belousov–Zhabotinsky chemical system. *Chaos Solitons Fractals* **2004**, *22*, 767–771. [CrossRef]
3. Uhlhaas, P.J.; Singer, W. Neural synchrony in brain disorders: Relevance for cognitive dysfunctions and pathophysiology. *Neuron* **2006**, *52*, 155–168. [CrossRef] [PubMed]
4. Carroll, T.L.; Perora, L.M. Synchronization in chaotic systems. *Phys. Rev. Lett.* **1990**, *64*, 821–824. [CrossRef]
5. Wang, P.; Wen, G.; Yu, X.; Yu, W.; Huang, T. Synchronization of multi-layer networks: From node-to-node synchronization to complete synchronization. *IEEE Trans Circuits Syst.* **2018**, *66*, 1141–1152. [CrossRef]
6. Fatemeh, P.; Mohadeseh, S.K.; Yaser, M.; Karthikeyan, R.; Sajad, J. Complete and partial synchronization in empirical brain networks. *Int. J. Electron. Commun. (AEU)* **2023**, *170*, 154863. [CrossRef]
7. Hoon, Y.; Mohd, R.M.; Mohd, Z.M.; Zawawi, M. Shunt active power filter: A review on phase synchronization control techniques. *Electronics* **2019**, *8*, 791. [CrossRef]
8. Sun, W.; Wang, S.; Wang, G.; Wu, Y. Lag synchronization via pinning control between two coupled networks. *Nonlinear Dyn.* **2015**, *79*, 2659–2666. [CrossRef]
9. Mihana, T.; Fujii, K.; Kanno, K.; Naruse, M.; Uchida, A. Laser network decision making by lag synchronization of chaos in a ring configuration. *Opt. Express* **2020**, *28*, 40112–40130. [CrossRef]
10. Long, C.; Zhang, G.; Zeng, Z.; Hu, J. Finite-time lag synchronization of inertial neural networks with mixed infinite time-varying delays and state-dependent switching. *Neurocomputing* **2021**, *433*, 50–58. [CrossRef]
11. Ibrahim, M.M.; Kamran, M.A.; Mannan, M.M.N.; Jung, I.H.; Kim, S. Lag synchronization of coupled time-delayed Fitzhugh–Nagumo neural networks via feedback control. *Sci. Rep.* **2021**, *11*, 1–15. [CrossRef] [PubMed]
12. Rasa, S.; Loreta, S.; Minvydas, R. Detection of lag synchronization based on matrices of delayed differences. *Commun. Nonlinear Sci. Numer. Simul.* **2023**, *116*, 106864. [CrossRef]
13. Ahmadlou, M.; Adeli, H. Visibility graph similarity: A new measure of generalized synchronization in coupled dynamic systems. *Physica D* **2012**, *241*, 326–332. [CrossRef]
14. Wu, Z.; Zhang, X.; Zhong, X. Generalized chaos synchronization circuit simulation and asymmetric image encryption. *IEEE Access* **2019**, *7*, 37989–38008. [CrossRef]
15. Ciszak, M.; Mayol, C.; Mirasso, C.R.; Toral, R. Anticipated synchronization in coupled complex Ginzburg-Landau systems. *Phys. Rev. E* **2015**, *92*, 032911. [CrossRef] [PubMed]
16. Emad, E.; Mahmoud, M.; Higazy, T.M.; Al, H. A New Nine-Dimensional Chaotic Lorenz System with Quaternion Variables: Complicated Dynamics, Electronic Circuit Design, Anti-Anticipating Synchronization, and Chaotic Masking Communication Application. *Mathematics* **2019**, *7*, 877. [CrossRef]
17. Srinivasan, K.; Sivaganesh, G.; Fonzin F.T.; Raja M.I. Analytical studies on complete, lag and anticipation synchronization in cascaded circuits with numerical and experimental confirmation. *Int. J. Electron. Commun. (AEU)* **2023**, *159*, 154491. [CrossRef]
18. Thang, M.H.; Sanjay, K.P.; Sayan, M.; Santo, B. Synchronization and secure communication in time delayed semiconductor laser systems. *Optik* **2016**, *127*, 10930–10947. [CrossRef]
19. Kamal, F.; Elsonbaty, A.; Elsaid, A. A novel fractional nonautonomous chaotic circuit model and its application to image encryption. *Chaos Solitons Fractals* **2021**, *144*, 110686. [CrossRef]
20. Voss, H.U. Anticipating chaotic synchronization. *Phys. Rev.* **2000**, *61*, 5115. [CrossRef]
21. Voss, H.U. Dynamic long-term anticipation of chaotic states. *Phys. Rev. Lett.* **2001**, *87*, 014102. [CrossRef] [PubMed]

22. Masoller, A.P.C. Anticipation in the synchronization of chaotic semiconductor lasers with optical feedback. *Phys. Rev. Lett.* **2001**, *86*, 2782. [CrossRef] [PubMed]
23. Pyragiene, T.; Pyragas, K. Anticipating synchronization in a chain of chaotic oscillators with switching parameters. *Phys. Lett.* **2015**, *379*, 3084–3088. [CrossRef]
24. Mousa, S.K.; Al Naimee, K.A. Experimental investigations of Synchronization in two optically coupled chaotic systems utilizing optical feedback and optical injection. *J. Univ. Anbar Pure Sci.* **2017**, *11*, 49–56. [CrossRef]
25. Blakely, J.N.; Stahl, M.T.; Corron, N.J. Time-shifted synchronization of chaotic oscillator chains without explicit coupling delays. *Chaos* **2009**, *19*, 043117. [CrossRef] [PubMed]
26. Corron, N.J.; Blakely, J.N.; Pethel, S.D. Lag and anticipating synchronization without time-delay coupling. *Chaos* **2005**, *15*, 023110. [CrossRef]
27. Smaoui, N.; Karouma, A.; Zribi, M. Secure communications based on the synchronization of the hyperchaotic Chen and the unified chaotic systems. *Commun. Nonlinear Sci. Numer. Simul.* **2011**, *16*, 3279–3293. [CrossRef]
28. Abarbanel, H.D.I.; Rulkov, N.F.; Suschik, M.M. Generalized synchronization of chaos: The auxiliary system approach. *Phys. Rev. E* **1996**, *53*, 4528–4535. [CrossRef]
29. Bin, Z.; Ran, L. The study of generalized synchronization between two identical neurons based on the Laplace transform method. *Appl. Sci.* **2021**, *11*, 11774. [CrossRef]
30. Nohel, J.A. Some problems in nonlinear Volterra integral equations. *Bull. Am. Math. Soc.* **1962**, *68*, 323–330. [CrossRef]
31. Roman, R.C.; Precup, R.E.; Petriu, E.M. Hybrid data-driven fuzzy active disturbance rejection control for tower crane systems. *Eur. J. Control* **2021**, *58*, 373–387. [CrossRef]
32. Chi, R.H.; Li, H.Y.; Shen, D.; Hou, Z.S.; Huang, B. Enhanced P-type control: Indirect adaptive learning rrom set-point updates. *IEEE Trans. Autom. Control* **2023**, *68*, 1600–1613. [CrossRef]

mathematics

MDPI

Article

# Improved Square-Root Cubature Kalman Filtering Algorithm for Nonlinear Systems with Dual Unknown Inputs

Zihao Lu [1,2], Na Wang [1,2,*] and Shigui Dong [1,2]

1 College of Automation, Qingdao University, Qingdao 266071, China; qduzhlu@163.com (Z.L.); dongshigui1999@163.com (S.D.)
2 Shandong Key Laboratory of Industrial Control Technology, Qingdao University, Qingdao 266071, China
* Correspondence: nwangqdu@qdu.edu.cn

**Abstract:** For nonlinear discrete systems with dual unknown inputs, there are many limitations regarding previous nonlinear filters. This paper proposes two new, improved square-root cubature Kalman filtering (ISRCKF) algorithms to estimate system states and dual unknown inputs. Improved square-root cubature Kalman filtering 1 (ISRCKF1) introduces an innovation that first obtains the unknown input estimates from the measurement equation, then updates the innovation to derive the unknown input estimates from the state equation, then uses the already obtained estimates of the dual unknown inputs to correct the one-step estimate of the state, and finally the minimum variance unbiased estimate of the state is obtained. Improved square-root cubature Kalman filtering 2 (ISRCKF2) builds a unified innovation feedback model, then applies the minimum variance unbiased estimation (MVUE) criterion to obtain the estimates of system states and dual unknown inputs, refining a more concise recursive filter but requiring stronger assumptions. Finally, simulation results demonstrate that the above two algorithms can achieve the optimal estimates of system states and dual unknown inputs simultaneously, and ISRCKF2 further enhances the accuracy of both state and dual unknown inputs estimation, which verifies the validity of the proposed algorithms.

**Keywords:** square-root cubature Kalman filter; dual unknown inputs; minimum variance unbiased estimation; state estimation; unknown input estimation

**MSC:** 93-10

## 1. Introduction

In practical engineering systems, different kinds of noise and various unmeasurable inputs or disturbances are often present in the system state equation and measurement equation due to environmental influences, improper selection of model parameters, equipment failures, etc. When dealing with the state estimation problems of the above systems, expert scholars usually refer to these disturbances or inputs that cannot be measured and for which no priori information is known collectively as unknown inputs [1–5]. The algorithms for state estimation of nonlinear systems containing unknown inputs are now widely used in many application fields such as autonomous navigation [6], target tracking [7], fault-tolerant control [8], and fault detection [9].

For nonlinear stochastic systems where the unknown input occurs only in the state equation, Ref. [10] presented an extended Kalman filter algorithm with unknown inputs for continuous-time systems that can identify the structural parameters and states of the system in real time. For discrete-time systems, Ref. [11] proposed an unknown input extended Kalman filter, which builds on the extended Kalman filter (EKF) [12] by completely decoupling the unknown inputs without forcing the measurement equation to be linear. However, the EKF algorithm requires linearization of the system, which inevitably introduces linearization errors and can lead to degraded filtering performance or even filter divergence when the system is strongly nonlinear. To this end, a nonlinear recursive filter is given in [13], which obtains unknown inputs for signals of arbitrary type via least-squares unbiased estimation and transforms the state estimation problem into a standard unscented Kalman filter (UKF) [14,15] problem. Since the unscented transform enables the posterior probability density function of Sigma points after propagation through a nonlinear function to reach at least the second-order term of the Taylor series expansion, its filtering accuracy is higher than that of the EKF, while overcoming the shortcoming that the EKF is only applicable to weakly nonlinear systems. In addition, Ref. [16] introduced statistical linearization and weighted least squares to estimate the unknown inputs and proposed a robust strong tracking unscented Kalman filtering algorithm with unknown inputs. However, UKF is prone to nonlocal effects of sampling when dealing with high-dimensional nonlinear systems, which leads to numerical instability in the filtering process and degradation of the filtering performance. Later, by improving the cubature Kalman filter (CKF) [17] algorithm, Ref. [18] designed a distributed filter that estimates the state and unknown inputs simultaneously. With further research, a nonlinear unknown input observer (NUIO) was proposed in [19] based on singular value decomposition-assisted dimensionality reduction CKF. The method preserves the third-order accuracy of Taylor expansion integral of the nonlinear state function by sampling the nonlinear part of the nonlinear state function instead of all of it.

However, the types of algorithms mentioned above only consider the case where the system state equation contain unknown inputs. With further research, some scholars have extended such filtering algorithms to the case where the measurement equation of the system model also contain unknown inputs. To address the problem of filter design for such direct feedthrough nonlinear systems, an unknown-input generalized extended Kalman filter is proposed in [20] for continuous-time systems. For discrete-time systems, an adaptive three-stage Kalman filter capable of tracking faults and unknown inputs is proposed in [21], which can be used in situations where faults and unknown inputs are not fully known, and its stability is demonstrated. Later, Ref. [22] proposed a robust EKF to estimate both the unknown inputs and the system states simultaneously. Ref. [23] proposed a novel adaptive three-level EKF for the problem of severe performance degradation of the conventional Kalman filter in handling unknown inputs, while a three-level UKF and robust three-level UKF were given in [24] to solve the unknown inputs and state estimation for nonlinear systems where the unknown input information is not completely known. In addition to this, many other scholars have considered the impact of uncertainties on system filtering caused by systems with both unknown inputs, missing measurements and multiplicative noise, and have carried out extensive research [25–29].

The above filtering algorithms all consider the case where the unknown input signals are the same in both state and measurement equations, but in practical application systems, the unknown input signals in the two equations are often different. For such systems, Ref. [30] designed a decoupling filter and an adaptive minimum upper filter to obtain optimal and suboptimal estimation of the state via introducing an adaptive adjustment factor, but this skips the estimates of the dual unknown inputs and does not give an explicit iterative formula for the estimates of the dual unknown inputs. Ref. [31] proposed a CKF-based NUIO method for robust sensor fault detection and demonstrated the superiority of this method compared to the NUIO methods of EKF and UKF. However, some scholars have found that constant rounding errors in CKF during iteration make the covariance matrix asymmetric or non-positive definite, which leads to degradation of filtering performance or even divergence. Later, Ref. [32] proposed a square-root cubature Kalman filter (SRCKF) based algorithm for battery charge state estimation. Since SRCKF directly propagates and updates the state covariance matrix square root by means of Cholesky decomposition, ensuring the non-negativity of the covariance matrix and avoiding the divergence of the filter.

To sum up, the research on state estimation for nonlinear systems containing unknown inputs has achieved the above fruitful results, and a SRCKF algorithm outperforms CKF and

UKF algorithms with regard to computational efficiency, filtering accuracy, and numerical stability, but the existing nonlinear filtering estimation algorithms with unknown inputs still have many limitations: (i) the traditional SRCKF algorithm has degraded filtering accuracy or even divergence when dealing with unknown inputs; (ii) existing nonlinear filtering algorithms consider unknown inputs, but assume that the unknown inputs in the state equation and measurement equation are the same, which narrows the applicability of the filter; (iii) only the estimation of the state can be achieved—the estimation of the unknown inputs cannot be obtained.

Based on the above analysis, this paper proposes two new, improved square-root cubature Kalman filter algorithms. Compared with the existing results, the main contributions have the following three aspects:

(1) When the state and measurement equations contain different unknown input signals without any prior information regarding the dual unknown inputs, according to the MVUE criterion, by designing the innovation to deal with the state and the dual unknown inputs, the traditional SRCKF is improved and extended to the nonlinear system with dual unknown inputs, which solves the problem that the performance of the SRCKF algorithm is seriously degraded or even cannot be applied in this case.
(2) The MVUE of the state and dual unknown inputs is obtained by minimizing the trace of the estimation error covariance matrix and then solving for it using Schur's complement lemma, provided that the undetermined gain matrix satisfies certain constraints.
(3) The ISRCKF does not require the unknown inputs in the state and measurement equations to be the same quantity, and the system under consideration is more in line with the practical application context, which is applied more loosely and with a wider range of applicability.

This paper is structured as follows. Section 2 gives a description of the research problem. In Section 3, an ISRCKF1 algorithm with dual unknown inputs is proposed. In Section 4, the simpler ISRCKF2 algorithm with dual unknown inputs is redesigned. In Section 5, we give a nonlinear example to study the validity of the two algorithms and compare the estimation results of the two algorithms with the SRCKF algorithm to prove the performance of the proposed algorithm. In Section 6, we provide our conclusion.

## 2. Problem Description

Consider the following direct feedthrough nonlinear discrete stochastic systems with dual unknown inputs:

$$x_k = f(x_{k-1}, u_{k-1}) + G_{k-1}d_{k-1} + w_{k-1} \tag{1}$$

$$y_k = \varphi(x_k) + H_k l_k + v_k \tag{2}$$

where $x_k \in \mathbb{R}^n$ and $y_k \in \mathbb{R}^p$ represent the state vector and the measured output vector of the system, respectively; $f(\cdot)$ and $\varphi(\cdot)$ are nonlinear transfer functions, $u_{k-1}$ is the known control input signal, $d_{k-1} \in \mathbb{R}^{m_1}$ and $l_k \in \mathbb{R}^{m_2}$ are unknown input vectors in the state and measurement equations, respectively; $w_{k-1} \in \mathbb{R}^n$ and $v_k \in \mathbb{R}^p$ are the process and measurement noise of the system, respectively, which are zero-mean Gaussian white noise signals that are not correlated with each other, and their nonsingular covariance matrices $Q_{k-1} = E(w_{k-1}w_{k-1}^T) \geq 0$ and $R_k = E(v_k v_k^T) > 0$ are known; $G_{k-1}$ and $H_k$ are determined matrices of suitable dimensional coefficients; the initial state value $x_0$ follows a Gaussian normal distribution and is independent of $w_{k-1}$ and $v_k$; the unbiased estimate $\hat{x}_0 = E(x_0)$ of $x_0$ and the initial covariance matrix $P_0^x = E[(x_0 - \hat{x}_0)(x_0 - \hat{x}_0)^T]$ are known; let $\hat{x}_{k|k}$ be an unbiased estimation of $x_k$, then $\hat{x}_{k|k-1}$ is an unbiased estimation of $f(x_{k-1})$, but a biased estimation of $x_k$ due to unknown inputs.

**Assumption 1.** *Assuming that the output, like most industrial systems, is linear and does not lose generality [19,31]. However, the EKF is a linearization of both state and measure equations* *using Jacobi matrices; in nonlinear situations, the output Function (2) can be linearized around the operating points as follows:*

$$y_k = C_k x_k + H_k l_k + v_k \tag{3}$$

*where* $C_k = \frac{\partial \varphi(x_k)}{\partial x_k}\Big|_{x_k = \hat{x}_{k|k-1}}$ *is the matrix of linearized measurement coefficients.*

**Remark 1.** *We generally default that systems (1)–(3) satisfy the state observability condition, and the specific research on parameter observability can refer to [33].*

**Assumption 2.** *The matrix* $D_k = \begin{bmatrix} C_k G_{k-1} & H_k \end{bmatrix}$ *has full rank, that is,* $rank(D_k) = m_1 + m_2$.

**Remark 2.** *Assumption 2 is the basic condition for decoupling the unknown inputs in the state and the measurement equations, requiring that the measurement vector dimension must be greater than the sum of the dual unknown input vectors, otherwise there is not enough information to estimate the dual unknown input vectors [11,13]. When Assumption 2 holds, the following four conclusions hold simultaneously.*

*(i.)* $rank(C_k G_{k-1}) = rank(G_{k-1}) = m_1$.
*(ii.)* $rank(H_k) = m_2$.
*(iii.)* There is no linear dependence between columns of $C_k G_{k-1}$ and columns of $H_k$.
*(iv.)* $p \geq m_1 + m_2$.

**Remark 3.** *The above optimal filtering problem for the system is to obtain stepwise recursive unbiased optimal filtering sequences* $\left\{\hat{l}_{0|0}, \cdots, \hat{l}_{k|k}\right\}$, $\left\{\hat{d}_{0|0}, \cdots, \hat{d}_{k-1|k}\right\}$ *and* $\left\{\hat{x}_{0|0}, \cdots, \hat{x}_{k|k}\right\}$ *for the dual unknown inputs and the system state based on the time series of measurements* $\{y_0, y_1, \cdots, y_k\}$, *provided that the initial state* $x_0$ *and its unbiased estimates* $\hat{x}_0$ *and covariance matrix* $P_0^x$ *are known.*

## 3. ISRCKF1

Consider the case where the unknown input signals in the equation of state and the measurement equation are different, i.e., $l_k \neq d_{k-1}$. For the system models (1)–(3), using the idea of innovation feedback, this section will give the ISRCKF filter for estimating the system state and the dual unknown inputs. The one-step estimate $\hat{x}_{k|k-1}$ of $x_k$, the prediction error covariance matrix $P_{k|k-1}^x$, and its square-root $S_{k|k-1}$ are first obtained from the known measurement sequence $\{y_0, y_1, \cdots, y_{k-1}\}$, the initial state estimate $\hat{x}_0$ and the covariance matrix $P_0^x$. The specific steps are given in Section 3.1; then solve for the undetermined gain matrix $N_k$ to obtain the estimate $\hat{l}_{k|k}$ of $l_k$. The specific expression and detailed derivation steps are given in Section 3.2. Based on the obtained unknown input estimate $\hat{l}_{k|k}$, design the estimator $\hat{d}_{k-1|k}$ for $d_{k-1}$. The specific expression for the filter gain $M_k$ to be determined and the detailed derivation steps are given in Section 3.3; further, based on the obtained dual unknown input estimates $\hat{l}_{k|k}$ and $\hat{d}_{k-1|k}$, the estimator $\hat{x}_{k|k}$ for $x_k$ is designed and the gain matrix $K_k$ scan be obtained by minimizing the trace of the covariance $P_{k|k}^x$ when performing the state update, as given in Section 3.4. Finally, update the square-root $S_{k|k}$ of $P_{k|k}^x$, given in Section 3.5.

### 3.1. Time Update

1. Factorize

$$S_{k-1|k-1} = \boldsymbol{chol}(P_{k-1|k-1}^x) \tag{4}$$

where $S_{k-1|k-1}$ denotes the square-root of $P_{k-1|k-1}^x$ and is obtained by performing Cholesky decomposition on $P_{k-1|k-1}^x$. It should be noted that in this paper, the above covariance is factorized only at the initial moment, i.e., the Cholesky decomposition is performed for the known initial covariance $P_0^x$ at $k = 1$.

2. Calculate the cubature points

$$\chi_{i,k-1|k-1} = S_{k-1|k-1}\xi_i + \hat{x}_{k-1|k-1},\ i = 1,2,\cdots,m \tag{5}$$

where $\chi_{i,k-1|k-1}$ is the cubature point and $\hat{x}_{k-1|k-1}$ is the optimal unbiased estimate of $x_{k-1}$ when the measurement time series $\{y_0, y_1, \cdots, y_{k-1}\}$ is known; $m$ is the number of cubature points and twice the dimension $n$ of the state $x$, that is, $m = 2n$; $\xi_i = \sqrt{m/2}\boldsymbol{I}_i$ is the set of $2n$ cubature points with the same weight based on the third-order spherical radial criterion,$\boldsymbol{I}_i$ is the $i$th column of the point set $\boldsymbol{I}$, and the point set $\boldsymbol{I}$ can be expressed as

$$\boldsymbol{I} = \left\{ \begin{pmatrix} 1 \\ 0 \\ \vdots \\ 0 \end{pmatrix}, \begin{pmatrix} 0 \\ 1 \\ \vdots \\ 0 \end{pmatrix}, \cdots, \begin{pmatrix} 0 \\ 0 \\ \vdots \\ 1 \end{pmatrix}, \begin{pmatrix} -1 \\ 0 \\ \vdots \\ 0 \end{pmatrix}, \begin{pmatrix} 0 \\ -1 \\ \vdots \\ 0 \end{pmatrix}, \cdots, \begin{pmatrix} 0 \\ 0 \\ \vdots \\ -1 \end{pmatrix} \right\} \tag{6}$$

3. Calculate the nonlinear propagated cubature points and the one-step state estimation

$$\chi^*_{i,k|k-1} = f(\chi_{i,k-1|k-1}, u_{k-1}) \tag{7}$$

$$\hat{x}_{k|k-1} = \frac{1}{m}\sum_{i=1}^{m} \chi^*_{i,k|k-1} \tag{8}$$

where $\chi^*_{i,k|k-1}$ denotes the updated cubature points; $\frac{1}{m} = \frac{1}{2n}$ is the weight corresponding to the cubature points.

4. Calculate the square-root of the prediction error covariance matrix

$$S_{k|k-1} = \boldsymbol{Tria}([\chi^*_{k|k-1}, S_{Q_{k-1}}]) \tag{9}$$

where $\boldsymbol{Tria}(\cdot)$ denotes triangular decomposition, $S_{k|k-1}$ is a lower triangular matrix, and $S_{Q_{k-1}}$ is the square-root of $Q_{k-1}$, which can be acquired by performing Cholesky decomposition on $Q_{k-1}$, i.e., $S_{Q_{k-1}} = \boldsymbol{chol}(Q_{k-1})$.

The central weighting matrix $\chi^*_{k|k-1}$ is given by

$$\chi^*_{k|k-1} = \frac{1}{\sqrt{m}}[\chi^*_{1,k|k-1} - \hat{x}_{k|k-1}, \chi^*_{2,k|k-1} - \hat{x}_{k|k-1}, \cdots, \chi^*_{m,k|k-1} - \hat{x}_{k|k-1}] \tag{10}$$

5. Evaluate the prediction error covariance matrix

$$\begin{aligned} P^x_{k|k-1} &= E[(f(x_{k-1}) - \hat{x}_{k|k-1} + w_{k-1})(f(x_{k-1}) - \hat{x}_{k|k-1} + w_{k-1})^T] \\ &= \tfrac{1}{m}\sum_{i=1}^{m} \chi^*_{i,k|k-1}\chi^{*T}_{i,k|k-1} - \hat{x}_{k|k-1}\hat{x}^T_{k|k-1} + Q_{k-1} \end{aligned} \tag{11}$$

where the expression for $\chi^*_{i,k|k-1}$ is given in Equation (7). The detailed derivation of Equation (11) can be found in [17]. Without affecting the calculation results, for ease of writing, we will simply abbreviate $f(x_{k-1}, u_{k-1})$ as $f(x_{k-1})$ from here on.

### 3.2. Estimation of Unknown Input $l_k$

Designing the innovation $\tilde{y}_k = y_k - C_k\hat{x}_{k|k-1}$, according to the system models (1)–(3), the innovation at moment k is expanded as follows:

$$\begin{aligned} \tilde{y}_k &= C_k x_k + H_k l_k + v_k - C_k\hat{x}_{k|k-1} \\ &= C_k[f(x_{k-1}) + G_{k-1}d_{k-1} + w_{k-1}] + v_k + H_k l_k - C_k\hat{x}_{k|k-1} \\ &= C_k[f(x_{k-1}) - \hat{x}_{k|k-1} + w_{k-1}] + v_k + C_k G_{k-1}d_{k-1} + H_k l_k \\ &= \Delta_k + D_k\begin{bmatrix} d_{k-1} \\ l_k \end{bmatrix} \end{aligned} \tag{12}$$

where $\boldsymbol{\Delta}_k$ and $\boldsymbol{D}_k$ are given by the following equations, respectively:

$$\boldsymbol{\Delta}_k = \boldsymbol{C}_k[\boldsymbol{f}(\boldsymbol{x}_{k-1}) - \hat{\boldsymbol{x}}_{k|k-1} + \boldsymbol{w}_{k-1}] + \boldsymbol{v}_k \tag{13}$$

$$\boldsymbol{D}_k = \begin{bmatrix} \boldsymbol{C}_k\boldsymbol{G}_{k-1} & \boldsymbol{H}_k \end{bmatrix} \tag{14}$$

From the known conditions we get

$$\boldsymbol{E}(\boldsymbol{\Delta}_k) = 0 \tag{15}$$

Using the innovation feedback $\widetilde{\boldsymbol{y}}_k$, the unbiased estimate $\hat{\boldsymbol{l}}_{k|k}$ for design $\boldsymbol{l}_k$ is as follows:

$$\hat{\boldsymbol{l}}_{k|k} = \boldsymbol{N}_k(\boldsymbol{y}_k - \boldsymbol{C}_k\hat{\boldsymbol{x}}_{k|k-1}) \tag{16}$$

The next step is to calculate the gain matrix $\boldsymbol{N}_k$. From Equations (12) and (16), we can obtain the estimation error of the unknown input $\boldsymbol{l}_k$:

$$\begin{aligned} \widetilde{\boldsymbol{l}}_{k|k} &= \boldsymbol{l}_k - \hat{\boldsymbol{l}}_{k|k} \\ &= (\boldsymbol{I}_{m_2} - \boldsymbol{N}_k\boldsymbol{H}_k)\boldsymbol{l}_k - \boldsymbol{N}_k\boldsymbol{C}_k\boldsymbol{G}_{k-1}\boldsymbol{d}_{k-1} - \boldsymbol{N}_k\boldsymbol{\Delta}_k \end{aligned} \tag{17}$$

**Theorem 1.** *If the system equation satisfies rank*$(\boldsymbol{D}_k) = m_1 + m_2$*, then for any* $\boldsymbol{d}_{k-1}$ *and* $\boldsymbol{l}_k$*, Equation (16) is the minimum variance unbiased estimator of* $\boldsymbol{l}_k$*, if and only if the following equation holds:*

$$\boldsymbol{N}_k = \begin{bmatrix} 0 & \boldsymbol{I}_{m_2} \end{bmatrix} (\boldsymbol{D}_k^T\widetilde{\boldsymbol{R}}_k^{-1}\boldsymbol{D}_k)^{-1}\boldsymbol{D}_k^T\widetilde{\boldsymbol{R}}_k^{-1} \tag{18}$$

*where* $\widetilde{\boldsymbol{R}}_k = \boldsymbol{C}_k\boldsymbol{P}^x_{k|k-1}\boldsymbol{C}_k^T + \boldsymbol{R}_k$.

**Proof.** First, we give the necessary conditions for Equation (18) to hold as follows:

$$\begin{bmatrix} 0 & \boldsymbol{I}_{m_2} \end{bmatrix} - \boldsymbol{N}_k\boldsymbol{D}_k = 0 \tag{19}$$

Equation (19) can be expanded to $\boldsymbol{I}_{m_2} - \boldsymbol{N}_k\boldsymbol{H}_k = 0$ and $-\boldsymbol{N}_k\boldsymbol{C}_k\boldsymbol{G}_{k-1} = 0$, at which point Equation (17) can be simplified as

$$\widetilde{\boldsymbol{l}}_{k|k} = -\boldsymbol{N}_k\boldsymbol{\Delta}_k \tag{20}$$

Thus combining Equations (15) and (20), regardless of the values taken by $\boldsymbol{d}_{k-1}$ and $\boldsymbol{l}_k$, $\hat{\boldsymbol{l}}_{k|k}$ is an unbiased estimate of the unknown input $\boldsymbol{l}_k$. Unbiasedness is proven.

Further based on Equations (11), (13), and (20), the covariance matrix $\boldsymbol{P}^l_{k|k} = \boldsymbol{E}[\widetilde{\boldsymbol{l}}_{k|k}\widetilde{\boldsymbol{l}}^T_{k|k}]$ of $\boldsymbol{l}_k$ can be approximated, that is,

$$\boldsymbol{P}^l_{k|k} = \boldsymbol{N}_k\widetilde{\boldsymbol{R}}_k\boldsymbol{N}_k^T \tag{21}$$

We choose the pending gain matrix $\boldsymbol{N}_k$ to minimize the $\boldsymbol{P}^l_{k|k}$ variance by satisfying the unbiased condition of Equation (19), so for solving $\boldsymbol{N}_k$ we can choose Equation (22) applied using the Lagrange multiplier method, which can be found in [1] for a more detailed description of the method.

$$\min \boldsymbol{tr}(\boldsymbol{P}^l_{k|k}) \quad \text{s.t.} \begin{bmatrix} 0 & \boldsymbol{I}_{m_2} \end{bmatrix} - \boldsymbol{N}_k\boldsymbol{D}_k = 0 \tag{22}$$

Similar to the proof of Gillijns in [34], the Lagrange multiplier method is adopted to solve the extreme value problem under this constraint. The Lagrangian is given by

$$\boldsymbol{F}_1 = \boldsymbol{tr}\left\{\boldsymbol{P}^l_{k|k}\right\} - 2\boldsymbol{tr}\{\Lambda_{1k}(\begin{bmatrix} 0 & \boldsymbol{I}_{m_2} \end{bmatrix} - \boldsymbol{N}_k\boldsymbol{D}_k)\} \tag{23}$$

where $\Lambda_{1k}$ is the Lagrange multiplier matrix, $\Lambda_{1k} \in R^{(m_1+m_2)\times m_2}$, and the factor "2" is intended to keep the calculation simple.

Taking the derivative of Equation (23) with respect to $N_k$ and making the resulting derivative zero, we get

$$N_k\widetilde{R}_k + \Lambda_{1k}^T D_k^T = 0 \tag{24}$$

The system of linear equations consisting of Equations (24) and (19) is as follows:

$$\begin{bmatrix} \widetilde{R}_k^T & D_k \\ D_k^T & 0 \end{bmatrix}\begin{bmatrix} N_k^T \\ \Lambda_{1k} \end{bmatrix} = \begin{bmatrix} 0 \\ \begin{bmatrix} 0 & I_{m_2} \end{bmatrix}^T \end{bmatrix} \tag{25}$$

According to the literature [34], the coefficient matrix of the system of equations (25) is nonsingular when $\widetilde{R}_k^T$ is invertible, at which point the linear system equation has a unique solution. Finally, using the Schur complement lemma, the gain matrix $N_k$ expression (18) is obtained. The proof is completed. □

### 3.3. *Estimation of Unknown Input* $d_{k-1}$

Using the estimate $\hat{l}_{k|k}$ of the unknown input $l_k$ obtained in the previous section, design a new innovation feedback as follows:

$$\widetilde{y}_k^{*1} = y_k - C_k\hat{x}_{k|k-1} - H_k\hat{l}_{k|k} \tag{26}$$

According to Equation (26), the unbiased estimate $\hat{d}_{k-1|k}$ of the unknown input $d_{k-1}$ is designed as follows:

$$\hat{d}_{k-1|k} = M_k\widetilde{y}_k^{*1} \tag{27}$$

Then, by combining Equations (12), (14), and (27), we can get the estimation error of the unknown input $d_{k-1}$:

$$\begin{aligned} \widetilde{d}_{k-1|k} &= d_{k-1} - \hat{d}_{k-1|k} \\ &= d_{k-1} - M_k(\widetilde{y}_k - H_k\hat{l}_{k|k}) \\ &= (I_{m_1} - M_kC_kG_{k-1})d_{k-1} - M_k\Delta_k - M_kH_k\widetilde{l}_{k|k} \end{aligned} \tag{28}$$

**Theorem 2.** *If the system equation satisfies* $rank(D_k) = m_1 + m_2$*, then for any* $d_{k-1}$ *and* $l_k$*, Equation (27) is the minimum variance unbiased estimator of* $d_{k-1}$*, if and only if the following equation holds:*

$$M_k = I_{m_1}\left[G_{k-1}^T C_k^T (\widetilde{R}_k^*)^{-1} C_k G_{k-1}\right]^{-1} G_{k-1}^T C_k^T (\widetilde{R}_k^*)^{-1} \tag{29}$$

*where* $\widetilde{R}_k^* = \widetilde{R}_k + H_kP_{k|k}^lH_k^T$.

**Proof.** First, we give the necessary conditions for Equation (29) to hold as follows:

$$I_{m_1} - M_kC_kG_{k-1} = 0 \tag{30}$$

From Equation (30), rewrite the expression for $\widetilde{d}_{k-1|k}$, that is,

$$\widetilde{d}_{k-1|k} = -M_k\Delta_k - M_kH_k\widetilde{l}_{k|k} \tag{31}$$

According to Equations (15) and (31), $E(\widetilde{d}_{k-1|k}) = 0$, so unbiasedness is proven.

Then, from Equations (11), (13), and (31), the covariance matrix $P_{k-1|k}^d = E(\widetilde{d}_{k-1|k}\widetilde{d}_{k-1|k}^T)$ of $d_{k-1}$ is obtained as

$$P_{k-1|k}^d = M_k\widetilde{R}_kM_k^T + M_kH_kP_{k|k}^lH_k^TM_k^T \tag{32}$$

Finally, by solving the following conditional extremum problem, Expression (29) of the undetermined gain matrix $M_k$, which holds Equation (30) and minimizes trace of $P^d_{k-1|k}$, is calculated, and then the minimum variance unbiased estimate $\hat{d}_{k-1|k}$ of $d_{k-1}$ and the corresponding minimum covariance $P^d_{k-1|k}$ are obtained:

$$\min tr(P^d_{k-1|k}) \quad \text{s.t. } I_{m_1} - M_k C_k G_{k-1} = 0 \tag{33}$$

This proof is similar to the proof of Theorem 1 above and therefore omitted. □

*3.4. Estimation of State $x_k$*

First, we consider updating the one-step estimate $\hat{x}_{k|k-1}$. Compensating for $\hat{x}_{k|k-1}$ by adding $\hat{d}_{k-1|k}$ obtained in the previous section, we can get the updated one-step estimate $\tilde{x}^*_{k|k-1}$ as follows:

$$\hat{x}^*_{k|k-1} = \hat{x}_{k|k-1} + G_{k-1}\hat{d}_{k-1|k} \tag{34}$$

From Equations (1) and (34), the estimation error of the updated one-step estimate $\tilde{x}^*_{k|k-1}$ is obtained:

$$\begin{aligned} \tilde{x}^*_{k|k-1} &= x_k - \hat{x}^*_{k|k-1} \\ &= f(x_{k-1}) - \hat{x}_{k|k-1} + w_{k-1} + G_{k-1}\tilde{d}_{k-1|k} \end{aligned} \tag{35}$$

Further, the covariance matrix is approximated from Equation (35):

$$\begin{aligned} P^{x^*}_{k|k-1} &= E[(x_k - \hat{x}^*_{k|k-1})(x_k - \hat{x}^*_{k|k-1})^T] \\ &= P^x_{k|k-1} + P^{xd*}_{k|k-1} + P^{d*x}_{k|k-1} + P^{d*d*}_{k|k-1} \end{aligned} \tag{36}$$

where the expression $P^x_{k|k-1}$ is given in Equation (11) above and the specific derivations for $P^{xd*}_{k|k-1}$, $P^{d*x}_{k|k-1}$, and $P^{d*d*}_{k|k-1}$ are given by Equations (38) and (39).

First, the unknown input prediction estimation error at moment $k$ is defined as

$$\tilde{d}^*_{k|k-1} = G_{k-1}(d_{k-1} - \hat{d}_{k-1|k}) \tag{37}$$

Combining Equations (11), (13), (20), (31), and (37), it is obtained that

$$\begin{aligned} P^{xd*}_{k|k-1} &= E(\tilde{x}_{k|k-1}\tilde{d}^{*T}_{k|k-1}) \\ &= E[(f(x_{k-1}) - \hat{x}_{k|k-1} + w_{k-1})(d_{k-1} - \hat{d}_{k-1|k})^T]G^T_{k-1} \\ &= E[(f(x_{k-1}) - \hat{x}_{k|k-1} + w_{k-1})\tilde{d}^T_{k-1|k}]G^T_{k-1} \\ &= E[(f(x_{k-1}) - \hat{x}_{k|k-1} + w_{k-1})(-M_k\Delta_k - M_kH_k\tilde{l}_{k|k})^T]G^T_{k-1} \\ &= E[(f(x_{k-1}) - \hat{x}_{k|k-1} + w_{k-1})(-M_k\Delta_k + M_kH_kN_k\Delta_k)^T]G^T_{k-1} \\ &= -E[(f(x_{k-1}) - \hat{x}_{k|k-1} + w_{k-1})\Delta^T_k](I_n - H_kN_k)^T M^T_k G^T_{k-1} \\ &= -E[(f(x_{k-1}) - \hat{x}_{k|k-1} + w_{k-1})(f(x_{k-1}) - \hat{x}_{k|k-1} + w_{k-1})^T]C^T_k(I_n - H_kN_k)^T M^T_k G^T_{k-1} \\ &= -P^x_{k|k-1}C^T_k(I_n - H_kN_k)^T M^T_k G^T_{k-1} \end{aligned} \tag{38}$$

where the expressions for $P^x_{k|k-1}$ and $M_k$ are given in Equations (11) and (30), respectively. Further, $P^{d*x}_{k|k-1} = (P^{xd*}_{k|k-1})^T$ is obtained.

From $P^{d*d*}_{k|k-1} = E[\tilde{d}^*_{k|k-1}\tilde{d}^{*T}_{k|k-1}]$, combined with Equation (37) we get

$$\begin{aligned} P^{d*d*}_{k|k-1} &= G_{k-1}E(\tilde{d}_{k-1|k}\tilde{d}^T_{k-1|k})G^T_{k-1} \\ &= G_{k-1}P^d_{k-1|k}G^T_{k-1} \end{aligned} \tag{39}$$

where $P^{d}_{k-1|k}$ is given in Equation (32) above. In summary, $P^{xd*}_{k|k-1}$, $P^{d*x}_{k|k-1}$, and $P^{d*d*}_{k|k-1}$ are derived.

Then, update the innovation based on the estimated values $\hat{d}_{k-1|k}$ and $\hat{l}_{k|k}$ of the unknown inputs $d_{k-1}$ and $l_k$ already obtained above:

$$\widetilde{y}^{*2}_{k} = y_k - C_k\hat{x}^{*}_{k|k-1} - H_k\hat{l}_{k|k} \tag{40}$$

Further, we can design the estimated value of state $x_k$ as follows:

$$\hat{x}_{k|k} = \hat{x}^{*}_{k|k-1} + K_k\widetilde{y}^{*2}_{k} \tag{41}$$

From Equations (3) and (41), the estimation error of state $x_k$ is obtained:

$$\begin{aligned} \widetilde{x}_{k|k} &= x_k - \hat{x}_{k|k} \\ &= \widetilde{x}^{*}_{k|k-1} - K_k(C_k\widetilde{x}^{*}_{k|k-1} + H_k\widetilde{l}_{k|k} + v_k) \end{aligned} \tag{42}$$

where $\widetilde{x}^{*}_{k|k-1}$ is given in equation (35) above.

**Remark 4.** *In Section 3.2 and Section 3.3, we have proved that $\hat{d}_{k-1|k}$ and $\hat{l}_{k|k}$ are both unbiased estimates. Combining Equations (42), (35), (13), and (15) yields $E(\widetilde{x}_{k|k}) = 0$, that is, the state estimate $\hat{x}_{k|k}$ obtained from Equation (41) is the unbiased estimate of $x_k$ for any value of $K_k$. Therefore, we only need to calculate the undetermined gain matrix $K_k$ next, which minimizes the trace of the state estimation error covariance matrix $P^{x}_{k|k}$. It should be noted that for state estimation, references* [35–37] *have also conducted research on state and parameter estimation for measurement scarcity and bilinear systems.*

The covariance matrix of the estimation error is given by the following equation, and the specific derivation process can be seen in Appendix A:

$$\begin{aligned} P^{x}_{k|k} = {} & (I_n - K_kC_k)P^{x^*}_{k|k-1}(I_n - C_k^TK_k^T) + K_kH_kP^{l}_{k|k}H_k^TK_k^T + K_kR_kK_k^T \\ & -(I_n - K_kC_k)[-(P^{x}_{k|k-1} + P^{d*x}_{k|k-1})C_k^TN_k^T + G_{k-1}M_kR_kN_k^T - G_{k-1}M_kH_kN_kR_kN_k^T]H_k^TK_k^T \\ & -(I_n - K_kC_k)(-G_{k-1}M_kR_k + G_{k-1}M_kH_kN_kR_k)K_k^T \\ & -K_kH_k[-N_kC_k(P^{x}_{k|k-1} + P^{xd*}_{k|k-1}) + N_kR_kM_k^TG_{k-1}^T - N_kR_kN_k^TH_k^TM_k^TG_{k-1}^T](I_n - C_k^TK_k^T) \\ & -K_kH_kN_kR_kK_k^T - K_k(-R_kM_k^TG_{k-1}^T + R_kN_k^TH_k^TM_k^TG_{k-1}^T)(I_n - C_k^TK_k^T) - K_kR_kN_k^TH_k^TK_k^T \end{aligned} \tag{43}$$

where $P^{x^*}_{k|k-1}$, $P^{l}_{k|k}$, and $P^{d*x}_{k|k-1} = (P^{xd*}_{k|k-1})^T$ are given by Equations (36), (21), and (38) above.

**Theorem 3.** *Under the condition that $(C_kP^{x^*}_{k|k-1}C_k^T + H_kP^{l}_{k|k}H_k^T)$ is positive definite, the undetermined gain matrix $K_k$ which minimizes the trace of $P^{x}_{k|k}$ is given by*

$$\begin{aligned} K_k = {} & [P^{x^*}_{k|k-1}C_k^T - (P^{x}_{k|k-1} + P^{d*x}_{k|k-1})C_k^TN_k^TH_k^T + G_{k-1}M_kR_kN_k^TH_k^T - G_{k-1}M_kR_k + G_{k-1}M_kH_kN_kR_k][C_kP^{x^*}_{k|k-1}C_k^T + H_kP^{l}_{k|k}H_k^T \\ & +R_k - C_k(P^{x}_{k|k-1} + P^{d*x}_{k|k-1})C_k^TN_k^TH_k^T - H_kN_kC_k(P^{x}_{k|k-1} + P^{xd*}_{k|k-1})C_k^T + C_kG_{k-1}M_kR_kN_k^TH_k^T + H_kN_kR_kM_k^TG_{k-1}^TC_k^T \\ & -C_kG_{k-1}M_kH_kN_kR_kN_k^TH_k^T - H_kN_kR_kN_k^TH_k^TM_k^TG_{k-1}^TC_k^T - C_kG_{k-1}M_kR_k - R_kM_k^TG_{k-1}^TC_k^T + C_kG_{k-1}M_kH_kN_kR_k \\ & +R_kN_k^TH_k^TM_k^TG_{k-1}^TC_k^T - H_kN_kR_k - R_kN_k^TH_k^T]^{-1} \end{aligned} \tag{44}$$

**Proof.** The gain matrix $K_k$ is obtained by minimizing the trace of $P^{x}_{k|k}$, which leads to the state estimation filter, that is, solving for the following equation:

$$\frac{\partial tr(P^{x}_{k|k})}{\partial K_k} = 0 \tag{45}$$

By substituting Equation (43) into Equation (45), and then combining the trace derivative rule of the matrix, we get

$$
\begin{aligned}
&-2P^{x^*}_{k|k-1}C_k^T + 2K_kC_kP^{x^*}_{k|k-1}C_k^T + 2K_kH_kP^{l}_{k|k}H_k^T + 2K_kR_k + 2(P^{x}_{k|k-1} + P^{d*x}_{k|k-1})C_k^TN_k^TH_k^T - 2G_{k-1}M_kR_kN_k^TH_k^T \\
&+2G_{k-1}M_kH_kN_kR_kN_k^TH_k^T - 2K_kC_k(P^{x}_{k|k-1} + P^{d*x}_{k|k-1})C_k^TN_k^TH_k^T - 2K_kH_kN_kC_k(P^{x}_{k|k-1} + P^{xd*}_{k|k-1})C_k^T \\
&+2K_kC_kG_{k-1}M_kR_kN_k^TH_k^T + 2K_kH_kN_kR_kM_k^TG_{k-1}^TC_k^T - 2K_kC_kG_{k-1}M_kH_kN_kR_kN_k^TH_k^T \\
&-2K_kH_kN_kR_kN_k^TH_k^TM_k^TG_{k-1}^TC_k^T + 2G_{k-1}M_kR_k - 2G_{k-1}M_kH_kN_kR_k - 2K_kC_kG_{k-1}M_kR_k - 2K_kR_kM_k^TG_{k-1}^TC_k^T \\
&+2K_kC_kG_{k-1}M_kH_kN_kR_k + 2K_kR_kN_k^TH_k^TM_k^TG_{k-1}^TC_k^T - 2K_kH_kN_kR_k - 2K_kR_kN_k^TH_k^T = 0
\end{aligned}
\tag{46}
$$

Finally, the collation leads to Equation (44). The proof is completed. □

### 3.5. Update the Square-Root $S_{k|k}$

This section calculates the square-root $S_{k|k}$ of $P^x_{k|k}$. Unlike the traditional SRCKF, the undetermined gain matrix $K_k$ in expression $S_{k|k}$ is obtained under the minimum variance unbiased estimation criterion, that is, $K_k$ in Section 3.4.

1. Calculate the cubature poialculate the cubature points

$$\chi_{i,k|k-1} = S_{k|k-1}\xi_i + \hat{x}_{k|k-1} \tag{47}$$

2. Calculate the nonlinear propagated cubature points and the predicted measurement

$$\gamma_{i,k|k-1} = C_k\chi_{i,k|k-1} \tag{48}$$

$$\hat{y}_{k|k-1} = C_k\hat{x}_{k|k-1} \tag{49}$$

3. Calculate the square-root of the estimation error covariance matrix

$$S_{k|k} = \boldsymbol{Tria}([\chi_{k|k-1} - K_k\gamma_{k|k-1}, K_kS_{R_k}]) \tag{50}$$

where the central weighting matrices $\chi_{k|k-1}$ and $\gamma_{k|k-1}$ are expressed as

$$\chi_{k|k-1} = \frac{1}{\sqrt{m}}[\chi_{1,k|k-1} - \hat{x}_{k|k-1}, \chi_{2,k|k-1} - \hat{x}_{k|k-1}, \cdots, \chi_{m,k|k-1} - \hat{x}_{k|k-1}] \tag{51}$$

$$\gamma_{k|k-1} = \frac{1}{\sqrt{m}}[\gamma_{1,k|k-1} - \hat{y}_{k|k-1}, \gamma_{2,k|k-1} - \hat{y}_{k|k-1}, \cdots, \gamma_{m,k|k-1} - \hat{y}_{k|k-1}] \tag{52}$$

### 3.6. Summary of ISRCKF1 Iteration Steps

To demonstrate the proposed filter design process more conveniently, the iterative steps of ISRCKF1 will be summarized in this section to give the dual unknown input and state estimation algorithm based on ISRCKF1 as follows:

**ISRCKF1 Iteration Steps**

Step 1: Initialization

$$
\begin{gathered}
\hat{x}_0 = E(x_0) \\
P_0^x = E[(x_0 - \hat{x}_0)(x_0 - \hat{x}_0)^T] \\
S_0 = \boldsymbol{chol}(P_0^x)
\end{gathered}
$$

where $S_0$ denotes the square-root factor of $P_0^x$ and is obtained by performing Cholesky decomposition on $P_0^x$. Set time $k = 1$.

Step 2: Time Update

$$\begin{aligned}
\chi_{i,k-1|k-1} &= S_{k-1|k-1}\xi_i + \hat{x}_{k-1|k-1} \\
\chi^*_{i,k|k-1} &= f(\chi_{i,k-1|k-1}, u_{k-1}) \\
\hat{x}_{k|k-1} &= \frac{1}{m}\sum_{i=1}^{m} \chi^*_{i,k|k-1} \\
S_{k|k-1} &= \boldsymbol{Tria}([\mathcal{X}^*_{k|k-1}, S_{Q_{k-1}}]) \\
P^x_{k|k-1} &= \frac{1}{m}\sum_{i=1}^{m} \chi^*_{i,k|k-1}\chi^{*T}_{i,k|k-1} - \hat{x}_{k|k-1}\hat{x}^T_{k|k-1} + Q_{k-1}
\end{aligned}$$

Step 3: Estimation of unknown input $\boldsymbol{l}_k$

$$\begin{aligned}
\widetilde{R}_k &= C_k P^x_{k|k-1} C^T_k + R_k \\
N_k &= \begin{bmatrix} 0 & I_{m_2} \end{bmatrix} (D^T_k \widetilde{R}^{-1}_k D_k)^{-1} D^T_k \widetilde{R}^{-1}_k \\
\hat{l}_{k|k} &= N_k (y_k - C_k \hat{x}_{k|k-1}) \\
P^l_{k|k} &= N_k \widetilde{R}_k N^T_k
\end{aligned}$$

Step 4: Estimation of unknown input $\boldsymbol{d}_{k-1}$

$$\begin{aligned}
\widetilde{R}^*_k &= \widetilde{R}_k + H_k P^l_{k|k} H^T_k \\
M_k &= I_{m_1}[G^T_{k-1} C^T_k (\widetilde{R}^*_k)^{-1} C_k G_{k-1}]^{-1} G^T_{k-1} C^T_k (\widetilde{R}^*_k)^{-1} \\
\widetilde{y}^{*_1}_k &= y_k - C_k \hat{x}_{k|k-1} - H_k \hat{l}_{k|k} \\
\hat{d}_{k-1|k} &= M_k \widetilde{y}^{*_1}_k \\
P^d_{k-1|k} &= M_k \widetilde{R}_k M^T_k + M_k H_k P^l_{k|k} H^T_k M^T_k
\end{aligned}$$

Step 5: Estimation of state $x_k$

$$\begin{aligned}
\hat{x}^*_{k|k-1} &= \hat{x}_{k|k-1} + G_{k-1}\hat{d}_{k-1|k} \\
P^{xd*}_{k|k-1} &= -P^x_{k|k-1} C^T_k (I_n - H_k N_k)^T M^T_k G^T_{k-1} \\
P^{d*x}_{k|k-1} &= (P^{xd*}_{k|k-1})^T \\
P^{d*d*}_{k|k-1} &= G_{k-1} P^d_{k-1|k} G^T_{k-1} \\
P^{x^*}_{k|k-1} &= P^x_{k|k-1} + P^{xd*}_{k|k-1} + P^{d*x}_{k|k-1} + P^{d*d*}_{k|k-1}
\end{aligned}$$

$$\begin{aligned}
K_k = {} & [P^{x^*}_{k|k-1} C^T_k - (P^x_{k|k-1} + P^{d*x}_{k|k-1}) C^T_k N^T_k H^T_k + G_{k-1} M_k R_k N^T_k H^T_k - G_{k-1} M_k R_k + G_{k-1} M_k H_k N_k R_k][C_k P^{x^*}_{k|k-1} C^T_k + H_k P^l_{k|k} H^T_k \\
& + R_k - C_k (P^x_{k|k-1} + P^{d*x}_{k|k-1}) C^T_k N^T_k H^T_k - H_k N_k C_k (P^x_{k|k-1} + P^{xd*}_{k|k-1}) C^T_k + C_k G_{k-1} M_k R_k N^T_k H^T_k + H_k N_k R_k M^T_k G^T_{k-1} C^T_k \\
& - C_k G_{k-1} M_k H_k N_k R_k N^T_k H^T_k - H_k N_k R_k N^T_k H^T_k M^T_k G^T_{k-1} C^T_k - C_k G_{k-1} M_k R_k - R_k M^T_k G^T_{k-1} C^T_k + C_k G_{k-1} M_k H_k N_k R_k \\
& + R_k N^T_k H^T_k M^T_k G^T_{k-1} C^T_k - H_k N_k R_k - R_k N^T_k H^T_k]^{-1}
\end{aligned}$$

$$\begin{aligned}
\widetilde{y}^{*_2}_k &= y_k - C_k \hat{x}^*_{k|k-1} - H_k \hat{l}_{k|k} \\
\hat{x}_{k|k} &= \hat{x}^*_{k|k-1} + K_k \widetilde{y}^{*_2}_k
\end{aligned}$$

Step 6: Update square-root $S_{k|k}$

$$\begin{aligned}
\chi_{i,k|k-1} &= S_{k|k-1}\xi_i + \hat{x}_{k|k-1} \\
\hat{y}_{k|k-1} &= C_k \hat{x}_{k|k-1} \\
\gamma_{i,k|k-1} &= C_k \chi_{i,k|k-1} \\
\mathcal{X}_{k|k-1} &= \frac{1}{\sqrt{m}}[\chi_{1,k|k-1} - \hat{x}_{k|k-1}, \chi_{2,k|k-1} - \hat{x}_{k|k-1}, \cdots, \chi_{m,k|k-1} - \hat{x}_{k|k-1}] \\
\Upsilon_{k|k-1} &= \frac{1}{\sqrt{m}}[\gamma_{1,k|k-1} - \hat{y}_{k|k-1}, \gamma_{2,k|k-1} - \hat{y}_{k|k-1}, \cdots, \gamma_{m,k|k-1} - \hat{y}_{k|k-1}] \\
S_{k|k} &= \boldsymbol{Tria}([\mathcal{X}_{k|k-1} - K_k \Upsilon_{k|k-1}, K_k S_{R_k}])
\end{aligned}$$

Step 7: Set time $k = k + 1$ and return to step 2.

## 4. ISRCKF2

In practice, the above ISRCKF1 needs to be updated gradually with innovation and the iterative process of the algorithm is cumbersome, making the solution process time-consuming and costly. In this section, we rederive a more concise filter for the system studied above, utilizing the same innovation feedback. Given $\{y_0, y_1, \cdots, y_{k-1}\}$, $\hat{x}_0$, and $P_0^x$, the filter first obtains $\hat{x}_{k|k-1}$, $P_{k|k-1}^x$, and $S_{k|k-1}$. The detailed derivation is the same as Section 3.1, and will not be repeated here; when the measurement is updated to step k, the filter estimate $\hat{x}_{k|k}$ of $x_k$ is obtained. The derivation and the specific expression for $L_k$ are given in Section 4.1; the next step is to obtain filter estimates $\hat{d}_{k-1|k}$ and $\hat{l}_{k|k}$ for the dual unknown inputs $d_{k-1}$ and $l_k$. The derivation and the specific expressions for $M_k$ and $N_k$ are given in Section 4.2; finally, the square-root $S_{k|k}$ of $P_{k|k}^x$ is updated, and its detailed derivation is the same as Section 3.5, which will not be repeated here.

### 4.1. Minimum Variance Unbiased Estimation of State $x_k$

This section calculates the filter undetermined gain matrix $L_k$ such that $\hat{x}_{k|k}$ is the minimum variance unbiased estimate of $x_k$.

First, the innovation feedback $y_k - C_k\hat{x}_{k|k-1}$ is used to correct the one-step estimate $\hat{x}_{k|k-1}$ to obtain the unbiased estimate $\hat{x}_{k|k}$ of $x_k$ and the corresponding covariance matrix. For systems (1)–(3), the innovation at time $k$ can be expanded as shown in Equation (12) above. Using the innovation feedback, the filtered estimate of $x_k$ can be obtained as follows:

$$\hat{x}_{k|k} = \hat{x}_{k|k-1} + L_k(y_k - C_k\hat{x}_{k|k-1}) \tag{53}$$

Then, calculate the gain matrix $L_k$. From Equations (1), (12), (14), and (53) we get

$$\begin{aligned}\tilde{x}_{k|k} &= x_k - \hat{x}_{k|k} \\ &= f(x_{k-1}) + G_{k-1}d_{k-1} + w_{k-1} - \hat{x}_{k|k-1} - L_k\Delta_k - L_kD_k\begin{bmatrix} d_{k-1} \\ l_k \end{bmatrix} \\ &= f(x_{k-1}) - \hat{x}_{k|k-1} + w_{k-1} - L_k\Delta_k + G_{k-1}d_{k-1} - L_kC_kG_{k-1}d_{k-1} - L_kH_kl_k\end{aligned} \tag{54}$$

According to Equation (13), Equation (54) can be transformed into

$$\begin{aligned}\tilde{x}_{k|k} &= f(x_{k-1}) - \hat{x}_{k|k-1} + w_{k-1} - L_k\Delta_k + G_{k-1}d_{k-1} - L_kC_kG_{k-1}d_{k-1} - L_kH_kl_k \\ &= C_k^{-1}(\Delta_k - v_k) - L_k\Delta_k + (G_{k-1} - L_kC_kG_{k-1})d_{k-1} - L_kH_kl_k \\ &= (C_k^{-1} - L_k)\Delta_k - C_k^{-1}v_k + (G_{k-1} - L_kC_kG_{k-1})d_{k-1} - L_kH_kl_k\end{aligned} \tag{55}$$

**Remark 5.** *It should be noted that the precondition for obtaining Equation (55) from Equation (54) is that the pseudoinverse matrix $C_k^{-1}$ exists, which is indeed a limitation of ISRCKF2 and a problem that we will address in future research.*

**Theorem 4.** *If the system equation satisfies $rank(D_k) = m_1 + m_2$, then for any $d_{k-1}$ and $l_k$, Equation (53) is the minimum variance unbiased estimator of all possible $x_k$, if and only if the following equation is true:*

$$L_k = [P_{k|k-1}^x C_k^T - ([G_{k-1} \quad 0] - P_{k|k-1}^x C_k^T \tilde{R}_k^{-1} D_k)(-D_k^T \tilde{R}_k^{-1} D_k)^{-1} D_k^T] \tilde{R}_k^{-1} \tag{56}$$

**Proof.** First, we give the necessary conditions for Equation (56) to hold as follows:

$$[G_{k-1} \quad 0] - L_kD_k = 0 \tag{57}$$

If Equation (57) holds, it is equivalent to $G_{k-1} - L_kC_kG_{k-1} = 0$ and $-L_kH_k = 0$. Substituting into Equation (54) gives

$$\tilde{x}_{k|k} = (C_k^{-1} - L_k)\Delta_k - C_k^{-1}v_k \tag{58}$$

Combining Equation (15) yields $E(\widetilde{x}_{k|k}) = 0$. Thus, the state estimate $\hat{x}_{k|k}$ obtained from Equation (53) is the unbiased estimate of $x_k$ for any value of $L_k$, regardless of the values of $d_{k-1}$ and $l_k$. Unbiasedness is proven.

Further from Equations (11), (13), and (58), $P^x_{k|k} = E(\widetilde{x}_{k|k}\widetilde{x}^T_{k|k})$ can be approximated, that is,

$$\begin{aligned} P^x_{k|k} &= E\left\{[(C_k^{-1} - L_k)\Delta_k - C_k^{-1}v_k][\Delta_k^T(C_k^{-1} - L_k)^T - v_k^T C_k^{-T}]\right\} \\ &= (C_k^{-1} - L_k)\widetilde{R}_k(C_k^{-1} - L_k)^T + C_k^{-1}R_k C_k^{-T} + E[(C_k^{-1} - L_k)\Delta_k(-v_k^T)C_k^{-T} - C_k^{-1}v_k\Delta_k^T(C_k^{-1} - L_k)^T] \\ &= (C_k^{-1} - L_k)\widetilde{R}_k(C_k^{-1} - L_k)^T + C_k^{-1}R_k C_k^{-T} + E[(C_k^{-1} - L_k)v_k(-v_k^T)C_k^{-T} - C_k^{-1}v_k v_k^T(C_k^{-1} - L_k)^T] \\ &= (C_k^{-1}\widetilde{R}_k - L_k\widetilde{R}_k)(C_k^{-T} - L_k^T) + C_k^{-1}R_k C_k^{-T} + [-(C_k^{-1} - L_k)R_k C_k^{-T} - C_k^{-1}R_k(C_k^{-T} - L_k^T)] \\ &= C_k^{-1}\widetilde{R}_k C_k^{-T} - C_k^{-1}\widetilde{R}_k L_k^T - L_k\widetilde{R}_k C_k^{-T} + L_k\widetilde{R}_k L_k^T + L_k R_k C_k^{-T} - C_k^{-1}R_k C_k^{-T} + C_k^{-1}R_k L_k^T \end{aligned} \tag{59}$$

Then, the undetermined gain matrix $L_k$ is solved by solving the following constrained optimization problem.

$$\min tr(P^x_{k|k}) \quad \text{s.t.} \begin{bmatrix} G_{k-1} & 0 \end{bmatrix} - L_k D_k = 0 \tag{60}$$

Let the Lagrangian be

$$F_2 = tr\left\{P^x_{k|k}\right\} - 2tr\{\Lambda_{2k}(\begin{bmatrix} G_{k-1} & 0 \end{bmatrix} - L_k D_k)\} \tag{61}$$

where $\Lambda_{2k}$ is the Lagrange multiplier matrix, $\Lambda_{2k} \in R^{(m_1+m_2)\times p}$, the factor "2" is intended to keep the calculation simple.

Taking the derivative of Equation (61) with respect to $L_k$, and making the resulting derivative zero, we get

$$L_k\widetilde{R}_k + \Lambda_{2k}^T D_k^T - P^x_{k|k-1}C_k^T = 0 \tag{62}$$

The system of linear equations consisting of Equations (62) and (57) is as follows:

$$\begin{bmatrix} \widetilde{R}_k^T & D_k \\ D_k^T & 0 \end{bmatrix}\begin{bmatrix} L_k^T \\ \Lambda_{2k} \end{bmatrix} = \begin{bmatrix} C_k P^x_{k|k-1} \\ \begin{bmatrix} G_{k-1} & 0 \end{bmatrix}^T \end{bmatrix} \tag{63}$$

Finally, using Schur's complement lemma to solve the above equations, Equation (56) is obtained. The proof is completed.□

### 4.2. Minimum Variance Unbiased Estimation of Unknown Inputs

This section considers the estimation of $d_{k-1}$ and $l_k$, where the estimation of $d_{k-1}$ is obtained in Section 4.2.1, and also determines the conditions that need to be satisfied for the undetermined gain matrix $M_k$. Section 4.2.2 obtains the estimation of $l_k$ and determines the conditions that need to be satisfied for the undetermined gain matrix $N_k$.

#### 4.2.1. Estimation of Unknown Input $d_{k-1}$

In this subsection, based on the innovation feedback $\widetilde{y}_k = y_k - C_k\hat{x}_{k|k-1}$ from the previous section, the gain matrix $M_k$ will be calculated.

First, using the innovation $\widetilde{y}_k$, design the unbiased estimate $\hat{d}_{k-1|k}$ of $d_{k-1}$ as follows:

$$\hat{d}_{k-1|k} = M_k(y_k - C_k\hat{x}_{k|k-1}) \tag{64}$$

Then, from Equations (12) and (64), the estimated error of the unknown input $d_{k-1}$ is obtained:

$$\begin{aligned} \widetilde{d}_{k-1} &= d_{k-1} - \hat{d}_{k-1|k} \\ &= (I_{m_1} - M_k C_k G_{k-1})d_{k-1} - M_k H_k l_k - M_k\Delta_k \end{aligned} \tag{65}$$

**Theorem 5.** *If the system equation satisfies* $\boldsymbol{rank}(\boldsymbol{D}_k) = \boldsymbol{m}_1 + \boldsymbol{m}_2$*, then for any* $\boldsymbol{d}_{k-1}$ *and* $\boldsymbol{l}_k$*, Equation (64) is the minimum variance unbiased estimator of* $\boldsymbol{d}_{k-1}$*, if and only if the following equation is true:*

$$\boldsymbol{M}_k = \begin{bmatrix} \boldsymbol{I}_{m_1} & 0 \end{bmatrix} \left(\boldsymbol{D}_k^T \widetilde{\boldsymbol{R}}_k^{-1} \boldsymbol{D}_k\right)^{-1} \boldsymbol{D}_k^T \widetilde{\boldsymbol{R}}_k^{-1} \tag{66}$$

**Proof.** First, we give the necessary conditions for Equation (66) to hold as follows:

$$\begin{bmatrix} \boldsymbol{I}_{m_1} & 0 \end{bmatrix} - \boldsymbol{M}_k \boldsymbol{D}_k = 0 \tag{67}$$

Equation (67) can be transformed into $\boldsymbol{I}_{m_1} - \boldsymbol{M}_k \boldsymbol{C}_k \boldsymbol{G}_{k-1} = 0$ and $-\boldsymbol{M}_k \boldsymbol{H}_k = 0$. At this point, Equation (65) can be simplified as

$$\widetilde{\boldsymbol{d}}_{k-1|k} = -\boldsymbol{M}_k \boldsymbol{\Delta}_k \tag{68}$$

Therefore, from Equations (15) and (68), we get $\boldsymbol{E}(\widetilde{\boldsymbol{d}}_{k-1|k}) = 0$. Unbiasedness is proven.

Further, from Equations (11), (13), and (68), the covariance matrix $\boldsymbol{P}_{k-1|k}^d = \boldsymbol{E}(\widetilde{\boldsymbol{d}}_{k-1|k} \widetilde{\boldsymbol{d}}_{k-1|k}^T)$ of $\boldsymbol{d}_{k-1}$ is calculated:

$$\boldsymbol{P}_{k-1|k}^d = \boldsymbol{M}_k \widetilde{\boldsymbol{R}}_k \boldsymbol{M}_k^T \tag{69}$$

Finally, solving the following constraint problem yields the undetermined gain matrix $\boldsymbol{M}_k$, which leads to the minimum variance unbiased estimate $\hat{\boldsymbol{d}}_{k-1|k}$ of $\boldsymbol{d}_{k-1}$ and the corresponding minimum variance matrix $\boldsymbol{P}_{k-1|k}^d$.

$$\min \boldsymbol{tr}(\boldsymbol{P}_{k-1|k}^d) \quad \text{s.t.} \begin{bmatrix} \boldsymbol{I}_{m_1} & 0 \end{bmatrix} - \boldsymbol{M}_k \boldsymbol{D}_k = 0 \tag{70}$$

Similar to the proof of Theorem 4, let the Lagrangian be

$$F_3 = \boldsymbol{tr}\left\{\boldsymbol{P}_{k-1|k}^d\right\} - 2\boldsymbol{tr}\left\{\Lambda_{3k}\left(\begin{bmatrix} \boldsymbol{I}_{m_1} & 0 \end{bmatrix} - \boldsymbol{M}_k \boldsymbol{D}_k\right)\right\} \tag{71}$$

where $\Lambda_{3k}$ is the Lagrange multiplier matrix, $\Lambda_{3k} \in \boldsymbol{R}^{(m_1+m_2)\times m_1}$, the factor "2" is intended to keep the calculation simple.

Taking the derivative of Equation (71) with respect to $\boldsymbol{M}_k$, and making the resulting derivative zero, we get

$$\boldsymbol{M}_k \widetilde{\boldsymbol{R}}_k + \Lambda_{3k}^T \boldsymbol{D}_k^T = 0 \tag{72}$$

The system of linear equations consisting of Equations (72) and (67) is as follows:

$$\begin{bmatrix} \widetilde{\boldsymbol{R}}_k^T & \boldsymbol{D}_k \\ \boldsymbol{D}_k^T & 0 \end{bmatrix} \begin{bmatrix} \boldsymbol{M}_k^T \\ \Lambda_{3k} \end{bmatrix} = \begin{bmatrix} 0 \\ \begin{bmatrix} \boldsymbol{I}_{m_1} & 0 \end{bmatrix}^T \end{bmatrix} \tag{73}$$

Using Schur's complement lemma to solve the above equations, Equation (66) is obtained. The proof is completed. □

4.2.2. Estimation of Unknown Input $\boldsymbol{l}_k$

Since the innovation feedback $\boldsymbol{y}_k - \boldsymbol{C}_k \hat{\boldsymbol{x}}_{k|k-1}$ used in ISRCKF2 is the same as the innovation $\widetilde{\boldsymbol{y}}_k$ designed for the estimation of $\boldsymbol{l}_k$ in Section 3.2, this subsection follows the same proof procedure as in Section 3.2 and will not be repeated here.

### 4.3. Summary of ISRCKF2 Iteration Steps

In order to clearly reflect the filter design process with simplified dual unknown inputs, this section will summarize the steps of ISRCKF2 and give a more concise iterative process as follows:

**ISRCKF2 Iteration Steps**

Step 1: Initialization

$$\hat{x}_0 = E(x_0)$$
$$P_0^x = E[(x_0 - \hat{x}_0)(x_0 - \hat{x}_0)^T]$$
$$S_0 = chol(P_0^x)$$

where $S_0$ denotes the square-root factor of $P_0^x$ and is obtained by performing Cholesky decomposition on $P_0^x$. Set time k = 1.

Step 2: Time Update

$$\chi_{i,k-1|k-1} = S_{k-1|k-1}\xi_i + \hat{x}_{k-1|k-1}$$
$$\chi^*_{i,k|k-1} = f(\chi_{i,k-1|k-1}, u_{k-1})$$
$$\hat{x}_{k|k-1} = \frac{1}{m}\sum_{i=1}^{m} \chi^*_{i,k|k-1}$$
$$S_{k|k-1} = Tria([\chi^*_{k|k-1}, S_{Q_{k-1}}])$$
$$P^x_{k|k-1} = \frac{1}{m}\sum_{i=1}^{m} \chi^*_{i,k|k-1}\chi^{*T}_{i,k|k-1} - \hat{x}_{k|k-1}\hat{x}^T_{k|k-1} + Q_{k-1}$$

Step 3: Estimation of state $x_k$

$$\widetilde{R}_k = C_k P^x_{k|k-1} C_k^T + R_k$$
$$L_k = [P^x_{k|k-1}C_k^T - (\begin{bmatrix} G_{k-1} & 0 \end{bmatrix} - P^x_{k|k-1}C_k^T\widetilde{R}_k^{-1}D_k)(-D_k^T\widetilde{R}_k^{-1}D_k)^{-1}D_k^T]\widetilde{R}_k^{-1}$$
$$\hat{x}_{k|k} = \hat{x}_{k|k-1} + L_k(y_k - C_k\hat{x}_{k|k-1})$$

Step 4: Estimation of unknown inputs $d_{k-1}$ and $l_k$

$$M_k = \begin{bmatrix} I_{m_1} & 0 \end{bmatrix}(D_k^T\widetilde{R}_k^{-1}D_k)^{-1}D_k^T\widetilde{R}_k^{-1}$$
$$N_k = \begin{bmatrix} 0 & I_{m_2} \end{bmatrix}(D_k^T\widetilde{R}_k^{-1}D_k)^{-1}D_k^T\widetilde{R}_k^{-1}$$
$$\hat{d}_{k-1|k} = M_k(y_k - C_k\hat{x}_{k|k-1})$$
$$\hat{l}_{k|k} = N_k(y_k - C_k\hat{x}_{k|k-1})$$

Step 5: Update of square-root $S_{k|k}$

$$\chi_{i,k|k-1} = S_{k|k-1}\xi_i + \hat{x}_{k|k-1}$$
$$\gamma_{i,k|k-1} = C_k\chi_{i,k|k-1}$$
$$\hat{y}_{k|k-1} = C_k\hat{x}_{k|k-1}$$
$$\chi_{k|k-1} = \frac{1}{\sqrt{m}}[\chi_{1,k|k-1} - \hat{x}_{k|k-1}, \chi_{2,k|k-1} - \hat{x}_{k|k-1}, \cdots, \chi_{m,k|k-1} - \hat{x}_{k|k-1}]$$
$$\gamma_{k|k-1} = \frac{1}{\sqrt{m}}[\gamma_{1,k|k-1} - \hat{y}_{k|k-1}, \gamma_{2,k|k-1} - \hat{y}_{k|k-1}, \cdots, \gamma_{m,k|k-1} - \hat{y}_{k|k-1}]$$
$$S_{k|k} = Tria([\chi_{k|k-1} - K_k\gamma_{k|k-1}, K_kS_{R_k}])$$

Step 6: Set time $k = k + 1$ and return to step 2.

## 5. Simulation Results

To demonstrate the effectiveness of the two filtering algorithms proposed in this paper, in this section, the ISRCKF algorithm is compared with the SRCKF algorithm proposed in [32]. First, the simulation uses a fifth-order two-phase nonlinear model of the induction motor [38,39], which can be described using the system models (1)–(3) as follows:

$$f(x_k,u_k) = \begin{bmatrix} x_{1,k-1} + h(-\gamma x_{1,k-1} + \frac{K}{T_r}x_{3,k-1} + Kpx_{5,k-1}x_{4,k-1} + \frac{1}{\sigma L_s}u_{1k}) \\ x_{2,k-1} + h(-\gamma x_{2,k-1} - Kpx_{5,k-1}x_{3,k-1} + \frac{K}{T_r}x_{4,k-1} + \frac{1}{\sigma L_s}u_{2,k}) \\ x_{3,k-1} + h(\frac{M}{T_r}x_{2,k-1} - \frac{1}{T_r}x_{3,k-1} - px_{5,k-1}x_{4,k-1}) \\ x_{4,k-1} + h(\frac{M}{T_r}x_{2,k-1} - \frac{1}{T_r}x_{4,k-1} + px_{5,k-1}x_{3,k-1}) \\ x_{5,k-1} + h(\frac{pM}{JT_r}(x_{3,k-1}x_{2,k-1} - x_{4,k-1}x_{1,k-1}) - \frac{T_L}{J}) \end{bmatrix}$$

where $x_{k-1}^T = [\, x_{1,k-1} \;\; x_{2,k-1} \;\; x_{3,k-1} \;\; x_{4,k-1} \;\; x_{5,k-1} \,] = [\, i_{sak-1} \;\; i_{sbk-1} \;\; \psi_{rak-1} \;\; \psi_{rbk-1} \;\; \omega_{k-1} \,]$ denotes the stator currents a and b, the rotor fluxes a and b and the angular speed, respectively, $u_k^T = [\, u_{1,k} \;\; u_{2,k} \,] = [\, u_{sak} \;\; u_{sbk} \,]$ represents the stator voltages control vector, $p$ is the number of pole pairs and $T_L$ is the load torque.

The rotor time constant $T_r$ and the parameters $\sigma$, $K$ and $\gamma$ are defined as follows:

$$T_r = \frac{L_r}{R_r},\ \sigma = 1 - \frac{M^2}{L_sL_r},\ K = \frac{M}{\sigma L_s L_r} \text{ and } \gamma = \frac{R_s}{\sigma L_s} + \frac{R_r M^2}{\sigma L_s L_r^2}.$$

where $R_s$ and $R_r$ are the per-phase resistances of the stator and rotor, $L_s$ and $L_r$ are the per-phase inductances of the stator and rotor, and $J$ is the rotor moment of inertia.

The simulations are carried out with the same numerical values as in [38], that is, $R_s = 0.18\ \Omega$, $R_r = 0.15\ \Omega$, $M = 0.068$ H, $p = 1$, $L_s = 0.0699$ H, $L_r = 0.0699$ H, $J = 0.0586$ kgm$^2$, $T_L = 10$ Nm and $h = 0.1$ ms.

The motor control input signals are $u_{1,k} = 350\cos(0.003k)$ and $u_{2,k} = 300\sin(0.003k)$. The system correlation coefficient matrix is as follows:

$$G_{k-1} = \begin{bmatrix} 1 & 0 \\ 0 & 1 \\ 0 & 0 \\ 0 & 0 \\ 0 & 0 \end{bmatrix}, C_k = \begin{bmatrix} 1 & 0 & 0 & 0 & 0 \\ 0 & 1 & 0 & 0 & 0 \\ 0 & 0 & 1 & 0 & 0 \\ 0 & 0 & 0 & 1 & 0 \\ 0 & 0 & 0 & 0 & 1 \end{bmatrix}, H_k = \begin{bmatrix} 0 & 0 \\ 0 & 0 \\ 1 & 0 \\ 0 & 1 \\ 0 & 0 \end{bmatrix}.$$

The covariance matrices for $w_{k-1}$ and $v_k$ are given by

$$Q_{k-1} = 10^{-4} \times \begin{bmatrix} 1 & 0 & 0 & 0 & 0 \\ 0 & 1 & 0 & 0 & 0 \\ 0 & 0 & 1 & 0 & 0 \\ 0 & 0 & 0 & 1 & 0 \\ 0 & 0 & 0 & 0 & 1 \end{bmatrix}, R_k = 10^{-2} \times \begin{bmatrix} 1 & 0 & 0 & 0 & 0 \\ 0 & 1 & 0 & 0 & 0 \\ 0 & 0 & 1 & 0 & 0 \\ 0 & 0 & 0 & 1 & 0 \\ 0 & 0 & 0 & 0 & 1 \end{bmatrix}.$$

ISRCKF1 and ISRCKF2 are set in the simulation the same system and filter initial conditions as the comparison algorithm, referring to [39], that is, $P_0^x = 100^2 I_5$, $x_{1,0} = x_{2,0} = x_{3,0} = x_{4,0} = x_{5,0} = 0$, $\hat{x}_{1,0} = 200$, $\hat{x}_{2,0} = 200$, $\hat{x}_{3,0} = 50$, $\hat{x}_{4,0} = 50$, $\hat{x}_{5,0} = 300$.

Setting the unknown input $d_{k-1}$ in the equation of state equal to $d_{k-1} = \begin{bmatrix} d_{1,k-1} & d_{2,k-1} \end{bmatrix}^T$, and the unknown input $l_k$ in the measurement equation equal to $l_k = \begin{bmatrix} l_{1k} & l_{2k} \end{bmatrix}^T$.

To quantitatively compare various nonlinear filtering algorithms, the root mean square error of $x_i$ at moment k is denoted by $RMSE_{i,k}$, which is defined as follows:

$$RMSE_{i,k} = \sqrt{\frac{1}{MC}\sum_{n=1}^{MC}\left(x_{i,k}^{(n)} - \hat{x}_{i,k|k}^{(n)}\right)^2},\ i = 1,2,3,4,5$$

where $MC$ denotes the number of Monte Carlo simulations and $MC = 100$ is set in the simulation; $x_{i,k}^{(n)}$ and $\hat{x}_{i,k|k}^{(n)}$ denote the actual and estimated values of the state $x_i$ at moment k under the nth Monte Carlo simulation, respectively. For unknown inputs $d_{k-1}$ and $l_k$, the above equation is also used as a quantitative evaluation index.

In the following, the validity of the proposed ISRCKF1 and ISRCKF2 for the system containing dual unknown inputs is verified, and to ensure the accuracy and generality of the experimental results, the number of sampling steps N is taken as large as possible, and the same as in [39] is set to be N = 1800.

Case 1: When the unknown inputs in the state and measurement equations are the same input signal, that is, $d_{k-1} = l_k \neq 0$, the state estimation performance difference of each filtering algorithm is considered. After 100 Monte Carlo simulations, the mean RMSE of ISRCKF1 and ISRCKF2 with dual unknown inputs and states between moments 0~1800 are shown in Tables 1–6. It should be added that "N/A" in the table means that the algorithm is not applicable. The time evolution curves of the state RMSE for the proposed filtering algorithms are shown in Appendix B.

**Table 1.** The mean RMSE of estimated values of the unknown input $d_{k-1}$.

| Unknown Input | ISRCKF1 | ISRCKF2 | SRCKF |
|---|---|---|---|
| $d_{1,k-1}$ | 0.6034 | 0.3962 | N/A |
| $d_{2,k-1}$ | 0.4944 | 0.3292 | N/A |

**Table 2.** The mean RMSE of estimated values of the unknown input $l_k$.

| Unknown Input | ISRCKF1 | ISRCKF2 | SRCKF |
|---|---|---|---|
| $l_{1k}$ | 0.1888 | 0.1360 | N/A |
| $l_{2k}$ | 0.3314 | 0.2216 | N/A |

**Table 3.** The mean RMSE of estimated values of the system states $x_k$.

| System State | ISRCKF1 | ISRCKF2 |
|---|---|---|
| $x_1$(A) | 0.0991 | 0.0985 |
| $x_2$(A) | 0.0983 | 0.1239 |
| $x_3$(Wb) | 0.1135 | 0.0841 |
| $x_4$(Wb) | 0.2169 | 0.1365 |
| $x_5$(Rad/s) | 0.6208 | 0.3631 |

**Table 4.** The mean RMSE of estimated values of the unknown input $d_{k-1}$.

| Unknown Input | ISRCKF1 | ISRCKF2 | SRCKF |
|---|---|---|---|
| $d_{1,k-1}$ | 0.7485 | 0.6686 | N/A |
| $d_{2,k-1}$ | 0.4927 | 0.3955 | N/A |

**Table 5.** The mean RMSE of estimated values of the unknown input $l_k$.

| Unknown Input | ISRCKF1 | ISRCKF2 | SRCKF |
|---|---|---|---|
| $l_{1k}$ | 0.2260 | 0.1816 | N/A |
| $l_{2k}$ | 0.3130 | 0.2121 | N/A |

**Table 6.** The mean RMSE of estimated values of the system states $x_k$.

| System State | ISRCKF1 | ISRCKF2 |
|---|---|---|
| $x_1$(A) | 0.0986 | 0.0989 |
| $x_2$(A) | 0.0983 | 0.0988 |
| $x_3$(Wb) | 0.1158 | 0.0810 |
| $x_4$(Wb) | 0.2197 | 0.1595 |
| $x_5$(Rad/s) | 0.7569 | 0.4841 |

Figure 1 and Table 1 show the estimation results of $d_{k-1}$ when considering the unknown input as sine and cosine electrical signals. It can be seen that the filtering algorithm given in this paper can track and estimate the unknown input $d_{k-1}$ better, and the ISRCKF2 estimation accuracy is higher.

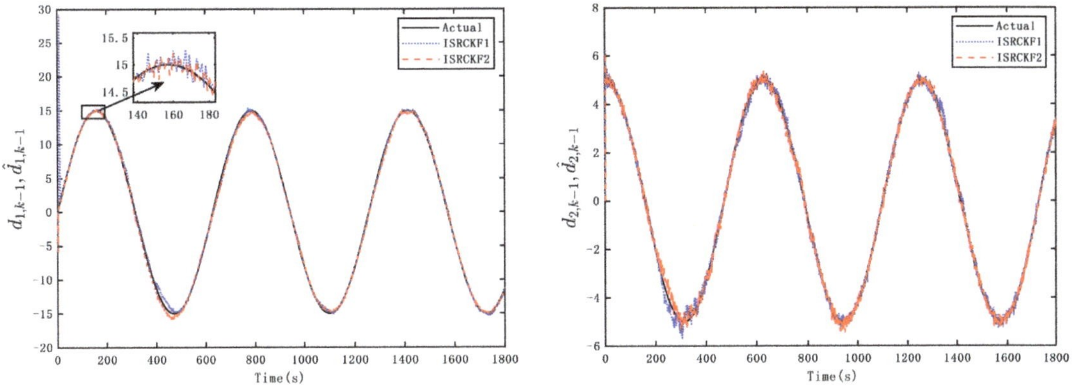


(a) (b)

**Figure 1.** Actual and estimated values of the unknown input $d_{k-1}$: (**a**) The unknown input component $d_{1,k-1}$: (**b**) The unknown input component $d_{2,k-1}$.

Figure 2 and Table 2 show the estimation results of $l_k$. We can also conclude that the filtering algorithm proposed can better track and estimate $l_k$, and the estimation of ISRCKF2 produces a smaller mean RMSE.

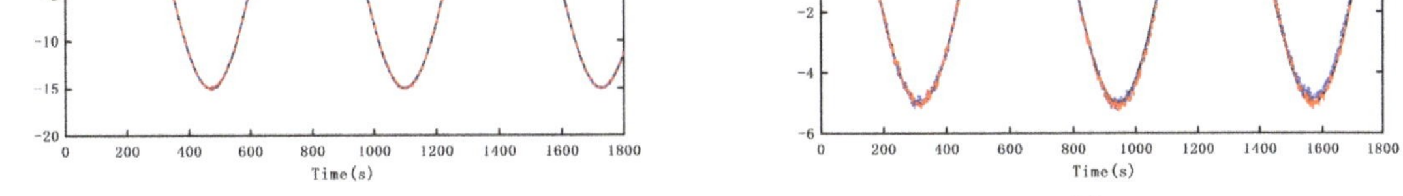


(a) (b)

**Figure 2.** Actual and estimated values of the unknown input $l_k$: (**a**) The unknown input component $l_{1k}$; (**b**) The unknown input component $l_{2k}$.

Figure 3 and Table 3 show the estimation results of the system states $x_k$. From Figure 3, we can see that the overall filtering performance of the proposed ISRCKF1 and ISRCKF2 is better than that of the SRCKF when considering the system with the same unknown inputs, proving that the ISRCKF improves the state estimation accuracy when dealing with nonlinear systems containing unknown inputs. Comparing the data results in Table 3, it can be seen that the estimation error performance of ISRCKF1 and ISRCKF2 is basically equivalent. However, since the simplified algorithm avoids the coupling error between the coupled terms and the cumulative error of the complex iteration steps, the ISRCKF2 estimation accuracy is more accurate.

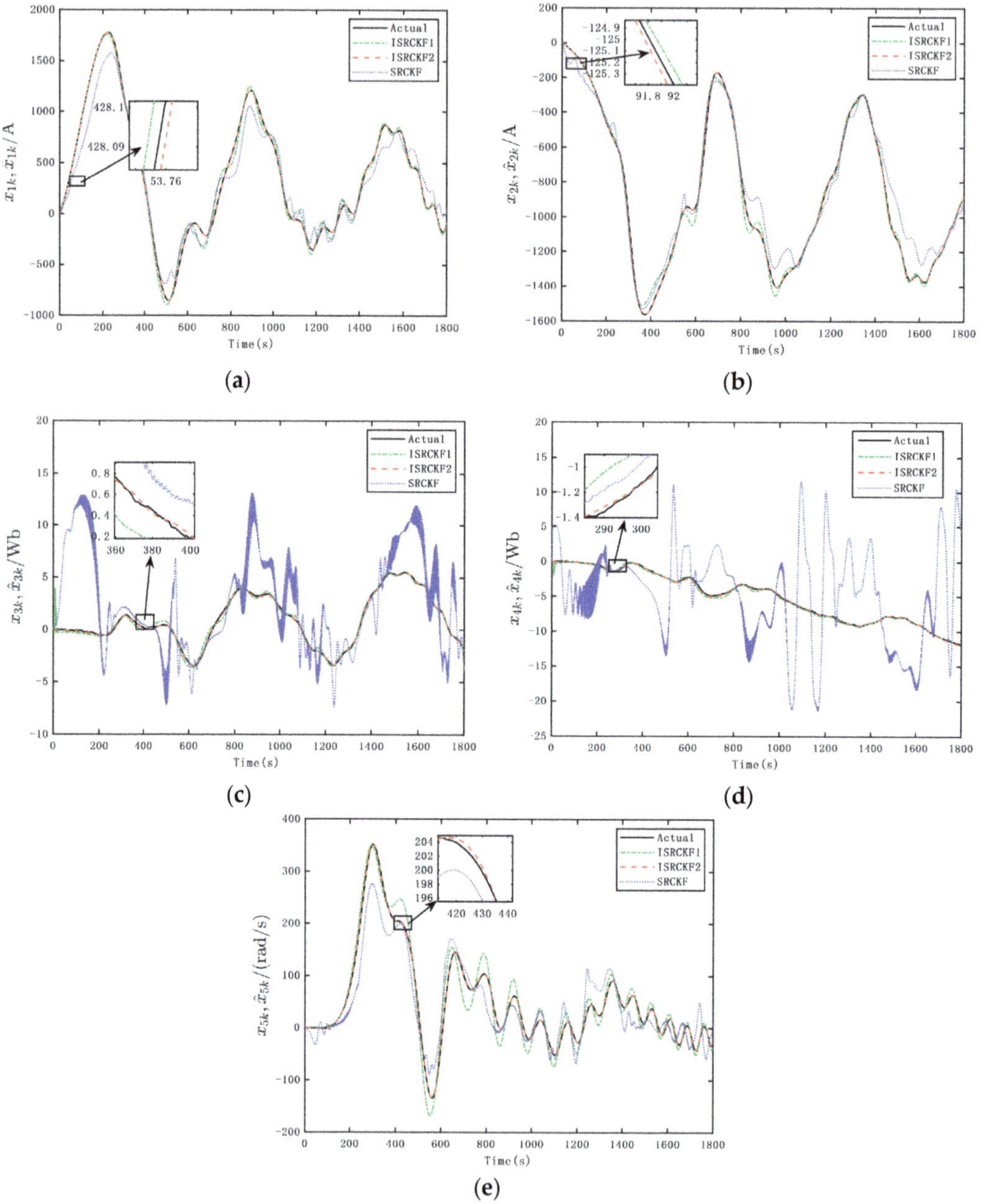


**Figure 3.** Estimation comparison of the system states $x_k$ under proposed algorithm and classical algorithm: (**a**) Stator current a; (**b**) Stator current b; (**c**) Rotor flux a; (**d**) Rotor flux b; (**e**) Angular speed.

Case 2: To further verify the validity of the proposed algorithm, the ISRCKF algorithm with dual unknown inputs is used for the direct feedthrough nonlinear discrete system with different unknown inputs in state and measurement equations, that is, $d_{k-1} \neq l_k \neq 0$, and the estimation of the simulation results of the dual unknown inputs and the state can be obtained as follows.

Figure 4 shows the true and estimated values of the unknown input $d_{k-1}$ for the two cases of sine and cosine signals, respectively, from which it can be seen that both algorithms can track and estimate the unknown inputs $d_{1,k-1}$ and $d_{2,k-1}$ better. Table 4 gives the root mean square error of $d_{k-1}$, from which we can conclude that the ISRCKF2 estimation is more accurate.

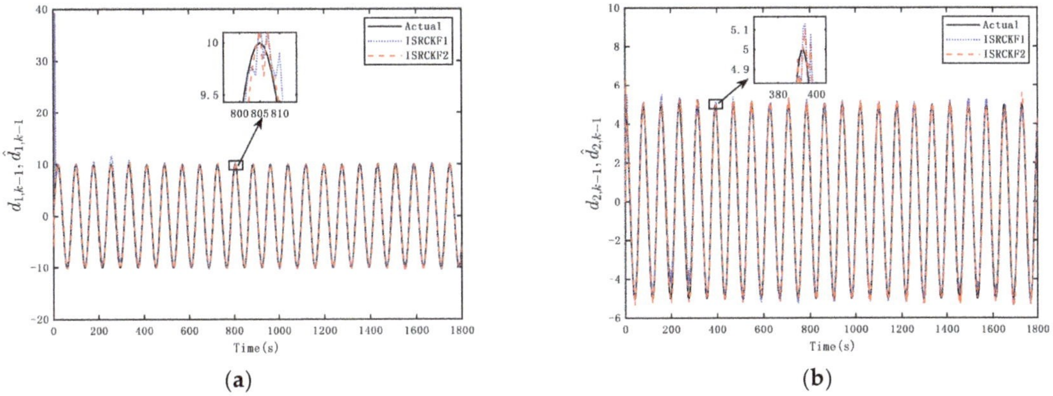


**Figure 4.** Actual and estimated values of the unknown input $d_{k-1}$: (**a**) The unknown input component $d_{1,k-1}$; (**b**) The unknown input component $d_{2,k-1}$.

Figure 5 and Table 5 show the estimation results when considering the unknown input $l_k$ as a step signal and a combination of constant, ramp, and step signals, respectively. We can see that both filtering algorithms presented in this paper can effectively track and estimate $l_{1k}$ and $l_{2k}$, and the mean RMSE of the estimation is not much different.

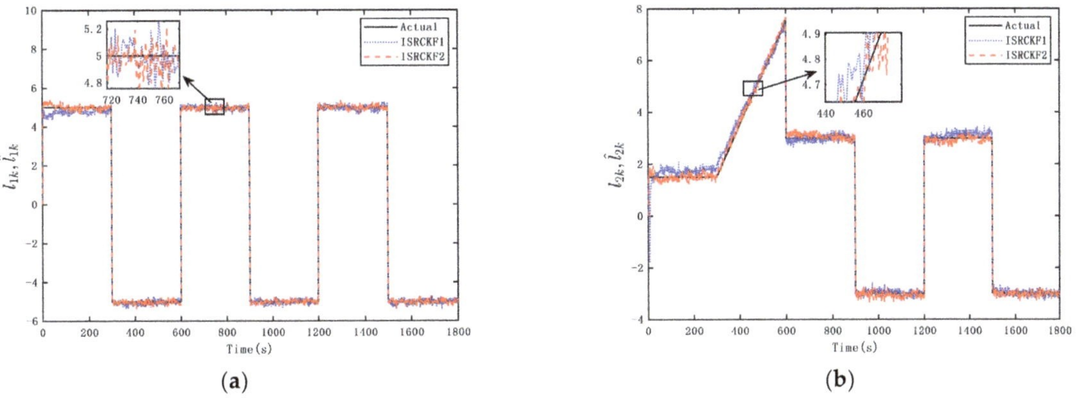


**Figure 5.** Actual and estimated values of the unknown input $l_k$: (**a**) The unknown input component $l_{1k}$; (**b**) The unknown input component $l_{2k}$.

Based on the above analysis, we can conclude that the two proposed algorithms can achieve the optimal estimation of the unknown inputs, regardless of whether the unknown inputs in the equation of state and the measurement equation are the same.

In engineering practice, the unknown inputs in the state and measurement equations are often different. Figure 6 shows the comparison of state estimation results in this case, and it can be seen that the two ISRCKF algorithms can completely track and estimate state $x_{1k} \sim x_{5k}$. By comparing the two sets of system states in Figures 3 and 6, it is intuitively clear that the system states change with the type of the unknown input itself.

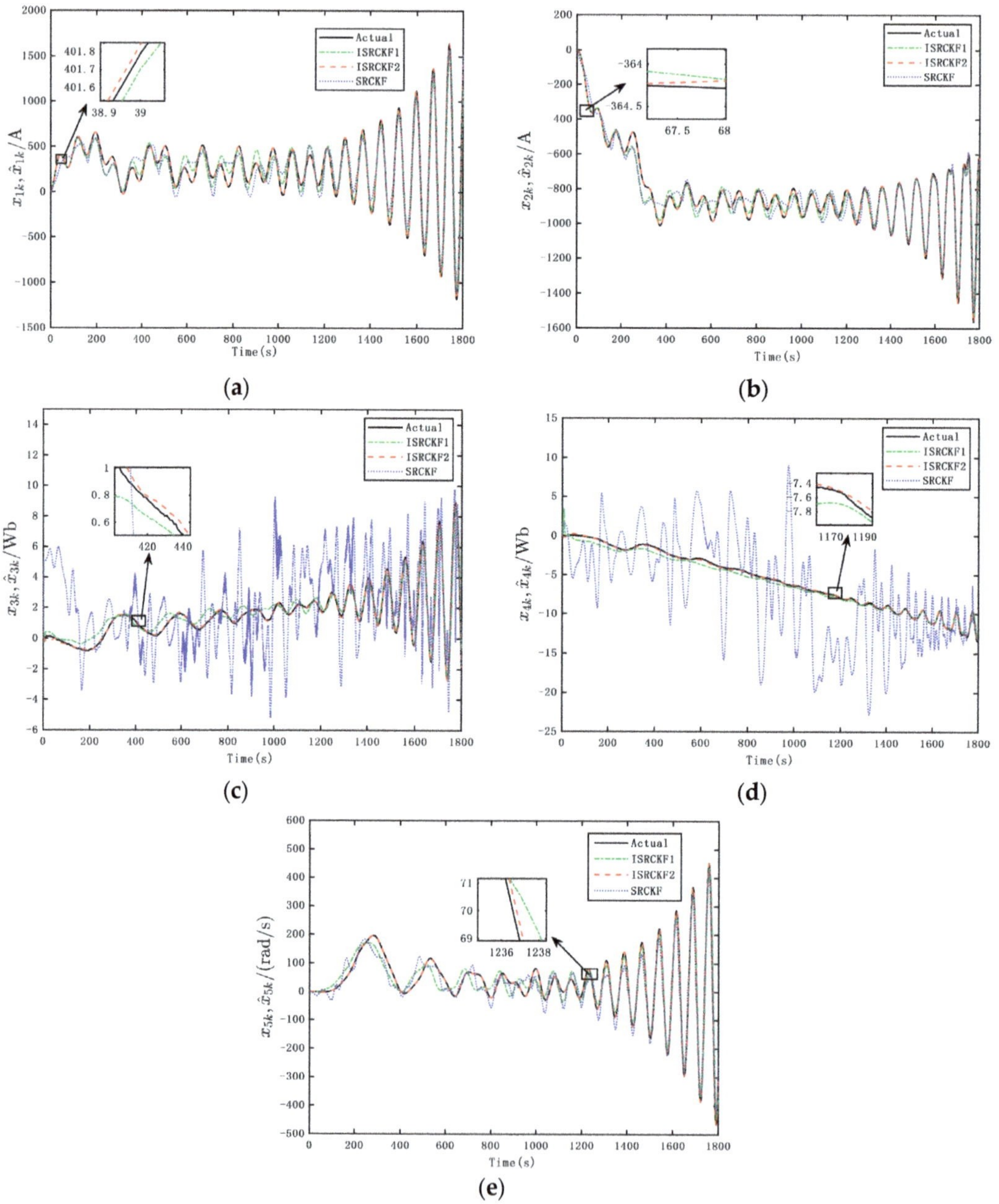


**Figure 6.** Estimation comparison of the system states $x_k$ under proposed algorithm and classical algorithm: (**a**) Stator current a; (**b**) Stator current b; (**c**) Rotor flux a; (**d**) Rotor flux b; (**e**) Angular speed.

The RMSE of the system state estimates obtained by applying the two proposed algorithms is shown in Table 6. Using numerical analysis of RMSE in Tables 4–6, it can be seen that for the direct feedthrough nonlinear discrete system with dual unknown inputs, both algorithms given in this paper can effectively achieve the estimation for the system states and the dual unknown inputs, which indicates that the ISRCKF algorithm is robust to the influence of unknown input signals, and ISRCKF2 avoids the cumulative error caused by the complex iterative steps as compared to ISRCKF1, so the state and the dual unknown inputs are estimated with higher accuracy. The above simulation examples show that the algorithm designed in this paper can still give higher accuracy state estimation results when dealing with nonlinear discrete systems affected by different unknown inputs, indicating that the system requirements for its application are more relaxed and the scope of application is broader.

## 6. Discussion

In this paper, two filtering algorithms are proposed to estimate the states and dual unknown inputs simultaneously. ISRCKF1 comprehensively considers the interaction between variables when dealing with nonlinear systems with dual unknown inputs and requires relatively weak preconditions and fewer constraints in its application, which makes it applicable to a wider range and is crucial for solving complex systems in the real world. However, due to the constant updating of the innovation, the complexity of the algorithm increases, resulting in a higher computation amount. In addition, the uncertainty of coupling terms between variables further challenges the algorithm, accumulating coupling term errors and affecting the estimation accuracy of the algorithm.

Based on ISRCKF1, we propose ISRCKF2. Compared with ISRCKF1, ISRCKF2 removes the coupling terms between the variables, but this requires additional constraints and certain assumptions in the application, i.e., the inverse of the matrix $\boldsymbol{C}_k$ is required to exist and to satisfy $\boldsymbol{C}_k\boldsymbol{C}_k^{-1} = \boldsymbol{I}_n$, which narrows down the applicability of ISRCKF2 to a certain extent, and it also requires future work to achieve our aim and solve our difficulties. Compared with ISRCKF1, ISRCKF2 has more concise iteration steps and less computation as it establishes a unified innovation feedback, while avoiding the coupling term error and improving in estimation accuracy.

It is assumed that the parameters of the system model are known in this paper. For the system with unknown parameters, the recursive generalized extended parameter estimation method is proposed in [40], and it is proposed in [41] that the model parameters can be obtained from the observed data by some identification methods such as the gradient algorithm and the method of least squares, which provides an important direction for us to generalize for the study of the system with unknown parameters in the future. The estimation of unknown inputs requires a more comprehensive consideration of their possible types of variation and the accuracy of the estimates, especially for the case of high-frequency unknown inputs. References [42,43] studies and discusses the impact of the characteristics of unknown inputs on the performance of estimators, using reduced order Das and Ghosal observer. In the case of unknown inputs, we will explore in more depth their types and limitations, as well as the relevant provisions for estimating the derivatives of unknown inputs in continuous systems. This will help to more accurately assess the performance of the estimator and to better cope with different types of unknown inputs.

## 7. Conclusions

In this paper, we focus on the problem of dual unknown input estimation and state estimation for direct feedthrough nonlinear discrete stochastic systems, extending the traditional SRCKF by proposing two new filtering algorithms to make the applied system more compatible with the practical context. The filtering algorithm is built on the minimum variance unbiased estimation criterion, and by designing innovation to deal with the dual unknown inputs, it can not only solve the problem that the traditional nonlinear filter filtering accuracy decreases or even cannot be used when the unknown inputs in the system state and measurement equations are different input signals, but also achieve the optimal estimation of the state and the unknown inputs simultaneously, even when the unknown inputs in the system state and measurement equations are the same input signals. Finally, through the simulation experiments on the fifth-order induction motor system, the designed filter can complete the estimation task better, which proves the effectiveness of the proposed algorithm.

It should be acknowledged that there are some limitations to this article. The strong assumption conditions satisfied by ISRCKF2 limit its application scope. In addition, this paper mainly focuses on certain common types of unknown input signals affecting state estimation. However, due to the complexity and uncertainty of unknown input signals, different unknown input signals may have different degrees of influence on the state estimation or even the state itself, which is still a difficult problem for the state estimation of nonlinear systems containing unknown inputs. Therefore, how to further weaken the assumptions of ISRCKF2 to expand the application scope, as well as how to design controllers to compensate the impact of the dual unknown inputs on the system state need to be further researched.

**Author Contributions:** Conceptualization, Z.L. and N.W.; methodology, Z.L. and N.W.; software, Z.L.; validation, Z.L., N.W. and S.D.; formal analysis, N.W.; investigation, Z.L.; resources, N.W.; data curation, Z.L.; writing—original draft preparation, Z.L.; writing—review and editing, Z.L. and N.W.; visualization, Z.L. and S.D.; supervision, N.W.; project administration, N.W.; funding acquisition, N.W. All authors have read and agreed to the published version of the manuscript.

**Funding:** This research was funded by the National Natural Science Foundation of China (61703221) and the Natural Science Foundation of Shandong Province (ZR2016FP10).

**Data Availability Statement:** The data presented in this study are available on request from the corresponding author.

**Conflicts of Interest:** The authors declare no conflicts of interest.

## Abbreviations

The following abbreviations are used in this manuscript:

| | |
|---|---|
| EKF | Extended Kalman filter |
| UKF | Unscented Kalman filter |
| CKF | Cubature Kalman filter |
| SRCKF | Square-root cubature Kalman filter |
| ISRCKF | Improved square-root cubature Kalman filter |
| MVUE | Minimum variance unbiased estimation |
| NUIO | Nonlinear unknown input observer |
| RMSE | Root mean square error |

## Appendix A

The specific derivation of Equation (43) is as follows:

$$\begin{aligned}
P^x_{k|k} &= E(\widetilde{x}_{k|k}\widetilde{x}^T_{k|k}) \\
&= E\Big\{[\widetilde{x}^*_{k|k-1} - K_k(C_k\widetilde{x}^*_{k|k-1} + H_k\widetilde{l}_{k|k} + v_k)][\widetilde{x}^*_{k|k-1} - K_k(C_k\widetilde{x}^*_{k|k-1} + H_k\widetilde{l}_{k|k} + v_k)]^T\Big\} \\
&= E\Big\{[(I_n - K_kC_k)\widetilde{x}^*_{k|k-1} - K_kH_k\widetilde{l}_{k|k} - K_kv_k][\widetilde{x}^{*T}_{k|k-1}(I_n - C^T_kK^T_k) - \widetilde{l}^T_{k|k}H^T_kK^T_k - v^T_kK^T_k]\Big\} \\
&= (I_n - K_kC_k)E(\widetilde{x}^*_{k|k-1}\widetilde{x}^{*T}_{k|k-1})(I_n - C^T_kK^T_k) - (I_n - K_kC_k)E(\widetilde{x}^*_{k|k-1}\widetilde{l}^T_{k|k})H^T_kK^T_k - (I_n - K_kC_k)E(\widetilde{x}^*_{k|k-1}v^T_k)K^T_k \\
&\quad - K_kH_kE(\widetilde{l}_{k|k}\widetilde{x}^{*T}_{k|k-1})(I_n - C^T_kK^T_k) + K_kH_kE(\widetilde{l}_{k|k}\widetilde{l}^T_{k|k})H^T_kK^T_k + K_kH_kE(\widetilde{l}_{k|k}v^T_k)K^T_k - K_kE(v_k\widetilde{x}^{*T}_{k|k-1})(I_n - C^T_kK^T_k) \\
&\quad + K_kE(v_k\widetilde{l}^T_{k|k})H^T_kK^T_k + K_kE(v_kv^T_k)K^T_k
\end{aligned} \tag{A1}$$

Combining Equations (36) and (21), and known conditions, substituting into Equation (A1) yields

$$P^x_{k|k} = (I_n - K_kC_k)P^{x*}_{k|k-1}(I_n - C_k^TK_k^T) + K_kH_kP^l_{k|k}H_k^TK_k^T + K_kR_kK_k^T - (I_n - K_kC_k)E(\widetilde{x}^*_{k|k-1}\widetilde{l}^T_{k|k})H_k^TK_k^T$$
$$-(I_n - K_kC_k)E(\widetilde{x}^*_{k|k-1}v_k^T)K_k^T - K_kH_kE(\widetilde{l}_{k|k}\widetilde{x}^{*T}_{k|k-1})(I_n - C_k^TK_k^T) + K_kH_kE(\widetilde{l}_{k|k}v_k^T)K_k^T \quad \text{(A2)}$$
$$-K_kE(v_k\widetilde{x}^{*T}_{k|k-1})(I_n - C_k^TK_k^T) + K_kE(v_k\widetilde{l}^T_{k|k})H_k^TK_k^T$$

Next calculate $E(\widetilde{x}^*_{k|k-1}\widetilde{l}^T_{k|k})$, $E(\widetilde{x}^*_{k|k-1}v_k^T)$, $E(\widetilde{l}_{k|k}\widetilde{x}^{*T}_{k|k-1})$, $E(\widetilde{l}_{k|k}v_k^T)$, $E(v_k\widetilde{x}^{*T}_{k|k-1})$ and $E(v_k\widetilde{l}^T_{k|k})$ to get $P^x_{k|k}$. According to Equations (20) and (13), we get

$$\begin{aligned}
&E(\widetilde{l}_{k|k}v_k^T)\\
&= -N_kE(\Delta_kv_k^T)\\
&= -N_kE\left\{[C_k(f(x_{k-1}) - \hat{x}_{k|k-1} + w_{k-1}) + v_k]v_k^T\right\}\\
&= -N_kR_k
\end{aligned} \quad \text{(A3)}$$

Similarly, it can be obtained that $E(v_k\widetilde{l}^T_{k|k}) = -R_kN_k^T$. According to Equations (20), (13), (11), (38), (35), (31), and (A3), it is obtained that

$$\begin{aligned}
&E(\widetilde{x}^*_{k|k-1}\widetilde{l}^T_{k|k})\\
&= -E[(f(x_{k-1}) - \hat{x}_{k|k-1} + w_{k-1} + G_{k-1}\widetilde{d}_{k-1|k})\Delta_k^T]N_k^T\\
&= -E\left\{[f(x_{k-1}) - \hat{x}_{k|k-1} + w_{k-1} + G_{k-1}\widetilde{d}_{k-1|k}][(f(x_{k-1}) - \hat{x}_{k|k-1} + w_{k-1})^TC_k^T + v_k^T]\right\}N_k^T\\
&= -E\left\{[f(x_{k-1}) - \hat{x}_{k|k-1} + w_{k-1}][f(x_{k-1}) - \hat{x}_{k|k-1} + w_{k-1}]^TC_k^TN_k^T\right\} - E\left\{G_{k-1}\widetilde{d}_{k-1|k}[f(x_{k-1}) - \hat{x}_{k|k-1} + w_{k-1}]^TC_k^TN_k^T\right\}\\
&-E(G_{k-1}\widetilde{d}_{k-1|k}v_k^TN_k^T)\\
&= -P^x_{k|k-1}C_k^TN_k^T - P^{d*x}_{k|k-1}C_k^TN_k^T + G_{k-1}E[(M_k\Delta_k + M_kH_k\widetilde{l}_{k|k})v_k^T]N_k^T\\
&= -(P^x_{k|k-1} + P^{d*x}_{k|k-1})C_k^TN_k^T + G_{k-1}M_kE(\Delta_kv_k^T)N_k^T + G_{k-1}M_kH_kE(\widetilde{l}_{k|k}v_k^T)N_k^T\\
&= -(P^x_{k|k-1} + P^{d*x}_{k|k-1})C_k^TN_k^T + G_{k-1}M_kR_kN_k^T - G_{k-1}M_kH_kN_kR_kN_k^T
\end{aligned} \quad \text{(A4)}$$

Similarly, we get $E(\widetilde{l}_{k|k}\widetilde{x}^{*T}_{k|k-1}) = -N_kC_k(P^x_{k|k-1} + P^{xd*}_{k|k-1}) + N_kR_kM_k^TG_{k-1}^T - N_kR_kN_k^TH_k^TM_k^TG_{k-1}^T$. According to equations (35), (31), (13), and (A3), it is obtained that

$$\begin{aligned}
&E(\widetilde{x}^*_{k|k-1}v_k^T)\\
&= E[(f(x_{k-1}) - \hat{x}_{k|k-1} + w_{k-1} + G_{k-1}\widetilde{d}_{k-1|k})v_k^T]\\
&= G_{k-1}E(\widetilde{d}_{k-1|k}v_k^T)\\
&= G_{k-1}E[(-M_k\Delta_k - M_kH_k\widetilde{l}_{k|k})v_k^T]\\
&= -G_{k-1}M_kE(\Delta_kv_k^T) - G_{k-1}M_kH_kE(\widetilde{l}_{k|k}v_k^T)\\
&= -G_{k-1}M_kR_k + G_{k-1}M_kH_kN_kR_k
\end{aligned} \quad \text{(A5)}$$

Similarly, we get $E(v_k\widetilde{x}^{*T}_{k|k-1}) = -R_kM_k^TG_{k-1}^T + R_kN_k^TH_k^TM_k^TG_{k-1}^T$. Finally, we substitute Equations (A3)–(A5) and the above results into Equation (A2) and then organize to obtain Equation (43). In summary, the derivation process of Equation (43) has been proved.

**Appendix B**

After performing 100 Monte Carlo simulation experiments, Figures A1 and A2 show the state RMSE time evolution curves of the proposed algorithms for the two cases.

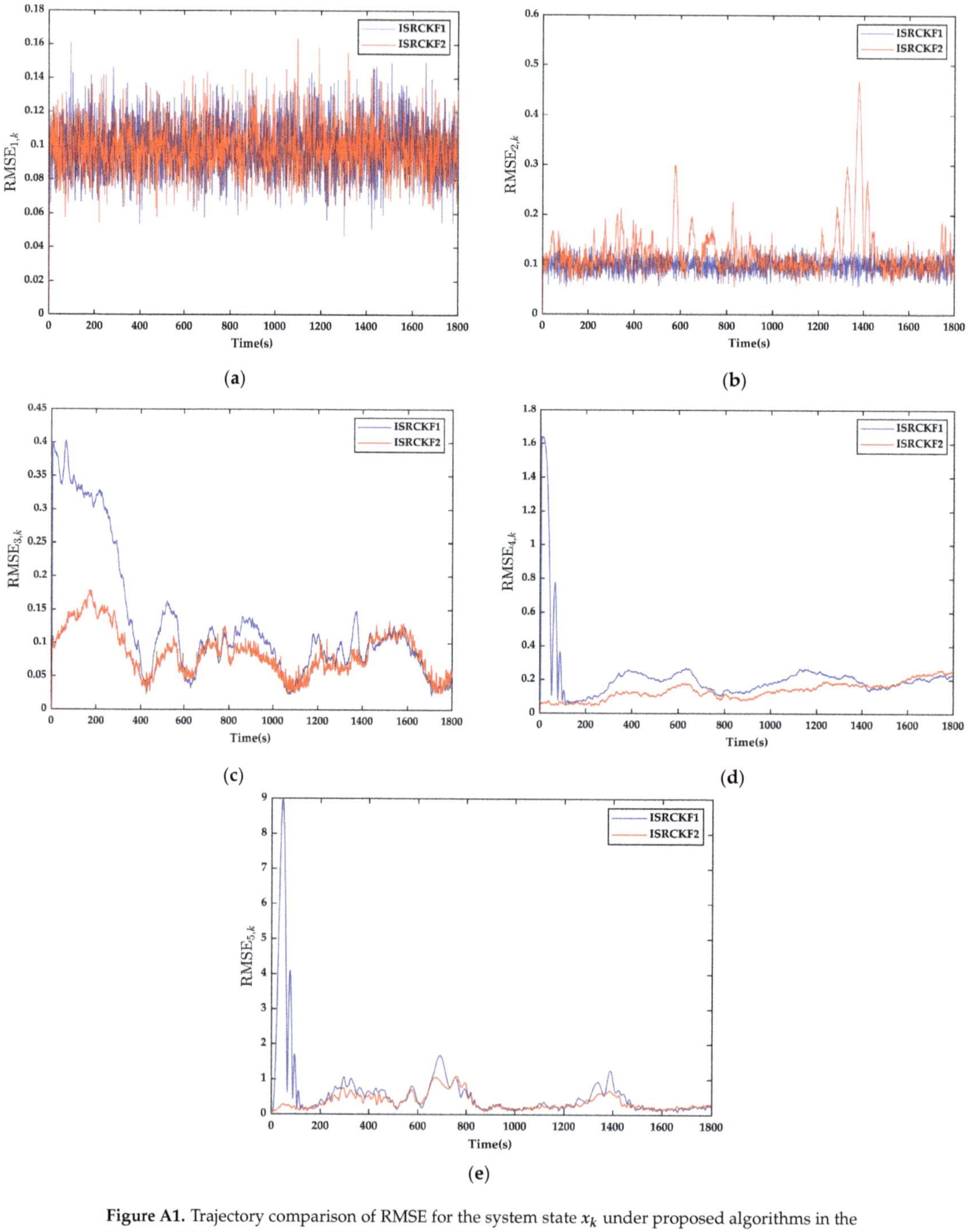


**Figure A1.** Trajectory comparison of RMSE for the system state $x_k$ under proposed algorithms in the first case: (**a**) Stator current a; (**b**) Stator current b; (**c**) Rotor flux a; (**d**) Rotor flux b; (**e**) Angular speed.

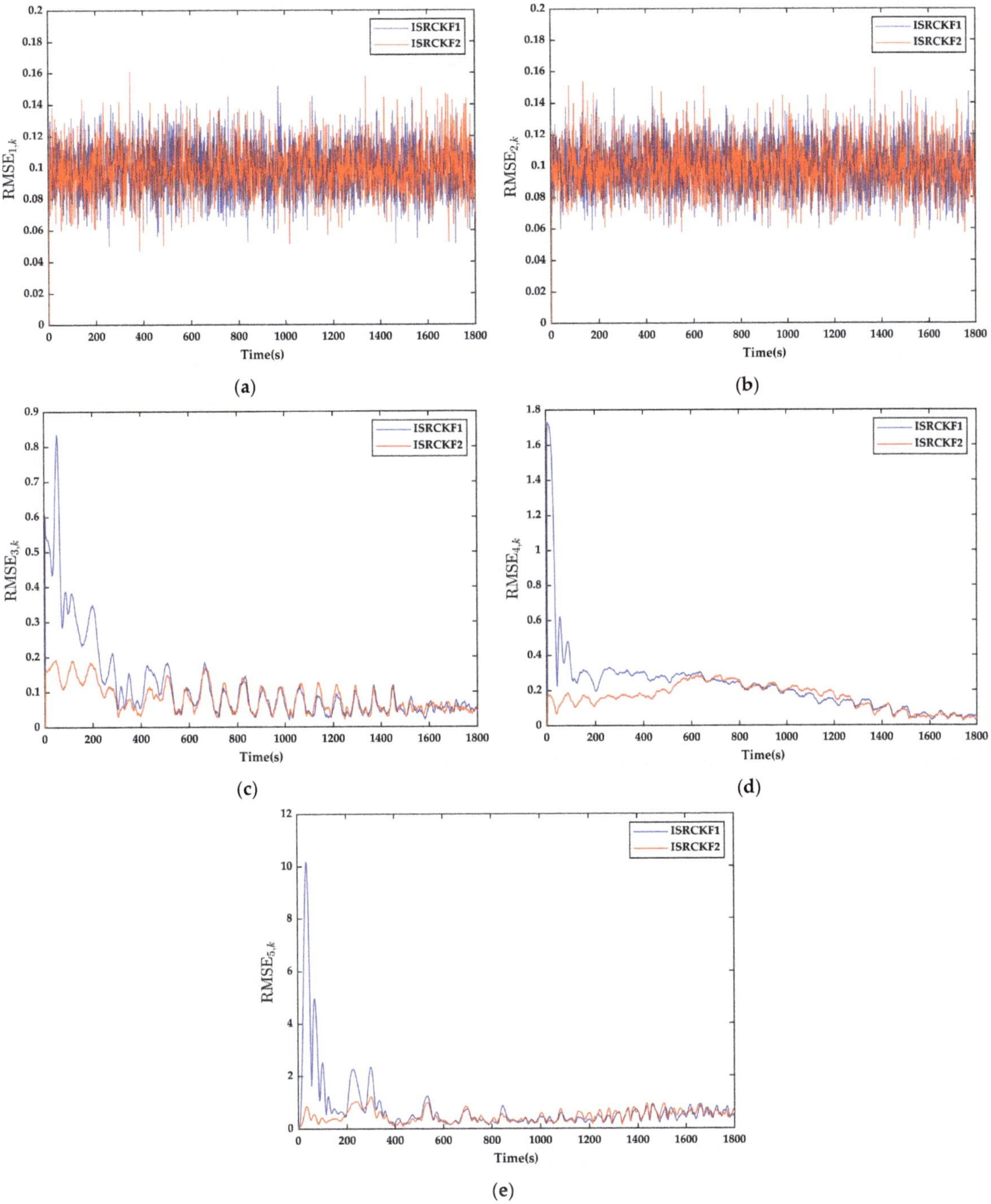


**Figure A2.** Trajectory comparison of RMSE for the system state $x_k$ under proposed algorithms in the second case: (**a**) Stator current a; (**b**) Stator current b; (**c**) Rotor flux a; (**d**) Rotor flux b; (**e**) Angular speed.

## References


1. Kitanidis, P.K. Unbiased minimum-variance linear state estimation. *Automatica* **1987**, *23*, 775–778. [CrossRef]
2. Darouach, M.; Zasadzinski, M. Unbiased minimum variance estimation for systems with unknown exogenous inputs. *Automatica* **1997**, *33*, 717–719. [CrossRef]
3. Gillijns, S.; De Moor, B. Unbiased minimum-variance input and state estimation for linear discrete-time systems. *Automatica* **2007**, *43*, 111–116. [CrossRef]
4. Lu, P.; Kampen, E.V.; Visser, C.D.; Chu, Q.P. Framework for state and unknown input estimation of linear time-varying systems. *Automatica* **2016**, *73*, 145–154. [CrossRef]
5. Delyon, B.; Zhang, Q.H. On the optimality of the Kitanidis filter for state estimation rejecting unknown inputs. *Automatica* **2021**, *132*, 109793. [CrossRef]
6. Xiao, M.L.; Zhang, Y.B.; Fu, H.M.; Wang, Z.H. Nonlinear unbiased minimum-variance filter for Mars entry autonomous navigation under large uncertainties and unknown measurement bias. *ISA Trans.* **2018**, *76*, 97–109. [CrossRef] [PubMed]
7. Asgari, M.; Khaloozadeh, H. Robust extended Kalman filtering for non-linear systems with unknown input: A UBB model approach. *IET Radar Sonar Navig.* **2020**, *14*, 1837–1844. [CrossRef]
8. Zhang, D.; Shen, Z.X.; Song, Y.D. Robust adaptive fault-tolerant control of nonlinear uncertain systems tracking uncertain target trajectory. *Inf. Sci.* **2017**, *415*, 446–460. [CrossRef]
9. Luo, Z.; Fang, H.J. Fault detection for non-linear system with unknown input and state constraints. *IET Signal Process.* **2013**, *7*, 800–806. [CrossRef]
10. Huang, J.S.; Li, X.Z.; Yang, X.J.; Zheng, Z.P.; Lei, Y. Experimental validation of the proposed extended Kalman filter with unknown inputs algorithm based on data fusion. *J. Low Freq. Noise Vib. Active Control* **2020**, *39*, 835–849. [CrossRef]
11. Meyer, L.; Ichalal, D.; Vigneron, V. An unknown input extended Kalman filter for nonlinear stochastic systems. *Eur. J. Control* **2020**, *56*, 51–61. [CrossRef]
12. Athans, M.; Wishner, R.P.; Bertolini, A. Suboptimal State Estimation for Continuous-time Nonlinear Systems from Discrete Noisy Measurements. *IEEE Trans. Autom. Control* **1968**, *13*, 504–514. [CrossRef]
13. Luo, Z.; Fang, H.J. Fault Detection for Nonlinear Systems with Unknown Input. *Asian J. Control* **2013**, *15*, 1503–1509. [CrossRef]
14. Julier, S.J.; Uhlmann, J.K.; Durrant-Whyte, H.F. A New Method for the Nonlinear Transformation of Means and Covariances in Filters and Estimators. *IEEE Trans. Autom. Control* **2000**, *45*, 477–482. [CrossRef]
15. Lefebvre, T.; Bruyninckx, H.; De Schutter, J. Comment on "A New Method for the Nonlinear Transformation of Means and Covariances in Filters and Estimators". *IEEE Trans. Autom. Control* **2002**, *47*, 1406–1409. [CrossRef]
16. Liu, X.H.; Guan, J.W.; Jiang, R.; Gao, X.; Chen, B.D.; Ge, S.S. Robust strong tracking unscented Kalman filter for non-linear systems with unknown inputs. *IET Signal Process.* **2022**, *16*, 351–365. [CrossRef]
17. Arasaratnam, I.; Haykin, S. Cubature Kalman Filters. *IEEE Trans. Autom. Control* **2009**, *54*, 1254–1269. [CrossRef]
18. Ding, J.L.; Xiao, J.; Zhang, Y.; Iqbal, M.N. Distributed State Estimation for Discrete-Time Nonlinear System with Unknown Inputs. *Circuits Syst. Signal Process.* **2014**, *33*, 3421–3441. [CrossRef]
19. Zhao, W.; Li, H.G.; Zou, L.Y.; Huang, W.J. Nonlinear Unknown Input Observer Based on Singular Value Decomposition Aided Reduced Dimension Cubature Kalman Filter. *Math. Probl. Eng.* **2017**, *2017*, 1267380. [CrossRef]
20. Pan, S.W.; Xiao, D.; Xing, S.T.; Law, S.S.; Du, P.Y.; Li, Y.J. A general extended Kalman filter for simultaneous estimation of system and unknown inputs. *Eng. Struct.* **2016**, *109*, 85–98. [CrossRef]
21. Xiao, Q.; Fu, H.M.; Wang, Z.H.; Zhang, Y.B.; Wu, Y.Z. The stability analysis of the adaptive three-stage Kalman filter. *Signal Process.* **2016**, *118*, 1–24. [CrossRef]
22. Jahanian, M.; Ramezani, A.; Moarefianpour, A.; Shoorehdeli, M.A. Robust extended Kalman filtering for nonlinear systems in the presence of unknown inputs and correlated noises. *Optim. Control Appl. Meth.* **2022**, *43*, 243–256. [CrossRef]
23. Xiao, M.L.; Zhang, Y.B.; Wang, Z.H.; Fu, H.M. An adaptive three-stage extended Kalman filter for nonlinear discrete-time system in presence of unknown inputs. *ISA Trans.* **2018**, *75*, 101–117. [CrossRef] [PubMed]
24. Xiao, M.L.; Zhang, Y.B.; Fu, H.M. Three-stage unscented Kalman filter for state and fault estimation of nonlinear system with unknown Input. *J. Frankl. Inst.* **2017**, *354*, 8421–8443. [CrossRef]
25. Lin, J.X.; Jiang, G.P.; Gao, Z.F.; Rong, L.N. State and input simultaneous estimation for discrete-time switched singular delay systems with missing measurements. *Int. J. Robust Nonlinear Control* **2017**, *27*, 2749–2772. [CrossRef]
26. Li, Y.Y.; Liu, S.; Zhong, M.Y.; Ding, S.X. State estimation for stochastic discrete-time systems with multiplicative noises and unknown inputs over fading channels. *Appl. Math. Comput.* **2018**, *320*, 116–130. [CrossRef]
27. Li, Z.J.; Sun, M.X.; Duan, Q.W.; Mao, Y. Robust State Estimation for Uncertain Discrete Linear Systems with Delayed Measurements. *Mathematics* **2022**, *10*, 1365. [CrossRef]
28. Fu, H.S.; Cui, B.T.; Zhuang, B.; Zhang, J.Z. State Estimation for a Class of Distributed Parameter Systems with Time-Varying Delay over Mobile Sensor-Actuator Networks with Missing Measurements. *Mathematics* **2021**, *9*, 661. [CrossRef]
29. Nejati, Z.; Abedi, M.; Faraji, A. Modified three stage Kalman filtering for stochastic non-linear systems with randomly occurring faults and intermittent measurements. *IET Control Theory Appl.* **2022**, *16*, 674–697. [CrossRef]
30. Feng, X.X.; Li, S.H.; Pan, F. State estimate for stochastic systems with dual unknown interference inputs. *Chin. J. Aeronaut.* **2020**, *33*, 2395–2407. [CrossRef]

31. Zarei, J.; Shokri, E. Robust sensor fault detection based on nonlinear unknown input observer. *Measurement* **2014**, *48*, 355–367. [CrossRef]
32. Cui, X.Y.; Jing, Z.; Luo, M.J.; Guo, Y.Z.; Qiao, H.M. A New Method for State of Charge Estimation of Lithium-Ion Batteries Using Square Root Cubature Kalman Filter. *Energies* **2018**, *11*, 209. [CrossRef]
33. Liu, J.B.; Gnanasekar, A.; Zhang, Y.; Bo, S.; Liu, J.F.; Hu, J.T.; Zou, T. Simultaneous state and parameter estimation: The role of sensitivity analysis. *Ind. Eng. Chem. Res.* **2021**, *60*, 2971–2982. [CrossRef]
34. Gillijns, S.; De Moor, B. Unbiased minimum-variance input and state estimation for linear discrete-time systems with direct feedthrough. *Automatica* **2007**, *43*, 934–937. [CrossRef]
35. Ding, F. State filtering and parameter estimation for state space systems with scarce measurements. *Signal Process.* **2014**, *104*, 369–380. [CrossRef]
36. Zhang, X.; Ding, F. Hierarchical parameter and state estimation for bilinear systems. *Int. J. Syst. Sci.* **2020**, *51*, 275–290. [CrossRef]
37. Zhang, X.; Ding, F.; Yang, E.F. State estimation for bilinear systems through minimizing the covariance matrix of the state estimation errors. *Int. J. Adaptive Control Signal Process.* **2019**, *33*, 1157–1173. [CrossRef]
38. Marino, R.; Peresada, S.; Valigi, P. Adaptive Input-output Linearizing Control of Induction Motors. *IEEE Trans. Autom. Control* **1993**, *38*, 208–221. [CrossRef]
39. Xiong, K.; Zhang, H.Y.; Chan, C.W. Performance evaluation of UKF-based nonlinear filtering. *Automatica* **2006**, *42*, 261–270. [CrossRef]
40. Ding, F.; Xu, L.; Zhang, X.; Zhou, Y.H. Filtered auxiliary model recursive generalized extended parameter estimation methods for Box-Jenkins systems by means of the filtering identification idea. *Int. J. Robust Nonlinear Control* **2023**, *33*, 5510–5535. [CrossRef]
41. Hu, C.; Ji, Y.; Ma, C.Q. Joint two-stage multi-innovation recursive least squares parameter and fractional-order estimation algorithm for the fractional-order input nonlinear output-error autoregressive model. *Int. J. Adapt. Control Signal Process.* **2023**, *37*, 1650–1670. [CrossRef]
42. Banerjee, A.; Das, G. Estimation of unknown input using reduced order Das and Ghosal observer. In Proceedings of the 2013 IEEE International Conference on Emerging Trends in Computing, Communication and Nanotechnology, Tirunelveli, India, 25–26 March 2013.
43. Banerjee, A.; Mondal, P.P.; Das, G. Construction of full order observer for linear time invariant systems using generalized matrix inverse. In Proceedings of the 2013 IEEE Conference on Information and Communication Technologies, Thuckalay, India, 11–12 April 2013.

applied sciences

MDPI

*Article*

# The Chinese Aviation Network: An Empirical Temporal Analysis on Its Structural Properties and Robustness

**Ruoshi Yang [1], Wei Sun [2], Meilong Le [1,*] and Hongyan Zhang [2]**

1 College of Civil Aviation, Nanjing University of Aeronautics and Astronautics, Nanjing 210016, China; yangruoshi@nuaa.edu.cn
2 School of Information Science and Technology, Hainan Normal University, Haikou 571158, China; cs.weisun@foxmail.com (W.S.); hongyan@hainnu.edu.cn (H.Z.)
* Correspondence: lemeilong@126.com; Tel.: +86-13918976990

**Abstract:** Complex networks have encouraged scholars to develop an effective method for abstracting and optimizing aviation networks. However, researchers often overlook the aviation network's temporal attribute and treat it as a static network. Aviation networks have strong temporal characteristics and the dynamic connection cannot be realistically described by a static network. It is necessary to more accurately and realistically represent these connections during the operation of an aviation network. This study explored temporal structures of the Chinese aviation temporal network (CATN) based on flight schedules and actual operational time data. Temporal networks based on time windows were represented to analyze the temporal topology features and robustness of the CATN. The results demonstrated the following: (1) based on the spatial-temporal aviation network, there is a morning departure peak (7:00–8:00) and an evening arrival peak at the airline hub (20:00–21:00); (2) examining the centrality of each airport in the CATN at various time intervals exposed fluctuations in their rankings, which could not be identified by a static network, and (3) the robustness of the CATN was found to be unaffected by time windows, but it displayed poor resilience against deliberate attacks, particularly when subjected to betweenness and closeness attacks, which target the network's shortest paths. For obtaining a greater understanding of the operating situation of civil aviation, displaying the topological features and robustness of the temporal network is of great importance.

**Keywords:** Chinese aviation temporal network; temporal network; topological structure; complex networks, robustness

## 1. Introduction

The aviation network functions as a typical complex system, with its subsystems that interact to generate functionality rather than being created by a specific subsystem [1–3]. Complex network theory enhances our understanding of the functionality and behavior of the aviation transportation system. The aviation network consists of nodes and edges, with nodes that represent airports or cities where airports are located and edges that represent connections between airports [4]. To model the US aviation network, Li-Ping [5] utilized a directed weighted network and analyzed its degree distribution and clustering coefficient. Li and Cai [6] analyzed the Chinese aviation network (CAN) with 128 airport nodes and 1165 flights, and found that its average shortest path was 2.067 and its clustering coefficient was 0.733, indicative of a small-world network classification. Barrat [7] examined the global aviation-weighted network and investigated the correlation between weight and the underlying topological structures of the network. Gautreau [8] analyzed the interaction between macroscopic and microscopic airport network characteristics in the US. Empirical studies demonstrate that both Chinese and global aviation networks have small-world properties, with short average shortest paths and high clustering coefficients. Both networks exhibit a double power-law distribution pattern in their respective degree distributions [9–12].

The majority of studies on aviation networks have concentrated on examining their static structures, overlooking the correlations present across various time periods. In static networks, the relationship between nodes depends only on the existence of a flight route, which disregards any temporal data such as departure and arrival times or delays. This oversimplified static abstraction leads to overestimating the connections among nodes and distorts the understanding of the system's structure. Temporal networks are characterized by the connections among nodes which vary over time periods [13]. Integration of the time dimension into the network model results in more complex interrelationships and presents novel challenges to static network theory.

Early researchers described temporal networks using weighted networks, with edge weight signifying the number of incidences over the entire time period. Nevertheless, this model considers an impractical assumption that the connections between pairs of nodes are randomly distributed within the time period [14]. To evaluate the presence of directed edges between nodes and to establish reachability graphs, Moody et al. [15] introduced the concept of temporal reachability. However, this model is only suitable for sparse networks. To overcome this drawback, several studies [16,17] discretized aggregated temporal networks into snapshot sequences using time windows, where the interactions between nodes within a time window belong to the same time window, and each snapshot within the time window is independent of the others. By correctly mapping temporal networks into static networks, the research characterized changes in network features such as the degree, betweenness, and clustering coefficient. Although these methods are broadly accepted in temporal network research, the selection of an appropriate time window size remains a crucial challenge. A small time window setting results in numerous invalid empty network structures, whereas using the entire observation window as the time window renders the temporal network static, leading to the loss of temporal information. Furthermore, these models ignore continuity in time when the interaction between nodes spans multiple time windows, resulting in inaccurate temporal information.

In recent years, there has been a growing trend among scholars to utilize temporal networks for the aviation transportation network. Ralvi et al. [18,19] captured the complexity within specific time windows and introduced corresponding complexity metrics to understand the dynamic evolution of complexity within these windows. Moreover, they introduced the concept of individual aircraft complexity to assess individual contributions and proposed a method for defining complex communities that evaluate the impact of various decisions on the overall system. To better represent the temporal relationship between flights, Mikko et al. [20] integrated temporal networks into the event graph framework to investigate the characteristics of these networks. Humberto and Lilian [21] analyzed the robustness of temporal networks in light of this framework.

This study aimed to model the CAN as a temporal network and to analyze its temporal topology to better understand the structural characteristics that determine the network's resistance to random and deliberate attacks. The researchers analyzed the robustness of the Chinese aviation temporal network (CATN) and provided insights into the obtained results. The remainder of the paper is organized as follows: Section 2 presents the methods used for building and researching the CATN. In Section 3, the theory of temporal networks was utilized to conduct an empirical analysis of the CATN. Section 4 further investigates the robustness of the temporal network model. The conclusions of the research are presented in Section 5.

## 2. Modeling and Research Methods

This study employed complex networks as a modeling tool to capture the dynamic behavior of the CATN and investigated its properties. Robustness measurements were utilized to quantify the network's structure and its ability to withstand potential disruptions.

### 2.1. Modeling the CATN

A temporal network refers to a system in which clear time evolution and temporal characteristics are evident [22,23]. In this system, nodes may emerge or vanish during the process of temporal changes, and interactions among nodes may appear or disappear with the progress of time. This study proposed an aggregated network $G = (V_G, E_G, T)$ to represent the temporal network. In this method, $V_G$ defines a set of nodes, and $E_G$ denotes a set of edges interacting within a finite time period of $[0, T]$. Each edge is represented with a quadruplet $e = (i, j, t, \Delta t)$, where $i$ and $j$ serve as node identifiers, $t$ stands for the time of interaction, and $\Delta t$ signifies the duration of connectivity. Although the implementation of static network visualization for the temporal network may yield improved visualization results, this approach does not fully incorporate topological and temporal information.

Ferreira [24] suggested the use of time windows to transform interactions that take place during the same time window into a static graph. A snapshot-based temporal graph [25] can map dynamic temporal aggregation networks ($G$) to a time sequence of static graphs $(G_1, G_2, \ldots, G_n)$. This method assists us in processing temporal information contained in the network's quadruplets. By defining time windows symbolized by $\omega$, a temporal network with $N = |V_G|$ nodes and duration $T$ can be efficiently divided into $L = \frac{T}{\omega}$ snapshots. In this scenario, $L$ signifies the number of snapshots obtained after the division. Each snapshot captures either all the events that occur within a specific time window or events that occur at a precise moment in time, with the collected information that characterizes these events.

In the framework of the CATN, a temporal network can be established by utilizing a triplet $e = (i, j, t)$ based on flight schedules. In this notation, $i$ and $j$ denote the departure and arrival airports of the flight, respectively, while $t$ indicates its departure time. Figure 1a shows a snapshot of the aggregation network, which consists of five airports and flights. Time information indicates that the departure time at $t$ is available on the edges. By utilizing a time window of $\omega = 1$ or $\omega = 2$, one can generate temporal snapshot sequences (see Figure 1b,c).

### 2.2. Temporally Topological Measures

The aggregated temporal network and snapshot-based temporal network are two methods for mapping dynamic temporal networks into static networks. Scholars have proposed numerous concepts and models for studying the topological properties of static networks. These methods can, to some extent, reflect the topological structure of the temporal network and have been extensively employed for various forms of air transportation networks [26–29]. A novel technique has been introduced for researching the structure of temporal topology, which is based on a static graph. A network is an undirected weighted graph $G(t)$ with nodes and edges varying across time periods of $t \in \{1, 2, \cdots, t, \cdots\}$. $V(t) = \{v_1(t), v_2(t), \cdots, v_n(t)\}$ represents the set of nodes in time period $t$, and each node represents an airport, with a total of $N = |V(t)|$. $E(t) = \{e_1(t), e_2(t), \cdots, e_m(t)\}$ represents the set of routes between airports, and there are $M = |E(t)|$ routes during $t$. Each edge $e_i(t)$ connects airports $a$ and $b$. Thus, an $N \times N$ adjacency matrix, $A(t)$, is associated with the undirected weighted graph, in which $A_{ij}$ is equal to 1 when there is a route between nodes $i$ and $j$; otherwise, $A_{ij}$ will be equal to 0.

#### 2.2.1. Node Degree and Distribution

The degree of each node $v(t) \in V(t)$, denoted as $d(t, v)$, is defined as the number of edges that connect to it. To calculate the degree of a given node $v(t)$, the adjacency matrix $A(t)$ can be utilized. If the $v$th row of $A(t)$ indicates the adjacent nodes to node $v$, then the degree of node $v$ can be obtained using the following equation:

$$d(t, v) = \sum_{i=1}^{n} A_{vi}(t) \tag{1}$$

where $A_{vj}(t)$ indicates whether there is at least one flight between nodes $v$ and $j$ at time $t$, and is equal to 0 if there are no flights that connect them.

The temporal average degree of a network refers to the average number of edges connecting each node in the network at a given time $t$. Higher average degrees in a network imply that nodes within this network are more likely to be interconnected, resulting in higher overall network connectivity. A mathematical formula for the temporal average degree is as follows:

$$\bar{d}(t) = \frac{1}{N}\sum_{i=1}^{n}\sum_{j=1}^{n} A_{ij} \quad (2)$$

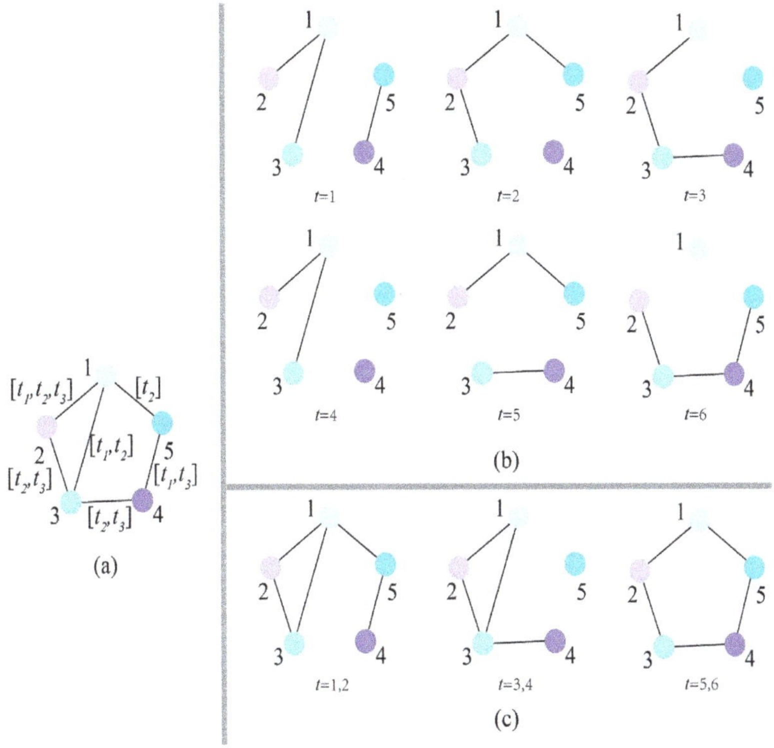


**Figure 1.** Different forms of instantaneous undirected networks: (**a**) an aggregated network, (**b**) a snapshot-based temporal network with a time window of 1 based on graph (**a**), and (**c**) a snapshot-based temporal network with a time window of 2 based on graph (**a**).

2.2.2. Temporal Average Path Length

The temporal average path length $L(t)$ refers to the average number of edges that connect any two nodes $i$ and $j$ in a network at a given time $t$. For pairs of nodes with no connection, their path length is defined as infinity. The following formula can be used to calculate $L(t)$:

$$L(t) = \frac{1}{N(N-1)}\sum_{i\neq j} l_{ij}(t) \quad (3)$$

where $l_{ij}(t)$ symbolizes the shortest path length between nodes $i$ and $j$ at time $t$. In an airline network, it represents the average number of transfers needed for any node to reach its final destination.

### 2.2.3. Clustering Coefficient

The clustering coefficient measures the possibility of two neighboring nodes being connected, and it is defined as the ratio of the actual edges between them to the number of all possible edges. The clustering coefficient quantifies the strength of intra-cluster connectivity and characterizes the closeness between nodes within a network. In particular, the clustering coefficient $C(t,v)$ of a given node $v(t)$ is computed using a function of the number of edges $n_e$ among its neighbors and $d(t,v)$, which is the degree of node $v(t)$. The function is defined as

$$C(t,v) = \begin{cases} \dfrac{2n_e}{d(t,v)(d(t,v)-1)} & \text{if } d(t,v) > 1 \\ 0 & \text{otherwise} \end{cases} \tag{4}$$

Note that, in the above equation, the condition $d(t,v) > 1$, indicates that the node $v(t)$ is connected to two or more neighboring nodes, while $\frac{d(t,v)(d(t,v)-1)}{2}$ represents the number of potential edges in the network. If the node $v(t)$ has less than or equal to one neighbor, then its clustering coefficient is equal to zero.

The average clustering coefficient $C(t)$ in a network is defined as the mean of all node clustering coefficients. It can be calculated as follows:

$$C(t) = \frac{1}{N}\sum_{i=1}^{n} C(t,i) \tag{5}$$

The clustering coefficient characterizes the density of a network and the inter-connection of nodes. A high clustering coefficient indicates a dense network with closely connected nodes, while a low clustering coefficient implies a sparser network with less connectivity between nodes.

### 2.2.4. Strength

During operational processes, aviation network traffic distribution is often non-uniform. The strength of a node is usually represented by the sum of the routes or flights. Weighted networks can be implemented to better represent and monitor such traffic. Temporal networks, likewise, can also be represented as undirected weighted graphs $G(t) = (V(t), E(t), w)$, where $V(t)$ and $E(t)$ have the same definitions as in undirected unweighted networks previously described. The function $w : E(t) \to \mathbb{R}$ maps the set of edges to the real number set $\mathbb{R}$ and denotes the weight of each edge. The adjacency matrix is defined by the following equation:

$$A_{ij}(t) = \begin{cases} w(i,j) & \text{if } (i,j) \in E(t) \\ 0 & \text{otherwise} \end{cases} \tag{6}$$

## 2.3. *Robustness*

The concept of network robustness is measured by the capability of a given network to continue to be connected, despite the failure or attack of a vertex [30]. Two main factors contributing to this problem are network attack strategies and the measurement of robustness.

### 2.3.1. Attack Strategies

Network attack methods can be categorized as either random failure or deliberate attack [31,32]. In aviation networks, random destruction resulting from non-subjective factors, such as the operation status of internal and external infrastructure within the aviation system and occasional extreme weather conditions, is classified as a random failure. The Rand function was utilized to generate random numbers, and the authors then applied the function that removes airport nodes and routes randomly. The network's performance was measured following the attack until the metric was smaller than or equal to the predetermined threshold. Exploring the effects of random failure and deliberate attacks is vital in understanding the resilience of a network and can guide the development of more efficient security mechanisms for network protection.

In addition to objective factors such as weather and equipment, the aviation network is also vulnerable to various forms of subjective interference, often in the form of biased attacks where critical nodes with higher importance are prioritized as preferred targets. The importance of a node depends on the extent of the decline in connectivity of the airline network following an attack. To assess node importance, centrality is utilized in deliberate attacks. This study employed four types of centrality, namely degree centrality (the number of connections between nodes), the nearest neighborhood degree centrality (local clustering measure), betweenness centrality (the extent of the node used as an intermediary between other nodes), and closeness centrality (the average distance between a node and all other nodes in the network). Formulas (7)–(10) show the computational methods for these centrality types, respectively:

$$C_{\mathrm{D}}(t,v) = d(t,v) \tag{7}$$

where $d(t,v)$ refers to the degree of node $v$ at time $t$, i.e., the number of edges connected to node $v$ at time $t$.

$$C_{\mathrm{D_{nn}}}(t,v) = \frac{\sum_{i \in N_i} d(t,i)}{d(t,v)} \tag{8}$$

where $N_i$ represents the set of neighboring nodes of node $i$.

$$C_{\mathrm{B}}(t,v) = \sum_{\substack{i=1 \\ i \neq v}}^{n} \sum_{\substack{j=1 \\ j \neq v}}^{n} \frac{\sigma_{ij}(t,v)}{\sigma_{ij}(t)} \tag{9}$$

where $\sigma_{ij}(t)$ stands for the number of shortest paths from node $i$ to node $j$ at time $t$, while $\sigma_{ij}(t,v)$ signifies the number of such paths that pass through node $v$.

$$C_{\mathrm{C}}(t,v) = \frac{N-1}{\sum_{\substack{i=1 \\ i \neq v}}^{n} l_{vi}(t)} \tag{10}$$

where $l_{vi}(t)$ is the length of the shortest path from node $i$ to node $v$ at time t.

### 2.3.2. Robustness Measurement

Initially, scholars typically used network performance parameters to evaluate network robustness. These parameters included the number of connected components, the maximum component size, the network diameter, the average shortest path length, the network efficiency, the clustering coefficient, and numerous centrality indicators. These indicators provide different perspectives for measuring network robustness and therefore necessitate tailored measures to be taken based on the specific problems that need to be addressed. Network efficiency and the maximum connected sub-graph proportion were used as measures of robustness.

In case a network is subjected to continuous attack, the size of the maximum connected sub-graph becomes an indispensable measure of network performance. It displays the maximum connected component that separates from the network after the attack and reflects the network connectivity. The size of the maximum connected sub-graph relative to the original network determines the proportion of retained nodes after an attack. A smaller value indicates more substantial damage to the network. The relative size can be evaluated using the following relation:

$$S(t) = \frac{N_{\mathrm{live}}}{N} \tag{11}$$

where $N_{\text{live}}(t)$ stands for the number of nodes preserved in the maximum connected subgraph following the network attack at time $t$, while $n$ symbolizes the total nodes in the original network at time $t$.

Latora [33] defined global efficiency, also known as network efficiency in this reference, as the average value of the reciprocal distances between all node pairs in a network. It indicates the convenience or difficulty of accessing one node from another, making it a fundamental measure of network connectivity. The calculation formula for global efficiency is as follows:

$$E(t) = \frac{1}{N(N-1)} \sum_{i \neq j} \frac{1}{l_{ij}(t)} \tag{12}$$

where $n$ refers to the number of nodes in the network at time $t$, and $l_{ij}(t)$ signifies the shortest distance between node $i$ and node $j$ at time $t$.

## 3. Empirical Analysis

### 3.1. Flight Data

Our research utilized data on all domestic flights in China for the year 2018, obtained from VariFlight, the leading aviation data service provider in China. Flights from Hong Kong, Macau, and Taiwan regions were excluded. These data included the most extensive information regarding the number of flights and airports in China during that year. The data consist of 11 valid attributes that are presented in Table 1. The redundancy of some redundant variables, which were not related to our research objectives, could not be used in our research and could potentially affect data processing speed. To address this issue, the researchers removed the delay information.

**Table 1.** Description of Data.

| Variable Type | Variable Name |
|---|---|
| Time Information | Departure time information: Scheduled departure time, Actual departure time |
| | Arrival time information: Scheduled arrival time, Actual arrival time |
| Delay Information | Departure delay in seconds |
| Airport Information | Landing airport information: Airport IATA code |
| | Take-off airport information: Airport IATA code |
| | Airport location information: Airport Longitude, Airport Latitude, Province, City |

During the process of collecting, storing, and retrieving data, missing values, outliers, or inconsistencies are often encountered. To ensure the accuracy and effectiveness of the data, it is necessary to analyze the missing values, outliers, and duplicates to determine the validity of the data. The authors dealt with the data that had undergone attribute processing. These data contained a total of 4,961,074 data entries, standing for the total number of flights in 2018. Of the 35,715 records, 0.07% had missing actual take-off and landing times. The purpose of processing the data was to create a temporal network. While constructing the network, this study had to consider the following constraints: the departure time of the flights is after 4:00 a.m. in the morning, and the landing time is before 4:00 a.m. the following day. Only data that met these conditions were defined as valid data.

A considerable amount of data in our available data was either missing, invalid, or redundant. Therefore, it was vital to conduct data preconditioning before data mining to improve data quality and satisfy the application requirements. The primary and most fundamental method to deal with missing data is to eliminate the samples that contain missed data. The existing data were mostly affected by missing values in actual takeoff and landing times. Upon comparison with actual operational data, it became evident that the presence of missing values was often observed in conjunction with distinct situations, such as diversion or midair returns. As a result, the missing data points were deemed invalid.

Apart from missing values, data validity needed to be assessed. Hence, to comply with the aforementioned constraints, the researchers had to remove data collected before 4:00 a.m. on 1 January 2018, because they are not classified as belonging to 2018. In addition, the data between 0:00 and 4:00 a.m. on 1 January 2019, were not available, so the researchers removed data between 4:00 a.m. on 30 December 2018 and 0:00 on 31 December 2018 to ensure data accuracy (Data from the year 2018 are inclusive of 1 January 2018, from 4:00 a.m. to 31 December 2018, until 4:00 a.m.). After the completion of the data preprocessing phase, a total of 4,915,350 valid data records were obtained. These records represent 99.08% of the initial dataset.

### 3.2. *Topological Properties of the CATN*

The CATN contains all flights that took place within China, excluding Hong Kong, Macao, and Taiwan, during the year of 2018. Each flight was defined by a quadruplet $(i, j, t, \Delta t)$, where $i$ and $j$ indicate the departure and arrival airports, respectively, $t$ denotes the departure time, and $\Delta t$ stands for the elapsed time from arrival to departure. By aggregating data from all the flights during 2018, the CATN could be expressed as a temporal network that included 230 airport nodes (i.e., $N = 230$) and a total of 3880 weighted edges that had time labels (i.e., $|E| = 3880$). Figure 2 exhibits an intuitive illustration of the CATN's configuration.

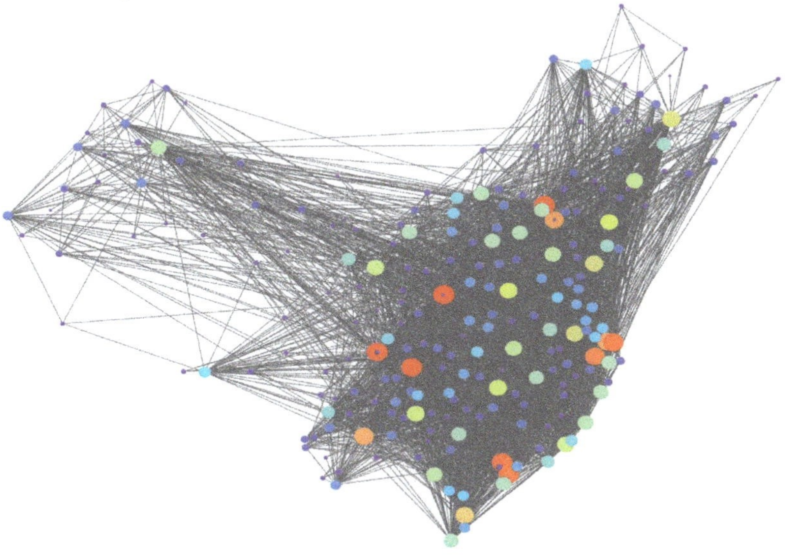

**Figure 2.** The CATN was constructed solely based on domestic civil aviation transport within China in 2018. The node's size and color stand for their respective degree, and the edges provide temporal information.

The presence of non-numerical time data stored at the edges of a temporal network makes it unsuitable for static network exploration. To avoid this situation, the authors used a snapshot-based temporal network model to investigate the topological features of dynamic networks. Our model partitioned time into hourly intervals, with a window of $\omega$ at one hour, and an overall aggregating time span of $T$ from 1 January 2018 at 4:00 a.m. to 31 December 2018 at 4:00 a.m., which resulted in the generation of 8760 snapshot networks. Our dynamic network model can thus be regarded as a sequence of static networks, in which each snapshot is analyzed to explore the topological and temporal characteristics of the CATN.

### 3.2.1. Temporal Attributes of Nodes and Edges

The temporal variation of China's airline routes throughout 2018 is displayed in Figure 3. The number of airlines in the Chinese airline network sharply increased at 6:00 and suddenly decreased at 10:00. The primary operational time of the overall network was from 6:00 to 24:00, such that 99% of flights in 97% of airports occurred during this time period. Peaks in flights occurred at 07:00 and 10:00, with each displaying an average number of flights of 510 and 530, respectively. These peaks signify a limitation in the operational capacity of the overall civil aviation network. The network was weighed based on the number of flights in each time period to investigate the changes in flight schedules, resulting in similar findings to those of the unweighted network. Unlike urban transportation, air transportation experiences long periods of sustained peak traffic [34].

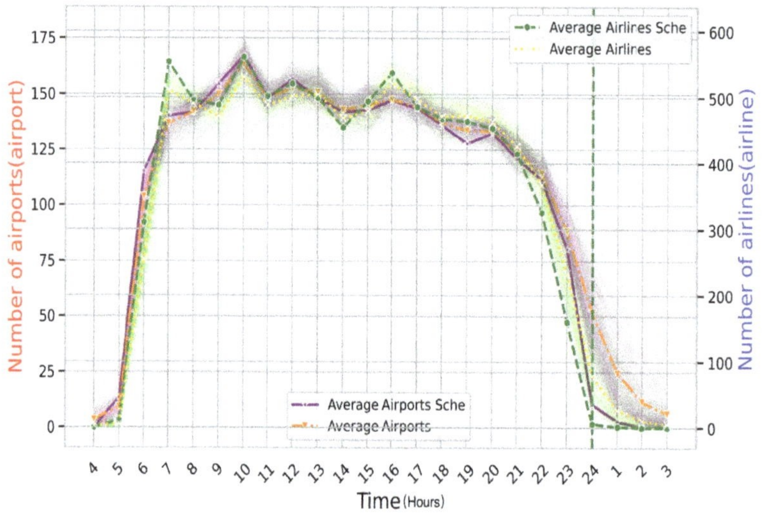


**Figure 3.** The presented graph depicts the temporal evolution of the number of airports and flight routes in the CATN. The yellow and green lines correspond to the average number of planned and actual flight routes per time period, whereas the purple and orange lines indicate the average number of planned and actual airports. The shaded regions of diverse colors illustrate the fluctuation of daily actual data around the mean value.

According to regulations of the Civil Aviation Administration, domestic airlines are advised to reduce the number of "red-eye flights", which are flights that depart and arrive after midnight. On average, there are fewer than 10 flights per day (0.1% of the total) that are scheduled to take off after midnight in the flight schedule, but in actual operations, an average of 94 flights per day (3% of the total) take off after midnight. To further explore this phenomenon, the authors analyzed the flight schedule by dividing time into hourly intervals based on scheduled departure time (Figure 4). The chart clearly shows that almost no flights are scheduled to take off after midnight in the Civil Aviation Administration's flight schedules. The researchers calculated the difference between the scheduled and actual number of flights for each hour of the year and found that there were discrepancies between 5:00 and 12:00 and between 15:00 and 16:00. This deviation led to flight delays and an increase in the number of "red-eye flights" in operation. However, during other times, the actual number of flights exceeded scheduled estimates, suggesting that airlines handled abnormal flights during those periods.

This study evaluated representative airports of various types, including domestic slot-coordinated airports such as Beijing Capital International and Xi'an Xianyang International, gateway airports such as Urumqi Diwopu International and Kunming Changshui

International, and provincial airports such as Lanzhou Zhongchuan International. Through examining the number of flights over time (Figure 5), the authors identified significant peak periods of outbound flights for Beijing, Xi'an, and Kunming, with values that dropped noticeably after 10:00. This suggested ample capacity at the operations, airspace, runways, and other support resources that could redistribute flights via ground wait and flow control after 10:00. However, Urumqi and Lanzhou had stable values, implying that their support resources such as runways and airspace were nearing saturation.

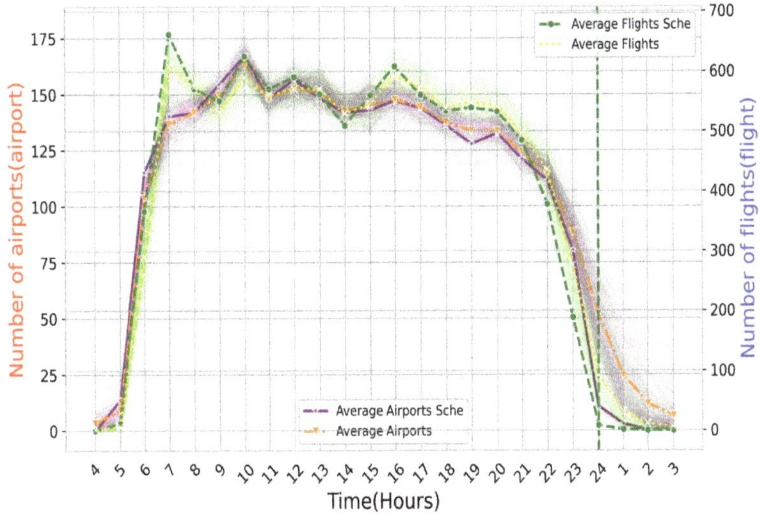


**Figure 4.** The presented figure showcases the changes in the hourly counts of airports and flights in the CATN. Specifically, the green and yellow lines correspond to the actual and planned average count of flights, respectively, for each time period. Equivalently, the purple and orange lines represent the actual and planned average count of airports, respectively, for each time period. Various shaded regions of divergent colors correspond to daily actual data that fluctuate around the mean.

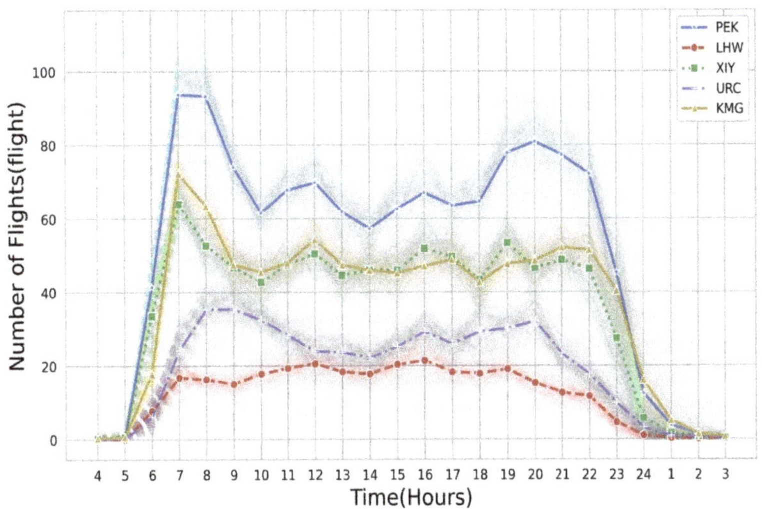


**Figure 5.** Flight volumes at specific airports over time.

Overall, the temporal variation of airline routes and flights in China highlights the challenges in the operational capacity of the CATN, particularly in handling peak traffic. These limitations lead to flight schedule delays, and airports resort to scheduling "red-eye flights" to accommodate all daily flights, consequently increasing the risk of aviation operations. Furthermore, evaluations of various representative airports reveal that they continue to encounter these constraints, although their operational capacities vary. Therefore, when developing flight optimization plans, it is crucial to optimize them based on the distinct operational capacities of different airports.

#### 3.2.2. Temporal Property of Node Degree

Node degree changes over time partially reflect the temporal characteristics of airport operation. Based on the CTAN constructed on 1 October 2018, the spatiotemporal characteristics of the air transportation network were illustrated at 6:00, 12:00, 18:00, and 24:00 by integrating the spatial information of airports' locations. The results revealed that, from 6:00 to 7:00 in the morning, routes were sparse with a small number of travelers, and most departures were from economically developed cities (Figure 6). After 8:00, a significant number of small and medium-sized airports joined the network and aggregated around the nearest hub airports with a central radiation structure. This gave rise to a rapid increase in the degree of hub nodes, and the entire civil aviation network entered its peak operation period. At noon, routes became denser, and there was a wider node distribution with the highest node degree value compared to other time periods. Although the air transportation network at 18:00 still contained some high-degree nodes, the number of such nodes gradually decreased due to the absence of takeoffs or landings at some small airports, leading to the disconnection of those airports from hub airports. At 24:00, the node distribution of the network was sparse with a small number of travelers, and most departures were from economically developed areas.

The six airports with the highest average node degree value in the air transportation network were Xi'an Xianyang, Beijing Capital, Chengdu Shuangliu, Chongqing Jiangbei, Shanghai Pudong, and Guangzhou Baiyun Airport, with values of 166, 165, 158, 154, 150, and 149, respectively. The high node degree of these airports indicated that they were the most closely connected airports in the air transportation network.

#### 3.2.3. Temporal Property of Average Shortest Path

The average path length of the aviation network can serve as a measure of the convenience of air travel for passengers. From a passenger's viewpoint, it is often impractical to take two flights within an hour. To evaluate this feature, the authors set the temporal window $\omega$ as one year. In the CATN network with a time window of one year, the shortest flight paths are summarized in Table 2. In summary, the average path length in the CATN is 1.97, suggesting that the majority of cities were accessible by changing two or fewer flights. The analysis revealed that 89.75% of city pairs could be reached either through direct flights or with just one flight change.

#### 3.2.4. Temporal Property of Clustering Coefficient

The clustering coefficient is a feature that can be applied to connected graphs. However, when time slicing was applied to the CATN, some graphs in certain time periods became disconnected, which largely consisted of a connected sub-graph and one to three small connected sub-graphs with no more than three airports, which were all branch airports. Hence, the small connected sub-graphs were ignored when calculating the clustering coefficient, and only the maximum connected sub-graph was taken into consideration.

The average clustering coefficient (C) of the network can reflect the overall network characteristics. Figure 7 exhibits that this topological feature demonstrates substantial fluctuations from 6:00 to 9:00 and after 22:00, signifying considerable alterations in the network structure during these periods. The clustering coefficient remained at a high and stable value from 9:00 to 21:00. When combined with the average shortest path, the

CATN demonstrated significant small-world characteristics with high clustering and short distances. The network structure in the early morning consisted of scattered flight routes that did not create a significantly connected sub-graph. As additional airport nodes and flight routes joined the network over time, its structure became increasingly complicated, which resulted in a rapid rise in the clustering coefficient between 5:00 and 7:00, and a subsequent sharp decrease from 7:00 to 9:00. This phenomenon can be attributed to a temporal asymmetry in the rate of growth between routes and airports within the network during distinct time intervals. Specifically, during the period from 5:00 to 7:00, the rate at which routes become part of the network surpasses the rate at which airports are incorporated. In contrast, between 7:00 and 9:00, the rate at which routes integrate into the network lags behind the rate at which airports join.

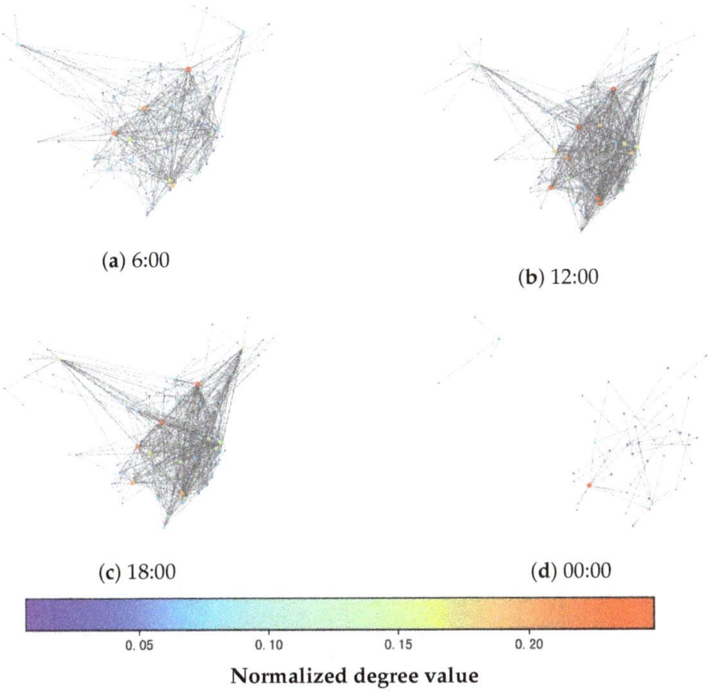


**Figure 6.** Spatiotemporal features of the aviation network at four specific time intervals. In the provided graph, the size of the nodes corresponds to the degree of those nodes. The color bar in the legend represents different value of normorlized degree. Each color corresponds to a specific range of values.

**Table 2.** Distribution of air routes based on the number of flights.

| Shortest Path | Number of Flights Needed to Change | Number of Paths | Percentage of Air Routes |
|---|---|---|---|
| 1 | 0 | 7156 | 13.59% |
| 2 | 1 | 40,118 | 76.17% |
| 3 | 2 | 5396 | 10.24% |

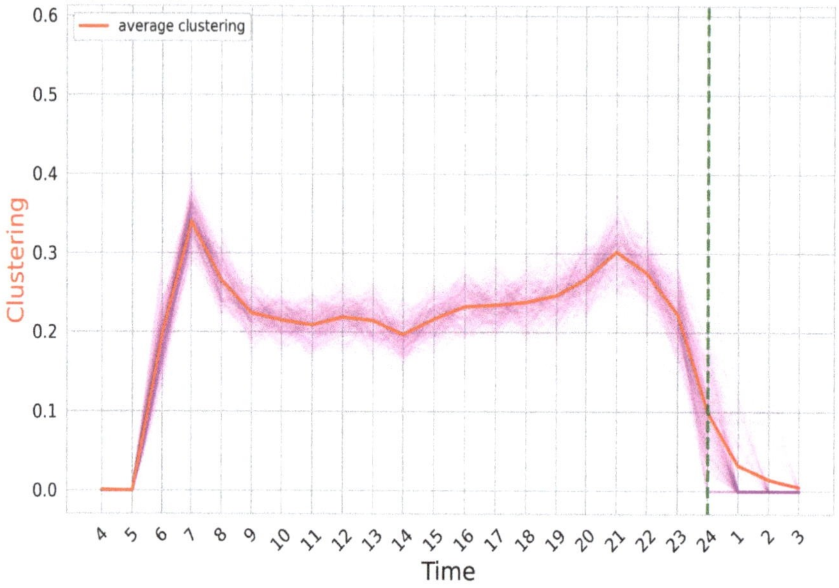


**Figure 7.** Clustering coefficient in 2018.

The findings of this study revealed the dynamic nature of the network through the examination of the degree and clustering coefficient. The analysis of the temporal and spatial variations in node degree illustrates the dynamic nature of airport operations and the fluctuating distribution of air traffic throughout the day. The fluctuation of the clustering coefficient reflects the changing patterns of connectivity and the growth of the network scale. This provides important data support for airport management and operational optimization, as well as information for the decision-making process of expanding routes and allocating airport resources to ensure effective connectivity and overall transportation performance.

## 4. Robustness of the CATN

This study investigated four distinct time-stamped versions of the CATN, corresponding to 6:00, 12:00, 18:00, and 24:00. In order to examine the centrality and robustness of the aviation network during these specific time intervals, all flight information from these four time periods in 2018 was combined.

### *4.1. Estimation of Attack Strategies*

Figure 8–11 illustrate the node centrality. The bar chart displays the centrality value of each airport, with the top 15 airports highlighted in blue. Degree centrality distribution was constant over time (Figure 8), which highlighted highly central airports located in hub airports such as Beijing and Shanghai. Airports with a higher nearest neighbor degree centrality are typically regional hub airports, such as provincial capital city airports or economically developed airports. This includes some hub airports with higher degree values, but the nearest neighbor degree of hub airports is generally smaller than that of regional hub airports. Furthermore, the overall nearest neighbor degree exhibits higher values compared to the node degree, implying a dense interconnection among airports and a preference for connecting with hub airports characterized by higher nearest neighbor degrees.

In contrast, the majority of the airports showed low betweenness centrality, with only a few displaying high centrality. These airports likely did not have high degrees, but their betweenness centrality played a significant role in maintaining network connectivity. Closeness centrality displayed a similar pattern to betweenness centrality, such that the top 15 airports with high betweenness centrality and closeness centrality displayed significant variations over time. In terms of time, the distribution of the CATN appeared to be unbalanced.

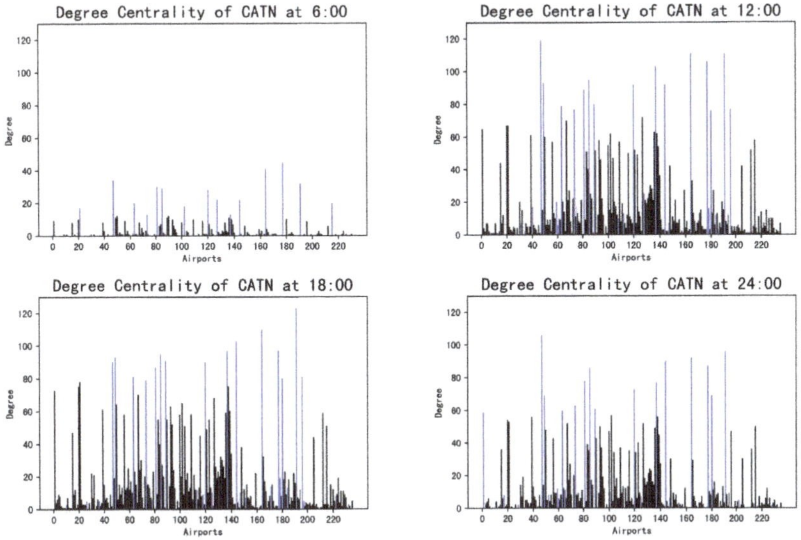


**Figure 8.** Temporal changes in degree centrality. The blue represent the top 15 airports with high degree centrality. The remaining airports are depicted in black.

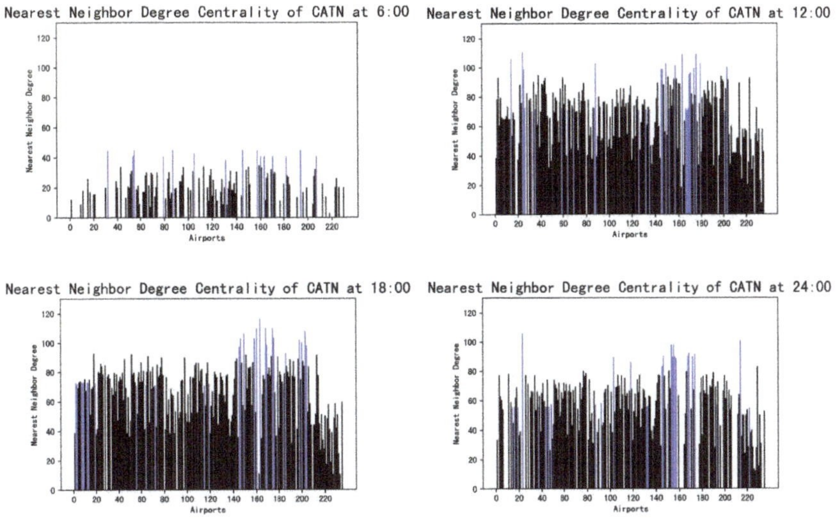


**Figure 9.** Temporal changes in nearest neighbor degree centrality. The blue represent the top 15 airports with high nearest neighbor degree centrality, while the remaining airports are depicted in black.

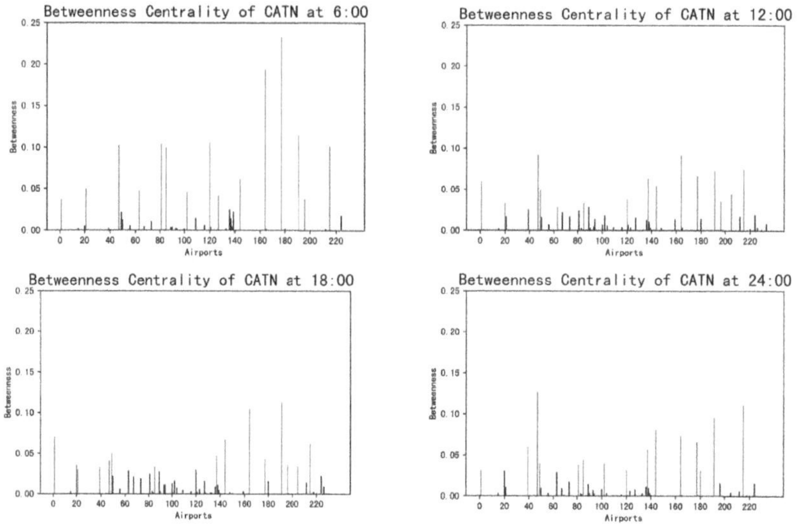


**Figure 10.** Temporal changes in betweenness centrality. The blue represent the top 15 airports with high nearest neighbor degree centrality, while the remaining airports are depicted in black.

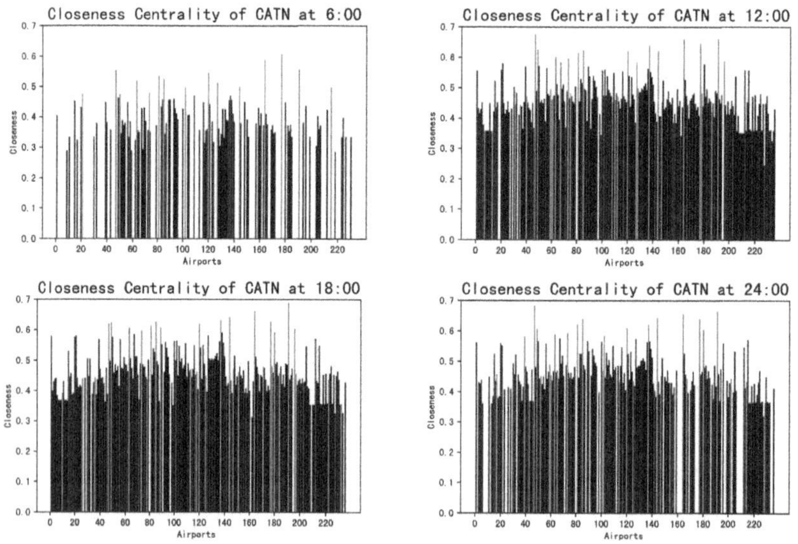


**Figure 11.** Temporal changes in closeness centrality. The blue represent the top 15 airports with high closeness centrality, while the remaining airports are depicted in black.

### 4.2. Influence of Structure on CTAN Robustness

This study performed attack experiments on four network snapshots employing centrality metrics as outlined in Section 2. In each experiment, only a single node was targeted, allowing us to investigate the alterations in the maximum connected component when employing both centrality-based and random attack strategies. As the number of deleted nodes increased, the proportion of nodes in the maximum connected component exhibited a significant downward trend (Figure 12). Random attacks were not effective in disrupting the connectivity of the aviation network in all four time intervals. On the contrary, targeted attacks substantially decreased the connectivity. For instance, when around 10% of the airports were removed, the size of the giant connected component shrunk by 30%, and the network gradually broke down after 20% of the airports were removed.

In comparison to the other four deliberate attack methods, degree-based attacks and nearest neighbor degree attacks had less impact on the robustness of the aviation network. Based on the results of the nearest neighbor degree attacks, an initial sharp decline was observed, indicating the ability of the nearest neighbor degree to rapidly identify crucial nodes in the aviation network, specifically hub airports. However, the subsequent attack effects were not significant and were comparable to random attacks. Referring to the findings in Figure 9, it can be inferred that the nearest neighbor degree attack primarily targets regional hub airports. Although these attacks do not result in a widespread paralysis of the aviation network, the transportation capacity of the network will diminish rapidly. Betweenness and closeness centrality-based attacks were more vulnerable. The uneven temporal distribution of the CATN led to frequent changes in the shortest paths. As both betweenness and closeness centrality indicators were related to the shortest path, safeguarding the hub airports located on the shortest path was necessary to ensure the connectivity of the aviation network.

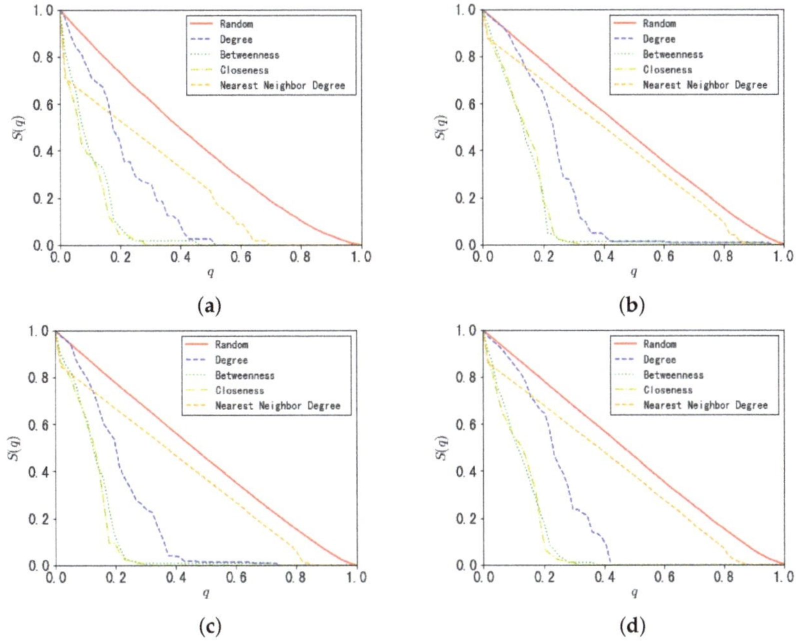


**Figure 12.** Temporal changes in the size of the maximum connected component in four specific time intervals: (**a**) the relationship between the relative size of the maximum connected component and the proportion of deleted nodes at 6:00, (**b**) the relationship between the relative size of the maximum connected component and the proportion of deleted nodes at 12:00, (**c**) the relationship between the relative size of the maximum connected component and the proportion of deleted nodes at 18:00, and (**d**) the relationship between the relative size of the maximum connected component and the proportion of deleted nodes at 24:00.

Network efficiency, represented by $E$, was another indicator for evaluating the robustness of the network under attack. Similar to the analysis conducted on giant components, attacks were carried out by targeting only one node at a time. Specifically, centrality and random attacks were performed on the CATN, and the alterations in CATN efficiency were observed across four time intervals. The attack results are exhibited in Figure 13. Random attacks were ineffective, so the efficiency did not decrease. Degree-based attacks and nearest neighbor degree attacks produced similar results with the size of the maximum connected component. Betweenness and closeness attacks, however, caused considerable damage to the aviation network. After removing 10% of the airports, the efficiency decreases by nearly 20% of the initial value, and the network gradually collapsed.

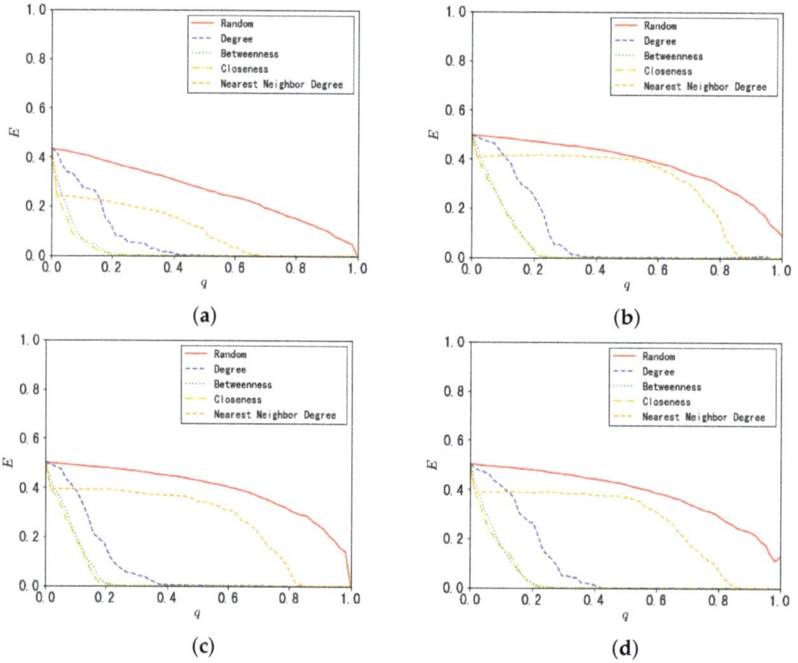


**Figure 13.** Temporal changes in network efficiency of four specific time intervals: (**a**) the relationship between weighted network efficiency and the proportion of node removal at 6:00, (**b**) the relationship between weighted network efficiency and the proportion of node removal at 12:00, (**c**) the relationship between weighted network efficiency and the proportion of node removal at 18:00, and (**d**) the relationship between weighted network efficiency and the proportion of node removal at 24:00.

By using the same attack strategy, it was observed that network efficiency was more substantially affected than the maximum connected component. This suggested that, although the aviation network could remain connected with a good network topology after an attack, its efficiency was greatly reduced. As a result, while the aviation network was possibly accessible topologically, it remained disconnected in a temporal sense.

In the aforementioned experiment, the authors employed a single-node attack approach. In order to investigate the network's robustness following attacks on multiple airports, the attack parameters were adjusted to target multiple nodes simultaneously. Specifically, the authors examined the effects of attacking 5 and 10 nodes at once, evaluating their performance based on the proportion of the maximum connected subgraph and network efficiency. The experimental findings, depicted in Figure 14, reveal that, when multiple airports are targeted, the CATN demonstrates substantial resilience against random attacks but exhibits weaker resistance against betweenness and closeness attacks. Consequently, the network's robustness against attack strategies remains comparable to that observed when targeting a single node. Nevertheless, it is notable that both network connectivity and efficiency deteriorate more rapidly.

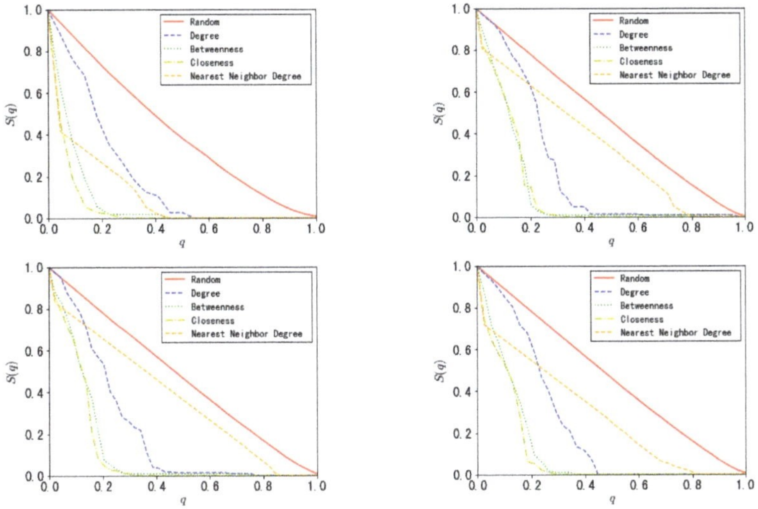


(**a**)

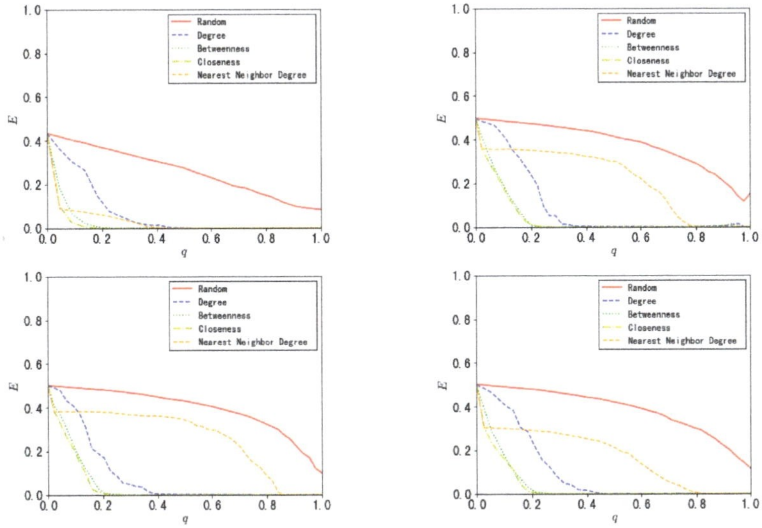


(**b**)

**Figure 14.** Temporal changes in the size of the maximum connected component and network efficiency of four specific time intervals: (**a**) attacking 5 nodes at once at 6:00, 12:00, 18:00, and 24:00; (**b**) attacking 10 nodes at once at 6:00, 12:00, 18:00, and 24:00.

## 5. Conclusions and Discussion

Complex network and temporal network theories were employed to explore the temporal and topological characteristics of the CATN. The network was defined as a dynamic temporal network and was mapped into a static network using time windows. The results indicated that the aviation network primarily operated from 6:00 to 24:00, with peak hours at 7:00 and 10:00. During these hours, the network exhibited an obvious scale-free characteristic and a small-world phenomenon, with the degree distribution rising rapidly in the initial stages before gradually stabilizing and finally rapidly dropping after the peak hours. Before final stabilization, the distribution of clustering coefficients initially increased and then decreased. The average shortest path showed that 80% of paths passed through hub airports, with an obvious central radiation structure. Furthermore, the robustness of the CATN at different times against random and deliberate attacks was explored using the proportion of maximum connected sub-graph nodes and network efficiency as robustness measures. The network was weak under deliberate attacks, particularly when using betweenness centrality and closeness centrality strategies. Further, it was observed that the CATN exhibited temporal unevenness in centrality, often leading to changes in the shortest path.

As a preliminary study, the feasibility of temporal network theory in CATN research was verified, and the law of robustness variation in the CATN was discussed. In our future research, an in-depth exploration will be conducted from two perspectives. First, the theory of temporal networks will be employed to investigate the intricate dynamics of networks in the CATN, aiming to provide practical solutions for optimizing flight schedules. This involves addressing cascading failures triggered by nodes or links within the network, as well as the dispersion and propagation of delays. Second, an exploration of more advanced techniques for temporal network modeling will be undertaken, aiming to accurately capture the dynamic structure of the CATN. Event-based temporal networks or dynamic graph neural networks can be utilized for this purpose.

**Author Contributions:** Conceptualization, R.Y. and W.S.; data curation, W.S.; formal analysis, R.Y.; funding acquisition, H.Z.; methodology, R.Y. and W.S.; resources, R.Y.; supervision, H.Z.; validation, R.Y., M.L. and H.Z.; visualization, W.S.; writing—original draft, R.Y. and W.S.; writing—review & editing, M.L. All authors have read and agreed to the published version of the manuscript.

**Funding:** This work was supported by the Hainan Provincial Natural Science Foundation of China (grant number 2019RC199) and the National Natural Science Foundation of China (grant number 62167003).

**Institutional Review Board Statement:** Not applicable.

**Informed Consent Statement:** Not applicable.

**Data Availability Statement:** The data and material used in this study were obtained from VariFlight. Due to the commercial nature of the data, access to the data is restricted by the provider. However, we have obtained the necessary permissions and licenses to use the data for our research study.

**Acknowledgments:** We gratefully acknowledge the financial support provided by the Hainan Provincial Natural Science Foundation of China (grant number 720RC616) and the National Natural Science Foundation of China (grant number 62167003).

**Conflicts of Interest:** The authors declare no conflict of interest.

## Abbreviations

The following abbreviations are used in this manuscript:

| | |
|---|---|
| CATN | Chinese aviation temporal network |
| CAN | Chinese temporal network |

## References

1. Cumelles, J.; Lordan, O.; Sallan, J.M. Cascading failures in airport networks. *J. Air Transp. Manag.* **2021**, *92*, 102026. [CrossRef]
2. Wang, J.; Bonilla, D.; Banister, D. Air deregulation in China and its impact on airline competition 1994–2012. *J. Transp. Geogr.* **2016**, *50*, 12–23. [CrossRef]
3. Sallan, J.M.; Lordan, O. *Air Route Networks through Complex Networks Theory*; Elsevier: Exeter Devon, UK, 2019.
4. Mou, J.-H. Analysis of Time Series Network Propagation Dynamics and Its Application in Aviation Networks. Master's Thesis, National University of Defense Technology, Changsha, China, 2017.
5. Chi, L.-P.; Wang, R.; Su, H.; Xu, X.-P.; Zhao, J.-S.; Li, W.; Cai, X. Structural properties of US flight network. *Chin. Phys. Lett.* **2003**, *20*, 1393.
6. Li, W.; Cai, X. Statistical analysis of airport network of China. *Phys. Rev. E* **2004**, *69*, 046106. [CrossRef]
7. Barrat, A.; Barthelemy, M.; Pastor-Satorras, R.; Vespignani, A. The architecture of complex weighted networks. *Proc. Natl. Acad. Sci. USA* **2004**, *101*, 3747–3752. [CrossRef]

8. Gautreau, A.; Barrat, A.; Barthélemy, M. Microdynamics in stationary complex networks. *Proc. Natl. Acad. Sci. USA* **2009**, *106*, 8847–8852. [CrossRef] [PubMed]
9. Zhang, L.; Du, H.; Zhao, Y.; Maeyer, P.D.; Zhang, X. Drawing topological properties from a multi-layered network: The case of an air transport network in "the belt and road" region. *Habitat Int.* **2019**, *93*, 102044. [CrossRef]
10. Lordan, O.; Sallan, J.M.; Simo, P.; Gonzalez-Prieto, D. Robustness of the air transport network. *Transp. Res. Part E Logist. Transp. Rev.* **2014**, *68*, 155–163. [CrossRef]
11. Cong, W.; Hu, M.; Dong, B.; Wang, Y.; Feng, C. Empirical analysis of airport network and critical airports. *Chin. J. Aeronaut.* **2016**, *29*, 512–519. [CrossRef]
12. Guida, M.; Maria, F. Topology of the Italian airport network: A scale-free small-world network with a fractal structure. *Chaos Solitons Fractals* **2007**, *31*, 527–536. [CrossRef]
13. Holme, P.; Saramäki, J. *Temporal Network Theory*; Springer: Cham, Switzerland, 2019; Volume 2.
14. Cheng, E.; Grossman, J.W.; Lipman, M.J. Time-stamped graphs and their associated influence digraphs. *Discret. Appl. Math.* **2003**, *128*, 317–335. [CrossRef]
15. Moody, J. The importance of relationship timing for diffusion. *Soc. Forces* **2002**, *81*, 25–56. [CrossRef]
16. Bassett, D.S.; Wymbs, N.F.; Porter, M.A.; Mucha, P.J.; Carlson, J.M.; Grafton, S.T. Dynamic reconfiguration of human brain networks during learning. *Proc. Natl. Acad. Sci. USA* **2011**, *108*, 7641–7646. [CrossRef] [PubMed]
17. Basu, P.; Bar-Noy, A.; Ramanathan, R.; Johnson, M.P. Modeling and analysis of time-varying graphs. *arXiv* **2010**, arXiv:1012.0260.
18. Isufaj, R.; Koca, T.; Piera, M.A. Spatiotemporal graph indicators for air traffic complexity analysis. *Aerospace* **2021**, *8*, 364. [CrossRef]
19. Isufaj, R.; Omeri, M.; Piera, M.A.; Valls, J.S.; Gallego, C.E.V. From Single Aircraft to Communities: A Neutral Interpretation of Air Traffic Complexity Dynamics. *Aerospace* **2022**, *9*, 613. [CrossRef]
20. Kivelä, M.; Cambe, J.; Saramäki, J.; Karsai, M. Mapping temporal-network percolation to weighted, zhestatic event graphs. *Sci. Rep.* **2018**, *8*, 12357 [CrossRef]
21. Sano, H.H.; Berton, L. A temporal event graph approach and robustness analysis for air transport network. *IEEE Trans. Netw. Sci. Eng.* 2021, **8**, 3453–3464. [CrossRef]
22. Holme, P.; Saramäki, J. A map of approaches to temporal networks. In *Temporal Network Theory*; Springer: Cham, Switzerland, 2019; pp. 1–24.
23. George, B.; Shekhar, S. Time-aggregated graphs for modeling spatio-temporal networks. *J. Data Semant.* **2008**, *XI*, 191–212.
24. Ferreira, A. Building a reference combinatorial model for manets. *IEEE Netw.* **2004**, *18*, 24–29. [CrossRef]
25. Hulovatyy, Y.; Chen, H.; Milenkovi'c, T. Exploring the structure and function of temporal networks with dynamic graphlets. *Bioinformatics* **2015**, *31*, 171–180. [CrossRef]
26. Zhang, J.; Cao, X.-B.; Du, W.-B.; Cai, K.-Q. Evolution of Chinese airport network. *Phys. A Stat. Mech. Appl.* **2010**, *389*, 3922–3931. [CrossRef] [PubMed]
27. Liu, H.-K.; Zhang, X.-L.; Zhou, T. Structure and external factors of Chinese city airline network. *Phys. Procedia* **2010**, *3*, 1781–1789. [CrossRef]
28. Lordan, O.; Sallan, J.M.; Simo, P.; Gonzalez-Prieto, D. Robustness of airline alliance route networks. *Commun. Nonlinear Sci. Numer. Simul.* **2015**, *22*, 587–595. [CrossRef]
29. Du, W.-B.; Zhou, X.-L.; Lordan, O.; Wang, Z.; Zhao, C.; Zhu, Y.-B. Analysis of the Chinese airline network as multi-layer networks. *Transp. Res. Part E Logist. Transp. Rev.* **2016**, *89*, 108–116. [CrossRef]
30. Wang, Y.; Zhan, J.; Xu, X.; Li, L.; Chen, P.; Hansen, M. Measuring the resilience of an airport network. *Chin. J. Aeronaut.* **2019**, *32*, 2694–2705. [CrossRef]
31. Morone, F.; Makse, H.A. Influence maximization in complex networks through optimal percolation. *Nature* **2015**, *524*, 65–68. [CrossRef]
32. Bai, B. Strategic business management for airport alliance: A complex network approach to simulation robustness analysis. *Phys. A Stat. Mech. Appl.* **2022**, *606*, 126682. [CrossRef]
33. Latora, V.; Marchiori, M. Efficient behavior of small-world networks. *Phys. Rev. Lett.* **2001**, *87*, 198701. [CrossRef]
34. Sun, L.; Axhausen, K.W.; Lee, D.H.; Huang, X. Understanding metropolitan patterns of daily encounters. *Proc. Natl. Acad. Sci. USA* **2013**, *110*, 13774–13779. [CrossRef]

symmetry MDPI

Article

# Maximum Convergence Rate Control of a Switched Electrical Network with Symmetries

Simone Fiori [1,*], Luca Ciaramicoli [2] and Giovanni Berluti [2]

[1] Department of Information Engineering, Marches Polytechnic University, I-60131 Ancona, Italy
[2] School of Electronics Engineering, Marches Polytechnic University, I-60131 Ancona, Italy
* Correspondence: s.fiori@staff.univpm.it

**Abstract:** The purpose of the present research endeavor is to propose a novel control strategy for a DC-DC electrical converter realized as a switched circuit. The present endeavor is based on an early work by Leonard and Krishnaprasad where a prototypical DC-DC converter was modeled as a state space dynamical system and controlled by an open-loop strategy based on Lie group theory. In this work, we shall rather introduce a closed-loop control strategy based on maximum convergence rate, study some features of the novel method, and illustrate its behavior by numerical simulations. A numerical comparison of the two methods complements the paper. The numerical comparison shows how the proposed feedback control method outperforms the static one in terms of convergence rate and resiliency against parameters mismatch.

**Keywords:** DC-DC electrical converter; open-loop control; feedback control; switched electrical network

## 1. Introduction

Switched electrical networks are systems that consist of interconnected subsystems or components that can be controlled or switched on and off. Switched systems appear in a wide range of applications, including power systems, telecommunications, and transportation systems [1]. In power systems, for example, switched networks can be used to model the operation of power generators, transmission lines, and loads. In telecommunications, switched networks are used to model the routing of data packets through a network of routers and switches. In transportation systems, switched networks can be used to model the operation of traffic lights and other control systems. Further applications are found in mechanical engineering and computer science [2], as well as other areas such as robotics, biomedical engineering, and environmental engineering.

A prominent instance of a switched system is a DC-voltage to DC-voltage (DC-DC) converter. DC-DC converters turn out to be essential components in a wide range of electronic systems, and they play a crucial role in converting sourced DC voltages to desired levels that suit the needs of different devices. There exist several different types of DC-DC converters, each with their own advantages and disadvantages, and the appropriate type depends on the specific application requirements. High-frequency DC-DC converters offer the advantage of smaller and lighter components but also present additional challenges, which can be addressed through different constructive topologies and control techniques. Modeling and analyzing DC-DC converters are of prime importance for understanding their behavior and predicting their performance.

There exist several methods for analyzing and synthesizing switched systems, including traditional methods such as Lyapunov stability analysis and modern methods such as hybrid system theory and nonlinear system theory [3]. Stability analysis is an important aspect of switched systems, as it determines whether a system is able to maintain its equilibrium or desired behavior over time. Control design is another important aspect of switched systems, as it involves the design of controllers or algorithms that can be used to steer the system towards a desired behavior. Switched systems with delays are also of particular interest, as some real-world systems exhibit non-negligible delays due to the inherent limitations of physical components or communication networks [4].

In the present paper, we shall take as a case study the prototypical model of a switched DC-DC converter studied in [5] and briefly revise its mathematical description as a first-order dynamical system (in three variables). In the first place, we shall revise some salient features of this electrical scheme that, being lossless, admits an invariant that shapes its state space to be a curved manifold. A lossless converter is one that presents no energy dissipation or, equivalently, a unitary conversion rate and represents henceforth an ideal system. Electrical energy conservation is an instance of symmetry, since state transition in such systems is modeled through mathematical operators that leave a quadratic form of the system's state unvaried.

We shall subsequently point out that the open-loop control strategy proposed in [5] is characterized by some advantages and some drawbacks that we propose to mitigate through a closed-loop control strategy that is much simpler to implement yet provides comparable performances. The design and implementation of the control algorithm makes explicit reference to manifold theory, a branch of mathematical analysis that, in the context of non-linear control, allows for dealing seamlessly with curved state spaces.

The present paper is organized as follows. Section 2 recalls the prototypical DC-DC converter examined by Leonard and Krishnaprasad in [5] along with its state space mathematical model and emphasizes some of its salient mathematical features. Section 3 summarizes a number of technical aspects related to the design of DC-DC converters. Section 4 reviews the Lie group-based control strategy deployed in [5] along with its salient features and drawbacks. Section 5 introduces a closed-loop control strategy to transfer the energy content from the input side of the converter to the output side based on maximum convergence rate. Section 6 illustrates the behavior of the introduced control strategy by a number of numerical tests, as well as a comparison of the introduced strategy with the original Leonard–Krishnaprasad method. Section 7 concludes the paper.

## 2. Electrical and Mathematical Model of a Prototypical DC-DC Converter

The present section summarizes the main features of DC-DC converters and illustrates the mathematical model of a prototypical converter.

### 2.1. Generality of DC-DC Converters

DC-DC converters are electronic circuits that convert direct current (DC) from one voltage level to another and constitute a viable counterpart to electrical transformers that may only be used in alternate current (AC) circuits. DC-DC converters are widely employed in a variety of applications, including power supplies for electronic devices, electric vehicles, and renewable energy systems [6–9]. There exist several types of DC-DC converters, including buck converters, boost converters, buck–boost converters, and Ćuk converters. Each type has its own advantages and disadvantages, and the appropriate type for a given application depends on the specific requirements and constraints.

Buck converters are used to step down the input voltage to a lower output voltage. They operate by switching the input voltage at a high frequency and using inductors and capacitors to filter the voltage ripple. The output voltage can be controlled by adjusting the duty cycle of the switching signal. Boost converters, on the other hand, are used to step up the input voltage to a higher output voltage.

Buck–boost converters can operate in either a buck mode or a boost mode, depending on the input and output voltage levels. They are often used in applications where the input voltage may be either higher or lower than the desired output voltage. Also, Ćuk converters are a hybrid of the buck and boost converters, and they can achieve a high conversion ratio, although their topology is generally more complex; a non-isolated Ćuk converter comprises two inductors, two capacitors, a switch and a diode. They are often used in high-power applications where a high conversion ratio is required. High-frequency DC-DC converters are those that operate at switching frequencies above a few hundred kilo-Hertz. They have the advantage of smaller and lighter components, but they also present the challenge of higher losses and electromagnetic interference (EMI) issues. Various topologies and control techniques have been developed to address these challenges and to enable the use of high-frequency DC-DC converters in a wide range of applications.

Modeling and analysis techniques for DC-DC converters are important for understanding their behavior and predicting their performance. Small-signal modeling and modal analysis are widely used approaches for this purpose. These methods involve linearizing the converter equations around the operating point and analyzing the resulting linear system to determine the dynamic response of the converter.

### 2.2. A Prototypical Converter Model and Its State Space Description

Direct current converters have been realized in various electrical configurations [10–12], which invariably contain an energy source (typically an electrical capacitor), an energy buffer (typically a capacitor), an energy-bridging element (typically an electrical inductor), and a switch that realizes a variable topology structure. Electronic switches in power applications are realized as (groups of) thyristors (for a review of high-power thyristors in power electronics applications, see, e.g., [13,14]), while MOSFET is the typical device in low-power applications.

As a prototypical model of a switched DC-DC converter, we adopted the circuital scheme studied in [5], as shown in Figure 1.

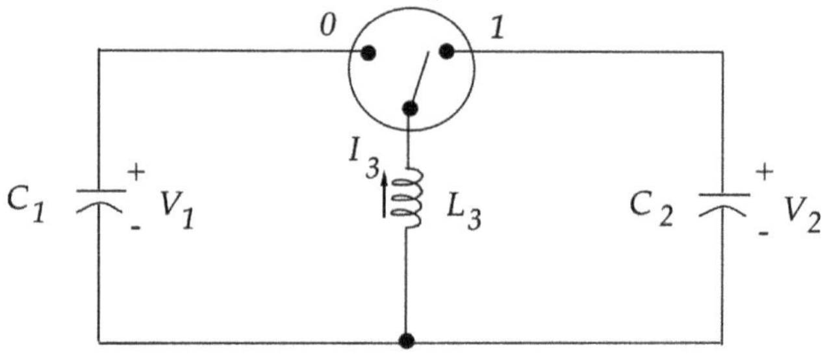


**Figure 1.** Prototypical circuital scheme of a DC-DC converter adapted from [5], where $C_1 > 0$, $C_2 > 0$, and $L_3 > 0$.

At the beginning of the energy transfer cycle, the capacitor $C_1$ is charged at a voltage $V_1(0)$, which constitutes the input to the converter, and the inductor current intensity $I_3(0) = 0$, while the capacitor $C_2$ contains no charges; hence, $V_2(0) = 0$. The control problem consists in transferring the electrical energy initially stored in $C_1$ to $C_2$, through the inductor $L_3$, by an appropriate sequence of commutations of the switch. In fact, the inductor serves as a temporary energy storage at every commutation.

By network analysis, it is found that the state space representation of the converter in Figure 1 is given by the following set of three differential equations:

$$\begin{cases} C_1 \dfrac{\mathrm{d}V_1}{\mathrm{d}t} = (1-u)I_3, \\ C_2 \dfrac{\mathrm{d}V_2}{\mathrm{d}t} = u\, I_3, \\ L_3 \dfrac{\mathrm{d}I_3}{\mathrm{d}t} = -(1-u)V_1 - u\, V_2, \end{cases} \tag{1}$$

where the variable $u \in \{0, 1\}$ constitutes the control input to the system and $t \in [0, t_f]$, where $t_f > 0$ denotes the total duration of a conversion cycle, namely, the total time allotted to the converter to transfer energy from the input capacitor to storage capacitor and, with reference to Figure 1, an input value $u = 1$ means that the switch is in the right-hand position.

Notice that the circuit in Figure 1 is devoid of dissipative elements, which corresponds to considering ideal (lossless) components. In addition, the energy buffer capacitor $C_2$ is not connected to any load. In the considered idealized model drawn from [5], the capacitor $C_1$ is substantially discharged into the capacitor $C_2$ by using an inductor as a temporary energy storage element and a switch to connect these three elements so that the energy transfer takes place. In normal operations, an electrical load might be a low-power DC motor [15] or a low-power AC motor connected to the DC supply through an inverter [10].

In the present idealized context, the energy content of the circuit stays unvaried over time, as shown in the following result.

**Theorem 1** (Energy conservation in the idealized DC-DC converter). *The total instantaneous electrical energy content of the circuit in Figure 1 is conserved over time (irrespective of the control input).*

**Proof.** The total instantaneous electrical energy content of the circuit in Figure 1 is defined as

$$E(V_1, V_2, I_3) := \tfrac{1}{2}C_1V_1^2 + \tfrac{1}{2}C_2V_2^2 + \tfrac{1}{2}L_3I_3^2. \quad (2)$$

It can be immediately verified that

$$\frac{\mathrm{d}E(V_1, V_2, I_3)}{\mathrm{d}t} = (1-u)V_1I_3 + u\,V_2I_3 - (1-u)I_3V_1 - u\,I_3V_2 = 0,\ \forall t \in [0, t_\mathrm{f}]. \quad (3)$$

Therefore, the energy content is conserved over time. □

The above theorem illustrates the existence of a symmetry in the system representing a lossless DC-DC energy converter.

At the end of the control cycle, whose duration is denoted as $t_\mathrm{f}$, it is expected that $V_1(t_\mathrm{f}) = 0$ and $I_3(t_\mathrm{f}) = 0$. Since the circuit is lossless, the following energy conservation property holds:

$$\tfrac{1}{2}C_1V_1^2(0) = \tfrac{1}{2}C_2V_2^2(t_\mathrm{f}). \quad (4)$$

Therefore, the output voltage of the converter will be

$$V_2(t_\mathrm{f}) = \pm\sqrt{\frac{C_1}{C_2}}V_1(0). \quad (5)$$

We therefore see that the circuital model under analysis may realize a buck–boost converter depending on the ratio $C_1/C_2$ between the capacitance values.

Since the electrical circuit is lossless, the state space of the circuit is not the entire flat space $\mathbb{R}^3$ but a curved space. To make such a conclusion more evident, it is convenient to introduce the new state variables

$$x^{(1)} := \sqrt{C_1}V_1,\ x^{(2)} := \sqrt{C_2}V_2,\ x^{(3)} := \sqrt{L_3}I_3 \quad (6)$$

as well as the state vector $x := \begin{pmatrix} x^{(1)} & x^{(2)} & x^{(3)} \end{pmatrix}^\top$. Notice that the states $x^{(1)}$ and $x^{(2)}$ are measured in $\mathrm{VF}^{1/2}$, while the state $x^{(3)}$ is measured in $\mathrm{AH}^{1/2}$. (V is the International System unit for Volt, F for Farad, and H for Henry.) The mathematical model (1) may then be recast as

$$\dot{x}(t) = (A + Bu(t))x(t),\ x(0) = \begin{pmatrix} \sqrt{2E_0} & 0 & 0 \end{pmatrix}^\top, \quad (7)$$

where $E_0 > 0$ denotes the energy content of the source (in joules) and, following [5], we have defined the constant matrices

$$A := \begin{pmatrix} 0 & 0 & \omega_1 \\ 0 & 0 & 0 \\ -\omega_1 & 0 & 0 \end{pmatrix}, \ B := \begin{pmatrix} 0 & 0 & -\omega_1 \\ 0 & 0 & \omega_2 \\ \omega_1 & -\omega_2 & 0 \end{pmatrix} \tag{8}$$

with $\omega_1 := 1/\sqrt{C_1 L_3}$ and $\omega_2 := 1/\sqrt{C_2 L_3}$. In terms of the new state variable vector, the conservation of energy reads

$$E(x^{(1)}, x^{(2)}, x^{(3)}) = \tfrac{1}{2} x^\top x = E_0, \tag{9}$$

which makes it apparent how the state space of the converter, in the normalized variables, is a three-dimensional sphere of radius $\sqrt{2E_0}$.

From a dynamical system perspective, energy conservation stems from the algebraic structure of the system (7), in which it can be immediately recognized that both matrices $A$ and $B$ are skew-symmetric, namely $A^\top = -A$ and $B^\top = -B$. As a consequence, it holds that $\dot{E} = x^\top \dot{x} = x^\top A x + (x^\top B x) u = 0$, irrespective of the values taken by the control input. In other terms, as the state $x$ evolves, the point of coordinates $x$ traces a curve laying on a spherical shell described by the equation $x^\top x = 2E_0$. In manifold calculus, such a surface is denoted as $\sqrt{2E_0}\mathbb{S}^2$, which represents a 3D sphere of radius $\sqrt{2E_0}$.

Generalizing from [5], we fix the target state to be $\bar{x}(t_f) = \begin{pmatrix} 0 & -\sqrt{2E_0} & 0 \end{pmatrix}^\top$. The objective of control design is to steer the state of the system from the initial state $x(0)$ to the desired final state $\bar{x}(t_f)$ under the constraint imposed by the dynamics (7), which, as we have just emphasized, hides a further constraint on the shape of the state space.

## 3. Technological Aspects of DC-DC Converters

In DC-DC converter modeling and design, there exist several key concerns related to performance and reliability, including noise levels, switching losses, component stress, and output voltage ripple. Here is a general overview of each of these concerns:

1. **Noise level:**
   DC-DC converters often involve rapid switching of currents and voltages, leading to electromagnetic interference. Electromagnetic interference can affect nearby electronic devices and may pose regulatory compliance challenges. Effective measures, such as shielding and filtering, are employed to mitigate EMI concerns [16]. In certain applications, the switching frequency or other factors can produce audible noise, which is undesirable. Design considerations include component and operational parameter selection to minimize audible noise.
2. **Switching losses:**
   DC-DC converters experience losses during state transitions, such as switching between on/off states of thyristors. Such losses result in heat generation and reduced efficiency. Design optimization aims to minimize switching losses through careful component selection and control strategies [17,18].
3. **Component stress:**
   *Voltage and Current Stress:* Components within DC-DC converters, such as diodes, capacitors, and inductors, endure high voltage and current stresses during operation. Component selection is crucial to ensure reliability by matching voltage and current ratings [19].
   *Thermal stress:* Elevated temperatures may arise due to high switching frequencies and load conditions in DC-DC converters. Effective thermal management techniques, heat sinks, and component selection are employed to manage thermal stress and prevent component failure [20].
4. **Output voltage ripple:**
   DC-DC converters are designed to provide a stable output voltage. However, due to switching and control dynamics, output voltage ripples may occur. Minimizing

output voltage ripple is crucial, especially in applications demanding precise voltage regulation [21]. There exist several strategies to mitigate output voltage ripple. The output voltage ripple is the ripple occurring across the capacitor $C_2$ and may be reduced, for a buck–boost converter, by an appropriate choice of the capacity $C_2$, considering the switching rate and the rated output power. Output voltage ripple might be reduced through the application of output filters, such as inductors and capacitors, which serve to smooth the output voltage and enhance regulation. However, additional filters increase cost, size, and circuit complexity and are not a preferred choice in practice.

5. **Overshoot and undershoot:**
Overshoot refers to a temporary increase in the output voltage or current above the desired steady-state value when there is a sudden increase in load or input voltage. Rapid load increases lead the converter to momentarily produce a higher output voltage to meet the higher demand [22]. Sudden changes in input voltage, like voltage spikes or rapid increases, can also induce overshoot as the converter adjusts. Overshoot can potentially exceed component or load limits, causing damage or instability and may affect the performance of devices powered by the converter, particularly in sensitive applications.
Undershoot refers to a temporary decrease in the output voltage or current below the desired steady-state value when there is a sudden decrease in load or input voltage. A sudden reduction in load leads to undershoot as the converter momentarily produces a lower output voltage. Rapid changes in input voltage, especially when the control loop reacts slowly, can also result in undershoot [23]. Undershoot can lead to temporary power loss or a voltage drop, potentially disrupting connected devices or systems. In applications sensitive to voltage fluctuations, undershoot can cause data loss or system instability.
To manage overshoot and undershoot in DC-DC converters, practitioners employ various strategies. Proper design and tuning of the control loop are critical for responding optimally to load or input changes while minimizing transient deviations, selection of components with appropriate voltage and current ratings helps avoid component stress and enhance converter performance, implementation of input voltage surge protection mechanisms can reduce the impact of rapid input voltage changes, and voltage and current regulation feedback loops help to maintain stable output values and minimize overshoot and undershoot. In practice, the specific behavior of overshoot and undershoot depends on factors like the converter topology, control strategy, load conditions, and input voltage characteristics. Practitioners aim to strike a balance between achieving a fast response to load or input changes and minimizing these transient deviations to ensure reliable operation in their specific applications.

6. **Stability issues:**
Stability issues in DC-DC converters refer to concerns regarding the converter's ability to maintain a consistent and controlled output voltage or current under different operational conditions. These issues are essential for ensuring that the converter operates reliably without undesirable oscillations or instability [24].
Several factors influence stability in DC-DC converters. The choice of control strategy and the adjustment of control loop parameters significantly impacts stability, different DC-DC converter designs may exhibit varying levels of stability, rapid changes in the input voltage can challenge the converter's ability to maintain a stable output, sudden variations in the connected load can affect the converter's capacity to sustain a stable output, and component tolerances and properties, such as capacitance and inductance values, can also influence stability.
Proper design considerations, including selecting suitable components, minimizing parasitic elements, and optimizing the control loop, are critical for ensuring stability. Furthermore, experimental validation through real-world testing is essential to confirm stability under practical conditions, including varying loads and input voltages.

Engineers may employ control and compensation techniques, such as PID control or lead–lag compensation, to enhance stability.
Ultimately, the effective management of stability issues in DC-DC converters is crucial for guaranteeing their reliable and predictable performance across a range of applications.

In addressing such technological issues, engineers and researchers employ modeling and simulation techniques to analyze DC-DC converter performance. Factors such as switching topology selection, control strategy design, component choice, and layout optimization may be considered to enhance converter performance while satisfying specific application requirements. Furthermore, advancements in semiconductor technology and power electronics have led to the development of more efficient and reliable DC-DC converter designs in recent years.

As a theoretic control paper concerned with the development of a closed-loop control strategy, the present document does not consider such technical concerns any further.

## 4. Review of Control Design by Averaging on Lie Groups

The fundamental observation behind the solution proposed in [5] is related to the action of a Lie group on a smooth manifold. In the present case of interest, it boils down to the fact that every solution of the system (7) may be written as $x(t) = \Phi(t)x(0)$, where $\Phi$ represents a matrix-valued function that takes values in the special orthogonal group SO(3) of three-dimensional rotations and obeys the differential equation $\dot{\Phi} = (A + Bu)\Phi$ with initial condition $\Phi(0) = I$ and $t \in [0,\ t_f]$.

A first design choice is that the input signal $u$ is taken as a concatenation of pulse trains. Each pulse switches between 0, in which it stays for $(1 - \mu)T$ secs, and 1, in which it stays for $\mu T$ secs, where $\mu$ denotes the duty cycle and $T$ the duration of each pulse (expressed in seconds). Each pulse train in the chain consists of a number of pulses, denoted as $q$. The design effort then consists in determining the total duration $q$ of each train and the duty cycle $\mu$ for each train. Each pulse train consists possibly of a different number of pulses, which, upon adjusting their duration, may even be fractional.

Since a pulse train is a fast-oscillating periodic signal, it is assumed that the net effect of a pulse train is equivalent to that of its average. In other words, the effect of a fast-switching signal is assumed to be equivalent to that of its average $\overline{u} \in [0,\ 1]$. Under this assumption, it is possible to determine an approximate solution to the equation $\dot{\Phi} = (A + Bu)\Phi$.

The next design choice is to fix the final time $t_f$ and to subdivide the interval $[0,\ t_f]$ into a number of sub-intervals, to which correspond some pre-specified points of the trajectory, which will then be followed by the system. By using the approximate solution to the equation $\dot{\Phi} = (A + Bu)\Phi$ within each sub-interval, it is therefore possible to make sure that the designed control $u$ steers the system from the given initial state to the given final state along a pre-fixed trajectory.

Such a control design possesses a number of advantages, which may be summarized as follows:

- The conversion time $t_f$ is pre-fixed and hence known in advance, which facilitates the integration of a converter into a more complex circuit.
- The trajectory over the state space may be optimized to fulfill further design constraints, such as keeping almost constant the current flowing through the inductor.
- Once the design of a converter is completed, its functioning is ensured by a solid theoretical background.

The method proposed in [5] clearly suffers from some drawbacks, which may be summarized as follows:

- The design is based on a number of arbitrary choices, such as the trajectory design.
- The actual computation of the pairs $(\mu, q)$ for each pulse train involves a non-linear optimization process that requires intensive computation (based on quaternions).

- The more accurate the averaging formulas required, the more complicated their implementation and usage.
- As an open-loop control algorithm was designed, it possesses no resiliency against parameter mismatch and modeling mismatch.

In the following, we shall propose a simpler, closed-loop control strategy that will be compared to the one devised by Leonard and Krishnaprasad.

## 5. Closed-Loop Control by a Criterion Function Optimization

Although buck, boost, buck–boost, and Ćuk converters were briefly revised, the present paper focuses on the buck–boost concept. As a matter of fact, Ćuk converters are of higher order compared to the other concepts [25], so a straightforward extension of the proposed method to control Ćuk converters would not be realistic.

In contrast to analog-type control based on high-frequency pulse-width modulation (PWM) actuation [12], in the present paper, digital control is envisaged. In fact, an assumption we make on the converter circuitry is that, once the switch has commuted from position 0 to 1 or from position 1 to 0, it cannot commute again before a given latency time, hereafter denoted as $h > 0$. Such an assumption implies that the signal $u(t)$ is piecewise constant to either 0 or 1 able to change its value at most once every $h$ seconds. Typically the value of $h$ is far less than the unity (namely, $h \ll 1$).

An important consequence of the above assumption is that we may associate with the continuous-time signal $u(t)$ a discrete-time sequence $u_n$ defined as $u_n := u(nh)$, where $n = 0, 1, 2, \ldots, N$ denotes a discrete-time index. In addition, since the input is piecewise constant, the state vector takes, within each interval $[nh, (n+1)h]$, a predictable value. In fact, within any such interval, the system (7) is of the type $\dot{x} = Fx$; hence, defining a discrete-time state vector sequence $x_n := x(nh)$, its solution reads $x_{n+1} = \exp(hF)x_n$ [26], where $F$ is either $A$ or $A + B$ and 'exp' denotes matrix exponential. In other terms, it holds that

$$x_{n+1} = \begin{cases} \exp(hA)x_n & \text{if } u_n = 0, \\ \exp(h(A+B))x_n & \text{if } u_n = 1, \end{cases} \tag{10}$$

where $\exp(\cdot)$ denotes the matrix exponential. Since the state transition matrix is either $\exp(hA)$ or $\exp(h(A+B))$, it is appropriate to define

$$R^h_A := \exp(hA) \text{ and } R^h_{AB} := \exp(h(A+B)), \tag{11}$$

so that the state transition iteration (10) recasts as

$$x_{n+1} = \begin{cases} R^h_A\, x_n & \text{if } u_n = 0, \\ R^h_{AB}\, x_n & \text{if } u_n = 1. \end{cases} \tag{12}$$

Notice that both $R^h_A$, $R^h_{AB}$ belong to the special orthogonal group SO(3) of three-dimensional rotations, and hence the state transition $x_n \to x_{n+1}$ represents a three-dimensional rotation that keeps unchanged the norm of each state vector instance. In fact, denoting by $R_n \in \{R^h_A, R^h_{AB}\}$ the $n$th transition matrix, the state transition from $x_0$ to $x_N$ is due to a sequence of pure rotations, namely

$$x_N = R_N\, R_{N-1}\, R_{N-2} \ldots R_2\, R_1\, R_0\, x_0, \tag{13}$$

where $N := \lfloor t_f/h \rfloor$ denotes the duration of the conversion cycle in terms of the number of discrete steps.

In order to design a control sequence $u_n$ that drives the state of the converter from $x_0$ to the desired state, which we shall denote as $x_f$, we shall make use of a method sometimes referred to as 'criterion function guidance', as explained, for example, in [27], based on the second method of Lyapunov. Such a control method is based on the definition of a criterion function $E$ as a weighted distance between the current state and the desired state, and in the development of an input control sequence that minimizes iteratively such a criterion function.

In the present work, we shall select as a criterion the function

$$E(x) := \tfrac{1}{2}(x - x_{\mathrm{f}})^\top P(x - x_{\mathrm{f}}), \tag{14}$$

where $P$ denotes a symmetric, semi-positive definite $3 \times 3$ matrix. The quadratic criterion function $E$ is defined as a squared weighted distance between the state and the desired state and has a measurement unit of $\mathrm{V}^2$. The time-derivative $q_u(x) := \frac{\mathrm{d}E}{\mathrm{d}t}$ of such a criterion function reads

$$\begin{aligned} q_u(x) &= (x - x_{\mathrm{f}})^\top P\dot{x} \\ &= (x - x_{\mathrm{f}})^\top PAx + u\,(x - x_{\mathrm{f}})^\top PBx. \end{aligned} \tag{15}$$

The rate of decrease depends on the input $u$ and the two possible decrease rates at state $x$ read

$$\begin{cases} q_0(x) = (x - x_{\mathrm{f}})^\top PAx, \\ q_1(x) = (x - x_{\mathrm{f}})^\top P(A + B)x. \end{cases} \tag{16}$$

On the basis of such decrease rates, we may define the first instance of the control strategy as

**(S1) if** $\min\{q_0, q_1\} < 0$, **then set** $u = 0$ **if** $q_0 \leqslant q_1$ **or set** $u = 1$ **if** $q_1 < q_0$.

Since it might happen that none of the control rates are negative, it is necessary to devise a second control strategy to apply in such event. We observe that during the first pulse, the switch must be in the position $u = 0$ to ensure that the inductor starts charging; during the conversion cycle, when there may be no energy transfer from the inductor to the storage capacitor, it is sensible to let the inductor charge by setting $u = 0$. Likewise, at the end of the conversion cycle, the switch must get back to the position $u = 0$ to ensure that the storage capacitor $C_2$ will not discharge into the inductor after having been charged to the fullest. On the basis of such reasoning, the second control strategy was devised as

**(S2) if** $\min\{q_0, q_1\} \geqslant 0$, **then set** $u = 0$.

The whole adopted control strategy is summarized in Figure 2.

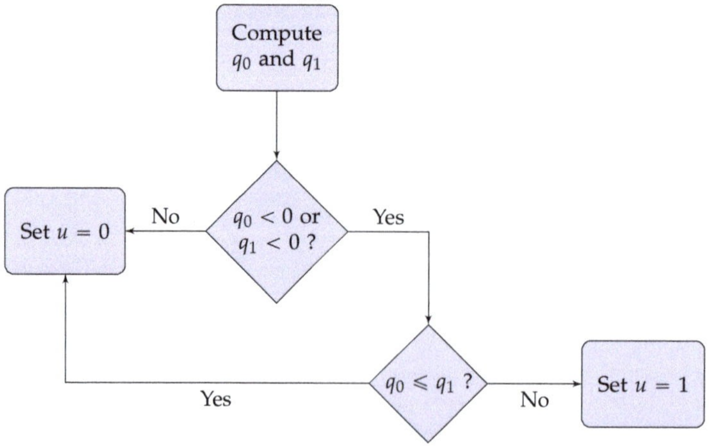


**Figure 2.** Flowchart detailing the devised control strategy **S1** + **S2** to transfer the whole energy content of the source capacitor $C_1$ to the target energy storage $C_2$.

We shall now assume the weighting matrix $P$ to be diagonal, namely that

$$P = \begin{pmatrix} p_1 & 0 & 0 \\ 0 & p_2 & 0 \\ 0 & 0 & p_3 \end{pmatrix}, \tag{17}$$

with $p_1, p_2, p_3 \geqslant 0$. Under this assumption, we were able to estimate the time $t^*$ at which the very first commutation takes place.

**Theorem 2** (First commutation time). *Define the unitless ratio* $\rho := \frac{p_2\omega_2}{(p_1-p_3)\omega_1}$. *Under the assumptions that* $p_3 < p_1$, $h\omega_1 < \pi$, *and* $\rho < 1$, *the exact time* $t^\star$ *at which the first commutation* $0 \to 1$ *of the control switch happens is given by*

$$\omega_1 t^\star = \cos^{-1}(\rho). \tag{18}$$

*In addition, the state* $x^\star$ *reached at the first commutation reads* $x^\star = \sqrt{2E_0}\begin{pmatrix}\rho & 0 & -\sqrt{1-\rho^2}\end{pmatrix}^\top$.

**Proof.** At the very beginning of the control cycle, it holds that

$$\begin{cases} q_0(x_0) = (x_0 - x_f)^\top PAx_0, \\ q_1(x_0) = (x_0 - x_f)^\top P(A+B)x_0. \end{cases} \tag{19}$$

Calculations reveal that

$$PAx_0 = (A+B)x_0 = 0, \tag{20}$$

which implies that $q_0(x_0) = q_1(x_0) = 0$. Therefore, the first switch state is determined by strategy **S2**.

The first commutation $0 \to 1$ of the input control $u$ will happen at a time $t^\star > 0$. We further notice that after the first pulse, the control variables take the values

$$\begin{cases} q_0(x(h)) = (R_A^h\, x_0 - x_f)^\top PAR_A^h\, x_0 = -(p_1 - p_3)\omega_1 \cos(h\omega_1)\sin(h\omega_1)\sqrt{2E_0}, \\ q_1(x(h)) = (R_A^h\, x_0 - x_f)^\top P(A+B)R_A^h\, x_0 = -p_2\omega_2 \sin(h\omega_1)\sqrt{2E_0}. \end{cases} \tag{21}$$

Hence, as long as $p_3 < p_1$ and $h\omega_1 < \pi$, it holds that $q_0(x(h)) < 0$ and $q_1(x(h)) < 0$. Hence, within the first pulse, the criterion function decreases (i.e., the converter state gets closer to the desired state), and hence $t^\star > h$.

The first commutation happens when the value of the function $q_0$ meets the value of the function $q_1$; hence, according to the devised strategy **S1**, the control input switches to 1. Such a condition corresponds to

$$\begin{aligned} 0 &= q_1(e^{At^\star}x_0) - q_0(e^{At^\star}x_0) \\ &= (e^{At^\star}x_0 - x_f)^\top PBe^{At^\star}x_0 \\ &= [(p_1 - p_3)\omega_1\cos(\omega_1 t^\star) - p_2\omega_2)]\sin(\omega_1 t^\star)\sqrt{2E_0}. \end{aligned} \tag{22}$$

The result (18) follows. The state $x^\star$ may be obtained by direct substitution of the value of $t^\star$ in the expression $\exp(At^\star)x_0$. □

Notice that before the first commutation happens, the energy content of the capacitor $C_2$ is still null; hence, the first pulse train serves to load the magnetic field of the inductor. In addition, the commutation time $t^\star$ does not depend on the energy content $E_0$.

It is instructive to observe that the ratio $\rho$ depends on the non-zero entries of the weighting matrix $P$ and on the ratio between the two capacitive elements, in fact $\omega_2/\omega_1 = \sqrt{C_1/C_2}$. In addition, it is worth observing that the weights $p_1$ and $p_3$ do not contribute independently of each other but only through their difference $p_1 - p_3$. For this reason, in the following, we shall always take $p_3 = 0$, which corresponds to leaving completely unconstrained the evolution of the current flowing through the inductor.

The advantages and drawbacks of the developed method in comparison to an open-loop method, recalled in Section 4, are summarized in Table 1 for the convenience of the reader.

**Table 1.** Advantages and drawbacks of the proposed control strategy compared to the LK one.

| Control Strategy | Advantages | Drawbacks |
|---|---|---|
| Leonard–Krishnaprasad Lie-group-averaging-based control strategy | Completes a conversion cycle in a pre-fixed time interval; specifies the trajectory in state space in advance (which may be optimized beforehand). | Assumes nominal values for the components; requires off-line computations to establish a control input sequence; requires re-calculation whenever changes in the conditions occur. |
| Proposed closed-loop feedback-type criterion-based control strategy | Deals with parameter mismatch as well as input/output voltage mismatch; requires no off-line computations to run; runs at lower switching frequency. | Does not guarantee a fixed conversion cycle duration; does not embody any limit on the intensity of the current flowing through the inductor. |

## 6. Numerical Experiments and Comparison

In order to exemplify the behavior of the devised control algorithm and to compare its performances to those exhibited by the original LK algorithm, we shall present two series of experiments. The first series was performed under ideal conditions, while the second series was performed under pre-set value mismatch.

### *6.1. Comparisons under Nominal Conditions*

In order to exemplify the performances of the devised control strategy, we shall repeat the numerical experiments conducted in [5] in order to be able to perform a comparison. The normalized values for the electrical components are $C_1 = 0.1$ Farad, $C_2 = 0.2$ Farad, $L_3 = 0.5$ Henry. In addition, we set $t_f = 1$ second and $E_0 = \frac{1}{2}$ joule. We shall mention, in our own interpretation of the choice of the authors of [5], that the considered circuit is 'normalized', namely the values of the capacitances and of the inductance are normalized by a factor several orders of magnitude less than the unit, since all that matters is ultimately the ratio $C_1/C_2$.

In order to run the control algorithm, we set the pulse duration to $h = 0.01$ s and the weights to be $p_1 = 2$, $p_2 = 1$. With these values, we have $\rho \approx 0.3536$ and $t^\star \approx 0.2704$ s.

The results obtained by the iteration (12) are displayed graphically in Figure 3.

The top panel of Figure 3 shows the sequence of pulse trains computed by the control algorithm of Figure 2. The control strategy resulting from the application of the devised algorithm is very clear. After a first long phase from $t = 0$ to $t = t^\star$, in which the inductor loads its magnetic field of electrical energy, the inductor drains energy from the capacitor $C_1$ for short time intervals and subsequently cedes part of its energy to the capacitor $C_2$, and this process repeats. The bottom panel of Figure 3 shows the evolution, over time, of the three state variables. In this panel, the vertical dashed line denotes the first commutation time $t^\star$, while the horizontal dashed line denotes the value of the ratio $\rho$ that the variable $x_1$ meets in correspondence of the first commutation time. The reached state at the end of the full conversion cycle was found to be $x_{100} = (0.0056 \; -0.9998 \; 0.0176)^\top$, namely, the desired state $x_f$ is reached with overall precision $\mathcal{O}(h)$. In fact, it is found that $\sqrt{(x_{100} - x_f)^\top P(x_{100} - x_f)} = \sqrt{2E(x_{100})} \approx 0.0068 \, \mathrm{J}^{\frac{1}{2}}$, which denotes an overall measure of conversion precision. (Here, J stands for joule.)

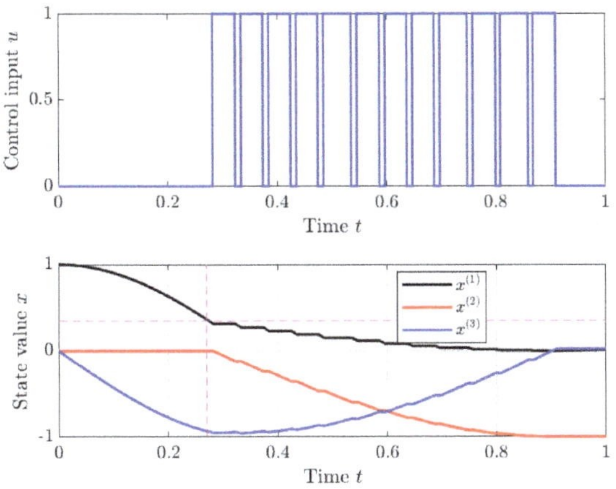


**Figure 3.** Result of application of the devised control algorithm to the experimental setting of paper [5] with $h = 0.01$ s: pulse trains and state variable values.

Figure 4 illustrates the values taken by the criterion function and by the control variables during the control cycle.

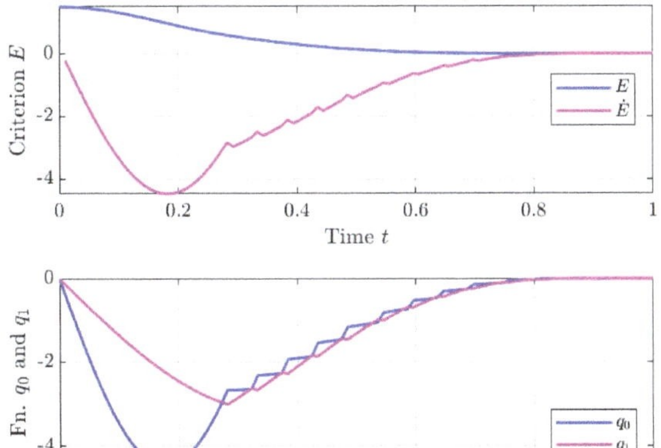


**Figure 4.** Result of application of the devised control algorithm to the experimental setting of paper [5] with $h = 0.01$ s: control variables used by the devised algorithm.

The top panel shows the values of the criterion function $E$ and of its derivative $\dot{E}$ during the control cycle. As expected, the criterion function $E$ decreases monotonically to zero. The bottom panel shows the values of the control variables $q_0$ and $q_1$, whose course meets the theoretical analysis.

In general, with the stated values for the electrical components of the converter, it is possible to predict the time $t^{\star}$ of the first $0 \rightarrow 1$ commutation as a function of the ratio $p_2/p_1$. This dependence is shown in Figure 5, where the open circle corresponds to the ratio $p_2/p_1 = 1/2$ chosen for this experiment. The graphical result shows that it is possible to control the first commutation time fairly easily, if desired, by selecting the ratio $p_2/p_1$.

By increasing the value of the latency time $h$, the precision of the conversion deteriorates, albeit not dramatically. Figure 6 illustrates the state trajectory of the converter when $h = 0.02$ s.

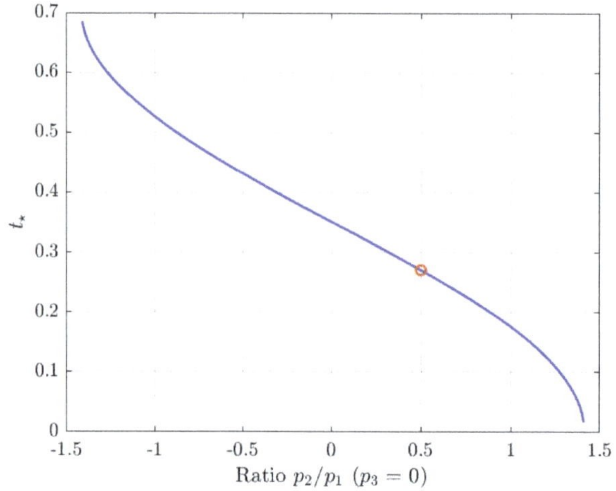


**Figure 5.** Dependence of the time $t^*$ of the first $0 \rightarrow 1$ commutation as a function of the ratio $p_2/p_1$.

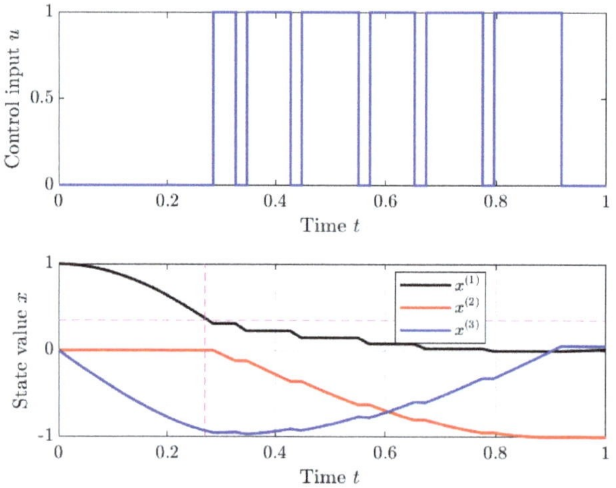


**Figure 6.** Result of application of the devised control algorithm to the experimental setting of paper [5] with $h = 0.02$ s.

In this experiment, the reached state is $x_{50} = (0.0187 \; -0.9986 \; 0.0503)^\top$, corresponding to an overall conversion precision of $\sqrt{2E(x_{50})} \approx 0.0201 \; \mathrm{J}^{\frac{1}{2}}$.

Given the linearity of the system (7) in the state variables, the result of conversion is independent of the initial condition. By this we mean that an initial state $(\sqrt{2E_0} \; 0 \; 0)^\top$ will result in a final state close to $(0 \; -\sqrt{2E_0} \; 0)^\top$, for every $E_0 > 0$. The performance of the algorithm instead depends noticeably on the value of the constant $h$. For example, for a value $h = 10^{-4}$ s, the precision of conversion turns out to be approximately $5.2019 \times 10^{-5} \; \mathrm{J}^{\frac{1}{2}}$.

Figure 7 illustrates the state trajectory and the switching sequence of the Leonard–Krishnaprasad method contrasted to the proposed control method. The state trajectories corresponding to the two methods differ substantially, especially the inductor current intensity. The overall evolution reveals comparable conversion performances obtained by a substantially different switching rate.

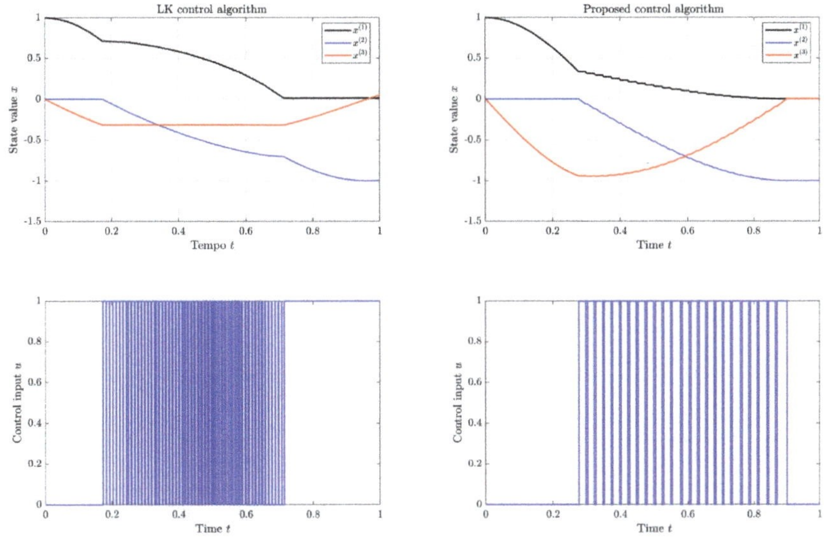


**Figure 7.** Comparison of the performances of the Leonard–Krishnaprasad (LK) and of the proposed methods on the experimental setting of paper [5] with $h = 0.005$ s.

As a further element of comparison, the number of $0 \to 1$ and $1 \to 0$ switchings were evaluated for both algorithms. These numbers, divided by the duration $t_f$ of the conversion cycle, give the switching rate corresponding to the compared control algorithms. The LK method results in a switching rate of 117 Hz, in contrast to the proposed method, which entails a switching rate of 48 Hz. Such rates appear to be in good accordance with the results displayed in the bottom panels of Figure 7. We shall note that, as the values of the parameters are not realistic, the values of the switching rates are not realistic either; however, they serve for comparison purposes.

### *6.2. Comparison under Nominal Value Mismatch*

In real-world settings, the values of the parameters of the electrical circuit in Figure 1 may not coincide with the nominal ones. In such an event, a closed-loop, feedback-driven algorithm is supposed to overperform a pre-set algorithm. The following numerical experiments aim to quantify the effects of parameter mismatch on the conversion performance and, ultimately, to assess the resilience of the considered control algorithms against parameter mismatch. The ideal values set in the tests were $C_1 = 0.1$ F, $C_2 = 0.2$ F, and $L_3 = 0.5$ H; therefore, $V_2(t_f) \approx \pm 0.7071 V_1(0)$.

#### 6.2.1. Mismatch in the Initial Source Voltage $V_1(0)$

The aim of the present set of experiments is to assess the entity of variation of the output voltage resulting from the variation of the source voltage. The results of three independent experiments are shown in Table 2. The percentage change in the output voltage as obtained by running both algorithms is less than 1% of its expected value, while being practically negligible for the proposed algorithm.

In addition, the adaptivity of the proposed algorithm allows us to attain conversion with fewer commutations of the switch (figures are omitted for brevity).

**Table 2.** Comparison of conversion results obtained by allowing a variation of the initial source voltage $V_1(0)$.

| Test Number | Initial Values | Obtained Results | | | |
|---|---|---|---|---|---|
| | | LK Algorithm | | Proposed Algorithm | |
| | | Final Values | Deviation of Voltage $V_2$ | Final Values | Deviation of Voltage $V_2$ |
| First test | $V_1(0) = 2$ V<br>$V_2(0) = 0$ V<br>$I_3(0) = 0$ A | $V_1(1) = +0.0229$ V<br>$V_2(1) = -1.4055$ V<br>$I_3(1) = +0.1030$ A | −0.6% | $V_1(1) = -0.0001$ V<br>$V_2(1) = -1.4142$ V<br>$I_3(1) = +0.0062$ A | ~0.00% |
| Second test | $V_1(0) = 5$ V<br>$V_2(0) = 0$ V<br>$I_3(0) = 0$ A | $V_1(1) = +0.0573$ V<br>$V_2(1) = -3.5138$ V<br>$I_3(1) = +0.2575$ A | −0.6% | $V_1(1) = -0.0003$ V<br>$V_2(1) = -3.5354$ V<br>$I_3(1) = +0.0155$ A | ~0.00% |
| Third test | $V_1(0) = 7$ V<br>$V_2(0) = 0$ V<br>$I_3(0) = 0$ A | $V_1(1) = +0.0802$ V<br>$V_2(1) = -4.9193$ V<br>$I_3(1) = +0.3604$ A | −0.6% | $V_1(1) = -0.0005$ V<br>$V_2(1) = -4.9496$ V<br>$I_3(1) = +0.0217$ A | ~0.00% |

#### 6.2.2. Mismatch in the Initial Output Voltage $V_2(0)$

In the present series of tests, the change in the $V_2$ output voltage, resulting from a non-zero initial voltage $V_2(0)$, is analyzed. Such tests aim at reproducing the presence of a residual voltage on the $C_2$ capacitor at the beginning of a conversion cycle. From the results of Table 3, it emerges that the percentage change of $V_2(1)$ pertaining to the proposed algorithm is lower.

**Table 3.** Comparison of conversion results obtained by allowing a variation of the initial output voltage $V_2(0)$.

| Test Number | Initial Values | Obtained Results | | | |
|---|---|---|---|---|---|
| | | LK Algorithm | | Proposed Algorithm | |
| | | Final Values | Deviation of Voltage $V_2$ | Final Values | Deviation of Voltage $V_2$ |
| First test | $V_1(0) = 1$ V<br>$V_2(0) = 0.5$ V<br>$I_3(0) = 0$ A | $V_1(1) = -0.4171$ V<br>$V_2(1) = -0.7521$ V<br>$I_3(1) = -0.1982$ A | 6.36% | $V_1(1) = -0.4139$ V<br>$V_2(1) = -0.7203$ V<br>$I_3(1) = -0.2413$ A | 1.87% |
| Second test | $V_1(0) = 1$ V<br>$V_2(0) = 1$ V<br>$I_3(0) = 0$ A | $V_1(1) = -0.8457$ V<br>$V_2(1) = -0.8014$ V<br>$I_3(1) = -0.4480$ A | 13.33% | $V_1(1) = -0.9751$ V<br>$V_2(1) = -0.7225$ V<br>$I_3(1) = -0.4484$ A | 2.18% |

#### 6.2.3. Mismatch in the Initial Current $I_3(0)$

The aim of the present series of simulations is to quantify the variation of the final output voltage caused by variations of the initial inductor current $I_3(0)$. The purpose of these experiments is to analyze the performance of both algorithms in the event of a residual current in the $L_3$ inductor at the beginning of a conversion cycle.

From the results of Table 4, it emerges how the percentage variation of the output voltage $V_2(1)$ is lower when the proposed control algorithm is applied.

#### 6.2.4. Large Mismatch in the Capacitance Values $C_1$ and $C_2$

The resilience of the analyzed control algorithms was assessed by recording the variation of the output voltage of the DC-DC converter resulting from large variations of the $C_1$ and $C_2$ values (far larger than commercial tolerance values). The purpose of this test is to analyze the performance of both algorithms in the event of a mistake in the choice of a single capacitor.

**Table 4.** Comparison of conversion results obtained by allowing a variation of the initial inductor current $I_3(0)$.

| Test Number | Initial Values | Obtained Results | | | |
|---|---|---|---|---|---|
| | | LK Algorithm | | Proposed Algorithm | |
| | | Final Values | Deviation of Voltage $V_2$ | Final Values | Deviation of Voltage $V_2$ |
| First test | $V_1(0) = 1$ V<br>$V_2(0) = 0$ V<br>$I_3(0) = 0.3$ A | $V_1(1) = -0.5453$ V<br>$V_2(1) = -0.7315$ V<br>$I_3(1) = -0.1982$ A | 3.45% | $V_1(1) = -0.4881$ V<br>$V_2(1) = -0.7209$ V<br>$I_3(1) = -0.1856$ A | 1.95% |
| Second test | $V_1(0) = 1$ V<br>$V_2(0) = 0$ V<br>$I_3(0) = 0.5$ A | $V_1(1) = -0.9012$ V<br>$V_2(1) = -0.7570$ V<br>$I_3(1) = -0.2502$ A | 7.06% | $V_1(1) = -0.9062$ V<br>$V_2(1) = -0.7291$ V<br>$I_3(1) = -0.4484$ A | 3.11% |
| Third test | $V_1(0) = 1$ V<br>$V_2(0) = 0$ V<br>$I_3(0) = -0.3$ A | $V_1(1) = -0.5224$ V<br>$V_2(1) = -0.6740$ V<br>$I_3(1) = +0.2325$ A | −4.68% | $V_1(1) = -0.5693$ V<br>$V_2(1) = -0.7208$ V<br>$I_3(1) = -0.1318$ A | 1.94% |

From the results displayed in Table 5, it appears that the effect of an incorrect choice in the capacitance value, when large, heavily affects the performance of the algorithms that control the DC-DC circuitry. Even in this case, however, the proposed algorithm shows a better resilience. In particular, the resilience against a large variation in the capacitance of the source capacitor $C_1$ is notable.

**Table 5.** Comparison of conversion results obtained by allowing a large variation of the capacitance of the capacitors $C_1$ and $C_2$. In these simulation experiments, $V_1(0) = 1$ V, $V_2(0) = 0$ V, and $I_3(0) = 0$ A.

| Test Number | Components Values | Obtained Results | | | |
|---|---|---|---|---|---|
| | | LK Algorithm | | Proposed Algorithm | |
| | | Final Values | Deviation of Voltage $V_2$ | Final Values | Deviation of Voltage $V_2$ |
| First test | $C_1 = 0.1$ F<br>$C_2 = 0.4$ F (+100%)<br>$L_3 = 0.5$ H | $V_1(1) = +0.0908$ V<br>$V_2(1) = -0.4659$ V<br>$I_3(1) = -0.1577$ A | −34.11% | $V_1(1) = +0.0453$ V<br>$V_2(1) = -0.4957$ V<br>$I_3(1) = -0.0551$ A | −29.89% |
| Second test | $C_1 = 0.2$ F (+100%)<br>$C_2 = 0.2$ F<br>$L_3 = 0.5$ H | $V_1(1) = +0.4312$ V<br>$V_2(1) = -0.9026$ V<br>$I_3(1) = -0.0051$ A | 27.65% | $V_1(1) = -0.4989$ V<br>$V_2(1) = -0.7085$ V<br>$I_3(1) = -0.3156$ A | 0.20% |

#### 6.2.5. Large Mismatch in the Values of the Inductance $L_3$

The aim of these further simulations is to assess the variation of the output voltage at the end of a conversion cycle resulting from a mismatch in the value of the inductance.

From the results of Table 6, it is clear that the percentage variation of the voltage $V_2(1)$ is lower with the proposed algorithm in all considered cases. The proposed closed-loop algorithm proved henceforth to be more resilient than the pre-fixed LK algorithm.

#### 6.2.6. Five Percent Tolerance in the Values of the Parameters of the Discrete Components

A five percent tolerance in the values of the parameters $C_1$, $C_2$, and $L_3$ may be expected in yield components, which would result in a degradation of conversion performances.

From the results of Table 7, it emerges that in response to a small mismatch of single components, the proposed algorithm works better in terms overall results compared to the LK one.

**Table 6.** Comparison of conversion results obtained by allowing a large variation of the inductance $L_3$ of the inductor. In these simulation experiments, $V_1(0) = 1$ V, $V_2(0) = 0$ V, and $I_3(0) = 0$ A.

| Test Number | Components Values | Obtained Results | | | |
|---|---|---|---|---|---|
| | | LK Algorithm | | Proposed Algorithm | |
| | | Final Values | Deviation of Voltage $V_2$ | Final Values | Deviation of Voltage $V_2$ |
| First test | $C_1 = 0.1$ F<br>$C_2 = 0.2$ F<br>$L_3 = 0.1$ H (−80%) | $V_1(1) = +0.6298$ V<br>$V_2(1) = +0.5262$ V<br>$I_3(1) = +0.2339$ A | −174.41% | $V_1(1) = +0.0106$ V<br>$V_2(1) = -0.7070$ V<br>$I_3(1) = -0.0149$ A | −0.01% |
| Second test | $C_1 = 0.1$ F<br>$C_2 = 0.2$ F<br>$L_3 = 0.8$ H (+60%) | $V_1(1) = -0.2614$ V<br>$V_2(1) = -0.6506$ V<br>$I_3(1) = -0.1036$ A | −7.99% | $V_1(1) = -0.0003$ V<br>$V_2(1) = -0.6891$ V<br>$I_3(1) = -0.0792$ A | −2.55% |
| Third test | $C_1 = 0.1$ F<br>$C_2 = 0.2$ F<br>$L_3 = 1$ H (+100%) | $V_1(1) = +0.3737$ V<br>$V_2(1) = -0.5874$ V<br>$I_3(1) = -0.1307$ A | −16.93% | $V_1(1) = -0.0179$ V<br>$V_2(1) = -0.6343$ V<br>$I_3(1) = -0.1397$ A | −10.30% |

**Table 7.** Comparison of conversion result degradation due to a mismatch in the values of the discrete components of ±5%. In these simulation experiments, $V_1(0) = 1$ V, $V_2(0) = 0$ V, and $I_3(0) = 0$ A.

| Test Number | Components Values | Obtained Results | | | |
|---|---|---|---|---|---|
| | | LK Algorithm | | Proposed Algorithm | |
| | | Final Values | Deviation of Voltage $V_2$ | Final Values | Deviation of Voltage $V_2$ |
| First test | $C_1 = 0.105$ F (+5%)<br>$C_2 = 0.2$ F<br>$L_3 = 0.5$ H | $V_1(1) = +0.0457$ V<br>$V_2(1) = -0.7205$ V<br>$I_3(1) = +0.0466$ A | 1.90% | $V_1(1) = -0.1643$ V<br>$V_2(1) = -0.7078$ V<br>$I_3(1) = -0.0628$ A | 0.10% |
| Second test | $C_1 = 0.095$ F (−5%)<br>$C_2 = 0.2$ F<br>$L_3 = 0.5$ H | $V_1(1) = -0.0251$ V<br>$V_2(1) = -0.6835$ V<br>$I_3(1) = +0.0567$ A | −3.34% | $V_1(1) = +0.0034$ V<br>$V_2(1) = -0.6892$ V<br>$I_3(1) = -0.0018$ A | −2.53% |
| Third test | $C_1 = 0.1$ F<br>$C_2 = 0.210$ F (+5%)<br>$L_3 = 0.5$ H | $V_1(1) = +0.0020$ V<br>$V_2(1) = -0.6884$ V<br>$I_3(1) = +0.0343$ A | −2.65% | $V_1(1) = -0.0163$ V<br>$V_2(1) = -0.6898$ V<br>$I_3(1) = +0.0108$ A | −2.45% |
| Fourth test | $C_1 = 0.1$ F<br>$C_2 = 0.190$ F (−5%)<br>$L_3 = 0.5$ H | $V_1(1) = +0.0218$ V<br>$V_2(1) = -0.7168$ V<br>$I_3(1) = +0.0699$ A | 1.37% | $V_1(1) = -0.1753$ V<br>$V_2(1) = -0.7091$ V<br>$I_3(1) = +0.0525$ A | 0.28% |
| Fifth test | $C_1 = 0.1$ F<br>$C_2 = 0.2$ F<br>$L_3 = 0.525$ H (+5%) | $V_1(1) = +0.0366$ V<br>$V_2(1) = -0.7056$ V<br>$I_3(1) = +0.0277$ A | −0.21% | $V_1(1) = +0.0088$ V<br>$V_2(1) = -0.7068$ V<br>$I_3(1) = +0.0123$ A | −0.04% |
| Sixth test | $C_1 = 0.1$ F<br>$C_2 = 0.2$ F<br>$L_3 = 0.475$ H (−5%) | $V_1(1) = -0.0143$ V<br>$V_2(1) = -0.6970$ V<br>$I_3(1) = +0.0786$ A | −1.42% | $V_1(1) = +0.0057$ V<br>$V_2(1) = -0.7070$ V<br>$I_3(1) = +0.0063$ A | −0.01% |

#### 6.2.7. Ten Percent Tolerance in the Values of the Parameters of the Discrete Components

Further to the previous section, a larger tolerance of ten percent in the values of the parameters of the discrete components, in line with commercial standards, may be taken as a basis for comparison of conversion performances.

The obtained results are displayed in Table 8, from which it emerges how in response to a higher mismatch of single components, the percentage variation of the final output voltage $V_2(1)$ stays proportional to the results obtained in Table 7. Again, it is clear how the proposed algorithm provides better results compared to the LK algorithm.

**Table 8.** Comparison of conversion result degradation due to a mismatch in the values of the discrete components of $\pm 10\%$. In these simulation experiments, $V_1(0) = 1$ V, $V_2(0) = 0$ V, and $I_3(0) = 0$ A.

| Test Number | Components Values | Obtained Results | | | |
|---|---|---|---|---|---|
| | | LK Algorithm | | Proposed Algorithm | |
| | | Final Values | Deviation of Voltage $V_2$ | Final Values | Deviation of Voltage $V_2$ |
| First test | $C_1 = 0.11$ F (+10%)<br>$C_2 = 0.2$ F<br>$L_3 = 0.5$ H | $V_1(1) = +0.0778$ V<br>$V_2(1) = -0.7368$ V<br>$I_3(1) = +0.0421$ A | 4.20% | $V_1(1) = -0.2178$ V<br>$V_2(1) = -0.7130$ V<br>$I_3(1) = -0.0789$ A | 0.83% |
| Second test | $C_1 = 0.09$ F (−10%)<br>$C_2 = 0.2$ F<br>$L_3 = 0.5$ H | $V_1(1) = -0.0640$ V<br>$V_2(1) = -0.6625$ V<br>$I_3(1) = +0.0624$ A | −6.31% | $V_1(1) = +0.0093$ V<br>$V_2(1) = -0.6708$ V<br>$I_3(1) = -0.0006$ A | −5.13% |
| Third test | $C_1 = 0.1$ F<br>$C_2 = 0.22$ F (+10%)<br>$L_3 = 0.5$ H | $V_1(1) = +0.0066$ V<br>$V_2(1) = -0.6740$ V<br>$I_3(1) = -0.0182$ A | −4.68% | $V_1(1) = +0.0163$ V<br>$V_2(1) = -0.6740$ V<br>$I_3(1) = +0.0074$ A | −4.68% |
| Fourth test | $C_1 = 0.1$ F<br>$C_2 = 0.18$ F (−10%)<br>$L_3 = 0.5$ H | $V_1(1) = +0.0333$ V<br>$V_2(1) = -0.7303$ V<br>$I_3(1) = +0.0895$ A | 3.28% | $V_1(1) = -0.2792$ V<br>$V_2(1) = -0.7102$ V<br>$I_3(1) = -0.0533$ A | 0.44% |
| Fifth test | $C_1 = 0.1$ F<br>$C_2 = 0.2$ F<br>$L_3 = 0.55$ H (+10%) | $V_1(1) = +0.0610$ V<br>$V_2(1) = -0.7061$ V<br>$I_3(1) = -0.0068$ A | −0.14% | $V_1(1) = +0.0079$ V<br>$V_2(1) = -0.7069$ V<br>$I_3(1) = +0.0089$ A | −0.03% |
| Sixth test | $C_1 = 0.1$ F<br>$C_2 = 0.2$ F<br>$L_3 = 0.45$ H (−10%) | $V_1(1) = +0.0404$ V<br>$V_2(1) = -0.6876$ V<br>$I_3(1) = -0.1096$ A | −2.76% | $V_1(1) = +0.0141$ V<br>$V_2(1) = -0.7069$ V<br>$I_3(1) = +0.0098$ A | −0.03% |

6.2.8. Combined Ten Percent Variation of the Values of the Parameters $C_1$ and $C_2$

In contrast to the case when just one of the components is subject to a mismatch in its nominal value, conversion performances are evaluated when the values of both capacitors are subjected to mismatch at the same time.

The results obtained by running both algorithms are displayed in Table 9, from which it is clear how the proposed algorithm yields better results than the LK one. When, in particular, the capacitance mismatch exhibits opposite signs (+10%/−10% or −10%/+10%), both algorithms decrease in performance. This is due to the fact that the final output value is strictly correlated to the capacitor ratio, based on Formula (5).

**Table 9.** Combined ten percent variation of the values of the parameters $C_1$ and $C_2$. In these simulation experiments, $V_1(0) = 1$ V, $V_2(0) = 0$ V, and $I_3(0) = 0$ A.

| Test Number | Components Values | Obtained Results | | | |
|---|---|---|---|---|---|
| | | LK Algorithm | | Proposed Algorithm | |
| | | Final Values | Deviation of Voltage $V_2$ | Final Values | Deviation of Voltage $V_2$ |
| First test | $C_1 = 0.11$ F (+10%)<br>$C_2 = 0.2$ F<br>$L_3 = 0.5$ H | $V_1(1) = +0.0778$ V<br>$V_2(1) = -0.7368$ V<br>$I_3(1) = +0.0421$ A | 4.20% | $V_1(1) = -0.2178$ V<br>$V_2(1) = -0.7130$ V<br>$I_3(1) = -0.0789$ A | 0.83% |
| First test | $C_1 = 0.11$ F (+10%)<br>$C_2 = 0.22$ F (+10%)<br>$L_3 = 0.5$ H | $V_1(1) = +0.0610$ V<br>$V_2(1) = -0.7061$ V<br>$I_3(1) = +0.0075$ A | −0.14% | $V_1(1) = +0.0079$ V<br>$V_2(1) = -0.7069$ V<br>$I_3(1) = -0.0098$ A | −0.03% |
| Second test | $C_1 = 0.11$ F (+10%)<br>$C_2 = 0.18$ F (−10%)<br>$L_3 = 0.5$ H | $V_1(1) = +0.0981$ V<br>$V_2(1) = -0.7664$ V<br>$I_3(1) = -0.0817$ A | 8.39% | $V_1(1) = -0.3667$ V<br>$V_2(1) = -0.7175$ V<br>$I_3(1) = -0.0714$ A | 1.47% |

**Table 9.** *Cont.*

| Test Number | Components Values | Obtained Results | | | |
|---|---|---|---|---|---|
| | | LK Algorithm | | Proposed Algorithm | |
| | | Final Values | Deviation of Voltage $V_2$ | Final Values | Deviation of Voltage $V_2$ |
| Third test | $C_1 = 0.09$ F ($-10\%$)<br>$C_2 = 0.22$ F ($+10\%$)<br>$L_3 = 0.5$ H | $V_1(1) = -0.0836$ V<br>$V_2(1) = -0.6361$ V<br>$I_3(1) = +0.0307$ A | $-10.04\%$ | $V_1(1) = +0.0232$ V<br>$V_2(1) = -0.6394$ V<br>$I_3(1) = -0.0044$ A | $-9.57\%$ |
| Fourth test | $C_1 = 0.09$ F ($-10\%$)<br>$C_2 = 0.18$ F ($-10\%$)<br>$L_3 = 0.5$ H | $V_1(1) = -0.0404$ V<br>$V_2(1) = -0.6876$ V<br>$I_3(1) = +0.0986$ A | $-2.76\%$ | $V_1(1) = +0.0141$ V<br>$V_2(1) = -0.7069$ V<br>$I_3(1) = +0.0088$ A | $-0.03\%$ |

#### 6.2.9. Combined Ten Percent Component Variation

The last series of simulated tests aim at assessing the resilience of the control algorithms when all discrete components are subjected to a ten percent change with respect to their nominal values. The obtained results are displayed Table 10.

**Table 10.** Combined ten percent variation of the values of all components. In these simulation experiments, $V_1(0) = 1$ V, $V_2(0) = 0$ V, and $I_3(0) = 0$ A.

| Test Number | Components Values | Obtained Results | | | |
|---|---|---|---|---|---|
| | | LK Algorithm | | Proposed Algorithm | |
| | | Final Values | Deviation of Voltage $V_2$ | Final Values | Deviation of Voltage $V_2$ |
| First test | $C_1 = 0.11$ F ($+10\%$)<br>$C_2 = 0.18$ F ($-10\%$)<br>$L_3 = 0.55$ H ($+10\%$) | $V_1(1) = +0.1435$ V<br>$V_2(1) = -0.7718$ V<br>$I_3(1) = +0.0337$ A | $9.15\%$ | $V_1(1) = -0.3522$ V<br>$V_2(1) = -0.7103$ V<br>$I_3(1) = -0.1003$ A | $0.45\%$ |
| Second test | $C_1 = 0.11$ F ($+10\%$)<br>$C_2 = 0.18$ F ($-10\%$)<br>$L_3 = 0.45$ H ($-10\%$) | $V_1(1) = +0.0503$ V<br>$V_2(1) = -0.7475$ V<br>$I_3(1) = +0.1436$ A | $5.71\%$ | $V_1(1) = +0.3987$ V<br>$V_2(1) = -0.7152$ V<br>$I_3(1) = -0.0308$ A | $1.15\%$ |
| Third test | $C_1 = 0.09$ F ($-10\%$)<br>$C_2 = 0.18$ F ($-10\%$)<br>$L_3 = 0.55$ H ($+10\%$) | $V_1(1) = +0.0064$ V<br>$V_2(1) = -0.7019$ V<br>$I_3(1) = +0.0510$ A | $-0.74\%$ | $V_1(1) = +0.0069$ V<br>$V_2(1) = -0.7070$ V<br>$I_3(1) = +0.0047$ A | $-0.01\%$ |
| Fourth test | $C_1 = 0.09$ F ($-10\%$)<br>$C_2 = 0.18$ F ($-10\%$)<br>$L_3 = 0.45$ H ($-10\%$) | $V_1(1) = -0.0877$ V<br>$V_2(1) = -0.6585$ V<br>$I_3(1) = +0.1591$ A | $-6.87\%$ | $V_1(1) = +0.0031$ V<br>$V_2(1) = -0.7071$ V<br>$I_3(1) = -0.0031$ A | ~0.00% |
| Fifth test | $C_1 = 0.09$ F ($-10\%$)<br>$C_2 = 0.22$ F ($+10\%$)<br>$L_3 = 0.55$ H ($+10\%$) | $V_1(1) = -0.0306$ V<br>$V_2(1) = -0.6394$ V<br>$I_3(1) = -0.0114$ A | $-9.57\%$ | $V_1(1) = +0.0122$ V<br>$V_2(1) = -0.6396$ V<br>$I_3(1) = +0.0000$ A | $-9.55\%$ |
| Sixth test | $C_1 = 0.09$ F ($-10\%$)<br>$C_2 = 0.22$ F ($+10\%$)<br>$L_3 = 0.45$ H ($-10\%$) | $V_1(1) = -0.1386$ V<br>$V_2(1) = -0.6219$ V<br>$I_3(1) = +0.0857$ A | $-12.04\%$ | $V_1(1) = +0.0207$ V<br>$V_2(1) = -0.6395$ V<br>$I_3(1) = +0.0018$ A | $-9.56\%$ |

From the obtained results it appears that the proposed algorithm shows better performance than the LK algorithm.

## 7. Conclusions and Devised Extensions

The purpose of the illustrated research was to propose a novel control strategy for a prototypical DC-DC electrical converter on the basis of an early work by Leonard and Krishnaprasad, where a DC-DC converter was modeled as a state space dynamical system and controlled by an open-loop strategy based on Lie group theory.

In the present research endeavor, we have introduced a closed-loop control strategy based on two criteria, namely maximum convergence rate and minimal energy exchange. A theoretical study of the time where the first commutation $0 \rightarrow 1$ of the switch happens was conducted. A series of numerical simulations were shown and examined to illustrate the devised control algorithm. The novel method was compared to the original one by Leonard and Krishnaprasad numerically.

We shall underline that the intention of the paper was mostly a theoretical speculation, without a direct connection to normal DC-DC converter operation, along the same lines as the reference paper [5]. In fact, an operation mode of a DC-DC converter does not imply that an input capacitor $C_1$ is discharged into an output capacitor $C_2$. In contrast, the design and operation (and control) of a converter is based on a practically constant voltage $V_1$, while the voltage $V_2$ exhibits a voltage ripple influenced by control and by the value of the capacitor $C_2$. Additionally, in the considered case study, no load is connected to the output of the converter, whereas in practical applications, an electrical load is certainly present.

A possible extension of the devised method would be that of a single source to multiple target storage unit (SS2MTU), namely to transfer the energy content of a source capacitor $C_s$ to a number of target capacitors $C_i$, with $i = 1, 2, \ldots, N$ with different capacitive elements so as to achieve different conversion ratios simultaneously.

A further extension, building on the SS2MSU system, would be a multiple DC source to a multiple DC storage unit (MSU2MTU), possibly with different capacitance values, to accommodate for a variety of concurring DC energy sources being converted to a multiple storage unit with possibly different conversion ratios.

**Author Contributions:** Conceptualization, S.F.; methodology, S.F., L.C. and G.B.; software, S.F., L.C. and G.B.; validation, L.C. and G.B.; formal analysis, S.F.; writing—original draft preparation, S.F.; writing—review and editing, S.F., L.C. and G.B.; visualization, L.C. and G.B.; supervision, S.F. All authors have read and agreed to the published version of the manuscript.

**Funding:** This research received no external funding.

**Data Availability Statement:** Data used within the paper are drawn from referenced literature.

**Acknowledgments:** The first author wishes to acknowledge the work of his former student, M.L. Bramucci, who re-implemented the LK method from scratch.

**Conflicts of Interest:** The authors declare no conflicts of interest.

## References


1. Li, Z.; Soh, Y.; Wen, C. (Eds.) *Switched and Impulsive Systems: Analysis, Design and Applications*; Springer: Berlin/Heidelberg, Germany, 2005.
2. Sun, Z.; Ge, S.S. (Eds.) *Switched Linear Systems: Control and Design*; Springer: Berlin/Heidelberg, Germany, 2005.
3. Liberzon, D. *Switching in Systems and Control*; Birkhäuser: Basel, Switzerland, 2006.
4. Mahmoud, M.S. *Switched Time-Delay Systems: Stability and Control*; Springer: Berlin/Heidelberg, Germany, 2010.
5. Leonard, N.; Krishnaprasad, P. Control of switched electrical networks using averaging on Lie groups. In Proceedings of 1994 33rd IEEE Conference on Decision and Control, Orlando, FL, USA, 14–16 December 1994; Volume 2, pp. 1919–1924. [CrossRef]
6. Corradini, L.; Maksimovic, D.; Mattavelli, P.; Zane, R. *Digital Control of High-Frequency Switched-Mode Power Converters*; John Wiley & Sons: Hoboken, NJ, USA, 2015.
7. Kursun, V.; Narendra, S.; De, V.; Friedman, E. Low-voltage-swing monolithic DC-DC conversion. *IEEE Trans. Circuits Syst. II Express Briefs* **2004**, *51*, 241–248. [CrossRef]
8. Mohan, N.; Undeland, T.M.; Robbins, W.P. *Power Electronics: Converters, Applications, and Design*, 3rd ed.; John Wiley & Sons: Hoboken, NJ, USA, 2005.
9. Shen, C.L.; Shen, Y.S.; Tsai, C.T. Isolated DC-DC converter for bidirectional power flow controlling with soft-switching feature and high step-up/down voltage conversion. *Energies* **2017**, *10*, 296. [CrossRef]
10. Drakunov, S.V.; Reyhanoglu, M.; Singh, B. Sliding mode control of DC-DC power converters. *IFAC Proc. Vol.* **2009**, *42*, 237–242. [CrossRef]
11. Sira-Ramirez, H.; Rios-Bolivar, M. Sliding mode control of DC-to-DC power converters via extended linearization. *IEEE Trans. Circuits Syst. I Fundam. Theory Appl.* **1994**, *41*, 652–661. [CrossRef]
12. Tan, S.C.; Lai, Y.; Cheung, M.; Tse, C. On the practical design of a sliding mode voltage controlled buck converter. *IEEE Trans. Power Electron.* **2005**, *20*, 425–437. [CrossRef]

13. Hingorani, N. Flexible AC transmission. *IEEE Spectr.* **1993**, *30*, 40–45. [CrossRef]
14. Rudervall, R.; Charpentier, J.P.; Sharma, R.Y. High voltage direct current (HVDC) transmission systems technology review paper. In Proceedings of the Energy Week, Washington, DC, USA, 7–8 March 2000.
15. Fiori, S.; Di Filippo, R. An improved chaotic optimization algorithm applied to a DC electrical motor modeling. *Entropy* **2017**, *19*, 665. [CrossRef]
16. Subramanian, A.; Govindarajan, U. Analysis and mitigation of conducted EMI in current mode controlled DC–DC converters. *IET Power Electron.* **2019**, *12*, 667–675. [CrossRef]
17. Mondzik, A.; Piróg, S.; Stala, R.; Penczek, A.; Gucwa, P. DC-DC boost converter with reduced switching losses in wide range of voltage gain. *Energies* **2023**, *16*, 4397. [CrossRef]
18. Zhang, K.; Shan, Z.; Jatskevich, J. Estimating switching loss and core loss in dual active bridge DC-DC converters. In Proceedings of the 2015 IEEE 16th Workshop on Control and Modeling for Power Electronics (COMPEL), Vancouver, BC, Canada, 12–15 July 2015; pp. 1–6. [CrossRef]
19. Gitau, M.N.; Masike, L.; Adams, G.P. A unified analysis of DC-DC converters's current stress. *Energies* **2023**, *16*, 3370. [CrossRef]
20. Kishor, Y.; Patel, R. Thermal modelling and reliability analysis of recently introduced high gain converters for PV application. *Clean. Energy Syst.* **2022**, *3*, 100016. [CrossRef]
21. Kuisma, M.; Jarvelainen, T.; Silventoinen, P. Analyzing voltage ripple in variable-frequency DC/DC boost converter. In Proceedings of the 2004 IEEE 35th Annual Power Electronics Specialists Conference (IEEE Cat. No.04CH37551), Aachen, Germany, 20–25 June 2004; Volume 2, pp. 1085–1089. [CrossRef]
22. Li, H.; Liu, Q.; Zhang, Z.; Jiang, X.; Cao, G.; Zhang, Y. A voltage vershoot design method for DC-DC converter based on state overshoot concept. In Proceedings of the IECON 2021—47th Annual Conference of the IEEE Industrial Electronics Society, Toronto, ON, Canada, 13–16 October 2021; pp. 1–6. [CrossRef]
23. Cao, J.; Chen, B.; Ye, H.; Feng, J. A peak-current mode boost converter with fast linear transient response. *Microelectron. J.* **2023**, *139*, 105897. [CrossRef]
24. Hamidi, S.A.; Nasiri, A. Stability analysis of a DC-DC converter for battery energy storage system feeding CPL. In Proceedings of the 2015 IEEE International Telecommunications Energy Conference (INTELEC), Osaka, Japan, 18–22 October 2015; pp. 1–5. [CrossRef]
25. Gholizadeh, H.; Gorji, S.A.; Afjei, E.; Sera, D. Design and implementation of a new Cuk-based step-up DC–DC converter. *Energies* **2021**, *14*, 6975. [CrossRef]
26. Moler, C.; Van Loan, C. Nineteen dubious ways to compute the exponential of a matrix, twenty-five years later. *SIAM Rev.* **2003**, *45*, 3–49. [CrossRef]
27. Lopez, I.; Mclnnes, C.R. Autonomous rendezvous using artificial potential function guidance. *J. Guid. Control. Dyn.* **1995**, *18*, 237–241. [CrossRef]

applied sciences

MDPI

Article

# Study on Stability of Elastic Compression Bending Bar in Viscoelastic Medium

Xiaochun Zhang [1,*], Jianhan Hu [2] and Shuyang Chen [2,3]

1 Intelligent Transport System Research Center, Southeast University, Nanjing 211189, China
2 School of Transportation, Southeast University, Nanjing 211189, China; 220213517@seu.edu.cn (J.H.); chenshuyang@sutpc.com (S.C.)
3 Shenzhen Urban Transport Planning Center, Shenzhen 518057, China
* Correspondence: 101008824@seu.edu.cn; Tel.: +86-13813802065

**Abstract:** In the southeastern coastal regions of China, thick layers of marine soft soil are widely distributed, exhibiting characteristics such as high compressibility, high porosity, low strength, high sensitivity, and easy thixotropy, and these viscoelastic behaviors of foundation soil have significant implications for elastic compression bending bar, as evidenced by issues such as post-construction settlement of roadbeds and long-term operation deviation of bridge pile foundations. In this study, a mechanical model of an elastic bar embedded in an elastic and viscoelastic medium, fixed at the base and free at the top, is established based on the Winkler foundation assumption. The deflection function of a bar subjected to both axial force and locally distributed horizontal load is derived using the Rayleigh-Ritz method. Utilizing the elastic-viscoelastic correspondence principle, the viscoelastic medium surrounding the bar is modeled as an elastic medium in which the ground reaction coefficient varies within phase space formulation. This study provides a robust theoretical foundation for soft soil foundation engineering projects and fills a significant gap in the literature by offering a comprehensive framework for understanding displacement in elastic bars within viscoelastic media. Drawing upon the derivation of the deformation function for elastic rods within a viscoelastic medium, the findings of this research hold significant applicability across a range of domains. These include, but are not limited to, the expansion of roadways in regions characterized by coastal soft soil, as well as the monitoring of deformation and lifespan in bridge pile foundations.

**Keywords:** viscoelastic mechanics; elastic compression bending bar; stability; temporal correlation

## 1. Introduction

The macroscopic study of soil rheology involves treating soil as a continuous and homogeneous medium and utilizing mathematical and mechanical methods to derive mathematical expressions that describe soil rheological phenomena based on macroscopic data. While this approach lacks the depth of microscopic studies in exploring the underlying mechanisms of soil rheology, it provides an intuitive depiction of the external characteristics of soil rheology. In particular, it enables the determination of rheological properties for soil samples from specific regions, thereby satisfying practical engineering requirements. As such, macroscopic study has long been a focus of soil rheology research. Common methods employed in macroscopic research include element theory and empirical theory. Element theory, also known as model theory, considers the rheological properties of soil to result from the combined action of elasticity, viscosity, and plasticity. These properties are simulated using spring, dashpot, and friction elements, respectively. Element theory is simple and intuitive, providing a comprehensive reflection of various rheological properties of media and is widely applied. Empirical theory, on the other hand, is based on specific soil qualities and conditions. It involves fitting a constitutive model to soil test results and has found some application in engineering.

In engineering practice, diseases such as post-construction settlement of roadbeds and long-term operation deviation of bridge pile foundations reflect the existence of rheological effects of foundation soil. In southeastern coastal areas of China, there are widely distributed thick layers of marine soft soil, which have characteristics such as high compressibility, high porosity, low strength, high sensitivity, and easy thixotropy. When carrying out roadbed, embankment filling and other earthwork operations in soft soil areas, soil consolidation settlement and long-term rheological effects are larger than those in non-soft soil areas, which will have a greater long-term impact on pile foundations of buildings or bridges in nearby areas. Therefore, when studying the impact of roadbed filling near existing bridges in soft soil areas on bridge piles, it is necessary to consider the long-term impact of soft soil rheology on bridge pile internal force state and geometric shape changes.

Notable examples of macroscopic research include the work of Taylor and Merchant [1], who considered the rheological properties of soil in their analysis of soil consolidation and employed the Kelvin model to describe the deformation of the soil skeleton. Tan [2,3] treated clay as a Maxwell body and, building on this theoretical foundation, studied the secondary time effect in the consolidation process of soft soil, proposing the Chen model. Gibson [4] used the Merchant model to fit soil rheological properties and conducted in-depth research on one-dimensional rheological consolidation problems.

In the field of engineering structures, a bar subjected to both axial force and lateral load is referred to as a compression-bending bar. Up to now, extensive investigations have been carried out into the influence and design of lateral braces on columns and beams, including the bracing stiffness, bracing strength requirements, and the corresponding buckling loads of members [5–11]. Winter [12] theoretically and experimentally developed the principles of minimum stiffness and minimum strength requirements for equally spaced braces by introducing a simplified rigid link model with fictitious hinges at the bracing points, which became the foundation of most of the practical bracing design formulations for braced columns. Timoshenko and Gere [13] derived the buckling loads and obtained the same values of full bracing stiffness as those by Winter for perfect straight columns, using a theoretical derivation based on the segment equilibrium and deformation compatibility. A typical example of such a bar is a pile foundation driven into soil, which often bears additional loads transmitted from the impact of pile loading construction in adjacent areas while simultaneously bearing axial loads. Several researchers have conducted forward-looking studies in this field and have made notable progress [14–17]. Dai [18] derived equations for calculating the internal force and deformation of horizontally loaded piles using the elastic foundation beam method in combination with the finite difference principle, matlock method (m method), two-parameter method, and p-y curve method. This research demonstrates the maturity of the finite difference method in its application to horizontally loaded piles.

Reese and Welch [19] established differential equations for the force of foundation piles under inclined eccentric load as early as 1975. Sastry et al. [20–22] later investigated the ultimate bearing capacity and force characteristics of piles under inclined loads in homogeneous sandy soils, homogeneous clay soils, and stratified soils through indoor model tests. Zhao [23] conducted an in-depth study of the ultimate bearing capacity of foundation piles under inclined load based on the principle of joint action of pile and soil and elasticity theory, using numerical methods and experimental means. By experiments and numerical simulations, some researchers have investigated the flexural behavior of the pile under combined axial and lateral loads. ZHUKOV and BALOV [24] studied pile behavior by in-situ tests, without considering the influence of the pile length above ground. ZHANG et al. [25] deduced the elastoplastic solutions for partially embedded piles, in which the coefficient of subgrade reaction was supposed to be a constant. By assuming that the coefficient of subgrade reaction increases linearly with depth, HAN and FROST [26] obtained the variational solutions for the transversely isotropic pile. The subgrade reaction method is widely used because of its simplicity and reasonable accuracy. However, the coefficient of subgrade reaction is not a fundamental soil property but a model parameter, and the soil resistance is modeled with independent springs without accounting for the continuity of soil [27]. POULOS and DAVIS [28] put forward the elasticity approach for single piles under lateral loads, in which the soil was assumed to be an ideal, homogeneous, isotropic, semi-infinite elastic material, and the differential solutions were obtained.

The research proposed a simplified analysis method for power series calculation based on the m method. However, this method has a large number of calculation parameters and its formulae are lengthy and complex, limiting its application to some extent. Due to the complexity of pile-soil interaction and the nonlinearity of pile perimeter resistance and displacement, single-parameter methods such as the m or c method often struggle to produce results that accurately reflect both pile displacement and internal force calculations. As a result, Wu [29] proposed a comprehensive stiffness principle and two-parameter method for calculating thrust piles, which established a power series solution. However, this solution does not account for the longitudinal and transverse bending that occurs in bridge piles or high bearing piles under horizontal and vertical (or inclined) loads, known as the gravity second-order effect. Obtaining an analytical solution for power series that accounts for longitudinal and transverse bending is challenging. In summary, there are relatively few research results available on elastic pile foundations in viscoelastic media.

In other area outside of soil mechanics, it is imperative to consider the post-buckling behavior when the supporting spring system exhibits nonlinearity or softening effects. In such scenarios, subcritical and unstable post-buckling behavior with localized failure modes may occur, rendering the failure mode highly sensitive to imperfections. Consequently, linear buckling analysis may provide limited guidance in predicting the ultimate load. An example of this phenomenon is the buckling of railways subjected to thermal strain [30–32]. And in other area outside of civil engineering, the viscoelastic-elastic complex system is still widely received by researchers [33–37].

In this paper, the comprehensive stiffness principle and two-parameter method are extended to the calculation of longitudinally and transversely bent piles, analogous to longitudinally and transversely bent beams in mechanics of materials [38,39]. The deflection differential equation of a composite loaded pile is established using this method and the corresponding finite difference equation is derived. The finite difference iterative format for pile deformation and internal force is then obtained based on the pile top load, pile end restraint conditions, and the continuity of deformation and internal force at the interface between the free and embedded solid sections of the pile. This study presents a time-history analysis of the displacement behavior of elastic compression bending bar embedded in viscoelastic media. The primary objective of this research is to enhance the long-term stability of elastic bars subjected to lateral disturbances in practical engineering applications. Additionally, this study addresses a significant gap in the existing literature by providing a comprehensive theoretical framework for understanding the temporal variation of displacement in elastic bars situated within viscoelastic media. The findings of this research have important implications for the design and construction of soft soil foundation engineering projects, offering a robust theoretical foundation for future work in this area.

## 2. Methods

### *2.1. Deflection Function for an Elastic Compression-Bending Bar within an Elastic Medium*

#### 2.1.1. Basic Model of Viscoelastic Mechanics

In the southeastern coastal region of China, where the soft soil layer is relatively thick, rock-socketed piles are commonly employed in the design of bridge pile foundations to ensure the stability of the superstructure. Pile lengths often exceed 50 m and can even reach 100 m in some areas. As such, the stability of single piles under these conditions can be approached as a slender bar stability problem. Unlike a compression bar model with a fixed bottom and free top, the pile body is enveloped by a thick layer of soft soil and cannot be simplified as a compression bar stability research model without lateral support. Similar to the support conditions of an elastic foundation beam, the pile side can be transformed into an elastic foundation support based on the Winkler foundation assumption. In actual working conditions, pile foundations bear not only axial loads transmitted from the bridge superstructure but also additional horizontal stresses caused by lateral deviation of piles, resulting in a state of compression-bending.

The constitutive equation of linear viscoelastic materials can be bifurcated into two primary categories: differential and integral types. The differential type is fundamentally premised on the utilization of ideal elements to construct a variety of models. These models subsequently facilitate the derivation of material constitutive equations, which are based on the inherent properties and combinational forms of the elements. This approach offers a level of convenience in resolving certain problems. As per the foundational theory of the differential type, it can be postulated that the one-dimensional differential constitutive equation of any given linear viscoelastic material can be articulated as follows:

$$P\sigma = Q\varepsilon \tag{1}$$

$$P = \sum_{k=0}^{n} p_k \frac{d^k}{dt^k},\ Q = \sum_{k=0}^{n} q_k \frac{d^k}{dt^k} \tag{2}$$

where: the differential operators $P$ and $Q$ are shown in Equation (2), where $p_k$ and $q_k$ are constants related to the elastic and viscous properties ($k$ is only a counting symbol, not the subgrade reaction coefficient mentioned below).

The constitutive equation of viscoelastic materials is derived from element models, which are formulated through the integration of elastic (spring) and viscous (dashpot) elements in a multitude of configurations. The spring elements adhere to the principles of Hooke's law.

$$\sigma = E\varepsilon,\ \tau = G\gamma \tag{3}$$

where $\sigma$ and $\tau$ are the normal stress and shear stress, respectively; $\varepsilon$ and $\gamma$ are the normal strain and shear strain, respectively; $E$ and $G$ are the tensile-compressive elastic modulus and shear elastic modulus.

The elements of a viscoelastic model, known as dashpots, adhere to Newton's law of viscosity, which is a fundamental principle in fluid mechanics as follow:

$$\sigma = \eta\dot{\varepsilon},\ \tau = \eta_1\dot{\gamma} \tag{4}$$

where $\eta$ and $\eta_1$ is the coefficient of viscosity, $\dot{\varepsilon}$ and $\dot{\gamma}$ are the normal strain and shear strain of viscosity. Figure 1 shows the mechanical model of Spring and Dashpot element.

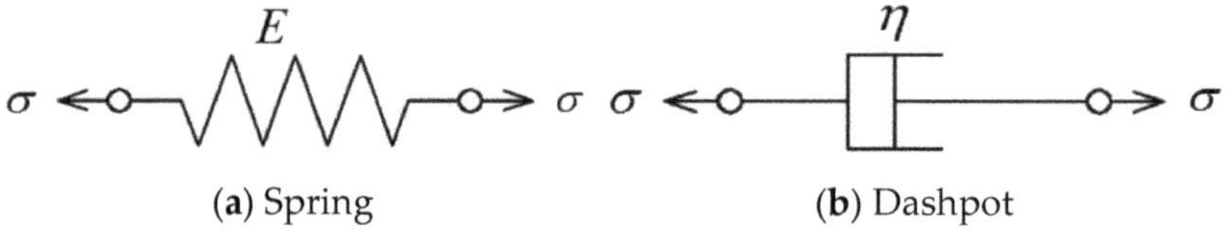


**Figure 1.** Spring elements and dashpot elements.

The Kelvin model is composed of a spring element and a dashpot element in parallel, as is shown in Figure 2. And the Van Der Poel model is formed by a spring element and a Kelvin model in series as is shown in Figure 3.

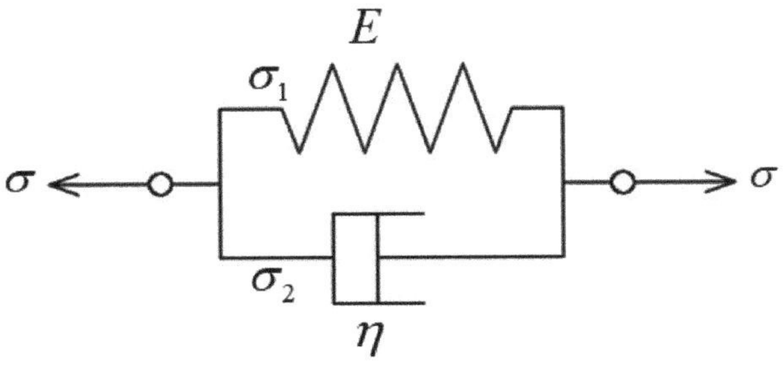


**Figure 2.** Kelvin model.

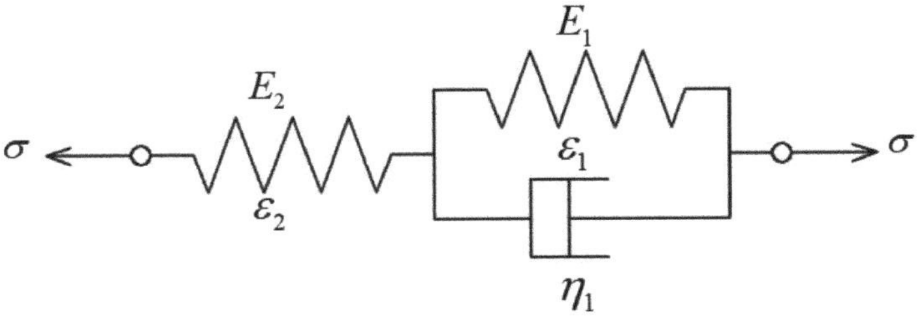


**Figure 3.** Van Der Poel model.

The constitutive equation of Van Der Poel is

$$\sigma + p_1\dot{\sigma} = q_0\varepsilon + q_1\dot{\varepsilon} \tag{5}$$

where:

$$p_1 = \frac{\eta_1}{E_1 + E_2},\ q_0 = \frac{E_1 E_2}{E_1 + E_2},\ q_1 = \frac{E_2\eta_1}{E_1 + E_2} \tag{6}$$

Given a constant stress $\sigma_0$, the creep equation of the Van Der Poel model is as follows:

$$\varepsilon(t) = \frac{\sigma_0}{E_2} + \frac{\sigma_0}{E_1}\left(1 - e^{-\frac{E_1}{\eta}t}\right) \tag{7}$$

2.1.2. Basic Physical Consumption of Elastic-Viscoelastic Media

The Winkler Foundation Model, also referred to as the Local Elastic Foundation Model, was a seminal concept introduced by the Czechoslovakian engineer Winkler in 1867. This model primarily postulates that the settlement of any given point on the foundation surface is directly proportional to the pressure exerted on the unit area at that point. In accordance with this model, the more rigid rock layer at the base of the foundation is considered a rigid layer, while the upper soil layer, which exhibits lesser stiffness, is discretized into independent springs on the rigid layer, as depicted in Figure 4.

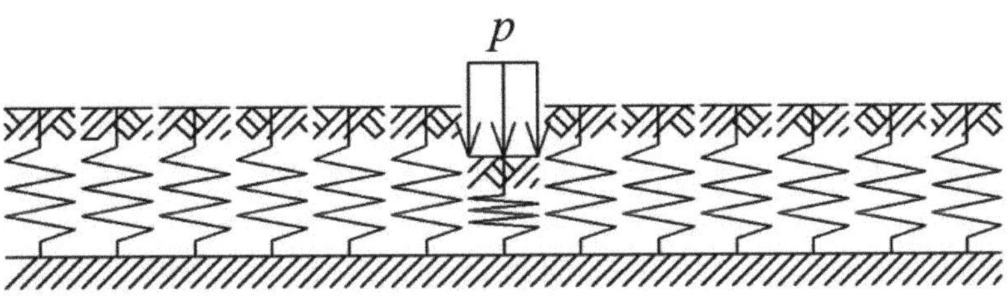


**Figure 4.** Winkler foundation model.

Viscoelastic mechanics encompasses two fundamental aspects: firstly, it provides a description of the viscoelastic properties of materials and expresses their constitutive relations; secondly, it involves the formulation and resolution of viscoelastic boundary value problems. As per the established principles of viscoelastic mechanics, the stress-strain relationship of elastic materials can be construed as a specific instance of viscoelasticity [40,41]. Consequently, a viscoelastic foundation can adhere to Winkler's assumption that views the foundation soil as discrete support points. By selecting an appropriate viscoelastic element model to replace the soil spring, we can apply the solution of an elastic beam on an elastic foundation to the principle of elastic-viscoelastic correspondence. Following a process of Laplace transform and inverse transformation, we can derive the solution for an elastic beam on a viscoelastic foundation.

#### 2.1.3. Mechanical Model of an Elastic Compression-Bending Bar in an Elastic Medium

Figure 5 depicts the mechanical model of an elastic compression-bending bar in an elastic medium. A constant load with a width of $L_p$ and magnitude of $p$ acts on the surface of the elastic medium at a distance of $d_p$ from one side of the bar to simulate subgrade filling on one side of the pile foundation; $q(x)$ represents the real passive load on the bar body caused by pile loading and soil self-weight; axial force $P$ acts on the end of the bar, inducing a small displacement $\Delta x$ in the vertical direction at the end of the bar. The bottom of the bar is fixed and the top is free; the length of bar is $L$, and $k(x)$ represents the ground reaction coefficient at different depths; $b_1$ denotes the calculated width of the elastic bar.

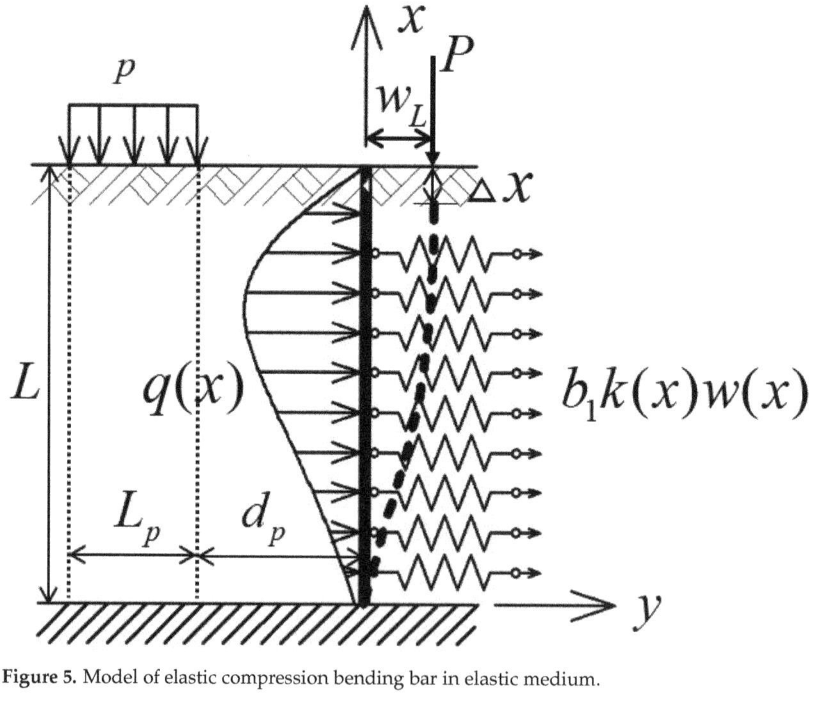


**Figure 5.** Model of elastic compression bending bar in elastic medium.

Drawing upon the relevant theories of elasticity and soil mechanics, it can be shown that, within an elastic medium, the horizontal additional stress $\sigma_x$ at a given point in the medium induced by a constant load $p$ of a certain width can be determined based on the Boussinesq solution and Flamant solution [42,43]. Subsequently, the real passive load $q(x)$

on the bar body can be obtained demonstrated in Equation (8) which is detailed introduced in Appendix A.

$$q(x) = \frac{b_1 p}{\pi}\left(\frac{(L-x)^2 \arctan\frac{L_p}{L-x}}{(L-x)^2 + (L_p + d_p)d_p} - \frac{(L_p + d_p)(L-x)}{(L-x)^2 + (L_p + d_p)^2} + \frac{(L-x)d_p}{(L-x)^2 + {d_p}^2}\right) \tag{8}$$

where: $q(x)$ is passive load on elastic bar caused by constant load and soil self-weight
$b_1$ is calculated width of the elastic bar
$L$ is length of bar
$L_p$ is width of constant load
$p$ is magnitude of constant load
$d_p$ is distance from the bar to constant load

The objective of this study is to investigate the displacement response of piles under the influence of ground load in the adjacent area. A crucial step in this analysis is to ascertain the additional pile load induced by the ground load. To this end, ABAQUS is employed as finite element model (FEM) solution software, pile material and geometric parameters are applied with roadbed filling area layout parameters, etc. that are specified at the outset of this section. Detailed parameters are shown in Table 1 and entire model are shown as Figure 6, which has a grey solid part for soil and a constant force adding in the yellow part.

**Table 1.** Parameters in ABAQUS model.

| Parameter | Magnitude | Unit |
|---|---|---|
| Width of bar/$b_1$ | 1 | m |
| Length of bar/$L$ | 50 | m |
| Elastic Modulus of bar/$E$ | $3 \times 10^4$ | MPa |
| Density of bar/$\rho$ | 2385 | kg/m$^3$ |
| Poisson's ratio/$\lambda$ | 0.2 | / |
| Distance from the bar to constant load/$d_p$ | 1 | m |
| Magnitude of constant force/$q$ | 50 | kPa |
| Length of constant force/$L_p$ | 20 | m |
| Reaction force coefficient/$k$ | 4500 | kN/m$^4$ |

**Figure 6.** FEM model of single bar with elastic medium.

As is shown in Figure 7, the numerical solution is in good agreement with the analytical solution. Therefore, the passive load of the pile body can be calculated by Equation (8).

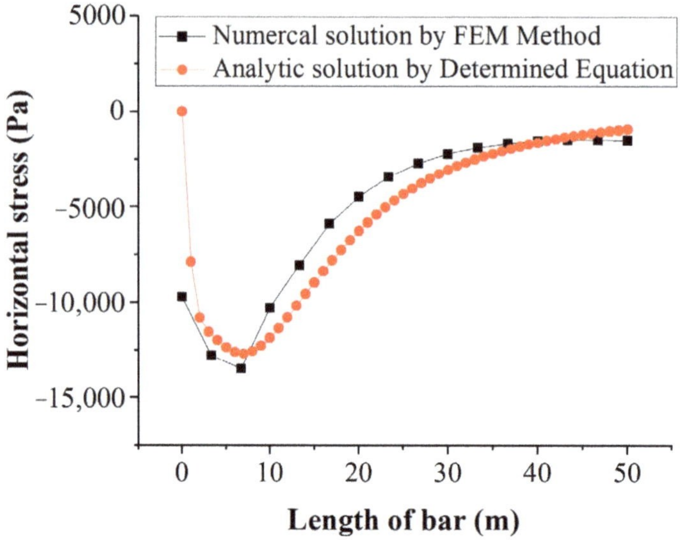


**Figure 7.** Comparison of numerical and analytical solutions for horizontal stress.

2.1.4. A Comprehensive Solution Formula for the Deflection of an Elastic Slender Bar within an Elastic Medium

Based on the law of conservation of energy, the total potential energy of both the bar and the surrounding soil $\Pi_p$ can be calculated when subjected to an axial force $P$ and a horizontal additional load $q(x)$ as shown in Equation (9) which is detailed introduced in Appendix B:

$$\Pi_p = \frac{EI}{2}\int_0^L (w'')^2 dx + \frac{b_1}{2}\int_0^L kw^2 dx - P\Delta x - b_1\int_0^L qwdx \tag{9}$$

where: $\Pi_p$ is total potential energy of both the bar and the surrounding soil

$E$ is elastic modulus of elastic bar

$I$ is moment of inertia of the cross section of elastic bar

$k$ is reaction force coefficient of elastic medium

$w$ is deflection of elastic bar as $w(x)$

$\Delta x$ is a small displacement caused by constant load $p$

$P$ is an axial force as shown in Figure 5

As per the mechanical model depicted in Figure 5, the boundary conditions for the compressed bar are defined as follows: $x = 0$, $w = 0$, $w' = 0$; $x = L$, $w'' = 0$. Given these boundary conditions, a test function $w(x)$ for the deflection curve of the micro-bent state bar can be selected in the form of a triangular series by the Laplace transform as:

$$w(x) = \sum_{n=1}^{i} w_n\left(1 - \cos\frac{(2n-1)\pi x}{2L}\right) \tag{10}$$

Substituting Equation (10) into Equation (9) yields:

$$\begin{aligned}\Pi_p &= \tfrac{EI}{2}\int_0^L\left(\sum_{n=1}^{i} w_n\left(\tfrac{(2n-1)\pi}{2L}\right)^2\cos\tfrac{(2n-1)\pi x}{2L}\right)^2 dx + \tfrac{b_1}{2}\int_0^L k\left(\sum_{n=1}^{i} w_n\left(1-\cos\tfrac{(2n-1)\pi x}{2L}\right)\right)^2 dx \\ &-\tfrac{P}{2}\int_0^L\left(\sum_{n=1}^{i} w_n\tfrac{(2n-1)\pi}{2L}\sin\tfrac{(2n-1)\pi x}{2L}\right)^2 dx - b_1\int_0^L q\sum_{n=1}^{i} w_n\left(1-\cos\tfrac{(2n-1)\pi x}{2L}\right)dx\end{aligned} \tag{11}$$

In accordance with the principle of minimum potential energy, for a system in a stable equilibrium state, $\Pi_p$ must be minimized, which means the first-order derivative of $\Pi_p$ is 0. This can be expressed as follows:

$$\begin{aligned}\frac{\partial \Pi_p}{\partial w_r} = & \sum_{n=1}^{i} w_n \int_0^L \Big( EI\frac{(2n-1)^2(2r-1)^2\pi^4}{16L^4} \cos\frac{(2n-1)\pi x}{2L} \cos\frac{(2r-1)\pi x}{2L} \\ & + b_1 k \Big(1-\cos\frac{(2n-1)\pi x}{2L}\Big)\Big(1-\cos\frac{(2r-1)\pi x}{2L}\Big) \\ & - P\frac{(2n-1)(2r-1)\pi^2}{4L^2} \sin\frac{(2n-1)\pi x}{2L} \sin\frac{(2r-1)\pi x}{2L}\Big) dx \\ & - b_1 \int_0^L q\Big(1-\cos\frac{(2r-1)\pi x}{2L}\Big) dx = 0, r = 1,2,3,\ldots,i\end{aligned} \tag{12}$$

Take:

$$\begin{aligned}W_{nr} = & \int_0^L \Big( EI\frac{(2n-1)^2(2r-1)^2\pi^4}{16L^4} \cos\frac{(2n-1)\pi x}{2L} \cos\frac{(2r-1)\pi x}{2L} \\ & + b_1 k \Big(1-\cos\frac{(2n-1)\pi x}{2L}\Big)\Big(1-\cos\frac{(2r-1)\pi x}{2L}\Big) \\ & - P\frac{(2n-1)(2r-1)\pi^2}{4L^2} \sin\frac{(2n-1)\pi x}{2L} \sin\frac{(2r-1)\pi x}{2L}\Big) dx\end{aligned} \tag{13}$$

$$Q_r = b_1 \int_0^L q\left(1-\cos\frac{(2r-1)\pi x}{2L}\right) dx \tag{14}$$

Then Equation (11) can be transformed into Equation (14):

$$\sum_{r=1}^{i}\sum_{n=1}^{i}(w_n W_{nr} - Q_r) = \begin{bmatrix} W_{11} & W_{12} & \cdots & W_{1i} \\ W_{21} & W_{12} & \cdots & W_{i2} \\ \vdots & \vdots & \cdots & \vdots \\ W_{i1} & W_{i2} & \cdots & W_{ii} \end{bmatrix} \begin{bmatrix} w_1 \\ w_2 \\ \vdots \\ w_i \end{bmatrix} - \begin{bmatrix} Q_1 \\ Q_2 \\ \vdots \\ Q_i \end{bmatrix} = 0 \tag{15}$$

Equation (15) delineates the phase space formulation representation of an elastic bar within an elastic medium by the Laplace transform, which is prepared for viscoelastic medium illustrated in Equation (23). The time-dependency and spatial-dependency of the viscoelastic medium are transposed into a phase space via the Laplace transform. This process effectively decouples the time-dependency from the deformation process inherent in the viscoelastic medium. Consequently, the deformation issue concerning the elastic bar within the viscoelastic medium, as referenced in this study, is reduced to a problem devoid of time-dependency. Upon completion of the solution, the Laplace inverse transformation is employed. This reconfigures the deformation function of the bar from the resolved phase space formulation back to a general solution exhibiting both time-dependency and spatial-dependency. The outcome is a universal function capable of predicting terminal deformations of the bar.

By solving Equation (15), the coefficients $w_n$ of the deflection curve test function for the bar body can be obtained. This yields an approximate solution $w(x)$ for the true displacement curve of the bar under varying axial loads $P$. For instance, when $i$ = 2 and $k = m(L-x)$, solving Equation (15) results in:

$$w_1 = \frac{288L^3\pi^2\left(81\pi^4(9EI\pi^2-4L^2P)Q_1 + 8L^5mb_1\left(\begin{array}{c}(27\pi^2-28)Q_1 \\ +2(62-9\pi^2)Q_2\end{array}\right)\right)}{\begin{array}{l}729\pi^8(EI\pi^2-4L^2P)(9EI\pi^2-4L^2P) \\ +16L^5mb_1\left(\begin{array}{c}41328L^2P\pi^4-180(511EI+27L^2P)\pi^6 \\ +9963EI\pi^8+4L^5mb_1(405\pi^4-3096\pi^2-8320)\end{array}\right)\end{array}} \tag{16}$$

$$w_2 = \frac{288L^3\pi^2\left(9\pi^4(EI\pi^2-4L^2P)Q_2 + 8L^5mb_1\left(\begin{array}{c}2(62-9\pi^2)Q_1 \\ +9(3\pi^2-28)Q_2\end{array}\right)\right)}{\begin{array}{l}729\pi^8(EI\pi^2-4L^2P)(9EI\pi^2-4L^2P) \\ +16L^5mb_1\left(\begin{array}{c}41328L^2P\pi^4-180(511EI+27L^2P)\pi^6 \\ +9963EI\pi^8+4L^5mb_1(405\pi^4-3096\pi^2-8320)\end{array}\right)\end{array}} \tag{17}$$

Thus, it can be inferred that within the context of a second-order triangular series, the deflection function for the bar body is:

$$w(x) = w_1\left(1 - \cos\frac{\pi x}{2L}\right) + w_2\left(1 - \cos\frac{3\pi x}{2L}\right) \tag{18}$$

*2.2. Deflection Function for an Elastic Compression-Bending Bar in a Viscoelastic Medium*

2.2.1. Mechanical Model of an Elastic Compression-Bending Bar in a Viscoelastic Medium

The interaction between elastic compression bending bar and the surrounding soil is complex and can be either static or dynamic. To simplify calculations, this interaction is often modeled using an elastic foundation beam. However, this approach has limitations as soil is a discrete, multiphase system with nonlinear and irrecoverable stress-strain characteristics under external forces. As such, modeling soil solely with springs may not accurately capture its behavior. Viscoelastic mechanics can better characterize the creep properties of soft soils by revealing their rheological properties. This typically involves establishing the viscoelastic instantaneous relationship of the material and determining boundary conditions and their solutions. By transforming the parameters characterizing the foundation soil's material properties in the elastic foundation model into time-dependent viscoelastic parameters, the springs in the elastic foundation can be replaced with a suitable viscoelastic model while maintaining the boundary conditions of the elastic foundation model. This allows for the solution of the beam on the elastic foundation to be transformed into a time-dependent viscoelastic solution.

In this part, adopts the method of the Winkler foundation hypothesis, selects a suitable viscoelastic constitutive model, and simplifies the soil around the slender bar shown in Figure 5 into discrete viscoelastic element supports. Taking the Van Der Poel model as an example, as shown in Figure 8.

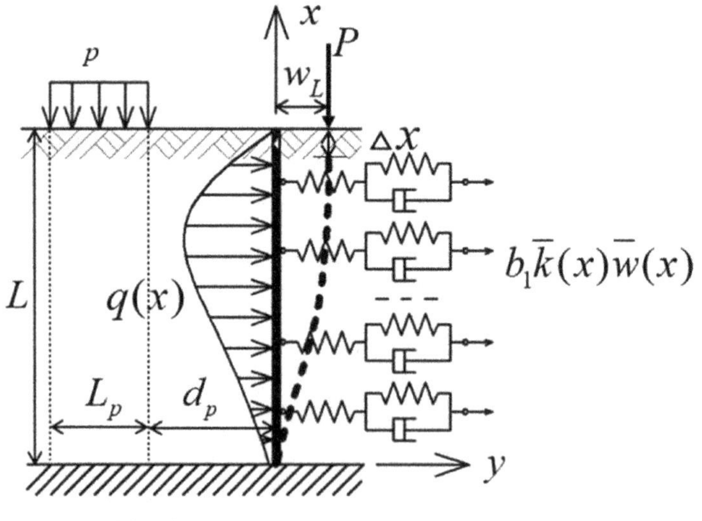


**Figure 8.** Model of elastic compression bending bar in viscoelastic medium.

2.2.2. A Comprehensive Solution Formula for the Deflection of an Elastic Slender Bar within a Viscoelastic Medium

In accordance with the principle of elastic-viscoelastic correspondence, the viscoelastic foundation elastic beam problem may be considered as an elastic foundation elastic beam problem with a reaction force coefficient of $\bar{k}(s, x)$. By analogy, utilizing the m-method to determine the value of the foundation reaction force coefficient, the foundation reaction force coefficient within the phase space formulation is established to be:

$$\overline{k}(s,x) = A\overline{k}_0(s)(L-x) \tag{19}$$

where: $\overline{k}$ is reaction force coefficient of viscoelastic medium in kelvin model

$A$ is constant

$s$ is displacement of elastic bar

Within Equation (19), $A$ represents a constant that governs the rate of change for the foundation reaction force coefficient $\overline{k}(s,x)$ from the ground surface to the depth of the soil. By substituting Equation (19) into Equation (13), the result is obtained:

$$\begin{aligned}\overline{W}_{nr} = & \int_0^L \left( EI\tfrac{(2n-1)^2(2r-1)^2\pi^4}{16L^4} \cos\tfrac{(2n-1)\pi x}{2L} \cos\tfrac{(2r-1)\pi x}{2L} \right. \\ & + A\overline{k}_0(s)(L-x)b_1\left(1-\cos\tfrac{(2n-1)\pi x}{2L}\right)\left(1-\cos\tfrac{(2r-1)\pi x}{2L}\right) \\ & \left. - P\tfrac{(2n-1)(2r-1)\pi^2}{4L^2} \sin\tfrac{(2n-1)\pi x}{2L} \sin\tfrac{(2r-1)\pi x}{2L} \right) dx, n = 1,2,\ldots; r = 1,2,\ldots \end{aligned} \tag{20}$$

In the event of an instantaneous increase in the side load of the bar, namely $q(t,x) = q(x)H(t)$, upon substitution into Equation (14), the result is obtained:

$$Q_r(t) = H(t)b_1 \int_0^L q\left(1-\cos\frac{(2r-1)\pi x}{2L}\right)dx = H(t)Q_r \tag{21}$$

where: $W$, $Q$ and $H$ are generalized function

Upon conducting Laplace transform on Equation (21), the result is obtained:

$$\overline{Q}_r = Q_r/s \tag{22}$$

Analogous to Equation (15), the subsequent matrix expression may be derived:

$$\sum_{r=1}^{i}\sum_{n=1}^{i}\left(\overline{w}_n(s)\overline{W}_{nr} - \overline{Q}_r\right) = \begin{bmatrix} \overline{W}_{11} & \overline{W}_{12} & \cdots & \overline{W}_{1i} \\ \overline{W}_{21} & \overline{W}_{12} & \cdots & \overline{W}_{i2} \\ \vdots & \vdots & \cdots & \vdots \\ \overline{W}_{i1} & \overline{W}_{i2} & \cdots & \overline{W}_{ii} \end{bmatrix} \begin{bmatrix} \overline{w}_1(s) \\ \overline{w}_2(s) \\ \vdots \\ \overline{w}_i(s) \end{bmatrix} - \begin{bmatrix} \overline{Q}_1 \\ \overline{Q}_2 \\ \vdots \\ \overline{Q}_i \end{bmatrix} = 0 \tag{23}$$

where: $\overline{W}$ and $\overline{Q}$ are generalized function under reaction force coefficient of viscoelastic medium

By solving Equation (23), the coefficients $\overline{w}_n(s)$ can be determined. Subsequently, the solution $w(s,x)$ for the deflection function of the elastic bar within the phase space formulation can be obtained. By performing an inverse Laplace transform on this solution, the deflection function $w(t,x)$ of the elastic bar within a viscoelastic medium can be derived.

2.2.3. Deflection Solution of an Elastic Slender Bar in a Kelvin Medium

The constitutive equation for a Kelvin body is given by $\sigma = q_0\varepsilon + q_1\dot{\varepsilon}$, where $q_0$ represents the Young's modulus ($E$) and $q_1$ represents the viscosity coefficient ($\eta$). By applying a Laplace transform to this constitutive equation and subsequently rearranging it, the reference value for the ground reaction coefficient $\overline{k}_0(s)$ can be determined:

$$\overline{k}_0(s) = \frac{\overline{\sigma}}{\overline{\varepsilon}} = E + \eta s \tag{24}$$

where: $\overline{k}_0$ is reference value for the ground reaction coefficient

$\overline{\sigma}$ is viscoelastic stress

$\overline{\varepsilon}$ is viscoelastic strain

Consequently, the ground reaction coefficient $\bar{k}(s,\ x)$ within the phase space formulation, as derived from the Kelvin model, is given by:

$$\bar{k}(s,x) = A(E+\eta s)(L-x) \tag{25}$$

Continuing with the example where $i = 2$, Equation (23) can be solved to determine the expression for the coefficient $\overline{w}_n(s)$, where $E_k$ denotes the elastic modulus of the spring element within the Kelvin model.

$$\overline{w}_1(s) = \frac{288L^3\pi^2\left(81\pi^4(9EI\pi^2-4L^2P)Q_1+8Ab_1L^5\begin{pmatrix}(27\pi^2-28)Q_1\\+2(62-9\pi^2)Q_2\end{pmatrix}(E_k+\eta s)\right)}{s\begin{pmatrix}729\pi^8(EI\pi^2-4L^2P)(9EI\pi^2-4L^2P)\\+144Ab_1L^5\pi^4(4592L^2P-20(511EI+27L^2P)\pi^2+1107EI\pi^4)(E_k+\eta s)\\+64A^2{b_1}^2L^{10}(405\pi^4-3096\pi^2-8320)(E_k+\eta s)^2\end{pmatrix}} \tag{26}$$

$$\overline{w}_2(s) = \frac{288L^3\pi^2\left(36L^2P\pi^4Q_2-9EI\pi^6Q_2+8Ab_1L^5\begin{pmatrix}2(9\pi^2-62)Q_1\\+9(28-3\pi^2)Q_2\end{pmatrix}(E_k+\eta s)\right)}{s\begin{pmatrix}256A^2{b_1}^2L^{10}(62-9\pi^2)^2(E_k+\eta s)^2\\-9\begin{pmatrix}EI\pi^6-4L^2P\pi^4+\\8Ab_1L^5(3\pi^2-28)(E_k+\eta s)\end{pmatrix}\begin{pmatrix}81(9EI\pi^6-4L^2P\pi^4)\\+8Ab_1L^5(27\pi^2-28)(E_k+\eta s)\end{pmatrix}\end{pmatrix}} \tag{27}$$

By applying an inverse Laplace transform to Equations (26) and (27), the two coefficients $w_1(t)$ and $w_2(t)$ for the deflection function $w(t,x)$ can be determined as:

$$w_1(t) = \frac{144e^{-N_7t}L^3\pi^2}{(N_3+E_k(N_4+E_kN_5))N_6}\begin{pmatrix}(e^{N_8t}-1)\begin{pmatrix}N_2(2N_3+E_kN_4)\\-N_1(N_4+2E_kN_5)\end{pmatrix}\eta\\+(N_1N_6+E_kN_2N_6)(2e^{N_7t}-e^{N_8t}-1)\end{pmatrix} \tag{28}$$

$$w_2(t) = R_1\left(R_{11}+\frac{e^{R_9t}(3R_{12}(R_2+E_kR_3)+R_{13}\eta)+e^{R_{10}t}(3R_{12}(R_2+E_kR_3)-R_{13}\eta)}{6R_{12}(9R_5R_7+E_k(9R_6R_7-E_kR_4+9R_8(R_5+E_kR_6)))}\right) \tag{29}$$

where:

$$\begin{aligned}
&N_1 = 81\pi^4(9EI\pi^2-4L^2P)Q_1;\ N_2 = 8Ab_1L^5((27\pi^2-28)Q_1+2(62-9\pi^2)Q_2)\\
&N_3 = 729\pi^8(EI\pi^2-4L^2P)(9EI\pi^2-4L^2P);\\
&N_4 = 144Ab_1L^5\pi^4(4592L^2P-20(511EI+27L^2P)\pi^2+1107EI\pi^4);\\
&N_5 = 64A^2{b_1}^2L^{10}(405\pi^4-3096\pi^2-8320);\\
&N_6 = \sqrt{({N_4}^2-4N_3N_5)\eta^2};\ N_7 = \tfrac{N_4\eta+2E_kN_5\eta+N_6}{2N_5\eta^2};\ N_8 = \tfrac{N_6}{N_5\eta^2};\\
&R_1 = 288L^3\pi^2;\ R_2 = 9\pi^4(4L^2P-EI\pi^2)Q_2;\\
&R_3 = 8Ab_1L^5(2(9\pi^2-62)Q_1+9(28-3\pi^2)Q_2);\\
&R_4 = 256A^2{b_1}^2L^{10}(62-9\pi^2)^2;\ R_5 = EI\pi^6-4L^2P\pi^4;\ R_6 = 8Ab_1L^5(3\pi^2-28)\\
&R_7 = 81(9EI\pi^6-4L^2P\pi^4);\\
&R_8 = 8Ab_1L^5(27\pi^2-28);\ R_9 = \tfrac{18E_kR_6R_8+9R_5R_8+9R_6R_7-2E_kR_4}{2(R_4-9R_6R_8)\eta}-\tfrac{3R_{12}}{2(R_4-9R_6R_8)\eta^2};\\
&R_{10} = \tfrac{18E_kR_6R_8+9R_5R_8+9R_6R_7-2E_kR_4}{2(R_4-9R_6R_8)\eta}+\tfrac{3R_{12}}{2(R_4-9R_6R_8)\eta^2};\\
&R_{11} = \tfrac{R_2+E_kR_3}{{E_k}^2R_4-9R_5R_7-9E_kR_6R_7-9E_kR_5R_8-9{E_k}^2R_6R_8};\\
&R_{12} = \sqrt{(4R_4R_5R_7+9{R_6}^2{R_7}^2-18R_5R_6R_7R_8+9{R_5}^2{R_8}^2)\eta^2}\\
&R_{13} = 9(R_2R_6R_7+R_2R_5R_8-2R_3R_5R_7)-9E_kR_3(R_6R_7+R_5R_8)-2E_kR_2(R_4-9R_6R_8)
\end{aligned}$$

As a result, the deflection function $w(t,x)$ for an elastic bar surrounded by a Kelvin body within a second-order triangular series has been determined. If the surrounding medium of the bar is either a Van Der Poel or Burgers body, the deflection function $w(t,x)$ for the elastic bar can also be calculated at varying levels of solution accuracy using the aforementioned method.

## 3. Results and Discussion

According to the rheological test data obtained from soft soil samples in the engineering site selection area, the Burgers model was employed as the viscoelastic constitutive model for the bar-side medium in theoretical calculations. The constitutive model parameters were determined to be $E_M$ = 116.2270 MPa, $E_K$ = 7.0203 MPa, $\eta_M$ = 511.5671 MPa·d, and $\eta_K$ = 8.6031 MPa·d. The soil sample utilized in this rheological test was extracted from a depth of 9–11 m below ground level, with a median value of 10 m being selected. Consequently, in the phase space formulation, $\overline{k}_0(s)$ represents the elastic ground reaction coefficient at a depth of 10 m below ground level. As per the definition of the ground reaction coefficient by the matlock method, the ground reaction coefficient increases linearly from 0 at ground level to the underground depth. Based on this definition, it was determined that the value of $A$ in Equation (25) is 0.1. All solution is solved by MATLAB based on extension of the Burgers model's solution in the Section 2 above, by transforming $K$ in Equation (25) into a basic burgers ground reaction coefficient which is treated under Laplace transform as $\overline{k}(s,x) = A\frac{E_1\eta_2(E_3+\eta_3 s)s}{\eta_2(E_3+\eta_3 s)s+E_1(E_3+(\eta_2+\eta_3)s)}(L-x)$. Detail is shown in Appendix C.

To enhance the accuracy of the calculation results, a solution accuracy of a seventh-order triangular series was employed. At this level of accuracy, the form of the coefficient $\overline{w}_n(s)$ solved according to Equation (23) is highly complex and obtaining an analytical solution for $w_n(t)$ through inverse Laplace transform is not straightforward. Abate J and Whitt W summarized existing research on Laplace numerical inverse transform in literature [44] and proposed a unified theoretical framework for Laplace inverse transformation numerical inversion. In this study, a MATLAB program plug-in developed by Tucker McClure based on this theoretical framework was utilized to write a calculation program and complete subsequent case calculations.

To investigate the compressive stability of longitudinal and transverse bending elastic bars in viscoelastic media, the value of bar end displacement $w(L)$ under long-term axial force and lateral loading was taken as a key parameter. This study examines the compressive stability of elastic compression bending bars in viscoelastic media.

### 3.1. Time Dependence of the Effect of Axial Force on the Lateral Deflection of the Bar Top

When an elastic bar is situated within a viscoelastic medium, its deflection over time can be determined using the deflection function Equation (10) of the longitudinal and transverse bending elastic bar. In this section, we investigate the compressive stability of the longitudinal and transverse bending elastic bar within the viscoelastic medium using the value of the bar end displacement $w(L)$ under long-term lateral pile load as a key parameter.

Taking a lateral pile load strength of 50 kPa and a bar length of 50 m as examples, and with a background of a typical bridge design life of 100 years in engineering, the total calculation time was set to 100 years. The change in transverse deflection, $w(L)$, of the bar top was investigated when different magnitudes of axial force, $P$, were applied to the bar top, as shown in Figure 9a. Analysis of Figure 9a reveals that within the 100-year calculation period, when $P$ is small (e.g., 50 kN, 100 kN, 500 kN, 1000 kN, 2500 kN), $w(L)$ maintains an approximate linear growth trend within the long-term load action time after the initial stage.

When $P$ is 5000 kN or 7500 kN, $w(L)$ exhibits a linear growth stage in the early period within the calculation period of 100 years, followed by a trend of nonlinear acceleration growth with time in the later period. When $P$ is 10,000 kN, $w(L)$ does not reach 100 years and tends towards $+\infty$ at a certain time point during this period. This indicates that when different axial forces $P$ are applied to the bar top and if the load action time is sufficiently large, they will all cause the bar top displacement $w(L)$ to tend towards $+\infty$.

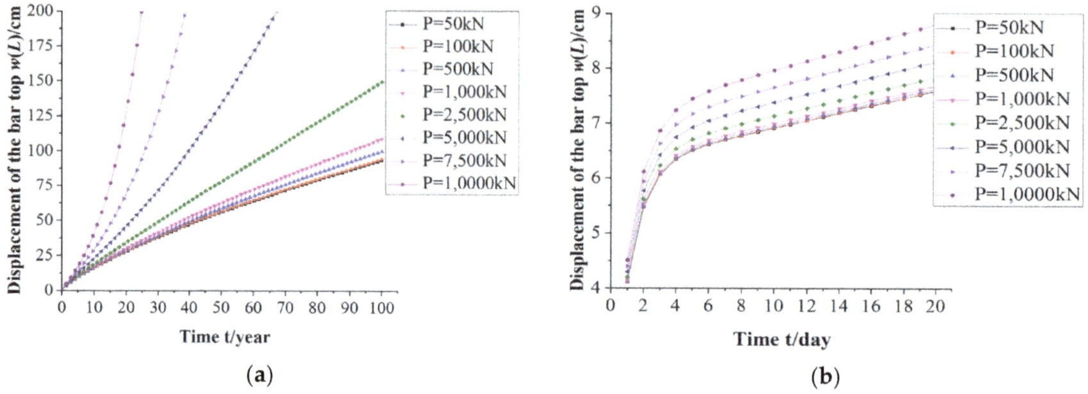


**Figure 9.** The time history of displacement variation when $L = 50$. (**a**) Total change duration of the lateral deflection $w(L)$; (**b**) The change duration of $w(L)$ during the initial stage of load application.

When $P$ is small, only its linear growth stage can be observed within the calculation period; when $P$ increases to a certain level, $w(L)$ will exhibit nonlinear acceleration growth within the calculation period but $w(L) \to +\infty$ has not yet occurred; when $P$ is large, $w(L) \to +\infty$ can be observed within the calculation period. By analogy with the definition of critical force $P_{cr}$ that causes simple compression bars to become unstable, we define the time that causes bars to become unstable as critical instability time $t_{cr}$.

The above results show that if $w(L) \to +\infty$ is used to define bar instability, different axial forces $P$ correspond to corresponding critical instability times $t_{cr}$; alternatively, specifications can be referred to and $w(L) = 10$ cm can be used as the basis for determining critical instability time $t_{cr}$.

In order to clarify the characteristics of bar displacement throughout the process and in conjunction with Burgers body's creep law, it is necessary to study the deformation morphology of long-term loaded bars. Figure 9b shows the $w(L)$-$t$ curve of bars subjected to axial force and horizontal additional load in their initial stage. Analysis of this figure reveals that instantaneous deformation occurs in bars at the moment of load action and that larger axial forces result in larger instantaneous deformation amounts.

As time progresses, $w(L)$ experiences rate of increase with time first increases and then decreases before finally entering a stable uniform growth stage. Larger axial forces result in larger growth amplitudes at each time point during nonlinear growth stage of $w(L)$ and also result in larger rates during its uniform growth stage. On the graph it shows that when axial force is large, $w(L)$-$t$ curve envelopes $w(L)$-$t$ curve when axial force is small from top to bottom.

In this section, considered a scenario as where the bar-side pile loading intensity is 50 kPa and the bar length is 50 m. The total calculation time is set to 100 years. We investigate the temporal evolution of lateral deflection $w(L)$ at the top of the bar under varying magnitudes of axial force $P$ as depicted in Figure 9.

Figure 9a,b illustrate that the bar undergoes instantaneous deformation upon the application of load. As time progresses, the rate of increase of $w(L)$ with respect to time initially increases before decreasing, ultimately reaching a stable and uniform growth phase. This displacement process closely resembles the three-stage creep characteristics of the Burgers model.

Over the course of the 100-year calculation period, when P is small, $w(L)$ exhibits an approximately linear growth trend during the long-term load application period following the initial stage. When $P$ is larger, such as 5000 kN or 7500 kN, $w(L)$ undergoes a linear growth phase in the early stage and exhibits a nonlinear acceleration trend with time in the later stage. When $P$ continues to increase to 10,000 kN, $w(L)$ does not reach 100 years and tends towards +∞ at some point during that period. This indicates that when different axial forces P are applied to the bar top and the load application time is sufficiently large, it will cause the bar top displacement $w(L)$ to tend towards +∞.

When $P$ is small, only its linear growth stage can be observed within the calculation period. When $P$ increases to a certain level, within the calculation period, $w(L)$ will exhibit nonlinear acceleration growth but has not yet shown the phenomenon of $w(L) \rightarrow +\infty$. When $P$ is large, $w(L) \rightarrow +\infty$ can be observed within the calculation period. Analogous to the definition of the critical force $P_{cr}$ that causes instability in a compressed bar, the time that causes bar instability is defined as the critical instability time $t_{cr}$.

These results demonstrate that if $w(L) \rightarrow +\infty$ is used to define bar instability, different axial forces $P$ correspond to their respective critical instability times $t_{cr}$. However, given that $w(L) \rightarrow +\infty$ is inconsistent with engineering practice and that $w(L) \rightarrow +\infty$ results cannot be obtained under common values of $P$ for vertical ultimate bearing capacity of pile foundations within a 100-year calculation period, it is recommended that $w(L)$ = 10 cm, as stipulated by specification [45], be utilized as the basis for determining the critical instability time $t_{cr}$.

Figure 10 illustrates that if $w(L)$ = 10 cm, as stipulated by the specification, is employed as the control condition for defining $t_{cr}$, the magnitude of $t_{cr}$ decreases as the axial force increases. This result suggests that when the bar length and bar-side pile loading intensity are held constant, an increase in axial force has a detrimental effect on the stability of the bar and reduces the time required for the bar displacement to exceed its limit.

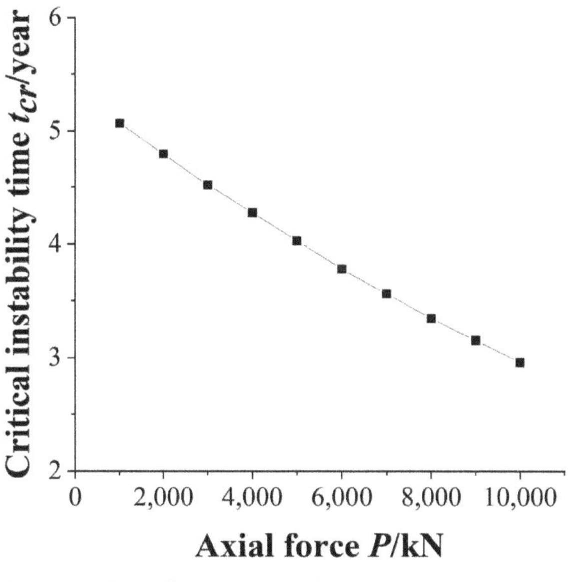


**Figure 10.** Relationship between axial force $P$ and critical instability time $t_{cr}$.

### *3.2. Time Dependence of the Effect of Bar Length on the Lateral Deflection of the Bar Top*

By maintaining the lateral pile load strength of the bar at a constant value of 50 kPa and using axial force values of $P$ = 2500 kN, 5000 kN, 7500 kN, and 10,000 kN, the transverse deflection $w(L)$ of the bar top over time was calculated for bar lengths ranging from 50 m to 140 m. Figure 11 illustrated the results obtained over a calculation period of 100 years. Analysis of the data reveals that when the axial force assumes smaller values (e.g., 2500 kN and 5000 kN), the correlation between bar length and the time course of bar displacement is minimal within the calculation period. Conversely, when the axial force assumes larger values (e.g., 7500 kN and 10,000 kN), an increase in bar length results in an initial slight decrease in the time course of bar instability, followed by an increase within the calculation period.

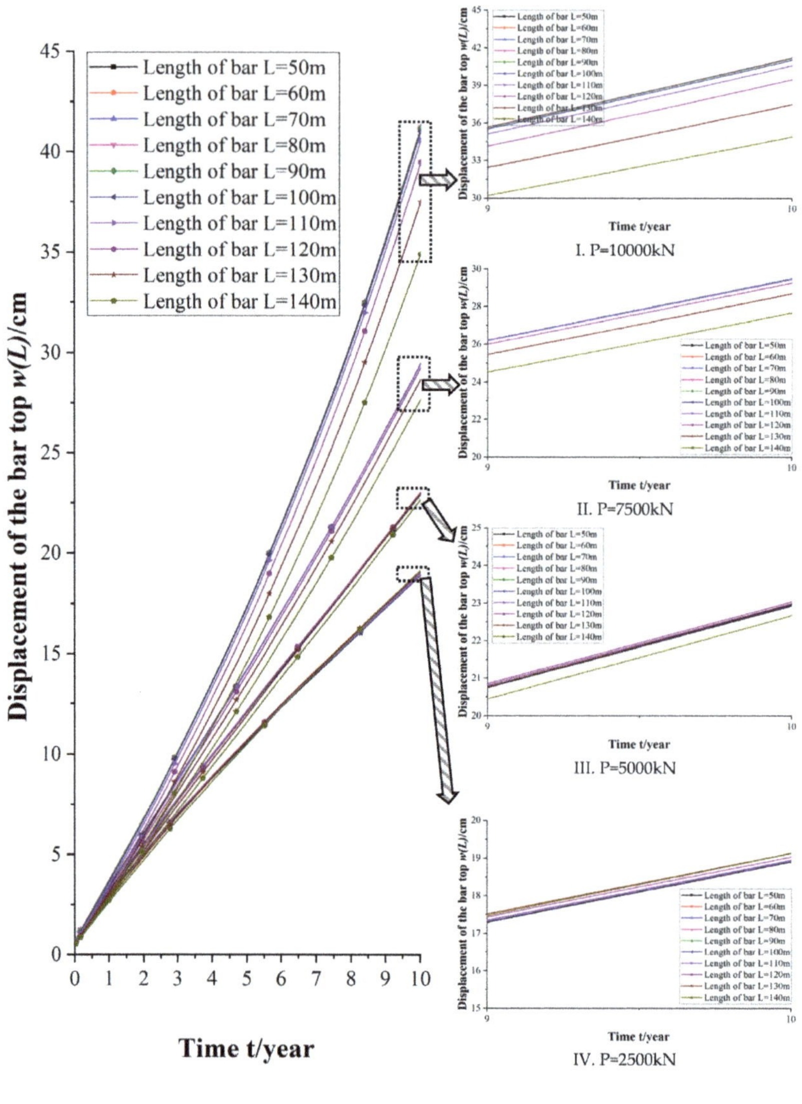


**Figure 11.** Relationship between axial force $P$ and $t_{cr}$ corresponding to different length of bar $L$. (**a**) Different axial force ($P$) series under variable length of bar ($L$); (**b**) Single axial force ($P$) series under variable length of bar ($L$).

By maintaining the loading intensity of the bar-side pile at 50 kPa and considering axial forces, denoted as P, of 2500 kN, 5000 kN, 7500 kN, and 10,000 kN for computing the temporal evolution of the lateral deflection, denoted as $w(L)$, at the top of the bar for bar lengths ranging from 50 m to 140 m over a period of 100 years. Based on the obtained results, when $w(L)$ reaches 10 cm, the relationship between bar length and the critical instability time, denoted as $t_c$ as shown in Figure 12.

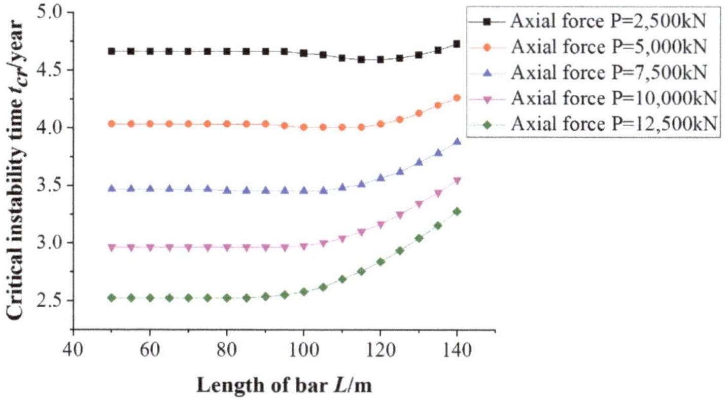


**Figure 12.** Relationship between $L$ and $t_{cr}$ under different axial forces.

As illustrated in Figure 12, the $t_{cr}$-$L$ curve remains relatively horizontal within the small bar length interval when the axial force is 2500 kN, 5000 kN, and 7500 kN. With an increase in $L$, $t_{cr}$ experiences a slight decrease before beginning to increase. For larger axial forces such as 10,000 kN and 12,500 kN, $t_{cr}$ remains relatively constant before continuously increasing with $L$. Given that pile foundation lengths between 50 m and 80 m are prevalent in deep soft foundation areas, it can be inferred that within this range of bar lengths, the instability time of bars is minimally impacted by bar length under equal pile loading intensity. However, for bar lengths exceeding 80 m, the relationship between bar length $L$ and bar instability time $t_{cr}$ under varying axial load levels should be carefully considered.

Figure 13 depicts the relationship between axial force $P$ and critical instability time $t_{cr}$ for different values of bar length $L$. It is evident that as axial force $P$ increases, the critical instability time $t_{cr}$ decreases more rapidly for smaller bar lengths $L$. Additionally, for bar lengths $L$ less than 80 m, there is a high degree of overlap in the $t_{cr}$-$L$ curve. This observation aligns with the conclusion that for bars with a length less than 80 m, bar length has minimal impact on its instability time.

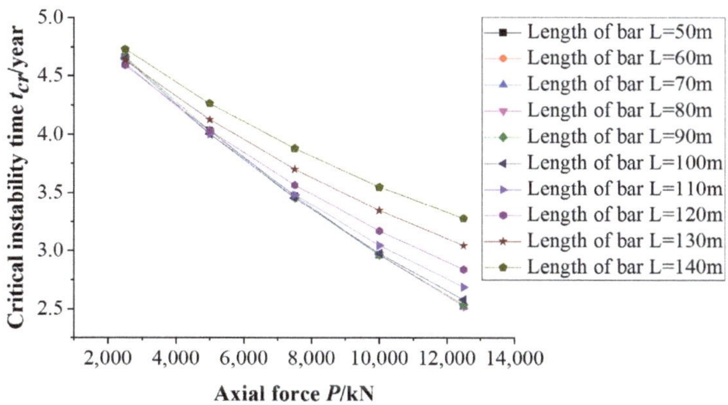


**Figure 13.** Relationship between $P$ and $t_{cr}$ corresponding to different $L$.

### 3.3. *Time Dependence of the Effect of Bar-Side Pile Loading Intensity on the Lateral Deflection of the Bar Top*

For a bar length $L$ of 50 m, the change in lateral deflection $w(L)$ of the bar top over time was calculated for axial forces of 2500 kN, 5000 kN, 7500 kN, and 10,000 kN respectively and pile loading intensities on the side of the bar of 10 kPa, 20 kPa, 30 kPa, 40 kPa, and 50 kPa. The calculation period spanned 100 years. Based on these results, the relationship between pile loading intensity and critical instability time $t_{cr}$ when $w(L)$ = 10 cm was determined as in Figure 13.

In order to accurately reflect the real-world deviation of pile foundations, the critical time $t_{cr}$ for bar instability is defined using the control condition $w(L)$ = 10. For instance, when P = 10,000 kN and the pile load intensity is 10 kPa, 20 kPa, 30 kPa, 40 kPa, and 50 kPa, the corresponding values of $t_{cr}$-$P$ when $w(L)$ reaches 10 cm are 11.4411 years, 6.81096 years, 4.81096 years, 3.67397 years, and 2.96164 years respectively. These results demonstrate that lower pile load intensities result in longer times for bars to exceed code requirements for deviation under the combined influence of axial force and horizontal additional load. As pile load intensity increases, this time decreases at a diminishing rate. Furthermore, $t_{cr}$-$P$ curves corresponding to lower pile load intensities exhibit larger average slopes within the same interval compared to those corresponding to higher pile load intensities. This indicates that as axial force increases, the time required for bars to exceed code requirements for deviation decreases more rapidly at lower pile load intensities.

As illustrated in Figure 14, the $t_{cr}$-$P$ curve is inversely related to pile loading intensity, with the magnitude of the downward shift decreasing continuously. For instance, when $P$ = 10,000 kN and pile loading intensity is 10 kPa, 20 kPa, 30 kPa, 40 kPa, and 50 kPa respectively, the time $t_{cr}$ at which $w(L)$ reaches 10 cm is 11.4411 years, 6.81096 years, 4.81096 years, 3.67397 years, and 2.96164 years. This observation suggests that for smaller pile loading intensities, the time required for a bar to produce an offset exceeding specification requirement under the combined action of axial force and additional horizontal load is longer. As pile loading intensity increases, this time decreases continuously with a decreasing rate of decrease. Furthermore, for $t_{cr}$-$P$ curves with smaller pile loading intensities, their average slope within the same interval is larger than that for larger pile loading intensities. This indicates that for smaller pile loading intensities, an increase in axial force results in a faster decrease in time for bars to produce an offset exceeding specification requirement.

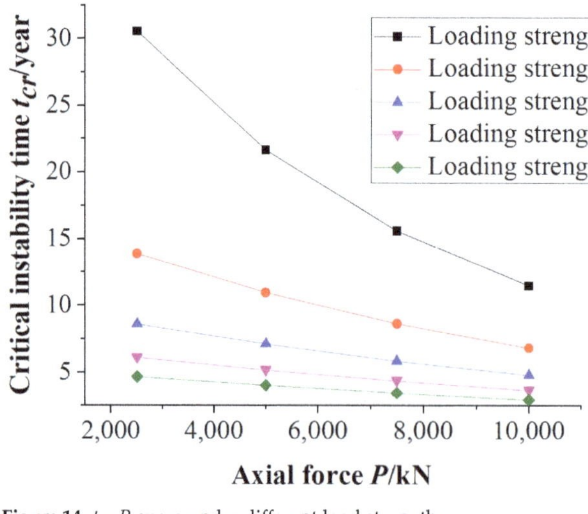


**Figure 14.** $t_{cr}$-$P$ curve under different load strength.

## 4. Conclusions

This study constructs a mechanical model of an elastic bar constrained by elastic and viscoelastic medium, fixed at the base and free at the top under Winkler foundation hypothesis, The Rayleigh-Ritz method is employed to derive the deflection function of the bar when subjected to both axial force and local horizontal distributed load. The primary conclusions are as follows:

1. A mechanical model of an elastic slender compression-bending bar, constrained by elastic-viscoelastic medium wrapping and fixed at the base while free at the top, was established. Utilizing the energy method criterion and the Rayleigh-Ritz method, an approximate solution for the deflection function of the bar body was derived when subjected to axial force $P$ at the top and horizontal additional load caused by pile loading at the side.
2. Employing the elastic-viscoelastic correspondence principle, the approximate solution for the deflection function of the bar body in an elastic medium was transformed into an approximate solution for the deflection function of an elastic compression-bending bar in a viscoelastic medium. Using a Kelvin body as an example for the side medium of the bar, a deflection function under a second-order triangular series was derived. MATLAB is applied in this study to compute the corresponding theoretical solutions under the burgers model.
3. The influence of the magnitude of axial force on the change of $w(L)$ over time under equal bar length and equal pile load strength is studied. The calculation results show that when there is constant load on the side of the bar, axial force acts on the top of the bar, and the larger the axial force, the faster the rate of change of $w(L)$ with time, and the earlier it develops towards instability. The influence of pile load strength on the change of $w(L)$ over time under equal bar length condition is studied. The results show that the change of bar length has little effect on the stability of elastic bar in viscoelastic medium, and the research should focus on the influence of axial force and pile load strength on the surface of side medium on bar stability.

This research presents the derivation of the deflection equation of an elastic rod in a viscoelastic medium and discusses the potential implications for the durability of a long rod under this assumption. However, the working condition assumed in this study has not been validated or compared with the actual engineering scenarios. In future research, more realistic engineering situations can be examined by incorporating boundary conditions, and contrasted with the actual engineering cases, to demonstrate the applicability and accuracy of this method in practical work.

**Author Contributions:** Methodology, X.Z.; Software, J.H.; Investigation, S.C.; Resources, S.C.; Writing—original draft, J.H.; Writing—review & editing, X.Z. All authors have read and agreed to the published version of the manuscript.

**Funding:** This research received no external funding.

**Institutional Review Board Statement:** Not applicable.

**Informed Consent Statement:** Not applicable.

**Data Availability Statement:** No new data were created or analyzed in this study. Data sharing is not applicable to this article.

**Conflicts of Interest:** The authors declare no conflict of interest.

## Appendix A

Here is a detail introduction of Equation (8): Flamant [43] illustrated a basic solution under polar coordinates, which is transferred into Cartesian coordinates [46] as shown in Figure A1 and Equation (A1):

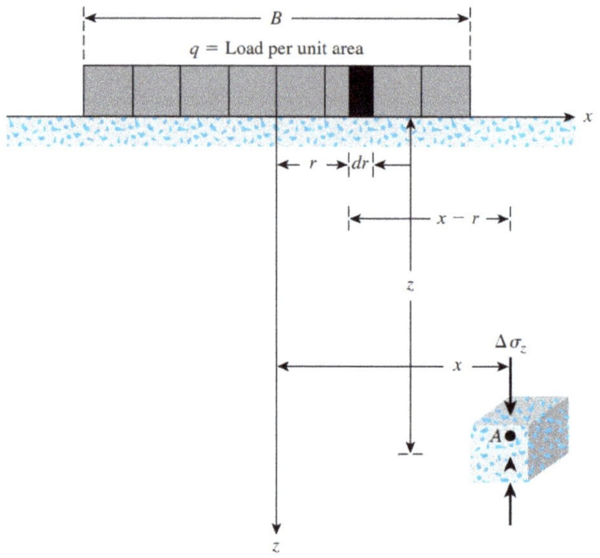


**Figure A1.** Vertical stress caused by a flexible strip load [46].

$$\sigma_z = \frac{q}{\pi}\left\{ \arctan\left[\frac{z}{x-\left(\frac{B}{2}\right)}\right] - \arctan\left[\frac{z}{x-\left(\frac{B}{2}\right)}\right] - \frac{Bz\left[x^2 - z^2 - \left(\frac{B^2}{4}\right)\right]}{\left[x^2 + z^2 - \left(\frac{B^2}{4}\right)\right] + B^2z^2} \right\} \tag{A1}$$

where: $q$ is the magnitude of constant load, as is load per unit area

$B$ is the total length of constant load

Then Equation (8) is transformed from Equation (A1) by converting normal coordinates into local coordinates as shown in Figure 5 according to the following principles: $B$ is converted to $L_p$, $x$ is converted to ($L_p + d_p$), and $z$ is converted to ($L - x$), then Equation (8) is evidently transformed.

## Appendix B

Here is a detail introduction of Equation (9) based on basic theory from Mechanics of Structure [47]: Under the action of the critical force, the bar body maintains a state of limit equilibrium within the plane Cartesian coordinate system shown in Figure 5. Assuming that the deflection curve of the rod element, represented by $w(x)$, undergoes slight bending in the critical state, the strain energy of the system in this state can be expressed by the following Equation (A2):

$$U = \frac{EI}{2}\int_0^L (w'')^2 dx + \frac{1}{2}\int_0^L b_1 k w^2 dx \tag{A2}$$

where: $q(x)$ is passive load on elastic bar caused by constant load and soil self-weight

$b_1$ is calculated width of the elastic bar

$L$ is length of bar

$L_p$ is width of constant load

$p$ is magnitude of constant load

$d_p$ is distance from the bar to constant load

When the critical force $P_{cr}$ is applied to the end of the rod, it induces a slight displacement $\Delta x$ in the vertical direction at the rod's end. The work performed by this displacement can be expressed as follows:

$$U_p = -P_{cr}\Delta x \tag{A3}$$

where: $P_{cr}$ is critical force

$\Delta x$ is a small displacement caused by constant load $p$

Based on the energy criterion, when the rod element is in a critical state as $U + U_p = 0$, then:

$$\frac{EI}{2}\int_0^L (w'')^2 dx + \frac{b_1}{2}\int_0^L kw^2 dx - P_{cr}\Delta x = 0 \tag{A4}$$

where: $\Pi_p$ is total potential energy of both the bar and the surrounding soil

$E$ is elastic modulus of elastic bar

$I$ is moment of inertia of the cross section of elastic bar

$k$ is reaction force coefficient of elastic medium

$w$ is deflection of elastic bar as $w(x)$

Then generalized function $R(w)$ is Solved from Equation (A4) as:

$$R(w) = \frac{EI\int_0^L (w'')^2 dx + b_1\int_0^L kw^2 dx}{2\Delta x} \tag{A5}$$

In Equation (A5):

$$\Delta x = \int_0^L \sqrt{1+(w')^2}dx - L \approx \frac{1}{2}\int_0^L (w')^2 dx \tag{A6}$$

Considering boundary conditions in Figure 5 as shown in follow:

$$x = 0,\ w = 0,\ w' = 0;\ x = L,\ w'' = 0 \tag{A7}$$

The deflection curve of the rod element in a state of slight bending can be approximated using a trial function in the form of a triangular series by Laplace transform as shown in Equation (A8):

$$w(x) = \sum_{n=1}^{i} w_n\left(1 - \cos\frac{(2n-1)\pi x}{2L}\right) \tag{A8}$$

Take Equations (A7) and (A8) into Equation (A6), the functional $R(w)$ could be expressed as:

$$R(w) = \frac{A(w_1, w_2, \ldots, w_i)}{B(w_1, w_2, \ldots, w_i)} \tag{A9}$$

In Equation (A9):

$$\begin{aligned} A(w_1, w_2, \ldots, w_i) = \ & EI\int_0^L \left(\sum_{n=1}^{i} w_n\left(\tfrac{(2n-1)\pi}{2L}\right)^2 \cos\tfrac{(2n-1)\pi x}{2L}\right)^2 dx \\ & + b_1\int_0^L k\left(\sum_{n=1}^{i} w_n\left(1-\cos\tfrac{(2n-1)\pi x}{2L}\right)\right)^2 dx \end{aligned} \tag{A10}$$

$$B(w_1, w_2, \ldots, w_i) = \int_0^L \left(\sum_{n=1}^{i} w_n\left(\frac{(2n-1)\pi}{2L}\right)\sin\frac{(2n-1)\pi x}{2L}\right)^2 dx \tag{A11}$$

when the minimum $R(w_1)$ is taken, $P_{cr}$ will be equal to $R(w)$, which means:

$$\frac{\partial R}{\partial w_r} = \frac{1}{B}\frac{\partial A}{\partial w_r} - \frac{A}{B^2}\frac{\partial B}{\partial w_r} = 0, r = 1,2,3,\ldots,i \tag{A12}$$

In Equation (A12):

$$\begin{aligned}\frac{\partial A}{\partial w_r} = & 2\sum_{n=1}^{i} w_n \int_0^L \left( EI\frac{(2n-1)^2(2r-1)^2\pi^4}{16L^4}\cos\frac{(2n-1)\pi x}{2L}\cos\frac{(2r-1)\pi x}{2L} \right. \\ & \left. +b_1 k\left(1-\cos\frac{(2n-1)\pi x}{2L}\right)\left(1-\cos\frac{(2r-1)\pi x}{2L}\right)\right)dx\end{aligned} \tag{A13}$$

$$\frac{\partial B}{\partial w_r} = 2\sum_{n=1}^{i} w_n \int_0^L \frac{(2n-1)(2r-1)\pi^2}{4L^2}\sin\frac{(2n-1)\pi x}{2L}\sin\frac{(2r-1)\pi x}{2L}dx \tag{A14}$$

Take the minimum $R(w)$ following principle of minimum potential energy as follow:

$$\frac{\partial A}{\partial w_r} - \frac{A}{B}\frac{\partial B}{\partial w_r} = \frac{\partial A}{\partial w_r} - P_{cr}\frac{\partial B}{\partial w_r} = 0 \tag{A15}$$

Then take Equations (A13) and (A14) into Equation (A15), the principle of minimum potential energy is list as:

$$\begin{aligned}& 2\sum_{n=1}^{i} w_n \int_0^L \left( EI\frac{(2n-1)^2(2r-1)^2\pi^4}{16L^4}\cos\frac{(2n-1)\pi x}{2L}\cos\frac{(2r-1)\pi x}{2L} \right. \\ & +b_1 k\left(1-\cos\frac{(2n-1)\pi x}{2L}\right)\left(1-\cos\frac{(2r-1)\pi x}{2L}\right) \\ & \left. -P_{cr}\frac{(2n-1)(2r-1)\pi^2}{4L^2}\sin\frac{(2n-1)\pi x}{2L}\sin\frac{(2r-1)\pi x}{2L}\right)dx = 0\end{aligned} \tag{A16}$$

Then defined:

$$\begin{aligned}W_{nr} = & 2\int_0^L \left( EI\frac{(2n-1)^2(2r-1)^2\pi^4}{16L^4}\cos\frac{(2n-1)\pi x}{2L}\cos\frac{(2r-1)\pi x}{2L} \right. \\ & +b_1 k\left(1-\cos\frac{(2n-1)\pi x}{2L}\right)\left(1-\cos\frac{(2r-1)\pi x}{2L}\right) \\ & \left. -P_{cr}\frac{(2n-1)(2r-1)\pi^2}{4L^2}\sin\frac{(2n-1)\pi x}{2L}\sin\frac{(2r-1)\pi x}{2L}\right)dx\end{aligned} \tag{A17}$$

By combining Equations (A16) and (A17), the matrix expression of phase space formulation is obtained as follow:

$$\sum_{r=1}^{i}\sum_{n=1}^{i} w_n W_{nr} = \begin{bmatrix} W_{11} & W_{12} & \cdots & W_{1i} \\ W_{21} & W_{12} & \cdots & W_{i2} \\ \vdots & \vdots & \cdots & \vdots \\ W_{i1} & W_{i2} & \cdots & W_{ii} \end{bmatrix} \begin{bmatrix} w_1 \\ w_2 \\ \vdots \\ w_i \end{bmatrix} = 0 \tag{A18}$$

Obviously, Equation (A18) have more than one solution, which means:

$$\begin{vmatrix} W_{11} & W_{12} & \cdots & W_{1i} \\ W_{21} & W_{12} & \cdots & W_{i2} \\ \vdots & \vdots & \cdots & \vdots \\ W_{i1} & W_{i2} & \cdots & W_{ii} \end{vmatrix} = 0 \tag{A19}$$

By selecting an appropriate number of expansion terms, denoted as $i$, the smallest positive root of Equation (A19) can be determined. This root represents the Euler formula for the critical force of a slender rod supported by lateral elasticity. For instance, when $i$ is equal to 2, the Euler formulas for the critical force, denoted as $P_{cr}$, can be derived for two commonly used values of the ground reaction coefficient.

When taken $k$ as constant, the solution is shown as follow:

$$\begin{aligned}P_{cr} = & \frac{135EI\pi^5+16b_1kL^4(45\pi-104)}{108L^2\pi^3} \\ & -\frac{\sqrt{729E^2I^2\pi^{10}-864b_1kEIL^4\pi^5(9\pi-28)+64{b_1}^2k^2L^8(3280-2232\pi+405\pi^2)}}{27L^2\pi^3}\end{aligned} \tag{A20}$$

when $k$ is increased linearly with depth, which means it follows Matlock method as:

$$k = m(L - x) \tag{A21}$$

Take Equation (A21) into Equation (A19), the solution is shown as follow:

$$\begin{aligned} P_{cr} &= \frac{405EI\pi^6+8b_1mL^5\left(135\pi^2-1148\right)}{324L^2\pi^4} \\ &- \frac{\sqrt{6561E^2I^2\pi^{12}-1296b_1mEIL^5\pi^6(27\pi^2-280)+16b_1{}^2m^2L^{10}\left(348196-70524\pi^2+3645\pi^4\right)}}{81L^2\pi^4} \end{aligned} \tag{A22}$$

## Appendix C

Here is a detail introduction of Burgers model and phase space formulation [48]:

The Burgers model is a typical generalized model consisting of a Maxwell model and a Kelvin model in series, as shown in Figure A2.

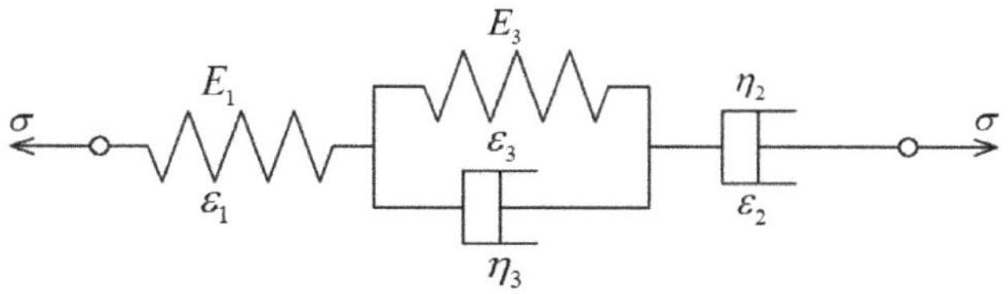


**Figure A2.** Burgers Model.

The Burgers model has the following constitutive relations as shown in Equation (A23)

$$\sigma + p_1\dot{\sigma} + p_2\ddot{\sigma} = q_1\dot{\varepsilon} + q_2\ddot{\varepsilon} \tag{A23}$$

where:

$$p_1 = \frac{\eta_2}{E_1} + \frac{\eta_2 + \eta_3}{E_3};\ p_2 = \frac{\eta_2\eta_3}{E_1E_3};\ q_1 = \eta_2;\ q_2 = \frac{\eta_2\eta_3}{E_3} \tag{A24}$$

Given a constant stress $\sigma_0$, the creep equation for the Burgers model is shown in Equation (A24):

$$\varepsilon(t) = \frac{\sigma_0}{E_1} + \frac{\sigma_0 t}{\eta_2} + \frac{\sigma_0}{E_3}\left(1 - e^{-\frac{E_3}{\eta_3}t}\right) \tag{A25}$$

Applying the Laplace transform to the generalization Equation (2) of the one-dimensional differential constitutive relations, then get Equation (A25):

$$\overline{P}(s) = \sum_{k=0}^{n} p_k s^k;\ \overline{Q}(s) = \sum_{k=0}^{n} q_k s^k \tag{A26}$$

Then the expression of Hooke's law for a similar uniaxial state in the phase space can be obtained as:

$$\overline{\sigma}(s) = \overline{E}(s)\overline{\varepsilon}(s) \tag{A27}$$

when applying the one-dimensional differential viscoelastic ontology to solve the related problems, it is only necessary to replace the material parameter of the corresponding elastic solution by $\overline{E}(s)$, to obtain the form of the elastic-like solution in the phase space, and then carry out the Laplace inverse transformation on it to obtain the corresponding viscoelastic solution. This is the principle of elastic-viscoelastic correspondence which applies the one-dimensional differential viscoelastic constitutive relations.

## References

1. Taylor, D.W.; Merchant, W. A theory of clay consolidation accounting for secondary compression. *Stud. Appl. Math.* **1940**, *19*, 167–185. [CrossRef]
2. Tan, T.K. One Dimensional problems of consolidation and secondary time effects. *Chin. Civ. Eng. J.* **1958**, *5*, 1–5.
3. Tan, T.K. Two Dimensional problems of settlements of clay layers due to consolidation and secondary time effects. *Acta Mech. Sin.* **1958**, *2*, 1–10.
4. Gibson, R.E. A one-dimensional consolidation problem with a moving boundary. *Q. Appl. Math.* **1960**, *18*, 123–129. [CrossRef]
5. Nethercot, D.A.; Rockey, K.C. The lateral buckling of beams having discrete intermediate restraints. *Struct. Eng.* **1972**, *50*, 391–403.
6. Mutton, B.R.; Trahair, N.S. Design requirements for column braces. *Civ. Eng. Trans.* **1975**, *CE17*, 30–35.
7. Yura, J.A. Winter's bracing approach revisited. *Eng. Struct.* **1996**, *18*, 821–825. [CrossRef]
8. Helwig, T.A.; Yura, J.A. Torsional bracing of columns. *ASCE J. Struct. Eng.* **1999**, *125*, 547–555. [CrossRef]
9. Gil, H.; Yura, J.A. Bracing requirements of inelastic columns. *J. Constr. Steel. Res.* **1999**, *51*, 1–19. [CrossRef]
10. McCann, F.; Wadee, M.A.; Gardner, L. Lateral stability of imperfect discretely braced steel beams. *J. Eng. Mech.* **2013**, *139*, 1341–1349. [CrossRef]
11. Foster, A.S.J.; Gardner, L. Ultimate behaviour of steel beams with discrete lateral restraints. *Thin Walled Struct.* **2013**, *72*, 88–101. [CrossRef]
12. Winter, G. Lateral bracing of columns and beams. *Trans. Am. Soc. Civ. Eng.* **1960**, *125*, 809–825. [CrossRef]
13. Timoshenko, S.P.; Gere, J.M. *Theory of Elastic Stability*, 2nd ed.; McGraw-Hill: New York, NY, USA, 1961.
14. Zhao, M.H. The Calculation of Piles under Simultaneous Axial and Lateral Loading. *J. Hunan Univ. Nat. Sci.* **1987**, *2*, 68–81. (In Chinese)
15. Zhou, J.K.; Du, Q.Q. Modified Winkler foundation model with horizontal force taken into account. *J. Hohai Univ. Nat. Sci.* **2002**, *32*, 669–673. (In Chinese)
16. Chen, L.J.; Yu, Q.D.; Dai, Z.H. Finite difference solution based on composite stiffness and bi-parameter method for calculating vertical and horizontal bending piles. *Chin. J. Rock Mech. Eng.* **2016**, *35*, 613–622. (In Chinese)
17. Feng, H.; Yang, Y.S.; Yu, H.T. Dynamic response of viscoelastic foundation beams under traveling wave effect. *Chin. J. Geotech. Eng.* **2020**, *42*, 126–132. (In Chinese)
18. Dai, Z.H.; Shen, P.S.; Zhang, J.W. Numerical solution of piles under lateral load of trapezoid-distribution by bi-parameter method. *Chin. J. Rock Mech. Eng.* **2004**, *23*, 2632–2638. (In Chinese)
19. Reese, L.C.; Welch, I.L.C. Lateral loading of deep foundations in stiff clay. *ASCE J. Soil Mech. Found. Div.* **1975**, *101*, 633–649. [CrossRef]
20. Sastry, V.V.R.N.; Meyerhof, G.G. Behaviour of flexible piles under inclined Loads. *Rev. Can. Geotech.* **1990**, *27*, 19–28. [CrossRef]
21. Meyerhof, G.G.; Yalcin, A.S. Behaviour of flexible batter piles under inclined loads in layered soil. *Rev. Can. Geotech.* **1993**, *30*, 247–256. [CrossRef]
22. Sastry, V.V.R.N.; Meyerhof, G.G. Behaviour of flexible piles in layered sands under eccentric and inclined loads. *Rev. Can. Geotech.* **1994**, *31*, 513–520. [CrossRef]
23. Zhao, M.H.; Hou, Y.Q.; Shan, Y.M. Calculation and model test study on the bridge piles under inclined loads. *J. Hunan Univ. Nat. Sci.* **1999**, *26*, 86–91. (In Chinese)
24. Zhukov, N.V.; Balov, I.L. Investigation of the effect of a vertical surcharge on horizontal displacements and resistance of pile columns to horizontal loads. *J. Geotech. Eng. ASCE* **1978**, *15*, 16–21.
25. Zhang, L.; Gong, X.N.; Yang, Z.X.; Yu, J.L. Elastoplastic solutions for single piles under combined vertical and lateral loads. *J. Cent. South Univ. Technol.* **2011**, *18*, 216–222. [CrossRef]
26. Han, J.; Frost, J.D. Load-deflection response of transversely isotropic piles under lateral loads. *Int. J. Numer. Anal. Methods Geomech.* **2000**, *24*, 509–529. [CrossRef]
27. Fan, C.C.; Long, J.H. Assessment of existing methods for predicting soil response of laterally loaded piles in sand. *Comput. Geotech.* **2005**, *32*, 274–289. [CrossRef]
28. Poulos, H.G.; Davis, E.H. *Pile Foundation Analysis and Design*; John Wiley & Sons: New York, NY, USA, 1980.
29. Wu, H.L. *The Composite Stiffness Principle and Bi-Parameter Method for Calculation of Laterally Loaded Piles*, 2nd ed.; China Communication Press: Beijing, China, 2000. (In Chinese)
30. Hunt, G.W.; Peletier, M.A.; Champneys, A.R.; Woods, P.D.; Ahmer Wadee, M.; Budd, C.J.; Lord, G.J. Cellular buckling in long structures. *Nonlinear Dyn.* **2000**, *21*, 3–29. [CrossRef]
31. Wadee, M.K.; Hunt, G.W.; Whiting, A.I.M. Asymptotic and Rayleigh–Ritz routes to localized buckling solutions in an elastic instability problem. *Proc. R. Soc. Lond. Ser. A* **1997**, *453*, 2085–2107. [CrossRef]
32. Hunt, G.W.; Wadee, M.K.; Shiacolas, N. Localized Elasticae for the Strut on the Linear Foundation. *J. Appl. Mech.* **1993**, *60*, 1033–1038. [CrossRef]
33. Groh, R.M.J.; Pirrera, A. Spatial chaos as a governing factor for imperfection sensitivity in shell buckling. *Phys. Rev. E* **2019**, *100*, 032205. [CrossRef]
34. Lugo, C.A.; Airoldi, C.; Chen, C.; Crosby, A.J.; Glover, B.J. Morpho elastic modelling of pattern development in the petal epidermal cell cuticle. *J. R. Soc. Interface* **2023**, *204*, 13.

35. Shen, J.J.; Pirrera, A.; Groh, R.M.J. Building blocks that govern spontaneous and programmed pattern formation in pre-compressed bilayers. *Proc. R. Soc. A* **2023**, *478*, 20220173. [CrossRef]
36. Xu, S.; Yan, Z.; Jang, K.I.; Huang, W.; Fu, H.; Kim, J.; Wei, Z.; Flavin, M.; McCracken, J.; Wang, R.; et al. Assembly of micro/nanomaterials into complex, three-dimensional architectures by compressive buckling. *Science* **2015**, *347*, 154–159. [CrossRef] [PubMed]
37. Yang, S.; Khare, K.; Lin, P.C. Harnessing Surface Wrinkle Patterns in Soft Matter. *Adv. Funct. Mater.* **2010**, *20*, 2550–2564. [CrossRef]
38. Sun, X.F.; Fang, X.S.; Guan, L.T. *Mechanics of Materials*, 5th ed.; Higher Education Press: Beijing, China, 2009. (In Chinese)
39. Liu, W. *The Program Design and Application of MATLABM*; Higher Education Press: Beijing, China, 2008. (In Chinese)
40. Kerr, A. On the Determination of Foundation Model Parameters. *J. Geotech. Geoenviron.* **1985**, *111*, 1334–1340. [CrossRef]
41. Kerr, A. Elastic and Viscoelastic Foundation Models. *ASME J. Appl. Mech.* **1964**, *31*, 491–498. [CrossRef]
42. Boussinesq, J. *Application des Potentiels à L'étude de L'équilibre et du Mouvement des Solides Élastiques: Principalement au Calcul des Déformations et des Pressions que Produisent, dans ces Solides, des Efforts Quelconques Exercés sur une Petite Partie de Leur Surface ou de leur Intérieur: Mémoire Suivi de Notes Étendues sur Divers Points de Physique, Mathematique et D'analyse*; Gauthier Villars: Paris, France, 1885.
43. Flamant, A. Sur la répartition des pressions dans un solide rectangulaire chargé transversalement. *CR Acad. Sci. Paris* **1892**, *114*, 1465–1468.
44. Abate, J.; Whitt, W. A Unified Framework for Numerically Inverting Laplace Transforms. *Inf. J. Comput.* **2006**, *18*, 408–421. [CrossRef]
45. Ministry of Housing and Urban-Rural Development of the People's Republic of China. *Technical Code for Building Pile Foundations (JGJ 94-2008)*; China Architecture & Building Press: Beijing, China, 2018.
46. Das, B.M. *Principles of Geotechnical Engineering*; Cengage Learning: Boston, MA, USA, 2021.
47. Krenk, S.; Høgsberg, J. *Statics and Mechanics of Structures*; Springer: Dordrecht, The Netherlands, 2013.
48. Christensen, R.M. *Theory of Viscoelasticity, an Introduction*; Elsevier: Amsterdam, The Netherlands, 1982.

mathematics

MDPI

Article

# Event-Triggered State Estimation for Uncertain Systems with Binary Encoding Transmission Scheme

**Zun Li, Binqiang Xue * and Youyuan Chen**

School of Automation, Qingdao University, Qingdao 266071, China
* Correspondence: xuebinqiang@qdu.edu.cn or xuebinqiang2005@163.com

**Abstract:** This paper proposes an event-triggered state estimation method for parameter-uncertain systems with a binary encoding transmission scheme. Firstly, a binary encoding transmission scheme is introduced between the state estimator and the system to improve the efficiency of network communication. Secondly, an event-triggering mechanism (ETM) is designed to ensure the accuracy of state estimation and reduce the computational burden of the state estimator. At the event-triggered moments, considering the uncertainty of the system, the binary encoding transmission scheme, and the ETM, a moving horizon estimator (MHER) is designed using the robust least squares optimization method to obtain optimal state estimation. At the no-event-triggered moments, the state estimation of the system is computed based on an open-loop state estimator (OLER). Furthermore, stability analysis showed that the state estimation error of the proposed method is bounded. Finally, the practical value of the proposed in this paper is confirmed through numerical simulation.

**Keywords:** moving horizon estimator (MHER); event-triggered mechanism (ETM); binary encoding transmission scheme; min–max problem

**MSC:** 93-10

## 1. Introduction

In recent years, state estimation theory has been extensively investigated [1–3]. Among the various state estimation methods, the moving horizon estimation (MHE) method continues to attract an increasing amount of attention because it can explicitly deal with constraints and limit the solutions of optimization problems to a fixed window length [4–6]. Using the moving horizon optimization principle, the authors of [7] designed a state estimation method with observable estimators for linear discrete systems. For distributed linear network systems, the authors of [8] proposed an MHE method with a consistent arrival cost. It is evident that the MHE method can provide solutions for state estimation problems in different systems.

Practical systems, such as robotic arms and unmanned aerial vehicles, are susceptible to external disturbances, which result in the uncertainty in the system model. However, state estimation for parameter-uncertain systems is challenging due to the uncertainty of the corresponding model parameters [9–11]. A regularized, robust method for dealing with uncertain data was proposed in [12], which effectively quantified the impact of uncertain parameters. The authors of [13] investigated the state estimation problem with respect to parameter-uncertain systems using min–max optimization theory. By designing a distributed moving horizon optimization algorithm for uncertain systems, the authors of [14] achieved good convergence of the state estimation error. The authors of [15] reduced the complexity of the state estimation scheme for distributed uncertain systems using a consistency algorithm.

At present, the time-triggered state estimation methods are mainly used for uncertain discrete systems, bringing about issues of limited communication bandwidth and energy shortages [16–20]. However, the event-triggered methods can effectively mitigate these issues [21–23]. For linear systems with an energy-harvesting sensor, the authors of [24] designed a dynamic event-triggered state estimation method to compute the real-time estimation of system states. The authors of [25] introduced a state estimation algorithm for linear systems based on an ETM that ensured the smooth operation of a system when affected by issues such as insufficient communication bandwidth. To meet the requirements of high-intensity state estimation for complex systems, two event-triggered state estimation methods were proposed in [26] for centralized systems and distributed systems, respectively. In general, there is usually a considerable distance between the systems and the state estimators, so it is necessary to ensure the remote and efficient transmission of information [27–29]. The authors of [30] proposed a state estimation method with unknown inputs. It allowed for the accurate transmission of useful information by quantifying the system measurements. Ref. [31] presented a binary encoding scheme for centralized systems and distributed systems that was capable of effectively transmitting the output information of sensors. Furthermore, in practical systems, system noise and random measurement noise are usually energy-limited [32], and it is difficult to accurately compute their variances, means, and other properties. However, a distinctive feature in this regard is that the noise from a system and random measurements is usually bounded. Therefore, in order to further progress research in this area, it is important to predefine the upper bounds of system noise and random measurement noise [33,34].

In summary, uncertain systems in industrial production usually need to carry out remote information transmission in a timely manner. At the same time, the uncertain parameters lead to an obvious increase in the computational burden of state estimation, which can severely disrupt the normal functioning of systems. Therefore, it is crucial to propose a state estimation method for uncertain systems that not only achieves efficient and reliable information transmission but also enables accurate state estimation in cases involving limited computational resources.

Based on the above analysis, this paper proposes an event-triggered state estimation method for parameter-uncertain systems with a binary encoding transmission scheme. The main points of innovation are as follows:

(1) By introducing a binary encoding transmission scheme, the efficient transmission of information between the state estimator and the system can be achieved.

(2) By designing an ETM, the state is estimated via the OLER when an event is not triggered. When an event is triggered, the MHER developed in this study enables optimal state estimation by solving a min–max problem using a robust least squares optimization method.

(3) Through stability analysis, the state estimation error is bounded when the noise remains bounded.

This paper consists of the following sections. Section 2 introduces the research subject and a flowchart of the designed state estimation method and presents the binary encoding transmission scheme and the ETM. In Section 3, the design process of the state estimator is described. The stability analysis is discussed in Section 4. Subsequently, Section 5 provides an example via a simulation. Finally, Section 6 provides a summary of the entire paper.

Given a symmetric positive definite matrix $K$, $\|\cdot\|_K^2$ is the square of the weighted Euclidean norm, $\|b\|_K^2 = b^T K b$. $\|\cdot\|$ represents the spectral norm of the matrix, and $I$ represents the identity matrix. $E(\cdot)$ represents the expectation of the vector parameters, and $\mathbf{R}^n$ stands for $n$-dimensional space. For the positive definite matrix $K$, $\lambda_{\max}(K)$ represents the maximum singular value of matrix K, and $\lambda_{\min}(K)$ represents the minimum singular value of matrix K. $col\{b_{t-N}, \ldots, b_t\}$ is denoted as $b_{t-N}^t$.

## 2. Problem Description

### 2.1. System Model

Consider the following class of parameter-uncertain linear discrete systems:

$$x_{t+1} = (A + \delta A_t)x_t + \omega_t,$$
$$y_t = Cx_t + v_t, \quad (1)$$

where $t \in \mathbf{N}_{\geq 0}$, $x_t \in \mathbf{X} \in \mathbf{R}^{n_x}$ represents the system state variable, $y_t \in \mathbf{Y} \in \mathbf{R}^{n_y}$ represents the measurement output, $\omega_t \in \mathbf{W} \subset \mathbf{R}^{n_x}$ denotes system noise, and $v_t \in \mathbf{V} \subset \mathbf{R}^{n_y}$ represents random measurement noise. $A$ and $C$ represent the matrices with appropriate dimensions. Additionally, the uncertainty of the model parameter is represented by $\delta A_t$, which is a matrix of the same type as $A$.

**Assumption 1.** *The matrix group $(C, A)$ is observable in N steps.*

**Assumption 2.** *Set $\mathbf{X}$, $\mathbf{W}$, and $\mathbf{V}$ such that they satisfy the following conditions:*

$$\begin{cases} \mathbf{X} = \{x_t : \|x_t\| \leqslant \rho_x\}, \\ \mathbf{W} = \{\omega_t : \|\omega_t\| \leqslant \rho_\omega\}, \\ \mathbf{V} = \{v_t : \|v_t\| \leqslant \rho_v\}, \end{cases}$$

*where $\rho_x$, $\rho_\omega$, and $\rho_v$ represent the upper bounds of the state variable, system noise, and random measurement noise, respectively.*

To cope with the complex network communication environment and ensure the efficient transmission of information, this paper introduces a binary encoding transmission scheme operating between the sensor and the state estimator. To reduce the estimation burden of the state estimator and ensure estimation accuracy, the ETM was designed. The state estimation method proposed in this paper is illustrated in Figure 1.

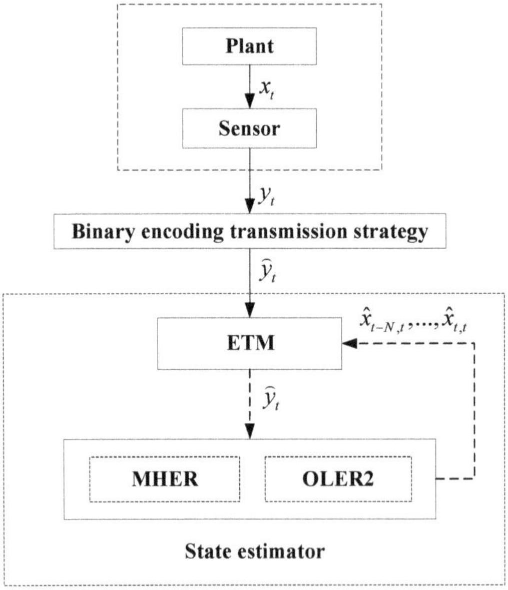


**Figure 1.** Event-triggered state estimation method.

*2.2. Design of Binary Encoding Transmission Scheme*

As shown in Figure 2, we designed a binary encoding transmission scheme for transmitting a sensor's measurement output to a state estimator.

**Figure 2.** Binary encoding transmission scheme.

In the binary encoding transmission scheme, a constant scalar $M_t$ within a certain range $[-U, U]$ is converted into a binary array with a length of $\Omega$ at time $t$. Then, $2^{\Omega}$ points are obtained from the binary array, and $\chi = \{l_1, \ldots, l_{2\Omega}\}$ are used to represent these evenly spaced points. There are a total of $2^{\Omega} - 1$ equally long intervals, each with a length of $\varsigma_l = 2U/2^{\Omega} - 1$. By utilizing a random truncation function, $M_t$ is processed to obtain the truncated output $\beta_t$.

When $l_i \leq M_t \leq l_{i+1}, i = 1, 2, \ldots, 2^{\Omega} - 1$, $\beta_t$ is generated via the following equation:

$$\begin{cases} \mathbf{P}\{\beta_t(M_t, \Omega) = l_i\} = 1 - p_t, \\ \mathbf{P}\{\beta_t(M_t, \Omega) = l_{i+1}\} = p_t, \end{cases} \tag{2}$$

where $0 \leq p_t \leq 1$, $p_t = (M_t - l_i)/\varsigma_l$. In addition, $\beta_t$ can be obtained based on

$$\beta_t(M_t, \Omega) = -U + \sum_{i=1}^{\Omega} o_{i,t} 2^{i-1} \varsigma_l. \tag{3}$$

Therefore, $M_t$ is encoded into a binary array: $O_t = \{o_{1,t}, o_{2,t}, \ldots, o_{\Omega,t}\}, o_{i,t} \in \{0, 1\}$.

To fulfill the requirement for transmitting the sensor measurement output, the designed encoding scheme is expanded into vector form: $M_t \in \mathbf{R}^{n_y}$, $\beta_t(M_t, \Omega) \in \mathbf{R}^{n_y}$.

To quantify the measurement output distortion caused by the binary encoding transmission scheme, the sensor measurement output recovered by the binary decoder is processed as follows:

$$\widehat{y}_i = \beta_t(y_i, \Omega), i = t - N, \ldots, t, \tag{4}$$

where $E\left\{\widehat{y}_i\right\} = y_i$, $N$ represents the window length of state estimation, and the truncation error is denoted as $\varepsilon_i = \widehat{y}_i - y_i$. $\Im$ represents the tight set that constrains the truncation error

$$\Im = \{\varepsilon_i \in \mathbf{R}^{n_y}, \|\varepsilon_{t-N}^t\| = \left\|\widehat{y}_{t-N}^t - y_{t-N}^t\right\| \leq \varpi, \varpi > 0\}. \tag{5}$$

**Remark 1.** *Different from the communication scheme designed in [31], the scheme designed in this paper only considers the error caused by the random truncation function in the communication process and restricts the truncation error. In actual information transmission, the probability of binary bit flipping in the binary coding transmission scheme is very low, and its error is negligible compared with the truncation error, so this problem is not studied in this paper.*

### 2.3. Design of ETM

The ETM is set on the estimator's side, and its configuration is illustrated in Figure 3.

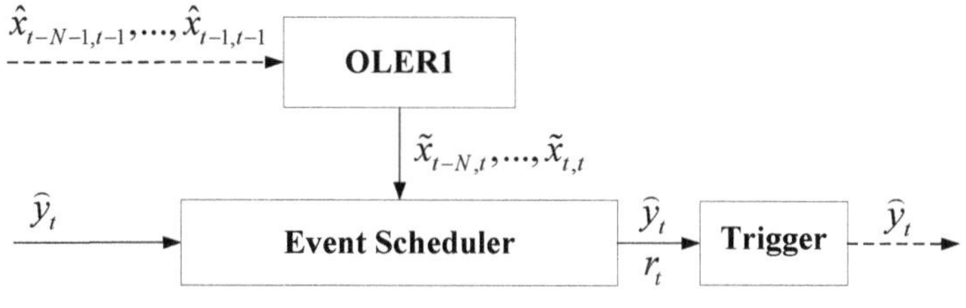


**Figure 3.** Schematic diagram of the ETM.

At $t > N$, the ETM obtains the output value after its transmission via the binary encoding transmission scheme, and OLER1 provides the current state estimation value; then, the two values are both transmitted to the event scheduler, which performs a calculation based on the following conditions:

$$r_t = \begin{cases} 0, \text{ if} \left\| \overset{\frown}{y}^{t}_{t-N} - \tilde{y}^{t,t}_{t-N,t} \right\| \leqslant \Delta, \\ 1, \text{ if} \left\| \overset{\frown}{y}^{t}_{t-N} - \tilde{y}^{t,t}_{t-N,t} \right\| > \Delta, \end{cases} \tag{6}$$

where $\Delta$ is the threshold to adjust the event trigger; $r_t = 1$ and $r_t = 0$ respectively indicate whether the event has been triggered; $\tilde{y}_{i,t}$ satisfies

$$\tilde{y}_{i,t} = C\tilde{x}_{i,t}, i = t - N + 1, \ldots, t, \tag{7}$$

$\tilde{x}_{i,t}$ represents the state estimation carried out by OLER1.

Based on the above analysis, at time $t$, the available information $\bar{y}_t$ of the state estimator is satisfied:

$$\begin{cases} \bar{y}_t = \overset{\frown}{y}_t, & \text{if } r_t = 1, \\ \bar{y}_t = \tilde{y}_{t,t}, & otherwise. \end{cases} \tag{8}$$

Then, the following equation is obtained

$$\left\| \overset{\frown}{y}^{t}_{t-N} - \bar{y}^{t}_{t-N} \right\| \leqslant \left\| \overset{\frown}{y}^{t}_{t-N} - \tilde{y}^{t,t}_{t-N,t} \right\| \leqslant \Delta. \tag{9}$$

To quantify the error caused by the ETM, we introduce a bounded vector $g_i$, $i = t - N, \ldots, t$, which satisfies the following relation:

$$\left\| g^{t}_{t-N} \right\| = \left\| \overset{\frown}{y}^{t}_{t-N} - \bar{y}^{t}_{t-N} \right\| \leqslant \left\| \overset{\frown}{y}^{t}_{t-N} - \tilde{y}^{t,t}_{t-N,t} \right\| \leqslant \Delta. \tag{10}$$

Thus, the relationship between the available information $\bar{y}_t$ regarding the state estimator and the system state variable $x_t$ can be expressed as follows:

$$\bar{y}_t = Cx_t + v_t + \varepsilon_t + g_t. \tag{11}$$

## 3. Design of State Estimator

The state estimator is composed of OLER2 and an MHER. When $r_t = 1$, an MHER is stimulated to obtain the optimal state estimation sequences within the current time window. When $r_t = 0$, OLER2 is used to obtain the open-loop state estimation sequences within the current time window.

### 3.1. Design of OLER

$\breve{x}_{i,t}$ represents the estimate of $x_{i,t}$. At $t > N$, the updated estimation sequences of state by the OLER1 are as follows:

$$\tilde{x}_{i+1,t} = A\tilde{x}_{i,t}, i = t - N, \ldots, t, \tag{12}$$

$$\tilde{x}_{t-N,t} = A\breve{x}_{t-N-1,t-1}, \tag{13}$$

$$\breve{x}_{t-N-1,t-1} = \begin{cases} \hat{x}^{*}_{t-N-1,t-1}, \text{if } r_{t-1} = 1, \\ \tilde{x}_{t-N-1,t-1}, \text{if } r_{t-1} = 0. \end{cases} \tag{14}$$

The OLER2 designed in the state estimator has the same structure as OLER1.

**Remark 2.** *OLER1 is designed to ensure that the ETM can operate properly when $t > N$. OLER2 is used to obtain the open-loop estimation of a state when an event has not been triggered, thus* *ensuring that the MHER can use the information from a previous moment when an event is triggered.*

### 3.2. Design of MHER

At $t > N$, based on the MHE method, the cost function is as follows:

$$J = \|\hat{x}_{t-N,t} - \bar{x}_{t-N}\|_G^2 + \sum_{i=t-N}^{t} \|y_i - C\hat{x}_{i,t}\|^2, \tag{15}$$

where $y_i = \bar{y}_i - \varepsilon_i - g_i$, and $G$ is the weighted positive-definite matrix. A vector $m_i = g_i + \varepsilon_i$ is introduced to represent the difference between the measurement output of the sensor and the available information on the state estimator, which satisfies the following relation:

$$\Psi = \{m_i \in \mathbf{R}^{n_y}, \|m_{t-N}^t\| = \|\varepsilon_{t-N}^t + g_{t-N}^t\| \leqslant \|\varepsilon_{t-N}^t\| + \|g_{t-N}^t\| \leqslant \varphi, \varphi > 0\}, \tag{16}$$

Therefore, Equation (8) can be written as

$$J = \|\hat{x}_{t-N,t} - \bar{x}_{t-N}\|_G^2 + \sum_{i=t-N}^{t} \|\bar{y}_i - C\hat{x}_{i,t} - m_i\|^2. \tag{17}$$

To account for the effect of the uncertain parameters $\delta A_i, i = t-N, \ldots, t-1$ and the output mismatch caused by $m_i$, this paper utilizes the robust least squares optimization method mentioned in [12,14]. Then, the following problems are introduced.

**Problem 1.** *$\bar{x}_{t-N}$ and $\bar{y}_{t-N}^t$ are known, and the optimal estimate of the state $\hat{x}_{t-N,t}^*$ satisfies*

$$\hat{x}_{t-N,t}^* = \arg\min_{\hat{x}_{t-N,t}} \max_{m_i, \delta A_i} J(\hat{x}_{t-N,t}, m_i, \delta A_i), i = t-N, \ldots, t, \tag{18}$$

*where $\hat{x}_{i,t}^* = A\hat{x}_{i-1,t}^*$; the prior estimate of state $x_{t-N}$ is represented as*

$$\bar{x}_{t-N} = \begin{cases} A\hat{x}_{t-N-1,t-1}^*, \text{if } r_{t-1} = 1, \\ A\tilde{x}_{t-N-1,t-1}, \text{if } r_{t-1} = 0, \end{cases} \tag{19}$$

*where $t > N$; and $\bar{x}_0$ represents the initial estimate of $x_0$.*

*To facilitate the subsequent computations, define the following terms in the corresponding manner:*

$$Q = \begin{bmatrix} C \\ CA \\ \vdots \\ CA^N \end{bmatrix}, \hat{Q}_t = \begin{bmatrix} C \\ C(A + \delta A_{t-N}) \\ \vdots \\ C\prod_{i=1}^{N}(A + \delta A_{t-i}) \end{bmatrix},$$

$$\Theta_t = \begin{bmatrix} 0 & \cdots & 0 & 0 \\ C & \cdots & 0 & 0 \\ \vdots & \vdots & \vdots & \vdots \\ C\prod_{i=1}^{N-1}(A + \delta A_{t-i}) & \cdots & C(A + \delta A_{t-i}) & C \end{bmatrix}.$$

*Therefore, we can obtain the following:*

$$\hat{x}_{t-N,t}^* = \arg\min_{\hat{x}_{t-N,t}} \max_{m_{t-N}^t, \delta A_{t-N}, \ldots \delta A_t} \left[\|\hat{x}_{t-N,t} - \bar{x}_{t-N}\|_G^2 + \|\bar{y}_{t-N}^t - \hat{Q}_t\hat{x}_{t-N,t} - m_{t-N}^t\|\right]. \tag{20}$$

By utilizing Equation (20) and Theorem 3.3 from [12], Problem 2 can be obtained.

**Problem 2.** *$\bar{x}_{t-N}$ and $\bar{y}_{t-N}^t$ are known. The optimal estimate of state $\hat{x}_{t-N,t}^*$ satisfies*

$$\hat{x}^*_{t-N,t} = \arg\min_{\hat{x}_{t-N,t}} \max_{\|f_t\| \leqslant \eta_1 \|\hat{x}_{t-N,t} - \bar{x}_{t-N,t}\| + \eta_2} \tilde{J}(\hat{x}_{t-N,t}, f_t), \tag{21}$$

*where*

$$\begin{aligned} \tilde{J}(\hat{x}_{t-N,t}, f_t) = \|\hat{x}_{t-N,t} - \bar{x}_{t-N}\|^2_G + \Big\| &(Q + \breve{Q}_t)(\hat{x}_{t-N,t} - \bar{x}_{t-N}) - \\ &(\bar{y}^t_{t-N} - Q\bar{x}_{t-N} - (\breve{Q}_t \bar{x}_{t-N} + m^t_{t-N}))\Big\|^2, \end{aligned} \tag{22}$$

$$\hat{x}^*_{i+1,t} = A\hat{x}^*_{i,t}, i = t-N, \ldots, t-1, \tag{23}$$

*where* $\breve{Q}_t = \hat{Q}_t - Q$, $\left\|\breve{Q}_t\right\| \leqslant \eta_1$, $\left\|\breve{Q}_t \bar{x}_{t-N} + m^t_{t-N}\right\| \leqslant \eta_2$ *and* $\|f_t\| \leqslant \eta_1 \|\hat{x}_{t-N,t} - \bar{x}_{t-N}\| + \eta_2$, *while* $\eta_1$ *and* $\eta_2$ *are two scalar parameters.*

Regarding Problem 2, considering the case where $\hat{x}^*_{t-N,t} \neq 0$ and introducing a scalar parameter $\theta_t$, $(\theta_t > 1)$, we can obtain the following expression:

$$\hat{x}^*_{t-N,t} = \left(\hat{G}_t + (\theta^*_t/\theta^*_t - 1)Q^T Q\right)^{-1} \left((\theta^*_t/\theta^*_t - 1)Q^T \bar{y}^t_{t-N} + \hat{G}_t \bar{x}_{t-N}\right), \tag{24}$$

where $\theta^*_t$ represents the solution that minimizes $\theta_t$, whose expression is as follows:

$$\theta^*_t = \arg\min_{\theta_t > 1} \Lambda(\theta_t). \tag{25}$$

The expression of $\Lambda(\theta_t)$ is as follows:

$$\begin{aligned} \Lambda(\theta_t) = &\|(\hat{x}_{t-N,t} - \bar{x}_{t-N,t})^*(\theta_t)\|^2_G + \theta_t S^2((\hat{x}_{t-N,t} - \bar{x}_{t-N,t})^*(\theta_t)) \\ &+ \|Q(\hat{x}_{t-N,t} - \bar{x}_{t-N,t})^*(\theta_t) - (\bar{y}^t_{t-N} - Q\bar{x}_{t-N})\|^2_{(\theta_t/\theta_t - 1)I}, \end{aligned} \tag{26}$$

where $S(\hat{x}_{t-N,t} - \bar{x}_{t-N,t}) = \eta_1 \|(\hat{x}_{t-N,t} - \bar{x}_{t-N,t})\| + \eta_2$.

In Equation (24), $\hat{G}_t$ satisfies $\hat{G}_t = G + \theta^*_t \eta_1(\eta_1 + \eta_2/\alpha^*_t(\theta^*_t))I$; here, $\alpha^*_t(\theta_t)$ is the unique solution of the following equation:

$$\begin{aligned} \alpha^2_t = &((\theta_t/\theta_t - 1)Q^T(\bar{y}^t_{t-N} - Q\bar{x}_{t-N}))^T [G + (\theta_t/\theta_t - 1)Q^T Q \\ &+ \theta_t \eta_1(\eta_1 + \eta_2/\alpha_t)I]^{-2}((\theta_t/\theta_t - 1)Q^T(\bar{y}^t_{t-N} - Q\bar{x}_{t-N})), \end{aligned} \tag{27}$$

where $\alpha_t = \|\hat{x}_{t-N,t} - \bar{x}_{t-N,t}\|$.

**Remark 3.** *Equation (24) is obtained by solving a min–max problem using the regularized robust least squares optimization method with bounded uncertainties. The specific solution process is detailed in [12].*

**Remark 4.** *As shown in Algorithm 1, the ETM only works when the time exceeds the length of the state estimation window N, and the MHER is always used for state estimation when the time is not exceeded. Unlike the methods in [25,26], this paper designs the ETM after remote network transmission has occurred, so the step involving the transmission of the estimated results to the plant side is omitted. The efficiency of state estimation is further improved.*

**Algorithm 1:** The state estimation algorithm designed in this paper.

**Step 1:** At $t \leq N$, execute the following steps:
**Step 1.1:** Based on binary encoding transmission scheme, $y_t$ is transmitted to obtain $\widehat{y}_t$.
**Step 1.2:** Based on the design of the MHER, $\hat{x}^*_{0,t}, \ldots, \hat{x}^*_{t,t}$ is obtained.

**Step 2:** At $t > N$, execute the following steps:
**Step 2.1:** Based on binary encoding transmission scheme, $y_t$ is transmitted to obtain $\widehat{y}_t$.
**Step 2.2:** Based on Equations (12)–(14), OLER1 calculates the value of $\tilde{x}_{t-N,t}, \ldots, \tilde{x}_{t,t}$.
**Step 2.3:** Based on Equation (6), the event scheduler calculates the value of $r_t$.
If $r_t = 1$, based on Equations (21)–(24), the MHER obtains $\hat{x}^*_{0,t}, \ldots, \hat{x}^*_{t,t}$ .
If $r_t = 0$, based on Equations (12)–(14), OLER2 obtains $\tilde{x}_{t-N,t}, \ldots, \tilde{x}_{t,t}$

**Step 3:** To update $\bar{y}^t_{t-N}$, let $t = t + 1$ and proceed to step 1.

## 4. Stability Analysis

The purpose of this section is to analyze whether the error of the designed event-triggered state estimation method is bounded under different event-triggered situations.

Referring to the relevant knowledge of linear matrix inequalities in [14], define the following as represented:

$$\begin{cases} \gamma_{\delta a} = \max\limits_{\delta A_i, i=t-N,\ldots,t-1} \|\delta A_i\|, \\ \gamma_{\breve{Q}} = \max\limits_{\delta A_i, i=t-N,\ldots,t-1} \left\|\breve{Q}_t\right\|, \\ \gamma_{\Theta} = \max\limits_{\delta A_i, i=t-N,\ldots,t-1} \|\Theta_t\|. \end{cases} \tag{28}$$

Based on the needs of stability analysis [35–37], define the following terms as they are shown:

$$\begin{cases} \underline{\sigma} = \lambda_{\min}(\hat{G}_t), \\ \bar{\sigma} = \lambda_{\max}(\hat{G}_t), \\ \underline{q} = \lambda_{\min}(Q^T Q), \\ \gamma_a = \|A\|, \\ \gamma_Q = \|Q\|. \end{cases} \tag{29}$$

Theorem 1 is introduced to analyze whether the state estimation error is bounded when OLER2 is used.

**Theorem 1.** *Under the conditions that Assumptions 1 and 2 are satisfied, if $r_t = 0$, the open-loop estimated value $\tilde{x}_{t-N,t}$ of the state satisfies*

$$\|x_{t-N} - \tilde{x}_{t-N,t}\| \leqslant \Upsilon, \tag{30}$$

*where* $\Upsilon = 1\Big/\sqrt{\underline{q}}(\varphi + \gamma_\Theta\sqrt{N}\rho_\omega + \sqrt{N+1}\rho_v + \gamma_{\breve{Q}}\rho_x)$.

**Proof of Theorem 1.** Based on Equations (1) and (7), we can determine that

$$y^t_{t-N} - \bar{y}^t_{t-N} = (Q + \breve{Q}_t)x_{t-N} + \Theta_t\omega^{t-1}_{t-N} + v^t_{t-N} - Q\tilde{x}_{t-N,t}, \tag{31}$$

Then, we can obtain the following:

$$Q(x_{t-N} - \tilde{x}_{t-N,t}) = (y^t_{t-N} - \bar{y}^t_{t-N}) - (\Theta_t\omega{}^{t-1}_{t-N} + v^t_{t-N} + \breve{Q}_t x_{t-N}). \tag{32}$$

If Assumptions 1 and 2 are satisfied, then $\underline{q} > 0$ holds; thus, it can be concluded that

$$\begin{aligned}\left\|x_{t-N}-\tilde{x}_{t-N,t}\right\| &\leqslant \frac{1}{\sqrt{q}}\Big(\|y_{t-N}^{t}-\tilde{y}_{t-N}^{t}\|+\|\Theta_t\|\left\|\omega_{t-N}^{t-1}\right\| \\ &\quad +\|v_{t-N}^{t}\|+\left\|\breve{Q}_t\right\|\|x_{t-N}\|\Big) \\ &\leqslant \frac{1}{\sqrt{q}}\Big(\varphi+\gamma_\Theta\sqrt{N}\rho_\omega+\sqrt{N+1}\rho_v+\gamma_{\breve{Q}}\rho_x\Big) \\ &=\Upsilon<+\infty.\end{aligned} \tag{33}$$

Thus, the proof of Theorem 1 is complete. □

Theorem 1 proves that the state estimation error of the system is bounded when an event is not triggered. To analyze whether the state estimation error of system is bounded when an event is triggered, Theorem 2 is introduced.

**Theorem 2.** *Under the conditions that Assumptions 1 and 2 are satisfied, if $r_t = 1$, the optimal estimate of the state obtained via the MHER satisfies*

$$\|x_{t-N}-\hat{x}_{t-N,t}^{*}\| < \Upsilon_{t-N}, \tag{34}$$

*where $\Upsilon_{t-N}$ is defined below.*

**Proof of Theorem 2.** Define $\hat{e}_{t-N,t} = x_{t-N} - \hat{x}_{t-N,t}^{*}$ and $\Xi(h) = \hat{G}_t(h) + (h/(h-1))Q^TQ$, $h > 1$. From Equation (11), we are left with

$$\bar{y}_{t-N}^{t} = \hat{Q}x_{t-N} + \Theta\omega_{t-N}^{t-1} + v_{t-N}^{t} + m_{t-N}^{t}. \tag{35}$$

Based on Equations (24) and (35), $\hat{e}_{t-N,t}$ is expressed as follows:

$$\begin{aligned}\hat{e}_{t-N,t} = \Xi(\theta_t^*)^{-1}[\hat{G}_t(x_{t-N}-\bar{x}_{t-N}) - (\theta_t^*/\theta_t^*-1)Q^T \\ (\breve{Q}x_{t-N}+\Theta\omega_{t-N}^{t-1}+v_{t-N}^{t}+m_{t-N}^{t})].\end{aligned} \tag{36}$$

At $t > N$, it can be derived from Equation (19) that

$$x_{t-N}-\bar{x}_{t-N} = \begin{cases} (A+\delta A_{t-N-1})x_{t-N-1} - A\hat{x}_{t-N-1,t-1}^{*} + \omega_{t-N-1}, \text{if } r_{t-1}=1, \\ (A+\delta A_{t-N-1})x_{t-N-1} - A\tilde{x}_{t-N-1,t-1} + \omega_{t-N-1}, \text{if } r_{t-1}=0. \end{cases} \tag{37}$$

Additionally, the state estimation error satisfies

$$e_{t-N-1,t-1} = \begin{cases} x_{t-N-1} - \hat{x}_{t-N-1,t-1}^{*}, \text{if } r_{t-1}=1, \\ x_{t-N-1} - \tilde{x}_{t-N-1,t-1}, \text{if } r_{t-1}=0. \end{cases} \tag{38}$$

By utilizing Equations (36)–(38), we can obtain the following:

$$\begin{aligned}\|\hat{e}_{t-N,t}\| \leqslant{}& \left\|\Xi(\theta_t^*)^{-1}\right\|\|\hat{G}_t\|\|A\|\|e_{t-N-1,t-1}\| + \left\|\Xi(\theta_t^*)^{-1}\right\|\|\hat{G}_t\|\|\delta A_{t-N-1}\|\|x_{t-N-1}\| \\ &+\left\|\Xi(\theta_t^*)^{-1}\right\|\|\hat{G}_t\|\|\omega_{t-N-1}\| + \Pi(\theta_t^*)\|Q^T\|\left\|\breve{Q}_t\right\|\|x_{t-N}\| \\ &+\Pi(\theta_t^*)\|Q^T\|\|\Theta_t\|\left\|\omega_{t-N}^{t-1}\right\| + \Pi(\theta_t^*)\|Q\|\|v_{t-N}^{t}\| \\ &+\Pi(\theta_t^*)\|Q\|\|m_{t-N}^{t}\|,\end{aligned} \tag{39}$$

where $\Pi(\theta_t^*) = (\theta_t^*/\theta_t^*-1)\left\|\Xi(\theta_t^*)^{-1}\right\|$.

When $h > 1$, $\left\|\Xi(\mathrm{h})^{-1}\right\| = (1/\lambda_{\min}(\Xi(\mathrm{h})))$ is satisfied, and $\lambda_{\min}(\hat{G}_t) + (h/h-1)\lambda_{\min}(Q^TQ) \leqslant \lambda_{\min}(\Xi(\mathrm{h}))$ is also satisfied. Then, for $\theta_t^* > 1$, we can determine that

$$\left\|\Xi(\theta_t^*)^{-1}\right\|\|\hat{G}_t\| \leqslant \left(\lambda_{\min}(\hat{G}_t) + \left(\frac{\theta_t^*}{\theta_t^*-1}\right)\lambda_{\min}\left(Q^TQ\right)\right)^{-1}\lambda_{\max}(\hat{G}_t). \tag{40}$$

If Assumptions 1 and 2 are satisfied, based on $\lambda_{\min}(Q^TQ) > 0$ and Equations (28), (29), and (40), we can conclude that

$$\left\|\Xi(\theta_t^*)^{-1}\right\|\|\hat{G}_t\|\|A\| < \frac{\lambda_{\max}(\hat{G}_t)\|A\|}{(\lambda_{\min}(\hat{G}_t) + \lambda_{\min}(Q^TQ))} = \frac{\bar{\sigma}\gamma_a}{\underline{\sigma}+\underline{q}}, \tag{41}$$

$$\left\|\Xi(\theta_t^*)^{-1}\right\|\|\hat{G}_t\|\|\delta A_{t-N-1}\|\|x_{t-N-1}\| < \frac{\lambda_{\max}(\hat{G}_t)\|\delta A_{t-N-1}\|\|x_{t-N-1}\|}{(\lambda_{\min}(\hat{G}_t) + \lambda_{\min}(Q^TQ))} = \frac{\bar{\sigma}\gamma_{\delta a}\rho_x}{\underline{\sigma}+\underline{q}}, \tag{42}$$

$$\left\|\Xi(\theta_t^*)^{-1}\right\|\|\hat{G}_t\|\|\omega_{t-N-1}\| < \frac{\lambda_{\max}(\hat{G}_t)\rho_\omega}{(\lambda_{\min}(\hat{G}_t) + \lambda_{\min}(Q^TQ))} = \frac{\bar{\sigma}\rho_\omega}{\underline{\sigma}+\underline{q}}. \tag{43}$$

The remaining terms in Equation (39) are processed as follows:

$$\begin{aligned}\Pi(\theta_t^*) &= \left(\tfrac{\theta_t^*}{\theta_t^*-1}\right)\left\|\Xi(\theta_t^*)^{-1}\right\| \leqslant \left(\tfrac{\theta_t^*}{\theta_t^*-1}\right) \times \frac{1}{\left(\lambda_{\min}(\hat{G}_t)+\left(\frac{\theta_t^*}{\theta_t^*-1}\right)\lambda_{\min}\left(Q^TQ\right)\right)} \\ &< \frac{1}{\lambda_{\min}\left(Q^TQ\right)} = \frac{1}{\underline{q}}.\end{aligned} \tag{44}$$

Based on Equations (28), (29), and (44), we can derive the following:

$$\Pi(\theta_t^*)\left\|Q^T\right\|\left\|\breve{Q}_t\right\|\|x_{t-N}\| < \frac{\gamma_Q\gamma_{\breve{Q}}\rho_x}{\underline{q}}, \tag{45}$$

$$\Pi(\theta_t^*)\left\|Q^T\right\|\|\Theta\|\left\|\omega_{t-N}^{t-1}\right\| < \frac{\gamma_Q\gamma_\Theta(\sqrt{N}\rho_\omega)}{\underline{q}}, \tag{46}$$

$$\Pi(\theta_t^*)\|Q\|\left\|v_{t-N}^t\right\| < \frac{\gamma_Q(\sqrt{N+1}\rho_v)}{\underline{q}}, \tag{47}$$

$$\Pi(\theta_t^*)\|Q\|\left\|m_{t-N}^t\right\| < \frac{\gamma_Q\varphi}{\underline{q}}. \tag{48}$$

If Assumptions 1 and 2 are satisfied, we can design a set of scalar parameters

$$\begin{cases} \Upsilon_0 = d_1\|x_0 - \bar{x}_0\| + d_2, \\ \Upsilon_{t-N} = d_3\|e_{t-N-1,t-1}\| + d_4 + d_5. \end{cases} \tag{49}$$

where

$$\begin{cases} d_1 = \dfrac{\bar{\sigma}}{\underline{\sigma} + \underline{q}}, \\ d_2 = \dfrac{(\gamma_Q(\gamma_{\underset{\smile}{Q}}(\sqrt{N}\rho_\omega) + \gamma_\Theta \rho_x + (\sqrt{N+1}\rho_v) + \varphi))}{\underline{q}}, \\ d_3 = \dfrac{\bar{\sigma}\gamma_a}{\underline{\sigma} + \underline{q}}, \\ d_4 = \dfrac{\bar{\sigma}(\gamma_{\delta a}\rho_x + \rho_\omega)}{\underline{\sigma} + \underline{q}}, \\ d_5 = \dfrac{\gamma_Q(\gamma_{\underset{\smile}{Q}}\rho_x + \gamma_\Theta(\sqrt{N}\rho_\omega) + (\sqrt{N+1}\rho_v) + \varphi)}{\underline{q}}. \end{cases} \tag{50}$$

By utilizing Equations (36), (41)–(43), and (45)–(50), we can transform Equation (39) into

$$\begin{aligned} \|\hat{e}_{t-N,t}\| &< \frac{\bar{\sigma}\gamma_a}{\underline{\sigma} + \underline{q}}\|e_{t-N-1,t-1}\| + \frac{\bar{\sigma}\gamma_{\delta a}\rho_x}{\underline{\sigma} + \underline{q}} + \frac{\bar{\sigma}\rho_\omega}{\underline{\sigma} + \underline{q}} + \frac{\gamma_Q\gamma_{\underset{\smile}{Q}}\rho_x}{\underline{q}} \\ &+ \frac{\gamma_Q\gamma_\Theta\left(\sqrt{N}\rho_\omega\right)}{\underline{q}} + \frac{\gamma_Q(\sqrt{N+1}\rho_v)}{\underline{q}} + \frac{\gamma_Q\varphi}{\underline{q}} \\ &= d_3\|e_{t-N-1,t-1}\| + d_4 + d_5 = \Upsilon_{t-N}. \end{aligned} \tag{51}$$

Similarly, at $t \leqslant N$, $\|e_{0,t}\|$ satisfies

$$\begin{aligned} \|e_{0,t}\| = \|x_0 - \hat{x}_{0,t}\| &< \frac{\bar{\sigma}}{\underline{\sigma} + \underline{q}}\|x_0 - \bar{x}_0\| + \frac{\gamma_Q\gamma_{\underset{\smile}{Q}}\left(\sqrt{N}\rho_\omega\right)}{\underline{q}} \\ &+ \frac{\gamma_Q\gamma_\Theta\rho_x}{\underline{q}} + \frac{\gamma_Q(\sqrt{N+1}\rho_v)}{\underline{q}} + \frac{\gamma_Q\varphi}{\underline{q}} \\ &= d_1\|x_0 - \bar{x}_0\| + d_2 = \Upsilon_0. \end{aligned} \tag{52}$$

In Equation (51), $\|e_{t-N-1,t-1}\|$ needs to be discussed. If $r_{t-1} = 0$, $\|e_{t-N-1,t-1}\| = \|x_{t-N-1} - \tilde{x}_{t-N-1}\| \leq \Upsilon$ is obtained from Theorem 1, then $\left\|x_{t-N} - \hat{x}^*_{t-N,t}\right\| = \|\hat{e}_{t-N,t}\| \leq d_3\Upsilon + d_4 + d_5$ can be obtained. If $r_{t-1} = 1$, $\|e_{t-N-1,t-1}\| = \left\|x_{t-N-1} - \hat{x}^*_{t-N-1,t-1}\right\| < \Upsilon_{t-N-1}$ is obtained from Equations (38) and (51); then, $\left\|x_{t-N} - \hat{x}^*_{t-N,t}\right\| = \|\hat{e}_{t-N,t}\| < d_3\Upsilon_{t-N-1} + d_4 + d_5$ can be obtained. $\hat{G}_t$ is selected such that $\left\|\Xi(\theta^*_t)^{-1}\right\|\|\hat{G}_t\|\|A\| < 1$ and $\left\|\Xi(\theta^*_t)^{-1}\right\|\|\hat{G}_t\| \neq 1, (\theta^*_t > 1)$, then $d_3 = \bar{\sigma}\gamma_a \Big/ (\underline{\sigma} + \underline{q}) \leqslant 1$ and $d_1 \neq 1$ can be obtained. At the same time, it can be surmised from Ref. [14] and Equation (50) that $d_1$, $d_2$, $d_4$, and $d_5$ are bounded parameters. Therefore, we ultimately arrive at the fact that $\Upsilon_{t-N}$ is bounded.

Thus, the proof of Theorem 2 is complete. □

Based on the Theorems 1 and 2 and Ref. [25], we can conclude that the state estimation error $e_{t-N,t}$ always satisfies $\|e_{t-N,t}\| \leqslant \max\{\Upsilon_{t-N}, \Upsilon, \Upsilon_{ss}\}$, regardless of whether the event at the previous time has been triggered. $\Upsilon_{ss}$ represents the steady-state value of $\Upsilon_{t-N}$,

$Y_{ss} = d_2 + d_3/1 - d_1$. Therefore, the stability of the event-triggered state estimation method proposed in this paper has been proven.

## 5. Simulation Example

A simulation example is introduced in this section to demonstrate that the proposed event state estimation method for uncertain systems with binary-coded communication schemes is feasible.

Consider the uncertain systems described by Equation (1) and let $A = \begin{bmatrix} 1.57 & 0.54 \\ -0.27 & 0.91 \end{bmatrix}$, $C = \begin{bmatrix} 0.9 & 0 \end{bmatrix}$, $\delta A_t = \text{diag}[0.06 \times \sin t \quad 0.09 \times \cos t]$, $\rho_x = 2$, $\rho_\omega = 0.3$, $\rho_v = 0.5$, and $G = \text{diag}\begin{bmatrix} 1.5 & 1.5 \end{bmatrix}$. The state estimation window length N is set to 4, the event-triggered threshold $\Delta$ is set to 0.21, and the simulation times are set to 150.

The event-triggered situations of the proposed method with $\Delta = 0.21$ are shown in Figure 4. As we can see from Figure 4, based on the ETM, the MHER and OLER2 in the state estimator operate alternately and efficiently, thus indicating that the efficiency of state estimation has been improved.

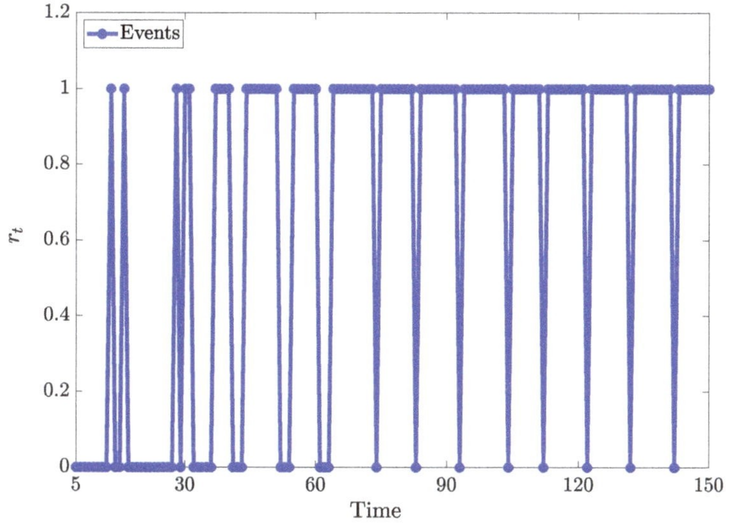


**Figure 4.** Event-triggered situations at $\Delta = 0.21$.

To demonstrate the good performance of the proposed method (EHME method), this paper compares the EMHE method with the $H_\infty$ Extended Kalman Filter method (HEKF method) proposed in [9] and the robust state estimation method (RBSE method) proposed in [3]. The state estimation results output by the EHME method, the HEKF method, and the RBSE method are shown in Figures 5 and 6, while the state estimation error of the three methods is presented in Figure 7. Figures 5–7 indicate that the EMHE method has a better state-tracking trajectory and a smaller state estimation error compared to the HEKF method and the RBSE method. Therefore, we can conclude that although the proposed method introduces the ETM and the binary encoding transmission scheme, it can still maintain good estimation performance.

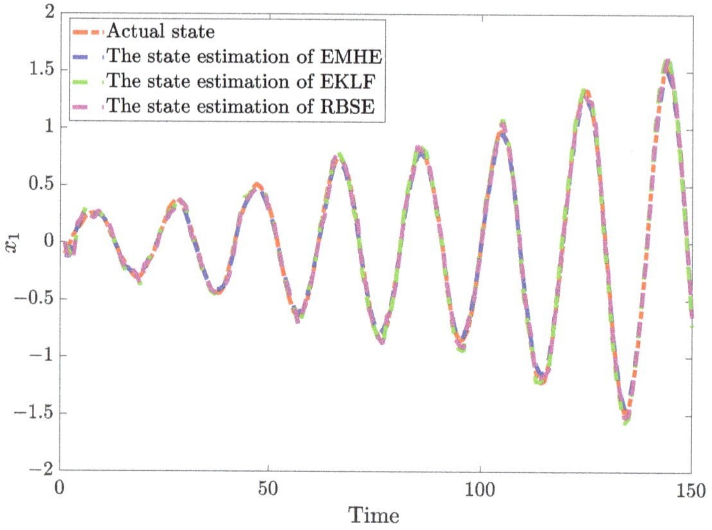


**Figure 5.** Estimation situations of the state $x_1$.

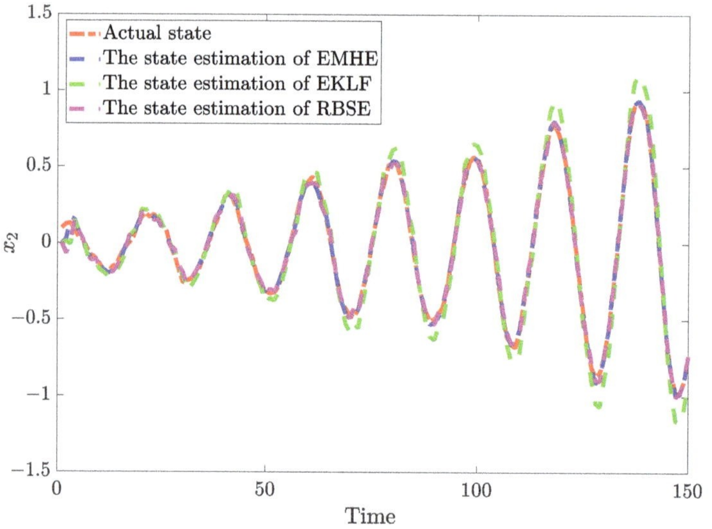


**Figure 6.** Estimation situations of the state $x_2$.

In order to further validate the practical application value of the method proposed in this paper, we calculated the times of the EMHE method, the method without an ETM, the HEKF method, and the RBSE method under the same simulation times and summarized them in Table 1. Since the method without an ETM always uses the MHER with a relatively complex algorithm to estimate the state, its calculation time is the longest. At the same time, since both the HEKF method and the RBSE method are designed on the basis of the Extended Kalman Filter method, the corresponding state estimation also takes a long time. However, the EMHE method reduces the use of the MHER and significantly shortens the computational time by introducing the an ETM. When the simulation times are set to 150, compared with the method without an ETM, the HEKF method, and the RBSE method, the EMHE method reduced the required time by 63.05% , 45.36%, and 51.69%, respectively.

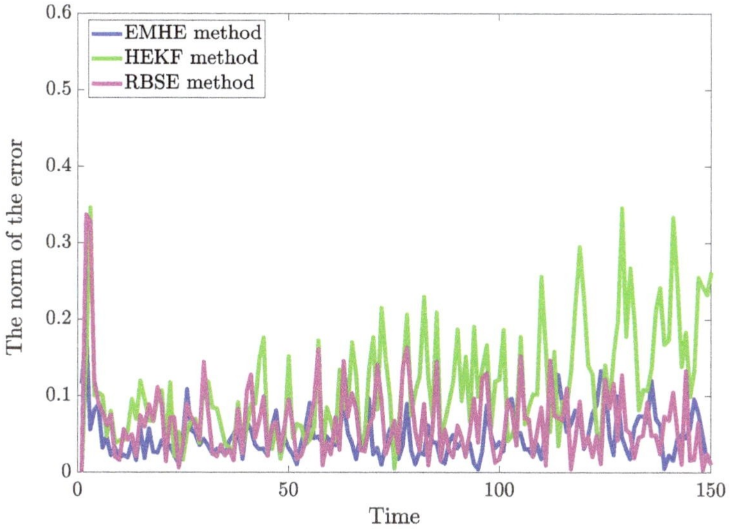


**Figure 7.** State estimation error of the three methods.

**Table 1.** Calculation time(s) of the four methods at different simulation times $j$.

| $j$ | 50 | 100 | 150 |
|---|---|---|---|
| EMHE method | 0.3572 | 1.0386 | 2.1563 |
| The method without an ETM | 1.4847 | 3.3549 | 5.8354 |
| HEKF method | 1.2131 | 2.5514 | 3.9462 |
| RBSE method | 1.3524 | 2.8647 | 4.4635 |

According to the above analysis, we can conclude that the method proposed in this paper saves network communication resources, reduces the computational burden of the estimator, ensures the stability of state estimation, and improves the state estimation performance of uncertain systems.

## 6. Conclusions

This paper proposes an event-triggered state estimation method for uncertain systems with a binary encoding transmission scheme. A binary encoding transmission scheme is introduced to ensure the remote and efficient transmission of information between a state estimator and a system. An ETM is incorporated to save network bandwidth and improve computational efficiency. an MHER is designed for optimal state estimation using the robust least squares optimization approach when an event is triggered, and the simpler OLER is used for state estimation when an event is not triggered. Stability analysis proved that the state estimation error is bounded. Finally, the proposed method's improvement in addressing estimation problems was validated through a simulation example.

Future work will investigate the design of an adaptive event-triggered mechanism for uncertain linear systems and distributed uncertain systems to further optimize state estimation methods.

**Author Contributions:** Conceptualization, Z.L. and B.X.; methodology, Z.L. and B.X.; software, Z.L.; data curation, Z.L.; writing—original draft preparation, Z.L. and Y.C.; writing—review and editing, Z.L. and Y.C.; funding acquisition, Z.L. All authors have read and agreed to the published version of the manuscript.

**Funding:** This work was supported in part by the National Natural Science Foundation of China under Grant 61603205, in part by the China Postdoctoral Science Foundation under Grant 2017M612205.

**Data Availability Statement:** Not applicable.

**Conflicts of Interest:** The authors declare no conflict of interest.

## References


1. Xue, B.Q.; Li, S.Q.; Zhu, Q.M. Moving Horizon State Estimation for Networked Control Systems with Multiple Packet Dropoutsin. *IEEE Trans. Autom. Control* **2012**, *57*, 2360–2366. [CrossRef]
2. Heo, S.; Park, C.G. Consistent EKF-based visual-inertial odometry on matrix Lie group. *IEEE Sens. J.* **2018**, *18*, 3780–3788. [CrossRef]
3. Wang, J.; Mao, Y.; Li, Z.Q.; Gao, J.W.; Liu H.B. Robust state estimation for uncertain linear discrete systems with d-step state delay. *IET Control Theory Appl.* **2021**, *15*, 1708–1723. [CrossRef]
4. Rengaswamy, R.; Narasimhan, S.; Kuppuraj, V. Receding-Horizon Nonlinear Kalman (RNK) Filter for State Estimation. *IEEE Trans. Autom. Control* **2013**, *58*, 2054–2059. [CrossRef]
5. Huang, Z.H.; Guo, Y.R.; Peng, H.; Xu, Y. Distributed Moving Horizon Estimation Over Wireless Sensor Networks: A Matrix-Weighted Consensus Approach . *IEEE Trans. Circuits Syst. II Express Briefs* **2023**, *70*, 1665–1669. [CrossRef]
6. Rao, C.; Rawlings, J.; Lee J. Constrained Linear State Estimation—A Moving Horizon Approach. *Automatica* **2001**, *37*, 1619–1628. [CrossRef]
7. Sui, D.; Johansen, A.; Feng, L. Linear Moving Horizon Estimation with Pre-Estimating Observer. *IEEE Trans. Autom. Control* **2010**, *55*, 2363–2368. [CrossRef]
8. Battistelli, G. Distributed Moving-Horizon Estimation with Arrival-Cost Consensus. *IEEE Trans. Autom. Control* **2019**, *64*, 3316–3323. [CrossRef]
9. Zhao, J. Dynamic State Estimation With Model Uncertainties Using $H_\infty$ Extended Kalman Filter. *IEEE Trans. Power Syst.* **2018**, *33*, 1099–1100. [CrossRef]
10. Qian, C.J. Semi-global stabilization of a class of uncertain nonlinear systems by linear output feedback. *IEEE Trans. Circuits Syst. II Express Briefs* **2005**, *52*, 218–222. [CrossRef]
11. Guo, G.; Wang, B.F. Robust kalman filtering for uncertain discrete-time systems with multiple packet dropouts. *Acta Autom. Sin.* **2010**, *36*, 767–772. [CrossRef]
12. Sayed, A.; Nascimento, V.; Cipparrone, F. A Regularized Robust Design Criterion for Uncertain Data. *SIAM J. Math. Anal.* **2002**, *23*, 1120–1142. [CrossRef]
13. Liu, S.; Zhao, G.R.; Zeng, B.; Gao, C. Moving Horizon Estimation for Uncertain Systems With Packet Dropouts and Quantization. *J. Syst. Eng. Electron.* **2020**, *42*, 912–918.
14. Alessandri, A.; Baglietto, M.; Battistelli, G. Robust Receding Horizon Estimation for Uncertain Discrete time Linear Systems. *IEEE Control Syst. Lett.* **2003**, *54*, 627–643. [CrossRef]
15. Wang, S.D.; Xue, B.Q. Distributed Moving Horizon Fusion Estimation for Nonlinear Constrained Uncertain Systems. *Mathematics* **2023**, *11*, 1507. [CrossRef]
16. Zhao, K.H. Stability of a Nonlinear Langevin System of ML-Type Fractional Derivative Affected by Time-Varying Delays and Differential Feedback Control. *Fractal Fract.* **2022**, *6*, 725. [CrossRef]
17. Xu, Y.; Zhou, J.Y.; Rao, H.X.; Lu, R.Q.; Xie, L.H. Reset Moving Horizon Estimation for Quantized Discrete Time Systems. *IEEE Trans. Autom. Control* **2021**, *66*, 4199–4205. [CrossRef]
18. Alessandri, A.; Baglietto, M.; Battistelli, G. Receding-horizon Estimation for Discrete Tme Linear Systems. *IEEE Trans. Autom. Control* **2003**, *48*, 473–478. [CrossRef]
19. Liu, A.; Yu, L.; Zhang, W.A.; Chen, M. Moving Horizon Estimation for Networked Systems with Quantized Measurements and Packet Dropouts. *IEEE Trans. Circuits Syst. II Express Briefs* **2012**, *60*, 1823–1834. [CrossRef]
20. Xue, B.Q.; Li, N.; Li, S.Y.; Zhu, Q.M. Moving Horizon Scheduling for Networked Control Systems with Communication Constraints. *IEEE Trans. Ind. Electron.* **2013**, *60*, 3318–3327. [CrossRef]
21. Zhou, H.D.; Sui, S.; Tong, S.C. Finite-Time Adaptive Fuzzy Prescribed Performance Formation Control for High-Order Nonlinear Multiagent Systems Based on Event-Triggered Mechanism. *IEEE Trans. Fuzzy Syst.* **2023**, *31*, 1229–1239. [CrossRef]
22. Wang, Y. An Event-Triggered Scheme for State Estimation of Preceding Vehicles under Connected Vehicle Environment. *IEEE Trans. Veh. Technol.* **2023**, *8*, 583–593.
23. Zhong, M.; Du, X.; Song, Y.; Xue, T.; Ding, S.X. Event-Triggered Parity Space Approach to Fault Detection for Linear Discrete-Time System. *IEEE Trans. Syst. Man Cybern. Syst.* **2022**, *52*, 4813–4822. [CrossRef]
24. Huang, J.R.; Shi, D.W.; Chen, T.W. Event-triggered State Estimation with an Energy Harvesting Sensor. *IEEE Trans. Autom. Control* **2017**, *62*, 4768–4775. [CrossRef]
25. Yin, X.Y.; Liu, J.F. Event-triggered State Estimation of Linear Systems Using Moving Horizon Estimation. *IEEE Trans. Control Syst. Technol.* **2021**, *29*, 901–909. [CrossRef]
26. Yin, X.Y.; Huang, B. Event-triggered Distributed Moving Horizon State Estimation of Linear Systems. *IEEE Trans. Syst. Man Cybern. Syst.* **2022**, *52*, 6439–6451. [CrossRef]
27. Alessandri, A.; Awawdeh, M. Moving-horizon estimation with guaranteed robustness for discrete-time linear systems and measurements subject to outliers. *Automatica* **2016**, *67*, 85–93. [CrossRef]

28. He, D.; Xu, C.; Zhu, J.; Du, H. Moving horizon $H_\infty$ estimation of constrained multisensor systems with uncertainties and fading channels. *IEEE Trans. Instrum. Meas.* **2021**, *70*, 1–12. [CrossRef]
29. Schiller, J.D.; Müller, M.A. Suboptimal Nonlinear Moving Horizon Estimation. *IEEE Trans. Autom. Control* **2023**, *68*, 2199–2214. [CrossRef]
30. Zou, L.; Wang, Z.D.; Hu, J.; Han, Q.L. Moving Horizon Estimation With Unknown Inputs under Dynamic Quantization Effects. *IEEE Trans. Autom. Control* **2020**, *65*, 5368–5375. [CrossRef]
31. Liu, Q.Y.; Wang, Z.D. Moving-horizon Estimation for Linear Dynamic Networks with Binary Encoding Schemes. *IEEE Trans. Autom. Control* **2021**, *66*, 1763–1770. [CrossRef]
32. Alessandri, A.; Baglietto, M.; Battistelli, G. Moving-horizon state estimation for nonlinear discrete-time systems: New stability results and approximation schemes. *Automatica* **2008**, *7*, 1753–1765. [CrossRef]
33. Farina, M.; Ferrari-Trecate, G.; Scattolini, R. Distributed Moving Horizon Estimation for Linear Constrained Systems. *IEEE Trans. Autom. Control* **2010**, *55*, 2462–2475. [CrossRef]
34. Gharbi, M.; Gharesifard, B.; Ebenbauer, C. Anytime Proximity Moving Horizon Estimation: Stability and Regret. *IEEE Trans. Autom. Control* **2023**, *68*, 3393–3408. [CrossRef]
35. Zhao, K.H. Solvability and GUH-stability of a nonlinear CF-fractional coupled Laplacian equations. *AIMS Math.* **2023**, *8*, 13351–13367. [CrossRef]
36. Zhao, K.H. Existence and Stability of a Nonlinear Distributed Delayed Periodic AG-Ecosystem with Competition on Time Scales. *Axioms* **2023**, *12*, 315. [CrossRef]
37. Ji, L.; Rawlings, B.J.; Hu, W.; Wynn, A.; Diehl, M. Robust Stability of Moving Horizon Estimation Under Bounded Disturbances *IEEE Trans. Autom. Control* **2016**, *61*, 3509–3514. [CrossRef]

applied sciences

MDPI

Article

# Dynamic Modeling of Motorized Spindle System with Unbalanced Mass and Spindle Inclination

**Dan Gu, Yufeng Xie, Shulin Liu *, Yuan Wei * and Jiayi Shen**

School of Mechatronic Engineering and Automation, Shanghai University, Shanghai 200444, China; gagarin1980@163.com (D.G.); xyf1178517671@163.com (Y.X.); s2414717277@shu.edu.cn (J.S.)
* Correspondence: lsl346@126.com (S.L.); ywei56@shu.edu.cn (Y.W.)

**Abstract:** The high-speed motorized spindle has been wildly used in the field of aerospace processing, due to its advantages such as high speed, high precision, and high efficiency. CNC machine tools used for processing aerospace products require high machining accuracy, and once the spindle fails, it will seriously affect the quality of product processing. Thus, it is important to study the faults of the spindle, especially the faults caused by subtle errors. In this work, a dynamic model of a spindle with unbalanced mass fault and spindle inclination fault is established, and the natural frequencies and mode shapes of the motorized spindle are calculated by using the whole transfer matrix method (WTMM). The deflections of the spindle initial end in the different situations are discussed when the two faults happen independently. The results show that the spindle end deflection of the same fault has different sensitivity at different speeds. At the third order of natural frequencies, the deflection of a motorized spindle is greatest regardless of the fault that occurs. Although the motorized spindle rotates at the same speed, different faults could cause different mode shapes. At the lower speed, when the unbalanced mass fault happens, the mode shape is in an arched shape, and while the spindle inclination fault happens, the mode shape is in a concave shape.

**Keywords:** motorized spindle; whole transfer matrix method; unbalanced mass; spindle inclination; dynamic modeling

## 1. Introduction

Aerospace products are mostly thin-walled parts with complex shapes, needed to suit the specific working environment and performance demands, and are typically processed using high-speed milling. In the field of aerospace processing, there are strict requirements for the precision of machine tool processing, in which the precision of the motorized spindle plays a vital role in the whole processing. The common faults of motorized spindles include spindle fault [1–3], bearing fault [4,5], lubrication system fault [6], stator and rotor fault [7,8], spindle housing fault [9], and cooling system fault [10]; the bearing fault is the most common and analyzed fault. However, compared with other faults, the spindle fault is more difficult to find and more likely to affect the progress of this machining. For aerospace equipment, a small error may lead to the failure of the whole mechanism, so the faults of motorized spindles cannot be ignored in the processing of aerospace equipment.

In recent years, the fault dynamics of motorized spindles have become an important issue. Wang et al. (2018) [11] established a dynamic model of a dual-rotor system to solve its unbalanced misalignment coupling faults. Xu et al. (2016) [12] obtained the unbalanced magnetic pull by numerical method and discussed the effects of initial static eccentricity on displacement spectra. Liu et al. (2019) [13] established a dynamic model of the permanent magnet synchronous motors' rotor-bearing system which considers unbalanced magnetic pull and the nonlinear restoring force. They concluded that unbalanced magnetic pull causes a negative stiffness effect on the natural frequencies. Werner (2017) [14] established an analytical rotor dynamic model which considers mass eccentricity, bent rotor deflection, and magnetic eccentricity. Zuo et al. (2018) [15] analyzed and modeled the longitudinal dynamics of the motorized spindle under the excitation of torque ripple. Based on the Jeffcott rotor model, Xiang et al. (2016) [16] analyzed the stiffness characteristics of the rotor system of the permanent magnet synchronous motors. Kumar (2017) [17] gave the reason of unbalanced magnetic pull. Based on the law of energy conservation, Bossmanns and Tu (2001) [18] presented a power flow model of high-speed motorized spindles. Li et al. (2004) [19,20] established a comprehensive integrated thermo-dynamic model for high-speed spindles. Xu et al. (2017) [21] built the dynamic model of the motorized spindle system. They used this dynamic model to show that the vibrations are proportional to the mass of the unbalance. Liu et al. (2016) [22] presented an improved dynamic model for unbalanced high-speed motorized spindles, and obtained the sensitivities of the motorized spindles.

Although a huge number of scholars have contributed to this research, the fault dynamics of motorized spindles still have some unsolved problems, such as the vibration of motorized spindles caused by the unbalanced mass and inclined axis. In practical engineering, the two kinds of faults will occur due to the uneven distribution of rotor material density during the manufacturing process or errors caused by the installation of a motorized spindle. Although the faults in these two cases can be quickly solved after discovery, they will still affect the normal operation of the motorized spindle before discovery. Moreover, it is also very important to judge whether these two faults occur. If they are found as early as possible, their influence on the normal operation of the whole machinery can be avoided. Therefore, it is necessary to analyze the influence of these two kinds of faults on the vibration of the spindle, so as to help the operator find out whether the faults happen as soon as possible.

In this work, a model of motorized spindles based on the whole transfer matrix method (WTMM) is established and a model of a spindle with fault caused by unbalanced mass or spindle inclination is given in Section 2. Then, a case study we conducted on a certain type of high-speed motorized spindle is described by using these models in Section 3. The deflection of the spindle end in different conditions is discussed and the mode shapes of the motorized spindle are presented considering the unbalanced mass fault and the spindle inclination fault, respectively. The research in this paper provides theoretical guidance for the design, operation, and maintenance of motorized spindles, which can effectively improve the accuracy and production efficiency of high-speed machining, and promote the development of a series of related technologies such as high-speed feed, high-performance tools, detection, and control.

## 2. Dynamic Modeling

The structure diagram of a high-speed motorized spindle is shown in Figure 1. Its main parts include a rotor spindle, shell, stator, and bearings. Figure 1b shows the structure diagram of a motorized spindle-bearing system.

Compared with other methods, the whole transfer matrix method (WTMM) has more efficient computing power and more extensive application capabilities [23,24]. So, in this paper, it is used to analyze the unbalanced mass fault and spindle inclination fault of a motorized spindle.

According to the WTMM, a motorized spindle is divided into subsegments. By lumped mass and lumped moment of inertia method, as shown in Figure 1c, every subsegment is simplified into two rigid discs which are connected by massless elastic shaft. The unit is formed by a disc and a shaft, and two units are connected by the node. Each node has four effective parameters: section deflection (radial displacement) $y$, section rotation angle $\theta$, section bending moment $M$, and section shear force $Q$. The vector composed of these four effective parameters is the cross-sectional state vector, which is recorded as

$$\mathbf{Z} = \begin{bmatrix} y & \theta & M & Q \end{bmatrix}^T \tag{1}$$

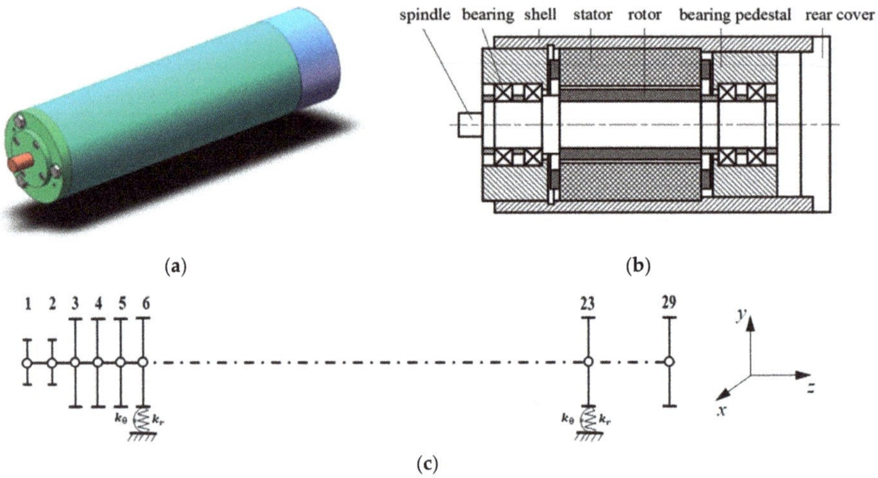


**Figure 1.** Schematic diagram of high-speed motorized spindle: (**a**) description of what is contained in the first panel; (**b**) structure of motorized spindle; (**c**) lumped mass model of high-speed motorized spindle.

### 2.1. Dynamic Modeling of Motorized Spindle without Fault

As shown in Figure 2a, the forces of the left and right sections of the rigid thin disc have a bending moment of $M^L$ and $M^R$ and a shear force of $Q^R$ and $Q^L$.

The equations of inertia force and moment of inertia are as follows [25]:

$$F = m\omega^2 y \tag{2}$$

$$M = \left(J_d - J_p\frac{\Omega}{\omega}\right)\omega^2\theta \tag{3}$$

where $y$ is the deflection of the disc, $\theta$ is the rotation angle of the disc, $\Omega$ is the rotation angular velocity of the rotor, and $\omega$ is the whirling angular velocity, $J_d$ and $J_p$ are the polar and equatorial moment of inertia, respectively. When the node is unsupported, the inertia force is zero. The D'Alembert equation can be established as follows:

$$\begin{cases} y^R = y^L \\ \theta^R = \theta^L \\ M^R = M^L - \left(J_d - J_p\frac{\Omega}{\omega}\right)\omega^2\theta + k_\theta\theta \\ Q^R = m\omega^2 y + Q^L - k_r y \end{cases} \tag{4}$$

where $k_r$ is the radial stiffness of the bearing and $k_\theta$ is the angular stiffness of the bearing.

Equation (4) is expressed in the form of a matrix to obtain

$$\begin{bmatrix} y \\ \theta \\ M \\ Q \end{bmatrix}_i^R = \begin{bmatrix} 1 & 0 & 0 & 0 \\ 0 & 1 & 0 & 0 \\ 0 & k_\theta - \left(J_d - J_p\frac{\Omega}{\omega}\right)\omega^2 & 1 & 0 \\ m\omega^2 - k_r & 0 & 0 & 1 \end{bmatrix}_i \begin{bmatrix} y \\ \theta \\ M \\ Q \end{bmatrix}_i^L \tag{5}$$

where $y$, $\theta$, $M$, and $Q$ are the deflection, rotation angle, bending moment, and shear force of the disc, respectively. The superscript $R$ or $L$ indicates the right side or left side of the disc. The subscript $i$ indicates the disc of the $i$th unit. Therefore, when the disc left side cross-sectional state of the $i$th unit is known, Equation (5) can be used to solve its right side cross-sectional state.

Similarly, as shown in Figure 2b, the left end of the elastic shaft connects the disc's right side of the same unit, and the right end of the shaft connects the left side of the disc which belongs to the next disc –shaft unit. Therefore, the shear force and bending moment at the left end of the shaft are superscripted as $R$ and subscripted as $i$. The shear force and bending moment at the right end are superscripted as $L$ and subscripted as $i+1$.

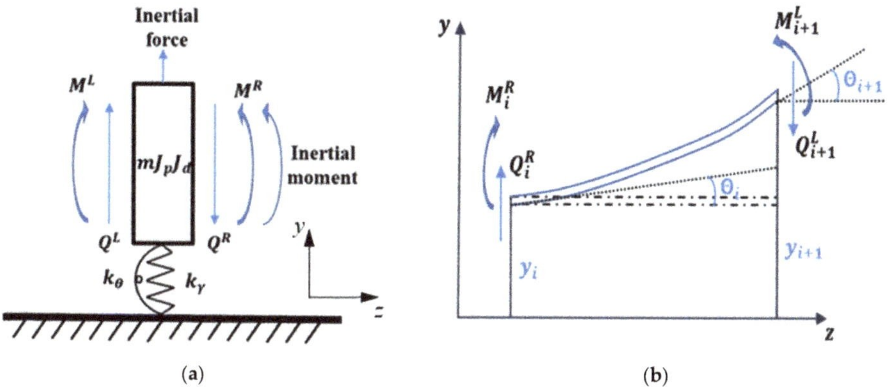


**Figure 2.** Stress diagram of unit: (**a**) stress diagram of unit; (**b**) stress diagram of massless elastic shaft.

It is also known from the bending deformation conditions of shafts in the mechanics of materials that

$$\begin{cases} y_{i+1} = y_i + \theta_i l_i + \frac{M_{i+1}^L l_i^2}{2E_i I_i} - Q_{i+1}^L \left( \frac{l_i^3}{3E_i I_i} + \frac{k_s l_i}{G_i A_i} \right) \\ \theta_{i+1} = \theta_i + \frac{M_{i+1}^L l_i}{E_i I_i} - Q_{i+1}^L \frac{l_i^2}{2E_i I_i} \\ M_{i+1}^L = M_i^R + Q_i^R l_i \\ Q_{i+1}^L = Q_i^R \end{cases} \tag{6}$$

where $k_s$ is the shape factor of the section, for the round solid shaft $k_s = 10/9$. $G_i$ is the shear modulus of the $i$th subsegment unit, $A_i$ is the sectional area of the $i$th subsegment unit, $E_i$ is the modulus of elongation of the subsegment, and $I_i$ is the inertia moment of the $i$th subsegment.

Similarly, by changing the above equation into matrix form, the following formula can be obtained:

$$\begin{bmatrix} y \\ \theta \\ M \\ Q \end{bmatrix}_{i+1}^{L} = \begin{bmatrix} 1 & l & \frac{l^2}{2EI} & \frac{l^3}{6EI}(1-v) \\ 0 & 1 & \frac{l}{EI} & \frac{l^2}{2EI} \\ 0 & 0 & 1 & l \\ 0 & 0 & 0 & 1 \end{bmatrix}_i \begin{bmatrix} y \\ \theta \\ M \\ Q \end{bmatrix}_i^{R} \tag{7}$$

where $v = k_s(6E_iI_i)/(G_iA_il_i^2)$ is called the shear effect coefficient. The vector on the left side of the equation is the cross-sectional state vector at the right end of the massless elastic shaft in the $i$th unit, that is, the cross-sectional state vector at the left side of the disc in the $(i+1)$th unit.

The transfer matrix of the whole unit can be obtained by combining the two equations of the same unit.

$$\begin{bmatrix} y \\ \theta \\ M \\ Q \end{bmatrix}_{i+1}^{L} = \begin{bmatrix} 1 & l & \frac{l^2}{2EI} & \frac{l^3}{6EI}(1-v) \\ 0 & 1 & \frac{l}{EI} & \frac{l^2}{2EI} \\ 0 & 0 & 1 & l \\ 0 & 0 & 0 & 1 \end{bmatrix}_i \times \begin{bmatrix} 1 & 0 & 0 & 0 \\ 0 & 1 & 0 & 0 \\ 0 & k_\theta - \left(J_d - J_p\frac{\Omega}{\omega}\right)\omega^2 & 1 & 0 \\ m\omega^2 - k_r & 0 & 0 & 1 \end{bmatrix}_i \begin{bmatrix} y \\ \theta \\ M \\ Q \end{bmatrix}_i^{L} \quad (8)$$

$$\begin{bmatrix} y \\ \theta \\ M \\ Q \end{bmatrix}_{i+1}^{L} = \begin{bmatrix} 1 + \frac{l^3}{6EI}(1-v)(m\omega^2 - k_r) & l + \frac{l^2}{2EI}\left(k_\theta - (J_d - J_p\frac{\Omega}{\omega})\omega^2\right) & \frac{l^2}{2EI} & \frac{l^3}{6EI}(1-v) \\ 1 \quad \frac{l^2}{2EI}(m\omega^2 - k_r) & 1 + \frac{l}{EI}\left(k_\theta - (J_d - J_p\frac{\Omega}{\omega})\omega^2\right) & \frac{l}{EI} & \frac{l^2}{2EI} \\ l(m\omega^2 - k_r) & k_\theta - (J_d - J_p\frac{\Omega}{\omega})\omega^2 & 1 & l \\ m\omega^2 - k_r & 0 & 0 & 1 \end{bmatrix}_i \begin{bmatrix} y \\ \theta \\ M \\ Q \end{bmatrix}_i^{L} \quad (9)$$

Equation (9) indicates that the cross-sectional state vector at the left end of the $i$th unit is transferred to the right end of the unit, that is, the left end of the next unit. For convenience of description, Equation (9), can be written as $\mathbf{Z}_{i+1}^L = \mathbf{T}_i \mathbf{Z}_i^L$, where $\mathbf{T}_i$ is the transfer matrix between units.

### 2.2. Dynamic Modeling of Motorized Spindle with Unbalanced Mass Fault

Unbalanced mass fault is the simplest and most common fault form of the motorized spindle. When an unbalanced mass fault occurs, the center line of the mass and the center line of the rotation are in the same plane, and the distance between them is $r$, which is called eccentricity. Therefore, the unbalanced force will be generated at this time:

$$F = U\omega^2 \quad (10)$$

where $U = mr$, m is the unbalanced mass.

Under the influence of the unbalanced force, the force state of the disc will change as shown in Figure 3. The D'Alembert equation is then re-established:

$$\begin{cases} M^R = M^L - \left(J_d - J_p\frac{\Omega}{\omega}\right)\omega^2\theta + k_\theta\theta + U\omega^2 \\ Q^R = m\omega^2 y + Q^L - k_r y \end{cases} \quad (11)$$

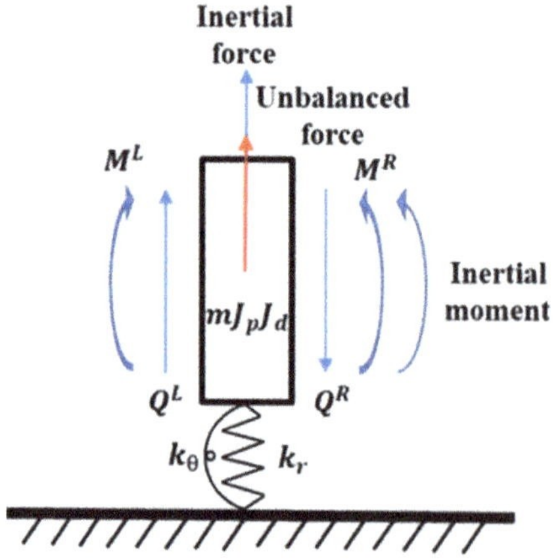


**Figure 3.** Stress diagram of rigid disc under unbalanced mass fault.

It is expressed in the form of a matrix to obtain

$$\begin{bmatrix} y \\ \theta \\ M \\ Q \end{bmatrix}_{i+1}^{L} = \begin{bmatrix} 1 & l & \frac{l^2}{2EI} & \frac{l^3}{6EI}(1-v) \\ 0 & 1 & \frac{l}{EI} & \frac{l^2}{2EI} \\ 0 & 0 & 1 & l \\ 0 & 0 & 0 & 1 \end{bmatrix}_i \times \begin{bmatrix} 1 & 0 & 0 & 0 \\ 0 & 1 & 0 & 0 \\ 0 & k_\theta - (J_d - J_p\frac{\Omega}{\omega})\omega^2 & 1 & 0 \\ m\omega^2 - k_r & 0 & 0 & 1 \end{bmatrix}_i \begin{bmatrix} y \\ \theta \\ M \\ Q \end{bmatrix}_i^L + \begin{bmatrix} 0 \\ 0 \\ 0 \\ U_i \end{bmatrix}_i \omega^2 \quad (12)$$

$$\begin{bmatrix} y \\ \theta \\ M \\ Q \end{bmatrix}_{i+1}^{L} = \begin{bmatrix} 1 + \frac{l^3}{6EI}(1-v)(m\omega^2 - k_r) & l + \frac{l^2}{2EI}\left(k_\theta - (J_d - J_p\frac{\Omega}{\omega})\omega^2\right) & \frac{l^2}{2EI} & \frac{l^3}{6EI}(1-v) \\ 1 \quad \frac{l^2}{2EI}(m\omega^2 - k_r) & 1 + \frac{l}{EI}\left(k_\theta - (J_d - J_p\frac{\Omega}{\omega})\omega^2\right) & \frac{l}{EI} & \frac{l^2}{2EI} \\ l(m\omega^2 - k_r) & k_\theta - (J_d - J_p\frac{\Omega}{\omega})\omega^2 & 1 & l \\ m\omega^2 - k_r & 0 & 0 & 1 \end{bmatrix}_i \begin{bmatrix} y \\ \theta \\ M \\ Q \end{bmatrix}_i^L + \begin{bmatrix} \alpha l \\ \alpha \\ 0 \\ 0 \end{bmatrix}_i \quad (13)$$

As the motorized spindle without fault, for convenience of description, Equation (13) can be written as $\mathbf{Z}_{i+1}^L = \mathbf{T}_i\mathbf{Z}_i^L + \mathbf{H}_i$, where $\mathbf{T}_i$ is the transfer matrix between units and the $\mathbf{H}_i$ is unbalance vector.

### 2.3. Dynamic Modeling of Motorized Spindle with Spindle Inclination Fault

As the inclination angle is relatively small in practical work, once it is ignored, it may lead to further serious failure problems such as the spindle hitting the film or misalignment. Therefore, it is necessary to study the influence of the inclination angle on the dynamic characteristics of a motorized spindle.

As shown in Figure 4, when the motorized spindle inclines, an inclination angle of magnitude is generated between the axis and the horizontal line. Because it is small,

$$\begin{cases} \sin\alpha = \alpha \\ \cos\alpha = 1 \end{cases} \quad (14)$$

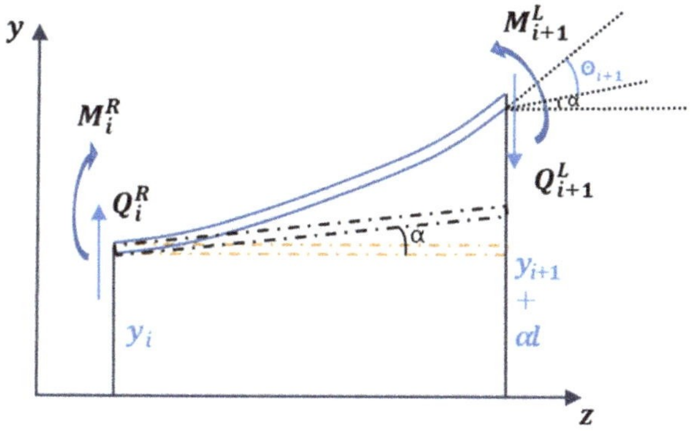


**Figure 4.** Stress diagram of massless elastic shaft under spindle inclination fault.

Due to the existence of the inclination angle, the actual value of the rotation angle at the right end of the spindle is the calculated rotation angle plus the inclination angle caused by the inclination, and the deflection will also be increased in the same way. The bending deformation analysis of the elastic shaft shows that

$$\begin{cases} y_{i+1} = y_i + \theta_i l_i + \frac{M_{i+1}^L l_i^2}{2E_iI_i} - Q_{i+1}^L\left(\frac{l_i^3}{3E_iI_i} + \frac{k_s l_i}{G_iA_i}\right) + \alpha l_i \\ \theta_{i+1} = \theta_i + \frac{M_{i+1}^L l_i}{E_iI_i} - Q_{i+1}^L\frac{l_i^2}{2E_iI_i} + \alpha \\ M_{i+1}^L = M_i^R + Q_i^R l_i \\ Q_{i+1}^L = Q_i^R \end{cases} \quad (15)$$

where $\alpha$ is the small angle at which the axis of the motorized spindle is inclined. The equation can be obtained by changing the above formula into matrix form:

$$\begin{bmatrix} y \\ \theta \\ M \\ Q \end{bmatrix}_{i+1}^{L} = \begin{bmatrix} 1 & l & \frac{l^2}{2EI} & \frac{l^3}{6EI}(1-v) \\ 0 & 1 & \frac{l}{EI} & \frac{l^2}{2EI} \\ 0 & 0 & 1 & l \\ 0 & 0 & 0 & 1 \end{bmatrix}_i \begin{bmatrix} y \\ \theta \\ M \\ Q \end{bmatrix}_i^{R} + \begin{bmatrix} \alpha l \\ \alpha \\ 0 \\ 0 \end{bmatrix}_i \tag{16}$$

Therefore, the transfer matrix of the whole unit is

$$\begin{bmatrix} y \\ \theta \\ M \\ Q \end{bmatrix}_{i+1}^{L} = \begin{bmatrix} 1+\frac{l^3}{6EI}(1-v)(m\omega^2-k_r) & l+\frac{l^2}{2EI}\left(k_\theta-(J_d-J_p\frac{\Omega}{\omega})\omega^2\right) & \frac{l^2}{2EI} & \frac{l^3}{6EI}(1-v) \\ \frac{l^2}{2EI}(m\omega^2-k_r) & 1+\frac{l}{EI}\left(k_\theta-(J_d-J_p\frac{\Omega}{\omega})\omega^2\right) & \frac{l}{EI} & \frac{l^2}{2EI} \\ l(m\omega^2-k_r) & k_\theta-(J_d-J_p\frac{\Omega}{\omega})\omega^2 & 1 & l \\ m\omega^2-k_r & 0 & 0 & 1 \end{bmatrix}_i \begin{bmatrix} y \\ \theta \\ M \\ Q \end{bmatrix}_i^{L} + \begin{bmatrix} \frac{l^3}{6EI}(1-v) \\ \frac{l^2}{2EI} \\ l \\ 1 \end{bmatrix}_i U_i\omega^2 \tag{17}$$

As with other conditions, Equation (17) is written as $\mathbf{Z}_{i+1}^{L} = \mathbf{T}_i\mathbf{Z}_i^{L} + \mathbf{B}_i$, where $\mathbf{T}_i$ is the transfer matrix between units and $\mathbf{B}_i$ is the unbalance vector.

## 3. Results and Discussion

The parameters of a certain high-speed motorized spindle used in a practical machining machine tool for aviation components are shown in Table 1.

**Table 1.** Parameters of high-speed motorized spindle.

| Parameter | Value |
|---|---|
| Spindle length/mm | 290 |
| Mass/kg | 9.87 |
| Current/A | 10 |
| Inner diameter of motor rotor/mm | 42 |
| Outer diameter of rotating spindle/mm | 40 |
| Outer diameter of motor rotor/mm | 100 |
| Type of bearing | 71,908 CE/HCP4A |
| Maximum speed/rpm | 58,000 |

### 3.1. Dynamic Characteristics of Motorized Spindle without Fault

From Section 2.1, the general equation from the first node to the last node can be written as

$$\mathbf{Z}_{n+1} = \mathbf{T}_n\mathbf{Z}_n = \mathbf{T}_n\mathbf{T}_{n-1}\mathbf{Z}_{n-1} = \cdots = \mathbf{T}_n\mathbf{T}_{n-1}\cdots\mathbf{T}_1\mathbf{Z}_1 = \mathbf{T}_{all}\mathbf{Z}_1 \tag{18}$$

where $\mathbf{Z}_{n+1}$ is the last subsegment state vector of the motorized spindle, $\mathbf{Z}_1$ is the initial subsegment state vector of the motorized spindle, and $\mathbf{T}_{all}$ is the total transfer matrix obtained by sequentially multiplying the transfer matrices of all units from the first to the last. Because both ends of the motorized spindle are free ends, we know that some parameters of the end, that is, the bending moment and shear force of these two sections, are zero. Therefore,

$$\Delta\omega = \begin{vmatrix} a_{31} & a_{32} \\ a_{41} & a_{42} \end{vmatrix} = \left(m\omega^2 - k_r\right)\left(k_\theta - \left(J_d - J_p\frac{\Omega}{\omega}\right)\omega^2\right) = 0 \tag{19}$$

where the critical angular velocity can be obtained by Ref. [26], $a_{31}$, $a_{32}$, $a_{41}$, and $a_{42}$ are the elements in the first or second columns of the third row and the first or second columns of the fourth row of the total transfer matrix, respectively, which contain the parameter angular velocity of $\omega$.

Figure 5 shows the curve of angular velocity with change and the iterative method is used to solve the problem in a small range of approximately 0. When $\Delta\omega = 0$, the value corresponding to the horizontal coordinate is the critical rotational speed. Within the specified operating speed range of the motorized spindle, there are four points satisfied as $\Delta\omega \approx 0$. This means that the system has four critical speeds, which are 16,255.16, 24,526.62, 36,430.34, and 53,907.12 rpm.

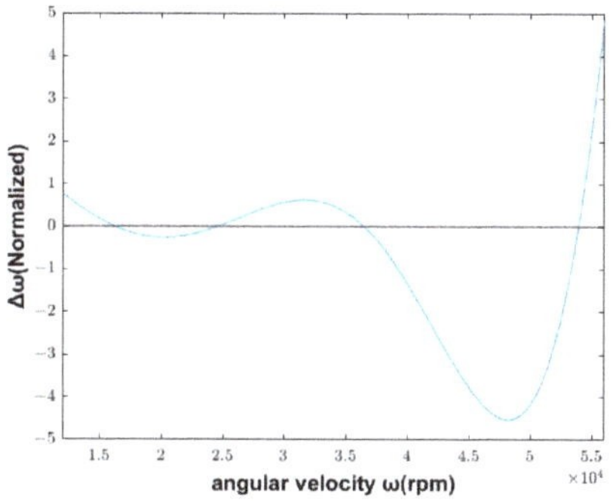


**Figure 5.** $\Delta\omega - \omega$ curve.

After the critical rotational speed is obtained, the relationship between the deflection of the initial end $y_1$ and the rotation angle of the initial end $\theta_1$ can be obtained by bringing the corresponding value of $\omega$ into Equation (18). For convenience of calculation, a dimensionless calculation is usually carried out, and the deflection of the initial end $y_1$ is set to 1. The angle $\theta_1$ can be obtained as follows:

$$\begin{pmatrix} y \\ \theta \end{pmatrix}^L = \begin{pmatrix} a_{31} & a_{32} \\ a_{41} & a_{42} \end{pmatrix} \begin{pmatrix} y \\ \theta \end{pmatrix}^R \Rightarrow \theta^L = a_{41}y^R + a_{42}\theta^R = 0 \Rightarrow \theta^R = -\frac{a_{41}}{a_{42}}y^R = \alpha y^R \\ \theta_1 = -\frac{a_{41}}{a_{42}}y_1 = \alpha y_1 \tag{20}$$

From this, the normalized state vector of the initial end at a certain critical speed can be obtained as follows:

$$\mathbf{Z}_1 = \begin{bmatrix} 1 & \alpha & 0 & 0 \end{bmatrix}^T \tag{21}$$

Equation (21) is multiplied by the element transfer matrix of each element in turn to obtain the cross-sectional section state vector of the corresponding node. At the different critical speeds, the deflection of each node can be calculated, and the mode shapes are shown in Figure 6.

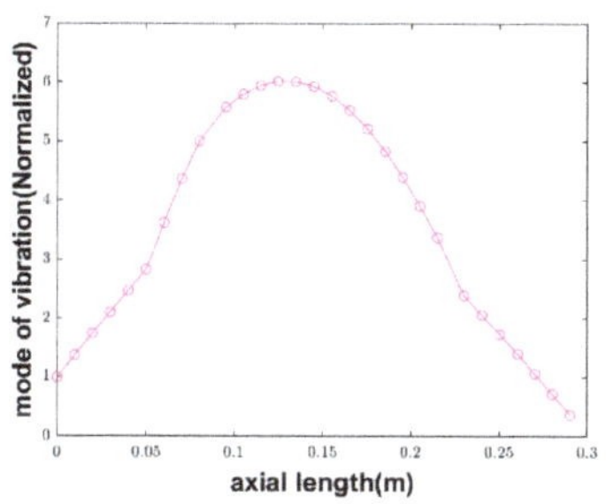


(**a**)

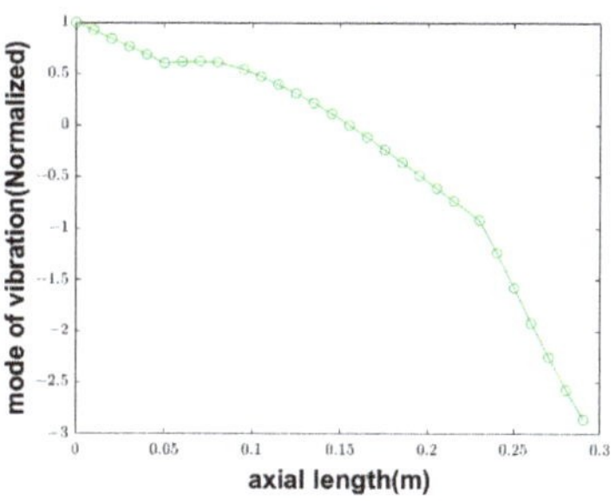


(**b**)

**Figure 6.** *Cont.*

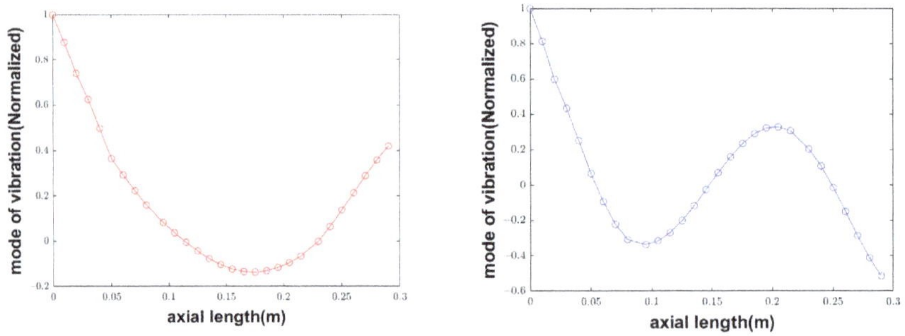


(c) (d)

**Figure 6.** The first fourth-order mode shapes of the motorized spindle: (**a**) first-order mode shape; (**b**) second-order mode shape; (**c**) third-order mode shape; (**d**) fourth-order mode shape.

From (a–d) in Figure 6, it can be seen that under different critical speeds, the radial displacement of the spindle changes with the axial direction of the motorized spindle, and the vibration of the motorized spindle is different at different critical speeds.

### *3.2. Dynamic Characteristics of Motorized Spindle with Unbalanced Mass Fault*

From Section 2.2, when the unbalanced mass fault happens, the general equation from the first node to the last node can be written as

$$\mathbf{Z}_{n+1} = \mathbf{T}_n\mathbf{Z}_n + \mathbf{H}_n = \mathbf{T}_n(\mathbf{T}_{n-1}\mathbf{Z}_{n-1} + \mathbf{H}_{n-1}) + \mathbf{H}_n = \cdots = \mathbf{T}_{all}\mathbf{Z}_1 + \sum_{k=1}^{n}(\prod_{j=k+1}^{n}\mathbf{T}_j\mathbf{H}_k) = \mathbf{T}_{all}\mathbf{Z}_1 + \mathbf{H}_{all} \quad (22)$$

where $\mathbf{H}_{all}$ is the sum of the unbalanced vector of all units from left to right. It is the same as the condition of the motorized spindle without fault; both ends of the spindle are free ends. The equation of angular velocity can be obtained:

$$\begin{bmatrix} y \\ \theta \end{bmatrix}_1 = -\begin{bmatrix} a_{31} & a_{32} \\ a_{41} & a_{42} \end{bmatrix}^{-1}\begin{bmatrix} h_3 \\ h_4 \end{bmatrix} \quad (23)$$

where $a_{31}$, $a_{32}$, $a_{41}$ and $a_{42}$ are the elements in the first or second columns of the third row and the first or second columns of the fourth row of the total transfer matrix, respectively. $h_3$ and $h_4$ are the third and fourth elements of $h_{all}$. All of them contain the parameter angular velocity of $\omega$. $y$ and $\theta$ indicate the section deflection and rotation angle of the initial end of the spindle, respectively.

Assuming that the amount of unbalance is 1–5 g·mm, the section deflection and rotation angle of different initial ends can be obtained under different angular velocities. Within the whole specified rotating speed, when the whole spindle is unbalanced, the deflection of the spindle initial end is shown in Figure 7a. Around the second-order critical rotating speed, the schematic diagram of the unbalanced mass fault influence on the deflection of the spindle intimal end when the amount of unbalance exists in different units is shown in Figure 7b.

As can be seen from Figure 7a, the deflection of the spindle initial end increases gradually at first when the unbalanced mass fault occurs, and reaches a local maximum when it reaches the first-order critical speed. At this time, the deflection direction of the shaft end suddenly changes. With the further increase in rotational speed, the deflection of the spindle initial end continues to increase along this direction and the direction suddenly changes again when the second-order critical rotational speed is reached. The same characteristics are shown at the third-order and fourth-order critical speeds. Moreover, the maximum value is reached at the third-order critical speed.

It can be seen from Figure 7b that the direction and magnitude of the spindle initial end deflection are different in different units when the unbalanced mass faults occur. The spindle initial end deflection is upward when a fault occurs in the 6th, 9th, 12th, or 15th unit. When the speed reaches the second-order critical speed, the direction suddenly changes, and the deflection value of the 12th unit is the largest, which may be more sensitive to the unbalanced fault. On the other hand, the spindle initial end deflection of the 18th, 21st, and 24th units are downward, and the direction suddenly changes when it reaches the second-order critical speed too.

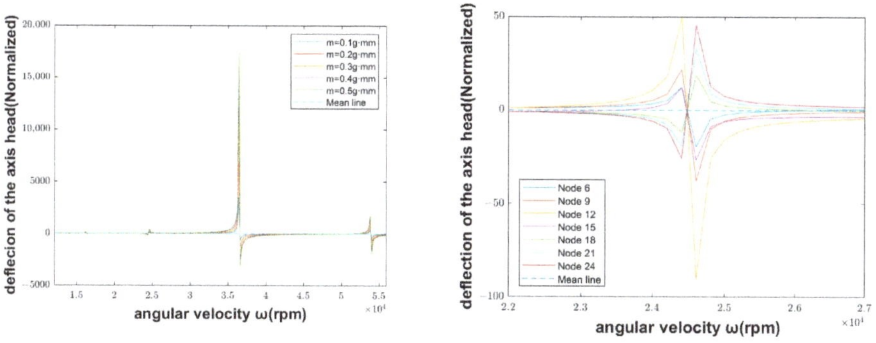


(a) (b)

**Figure 7.** Deflection of the spindle initial end in the case of unbalanced mass fault: (**a**) deflection of the spindle initial end; (**b**) spindle initial end deflection when fault occurs at different units.

The vibration mode of the whole motorized spindle can be further solved when the unbalanced mass fault occurs, as shown in Figure 8a. Around the first-order critical speed, the influence of the unbalanced mass fault on the middle part of the spindle is slightly greater than that on the spindle ends. While with the increase in speed, the influence of the unbalanced mass fault on the spindle ends begins to be greater than that on the middle part of the spindle around the second-order and third-order critical speeds. Around the third-order critical speed, the influence on the spindle end reaches the maximum value. After the rotating speed continues to increase, the degree of influence between the spindle end and the middle part of the spindle caused by the unbalanced mass fault gradually decreases.

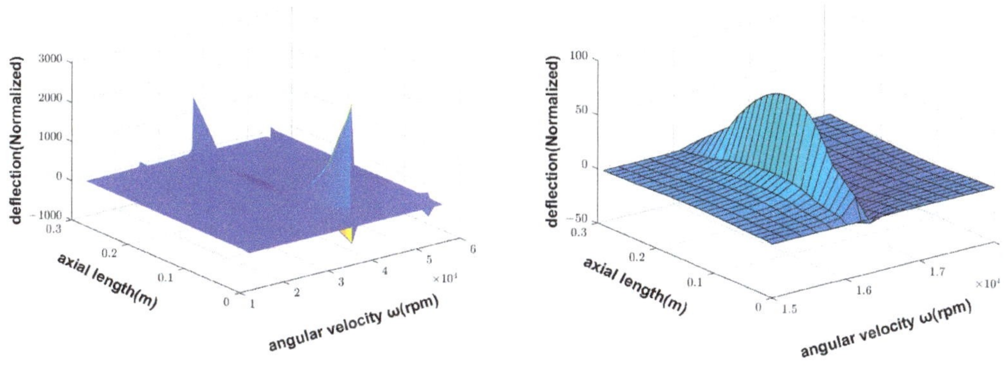


(a) (b)

**Figure 8.** Mode shape in the case of unbalanced mass fault: (**a**) mode shape in working speed range; (**b**) mode shape around the first-order critical speed.

The vibration mode of the whole motorized spindle under the unbalanced mass fault is further analyzed around the first-order critical speed, as shown in Figure 8b. Before reaching the first-order critical speed, the middle part of the spindle deflection is upward and greater than the spindle end deflection. Therefore, at this time, the motorized spindle is arched upward. However, when the first-order critical speed is reached, the middle part of the spindle deflection begins to change to the opposite direction, so the whole spindle is in a concave shape.

It can be seen that the dynamic characteristics of the whole spindle cannot be ignored when the unbalanced mass fault occurs, especially when the speed is close to the critical speed, and different parts of the spindle have a different sensitivity to unbalanced mass faults.

### 3.3. Dynamic Characteristics of Motorized Spindle with Spindle Inclination Fault

From Section 2.3, when the spindle inclination fault occurs, the general equation from the first node to the last node can be written as

$$\mathbf{Z}_{n+1} = \mathbf{T}_n\mathbf{Z}_n + \mathbf{B}_n = \mathbf{T}_n(\mathbf{T}_{n-1}\mathbf{Z}_{n-1} + \mathbf{B}_{n-1}) + \mathbf{B}_n = \cdots = \mathbf{T}_{all}\mathbf{Z}_1 + \sum_{k=1}^{n} \left( \prod_{j=k+1}^{n} \mathbf{T}_j\mathbf{B}_k \right) = \mathbf{T}_{all}\mathbf{Z}_1 + \mathbf{B}_{all} \tag{24}$$

where $\mathbf{B}_{all}$ is the sum of the unbalance vector of all units from left to right. Both ends of the spindle are free ends. The equation of angular velocity is

$$\begin{bmatrix} y \\ \theta \end{bmatrix}_1 = -\begin{bmatrix} a_{31} & a_{32} \\ a_{41} & a_{42} \end{bmatrix}^{-1} \begin{bmatrix} b_3 \\ b_4 \end{bmatrix} \tag{25}$$

where $b_3$ and $b_4$ are the third and fourth elements of $b_{all}$. All of them contain a parameter angular velocity of $\omega$. $y$, $\theta$ indicate the section deflection and rotation angle of the initial end of the spindle.

Assuming that the inclination angles are 0.1–0.5°, the section de-flection and rotation angle of different initial ends can be obtained under different angular velocities. Within the whole specified rotating speed, when the spindle inclines, the deflection of the spindle initial end is shown in Figure 9a. Around the second-order critical rotating speed, the spindle inclination fault influence on the deflection of the spindle initial end when the spindle incline at different inclination angles is shown in the schematic diagram, Figure 9b.

As can be seen from Figure 9a, the deflection of the initial spindle end increases gradually at first when the spindle inclination fault occurs, then the deflection suddenly changes direction and increases gradually along this direction, when approaching the first-order critical speed. At the first-order critical speed, the deflection decreases suddenly. Until the deflection equals zero, the deflection increases gradually along the opposite direction. The same characteristics are shown at the second-order and fourth-order critical speeds. However, at the fourth-order critical speed, the local maximum is smaller than it is at the second-order critical speed. Differently, when the rotational speed reaches the third-order critical speed, the deflection suddenly jumps to the maximum value in the upward direction. Then the direction suddenly changes, and decreases gradually.

As shown in Figure 9b, the bigger the angle, the higher the deflection of the spindle initial end. However, the changing trend of the motorized spindle initial end deflection is the same at different angles. That is to say, in the case of spindle inclination fault at the small angle, the inclination angle only changes the size of the deflection, but does not influence the trend of the deflection changing with the rotating speed.

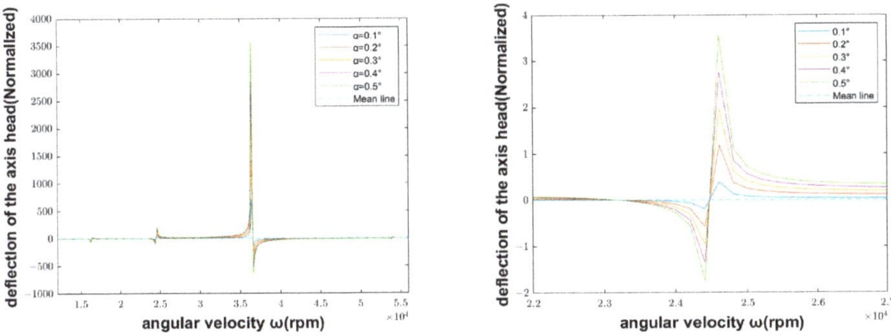


(**a**) (**b**)

**Figure 9.** Deflection of the spindle initial end in the case of spindle inclination fault: (**a**) deflection of the spindle initial end; (**b**) spindle intimal end deflection when fault occurs at different inclinations.

As shown in Figure 10a, different from the unbalanced mass fault, the spindle inclination fault affects the whole motorized spindle around the first-order critical speed, while with the increase in the speed, the influence of the spindle inclination fault on the spindle initial end begins to be greater than that on the middle part of the spindle, around the second-order and third-order critical speed. In the third order, the influence on the spindle initial end reaches the maximum value. After the speed continues to increase, the influence of fault on the motorized spindle begins to increase. At the fourth-order critical speed, the deflection of the spindle end directly shows an up direction and the other end shows a down direction. Although the deflection at this time is small, if we do not pay attention to this situation, the spindle inclination fault may further develop into a membrane collision fault, which will cause serious damage to the motorized spindle.

The vibration mode of the motorized spindle under spindle inclination fault is further analyzed around the first-order critical speed, as shown in Figure 10b. Before reaching the first-order critical speed, the spindle deflection is downward, and the middle part of the spindle deflection is greater than the spindle end. Therefore, the motorized spindle is a concave type. However, after reaching the first-order critical speed, it is still in a concave shape, but the deflection is smaller than that before the first-order speed.

As can be seen from Figure 10a, the deflection of the spindle initial end increases gradually at first when unbalanced mass fault occurs, and reaches a local maximum when it reaches the first-order critical speed. At this time, the deflection direction of the shaft end suddenly changes. With the further increase in rotational speed, the deflection of spindle initial end continues to increase along this direction, and the direction suddenly changes again when the second-order critical rotational speed is reached. The same characteristics are shown at the third-order and fourth-order critical speeds. Moreover, the maximum value is reached at the third-order critical speed.

It can be observed that when the spindle inclination fault occurs, the dynamic characteristics of the spindle change with the increase in rotating speed. At a low-rotating speed, the influence of elastic deformation on the spindle deflection is greater than the inclination on it, while with the increase in rotating speed, the influence of inclination gradually becomes dominant.

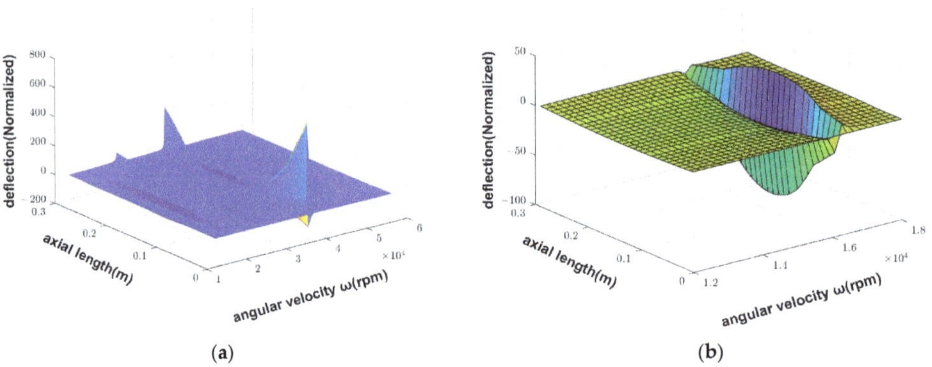


**Figure 10.** Mode shape in the case of spindle inclination fault: (**a**) mode shape in working speed range; (**b**) mode shape around the first-order critical speed.

## 4. Conclusions

In this work, focusing on the unbalanced mass fault and inclined axis fault, we improved the traditional WTMM. Then we used it to analyze the mode shape of motorized spindle under the fault at different rotation speeds. We also studied which part of the motorized spindle is more sensitive to the unbalanced mass fault, and the influence of different tilt angles on vibration of motorized spindle. The findings in this work will have different applications in many different types of machinery in which there are rotating cylinders, shafts, or spindles. The main conclusions are as follows:

(1) If the unbalanced mass fault occurs, when the spindle speed is close to the first-order or second-order critical speed, the deflection at the initial end of the spindle is small; when the spindle speed is close to the fourth-order critical speed, the deflection is large. The opposite is true when the spindle inclination fault occurs. It can be seen that the motorized spindle is more sensitive at high rotating speeds when the unbalanced mass fault occurs, and it is more sensitive at low rotating speed when the spindle inclination fault occurs.

(2) CNC machine tools are vibration systems with high stiffness, and the support system is flexible or nearly flexible. The rotor may exhibit rigid behavior, manifested as a smaller rotor vibration and larger bearing seat vibration. So, regardless of whether the spindle has an unbalanced mass fault or spindle inclination fault, when the spindle speed reaches the third-order critical speed, the deflection at the initial end of the spindle is the largest.

(3) When the spindle speed approaches the first-order critical speed, an unbalanced mass fault will make the middle part of the spindle bend upward, while the spindle inclination fault will make the middle part of the spindle bend downward.

(4) When the unbalanced mass fault occurs, the deflection will be increased first, then suddenly decrease until the direction change, and the deflection will continue to increase along this direction. When the spindle inclination fault occurs, the deflection will be increased first in the opposite direction, then gradually decreases after reaching the maximum value, and then gradually continues to increase along the original direction when it returns to the original direction.

**Author Contributions:** Conceptualization, S.L. and Y.W.; methodology, D.G. and Y.X.; software, Y.X.; validation, J.S.; formal analysis, D.G. and Y.X.; investigation, D.G. and Y.X.; resources, Y.W.; data curation, Y.X.; writing—original draft preparation, D.G. and Y.X.; writing—review and editing, S.L. and Y.W.; supervision, S.L. and Y.W; project administration, S.L.; funding acquisition, S.L. All authors have read and agreed to the published version of the manuscript.

**Funding:** This research was funded by the National Natural Science Foundation of China (No. 11802168, No. 52075310).

**Institutional Review Board Statement:** Not applicable.

**Informed Consent Statement:** Not applicable.

**Data Availability Statement:** The data presented in this study are available upon request from the corresponding author.

**Conflicts of Interest:** The authors declare no conflict of interest.

## References

1. Lin, C.-W.; Lin, Y.-K.; Chu, C.-H. Dynamic models and design of spindle-bearing systems of machine tools: A review. *Int. J. Precis. Eng. Manuf.* **2013**, *14*, 513–521. [CrossRef]
2. Gu, D.; Wei, Y.; Xiong, B.; Liu, S.; Zhao, D.; Wang, B. Three degrees of freedom chatter stability prediction in the milling process. *J. Mech. Sci. Technol.* **2020**, *34*, 3489–3496. [CrossRef]
3. Xiong, B.; Wei, Y.; Gu, D.; Zhao, D.; Wang, B.; Liu, S. Chatter stability analysis of variable speed milling with helix angled cutters. *Proc. Inst. Mech. Eng. Part B J. Eng. Manuf.* **2021**, *235*, 850–861. [CrossRef]
4. Appana, D.K.; Prosvirin, A.; Kim, J.-M. Reliable fault diagnosis of bearings with varying rotational speeds using envelope spectrum and convolution neural networks. *Soft Comput.* **2018**, *22*, 6719–6729. [CrossRef]
5. Tra, V.; Kim, J.; Kim, J.-M. Fault Diagnosis of Bearings with Variable Rotational Speeds Using Convolutional Neural Networks. In *Advances in Computer Communication and Computational Sciences*; Springer: Singapore, 2019; pp. 71–81.
6. Feng, H.; Jiang, S. Dynamics of a motorized spindle supported on water-lubricated bearings. *Proc. Inst. Mech. Eng. Part C J. Mech. Eng. Sci.* **2017**, *231*, 459–472. [CrossRef]
7. Ma, J.; Wu, L.; Zhu, Z.Q. Effect of magnet thickness on electromagnetic performance of high speed permanent magnet machines. In Proceedings of the 2017 IEEE International Electric Machines and Drives Conference (IEMDC), Miami, FL, USA, 21–24 May 2017; pp. 1–7.
8. Wang, B.; Wei, Y.; Liu, S.; Gu, D.; Zhao, D. Intelligent chatter detection for CNC machine based on RFE multi-feature selection strategy. *Meas. Sci. Technol.* **2021**, *32*, 095904. [CrossRef]
9. Shan, W.; Chen, X.; He, Y.; Zhou, J. A novel experimental research on vibration characteristics of the running high-speed motorized spindles. *J. Mech. Sci. Technol.* **2013**, *27*, 2245–2252. [CrossRef]
10. Fan, K.; Xiao, J.; Wang, R.; Gao, R. Thermoelectric-based cooling system for high-speed motorized spindle I: Design and control mechanism. *Int. J. Adv. Manuf. Technol.* **2022**, *121*, 3787–3800. [CrossRef]
11. Wang, N.; Jiang, D. Vibration response characteristics of a dual-rotor with unbalance-misalignment coupling faults: Theoretical analysis and experimental study. *Mech. Mach. Theory* **2018**, *125*, 207–219. [CrossRef]
12. Xu, X.; Han, Q.; Chu, F. Nonlinear vibration of a generator rotor with unbalanced magnetic pull considering both dynamic and static eccentricities. *Arch. Appl. Mech.* **2016**, *86*, 1521–1536. [CrossRef]
13. Liu, H.; Wu, Y.; Wang, X.; Yan, P.; Zhang, X. Nonlinear normal modes and primary resonance for permanent magnet synchronous motors with a nonlinear restoring force and an unbalanced magnetic pull. *Nonlinear Dyn.* **2019**, *97*, 1197–1213. [CrossRef]
14. Werner, U. Influence of electromagnetic field damping on forced vibrations of induction rotors caused by dynamic rotor eccentricity. *ZAMM—J. Appl. Math. Mech./Z. Für Angew. Math. Und Mech.* **2017**, *97*, 38–59. [CrossRef]
15. Zuo, S.; Li, D.; Mao, Y.; Deng, W. Longitudinal vibration analysis and suppression of electric wheel system driven by in-wheel motor considering unbalanced magnetic pull. *Proc. Inst. Mech. Eng. Part D J. Automob. Eng.* **2019**, *233*, 2729–2745. [CrossRef]
16. Xiang, C.; Liu, F.; Liu, H.; Han, L.; Zhang, X. Nonlinear dynamic behaviors of permanent magnet synchronous motors in electric vehicles caused by unbalanced magnetic pull. *J. Sound Vib.* **2016**, *371*, 277–294. [CrossRef]
17. Kumar, G.; Kalita, K. Vibration control using BCW induction motor. *Procedia Eng.* **2016**, *144*, 94–101. [CrossRef]
18. Bossmanns, B.; Tu, J.F. A power flow model for high speed motorized spindles—Heat generation characterization. *J. Manuf. Sci. Eng.* **2000**, *123*, 494–505. [CrossRef]
19. Li, H.; Shin, Y.C. Integrated dynamic thermo-mechanical modeling of high speed spindles, Part 1: Model development. *J. Manuf. Sci. Eng.* **2004**, *126*, 148–158. [CrossRef]
20. Li, H.; Shin, Y.C. Integrated dynamic thermo-mechanical modeling of high speed spindles, Part 2: Solution procedure and validations. *J. Manuf. Sci. Eng.* **2004**, *126*, 159–168. [CrossRef]
21. Xul, J.; Zheng, X.; Zhang, J.; Liu, X. Vibration characteristics of unbalance response for motorized spindle system. *Procedia Eng.* **2017**, *174*, 331–340. [CrossRef]
22. Liu, J.; Lai, T.; Chen, X. Dynamics analysis of unbalanced motorized spindles supported on ball bearings. *Shock Vib.* **2016**, *2016*, 2787524. [CrossRef]
23. Su, X.; Lu, H.; Zhang, X.; Fan, W.; Zhang, Y. Analysis of dynamic characteristic for misalignment-spline gear shaft based on whole transfer matrix method. *J. Vibroeng.* **2018**, *20*, 1392–1408. [CrossRef]
24. Huang, S.; Chi, X.; Xu, Y.; Sun, Y. Dynamic characteristics of coupling shaft system in tufting machine based on the Riccati whole transfer matrix method. *Proc. Inst. Mech. Eng. Part C J. Mech. Eng. Sci.* **2017**, *231*, 356–371. [CrossRef]

25. Jiang, S.; Zheng, S. A modeling approach for analysis and improvement of spindle-drawbar-bearing assembly dynamics. *Int. J. Mach. Tools Manuf.* **2010**, *50*, 131–142. [CrossRef]
26. Zhong, Y.; He, Y.; Wang, Z.; Li, F. *Rotor Dynamics*; Tsinghua University Press: Beijing, China, 1987.

applied sciences

MDPI

*Article*

# Generalized Function Projective Synchronization of Two Different Chaotic Systems with Uncertain Parameters

**Bin Zhen * and Yu Zhang**

School of Environment and Architecture, University of Shanghai for Science and Technology, Shanghai 200093, China; zhangyu980831@163.com
* Correspondence: zhenbin80@163.com

**Abstract:** This study proposes a new approach to realize generalized function projective synchronization (GFPS) between two different chaotic systems with uncertain parameters. The GFPS condition is derived by converting the differential equations describing the synchronization error systems into a series of Volterra integral equations on the basis of the Laplace transform method and convolution theorem, which are solved by applying the successive approximation method in the theory of integral equations. Compared with the results obtained by constructing Lyapunov functions or calculating the conditional Lyapunov exponents, the uncertain parameters and the scaling function factors considered in this paper have fewer restrictions, and the parameter update laws designed for the estimation of the uncertain parameters are simpler and easier to realize physically.

**Keywords:** generalized function projective synchronization; chaotic systems; Volterra integral equations; the Laplace transform method

## 1. Introduction

Since Pecora and Carroll realized synchronization between two chaotic systems in 1990 [1,2], chaos synchronization has been widely explored and studied due to its potential applications in vast areas of physics and engineering science [3–6]. Synchronization is a process wherein two or more systems adjust a given property of their motion. Different types of synchronization phenomena have been numerically observed and experimentally verified in a variety of chaotic systems, such as complete synchronization [1,7], phase synchronization [8,9], anti-phase synchronization [10,11], lag synchronization [12,13], generalized synchronization [14,15] and projective synchronization [16,17].

Complete synchronization (CS) is characterized by a coincidence of states of two chaotic systems while evolving in time. CS only appears when two interacting systems are identical. However, it is difficult to construct two absolutely identical systems in the real world. Generalized synchronization (GS) refers to a functional relation of states of two different chaotic systems, which is an extension of CS. Projective synchronization (PS) is more complex than CS but simpler than GS. Additionally, PS is useful in extending binary digital to M-nary digital communication to achieve faster communication [18–20].

PS was first reported by Mainieri et al. [16] in a class of systems with partial linearity in which the drive and response vectors evolve in a proportional scale. Xu [21] showed that the scaling factor of PS in coupled partially linear systems is unpredictable and can be arbitrarily maneuvered by introducing a feedback control to master systems. The PS between two chaotic discrete dynamical systems was achieved by Xin and Wu [22] via linear state error feedback control. Wen and Xu [23] and Yan and Li [24] extended the projective synchronization feature to general nonlinear systems, including non-partially linear chaotic systems, by applying controllers to response systems, which is called generalized projective synchronization (GPS) and was studied later [25–27]. CS and anti-synchronization are particular cases of GPS. As an extension of GPS, modified projective synchronization (MPS) was introduced by Li et al. [28] and He et al. [29], where the responses of synchronized dynamical states can be synchronized up to a constant matrix. The combined projective synchronization (CPS) among different nonlinear systems has been reported by Feng et al. [30] by means of the active backstepping design method. Chen et al. [31,32] considered two chaotic systems synchronized up to a scaling function factor, which is called function projective synchronization (FPS). The FPS between two novel chaotic systems has been investigated by El Dessoky et al. [33] and Bekiros et al. [34] using the Lyapunov method of stability. Du et al. [35] and Srisuntorn et al. [36] discussed modified function projective synchronization (MFPS), in which two chaotic systems can be synchronized up to a desired scaling function matrix. As another extension of MPS and FPS, in generalized function projective synchronization (GFPS), the scaling function matrix is more generalized than that of MFPS, as has been discussed by Yu and Li [37], Li and Zhao [38], Wu et al. [39] and Al-Azzawi [40].

Most of the previous works have focused on the constant scaling factor and mainly on chaotic systems with exactly known parameters. However, in many practical situations, GFPS needs to be investigated while some parameters are uncertain. It is easy to understand that GFPS may be greatly affected by these uncertainties. Yu and Li [37] designed an adaptive controller to achieve GFPS between two different chaotic systems and gain some parameter update laws to estimate the unknown parameters of chaotic systems. Li and Zhao [38] firstly discussed the GFPS conditions of two different uncertain hyperchaotic systems when the scaling function factors were periodic functions and polynomial functions. However, only a few theoretical results about GFPS between two different uncertain chaotic systems have been obtained [41–45]. Furthermore, the aforementioned controllers all are designed through the construction of proper Lyapunov functions. The added controllers therefore are often very specific and may be too difficult to realize physically.

In this paper, an effective approach is proposed to design an adaptive controller to realize GFPS between two different uncertain chaotic systems. Moreover, the unknown parameters in the coupled system can be estimated by given parameter update laws. The key to realizing GFPS between two chaotic systems is the determination of the stability of the synchronization error system. In previous works, this problem usually was solved by using the Lyapunov method. However, there are two main difficulties in this method: the first one is that it is challenging to construct a proper Lyapunov function; the second is that the GFPS conditions derived using the Lyapunov method are usually sufficient and highly conserved. The controllers and parameter update laws designed based on such conditions are often very specific. In fact, GFPS can be viewed as a unique case of GS, which can be investigated using the auxiliary system approach [46]. To avoid the construction of Lyapunov functions, the approach presented in this paper converts the differential equations describing the synchronization error system between the auxiliary system and the response system into a series of Volterra integral equations by using the Laplace transform method. Then, the dynamical behavior of the error system near its origin can be analyzed by employing the successive approximation method [47] . In the theory of integral equations, the significance of the Laplace transform method is that the GFPS condition can be obtained without the construction of Lyapunov functions, even if the scaling function factors are periodic functions, polynomial functions or any other complex functions. Compared to previous works, our approach allows the easier design of controllers and simpler parameter update laws and removes some restrictions on the uncertain parameters. The feasibility and effectiveness of the approach are illustrated by several numerical simulations.

The rest of the paper is organized as follows. In Section 2, the GFPS scheme for two different uncertain chaotic systems is introduced. In Section 3, the GFPS condition is investigated using the Laplace transform method combined with the successive approximation method in the theory of Volterra integral equations. A numerical example is provided in Section 4 to demonstrate the effectiveness of our GFPS scheme. In Section 5, simple parameter update laws are designed according to the GFPS condition derived in Section 3 to estimate the uncertain parameters in chaotic systems, the validity of which are illustrated by numerical simulations. Finally, conclusions are drawn in Section 6.

## 2. GFPS Scheme between Two Different Uncertain Chaotic Systems

Consider two chaotic systems (drive–response systems) with uncertain parameters given in the following form:

$$\begin{aligned} \dot{x} &= f_1(x, \phi), \\ \dot{y} &= f_2(y, \psi) + u, \end{aligned} \tag{1}$$

where $x, y \in R^n$; $\phi \in R^m, \psi \in R^l$ are uncertain parameter vectors, and $u$ is the controller to be designed. For simplicity, assume that $f_1(0, \phi) = f_2(0, \psi) = 0$ hold for any values of $\phi$ and $\psi$. If there exists a scaling function matrix $h = diag\{h_1(x), h_2(x), \cdots, h_n(x)\}$, such that

$$\lim_{t \to \infty} ||y - h(x)x|| = 0, \tag{2}$$

then the two chaotic systems in system (1) are said to be GFPS with respect to the scaling function matrix $h$ [37,40]. The two chaotic systems are said to be FPS if

$$h_1(x) = h_2(x) = \cdots = h_n(x)$$

in Equation (2). Moreover, we find MPS between the two chaotic systems if $h = diag\{h_1, h_2, \cdots, h_n\}$, $h_i (i = 1, 2, \cdots, n)$ are real constants in Equation (2). Obviously, both FPS and MPS are specific cases of GFPS.

The controller $u$ and the parameter update laws in system (1) are designed as below:

$$\begin{aligned} u &= J(x) f_1(x, \hat{\phi}) - f_2(y, \hat{\psi}) + k\delta_1, \\ \dot{\hat{\phi}} &= g_1(\hat{\phi} - \phi, \delta_1), \\ \dot{\hat{\psi}} &= g_2(\hat{\psi} - \psi, \delta_1), \end{aligned} \tag{3}$$

where $J(x) = d(h(x)x)/dx$, $k$ is the control parameter, and $\delta_1 = y - h(x)x$ is the synchronization error; $\hat{\phi}$ and $\hat{\psi}$ are the estimated values of uncertain parameter vectors $\phi$ and $\psi$, respectively. In addition, $g_1(\hat{\phi} - \phi, 0) = g_2(\hat{\psi} - \psi, 0) = 0$ hold if

$$\hat{\phi} = \phi \quad and \quad \hat{\psi} = \psi. \tag{4}$$

The GFPS occurring in system (1) can be regarded as a specific case of GS between the two chaotic systems, which can be detected using the auxiliary system approach [46]. Consider the following identical copy of the response system driven by the same driving signal $x$,

$$\begin{aligned} \dot{z} &= f_2(z, \psi) + \bar{u}, \\ \bar{u} &= J(x) f_1(x, \bar{\phi}) - f_2(z, \bar{\psi}) + k\delta_2, \\ \dot{\bar{\phi}} &= g_1(\bar{\phi} - \phi, \delta_2), \\ \dot{\bar{\psi}} &= g_2(\bar{\psi} - \psi, \delta_2), \end{aligned} \tag{5}$$

in which $z \in R^n$ and $\delta_2 = z - h(x)x$. $\bar{\phi}$ and $\bar{\psi}$ are the estimated values of uncertain parameter vectors $\phi$ and $\psi$, respectively. Additionally, $g_1(\bar{\phi} - \phi, 0) = g_2(\bar{\psi} - \psi, 0) = 0$ hold for

$$\bar{\phi} = \phi \quad and \quad \bar{\psi} = \psi. \tag{6}$$

Then, GFPS conditions (2) and (4) lead to

$$\lim_{t \to \infty} ||y - z|| = 0, \quad \lim_{t \to \infty} ||\hat{\phi} - \bar{\phi}|| = 0, \quad \lim_{t \to \infty} ||\hat{\psi} - \bar{\psi}|| = 0. \tag{7}$$

## 3. Detection of GFPS Condition Based on the Laplace Transform

By letting

$$e_1 = \frac{y-z}{2}, \qquad e_2 = \frac{\hat{\phi}-\bar{\phi}}{2}, \qquad e_3 = \frac{\hat{\psi}-\bar{\psi}}{2}$$
$$e_4 = \frac{y+z}{2}, \qquad e_5 = \frac{\hat{\phi}+\bar{\phi}}{2}, \qquad e_6 = \frac{\hat{\psi}+\bar{\psi}}{2}$$

systems (1), (3) and (5) become

$$\begin{aligned}
\dot{e}_1 &= ke_1 + \frac{J(x)}{2}[f_1(x, e_2+e_5) - f_1(x, e_5-e_2)] + \frac{1}{2}[f_2(e_1+e_4, \psi) - f_2(e_4-e_1, \psi)] \\
&\quad + \frac{1}{2}[f_2(e_4-e_1, e_6-e_3) - f_2(e_4+e_1, e_6+e_3)], \\
\dot{e}_2 &= \frac{1}{2}[g_1(e_2+e_5-\phi, e_1+e_4-h(x)x) - g_1(e_5-e_2-\phi, e_4-e_1-h(x)x)], \\
\dot{e}_3 &= \frac{1}{2}[g_2(e_3+e_6-\psi, e_1+e_4-h(x)x) - g_2(e_6-e_3-\psi, e_4-e_1-h(x)x)], \\
\dot{e}_4 &= ke_4 + \frac{J}{2}[f_1(x, e_2+e_5) + f_1(x, e_5-e_2)] + \frac{1}{2}[f_2(e_1+e_4, \psi) + f_2(e_4-e_1, \psi)] \\
&\quad - \frac{1}{2}[f_2(e_4-e_1, e_6-e_3) + f_2(e_4+e_1, e_6+e_3)], \\
\dot{e}_5 &= \frac{1}{2}[g_1(e_2+e_5-\phi, e_1+e_4-h(x)x) + g_1(e_5-e_2-\phi, e_4-e_1-h(x)x)], \\
\dot{e}_6 &= \frac{1}{2}[g_2(e_3+e_6-\psi, e_1+e_4-h(x)x) + g_2(e_6-e_3-\psi, e_4-e_1-h(x)x)], \\
\dot{x} &= f_1(x, \phi).
\end{aligned} \tag{8}$$

GFPS condition (7) can be written as

$$\lim_{t\to\infty} ||e_i|| = 0, \quad i = 1, 2, 3. \tag{9}$$

For sufficiently small $e_{1,2,3}$ near $(0,0,0)$, the right-hand side of the first three equations in system (8) can be expanded as

$$\begin{aligned}
\dot{e}_1 &= ke_1 + F_1(x, e_2, e_3), \\
\dot{e}_2 &= \alpha_1 e_1 + \alpha_2 e_2 + F_2(e_1, e_2), \\
\dot{e}_3 &= \alpha_3 e_1 + \alpha_4 e_3 + F_3(e_1, e_3).
\end{aligned} \tag{10}$$

where

$$F_1(x, e_2, e_3) = J(x) \sum_{n=1}^{\infty} \frac{1}{(2n-1)!} \frac{\partial^{(2n-1)} f_1(x, \phi)}{\partial \phi^{(2n-1)}} e_2^{(2n-1)} - \sum_{n=1}^{\infty} \frac{1}{(2n-1)!} \frac{\partial^{(2n-1)} f_2(h(x)x, \psi)}{\partial \psi^{(2n-1)}} e_3^{(2n-1)},$$

$$F_2(e_1, e_2) = \sum_{n=1}^{\infty} \frac{1}{(2n+1)!} [\frac{\partial^{(2n+1)} g_1(0,0)}{\partial \delta_1^{(2n+1)}} e_1^{(2n+1)} + \frac{\partial^{(2n+1)} g_1(0,0)}{\partial(\hat{\phi}-\phi)^{(2n+1)}} e_2^{(2n+1)}],$$

$$F_3(e_1, e_3) = \sum_{n=1}^{\infty} \frac{1}{(2n+1)!} [\frac{\partial^{(2n+1)} g_2(0,0)}{\partial \delta_1^{(2n+1)}} e_1^{(2n+1)} + \frac{\partial^{(2n+1)} g_2(0,0)}{\partial(\hat{\psi}-\psi)^{(2n+1)}} e_3^{(2n+1)}],$$

$$\alpha_1 = \frac{\partial g_1(0,0)}{\partial \delta_1}, \quad \alpha_2 = \frac{\partial g_1(0,0)}{\partial(\hat{\phi}-\phi)}, \quad \alpha_3 = \frac{\partial g_2(0,0)}{\partial \delta_1}, \quad \alpha_4 = \frac{\partial g_2(0,0)}{\partial(\hat{\psi}-\psi)}.$$

Consider the Laplace transform, defined as follows:

$$\begin{aligned} E_j(s) &= L[e_j] = \int_0^{+\infty} e_j(t)e^{-st}dt, \\ e_j(t) &= L^{-1}[E_j] = \frac{1}{2\pi i}\int_{\sigma-i\infty}^{\sigma+i\infty} E_j(s)e^{st}ds, \qquad j = 1,2,3. \end{aligned} \tag{11}$$

Both $e_j$ and $E_j$ are vectors with the same dimensions for $j = 1,2,3$. Taking the Laplace transform of $e_{1,2,3}$ on both sides of system (10) yields

$$(S - M)[E_1, E_2, E_3]^T = [e_{10}, e_{20}, e_{30}]^T + [\hat{F}_1, \hat{F}_2, \hat{F}_3]^T, \tag{12}$$

where

$$S = \begin{bmatrix} sI_n & 0 & 0 \\ 0 & sI_m & 0 \\ 0 & 0 & sI_l \end{bmatrix}, \qquad M = \begin{bmatrix} kI_n & 0 & 0 \\ \alpha_1 & \alpha_2 & 0 \\ \alpha_3 & 0 & \alpha_4 \end{bmatrix}, \tag{13}$$

$I_n$, $I_m$, $I_l$ represent the $n \times n$, $m \times m$ and $l \times l$ real identity matrices, respectively. $T$ denotes the transpose of vectors. $e_{j0}, j = 1,2,3$, are given initial values of system (10) and $\hat{F}_{1,2,3}$ are the Laplace transforms of $F_{1,2,3}$, respectively, i.e.,

$$\hat{F}_j = L[F_j] = \int_0^{+\infty} F_j(t)e^{-st}dt, \quad j = 1,2,3.$$

The solution to Equation (12) can be derived by using Cramer's rule [48]. Take, for example, the expression of $E_1$, which is an $n$-dimensional vector and always can be given by the form below:

$$E_1 = \begin{bmatrix} E_{11} \\ E_{12} \\ \vdots \\ E_{1n} \end{bmatrix} = \frac{1}{D(s)} \begin{bmatrix} w_{11}(s) \\ w_{12}(s) \\ \vdots \\ w_{1n}(s) \end{bmatrix} + \frac{1}{D(s)} \begin{bmatrix} w_{21}(s)\hat{F}_{11}(s) \\ w_{22}(s)\hat{F}_{12}(s) \\ \vdots \\ w_{2n}(s)\hat{F}_{1n}(s) \end{bmatrix}, \tag{14}$$

where $(E_{11}, E_{12}, \cdots, E_{1n})$ and $(\hat{F}_{11}, \hat{F}_{12}, \cdots, \hat{F}_{1n})$ are the $n$ components of the vectors $E_1$ and $\hat{F}_1$, respectively. $D(s)$ is the determinant of matrix $(S - M)$. $w_{1j}, w_{2j}, j = 1, \cdots, n$, are polynomials of $s$, and their highest powers are less than that of $D(s)$. Taking the inverse Laplace transform on Equation (14) and applying the convolution theorem, one has

$$e_{1j} = L^{-1}[E_{1j}] = L^{-1}[\frac{w_{1j}(s)}{D(s)}] + L^{-1}[\frac{w_{2j}(s)}{D(s)}] * F_{1j}(t), \quad j = 1, \cdots, n. \tag{15}$$

where $(e_{11}, e_{12}, \cdots, e_{1n})$ and $(F_{11}, F_{12}, \cdots, F_{1n})$ are the $n$ components of the vectors $e_1$ and $F_1$, respectively. $*$ denotes the convolution operation. First, we introduce the following theorem.

**Theorem 1.** *The necessary condition for $e_{1j} \to 0$ with $t \to \infty$, $j = 1,2,\cdots,n$, in Equation (15) is that all eigenvalues of matrix $M$ defined in Equation (13) have negative real parts.*

**Proof.** Without loss of generality, for any $i = 1,2$ and $j = 1,2,\cdots,n$, $w_{ij}(s)/D(s)$ have the following form

$$\frac{w_{ij}(s)}{D(s)} = \frac{\sum\limits_{p=0}^{v} a_p s^p}{\sum\limits_{p=0}^{n+m+l} b_p s^p}, \tag{16}$$

where $v < n + m + l$, $a_p$ and $b_p$ are constants. Consider the following three cases of the inverse Laplace transforms for Equation (16):

- $D(s)$ has $n+m+l$ single roots $s_1, s_2, \cdots, s_{(n+m+l)}$
  $\frac{w_{ij}(s)}{D(s)} = \frac{q_1}{s-s_1} + \frac{q_2}{s-s_2} + \cdots + \frac{q_{(n+m+l)}}{s-s_{(n+m+l)}}$,
  where $q_p = \frac{w_{ij}(s)}{D(s)}(s-s_p)\Big|_{s=s_p}, \quad p = 1,2,\cdots,n+m+l.$
  $L^{-1}[\frac{w_{ij}(s)}{D(s)}] = q_1 e^{s_1 t} + q_2 e^{s_2 t} + \cdots + q_{(n+m+l)} e^{s_{(n+m+l)} t}$
- $D(s)$ has a pair of conjugate complex roots $s_{1,2} = \beta \pm j\omega$
  $\frac{w_{ij}(s)}{D(s)} = \frac{w_{ij}(s)}{(s-\beta-j\omega)(s-\beta+j\omega)Q(s)} = \frac{q_1}{s-\beta-j\omega} + \frac{q_2}{s-\beta+j\omega} + \frac{P(s)}{Q(s)}$,
  where $q_1 = \frac{w_{ij}(s)}{D'(s)}\Big|_{s=\beta+j\omega}, \quad q_2 = \frac{w_{ij}(s)}{D'(s)}\Big|_{s=\beta-j\omega}$,
  $L^{-1}[\frac{w_{ij}(s)}{D(s)}] = q_1 e^{(\beta+j\omega)t} + q_2 e^{(\beta-j\omega)t} + L^{-1}[\frac{P(s)}{Q(s)}]$
- $D(s)$ has $(n+m+l)$-repeated real roots $s = s_0$
  $\frac{w_{ij}(s)}{D(s)} = \frac{q_1}{s-s_0} + \frac{q_2}{(s-s_0)^2} + \cdots + \frac{q_{(n+m+l)}}{(s-s_0)^{(n+m+l)}}$,
  where $q_{(n+m+l-p)} = \frac{1}{p!}\frac{d^p}{ds^p}[(s-s_0)^{(n+m+l)}\frac{w_{ij}(s)}{D(s)}]\Big|_{s=s_0}, \quad p = 1,2,\cdots,n+m+l-1,$
  $q_{(n+m+l)} = [(s-s_0)^{(n+m+l)}\frac{w_{ij}(s)}{D(s)}]\Big|_{s=s_0}$,
  $L^{-1}[\frac{w_{ij}(s)}{D(s)}] = (q_1 + q_2 t + \cdots + q_{(n+m+l)} t^{(n+m+l-1)}) e^{s_0 t}$

From the analysis of the inverse Laplace transform of Equation (16), it is clear that the inverse Laplace transform of $w_{ij}(s)/D(s)$, $i = 1,2$, in Equation (15) must be a sum of exponential functions. When $t \to \infty$, all roots of $D(s)$ must have negative real parts if

$$L^{-1}[\frac{w_{1j}(s)}{D(s)}], L^{-1}[\frac{w_{2j}(s)}{D(s)}] \to 0, \quad j = 1,2,\cdots,n$$

□

Assume that matrix $M$ only has eigenvalues with negative real parts; from Equation (15), when $t \to +\infty$, $e_{1j}$ can be written in the following form:

$$e_{1j} = \int_0^t R_{1j}(t-\tau)F_{1j}(\tau)d\tau, \quad j = 1,2,\cdots,n \tag{17}$$

where

$$R_{1j} = L^{-1}[\frac{w_{2j}(s)}{D(s)}], \quad j = 1,2,\cdots,n$$

One can easily confirm that the condition that matrix $M$ only has eigenvalues with negative real parts also is necessary for $e_{2j} \to 0, j = 1,2,\cdots,m$ and $e_{3j} \to 0, j = 1,2,\cdots,l$. Similarly, under such conditions, when $t \to +\infty$, $e_{2j}$ and $e_{3j}$ have the following form:

$$\begin{aligned} e_{2j} &= \int_0^t R_{2j}(t-\tau)F_{1j}(\tau)d\tau + \int_0^t R_{3j}(t-\tau)F_{2j}(\tau)d\tau, \quad j = 1,2,\cdots,m \\ e_{3j} &= \int_0^t R_{4j}(t-\tau)F_{1j}(\tau)d\tau + \int_0^t R_{5j}(t-\tau)F_{2j}(\tau)d\tau + \int_0^t R_{6j}(t-\tau)F_{3j}(\tau)d\tau, \\ &\qquad j = 1,2,\cdots,l \end{aligned} \tag{18}$$

in which $R_{ij} \to 0$ for $t \to +\infty, i = 2,3,\cdots,6$. $(F_{21}, F_{22}, \cdots, F_{2m})$ and $(F_{31}, F_{32}, \cdots, F_{3l})$ are the $m$ and $l$ components of the vectors $F_2$ and $F_3$, respectively, which are defined in Equation (10).

**Theorem 2.** *$e_{1j} = 0, j = 1,2,\cdots,n$, $e_{2j} = 0, j = 1,2,\cdots,m$, $e_{3j} = 0, j = 1,2,\cdots,l$, are the unique continuous solutions of Equations (17) and (18).*

**Proof.** In fact, Equations (17) and (18) are a series of Volterra integral equations, which can be solved by using the successive approximation method [47] in the theory of integral equations. Consider the Volterra integral equation as below:

$$A(t) = B(t) + \int_0^t T(t-\tau)H(\tau, A(\tau))d\tau, \tag{19}$$

where $A(t) \in R^n$ and $|A(t)| < \infty$, $T(t)$ is a $n \times n$ matrix, and $|T(t)| \in L[0, t_0]$ for any $0 < t_0 < \infty$. Both $B(t)$ and $H(t, A(t))$ are vectors with $n$ continuous components. Furthermore, for any $\eta > 0$, there must exist a constant $\rho(\eta) > 0$ such that

$$|H(t, A_1) - H(t, A_2)| \leq \rho(\eta)|A_1 - A_2|, \quad (|A_1|, |A_2| \leq \eta) \tag{20}$$

From the results given by Nohel [47], if the above conditions are satisfied, the successive approximations

$$\omega_0(t) = 0, \quad \omega_{n+1}(t) = B(t) + \int_0^t T(t-\tau)H(\tau, \omega_n(\tau))d\tau, \quad n = 0, 1, 2, \cdots \tag{21}$$

will uniformly converge to the unique continuous solution $A(t) = \omega(t)$ of Equation (19).

Comparing Equations (17)–(19), it is easy to check that $F_{1,2,3}$, which are defined in Equation (10), are smooth enough to satisfy condition (20). From Equation (21), $e_{1j} = 0$, $j = 1, 2, \cdots, n$, $e_{2j} = 0$, $j = 1, 2, \cdots, m$, $e_{3j} = 0$, $j = 1, 2, \cdots, l$, are the unique continuous solutions of Equations (17) and (18). □

According to Theorems 1 and 2, we have the following result.

**Theorem 3.** *GFPS in systems (1) and (3) can be achieved if the matrix M defined in Equation (13) only has eigenvalues with negative real parts.* □

## 4. Discussion of GFPS Scheme

To illustrate the effectiveness of the GFPS scheme proposed in the previous sections on the basis of the Laplace transform method, the following chaotic systems with uncertain parameters, which have been considered in [37,38], are discussed:

$$\begin{aligned} \dot{x} &= f(x) + \Phi(x)\Theta, \\ \dot{y} &= g(y) + \Psi(y)\Omega + U, \end{aligned} \tag{22}$$

where $x, y \in R^n$ are the state vectors; $\Phi(x) : R^n \to R^{n\times p}$, $\Psi(y) : R^n \to R^{n\times q}$, $\Theta \in R^p$ and $\Omega \in R^q$ represent the vectors of uncertain parameters; and $U \in R^n$ is a controller. For simplicity, $f(0) + \Phi(0)\Theta = g(0) + \Psi(0)\Omega = 0$ hold for any values of $\Theta$ and $\Omega$. Define the error vector of GFPS as

$$e = y - h(x)x,$$

where $h(x) = diag\{h_1(x), h_2(x), \cdots, h_n(x)\}$, $h_i(x) : R^n \to R$ $(i = 1, 2, \cdots, n)$ are scaling function factors that compose the scaling function matrix $h(x)$. According to [38], GFPS in (22) can be achieved by the controller and the parameter update laws as follows:

$$\begin{aligned} U &= J(f(x) + \Phi(x)\hat{\Theta}) - g(y) - \Psi(y)\hat{\Omega} + Ke, \\ \dot{\hat{\Theta}} &= -(J\Phi(x))^T e - (\hat{\Theta} - \Theta), \\ \dot{\hat{\Omega}} &= \Psi^T(y)e - (\hat{\Omega} - \Omega), \end{aligned} \tag{23}$$

where $J = d(h(x)x)/dx$, $\hat{\Theta}$ and $\hat{\Omega}$ are estimated values of the uncertain parameters in Equation (22); $K$ is a $n \times n$ matrix, which can be decomposed into $K = K_1 + K_2$, in which $K_1$ and $K_2$ satisfy the following assumptions:

$$\begin{aligned} &(i) \quad K_1^T = -K_1, \\ &(ii) \quad K_2 = diag\{b_1, b_2, \cdots, b_n\},\ b_i < 0,\ (i = 1, 2, \cdots, n). \end{aligned} \tag{24}$$

From the first condition in condition (24), all real parts of the eigenvalues of $K_1$ are zero since it is a real antisymmetric matrix. Then, all eigenvalues of matrix $K$ ($K = K_1 + K_2$) must have negative real parts if condition (24) is satisfied.

From Theorem 3 in Section 3, GFPS in system (22) appears due to the controller and parameter update laws presented in Equation (23) if all eigenvalues of the matrix

$$\begin{bmatrix} K & 0 & 0 \\ 0 & -I_p & 0 \\ 0 & 0 & -I_q \end{bmatrix}, \tag{25}$$

have negative real parts, in which $I_p$ and $I_q$ are the $p \times p$ and $q \times q$ real identity matrices, respectively. Obviously, the above GFPS condition implies that all eigenvalues of matrix $K$ should have negative real parts.

**Remark 1.** *The GFPS condition derived in [38] is a special condition of Theorem 3 in Section 3.*

An example is given below to demonstrate that the condition (24) is not necessary for GFPS occurring in system (22) under the controller and parameter update laws in (23). Consider the hyperchaotic Chen system and hyperchaotic Lorenz system as the drive and response systems, respectively [38].

$$\begin{aligned} \dot{x}_1 &= -ax_1 + ax_2 + x_4, \\ \dot{x}_2 &= dx_1 + cx_2 - x_1x_3, \\ \dot{x}_3 &= -bx_3 + x_1x_2, \\ \dot{x}_4 &= rx_4 + x_2x_3, \end{aligned} \tag{26}$$

and

$$\begin{aligned} \dot{y}_1 &= -a_1y_1 + a_1y_2 + y_4 + u_1, \\ \dot{y}_2 &= b_1y_1 - y_2 - y_1y_3 + u_2, \\ \dot{y}_3 &= -c_1y_3 + y_1y_2 + u_3, \\ \dot{y}_4 &= d_1y_4 - y_1y_3 + u_4, \end{aligned} \tag{27}$$

where $a, b, c, d, r, a_1, b_1, c_1$ and $d_1$ are uncertain parameters to be identified. The true values of the uncertain parameters of systems (26) and (27) are chosen as $a = 35$, $b = 3$, $c = 12$, $d = 7$, $r = 0.08$, $a_1 = 10$, $b_1 = 28$, $c_1 = 8/3$ and $d_1 = 1.3$. In addition, we choose the scaling function matrix as $h(x) = diag\{h_1(x_1), h_2(x_2), h_3(x_3), h_4(x_4)\}$, where

$$\begin{aligned} h_1(x_1) &= 2\sin(x_1 + 1) + 1, \\ h_2(x_2) &= 3\sin(-2x_2 + 3) + 1, \\ h_3(x_3) &= -\sin(3x_3 - 1) + 1, \\ h_4(x_4) &= -2\sin(-x_4 + 4) + 1, \end{aligned} \tag{28}$$

Then, one can find that $J = d(h(x)x)/dx = diag\{J_1, J_2, J_3, J_4\}$, where

$$\begin{aligned} J_1 &= 2\cos(x_1+1)x_1 + 2\sin(x_1+1) + 1, \\ J_2 &= -6\cos(2x_2-3)x_2 - 3\sin(2x_2-3) + 1, \\ J_3 &= -3\cos(3x_3-1)x_3 - \sin(3x_3-1) + 1, \\ J_4 &= 2\cos(x_4-4)x_4 + 2\sin(x_4-4) + 1, \end{aligned} \tag{29}$$

According to the theoretical analysis, the controller and parameter update laws presented in (23) have the following form:

$$\begin{aligned} u_1 &= J_1x_4 - y_4 + \hat{a}_1(y_1-y_2) - \hat{a}J_1(x_1-x_2) + \sum_{i=1}^{4} k_{1i}e_i, \\ u_2 &= y_2 + y_1y_3 - J_2x_1x_3 - \hat{b}_1y_1 + \hat{d}J_2x_1 + \hat{c}J_2x_2 + \sum_{i=1}^{4} k_{2i}e_i, \\ u_3 &= J_3x_1x_2 - y_1y_2 + \hat{c}_1y_3 - \hat{b}J_3x_3 + \sum_{i=1}^{4} k_{3i}e_i, \\ u_4 &= y_1y_3 + J_4x_2x_3 - \hat{d}_1y_4 + \hat{r}J_4x_4 + \sum_{i=1}^{4} k_{4i}e_i, \end{aligned} \tag{30}$$

$$\begin{aligned} \dot{\hat{a}} &= (x_1-x_2)J_1e_1 + (a-\hat{a}), \\ \dot{\hat{b}} &= x_3J_3e_3 + (b-\hat{b}), \\ \dot{\hat{c}} &= -x_2J_2e_2 + (c-\hat{c}), \\ \dot{\hat{d}} &= -x_1J_2e_2 + (d-\hat{d}), \\ \dot{\hat{r}} &= -x_4J_4e_4 + (r-\hat{r}), \end{aligned} \tag{31}$$

$$\begin{aligned} \dot{\hat{a}}_1 &= (y_2-y_1)e_1 + (a_1-\hat{a}_1), \\ \dot{\hat{b}}_1 &= y_1e_2 + (b_1-\hat{b}_1), \\ \dot{\hat{c}}_1 &= -y_3e_3 + (c_1-\hat{c}_1), \\ \dot{\hat{d}}_1 &= y_4e_4 + (d_1-\hat{d}_1), \end{aligned} \tag{32}$$

where $k_{ij}$ $(i = 1,2,3,4; j = 1,2,3,4)$ are real constants that compose the matrix $K$; $e_i = (y_i - h_ix_i)$, $i = 1,2,3,4$; $\hat{a}$, $\hat{b}$, $\hat{c}$, $\hat{d}$, $\hat{r}$, $\hat{a}_1$, $\hat{b}_1$, $\hat{c}_1$ and $\hat{d}_1$ are estimated values of the unknown parameters in systems (26) and (27). From the analytical result given in [38], GFPS between systems (26) and (27) will be achieved if the $K$ matrix is chosen as

$$K = \begin{bmatrix} -1 & 1 & 0 & 0 \\ -1 & -1 & 1 & 0 \\ 0 & -1 & -1 & 1 \\ 0 & 0 & -1 & -1 \end{bmatrix}, \tag{33}$$

which satisfies condition (24). However, the GFPS conditions obtained in this paper for systems (26) and (27) only require that all eigenvalues of $K$ matrix have negative real parts. Numerical simulations for systems (26), (27) are carried out below by selecting

$$K = \begin{bmatrix} -1 & 1 & 0 & 0 \\ -2 & -1 & 1 & 0 \\ 0 & -1 & -1 & 1 \\ 0 & 0 & -1 & -1 \end{bmatrix}. \tag{34}$$

**Remark 2.** *The $K$ matrix (34) is not antisymmetric, which does not satisfy the first condition in (24).*

The four eigenvalues of $K$ matrix (34) are $-1 \pm 1.85i$ and $-1 \pm 0.77i$. The numerical results are shown in Figures 1 and 2, in which the initial values of the systems (26) and (27) are taken as $x_1(0) = 3$, $x_2(0) = 5$, $x_3(0) = 7$, $x_4(0) = 8$, $y_1(0) = -12$, $y_2(0) = 2$, $y_3(0) = 3$, $y_4(0) = -5$ and the estimated parameters have initial conditions $\hat{a}(0) = 4$, $\hat{b}(0) = -6$, $\hat{c}(0) = -2$, $\hat{d}(0) = 1$, $\hat{r}(0) = -5$, $\hat{a}_1(0) = 1$, $\hat{b}_1(0) = -3$, $\hat{c}_1(0) = 0.3$ and $\hat{d}_1(0) = 0.1$. Although $K$ matrix (34) does not satisfy condition (24), Figure 1 shows that GFPS between systems (26) and (27) is achieved with the scaling function factors (28). Figure 2 illustrates that the estimated parameters $\hat{a}$, $\hat{b}$, $\hat{c}$, $\hat{d}$, $\hat{r}$, $\hat{a}_1$, $\hat{b}_1$, $\hat{c}_1$, $\hat{d}_1$ in systems (26) and (27) adapt themselves to the true values $a = 35$, $b = 3$, $c = 12$, $d = 7$, $r = 0.08$, $a_1 = 10$, $b_1 = 28$, $c_1 = 8/3$ and $d_1 = 1.3$, respectively. The numerical results in Figures 1 and 2 demonstrate that GFPS between systems (26) and (27) can be achieved with the scaling function factors (28) and matrix (34), and the parameter estimate errors converge to zero.

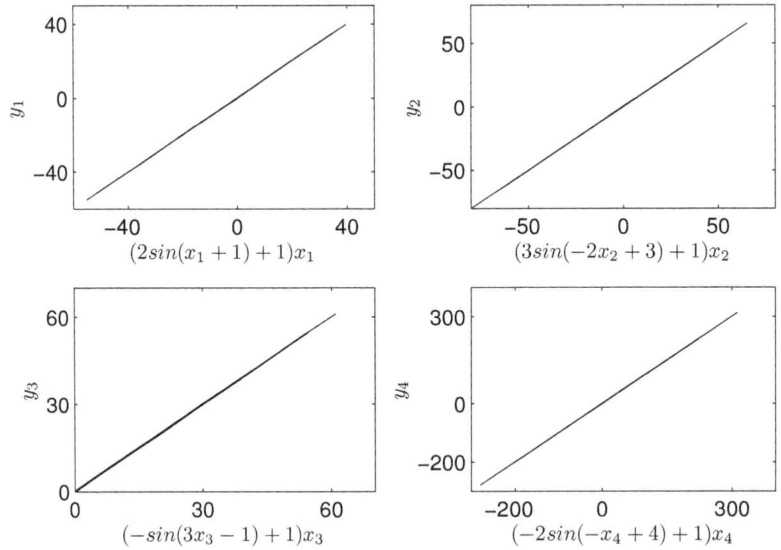


**Figure 1.** GFPS between systems (26) and (27) is achieved with the scaling function matrix $h(x) = diag\{2sin(x_1 + 1) + 1, 3sin(-2x_2 + 3) + 1, -sin(3x_3 - 1) + 1, -2sin(-x_4 + 4) + 1\}$ and $K$ matrix (34).

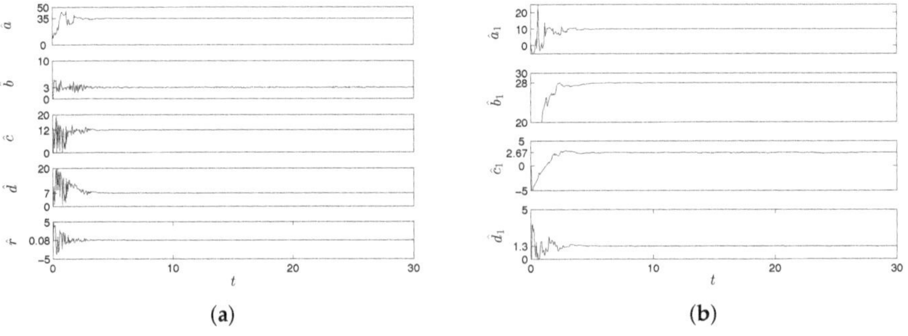


**Figure 2.** The estimates of the uncertain parameters of systems (26) and (27) for GFPS with the scaling function matrix $h(x) = diag\{2sin(x_1 + 1) + 1, 3sin(-2x_2 + 3) + 1, -sin(3x_3 - 1) + 1, -2sin(-x_4 + 4) + 1\}$ and $K$ matrix (34). (**a**) uncertain parameters in system (26) (**b**) uncertain parameters in system (27).

**Remark 3.** *The GFPS condition derived using the Laplace transform method is more generalized than that proposed in [38], which is obtained using the Lyapunov method.*

To further verify the theory's parts in more detail, another $K$ matrix,

$$K = \begin{bmatrix} -1 & 2 & 3 & 4 \\ -2 & 1 & 2 & 3 \\ 2 & -3 & -4 & -4 \\ -2 & -3 & -4 & -4 \end{bmatrix}, \tag{35}$$

is chosen to perform the numerical simulations for systems (26) and (27) with scaling function factors (28), in which all the initial conditions remain the same.

**Remark 4.** *For matrix (35), neither of the two conditions in (24) is satisfied.*

Matrix (35) has four eigenvalues with negative real parts, $-3.4925 \pm 0.6286i$, $-0.5075 \pm 0.2452i$, which satisfies the condition given in Theorem 1 in Section 3. The numerical results are presented in Figures 3 and 4. Figure 3 illustrates that GFPS between systems (26) and (27) occurs with the scaling function factors (28). Figure 4 shows that the estimated parameters $\hat{a}$, $\hat{b}$, $\hat{c}$, $\hat{d}$, $\hat{r}$, $\hat{a}_1$, $\hat{b}_1$, $\hat{c}_1$, $\hat{d}_1$ in systems (26) and (27) also can converge to their true values $a = 35$, $b = 3$, $c = 12$, $d = 7$, $r = 0.08$, $a_1 = 10$, $b_1 = 28$, $c_1 = 8/3$ and $d_1 = 1.3$, respectively. The numerical results in Figures 1–4 demonstrate that GFPS between systems (26) and (27) can be achieved with the scaling function factors (28), and the parameter estimate errors converge to zero as long as all eigenvalues of $K$ matrix have negative real parts, which verifies the correctness of our theoretical results in Section 3.

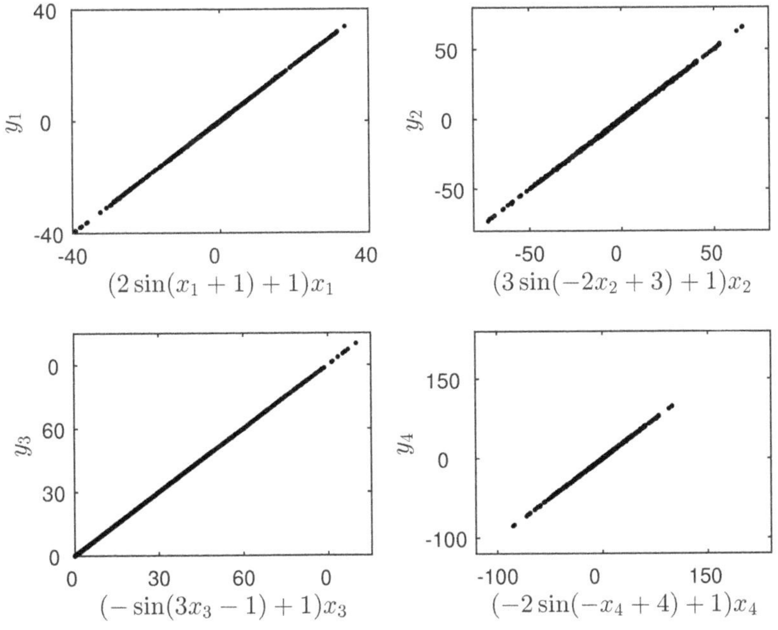


**Figure 3.** GFPS between systems (26) and (27) is achieved with the scaling function matrix $h(x) = diag\{2sin(x_1+1)+1, 3sin(-2x_2+3)+1, -sin(3x_3-1)+1, -2sin(-x_4+4)+1\}$ and $K$ matrix (35).

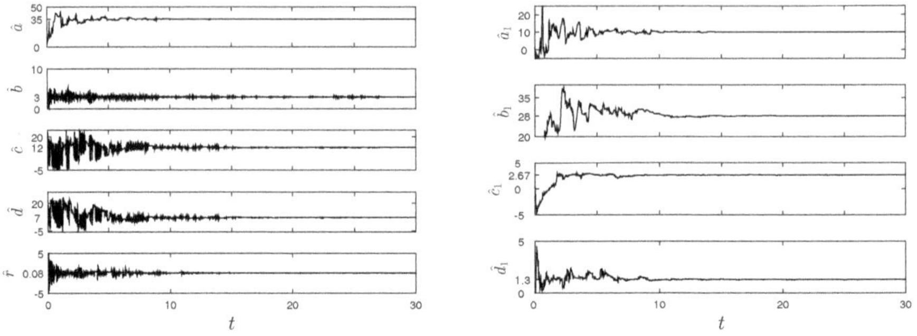


(**a**) (**b**)

**Figure 4.** The estimates of the uncertain parameters of systems (26) and (27) for GFPS with the scaling function matrix $h(x) = diag\{2sin(x_1+1)+1, 3sin(-2x_2+3)+1, -sin(3x_3-1)+1, -2sin(-x_4+4)+1\}$ and $K$ matrix (35). (**a**) uncertain parameters in system (26) (**b**) uncertain parameters in system (27).

## 5. The Design of Simple Parameter Update Laws

According to the GFPS condition given in Theorem 1 in Section 3, simple parameter update laws can be designed for the estimation of the uncertain parameters in system (1). From $M$ matrix defined in Equation (13), $g_1(\hat{\phi}-\phi, e_1)$ and $g_2(\hat{\psi}-\psi, e_1)$ in Equation (3) can be selected so that the eigenvalues of matrix $M$ are independent of $\phi$ and $\psi$. Consider that $g_1(\hat{\phi}-\phi, e_1)$ and $g_2(\hat{\psi}-\psi, e_1)$ in Equation (3) have the following form:

$$\begin{aligned} g_1(\hat{\phi}-\phi, \delta_1) &= \gamma_{11}(\hat{\phi}-\phi) + \gamma_{12}\Gamma_1(\delta_1), \\ g_2(\hat{\psi}-\psi, e_1) &= \gamma_{21}(\hat{\psi}-\psi) + \gamma_{22}\Gamma_2(\delta_1), \end{aligned} \tag{36}$$

where $\gamma_{11}, \gamma_{12}, \gamma_{21}, \gamma_{22}$ are control parameters; $\Gamma_{1,2}(\delta_1)$ are the functions of $\delta_1$, $\left.\frac{\partial \Gamma_{1,2}(\delta_1)}{\partial \delta_1}\right|_{\delta_1=0} = 0$. With parameter update laws (36), matrix $M$ in Equation (13) becomes

$$M = \begin{bmatrix} kI_n & 0 & 0 \\ 0 & \gamma_{11}I_m & 0 \\ 0 & 0 & \gamma_{21}I_l \end{bmatrix}, \tag{37}$$

where $I_n$, $I_m$ and $I_l$ represent $n \times n$, $m \times m$ and $l \times l$ real identity matrices, respectively. All eigenvalues of matrix (37) have negative real parts as long as matrices $k$, $\gamma_{11}$ and $\gamma_{21}$ have no eigenvalues with non-negative real parts. For the sake of simplicity, one can choose all $k$, $\gamma_{11}$ and $\gamma_{21}$ to be scalars. Therefore, GFPS in system (1) can be achieved by the controller and parameter update laws presented in Equations (3) and (36) if $k, \gamma_{11}, \gamma_{21} < 0$.

To illustrate the effectiveness of the parameter update laws (36), drive–response systems constructed with the Lorenz and Chen systems are selected to perform numerical simulations. The Lorenz chaotic system [49], as the drive system, can be described by the following equations:

$$\begin{aligned} \dot{x}_1 &= \sigma(x_2 - x_1), \\ \dot{x}_2 &= rx_1 - x_2 - x_1x_3, \\ \dot{x}_3 &= x_1x_2 - \beta x_3, \end{aligned} \tag{38}$$

where $x_1, x_2, x_3$ are state variables, and $\sigma, r, \beta$ are uncertain parameters to be identified. In particular, when $\sigma = 10$, $r = 28$ and $\beta = 8/3$, system (38) displays a chaotic attractor.

The Chen chaotic system [50], as the response system, is given as below:

$$\begin{aligned}\dot{y}_1 &= a(y_2 - y_1) + u_1,\\ \dot{y}_2 &= (c-a)y_1 - y_1y_3 + cy_2 + u_2,\\ \dot{y}_3 &= y_1y_2 - by_3 + u_3,\end{aligned} \tag{39}$$

where $y_1, y_2, y_3$ are state variables, $a, b, c$ are uncertain parameters to be estimated and $u_1, u_2, u_3$ are the control laws to be designed. System (39) exhibits chaotic dynamics when $a = 35$, $b = 3$ and $c = 28$. The scaling function matrix between systems (38) and (39) is chosen as $h(x) = diag\{2sin(x_1+1)+1, 3sin(-2x_2+3)+1, -sin(3x_3-1)+1)\}$, and one has

$$\begin{aligned}J = \frac{d(h(x)x)}{dx} = diag\{&2\cos(x_1+1)x_1 + 2\sin(x_1+1)+1,\\ &-6\cos(2x_2-3)x_2 - 3\sin(2x_2-3)+1,\\ &-3\cos(3x_3-1)x_3 - \sin(3x_3-1)+1\}.\end{aligned}$$

According to Equations (3) and (36), the controller $u_1, u_2, u_3$ and the parameter update laws can be given by

$$\begin{aligned}u_1 =& (2\cos(x_1+1)x_1 + 2\sin(x_1+1)+1)\hat{\sigma}(x_2 - x_1)\\ &+ k(y_1 - (2\sin(x_1+1)+1)x_1) + \hat{a}(y_1 - y_2),\\ u_2 =& (-6\cos(2x_2-3)x_2 - 3\sin(2x_2-3)+1)(\hat{r}x_1 - x_2 - x_1x_3)\\ &+ k(y_2 - (3\sin(-2x_2+3)+1)x_2) - (\hat{c}-\hat{a})y_1 + y_1y_3 - \hat{c}y_2,\\ u_3 =& (-3\cos(3x_3-1)x_3 - \sin(3x_3-1)+1)(x_1x_2 - \hat{\beta}x_3)\\ &+ k(y_3 - (-\sin(3x_3-1)+1)x_3) - y_1y_2 + \hat{b}y_3,\end{aligned} \tag{40}$$

$$\begin{aligned}\dot{\hat{\sigma}} &= \gamma_{11}(\hat{\sigma} - \sigma) + \gamma_{12}(1 - cos(y_1 - (2sin(x_1+1)+1)x_1)),\\ \dot{\hat{r}} &= \gamma_{11}(\hat{r} - r) + \gamma_{12}(1 - cos(y_2 - (3sin(-2x_2+3)+1)x_2)),\\ \dot{\hat{\beta}} &= \gamma_{11}(\hat{\beta} - \beta) + \gamma_{12}(1 - cos(y_3 - (-sin(3x_3-1)+1)x_3)),\end{aligned} \tag{41}$$

$$\begin{aligned}\dot{\hat{a}} &= \gamma_{21}(\hat{a} - a) + \gamma_{22}(1 - cos(y_1 - (2sin(x_1+1)+1)x_1)),\\ \dot{\hat{b}} &= \gamma_{21}(\hat{b} - b) + \gamma_{22}(1 - cos(y_2 - (3sin(-2x_2+3)+1)x_2)),\\ \dot{\hat{c}} &= \gamma_{21}(\hat{c} - c) + \gamma_{22}(1 - cos(y_3 - (-sin(3x_3-1)+1)x_3)),\end{aligned} \tag{42}$$

where $\hat{\sigma}, \hat{r}, \hat{\beta}, \hat{a}, \hat{b}$ and $\hat{c}$ are estimated values of the uncertain parameters in systems (38) and (39). The true values of the uncertain parameters of systems (38) and (39) are chosen as $\sigma = 10$, $r = 28$, $\beta = 8/3$ and $a = 35$, $b = 3$, $c = 28$. According to the previous analysis results in the section, if $k, \gamma_{11}, \gamma_{21}, \gamma_{12}, \gamma_{22}$ are chosen as $k = \gamma_{11} = \gamma_{21} = -1$ and $\gamma_{12} = \gamma_{22} = 0.01$, GFPS between uncertain Lorenz and Chen chaotic systems will be achieved under the control of the controller (40), and the uncertain parameters will be estimated using the parameter update laws (41) and (42). The initial conditions are given by $x_1(0) = 0.1$, $x_2(0) = 0.2$, $x_3(0) = 0.3$, $y_1(0) = 0.4$, $y_2(0) = 0.5$, $y_3(0) = 0.6$, $\hat{\sigma} = 9$, $\hat{r} = 25$, $\hat{\beta} = 3$, $\hat{a} = 33$, $\hat{b} = 2$ and $\hat{c} = 29$ to perform the numerical simulations. As shown in Figures 5 and 6, GFPS between systems (38) and (39) is achieved, and the uncertain parameters are also identified.

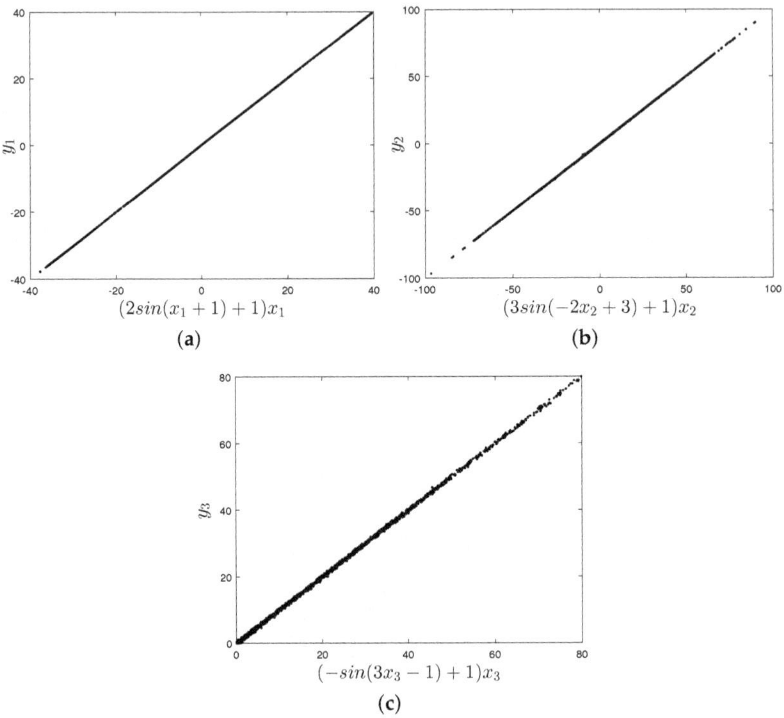


**Figure 5.** GFPS between systems (38) and (39) is achieved with the scaling function matrix $h(x) = diag\{2sin(x_1+1)+1, 3sin(-2x_2+3)+1, -sin(3x_3-1)+1)\}$. (**a**) $(2sin(x_1+1)+1)x_1$ vs. $y_1$ (**b**) $(3sin(-2x_2+3)+1)x_2$ vs. $y_2$ (**c**) $(-sin(3x_3-1)+1))x_3$ vs. $y_3$.

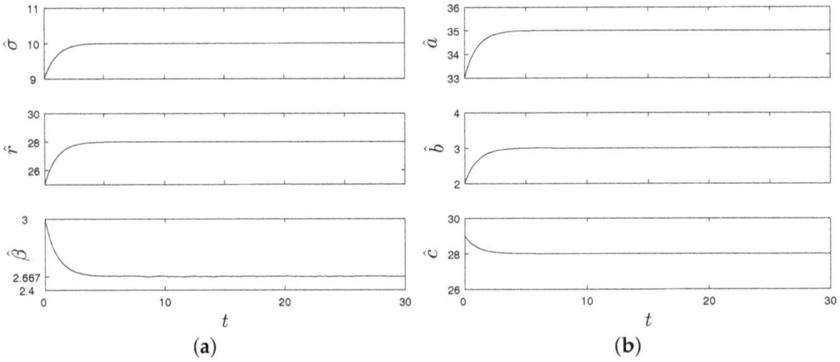


**Figure 6.** The estimates of the uncertain parameters of systems (38) and (39) for GFPS with the scaling function matrix $h(x) = diag\{2sin(x_1+1)+1, 3sin(-2x_2+3)+1, -sin(3x_3-1)+1)\}$. (**a**) uncertain parameters in system (38) (**b**) uncertain parameters in system (39).

If the scaling function matrix is taken as $h(x) = diag\{0.02x_1^2 + 0.2x_1 + 1, -0.01x_2^2 + 0.1x_2 - 1, 0.03x_3^2 - 0.1x_3 + 0.5\}$ and all other parameter conditions in systems (38) and (39) remain the same, GFPS between systems (38) and (39) still can be achieved and the uncertain parameters are also identified under the controller and parameter update laws given by Equations (3) and (36), respectively. The simulation results are shown in Figures 7 and 8, which illustrate the validity of the parameter update laws (36) proposed in the section. It is worth pointing out that the parameter update laws (36) are still valid even if the scaling function factors are other complex functions.

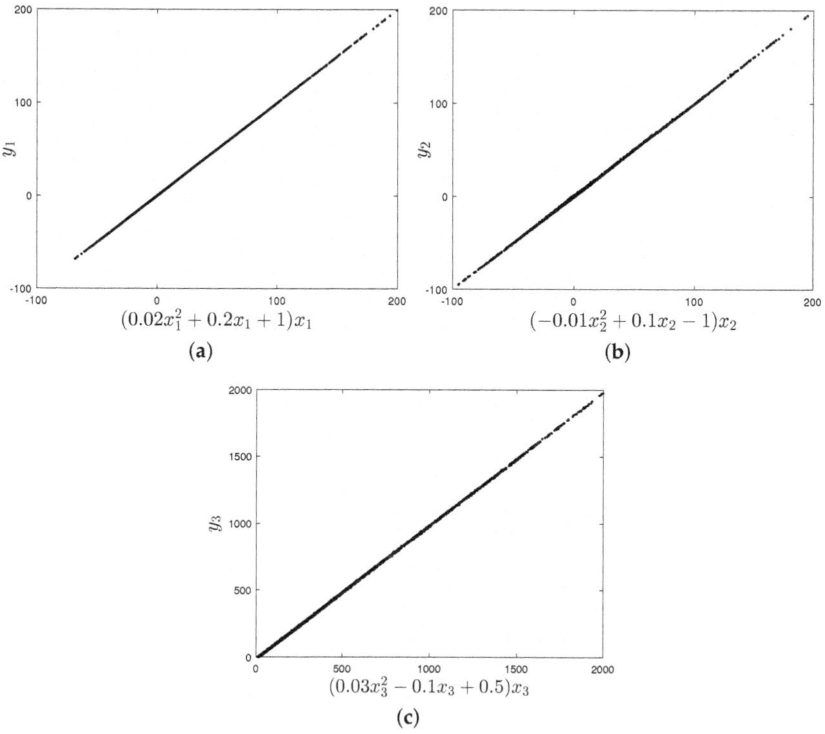

Figure 7. GFPS between systems (38) and (39) is achieved with the scaling function matrix $h(x) = diag\{0.02x_1^2 + 0.2x_1 + 1, -0.01x_2^2 + 0.1x_2 - 1, 0.03x_3^2 - 0.1x_3 + 0.5)\}$. (a) $(0.02x_1^2 + 0.2x_1 + 1)x_1$ vs. $y_1$ (b) $(-0.01x_2^2 + 0.1x_2 - 1)x_2$ vs. $y_2$ (c) $(0.03x_3^2 - 0.1x_3 + 0.5)x_3$ vs. $y_3$.

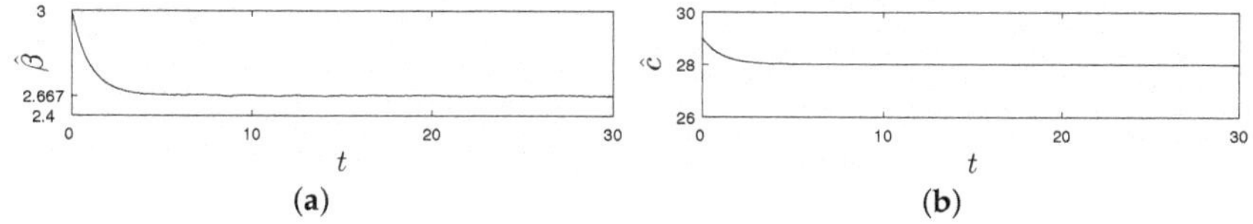

Figure 8. The estimates of the uncertain parameters of systems (38) and (39) for GFPS with the scaling function matrix $h(x) = diag\{0.02x_1^2 + 0.2x_1 + 1, -0.01x_2^2 + 0.1x_2 - 1, 0.03x_3^2 - 0.1x_3 + 0.5\}$. (a) uncertain parameters in system (38) (b) uncertain parameters in system (39).

## 6. Conclusions

In this paper, an adaptive controller is proposed to realize GFPS between two different chaotic systems (drive–response systems) with uncertain parameters based on the Laplace transform method. Compared to previous research, first, the uncertain chaotic systems considered in this paper are more generalized and have fewer restrictions on uncertain parameters; second, the GFPS scheme proposed in this paper is more generalized, and it can be used even when the scaling function factors are periodic functions, polynomial functions or other complex functions. Third, the parameter update laws designed for the estimation of uncertain parameters based on the GFPS condition derived using the Laplace transform method are simpler.

The greatest advantage of the GFPS scheme based on the Laplace transform method is that it can be used to design simple controllers and parameter update laws without constructing Lyapunov functions. The added controllers designed by this approach are more generalized and easier to realize physically than those designed through the Lyapunov function method.

It should be noted that we only consider the local stability of the origin in the GFPS error system between two chaotic systems in this paper. Moreover, to obtain the GFPS conditions, we require the smoothness of the trajectories in two chaotic systems to guarantee the existence, uniqueness and continuousness of solutions in a class of Volterra integral equations. According to the works in the literature [51,52], the global stability of the GFPS error system between two non-smooth chaotic systems is worthy of discussion, and this will be one of our next research points.

**Author Contributions:** Conceptualization, B.Z.; methodology, B.Z. and Y.Z.; software, Y.Z.; validation, Y.Z.; writing—original draft preparation, B.Z. and Y.Z.; writing—review and editing, B.Z.; funding acquisition, B.Z. All authors have read and agreed to the published version of the manuscript.

**Funding:** This research was funded by the National Natural Science Foundation of China, grant number 11672185.

**Institutional Review Board Statement:** Not applicable.

**Informed Consent Statement:** Not applicable.

**Data Availability Statement:** The study did not report any data.

**Acknowledgments:** The authors greatly appreciate the anonymous referees for their valuable suggestions and questions.

**Conflicts of Interest:** The authors declare no conflict of interest.

## References

1. Pecora, L.M.; Carroll, T.L. Synchronization in chaotic systems. *Phys. Rev. Lett.* **1990**, *64*, 821–824. [CrossRef] [PubMed]
2. Carroll, T.L.; Pecora, L.M. Synchronizing chaotic circuits. *IEEE Trans. Circ. Syst. I* **1991**, *38*, 453–456. [CrossRef]
3. Chen, G.; Dong, X. *From Chaos to Order: Methodologies, Perspectives and Applications*; World Scientific: Singapore, 1998;
4. Klug, F. Synchronization transitions in supply chain networks. *Int. J. Syst.-Sci.-Oper. Logist.* **2023**, *10*, 2224104.
5. Zhong, X.H.; Jia, Z.W.; Li, Q.T.; Wang, L.S.; Guo, Y.Y.; Sun, Y.H.; Wang, Y.C.; Wang, A.B. Multi-wavelength broadband chaos generation and synchronization using long-cavity FP lasers. *IEEE J. Sel. Top. Quantum Electron.* **2023**, *29*, 0600207. [CrossRef]
6. Khezeli, M.; Najafi, E.; Molana, M.H.; Seidi, M. A sustainable and resilient supply chain (RS-SCM) by using synchronisation and load-sharing approach: application in the oil and gas refinery. *Int. J. Syst.-Sci.-Oper. Logist.* **2023**, *10*, 2198055. [CrossRef]
7. Li, J.; Li, H.L.; Yang, J.P.; Yang, J.K.; Zhang, L. Hybrid control-based synchronization of fractional-order delayed complex-valued fuzzy neural networks. *Comput. Appl. Math.* **2023**, *42*, 154. [CrossRef]
8. Rosenblum, M.G.; Pikovsky, A.S.; Kurths, J. Phase synchronization of chaotic oscillators. *Phys. Rev. Lett.* **1996**, *76*, 1804–1807. [CrossRef]
9. Haerter, P.; Viana, R.L. Synchronization of phase oscillators due to nonlocal coupling mediated by the slow diffusion of a substance. *Braz. J. Phys.* **2023**, *53*, 114. [CrossRef]
10. Cao, L.; Lai, Y. Antiphase synchronism in chaotic systems. *Phys. Rev. E* **1998**, *58*, 382–386. [CrossRef]
11. Thazhathethil, R.; Pallimanhiyil, A.S. In-phase and anti-phase bursting dynamics and synchronisation scenario in neural network by varying coupling phase. *J. Biol. Phys.* **2023**. [CrossRef]
12. Rosenblum, M.G.; Pikovsky, A.S.; Kurths, J. From phase to lag synchronization in coupled chaotic oscillators. *Phys. Rev. Lett.* **1997**, *78*, 4193–4296. [CrossRef]
13. Kumar, V.; Heiland, J.; Benner, P. Exponential lag synchronization of Cohen-Grossberg neural networks with discrete and distributed delays on time scales. *Neural Process. Lett.* **2023**. [CrossRef]

14. Kocarev, L.; Parlitz, U. Generalized synchronization, predictability, and equivalence of unidirectionally coupled dynamical systems. *Phys. Rev. Lett.* **1996**, *76*, 181–1819. [CrossRef] [PubMed]
15. Xing, Y.A.; Dong, W.J.; Zeng, J.; Guo, P.T.; Zhang, J.; Ding, Q. Study of generalized chaotic synchronization method incorporating error-feedback coefficients. *Entropy* **2023**, *25*, 818. [CrossRef]
16. Mainieri, R.; Rehacek, J. Projective synchronization in the threedimensional chaotic systems. *Phys. Rev. Lett.* **1999**, *82*, 3042–3045. [CrossRef]
17. Han, J.; Chen, G.C.; Wang, L.M.; Zhang, G.D.; Hu, J.H. Direct approach on fixed-time stabilization and projective synchronization of inertial neural networks with mixed delays. *Neurocomputing* **2023**, *535*, 97–106. [CrossRef]
18. Chee, C.Y.; Xu, D.L. Chaos-based M-nary digital communication technique using controlled projective synchronization. *IEEE Proc. Circuits Devices Syst.* **2006**, *153*, 357–360. [CrossRef]
19. Zhang, F.F.; Leng, S.; Li, Z.F.; Jiang, C.M. Time delay complex Chen chaotic system and secure communication scheme for wireless body area networks. *Entropy* **2020**, *22*, 1420. [CrossRef]
20. Baluni, S.; Yadav, V.K.; Das, S. Quasi projective synchronization of time varying delayed complex valued Cohen-Grossberg neural networks. *Inf. Sci.* **2022**, *612*, 231–240. [CrossRef]
21. Xu, D.L. Control of projective synchronization in chaotic systems. *Phys. Rev. E.* **2001**, *63*, 027201. [CrossRef]
22. Xin, B.G.; Wu, Z.H. Projective synchronization of chaotic discrete dynamical systems via linear state error feedback control. *Entropy* **2015**, *17*, 2677–2687. [CrossRef]
23. Wen, G.; Xu, D. Nonlinear observer control for full-state projective synchronization in chaotic continuous-time system. *Chaos Solitons Fractals* **2005**, *26*, 71–77. [CrossRef]
24. Yan, J.; Li, C. Generalized projective synchronization of a unified chaotic system. *Chaos Solitons Fractals* **2005**, *26*, 1119–1124. [CrossRef]
25. Zhou, Y.; Wang, H.; Liu, H. Generalized function projective synchronization of incommensurate fractional-order chaotic systems with inputs saturation. *Int. J. Fuzzy Syst.* **2019**, *21*, 823–836. [CrossRef]
26. Zhang, L.; Wang, Y.; Wang, Q.; Qiao, S.; Wang, F. Generalized projective synchronization for networks with one crucial node and different dimensional nodes via a single controller. *Asian J. Control* **2020**, *24*, 1471–1483. [CrossRef]
27. Li, X.; Zhao, X.; Lu, Y.; Xu, T.; Hou, P.; Cui, X. Generalized projection synchronization and generalized dislocation synchronization of fractional-order financial chaotic systems with different orders. In Proceedings of the International Symposium on Electrical, Electronics and Information Engineering, Soul, Republic of Korea, 19–21 February 2021; pp. 635–638.
28. Li, G.H. Modified projective synchronization of chaotic system. *Chaos Solitons Fractals* **2007**, *32*, 1786–1790. [CrossRef]
29. He, J.M.; Lei, T.F.; Chen, F.Q. Global matrix projective synchronization of delayed fractional-order neural networks. *Soft Comput.* **2023**, *27*, 8991–9000. [CrossRef]
30. Feng, C.F.; Tan, Y.R.; Wang, Y.H.; Yang, H.J. Active backstepping control of combined projective synchronization among different nonlinear systems. *Automatika* **2017**, *58*, 295–301. [CrossRef]
31. Chen, Y.; Li, X. Function projective synchronization between two identical chaotic systems. *Int. J. Mod. Phys. C* **2007**, *18*, 883–888. [CrossRef]
32. Chen, Y.; An, H.L. The function cascade synchronization approach with uncertain parameters or not for hyperchaotic systems. *Appl. Math. Comput.* **2008**, *197*, 96–110. [CrossRef]
33. El-Dessoky, M.M.; Alzahrani, E.O.; Almohammadi, N.A. Chaos control and function projective synchronization of noval chaotic dynamical system. *J. Comput. Anal. Appl.* **2019**, *26*, 1.
34. Bekiros, S.; Yao, Q.J.; Mou, J.; Alkhateeb, A.F.; Jahanshahi, H. Adaptive fixed-time robust control for function projective 332 synchronization of hyperchaotic economic systems with external perturbations. *Chaos Solitons Fractals* **2023**, *172*, 113609. [CrossRef]
35. Du, H.Y.; Zeng, Q.S.; Wang, C.H. Modified function projective synchronization of chaotic system. *Chaos Solitons Fractals* **2009**, *197*, 2399–2404. [CrossRef]
36. Srisuntorn, R.; Weera, W.; Botmart, T. Modified function projective synchronization of master-slave neural networks with mixed interval time-varying delays via intermittent feedback control. *AIMS Math.* **2022**, *7*, 18632–18661. [CrossRef]
37. Yu, Y.; Li, H.X. Adaptive generalized function projective synchronization of uncertain chaotic systems. *Nonlinear Anal. RWA* **2010**, *11*, 2456–2464. [CrossRef]
38. Li, Z.B.; Zhao, X.S. Generalized function projective synchronization of two different hyperchaotic systems with unknown parameters. *Nonlinear Anal. RWA* **2011**, *12*, 2607–2615. [CrossRef]
39. Wu, X.J.; Fu, Z.Y.; Kurths, J. A secure communication scheme based generalized function projective synchronization of a new 5D hyperchaotic system. *Phys. Scr.* **2015**, *90*, 045210. [CrossRef]
40. Al-Azzawi, S.F., Al-Talib, Z.S. Generalized function projective synchronization via nonlinear controller strategy. *J. Interdiscip. Math.* **2022**, *25*, 1753-1765. [CrossRef]
41. Zhou, X.B.; Xiong, L.L.; Cai, X.M. Adaptive Switched Generalized Function Projective Synchronization between Two Hyperchaotic Systems with Unknown Parameters. *Entropy* **2013**, *16*, 377–388. [CrossRef]
42. Sun, J.W.; Guo, J.C.; Zhang, X.C. Adaptive generalized hybrid function projective dislocated synchronization of new four-dimensional uncertain chaotic systems. *Appl. Math. Comput.* **2015**, *252*, 304–314. [CrossRef]

43. Hamel, S.; Boulkroune, A. A generalized function projective synchronization sheme for uncertain chaotic systems subject to input nonlinearities. *Int. J. Gen. Syst.* **2016**, *45*, 689–710. [CrossRef]
44. Li, Z.B.; Tang, J.S. Generalized binary function projective synchronization of chaotic systems with unknown parameters. *Optik* **2017**, *137*, 101–107. [CrossRef]
45. Li, Q.P.; Liu, S.Y.; Chen, Y.G. Finite-time adaptive modified function projective multi-lag generalized compound synchronization for multiple uncertain chaotic systems. *Int. J. Appl. Math. Comput. Sci.* **2018**, *28*, 613–624. [CrossRef]
46. Abarbanel, H.D.I.; Rulkov, N.F.; Sushchik, M.M. Gneralized synchronization of chaos: The auxiliary system approach. *Phys. Rev. E* **1996**, *53*, 4528–4535. [CrossRef]
47. Nohel, J.A. Some problems in nonlinear Volterra integral equations. *Bull. Am. Math. Soc.* **1962**, *68*, 323–330. [CrossRef]
48. Kyrchei, I. Cramer's rule for generalized inverse solutions. In *Advances in Linear Algebra Research*; Kyrchei, I., Ed.; Nova Science Publishers: Hauppauge, NY, USA, 2015, pp. 79–132.
49. Lorenz, E.N. Deterministic nonperiodic flow. *J. Atmos. Sci.***1963**, *20*, 130–141. [CrossRef]
50. Chen, G.; Ueta, T. Yet another chaotic attrator. *Int. J. Bifurc. Chaos* **1999**, *9*, 1465–1466. [CrossRef]
51. Zhang, T.W.; Xiong, L.L. Periodic motion for impulsive fractional functional differential equations with piecewise Caputo derivative. *Appl. Math. Lett.* **2020**, *101*, 106072. [CrossRef]
52. Liu, Y.T.Global exponential stability and synchronization of discrete-time fuzzy bidirectional associative memory neural networks via Mittag-Leffler difference approach. *Int. J. Fuzzy Syst.* **2023**, *25*, 1922–1934. [CrossRef]

applied sciences

MDPI

Article

# Effect of the Dynamic Froude–Krylov Force on Energy Extraction from a Point Absorber Wave Energy Converter with an Hourglass-Shaped Buoy

**Houssein Yassin [1,†], Tania Demonte Gonzalez [1,†], Gordon Parker [1,*,†] and David Wilson [2,†]**

1 Mechanical Engineering–Engineering Mechanics Department, Michigan Technological University, Houghton, MI 49931, USA
2 Sandia National Laboratories, Albuquerque, NM 87185, USA
* Correspondence: ggparker@mtu.edu; Tel.: +1-(906)-370-1341
† These authors contributed equally to this work.

**Abstract:** Point absorber wave energy converter (WEC) control strategies often require accurate models for maximum energy extraction. While linear models are suitable for small motions, the focus is on the nonlinear model of an hour-glass shaped buoy undergoing large vertical displacements. Closed-form expressions for the static and dynamic Froude–Krylov forces are developed. It is shown that, in general, the dynamic and static forces are of similar magnitude, which is not the case for a spherical buoy. While the dynamic force reduces the amplitude of the net buoy force, its shape predicts a larger buoy response than if neglected, causing the nonlinear terms to have an even more significant effect. An input-state feedback linearizing controller is developed to show how the nonlinear model can be used in a control law. A 2.5 m buoy example is simulated to illustrate the approach of tracking an arbitrary displacement reference. For the case considered, the extracted power is 30% larger when the nonlinear dynamic FK force is used in the control law. The hourglass buoy is also compared to a spherical buoy to illustrate differences in their response to regular waves and energy extraction when using the same control laws. A spherical buoy diameter of 7.5 m was required to obtain the same power output as a 5 m tall hourglass buoy. A power-force-amplitude (PFA) metric is introduced to compare energy extraction performance and power take-off requirements. The hourglass buoy's PFA was 13% larger than the spherical buoy implying that it can produce similar power but with less control effort.

**Keywords:** wave energy; nonlinear model; Froude–Krylov force; input-state feedback linearization; power-force-amplitude

## 1. Introduction

Sea waves are an attractive source of energy due to their relatively large energy density as compared to solar and wind resources [1]. This same high-energy sea environment conspires against the widespread use of wave energy converters (WECs) due to engineering challenges, including extreme event survivability and maintenance [2–5]. Extracting energy and transferring it to a load efficiently is a WEC's typical objective. Harmonizing its mechanical, power take-off (PTO), and control strategy designs may lead to the practical application of WECs for terrestrial and at-sea energy grids.

A point absorber WEC's PTO extracts energy from the relative velocity between its floating buoy and fixed base. By definition, maximum energy extraction is achieved by controlling its PTO force such that the integrated product of the force and relative velocity is maximized subject to the PTO's force and stroke limits. At times the PTO will do work on the buoy, momentarily sacrificing energy extraction to produce a buoy response leading to greater energy extraction later. Energy optimal control has been studied extensively for assumed buoy geometries, including quantifying upper bounds [1,6,7], impedance matching [7–11], and both closed-form [12] and numerical optimal control solutions [13] such as with model predictive control [14–16]. Early efforts focused on buoy response regimes that can be modeled using linear differential equations considering a buoy's nonlinear response a more recent focus [17–19].

While this paper investigates a particular buoy geometry, the hourglass shape of Wilson et al. [17,20,21], it is important to note that understanding mechanical energy extraction is only part of the more complex problem of energy production. Addressing this system-level objective requires simultaneous consideration of the entire process, including the design of the WEC, its PTO, and electrical energy transmission power electronics [22].

Since a point absorber's buoy shape and mass affects its dynamics response, the shape is a crucial design consideration and should be exploited. Garcia et al. [23,24] quantified the benefit of the simultaneous design of a point absorber's buoy geometry and control law. Adaptive shape modification is another method for expanding the performance of a WEC operating in off-nominal sea conditions [25]. The review article by Garcia and Forehand provides a comprehensive accounting of past and current efforts in shape optimization, clearly illustrating the importance of geometry for a broad range of WEC types.

While buoy shape affects a WEC's model form when operating in a linear regime, it can profoundly affect its nonlinear model and response. Computational fluid dynamics (CFD) models can capture these effects but are impractical for real-time, model-based control laws [26]. Closed-form nonlinear Froude–Krylov (FK) force models have been developed in [27–29] to incorporate buoy shape effects into model-based control laws for large-motion, nonlinear operating regimes. Giorgi et al. [27] present an analytical approach with 2% error to derive the nonlinear FK forces using Airy's wave theory for deep water waves to represent the static and dynamic pressure forces acting on the buoy.

Researchers have proposed various control strategies that incorporate the analytical nonlinear FK forces. The authors in [30,31] developed a sliding mode controller for a heaving point absorber that demonstrated the robustness and energy maximization of the WEC. In [32], an optimal model-based control was developed for a data-based modeled heaving point absorber. The latching control is a well-established technique that has also been used in WECs, including those that utilize the computationally efficient closed-form nonlinear FK forces [26,33,34]. All of these papers have used conventional buoy shapes such as spheres and cylinders, which have well-studied hydrodynamics. In this study, a state feedback linearization control scheme is proposed for an unconventional hourglass-shaped heaving point absorber.

An hourglass-shaped, point absorber buoy was considered in three recent articles by Wilson et al. [17,20,21] and is the motivation for this paper. In [20], the authors developed the hourglass-shaped buoy's nonlinear hydrostatic force term and compared its energy extraction performance to a cylindrical buoy. Using the results from [35], where a cubic term was inserted into the cylindrical buoy's control strategy, the authors recreated the effect naturally through the hourglass shape of the buoy. They showed that the hourglass buoy provided the same reactive power management as the nonlinear control of [35] using a velocity feedback control law. In [17], the hourglass buoy control was extended further using the Hamiltonian surface shaping and power flow control (HSSPFC) technique. The model for the hourglass buoy was the same as [20] where again, the dynamic Froude–Krylov force was not included. The energy extraction performance of the hourglass buoy was compared to a cylindrical buoy with an artificially applied cubic nonlinearity. A key result was that the hourglass buoy, containing the natural cubic nonlinearity, outperformed the cylindrical buoy with the artificial cubic term. In [21], a rate feedback control strategy was used for an hourglass-shaped buoy point absorber capable of adapting its cone angle based on incident wave conditions. Using an irregular wave example, it was shown that the energy extraction increased by 50% when the cone angle was varied. The model used for the simulation was the same as described previously.

The goal of this paper is to explore the effect of the nonlinear dynamic Froude–Krylov force, derived using the approach of Giorgi and Ringwood [27], on an hourglass-shaped buoy point absorber. We first examine the buoy's simulated response to regular waves with and without the dynamic FK force in the model. It is shown that neglecting this term can lead to underprediction of the buoy motion in some cases. Thus, it should be included in the analysis and model-based control system design. Adjusting the draft could potentially increase the energy extracted, as shown on a spherical buoy in [36]. However, for the hourglass-shaped buoy, varying the draft leads to additional terms in the static and dynamic Froude–Krylov force expressions. Therefore, in this study, the considered buoy's draft is at its vertical midpoint in all cases. A feedback linearizing controller is used to investigate further the effect of including the dynamic FK force term in the hourglass-shaped buoy's PTO control law. It should be noted that this control strategy is provided merely as an example of how the nonlinear model can be used in a model-based control strategy. A sinusoidal reference trajectory is used but is not presumed to be optimal from an energy extraction perspective. Finally, the closed-loop energy extraction performance of hourglass and spherical buoys are compared using the previously mentioned feedback linearizing controller, where the nonlinear models are used for each of their control laws. This is performed to illustrate the unique behavior of the hourglass shape to help explain why it extracts greater energy than a similarly sized spherical buoy.

This article has three contributions. First, the closed-form dynamic Froude–Krylov force expression for an hourglass-shaped buoy was developed. Second, we show that the dynamic Froude–Krylov force of an hourglass-shaped buoy is not negligible and should be considered during analysis and control system design. Third, the closed-loop energy extraction of an hourglass point absorber is greater than a similarly sized spherical device.

The remainder of this paper is organized as follows. Section 2 describes the model used by the simulation and the control law, including developing the hourglass buoy's FK forces. Section 3 shows the contributions of the static and dynamic FK forces to the buoy's free response. Section 4 develops the feedback linearization control law used for evaluating the performance of the hourglass and spherical buoys whose FK force model is described in Section 5. The results of a simulated case study are provided in Section 6 with concluding remarks in Section 7.

## 2. Hourglass Buoy Model

The buoy of mass $m$ and weight $F_g$ has the geometry shown in Figure 1 where its dynamic model of Equation (1) is used for simulation, assessing energy extraction performance, and by the model-based control law.

$$m\ddot{\zeta} = F_g + F_{fk,s} + F_{fk,d} + F_r + F_c \tag{1}$$

The buoy is constrained to heave with potentially large vertical displacement $\zeta$ defined relative to the still water line. This assumption simplifies the model considerably compared to general motion, and state space models such as used in [37]. The nonlinear static and dynamic Froude–Krylov forces, $F_{fk,s}$ and $F_{fk,d}$, are the main focus of this work and are developed in Section 2.1. The radiation force model, $F_r$, is described in Section 2.2, and the power take-off (PTO) device is assumed to be ideal and applies $F_c$ to the buoy according to the control law of Section 4.

It is assumed that the waves are regular with their elevation given by Equation (2). The waves are also assumed to be linear with steepness, $S_w = 2A/\lambda \leq 0.018$, and wavelength, $\lambda$, much greater than the buoy's maximum diameter or $\lambda \gg 2h\tan(\alpha)$ and are the justification for neglecting the diffraction forces [8].

$$\eta = A\cos(\omega t) \tag{2}$$

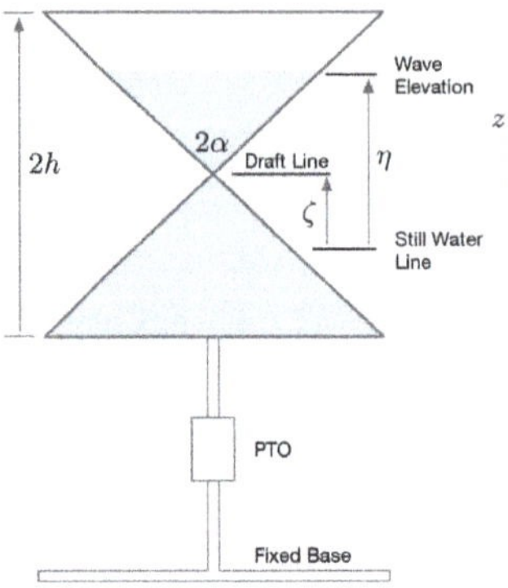


**Figure 1.** Hourglass buoy, notation, and coordinate system. Blue denotes the submerged volume.

*2.1. Froude–Krylov Force*

The Froude–Krylov (FK) force, $F_{fk}$ of Equation (3), is the integration of the pressure field forces, $\vec{P}(t)$, over the instantaneous wetted surface of the buoy, $S$ [27]

$$F_{fk} = -\int\int \vec{P}(t) \cdot \hat{n}\, dS \tag{3}$$

where $\hat{n}$ is the unit vector pointing outwards from the buoy's surface. The pressure will be approximated using Airy's Wave Theory shown in Equation (4)

$$P(t) = \rho g e^{\chi z} A \cos(\omega t) - \rho g z \tag{4}$$

where $\rho$ is the density of water, $g$ is the gravity acceleration constant, and $\chi$ is the wave number. The two terms of $\vec{P}(t)$, after substitution into Equation (3), give the dynamic and static FK forces used in Equation (1).

$$F_{fk} = F_{fk,d} + F_{fk,s} \tag{5}$$

The approach of Giorgi and Ringwood [27] is used to obtain closed-form expressions for the FK forces shown in Equation (6) where cylindrical coordinates are used when integrating Equation (3).

$$\begin{aligned} F_{fk,s} &= -\rho g \int_0^{2\pi} \int_{z_1}^{z_2} z f(z) f'(z) dz\, d\theta \\ F_{fk,d} &= -\rho g A \cos(\omega t) \int_0^{2\pi} \int_{z_1}^{z_2} e^{\chi z} f(z) f'(z) dz\, d\theta \end{aligned} \tag{6}$$

The revolution profile for the hourglass buoy geometry of Figure 1 is $f(z) = \overline{\alpha}(z+h)$ for the lower cone and $f(z) = \overline{\alpha}(z - \zeta + \eta)$ for the upper cone with $\overline{\alpha} = \tan(\alpha)$. The cylindrical coordinate rotation limits, $[0,\ 2\pi]$, and the vertical limits, $[z_1,\ z_2]$, are based on the instantaneous wetted surface with again with two parts: (1) the immersed portion of the lower cone with vertical limits $[-h,\ 0]$ and the immersed part of the upper cone with limits $[\zeta - \eta,\ 0]$. The resulting FK force terms are given in Equation (7) where the first term of $F_{fk,s}$ is the weight of the buoy, $F_g$.

$$\begin{aligned} F_{fk,s} &= \frac{1}{3}\pi \rho g \overline{\alpha}^2 \left[h^3 + (\eta - \zeta)^3\right] \\ F_{fk,d} &= \frac{2\pi}{\chi^2} \rho g \overline{\alpha}^2 \eta \left[2 - e^{-\chi h} - e^{-\chi(\eta - \zeta)} - \chi(h + \eta - \zeta)\right] \end{aligned} \tag{7}$$

Rewriting Equation (7) for substitution into Equation (1) yields Equation (8) where the first equation is the buoyancy force and is the same as developed in [17] using a geometric approach.

$$\begin{aligned} F_b = F_{fk,s} + F_g &= \frac{1}{3}\pi\rho g\overline{\alpha}^2(\eta - \zeta)^3 \\ F_{fk,d} &= \frac{2\pi}{\chi^2}\rho g\overline{\alpha}^2\eta\left[2 - e^{-\chi h} - e^{-\chi(\eta-\zeta)} - \chi(h + \eta - \zeta)\right] \end{aligned} \tag{8}$$

## 2.2. Radiation Force

Although not entirely consistent with the large motion assumption mentioned earlier, a linear radiation force model is used here to keep the focus on the effect of the FK forces. The radiation forces are split into two components, one related to velocity and the other to acceleration, shown in Equation (9) [30],

$$F_r = \int_0^\infty h_r(\tau)\dot{\zeta}(t - \tau)d\tau - m_a\ddot{\zeta} \tag{9}$$

where $h_r$ is the radiation impulse response function and $m_a$ is the added mass. However, in our analysis, we consider regular harmonic waves of Equation (2) where the integral of Equation (9) becomes Equation (10).

$$F_r = -b\dot{\zeta} - m_a\ddot{\zeta} \tag{10}$$

where $b$ is the linear damping constant specific to the regular wave frequency. Substituting Equation (10) and Equation (8) into Equation (1), and neglecting the diffraction force yields the nonlinear model we will use for the rest of the analysis.

$$M\ddot{\zeta} + b\dot{\zeta} - \frac{1}{3}\pi\rho g\overline{\alpha}^2(\eta - \zeta)^3 - \frac{2\pi}{\chi^2}\rho g\overline{\alpha}^2\eta\left[2 - e^{-\chi h} - e^{-\chi(\eta-\zeta)} - \chi(h + \eta - \zeta)\right] = F_c \tag{11}$$

where $M = m + m_a$.

A point absorber is often modeled using the Cummins equation shown in Equation (12) [8,38,39].

$$M\ddot{\zeta} + \int_0^\infty h_r(\tau)\dot{\zeta}(t - \tau)d\tau + k\zeta = f_f \tag{12}$$

where $f_f$ contains all external forces applied to the buoy, including diffraction, and those supplied by the power take-off. It is instructive to compare Equation (12) to the nonlinear model of Equation (11). The first two terms were introduced in Section 2.2. The third term of Equation (12), $k\zeta$, models the linear hydrostatic force and is valid for small motions about an equilibrium. This is replaced with a cubic term in Equation (11) appropriate for potentially large vertical displacement. The fourth term on the right of Equation (11) is the dynamic Froude–Krylov force and could be included in the $f_f$ of Equation (12) as it can be classified as an external force. Typically, the $f_f$ includes the linear diffraction force, which we have neglected as mentioned previously, and the PTO control force, $F_c$ of Equation (11).

# 3. Contribution of Froude–Krylov Forces to the Free Response

Before exploring the control system design of Section 4, it is helpful to examine the relative effects of the terms in Equation (11). Instead of a nondimensional assessment, we will use the example buoy used for the control system design and rewrite Equation (11) as

$$M\ddot{\zeta} + b\dot{\zeta} - F_b - F_{fk,d} = F_c \tag{13}$$

The buoy characteristics are given in Table 1, where $m_a$ and $b$ were obtained using WAMIT for the buoy geometry of Figure 1. The incident regular, linear wave has a period of 18 s and an amplitude of $A = 0.35$ m. The amplitude was selected to yield maximum buoy displacement such that it nearly leaves the water. While this is not physically possible for a free-floating buoy, it provides a limiting scenario. Assuming the water depth, $h_w$, is large such that $\chi h_w \gg 1$, the dispersion equation.

$$\omega^2 = g\chi \tanh(\chi h_w) \approx g\chi \tag{14}$$

gives $\chi \approx \frac{\omega^2}{g} = 0.0125$ with wavelength $\lambda = \frac{2\pi}{\chi} = 502.65$m.

**Table 1.** Hourglass buoy parameters.

| Feature | Symbol | Value | Units |
|---|---|---|---|
| Height | $2h$ | 5 | m |
| Cone Angle | $\alpha$ | 60 | degree |
| Draft | $h_0$ | 2.5 | m |
| Mass | $m$ | 50,376 | kg |
| Density | $\rho_b$ | 1000 | kg/m$^3$ |
| Added Mass | $m_a$ | 59,250 | kg |
| Damping | $b$ | 20,000 | N/(m/s) |

It should be noted that under some conditions, the dynamic FK force is small compared to $F_b$ and could be neglected. For example, for large wavelength, small wave amplitudes, small buoy displacements, and small buoy size such that $\chi\eta \ll 1$, $\chi\zeta \ll 1$ and, $\chi h \ll 1$, the Taylor series expansion of Equation (8) gives $F_{fk,d} \approx 0$.

A Simulink simulation using a fixed-step solver, ODE3, with a 0.01 s time step was created using Equation (13) where the dynamic FK force could be selectively included. The model parameters are provided in Table 1 where the hydrodynamic coefficients $m_a$ and $b$ were calculated using WAMIT's boundary-element-method solver [40]. The hourglass buoy geometry was discretized by a 221-panel mesh, and the solution was calculated over the angular frequency range of 0.1–12 rad/s with 0.01 rad/s increments. The equilibrium configuration was with the buoy submerged to its halfway point. To check convergence, the mesh was increased to 349 panels yielding nearly identical results as the 221-panel model. Figure 2 compares the three noninertial forces of Equation (11), as well as their sum, for the regular wave described above with $F_c = 0$. The buoyancy and dynamic FK forces, $F_b$ and $F_{fk,d}$, are of similar amplitude but phase shifted nearly 180°. The radiation damping force, $b\dot{\zeta}$, is small in comparison. The force sum is also shown and illustrates an interesting result arising from the combination of $F_b$ and $F_{fk,d}$. Namely, the force sum changes sign often, which is not the case if $F_{fk,d}$ is omitted. This has a dramatic effect on the buoy motion.

To further investigate the effect of the combination of $F_b$ and $F_{fk,d}$ a simulation was performed with $F_{fk,d} = 0$ with the results shown in Figure 3. For these conditions, the buoy behaves as a wave-follower with very little vertical motion and thus very low $F_b$, about 15 kN compared to the 600 kN of Figure 2. When $F_{fk,d}$ is omitted, the buoy response cannot fully excite its buoyancy force's nonlinearity. In contrast, when $F_{fk,d}$ is included, the buoy motion is excited, effectively turning on the nonlinearity that leads to even greater motion.

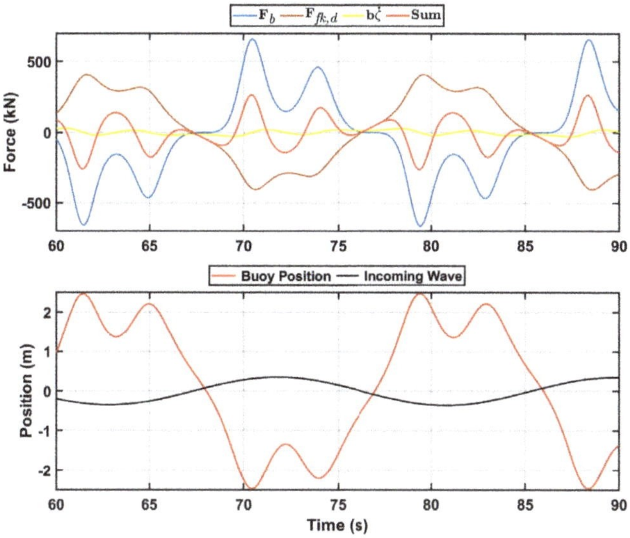


**Figure 2.** Simulated steady state results comparing the three noninertial terms of Equation (11): $F_b = F_{fk,s} + F_g$, $F_{fk,d}$, $b\dot{\zeta}$ and their sum (upper plot) and the buoy displacement and wave elevation (lower plot) for a deep water wave with $T = 18$ s and the buoy parameters shown in Table 1.

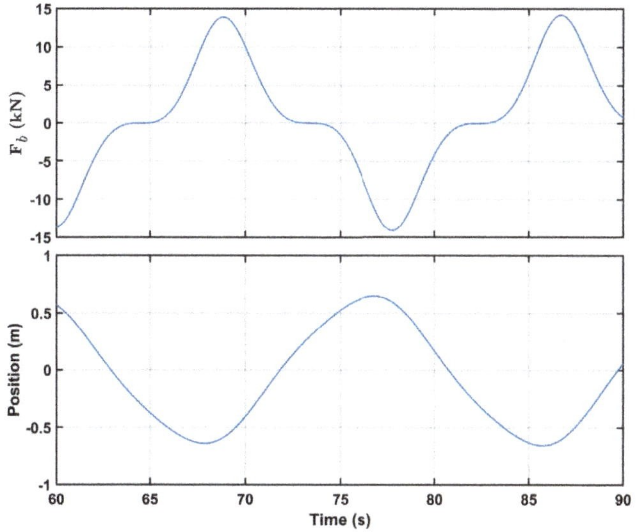


**Figure 3.** Simulated steady state results of a model without $F_{fk,d}$. The upper plot is $F_b = F_{fk,s} + F_g$ force and the lower plot is the buoy displacement.

## 4. Control Design

A general control architecture for extracting energy is shown in Figure 4, where it is assumed the instantaneous displacement states, wave elevation $\eta$, and the reference displacement states, $\zeta_r$, $\dot{\zeta}_r$ and $\ddot{\zeta}_r$ are known. A variety of control laws could be used with this architecture, where we will consider feedback linearization with a sinusoidal reference to illustrate a model-based approach. It should be noted that this results in the cancellation of all the nonlinear terms in Equation (13). Such cancellation is avoided by including these terms in the reference command leading to a control solution that does not require a reference and relies on exploiting the buoy's nonlinear response to extract energy.

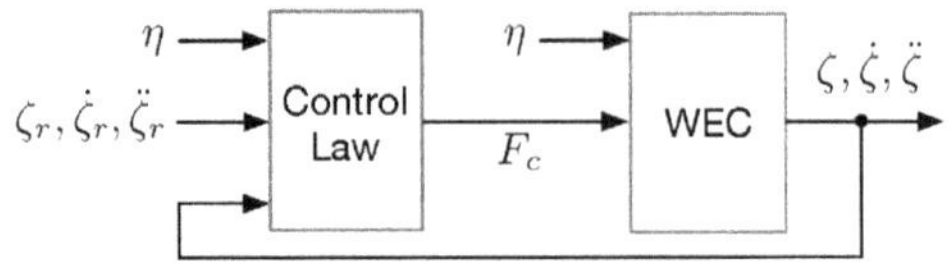


**Figure 4.** Control architecture illustrating its access to wave elevation, buoy states, and reference states.

The error states are defined by Equation (15)

$$\begin{aligned} e_1 &= \zeta - \zeta_r \\ e_2 &= \dot{\zeta} - \dot{\zeta}_r \end{aligned} \tag{15}$$

and then differentiated to form Equation (16).

$$\begin{aligned} \dot{e_1} &= e_2 \\ \dot{e_2} &= \ddot{\zeta} - \ddot{\zeta}_r \end{aligned} \tag{16}$$

Equation (11) can be solved for $\ddot{\zeta}$

$$\begin{aligned} \ddot{\zeta} = M^{-1}\Big(F_c - b\dot{\zeta} + \frac{1}{3}\pi\rho g \overline{\alpha}^2 (\eta - \zeta)^3 \\ + \frac{2\pi}{\chi^2}\rho g \overline{\alpha}^2 \eta \Big[2 - e^{-\chi h} - e^{-\chi(\eta - \zeta)} - \chi(h + \eta - \zeta)\Big]\Big) \end{aligned} \tag{17}$$

and substituted into $\dot{e}_2$ of Equation (16) to create Equation (18)

$$\dot{e_2} = \nu - \frac{b}{M} e_2 \tag{18}$$

where $\nu$ is:

$$\begin{aligned} \nu = M^{-1}\Big(F_c - b\dot{\zeta}_r + \frac{1}{3}\pi\rho g \overline{\alpha}^2 (\eta - \zeta)^3 \\ + \frac{2\pi}{\chi^2}\rho g \overline{\alpha}^2 \eta \Big[2 - e^{-\chi h} - e^{-\chi(\eta - \zeta)} - \chi(h + \eta - \zeta)\Big]\Big) - \ddot{\zeta}_r \end{aligned} \tag{19}$$

and the error state system of Equation (16) is written compactly in Equation (20).

$$\begin{aligned} \dot{e_1} &= e_2 \\ \dot{e_2} &= -\frac{b}{M} e_2 + \nu \end{aligned} \tag{20}$$

State feedback control

$$\nu = -\vec{k}^T \vec{e} \tag{21}$$

is applied to $\nu$ and the control law is Equation (22)

$$\begin{aligned} F_c = -M\vec{k}^T\vec{e} + M\ddot{\zeta}_r + b\dot{\zeta}_r - \frac{1}{3}\pi\rho g \overline{\alpha}^2 (\eta - \zeta)^3 \\ - \frac{2\pi}{\chi^2}\rho g \overline{\alpha}^2 \eta \Big[2 - e^{-\chi h} - e^{-\chi(\eta - \zeta)} - \chi(h + \eta - \zeta)\Big] \end{aligned} \tag{22}$$

which is simply Equation (23).

$$F_c = -M\vec{k}^T\vec{e} + M\ddot{\zeta}_r + b\dot{\zeta}_r - F_b - F_{fk,d} \tag{23}$$

A variety of methods can be used to create the gain vector $\vec{k}$, where the infinite horizon, linear quadratic regulator approach is used, which minimizes the cost function of Equation (24) [41]

$$J = \int_0^{\infty} (\vec{e}^T Q \vec{e} + v^T R v) dt \tag{24}$$

where in our case

$$Q = \begin{bmatrix} 10 & 0 \\ 0 & 1 \end{bmatrix} \quad \text{and} \quad R = 1 \tag{25}$$

yielding $\vec{k}^T = \begin{bmatrix} 3.16 & 2.53 \end{bmatrix}$. The effect of varying the $Q_{11}$ gain value, nominally set to 10, will be examined in Section 6.3.

The control law shown in the block diagram of Figure 4 can be summarized by Equation (23), the error vector of Equation (15), the hourglass buoy parameters of Table 1, the Froude–Krylov terms of Equation (8) and the gains of $\vec{k}$ above. It requires the reference trajectory and measurements of the wave elevation and the buoy displacement states.

## 5. Spherical Buoy Model

In Section 6, the energy extraction results of applying the control law of Section 4 to the hourglass buoy and the spherical buoy of [30] are discussed. A brief description of the spherical buoy model is provided here.

The spherical buoy model form is identical to that of the hourglass buoy of Equation (13) but with different expressions for the buoyancy and dynamic FK forces shown in Equation (26)).

$$\begin{aligned} F_b &= \frac{1}{3}\pi\rho g(\eta - \zeta)\left[3R^2 - (\eta - \zeta)^2\right] \\ F_{fk,d} &= \frac{2\pi}{\chi^2}\rho g\eta\left[1 - e^{-\chi(\eta+R-\zeta)} - \chi\left(\eta - \zeta + Re^{-\chi(\eta+R-\zeta)}\right)\right] \end{aligned} \tag{26}$$

The control law form is the same as given in Equation (23) with $F_b$ and $F_{fk,d}$ of Equation (26) and the spherical buoy parameters provided above. The LQR gains are the same as those used for the hourglass buoy.

## 6. Results

The hourglass and spherical buoys described above were modeled in Simulink and simulated using a fixed-step solver, ODE3, with a 0.01 s time step. The Simulink model is shown in Figure 5 where a MATLAB Function was used to implement the buoy dynamic model. The control law was organized into a subsystem. The incident regular wave elevation of Equation (2) was $A = 0.5$ m with a period of six seconds, $\omega = \frac{\pi}{3}$ matching the conditions of [30]. The reference displacement, $\zeta_r$ of Equation (27), used by the tracking control law, had the same period but with an amplitude of two meters shown in Equation (27).

$$\zeta_r = 2\sin\left(\frac{\pi}{3}t\right) \tag{27}$$

The absorbed energy, $E$, was calculated using Equation (28).

$$E = \int_0^t F_c \dot{\zeta}\, dt \tag{28}$$

Two closed-loop cases are explored below: (1) the effect of including the nonlinear FK force in the control law and (2) a comparison between the hourglass and spherical buoys.

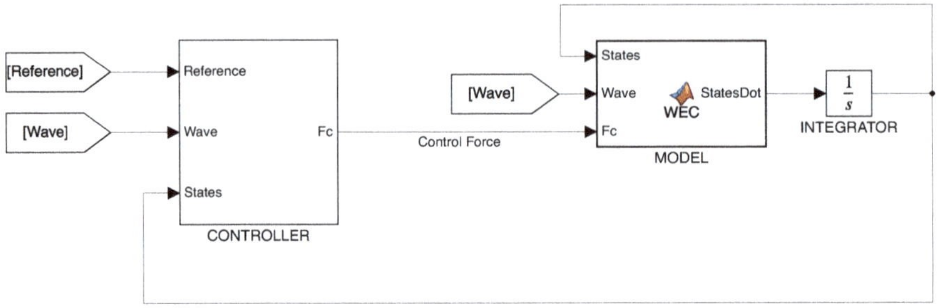


**Figure 5.** Simulink model of the closed loop system. The plant model for both the hourglass and spherical buoys was implemented using a MATLAB Function block and an integrator.

### 6.1. Effect of the Nonlinear FK Force

Calculating $F_{fk,d}$ in the control law requires knowledge of the wave number and hence more information about the incident waves. If $F_{fk,d}$ has negligible performance on the closed-loop energy extraction, then a model-based control strategy would be easier to implement.

Figure 6 compares the energy extraction performance of the hourglass buoy using the control law of Equation (23) with and without $F_{fk,d}$. The power is about 30% larger for the case where $F_{fk,d}$ is used in the control law. As expected, the closed-loop tracking performance suffered due to the model uncertainty, as illustrated in the last two plots of Figure 6. The last plot shows the error between the reference trajectory and the buoy displacement response using $e_1$ of Equation (15). For the case where the dynamic Froude–Krylov force is included in the control law, the error is nearly zero. In contrast, the error is about 0.8 m when the dynamic term is neglected by the control law. This same behavior is illustrated in the velocity response in the third plot. The PTO force increased slightly as the control law compensated for the model uncertainty through its error correction term. The position tracking error, shown in the last subplot, could be improved by increasing the $\vec{k}$ of Equation (21) but was not performed here.

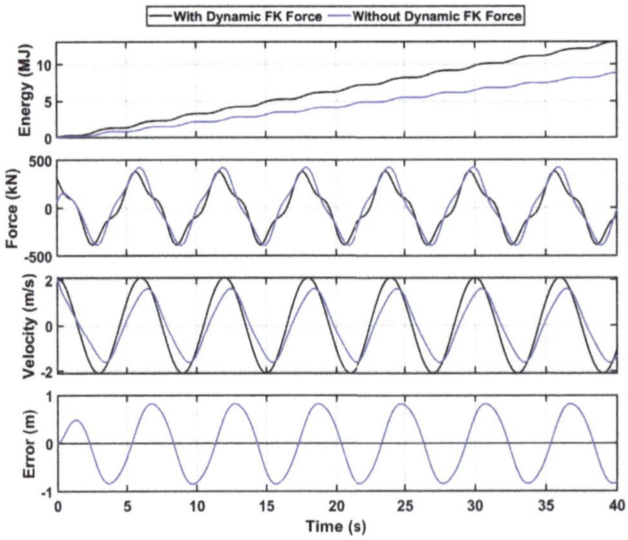


**Figure 6.** Closed–loop hourglass buoy response comparing the effect of including $F_{fk,d}$ in the control law.

### 6.2. Hourglass and Spherical Buoy Comparison

The free-response results of Section 3 indicated that the hourglass buoy possessed synergistic behavior between its buoyancy and dynamic FK forces. In this section, we compare the closed-loop response of the hourglass buoy to a spherical one using the control law of Equation (23).

Two cases are considered: (1) a diameter of 5 m compared to the hourglass buoy whose total height was also $2h = 5$ m, and (2) a diameter of 7.5 m so that the power generated for both buoys was approximately equal. The properties of the two spherical buoys are given in Table 2 where the added mass and damping were estimated from [42].

**Table 2.** Spherical buoy parameters.

| Feature | Symbol | Case 1 | Case 2 | Units |
|---|---|---|---|---|
| Radius | $R$ | 2.5 | 3.75 | m |
| Draft | $h_d$ | 2.5 | 3.75 | m |
| Mass | $m$ | 32,725 | 110,446 | kg |
| Density | $\rho_b$ | 500 | 500 | kg/m$^3$ |
| Added Mass | $m_a$ | 14,019 | 47,492 | kg |
| Damping | $b$ | 11,208 | 18,665 | N/(m/s) |

Figure 7 shows the energy extracted, control force, velocity, and position error, $\zeta_r - \zeta$. The energy and control force plots show the results of both buoys. Since the tracking response was nearly identical for both buoys, the velocity and position error plots are shown for only the hourglass buoy.

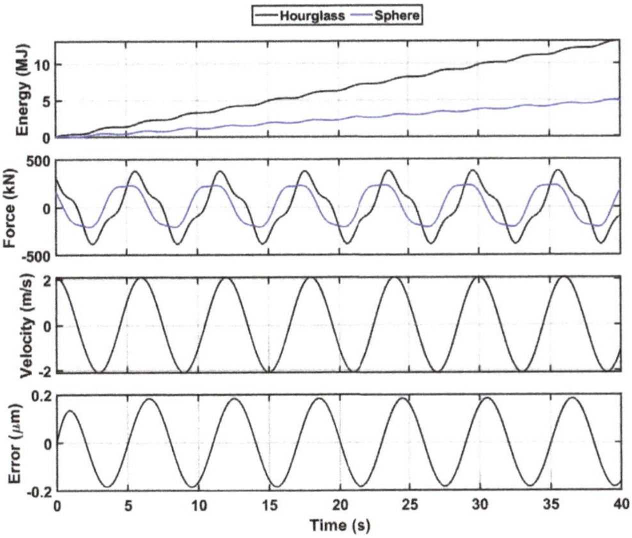


**Figure 7.** Closed–loop response and performance comparing the hourglass and spherical buoys. The hourglass buoy height and the spherical buoy diameter are both 5 m.

While the hourglass-shaped buoy shows about 160% increased energy extraction over the sphere, it should not be concluded that the hourglass shape is better than the spherical one for all conditions. An examination of the control force and velocity plots shows that the hourglass force is better aligned with the buoy velocity peaks. Thus, more energy is extracted. This could be due to differences in the mass properties of the two buoys and the effect this has on their responses resulting from their different nonlinear Froude–Krylov force expressions. Furthermore, the hourglass buoy control amplitude is near twice that of the sphere.

To explore this further, the radius of the spherical buoy was increased to 3.75 m to yield similar energy extraction as the hourglass shown in Figure 8. This resulted in larger control forces, but the phasing between the control force and velocity was not as efficient as with the hourglass buoy for these conditions.

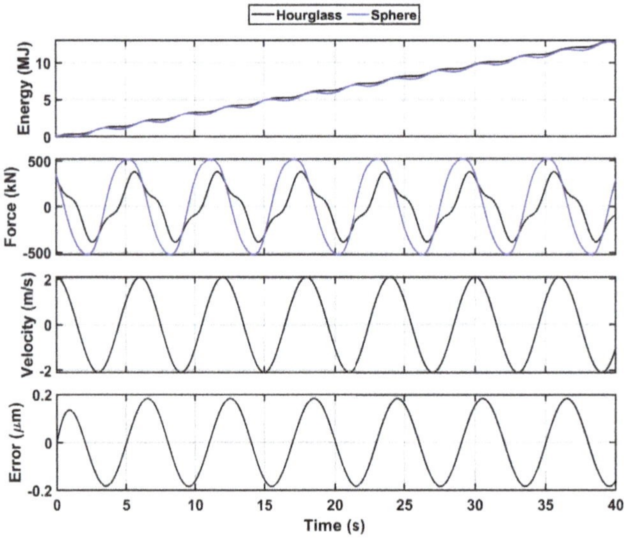


**Figure 8.** Closed–loop response and performance comparing the hourglass and spherical buoys. The hourglass buoy's height is 5 m, and the spherical buoy diameter is 7.5 m so as to yield a similar energy extraction performance as the hourglass buoy.

A comparison of the power used in the Figures above gives a partial indicator of the WEC's performance. As a performance metric, extracted power ignores the requirements placed on the PTO in terms of force and stroke. Equation (29) was used to capture this and allow comparison across different geometries. This power, force, amplitude (PFA) metric reduces commensurately with the PTO size in terms of maximum force and stroke.

$$PFA = \frac{A_P}{A_F A} \tag{29}$$

where $A_P$ is the amplitude of the real power, $A_F$ is the amplitude of the control force and $A$ is the amplitude of the buoy position.

The PFA for the hourglass is 0.99 compared to 0.88 for the 3.75-m diameter sphere, about a 13% improvement for the hourglass buoy. Since, in both cases, the tracking control was nearly perfect, the buoy position had no effect.

### 6.3. Sensitivity Analysis

To observe the robustness of the controller, a sensitivity analysis was performed for the hourglass-shaped buoy by simultaneously increasing the added mass, $m_a$, and the wave number, $\chi$, used by the controller by 50%. The model used $m_a = 59,250$ kg and $\chi = 0.0125$ as described in Section 3. When these values are used by the controller, the resulting response is referred to as "Exact" whereas the "Perturbed" values were increased to $m_a = 88,875$ kg and $\chi = 0.0188$.

Figure 9 shows the results where the LQR gain vector is unchanged from Section 4. The last plot shows that the tracking performance contains some errors for the perturbed case. However, this error is sufficiently small that it did not affect the energy production or the actuator force. Table 3 shows that the displacement tracking error is reduced as the $Q_{11}$ gain of the LQR controller is increased.

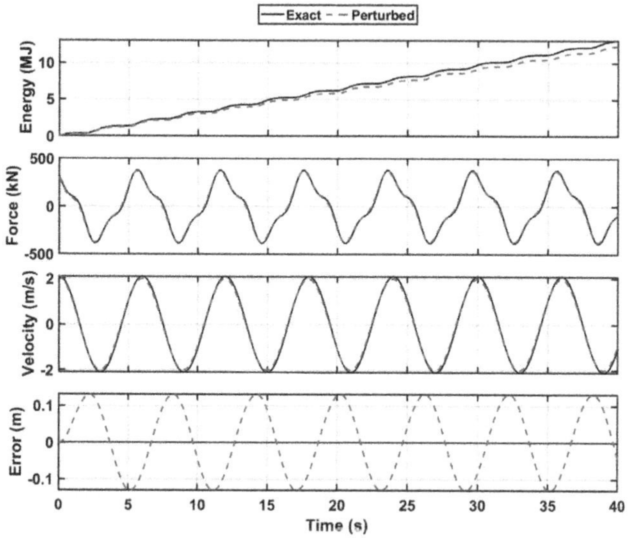


**Figure 9.** Comparison of the controller performance when using the exact model parameters and perturbed values.

**Table 3.** Displacement tracking error due to $Q_{11}$ LQR gain value.

| First State Gain | Displacement Error (m) |
|---|---|
| 10 | 0.130 |
| 20 | 0.098 |
| 50 | 0.064 |
| 100 | 0.015 |

## 7. Conclusions and Future Work

Buoy geometry can significantly impact a point absorber's dynamic response due to the static and dynamic Froude–Krylov forces. The hourglass-shaped buoy's static and dynamic force expressions have nonlinearities of the form $(\eta-\zeta)^3$ and -$\eta\left[e^{-\chi(\eta-\zeta)}+\chi(\eta-\zeta)\right]$, respectively, as seen in Equation (8). Although the nonlinear dynamic forces are less than the static, their shape results in a total force that excites the buoy's vertical motion more than the static force alone. This additional motion results in the nonlinear terms, both static and dynamic, having an even more significant effect. Exploiting buoy shapes to create synergy between the static and dynamic forces can be used to increase the efficiency of a point absorber. While this effect is evident in the hourglass buoy, it may be possible to find shapes that are easier to implement physically, with similar synergy between their static and dynamic forces.

Model-based control usually requires reduced-order models that can execute in real time. The nonlinear dynamic forces, though less than the static, were found to impact the energy extraction of the closed-loop system. This effect will vary under different wave conditions, but they should be included when possible.

Experimental validation of static and dynamic Froude–Krylov force expressions for the hourglass buoy is one of the next steps. Incorporating the effects of significant motion on added mass and radiation damping into the control law should be explored. Shape adaptation, and its effect on the Froude–Krylov forces, is yet another feature that could be exploited to increase point absorber efficiency.

**Author Contributions:** Conceptualization, H.Y., G.P. and D.W.; Methodology, H.Y.; Validation, G.P.; Formal analysis, H.Y. and T.D.G.; Writing—original draft, H.Y.; Writing—review & editing, T.D.G., G.P. and D.W. All authors have read and agreed to the published version of the manuscript.

**Funding:** This research received no external funding.

**Data Availability Statement:** The data presented in this study are available on request from the corresponding author.

**Acknowledgments:** Sandia National Laboratories is a multimission laboratory managed and operated by National Technology & Engineering Solutions of Sandia, LLC, a wholly owned subsidiary of Honeywell International Inc., for the U.S. Department of Energy's National Nuclear Security Administration under contract DE-NA0003525. This paper describes objective technical results and analysis. Any subjective views or opinions that might be expressed in the paper do not necessarily represent the views of the U.S. Department of Energy or the United States Government.

**Conflicts of Interest:** The authors declare no conflict of interest.

## References

1. Falnes, J. A review of wave-energy extraction. *Mar. Struct.* **2007**, *20*, 185–201. [CrossRef]
2. Rémouit, F.; Chatzigiannakou, M.A.; Bender, A.; Temiz, I.; Sundberg, J.; Engström, J. Deployment and maintenance of wave energy converters at the Lysekil research site: A comparative study on the use of divers and remotely-operated vehicles. *J. Mar. Sci. Eng.* **2018**, *6*, 39. [CrossRef]
3. Henry, A.; Doherty, K.; Cameron, L.; Whittaker, T.; Doherty, R. *Advances in the Design of the Oyster Wave Energy Converter*; RINA Marine and Offshore Renewable Energy: London, UK, 2010.
4. Lehmann, M.; Karimpour, F.; Goudey, C.A.; Jacobson, P.T.; Alam, M.R. Ocean wave energy in the United States: Current status and future perspectives. *Renew. Sustain. Energy Rev.* **2017**, *74*, 1300–1313. [CrossRef]
5. Drew, B.; Plummer, A.R.; Sahinkaya, M.N. A review of wave energy converter technology. *Proc. Inst. Mech. Eng. Part A J. Power Energy* **2009**, *223*, 887–902. [CrossRef]
6. Budar, K.; Falnes, J. A resonant point absorber of ocean-wave power. *Nature* **1975**, *256*, 478–479. [CrossRef]
7. Falnes, J.; Kurniawan, A. *Ocean Waves and Oscillating Systems: Linear Interactions Including Wave-Energy Extraction*; Cambridge University Press: Cambridge, UK, 2020; Volume 8.
8. Korde, U.A.; Ringwood, J. *Hydrodynamic Control of Wave Energy Devices*; Cambridge University Press: Cambridge, UK, 2016.
9. Faedo, N.; Pasta, E.; Carapellese, F.; Orlando, V.; Pizzirusso, D.; Basile, D.; Sirigu, S.A. Energy-maximising experimental control synthesis via impedance-matching for a multi degree-of-freedom wave energy converter. *IFAC-PapersOnLine* **2022**, *55*, 345–350. [CrossRef]
10. Lattanzio, S.; Scruggs, J. Maximum power generation of a wave energy converter in a stochastic environment. In Proceedings of the 2011 IEEE International Conference on Control Applications (CCA), Denver, CO, USA, 28–30 September 2011; pp. 1125–1130.
11. Bacelli, G.; Nevarez, V.; Coe, R.G.; Wilson, D.G. Feedback resonating control for a wave energy converter. *IEEE Trans. Ind. Appl.* **2019**, *56*, 1862–1868. [CrossRef]
12. Zou, S.; Abdelkhalik, O.; Robinett, R.; Bacelli, G.; Wilson, D. Optimal control of wave energy converters. *Renew. Energy* **2017**, *103*, 217–225. [CrossRef]
13. Abdelkhalik, O.; Darani, S. Optimization of nonlinear wave energy converters. *Ocean. Eng.* **2018**, *162*, 187–195. [CrossRef]
14. Li, G.; Belmont, M.R. Model predictive control of sea wave energy converters—Part I: A convex approach for the case of a single device. *Renew. Energy* **2014**, *69*, 453–463. [CrossRef]
15. Brekken, T.K. On model predictive control for a point absorber wave energy converter. In Proceedings of the 2011 IEEE Trondheim PowerTech, Trondheim, Norway, 19–23 June 2011; pp. 1–8.
16. Zou, S.; Abdelkhalik, O.; Robinett, R.; Korde, U.; Bacelli, G.; Wilson, D.; Coe, R. Model Predictive Control of parametric excited pitch-surge modes in wave energy converters. *Int. J. Mar. Energy* **2017**, *19*, 32–46. [CrossRef]
17. Wilson, D.G.; Robinett, R.D., III; Bacelli, G.; Abdelkhalik, O.; Coe, R.G. Extending Complex Conjugate Control to Nonlinear Wave Energy Converters. *J. Mar. Sci. Eng.* **2020**, *8*, 84. [CrossRef]
18. Richter, M.; Magana, M.E.; Sawodny, O.; Brekken, T.K. Nonlinear model predictive control of a point absorber wave energy converter. *IEEE Trans. Sustain. Energy* **2012**, *4*, 118–126. [CrossRef]
19. Darani, S.; Abdelkhalik, O.; Robinett, R.D.; Wilson, D. A hamiltonian surface-shaping approach for control system analysis and the design of nonlinear wave energy converters. *J. Mar. Sci. Eng.* **2019**, *7*, 48. [CrossRef]
20. Wilson, D.G.; Robinett, R.D.; Bacelli, G.; Abdelkhalik, O.; Weaver, W.W.; Coe, R. Nonlinear WEC optimized geometric buoy design for efficient reactive power requirements. In Proceedings of the OCEANS 2019 MTS/IEEE, Seattle, WA, USA, 27–31 October 2019; pp. 1–6.

21. Wilson, D.G.; Robinett, R.D.; Weaver, W.W.; Glover, S.F. Wave Energy Converter Buoy with Variable Geometry that Improves Energy and Power Capture for Changing Sea States. In Proceedings of the 2022 International Symposium on Power Electronics, Electrical Drives, Automation and Motion (SPEEDAM), Sorrento, Italy, 22–24 June 2022; pp. 918–923.
22. Bacelli, G.; Coe, R.G. Comments on control of wave energy converters. *IEEE Trans. Control Syst. Technol.* **2020**, *29*, 478–481. [CrossRef]
23. Garcia-Rosa, P.B.; Bacelli, G.; Ringwood, J.V. Control-informed geometric optimization of wave energy converters: The impact of device motion and force constraints. *Energies* **2015**, *8*, 13672–13687. [CrossRef]
24. Garcia-Rosa, P.B.; Ringwood, J.V. On the sensitivity of optimal wave energy device geometry to the energy maximizing control system. *IEEE Trans. Sustain. Energy* **2015**, *7*, 419–426. [CrossRef]
25. Abdelkhalik, O.; Coe, R.G.; Bacelli, G.; Wilson, D.G. WEC geometry optimization with advanced control. In Proceedings of the ASME International Conference on Offshore Mechanics and Arctic Engineering, Trondheim, Norway, 25–30 June 2017; Volume 57786, p. V010T09A031.
26. Giorgi, G.; Ringwood, J.V. Implementation of latching control in a numerical wave tank with regular waves. *J. Ocean. Eng. Mar. Energy* **2016**, *2*, 211–226. [CrossRef]
27. Giorgi, G.; Ringwood, J.V. Computationally efficient nonlinear Froude–Krylov force calculations for heaving axisymmetric wave energy point absorbers. *J. Ocean. Eng. Mar. Energy* **2017**, *3*, 21–33. [CrossRef]
28. Giorgi, G.; Ringwood, J.V. A Compact 6-DoF Nonlinear Wave Energy Device Model for Power Assessment and Control Investigations. *IEEE Trans. Sustain. Energy* **2019**, *10*, 119–126. [CrossRef]
29. Giorgi, G.; Sirigu, S.; Bonfanti, M.; Bracco, G.; Mattiazzo, G. Fast nonlinear Froude–Krylov force calculation for prismatic floating platforms: A wave energy conversion application case. *J. Ocean. Eng. Mar. Energy* **2021**, *7*, 439–457. [CrossRef]
30. Demonte Gonzalez, T.; Parker, G.G.; Anderlini, E.; Weaver, W.W. Sliding mode control of a nonlinear wave energy converter model. *J. Mar. Sci. Eng.* **2021**, *9*, 951. [CrossRef]
31. Zou, S.; Song, J.; Abdelkhalik, O. A sliding mode control for wave energy converters in presence of unknown noise and nonlinearities. *Renew. Energy* **2023**, *202*, 432–441. [CrossRef]
32. Faedo, N.; Giorgi, G.; Ringwood, J.V.; Mattiazzo, G. Optimal control of wave energy systems considering nonlinear Froude–Krylov effects: Control-oriented modelling and moment-based control. *Nonlinear Dyn.* **2022**, *109*, 1777–1804. [CrossRef]
33. Penalba, M.; Giorgi, G.; Ringwood, J.V. Mathematical modelling of wave energy converters: A review of nonlinear approaches. *Renew. Sustain. Energy Rev.* **2017**, *78*, 1188–1207. [CrossRef]
34. Giorgi, G.; Ringwood, J.V. Comparing nonlinear hydrodynamic forces in heaving point absorbers and oscillating wave surge converters. *J. Ocean. Eng. Mar. Energy* **2018**, *4*, 25–35. [CrossRef]
35. Wilson, D.G.; Robinett, R., III; Abdelkhalik, O.; Bacelli, G. *Nonlinear Control Design for Nonlinear Wave Energy Converters;* Technical Report SAND2018-5214C; Sandia National Lab. (SNL-NM): Albuquerque, NM, USA, 2018.
36. Tan, J.; Polinder, H.; Laguna, A.J.; Miedema, S. A numerical study on the performance of the point absorber Wave Energy Converter integrated with an adjustable draft system. *Ocean. Eng.* **2022**, *254*, 111347. [CrossRef]
37. Forehand, D.I.; Kiprakis, A.E.; Nambiar, A.J.; Wallace, A.R. A fully coupled wave-to-wire model of an array of wave energy converters. *IEEE Trans. Sustain. Energy* **2015**, *7*, 118–128. [CrossRef]
38. Cummins, W. The Impulse Response Function and Ship Motions. *Schiffstechnik* **1962**, *47*, 101–109.
39. Zou, M.; Chen, M.; Zhu, L.; Li, L.; Zhao, W. A constant parameter time domain model for dynamic modelling of multi-body system with strong hydrodynamic interactions. *Ocean. Eng.* **2023**, *268*, 113376. [CrossRef]
40. WAMIT Inc. *The State of the Art in Wave Interaction Analysis;* WAMIT Inc.: Chestnut Hill, MA, USA, 2023.
41. Hendricks, E.; Jannerup, O.; Sørensen, P.H. *Linear Systems Control: Deterministic and Stochastic Methods;* Springer: Berlin/Heidelberg, Germany, 2008.
42. Havelock, T.H. Waves due to a floating sphere making periodic heaving oscillations. *Proc. R. Soc. Lond. Ser. A Math. Phys. Sci.* **1955**, *231*, 1–7.

applied sciences

MDPI

Article

# Dynamics Analysis and Chaos Identification of Compound Pendulum Jaw Crusher with Joint Clearance

**Shenpeng Wang, Yan Cui * and Chune Wang**

College of Mechanical and Automotive Engineering, Shanghai University of Engineering and Science, Shanghai 201620, China
* Correspondence: cuiyan0312@126.com; Tel.: +86-13917337356

**Abstract:** A dynamic model of the working mechanism of a compound pendulum jaw crusher with clearance is established by the Lagrange multiplier method (LMD) based on the L-N contact force model and modified Coulomb friction force model. The correctness of the dynamic model is verified by MATLAB and Adams comparison simulation, and the best selection range of driving speed and clearance size is determined according to the laminating crushing theory (LCT). The effects of driving speed and clearance size on the shock and chaos phenomenon in the clearance mechanism are analyzed, as well as the vibration force applied to the foundation. The results show that, under the premise of meeting the requirements of crusher production capacity, increasing the driving speed and reducing the clearance size can weaken the shock and chaos phenomenon during the operation of the mechanism, improve the dynamic performance and enhance the operation stability, and the service life of foundation is prolonged with the stability of vibration force. Furthermore, the optimal combination of driving speed and clearance size is determined, which provides a theoretical basis for the optimization of the vibration force of the jaw crusher.

**Keywords:** clearance; crusher; chaos; shock; vibration

## 1. Introduction

Jaw crushers are utilized extensively in the industries of mining, metallurgy, building materials, ceramics, and engineering of massive minerals with small and medium particle size crushing [1–3]. There are now more demands on crushers in terms of efficiency, energy use, and cost due to new issues including the rise in low-quality ore, energy shortages, and the rising demand for building materials [4]. The frame will be subjected to the inertia force produced by the eccentric shaft, elbow plate, and other components when the crusher is operating, which will cause the crusher to vibrate and transfer to the fixed foundation of the crusher, reducing the service life of the foundation [5].

Research on the optimization of the crusher's vibration force must be conducted in order to lower the failure rate of the machine and avoid equipment damage. Yang [6] established a parametric model of a jaw crusher using the ANSYS Parametric Design Language (APDL) and optimized the design with the minimum weight of a movable jaw as the objective function. This reduced the vibration caused by the inertia force of the working mechanism, improved the mechanism's dynamic performance and extended the service life of the product. Li [7] proposed a vibrator–pendulum coupling system in which the pendulum is driven by the synchronous motion of the two vibrators directly. The synchronization criterion of the two vibrators and the stability criterion for the synchronous states were derived, allowing a new kind of vibration jaw crusher to be designed to work with its swing motion trajectory under the premise of stable zero-phase difference between the two vibrators. In order to create a more efficient dynamic balance mechanism, Cheng [8] established a real-time dynamic model of a crusher based on multi-body dynamics (MBD) and the discrete element method (DEM). Using the response surface method (RSM), the predictive regression model of crusher performance was constructed. The amplitude of vibration force was greatly decreased by increasing the mass of the fixed cone for optimization.

Current research on jaw crushers has concentrated on the design and optimization of the mechanism structure [9], ignoring the existence of joint clearance. When the crusher is operating at a high speed and under a heavy load, the revolute joint's elements will lose contact with one another. When they come into contact again owing to constraint and inertia, a collision will happen. At this point, the revolute joint's response force can be several times or even a dozen times as great as when there is no clearance [10], which invariably has an effect on the assessment of crusher vibration.

The dynamic performance of the clearance mechanism system can be better described by chaos, and the stability of the mechanism operation can also be assessed by chaos [11]. Wu [12] described the system dynamics for the planar slider-crank mechanism with multiple clearance joints by the correlation dimension and bifurcation actions. The dynamic equations of motion were derived using generalized coordinates and the Lagrange method, and the effects of clearance size and driving speed on system bifurcation were investigated. Chen [13] utilized the KED method with the Lagrange method to establish a nonlinear dynamic model of a spatially parallel mechanism. The study showed the influence of various clearances on the parallel mechanism dynamics behavior, including chaos and bifurcation phenomenon, which provided a methodology for forecasting the dynamics behavior of parallel mechanisms with clearance and flexible links.

By summarizing the above-mentioned research, one challenging issue exists in the jaw crusher. The presented idealized model does not accurately describe the dynamics of the crusher, which greatly affects the evaluation of the dynamic performance. Specifically, the existing research does not consider the effect of revolute joint with clearance and lacks comparative analysis of the dynamic characteristics under different working conditions. There is also a lack of specific analysis of the key parameters affecting the vibration force. Numerous scholars have used nonlinear dynamics techniques such as chaos to accomplish research and analysis in the field of mechanism with clearance, and they have produced a wealth of fruitful findings. In these backgrounds, the PE250 × 400 compound pendulum jaw crusher manufactured by Shanghai Clirik Machinery Co., Ltd. in China is used as the research object. Chaos is incorporated into the vibration force assessment process to examine the best combination of driving speed and clearance size after taking the clearance factor into account. In Section 2, the model of the revolute joint with clearance, the contact force model, and the friction force model is introduced. In Section 3, the dynamic model of the crusher working mechanism with clearance is established by the Lagrange multiplier method (LMD). In Section 4, the dynamic equation is solved by the variable step fourth-order Runge–Kutta method. Comparative simulations between MATLAB and Adams are used to confirm the correctness of the theoretical model, and the range of driving speed and clearance size is determined by the laminar crushing theory (LCT) without affecting the production capacity. In Section 5, the shock and chaos phenomenon of the clearance mechanism under the situations of a single driving speed, a single clearance size, and a combination of driving speed and clearance size are analyzed, respectively. The effects of clearance on the vibration force applied to the foundation are investigated. Finally, the optimal combination of parameters is given.

## 2. Clearance Model

### 2.1. Revolute Joint Model with Clearance

The shaft and bearing of the revolute joint are viewed as independent parts, and Figure 1 depicts their model [14]. Where $i$, $j$ stand for the bearing and shaft, $O_i$, $O_j$ for the center of mass, $Q_i$, $Q_j$ for the collision point at impact, $\boldsymbol{r}_i^O$, $\boldsymbol{r}_j^O$ for the center of mass' position vector, $\boldsymbol{r}_i^Q$, $\boldsymbol{r}_j^Q$ for the position vector of collision point, and $F_n$, $F_t$ for the normal contact force and tangential friction force at impact.

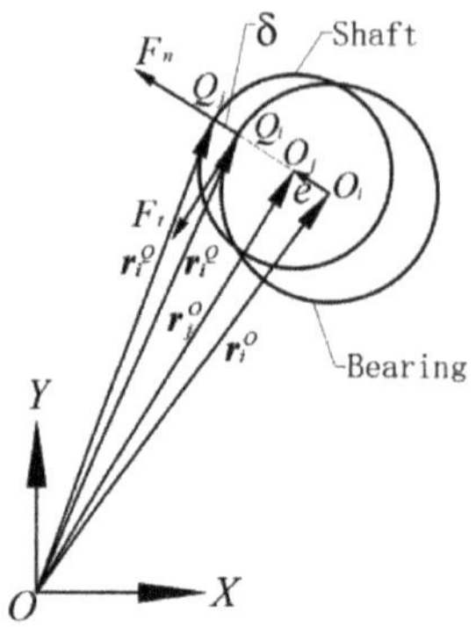


**Figure 1.** Revolute joint model with clearance.

The eccentric vector between the bearing and shaft can be defined as follows:

$$e = r_j^O - r_i^O \tag{1}$$

Its unit normal vector can be expressed as follows:

$$n = e/e \tag{2}$$

where $e$ is the amplitude of the eccentric vector, $e = \sqrt{e^T e}$.

The angle between the eccentric vector and the positive direction of the X-axis can be defined as follows:

$$\varphi = \cos^{-1}(n \cdot I) \tag{3}$$

where $I$ is the unit vector in the positive direction of the X-axis.

The position vector of the collision point can be defined as follows:

$$r_k^Q = r_k^O + R_k n \quad (k = i, j) \tag{4}$$

where $r_i$, $r_j$ are the radius of the bearing and shaft, respectively.

The velocity vector at the collision point is produced by taking the derivative of Equation (4) with respect to time, which can be stated as follows:

$$\dot{r}_k^Q = \dot{r}_k^O + R_k \dot{n} \quad (k = i, j) \tag{5}$$

The velocity vector at the collision point is projected onto the collision plane in the normal and tangential directions, then the relative normal and tangential velocities can be defined as the following:

$$\begin{cases} v_n = \left(\dot{r}_i^Q - \dot{r}_j^Q\right)^{\mathrm{T}} n \\ v_t = \left(\dot{r}_i^Q - \dot{r}_j^Q\right)^{\mathrm{T}} t \end{cases} \tag{6}$$

where $t$ is the tangential vector, obtained by $n$ rotating 90° counterclockwise.

The embedding depth between the bearing and shaft can be expressed as follows:

$$\delta = e - c \tag{7}$$

where $c$ is clearance size, $c = R_i - R_j$.

## 2.2. Contact Force Model

In the impact process, the normal contact force will cause certain shock and vibration, which greatly affects the performance of the mechanism, so the selection of the normal contact force model is an important step in the dynamic modeling process. In this study, the L-N contact force model [15] will be used, which is based on the spring damping model with the addition of energy dissipation, material properties, elastic deformation, and collision velocity to more closely match the actual situation, and it can be expressed as follows:

$$F_n = K\delta^m\left(1 + \frac{3\left(1 - c_e^2\right)}{4}\frac{\dot{\delta}}{\dot{\delta}^{(-)}}\right) \tag{8}$$

where $K$ is the stiffness coefficient, which can be defined as the following:

$$K = \frac{4}{3\left(h_i + h_j\right)}\sqrt{\frac{R_i R_j}{R_i - R_j}} \tag{9}$$

where $h_k$ is combined constant, which can be defined as the following:

$$h_k = \frac{1 - \nu_k^2}{E_k}(k = i, j) \tag{10}$$

where $m$ is the power exponent of the metallic surface, $c_e$ is material recovery coefficient, $\dot{\delta}^{(-)}$ is normal relative initial impact velocity, $\nu_k$ is Poisson's ratio, and $E_k$ is Elastic modulus.

### 2.3. Friction Force Model

In order to create dynamic friction, Ambrósio added the dynamic correction factor $c_d$ to the Coulomb friction model [16]. The model is capable of effectively describing the dry friction behavior between colliding entities as well as the viscous and micro-slip phenomenon that occur at very low velocities, while preventing numerical operations from becoming unstable owing to abrupt changes in the tangential velocity direction, hence increasing the stability of the numerical integration. It can be expressed as follows:

$$F_t = -c_f c_d F_n \frac{\nu_t}{|\nu_t|} \tag{11}$$

where $c_d$ is the dynamic correction factor, which can be defined as the following:

$$c_d = \begin{cases} 0 & |\nu_t| < v_0 \\ \frac{|\nu_t| - v_0}{v_1 - v_0} & v_0 \leqslant |\nu_t| \leqslant v_1 \\ 1 & |\nu_t| > v_1 \end{cases} \tag{12}$$

where $c_f$ is coefficient of sliding friction, $\nu_t$ is relative sliding velocity, $v_0$, $v_1$ are the static and dynamic friction speed limits, respectively.

## 3. Dynamic Modeling

Figure 2 depicts the working mechanism of the jaw crusher with clearance in the revolute joint, which includes frame 0, eccentric shaft 1, moveable jaw 2, and elbow plate 3. The driving part is the eccentric shaft 1, which provides power to the moveable jaw 2 via the revolute joint A to accomplish the crushing operation. The components are all hard and difficult to bend, so the influence of revolute joint A on the mechanism is greater and the existence of its clearance must be taken into account [17], which has been represented in Figure 2a in an exaggerated manner.

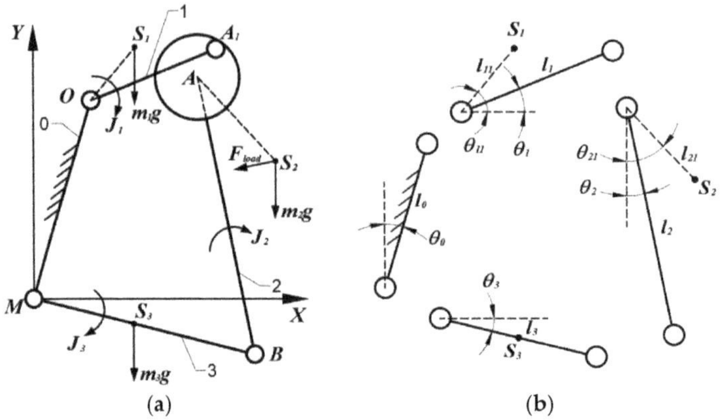


**Figure 2.** Working mechanism of the jaw crusher with clearance. (**a**) Demonstration of mass, rotational inertia and load; (**b**) Demonstration of length and angle.

The system generalized coordinates are $q = [x_1, y_1, \theta_1, x_2, y_2, \theta_2, x_3, y_3, \theta_3]^T$. Ideally, eight constraint equations can be introduced for the four moving subsets; however, the existence of the clearance causes the constraint to fail at revolute joint A. The constraint equation can be shown as follows:

$$\boldsymbol{\Phi}(q,t) = \begin{bmatrix} x_1 - l_0\sin\theta_0 - l_{11}\cos\theta_{11} \\ y_1 - l_0\cos\theta_0 - l_{11}\sin\theta_{11} \\ x_2 - l_3\cos\theta_3 + l_2\sin\theta_2 - l_{21}\sin\theta_{21} \\ y_2 + l_3\sin\theta_3 - l_2\cos\theta_2 + l_{21}\cos\theta_{21} \\ x_3 - (l_3/2)\cos\theta_3 \\ y_3 + (l_3/2)\sin\theta_3 \end{bmatrix} = 0 \tag{13}$$

where $x_1, y_1, x_2, y_2, x_3, y_3$ represent the horizontal and vertical coordinates of the centers of mass $S_1, S_2, S_3$, respectively. $\theta_0, \theta_1, \theta_2, \theta_3$ represent the angle between each linkage and the coordinate axis, respectively. $l_0, l_1, l_2, l_3$ represent the length of each linkage, respectively. $l_{11}, l_{21}$ represent the length of the center of mass $S_1, S_2$ to their left endpoints, respectively. $\theta_{11}, \theta_{21}$ represent the angle between the line of the center of mass and the left endpoint and the coordinate axis, respectively, $\theta_{11} = \theta_1 + 30°, \theta_{21} = \theta_2 + 15°$.

By taking the first-order derivative of Equation (13) with respect to time, the velocity constraint equation is stated as follows:

$$\boldsymbol{\Phi}_q\dot{q} = -\boldsymbol{\Phi}_t \equiv v \tag{14}$$

where $\boldsymbol{\Phi}_q$ is the Jacobian matrix of the constraint equation. $\boldsymbol{\Phi}_t$ is the first-order derivative of the constraint equation with respect to time.

By taking the second-order derivative of Equation (13) with respect to time, the acceleration constraint equation is stated as follows:

$$\boldsymbol{\Phi}_q\ddot{q} = -\left(\boldsymbol{\Phi}_q\dot{q}\right)_q\dot{q} - 2\boldsymbol{\Phi}_{qt}\dot{q} - \boldsymbol{\Phi}_{tt} \equiv \gamma \tag{15}$$

where $\boldsymbol{\Phi}_{qt}$ is the time partial derivative of the Jacobian matrix. $\boldsymbol{\Phi}_{tt}$ is the second-order partial derivative of the constraint equation with respect to time.

The dynamic equation based on the Lagrange multiplier method (LMD) [18] can be defined as the following:

$$\boldsymbol{M}\ddot{q} + \boldsymbol{\Phi}_q^{\mathrm{T}}\lambda = \boldsymbol{Q} - \boldsymbol{F} \tag{16}$$

where $\boldsymbol{M}$ is the generalized quality matrix, which can be shown as follows:

$$\boldsymbol{M} = diag[m_1, m_1, J_1, m_2, m_2, J_2, m_3, m_3, J_3] \tag{17}$$

where $\boldsymbol{Q}$ is the system generalized external forces, which can be shown as follows:

$$\boldsymbol{Q} = [0, -m_1 g, M_0, F_{load}\sin\theta_2, -m_2 g - F_{load}\cos\theta_2, 0, 0, -m_3 g, 0]^T \tag{18}$$

where $\boldsymbol{F}$ is the collision force matrix, which can be shown as follows:

$$F = \begin{cases} \begin{bmatrix} -F_n\cos\varphi + F_t\sin\varphi, F_n\sin\varphi + F_t\cos\varphi, 0, \\ F_n\cos\varphi - F_t\sin\varphi, -F_n\sin\varphi - F_t\cos\varphi, 0, \\ 0, 0, 0 \end{bmatrix}^T & \delta > 0 \\ 0 & \delta \leq 0 \end{cases} \tag{19}$$

where $\lambda$ is the Lagrange multiplier array. $m_1, m_2, m_3$ are the masses of the eccentric shaft, moveable jaw, and elbow plate, respectively; $J_1, J_2, J_3$ are the rotational inertia, respectively; $M_0$ is the driving torque; $F_{load}$ is the fixed load.

The displacement constraint equation and the velocity constraint equation are inserted into the system dynamic equation through Baumgarte's default stabilization procedure [19], and the dynamic equation of the mechanism with clearance in the revolute joint is derived as follows:

$$\begin{pmatrix} \boldsymbol{M} & \boldsymbol{\Phi}_q^{\mathrm{T}} \\ \boldsymbol{\Phi}_q & 0 \end{pmatrix} \begin{pmatrix} \ddot{q} \\ \lambda \end{pmatrix} = \begin{pmatrix} \boldsymbol{Q} - \boldsymbol{F} \\ \gamma - 2\alpha\dot{\boldsymbol{\Phi}} - \beta^2\boldsymbol{\Phi} \end{pmatrix} \tag{20}$$

where $\alpha, \beta$ are Baumgarte stability factors, respectively.

## 4. Validation Analysis

The structural parameters of the working mechanism of the jaw crusher are shown in Table 1. The dynamic simulation parameters are shown in Table 2.

**Table 1.** Structural parameters of the jaw crusher mechanism.

| Components * | Rotational Inertia $J$ | Quality $m$ | Length $L$ |
|---|---|---|---|
| frame 0 | — | 2597 | 0.64 |
| eccentric shaft 1 | 0.2 | 95 | 0.01 |
| moveable jaw 2 | 128.48 | 524 | 0.695 |
| elbow plate 3 | 0.7154 | 22 | 0.3 |

* The units of each parameter are kg·m$^2$, kg, m, respectively.

**Table 2.** Dynamic simulation parameters.

| Parameter | Numerical Value | Parameter | Numerical Value |
|---|---|---|---|
| $m$ | 1.5 | $E_k$ | 207 Gpa |
| $c_e$ | 0.9 | $v_0$ | 0.1 m/s |
| $v_k$ | 0.29 | $v_1$ | 1 m/s |
| $c_f$ | 0.1 | $F_{load}$ | 8.8 KN |
| $\alpha$ | 20 | $\dot{\delta}^{(-)}$ | 0.5 m/s |
| $\beta$ | 20 | $\theta_0$ | 22.5° |
| $R_i$ | 0.13 m | $\theta_1$ | 0° |
| $l_{21}$ | 0.006 m | $l_{21}$ | 0.3 m |

As can be observed from Equation (20), the dynamic equation is coupled and extremely nonlinear, hence it cannot be solved directly to produce a stable solution. Therefore, in this study, the equation is iteratively solved by the variable step fourth-order Runge–Kutta method [20] to minimize calculation error. The initial condition is set to the ideal state, and eccentric shaft 1 is parallel to the X-axis. The solution process is shown in Figure 3.

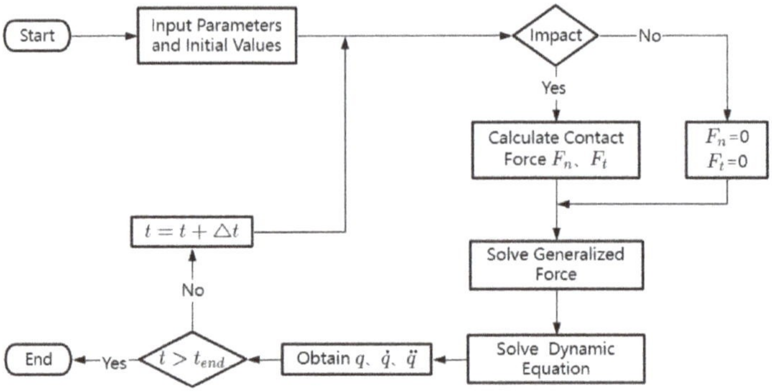


**Figure 3.** Flow chart for solving dynamic equation.

### *4.1. Validation of Dynamical Model*

The primary components of the PE250 × 400 crusher are imported into Adams software for virtual prototype simulation in order to confirm the correctness of the dynamic model of the working mechanism of the crusher with clearance. The simulation interface is shown in Figure 4, and the results obtained are compared with those of MATLAB. The simulation parameters are the same in MATLAB and Adams, where the clearance size at the revolute joint A is 0.06 mm and the system driving speed is 30 rad/s.

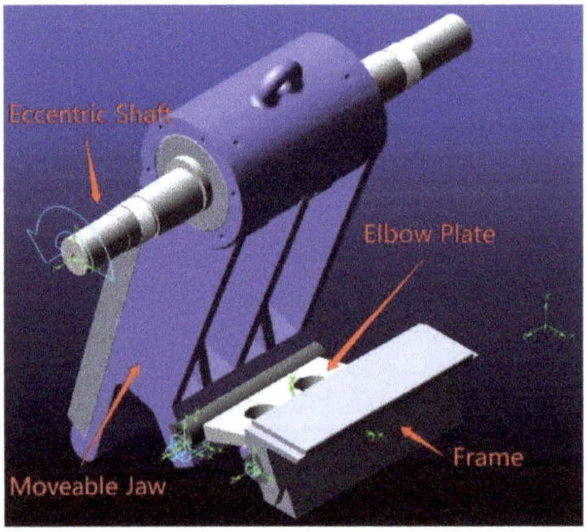


**Figure 4.** Adams virtual simulation interface.

After the mechanism has stabilized in operation, the displacement and velocity curves of the moveable jaw are shown in Figure 5a,b, respectively, and the collision force curve at the revolute joint A is shown in Figure 5c,d. Analysis of Figure 5a shows that the displacement curves obtained by Adams and those obtained by MATLAB are basically the same. A closer look at Figure 5b reveals that during motion, the velocity curve will exhibit noticeable fluctuations at the crest and trough. The fluctuation value of MATLAB is slightly larger, and the two curves essentially continue to overlap after the fluctuation. Analysis of Figure 5c,d shows that the collision forces at the revolute joint A obtained by Adams and MATLAB maintains the same waveform trend overall, although they exhibit complex nonlinear oscillations. There are slight differences in the curve fluctuations and peak magnitudes. In the time interval [12, 12.4], 12.099 s for the first time to reach the peak collision force of 37,796 N for Adams and 12.106 s for the first time to reach the peak collision force of 37,578 N for MATLAB.

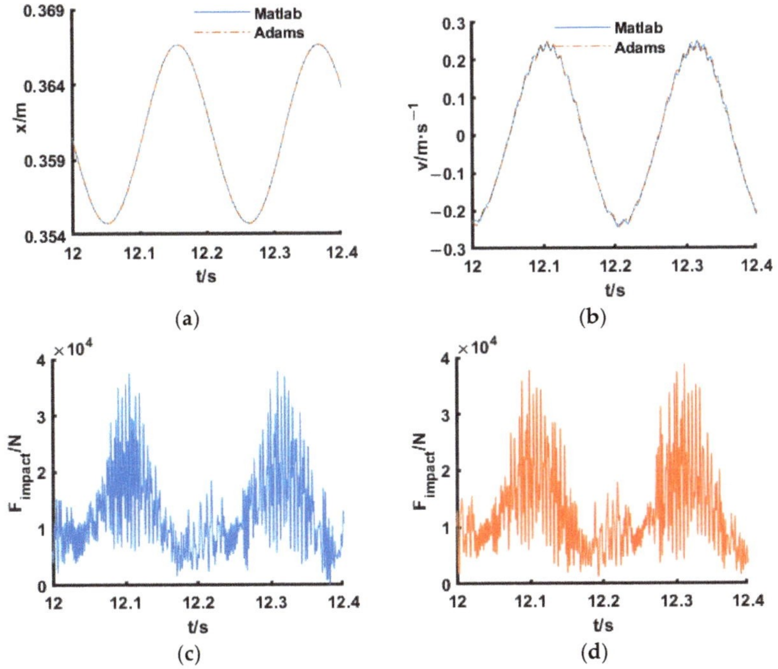


**Figure 5.** Comparison of results from Adams and MATLAB: (**a**) Moveable jaw displacement; (**b**) Moveable jaw velocity; (**c**,**d**) Collision force at the revolute joint A.

In summary, the results from Adams and MATLAB are similar in terms of trends, which verifies the correctness of the dynamic model of the working mechanism of the crusher with clearance. The slight difference in the peak is analyzed for possible reasons [21]:

(1) Errors in the structural parameters of the virtual prototype. Adams is deficient in modeling capability, so a 3D model of the crusher is built in SOLIDWORKS and imported into Adams, while some structural simplifications are made to the model, such as merging the bearing rollers of the eccentric shaft and thickening the revolute joint where the elbow plate is connected to the moveable jaw, resulting in minor changes in mass, volume and other parameters.

(2) The difference between the two solution methods. The normal contact force model and the tangential friction force model in the dynamic model are not fully consistent with the mechanical model in Adams. In Adams, the collision forces between the components are defined using the Contact tool. The normal contact force is calculated by the Impact function in the function library, and the tangential friction force is calculated by the Coulomb friction model.

### *4.2. Validation of Influence Parameters*

Crusher production capacity is defined as the volume of material that a crusher can crush per unit of time, which is an important indicator of crusher performance [22]. According to the lamination crushing theory (LCT), the movable jaw follows the eccentric shaft to move reciprocally once, and the ore will be crushed and moved down layer by layer until they are discharged from the crushing chamber. The specific discharge volume per unit of time is obtained by calculating the discharge volume of a single layer. The cross-section diagram of the broken layer is shown in Figure 6. $B$ and $b$ are the inlet size and discharge size when the movable jaw moves to the loose edge position, $a$ is the engagement angle of the crushing chamber, $H$ is the height of the crushing chamber, $s$ is the horizontal travel of the lower end point of the movable jaw, $AA_1BB_1$ is the trapezoidal cross-section of the discharge prism during one movement cycle, and $h$ is the height of the cross-section.

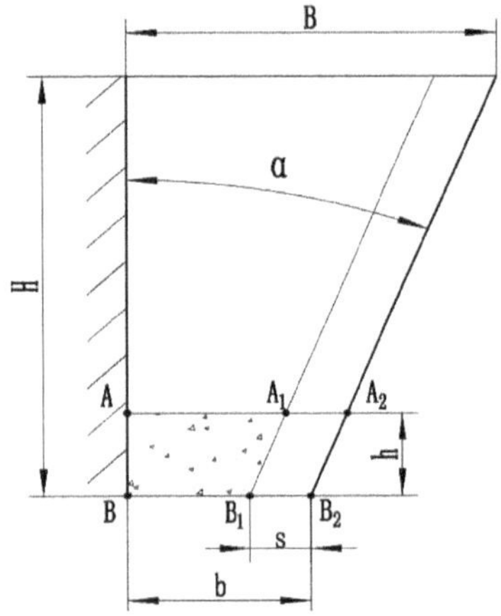


**Figure 6.** Cross-section diagram of the broken layer.

The crusher production capacity can be expressed as follows:

$$Q = 60nV\mu \tag{21}$$

where $V$ is the volume of material discharged once by the reciprocating motion of the moveable jaw, which can be defined as the following:

$$V = S \cdot L \tag{22}$$

where $S$ is the area of the trapezoidal section of the prism, which can be defined as follows:

$$S = \frac{2b - s}{2}h \tag{23}$$

when the material breaks away from the moveable jaw, it does free fall motion, so the height of the cross-section can be shown as follows:

$$h = \frac{1}{2}gt^2 \tag{24}$$

In a motion cycle, the discharge time occupies half of the time, so the discharge time can be expressed as follows:

$$t = \frac{30}{n} \tag{25}$$

where $n$ is the eccentric shaft speed, $g$ is the acceleration of gravity, $L$ is the feed opening length, and $\mu$ is the loosening factor of the mineral.

Combined with the above equations, it can be seen that the production capacity of the crusher is determined by the eccentric shaft speed, the horizontal travel of the lower end point of the movable jaw and the discharge size. While the driving speed and the clearance size are important influencing factors of the shock and chaos phenomenon in the mechanism [23], if these two parameters are analyzed, their influence on the production capacity also needs to be considered. The driving speed corresponds to the eccentric shaft speed, and the clearance size corresponds to the horizontal travel of the lower end point of the movable jaw and the discharge size, i.e.,

Given a driving speed of 30 rad/s and common fit tolerances of 0.06 and 0.12 mm for the revolute joint A, a comparison of the displacement of the revolute joint B under different clearance size conditions is shown in Figure 7a. It can be seen that the displacement curves basically overlap, indicating that the change in clearance size will not affect the horizontal travel of the lower end point of the movable jaw and the discharge size, which will not affect the production capacity.

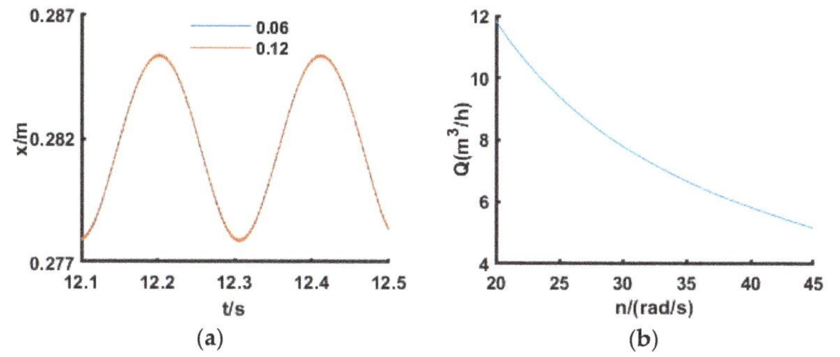


**Figure 7.** Actual results of influence parameters: (**a**) Displacement of revolute joint B at different clearance sizes; (**b**) Relationship between eccentric shaft speed and production capacity.

Taking granite ore crushing as an example, the given feed opening length is 0.4 m, the crushing chamber height is 0.6 m and the loosening factor is 0.7. According to Equation (21), the relationship between driving speed and production capacity of a PE250 $\times$ 400 compound pendulum jaw crusher can be obtained, as shown in Figure 7b. As can be seen from the figure, the production capacity decreases as the driving speed increases. The single-layer material fall height $h$ should be less than the crushing chamber height $H$. Combining with Equations (24) and (25), the driving speed should be greater than 25 rad/s. Meanwhile, the maximum production capacity of the crusher is calculated as 9.41 $m^3/h$. According to the definition of crusher production capacity affluence factor [24], the actual production capacity multiplied by the affluence factor over the maximum production capacity can meet the production demand and the general affluence factor $k$ = 1.67. Therefore, the minimum production capacity allowed for the crusher is 5.63 $m^3/h$, which corresponds to a maximum driving speed of 41.45 rad/s. For the convenience of the subsequent parameter analysis, the maximum driving speed is chosen to be 40 rad/s.

In summary, the influence of driving speed and clearance size on the working mechanism of the crusher should be studied without affecting production capacity. The range of driving speed and clearance size can be chosen as [25, 40] and [0.03, 0.12], respectively.

## 5. Analysis of Dynamic Performance

### 5.1. Influence of Driving Speed

Given the clearance size of 0.06 mm and the driving speed of 25, 30, 35, and 40 rad/s, Figures 8–10 represent the characteristic images of the axial trajectory of the shaft, the vibration force applied to the foundation, and the Poincare mapping at different driving speeds, respectively.

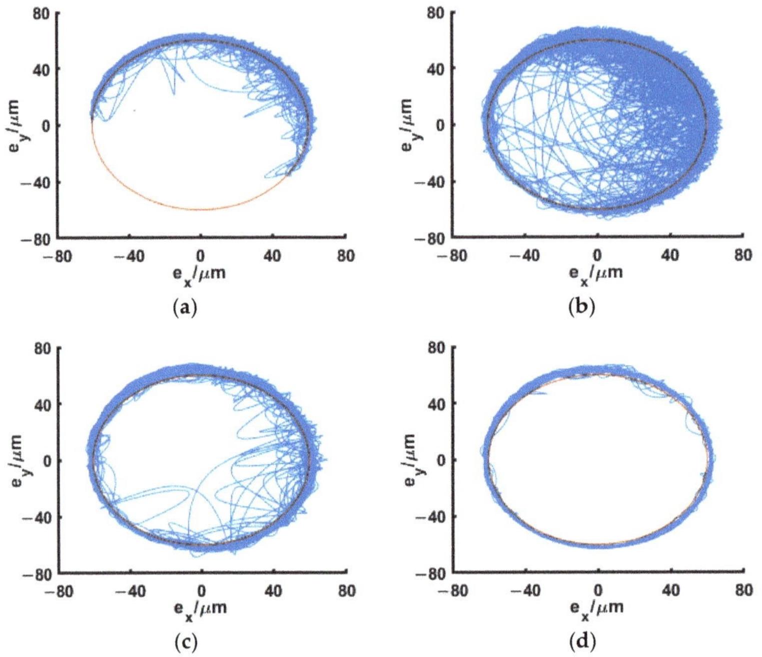


**Figure 8.** Axis trajectory of shaft at different driving speeds. (**a**) $\omega = 25$. (**b**) $\omega = 30$. (**c**) $\omega = 35$. (**d**) $\omega = 40$.

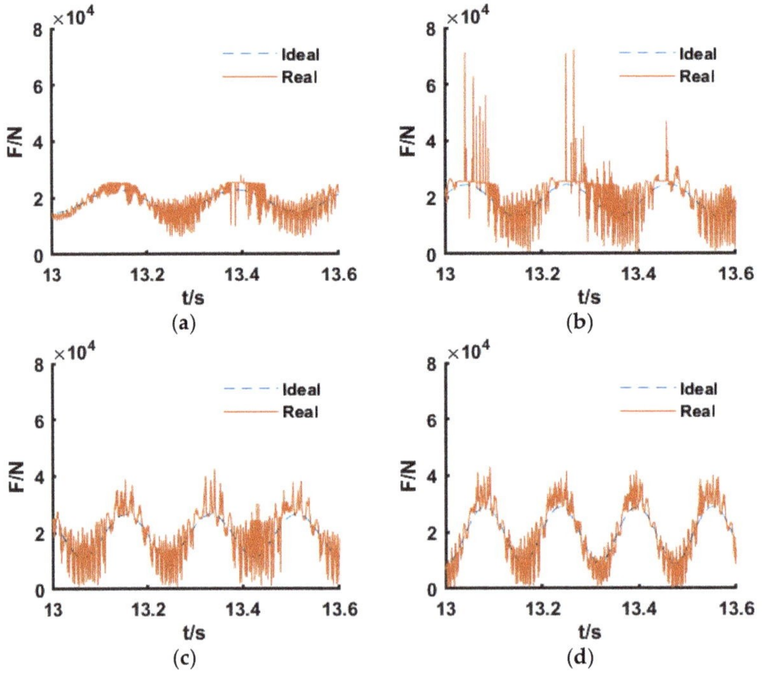


**Figure 9.** Vibration force on the foundation at different driving speeds. (**a**) $\omega = 25$. (**b**) $\omega = 30$. (**c**) $\omega = 35$. (**d**) $\omega = 40$.

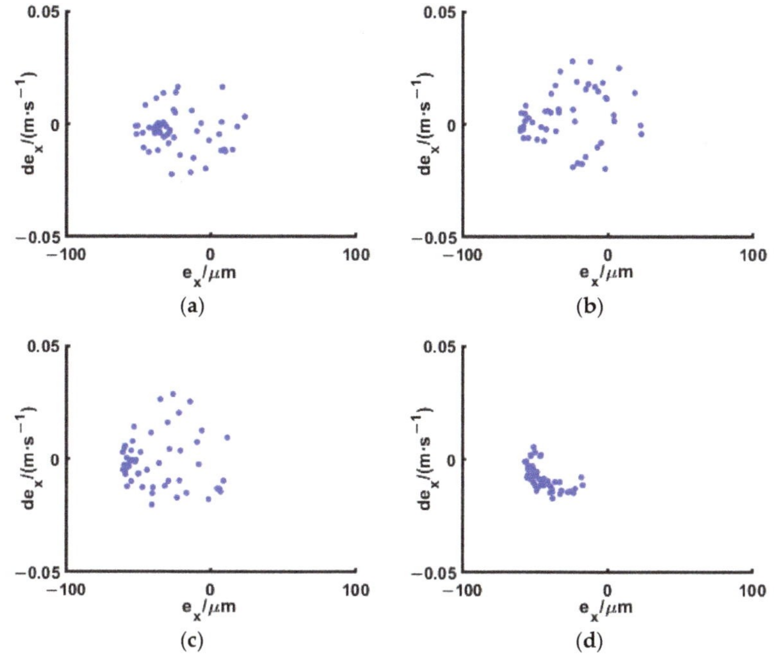


**Figure 10.** Poincare mapping at different driving speeds. (**a**) $\omega = 25$. (**b**) $\omega = 30$. (**c**) $\omega = 35$. (**d**) $\omega = 40$.

In Figure 8, the solid circle, which has the origin as its center and the clearance size as its radius, stands for the axial motion boundary of the shaft. The distribution of the collision locations is formed by the contact between the motion trajectory and the solid circle. According to an analysis of Figure 8, the collision points are concentrated in the upper right corner of the boundary at 25 rad/s, chaotically distributed at 30 rad/s and 35 rad/s, and steadily embedded at 40 rad/s. It shows that as the driving speed increases, the shock phenomenon in the mechanism first increases and then decreases. The reason is that when the eccentric shaft is rotating at low speed, the moveable jaw is affected by gravity and fixed load, and the collision occurs mostly on the upper right side of the bearing and shaft, which is easy to cause irregular wear of the revolute joint. As the driving speed increases, the collision forces increase accordingly and the motion becomes chaotic. If the driving speed is further increased, it is difficult for the bearing to make a large movement in a very short period due to the inertia force, and it can only make a stable movement along the axis motion boundary.

According to an analysis of Figure 9, compared to the case when there is no clearance, the existence of clearance causes nonlinear oscillations in the vibration forces applied to the foundation. The peak vibration forces at 25 rad/s, 30 rad/s, 35 rad/s and 40 rad/s are $2.3 \times 10^4$ N, $7.4 \times 10^4$ N, $4.1 \times 10^4$ N and $4 \times 10^4$ N, which are consistent with the variation in the shock phenomenon. The amplitude of the oscillation is greatest at a driving speed of 30 rad/s, with a peak increase of approximately 180%. It shows that as the driving speed increases, the influence on the vibration force applied to the foundation and the foundation service life first increases and then decreases.

The Poincare mapping technique [25] is a qualitative approach to identifying the chaos phenomenon. When the discrete points show one or fewer, closed curves, hierarchical and banded distributions, it means that the system is in a state of periodic, quasi-periodic, and chaotic motion. Figure 10 shows that the Poincare mapping points display a band distribution and no association between neighboring mapping points as the driving speed increases, but the distribution range widens. At the driving speed of 40 rad/s, the distribution range of the Poincare mapping points shrinks and moves in the direction of a point-like trend. It shows that as the driving speed increases, the chaos phenomenon decreases and the stability of the mechanism increases. It is shown that as the driving speed increases, the chaos phenomenon, like the shock phenomenon, first increases and then decreases, i.e., the operating stability of the mechanism first decreases and then increases.

In summary, for a clearance size of 0.06 mm and a driving speed range of 25 rad/s to 40 rad/s, the driving speed of 40 rad/s is chosen as the best considering the wear problem. Therefore, in the case of a specific clearance size, the maximum value within the driving speed selection range should be selected, which can weaken the chaos and shock phenomenon in the mechanism, enhance the stability of the mechanism operation, weaken the influence of the clearance on the vibration force applied to the foundation, and prolong the service life of the foundation.

### 5.2. Influence of Clearance Size

Given the driving speed of 30 rad/s and the clearance size of 0.03, 0.06, 0.09, 0.12 mm, Figures 11–13 represent the characteristic images of the axial trajectory of the shaft, the vibration force applied to the foundation, and the Poincare mapping at different driving speeds, respectively.

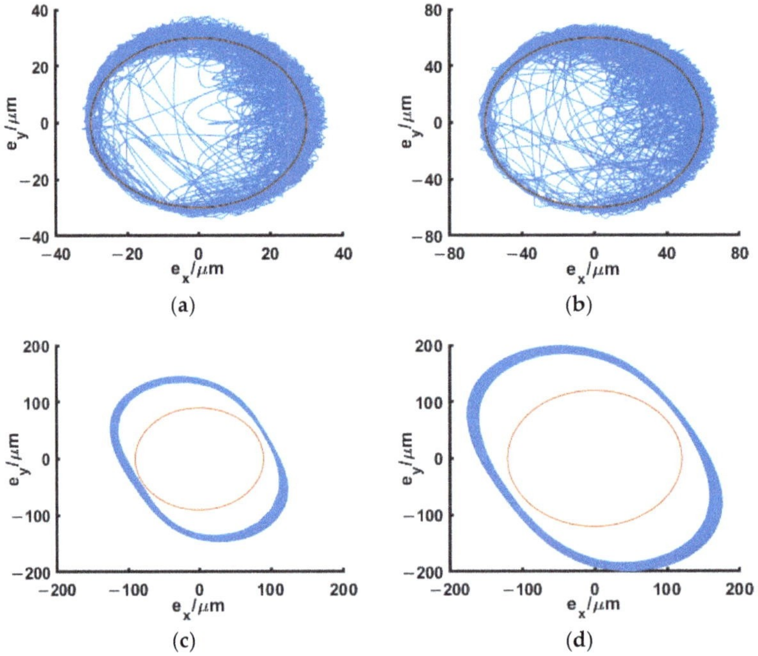


**Figure 11.** Axis trajectory of shaft at different clearance sizes. (**a**) $r = 0.03$. (**b**) $r = 0.06$. (**c**) $r = 0.09$. (**d**) $r = 0.12$.

Analysis of Figure 11 shows that when the clearance size is 0.03 mm or 0.06 mm, the collision point distribution is chaotic. After increasing to 0.09 mm, the axial trajectory is distributed in an elliptical band, and the clearance joint elements contact stably. After increasing to 0.12 mm, the state remains stable and the embedding depth increases. It visually demonstrates the change process of the shock phenomenon that first enhances and then disappears as the clearance size increases. This is because the limited clearance sizes reduce the eccentric shaft's range of motion and increase the likelihood of collisions.

As the clearance size increases, the collision forces increase accordingly and the motion becomes chaotic. When the clearance size is increased to a certain level, the movable jaw is able to maintain contact with the eccentric shaft due to the inertial force and make a stable motion. This is consistent with the scenario of the "momentum exchange method" and the "massless linkage method" proposed by Flores [26].

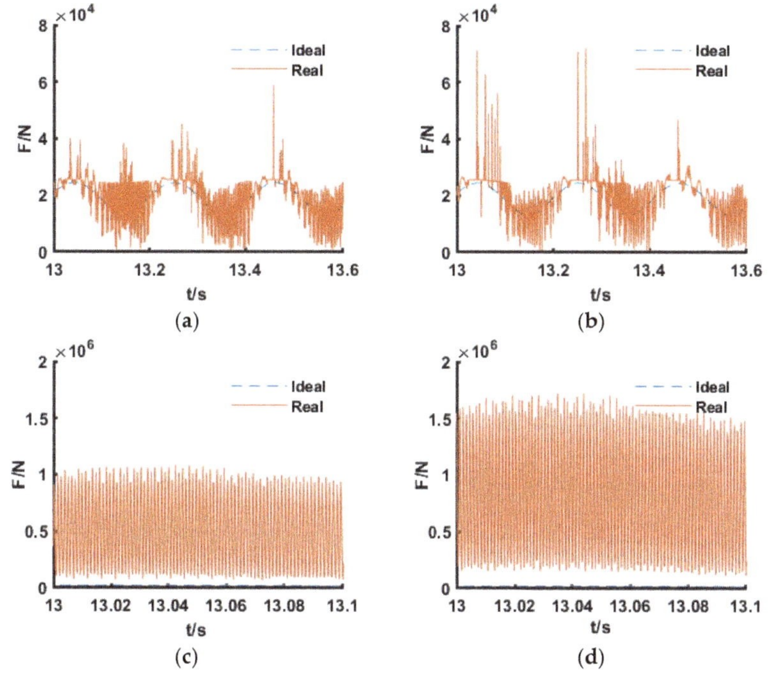


**Figure 12.** Vibration force on the foundation at different clearance sizes. (**a**) $r = 0.03$. (**b**) $r = 0.06$. (**c**) $r = 0.09$. (**d**) $r = 0.12$.

Analysis of Figure 12 shows that the vibration force on the foundation is influenced by changes in the shock phenomenon and exhibits the same dynamic characteristics. When the clearance size is 0.03 mm and 0.06 mm, the peak vibration force is $6 \times 10^4$ N and $7.4 \times 10^4$ N, respectively, when the clearance size increases to 0.09 mm and 0.12mm, the change curve smooths out and the peak vibration force is $1.1 \times 10^6$ N and $1.7 \times 10^6$ N, respectively. However, at 0.12 mm, the mean vibration force with clearance is $8 \times 10^5$ N and the mean vibration force without clearance is $1.8 \times 10^4$ N, showing an increase of nearly 45 times in comparison, far exceeding the amplitude of oscillation at 0.06 mm. It is shown that the smaller the clearance size, the less the influence on the vibration force on the foundation and the less the influence on the service life of the foundation.

Analysis of Figure 13 shows that for the clearance size of 0.03 mm, there is no correlation between adjacent Poincare mapping points. After increasing to 0.06 mm, the distribution of mapping points expands. After increasing to 0.09 mm and 0.12 mm, the mapping points form a closed curve. It is shown that as the clearance size increases, the chaotic motion of the system gradually strengthens and then enters the state of quasi-periodic motion, where the stability of the mechanism operation first weakens and then strengthens.

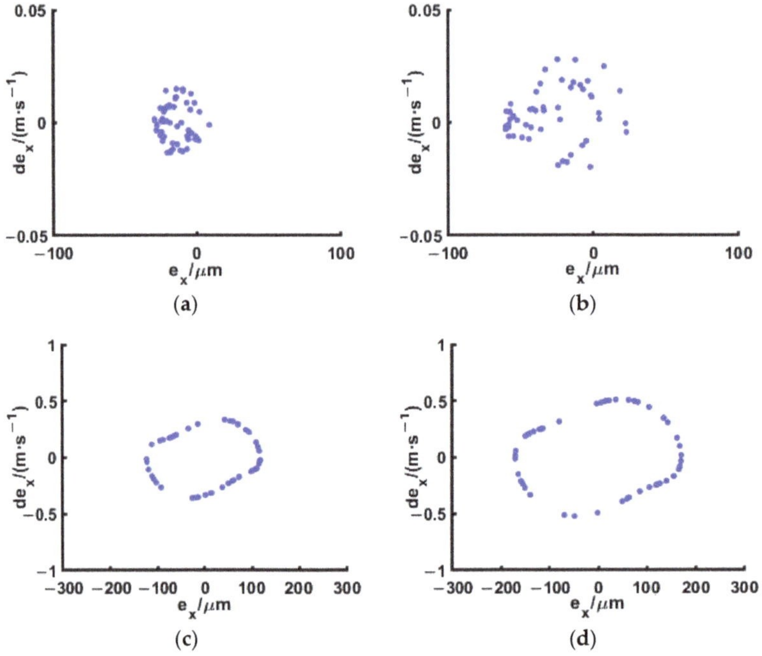


**Figure 13.** Poincare mapping at different clearance sizes. (**a**) $r = 0.03$. (**b**) $r = 0.06$. (**c**) $r = 0.09$. (**d**) $r = 0.12$.

In summary, for a driving speed of 30 rad/s and a clearance size range of 0.03 mm to 0.12 mm, the clearance size of 0.03 mm is chosen as the best considering the magnitude of the vibration force. Therefore, in the case of a specific driving speed, the minimum value within the clearance size selection range should be selected, which can weaken the chaos and shock phenomenon in the mechanism, enhance the stability of the mechanism operation, weaken the influence of the clearance on the vibration force applied to the foundation, and prolong the service life of the foundation.

### *5.3. Influence of Driving Speed and Clearance Size*

Figure 14a shows the root-mean-square values of the vibration force on the foundation for a driving speed range of [25, 40] and a clearance size range of [0.03, 0.12]. The three-dimensional surface is able to show the magnitude of $F_{rms}$ corresponding to different $(r, \omega)$. The contour plots of $F_{rms}$ are shown in Figure 14b, which can graphically express the effects of the driving speed and the clearance size together on the vibration force applied to the foundation.

Analysis of Figure 14a,b shows that the contours are densely distributed on the right side, and the value of each contour increases tenfold, from $1 \times 10^4$ N all the way to $9 \times 10^4$ N. It indicates that the vibration force increases sharply after the clearance size increases to a certain degree, which verifies the analysis of the vibration force in Section 5.2. The combination of driving speed and clearance size should be chosen from the darkest color part since its vibration force is much lower than that of the other parts and its influence on the service life of the foundation is likewise much smaller.

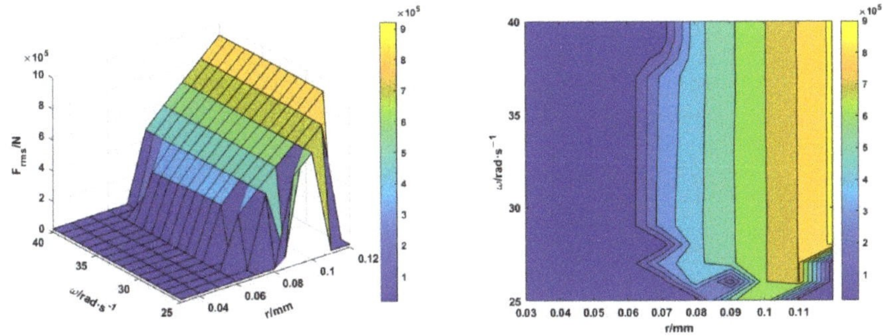


(**a**) (**b**)

**Figure 14.** Root-mean-square values of the vibration force on the foundation: (**a**) $r - \omega - F_{rms}$ three-dimensional surface diagram; (**b**) $r - \omega - F_{rms}$ contour map.

According to the previous analysis, the peak is able to reflect the shock phenomenon and the operational stability of the mechanism. Based on the conclusion in Section 5.2 on the selection of clearance size, the local surface and contour maps of the peak vibration force are plotted based on the darkest color part in Figure 14b with $r$ = 0.06 mm as the dividing line. Figure 15a shows the peak vibration forces on the foundation for a driving speed range of [25, 40] and a clearance size range of [0.03, 0.06]. The three-dimensional surface is able to show the magnitude of $F_{max}$ corresponding to different $(r, \omega)$. The contour plots of $F_{max}$ are shown in Figure 15b.

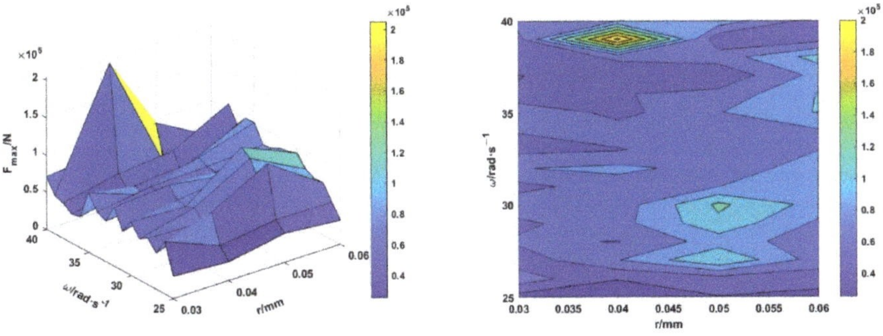


(**a**) (**b**)

**Figure 15.** Peak vibration forces on the foundation: (**a**) $r - \omega - F_{max}$ three-dimensional surface diagram; (**b**) $r - \omega - F_{max}$ contour map.

Analysis of Figure 15a,b shows that the contour distribution is complex, increasing from $2 \times 10^4$ N all the way to $2 \times 10^5$ N, where the darkest color part is distributed near the driving speed of 25 rad/s. Considering that a smaller driving speed can lead to irregular wear of the revolute joint, the combination solution is chosen from the part with the second darkest color. Combining the conclusions on the driving speed and clearance size selection in Sections 5.1 and 5.2, the optimal combination solution is 39 rad/s and 0.03 mm, with a mean vibration force of $2.19 \times 10^4$ N and a peak vibration force of $3.8 \times 10^4$ N.

Bo, etc., [27] analyzed the effect of the revolute joint with clearance on the dynamic characteristics of the PE250 × 400 compound pendulum jaw crusher in terms of jaw speed, acceleration and vibration force at a driving speed of 30 rad/s and a clearance size of 0.06 mm. Some of the characteristic images are shown in Figures 8b, 9b and 10b. Figure 16 shows the axis trajectory of the shaft, the vibration force applied to the foundation, and the Poincare mapping of the crusher with clearance when the optimal driving speed and clearance size suggested in this study are utilized.

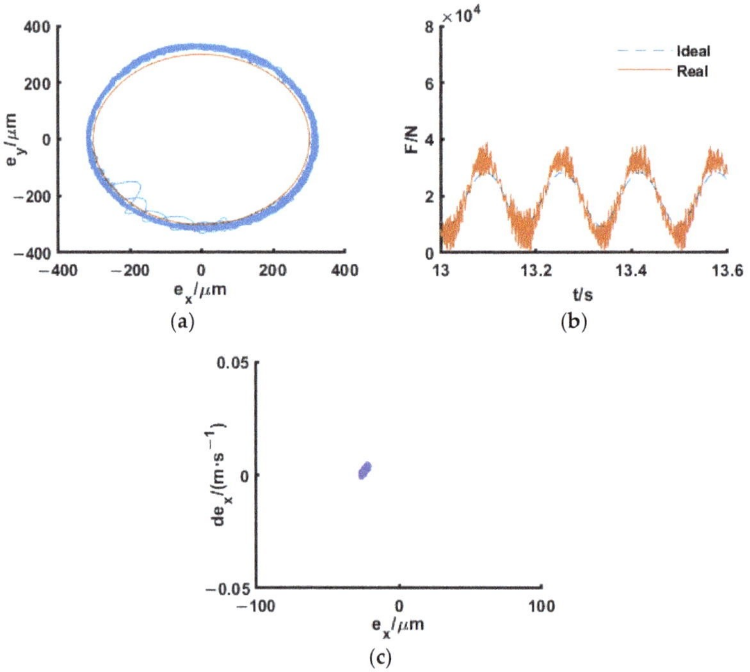


**Figure 16.** Optimized dynamic characteristic images: (**a**) the axis trajectory of shaft; (**b**) the vibration force applied to the foundation; (**c**) the Poincare mapping.

Comparing Figure 8b with Figure 16a, Figure 9b with Figure 16b, and Figure 10b with Figure 16c, it is clear that the optimal combination solution significantly reduces the shock and chaos phenomenon in the mechanism, maintains the stable contact between the revolute joint's elements, and reduces the peak vibration force applied to the foundation from $7.4 \times 10^4$ N to $3.8 \times 10^4$ N, or about 48.6%, with a smooth change curve.

## 6. Conclusions

Dynamics analysis and chaos identification of the compound pendulum jaw crusher with joint clearance is conducted in this paper. Because the existing research only focuses on the optimal design of the crusher structure and the existence of clearance is neglected, this study introduces chaos into the dynamic characteristic evaluation system and analyses the effect of clearance on the crusher under different driving speeds and clearance sizes. On the basis of the proposed theoretical investigations, the obtained major conclusions can be drawn:

- Comparative simulations in MATLAB and Adams verify the correctness of the clearance mechanism modeling method and solution method, which provides a reference for the research on the dynamics of crushers with clearance.
- The dynamic simulation analysis verifies that the existence of clearance leads to significant oscillations in the vibration forces, which seriously affect the service life of the foundations. A larger driving speed and a smaller clearance size can be beneficial to reduce the shock and chaos phenomenon during the operation of the mechanism, reduce the influence of the clearance on the vibration force applied to the foundation, and improve the stability of the mechanism operation.

- Analysis of the combined effect of driving speed and clearance size shows that the optimal combination of 39 rad/s and 0.03 mm reduces the peak vibration force by 48.6% while meeting the crusher production capacity.

**Author Contributions:** Conceptualization, Y.C. and S.W.; methodology, S.W.; software, S.W.; validation, Y.C., S.W. and C.W.; formal analysis, S.W.; investigation, S.W.; resources, Y.C.; data curation, S.W.; writing—original draft preparation, S.W.; writing—review and editing, S.W.; visualization, Y.C.; supervision, C.W.; project administration, C.W.; funding acquisition, Y.C. All authors have read and agreed to the published version of the manuscript.

**Funding:** This research was funded by National Natural Science Foundation of China (12272217).

**Institutional Review Board Statement:** Not applicable.

**Informed Consent Statement:** Not applicable.

**Data Availability Statement:** The data are available from the corresponding author upon reasonable request.

**Conflicts of Interest:** The authors declare no conflict of interest.

## References

1. Barrios, G.K.; Jiménez-Herrera, N.; Fuentes-Torres, S.N.; Tavares, L.M. DEM Simulation of Laboratory-Scale Jaw Crushing of a Gold-Bearing Ore Using a Particle Replacement Model. *Minerals* **2020**, *10*, 717. [CrossRef]
2. Deniz, V. A Study on the Effects of Coal Feed Size and Coal Type on the Performance of a Laboratory Jaw Crusher. *Energy Sources Part A* **2014**, *36*, 1249–1255. [CrossRef]
3. Ulsen, C.; Tseng, E.; Angulo, S.C.; Landmann, M.; Contessotto, R.; Balbo, J.T.; Kahn, H. Concrete Aggregates Properties Crushed by Jaw and Impact Secondary Crushing. *J. Mater. Res. Technol.* **2019**, *8*, 494–502. [CrossRef]
4. Osanloo, M.; Paricheh, M. In-pit Crushing and Conveying Technology in Open-pit Mining Operations: A Literature Review and Research Agenda. *Int. J. Min. Reclam. Environ.* **2020**, *34*, 430–457. [CrossRef]
5. Lapin, S.K. Vibration Insulation for Jaw Crushers—An Effective Means of Reducing Bed and Foundation Deformations. *Soil Mech. Found. Eng.* **2000**, *37*, 198–201. [CrossRef]
6. Yang, C.L.; Huang, D.M.; Xiong, J.F.; Liu, X.H. The Weight Reduction Optimization of Jaw Crusher about Moving Jaw Based on Parametric Finite Element Analysis. In Proceedings of the 1st International Workshop on Hydraulic Equipment and Support Systems for Mining (IWHEM 2012), Huludao, China, 17–18 August 2012; p. 275.
7. Li, Z.; Chen, W.; Zhang, W.; Zhang, X.; Wen, B. Theoretical, Numerical, and Experimental Study on the Synchronization in a Vibrator–pendulum Coupling System. *Arch. Civ. Mech. Eng.* **2022**, *22*, 157. [CrossRef]
8. Cheng, J.; Ren, T.; Zhang, Z.; Jin, X.; Liu, D. Influence of Two Mass Variables on Inertia Cone Crusher Performance and Optimization of Dynamic Balance. *Minerals* **2021**, *11*, 163. [CrossRef]
9. Zhao, Z.; Li, Y.; Li, W.; Zhan, X.; Zhu, X.; Zhong, J. Research on the Biaxial Compound Pendulum Jaw Crusher Based on Seven-bar Mechanism. *Proc. Inst. Mech. Eng. Part C J. Mech. Eng. Sci.* **2015**, *230*, 1876–1889.
10. Wang, X.; Liu, G.; Ma, S. Dynamic Analysis of Planar Mechanical Systems with Clearance Joints Using a New Nonlinear Contact Force Model. *J. Mech. Sci. Technol.* **2016**, *30*, 1537–1545. [CrossRef]
11. Chen, X.; Jiang, S.; Deng, Y.; Wang, Q. Dynamics analysis of 2-DOF complex planar mechanical system with joint clearance and flexible links. *Nonlinear Dyn.* **2018**, *93*, 1009–1034. [CrossRef]
12. Wu, X.; Sun, Y.; Wang, Y.; Chen, Y. Correlation Dimension and Bifurcation Analysis for the Planar Slider-crank Mechanism with Multiple Clearance Joints. *Multibody Syst. Dyn.* **2021**, *52*, 95–116. [CrossRef]
13. Chen, X.; Jia, Y.; Deng, Y.; Wang, Q. Dynamics Behavior Analysis of Parallel Mechanism with Joint Clearance and Flexible Links. *Shock Vib.* **2018**, *2018*, 9430267. [CrossRef]
14. Farahan, S.B.; Ghazavi, M.R.; Rahmanian, S. Bifurcation in a Planar Four-bar Mechanism with Revolute Clearance Joint. *Nonlinear Dyn.* **2017**, *87*, 955–973. [CrossRef]
15. Lankarani, H.M.; Nikravesh, P.E. A Contact Force Model With Hysteresis Damping for Impact Analysis of Multibody Systems. *J. Mech. Des.* **1990**, *112*, 369–376. [CrossRef]
16. Ambrósio, J.A.C. Dynamics of Structures Undergoing Gross Motion and Nonlinear Deformations: A Multibody Approach. *Comput. Struct.* **1996**, *59*, 1001–1012. [CrossRef]
17. Chen, X.; Wang, J. Nonlinear Dynamic Analysis and Experimental Study of Multi-link Press with Dry Friction Clearances of Revolute Joints. *Meccanica* **2022**, *57*, 2627–2652. [CrossRef]
18. Chen, X.; Guo, J. Effects of Spherical Clearance Joint on Dynamics of Redundant Driving Spatial Parallel Mechanism. *Robotica* **2021**, *39*, 1064–1080. [CrossRef]
19. Baumgarte, J. Stabilization of Constraints and Integrals of Motion in Dynamical Systems. *Comput. Methods Appl. Mech. Eng.* **1972**, *1*, 1–16. [CrossRef]

20. Bai, Z.F.; Zhao, Y.; Chen, J. Dynamics Analysis of Planar Mechanical System Considering Revolute Clearance Joint Wear. *Tribol. Int.* **2013**, *64*, 85–95. [CrossRef]
21. Chen, X.; Jiang, D. Design, Kinematics, and Statics of a Novel Wave Energy Converter with Parallel Mechanism. *Int. J. Adv. Robot. Syst.* **2019**, *16*, 1729881419876214. [CrossRef]
22. Zhong, X.; Niu, X.; Ji, Q.; Shen, X. Simulation Analysis of Cavity Shape of Compound Pendulum Jaw Crusher. *J. Phys. Conf. Ser.* **2020**, *1637*, 012130. [CrossRef]
23. Tang, Y.; Chang, Z.; Dong, X.; Hu, Y.; Yu, Z. Nonlinear Dynamics and Analysis of a Four-bar Linkage with Clearance. *Front. Mech. Eng.* **2013**, *8*, 160–168. [CrossRef]
24. Chen, Z.; Wang, G.; Xue, D.; Bi, Q. Simulation and Optimization of Gyratory Crusher Performance Based on the Discrete Element Method. *Powder Technol.* **2020**, *376*, 93–103. [CrossRef]
25. Javanfar, A.; Bamdad, M. Effect of Novel Continuous Friction Model on Nonlinear Dynamics of the Mechanisms with Clearance Joint. *Proc. Inst. Mech. Eng. Part C J. Mech. Eng. Sci.* **2021**, *236*, 6040–6052. [CrossRef]
26. Flores, P.; Ambrosio, J. Revolute Joints with Clearance in Multibody Systems. *Comput. Struct.* **2004**, *82*, 1359–1369. [CrossRef]
27. Bo, S.J.; Guo, R.B.; Zhu, H. Dynamic Study for Working Device of PE-250*400 Crusher. *Adv. Mater. Res.* **2013**, *753–755*, 1699–1702. [CrossRef]

*entropy*

MDPI

*Article*

# Decomposition–Coordination of Double-Layer MPC for Constrained Systems

**Hongrui Wang [1,2,3,4], Pengbin Zhang [1,2,3,4], Zhijia Yang [1,2,3] and Tao Zou [5,*]**

[1] State Key Laboratory of Robotics, Shenyang Institute of Automation, Chinese Academy of Sciences, Shenyang 110016, China
[2] Key Laboratory of Networked Control Systems, Chinese Academy of Sciences, Shenyang 110016, China
[3] Institutes for Robotics and Intelligent Manufacturing, Chinese Academy of Sciences, Shenyang 110169, China
[4] University of Chinese Academy of Sciences, Beijing 100049, China
[5] School of Mechanical and Electrical Engineering, Guangzhou University, Guangzhou 510006, China
* Correspondence: tzou@gzhu.edu.cn

**Abstract:** Large-scale industrial processes usually adopt centralized control and optimization methods. However, with the growth of the scale of industrial processes leading to increasing computational complexity, the online optimization capability of the double-layer model predictive control algorithm is challenged, exacerbating the difficulty of the widespread implementation of this algorithm in the industry. This paper proposes a distributed double-layer model predictive control algorithm based on dual decomposition for multivariate constrained systems to reduce the computational complexity of process control. Firstly, to solve the problem that the original dual decomposition method does not apply to constrained systems, two improved dual decomposition model prediction control methods are proposed: the dual decomposition method based on the quadratic programming in the subsystem and the dual decomposition method based on constraint zones, respectively. It is proved that the latter will certainly converge to the constraint boundaries with appropriate convergence factors for the controlled variables. The online optimization ability of the proposed two methods is compared in discussion and simulation, concluding that the dual decomposition method based on the constraint zones exhibits superior online optimization ability. Further, a distributed double-layer model predictive control algorithm with dual decomposition based on constraint zones is proposed. Different from the objective function of the original dual decomposition model predictive control, the proposed algorithm's dynamic control-layer objective function simultaneously tracks the steady-state optimization values of the controlled and manipulated variables, giving the optimal solution formulation of the optimization problem consisting of this objective function and the constraints. The algorithm proposed in this paper achieves the control goals while significantly reducing the computational complexity and has research significance for promoting the industrial implementation of double-layer model predictive control.

**Keywords:** double-layer; dual decomposition; model predictive control

## 1. Introduction

Modern industrial processes are characterized by large-scale components, extensive spatial structure, and strong sub-unit coupling [1]. The hierarchical optimization and control structure is often used for high-dimensional complex systems. It contains a planning and scheduling layer, real-time optimization (RTO) layer, advanced process control (APC) layer, and a regular control layer [2,3]. Model predictive control (MPC) is a type of APC algorithm that deals with multi-input multi-output constraint systems. The successful implementation of MPC can generate considerable revenues for companies, but its practical application is not universal for smaller companies with insufficient capacities. Due to the time scale between layers being different, disturbances entering into the process during any control period will result in a shift in the static operating point, and dynamic control cannot work on the optimal point, thus affecting economic efficiency. The optimal results of the RTO layer may not be suitable for current operating conditions. A two-stage MPC structure is commonly used in the industry [4–7], so many scholars proposed double-layer model predictive control (DLMPC). The steady-state optimization (SSO) layer of the DLMPC is able to adjust constraints and optimize steady-state operating points based on the steady-state model, while the dynamic control layer accomplishes the tracking of set values based on the dynamic control (DC) model [8–11].

The premise of implementing hierarchical optimization and control structure is based on a decentralized control system, realizing the coordination optimization among subsystems with control algorithms. Compared to centralized control systems, decentralized control systems can reduce computational complexity. Tamás Keviczky et al., facing a class dynamic decoupling systems' optimal control problem, designed a decentralized rolling horizon control (RHC) scheme, describing the coupling between systems through diagrams, which decomposes centralized RHC controllers into some small RHC controllers to reduce the complexity of the problem and analyzes the sufficient stability conditions for the proposed solution [12]. For extensive dynamic processes with input constraints, Alessandro Alessio proposed decentralized model predictive control that approximated the global model by decomposing it into several smaller models for local predictions, giving sufficient conditions for asymptotic closed-loop stability in the absence of intermittent communication of measurement data [13]. Focusing on the lack of intermittent measurement data communication between decentralized model prediction controllers, Davide Barcelli gave sufficient criteria for the asymptotic tracking of output setpoints and rejection of constant measurement disturbances, and proposed a decentralized model predictive control method for extensive process setpoint tracking [14].

Although powerful algorithms quantify the interactions of decentralized control loops, decentralized control systems still lack the coupling between multiple variables [15]. In contrast to decentralized control systems, distributed control methods enable controllers to communicate with each other to cooperate while operating individually. Each controller considers the dynamic interactions among systems [16]. Aiming at the internet multi-objective optimal problem, Andrea Camisa proposed a distributed model predictive control (DMPC) calculation scheme of negotiation among agents based on cooperative game theory [17]. M. Francisco introduced a new fuzzy inference system into the cooperative game algorithm to propose multi-agent fuzzy negotiation distributed model predictive control considering economic criteria and process constraints on the negotiation process of the agents [18]. For the control of large urban traffic networks, Zhao Zhou et al. proposed a two-level hierarchical control framework, which uses the MPC method, with the upper level solving higher-level optimization problems for traffic-demand balance and the lower level adopting distributed control schemes within each sub-network to reduce the computational complexity [19]. A negotiation strategy among agents was designed using fuzzy rules under a hierarchical DMPC control architecture. The agents negotiate in pairs at the lower layers based on coupling and communication networks, with the lower-level negotiation process avoiding a portfolio explosion [20]. Under non-iterative, non-cooperative MPC algorithms, Marcello Farina proposed a DMPC algorithm that requires only partially connected communication networks and structural information [21]. In terms of non-cooperative strategies, Haimin Hu proposed new distributed iterative learning model predictive control that uses the local states and input trajectories of the previous iteration to construct time-varying safety sets and terminal cost functions [22]. For multi-area interconnected power systems, an algorithm based on Laguerre functions for DMPC containing the game-theoretic Nash equilibrium was proposed, with each regional DMPC controller coordinating with other controllers to find Nash equilibrium points [23].

Many scholars have proposed distributed or decentralized MPC algorithms based on dual decomposition. Joseph J. Yame et al. proposed a new DMPC containing both decomposition subsystem and integrated coordination sub-controller stages [24]. Takumi

Namba et al. proposed a dual decomposition DMPC and applied it to a microgrid with the large-scale introduction of PV power [25]. Yuji Wakasa et al. proposed a decentralized model predictive control algorithm based on dual decomposition, which enables the decentralized control approach to solve the original optimization problem accurately using iterations [26]. Xi et al. proposed a decomposed–coordinated model predictive control (DCMPC) algorithm based on the theory of dual decomposition, but the algorithm is oriented towards DMC with only equation constraints without considering the case with constraints [2]. The authors used hierarchical distributed predictive control as a key search term to review the relevant literature [27–31]. At present, there are few research results on distributed control algorithms in the DLMPC structure. Yang Kai et al. proposed an integrated algorithm for real-time optimization and distributed control, with an overall economic optimization model for the upper layer and a distributed dynamic control structure for the lower layer [32]. Shi et al. proposed a distributed two-layer structure strategy for large-scale systems, with an online adaptive constraint adjustment scheme for the upper layer considering the possible constraints and priority order in the process. Based on the Pareto optimal algorithm, the lower layer proposed a new cooperative distributed dynamic matrix control based on a Jacobi-type iterative cooperation approach to achieve a globally optimal solution [33].

Centralized control optimization is often employed for industrial process control. However, the high computational effort of centralized control optimization challenges the MPC's online optimization capabilities, exacerbating the difficulty of the widespread implementation of DLMPC algorithms in the industry. Reducing the more considerable computational complexity with a bit of sacrifice of control performance under satisfying the control goals is of greater significance in promoting the popularization of the industrial implementation of DLMPC algorithms. To reduce the computational complexity of industrial process control, research in this paper is oriented towards constrained multivariable distributed control systems. Based on decomposition–coordination MPC, we propose two strategies for adding constraints, and based on one of them we propose decomposition–coordination of DLMPC for constrained systems, with the following main contributions: Firstly, two methods are proposed to add variable constraints based on the original decomposition–coordination (dual decomposition method) MPC algorithm to solve the problem that the original method is not applicable to multivariable systems with constraints. In the first method, based on the dual decomposition method, the suboptimization problem with constraints for each subsystem forms a quadratic programming (QP) problem. The second method is the dual decomposition method based on the constrained zones, which analyzes the convergence relationship between the variables and the constraints. It proves that if the convergence factor is small enough, the solutions will eventually definitely converge to the boundaries of the constraints. Then, both proposed methods are discussed and analyzed based on their performances, concluding that the second method has superior online optimization capabilities, which are validated by Simulation 1. Further, a distributed DLMPC algorithm based on the dual decomposition of constraint zones is proposed, where the decomposition–coordinated dynamic control layer simultaneously tracks the steady-state optimized values of the controlled variable (CV) and manipulated variable (MV), which is also different from the original decomposition–coordinated dynamic control objective function, giving an optimal solution expression that added the tracking of the steady-state optimized values of the MV, and proving the effectiveness of the proposed algorithm through Simulation 2. The improved DLMPC algorithm in this paper satisfies the control goals and constraints while greatly reducing the computational complexity of the dynamic control layer, thus improving the online optimization capability of the algorithm. It is of interest and value to provide a fundamental theoretical study for the industrial implementation of distributed DLMPC.

This paper is arranged as follows: Section 1 is the Introduction. Section 2 provides an overview of DLMPC and DCMPC as a foundation for the later paper. Section 3 proposes two improved dual decomposition methods, namely, the dual decomposition method based on subsystem QP and the dual decomposition method based on the constraint zones, discussing and analyzing the performance of the two methods. Section 4 proposes a new DLMPC algorithm based on the dual decomposition method of the constraint zone. Based on the original DCMPC, which only tracks the external targets of the controlled variables, it adds the ability to track the external targets of the manipulated variable, to track the steady-state optimized value of the MV, and to give a characterization of the optimal solution under such an objective function. Section 5 simulates and validates the algorithms proposed in Sections 3 and 4, respectively, employing the Shell heavy oil fractionation model. Section 6 is conclusions.

Partial abbreviations and notations are shown in Table 1.

**Table 1.** Main abbreviations or notations.

| Abbreviations/Notations | Meaning | Notations | Meaning |
|---|---|---|---|
| CV | Controlled variable | $\underline{\mathbf{U}}$ | MV constraint lower limits |
| MV | Manipulated variable | $\underline{\mathbf{Y}}$ | CV constraint lower limits |
| $\mathbf{U}$ | MV | $\overline{\mathbf{U}}$ | MV constraint upper limits |
| $\mathbf{Y}$ | CV | $\overline{\mathbf{Y}}$ | MV constraint upper limits |
| Subscript T | Tracking targets | Subscript $^{ss}$ | Steady-state optimization layer |
| $\Delta$ | Increment | $\widetilde{y}_{PM}(k)$ | The output predicted value of $P$ time domain in the future under $M$ control law at the time $k$ |
| $\mathbf{S}$ | Shift matrix | $\widetilde{y}_{P0}(k)$ | The initial output predicted value of the future $P$ time domains at the time $k$ |
| $\mathbf{h}$ | Correction matrix | $\mathbf{w}(k)$ | The reference trajectory of the controlled variable at the time $k$ |
| $N$ | Model length | $u_M(k)$ | $M$ control law increments at the time $k$ |
| $P$ | Prediction time horizon | $\mathbf{G}_{ss}$ | Steady-state transfer function matrix |
| $M$ | Control time horizon | $\widetilde{y}_{cor}$ | Corrected predicted values of the controlled variables |

## 2. Preliminary Knowledge

### 2.1. Double-Layer Model Predictive Control

DLMPC contains a steady-state optimization layer and dynamic control layer, and the structure is shown in Figure 1.

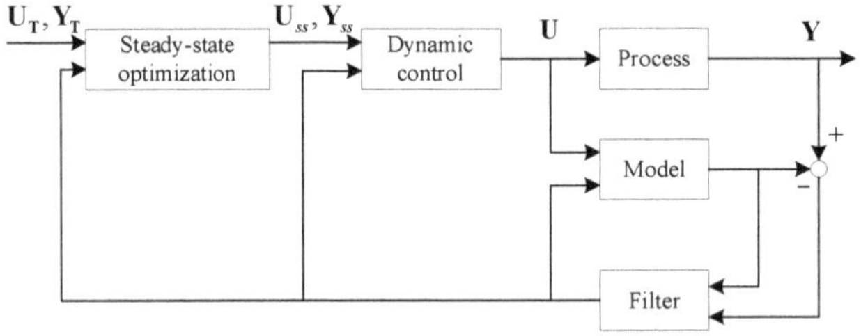


**Figure 1.** DLMPC structure.

The steady-state optimization layer calculates the steady-state target through the steady model of plants, and the steady-state targets are tracked in the dynamic control layer. There are two modes in the steady-state optimization layer: economic optimization (EO) and target tracking (TT). The economic optimization mode characterizes the benefit maximization or energy consumption minimization as a linear programming problem with the manipulated variables, as shown in Equation (1).

$$\min_{\Delta \mathbf{U}_{ss}(k)} \Xi = \mathbf{C}^T * \Delta \boldsymbol{U}_{ss}(k) \tag{1}$$

where **C** is the economic cost coefficient. The purpose of the steady-state target tracking model is to find the optimal steady-state target based on a given RTO setpoint. Its optimization problem can be characterized as a quadratic programming problem, as shown in Equation (2).

$$\min_{\Delta U_{ss}(k)} \Xi = \|U_{ss}(k) - \mathbf{U}_T\|^2_{\mathbf{R}_{ss}} + \|U_{ss}(k) - \mathbf{Y}_T\|^2_{\mathbf{Q}_{ss}} + \|\Delta U_{ss}(k)\|^2_{\mathbf{O}_{ss}} \tag{2}$$

The optimal steady-state operating point calculated by the steady-state optimization layer is used as the setpoint for tracking by the dynamic control layer, which forms the online optimization problem shown in Equation (3).

$$\min_{\Delta u_M(k)} J(k) = \|Y_{ss}(k) - \tilde{y}_{PM}(k)\|^2_{\mathbf{Q}} + \|U_{ss}(k) - u_M(k)\|^2_{\mathbf{R}} + \|\Delta u_M(k)\|^2_{\mathbf{O}} \tag{3}$$

where $M$ is the control time horizon, and $P$ is the prediction time horizon. The error power matrix and control power matrix are shown in Equation (4).

$$\begin{aligned} \mathbf{Q} &= \text{block} - \text{diag}(\mathbf{Q}_1, \cdots, \mathbf{Q}_p), \mathbf{Q}_i = \text{diag}[\mathbf{q}_i(1), \cdots, \mathbf{q}_i(\mathbf{P})] \\ \mathbf{R} &= \text{block} - \text{diag}(\mathbf{R}_1, \cdots, \mathbf{R}_m), \mathbf{R}_j = \text{diag}[\mathbf{r}_j(1), \cdots, \mathbf{r}_j(\mathbf{M})] \\ \mathbf{O} &= \text{block} - \text{diag}(\mathbf{O}_1, \cdots, \mathbf{O}_m), \mathbf{O}_j = \text{diag}[\mathbf{o}_j(1), \cdots, \mathbf{o}_j(\mathbf{M})] \end{aligned} \tag{4}$$

where block − diag means block diagonal matrix and diag means diagonal matrix.

The dynamic prediction model **A** (consisting of $p \times m$ block of matrices $P \times M$) is used to calculate the prediction value $\tilde{y}_{PM}(k)$ at time $k$ as shown in Equation (5).

$$\tilde{y}_{PM}(k) = \tilde{y}_{P0}(k) + \mathbf{A}\Delta u_M(k) \tag{5}$$

where $\tilde{y}_{P0}(k)$ is the initial predicted value, consisting of the predicted value at time $k-1$ and the prediction error. Using the optimal solution $\Delta u^*_M(k)$ of Equation (3) to calculate the manipulated variables at time $k$ is shown in Equation (6).

$$u_M(k) = u_M(k-1) + \Delta u^*_M(k) \tag{6}$$

### 2.2. Decomposed–Coordinated MPC

The online optimization problem for an unconstrained multi-input multi-output (MIMO) system is shown in Equation (7).

$$\begin{aligned} &\min_{\Delta u_M(k)} J(k) = \|\mathbf{w}(k) - \tilde{y}_{PM}(k)\|^2_{\mathbf{Q}} + \|\Delta u_M(k)\|^2_{\mathbf{R}} \\ &\text{s.t.} \quad \tilde{y}_{PM}(k) = \tilde{y}_{P0}(k) + \mathbf{A}\Delta u_M(k) \end{aligned} \tag{7}$$

where we suppose the system with $m$ inputs and $m$ outputs and decompose the original system into $m$ single-input single-output (SISO) subsystems. Due to the linear additivity among the subsystems, the original online optimization problem equals the sum of the predictive control sub-optimization problems, as shown in Equation (8).

$$\begin{aligned} &\min_{\Delta u_M(k)} J(k) = \sum_{i=1}^{m} \left\{ \|\mathbf{w}_i(k) - \tilde{y}_{i,PM}(k)\|^2_{\mathbf{Q}_i} + \|\Delta u_{i,M}(k)\|^2_{\mathbf{R}_i} \right\} \\ &\text{s.t.} \quad \tilde{y}_{i,PM}(k) = \tilde{y}_{i,P0}(k) + \sum_{j=1}^{m} \mathbf{A}_{ij}\Delta u_{j,M}(k),\ i = 1, \ldots, m \end{aligned} \tag{8}$$

Decomposed–Coordinated MPC originated from large-system control theory [34], a method that regards the association constraint $\tilde{y}_{PM}(k)$ in a coupled system as an independent variable with the same position as $\Delta u_M(k)$. Introducing the coordination factor $\boldsymbol{\lambda}$ calculates $\Delta u_{i,M}(k)$ and $\tilde{y}_{i,PM}(k)$ separately for each subsystem. By iterating and updating the coordination factor, so that $\widetilde{\boldsymbol{y}}_{PM}(k)$ and $\Delta\boldsymbol{u}_M(k)$ are associated in equilibrium, the optimal solution $\Delta u_M(k)$ is obtained as equivalent to the original problem.

Firstly, by introducing Lagrange multipliers $\boldsymbol{\lambda}$, the Lagrange function of Equation (8) is formed, as shown in Equation (9).

$$L(\Delta\boldsymbol{u}_M(k),\widetilde{\boldsymbol{y}}_{PM}(k),\boldsymbol{\lambda}(k)) \triangleq J(k) + \sum_{i=1}^{m} \lambda_i^{\mathrm{T}}(k)\left(\widetilde{\boldsymbol{y}}_{i,PM}(k) - \widetilde{\boldsymbol{y}}_{i,P0}(k) - \sum_{j=1}^{m} \mathbf{A}_{ij}\Delta\boldsymbol{u}_{j,M}(k)\right) \quad (9)$$

where $\boldsymbol{\lambda}(k) = [\ \boldsymbol{\lambda}_1^{\mathrm{T}}(k) \ \cdots \ \boldsymbol{\lambda}_m^{\mathrm{T}}(k)\ ]^{\mathrm{T}}$ and $\boldsymbol{\lambda}_i^{\mathrm{T}}(k) = [\ \lambda_i(k+1) \ \cdots \ \lambda_i(k+P)\ ]$, $i = 1,\ldots,m$. Then, the dual problem is solved, as shown in Equation (10), and the obtained unconstrained optimal solution is that of the original optimization problem.

$$\max_{\boldsymbol{\lambda}(k)} \min_{\Delta\boldsymbol{u}_M(k),\widetilde{\boldsymbol{y}}_{PM}(k)} L(\Delta\boldsymbol{u}_M(k),\widetilde{\boldsymbol{y}}_{PM}(k),\boldsymbol{\lambda}(k)) \quad (10)$$

The whole problem is decomposed into $m$ sub-optimization problems by solving the sub-optimization problem for $\Delta\boldsymbol{u}_{i,M}(k)$ and $\widetilde{y}_{i,PM}(k)$, respectively, and then updating the coordination factor $\boldsymbol{\lambda}(k)$ and iterating until the stopping condition. The procedure can be depicted as a two-stage optimization algorithm.

In the first stage, the Lagrange function of the original problem is minimized for a given coordination factor $\hat{\boldsymbol{\lambda}}(k)$ (Equation (9)), which can be expressed as the sum of the Lagrange functions of multiple sub-optimization problems, as shown in Equation (11).

$$\min_{\Delta\boldsymbol{u}_M(k),\widetilde{\boldsymbol{y}}_{PM}(k)} L\left(\Delta\boldsymbol{u}_M(k),\widetilde{\boldsymbol{y}}_{PM}(k),\hat{\boldsymbol{\lambda}}(k)\right) = \sum_{i=1}^{m} L_i\left(\Delta\boldsymbol{u}_{i,M}(k),\widetilde{\boldsymbol{y}}_{i,PM}(k),\hat{\boldsymbol{\lambda}}(k)\right) \quad (11)$$

where $L_i$ is shown in Equation (12).

$$\begin{aligned} &L_i(\Delta\boldsymbol{u}_{i,M}(k),\widetilde{y}_{i,PM}(k),\hat{\lambda}(k)) \\ &\triangleq \|\mathbf{w}_i(k) - \widetilde{\boldsymbol{y}}_{i,PM}(k)\|_{\mathbf{Q}_i}^2 + \|\Delta\boldsymbol{u}_{i,M}(k)\|_{\mathbf{R}_i}^2 \\ &+\hat{\lambda}_i^{\mathrm{T}}(k)(\widetilde{\boldsymbol{y}}_{i,PM}(k) - \widetilde{\boldsymbol{y}}_{i,P0}(k)) - \sum_{j=1}^{m}\left(\hat{\lambda}_j^{\mathrm{T}}(k)\mathbf{A}_{ji}\right)\Delta\boldsymbol{u}_{i,M}(k) \end{aligned} \quad (12)$$

According to the extreme value necessary condition, we can obtain $\widetilde{\boldsymbol{y}}_{i,PM}^*(k)$ and $\Delta\boldsymbol{u}_{i,M}^*(k)$, as shown in Equations (13) and (14).

$$\widetilde{\boldsymbol{y}}_{i,PM}^*(k) = \mathbf{w}_i(k) - 0.5\mathbf{Q}_i^{-1}\hat{\lambda}_i(k) \quad (13)$$

$$\Delta\boldsymbol{u}_{i,M}^*(k) = 0.5\mathbf{R}_i^{-1}\sum_{j=1}^{m}\left(\mathbf{A}_{ji}^{\mathrm{T}}\hat{\lambda}_j(k)\right) \quad (14)$$

In the second stage, the coordination factor $\hat{\boldsymbol{\lambda}}(k)$ is updated according to the solution of $\max_{\hat{\boldsymbol{\lambda}}(k)}\varphi(\hat{\boldsymbol{\lambda}}(k))$ using $\widetilde{\boldsymbol{y}}_{i,PM}^*(k)$ and $\Delta\boldsymbol{u}_{i,M}^*(k)$, where $\varphi(\hat{\boldsymbol{\lambda}}(k))$ is shown in Equation (15).

$$\varphi(\hat{\boldsymbol{\lambda}}(k)) \triangleq L\left(\Delta\boldsymbol{u}_M^*(k),\widetilde{\boldsymbol{y}}_{PM}^*(k),\hat{\boldsymbol{\lambda}}(k)\right) = \sum_{i=1}^{m} L_i\left(\Delta\boldsymbol{u}_{i,M}^*(k),\widetilde{\boldsymbol{y}}_{i,PM}^*(k),\hat{\boldsymbol{\lambda}}(k)\right) \quad (15)$$

$\hat{\lambda}(k)$ is modified by the gradient algorithm, as shown in Equation (16), where $l$ is the number of iterations and $\alpha(k)$ is the iteration step size.

$$\begin{cases} \hat{\lambda}_i^{l+1}(k) = \hat{\lambda}_i^{l}(k) + \alpha(k)\nabla\gamma(\hat{\lambda}_i^{l}) \\ \nabla\gamma(\hat{\lambda}_i^{l}) \triangleq \frac{\partial\varphi(\hat{\lambda}(k))}{\partial\hat{\lambda}_i^{l}(k)} = \widetilde{\boldsymbol{y}}_{i,PM}^{l}(k) - \widetilde{\boldsymbol{y}}_{i,P0}(k) - \sum_{j=1}^{m}\mathbf{A}_{ij}\Delta\boldsymbol{u}_{j,M}^{l}(k) \end{cases} \quad (16)$$

Once the difference between the two neighboring iterations' coordination factors is sufficiently small, it means that $\widetilde{y}_{PM}(k)$ and $\Delta u_M(k)$ have reached a state of associative equilibrium. The stopping condition of the iteration is shown in Equation (17).

$$\|\hat{\lambda}_i^{l+1}(k) - \hat{\lambda}_i^{l}(k)\| < \varepsilon,\ i = 1, \ldots, m \tag{17}$$

The relation between the solution of the decomposition–coordination method and the optimal solution of the original problem has been shown in [2], where as long as the iterative process converges, the final solution $\Delta u_M^*(k)$ is the optimal solution of the centralized solution of the original problem, as demonstrated in detail in [2].

## 3. Constrained Decomposition–Coordination Strategy

The original DCMPC is oriented towards multivariable systems with no constraints (only one equation constraint). However, most systems have constraint requirements on the controlled and manipulated variables. This paper proposes an improved DCMPC method for constrained multivariable systems, incorporating the simple handling of inequality constraints to the original method. The improved DCMPC can meet the control requirements of systems with inequality constraints.

### *3.1. Problem Description*

Suppose a constrained MIMO system, with $m$ inputs and $m$ outputs, has additional constraints on the CV, MV, and MV increments compared to the unconstrained system. The online optimization problem formed using a centralized optimization approach is shown in Equation (18).

$$\begin{aligned}
\min_{\Delta u_M(k)} \ & J(k) = \|\mathbf{w}(k) - \widetilde{y}_{PM}(k)\|_{\mathbf{Q}}^2 + \|\Delta u_M(k)\|_{\mathbf{R}}^2 \\
\text{s.t.}\ \ & \widetilde{y}_{PM}(k) = \widetilde{y}_{P0}(k) + \mathbf{A}\Delta u_M(k) \\
& u_M(k) = u_M(k-1) + \Delta u_M(k) \\
& \underline{\mathbf{Y}} \le \widetilde{y}_{PM}(k) \le \overline{\mathbf{Y}} \\
& \underline{\mathbf{U}} \le u_M(k) \le \overline{\mathbf{U}} \\
& \Delta\underline{\mathbf{U}} \le \Delta u_M(k) \le \Delta\overline{\mathbf{U}}
\end{aligned} \tag{18}$$

Decomposing the original system into $m$ SISO subsystems with linear additivity between the subsystems, the original centralized optimization problem is rewritten, as shown in Equation (19).

$$\begin{aligned}
\min_{\Delta u_M(k)} \ & J(k) = \sum_{i=1}^{m} \left\{ \|\mathbf{w}_i(k) - \widetilde{y}_{i,PM}(k)\|_{\mathbf{Q}_i}^2 + \|\Delta u_{i,M}(k)\|_{\mathbf{R}_i}^2 \right\} \\
\text{s.t.}\ \ & \widetilde{y}_{i,PM}(k) = \widetilde{y}_{i,P0}(k) + \sum_{j=1}^{m} \mathbf{A}_{ij}\Delta u_{j,M}(k),\ i = 1, \ldots, m \\
& u_{i,M}(k) = u_{i,M}(k-1) + \Delta u_{i,M}(k) \\
& \underline{\mathbf{Y}}_i \le \widetilde{y}_{i,PM}(k) \le \overline{\mathbf{Y}}_i \\
& \underline{\mathbf{U}}_i \le u_{i,M}(k) \le \overline{\mathbf{U}}_i \\
& \Delta\underline{\mathbf{U}}_i \le \Delta u_{i,M}(k) \le \Delta\overline{\mathbf{U}}_i
\end{aligned} \tag{19}$$

### *3.2. The Dual Decomposition Method Based on Subsystem QP*

The first improved dual decomposition method proposed in this paper for constrained MIMO distributed systems introduces CV, MV, and MV incremental constraints into the solving process of each distributed system to form the subsystem QP problem.

The equation constraint that contains the association of $\widetilde{y}_{i,PM}(k)$ and $\Delta u_{i,M}(k)$ is first introduced into the objective function via a Lagrange multiplier to form the dual problem for $m$ subsystems, as shown in Equation (20).

$$\begin{aligned} \min \quad & L_i(\Delta u_{i,M}(k), \widetilde{y}_{i,PM}(k), \lambda) \\ \text{s.t.} \quad & u_{i,M}(k) = u_{i,M}(k-1) + \Delta u_{i,M}(k) \\ & \underline{\mathbf{Y}}_i \leq \widetilde{y}_{i,PM}(k) \leq \overline{\mathbf{Y}}_i \\ & \underline{\mathbf{U}}_i \leq u_{i,M}(k) \leq \overline{\mathbf{U}}_i \\ & \Delta\underline{\mathbf{U}}_i \leq \Delta u_{i,M}(k) \leq \Delta\overline{\mathbf{U}}_i \end{aligned} \tag{20}$$

Therefore, the association of $\widetilde{y}_{i,PM}(k)$ and $\Delta u_{i,M}(k)$ in the constraints no longer remains. The constraints on $\widetilde{y}_{i,PM}(k)$ and $\Delta u_{i,M}(k)$ are already independently separable, so the optimization problem of Equation (20) is further decomposed into the QP problems corresponding to $\widetilde{y}_{i,PM}(k)$ and $\Delta u_{i,M}(k)$, as shown in Equations (21) and (22).

$$\begin{aligned} \min_{\widetilde{y}_{i,PM}(k)} \quad & \|\mathbf{w}_i(k) - \widetilde{y}_{i,PM}(k)\|^2_{\mathbf{Q}_i} + \hat{\lambda}_i^{\mathrm{T}}(k)(\widetilde{y}_{i,PM}(k) - \widetilde{y}_{i,P0}(k)) \\ \text{s.t.} \quad & \underline{\mathbf{Y}}_i \leq \widetilde{y}_{i,PM}(k) \leq \overline{\mathbf{Y}}_i \end{aligned} \tag{21}$$

$$\begin{aligned} \min_{\Delta u_{i,M}(k)} \quad & \|\Delta u_{i,M}(k)\|^2_{\mathbf{R}_i} - \sum_{j=1}^{m} \left(\hat{\lambda}_j^{\mathrm{T}}(k)\mathbf{A}_{ji}\right) \Delta u_{i,M}(k) \\ \text{s.t.} \quad & u_{i,M}(k) = u_{i,M}(k-1) + \Delta u_{i,M}(k) \\ & \underline{\mathbf{U}}_i \leq u_{i,M}(k) \leq \overline{\mathbf{U}}_i \\ & \Delta\underline{\mathbf{U}}_i \leq \Delta u_{i,M}(k) \leq \Delta\overline{\mathbf{U}}_i \end{aligned} \tag{22}$$

Using the QP solution method, $\widetilde{y}_{i,PM}(k)$ and $\Delta u_{i,M}(k)$ can be obtained and brought into Equation (16) to update the coordination factor $\boldsymbol{\lambda}$. The remaining steps are the same as the decomposition–coordination method and will not be repeated here.

### *3.3. The Dual Decomposition Method Based on the Subsystem Constrained Zone*

The second improved dual decomposition method proposed in this paper for constrained MIMO distributed systems introduces CV, MV, and MV incremental constraints into each distributed system after the solution has been solved, and the zones formed by the constraints limit the solution.

Firstly, the problem is dealt with as an optimization problem with equality constraints, equivalent to Equation (9). Then, the steps of forming the Lagrange function are the same as the original method, as shown in Equations (10)–(12). In solving the optimal of the sub-problem stage, the proposed strategy adds the inequality constraints after obtaining $\widetilde{y}^*_{i,PM}(k)$ and $\Delta u^*_{i,M}(k)$, as shown in Equations (23) and (24).

$$\begin{aligned} & K\Delta u^*_{i,M}(k) = \mathrm{Sat}_1[0.5\mathbf{R}_i^{-1}\sum_{j=1}^{m} A_{ji}^T \hat{\lambda}_j(k)] \\ & \mathrm{Sat}_1(\alpha) = \begin{cases} \overline{\mathbf{U}}_{i,M} - u_{i,M}(k-1) & \alpha > \overline{\mathbf{U}}_{i,M} \\ \alpha & \underline{\mathbf{U}}_{i,M} \leq \alpha \leq \overline{\mathbf{U}}_{i,M} \\ \underline{\mathbf{U}}_{i,M} - u_{i,M}(k-1) & \alpha < \underline{\mathbf{U}}_{i,M} \end{cases} \end{aligned} \tag{23}$$

$$\begin{aligned} & \widetilde{y}^*_{i,PM}(k) = \mathrm{Sat}_2[\mathbf{w}_i(k) - 0.5\mathbf{Q}_i^{-1}\hat{\lambda}_j(k)] \\ & \mathrm{Sat}_2(\beta) = \begin{cases} \overline{\mathbf{Y}}_{i,PM} & \beta > \overline{\mathbf{Y}}_{i,PM} \\ \beta & \underline{\mathbf{Y}}_{i,PM} \leq \beta \leq \overline{\mathbf{Y}}_{i,PM} \\ \underline{\mathbf{Y}}_{i,PM} & \beta < \underline{\mathbf{Y}}_{i,PM} \end{cases} \end{aligned} \tag{24}$$

where $K$ is the lower-unit triangular matrix of $M \times M$. The above constraint limits are only for the case of containing $\underline{\mathbf{U}}_i \leq \boldsymbol{u}_{i,M}(k) \leq \overline{\mathbf{U}}_i$, when simultaneously containing increments of $\Delta \boldsymbol{u}_{i,M}(k)$ are considered, as shown in Equation (25).

$$\Delta \boldsymbol{u}^*_{i,M}(k) = \max(\min(\Delta\underline{\mathbf{U}}_{i,M}, \mathrm{K}^{-1}\mathrm{Sat}_1[0.5\mathbf{R}_i^{-1}\sum_{j=1}^{m} \mathbf{A}_{ji}^T \hat{\boldsymbol{\lambda}}_j(k)]), \Delta\overline{\mathbf{U}}_{i,M}) \quad (25)$$

In the second stage, updating the coordination factor is the same as the original method based on calculated $\widetilde{\boldsymbol{y}}^*_{i,PM}(k)$ and $\Delta \boldsymbol{u}^*_{i,M}(k)$, as shown in Equations (15)–(17). Equations (23) and (25) indicate that the manipulated variables must exist within the constraints. Whether Equation (25) can make controlled variables exist within the constraints needs further analysis.

Suppose $\beta > \overline{\mathbf{Y}}_{i,PM}$ exists when $\widetilde{\boldsymbol{y}}^*_{i,PM}(k)$ is calculated iteratively in $l$ times, then $\widetilde{\boldsymbol{y}}^*_{i,PM}(k) = \overline{\mathbf{Y}}_{i,PM}$ exists. According to Equation (17), when the iteration stop condition is satisfied, it approximately means $\nabla\gamma(\hat{\boldsymbol{\lambda}}^l_i) = 0, i = 1, \ldots, m$, as shown in Equation (26).

$$\widetilde{\boldsymbol{y}}^*_{i,PM}(k) - \widetilde{\boldsymbol{y}}_{i,P0}(k) - \sum_{j=1}^{m} \mathbf{A}_{ij}\Delta \boldsymbol{u}^*_{j,M}(k) = 0, i = 1, \ldots, m \quad (26)$$

When $\widetilde{\boldsymbol{y}}^*_{i,PM}(k)$ surpasses the upper limit of the constraint, then $\widetilde{\boldsymbol{y}}^*_{i,PM}(k) = \overline{\mathbf{Y}}_{i,PM}$ holds through the constraint zone, and Equation (26) is rewritten as Equation (27).

$$\overline{\mathbf{Y}}_{i,PM} - \widetilde{\boldsymbol{y}}_{i,P0}(k) - \sum_{j=1}^{m} \mathbf{A}_{ij}\Delta \boldsymbol{u}^*_{j,M}(k) = 0, i = 1, \ldots, m \quad (27)$$

When we substitute the equality constraints of Equations (9) and (14) into Equation (27), we obtain that seen in Equation (28).

$$\overline{\mathbf{Y}}_{i,PM} - \underbrace{\left(\widetilde{\boldsymbol{y}}_{i,P0}(k) + \sum_{j=1}^{m} \mathbf{A}_{ij} 0.5\mathbf{R}_i^{-1} \sum_{j=1}^{m}\left(\mathbf{A}_{ji}^{\mathrm{T}}\hat{\boldsymbol{\lambda}}_j(k)\right)\right)}_{\widetilde{\boldsymbol{y}}_{i,PM}(k)} = 0, i = 1, \ldots, m \quad (28)$$

We can notice that the sum in parentheses in Equation (28) is the predicted value of the controlled variable in the future $P$ time horizon generated by the control law $\Delta \boldsymbol{u}_M(k)$. As long as the predetermined $\varepsilon$ is a small-enough positive number, and the coordination factor is continuously updated to the iteration stop condition, it can be understood from Equation (17) that $\overline{\mathbf{Y}}_{i,PM}$ and $\Delta \boldsymbol{u}_M(k)$ can reach an associated equilibrium state. In the continuous iterative process, the result $\widetilde{\boldsymbol{y}}_{i,PM}(k)$ generated by the control law $\Delta \boldsymbol{u}_M(k)$ will eventually converge to $\overline{\mathbf{Y}}_{i,PM}$. Similarly, $\beta < \underline{\mathbf{Y}}_{i,PM}$ converges to $\underline{\mathbf{Y}}_{i,PM}$ when $\beta < \underline{\mathbf{Y}}_{i,PM}$. It is proved that the added inequality constraint strategy is also effective for the controlled variables.

### *3.4. Performance Comparison of the Proposed Algorithms*

This paper proposes two improved methods based on the decomposition–coordination method for constrained distributed MPC systems. As demonstrated by the simulations in Section 5.1, both methods can satisfy the demand for constraints. However, the first method has greater computational effort than the second.

The first method introduces the equation constraint containing the association of $\widetilde{\boldsymbol{y}}_{i,PM}(k)$ and $\Delta \boldsymbol{u}_{i,M}(k)$ into the dual problem $L_i$ by Lagrange multipliers, so that $L_i$ is only related to $\widetilde{\boldsymbol{y}}_{i,PM}(k)$ and $\Delta \boldsymbol{u}_{i,M}(k)$, rather than to the other subsystem variables, thus decomposing the coupling between the subsystems. At the same time, $\widetilde{\boldsymbol{y}}_{i,PM}(k)$ and $\Delta \boldsymbol{u}_{i,M}(k)$ are independently separable in the constraints, so that $L_i$ is further decomposed into the two QP problems $\widetilde{\boldsymbol{y}}_{i,PM}(k)$ and $\Delta \boldsymbol{u}_{i,M}(k)$, which is equivalent to dividing the original centralized optimization problem into sub-optimization problems. For QP calculations containing a large matrix, this method can significantly reduce the computational scale. However, as suboptimization problems become more extensive, the computation time is not advantageous for some small and medium-sized computation scales. The second method is based on the original decomposition–coordination method, without adding the QP calculation, which is only through the logical determination of MV and CV constraints, so its computation time has a significant advantage compared with the first method. Compared with the centralized optimization method, the second method uses distributed parallel online iterative operations to avoid the high-dimensional matrix inverse calculation, so it has more superiority in computing control laws. The specific algorithm comparison analysis is detailed in Section 5.1.

## 4. PDLMPC Algorithm

This paper proposes two improved dual decomposition methods for constrained multivariate systems. After the above discussion and analysis and subsequent validation, the online optimization performance of the dual decomposition method based on the constrained zone is more advantageous, so an improved PDLMPC method is introduced into the DLMPC architecture. In this method, the steady-state optimization layer adopts the centralized optimization method, and the dynamic control layer adopts the constrained decomposition–coordination method. The structure is shown in Figure 2.

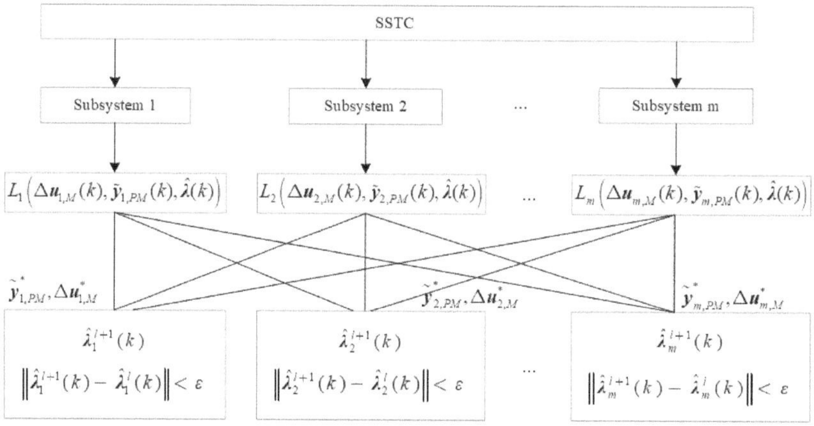


**Figure 2.** PDLMPC structure.

Different from the matrix dimension of the dynamic model, the steady-state model has a smaller one, and it is a one-step optimization. The prediction and control horizon are both one, so the centralized optimization will not generate a large computational burden. The centralized optimization problem for the steady-state layer of a constrained multivariable system is shown in Equation (29).

$$\begin{aligned}
\min_{\Delta \boldsymbol{U}_{\mathrm{ss}}(\mathrm{k})} \Xi = {} & \|\boldsymbol{U}_{\mathrm{ss}}(\mathrm{k}) - \mathbf{U}_{\mathrm{T}}\|^2_{\mathbf{R}_{\mathrm{ss}}} + \|\mathbf{Y}_{\mathrm{ss}}(\mathrm{k}) - \mathbf{Y}_{\mathrm{T}}\|^2_{\mathbf{Q}_{\mathrm{ss}}} + \|\Delta \boldsymbol{U}_{\mathrm{ss}}(\mathrm{k})\|^2_{\mathbf{O}_{\mathrm{ss}}} \\
\text{s.t.} \quad & \boldsymbol{U}_{\mathrm{ss}}(\mathrm{k}) = \boldsymbol{U}_{\mathrm{ss}}(\mathrm{k}-1) + \Delta \boldsymbol{U}_{\mathrm{ss}}(\mathrm{k}) \\
& \mathbf{Y}_{\mathrm{ss}}(\mathrm{k}) = \mathbf{Y}_{\mathrm{ss}}(k-1) + \mathbf{G}_{\mathrm{ss}} \Delta \boldsymbol{U}_{\mathrm{ss}}(\mathrm{k}) + \mathbf{e}(\mathrm{k}) \\
& \underline{\mathbf{Y}}_{\mathrm{ss}} \le \mathbf{Y}_{\mathrm{ss}}(\mathrm{k}) \le \overline{\mathbf{Y}}_{\mathrm{ss}} \\
& \underline{\mathbf{U}}_{\mathrm{ss}} \le \boldsymbol{U}_{\mathrm{ss}}(\mathrm{k}) \le \overline{\mathbf{U}}_{\mathrm{ss}}
\end{aligned} \tag{29}$$

where $\mathbf{e}(\mathrm{k})$ is the prediction error passed by the dynamic control layer. Solving Equation (24) to obtain the steady-state target values $\mathbf{Y}_{ss}(k)$ and $\boldsymbol{U}_{ss}(k)$, the online optimization problem of the dynamic control layer at time $k$ is shown in Equation (30).

$$\begin{aligned}
\min_{\Delta \boldsymbol{u}_M(k)} \ & J(k) = \|\mathbf{Y}_{ss}(k) - \widetilde{\boldsymbol{y}}_{PM}(k)\|_Q^2 + \|\boldsymbol{U}_{ss}(k) - \boldsymbol{u}_M(k)\|_O^2 + \|\Delta \boldsymbol{u}_M(k)\|_R^2 \\
\text{s.t.} \ & \widetilde{\boldsymbol{y}}_{PM}(k) = \widetilde{\boldsymbol{y}}_{P0}(k) + \mathbf{A}\Delta \boldsymbol{u}_M(k) \\
& \boldsymbol{u}_M(k) = \boldsymbol{u}_M(k-1) + \Delta \boldsymbol{u}_M(k) \\
& \underline{\mathbf{Y}} \leq \widetilde{\boldsymbol{y}}_{PM}(k) \leq \overline{\mathbf{Y}} \\
& \underline{\mathbf{U}} \leq \boldsymbol{u}_M(k) \leq \overline{\mathbf{U}} \\
& \Delta\underline{\mathbf{U}} \leq \Delta \boldsymbol{u}_M(k) \leq \Delta\overline{\mathbf{U}}
\end{aligned} \tag{30}$$

Firstly, the inequality constraints are ignored, and only the dual problem under equality constraints is considered, as shown in Equation (31).

$$\begin{aligned}
& L\left(\Delta \boldsymbol{u}_M(k), \widetilde{\boldsymbol{y}}_{PM}(k), \hat{\boldsymbol{\lambda}}(k)\right) \\
&= \sum_{i=1}^{m}\left\{\|\mathbf{Y}_{i,ss}(k) - \widetilde{\boldsymbol{y}}_{i,PM}(k)\|_{\mathbf{Q}_i}^2 + \|\boldsymbol{U}_{i,ss}(k) - \boldsymbol{u}_{i,M}(k)\|_{\mathbf{O}_i}^2 + \|\Delta \boldsymbol{u}_{i,M}(k)\|_{\mathbf{R}_i}^2\right\} + \sum_{i=1}^{m} \hat{\lambda}_i^{\mathrm{T}}(k)\left(\widetilde{\boldsymbol{y}}_{i,PM}(k) - \widetilde{\boldsymbol{y}}_{i,P0}(k) - \sum_{j=1}^{m} \mathbf{A}_{ij}\Delta \boldsymbol{u}_{j,M}(k)\right) \\
&= \sum_{i=1}^{m}\left\{\|\mathbf{w}_i(k) - \widetilde{\boldsymbol{y}}_{i,PM}(k)\|_{\mathbf{Q}_i}^2 + \|\boldsymbol{U}_{i,ss}(k) - \boldsymbol{u}_{i,M}(k)\|_{\mathbf{O}_i}^2 + \|\Delta \boldsymbol{u}_{i,M}(k)\|_{\mathbf{R}_i}^2 + \hat{\lambda}_i^{\mathrm{T}}(k)(\widetilde{\boldsymbol{y}}_{i,PM}(k) - \widetilde{\boldsymbol{y}}_{i,P0}(k))\right\} - \sum_{j=1}^{m}\sum_{i=1}^{m}(\hat{\lambda}_i^{\mathrm{T}}(k)\mathbf{A}_{ij}\Delta \boldsymbol{u}_{j,M}(k)) \\
&= \sum_{i=1}^{m} L_i\left(\Delta \boldsymbol{u}_{i,M}(k), \widetilde{\boldsymbol{y}}_{i,PM}(k), \hat{\boldsymbol{\lambda}}(k)\right)
\end{aligned} \tag{31}$$

At the first level, the dual problem $\min\limits_{\Delta \boldsymbol{u}_M(k), \widetilde{\boldsymbol{y}}_{PM}(k)} L\left(\Delta \boldsymbol{u}_M(k), \widetilde{\boldsymbol{y}}_{PM}(k), \hat{\boldsymbol{\lambda}}(k)\right)$ is minimized to solve $\Delta \boldsymbol{u}_{i,M}^*(k)$ and $\widetilde{\boldsymbol{y}}_{i,PM}^*(k)$, as shown in Equations (32) and (33).

$$\begin{aligned}
& \Delta \boldsymbol{u}_{i,M}^*(k) = \max(\min(\underline{\mathbf{U}}_i, \Delta\mathrm{Sat}_1[0.5(\mathbf{R}_i + \mathbf{O}_i)^{-1}(\sum_{j=1}^{m} \mathbf{A}_{ji}^T \hat{\boldsymbol{\lambda}}_j(k) + 2(\boldsymbol{U}_{i,ss}(k) - \boldsymbol{u}_{i,M}(k-1))]), \Delta\overline{\mathbf{U}}_i) \\
& \mathrm{Sat}_1(\alpha) = \begin{cases} \overline{\mathbf{U}}_{i,M} - \boldsymbol{u}_{i,M}(k-1) & \alpha > \overline{\mathbf{U}}_{i,M} \\ \alpha & \underline{\mathbf{U}}_{i,M} \leq \alpha \leq \overline{\mathbf{U}}_{i,M} \\ \underline{\mathbf{U}}_{i,M} - \boldsymbol{u}_{i,M}(k-1) & \alpha < \underline{\mathbf{U}}_{i,M} \end{cases}
\end{aligned} \tag{32}$$

$$\begin{aligned}
& \widetilde{\boldsymbol{y}}_{i,PM}^*(k) = \mathrm{Sat}_2[\mathbf{Y}_{i,ss}(k) - 0.5\mathbf{Q}_i^{-1}\hat{\boldsymbol{\lambda}}_j(k)] \\
& \mathrm{Sat}_2(\beta) = \begin{cases} \overline{\mathbf{Y}}_{i,PM} & \beta > \overline{\mathbf{Y}}_{i,PM} \\ \beta & \underline{\mathbf{Y}}_{i,PM} \leq \beta \leq \overline{\mathbf{Y}}_{i,PM} \\ \underline{\mathbf{Y}}_{i,PM} & \beta < \underline{\mathbf{Y}}_{i,PM} \end{cases}
\end{aligned} \tag{33}$$

The steps for updating the coordination factor in the second stage are the same as the original method. That is, Equations (15)–(17) are performed. When the iteration is stopped, the optimal manipulated variable is applied to the controlled process. The deviation between the sampling result $\boldsymbol{y}(k+1)$ at time $k+1$ and the predicted value $\widetilde{\boldsymbol{y}}_N(k+1|k)$ at time $k+1$ is taken as the prediction error $\mathbf{e}(k+1)$. The prediction error is transmitted to the steady-state optimization layer and the feedback correction module. The predicted initial value $\widetilde{\boldsymbol{y}}_{N0}(k+1)$ is obtained after feedback correction and shifting, as shown in Equation (34).

$$\begin{cases} \widetilde{\boldsymbol{y}}_{\mathrm{cor}}(k+1) = \widetilde{\boldsymbol{y}}_N(k) + \mathbf{he}(k+1) \\ \widetilde{\boldsymbol{y}}_{N0}(k+1) = \mathbf{S}\widetilde{\boldsymbol{y}}_{\mathrm{cor}}(k+1) \end{cases} \tag{34}$$

In summary, the improved DLMPC algorithm based on the decomposition–coordination method for constrained multivariable systems is complete. The algorithm adopts the overall optimization mode in the steady-state layer, which can give a more comprehensive steady-state target value. Taking the steady-state target value as the setting point of the dynamic control layer can provide more global information for each subsystem and more reasonable tracking targets. The distributed architecture of the dynamic control layer ensures the information transmission between subsystems through decomposition and coordination.

At the same time, a simple constraint method is proposed. Through theoretical analysis, it is proved that the method can ensure that MV and CV run in the constraint conditions. However, this method is also incomplete and needs specific skills and experience when setting coefficients such as weight of control, error weight, and iteration stop accuracy.

## 5. Simulation

The simulation object used in this paper is the Shell heavy oil fractionation tower model, which is a typical, large, constrained, multivariable process, the structure of which is shown in Figure 3. After simplification, it can be regarded as a linear model with three inputs and three outputs, as shown in Equation (35). The controlled variables are product draw $y_1$, side product draw $y_2$, and bottom reflux temperature $y_3$. The manipulated variables are the top flow $u_1$, side flow $u_2$, and bottom reflux heat transmission rate $u_3$.

$$\begin{bmatrix} y_1 \\ y_2 \\ y_3 \end{bmatrix} = \begin{bmatrix} \frac{4.05e^{-27s}}{50s+1} & \frac{1.77e^{-28s}}{60s+1} & \frac{5.88e^{-27s}}{50s+1} \\ \frac{5.39e^{-18s}}{50s+1} & \frac{5.72e^{-14s}}{60s+1} & \frac{6.90e^{-15s}}{40s+1} \\ \frac{4.38e^{-20s}}{33s+1} & \frac{4.42e^{-22s}}{44s+1} & \frac{7.20}{19s+1} \end{bmatrix} \cdot \begin{bmatrix} u_1 \\ u_2 \\ u_3 \end{bmatrix} \tag{35}$$

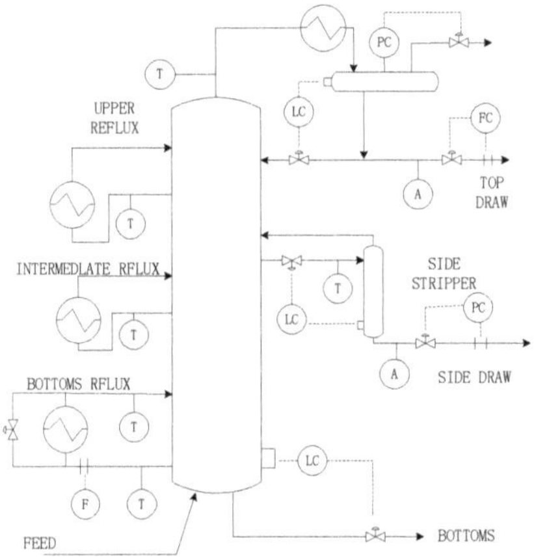


**Figure 3.** Shell heavy oil fractionation structure.

### 5.1. *PDD1 and PDD2 Validation of the Static Model*

To verify the performance and effectiveness of the two algorithms proposed in Section 4, the static model of Shell heavy oil fractionation with both a prediction and control time horizon of 1 is employed, as shown in Equation (36). The optimization problem adopts the form of Equation (18), which tracks only the external target of the controlled variable. This optimization problem is required to satisfy the CV and MV constraints, as shown in Table 2. The comparison algorithm is the original centralized optimization method (OC), verifying that the optimal solution sought by the improved dual decomposition method based on the subsystem QP (PDD1) and the improved dual decomposition method based on the subsystem constraint zone (PDD2) is consistent with the optimal solution of the CMPC. The control goal is to achieve a given target value $\mathbf{Y}_\mathrm{T} = \begin{bmatrix} 0.9 & 0.9 & 0.9 \end{bmatrix}^T$ for the CV while satisfying the constraints (as shown in Table 2). The simulation step size is 100.

$$\begin{bmatrix} y_1 \\ y_2 \\ y_3 \end{bmatrix} = \begin{bmatrix} 4.05 & 1.77 & 5.88 \\ 5.39 & 5.72 & 6.90 \\ 4.38 & 4.42 & 7.2 \end{bmatrix} \cdot \begin{bmatrix} u_1 \\ u_2 \\ u_3 \end{bmatrix} \tag{36}$$

**Table 2.** Parameter setting.

| Parameter Name | Parameter Value | Parameter Name | Parameter Value |
|---|---|---|---|
| CV constraint upper limit | $\overline{\mathbf{Y}} = \begin{bmatrix} 1 & 1 & 1 \end{bmatrix}^T$ | CV constraint lower limit | $\underline{\mathbf{Y}} = \begin{bmatrix} -1 & -1 & -1 \end{bmatrix}^T$ |
| MV constraint upper limit | $\overline{\mathbf{U}} = \begin{bmatrix} 0.1 & 0.1 & 0.1 \end{bmatrix}^T$ | MV constraint lower limit | $\underline{\mathbf{U}} = \begin{bmatrix} -0.1 & -0.1 & -0.1 \end{bmatrix}^T$ |

Simulations are first performed in the unconstrained case to ensure that the constraints in Table 2 are positive, comparing the algorithms to the original decomposition–coordination method (DD) and OC. The simulation results are shown in Figure 4. The optimal solution trajectories of OC and DD can be seen to overlap almost exactly, with $u_1$ and $u_3$ in Figure 4a both exceeding the constraint upper bound of 0.1, proving that the added constraint is positive.

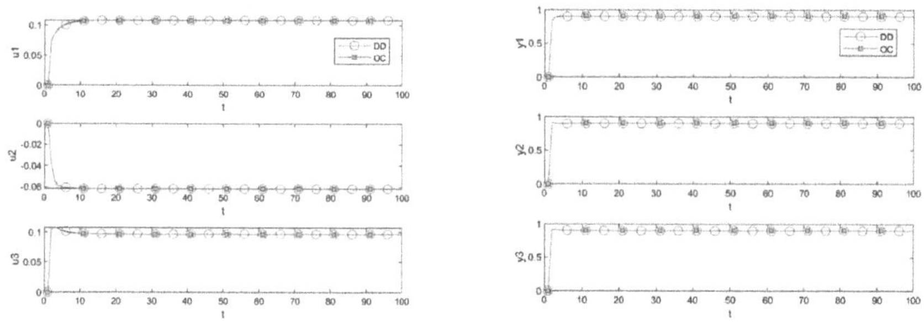


(a) (b)

**Figure 4.** Comparisons for the unconstrained case. (**a**) MV comparison; (**b**) CV comparison.

Comparing the centralized optimization method with constraints, the PDD1 and PDD2 are proposed in this paper. The results show that the three algorithms' optimal solutions are consistent in the presence of constraints, as shown in Figure 5. Comparing the optimal solution trajectory of $u_1$ and $u_3$ in Figures 4a and 5a, $u_1$ and $u_3$ are obviously constrained by the constraints. In Figure 5b, the CVs of the proposed algorithms meet the external target values required for control, and the trajectories of the CVs can track the EVs without zero deviation. The constraints on the CVs show that the CVs have residual degrees of freedom, which means that the CVs track without zero deviation not due to the positive constraints that passively form the control trajectories, but rather due to the naturally occurring control trajectory formed in response to the control law.

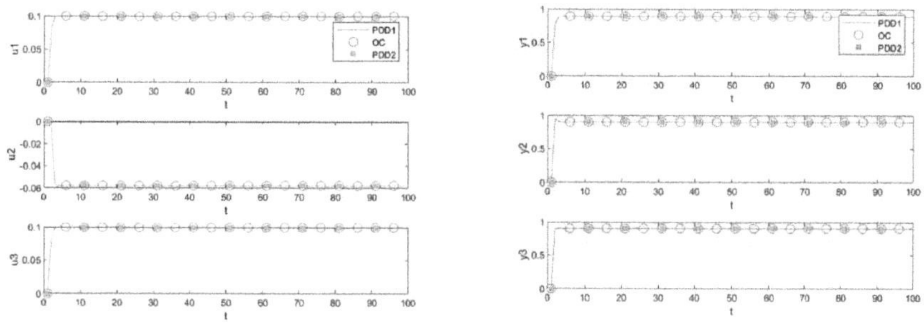


(a) (b)

**Figure 5.** Comparisons for the constrained case. (**a**) MV comparison; (**b**) CV comparison.

As for the algorithm running time, PDD1 took about 14.68 s, OC took about 0.522 s, and PDD2 took about 0.0259 s. We noticed that PDD1 takes the longest time and PDD2 has an obvious advantage in reducing the amount of computation and shortening the running time, which verifies the conjecture in Section 3.4.

In summary, Simulation 1 demonstrates that the constrained decomposition–coordination method proposed in Section 4 can ensure that the variables are constrained. At the same time, the control goal can be met under positive constraints. The resulting optimal solution trajectory is consistent with the constrained CMPC algorithm, verifying the effectiveness of the proposed algorithm and laying the foundation for the introduction of DLMPC.

The above simulation results show the effectiveness of the two improved dual decomposition methods proposed in this paper. Both methods can ensure that the variables are constrained and the control goal can be achieved under positive constraints. The optimal solution trajectory is consistent with the constrained OC algorithm. At the same time, the calculation time of the three algorithms is compared. PDD2 takes the shortest time and shows excellent online optimization performance.

### *5.2. PDLMPC Algorithm Verification Based on PDD2*

Although PDD1 achieves results consistent with OC, it takes longer. It is not advantageous for small- to medium-sized solutions, so we use a modified double-layer model predictive control algorithm (PDLMPC) based on PDD2 for subsequent simulation validation and a comparison algorithm using the original double-layer model predictive control algorithm (ODLMPC) with centralized optimization, using the Shell heavy oil fractionation model as shown in Equation (35). The steady-state optimization layer is required to track the external value for a given MV and CV and constitutes the steady-state optimization problem, as shown in Equation (18). The dynamic control layer tracks the steady-state targets $\mathbf{Y}_{ss}$, as shown in Equation (25). The control goal is to bring the CV to the given target value $\mathbf{Y}_{\mathrm{T}} = \begin{bmatrix} 0.5 & 0.5 & 0.5 \end{bmatrix}^T$ while satisfying the constraints in Table 3.

**Table 3.** Parameter setting.

| Parameter Name | Parameter Value | Parameter Name | Parameter Value |
|---|---|---|---|
| CV constraint upper limit | $\overline{\mathbf{Y}} = \begin{bmatrix} 1 & 1 & 1 \end{bmatrix}^T$ | CV constraint lower limit | $\underline{\mathbf{Y}} = \begin{bmatrix} -1 & -1 & -1 \end{bmatrix}^T$ |
| MV constraint upper limit | $\overline{\mathbf{U}} = \begin{bmatrix} 0.1 & 0.1 & 0.1 \end{bmatrix}^T$ | MV constraint lower limit | $\underline{\mathbf{U}} = \begin{bmatrix} -0.1 & -0.1 & -0.1 \end{bmatrix}^T$ |
| MV incremental constraint lower limit | $\Delta\overline{\mathbf{U}} = \begin{bmatrix} 0.05 & 0.05 & 0.05 \end{bmatrix}^T$ | MV incremental constraint lower limit | $\Delta\underline{\mathbf{U}} = \begin{bmatrix} -0.05 & -0.05 & -0.05 \end{bmatrix}^T$ |

The ODLMPC and DCDLMPC (centralized steady-state optimization for the upper layer and decomposed coordinated control for the lower layer) algorithms were first run without constraints to prove that the constraints in Table 3 are positive, and the simulation results are shown in Figure 6. The MVs of both ODLMPC and DCDLMPC algorithms exceeded the constraint limits in the unconstrained case, proving that the constraints in Table 3 are positive constraints. Ensuring that the controlled variable eventually reaches the control target is not the constraint forcing the trajectory to form so that the constraint on the controlled variable is greater than the control target value. It is worth noting that in Figure 6 the results of the two control algorithms are inconsistent, with the control results of DCDLMPC being more advantageous. This situation is because the control and error weight parameters of DLMPC are not set correctly.

Results from the ODLMPC and the PDLMPC algorithms presented in Section 5, running with constraints, are shown in Figure 7. Figure 7a clearly shows that the MV under the PDLMPC algorithm is restricted by the same constraints as the ODLMPC and that both the CV and MV are able to track the steady-state optimization trajectory given by the SSTC layer to realize the control goal. The average time taken for five runs of the DLMPC algorithm was 4.87 s, and the average time taken for five runs of the PDLMPC algorithm was 2.58 s, where the controllers of the three subsystems were not run in parallel but in series, which would have been smaller if they had been run in parallel. The PDLMPC algorithm reduces the computational complexity with less computational burden by giving up some of the tracking performance, and this sub-optimal result is acceptable.

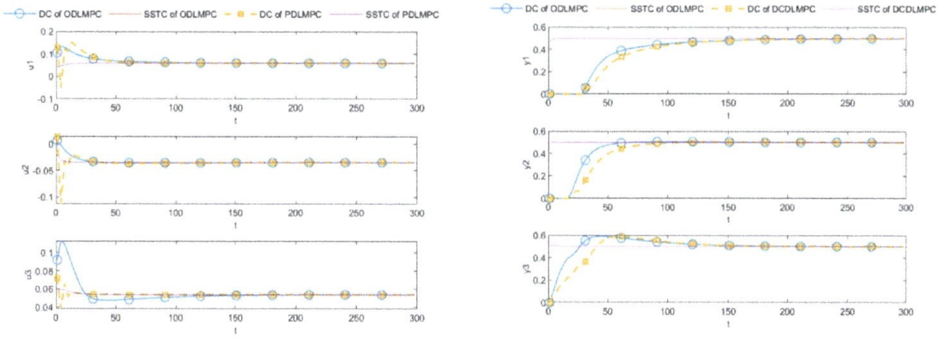


(a) (b)

**Figure 6.** Comparisons of DCDLMPC and ODLMPC. (**a**) MV comparison; (**b**) CV comparison.

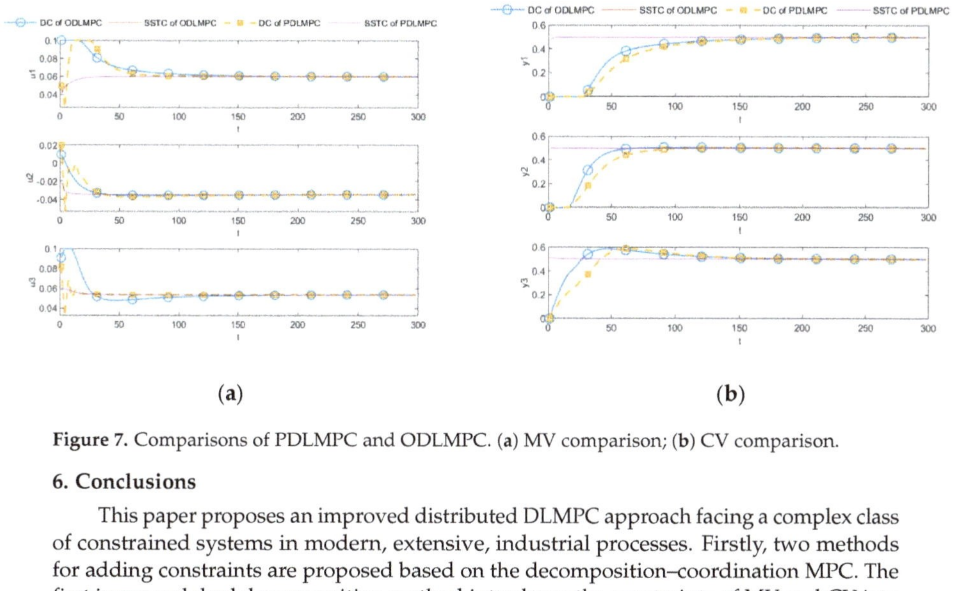


(a) (b)

**Figure 7.** Comparisons of PDLMPC and ODLMPC. (**a**) MV comparison; (**b**) CV comparison.

## 6. Conclusions

This paper proposes an improved distributed DLMPC approach facing a complex class of constrained systems in modern, extensive, industrial processes. Firstly, two methods for adding constraints are proposed based on the decomposition–coordination MPC. The first improved dual decomposition method introduces the constraints of MV and CV into solving the distributed subsystem, forming the QP problem with the dual function of the subsystem to solve it. The second is a dual decomposition method based on constrained zones, which introduces the constraints of MV and CV into the distributed subsystem after solving $\tilde{\boldsymbol{y}}^*_{i,PM}(k)$ and $\Delta \boldsymbol{u}^*_{i,M}(k)$. Furthermore, the convergence of the method is analyzed: as long as the parameters are set appropriately, the CV will eventually converge to the constraint boundary. The online optimization capabilities of the two proposed methods are discussed and compared, concluding that the second proposed method, the pairwise decomposition method based on constraint zones, has superior online optimization capability. This guess is proved by comparing the two proposed methods and the centralized optimization method through simulation. Based on the above work, an improved distributed DLMPC algorithm based on the pairwise decomposition method with constrained zones is proposed. Different from the objective function in the original decomposition–coordination method, the objective function used in the dynamic control layer of the improved distributed DLMPC algorithm tracks both the steady-state optimized values of MV and CV and the dynamic control layer of the decomposition–coordination with constraints designed for this objective function, which gives the characterization of the optimal solution as well as the strategy for processing the constraints. The proposed algorithm can reduce the computational complexity while achieving the control goals. The effectiveness and rationality of the proposed algorithm are validated through simulations and compared to the simulation results of DCDLMPC without constraints. It is evident that PDLMPC can make the manipulated variables constrained. Compared with ODLMPC, PDLMPC uses less running time and the control effect is similar, so it can meet the control requirements and achieve the control goals. Of course, the PDLMPC algorithm also has shortcomings; it has specific requirements for parameter settings, and further research will be carried out in the future to optimize the parameters of this algorithm.

**Author Contributions:** Original draft preparation, H.W.; methodology, H.W. and P.Z.; writing—review and editing, Z.Y. and T.Z. All authors have read and agreed to the published version of the manuscript.

**Funding:** This research was funded by the National Natural Science Foundation of China, grant number 52171331, the Science and Technology Planning Project of Guangzhou City, grant number 202102010411, and the National Key R&D Program of China, grant number 2017YFA0700303.

**Institutional Review Board Statement:** Not applicable.

**Data Availability Statement:** Not applicable.

**Conflicts of Interest:** The authors declare no conflict of interest.

## References

1. Tatjewski, P. Advanced control and on-line process optimization in multilayer structures. *Annu. Rev. Control* **2008**, *32*, 71–85. [CrossRef]
2. Xi, Y.G.; Li, D.W. *Predictive Control*; National Defense Industry Press: Beijing, China, 2013.
3. Rawlings, J.B.; Mayne, D.Q. *Model Predictive Control: Theory and Design*; NOB HILL PUBLISHING, LLC: Santa Barbara, CA, USA, 2009.
4. Sorensen, R.C.; Cutler, C.R. Lp integrates economics into dynamic matrix control: Process control and information systems: A special report. *Hydrocarb. Process.* **1998**, *77*, 57–65.
5. Brosilow, C.; Zhao, G. A Linear Programming Approach to Constrained Multivariable Process Control. In *System Identification and Adaptive Control*; Part 3 of 3, Volume 27 of Control and Dynamic Systems; Leondes, C.T., Ed.; Academic Press: Cambridge, MA, USA, 1988; pp. 141–181. [CrossRef]
6. Morshedi, A.M.; Cutler, C.R.; Skrovanek, T.A. Optimal solution of dynamic matrix control with linear programing techniques (ldmc). In Proceedings of the 1985 American Control Conference, Boston, MA, USA, 19–21 June 1985; pp. 199–208. [CrossRef]
7. Qin, S.; Badgwell, T.A. A survey of industrial model predictive control technology. *Control Eng. Pract.* **2003**, *11*, 733–764. [CrossRef]
8. Zou, T.; Ding, B.C.; Zhang, D. *MPC: An Introducion to Indus-Trial Applications*; Chemical Industry Press: Beijing, China, 2010.
9. Yang, Y.; Ding, B. Two-layer model predictive control for chemical process model with integrating controlled variables. *Can. J. Chem. Eng.* **2020**, *98*, 237–253. [CrossRef]
10. Liu, J.; Sun, H.; Lu, Y.; Hu, J.; Zou, T. A weighted local steady-state determination approach based on the globally optimal economic steady-states. *Can. J. Chem. Eng.* **2021**, *99*, 2229–2240. [CrossRef]
11. Liu, J.; Sun, H.; Zhang, Y.; Hu, J.; Zou, T. Steady-state sequence optimization with incremental input constraints in two-layer model predictive control. *ISA Trans.* **2022**, *128*, 144–158. [CrossRef]
12. Keviczky, T.; Borrelli, F.; Balas, G.J. Decentralized receding horizon control for large scale dynamically decoupled systems. *Automatica* **2006**, *42*, 2105–2115. [CrossRef]
13. Alessio, A.; Barcelli, D.; Bemporad, A. Decentralized model predictive control of dynamically coupled linear systems. *J. Process. Control* **2011**, *21*, 705–714. [CrossRef]
14. Barcelli, D.; Bemporad, A. Decentralized Model Predictive Control of Dynamically-Coupled Linear Systems: Tracking under Packet Loss. *IFAC Proc. Vol.* **2009**, *42*, 204–209. [CrossRef]
15. Christofides, P.D.; Scattolini, R.; de la Peña, D.M.; Liu, J. Distributed model predictive control: A tutorial review and future research directions. *Comput. Chem. Eng.* **2013**, *51*, 21–41. [CrossRef]
16. Negenborn, R.; Maestre, J. Distributed Model Predictive Control: An Overview and Roadmap of Future Research Opportunities. *IEEE Control Syst.* **2014**, *34*, 87–97. [CrossRef]

17. Camisa, A.; Köhler, P.N.; Müller, M.A.; Notarstefano, G.; Allgöwer, F. A distributed optimization algorithm for Nash bargaining in multi-agent systems. *IFAC-PapersOnLine* **2020**, *53*, 2684–2689. [CrossRef]
18. Francisco, M.; Mezquita, Y.; Revollar, S.; Vega, P.; De Paz, J.F. Multi-agent distributed model predictive control with fuzzy negotiation. *Expert Syst. Appl.* **2019**, *129*, 68–83. [CrossRef]
19. Zhou, Z.; De Schutter, B.; Lin, S.; Xi, Y. Two-Level Hierarchical Model-Based Predictive Control for Large-Scale Urban Traffic Networks. *IEEE Trans. Control Syst. Technol.* **2017**, *25*, 496–508. [CrossRef]
20. Masero, E.; Francisco, M.; Maestre, J.M.; Revollar, S.; Vega, P. Hierarchical distributed model predictive control based on fuzzy negotiation. *Expert Syst. Appl.* **2021**, *176*, 114836. [CrossRef]
21. Farina, M.; Scattolini, R. Distributed non-cooperative MPC with neighbor-to-neighbor communication. *IFAC Proc. Vol.* **2011**, *44*, 404–409. [CrossRef]
22. Hu, H.; Gatsis, K.; Morari, M.; Pappas, G.J. Non-Cooperative Distributed MPC with Iterative Learning. *IFAC-PapersOnLine* **2020**, *53*, 5225–5232. [CrossRef]
23. Sun, B.; Tang, Y.; Ye, L.; Chen, C.; Zhang, C.; Zhong, W. A Frequency Control Strategy Considering Large Scale Wind Power Cluster Integration Based on Distributed Model Predictive Control. *Energies* **2018**, *11*, 1600. [CrossRef]
24. Yame, J.J.; Gabsi, F.; Darure, T.; Jain, T.; Hamelin, F.; Sauer, N. Optimality Condition Decomposition Approach to Distributed Model Predictive Control. In Proceedings of the 2019 American Control Conference (ACC), Philadelphia, PA, USA, 10–12 July 2019; pp. 742–747. [CrossRef]
25. Namba, T.; Takeda, K.; Takaba, K. Dual Decomposition-Based Distributed Microgrid Managament with PV Prediction. In Proceedings of the 2018 57th Annual Conference of the Society of Instrument and Control Engineers of Japan (SICE), Nara, Japan, 11–14 September 2018; pp. 964–970. [CrossRef]
26. Wakasa, Y.; Arakawa, M.; Tanaka, K.; Akashi, T. Decentralized model predictive control via dual decomposition. In Proceedings of the 2008 47th IEEE Conference on Decision and Control, Cancún, Mexico, 9–11 December 2008; pp. 381–386. [CrossRef]
27. Zafra-Cabeza, A.; Maestre, J.M. *A Hierarchical Distributed MPC Approach: A Practical Implementation*; Springer: Dordrecht, The Netherlands, 2013; pp. 451–464. ISBN 978-94-007-7006-5. [CrossRef]
28. Nabais, J.L.; Negenborn, R.R.; Carmona-Ben ́ıtez, R.B.; Mendon ̧ca, L.F.; Botto, M.A. *Hierarchical MPC for Multiple Commodity Transportation Networks*; Springer: Dordrecht, The Netherlands, 2014; pp. 535–552. ISBN 978-94-007-7006-5. [CrossRef]
29. Kozma, A.; Savorgnan, C.; Diehl, M. *Distributed Multiple Shooting for Large Scale Nonlinear Systems*; Springer: Dordrecht, The Netherlands, 2013; pp. 327–340. ISBN 978-94-007-7006-5. [CrossRef]
30. Kang, W.; Li, Q.; Chen, M.; Peng, C.; Chen, F. A two-layer distributed control method for islanded microgrids by using multi-agent systems. *Zhongguo Dianji Gongcheng Xuebao/Proc. Chin. Soc. Electr. Eng.* **2018**, *38*, 770–781. [CrossRef]
31. Tilli, A.; Garone, E.; Conficoni, C.; Cacciari, M.; Bosso, A.; Bartolini, A. A two-layer distributed MPC approach to thermal control of Multiprocessor Systems-on-Chip. *Control Eng. Pract.* **2022**, *122*, 105099. [CrossRef]
32. Yang, K.; Li, L.; Xue, F. Real-time optimization and distributed control integration algorithm. *Comput. Meas. Control.* **2016**, *24*, 74–78. [CrossRef]
33. Shi, Y.; Zhang, Z.; Sun, P.; Xie, L.; Chen, Q.; Su, H.; Chen, X. Two-layer structure strategy for large-scale systems integrating online adaptive constraints adjustment method and cooperative distributed DMC algorithm. *Control. Eng. Pract.* **2021**, *116*, 104932. [CrossRef]
34. Da, Q.; He, J. *Large Systems Theory and Methods*; Southeast University Press: Nanjing, China, 1989.

mathematics

MDPI

Article

# Dynamical Analysis of a One- and Two-Scroll Chaotic System

Meng Liu [1], Zhaoyan Wu [2,3,*] and Xinchu Fu [4]

1 Jiaxing Nanyang Polytechnic Institute, Jiaxing 314003, China
2 School of Mathematics and Statistic, Jiangxi Normal University, Nanchang 330022, China
3 Jiangxi Provincial Center of Applied Mathematics, Jiangxi Normal University, Nanchang 330022, China
4 Department of Mathematics, Shanghai University, Shanghai 200072, China
* Correspondence: zhywu@jxnu.edu.cn; Tel.: +86-791-8812-0367

**Abstract:** In this paper, a three-dimensional (3D) autonomous chaotic system is introduced and analyzed. In the system, each equation contains a quadratic crossproduct. The system possesses a chaotic attractor with a large chaotic region. Importantly, the system can generate both one- and two-scroll chaotic attractors by choosing appropriate parameters. Some of its basic dynamical properties, such as the Lyapunov exponents, Lyapunov dimension, Poincaré maps, bifurcation diagram, and the chaotic dynamical behavior are studied by adjusting different parameters. Further, an equivalent electronic circuit for the proposed chaotic system is designed according to Kirchhoff's Law, and a corresponding response electronic circuit is also designed for identifying the unknown parameters or monitoring the changes in the system parameters. Moreover, numerical simulations are presented to perform and complement the theoretical results.

**Keywords:** chaotic system; Lyapunov exponent; bifurcation; electronic circuit; parameter identification

**MSC:** 34C28

## 1. Introduction

Since the first discovery of chaotic attractors by Lorenz in 1963 [1], chaos has attracted attention and interest for its useful speciality and application in information and computer science [2]. The proposals of new chaotic systems have been extensively studied by scientists in the past decades. In 1976, Rössler found a new simple 3D quadratic autonomous chaotic system with only one quadratic nonlinearity on the right-hand side [3].

In 1999, Chen found another chaotic attractor [4]. Recently, Lü and Chen further found a new chaotic system, which represented the transition between the Lorenz and the Chen system [5]. Moreover, Liu and Chen introduced a new chaotic system with three quadratic nonlinearities on the right-hand side in 2003 [6], which displayed two- and four-scroll attractors for different parameters. Then, Lü and Chen constructed another simple 3D system, which displayed two chaotic attractors simultaneously [7].

During the past few years, some new 3D chaotic systems have been analyzed [8–20]. To classify these 3D autonomous chaotic systems, Vaněček and Čelikoshý [21] gave a divertive classification by separating the linear and quadratic parts of a 3D autonomous system. The linear part was described by a constant matrix $A = [a_{ij}]_{3\times 3}$. The Lorenz system satisfied $a_{12}a_{21} > 0$, the Chen system satisfied $a_{12}a_{21} < 0$, and the Lü system satisfied $a_{12}a_{21} = 0$. As is known, the Lorenz system and Chen system display a two-scroll chaotic attractor separately. In this paper, we introduce a 3D autonomous system, in which each equation contains a quadratic crossproduct, and the constant matrix of the linear part satisfies $a_{12}a_{21} = 0$. Different from Lorenz-like systems, the proposed system can display different numbers of scroll chaotic attractors simultaneously. The system is described by:

$$\begin{cases} \dot{x} = ay - yz, \\ \dot{y} = -bz + zx, \\ \dot{z} = cx + xy - dz, \end{cases} \tag{1}$$

in which $(x, y, z)^T \in \Re^3$, and $a, b, c$, and $d$ are real parameters. Though system (1) has three quadratic nonlinearities on the right-hand side, it can display only one-scroll attractor in contrast to the Rössler attractor and Sprott's attractor [22,23]. Simultaneously, with the appropriate parameters, the system (1) can display a two-scroll attractor in contrast to the famous Lorenz attractor. This system is a supplement to the discovery of two-scroll band structure attractors.

Further, according to Kirchhoff's law, we design an equivalent electronic circuit for the proposed chaotic system to show its practical applications. The system parameters of an electronic circuit maybe unknown or uncertain. Thus, based on the parameter identification and adaptive synchronization of drive–response systems, we design a corresponding response electronic circuit to identify the unknown parameters or monitor the changes in the system parameters.

The outline of this paper is as follows. In Section 2, the basic dynamical behavior in the parameter space is discussed, and some parameter examples for generating chaos are given. In Section 3, bifurcation analysis and the simulation results of the chaotic system are presented. In Section 4, a vector map is employed to generalized different attractors with the same parameters in the system. In Section 5, the adaptive synchronization problem between the drive–response systems with fully unknown parameters is studied. Finally, conclusions are drawn in Section 6.

## 2. Basic Dynamical Behavior of the System

The divergence of system (1) is

$$\nabla \dot{V} = \frac{\partial \dot{x}}{\partial x} + \frac{\partial \dot{y}}{\partial y} + \frac{\partial \dot{z}}{\partial z} = 0 + 0 - d = -d. \tag{2}$$

Therefore, when parameter $d$ is positive, system (1) is dissipative.

The equilibria of system (1) can be obtained by solving the following algebraic equations:

$$ay - yz = 0, \ -bz + xz = 0, \ cx + xy - dz = 0.$$

When $bd \neq 0$, the system has three equilibria:

$$S_1 = (0, 0, 0)^T, \ S_2 = (b, \frac{ad - bc}{b}, a)^T, \ S_3 = (b, 0, \frac{bc}{d})^T.$$

In addition, under the condition $b = 0$ (or $d = 0$), the system has a unique equilibrium $S_{01}^* = (0, 0, 0)^T$ (or $S_{02}^* = (b, -c, a)^T$). In the following, we let $b \neq 0$ and $d > 0$. The Jacobian matrix of system (1) at the three equilibria $S_1$, $S_2$, and $S_3$ are

$$J_1 = \begin{pmatrix} 0 & a & 0 \\ 0 & 0 & -b \\ c & 0 & -d \end{pmatrix}, J_2 = \begin{pmatrix} 0 & 0 & \frac{bc-ad}{b} \\ a & 0 & 0 \\ \frac{ad}{b} & b & -d \end{pmatrix},$$

$$J_3 = \begin{pmatrix} 0 & \frac{ad-bc}{d} & 0 \\ \frac{bc}{d} & 0 & 0 \\ c & b & -d \end{pmatrix}.$$

The characteristic equations of $J_1$, $J_2$, and $J_3$ are

$$\lambda^3 + d\lambda^2 + abc = 0, \tag{3}$$

$$\lambda^3 + d\lambda^2 + \frac{a^2d^2 - abcd}{b^2}\lambda + a^2d - abc = 0, \tag{4}$$

$$(\lambda + d)(\lambda^2 + \frac{b^2c^2 - abcd}{d^2}) = 0. \tag{5}$$

Obviously, from Equation (5), the equilibrium $S_3$ is a saddle for $bc(bc - ad) < 0$ and is a center for $bc(bc - ad) > 0$.

According to the Routh–Hurwitz criterion [24,25], for a cubic characteristic equation

$$\lambda^3 + a_1\lambda^2 + a_2\lambda + a_3 = 0, \tag{6}$$

the real part of the roots of the cubic Equation (6) is negative if and only if $a_1 > 0$, $a_3 > 0$, $a_1a_2 - a_3 > 0$, i.e., (6) satisfies the condition $|arg(\lambda)| > \pi/2$. Then, the equilibrium point of system (1) is locally asymptotically stable.

Comparing Equation (3) with Equation (6), it is impossible to satisfy the conditions $abc > 0$ and $-abc > 0$ simultaneously, i.e., when $abc > 0$ (or $abc < 0$) and $d > 0$, the equilibrium $S_1$ is unstable. For instance, when $a = 0.4$, $b = 60$, $c = 16$, and $d = 10$, the three eigenvalues corresponding to $S_1 = (0, 0, 0)$ are $\lambda_1 = -12.4696$ and $\lambda_{2,3} = 1.2348 \pm 5.4102i$, and the system has a chaotic attractor at the unstable equilibrium $S_1$ for the initial value $(0.01, 0.01, 0.01)^T$ as shown in Figure 1.

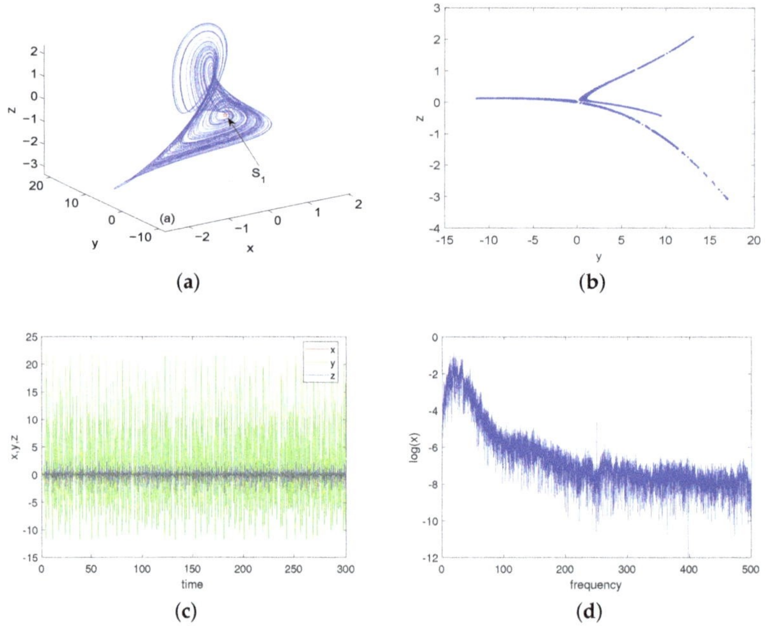


**Figure 1.** (**a**) Chaotic attractor of system (1) with $(a, b, c, d) = (0.4, 60, 16, 10)$ at the initial values $(0.01, 0.01, 0.01)^T$. (**b**) The corresponding Poincaré map on plane $x = 0$. (**c**) The time series of the $x, y, z$ states. (**d**) The power spectrum of the $x$ state.

Similarly, comparing Equation (4) with Equation (6), when one or two of the three conditions (i.e., $a(ad - bc) > 0$, $d^2 - b^2 > 0$, and $d > 0$) are not satisfied, the equilibrium $S_2$ is unstable, and system (1) can generate chaos at $S_1$ and $S_2$. For instance, when $a = 5$, $b = 50$, $c = -6$, and $d = 13$, the three eigenvalues corresponding to $S_2 = (50, 7.3, 5)^T$ are $\lambda_1 = -18.0662$ and $\lambda_{2,3} = 2.5331 \pm 9.7263i$, and the three eigenvalues corresponding to

$S_3 = (50, 0, -23.0769)^T$ are $\lambda_1 = -13$ and $\lambda_{2,3} = \pm 25.4544i$. The system has a chaotic attractor at unstable equilibrium $S_2$, as shown in Figure 2.

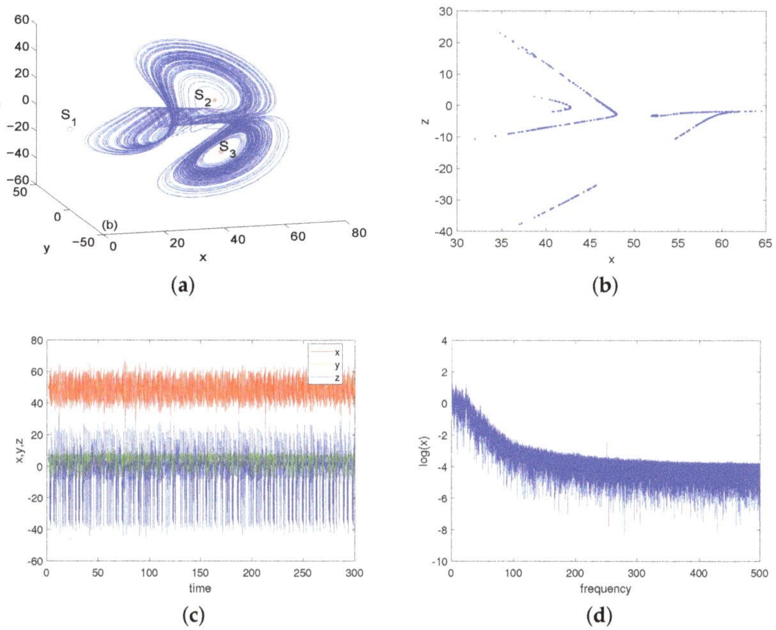


**Figure 2.** (**a**) Chaotic attractor of system (1) with $(a, b, c, d) = (5, 50, -6, 13)$ at the initial values $(50.01, 7.31, 5.01)^T$. (**b**) The corresponding Poincaré map on plane $y = 5$. (**c**) The time series of the $x, y, z$ states. (**d**) The power spectrum of the $x$ state.

## 3. Bifurcation Analysis

As is known, for a 3D autonomous system, its three Lyapunov exponents $L_1$, $L_2$, and $L_3$ can be obtained by using the Wolf algorithm [26]. For the equilibrium points, $L_3 < L_2 < L_1 < 0$, for the periodic orbits, $L_3 < L_2 < 0$, $L_1 = 0$, and for the chaotic attractor, $L_3 < 0, L_2 = 0$, $L_1 > 0$. In the following, the Lyapunov exponent spectrum and the corresponding bifurcation diagram of state variable $x$ with respect to different parameters are shown, and the basic dynamics of the chaotic system (1) are summarized as follows. In addition, the Lyapunov exponents $L_i$ and the Lyapunov dimension $D_L$ are listed, in which the Lyapunov dimension of chaos attractors is a fractional dimension, described as:

$$D_L = j + \frac{1}{\mid L_{j+1} \mid} \sum_{i=1}^{j} L_i = 2 + \frac{L_1 + L_2}{\mid L_3 \mid}. \tag{7}$$

In this section, system (1) is investigated under the condition that the four parameters are all positive, as shown in Table 1. Some examples according to different conditions of the parameters are shown in Tables 1 and 2, which cause system (1) to display chaotic attractors at $S_1$ and $S_2$, respectively.

**Table 1.** Parameter examples and Lyapunov exponents for chaotic system (1) that generate chaos at the equilibrium $S_1$.

| $abc > 0$ and $d > 0$ | Parameter Examples | Lyapunov Exponents | $D_L$ | Initial Values |
|---|---|---|---|---|
| $a > 0,\ b > 0,\ c > 0,$ | $a = 0.3,\ b = 60,\ c = 16,\ d = 10$ | $L_1 = 0.422,\ L_2 = 0,\ L_3 = -10.419$ | 2.0404 | $(0.01, 0.01, 0.01)^T$ |
| | $a = 0.45,\ b = 62,\ c = 16.1,\ d = 9$ | $L_1 = 0.883,\ L_2 = 0,\ L_3 = -9.883$ | 2.0897 | $(0.01, 0.01, 0.01)^T$ |
| | $a = 1.2,\ b = 18,\ c = 15,\ d = 8$ | $L_1 = 0.776,\ L_2 = 0,\ L_3 = -8.775$ | 2.0885 | $(0.01, 0.01, 0.01)^T$ |
| $a > 0,\ b < 0,\ c < 0,$ | $a = 0.85,\ b = -20,\ c = -16,\ d = 10$ | $L_1 = 0.314,\ L_2 = 0,\ L_3 = -10.314$ | 2.0306 | $(0.01, 0.01, 0.01)^T$ |
| | $a = 0.85,\ b = -30,\ c = -19,\ d = 1$ | $L_1 = 0.895,\ L_2 = 0,\ L_3 = -10.891$ | 2.0821 | $(0.01, 0.01, 0.01)^T$ |
| | $a = 0.65,\ b = -30,\ c = -16,\ d = 10$ | $L_1 = 0.443,\ L_2 = 0,\ L_3 = -10.441$ | 2.0424 | $(0.01, 0.01, 0.01)^T$ |
| $a < 0,\ b < 0,\ c > 0,$ | $a = -2,\ b = -21,\ c = 16,\ d = 10$ | $L_1 = 1.022,\ L_2 = 0,\ L_3 = -11.020$ | 2.0931 | $(0.01, 0.01, 0.01)^T$ |
| | $a = -2,\ b = -15,\ c = 16,\ d = 9$ | $L_1 = 0.881,\ L_2 = 0,\ L_3 = -9.882$ | 2.0896 | $(0.01, 0.01, 0.01)^T$ |
| | $a = -10,\ b = -8,\ c = 9,\ d = 8$ | $L_1 = 1.115,\ L_2 = 0,\ L_3 = -9.115$ | 2.1236 | $(0.01, 0.01, 0.01)^T$ |
| $a < 0,\ b > 0,\ c < 0,$ | $a = -1,\ b = 40,\ c = -18,\ d = 12$ | $L_1 = 0.924,\ L_2 = 0,\ L_3 = -12.918$ | 2.0715 | $(0.01, 0.01, 0.01)^T$ |
| | $a = -0.7,\ b = 30,\ c = -16,\ d = 10$ | $L_1 = 0.570,\ L_2 = 0,\ L_3 = -10.568$ | 2.0539 | $(0.01, 0.01, 0.01)^T$ |
| | $a = -1.2,\ b = 20,\ c = -16,\ d = 9$ | $L_1 = 0.743,\ L_2 = 0,\ L_3 = -9.742$ | 2.0763 | $(0.01, 0.01, 0.01)^T$ |

**Table 2.** Parameter examples and Lyapunov exponents for chaotic system (1) that generate chaos at the equilibrium $S_1$.

| $abc < 0$ and $d > 0$ | Parameter Examples | Lyapunov Exponents | $D_L$ | Initial Values |
|---|---|---|---|---|
| $a(ad - bc) > 0,$ $d^2 - b^2 < 0,\ b > 0$ | $a = 5,\ b = 50,\ c = -6,\ d = 13$ | $L_1 = 1.283,\ L_2 = 0,\ L_3 = -14.261$ | 2.0906 | $(50.01, 7.31, 5.01)^T$ |
| | $a = 12,\ b = 50,\ c = -6.1,\ d = 13$ | $L_1 = 1.586,\ L_2 = 0,\ L_3 = -14.574$ | 2.1093 | $(50.01, 7.31, 5.01)^T$ |
| | $a = -16,\ b = 60,\ c = 3,\ d = 9$ | $L_1 = 1.328,\ L_2 = 0,\ L_3 = -10.328$ | 2.1290 | $(50.01, 7.31, 5.01)^T$ |
| $a(ad - bc) > 0,$ $d^2 - b^2 < 0,\ b < 0$ | $a = 16,\ b = -60,\ c = 3,\ d = 9$ | $L_1 = 1.365,\ L_2 = 0,\ L_3 = -10.367$ | 2.1277 | $(-50.01, 7.31, 5.01)^T$ |
| | $a = 10,\ b = -50,\ c = 3,\ d = 10$ | $L_1 = 1.376,\ L_2 = 0,\ L_3 = -11.366$ | 2.1212 | $(-50.01, 7.31, 5.01)^T$ |
| | $a = 20,\ b = -70,\ c = 3,\ d = 10$ | $L_1 = 1.383,\ L_2 = 0,\ L_3 = -11.379$ | 2.1216 | $(-50.01, 7.31, 5.01)^T$ |

We fixed $b = 60$, $c = 16$, and $d = 10$, and the Lyapunov exponent spectrum with respect to $a$ is shown in Figure 3. When the parameter $a$ varied in the small interval $(0, 0.66)$, system (1) had very rich dynamical behaviors, i.e., when $a \in (0, 0.171) \cup (0.219, 0.231) \cup (0.257, 0.269) \cup (0.473, 0.526)$, the maximum Lyapunov exponent equaled zero, and system (1) had periodic orbits, and when $a \in (0.171, 0.219) \cup (0.231, 0.257) \cup (0.269, 0.473) \cup (0.526, 0.66)$, there was one positive Lyapunov exponent, and system (1) was chaotic.

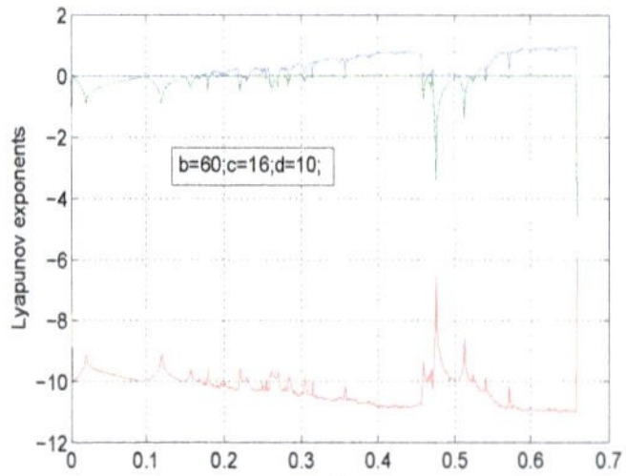


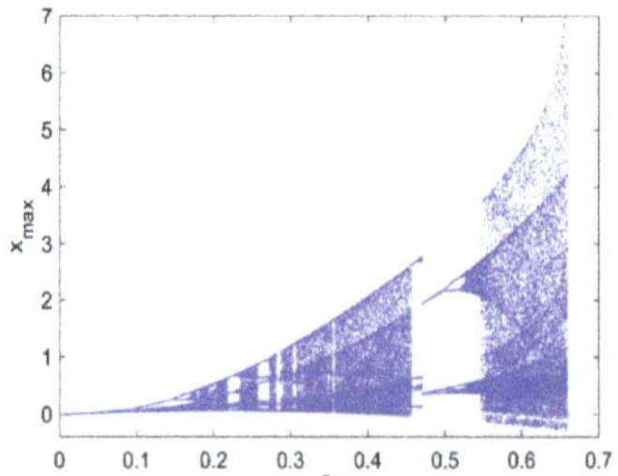


**Figure 3.** The Lyapunov exponents spectrum and the bifurcation diagram of system (1) with $(b, c, d) = (60, 16, 10)$ and $0 < a < 0.66$ at the initial values $(0.01, 0.01, 0.01)^T$: (**a**) Lyapunov exponents; (**b**) bifurcation diagram.

We fixed $a$, $c$, and $d$, and the Lyapunov exponent spectrum with respect to $b$ is shown in Figures 4 and 5. We fixed $a = 0.4$, $c = 16$, and $d = 10$; when $b$ varied in the interval $(0, 97)$, system (1) had very rich dynamical behaviors at the initial values $(0.01, 0.01, 0.01)^T$, i.e., when $b \in (0, 25.7) \cup (29.3, 30.2) \cup (32.9, 35) \cup (39.3, 41) \cup (68.5, 78.5)$, the maximum Lyapunov exponent equaled zero, and system (1) had periodic orbits, and when $b \in (25.7, 29.3) \cup (30.2, 32.9) \cup (35, 39.3) \cup (41, 68.5) \cup (78.5, 97)$, there was one positive Lyapunov exponent, and system (1) was chaotic. On the other hand, we fixed $a = 5, c = -6$, and $d = 13$; when $b$ varied in the interval $(45, 60)$, system (1) had very rich dynamical behaviors at the initial values $(50.01, 7.31, 5.01)^T$ as well, i.e., when $b \in (45.6, 46.9) \cup (50.3, 50.5) \cup (54.6, 55.2)$, the maximum Lyapunov exponent equaled zero, and system (1) had periodic orbits, and when $b \in (45, 45.6) \cup (46.9, 50.3) \cup (50.5, 54.6) \cup (55.2, 56.82)$, there was one positive Lyapunov exponent, and system (1) was chaotic.

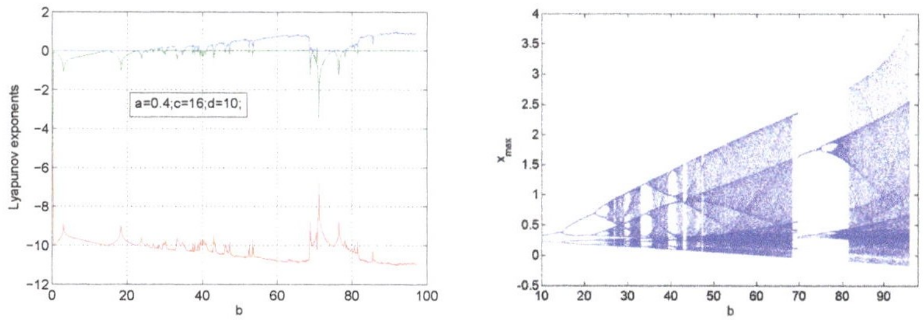


(a) (b)

**Figure 4.** The Lyapunov exponents spectrum and the bifurcation diagram of system (1) with $(a, c, d) = (0.4, 16, 10)$ and $0 < b < 97$ at the initial values $(0.01, 0.01, 0.01)^T$: (**a**) Lyapunov exponents; (**b**) bifurcation diagram.

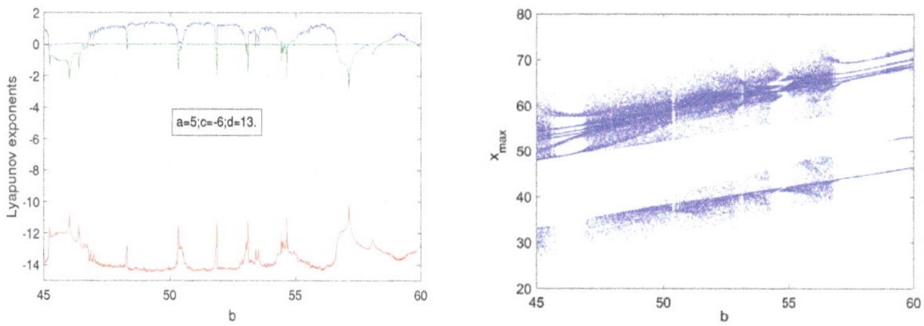


(a) (b)

**Figure 5.** The Lyapunov exponents spectrum and the bifurcation diagram of system (1) with $(a, c, d) = (5, -6, 13)$ and $45 < b < 60$ at the initial values $(50.01, 7.31, 5.01)^T$: (**a**) Lyapunov exponents; (**b**) bifurcation diagram.

We fixed $a = 0.4, b = 60$, and $d = 10$; when $c$ varied in the interval $(2.7, 20.2)$, system (1) had very rich dynamical behaviors, i.e., when $c \in (2.7, 9.1) \cup (10.12, 10.21) \cup (11.7, 12.1) \cup (17.38, 17.77)$, the maximum Lyapunov exponent equaled zero, and system (1) had periodic orbits, and when $c \in (9.1, 10.12) \cup (10.21, 11.7) \cup (12.1, 17.38) \cup (17.77, 20.2)$, there was one positive Lyapunov exponent, and system (1) was chaotic. The corresponding Lyapunov exponent and bifurcation diagram are shown in Figure 6.

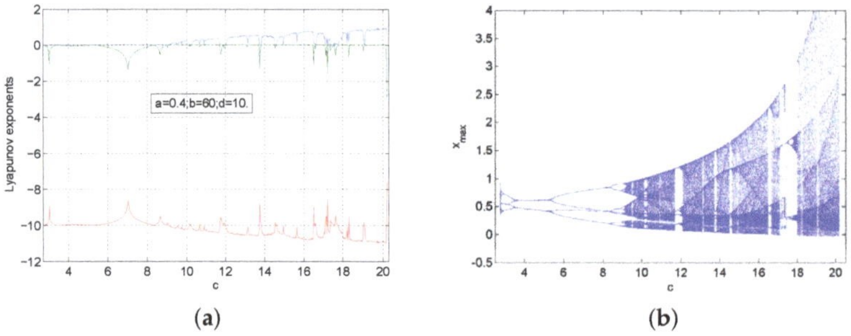


**Figure 6.** The Lyapunov exponents spectrum and the bifurcation diagram of system (1) with $(a, b, d) = (0.4, 60, 10)$ and $2.7 < c < 20.2$ at the initial values $(0.01, 0.01, 0.01)^T$: (**a**) Lyapunov exponents; (**b**) bifurcation diagram.

We fixed $a = 0.4, b = 60$, and $c = 16$; when $d$ varied in the interval $(8.49, 20)$, system (1) had very rich dynamical behaviors at the initial values $(0.01, 0.01, 0.01)^T$, i.e., when $d \in (9.35, 9.65) \cup (11.58, 20)$, the maximum Lyapunov exponent equaled zero, and system (1) had periodic orbits, and when $d \in (8.49, 9.35) \cup (9.65, 11.58)$, there was one positive Lyapunov exponent. and system (1) was chaotic. The corresponding Lyapunov exponent and bifurcation diagram are shown in Figures 7 and 8. On the other hand, we fixed $a = 5, b = 50$, and $c = -6$; when $d$ varied in the interval $(5, 30)$, system (1) had very rich dynamical behaviors at the initial values $(50.01, 7.31, 5.01)^T$ as well, i.e., when $b \in (5, 5.4) \cup (11.9, 12.2) \cup (13.3, 13.6) \cup (13.8, 14.1) \cup (14.8, 15.1) \cup (16.6, 16.8) \cup (17.8, 23.1) \cup (24.9, 30)$, the maximum Lyapunov exponent equaled zero, and system (1) had periodic orbits, and when $b \in (5.4, 11.9) \cup (12.2, 13.3) \cup (13.6, 13.8) \cup (14.1, 14.9) \cup (15.1, 16.6) \cup (16.8, 17.8) \cup (23.1, 24.9)$, there was one positive Lyapunov exponent, and system (1) was chaotic.

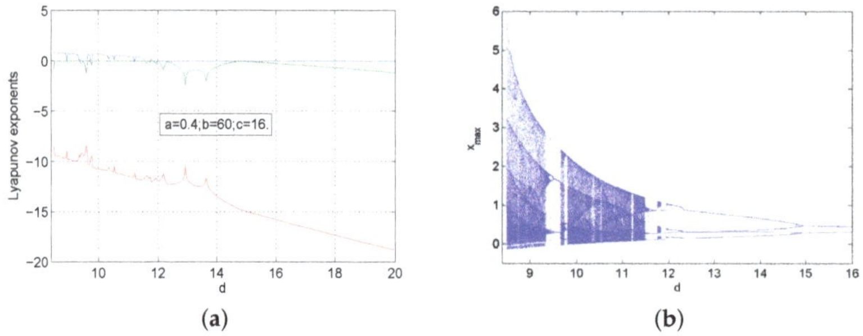


**Figure 7.** The Lyapunov exponents spectrum and the bifurcation diagram of system (1) with $(a, b, c) = (0.4, 60, 16)$ and $8.49 < d < 20$ at the initial values $(0.01, 0.01, 0.01)^T$: (**a**) Lyapunov exponents; (**b**) bifurcation diagram.

In Figure 9, some simulation results of system (1) with different parameter values are given in the $x - y - z$ space.

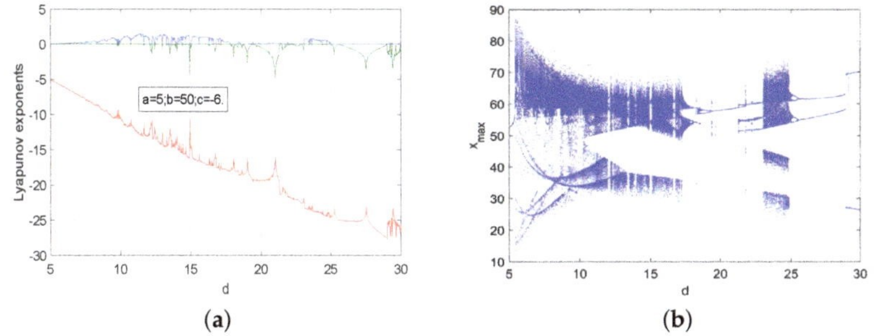


**Figure 8.** The Lyapunov exponents spectrum and the bifurcation diagram of system (1) with $(a, b, c) = (5, 50, -6)$ and $5 < d < 30$ at the initial values $(50.01, 7.31, 5.01)^T$: (**a**) Lyapunov exponents; (**b**) bifurcation diagram.

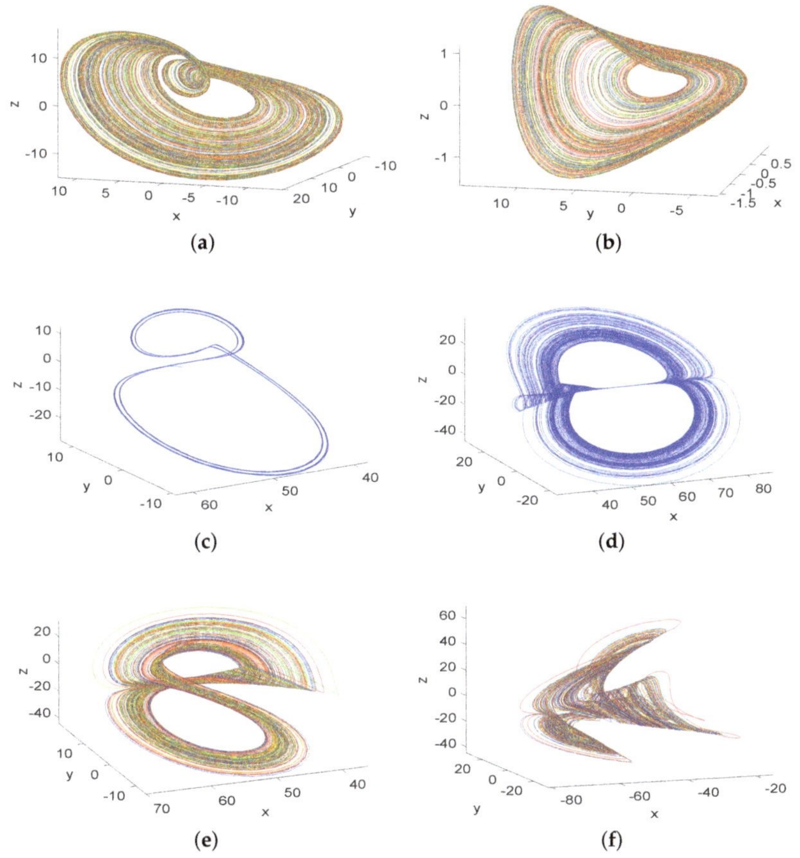


**Figure 9.** The phase portrait of system (1) with different parameter values: (**a**) $a = 5.8, b = 9, c = 15, d = 11$; (**b**) $a = 0.4, b = 60, c = 10, d = 10$; (**c**) $a = 5, b = 50, c = -6, d = 25.5$; (**d**) $a = 5, b = 60, c = -6, d = 34$; (**e**) $a = 5, b = 55, c = -6, d = 16$; (**f**) $a = 16, b = -60, c = 3, d = 9$.

## 4. Chaotic Systems Generalized by a Vector Map

In this section, we introduce several chaotic systems generalized by a vector map. Firstly, system (1) can be written as:

$$\begin{pmatrix} \dot{x} \\ \dot{y} \\ \dot{z} \end{pmatrix} = \begin{pmatrix} 0 & y & 0 \\ 0 & 0 & z \\ x & 0 & 0 \end{pmatrix} A + \begin{pmatrix} x & 0 & 0 \\ 0 & y & 0 \\ 0 & 0 & z \end{pmatrix} B + \begin{pmatrix} yz & 0 & 0 \\ 0 & xz & 0 \\ 0 & 0 & xy \end{pmatrix} C, \tag{8}$$

where $A = (c, a, -b)^T$, $B = (0, 0, -d)^T$, and $C = (-1, 1, 1)^T$ are the parameter vectors. Now, we define the following vector maps $\phi_i : \Re^3 \to \Re^3, i = 1, 2:$

$$\begin{aligned} \phi_1((x_1, x_2, x_3)^T) &= (x_2, x_3, x_1)^T, \\ \phi_2((x_1, x_2, x_3)^T) &= (x_3, x_1, x_2)^T. \end{aligned} \tag{9}$$

Then, system (10) is chaotic for the same parameters as (8):

$$\begin{aligned} \begin{pmatrix} \dot{x} \\ \dot{y} \\ \dot{z} \end{pmatrix} = \begin{pmatrix} 0 & y & 0 \\ 0 & 0 & z \\ x & 0 & 0 \end{pmatrix} \phi_i(A) + \begin{pmatrix} x & 0 & 0 \\ 0 & y & 0 \\ 0 & 0 & z \end{pmatrix} \phi_i(B) \\ + \begin{pmatrix} yz & 0 & 0 \\ 0 & xz & 0 \\ 0 & 0 & xy \end{pmatrix} \phi_i(C). \end{aligned} \tag{10}$$

Substituting Equation (9) into (10) yields two chaotic systems (11) and (12) with $a = 1, b = 22, c = 16$, and $d = 10$ and the initial value $(0.1, 0.1, 0.1)^T$. The two chaotic attractors are shown in Figure 10a.

$$\begin{cases} \dot{x} = -by + yz, \\ \dot{y} = cz + zx - dy, \\ \dot{z} = ax - xy, \end{cases} \tag{11}$$

$$\begin{cases} \dot{x} = cy + yz - dx, \\ \dot{y} = az - zx, \\ \dot{z} = -bx + xy. \end{cases} \tag{12}$$

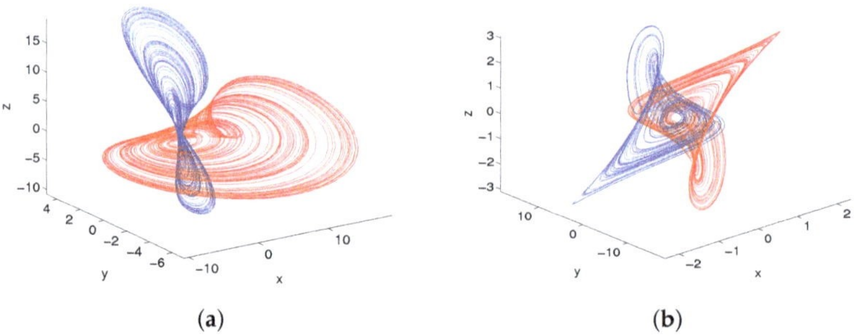


**Figure 10.** (**a**) The red plot corresponds to system (11), and the blue plot corresponds to system (12), with (a,b,c,d)=(1,22,16,10). (**b**) The red plot corresponds to system (13), and the blue plot corresponds to system (1), with $(a, b, c, d) = (0.4, 50, 16, 10)$.

In addition, if we only revise $C = (1, -1, -1)^T$ in system (8), system (13) is chaotic for the same parameters as system (1), when we have the parameters $a = 0.4, b = 50, c = 16,$

and $d = 10$ and the initial value $(0.1, 0.1, 0.1)^T$. The two chaotic attractors are shown in Figure 10b.

$$\begin{cases} \dot{x} = ay + yz, \\ \dot{y} = -bz - xz, \\ \dot{z} = cx - xy - dz. \end{cases} \tag{13}$$

## 5. Electronic Circuit Design

In this section, we present an equivalent electronic circuit for the proposed chaotic system (1). The circuit implementation shows that it can be practically used in technological applications. In order to implement the equations, we considered the analog circuit design using Multisim software, as depicted in Figure 11, with AD712KN operational amplifiers and AD633 analog multipliers all powered by ±15 V symmetric voltages.

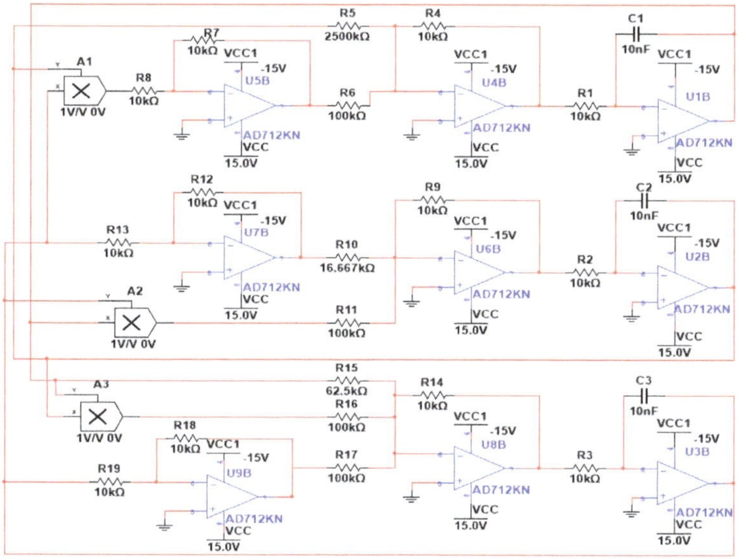


**Figure 11.** Electronic circuit schematics of the chaotic system (1).

Using the Kirchhoff Law for the analog circuit, the generated nonlinear equations are described as

$$\begin{cases} \dot{x} = \frac{R_4}{R_1 R_5 C_1} y + \frac{R_4 R_7}{R_1 R_6 R_8 C_1} yz, \\ \dot{y} = -\frac{R_9 R_{12}}{R_2 R_{10} R_{13} C_2} z + \frac{R_9}{R_2 R_{11} C_2} zx, \\ \dot{z} = \frac{R_{14}}{R_3 R_{15} C_3} x + \frac{R_{14}}{R_3 R_{16} C_3} xy - \frac{R_{14} R_{18}}{R_3 R_{17} C_3} z. \end{cases} \tag{14}$$

Comparing Equation (14) with Equation (1), the common circuital component values were selected as $C_1 = C_2 = C_3 = 10$ nF, $R_i = 10$ kΩ $(i = 1, 2, 3, 4, 9, 14)$, $R_i = 100$ kΩ $(i = 6, 11, 16)$, and $R_i = 10$ kΩ $(i = 7, 8, 12, 13, 18, 19)$. When we chose $R_5 = 2500$ kΩ, $R_{10} = 16.67$ kΩ, $R_{15} = 62.5$ kΩ, and $R_{17} = 100$ kΩ, we obtained a one-scoll chaotic attractor similar to the one obtained by numerical simulation with $(a, b, c, d) = (0.4, 60, 16, 10)$, and the Multisim results on oscilloscope are shown as Figure 12a,b. When we chose $R_5 = 200$ kΩ, $R_{10} = 20$ kΩ, $R_{15} = 166.67$ kΩ, and $R_{17} = 76.92$ kΩ, and modified the connection $R_{15}$ to $R_1$, we obtained a two-scoll chaotic attractor similar to the one obtained by numerical simulation with $(a, b, c, d) = (5, 50, -6, 13)$, and the Multisim results on the oscilloscope are shown as Figure 12c,d.

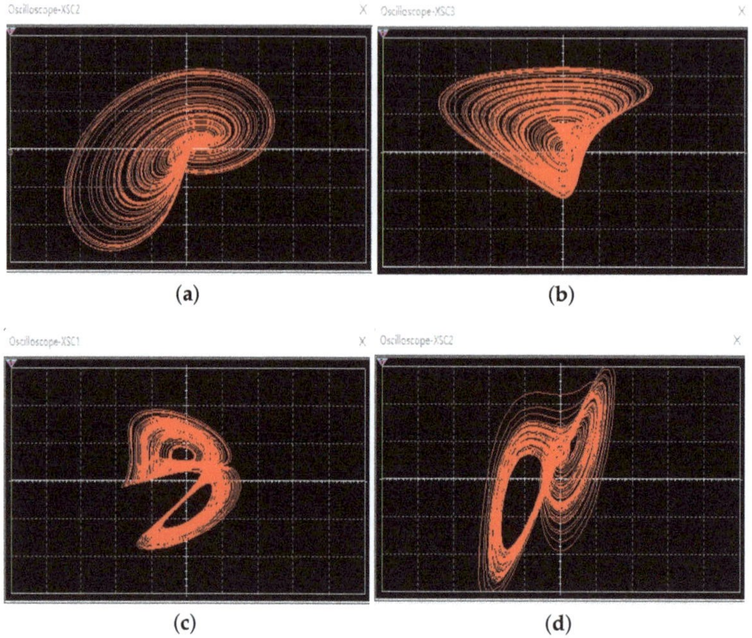


**Figure 12.** Phase spaces of the chaotic system (1) on an oscilloscope obtained from the analog circuit: (**a**) x-z; (**b**) y-z; (**c**) x-z; (**d**) y-z.

## 6. Parameter Identification

In this section, we supposed that the parameters $a,\ b,\ c,$ and $d$ of system (1) were unknown and needed to be identified. We regarded system (1) as the drive system. The response system with adaptive controllers and updating laws was designed as:

$$\begin{cases} \dot{x}_1 = a_1 y_1 - y_1 z_1 + u_1, \\ \dot{y}_1 = -b_1 z_1 + z_1 x_1 + u_2, \\ \dot{z}_1 = c_1 x_1 + x_1 y_1 - d_1 z_1 + u_3, \end{cases} \tag{15}$$

where $a_1,\ b_1,\ c_1,$ and $d_1$ were the estimations of $a,\ b,\ c,$ and $d$, and $u_1,\ u_2,$ and $u_3$ were controllers to be designed.

**Theorem 1.** *If we design the controllers $u_1,\ u_2,$ and $u_3$ in (15) as*

$$\begin{cases} u_1 = -k_1 e_x, \\ u_2 = -k_2 e_y - x_1 e_z, \\ u_3 = -k_3 e_z - x e_y, \\ \dot{k}_1 = \alpha_1 e_x^2, \\ \dot{k}_2 = \alpha_2 e_y^2, \\ \dot{k}_3 = \alpha_3 e_z^2, \end{cases} \tag{16}$$

*where $\alpha_1,\ \alpha_2,$ and $\alpha_3$ are positive constants, and the updating laws of $a_1,\ b_1,\ c_1,$ and $d_1$ as*

$$\begin{cases} \dot{a}_1 = -\theta_1 y_1 e_x, \\ \dot{b}_1 = \theta_2 z_1 e_y, \\ \dot{c}_1 = -\theta_3 x_1 e_z, \\ \dot{d}_1 = \theta_4 z_1 e_z, \end{cases} \tag{17}$$

*where $\theta_1$, $\theta_2$, $\theta_3$, and $\theta_4$ are positive constants, then the adaptive synchronization between the drive–response systems (1) and (15) is achieved, and the unknown parameters $a$, $b$, $c$, and $d$ in (1) are identified by $a_1$, $b_1$, $c_1$, and $d_1$ in (15), with controllers (16) and updating laws (17).*

**Proof.** Let $e_x = x_1 - x, e_y = y_1 - y$, and $e_z = z_1 - z$; then, one has

$$\begin{cases} \dot{e_x} = (a_1 - a)y_1 + ae_y - y_1 e_z - e_y z + u_1, \\ \dot{e_y} = -(b_1 - b)z_1 - be_z + x_1 e_z + e_x z + u_2, \\ \dot{e_z} = (c_1 - c)x_1 + ce_x + e_x y_1 + xe_y - (d_1 - d)z_1 - de_z + u_3. \end{cases} \tag{18}$$

We consider the following Lyapunov function

$$\begin{aligned} V(t) =& \frac{e_x^2 + e_y^2 + e_z^2}{2} + \frac{(a_1 - a)^2}{2\theta_1} + \frac{(b_1 - b)^2}{2\theta_2} \\ &+ \frac{(c_1 - c)^2}{2\theta_3} + \frac{(d_1 - d)^2}{2\theta_4} + \sum_{i=1}^{3} \frac{(k_i - k_i^*)^2}{2\alpha_i}, \end{aligned} \tag{19}$$

where $k_1^*$, $k_2^*$, and $k_3^*$ are arbitrary positive constants to be determined.

Then, the derivative of $V(t)$ along the trajectories of (18) gives

$$\begin{aligned} \dot{V}(t) =& e_x \dot{e}_x + e_y \dot{e}_y + e_z \dot{e}_z + \frac{(a_1 - a)\dot{a}_1}{\theta_1} + \frac{(b_1 - b)\dot{b}_1}{\theta_2} \\ &+ \frac{(c_1 - c)\dot{c}_1}{\theta_3} + \frac{(d_1 - d)\dot{d}_1}{\theta_4} + \sum_{i=1}^{3} \frac{(k_i - k_i^*)\dot{k}_i}{\alpha_i} \\ =& e^T (P - K) e, \end{aligned}$$

where $e = (e_x, e_y, e_z)^T$,

$$P = \begin{pmatrix} 0 & \frac{a}{2} & \frac{c}{2} \\ \frac{a}{2} & 0 & -\frac{b}{2} \\ \frac{c}{2} & -\frac{b}{2} & -d \end{pmatrix}, \; K = \begin{pmatrix} k_1^* & 0 & 0 \\ 0 & k_2^* & 0 \\ 0 & 0 & k_3^* \end{pmatrix}.$$

Then, one can choose $k_1^*$, $k_2^*$, and $k_3^*$ large enough such that $P - K < 0$, i.e., $\dot{V}(t) < 0$, which implies that the adaptive synchronization is achieved, and the unknown parameters $a$, $b$, $c$, and $d$ are identified by $a_1$, $b_1$, $c_1$, and $d_1$. Thus, the proof is complete. □

In the simulations, we designed the corresponding electronic circuit with controllers (16) and updating laws (17) to identify the unknown parameters using Multisim software. Figure 13 shows the electronic circuit design with the AD712KN operational amplifiers and AD633 analog multipliers all powered by $\pm 15$ V symmetric voltages. We supposed that the resistances $R_{22}, R_{12}, R_{14}$, and $R_{17}$ of the drive system corresponding to the parameters $a = 0.4$, $b = 60$, $c = 16$, and $d = 10$ of system (1) were unknown and needed to be identified. We chose the following resistances of the updating laws $R_a = R_b = R_c = R_d = 50$ kΩ and the following resistances of the controllers $R_{K_1} = R_{K_2} = R_{K_3} = 100$ kΩ. The Multisim results on the oscilloscope are shown as Figure 14a,b. Clearly, the unknown parameters $a$, $b$, $c$, and $d$ were well identified by $a_1$, $b_1$, $c_1$, and $d_1$.

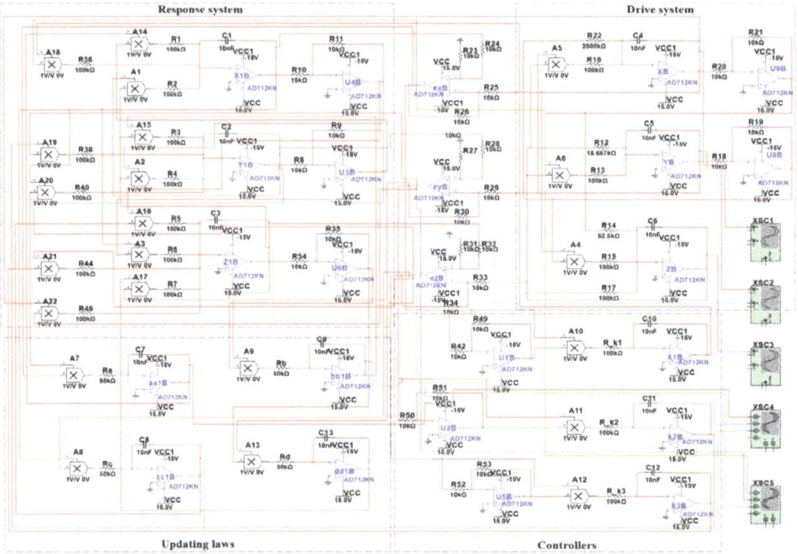


**Figure 13.** Electronic circuit schematics of the parameter identification.

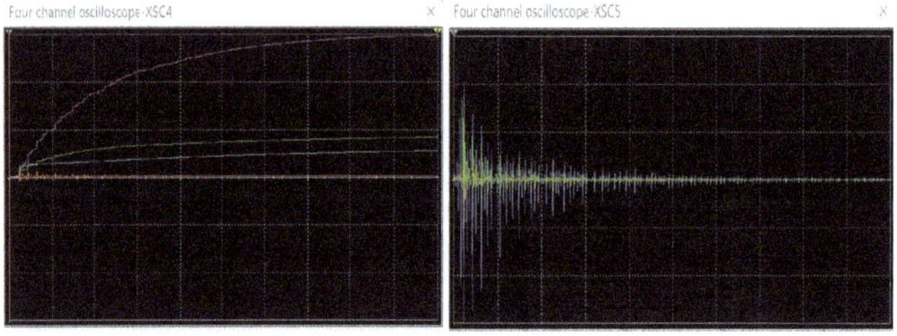


(a) (b)

**Figure 14.** Identification of $a_1, b_1, c_1$, and $d_1$ by an electronic circuit.

## 7. Conclusions

In this paper, we introduced and studied a new 3D autonomous chaotic system, which could generate one-scroll and two-scroll chaotic attractors with different parameters. The dynamical behaviors and properties of this chaotic system were investigated both theoretically and numerically. The Lyapunov exponent spectrum and the corresponding bifurcation diagram, with respect to different parameters, were presented, and these validated the correctness of our results. Spectral analysis showed that the system had a large chaos region. Moreover, a vector map was employed to the generalized chaotic system. Compared with the famous Rössler, Sprott, and Lorenz attractors, this system is a supplement to the discovery of one-scroll and two-scroll attractors. Further, we designed an equivalent electronic circuit for the proposed chaotic system based on Kirchhoff's Law to show its practical applications. We designed a corresponding response electronic circuit to identify the unknown parameters or monitor the changes in the system parameters as well. Finally, numerical simulations were presented to perform and complement the theoretical results.

**Author Contributions:** M.L. wrote the original draft, Z.W. and X.F. reviewed and edited the paper. All authors have read and agreed to the published version of the manuscript.

**Funding:** This work was jointly supported by the scientific research project of the Education Department of Zhejiang Province (No.Y202044362) and the NSFC (No. 61963019).

**Institutional Review Board Statement:** Not applicable.

**Informed Consent Statement:** Not applicable.

**Data Availability Statement:** The data are presented in the manuscript.

**Conflicts of Interest:** The authors declare no conflict of interest.

## References

1. Lorenz, E.N. Deterministic Nonperiodic Flow. *J. Atmos. Sci.* **1963**, *20*, 130–141. . [CrossRef]
2. Fallahi, K.; Leung, H. A chaos secure communication scheme based on multiplication modulation. *Commun. Nonlinear Sci. Numer. Simul.* **2010**, *15*, 368–383. . [CrossRef]
3. Rössler, O.E. An equation for continuous chaos. *Phys. Lett. A* **1976**, *57*, 397–398. . [CrossRef]
4. Chen, G.; Ueta, T. Yet another chaotic attractor. *Int. J. Bifurc. Chaos* **1999**, *9*, 1465–1466. . [CrossRef]
5. Lü, J.; Chen, G.; Cheng, D.; Celikovsky, S. Bridge the gap between the Lorenz system and the Chen system. *Int. J. Bifurc. Chaos Appl. Sci. Eng.* **2002**, *12*, 2917–2926. . [CrossRef]
6. Liu, W.; Chen, G. A new chaotic system and its generation. *Int. J. Bifurc. Chaos Appl. Sci. Eng.* **2003**, *13*, 261–267. . [CrossRef]
7. Lü, J.; Chen, G.; Cheng, D. A New Chaotic System and Beyond: The Generalized Lorenz-like System. *Int. J. Bifurc. Chaos* **2004**, *14*, 1507–1537. [CrossRef]
8. Dong, G.; Du, R.; Tian, L.; Jia, Q. A novel 3D autonomous system with different multilayer chaotic attractors. *Phys. Lett. Sect. A: Gen. At. Solid State Phys.* **2009**, *373*, 3838–3845. . [CrossRef]
9. Wang, Z.; Qi, G.; Sun, Y.; Van Wyk, B.J.; Van Wyk, M.A. A new type of four-wing chaotic attractors in 3-D quadratic autonomous systems. *Nonlinear Dyn.* **2010**, *60*, 443–457. . [CrossRef]
10. Wei, Z.; Yang, Q. Dynamical analysis of a new autonomous 3-D chaotic system only with stable equilibria. *Nonlinear Anal. Real World Appl.* **2011**, *12*, 106–118. . [CrossRef]
11. Liu, Y. Analysis of global dynamics in an unusual 3D chaotic system. *Nonlinear Dyn.* **2012**, *70*, 2203–2212. [CrossRef]
12. Wang, X.; Chen, G. Constructing a chaotic system with any number of equilibria. *Nonlinear Dyn.* **2013**, *71*, 429–436. . [CrossRef]
13. Li, C.; Sprott, J.C.; Xing, H. Constructing chaotic systems with conditional symmetry. *Nonlinear Dyn.* **2017**, *87*, 1351–1358. . [CrossRef]
14. Wang, G.; Yuan, F.; Chen, G.; Zhang, Y. Coexisting multiple attractors and riddled basins of a memristive system. *Chaos* **2018**, *28*, 013125. . [CrossRef]
15. Zhang, X.; Wang, C.; Yao, W.; Lin, H. Chaotic system with bondorbital attractors. *Nonlinear Dyn.* **2019**, *97*, 2159–2174. . [CrossRef]
16. Xie, Q.; Zeng, Y. Generating different types of multi-double-scroll and multi-double-wing hidden attractors. *Eur. Phys. J. Spec. Top.* **2020**, *229*, 1361–1371. . [CrossRef]
17. Gong, L.; Wu, R.; Zhou, N. A New 4D Chaotic System with Coexisting Hidden Chaotic Attractors. *Int. J. Bifurc. Chaos* **2020**, *30*, 2050142. . [CrossRef]
18. Zhou, L.; You, Z.; Tang, Y. A new chaotic system with nested coexisting multiple attractors and riddled basins. *Chaos Solitons Fractals* **2021**, *148*, 111057. . [CrossRef]
19. Ma, C.; Mou, J.; Xiong, L.; Banerjee, S.; Liu, T.; Han, X. Dynamical analysis of a new chaotic system: Asymmetric multistability, offset boosting control and circuit realization. *Nonlinear Dyn.* **2021**, *103*, 2867–2880. . [CrossRef]
20. Wang, R.; Li, C.; Kong, S.; Jiang, Y.; Lei, T. A 3D memristive chaotic system with conditional symmetry. *Chaos Soliton Fractal* **2022**, *158*, 111992. [CrossRef]
21. Vanecek, A., Celikoshy, S. *Control Systems: From Linear Analysis to Synthesis of Chaos*; Prentice-Hall: London, UK, 1996.
22. Sprott, J.C. Some simple chaotic flows. *Phys. Rev. E* **1994**, *50*, R647–R650. . [CrossRef] [PubMed]
23. Sprott, J.C. Simplest dissipative chaotic flow. *Phys. Lett. Sect. A: Gen. At. Solid State Phys.* **1997**, *228*, 271–274. . [CrossRef]
24. Huang, K.; Yang, Q. Stability and Hopf bifurcation analysis of a new system. *Chaos Solitons Fractals* **2009**, *39*, 567–578. . [CrossRef]
25. Zheng, S.; Dong, G.; Bi, Q. A new hyperchaotic system and its synchronization. *Appl. Math. Comput.* **2010**, *215*, 3192–3200. . [CrossRef]
26. Wolf, A.; Swift, J.B.; Swinney, H.L.; Vastano, J.A. Determining Lyapunov exponents from a time series. *Phys. D: Nonlinear Phenom.* **1985**, *16*, 285–317. . [CrossRef]

*axioms*

MDPI

*Article*

# A PDE Model Approach to Formation Control of Large-Scale Mobile Sensor Networks with Boundary Uncertainties

**Xueming Qian [1,2,*] and Baotong Cui [2]**

[1] School of Internet of Things, Wuxi Vocational College of Science and Technology, Wuxi 214028, China
[2] School of Internet of Things Engineering, Jiangnan University, Wuxi 214122, China
* Correspondence: xmqian1981@gmail.com

**Abstract:** This paper investigates the formation problem of an array of large-scale mobile sensor networks. A new framework for the dynamic of mobile sensors as a continuum described by the parabolic system with boundary disturbance is proposed. The communication topology of agents is a chain graph and fixed. Leader feedback laws which are designed in a manner to the boundary control of large-scale mobile sensor networks allow the mobile sensors to achieve the formation steadily. By referring to the Lyapunov functional method and employing a boundary control approach, a new protocol is established to deal with a formation problem for the large-scale mobile sensor networks. Finally, numerical examples are given to illustrate the usefulness of the results.

**Keywords:** boundary control; formation control; mobile sensors; parabolic systems; boundary disturbance; consensus

**MSC:** 93C20; 93A15; 93A16; 93D50; 93D05

## 1. Introduction

During the last few decades, wireless sensor networks have been intensively explored and applied in a wide variety of fields [1,2], including military, surroundings, health, and home. A collection of sensing devices which are connected by wireless communication can be built into a wireless sensor network in engineering. In order to adapt to the requirement of flexibility in applications, the sensing nodes of wireless sensor networks can be endowed mobility using mobile devices. Mobile sensing devices reduce power consumption and costs while also improving performance. In the last few years, mobile sensor networks have attracted a lot more interest because they have been successful in a wide range of applications including formation control [3–5], target tracking [6–8], environmental monitoring [9,10], and other areas. The study of cooperative control of mobile sensing networks has been viewed as a crucial issue.

Ref. [11] presents a robust control system for mobile sensors moving collaboratively in a distributed environment. A gradient climbing task in which mobile sensor networks can adapt its configuration in response to the measured environment. In [12], three distributed stable deployment algorithms based on flocking in mobile sensor networks are addressed. In addition, from the view of swarm, the control problem of wireless sensor networks with mobile multi-robots is discussed in [13,14]. Moreover, formation control of mobile sensor networks has been investigated intensively. A distributed control model is given to form the desired formation in [15]. The distributed observers are studied to achieve the formation in [16]. Stabilization of geometric pattern based on Laplacian are considered in [8,9,17]. Deployment onto a desired planar curve is investigated in [18] and [19] using coverage control algorithms and Lie group setting, respectively.

In general, the model suggested in [20] or versions of these models are used in the majority of studies concerning mobile sensor networks models. The model presented by Reynolds contains three heuristic rules. They are alignment, cohesion, and separation. Some additional rules such as obstacle avoidance capabilities have been suggested ever since. It is not difficult to see that these models are the particle system essentially. The discrete dynamic characteristics are exhibited in these models. Nevertheless, an increasing number of studies have found that continuity and layered property exist in biological systems [21], social systems and so on. In [21], the authors proposed the starling swarm coordination for studying the behavior inertia and information transfer of the birds, which could study the cooperative flight of unmanned aerial vehicle. These phenomena show us that the system should be interpreted as a continuum when it is made up of a significant number of mobile sensing devices. That is to say, large-scale mobile sensor networks(LMSNs) is a fluid instead of a particle system. Therefore, the large-scale mobile sensor networks should be modeled as a partial differential equation (PDE).

For the past few years, a partial differential equation has been an effective tool that begins to analyze the multi-agent system [22–24]. In particular, [22] used the LWR equation for crowd evacuation, Ref. [23] analyzes the stability of vehicular platoons based on an unstable hyperbolic system, and Ref. [24] models multi-agents as reaction–advection–diffusion equations for deployment onto spatial curves. Furthermore, growing results of experiments suggest that the partial differential equation has gradually become an important way of system modeling. However, it should be pointed out that a PDE-based model for formation control of large-scale mobile sensor networks still has little research attention.

Therefore, the major purpose of this research is to investigate the formation issue for an array of large-scale mobile sensor networks based on parabolic systems with boundary disturbance. For the PDE model of mobile sensor networks, the boundary control approach is addressed. Such a framework which implies the communication topology of mobile sensors is a chain graph and fixed. Hence, stable formation onto planar curves can be achieved by leaders, i.e., the boundary control of the leader such that mobile sensors will reach consensus. By constructing proper Lyapunov functional and employing the boundary control approach, some sufficient conditions have been established to cope with the formation issue for large group mobile sensing devices.

This study mainly contributes the following innovative points: (1) a framework for the dynamic of large-scale mobile sensors as a continuum and modeled by parabolic system is presented; (2) the design step about formation of large-scale mobile sensor networks is proposed; and (3) boundary disturbance in the larger group mobile sensing devices is considered. In order to exhibit the utility of the proposed boundary control strategy, numerical simulations are provided.

## 2. Problem Formulation and Preliminaries

In this study, a new framework is proposed for the dynamics of a large group of mobile sensors as a continuum by the parabolic system. Large-scale mobile sensor networks are moving in an obstacle-free two-dimensional plane. Energy constraints at the level of network nodes are not considered here. The dynamics of mobile sensor networks are described by the following integrator:

$$\frac{\partial \boldsymbol{x}(\theta,t)}{\partial t} = \boldsymbol{v}(\theta,t), \tag{1}$$

where $\boldsymbol{x}(\theta,t) = [x_1(\theta,t), x_2(\theta,t)]^T$ denotes the position of agent $\theta$ at time $t$. The parameter $\theta$ as the identity of each mobile sensor in a large mobile sensor network, and $\theta \in [0,1]$. Namely, $\theta$ is a mobile sensor's index code and as the spatial variable of mobile sensor network for the group's collective dynamics. $\boldsymbol{v}(\theta,t) = [v_1(\theta,t), v_2(\theta,t)]^T$ is the velocity for the agent $\theta$, and the control input $\boldsymbol{v}(\theta,t)$ is determined by

$$\boldsymbol{v}(\theta,t) = D\frac{\partial^2 \boldsymbol{x}(\theta,t)}{\partial \theta^2}, \tag{2}$$

with diffusion operator $D = \text{diag}[d_1, d_2]$, constants $d_1, d_2$ are both positive.

The velocity input of diffusion-based feedback indicates all the agents utilize only local information from the nearest-neighbor. A distributed communication topology is given, which is a chain graph.

**Remark 1.** *It is worth mentioning that, in this paper, the input control (2) is given for the first-order integrator, which implies that this paper is studying the consensus of large-scale mobile sensor networks in the framework of the PDE model. In other words, this paper is concerned with the theoretical analysis of the collaborative dynamics of the system as a whole when the large-scale mobile nodes are viewed as a continuum (PDE model). The object of study here is the continuum, not the network.*

*The PDE model approach has natural advantages for the study of continuum models. However, the control inputs proposed in (2) in the paper also bring some limitations to the system model. In addition, (2) only considers the distributed chain communication topology of the system but does not describe phenomena such as output data missing, sensor saturation, or even node failure. These are currently being considered by the authors.*

**Remark 2.** *In recent years, several important results about multi-agent system based on individual have been obtained. First-order integrator dynamics often can be presented as the model of multi-agent systems. The agent dynamics is expressed in the following:*

$$\dot{x}_i(t) = \sum_{j \in \mathcal{N}_i} (x_j(t) - x_i(t)), \quad i = 1, 2, \cdots, n,$$

*where $\mathcal{N}_i$ is the set of neighbors that communicate with agent i. It is not difficult to obtain an associated continuum model by continuous approximation as follows:*

$$\frac{\partial x(\theta, t)}{\partial t} = \frac{\partial^2 x(\theta, t)}{\partial \theta^2},$$

*where the velocity feedback of each agent(mobile sensor) employs the diffusion term.*

The initial condition associated with (1) is given by

$$\boldsymbol{x}(\theta, 0) = \boldsymbol{x}_0(\theta), \tag{3}$$

where $\boldsymbol{x}(\theta, 0) = [x_1(\theta, 0), x_2(\theta, 0)]^T, \boldsymbol{x}_0(\theta) = [x_{10}(\theta), x_{20}(\theta)]^T$, and have the mixed boundary condition

$$\boldsymbol{x}(0, t) = [0, 0]^T, \qquad \frac{\partial \boldsymbol{x}(1, t)}{\partial \theta} = \boldsymbol{u}(t) + \boldsymbol{w}(t), \tag{4}$$

where $\boldsymbol{u}(t) = [u_1(t), u_2(t)]^T$ is the controls to be designed, $\boldsymbol{w}(t) = [w_1(t), w_2(t)]^T$ is an boundary disturbance, which assumes the following:

$$\int_0^{+\infty} \boldsymbol{w}^T(t)\boldsymbol{w}(t)dt < +\infty. \tag{5}$$

It is worth mentioning that the initial condition should be compatible with the boundary condition, i.e.,

$$\left.\frac{d\boldsymbol{x}_0(\theta)}{d\theta}\right|_{\theta=0} = \boldsymbol{0}, \qquad \left.\frac{d\boldsymbol{x}_0(\theta)}{d\theta}\right|_{\theta=1} = \boldsymbol{u}(0) + \boldsymbol{w}(0).$$

A large group of mobile sensors as a continuum by system (1), index $\theta = 1$ is the leader sensor, and $\theta = 0$ is the anchor sensor, the rest of index $0 < \theta < 1$ are the follower sensors. When proper boundary control law is designed such that $\lim_{t \to +\infty} \boldsymbol{x}(\theta, t) = \bar{\boldsymbol{x}}(\theta)$ for all $\theta \in [0, 1]$, then mobile sensors are said to achieve formation consensus. Meanwhile, a large group of mobile sensors is also said to achieve consensus if $\boldsymbol{u}(t) = \boldsymbol{0}$ for the leader.

Our main purpose in this paper is to accomplish the desired formation for an array of large mobile sensor networks based on heat equation with boundary disturbance by using the boundary control approach. Clearly, the 2D formation problem can be discussed in two aspects. On the one hand, what kind of formation can be reached by a continuum of mobile sensors under diffusion-based feedback? On the other hand, how can the boundary controller be designed such that mobile sensors steadily reach the formation?

## 3. Formation Control Scheme Design

First of all, we consider the diffusion-based feedback as velocity control input in continuum of mobile sensors. According to (1) and (2), the dynamics of mobile sensors are governed by heat equation

$$\frac{\partial x(\theta,t)}{\partial t} = D\frac{\partial^2 x(\theta,t)}{\partial \theta^2}. \tag{6}$$

The equilibrium equation of (6) is a second ordinary differential equation

$$\frac{d^2\bar{x}(\theta)}{d\theta^2} = \mathbf{0}. \tag{7}$$

It is not difficult to solve the solution of (7) being a family of linear functions. Those functions can be decomposed to vertical and horizontal directions, respectively. The compact form of linear functions can be described as

$$\bar{x}(\theta) = \begin{bmatrix} \bar{x}_1(\theta) \\ \bar{x}_2(\theta) \end{bmatrix} = \begin{bmatrix} a_1 & a_2 \\ a_3 & a_4 \end{bmatrix}\begin{bmatrix} 1 \\ \theta \end{bmatrix}, \tag{8}$$

where constants $a_1, a_2, a_3$ and $a_4$ are formation coefficients, which are picked by users to generate the shape of the target formation. The type of formation curve is determined by the basis $(1,\theta)$, which is the solution of (7). Therefore, mobile sensors which are governed by (6) have the capacity of achieving linear formations or rendezvousing at the origin.

Next, in general, the boundary control approach is an effective way for large mobile sensor networks described by parabolic systems. Thus, we consider the leader agent($\theta = 1$) as boundary control input enabling large mobile sensors reach the desired formation stably. The control law of leader can be designed as follows:

$$u(t) = -Kx(1,t) + K\bar{x}(1) + \frac{d\bar{x}(1)}{d\theta}, \tag{9}$$

where $x(1,t) = [x_1(1,t), x_2(1,t)]^T$ and $\frac{d\bar{x}(1)}{d\theta}$ means $\frac{d\bar{x}(\theta)}{d\theta}|_{\theta=1}$.

To have mobile sensors successfully reach the formation, we shift the equilibrium $\bar{x}(\theta)$ to the origin by transformation $y(\theta,t) = x(\theta,t) - \bar{x}(\theta)$, and the formation error dynamics for mobile sensors can be obtained from (3), (4), and (6) as follows:

$$\frac{\partial y(\theta,t)}{\partial t} = D\frac{\partial^2 y(\theta,t)}{\partial \theta^2}, \tag{10}$$

$$y(\theta,0) = y_0(\theta), \tag{11}$$

$$y(0,t) = \mathbf{0}, \quad \frac{\partial y(1,t)}{\partial \theta} = U(t) + w(t), \tag{12}$$

where $y(\theta,t) = [y_1(\theta,t), y_2(\theta,t)]^T$.

Substituting error $y(\theta,t)$ into (4) and (8) leads to

$$U(t) = -Ky(1,t). \tag{13}$$

In brief, the design procedure of formation deployment of large mobile sensors is listed in the following:

i. Find the equilibrium equation of dynamics model;

ii. Analyze the basis of equilibrium equation to determine the type of equilibrium curve component;
iii. Select the formation coefficients to determine the shape of equilibrium curve component;
iv. Orthogonal synthesis of the equilibrium curve of two coordinate axes to determine a target formation;
v. Implement the control law for leader mobile sensor to achieve the formation.

## 4. Stability Analysis of the Closed-Loop System

To give our main results, we introduce the following definition and lemmas:

**Definition 1** ([25]). *A parabolic system is said to be $L_2[0,1]$ stable, if*

$$\lim_{t\to+\infty}\int_0^1 \boldsymbol{y}^T(\theta,t)\boldsymbol{y}(\theta,t)d\theta = \mathbf{0},$$

*holds.*

Definition 1 implies the closed-loop system state $\boldsymbol{x}(\theta,t)$ is $L_2[0,1]$ stable at the equilibrium curve $\bar{\boldsymbol{x}}(\theta)$.

**Definition 2** ([26]). *A parabolic system is said to be globally asymptotically stable, if*

$$\lim_{t\to+\infty}\boldsymbol{y}(\theta,t) = \mathbf{0},$$

*holds.*

Definition 2 implies $\lim_{t\to+\infty}\boldsymbol{x}(\theta,t) = \bar{\boldsymbol{x}}(\theta)$, that is to say, a large group of mobile sensors will converge into the formation $\bar{\boldsymbol{x}}(\theta)$.

**Lemma 1** ([27]). *Let $z \in W^{1,2}(a,b)$ be a scalar function with $z(a) = 0$. Then,*

$$\int_a^b z^2(\xi)d\xi \le \frac{(b-a)^2}{2}\int_a^b \left(\frac{dz(\xi)}{d\xi}\right)^2 d\xi,$$

$$\max_{\xi\in[a,b]} z^2(\xi) \le (b-a)\int_a^b \left(\frac{dz(\xi)}{d\xi}\right)^2 d\xi.$$

**Lemma 2** ([25]). *(The special case of Barbalat's Lemma) Let $f(t)$ be a non-negative function defined on $[0,+\infty)$. If $f(t)$ is differentiable on $[0,+\infty)$ and $\sup_{t\ge 0}\frac{df(t)}{dt} < +\infty$, and $\int_0^{+\infty} f(t)dt < +\infty$. Then, there holds*

$$\lim_{t\to+\infty} f(t) = 0.$$

The following theorems present the main outcomes of the study.

### 4.1. Boundedness

**Theorem 1.** *Under the control scheme (13), the system (10) with boundary disturbance $\boldsymbol{w}(t)$ satisfying (5) and the initial condition $\boldsymbol{y}_0(\theta)$ satisfying $\int_0^{+\infty} \boldsymbol{y}_0^T(\theta)\boldsymbol{y}_0(\theta)d\theta < +\infty$ is uniformly square integrable, i.e.,*

$$\sup_{\theta\in[0,1]}\int_0^{+\infty} \boldsymbol{y}^T(\theta,t)\boldsymbol{y}(\theta,t)dt < +\infty,$$

*and $\int_0^1 \boldsymbol{y}^T(\theta,t)\boldsymbol{y}(\theta,t)d\theta$ is bounded on $[0,+\infty)$ if $\sigma_2 < \sigma_1$.*

**Proof.** Consider a Lyapunov functional as

$$V(t) = \frac{1}{2}\int_0^1 \boldsymbol{y}^T(\theta,t)\boldsymbol{y}(\theta,t)d\theta. \quad (14)$$

For the system (10), the time derivative of $V(t)$ can be computed as follows:

$$\begin{aligned}\frac{dV(t)}{dt} &= \int_0^1 \boldsymbol{y}^T(\theta,t)\frac{\partial \boldsymbol{y}(\theta,t)}{\partial t}d\theta \\ &= \int_0^1 \boldsymbol{y}^T(\theta,t)D\frac{\partial^2 \boldsymbol{y}(\theta,t)}{\partial \theta^2}d\theta \\ &= \boldsymbol{y}^T(\theta,t)D\frac{\partial \boldsymbol{y}(\theta,t)}{\partial \theta}\bigg|_0^1 \\ &\quad - \int_0^1 \left(\frac{\partial \boldsymbol{y}(\theta,t)}{\partial \theta}\right)^T D\frac{\partial \boldsymbol{y}(\theta,t)}{\partial \theta}d\theta \\ &\le \boldsymbol{y}^T(1,t)D\frac{\partial \boldsymbol{y}(1,t)}{\partial \theta} \\ &\quad - \int_0^1 \left(\frac{\partial \boldsymbol{y}(\theta,t)}{\partial \theta}\right)^T D\frac{\partial \boldsymbol{y}(\theta,t)}{\partial \theta}d\theta.\end{aligned}$$

By employing boundary control scheme (13), we have

$$\begin{aligned}\frac{dV(t)}{dt} \le &-\boldsymbol{y}^T(1,t)DK\boldsymbol{y}(1,t) + \boldsymbol{y}^T(1,t)D\boldsymbol{w}(t) \\ &- \int_0^1 \left(\frac{\partial \boldsymbol{y}(\theta,t)}{\partial \theta}\right)^T D\frac{\partial \boldsymbol{y}(\theta,t)}{\partial \theta}d\theta \\ \le &(-\sigma_1+\sigma_2)\boldsymbol{y}^T(1,t)\boldsymbol{y}(1,t) \\ &- \sigma_3\int_0^1 \left(\frac{\partial \boldsymbol{y}(\theta,t)}{\partial \theta}\right)^T \frac{\partial \boldsymbol{y}(\theta,t)}{\partial \theta}d\theta \\ &+ \frac{1}{4}\boldsymbol{w}^T(t)\boldsymbol{w}(t),\end{aligned} \quad (15)$$

where $\sigma_1 = \lambda_{\min}(DK)$, $\sigma_2 = \lambda_{\max}(D^2)$ and $\sigma_3 = \lambda_{\min}(D)$.
Integrating (15) over $[0,t]$, we obtain

$$\begin{aligned}&\int_0^1 \boldsymbol{y}^T(\theta,t)\boldsymbol{y}(\theta,t)d\theta \\ \le &\frac{1}{2}\int_0^t \boldsymbol{w}^T(\tau)\boldsymbol{w}(\tau)d\tau + \int_0^1 \boldsymbol{y}_0^T(\theta)\boldsymbol{y}_0(\theta)d\theta.\end{aligned}$$

In addition, integrating (15) over $[0,+\infty]$, then

$$\begin{aligned}&\int_0^{+\infty}\Bigg[(\sigma_1-\sigma_2)\boldsymbol{y}^T(1,t)\boldsymbol{y}(1,t) \\ &+ \sigma_3\int_0^1 \left(\frac{\partial \boldsymbol{y}(\theta,t)}{\partial \theta}\right)^T \frac{\partial \boldsymbol{y}(\theta,t)}{\partial \theta}\Bigg]d\theta \\ \le &\frac{1}{4}\int_0^{+\infty} \boldsymbol{w}^T(t)\boldsymbol{w}(t)dt + \int_0^1 \boldsymbol{y}_0^T(\theta)\boldsymbol{y}_0(\theta)d\theta.\end{aligned}$$

Considering (5), it implies that $\int_0^1 \boldsymbol{y}^T(\theta,t)\boldsymbol{y}(\theta,t)d\theta$, $\int_0^{+\infty} \boldsymbol{y}^T(1,t)\boldsymbol{y}(1,t)dt$ and $\int_0^{+\infty}\int_0^1 \left(\frac{\partial \boldsymbol{y}(\theta,t)}{\partial \theta}\right)^T \frac{\partial \boldsymbol{y}(\theta,t)}{\partial \theta}d\theta dt$ are bounded.

By using Lemma 1, we have

$$\begin{aligned}&\int_0^{+\infty}\int_0^1 \boldsymbol{y}^T(\theta,t)\boldsymbol{y}(\theta,t)d\theta dt\\ &\leq \frac{1}{2}\int_0^{+\infty}\int_0^1 \left(\frac{\partial \boldsymbol{y}(\theta,t)}{\partial\theta}\right)^T \frac{\partial \boldsymbol{y}(\theta,t)}{\partial\theta} d\theta dt,\end{aligned}$$

and

$$\begin{aligned}&\int_0^{+\infty} \boldsymbol{y}^T(\theta,t)\boldsymbol{y}(\theta,t)dt\\ &\leq \int_0^{+\infty}\int_0^1 \left(\frac{\partial \boldsymbol{y}(\theta,t)}{\partial\theta}\right)^T \frac{\partial \boldsymbol{y}(\theta,t)}{\partial\theta} d\theta dt,\end{aligned}$$

which implies that $\boldsymbol{y}(\theta,t)$ is uniformly square integrable. The proof of Theorem 1 is completed. □

### 4.2. $L_2[0,1]$ Stability

**Theorem 2.** *Under the control scheme (13), and boundary disturbance $\boldsymbol{w}(t)$ is bounded on $[0,+\infty)$, the closed-loop system (11) is $L_2[0,1]$ stable, i.e.,*

$$\lim_{t\to+\infty}\int_0^1 \boldsymbol{y}^T(\theta,t)\boldsymbol{y}(\theta,t)d\theta = \mathbf{0},$$

*if $\sigma_2 < \sigma_1$.*

**Proof.** In this part, the $L_2[0,1]$ stability of the closed-loop system (10) is analyzed. From (15) and Lemma 1, it is easy to see that

$$\frac{1}{2}\frac{d}{dt}\left(\int_0^1 \boldsymbol{y}^T(\theta,t)\boldsymbol{y}(\theta,t)d\theta\right) \leq \frac{1}{4}\boldsymbol{w}^T(t)\boldsymbol{w}(t).$$

On account of boundary disturbance $\boldsymbol{w}(t)$ is bounded on $[0,+\infty)$, there exists a constant $M$, such that

$$\frac{d}{dt}\left(\int_0^1 \boldsymbol{y}^T(\theta,t)\boldsymbol{y}(\theta,t)d\theta\right) \leq M, \tag{16}$$

holds.

By the Invariance Principle, (16) implies that the state of closed-loop system (10) is $L_2[0,1]$ stable. This completes the proof of Theorem 2. □

That is to say, a large mobile sensor network is $L_2[0,1]$ stable reaches the formation $\bar{\boldsymbol{x}}(\theta)$ under the guidance of our leader.

In the following, two valuable special scenarios are considered.

### 4.3. Special Scenarios

#### 4.3.1. Scenario 1

When the boundary disturbance $\boldsymbol{w}(t)$ vanishes in the closed-loop system (10), we have the following result easily:

**Theorem 3.** *Under the control scheme (13), and boundary disturbance $\boldsymbol{w}(t) = \mathbf{0}$ and $\int_0^1 \left(\frac{\partial \boldsymbol{y}_0(\theta)}{\partial\theta}\right)^T \frac{\partial \boldsymbol{y}_0(\theta)}{\partial\theta} d\theta$ is bounded, the closed-loop system (10) with auxiliary boundary condition $\frac{\partial \boldsymbol{y}(0,t)}{\partial\theta} = \mathbf{0}$ is globally asymptotically stable, i.e.,*

$$\lim_{t\to+\infty} \boldsymbol{y}(\theta,t) = \mathbf{0},$$

*if* $\sigma_2 < \sigma_1$.

**Proof.** To discuss the globally asymptotically stability of the closed-loop system (10), we introduce the auxiliary function as follows:

$$\begin{aligned} g(t) = & \frac{1}{2}\int_0^1 \left(\frac{\partial \boldsymbol{y}(\theta,t)}{\partial \theta}\right)^T \frac{\partial \boldsymbol{y}(\theta,t)}{\partial \theta} d\theta \\ & + \frac{1}{2} K \boldsymbol{y}^T(1,t)\boldsymbol{y}(1,t). \end{aligned} \tag{17}$$

Calculating the time derivative of $g(t)$, we obtain

$$\begin{aligned} \frac{dg(t)}{dt} \leq & \int_0^1 \left(\frac{\partial \boldsymbol{y}(\theta,t)}{\partial \theta}\right)^T \frac{\partial^2 \boldsymbol{y}(\theta,t)}{\partial \theta \partial t} d\theta \\ & + K\boldsymbol{y}^T(1,t)\frac{\partial \boldsymbol{y}(1,t)}{\partial t} \\ \leq & \left(\frac{\partial \boldsymbol{y}(1,t)}{\partial \theta}\right)^T \frac{\partial \boldsymbol{y}(1,t)}{\partial t} \\ & - \int_0^1 \left(\frac{\partial^2 \boldsymbol{y}(\theta,t)}{\partial \theta^2}\right)^T \frac{\partial^2 \boldsymbol{y}(\theta,t)}{\partial \theta^2} d\theta \\ \leq & - \int_0^1 \left(\frac{\partial^2 \boldsymbol{y}(\theta,t)}{\partial \theta^2}\right)^T \frac{\partial^2 \boldsymbol{y}(\theta,t)}{\partial \theta^2} d\theta \leq \boldsymbol{0}. \end{aligned} \tag{18}$$

Integrating (18) over $[0,1]$, we derive

$$\begin{aligned} g(t) \leq & \, g(0) \\ = & \frac{1}{2}\int_0^1 \left(\frac{\partial \boldsymbol{y}_0(\theta)}{\partial \theta}\right)^T \frac{\partial \boldsymbol{y}_0(\theta)}{\partial \theta} d\theta \\ & + \frac{1}{2} K \boldsymbol{y}_0^T(1)\boldsymbol{y}_0(1), \end{aligned}$$

which implies that $g(t)$ is bounded on $[0,+\infty)$.

Accordingly, $\int_0^1 \left(\frac{\partial \boldsymbol{y}(\theta,t)}{\partial \theta}\right)^T \frac{\partial \boldsymbol{y}(\theta,t)}{\partial \theta} d\theta$ and $\boldsymbol{y}(1,t)$ are bounded on $[0,+\infty)$.

Moreover, by using Lemma 1, we have

$$\max_{\theta \in [0,1]} |\boldsymbol{y}(\theta,t)| \leq \sqrt{\int_0^1 \left(\frac{\partial \boldsymbol{y}(\theta,t)}{\partial \theta}\right)^T \frac{\partial \boldsymbol{y}(\theta,t)}{\partial \theta} d\theta},$$

and thus $\boldsymbol{y}(\theta,t)$ is uniformly bounded.

Noting that $\boldsymbol{y}(\theta,t)$ is once continuously differentiable for any $t \in [0,+\infty)$, we can conclude from Lemma 2 that $\lim_{t\to+\infty} \boldsymbol{y}(\theta,t) = \boldsymbol{0}$. □

In other words, the state $\boldsymbol{y}(\theta,t)$ of closed-loop system (10) converges to zero. In this case, a large mobile sensor network has global asymptotic stability reaching the formation $\bar{x}(\theta)$ under the guidance of our leader.

4.3.2. Scenario 2

In the following, we consider the case when system (10) has no control law of leader, i.e., the boundary condition of system (10) satisfies the Dirichlet boundary condition $\boldsymbol{y}(0,t) = \boldsymbol{y}(1,t) = \boldsymbol{0}$. From what has been discussed above, the corollary is easily obtained from Theorem 3.

**Corollary 1.** *Under boundary disturbance* $\boldsymbol{w}(t) = \mathbf{0}$ *and with* $\int_0^1 \left(\frac{\partial y_0(\theta)}{\partial \theta}\right)^T \frac{\partial y_0(\theta)}{\partial \theta} d\theta$ *being bounded, the closed-loop system (10) with Dirichlet boundary condition is globally asymptotically stable, i.e.,*

$$\lim_{t \to +\infty} \boldsymbol{y}(\theta, t) = 0,$$

*if* $\sigma_2 < \sigma_1$.

That is, when no leader is in a large mobile sensor network, mobile sensors will reach consensus at zero globally asymptotically stable under the condition of Corollary 1.

## 5. Simulation Example

In this section, we present the simulation results to justify the Theorem 2 and Corollary 1 obtained above. In the following examples, we consider the mobile sensor networks including 10 mobile sensors to deployment. It is worth noting that simulating large-scale agents in formation employing the PDE approach can in fact illustrate the deployment of any finite number of mobile agents. Thus, unlike the formation control of discrete individual systems, even though the "10 agents" are not truly "large-scale", their simulation results in the PDE model are the same as those of the large-scale agents.

The boundary control law (13) is implemented by the leader sensor. In this manner, the mobile sensor $\theta$ can stably move to the desired formation rapidly.

In simulation, the initial positions of mobile sensors are sampled from the Gaussian distribution $\mathcal{N}(0,1)$ and boundary disturbance $\boldsymbol{w}(t) = [e^{-\frac{t}{2}}, \frac{1}{2t+1}]^T$. The coefficient of diffusion operator $D = 0.015I$.

Our first scenario focuses on the mobile sensors in a straight line under the control law of the leader. The feedback gain of boundary controller is considered as follows: $K = \text{diag}[5, 8]$. Figure 1 depicts the spatio-temporal trajectories of the mobile sensors, which show that 10 agents move onto a desired straight line in the time interval $[0, 40]$. Figure 1 is projected onto the plane $x_1x_2$ as Figure 2, where the motion of the mobile sensors on 2D can be clearly shown. The simulations above verify Theorem 2.

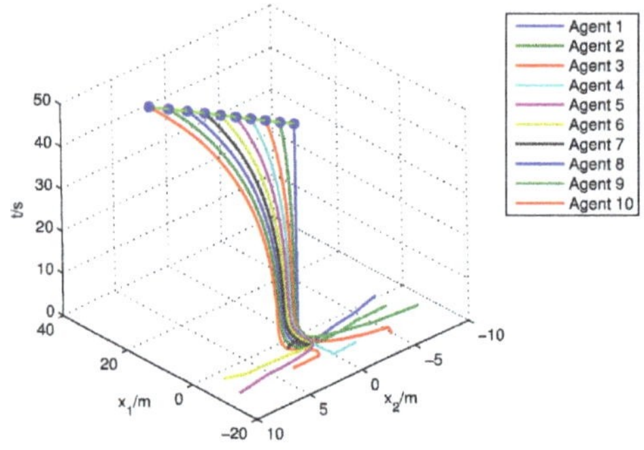


**Figure 1.** The spatio-temporal trajectory of mobile sensors.

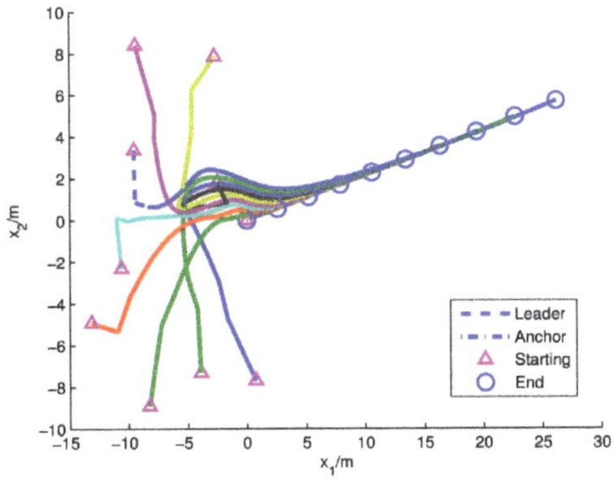


**Figure 2.** 2D trajectory of mobile sensors.

Next, Figure 3 is formed using parameter $K = \mathbf{0}$, which can verify Corollary 1. It is also clear that the distributed parameter multi-agent system under feedback input $v(\theta,t) = \frac{\partial^2 \mathbf{x}(\theta,t)}{\partial \theta^2}$ will rendezvous at the origin if no boundary controller is implemented by the leader. Figure 4 demonstrates the 2D trajectory of mobile sensors. As can be seen from it, 10 agents rendezvous at the origin on the projection plane $x_1x_2$.

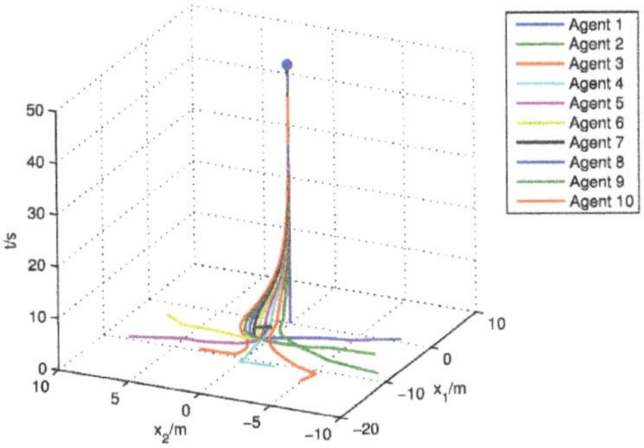


**Figure 3.** The spatio-temporal trajectory of mobile sensors.

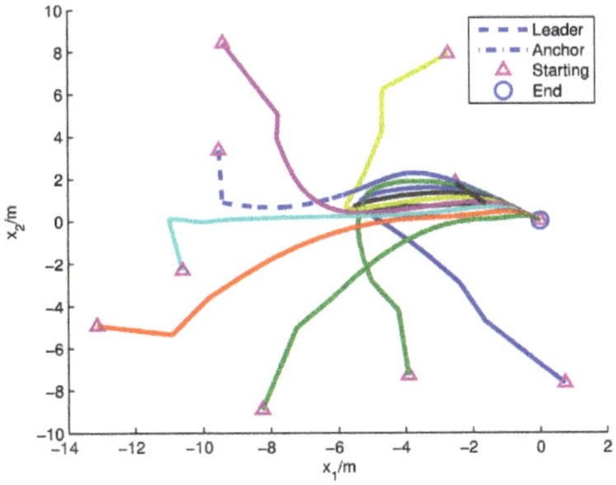


**Figure 4.** 2D trajectory of mobile sensors.

## 6. Conclusions

This paper has studied the formation problem for an array of large mobile sensor networks. A new framework of mobile sensor networks has been introduced by modeling the dynamic of mobile sensors as a continuum which is described by the parabolic system with boundary disturbance such that the model of the system conforms more to the realities of the situation in nature. In this model, the communication topology of agents is a chain graph and fixed. By utilizing the Lyapunov functional method and boundary control approach, leader feedback laws which are designed in a manner for the boundary control of large mobile sensor networks allow the mobile sensors to stably achieve the formation. In fact, the large mobile sensor network addressed can be used for agricultural applications, the formation of unmanned aerial systems, and so on. Simulation results further illustrated the effectiveness of our results.

**Author Contributions:** Conceptualization, X.Q. and B.C.; methodology, X.Q.; software, X.Q.; formal analysis, X.Q.; resources, B.C.; writing—original draft preparation, X.Q.; writing—review and editing, X.Q.; supervision, B.C.; project administration, B.C.; funding acquisition, X.Q. All authors have read and agreed to the published version of the manuscript.

**Funding:** This research was funded by the Natural Science Foundation of the Jiangsu Higher Education Institutions of China (No. 17KJB510051), the Qing Lan Project of the Jiangsu Higher Education Institutions (Young and Middle-aged Academic Leader (2022), the Excellent Teaching Team (2020)), the Soft Science Research Project of Wuxi (No. KX-22-B60), and the Advanced Research and Study Project for Academic Leaders of Jiangsu Higher Vocational Colleges (No. 2021GRFX068).

**Institutional Review Board Statement:** Not applicable.

**Informed Consent Statement:** Not applicable.

**Data Availability Statement:** Not applicable.

**Conflicts of Interest:** The authors declare no conflict of interest.

### Abbreviations

The following abbreviations are used in this manuscript:

| | |
|---|---|
| 2D | Two-dimensional |
| LMSNs | Large mobile sensor networks |
| LWR equation | Lighthill–Whitham–Richards equation |
| PDE | Partial differential equation |

## References

1. Ma, Z.; Sun, Y.; Mei, T. A review of wireless sensor networks. *J. Commun.* **2004**, *4*, 114–124.
2. Xu, Z.; Zhang, J. Analysis of wireless sensor network technology and its application technology bottleneck in Internet of Things. *Internet Things Technol.* **2011**, *1*, 79–81.
3. Gharajeh, M.S.; Jond, H.B. An intelligent approach for autonomous mobile robots path planning based on adaptive neuro-fuzzy inference system. *Ain Shams Eng. J.* **2022**, *13*, 101491. [CrossRef]
4. Chen, L.; Wang, Y.; Zhu, Q. Swarm formation control of multiple mobile robots based agent. *Syst. Eng. Electron.* **2006**, *5*, 731–735.
5. Heidari, R.; Khajehzadeh, A.; Eslami, M. A consensus event-triggered control of networked multi-agent systems with non-ideal communication network. *Ain Shams Eng. J.* **2021**, *12*, 3783–3790. [CrossRef]
6. Tu, Z.; Wang, Q.; Shen, Y. Simultaneous optimization of target tracking and monitoring in mobile sensor networks. *J. Autom.* **2012**, *38*, 452–461.
7. Wu, M.; Li, L.; Wei, Z.; Wang, H. A consistent observation method for target tracking heterogeneous sensors in dynamic environment of mobile robots. *J. Opt.* **2015**, *35*, 202–210.
8. Mao, W.; Zhang, W.; Qiu, X. Robot trajectory tracking control based on multi-sensor fusion. *Control Eng.* **2020**, *27*, 1125–1130.
9. Liu, H.; Yang, H. Application of wireless sensing technology in building environment monitoring. *Chem. Autom. Instrum.* **2011**, *38*, 1413–1416.
10. Sheng, X. A regional environmental monitoring system based on quadrotor UAVs. *Internet Things Technol.* **2021**, *11*, 25–26.
11. Jia, S.; Gao, X.; Lu, Q. Distributed dynamic path planning algorithms in wireless sensor networks. *J. Sens. Technol.* **2013**, *5*, 695–700.

12. Xiao, Q.; Ning, D. A swarming control model for mobile sensor networks based on tracking a single moving target. *Comput. Knowl. Technol.* **2014**, *10*, 1397–1400.
13. Zhou, H.; Xu, C.; Li, D.; Shao, S. Application of cluster intelligence algorithm in WSN node localization. *Comput. Eng.* **2008**, *20*, 18–20.
14. Fang, Z.; Gu, A.; Wang, L.; Zhang, X. Design of mobile robot system based on whisker sensor. *China Instrum.* **2009**, *1*, 82–84.
15. Wang, J. Distributed target tracking based on sensing agents and movement control laws. *J. Electron. Meas. Instrum.* **2019**, *33*, 156–163.
16. Zhong, L.; Lu, D.; Ye, J.; Qi, T. Distributed asynchronous measurement sensor network for robot control. *Autom. Instrum.* **2011**, *26*, 1–4.
17. Deng, Y. Research on adaptive control of ship formation collision avoidance. *Ship Sci. Technol.* **2017**, *39*, 31–33.
18. Wang, Y.; Hong, B. Fast convergence of distributed control algorithms for arbitrary formation of robot troops. *Comput. Res. Dev.* **2002**, *10*, 1274–1280.
19. Li, J.; Peng, S.; An H.; Xiang, X.; Shen, L. UAV formation aggregation method based on differential geometry and Lie group. *J. NUDT* **2013**, *35*, 157–164.
20. Wu, Z.; Yu, H.; Wang, R. Flocking control of dynamic topology-based multiple motion agents with a leader. *J. HUST (Nat. Sci. Ed.)* **2008**, *10*, 29–31.
21. Attanasi, A.; Cavagna, A.; Del Castello, L.; Giardina, I.; Grigera, T.S.; Jelić, A.; Melillo, S.; Parisi, L.; Pohl, O.; Shen, E.; et al. Information transfer and behavioural inertia in starling flocks. *Nat. Phys.* **2014**, *10*, 691–696. [CrossRef]
22. Qin, W.; Zhuang, B.; Cui, B. Boundary control of crowd evacuation system based on continuum model. *Control Decis.* **2018**, *33*, 2073–2079.
23. Qi, J.; Zhang, J. Boundary control for an unstable wave equation in annular with symmetrical initial data. *J. DU (Eng. Ed.)* **2016**, *33*, 467–472.
24. Qi, J.; Wang, C.; Pan, F. Boundary observation and output feedback control for spherically symmetric n-dimensional reaction diffusion equations. *J. Autom.* **2015**, *41*, 1356–1364.
25. Bao, L.; Huang, P. $H_\infty$ control of delayed distributed parameter switching systems. *Sci. Technol. Eng.* **2019**, *19*, 28–33.
26. Qian, X. Global synchronization in an array of coupled neural networks with impulse and time-varying delayed coupling. *J. YU (Nat. Sci. Ed.)* **2011**, *14*, 21–26.
27. Luo, S.; Chen, W. Periodic intermittent sampling control of a class of stochastic reaction diffusion neural networks. *Syst. Sci. Math.* **2017**, *37*, 652–664.

*entropy*

MDPI

*Article*

# Online Tracking of Maneuvering Target Trajectory Based on Chaotic Time Series Prediction

**Qian Wei [1], Peng Su [1], Lin Zhou [1,*] and Wentao Shi [2,*]**

[1] School of Artificial Intelligence, Henan University, Zhengzhou 450046, China
[2] School of Marine Science and Technology, Northwestern Polytechnical University, Xi'an 710072, China
* Correspondence: zhoulin@henu.edu.cn (L.Z.); swt@nwpu.edu.cn (W.S.)

**Abstract:** Online prediction of maneuvering target trajectory is one of the most popular research directions at present. Specifically, the primary factors balancing, between prediction accuracy and response time, will give the research substance. This paper presents an online trajectory prediction algorithm based on small sample chaotic time series (OTP-SSCT). First, we optimize in terms of data breadth. The dynamic split window is built according to the motion characteristics of the maneuvering target, thus realizing trajectory segmentation and constructing a small sample chaotic time series prediction set. Second, since fully considering the motion patterns of maneuvering targets, we introduce the spatiotemporal features into the particle swarm optimization (PSO) model identification algorithm, which improves the identification sensitivity of key trajectory data points. Furthermore, we propose a feedback optimization strategy of residual compensation to correct the trajectory prediction values to improve the prediction accuracy. For the initial value sensitivity problem of the PSO model identification algorithm, we propose a new initial population strategy, which improves the effectiveness of initial parameters on model identification. Through simulation experiment analysis, it is verified that the proposed OTP-SSCT algorithm achieves better prediction accuracy and faster response time.

**Keywords:** online prediction; trajectory segmentation; PSO model identification; feedback optimization; initial value sensitivity

## 1. Introduction

Trajectory prediction combines trajectory information as time series, and uses prediction algorithms to reasonably predict the future movement trend of maneuvering target. It is widely used in unmanned aerial vehicle (UAV) trajectory planning, intelligent transportation system, pedestrian trajectory tracking and other fields [1–3]. In recent years, taking advantage of sensor network technical development, the located target has higher mobility and is nonlinear and time-varying. The classical trajectory prediction algorithm with a pre-set motion model usually cannot dynamically adjust the model parameters. Hence, the phenomenon of model mismatch, which leads to inaccurate results and occurs in the trajectory prediction process of strong maneuverability targets. In order to solve the insufficient dynamics problem, many scholars began to study online trajectory prediction methods based on dynamic learning; however, those methods need to consider both prediction accuracy and prediction response time. So, the establishment has processed ultra-big data and the high-speed calculation capacity prediction system—the key question is what kind of method is appropriate.

Generally, online trajectory prediction methods are mainly divided into two main categories. One is the prediction algorithm with pre-set motion model. These methods mainly establish the motion model in advance through the laws of dynamics or physics, and then adjust the model parameters according to the motion state to achieve an online prediction of maneuvering targets [4]. Zhang et al. proposed an online four-dimensional trajectory prediction method [5]. This method improved the accuracy of trajectory prediction by updating the flight intention with the horizontal and vertical intention models of the vehicle, while incorporating the current state. A multimodel-based extended Kalman filter (EKF) was proposed in [6]. The method combined the EKF with the interacting multiple model (IMM) framework to predict the vehicle trajectory by selecting different EKF models. IMM is an adaptive algorithm that can effectively adjust the probability of each model, and is especially suitable for the localization tracking of maneuvering targets [7]. Wu et al. proposed an adaptive trajectory prediction algorithm [8]. By integrating two motion models and adjusting the model parameters adaptively, the algorithm could accurately predict the trajectory of pedestrians. Xie et al. proposed a vehicle trajectory prediction algorithm, which considered both dynamic motion with physical laws in the short term and advanced driving patterns with maneuverability estimation in the long term [9]; thus, the long-term prediction of vehicle trajectories was achieved.

The above methods are suited for maneuvering targets with low motion complexity. For moving targets with high maneuverability, it is difficult to directly establish a prediction model to adapt to the nonlinearity shifts in the track process. For this reason, the complexity of the model will increase sharply, and the robustness of the model will deteriorate. So, the prediction algorithm with a parser motion model can hardly provide good performance for high maneuverability targets.

The other category is data-driven trajectory prediction algorithms, which are divided into two methods, classical prediction methods based on regression theory and model identification methods using artificial intelligence data mining [10,11]. Regression prediction methods use a large number of trajectory data to build regression equations and achieve online prediction through iterative regression [12]. These algorithms include linear regression, logistic regression, ridge regression, autoregressive moving average, etc. Model identification methods focus on mining trajectory information deeply and identify target tracking models by intelligent algorithms; therefore, these methods are enable online trajectory prediction with high trajectory complexity. Liu et al. proposed a trajectory estimation framework for multi-UAV based on the prediction of user's movement information [13]. The method trains the parameters of the echo state network model by the true trajectory data set, thereby establishing a tracking model to predict the trajectory of the UAV. Shi et al. proposed a flight trajectory prediction model based on a long short-term memory (LSTM) network [14]. The sliding window method was introduced to avoid the dynamic dependence of adjacent states in long-term sequence, which improves the accuracy of online prediction. Ma et al. proposed a deep learning-based four-dimensional trajectory prediction algorithm, in which the flight trajectory features in the spatial and temporal dimensions were extracted by convolutional neural networks and LSTM, respectively [15]. Han et al. proposed a short-term online trajectory prediction model based on the gated cyclic neural network, which was trained and updated with model parameters by the batch method to achieve an online prediction of short-term flight trajectories [16].

Generally, unlike the prediction algorithm of pre-set motion model, the above methods learn the target motion law by mining the trajectory information, and thus achieve online trajectory prediction. By avoiding the physical modeling problem of complex motion, such algorithms are able to track highly maneuverable targets with surprising accuracy. However, the accuracy of the model identification is often related to the amount of training data. Such algorithms need the support of large-scale computation and real time prediction; therefore, processing the data effectively and establishing an accurate model is the bridge between the two aspects.

Due to the highly nonlinear and time-varying characteristics of maneuver target trajectory, a simple time series analysis is difficult to use to analyze the comprehensive reflect target's motion laws. In recent years, the prediction of chaotic time series has become a popular research direction regarding the aspect of trajectory prediction [17,18]. To analyze the deep behavioral characteristics of maneuvering targets, many methods perform prediction model identification on the basis of chaos theory. Liu et al. proposed a ship trajectory prediction model, which enhanced the optimization effect of the differential evolution algorithm on the model parameters by introducing chaos theory [19]. Hong et al. proposed a motion prediction model of floating platform, which optimized the model parameters by a chaotic efficient bat algorithm, thus achieving complex nonlinear trajectory prediction [20]. Recently, the Volterra series model, as a common model, has been used to predict chaotic time series, which can fit nonlinear data for accurate prediction [21]. Han et al. proposed a local Volterra prediction algorithm based on the clustering of phase points [22]. The method compared the similarity of predicted phase points and observation points in the clustering, and the optimal adjacent phase points were used to train the prediction model. Qiao et al. proposed a novel hybrid prediction model based on the Volterra filter [23]. The model parameters were optimized by an improved whale optimization algorithm. Lv et al. proposed a hybrid dynamic prediction algorithm based on the Volterra prediction model [24]. By introducing the sliding window method, the algorithm realized the online identification of the Volterra model parameters, and the dynamic relations were constructed from the high-dimensional phase space to realize the trajectory prediction of complex systems. Therefore, combining the prediction method with the chaos theory method, the algorithm achieves a better trajectory prediction effect; however, most algorithms based on chaos theory are only suitable for off-line trajectory prediction. For the complex motion model, how to extract the trajectory features in real time and build the prediction model dynamically are still two major problems to be solved. At present, the online trajectory prediction for chaotic time series has some deficiency in dynamic construction of models.

To summarize: in order to achieve stable online prediction of maneuvering target trajectories, this paper proposes an online trajectory prediction algorithm based on small sample chaotic time series (OTP-SSCT). Firstly, a small sample chaotic time series prediction set is constructed using a dynamic sliding window. Secondly, combining the spatiotemporal characteristics of maneuvering targets, the online trajectory prediction model is constructed using the improved PSO algorithm. Finally, the feedback optimization strategy is adopted to further improve the accuracy of the online trajectory prediction. The main contributions of this paper are as follows:

(1) In this paper, an online trajectory prediction method of maneuvering targets is proposed according to chaotic time series analysis. Thus, considering the motion characteristics of maneuvering targets, we construct a small sample chaotic time series prediction set by an innovative trajectory segmentation method. It realizes dynamic segmentation for trajectory data. Since fully considering nonlinear maneuver characteristics of trajectory data, spatiotemporal features are introduced in the improved PSO algorithm to increase the identification sensitivity of key trajectory data points. For the superposition of prediction errors caused by model identification, the trajectory predicted values are corrected by the feedback optimization strategy to ensure the prediction accuracy.

(2) For the initial value sensitivity problem of PSO model identification algorithm, a new initial population strategy is proposed in this paper. According to the spatial distribution of candidate populations, the strategy selects some improvement positions to compensate the uniformity of the initial population distribution, which avoids the algorithm from falling into local traps.

The content of this paper is organized as follows. Section 2 introduces the Volterra prediction model based on phase space reconstruction of chaotic time series. Section 3 elaborates principle and implementation process of the proposed OTP-SSCT algorithm. Further, a new initial population strategy of the PSO model identification algorithm is presented in Section 4. Section 5 gives the pseudo-code of our algorithm. Section 6 verifies the feasibility and effectiveness of the proposed algorithm. Section 7 provides the conclusions of this paper.

## 2. Volterra Prediction Model Based on Phase Space Reconstruction

The trajectory of maneuvering target is highly nonlinear and time-varying, which increases the difficulty of the prediction model identification. In this paper, an online prediction method of maneuver trajectory is proposed according to chaotic time series analysis. In order to obtain more motion laws of the target, the trajectory data can be reconstructed into the phase space by chaos theory [25]. It is easier to fit the evolution law between phase points by the nonlinear prediction model in the high-dimensional phase space. Thus, accurate prediction of maneuvering targets can be achieved. The trajectory prediction process of Volterra model is shown in Figure 1.

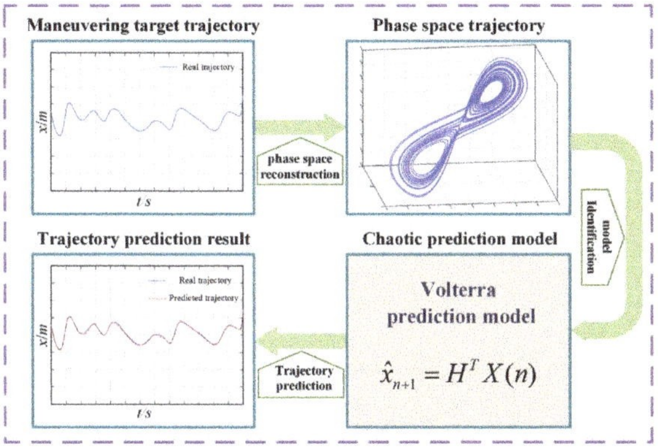


**Figure 1.** Trajectory prediction process of Volterra model.

Assuming that the observed trajectory one-dimensional time series is $x(n)$, $n = 1, 2, \cdots, N$. According to Takens embedding theorem, the phase space reconstruction of the original sequence does not change the evolutionary information of the system [26]. The reconstructed multidimensional time series is represented by phase points as follows:

$$X_l = [x(l), x(l+\tau), \cdots, x(l+(m-1)\tau)]^{\mathrm{T}} \qquad l = 1, 2, \cdots, L \tag{1}$$

where $X_l$ is the $l$-th phase point in phase space. $m$ is the embedding dimensions. $\tau$ is the delay time. $L = N - (m-1)\tau$ is the number of phase points. $N$ is the number of trajectory data.

In the phase space, there is a mapping $F : R^m \to R^m$ to represent the evolution law between phase points. Thus, there exists a new mapping $f : R^m \to R$ to represent the evolutionary law between trajectory data and phase points, as shown below:

$$x(l+(m-1)\tau+1) = f(X_l) \qquad l = 1, 2, \cdots, L \tag{2}$$

where $x(l+(m-1)\tau+1)$ represents the $m$-th dimensional component of $X_{l+1}$.

The Volterra model can characterize arbitrary nonlinear functions and has good scalability. In this paper, the Volterra model is chosen as the mapping relationship $f$ in Equation (2). The $p$-order discrete Volterra model with memory length $m$ is expressed as follows:

$$y(n) = \sum_{k=1}^{p} \sum_{i_1, \cdots, i_k = 0}^{m-1} h_k(i_1, i_2, \cdots, i_k) \prod_{j=1}^{k} x(n - i_j) \qquad n = 1, 2, \cdots, N \tag{3}$$

where $x(n)$ and $y(n)$ is the input and the output, respectively. $p$ is the order of Volterra prediction model. $m$ is the embedding dimensions (memory length). $h_k(i_1, i_2, \cdots, i_k)$ is the $k$-th kernel parameter value, which can be expressed in the form of kernel vector $\boldsymbol{H} = [h_1(0), h_1(1), \cdots, h_1(m-1), h_2(0,0), h_2(0,1), \cdots, h_2(0, m-1), \cdots, h_p(m-1, m-1)]$.

Combining Equations (2) and (3), the prediction values $\hat{x}(n+1)$ are obtained from the Volterra prediction model as follows:

$$\hat{x}(n+1) = \sum_{k=1}^{p} \sum_{i_1,\cdots,i_k=0}^{m-1} h_k(i_1, i_2, \cdots, i_k) \prod_{j=1}^{k} x(n - i_j \times \tau) \qquad n = 1, 2, \cdots, N \tag{4}$$

Determining the kernel parameters value of the Volterra prediction model is a typical nonlinear system identification problem. The OTP-SSCT algorithm is proposed in Section 3. By constructing a small sample chaotic time series prediction set, the parameters of Volterra model are dynamically identified. Thus, online prediction of maneuvering target trajectories is achieved.

## 3. Online Trajectory Prediction Algorithm Based on Small Sample Chaotic Time Series

Traditional prediction algorithms with pre-set motion model will rarely change the model parameters, resulting in insufficient dynamic capabilities. In contrast, data-fitting based prediction algorithms require large-scale training data support, resulting in poor online performance. To overcome these problems, this paper proposes the OTP-SSCT algorithm, which can dynamically construct the prediction model and realize the online prediction of maneuvering trajectory.

### 3.1. Algorithm Model Framework

The framework diagram of the proposed algorithm is shown in Figure 2. To begin with, the input trajectory data are pre-processed by the presented trajectory segmentation method. The dynamic split window is built according to the motion characteristics of maneuvering targets, thus realizing trajectory segmentation and constructing a small sample chaotic time series prediction set. Secondly, in combination with that prediction set, we construct an online trajectory prediction model through improving the PSO model identification algorithm. In addition, we propose an initial population strategy to improve the effectiveness of initial parameters on model identification. Finally, while outputting the trajectory prediction values, we also compensate them by the feedback optimization strategy, so as to improve the prediction accuracy.

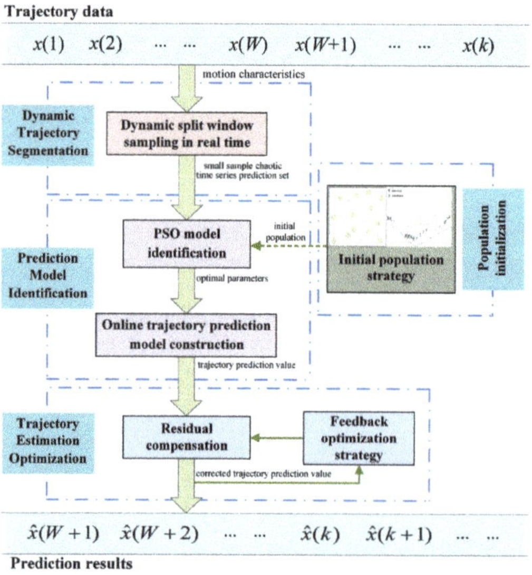


**Figure 2.** Framework diagram of the proposed algorithm.

### 3.2. Dynamic Trajectory Segmentation

The long-term movement of the maneuvering target generates a large amount of redundant historical trajectory information, some of which is no longer valid or even interferes with the construction of the prediction model. In this section, we propose an innovative trajectory segmentation method, which is used to construct a small sample chaotic time series prediction set of the target. The method achieves trajectory segmentation through the dynamic split window. Firstly, we design a similarity function to measure the dynamic changes of the trajectory data, and also define the magnitude of the similarity value by the threshold method. Secondly, by extracting the motion characteristics of the maneuvering target, we give the adjustment rules for the length of dynamic split window. Finally, the trajectory segmentation is performed through the dynamic split window, and a small sample chaotic time series prediction set is constructed for the subsequent model identification. The flow chart of trajectory segmentation with dynamic split window is shown in Figure 3.

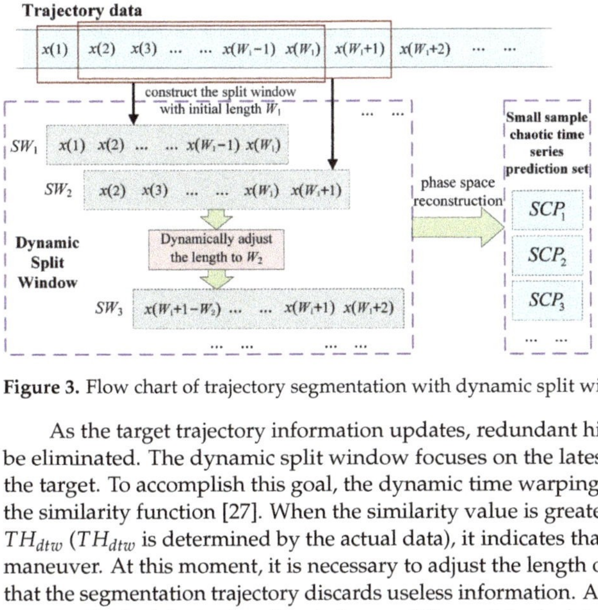


**Figure 3.** Flow chart of trajectory segmentation with dynamic split window.

As the target trajectory information updates, redundant historical trajectories need to be eliminated. The dynamic split window focuses on the latest trajectory information of the target. To accomplish this goal, the dynamic time warping (DTW) distance is used as the similarity function [27]. When the similarity value is greater than the preset threshold $TH_{dtw}$ ($TH_{dtw}$ is determined by the actual data), it indicates that the target produces a new maneuver. At this moment, it is necessary to adjust the length of dynamic split window so that the segmentation trajectory discards useless information. After each length adjustment, we save the first dynamic split window as $SW_{old}$, while the latest dynamic split window is $SW_{new}$. The similarity function is as follows:

$$DTW(SW_{new}, SW_{old}) = \underset{P=p_1,\cdots,p_k,\cdots,p_K}{\arg\min} \sqrt{\sum_{k=1}^{K} [p_k(i,j)]^2} \tag{5}$$

where

$$p_k(i,j) = (x(i) - x(j))^2 + \min\{p_k(i-1,j-1), p_k(i-1,j), p_k(i,j-1)\}$$

$P$ is the warping path between $SW_{new}$ and $SW_{old}$. $K \in [W, 2W-1)$ is the length of the warping path. $W$ is the length of dynamic split window. $x(i) \in SW_{new}$ and $x(j) \in SW_{old}$ are the trajectory data, $1 \le i, j \le W$. More details can be found in [27].

The length of the dynamic split window is determined by the motion characteristics of the maneuvering target. At first, we define that the motion characteristics consist of two components, trajectory data mean volatility ratio $R_V$ and tracking model error volatility ratio $R_E$. The calculation methods are shown below, respectively:

$$R_V = \frac{V_{new} - V_{old}}{V_{old}} \tag{6}$$

$$R_E = \frac{E_{new} - E_{old}}{E_{old}} \tag{7}$$

where

$$V_{new} = \frac{\mathrm{Var}(SW_{new})}{\mathrm{Max}(SW_{new}) - \mathrm{Min}(SW_{new})} \tag{8}$$

$$E_{new} = \frac{1}{W}\sum_{l=1}^{W} |x(l) - \hat{x}(l)| \qquad x(l) \in SW_{new} \tag{9}$$

$V_{old}$ and $E_{old}$ are calculated in the same way as Equations (8) and (9). $V_{new}$ and $V_{old}$ are the fluctuation values of $SW_{new}$ and $SW_{old}$, respectively, indicating the intensity of the target motion. Var denotes calculation of variance. Max and Min denote calculation of the maximum and minimum values, respectively. $E_{new}$ and $E_{old}$ are the average prediction errors of $SW_{new}$ and $SW_{old}$, respectively, indicating the accuracy of the prediction model. $\hat{x}(l)$ and $x(l)$ denote the prediction value and true value of the trajectory, respectively.

The dynamic split window length $W$ can be dynamically adjusted according to the motion characteristics of Equations (6) and (7), as follows:

$$W = w + \Delta w \tag{10}$$

where $w$ is the fixed length, determined by the actual data. $\Delta w$ is the dynamic length, defined by the motion characteristics, and the calculation method is as follows:

$$\Delta w = (\lambda \times R_V + \mu \times R_E) \times \frac{w}{2} \tag{11}$$

$$\lambda = \begin{cases} 0 & 0 \le |R_V| < \delta \\ 1 & \delta \le |R_V| \end{cases} \tag{12}$$

$$\mu = \begin{cases} 0 & 0 \le |R_E| < \varepsilon \\ 1 & \varepsilon \le |R_E| \end{cases} \tag{13}$$

where $\lambda$ and $\mu$ are the selection factors. $0 \le \delta \le 1$ and $0 \le \varepsilon \le 1$ are the regulation thresholds.

After segmenting the trajectory data through the dynamic split window, the trajectory data $SW_k$ at the $k$-th moment is reconstructed according to Equation (4). Thereby, a small sample chaotic time series prediction set is constructed as follows:

$$SCP_k = [X(1), X(2), \cdots, X(l), \cdots, X(L)] \qquad L = W - (m-1)\tau \tag{14}$$

where $X(l) = [\, x(l+(m-1)\tau), \cdots, x(l), x^2(l+(m-1)\tau), \cdots, x^2(l), \cdots, x^p(l+(m-1)\tau), \cdots, x^p(l)\,]^{\mathrm{T}}$. $L$ is the number of phase points. $X(l)$ is a $D$-dimensional phase points, $D = \sum_{k=0}^{p} C_{k+m-1}^{k}$. $x(l) \in SW_k$ is the trajectory data. $m$ is the embedding dimensions. $\tau$ is the delay time. $p$ is the order of Volterra prediction model.

### 3.3. Prediction Model Construction

Since the dynamic split window focuses on the latest trajectory data, the small sample chaotic time series prediction set we constructed is optimized in terms of data breadth; however, the spatiotemporal features of the trajectories themselves cannot be ignored. To further explore the motion patterns of maneuvering targets, we improve the sensitivity of the algorithm to identify key points by extracting the spatiotemporal features of trajectory data, which is optimized in terms of data depth—see Figure 4.

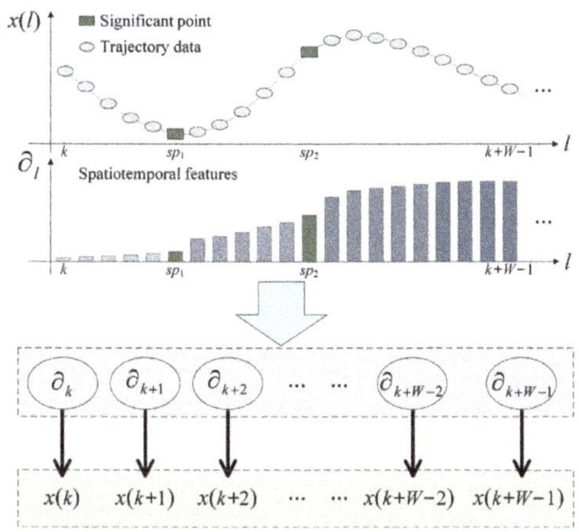


**Figure 4.** Extraction diagram of trajectory spatiotemporal features.

The movement trajectory of a maneuvering target may seem haphazard, but there are precursors before the maneuver occurs. In order to visualize the change pattern of target maneuvers, the trajectory data are segregated by significant points. The spatiotemporal features of the trajectory are extracted as follows:

$$\frac{|x(j) - x(i)|}{(|x(j) + x(i)|)/2} \geq \rho \qquad j - i \geq \sigma \tag{15}$$

where $x(i)$ and $x(j)$ are the trajectory data, $1 \leq i < j$. $\sigma$ and $\rho$ are the distance threshold and the position threshold, respectively.

The significant point in the dynamic split window $SW_k$ of the $k$-th moment is measured by Equation (15), which is recorded as $SP_k$:

$$SP_k = [x(sp_1), \cdots, x(sp_q), \cdots, x(sp_Q)] \qquad sp_q \in \{1, 2, \cdots, W\} \tag{16}$$

where $x(sp_q)$ is the $q$-th significant point, $sp_q$ is the position in $SW_k$. $W$ is the length of dynamic split window.

The significant points spatially represent the different trends of the trajectories, while the trajectory data follow a temporal order. Therefore, combined with the characteristics of sigmoid function, the spatiotemporal features $\partial_l$ of of the trajectory data are mined:

$$\partial_l = \begin{cases} a_l/(Q+1) & k \leq l < sp_1 \\ 2 \times a_l/(Q+1) & sp_1 \leq l < sp_2 \\ \vdots & \\ (q+1) \times a_l/(Q+1) & sp_q \leq l < sp_{q+1} \\ \vdots & \\ a_l & sp_Q \leq l \leq k + W - 1 \end{cases} \tag{17}$$

where

$$a_l = 1 \Big/ \left[1 + e^{(-10 \times l + 5)}\right] \qquad l \in [k, k + W - 1]$$

The PSO model identification algorithm is a problem-oriented metaheuristic that gradually approximates the optimal solution through the update iteration of particle swarm. The fitness function is the central part of PSO model identification algorithm, which controls the direction of population optimization. For the particle $\hat{H}_i$, the fitness function of the original algorithm as follows:

$$\begin{aligned} f_e(e_i) &= \frac{1}{L}\sum_{l=1}^{L}(x(l)-\hat{x}(l))^2 \\ &= \frac{1}{L}\sum_{l=1}^{L}\left(x(l)-\hat{H}_i\times X(l)\right)^2 \end{aligned} \quad (18)$$

where $\hat{H}_i$ is the $i$-th particle. $L$ is the number of phase points in the small sample chaotic time series prediction set. $x(l)$ is the trajectory data and $x(l)\in SW_k$. $X(l)$ is a D-dimensional phase points.

The velocity and position evolution rules of the $i$-th particle $\hat{H}_i$ are as follows:

$$v_i(t+1)=\omega\times v_i(t)+c_1\times r_1\times\left(P_i(t)-\hat{H}_i(t)\right)+c_2\times r_2\times\left(P_g(t)-\hat{H}_i(t)\right) \quad (19)$$

$$\hat{H}_i(t+1)=\hat{H}_i(t)+v_i(t+1) \quad (20)$$

where $t$ is the iteration number of the PSO algorithm. $v_i$ is the velocity. $\omega$ is the inertia weight. $c_1$ and $c_2$ are learning factors. $r_1$ and $r_2$ are random numbers between the interval [0, 1]. $P_i(t)$ is the individual historical optimal position. $P_g(t)$ is the overall historical optimal position.

In this section, we design an improved fitness function by incorporating spatiotemporal features into a small sample chaotic temporal prediction set. By improving the sensitivity of the algorithm to identify key points in the trajectory, the population is led to evolve toward the optimal solution. The improved fitness function is as follows:

$$f_e(e_i)=\frac{1}{L}\sum_{l=1}^{L}\left[\beta_l\times\left(x(l)-\hat{H}_i\times X(l)\right)^2\right] \quad (21)$$

$$\beta_l=b_l\bigg/\sum_{i=1}^{M}b_i \qquad l=1,2,\cdots,L \quad (22)$$

where

$$b_l=\partial_{l+(m-1)\tau}+\cdots+\partial_l+\partial^2_{l+(m-1)\tau}+\cdots+\partial^2_l+\cdots+\partial^p_{l+(m-1)\tau}+\cdots+\partial^p_l$$

$\beta_l$ is the normalized effect factor, and $\partial_l$ is the trajectory spatiotemporal feature.

The PSO algorithm finds the global optimal position with the smallest fitness value, which is the optimal parameter value of Volterra prediction model. The first model identification applies the initial population strategy proposed in Section 4. As the dynamic split window is updated online, the saved optimal values are utilized to initialize the particle swarm to decrease the identification time. Ultimately, the identified model enables online trajectory prediction of the maneuvering target.

*3.4. Trajectory Estimation Optimization*

Since the trajectory segmentation method is adopted to accomplish online identification, there may be superposition of prediction errors. Thereby, leading to the degradation in prediction accuracy of the algorithm. To improve the prediction performance, we propose a feedback optimization strategy of residual compensation to correct the predicted values. The feedback optimization strategy of residual compensation is shown in Figure 5. The calculation method is as follows:

$$\hat{x}_{adj}(k+1)=\hat{x}(k+1)+resi(k+1) \quad (23)$$

$$resi(k+1) = SW_k \times \boldsymbol{\theta} \quad (24)$$

where $\hat{x}_{adj}(k+1)$ and $\hat{x}(k+1)$ represent the corrected prediction and model prediction, respectively. $resi(k+1)$ is the residual compensation. $SW_k$ is the dynamic split window. $\boldsymbol{\theta}$ is the compensation coefficient vector.

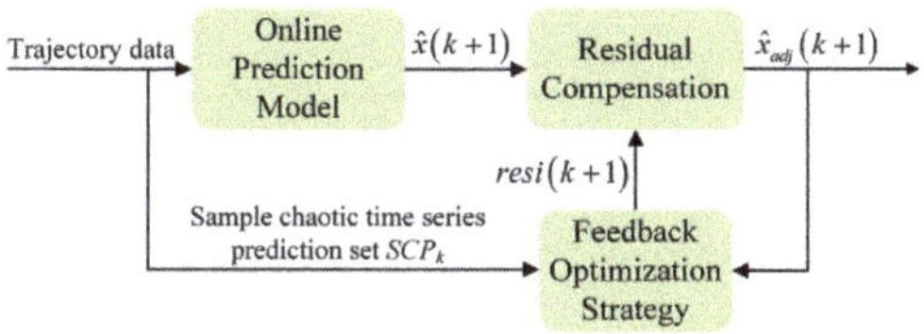


**Figure 5.** Feedback optimization strategy of residual compensation.

The least square (LS) algorithm is a simple linear recursive estimator with the advantages of small memory and fast calculation speed. In this paper, the LS algorithm is used to solve the compensation coefficient vector $\boldsymbol{\theta}$. The input item is $SCP_k$, and the expected output is $\Delta E$:

$$\Delta E = [e(k-W+1), e(k-W+2), \cdots, e(k)] \quad (25)$$

where

$$e(k) = x(k) - \hat{x}(k)$$

$e(k)$ is the error between the true value and the predicted value. $\Delta E$ is the error vector of $SW_k$. $W$ is the dynamic split window length.

## 4. Initial Population Strategy for PSO Model Identification Algorithm

The PSO model identification algorithm is an intelligent optimization algorithm with simpler structure, easier convergence and stronger robustness. However, the method suffers from the initial value sensitivity problem, and it is prone to fall into the local optimum trap in the process of seeking the optimal. To avoid this problem, on the premise of ensuring randomness, we propose a new initial population strategy to improve the optimization performance of the identification algorithm.

### 4.1. Chaotic Population Initialization

Since chaotic motion has obvious characteristics of nonlinearity, randomness and ergodicity, initializing the particle swarm with the idea of chaos can improve the quality of the population. The chaotic initialization method utilizes chaotic mapping and obtains the initial population through iterative means; however, it is difficult to traverse the entire search space when the number of iterations is insufficient. Reverse learning is an advancement of the traditional iterative approach, the idea of which is considering the opposite direction when generating a feasible solution. The schematic diagram is shown in Figure 6. Thus, it is combined with the reverse learning method to generate chaotic initial populations.

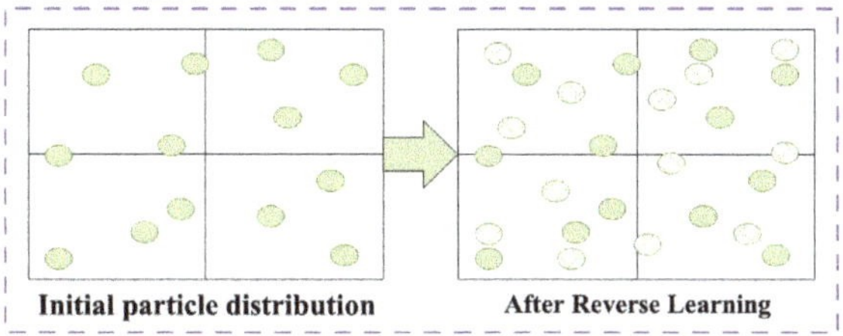


**Figure 6.** Reverse optimization of initial population.

A tent map is a chaotic map in its parameter range, with uniform distribution function and good correlation. The tent chaotic map model is as follows:

$$T_{k+1} = \begin{cases} T_k/\nu & T_k \in [0,\nu) \\ (1-T_k)/(1-\nu) & T_k \in [\nu,1] \end{cases} \tag{26}$$

where $\nu \in (0,1)$ is the model parameter. When $T_k \in [0,1]$ and $T_k \neq \nu$, the above mapping is in a chaotic state.

We assume that the population size is $N$, and the search space is $D$-dimensional. The upper and lower search limits for each dimension are $x^d_{\text{min}}$ and $x^d_{\text{max}}$, respectively.

Firstly, it generates a $D$-dimensional random vector $\mathbf{T}_1$:

$$\mathbf{T}_1 = \left[t^1_1, t^2_1, \cdots, t^d_1, \cdots, t^D_1\right] \qquad t^d_1 \in (0,1) \tag{27}$$

Secondly, each dimension of $\mathbf{T}_1$ is substituted into Equation (26) for iteration. After $N/2$ iterations, a group of $D$-dimensional chaotic vectors are generated:

$$\mathbf{T}_n = \left[t^1_n, t^2_n, \cdots, t^d_n, \cdots, t^D_n\right] \qquad n = 1,2,\cdots,N/2 \tag{28}$$

where the value of each vector $t^d_n \in (0,1)$, and $d = 1,2,\cdots,D$ is the dimension of the search space.

Then, all chaotic vectors are mapped to the search space to generate the candidate population set $\mathbf{C}_1 = \{\mathbf{CX}_n\}$:

$$\mathbf{CX}_n = \left[cx^1_n, cx^2_n, \cdots, cx^d_n, \cdots, cx^D_n\right] \tag{29}$$

where

$$cx^d_n = \left(x^d_{\text{max}} - x^d_{\text{min}}\right) \times t^d_n + x^d_{\text{min}}$$

Finally, the reverse learning method is applied to generate the candidate population set $\mathbf{C}_2 = \{\mathbf{RX}_n\}$:

$$\mathbf{RX}_n = \left[rx^1_n, rx^2_n, \cdots, rx^d_n, \cdots, rx^D_n\right] \tag{30}$$

where

$$rx^d_n = x^d_{\text{max}} - x^d_{\text{min}} + cx^d_n$$

The chaotic candidate population set $\{\mathbf{C}_1 \cup \mathbf{C}_2\}$ generated by the above method take into account both diversity and ergodicity. However, due to the randomness of the chaotic mapping, the distribution of candidate populations in the search space is not uniform. Thus, the algorithm is prone to fall into the local optimum trap.

*4.2. Spatial Distribution Compensation*

To ensure the homogeneity of the initial population, we propose a spatial distribution compensation method to optimize the chaotic initialization, which makes the diversity and homogeneity of the initial population guaranteed.

All dimensions of the search space are divided equally, and interval values $I^d_k$ and spatial interval $\mathbf{S}^d_j$ are generated in each dimension, as follows:

$$I^d_k = x^d_{\text{min}} + (k-1) \times \left(x^d_{\text{max}} - x^d_{\text{min}}\right)/Q \qquad k = 1,2,\cdots,Q+1 \tag{31}$$

$$\mathbf{S}^d_j = \left[I^d_j, I^d_{j+1}\right] \qquad j = 1,2,\cdots,Q \tag{32}$$

where $Q$ is the average number of divisions. $d = 1,2,\cdots,D$ is the dimension of the search space.

Spatial interval $\mathbf{S}_j^d$ divides each dimension of the search space equally into discrete grid states, as shown in Figure 7, where each grid represents an interval.

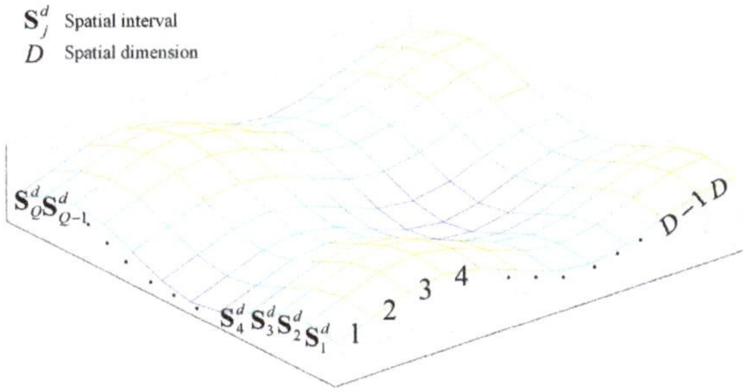


**Figure 7.** Discrete gridding of the search space.

From the distribution of the chaotic candidate population set in the search space, defining the following sparsity function $f_s\left(\mathbf{S}_j^d\right)$:

$$f_s\left(\mathbf{S}_j^d\right) = 1 - r_j^d \times c_j^d \Big/ N \tag{33}$$

where $r_j^d \in [0,1]$ is a randomly generated value. $c_j^d$ is the number of the candidate population in the $j$-th interval of $d$-th dimension. $N$ is the population size.

The larger the sparsity value, the fewer candidate populations fall in the interval; therefore, calculate the sparsity values of each interval and arrange in descending order, and select the median of the $M$ intervals with the greatest sparsity as the new candidate populations.

$$sx_i^d = \left(I_{i+1}^d - I_i^d\right) \Big/ 2 \qquad i = 1, 2, \cdots, M \tag{34}$$

where $M$ $(M < Q)$ is determined by the actual distribution. $I_i^d$ is the $i$-th interval value after sorting according to sparsity.

Combining the interval median values calculated by Equation (34), it generates the candidate population position set $\mathbf{C}_3 = \{\mathbf{SX}_i\}$:

$$\mathbf{SX}_i = \left[sx_i^1, sx_i^2, \cdots, sx_i^d, \cdots, sx_i^D\right] \tag{35}$$

In summary, there are generated $(N + M)$ groups of candidate population set $\mathbf{C}$:

$$\mathbf{C} = \{\mathbf{C}_1 \cup \mathbf{C}_2 \cup \mathbf{C}_2\} = \{\mathbf{CX}_n \cup \mathbf{RX}_n \cup \mathbf{SX}_i\} \quad n = 1, 2, \cdots, N/2 \quad i = 1, 2, \cdots, M \tag{36}$$

We select $N$ groups of candidate populations with the best fitness as the initial populations of the PSO model identification algorithm. The proposed initial population strategy avoids the algorithm from falling into local traps by increasing the diversity and homogeneity of the initial population. Meanwhile, it improves the efficiency of the algorithm in finding the optimum.

## 5. Algorithm Implementation

For the online prediction scenario of maneuvering target trajectory, the OTP-SSCT algorithm is proposed in this paper. Firstly, we initialize the algorithm parameters and use the proposed initial population strategy to obtain the initial population of the PSO model identification algorithm. Secondly, the split window length is dynamically adjusted by the motion characteristics of the maneuvering target, and a small sample chaotic time series prediction set is constructed. Then, we use the improved PSO algorithm to identify the trajectory prediction model. Finally, the prediction results are obtained, while the feedback optimization strategy is applied to further improve the prediction accuracy. According to the above analysis, more details related to the proposed OTP-SSCT algorithm, and the pseudo-code is shown in Algorithm 1.

**Algorithm 1:** OTP-SSCT algorithm.

**Input:** Trajectory data $x(n), n = 1, 2, \cdots, N$
**Output:** Prediction results $\hat{x}(n), n = W, W+1, \cdots, N, \cdots$

Initializing the dynamic split window length $W$, embedding dimension $m$ and delay time $\tau$ based on the historical trajectory data;
Obtaining the initial population according to (26)–(36);
**for** $k = 1 : N$ **do**
    Calculating the similarity distance $DTW$ according to (5);
    **if** $DTW > TH_{dtw}$ **then**
        Adjusting the dynamic split window length $W$ according to (10)–(13);
    **end if**
    Segmenting the trajectory data $x(n)$ according to the dynamic split window;
    Constructing the small sample chaotic time series prediction set $SCP_k$ according to (14);
    Extracting trajectories spatiotemporal features $\partial_n$ according to (17);
    Identifying the model parameters $\boldsymbol{H}$ by the improved PSO model identification algorithm according to (19)–(22);
    Constructing the trajectory prediction model according to (4);
    Obtaining the prediction results $\hat{x}(n)$ and performing the feedback optimization strategy according to (23)–(25);
**end for**

## 6. Simulation and Analysis

To address the problem of stable online prediction of maneuvering target trajectories, this paper proposes the OTP-SSCT algorithm. Considering the motion characteristics of trajectory data, the improved PSO model identification algorithm is used to identify the parameters of Volterra model. Finally, the online trajectory prediction of maneuvering target is realized.

In this experiment, considering the maneuvering characteristics of the UAV trajectory, we choose the 3rd-order Volterra prediction model as the structural basis of the identification model, as follows:

$$
\begin{aligned}
\hat{x}(n+1) &= \sum_{k=1}^{3} \sum_{i_1, i_2, i_3=0}^{m-1} h_k(i_1, i_2, i_3) \prod_{j=1}^{k} x(n - i_j \times \tau) \\
&= \sum_{i_1=0}^{m-1} h_1(i_1) x(n - i_1 \times \tau) + \sum_{i_1=0}^{m-1} \sum_{i_2=i_1}^{m-1} h_2(i_1, i_2) x(n - i_1 \times \tau) x(n - i_2 \times \tau) \\
&+ \sum_{i_1=0}^{m-1} \sum_{i_2=i_1}^{m-1} \sum_{i_3=i_2}^{m-1} h_3(i_1, i_2, i_3) x(n - i_1 \times \tau) x(n - i_2 \times \tau) x(n - i_3 \times \tau)
\end{aligned}
\tag{37}
$$

where $\hat{x}(n+1)$ and $x(n)$ represent the prediction result and the trajectory data, respectively. $m$ and $\tau$ represent the memory length of the model and the delay time, respectively. $h_k(i_1, i_2, i_3)$ is the kernel parameter.

To verify the performance of our OTP-SSCT algorithm, the following simulation experiments are designed in three directions.

(1) For the initial population setting problem of the PSO algorithm, different population selection strategies are used for particle swarm initialization and model parameter identification. The performance of the proposed population initialization strategy is verified by comparing the identification accuracy.

(2) The model identification of chaotic time series in Lorenz-X dimension is performed with different optimization algorithms. In comparison with several performance metrics, it tests the identification capability of the proposed improved PSO model identification algorithm.

(3) Aiming at online trajectory prediction scenarios of maneuvering targets, different prediction algorithms are adopted for online trajectory prediction of UAV three-degree-of-freedom maneuvering target. By analyzing the precision metrics such as prediction accuracy and response time, it verifies the performance of the proposed OTP-SSCT algorithm.

This paper uses four precision metrics for the quantitative evaluation: mean square error (MES), root mean square error (RMSE), mean absolute percentage error (MAPE) and coefficient of determination $R^2$. The software used for the simulation experiment is MATLAB2019a. The simulation environment: CPU is Intel Core i5-6500 3.20 GHz, memory is 8GB, and the operating system is Microsoft Windows 10.

### 6.1. Parameter Identification Experiment with Different Initialization Methods

Initial population directly affects the optimization performance of PSO algorithm; therefore, given the real parameters, the performance of our proposed initialization strategy is tested by comparing the identification errors of different initialization methods. The experimental process is shown in Figure 8.

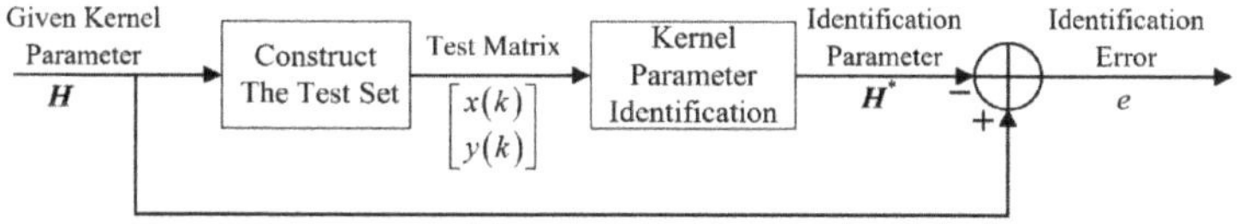


**Figure 8.** Flowchart of *experiment* 1.

The comparison methods are random initialization method, chaotic initialization method and sequence initialization method [28]. Identifying a 3rd-order Volterra model with a memory length of 3. The given kernel parameter vector is as follows:

$$\boldsymbol{H} = [1.54, -1, 0.56, 0.87, 0, 0, 1.13, 1.21, -1.29, 1.45, 0, 0, 0.88, 0, 0, -0.66, 0, 0, 1.06]^{\mathrm{T}}$$

For the Volterra model with given kernel parameters $\boldsymbol{H}$, the input signal $x(k) \in [0,1]$ is uniform white noise, and the output signal is $y(k)$. $x(k)$ and $y(k)$ constitute the test set, and the PSO algorithm with four initialization methods is used for parameter identification, respectively. Setting the population size to 100 and the number of iterations to 500. Each method is simulated for 50 times, and the identification results are averaged.

It can be seen from Figure 9 and Table 1 that the PSO algorithm using four different initialization methods can identify approximate values close to the true values of the kernel parameters. By comparing the absolute error between the identification results and the true values, the minimum error of our method is about 0.0019, the maximum error is about 0.0143, and the average error is about 0.0053. It can be seen that the average error of our method is 73.3%, 65.6% and 36.9% smaller than that of the random initialization method, chaotic initialization method and sequence initialization method, respectively; therefore, the initial population strategy proposed in this paper significantly improves the identification effect of the PSO model identification algorithm.

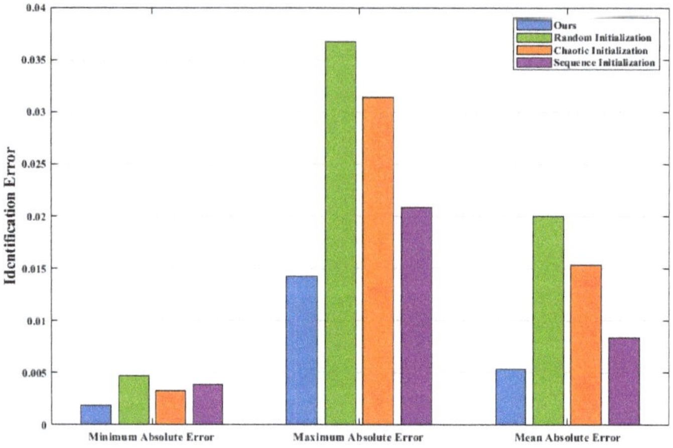


**Figure 9.** Comparison of parameter identification error.

**Table 1.** Kernel parameter identification results.

| *H* [1] | T-V [2] | Ours | | R-I [3] | | C-I [4] | | S-I [5] | |
|---|---|---|---|---|---|---|---|---|---|
| | | $H^*$ [6] | *E* [7] | $H^*$ | *E* | $H^*$ | *E* | $H^*$ | *E* |
| $h_1(0)$ | 1.54 | 1.5374 | 0.0026 | 1.5096 | 0.0304 | 1.5367 | 0.0033 | 1.5354 | 0.0046 |
| $h_1(1)$ | −1 | −0.9935 | 0.0065 | −0.9633 | 0.0367 | −0.9735 | 0.0265 | −0.9792 | 0.0208 |
| $h_1(2)$ | 0.56 | 0.5554 | 0.0046 | 0.5301 | 0.0299 | 0.5473 | 0.0127 | 0.5503 | 0.0097 |
| $h_2(0,0)$ | 0.87 | 0.8738 | 0.0038 | 0.8752 | 0.0052 | 0.8734 | 0.0034 | 0.8750 | 0.0050 |
| $h_2(0,1)$ | 0 | $-5.6 \times 10^{-4}$ | $5.6 \times 10^{-4}$ | −0.0074 | 0.0074 | −0.0098 | 0.0098 | −0.0060 | 0.0060 |
| $h_2(0,2)$ | 0 | $-8.6 \times 10^{-4}$ | $8.6 \times 10^{-4}$ | 0.0086 | 0.0086 | 0.0093 | 0.0093 | 0.0074 | 0.0074 |
| $h_2(1,1)$ | 1.13 | 1.1377 | 0.0077 | 1.0990 | 0.0310 | 1.0986 | 0.0314 | 1.1193 | 0.0107 |
| $h_2(1,2)$ | 1.21 | 1.2081 | 0.0019 | 1.2053 | 0.0047 | 1.2028 | 0.0072 | 1.2061 | 0.0039 |
| $h_2(2,2)$ | −1.29 | −1.2833 | 0.0067 | −1.2636 | 0.0264 | −1.2718 | 0.0182 | −1.2807 | 0.0093 |
| $h_3(0,0,0)$ | 1. 45 | 1.4524 | 0.0024 | 1.4567 | 0.0067 | 1.4395 | 0.0105 | 1.4436 | 0.0064 |
| $h_3(0,0,1)$ | 0 | $-4.3 \times 10^{-5}$ | $4.3 \times 10^{-5}$ | −0.0336 | 0.0336 | 0.0078 | 0.0078 | 0.0160 | 0.0160 |
| $h_3(0,0,2)$ | 0 | $7.2 \times 10^{-4}$ | $7.2 \times 10^{-4}$ | 0.0011 | 0.0011 | 0.0441 | 0.0441 | 0.0051 | 0.0051 |
| $h_3(0,1,1)$ | 0.88 | 0.8850 | 0.0050 | 0.8971 | 0.0171 | 0.8554 | 0.0246 | 0.8862 | 0.0062 |
| $h_3(0,1,2)$ | 0 | −0.0079 | 0.0079 | 0.0255 | 0.0255 | 0.0028 | 0.0275 | −0.0466 | 0.0466 |
| $h_3(0,2,2)$ | 0 | 0.0045 | 0.0045 | −0.0121 | 0.0121 | 0.0516 | 0.0516 | 0.0162 | 0.0162 |
| $h_3(1,1,1)$ | −0.66 | −0.6569 | 0.0031 | −0.6408 | 0.0192 | −0.6353 | 0.0247 | −0.6682 | 0.0082 |
| $h_3(1,1,2)$ | 0 | −0.0065 | 0.0065 | −0.0910 | 0.0910 | 0.0402 | 0.0402 | 0.0202 | 0.0202 |
| $h_3(1,2,2)$ | 0 | 0.0105 | 0.0105 | 0.0765 | 0.0765 | −0.0085 | 0.0085 | 0.0040 | 0.0040 |
| $h_3(2,2,2)$ | 1.06 | 1.0457 | 0.0143 | 1.0474 | 0.0126 | 1.0534 | 0.0066 | 1.0526 | 0.0074 |

Explain: [1] ***H*** represents the kernel parameters; [2] **T-V** represents true value; [3] **R-I** represents random initialization; [4] **C-I** represents chaotic initialization; [5] **S-I** represents sequence initialization; [6] ***H**** represents the identification result; [7] ***E*** represents the identification error.

### *6.2. Model Identification Experiment with Different Optimization Algorithms*

To further evaluate the performance of our algorithm in identifying the model parameters. Firstly, the X-dimensional time series generated by the Lorenz equation is used as validation data, and it is divided into historical training set and online prediction set. Secondly, the algorithm parameters are determined and the training set is used to identify the model. Finally, the prediction set is sampled by the dynamic split window, identifies the prediction model online and performs single-step prediction. In addition, improved particle swarm optimization algorithm (IPSO) [29], genetic algorithm (GA) [30] and beetle antennae search algorithm (BAS) [31] are used for comparative analysis. The parameter of the algorithms are shown in Table 2.

**Table 2.** Algorithm parameter setting of *experiment* 2.

| Algorithm | Parameter Settings |
|---|---|
| Ours | $Q = 29, w = 0.8, c_1 = c_2 = 2, r_1 = rand, r_2 = rand$ |
| IPSO | $S = 5, \lambda_1 = \lambda_2 = \lambda_3 = 2, \alpha_1 = rand, \alpha_2 = rand, \alpha_2 = rand$ |
| GA | $p_c = 0.8, p_m = 0.01$ |
| BAS | $\eta = 0.95, \delta = 2$ |

The Lorenz equation is a classical chaotic system, and the generated chaotic time series are often used for algorithm verification. The mathematical expression of the Lorenz equation is as follows:

$$\begin{cases} x' = & -\sigma(x-y) \\ y' = & (r-z)x - y \\ z' = & xy - bz \end{cases}$$

Based on the above Lorenz equation, we set the initial value to (1,0,0), $\sigma = 10$, $r = 28$ and $b = \frac{8}{3}$. The fourth-order Runge–Kutta method is used for sampling, with an integration step of 0.02. In the generated chaotic time series, 700 sample points of the Lorenz-X dimensional are selected as the experimental data. The initial length of dynamic split window is set to 200, and the last 500 sample points are used as the test set. The delay time $\tau = 8$ and the embedding dimension $m = 5$ are determined by the historical trajectory. To ensure the objectivity of comparison experiment, the population size of each algorithm is 100 and the maximum number of iterations is 500.

Figure 10 shows the trajectory prediction of Lorenz-X dimensional by four model identification algorithms. It can be seen that the models identified by the four algorithms can achieve nonlinear trajectory prediction, and the predicted trajectory of our algorithm has best fit with the true trajectory, while the BAS algorithm has the most deviation points. Figure 11 shows the trajectory prediction errors of the four algorithms, where $Err = |x(n) - \hat{x}(n)|$. It can be found that our algorithm has only 7.2% of prediction errors greater than 0.2, compared to 21.4%, 42% and 38.4% for IPSO, GA and BAS algorithms, respectively. Therefore, the identification model of our algorithm can control the prediction error within a low fluctuation range, which indicates that the model identification performance is much higher than the other three algorithms.

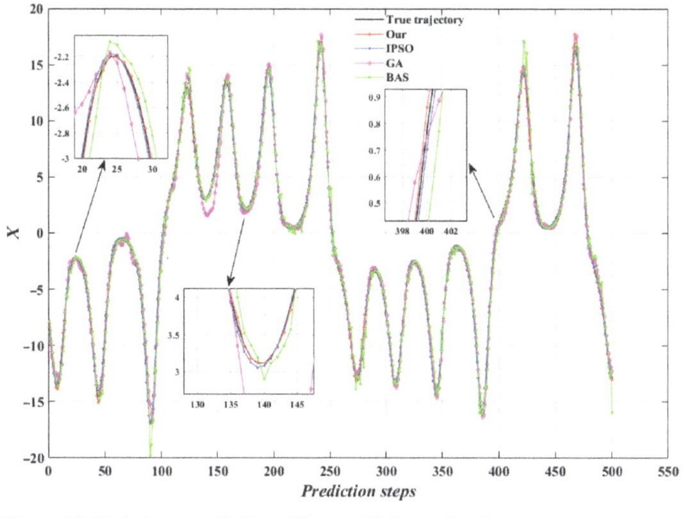


**Figure 10.** Trajectory prediction of Lorenz-X dimensional.

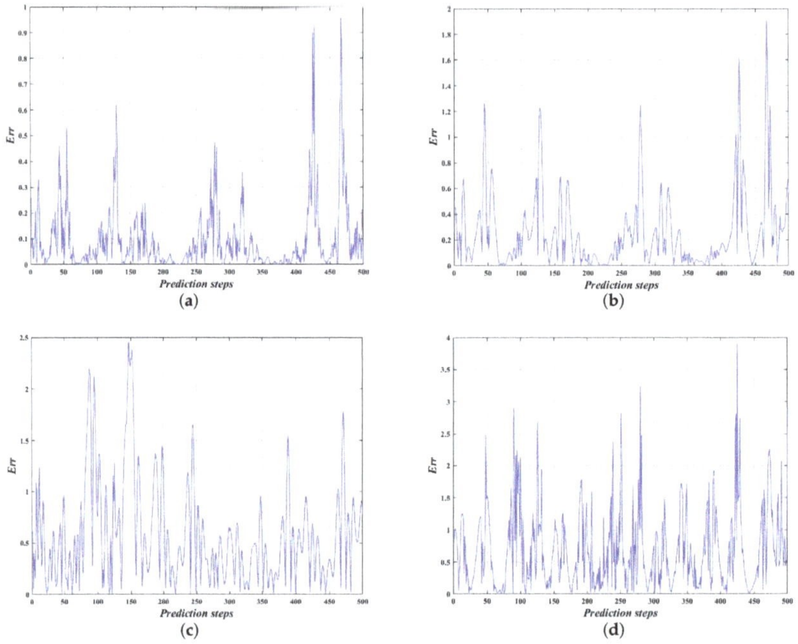


**Figure 11.** Trajectory prediction errors of the four algorithms. (**a**) Our (OTP-SSCT) algorithm; (**b**) the IPSO algorithm; (**c**) the GA algorithm; (**d**) the BAS algorithm.

From Figure 12, the average prediction error of our algorithm are only 35.9%, 16.2%, and 13.6% of IPSO, GA and BAS algorithms, respectively, and the maximum prediction error are only 50.1%, 38.9% and 24.5%, respectively. Obviously, the prediction accuracy of the model identified by our algorithm is higher than the other three optimization algorithms. According to the maneuver of the Lorenz-X dimensional trajectory, it is calculated the average prediction error at each 50 steps, as shown in Figure 13. It is shown that our algorithm can maintain a high prediction accuracy for strongly nonlinear trajectory over time, while the other three algorithms are affected by the fluctuations of trajectories. Meanwhile, it indicates that the proposed algorithm is highly robust and can achieve model identification of strongly nonlinear trajectories.

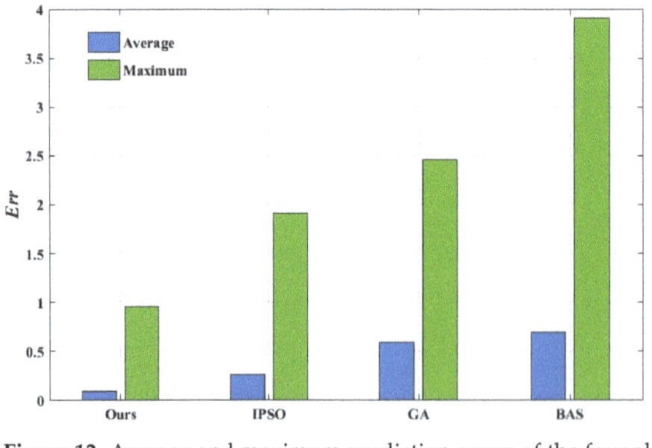


**Figure 12.** Average and maximum prediction errors of the four algorithms.

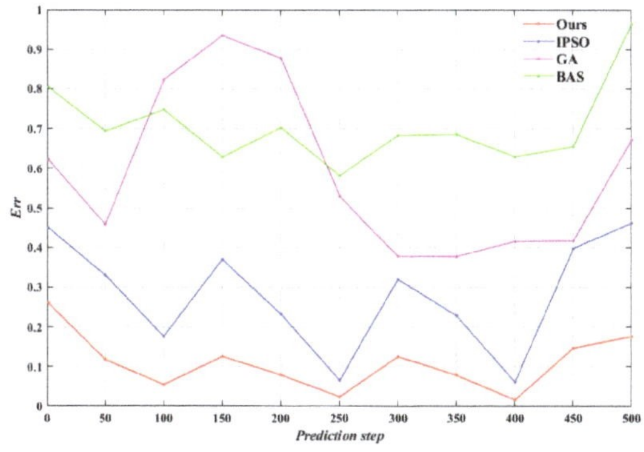


**Figure 13.** Average prediction error of the prediction process in *experiment* 2.

Table 3 shows that our algorithm has the smallest values of both MSE and RMSE, and $R^2$ is closest to 1. It shows that our algorithm has a significant advantage in nonlinear trajectory prediction. Compared with the other three algorithms, the MAPE value of our algorithm is the lowest, indicating that the OTP-SSCT algorithm has the best stability. The statistics in Table 3 show the average results from 50 independent replicate experiments for each algorithm. In summary, the proposed algorithm has certain advantages in terms of prediction accuracy and stability.

**Table 3.** Comparison of precision metrics of *experiment* 2.

| Trajectory | Algorithm | Precision Metrics | | | |
|---|---|---|---|---|---|
| | | MSE | RMSE | MAPE | $R^2$ |
| Lorenz-X | Ours | 0.0262 | 0.1618 | 0.0235 | 0.9996 |
| | IPSO | 0.1575 | 0.3969 | 0.0702 | 0.9975 |
| | GA | 0.5872 | 0.7663 | 0.1844 | 0.9907 |
| | BAS | 0.8765 | 0.9362 | 0.2098 | 0.9861 |

### 6.3. Online Trajectory Prediction Experiment for Maneuvering Targets

In order to test the performance of the OTP-SSCT algorithm for the online prediction of maneuvering target trajectories, the mapping $f$ in Equation (2) is replaced by LSTM algorithm, kernel least mean squares (KLMS) algorithm, kernel recursive least squares (KRLS) algorithm and support vector regression (SVR) algorithm. We perform online single-step prediction of maneuvering target trajectories using the five algorithms separately, and a comparative analysis is given by precision metrics.

The trajectory maneuver can be divided into horizontal maneuvers, vertical maneuvers and spatial combination maneuvers. We adopt the UAV three-degree-of-freedom model to simulate the generation of real-time trajectory data. Considering the UAV as a mass point and the ground coordinate system as an inertial coordinate system, the motion equations are as follows:

$$\begin{cases} \dot{x}_t &= v_t \cos\theta_t \cos\psi_t \\ \dot{y}_t &= v_t \cos\theta_t \sin\psi_t \\ \dot{z}_t &= v_t \cos\theta_t \\ \dot{v}_t &= g(F_t - \sin\theta_t) \\ \dot{\theta}_t &= \frac{g}{v_t}(T_t \cos\phi_t - \cos\theta_t) \\ \dot{\psi}_t &= gT_t \sin\phi_t / (v_t \cos\theta_t) \end{cases} \tag{38}$$

where $x_t$ , $y_t$ and $z_t$ represent the horizontal and height coordinates of the UAV in the inertial coordinate system, respectively. $v_t$, $\theta_t$ and $\psi_t$ represent the speed, climb angle and heading angle of the UAV, respectively. $\phi_t$ is the roll angle. $g$ is the gravitational acceleration. $F_t$ and $T_t$ represent the horizontal overload and the longitudinal overload, respectively. $[x_t, y_t, z_t, v_t, \theta_t, \psi_t]^T$ and $[F_t, T_t, \phi_t]^T$ are the state variables and control variables of the UAV, respectively.

A randomly generated maneuvering trajectory by Equation (38) was used as the experimental data. The number of sensor samples is 300 and the sampling interval is 0.2 s. The initial length of dynamic split window is set to 20, and predict the latter 280 trajectory data online. Adopt three-dimensional coordinate independent prediction to improve the prediction accuracy. The delay time $\tau$ and the embedding dimension $m$ of each dimension of the trajectory data are determined by the historical trajectory. X-dimension: $\tau = 5, m = 2$. Y-dimension: $\tau = 6, m = 2$. Z-dimension: $\tau = 6, m = 2$.

As can be seen from Figure 14, the five prediction trajectories roughly match the true trajectories. Among them, our algorithm is closest to the true trajectory, while the other four algorithms all appear a lot of deviation points. Figure 15 shows the online trajectory prediction errors for each dimension. From Figure 15, the farthest distance in space between the predicted trajectory and the true trajectory can be calculated, where ours, LSTM, KLMS, KRLS and SVR algorithms are about 1.34 m, 8.57 m, 9.54 m, 2.55 m and 8.93 m, respectively. Meanwhile, the average prediction error of our algorithm improves approximately 87.6%, 87.3%, 45.9% and 78.5% over LSTM, KLMS, KRLS and SVR algorithms, respectively. Obviously, our algorithm has the smallest error fluctuation range and the strongest robustness performance. The mean absolute error is calculated for every 20 steps, as shown in Figure 16. It can be found that the prediction accuracy of our algorithm for maneuvering target trajectory is better than the other four algorithms, which can achieve high accuracy online trajectory prediction.

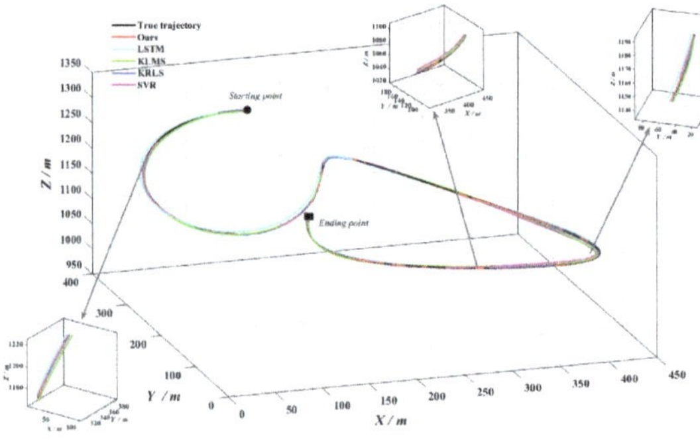


**Figure 14.** Online trajectory prediction of maneuvering target.

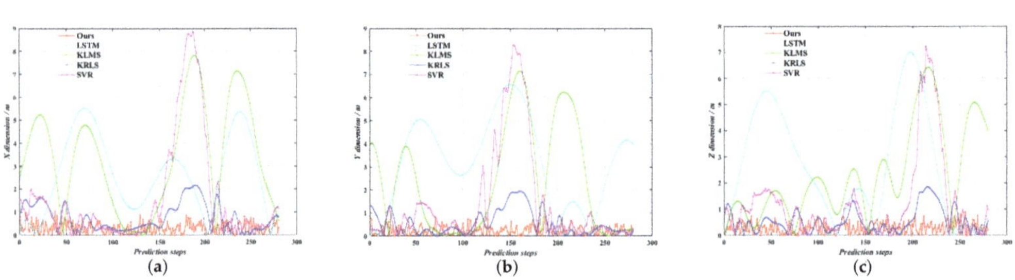


**Figure 15.** Online trajectory prediction errors of five algorithms. (**a**) X-dimension prediction errors; (**b**) Y-dimension prediction errors; (**c**) Z-dimension prediction errors.

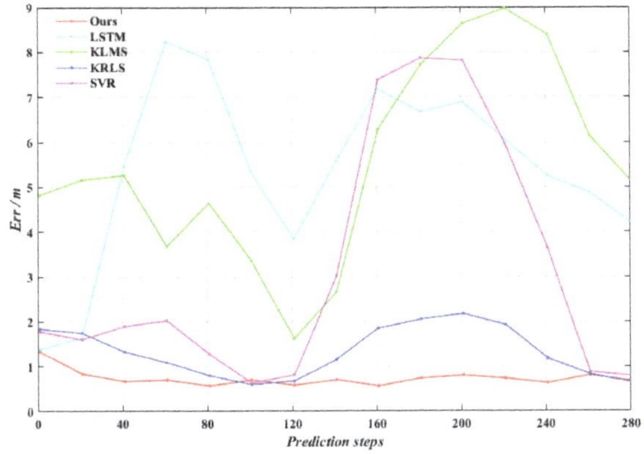


**Figure 16.** Average prediction error of the prediction process in *experiment* 3.

From the precision metrics in Table 4, it can be seen that our algorithm has optimal trajectory prediction results for the maneuvering target and it can accurately track the true trajectory in the whole stage. The LSTM algorithm uses gradient descent algorithm to learn the long-term dependencies of data through three gating units. The network parameters are trained by the dynamic split window in this experiment. Since the algorithm is prone to falling into local extremes or overfitting, and the insufficient amount of training data can also lead to model training failure, which affects the prediction performance. The KLMS and KRLS algorithms are adaptive filtering algorithms based on kernel method. In this experiment, the Gaussian kernel function is chosen as the kernel function. Both algorithms require only a small amount of computation for filtering estimation and weight update, but memory consumption will increase linearly with the number of iterations. Although there are some sparse methods such as novel criterion and approximate linear correlation, they still cannot maintain the prediction accuracy of the algorithm. The SVR algorithm is used to find a minimum interval band to wrap all the training samples. In this experiment, we use the dynamic split window to segment the training and prediction sets. It can be found that the SVR algorithm has a large prediction error at maneuvering turns, which is due to the poor preprocessing of the training data thus leading to parameter identification failure.

**Table 4.** Comparison of precision metrics of *experiment* 3.

| Prediction Dimension | Prediction Algorithm | Precision Metrics | | | |
|---|---|---|---|---|---|
| | | MSE/m$^2$ | RMSE/m | MAPE | $R^2$ |
| X | Ours | 0.3521 | 0.5934 | 0.0059 | 0.9998 |
| | LSTM | 14.943 | 3.8655 | 0.0388 | 0.9994 |
| | KLMS | 26.065 | 5.1054 | 0.0422 | 0.9988 |
| | KRLS | 1.5008 | 1.2251 | 0.0111 | 0.9997 |
| | SVR | 12.407 | 3.5224 | 0.0146 | 0.9993 |
| Y | Ours | 0.2383 | 0.4881 | 0.0053 | 0.9998 |
| | LSTM | 21.746 | 4.6633 | 0.0563 | 0.9985 |
| | KLMS | 17.012 | 4.1245 | 0.0504 | 0.9989 |
| | KRLS | 0.9433 | 0.9712 | 0.0126 | 0.9996 |
| | SVR | 10.144 | 3.1851 | 0.0338 | 0.9993 |
| Z | Ours | 0.2848 | 0.5337 | $4.7 \times 10^{-4}$ | 0.9999 |
| | LSTM | 17.764 | 4.2148 | 0.0036 | 0.9987 |
| | KLMS | 12.886 | 3.5897 | 0.0032 | 0.9991 |
| | KRLS | 0.7959 | 0.8921 | $7.6 \times 10^{-4}$ | 0.9997 |
| | SVR | 6.3458 | 2.5191 | 0.0017 | 0.9995 |

Figure 17 shows the single-step predicted response time of the five algorithms for 50 independent replicate experiments, in which the average single-step prediction time of our algorithm, LSTM, KLMS, KRLS and SVR algorithms are about 0.03352, 0.04344, 0.03257, 0.03604 and 0.04102, respectively. Among them, the single-step prediction time of our algorithm is about 89.8% of five algorithms on average, indicating that our algorithm has an outstanding position in terms of response time. Notice that the single-step prediction time of our algorithm is slightly longer than that of the KLMS algorithm. This is due to the first model identification adopting the improved initial population strategy, which consumes more time. Although the saved optimal values are later used for population initialization, maneuvering changes of the target still increase the identification time. However, it is undeniable that our algorithm outperforms the KLMS algorithm in terms of prediction accuracy.

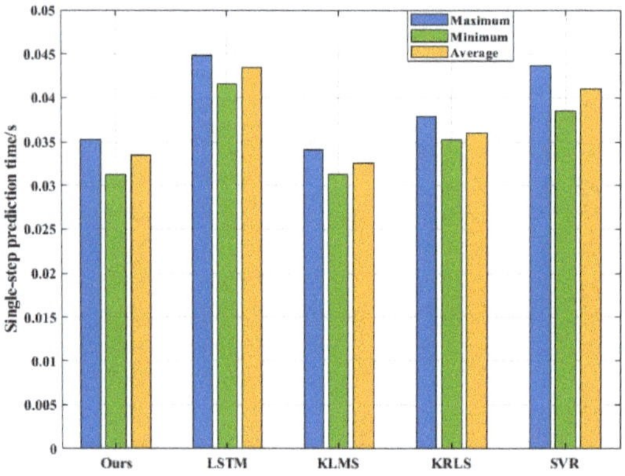


**Figure 17.** Comparison of single-step prediction times.

## 7. Conclusions

In this paper, we propose the OTP-SSCT algorithm to achieve a stable online prediction of maneuvering target trajectories. The algorithm has the following advantages: (1) We build an online prediction model of maneuver trajectory according to chaotic time series analysis. Thus, a small sample chaotic time series prediction set that characterizes the target motion characteristics is constructed through the trajectory segmentation method. By fully considering the nonlinear maneuver characteristics of trajectory data, we introduce the spatiotemporal features into the PSO model identification algorithm, thus improving the identification sensitivity of key points. Meanwhile, the feedback optimization strategy is used to avoid the superposition of prediction errors and improve the prediction accuracy. (2) We propose a new initial population strategy to solve the initial value sensitivity problem of PSO model identification algorithm, thus improving the effectiveness of the initial parameters on model identification. In our future work, it will be interesting to further improve the speediness of the algorithm.

**Author Contributions:** Conceptualization, Q.W.; methodology, Q.W. and P.S.; validation, P.S.; formal analysis, L.Z.; writing—original draft preparation, P.S.; writing—review and editing, W.S.; funding acquisition, Q.W. All authors have read and agreed to the published version of the manuscript.

**Funding:** This research was funded by the National Science Foundation Council of China (61976080, 61771006), the Key Research Projects of University in Henan Province, China (21A413002, 20B510001), the Programs for Science and Technology Development of Henan Province, China (212102310298, 222102210002, 222102210004, 222102220028, 222102210088), Graduate Education Innovation and Quality Improvement Action Plan project (SYLKC2022013).

**Institutional Review Board Statement:** Not applicable.

**Informed Consent Statement:** Not applicable.

**Data Availability Statement:** Not applicable.

**Conflicts of Interest:** The authors declare no conflict of interest.

## References


1. Lin, Y.; Saripalli, S. Sampling-based path planning for UAV collision avoidance. *IEEE Trans. Intell. Transp. Syst.* **2017**, *18*, 3179–3192. [CrossRef]
2. Huang, Z.; Xia, J.; Li, F.; Li, Z.; Li, Q. A peak traffic congestion prediction method based on bus driving time. *Entropy* **2019**, *21*, 709. [CrossRef] [PubMed]
3. Alahi, A.; Goel, K.; Ramanathan, V.; Robicquet, A.; Fei-Fei, L.; Savarese, S. Social lstm: Human trajectory prediction in crowded spaces. In Proceedings of the 2016 IEEE Conference on Computer Vision and Pattern Recognition, Las Vegas, NV, USA, 27–30 June 2016; pp. 961–971.
4. Rudenko, A.; Palmieri, L.; Herman, M.; Kitani, K.M.; Gavrila, D.M.; O Arras, K. Human motion trajectory prediction: A survey. *Int. J. Robot. Res.* **2020**, *39*, 895–935. [CrossRef]
5. Zhang, J.; Liu, J.; Hu, R.; Zhu, H. Online four dimensional trajectory prediction method based on aircraft intent updating. *Aerosp. Sci. Technol.* **2018**, *77*, 774–787. [CrossRef]
6. Abbas, M.T.; Jibran, M.A.; Afaq, M.; Song, W. An adaptive approach to vehicle trajectory prediction using multimodel Kalman filter. *Trans. Emerg. Telecommun. Technol.* **2020**, *31*, e3734. [CrossRef]
7. Hwang, I.; Seah, C.E.; Lee, S. A study on stability of the interacting multiple model algorithm. *IEEE Trans. Autom. Control.* **2016**, *62*, 901–906. [CrossRef]
8. Wu, H.; Xu, W.; Yao, B.; Hu, Y.; Feng, H. Interacting multiple model-based adaptive trajectory prediction for anticipative human following of mobile industrial robot. *Procedia Comput. Sci.* **2020**, *176*, 3692–3701. [CrossRef]
9. Xie, G.; Gao, H.; Qian, L.; Huang, B.; Li, K.; Wang, J. Vehicle trajectory prediction by integrating physics-and maneuver-based approaches using interactive multiple models. *IEEE Trans. Ind. Electron.* **2017**, *65*, 5999–6008. [CrossRef]
10. Zhang, X.; Zhao, Z.; Zheng, Y.; Li, J. Prediction of taxi destinations using a novel data embedding method and ensemble learning. *IEEE Trans. Intell. Transp. Syst.* **2019**, *21*, 68–78. [CrossRef]
11. Heravi, E.J.; Khanmohammadi, S. Long term trajectory prediction of moving objects using gaussian process. In Proceedings of the 2011 First International Conference on Robot, Vision and Signal Processing, Kaohsiung, Taiwan, 21–23 November 2011; pp. 228–232.
12. Tran, Q.; Firl, J. Online maneuver recognition and multimodal trajectory prediction for intersection assistance using non-parametric regression. In Proceedings of the 2014 IEEE Intelligent Vehicles Symposium proceedings, Dearborn, MI, USA, 8–11 June 2014; pp. 918–923.
13. Liu, X.; Liu, Y.; Chen, Y.; Hanzo, L. Trajectory design and power control for multi-UAV assisted wireless networks: A machine learning approach. *IEEE Trans. Veh. Technol.* **2019**, *68*, 7957–7969. [CrossRef]
14. Shi, Z.; Xu, M.; Pan, Q.; Yan, B.; Zhang, H. LSTM-based flight trajectory prediction. In Proceedings of the 2018 International Joint Conference on Neural Networks, Rio de Janeiro, Brazil, 8–13 July 2018; pp. 1–8.
15. Ma, L.; Tian, S. A hybrid CNN-LSTM model for aircraft 4D trajectory prediction. *IEEE Access* **2020**, *8*, 134668–134680. [CrossRef]
16. Han, P.; Wang, W.; Shi, Q.; Yang, J. Real-time short-term trajectory prediction based on GRU neural network. In Proceedings of the 2019 IEEE/AIAA 38th Digital Avionics Systems Conference, San Diego, CA, USA, 8–12 September 2019; pp. 1–8.
17. Adewumi, A.; Kagamba, J.; Alochukwu, A. Application of chaos theory in the prediction of motorised traffic flows on urban networks. *Math. Probl. Eng.* **2016**, *2016*, 5656734. [CrossRef]
18. Jiang, P.; Wang, B.; Li, H.; Lu, H. Modeling for chaotic time series based on linear and nonlinear framework: Application to wind speed forecasting. *Energy* **2019**, *173*, 468–482. [CrossRef]
19. Liu, J.; Shi, G.; Zhu, K. Vessel trajectory prediction model based on AIS sensor data and adaptive chaos differential evolution support vector regression (ACDE-SVR). *Appl. Sci.* **2019**, *9*, 2983. [CrossRef]
20. Hong, W.C.; Li, M.W.; Geng, J.; Zhang, Y. Novel chaotic bat algorithm for forecasting complex motion of floating platforms. *Appl. Math. Model.* **2019**, *72*, 425–443. [CrossRef]
21. Chatterjee, A. Parameter estimation of Duffing oscillator using Volterra series and multi-tone excitation. *Int. J. Mech. Sci.* **2010**, *52*, 1716–1722. [CrossRef]
22. Han, F.; Yang, S.; Song, S. Local Volterra multivariable chaotic time series multi-step prediction based on phase points clustering. *J. Vibroeng.* **2018**, *20*, 2486–2503. [CrossRef]
23. Qiao, W.; Yang, Z.; Kang, Z.; Pan, Z. Short-term natural gas consumption prediction based on Volterra adaptive filter and improved whale optimization algorithm. *Eng. Appl. Artif. Intell.* **2020**, *87*, 103323. [CrossRef]
24. Lv, M.; Zhang, X.; Chen, H.; Ling, C.; Li, J. An accurate online prediction model for kiln head temperature chaotic time series. *IEEE Access* **2020**, *8*, 44288–44299. [CrossRef]
25. Ali, S.; Basharat, A.; Shah, M. Chaotic invariants for human action recognition. In Proceedings of the 2007 IEEE 11th International Conference on Computer Vision, Rio de Janeiro, Brazil, 14–21 October 2007; pp. 1–8.

26. Li, Y.; Song, Y.; Li, C. Selection of parameters for phase space reconstruction of chaotic time series. In Proceedings of the 2010 IEEE Fifth International Conference on Bio-Inspired Computing: Theories and Applications, Changsha, China, 23–26 September 2010; pp. 30–33.
27. Kate, R.J. Using dynamic time warping distances as features for improved time series classification. *Data Min. Knowl. Discov.* **2016**, *30*, 283–312. [CrossRef]
28. Elsayed, S.; Sarker, R.; Coello, C.A.C. Sequence-based deterministic initialization for evolutionary algorithms. *IEEE Trans. Cybern.* **2017**, *47*, 2911–2923. [CrossRef] [PubMed]
29. Chang, W.D. Volterra filter modeling of nonlinear discrete-time system using improved particle swarm optimization. *Digit. Signal Process.* **2012**, *22*, 1056–1062. [CrossRef]
30. Tang, H.; Liao, Y.H.; Cao, J.Y.; Xie, H. Fault diagnosis approach based on Volterra models. *Mech. Syst. Signal Process.* **2010**, *24*, 1099–1113. [CrossRef]
31. Wu, Q.; Shen, X.; Jin, Y.; Chen, Z.; Li, S.; Khan, A.H.; Chen, D. Intelligent beetle antennae search for UAV sensing and avoidance of obstacles. *Sensors* **2019**, *19*, 1758. [CrossRef] [PubMed]

mathematics

MDPI

Article

# Numerical Solution of Finite Kuramoto Model with Time-Dependent Coupling Strength: Addressing Synchronization Events of Nature

**Dharma Raj Khatiwada**

School of STEM, College of Agriculture, Community and Sciences, Kentucky State University, 400 E Main St., Frankfort, KY 40601, USA; dharma.khatiwada@kysu.edu

**Abstract:** The synchronization of an ensemble of oscillators is a phenomenon present in systems of different fields, ranging from social to physical and biological systems. This phenomenon is often described mathematically by the Kuramoto model, which assumes oscillators of fixed natural frequencies connected by an equal and uniform coupling strength, with an analytical solution possible only for an infinite number of oscillators. However, most real-life synchronization systems consist of a finite number of oscillators and are often perturbed by external fields. This paper accommodates the perturbation using a time-dependent coupling strength $K(t)$ in the form of a sinusoidal function and a step function using 32 oscillators that serve as a representative of finite oscillators. The temporal evolution of order parameter $r(t)$ and phases $\theta_j(t)$, key indicators of synchronization, are compared between the uniform and time-dependent cases. The identical trends observed in the two cases give an important indication that the synchrony persists even under the influence of external factors, something very plausible in the context of real-life synchronization events. The occasional boosting of coupling strength is also enough to keep the assembly of oscillators in a synchronized state persistently.

**Keywords:** Kuramoto model; finite oscillators; synchronization

**MSC:** 93C10; 37M05

## 1. Introduction

The phenomenon by which a large population of elements oscillating with different natural frequencies lock to a common frequency is called synchronization. Such phenomena have been observed in many real-world systems, such as rhythmic response patterns of pacemaker cells in the heart [1], a synchronously flashing group of fireflies [2,3], crickets that chirp in unison [4], etc. Arrays of lasers and superconducting Josephson junctions are examples in physics that obey collective synchronization [5,6].

One of the most successful approaches for exploring and analyzing synchronization phenomenon for the population of elements as an ensemble of phase oscillators is given by Kuramoto [7], which treats each oscillator only by their phases [8]. The oscillators proposed in his model are assumed to be non-linear, and there is a weak but uniform coupling between every oscillator (all-to-all coupling).

There is a threshold $K_c$ for the coupling strength $K$, which characterizes the interaction term in the Kuramoto mathematical model, above which a phase transition analogous to magnetic phase transition occurs [9]. In other words, the system shifts from an incoherent state to a partially synchronized state and ultimately to the fully synchronized one with $K >> K_c$. Kuramoto's original model can be considered a basic model in the sense that it describes oscillators of the fixed natural frequencies $w_j$, connected via 'all-to-all' uniform $K$ [10] and solvable only in the thermodynamic limit ($N \to \infty$). However, the real-world systems that exhibit synchronous behavior consist of a finite number of oscillators, and the coupling is expected to be perturbed by time-dependent external factors. Some of the promising examples include the natural firing rate of neurons in the brain, which correspond to the natural frequencies, and the degree of excitation/inhibition from neurons to stimulate a neuron via a dendrite (modeled as time-dependent coupling strength) [10–12].

To address the inevitable influence of external fields, this paper incorporates a time-dependent coupling strength of the form $K(t) = K_1(1 + A\sin(\Omega t))$ as suggested in [10,11,13] and also mentioned in [14] with a slightly different scope. However, the term $K_1$ plays the same role as the uniform coupling strength of Kuramoto's original model, a different approach from [10,11,13]. The influence of the external field can be oscillatory, so the choice of the term $A\sin(\Omega t)$ is justifiable.

The above form of $K(t)$ is tested here for an 32 oscillators system, an even power of 2 oscillators and a representative of a small number of finite oscillators such as synchronously clapping audiences at an orchestra.

This paper also intends to use coupling strength $K(t)$ in the form of a step function. Unlike the trigonometric function, $A\sin(\Omega t)$, which acts continuously in the system of oscillators, a different (higher) value of coupling strength $K$ acts for only a certain time interval where the order parameter $r(t)$, a key indicator that reflects the synchronous behavior of oscillators (Section 2), goes beyond its pre-defined threshold value. It is assumed that the system of such oscillators' assembly preserves its autonomy except at some intervals of time where $r$ approaches some threshold value. $K(t)$ serving as a step-function has been tested in various articles, notably in [15], which aims to implement this approach for two oscillators' assembly. However, the approach implemented here is unique in the sense that $K(t)$ only takes a higher value if the order parameter $r(t)$ goes beyond a certain threshold. Additionally, the step-function model of $K(t)$ tested here is for 32 oscillators, which can serve as a representative for a wide range of finite oscillators exhibiting synchronous behavior in nature.

As mentioned earlier, there is a phase transition from an incoherent state to being partially synchronized in the coupling strength $K \geq K_c$ and finally to the fully synchronized state for $K >> K_c$. The testing of synchronized states at some value of $K$ greatly depends on the critical value $K_c$ for the given oscillators' assembly. However, such critical value has only been found analytically in the infinite oscillators' assembly so far. Still, this paper picks the value of $K_c$ equal to the critical value associated with infinite oscillators that comes from Equation (8). The synchronization pattern, if observed in the vicinity of such analytical value, will help us to locate the respective $K_c$ of finite oscillators more precisely.

The following two methods will be followed to observe the impact of external fields,

- Step function: $K < K_c$ can be used except for the order parameter $r(t) \leq r_{th}$, where a higher $K > K_c$ is applied to elevate oscillators so that they maintain their synchronized state;
- Trigonometric function: $K_1$ can be set in the vicinity of $K_c$ to observe whether the perturbation part $A\sin(\Omega t)$ can make an impact to elevate the oscillators into the synchronized state.

The synchronization events observed in nature have displayed their synchronous behavior persistently, despite the presence of external fields, so the results from this proposed work are expected to find a close match between time-dependent and uniform parameters. In the case of a step function, its usefulness lies in whether the oscillators maintain the synchrony with the boost of a higher coupling strength $K$ value for $r \leq r_{th}$.

## 2. Kuramoto Model with the Infinite Number of Oscillators

In a system of $N$ phase oscillators, each oscillating with natural frequencies $w_j$, the time evolution of phase $\theta_j$ of the j$^{th}$ oscillator is given by following Kuramoto [7] equation:

$$\dot{\theta}_j(t) = \omega_j + \frac{K}{N}\sum_{k=1}^{N} \sin(\theta_k(t) - \theta_j(t)), \quad i = 1, ..., N \tag{1}$$

where $K \geq 0$, is the homogeneous all-to-all coupling strength. The factor $1/N$ ensures that the system converges in the thermodynamic limit (as $N \to \infty$). The frequencies $w_j$ are distributed according to some probability density $g(w)$, where $g(w)$ is assumed to be unimodal and symmetric about the mean frequency $\nu$ i.e., $g(\nu + \omega) = g(\nu - \omega)$ [6,16]. For example, $g(w)$ can be a Gaussian distribution or a Cauchy–Lorentzian distribution. Non-identical values of the frequency of each individual, especially in the context of biological oscillators such as fireflies, reflect their inevitable variability found in nature [17].

Equation (1) can be related to a physically sensible quantity called order parameter $r(t)$, which is simply an average (centroid vector of the phase distribution) of the complex number (Z) that represents the phase of the oscillators with each moving in a unit circle.

$$Z = r(t)e^{i\psi(t)} = \frac{1}{N}\sum_{k=1}^{N} e^{i\theta_k(t)}, \text{with}, r(t) = |Z| \tag{2}$$

Multiplying Equation (2) by $e^{-i\theta_j(t)}$, the following relation is obtained.

$$r(t)e^{i(\psi(t)-\theta_j(t))} = \frac{1}{N}\sum_{k=1}^{N} e^{i(\theta_k(t)-\theta_j(t))} \tag{3}$$

Now, comparing the imaginary part of the above relation and making the substitution in Equation (1), we find:

$$\dot{\theta}_j(t) = w_j + Kr(t)\sin(\psi(t) - \theta_j(t)), \quad i = 1, ..., N \tag{4}$$

- $K \to 0 : \theta_j(t) = w_j + \theta(0)$
- $K \to \infty : \frac{\dot{\theta}_j(t) - w_j}{Kr(t)} = \sin(\psi(t) - \theta_j(t)) \implies \theta_j(t) = \psi(t)$

In the thermodynamic limit, oscillators are expected to be distributed with a probability density $\rho(\theta, \omega, t)$. Now, the arithmetic mean in Equation (2) becomes an average over phase and frequency:

$$r(t)e^{i\psi(t)} = \int_{-\pi}^{\pi}\int_{-\infty}^{\infty} e^{i\theta}\rho(\theta, \omega, t)g(\omega)d\theta d\omega \tag{5}$$

For $K \to 0$, Equation (5) becomes

$$r(t)e^{i\psi(t)} = \int_{-\infty}^{\infty}\left(\int_{-\pi}^{\pi} e^{i(\omega t+\theta_0)}\rho(\theta, \omega, t)d\theta\right)g(\omega)d\omega = 0 \tag{6}$$

Therefore, $r(t) \to 0$ as $t \to \infty$ and the oscillators are out of synchrony.
However for $K \to \infty$, $\theta_j(t) = \psi(t)$

$$r(t)e^{i\psi(t)} = e^{i\psi(t)}\int_{-\pi}^{\pi}\int_{-\infty}^{\infty} \rho(\theta, \omega, t)g(\omega)d\theta d\omega \tag{7}$$

These integrals are equal to 1 as they are the integral of densities over the entire range. It means $r(t) \to 1$ and there is perfect synchrony among the oscillators [7,16]. For the intermediate range of coupling strength, $K_c < K < \infty$ (with $K_c$, critical coupling strength; defined in Equation (8)), some of the oscillators are phase-locked ($\dot{\theta}_j = 0$) and others rotate out of synchrony with the locked oscillators [8].

Kuramoto obtained the exact solution of $K_c$ after evaluating the contribution of both drifting and locked oscillators (Equation (5)) as

$$K_c = \frac{2}{\pi g(0)} \tag{8}$$

where g(0) is the frequency distribution of interest, evaluated at $\omega = 0$.

For the Gaussian distribution of frequencies that is used in this research, the approximate solution or $r(t)$ near the bifurcating point $K_c$ is given by [8]

$$r \approx \sqrt{\frac{8(K - K_c)}{-K_c^3 g''(0)}} \tag{9}$$

where $g''(0)$ is the second derivative of $g(0)$ evaluated at $\omega = 0$.

### 2.1. Synchronization Visualization

The exact solution for the critical coupling strength $K_c$ given in Equation (8) is achievable as $N \to \infty$. The numerical solution of $N = 32$ oscillators for this work will pick the value of $K_c$ that comes from Equation (8).

Figure 1 compares the behavior of $r(t)$ versus $K$ for four different assemblies of oscillators, with numbers ranging from 32 to 256 near the onset of the synchronization. The natural frequencies $w_j$ are drawn from the Gaussian. The number of oscillators used in four cases replicates the number in various synchronized events of nature, such as fireflies displaying synchrony. As $N$ increases, the bifurcation point approaches its counterpart for $N \to \infty$. In the case of 32 oscillators, however, there is a larger discrepancy.

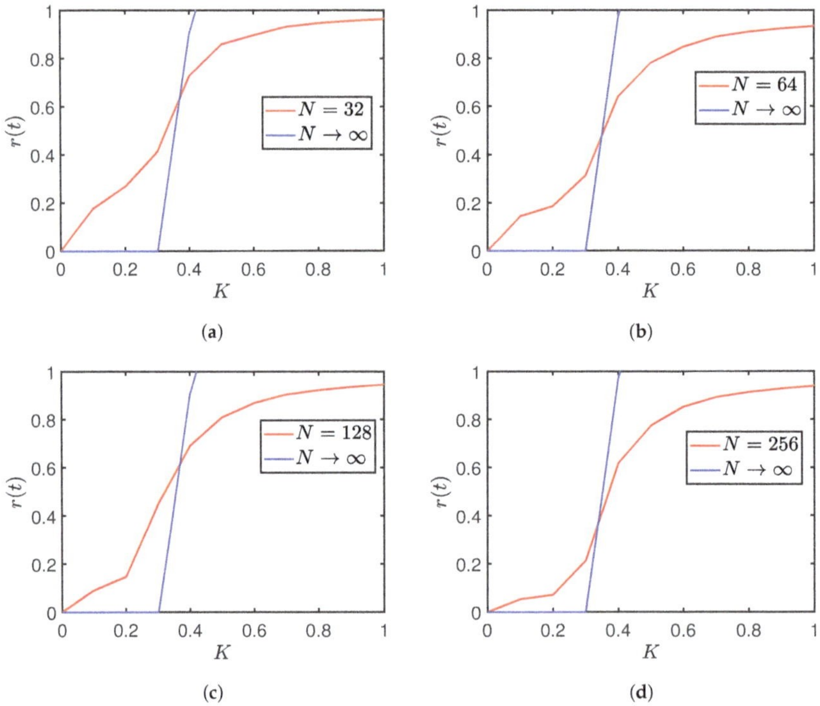


**Figure 1.** $r(t)$ vs. $K$ plots of (**a**) 32, (**b**) 64, (**c**) 128, and (**d**) 256 oscillators: simulation versus approximated solution (Equation (9)) near critical coupling strength $K_c$. $w_j$ are drawn from the Gaussian with mean $\mu = 0$ & standard deviation $\sigma = 0.2$. $K_c$ approaches its analytical counterpart (evaluated for $N \to \infty$) as the number of oscillators increases from 32 to 256.

Figure 2a displays the finite size effect on the temporal evolution of phase angle $\theta_j(t)$ and the order parameter $r(t)$ of the 32 oscillators. The use of $K_c = 0.32$ for dealing with 32 oscillators cannot be justified; however, $K = 0.5 > K_c$ keeps the finite size effect on the sideline. The phases of oscillators start gathering coherently for $K > K_c$. The fluctuation

($\approx N^{-\frac{1}{2}}$) of $r(t)$ is significantly higher in the 32 oscillator case compared to Figure 2b, which uses 256 oscillators.

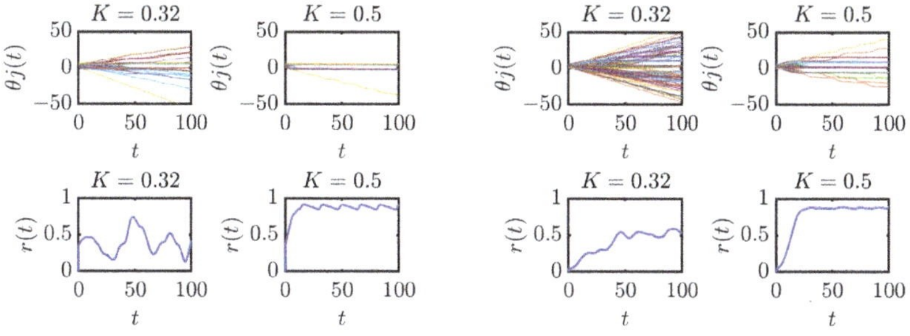


(a) (b)

**Figure 2.** Synchronization pattern: 32 vs. 256 oscillators. The critical value $K_c$ is expected to reach closer to 0.32 (exact value in the thermodynamic limit, $N \to \infty$) with the increase in the number of oscillators. Two partially synchronized groups are visible in the 32 oscillator case (**a**), while there are many of them in the case of 256 oscillators (**b**). Additionally, the fluctuation in the typical $r(t)$ vs. $t$ plot, which is proportional to $N^{-\frac{1}{2}}$, is less prominent in the case of the 256 oscillator system compared to 32 oscillators.

### 2.2. *Time-Dependent Parameters*

The inherent time-variability of oscillators can be addressed by incorporating time-dependent coupling strength $K(t)$ in Kuramoto's original model, which assumes all-to-all homogeneous coupling $K$.

In this research work, the following form of $K(t)$ has been used:

$$K(t) = K_1 + A\sin(\Omega t) \tag{10}$$

where $K_1$ is the uniform coupling strength with the value chosen in the vicinity of $K_c$). Other parameters, $A$ and $\Omega$, represent the amplitude and frequency of the perturbation term, respectively.

Now, we introduce $K(t)$ for the system of $N = 32$ oscillators as:

$$\dot{\theta}_j(t) = \omega_j + \frac{(K_1 + A\sin(\Omega t))}{N} \sum_{k=1}^{N} \sin(\theta_k(t) - \theta_j(t)) \quad i = 1, \ldots.., N \tag{11}$$

The natural frequencies $w_j$ are chosen randomly from the standard Gaussian density function. The amplitude $A$ and frequency $\Omega$ of the perturbation term have also been treated as variables separately to see their impact on the synchronization pattern in the $r(t)$ vs. $t$ graph using Equation (2).

## 3. Results

As discussed in Section 2, the main objective of this work is to check for the similarity in $r(t)$ and $\theta(t)$ vs. $t$ graphs using the time-dependent and uniform coupling strengths in the vicinity of $K_c$.

Numerical integration of Equation (11) was carried out using the Runge–Kutta 4th order method, with a time step of $dt = 0.01$ throughout 100 (some unit of time). The values of $K_1 = K$ have been chosen in the vicinity of critical coupling strength $K_c$ based on Equation (8). The natural frequencies of the oscillators were randomly chosen to be from standard Gaussian distribution with mean $\mu = 0$ and standard deviation $\sigma = 0.2$. Initial phases $\theta_j(0)$ were also randomly chosen in the range of $[0,2\pi]$. A default setting of the random number generation was used to ensure the same numbers were being used in the successive iteration.

Now, the standard Gaussian distribution of natural frequencies given by

$$g(w) = \frac{1}{\sqrt{2\pi\sigma}} e^{-\frac{1}{2}(\frac{w-\mu}{\sigma})^2} \quad (12)$$

From the above equation $g(0) \approx 1.995$ for $\mu = 0$ and $\sigma = 0.2$.
Critical coupling strength $K_c$ based on Equation (8) equals to

$$K_c = \frac{2}{\pi g(0)} \approx 0.32 \quad (13)$$

### 3.1. Using Periodic Function as Perturbation

The comparisons were made in the $r(t)$ vs. $t$ graphs using the uniform and time-varying coupling strengths. Additionally, the response of $K(t)$ was observed more closely in $r(t)$ vs. $t$ by varying amplitude $A$ and frequency $\Omega$ of the perturbation term separately.

The temporal evolution of $r(t)$ in Figure 3a for the $K = 0.5$ & $K(t) = K_1 + A\sin(\Omega t)$ with $K_1 = 0.5$ are similar from the perspectives of their synchronization pattern. The extra oscillatory behavior in the graph is simply due to the sinusoidal term embedded in $K(t)$. Similar trends are observed even for the $K = 0.2$ & $K(t) = K_1 + A\sin(\Omega t)$ with $K_1 = 0.2$ in Figure 3b, except for the fact that they no longer exhibit the coherent behavior.

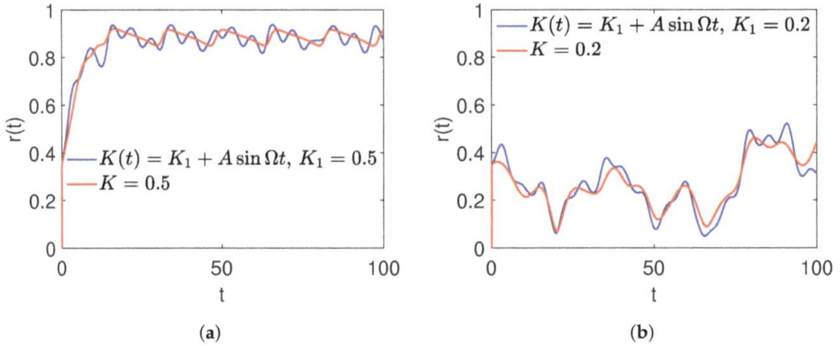


**Figure 3.** Comparison of $r(t)$ vs. $t$ graphs for $K = K_1 < K_c$ and $K = K_1 > K_c$ $A = 0.2$, $\Omega = 1$. (**a**): Temporal evolution of $r(t)$ for $K = 0.5 = K_1 > K_c$. Results are identical for the time-dependent and uniform coupling strength. (**b**): $r(t)$ vs. $t$ comparison using $K = 0.2$ & $K(t)$ with $K_1 = 0.2$. These graphs are fairly similar.

Figure 4a,c display the time evolution of $\theta(t)$ and $r(t)$ with variable $A$ for $K(t) = K_1 + A\sin(\Omega t)$ with $K_1 = 0.2$. The oscillators remain out of synchrony except for some periodic peaks at which $K(t) \to K_c$. It is usually seen if amplitude $A$ goes on increasing.

Similarly, Figure 4b,d show $\theta_j(t)$ and $r(t)$ vs. $t$ plots for $K(t) = K_1 + A\sin(\Omega t)$ with $K_1 = 0.5$, using different A values. As $K_1 = 0.5$, which is well above $K_c$, the oscillators display synchronous behavior. However, oscillators start losing their rhythmic pattern when A increases significantly. The impact of various values $\Omega$ in $K(t)$ was not significant in both regions, mainly from the perspective of reshaping the synchronization pattern.

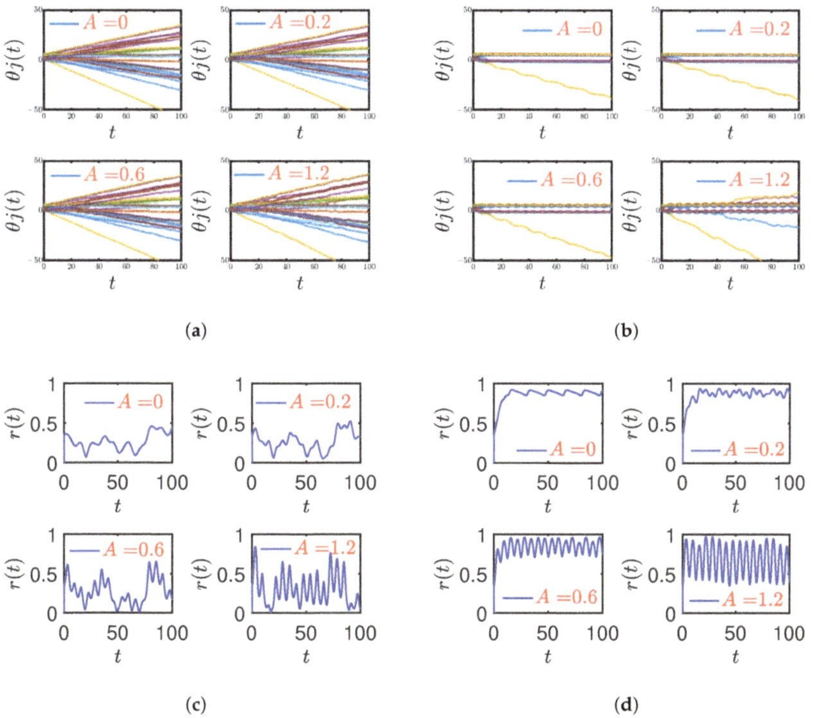


**Figure 4.** $\theta(t)$ and $r(t)$ vs. $t$ graphs for different values of $A$, $\Omega = 1$. Two regions, $K_1 < K_c$ and $K_1 > K_c$, are tested. (**a**): $K_1 = 0.2 < K_c$ forces the oscillators to remain out of synchrony except for some with higher values of $A$. (**b**): Drifting of some oscillators are associated with the higher values of $A$ (=1.2) even for $K_1 = 0.5 > K_c$. It gives the important indication that the higher $A$ values are destabilizing the synchronization pattern. (**c**,**d**) are the respective $r(t)$ vs. $t$ graphs for $K_1 < K_c$ and $K_1 > K_c$.

### 3.2. Using Step Function as a Perturbation

An alternative approach that essentially replaces the sinusoidal function as a perturbation term is to provide a uniform but higher value of coupling strength $K(t)$ for the time at which order parameter $r(t)$ goes below some threshold value $r_{th}$. A step function may fulfill such requirements and thus keep oscillators in their synchronized state persistently. A system of such oscillators displays autonomous behavior, except for $r(t) \leq r_{th}$, where a uniform but higher value of $K(t)$ is applied to maintain the synchronized state of oscillators.

One important assumption is made in the numerical solution using the additional coupling strength $K(t)$ in the form of a step function.

- The threshold value $r_{th}$ of the order parameter selected in the $r(t)$ vs. $t$ graph, below which an additional value of coupling strength ($K_{add}$) is applied, is a tentative mid-value ($0 \leq r(t) \leq 1$). Oscillations observed in the $r(t)$ vs. $t$ graph above $r_{th}$ are assumed to be holding the synchronous behavior of oscillators.

The order parameter $r(t)$ vs. $t$ graph (Figure 5a) looks very random in the case of 32 oscillators. Additionally, the lack of precise information regarding the critical coupling strength $K_c$ gives the real difficulty in the numerical solution. The extrapolation technique by which special time points $t_2$, which correspond to the $r_{th}$, are estimated, fails to reflect the closer values. There is a discrepancy between actual time points corresponding to $r_{th}$ and the estimated $t_2$ (represented by red asterisk in Figure 5a). Figure 5c (corresponding $K$ vs. $t$ plot in Figure 5b) reflects such discrepancy, as $r(t)$ occasionally goes below its threshold value $r_{th} = 0.2$. However, such occasional drifting can be eliminated if the tolerance value is increased. The higher the tolerance, the more of these $r(t) \leq r_{th}$ points are boosted with the additional coupling strength $K_{add}$, thus ultimately, preserving the synchronous behavior. This will also increase the average value of the coupling strength $K_{ave}$, as $K_{add}$ will be applied for a longer interval of time. Figure 5d, which is generated using $t_{tol} = t_2 \pm 4$, does not include such occasional drifting of the oscillators or, to be more precise, the order parameter $r(t)$ at least does not go beyond $r_{th} = 0.25$.

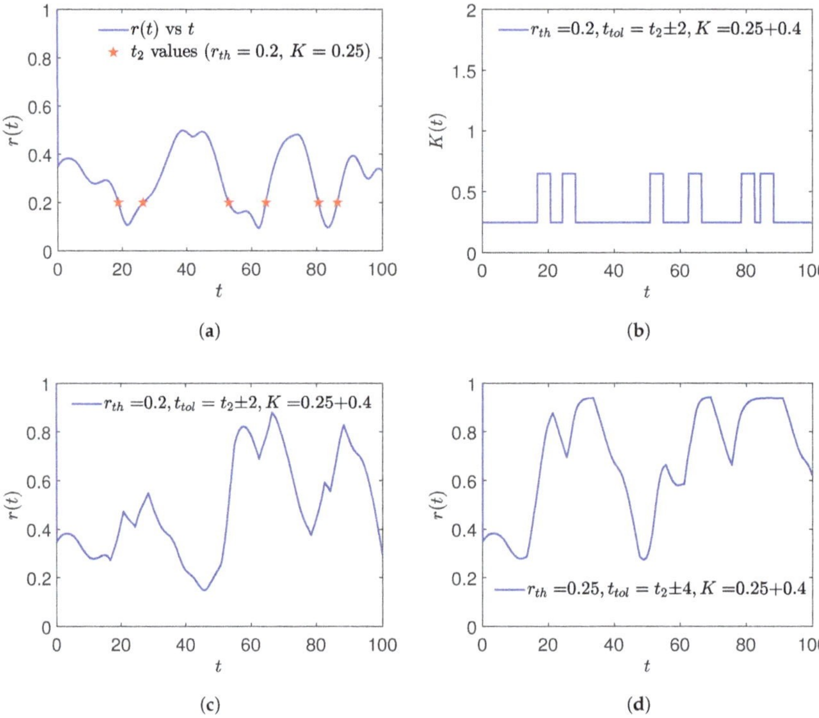


**Figure 5.** (**a**): Estimating the $t_2$ values at $r = r_{th}$ for a 32 oscillator system. These points lack periodicity and are located at random distances in the $r(t)$ vs. $t$ curve. (**b**): $K$ vs. $t$ graph for 32 oscillator system after giving $K_{add}$ in the vicinity of $r = r_{th} = 0.2$. The higher value of $K$ is applied in a narrow interval, causing no noticeable increase in the $K_{ave}$. (**c**): Respective $r(t)$ vs. $t$ graph for a 32 oscillator system after giving $K_{add}$ in the vicinity of $r = r_{th} = 0.2$. Finally, the higher tolerance value $\pm 4$ is used in (**d**) of $r(t)$ vs. $t$ graphs. Now, every point in the graph lies above the $r_{th}$ due to the increased value of tolerance.

## 4. Conclusions and Discussions

This paper was focused on the synchronization of finite phase oscillators using time-dependent coupling strength $K(t)$. In contrast, the original Kuramoto model [7] was formulated for infinite oscillators using a homogeneous all-to-all coupling strength, $K$.

Keeping synchronization phenomena exhibited by different real-world examples, such as periodic flashing of fireflies, in consideration, the choice of using finite oscillators is a necessity. Additionally, considering the influence of external fields on the behavior of each oscillator in the form of a perturbation term makes the original Kuramoto model more realistic. Finding a good match between time-dependent and uniform cases is a required outcome, as real-life synchronization events have been in existence despite the influence of external factors.

A similar pattern observed in $r(t)$ and $\theta(t)$ vs. $t$ graphs for uniform and time-dependent coupling strengths have a broad indication in nature: time variability is expected but well-adjusted by the oscillators displaying the synchronous behavior. $A$ and $\Omega$ of the perturbation part had little impact on the $r(t)$ vs. $t$ graphs. However, the drifting of oscillators even with $K = 0.5 > K_c$ was noticeable for higher values of A (Figure 4b).

The numerical solution that involved the step function (Section 3.2) was carried out with the assumption that the persistence of such synchronized states can be achieved by supplying higher coupling strength $K(t) = K_{initial} + K_{add}$ in the system of oscillators if the order parameter $r(t)$ goes beyond some threshold $r_{th}$. Such thresholds are user-defined, which means a limit can be set for the oscillator's phase difference from each other that can still be included in the group of oscillators that are displaying the synchronous behavior. For example, a firefly (oscillator) flashing with some period plus some margin can still be included in the main group of fireflies that are exhibiting synchronous behavior. Mathematically, they can be kept in one group if their frequencies are within some value of $\sigma$. The synchronous behavior of oscillators is always ensured for $K(t) \geq K_c$ if $K_c$ is precisely known. Thus, Section 3.2 is simply trying to figure out whether such synchronized states are achievable for the coupling strength $K(t)$ which is, in general, kept less than $K_c$, except for the occasional cases ($r(t) \to r_{th}$), where a higher $K(t)$ acts for a brief period.

This approach, however, does not provide a conclusive picture. This is mainly due to the uncertainties associated with $K_c$ for 32 oscillators. The other parameters such as $K_{initial}$, $K_{add}$, $r_{th}$, and $t_{tol}$ also come with some degree of uncertainties because of $K_c$. However, such a limitation was essentially overcome using the higher tolerance values as in Figure 5d.

**Funding:** National Science Foundation (NSF)—HBCU-UP-Implementation Grant: Preparing the Pipeline of Next Generation STEM Professionals. (Award # HRD 2011917).

**Data Availability Statement:** Not applicable.

**Acknowledgments:** The author gratefully acknowledges the support of the National Science Foundation (NSF) – HBCU-UP- Implementation Grant: Preparing the Pipeline of Next Generation STEM Professionals. (Award # HRD 2011917).

**Conflicts of Interest:** The author declares no conflict of interest.

## References

1. Peskin, C.S. *Mathematical Aspects of Heart Physiology*; Courant Institute of Mathematical Science Publication: New York, NY, USA, 1975; pp. 268–278.
2. Choi, H.H.; Lee, J.R. Principles, Applications, and Challenges of Synchronization in Nature for Future Mobile Communication Systems. *Mob. Inf. Syst.* **2017**, *2017*, 8932631. [CrossRef]
3. Buck, J. Synchronous Rhythmic Flashing of Fireflies. II. *Q. Rev. Biol.* **1988**, *63*, 256. [CrossRef] [PubMed]
4. Walker, T.J. Acoustic synchrony: Two mechanisms in the snowy tree cricket. *Science* **1969**, *166*, 891–894. [CrossRef] [PubMed]
5. Ochab, J.; Gora, P.F. Synchronization of Coupled Oscillators in a Local One-Dimensional Kuramoto Model. *Acta Phys. Pol. B* **2010**, *3*, 453.
6. Chopra, N.; Spong, M.W. On Synchronization of Kuramoto Oscillators. In Proceedings of the 44th IEEE Conference on Decision and Control, Seville, Spain, 15 December 2005; pp. 3916–3922.
7. Kuramoto, Y. *Chemical Oscillations, Waves, and Turbulence*; Springer: Berlin, Germany, 1984.
8. Acebron, J.A.; Bonilla, L.L.; Vicente, C.J.P.; Ritort, F.; Spigler, R. The Kuramoto model: A simple paradigm for synchronization phenomena. *Rev. Mod. Phys.* **2005**, *77*, 137–185. [CrossRef]
9. Sakaguchi, H. Synchronization in coupled phase oscillators. *arXiv* **2008**, arXiv:0708.2460.
10. Cumin, D.; Unsworth, C.P. Generalising the Kuramoto Model for the Study of Neuronal Synchronization in the Brain. *Phys. D Nonlinear Phenom.* **2007**, *226*, 181–196. [CrossRef]
11. Petkoski, S.; Stefanovska, A. Kuramoto model with time-varying parameters. *Phys. Rev. E* **2012**, *86*, 046212. [CrossRef] [PubMed]
12. Moreira, C.A.; de Aguiar, M.A.M. Global synchronization of partially forced Kuramoto oscillators on networks. *Phys. A Stat. Mech. Its Appl.* **2019**, *514*, 487–496. [CrossRef]
13. Childs, L.M.; Strogatz, S.H. Stability diagram for the forced Kuramoto model. *Chaos Interdiscip. J. Nonlinear Sci.* **2008**, *18*, 043128. [CrossRef] [PubMed]

14. Wright, E.A.P.; Yoon, S.; Mendes, J.F.F.; Goltsev, A.V. Topological phase transition in the periodically forced Kuramoto model. *Chaos Solitons Fractals* **2021**, *145*, 110816. Available online: https://www.sciencedirect.com/science/article/pii/S0960077921001685 (accessed on 6 September 2022). [CrossRef]
15. Savostyanov, A.; Shapoval, A.; ShnirmanS, M. The inverse problem for the Kuramoto model of two nonlinear coupled oscillators driven by applications to solar activity. *Phys. D Nonlinear Phenom.* **2020**, *401*, 132160. [CrossRef]
16. Strogatz, S.H. From Kuramoto to Crawford:exploring the onset of synchronization in populations of coupled oscillators. *Phys. D* **2000**, *143*, 1–20. [CrossRef]
17. English, L.Q. Synchronization of oscillators: An ideal introduction to phase transitions. *Eur. J. Phys.* **2008**, *29*, 143–153. [CrossRef]

applied sciences

MDPI

*Article*

# Habitat Destruction and Restoration in Relation to Extinction and Survival of Species in Competitive Communities

**Kornphong Chonsiripong [1], Eakkapong Duangdai [2], Radom Pongvuthithum [3,4] and Chulin Likasiri [4,*]**

[1] Faculty of Science at Sriracha, Kasetsart University, Chon Buri 20230, Thailand
[2] Department of Mathematics and Statistics, Faculty of Science and Technology, Chiang Mai Rajabhat University, Chiang Mai 50300, Thailand
[3] Department of Mechanical Engineering, Chiang Mai University, Chiang Mai 50200, Thailand
[4] Research Group in Mathematics and Applied Mathematics, Department of Mathematics, Faculty of Science, Chiang Mai University, Chiang Mai 50200, Thailand
* Correspondence: chulin.l@cmu.ac.th

**Abstract:** We studied the N-species competitive coexistence model with direct effect on habitat destruction to analyze the behaviors of abundant and extinct species in the system caused by habitat loss. The nontrivial equilibrium points of the system are determined for a general habitat destruction function. For the trivial equilibrium, species that survived the habitat destruction are identified using eigenvalues of the Jacobian matrix. Solutions of the system are also presented using the recursive method. Three special cases of habitat destruction functions are addressed: continuous destruction, which is a typical habitat destruction; sudden habitat destruction, which is similar to natural phenomena such as earthquakes or floods; and sudden habitat destruction with aftershocks. The proportional abundances of 50 species are numerically portrayed in each case. We found that the survival of a species is guaranteed if its corresponding eigenvalue is positive. However, the fact that a species has negative corresponding eigenvalue does not guarantee its extinction, as this also depends on the initial number of that species.

**Keywords:** mathematical model; eigenvalue; equilibrium analysis; recursive method; habitat destruction; species extinction

## 1. Introduction

Habitat destruction is a known cause of species extinction over the past billion years. Dominant species are not endangered if their habitat is not completely destroyed, as their territory might be randomly spread across the whole habitat and, hence, they can survive in undisturbed areas. Consequently, extinctions of rare species are assumed to be biased. Early work in this area by Nee and May [1] analyzed the effect of habitat removal of two species that are regionally abundant. Later, Tilman et al. [2] developed the coexistence model, which became widely applied to competitive multi-species' habitat loss. In the Tilman study, extinction in abundant species may be due to habitat loss with time delay. In a subsequent work by Tilman et al. [3], an analytical model of competitive coexistence in spatial habitats is introduced. Here, the extinct species with the poorest distribution are the best competitors. However, since the model assumptions are on dispersal ability and mortality rates, this extinction model is considerably biased.

Tilman's models have been widely studied over the past three decades. A wider range of parameters is investigated by Klausmeier [4], and a critical habitat size for a single species was also examined using a spatially explicit model. A prey–predator metapopulation model in Bascompte and Solé [5] indicates that dispersal rate is crucial for habitat destruction. The extension of this model by Swihart et al. [6] shows that the effect of habitat destruction on the coexistence of prey and predator depends on the invasion or colonization rate of prey and predator. More works related to space utilization on prey-predator interaction model can be found in [7–9].

Accounting for spatial structure and heterogeneity, the work of Prakash and de Roos [10] demonstrates that a predator–prey metapopulation is affected by habitat destruction. Their model is similar to the NMH model of Nee et al. [11]. However, in their model, prey are dispersed in inhabited patches of both predator and prey. The single-species argument NMH model is generally not enough to explain the effects of predator–prey interaction since it allowed the feedback loop into the model. In a later work by Prakash and de Roos [12] on metacommunities of mutualistic species, they vary the strength of mutualistic interaction and fraction of suitable habitat.

In Ovaskainen [13], an eigenvalue–eigenvector relationship was analytically studied to find the effects of habitat destruction and restoration in a metapopulation model using simple models. Earlier, Mean-field theory (MFT) had been used by Nakagiri et al. [14] to develop a model of ecosystems containing two or three species and study the relationships between habitat destruction and species extinction. A nonlinear n-population metapopulation model was also improved from Tilman's model by Lin et al. [15], whose work showed that the extinction number of superior competitors is related to the disharmony exponent of metapopulation. Lin and Zhen-Shan [16] demonstrated the odd-even evolution rule to show that the extinctions of some species are biased. Nakagiri et al. [17] used MFT to simulate and showed that a parity law reveals even-odd species different behaviors.

Species–area and endemics–area relationships (SAR and EAR) models have also been used to predict extinctions. Rybicki et al. [18] showed that EAR underestimated extinctions due to habitat loss while SAR gave a good approximation of short-term extinction unless the remaining habitat is so highly fragmented that SAR also underestimated the extinctions. Rompre et al. [19] used SAR model to predict the rates of bird species loss in fragmented forests and the results showed that SAR underestimated actual losses in the Panama Canal region due to nonrandom patterns of species distributions. Due to a lack of temporal data, long-term losses have not been widely investigated compared to short-terms losses. However, a recent study by Dri [20] showed that habitat losses reduced bird species diversity in urban South America over a time span of only 10 years.

More works on metapopulation model have been undertaken by Xu et al. [21] and Morozov and Li [22]. Through existence and equilibrium stability, the work by Xu et al., showed that the condition of extinction depended on habitat destruction and creation. Based on the original deterministic model, their model is a stochastic differential equations model. Another extension to Tilman's model was carried out by Morozov and Li [22]. In their work, three scenarios on the abundant species are studied: the most abundant species being the best competitors; equally abundant species; and the most abundant species being the poorest competitors. Explicit expressions for species equilibrium in these different scenarios are derived in their study.

Tilman's model was further modified by Liu et al. [23]. In this work, the rates of temporal heterogeneity are introduced via H(t) function. The parameters used are similar to Tilman's input parameters and three types of destruction were simulated: instantaneous destruction, continuous-complete destruction, and continuous-partial destruction. With a slight modification to the model in Tilman et al. [2], the direct effects of habitat destruction on species abundance are captured. Weak and strong Al-lee-like effects are studied in Chen et al. [24] whose work is also based on Tilman's model with a drastically shortened time delay of the species extinction.

In Pongvuthithum and Likasiri [25], the metapopulation model was improved by approximating the direct effect rate of habitat destruction using differentiable function. This approximation allowed analytical discussion on the equilibrium and stability of the model. Various habitat destruction functions were also studied and discussed including cyclical habitat destruction. Due to the superiority of Pongvuthithum and Likasiri's model to Liu et al., and Tilman's model, we are interested in extending this model and predicting the survival species. We also aim to analyze the equilibrium and stability of the model.

Further, due to the model's specific form, we are able to present a recursive method to find the model solution. Several habitat-loss behaviors will also be studied numerically.

**2. The Models and Analyses**

The model in Pongvuthithum and Likasiri [25] describing the habitat proportion occupied by species $i$, $p_i$, can be written as follows:

$$\begin{aligned} F_1 &= \tfrac{dp_1}{dt} = c_1 p_1 (h(t) - p_1) - m_1 p_1 + f(t) p_1 \\ F_i &= \tfrac{dp_i}{dt} = c_i p_i \left( h(t) - \sum_{j=1}^{i} p_j \right) - m_i p_i - \sum_{j=1}^{i-1} c_j p_i p_j + f(t) p_i, i = 2, 3, \ldots, n \end{aligned} \quad (1)$$

where $c_i$ is the colonization or invasion rate of $ith$ species. In this study, the $(i+1)^{th}$ species is a better colonization or invasion species compared to the $ith$ species,

$m_i$ is the mortality rate of $ith$ species,
$h(t)$ represents the remaining habitat,
$f(t)p_i$ is the direct effect of habitat destruction where $f(t) = \frac{h'(t)}{h(t)}$ is the ratio between the rate of change of the remaining habitat and the remaining habitat; in other words, $f(t)$ is the balance between the two figures. If $f(t)$ is a decreasing function, the direct effect of the habitat loss will be decreasing. Otherwise, it will be increasing over time.

The Jacobian of the system (1) can be expressed as a lower triangular matrix:

$$J = \begin{bmatrix} c_1 h - 2c_1 p_1 - m_1 + f & 0 & \cdots & \cdots & 0 \\ -(c_2 + c_1) p_2 & c_2 h - (c_2 + c_1) p_1 - 2c_2 p_2 - m_2 + f & 0 & \cdots & 0 \\ -(c_3 + c_1) p_3 & -(c_3 + c_2) p_3 & \frac{\partial F_3}{\partial p_3} & \cdots & 0 \\ \vdots & \vdots & \cdots & \cdots & \vdots \\ -(c_n + c_1) p_n & -(c_n + c_2) p_n & \cdots & -c_n p_n - c_{n-1} p_n & \frac{\partial F_n}{\partial p_n} \end{bmatrix}$$

where $\frac{\partial F_j}{\partial P_j}$, the $jth$ diagonal element is $c_j h - c_j \sum_{i=1}^{j-1} p_i - \sum_{i=1}^{j-1} c_i p_i - 2c_j p_j - m_j + f$.

System (1) has the trivial equilibrium point at $p_i = 0, \forall i$ with the Jacobian at the equilibrium:

$$J_0 = \begin{bmatrix} c_1 h - m_1 + f & 0 & \cdots & 0 \\ 0 & c_2 h - m_2 + f & & \vdots \\ \vdots & & \ddots & 0 \\ 0 & \cdots & 0 & c_n h - m_n + f \end{bmatrix}$$

From the structure of $J_o$, the linearized system is decoupled. Therefore, to guarantee the survival of the $ith$ species, it is sufficient to show that there exists a time $T \geq 0$ in which the $ith$ term of the diagonal of $J_o$ (or the $ith$ eigenvalue of $J_o$) is positive for all $t \geq T$. However, it does not mean that the species having negative eigenvalues for all $t \geq 0$ will be extinct since the associated solution of (1) might not converge to the origin depending on the initial condition. There are several special cases worth exploiting explicitly for the trivial equilibrium case:

Case 1: $c_i h - m_i + f(t) = \lambda_i$ case when $\lambda_i$ are constant and $h(t)$ is a solution to the equation:

$$h' + k_i h = c_i h^2$$

where $k_i = \lambda_i + m_i$ are constant.

Solving the above differential equation yields

$$h = k_i / \left( c_i + \hat{c} k_i e^{-k_i t} \right),$$

where $\hat{c} = \frac{1}{h(0)} - \frac{c_i}{k_i}$. Hence, in order for $h = \frac{k_i}{c_i + \hat{c} k_i e^{-k_i t}}, \forall i$, we must have $k_i = k$ and $c_i = c$ for some constants $k$ and $c$. Thus, in this case the habitat is equally distributed among species and $\lambda_i = k - m_i$. Hence, for any $i$, if $\lambda_i > 0$, the *ith* species survives the habitat destruction.

**Remark 1.** *For constant h, i.e., $f(t) = 0$, we have that $\lambda_i = c_i h - m_i$ is constant and is a special case to this case.*

Case 2: A more general $c_i h - m_i + f(t)$ case. In order for the *ith* species to survive the habitat loss, $c_i h - m_i + f(t)$ needs to be positive, $\forall t > T$, for some large enough T.

**Remark 2.** *We assume in this study that $c_{i+1} > c_i$, $\forall i \in \{1, 2, \ldots, n\}$. Consider the ith and $(i+1)^{th}$ diagonal elements in the Jacobian $J_0$, corresponding to the two adjacent species ith and $(i+1)^{th}$,*

$$c_{i+1} h - m_{i+1} + f(t) > c_i h - m_i + f(t) \text{ iff } h > \frac{m_{i+1} - m_i}{c_{i+1} - c_i}$$

Hence, if $c_i h - m_i + f(t) > 0, \forall t > \text{T}$ and $h > \frac{m_{i+1} - m_i}{c_{i+1} - c_i}$, the $(i+1)th$ species will survive as well as the *ith* species.

A special case when $m_i = m$, $\forall i \in \{1, 2, \ldots, n\}$ studied in Tilman et al. [2,3] and Liu et al. [23], if the *ith* species survives, all other higher ranked species will also survive the habitat destruction.

Case 3: $h(t)$ is a monotone decreasing function. If the *ith* diagonal element of the Jacobian matrix $J_0$, $c_i h - m_i - f > 0$, the *ith* species survive the habitat loss. Since $h(t) > 0$, we have $h' < m_i h - c_i h^2$. While $h$ is decreasing as the habitat is being destroyed, we have that $h' < 0$. As a consequence, the *ith* species will survive the habitat loss if $h < \frac{c_i}{m_i}, \forall t > T$, for some large enough T. In particular, if $\lim_{t \to \infty} h' = 0$ and $\lim_{t \to \infty} h = h_\infty$ we have that the *ith* species will survive so long as $h_\infty < \lim_{t \to \infty} \frac{c_i}{m_i}$. We will illustrate this case in the simulation.

### 2.1. Finding a Nontrivial Equilibrium Point

For Complete Survival Equilibrium: $p_i^* \neq 0, \forall i$ with $f(t) = 0$, the abundance equilibrium in a recursive form shown in Tilman et al., (1994) [2] is

$$p_i^* = h - \frac{m_i}{c_i} - \sum_{j=1}^{i-1} p_j^* \cdot \left( 1 + \frac{c_j}{c_i} \right), i = 2, \ldots, n,$$

where $p_1^* = h - \frac{m_1}{c_1}$,

In general, if $f(t) \neq 0$, the equilibrium, $p_{fi}^*$, can be written as

$$p_{fi}^* = p_i^* + (-1)^i f(t) \prod_{j=1}^{\lceil \frac{i}{2} \rceil} c_{2j-2} / \prod_{j=1}^{\lceil \frac{i}{2} \rceil} c_{2j-1}, \quad i = 1, \ldots, n \tag{2}$$

where $p_i^*$ is the equilibrium when $f(t) = 0$, written explicitly in Pongvuthithum and Likasiri [25], are governed by the following equations:

$$p_{2i}^* = (-1)^{2i+1} h \frac{\prod_{j=1}^{2i-1} c_j}{\prod_{j=2}^{2i} c_j} + \sum_{k=1}^{i} (m_{2k-1} - m_{2k}) \frac{\prod_{j=2k+1}^{2i-1} c_j}{\prod_{j=2k}^{2i} c_j} + \sum_{k=1}^{i} (m_{2k-1} - m_{2k}) \frac{\prod_{j=2k}^{2i-2} c_j}{\prod_{j=2k-1}^{2i-1} c_j}, \quad (3)$$

and

$$p_{2i+1}^* = h \frac{\prod_{j=1}^{2i-1} c_j}{\prod_{j=2}^{2i} c_j} + \sum_{k=1}^{i} (-m_{2k-1} + m_{2k}) \frac{\prod_{j=2k+1}^{2i-1} c_j}{\prod_{j=2k}^{2i} c_j} + \sum_{k=1}^{i} (-m_{2k-1} + m_{2k}) \frac{\prod_{j=2k}^{2i} c_j}{\prod_{j=2k-1}^{2i+1} c_j}, \quad (4)$$

where $p_1^* = h - \frac{m_1}{c_1}$.

This equilibrium point is useful in determining habitat abundance. It has been known that the equilibrium is locally stable if the eigenvalues of the Jacobian matrix $\lambda_i = c_i h - c_i \sum_{j=1}^{i-1} p_{fj}{}^* - \sum_{j=1}^{i-1} c_j p_{fj}{}^* - 2c_i p_{fi}{}^* - m_i + f < 0, \forall i$. All species will converge to the equilibrium $p_{fi}^* \neq 0$, if the initial conditions are close enough to $p_{fi}^*$. Unfortunately, finding the region of attraction of an equilibrium point for a nonlinear system is very difficult and usually impossible. As a result, we generally cannot conclude where the solution will converge analytically. However, due to the special form of the model, an analytic solution can be written explicitly, as in the next section.

### 2.2. Finding a Solution

The system (1) can be rewritten as

$$\frac{dp_1}{dt} = (c_1 h(t) - m_1 + f)p_1 - c_1 p_1^2,$$

$$\frac{dp_2}{dt} = (c_2 h(t) - (c_1 + c_2)p_1 - m_2 + f)p_2 - c_2 p_2^2,$$

$$\frac{dp_3}{dt} = (c_3 h(t) - (c_1 + c_3)p_1 - (c_2 + c_3)p_2 - m_3 + f)p_3 - c_3 p_3^2, \ldots,$$

$$\frac{dp_n}{dt} = \left( c_n h(t) - \sum_{i=1}^{n-1} (c_i + c_n)p_i - m_n + f \right) p_n - c_n p_n^2$$

Since $A_n(t) = c_n h(t) - \sum_{i=1}^{n-1} (c_i + c_n)p_i - m_n + f$ is a function of $t$, it can be viewed as a special case of the Bernoulli equation in the following form:

$$\frac{dp_n}{dt} = A_n(t)p_n - c_n p_n^2, \forall n$$

where $A_1(t) = c_1 h(t) - m_1 + f$ and $A_n(t) = c_n h(t) - \sum_{i=1}^{n-1} (c_i + c_n)p_i - m_n + f, \forall n = 2, \ldots,$ whose solution can be solved analytically and recursively as

$$\frac{1}{P_n(t)} = \frac{1}{P_n(0)} \exp\left( -\int_0^t A_n(\tau)d\tau \right) + c_n \int_0^t \exp\left( \int_t^{\hat{\tau}} A_n(\tau)d\tau \right) d\hat{\tau}, \forall n \quad (5)$$

## 3. Varieties of Habitat Destruction and Their Simulation Results

Three special cases of $h(t)$ are considered in this study.

### 3.1. Basic Continuous Destruction

To illustrate the case when $h(t)$ is a decreasing function or the habitat is continuously being destroyed, we use the exponentially decay function for the remaining habitat, which is defined as $h(t) = h_\infty - (h_\infty - 1) \cdot e^{-t}$. Shown in Figure 1, the final ratio of habitat destruction, $h_\infty = \lim_{t\to\infty} h(t)$, is equal to 0.20, which is to say 20% of the habitat remains after the destruction.

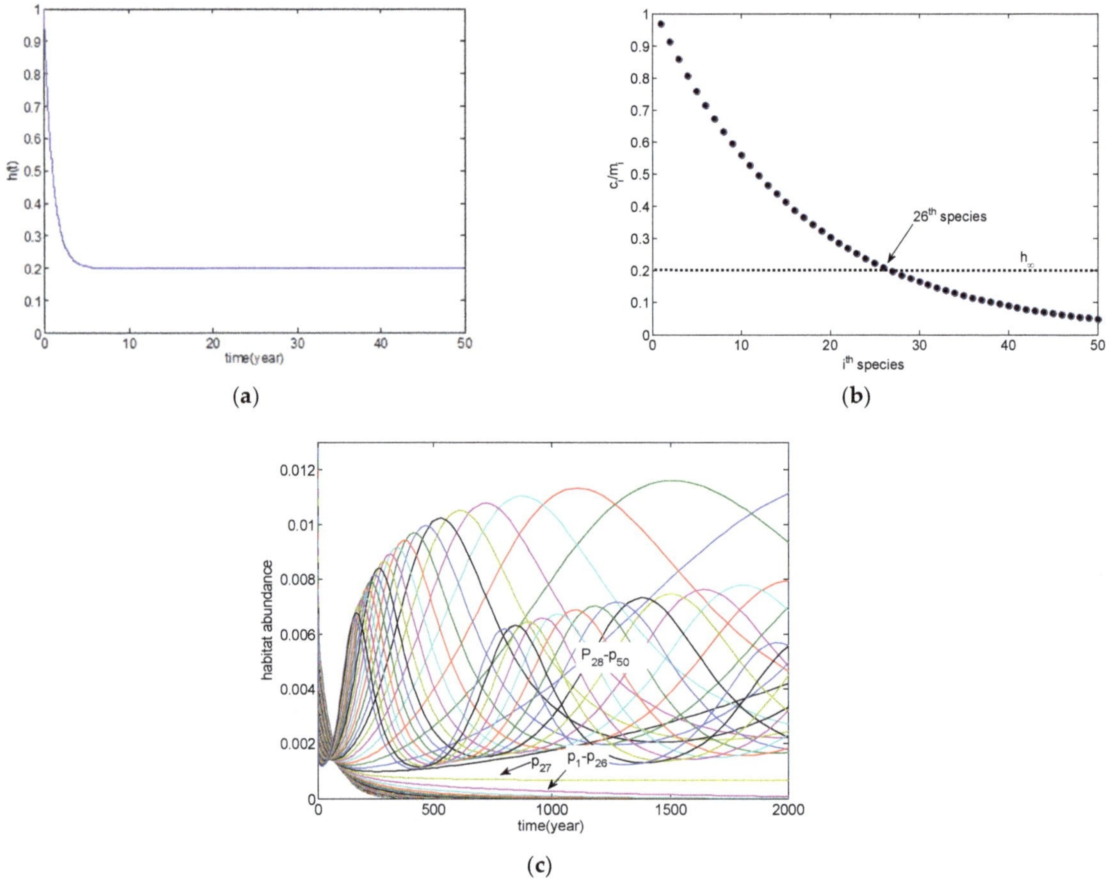


**Figure 1.** (**a**) shows the habitat destruction function when $h(t) = \mathrm{e}^{-\mathrm{t}}$ and $h_\infty = 0.20$. (**b**) compares $h_\infty = \lim\limits_{t\to\infty} h(t)$ and the ratio $\frac{c_i}{m_i}$ of species $i$. In this particular example, the ratio $\frac{c_i}{m_i} > h_\infty$ for all $i \geq 27$. The ratio (**c**) shows the proportional abundance of all 50 species. The first 26 species become extinct whereas the remaining 24 species survive the habitat loss.

### 3.2. Sudden Habitat Destruction

Habitat destruction such as flooding or earthquake can be described using a unimodal function where the remaining habitat shown in Figure 2 can be written as $h(t) = 1 - 0.94(4)\left[e^{-t} - e^{-2t}\right] - 0.1\left[1 - e^{-t}\right]$. We assume that the habitat is destroyed by a natural disaster until there is only 1% of the habitat remaining, and then it is restored back to 90% of the original level.

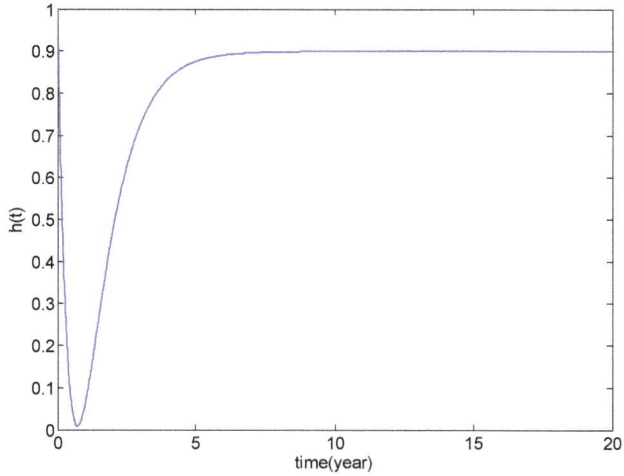


**Figure 2.** The function of the remaining habitat governed by $h(t) = 1 - 0.94(4)\left[e^{-t} - e^{-2t}\right] - 0.1\left[1 - e^{-t}\right]$ when $h_\infty = 0.90$.

Using the parameters given in Tilman et al., (1994) [2], we consider the case where all species experience the same loss, i.e., $m_i = m = 0.02, \forall i$, and $c_i = m/(1-q)^{2i-1}$, where $q = 0.03$. These parameters will ensure the competitiveness assumed in this study, that is, the $(i+1)th$ species has a better ratio of colonization compared to the $ith$ species.

In this case, $\lim_{t\to\infty} h = h_\infty = 0.90$. The analysis shows that the $ith$ species will survive so long as $h^* > \lim_{t\to\infty} \frac{c_i}{m_i}$.

To capture this case, we then simulate the proportional abundance of species 1–50 using a unimodal function giving $h(t) = 1 - 0.94(4)\left[e^{-t} - e^{-2t}\right] - 0.1\left[1 - e^{-t}\right]$ as shown in Figure 2 and $h_\infty = 0.90$ or 10% of the habitat has been completely destroyed. Simulations of Equation (1) in this case are shown in the Figure 3a–d.

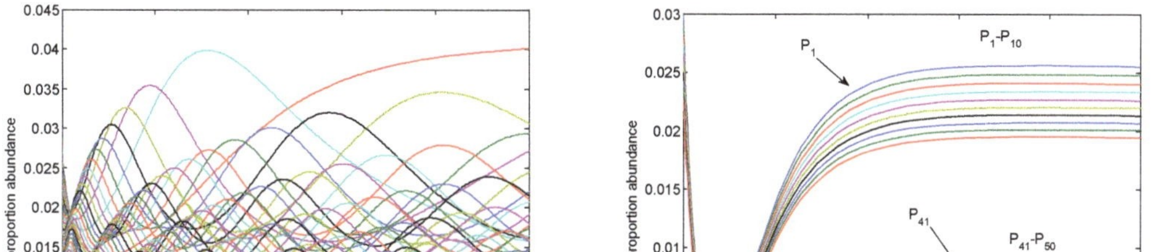


(**a**) (**b**)

**Figure 3.** *Cont.*

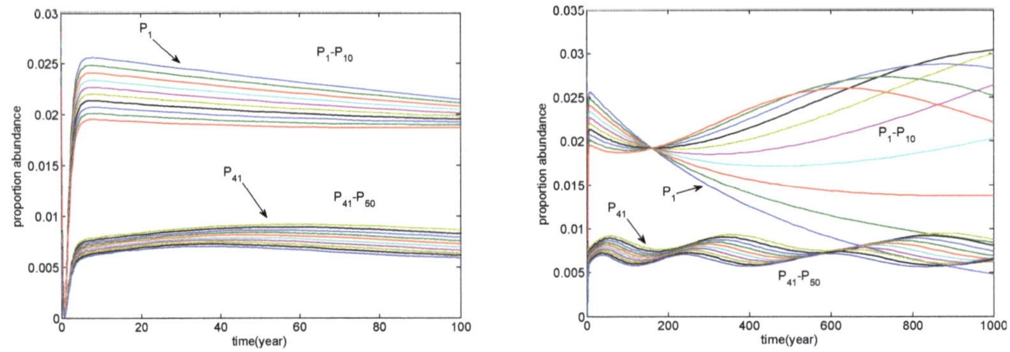


(c) (d)

**Figure 3.** Proportional abundance of 50 species when $h(t) = 1 - 0.94(4)\left[e^{-t} - e^{-2t}\right] - 0.1\left[1 - e^{-t}\right]$, 10% of the habitat is completely destroyed. The simulation shows 48 out of 50 species converging to the second equilibrium (2.2) in (**a**). Species 1 and 2 converge to the trivial equilibrium; therefore, they both will become extinct while the remaining 48 species will survive the habitat loss. (**b–d**) shows the proportional abundance of species 1–10 and 41–50 of the first 10 years, 100 years, and 1000 years, respectively. Better competitors approach their equilibria faster and they oscillate closer around the equilibria. The simulation closely shows how the proportional abundances of species 41–50 oscillate around their equilibriums.

### *3.3. Sudden Habitat Destruction with Aftershock*

Habitat destruction with aftershock such as flooding or earthquake with a smaller amplitude aftershock can be described using a bimodal function. Here, the remaining habitat shown in Figure 4 is governed by $h(t) = 1 - 0.935(4)\left(e^{-t} - e^{-2t}\right) - \frac{0.375}{\left(1+5(t-5)^2\right)} - 0.1\left(1 - e^{-t}\right)$ where we assume that the habitat is first reduced to only 1% and then naturally restored. There is one aftershock, which destroys the habitat to 50%, which happens as the habitat is being restored to 80%. Eventually, only 10% of the habitat is completely destroyed.

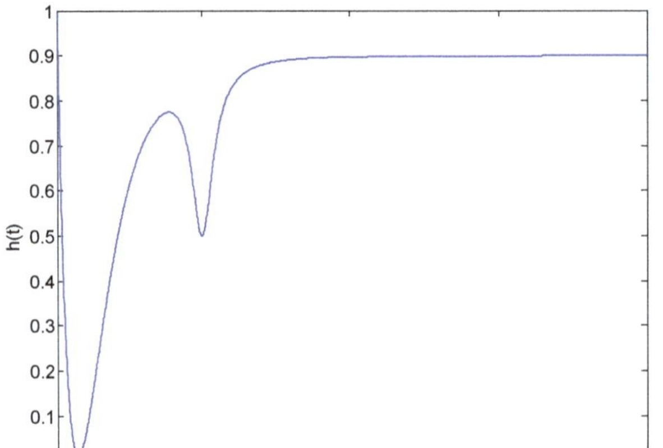


**Figure 4.** The remaining habitat function of sudden habitat destruction with one after shock governed by $h(t) = 1 - 0.935(4)\left(e^{-t} - e^{-2t}\right) - \frac{0.375}{\left(1+5(t-5)^2\right)} - 0.1\left(1 - e^{-t}\right)$ where $h_\infty = 0.90$.

To capture this case, we simulate the proportional abundance of species 1–50 using $h(t)$ as shown in Figure 4 and $h_\infty = 0.90$ or 10% of the habitat having been completely destroyed. Simulations of Equation (1) in this case are shown in Figures 5 and 6. While Figure 5a–d show 10,000-year simulations of specific groups of species, Figure 6a–d capture the first 10, 100, 1000 and 2000 years of species 1–10 and 41–50, respectively. Note that all other parameters are also taken from Tilman et al., (1994) [2].

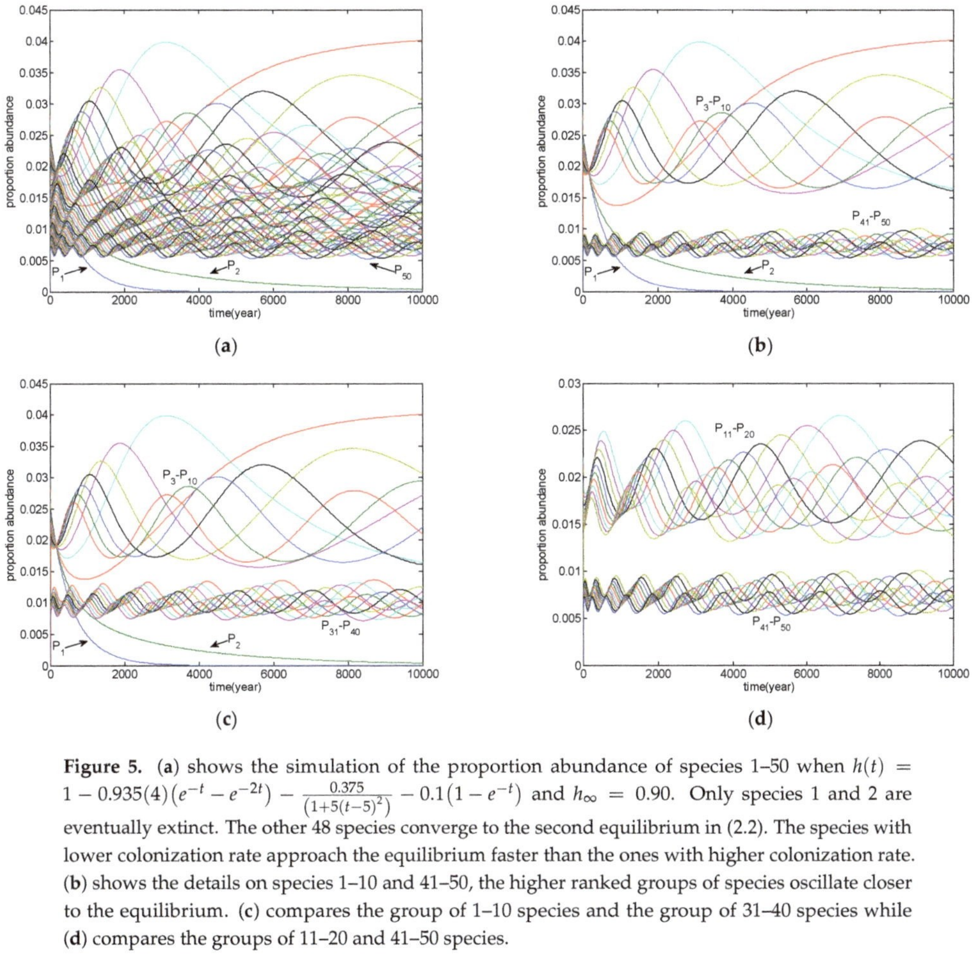


**Figure 5.** (**a**) shows the simulation of the proportion abundance of species 1–50 when $h(t) = 1 - 0.935(4)(e^{-t} - e^{-2t}) - \frac{0.375}{(1+5(t-5)^2)} - 0.1(1 - e^{-t})$ and $h_\infty = 0.90$. Only species 1 and 2 are eventually extinct. The other 48 species converge to the second equilibrium in (2.2). The species with lower colonization rate approach the equilibrium faster than the ones with higher colonization rate. (**b**) shows the details on species 1–10 and 41–50, the higher ranked groups of species oscillate closer to the equilibrium. (**c**) compares the group of 1–10 species and the group of 31–40 species while (**d**) compares the groups of 11–20 and 41–50 species.

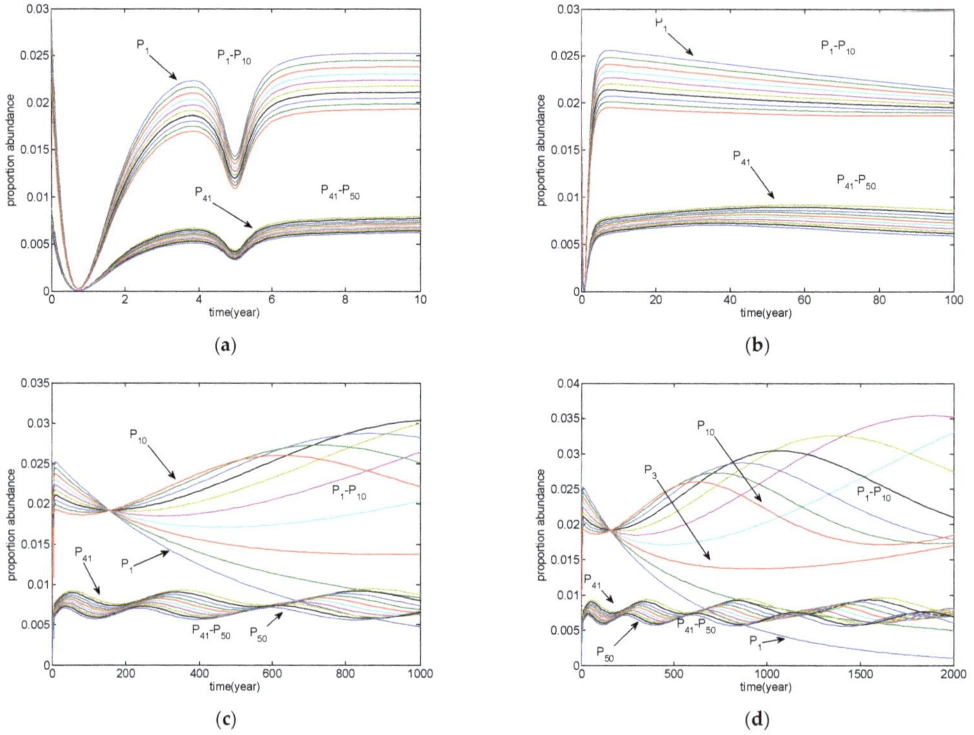


**Figure 6.** Proportional Abundance of 50 species when $h(t) = 1 - 0.935(4)\left(e^{-t} - e^{-2t}\right) - \frac{0.375}{\left(1+5(t-5)^2\right)} - 0.1\left(1 - e^{-t}\right)$ and $h_\infty = 0.90$, where 10% of the habitat cannot be restored. The 10-year simulation is shown in (**a**), where we can see species 1 has the maximum occupancy rate. (**b**) shows a 100-year simulation, where lower ranked species still show high occupancy rate, especially species 1 and 2, appearing in the top two. (**c**) shows 1000-year simulation where we start to see species 1 and 2 dropping from the top two and species 1 eventually becoming minimal near the end of the simulation. (**d**) shows 2000 years simulation; here species 1 and 2 are approaching their zero equilibrium.

In the analyses, we show in Remark 2 that the survival of the *ith* species can be guaranteed by the sign of its corresponding eigenvalue. That is, it is sufficient for a species to survive if there exists a time $T \geq 0$ in which its corresponding eigenvalue is positive for all $t \geq T$. We have plotted the eigenvalues of the system in the previous simulation to illustrate this. Figure 7a shows the corresponding eigenvalues of all 50 species over 10 years. A closer look at Figure 7b shows that the eigenvalues of species 1–3 are negative for some time $T \geq 0$. However, species 3 does not become extinct. As mentioned earlier in the analysis, having a negative eigenvalue for all $t \geq 0$ does not necessarily mean that the species will become extinct since the associated solution of (1) might not converge to the origin depending on the initial condition.

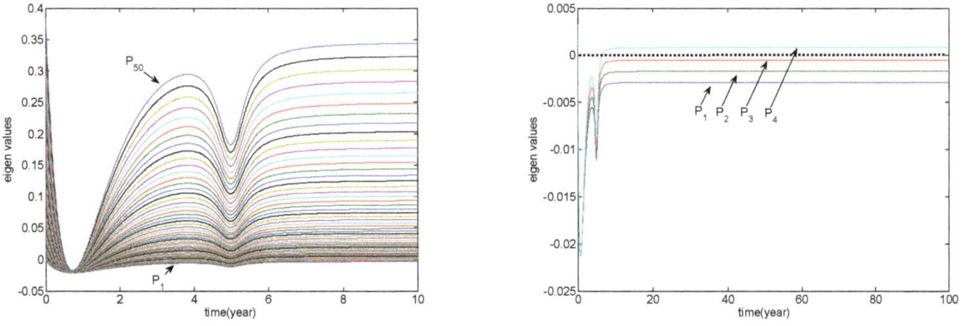


(**a**) (**b**)

**Figure 7.** (**a**) shows corresponding eigenvalues for all 50 species over 10 years. (**b**) shows the eigenvalues of species 1-4 over 100 years. We can see that species 1-3 have negative eigenvalues for all $t \geq 0$.

## 4. Summary and Discussion

We study the generalized N-species competitive coexistence model proposed in Pongvuthithum and Likasiri (2010) [25]. Due to the specific form of the model, each equation can be reduced to the Bernoulli equation where we show how to construct a solution in integral form recursively. The model considered is a smooth direct effect on habitat destruction on the abundance and extinction of species in the system. We analyze the behaviors of survival and extinction species and predict the survival species by using the corresponding eigenvalues of the Jacobian matrix, and the equilibrium and stability of the model.

Special cases of continuous habitat destruction and sudden habitat destruction, such as flooding and earthquakes with and without aftershocks, are discussed numerically. In the simulations, we can see that we can predict the survival of a species by its corresponding eigenvalue. If there exists a time $T \geq 0$ in which its corresponding eigenvalue is positive for all $t \geq T$, that species' survival of the habitat loss is abundant. However, corresponding negative eigenvalues for all $t \geq 0$ of a specific species does not predict the extinction of a species since the solution of the system may or may not converge to its equilibrium depending on the initial condition.

**Author Contributions:** Conceptualization, R.P. and C.L.; methodology, K.C. and E.D.; software, R.P.; validation, E.D. and K.C.; formal analysis, R.P.; investigation, E.D.; resources, E.D.; data curation, K.C.; writing—original draft preparation, C.L.; writing—review and editing, C.L.; visualization, R.P.; supervision, C.L.; project administration, C.L.; funding acquisition, K.C. All authors have read and agreed to the published version of the manuscript.

**Funding:** This research funded by Research Group in Mathematics and Applied Mathematics, Department of Mathematics, Faculty of Science, Chiang Mai University.

**Institutional Review Board Statement:** Not applicable.

**Informed Consent Statement:** Not applicable.

**Data Availability Statement:** Not applicable.

**Acknowledgments:** This research is supported by the Research Group in Mathematics and Applied Mathematics, Department of Mathematics, Faculty of Science, Chiang Mai University; and School of Science, King Mongkut's Institute of Technology Ladkrabang. The authors would like to thank Wiriya Sungkhaniyom for her proofreading help.

**Conflicts of Interest:** The authors declare no conflict of interest.

## References


1. Nee, S.; May, R.M. Dynamics of metapopulations: Habitat destruction and competitive coexistence. *J. Anim. Ecol.* **1992**, *61*, 37–40. [CrossRef]
2. Tilman, D.; May, R.M.; Lehman, C.L.; Nowak, M.A. Habitat destruction and the extinction debt. *Nature* **1994**, *371*, 65–66. [CrossRef]
3. Tilman, D.; Lehman, C.L.; Yin, C. Habitat destruction, dispersal, and deterministic extinction in competitive communities. *Am. Nat.* **1997**, *149*, 407–435. [CrossRef]
4. Klausmeier, C.A. Extinction in multispecies and spatially explicit models of habitat destruction. *Am. Nat.* **1998**, *152*, 303–310. [CrossRef] [PubMed]
5. Bascompte, J.; Solé, R.V. Effects of habitat destruction in a prey–predator metapopulation model. *J. Theor. Biol.* **1998**, *195*, 383–393. [CrossRef] [PubMed]
6. Swihart, R.K.; Feng, Z.; Slade, N.A.; Mason, D.M.; Gehring, T.M. Effects of habitat destruction and resource supplementation in a predator–prey metapopulation model. *J. Theor. Biol.* **2001**, *210*, 287–303. [CrossRef] [PubMed]
7. Keitt, T.H.; Johnson, A.R. Spatial heterogeneity and anomalous kinetics: Emergent patterns in diffusion-limited predatory-prey interaction. *J. Theor. Biol.* **1995**, *172*, 127–139. [CrossRef]
8. Hanski, I.; KorpimÄki, E. Microtine rodent dynamics in northern Europe: Parameterized models for the predator-prey interaction. *Ecology* **1995**, *76*, 840–850. [CrossRef]
9. De Roos, A.M.; McCauley, E.; Wilson, W.G. Pattern formation and the spatial scale of interaction between predators and their prey. *Theor. Popul. Biol.* **1998**, *53*, 108–130. [CrossRef]
10. Prakash, S.; de Roos, A.M. Habitat destruction in a simple predator–prey patch model: How predators enhance prey persistence and abundance. *Theor. Popul. Biol.* **2002**, *62*, 231–249. [CrossRef]
11. Nee, S.; Hassell, M.P.; May, R.M. Two-species metapopulation models. In *Metapopulation Biology*; Elsevier: Amsterdam, The Netherlands, 1997; pp. 123–147.
12. Prakash, S.; de Roos, A.M. Habitat destruction in mutualistic metacommunities. *Theor. Popul. Biol.* **2004**, *65*, 153–163. [CrossRef] [PubMed]
13. Ovaskainen, O. Habitat destruction, habitat restoration and eigenvector–eigenvalue relations. *Math. Biosci.* **2003**, *181*, 165–176. [CrossRef]
14. Nakagiri, N.; Tainaka, K.-I.; Tao, T. Indirect relation between species extinction and habitat destruction. *Ecol. Model.* **2001**, *137*, 109–118. [CrossRef]
15. Lin, Z.-S.; Qi, X.-Z.; Li, B.-L. Can best competitors avoid extinction as habitat destruction? *Ecol. Model.* **2005**, *182*, 107–112. [CrossRef]
16. Lin, Z.-S. The ecological order of persisting species during habitat destruction. *Ecol. Model.* **2005**, *184*, 249–256. [CrossRef]
17. Nakagiri, N.; Tainaka, K.-I. Indirect effects of habitat destruction in model ecosystems. *Ecol. Model.* **2004**, *174*, 103–114. [CrossRef]
18. Rybicki, J.; Hanski, I. Species–area relationships and extinctions caused by habitat loss and fragmentation. *Ecol. Lett.* **2013**, *16*, 27–38. [CrossRef]
19. Rompré, G.; Robinson, W.D.; Desrochers, A.; Angehr, G. Predicting declines in avian species richness under nonrandom patterns of habitat loss in a Neotropical landscape. *Ecol. Appl.* **2009**, *19*, 1614–1627. [CrossRef]
20. Dri, G.F.; Fontana, C.S.; de Sales Dambros, C. Estimating the impacts of habitat loss induced by urbanization on bird local extinctions. *Biol. Conserv.* **2021**, *256*, 109064. [CrossRef]
21. Xu, D.; Feng, Z.; Allen, L.J.; Swihart, R.K. A spatially structured metapopulation model with patch dynamics. *J. Theor. Biol.* **2006**, *239*, 469–481. [CrossRef]
22. Morozov, A.; Li, B.-L. Abundance patterns in multi-species communities exposed to habitat destruction. *J. Theor. Biol.* **2008**, *251*, 593–605. [CrossRef]
23. Liu, H.; Lin, J.; Zhang, M.; Lin, Z.; Wen, T. Extinction of poorest competitors and temporal heterogeneity of habitat destruction. *Ecol. Model.* **2008**, *219*, 30–38. [CrossRef]
24. Chen, L.-L.; Hui, C. Habitat destruction and the extinction debt revisited: The Allee effect. *Math. Biosci.* **2009**, *221*, 26–32. [CrossRef] [PubMed]
25. Pongvuthithum, R.; Likasiri, C. Analytical discussions on species extinction in competitive communities due to habitat destruction. *Ecol. Model.* **2010**, *221*, 2634–2641. [CrossRef]

entropy

MDPI

Article

# Dynamics of Information Flow between the Chinese A-Share Market and the U.S. Stock Market: From the 2008 Crisis to the COVID-19 Pandemic Period

**Chun-Xiao Nie [1,2,*] and Jing Xiao [1]**

[1] School of Statistics and Mathematics, Zhejiang Gongshang University, Hangzhou 310018, China
[2] Collaborative Innovation Center of Statistical Data Engineering, Technology & Application, Zhejiang Gongshang University, Hangzhou 310018, China
* Correspondence: niechunxiao2009@163.com

**Abstract:** The relationship between the Chinese market and the US market is widely concerned by researchers and investors. This paper uses transfer entropy and local random permutation (*LRP*) surrogates to detect the information flow dynamics between two markets. We provide a detailed analysis of the relationship between the two markets using long-term daily and weekly data. Calculations show that there is an asymmetric information flow between the two markets, in which the US market significantly affects the Chinese market. Dynamic analysis based on weekly data shows that the information flow evolves, and includes three significant periods between 2004 and 2021. We also used daily data to analyze the dynamics of information flow in detail over the three periods and found that changes in the intensity of information flow were accompanied by major events affecting the market, such as the 2008 financial crisis and the COVID-19 pandemic period. In particular, we analyzed the impact of the S&P500 index on different industry indices in the Chinese market and found that the dynamics of information flow exhibit multiple patterns. This study reveals the complex information flow between two markets from the perspective of nonlinear dynamics, thereby helping to analyze the impact of major events and providing quantitative analysis tools for investment practice.

**Keywords:** stock market; transfer entropy; information flow; surrogates

## 1. Introduction

Both China and the United States are large economies with high influence, so analyzing the relationship between the two stock markets is of great significance for investment practice and risk management. Currently, there are many studies analyzing the relationship between the two markets based on different datasets, in which multiple empirical tools are used. For example, Goh et al. found that some economic variables in the United States can improve the forecast performance of the Chinese stock market [1]. Jian Chen et al. found that US economic variables can help predict the monthly volatility of the Chinese market [2]. In addition, previous studies have also shown that US economic policy uncertainty can significantly explain the returns of China's A-share market [3]. Researchers not only pay attention to how economic variables and policies in the US market affect the Chinese market but also directly discuss the relationship between the returns of the two markets. These studies focus on the impact of special events on the relationship between markets, such as the 2007–2009 crisis [4–10], the Eurozone crisis [11], and the COVID-19 pandemic [12–14]. Empirical analysis shows that the financial crisis has had a significant impact on the BRICS and emerging markets, such as leading to significant risk spillover effects and risk contagion [5–7]. In particular, the Chinese market was significantly affected by the US market during the crisis [4–6,8,10,11]. These studies used many types of quantitative tools, such as multivariate GARCH models [4,6,10], Granger causality test [5], latent factor models [9], correlation contagion tests [8], etc. In particular, recently, the COVID-19 pandemic has significantly impacted the Chinese market and the US market [12]. For example, research reveals that China's stock market and the US stock market have significant spillover effects during the pandemic [13]. In addition, the pandemic has also impacted the correlation structure between market indexes [14].

Existing studies have revealed that there are extensive links between markets, and the Chinese market is significantly affected by the US market, especially under the impact of major events. However, previous studies have not fully analyzed the long-term relationship between the two markets. For example, most studies only analyze the static relationship in a specified period but do not focus on dynamics. This study analyzes the long-term relationship between the two markets from the perspective of nonlinear information flow dynamics. We use transfer entropy ($TE$) to quantitatively characterize nonlinear relationships [15], and analyze the dynamics in detail by surrogate time series [16]. Surrogates is an effective technique for analyzing nonlinear time series, which generates simulated time series by keeping some characteristics of the original time series [17–19].

At present, transfer entropy has expanded into a "toolbox" that includes multiple tools. For example, Marschinski et al. introduced effective transfer entropy by shuffling the source time series, where the shuffling process is used to destroy the correlation [20]. Further, the group transfer entropy and the effective group transfer entropy can be constructed by considering multiple time series [21]. Jizba et al. constructed the Rényian transfer entropies by extending Shannon entropy to Rényi entropy [22]. Papana et al. extended the transfer entropy method to analyze non-stationary time series [23]. Recently, Nie proposed a method to detect local information flow [16]. This method obtains the baseline $TE$ value from local surrogates so that it is possible to observe the contribution of observations within a small computational window to the original $TE$ value. An advantage of this method is that changes in the intensity of the information flow can be clearly observed, so that we can identify the impact of events on the information flow [16].

Since transfer entropy is a non-parametric method that effectively extracts the causal relationship between variables, it is suitable for analyzing high-complexity financial time series. For example, methods based on transfer entropy have been widely used to analyze financial time series, such as the relationship between stock markets [24–28], the price—volume relationship [16,29–31], and the foreign exchange market [32]. In particular, early empirical research shows that the US market is a core source of information for global financial markets [28]. Kim et al. considered 10 important indexes and found that the structure of the information transfer network was affected by the crisis [33].

This study focuses on analyzing the evolutionary characteristics of the information flow between China's A-share market and the US market. We use nonlinear analysis methods to characterize the complex information flow dynamics between the two markets and, in particular, identify periods of high information flow intensity by $TE$ and $LRP$. This study reveals the details of information transfer between markets based on long-term data and from multiple time scales. In the following, we first describe the data set and analyze basic descriptive statistics. Secondly, we review the concept and calculation method of transfer entropy, and the steps of $LRP$-based analysis. Third, we use weekly data and daily data to analyze the information flow between the two markets. Finally, we analyze the information flow between the S&P500 index and the industry indices of the Chinese A-share market.

## 2. Data

This paper uses three datasets, which include closing indices for two market composite indices and some industry indices. The composite index data includes daily and weekly closing data of S&P500 and CSI300 indices, as well as weekly closing data of some industry indices in the Chinese A-share market. All data are extracted from the iFinD database of Tonghuashun company. For the original closing index time series $P = \{P(t)\}$, we calculate the series $R = \{R(t)\}$, where $R(t) = log(P(t+1)) - log(P(t))$.

For weekly data, we use $R_{us}$ and $R_{cn}$ to represent the return series of the S&P500 index and the CSI300 Index, respectively. We considered the data from 31 December 2004 to 17 June 2022, a total of 890 observations, so that each preprocessed series includes 889 observations. Table 1 lists some descriptive statistics of the two series. In addition, the Pearson correlation coefficient of the two-time series is 0.2118, which implies a weak correlation.

**Table 1.** Descriptive statistics of preprocessed time series.

| Series | Average | Std | Skewness | Kurtosis |
|---|---|---|---|---|
| $R_{us}$ | 0.0012 | 0.0261 | −2.0557 | 23.1000 |
| $R_{cn}$ | 0.0016 | 0.0359 | −0.2955 | 2.2400 |
| $R^1_{us}$ | $-1.2752 \times 10^{-4}$ | 0.0179 | −0.4871 | 9.0371 |
| $R^1_{cn}$ | $4.3938 \times 10^{-4}$ | 0.0242 | −0.5126 | 2.0974 |
| $R^2_{us}$ | $3.9919 \times 10^{-4}$ | 0.0077 | −0.2361 | 4.2776 |
| $R^2_{cn}$ | $5.8243 \times 10^{-4}$ | 0.0161 | −0.9794 | 6.7738 |
| $R^3_{us}$ | $6.0558 \times 10^{-4}$ | 0.0138 | −1.0405 | 17.2147 |
| $R^3_{cn}$ | $2.0142 \times 10^{-4}$ | 0.0134 | −0.3052 | 3.6399 |
| $R^1_{ind}$ | 0.0007 | 0.0437 | −0.2044 | 2.0224 |
| $R^2_{ind}$ | 0.0016 | 0.0460 | −0.3658 | 2.0825 |
| $R^3_{ind}$ | 0.0007 | 0.0391 | −0.3157 | 1.7477 |

In addition to the weekly data, we also analyzed the daily data of the two indexes. To clearly analyze the impact of major events, we divide the daily data into three periods. For the preprocessed index data, we use $R^1_{cn}$ and $R^1_{us}$ to represent the data of the first period and use similar symbols for other periods. The time intervals corresponding to the three data sets are 4 January 2007–31 December 2010, 2 January 2014–29 December 2017, and 2 January 2018–31 December 2021 respectively. Since the trading days of the two markets are not synchronized, we remove the missing values in the two series, leaving observations with the same time stamp. The time series of the three periods without missing values include 944, 948, and 942 observations, respectively. Table 1 lists the basic descriptive statistics of each return series. Here, we always use the rounding method to keep four decimal places.

We can find from Table 1 that all the averages are close to zero. In three of the four pairs of time series, the standard deviation of the series in the Chinese market is greater than that in the US market. In addition, all skewness values are less than zero, and in particular, all excess kurtosis values are greater than zero. Here, we report the excess kurtosis obtained by subtracting 3 from the original kurtosis value. The kurtosis values imply that all distributions have peak characteristics and are not normal.

In addition to the market composite index, we used 31 industry indices compiled by SWS Research Co., Ltd. Here, we used three of them for dynamic analysis. Table 1 lists the descriptive statistics of the return series for the three indices. The symbols $R^1_{ind}$, $R^2_{ind}$ and $R^3_{ind}$ correspond to the industry index codes, 801040.SL (Steel industry), 801180.SL (Real estate) and 801960.SL (Petrochemical industry), respectively. It can be found that the average of $R^2_{ind}$ is significantly larger than the other two indices, while the difference between the standard deviations is small. In particular, all skewness values are less than 0, and kurtosis values are greater than 0.

## 3. Method

### 3.1. Transfer Entropy

We consider two random variables $I$ and $J$, the marginal probability distribution is $p_I(\cdot)$ is $p_J(\cdot)$, and the joint distribution is $p_{I,J}(\cdot)$. We assume that the underlying dynamic structure conforms to a stationary Markov process, where $I$ and $J$ correspond to orders $k$ and $l$, respectively. For $I$, the conditional distribution of state at time $t+1$ ($i_{t+1}$) is independent of state in $t-k$ ($i_{t-k}$), that is, $p_I(i_{t+1}|i_t, \cdots, i_{n-k+1}) = p_I(i_{t+1}|i_t, \cdots, i_{n-k})$. Similarly, for $J$, $p_J(j_{t+1}|j_t, \cdots, j_{n-k+1}) = p_J(j_{t+1}|j_t, \cdots, j_{n-k})$, where $j_t$ represents the state at time $t$. The transfer entropy of $J \to I$ is defined as Equation (1) [15], where the symbols $i_t^k = (i_t, \cdots, i_{n-k+1})$ and $j_t^k = (j_t, \cdots, j_{n-k+1})$. In this article, we set $l = k = 1$.

$$TE_{J\to I} = \sum p(i_{t+1}, i_t^k, j_t^l) log \frac{p(i_{t+1}|i_t^k, j_t^l)}{p(i_{t+1}|i_t^k)} \tag{1}$$

### 3.2. The Method Used to Estimate the Transfer Entropy

For a time series, we can encode it into a time series including states and then estimate the transfer entropy. Here, we use a method proposed in previous studies to encode the time series [25,26,34], which encodes the original time series through a group of quantiles. One of the advantages of this method is that a large weight can be assigned to the tail of the distribution, thereby focusing on the influence of extreme values in the financial market [25,26]. In addition, we use the R package RTransferEntropy developed by Behrendt et al. [34].

If the time series $s = \{s(t)\}$ is coded into $n$ states, $n-1$ quantiles $Q = \{q_1, q_2, \cdots, q_{n-1}\}$ are required, as shown in Equation (2). For example, if the observation $s(t)$ is between the quantiles $q_2$ and $q_3$ of the distribution, then $s'(t) = 2$. The converted time series $s' = \{s'(t)\}$ includes $n$ states, and can be used for conditional distribution and joint distribution.

$$s'(t) = \begin{cases} 1 & s(t) \leq q_1 \\ 2 & s(t) \in (q_2, q_3) \\ \cdots & \\ n-1 & s(t) \in (q_{n-2}, q_{n-1}) \\ n & s(t) \geq q_{n-1} \end{cases} \tag{2}$$

In order to comprehensively analyze the significance of information flow at different quantile parameters, we set the following five sets of parameters: $Q_1 = \{q(0.05), q(0.5), q(0.95)\}$, $Q_2 = \{q(0.10), q(0.5), q(0.90)\}$, $Q_3 = \{q(0.15), q(0.5), q(0.85)\}$, $Q_4 = \{q(0.20), q(0.5), q(0.80)\}$ and $Q_5 = \{q(0.25), q(0.5), q(0.75)\}$, where the symbol $q(\alpha)$ represents $\alpha$-quantile. From the first to the fifth set of parameters, the weight of the extreme values of the tail decreases. In the first group, extreme values are given the greatest weight, and the distribution in the fifth group is equally divided into four parts.

Here, we use the method proposed by Dimpfl et al. to test the significance of $TE$ [25,34]. The method first estimates the dynamics between two variables through the Markov process. Secondly, based on the Markov process, a pair of simulated time series whose dependence has been eliminated can be generated, and the $TE$ value can be estimated. Third, we repeatedly generate the simulated time series $N_{boot}$ times to obtain a $TE$ value distribution for comparison. Finally, the $p$-value is generated by comparing the original $TE$ value with the quantile of the baseline distribution. In this article, we always set $N_{boot} = 1000$.

### 3.3. LRP-Based Analysis

The local random permutation ($LRP$) method uses $RP$ surrogates locally to estimate the intensity of the information flow within a period [16]. Previous studies have confirmed that this method can effectively describe changes in information flow through toy models [16]. In particular, to maintain the correlation between series, $LRP$ provides surrogates that keep the Pearson correlation unchanged.

We consider the information flow $s_1 \to s_2$ between the series $s_1 = \{s_1(t)\}$ and $s_2 = \{s_2(t)\}$, where each series includes $N$ observations. We assume that the length of the calculation window for constructing local $RP$ surrogates is $L_w$, and the $TE_{s_1 \to s_2}$ value is significant.

1 We extract the observations in the time interval $[1, L_w]$ to obtain the sub-time series $s_1^{[1,L_w]} = \{s_1(1), \cdots, s_1(L_w)\}$ and $s_2^{[1,L_w]} = \{s_2(1), \cdots, s_2(L_w)\}$, and shuffle the observations of the sub-sequences $s_1^{[1,L_w]}$ and $s_2^{[1,L_w]}$. In the scrambling step, the cor-

respondence between the timestamps of the two series does not change, that is, the series $\{(s_1(k), s_2(k))\}$ composed of pairs is scrambled.

2 We calculate the transfer entropy value between the time series obtained by local scrambling in step 1, and denoted it as $TE^{[1,L_w]}_{s_1 \to s_2}(1)$.

3 We repeat steps 1 and 2 a total of $M$ times to obtain a distribution of $TE$ values used as a benchmark ($TE^1_{ben} = \{TE^{[1,L_w]}_{s_1 \to s_2}(k) | k = 1, 2, \cdots, M\}$).

4 We calculate the Z-score $Z_{[1,L_w]} = \frac{TE - m_{[1,L_w]}}{s_{[1,L_w]}}$.

5 We move the calculation window one observation each time, and repeat steps 1–4 to get the distribution $TE^i_{ben} = \{TE^{[i,L_w]}_{s_1 \to s_2}(k) | k = 1, 2, \cdots, M\}$ and the Z-score. For example, the second Z-value is obtained by scrambling all observations in the interval $[2, L_w + 1]$. Finally, we get the Z-score sequence $\{Z_{[i,i+L_w-1]} | i = 1, 2, \cdots, N - L_w + 1\}$.

The estimation of the $TE$ value requires setting the parameter $Q_k$, so the parameter may affect the calculation result. Here, we use the following method to synthesize the calculation results of different parameters. For a sequence of Z-scores ($\{z_k(t)\}$) generated by a parameter $Q_k$, we construct a 0–1 sequence using Equation (3), where $Z_{th}$ is the threshold. We set the threshold $Z_{th} = 1.645$ here. In this way, we obtain a series $I_k = \{I_k(t) | t = 1, 2, \cdots, N - L_w + 1\}$ based on Z-scores, in which observations only take values in $\{0, 1\}$. If $TE^i_{ben}$ is a normal distribution, then the original $TE$ value is greater than most of the values (95%) in $TE^i_{ben}$, which implies that there is a significant information flow in the period $[i, i + L_w - 1]$. In this article, we always set $M = 1000$. Then, we calculate the series $I = \{I(t)\}$ ($I(t) = \sum_k I_k(t)$).

$$I_k(t) = \begin{cases} 1, & z_k(t) \geq z_{th} \\ 0, & z_k(t) < z_{th} \end{cases} \tag{3}$$

Since each Z-score corresponds to an interval, in this article, we use the following convention to plot the figure: the right end of the interval corresponding to the value is used as the time stamp in the figure. For example, for the value $Z_{[1,L_w]}$, the time label in the figure is the timestamp corresponding to $L_w$. Similarly, we plot the time series $\{I(t)\}$.

## 4. Results

### 4.1. Information Flow Analysis Based on Weekly Data

#### 4.1.1. $TE$ Values of Different Parameters

We use a group of parameters $Q_k$ ($k = 1, \cdots, 5$) to calculate the information flow between $R_{us}$ and $R_{cn}$. Table 2 shows the calculated results, all $TE_{us \to cn}$ values are significant at the significance level $\alpha = 0.05$, and three of the $p$-values are less than 0.01. In addition, all $TE_{cn \to us}$ values are not significant. Table 2 suggests that there is an asymmetric information flow between the Chinese market and the US market, and the US market is a dominant information source.

**Table 2.** Information flow analysis of long-term weekly data.

| $Q_k$ | $TE_{cn \to us}$ | $p_{cn \to us}$ | $TE_{us \to cn}$ | $p_{us \to cn}$ |
|---|---|---|---|---|
| $Q_1$ | 0.040537 | 0.054 | 0.048419 | 0.006 |
| $Q_2$ | 0.0352622 | 0.261 | 0.0519378 | 0.005 |
| $Q_3$ | 0.0381743 | 0.141 | 0.0449418 | 0.039 |
| $Q_4$ | 0.0357750 | 0.183 | 0.0492554 | 0.008 |
| $Q_5$ | 0.047230 | 0.014 | 0.044303 | 0.031 |

The minimum $p$-value of 0.005 corresponds to the parameter $Q_2$, which means that significant information flow can be detected when a large weight is assigned to extreme values. In addition, the $p$-value corresponding to the equally divided quantile interval is also less than 0.05. This means that significant $us \to cn$ information flow can be detected regardless of whether a greater weight is given to the extreme values.

In summary, there is sufficient evidence to support the existence of information flow $us \to cn$. The change of the $p$-value with the change of the parameter implies that the analysis of the information flow depends on the choice of the parameter, and thus the calculation results of multiple parameters need to be integrated.

#### 4.1.2. Information Flow Dynamics for Weekly Data

In the previous section, we analyzed the information flow between the two markets globally. However, the static analysis only shows the existence of information flow in the considered period, thus lacking local dynamic analysis. Below, we use $LRP$ to analyze the dynamics of information flow, in particular, to detect periods with localized strong information flow.

We set five sets of parameters $Q_k$ $(k = 1, \cdots, 5)$ and $L_w = 96$, where the calculation window corresponds to two years of trading weeks. The five subfigures of Figure 1 show the $Z$-score time series. We observe that all subfigures include local peaks larger than 1.645, implying non-trivial information flow dynamics between series. These peaks correspond to periods of localized strong information flow. For example, for Figure 1a, it can be seen that the $Z$-score increased rapidly in 2008 and dropped significantly thereafter. Furthermore, we can also observe that the number of peaks in all $Z$-score series is greater than 1. For example, three localized peaks can be observed in Figure 1b, implying the existence of three periods of strong information flow.

The numbers of local peaks for the five $Z$-score series are 2, 3, 2, 3, and 3, respectively. Table 3 lists the periods corresponding to these peaks. We use $T_1$, $T_2$, and $T_3$ to represent the time intervals in three periods, where each period corresponds to two endpoints. For example, the starting time corresponding to $T_1$ $(Q_1)$ is 8 June 2007, and the closing time is 24 April 2009. The first line indicates that there is a significant flow of information in the period 8 June 2007–24 April 2009. We list the $Z$-score at the corresponding time interval in the fourth column. The maximum value corresponds to the $T_1$ period, while the minimum value corresponds to the $T_2$ period.

**Table 3.** Three significant periods of the information flow of parameter.

| $Q$ | $T_i$ | $t$ | $Z$ | **Period** |
|---|---|---|---|---|
| $Q_1$ | $T_1$ | 24 April 2009 | 4.0550 | 8 June 2007–24 April 2009 |
| $Q_1$ | $T_2$ | 17 November 2017 | 2.2084 | 1 January 2016–17 November 2017 |
| $Q_2$ | $T_1$ | 9 January 2009 | 3.0700 | 2 March 2007–19 January 2009 |
| $Q_2$ | $T_2$ | 15 May 2015 | 2.5980 | 19 July 2013–15 May 2015 |
| $Q_2$ | $T_3$ | 12 October 2018 | 2.4534 | 2 December 2016–12 October 2018 |
| $Q_3$ | $T_1$ | 8 May 2009 | 2.9067 | 22 June 2007–8 May 2009 |
| $Q_3$ | $T_2$ | 17 April 2015 | 2.9681 | 21 June 2013–17 April 2015 |
| $Q_4$ | $T_1$ | 2 January 2009 | 3.6279 | 16 February 2007–2 January 2009 |
| $Q_4$ | $T_2$ | 8 November 2013 | 1.8349 | 23 December 2011–8 November 2013 |
| $Q_4$ | $T_3$ | 28 February 2020 | 3.2564 | 13 April 2018–28 February 2020 |
| $Q_5$ | $T_1$ | 9 April 2010 | 3.0202 | 23 May 2008–9 April 2010 |
| $Q_5$ | $T_2$ | 21 November 2014 | 1.8017 | 18 March 2013–21 November 2014 |
| $Q_5$ | $T_3$ | 13 March 2020 | 2.0941 | 27 April 2018–13 March 2020 |

We find that most of the local peaks correspond to periods that are close to each other, such as the $T_1$ period of parameter $Q_1$ and the $T_1$ period of $Q_2$ are 8 June 2007–24 April 2009 and 2 March 2007–19 January 2009 respectively. Table 3 shows that the 2008 crisis significantly affected the information flow of $us \to cn$. In addition, the periods corresponding to several local peaks also include major events. For example, the third peak of $Q_5$ includes the oil crisis and the collapse of the US stock market. In summary, the $LRP$-based method identifies some local peaks, indicating that the dynamics of information flow are non-trivial.

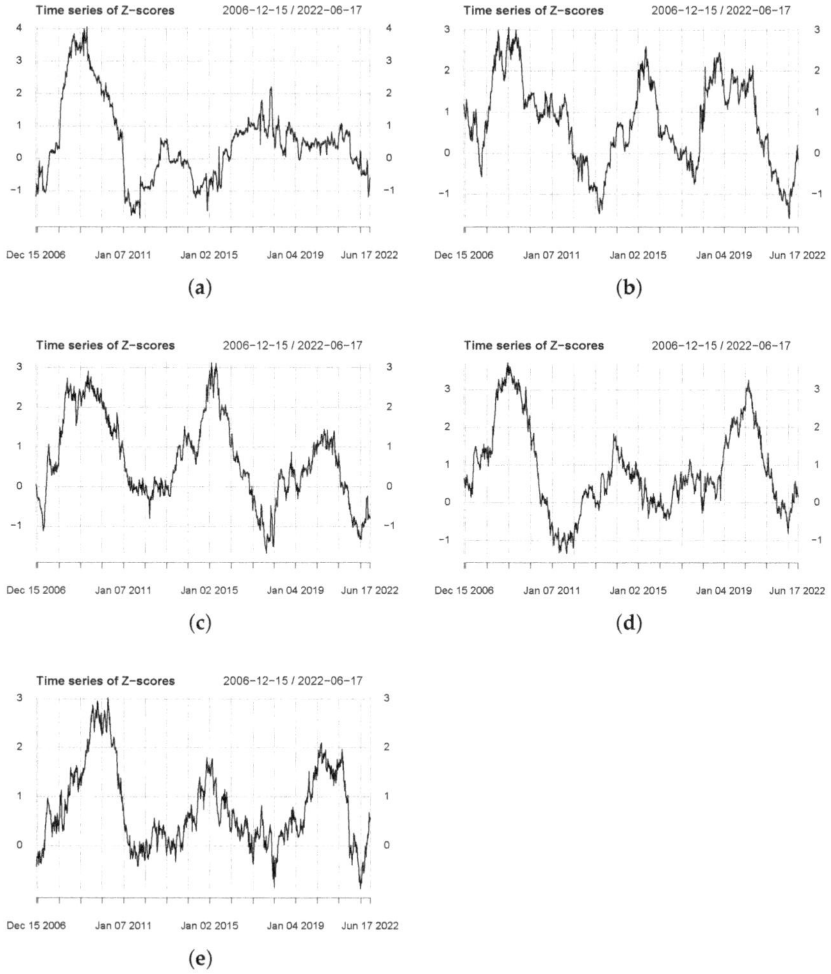


**Figure 1.** The figure shows the $Z$-scores for different parameters $Q$, where (**a–e**) correspond to $Q_1$–$Q_5$, respectively.

### 4.1.3. Comprehensive Analysis of Z-Score Series

We roughly identified three important periods by $Z$-scores in the previous subsection. Below, we calculate the $I$ series from the $I_k$ series of five $Z$-score series. The series $I$ takes the value in the set $\{0, 1, 2, 3, 4, 5\}$. If $I_k(t) = 0$, the observations in the time period $[t - L_w + 1, t]$ do not significantly affect the analysis results of the information flow. On the contrary, if $I(t) = 5$, it implies that the information flow within the period $[t - L_w + 1, t]$ is significant and does not depend on the choice of parameters. A large $I(t)$ value implies that there is information flow in multiple quantile intervals $\{Q_k | k = 1, \cdots, 5.\}$.

Figure 2 shows the $I$ series. There are three periods where most observations are greater than or equal to 1. In particular, the analysis shows that there is a period where most observations are greater than or equal to 3, that is, 8 February 2008–24 September 2010. During the 2008–2009 crisis in the US market, there was strong information flow $us \to cn$. We use $N_i$ to represent the number of observations equal to $i$ in the $I$ series, as shown in Table 4. It can be found that if the calculation results of 5 parameters are integrated, nearly half of the observations are greater than or equal to 1. This implies that information flows widely exist between the two markets during the period under consideration. Table 5 lists the three periods identified by Figure 2. Each period includes some major events that affect the market. The first period corresponds to the 2007–2009 crisis in the US market. The second period includes the crisis in the Chinese stock market in 2015. In addition, before the stock market crash in 2015, the bubble was generated in the second half of 2014. The second period not only includes the breakage period of the bubble, but also the generation period. The third period includes a series of major events, such as the circuit breaker in the US market in March 2020, and the COVID-19 pandemic. Based on this table, we divide three periods for the analysis of daily data in the next section: 2007–2010, 2014–2017, and 2008–2021.

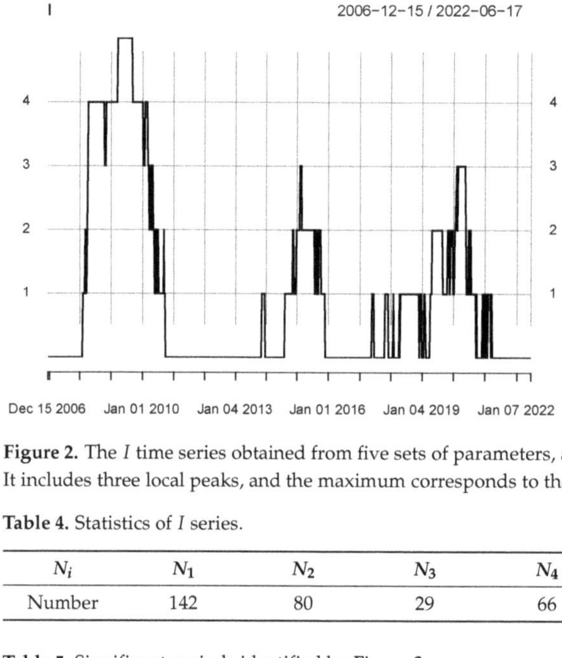


**Figure 2.** The $I$ time series obtained from five sets of parameters, and takes values in $\{0,1,2,3,4,5\}$. It includes three local peaks, and the maximum corresponds to the 2008 crisis.

**Table 4.** Statistics of $I$ series.

| $N_i$ | $N_1$ | $N_2$ | $N_3$ | $N_4$ | $N_5$ | **Sum** |
|---|---|---|---|---|---|---|
| Number | 142 | 80 | 29 | 66 | 24 | 341 |

**Table 5.** Significant periods identified by Figure 2.

| $T_i$ | **Period** |
|---|---|
| $T_1$ | 8 February 2008–24 September 2010 |
| $T_2$ | 15 August 2014–20 November 2015 |
| $T_3$ | 13 April 2018–19 March 2021 |

### 4.2. *Information Flow Analysis of Daily Data*

#### 4.2.1. Global Analysis of Daily Data

In this section, we analyze the relationship between the daily data of the two indices. Table 6 lists the transfer entropy values $TE_{us \to cn}$ of different parameters. For the first and third periods, all $TE$ values are significant at significance level $\alpha = 0.05$. For the second period, the $TE$ values of parameters $Q_1$ and $Q_2$ were not significant. However, the $p$-value for the $TE$ value of $Q_4$ was less than 0.01. This implies that extreme values have little effect on information flow in the second period. In summary, we still found significant information flows in the three periods of daily data. Below, we use $LRP$-based analysis to discuss the dynamics of information flow in each period.

**Table 6.** Analysis of $TE$ value of daily data.

| Period | 2007–2010 | 2014–2017 | 2018–2021 |
|---|---|---|---|
| $Q_1$ | 0.0573 (<0.001) | 0.0322 (0.262) | 0.0587 (<0.001) |
| $Q_2$ | 0.0490 (0.006) | 0.0286 (0.529) | 0.0630 (<0.001) |
| $Q_3$ | 0.0510 (0.002) | 0.0505 (0.002) | 0.0545 (<0.001) |
| $Q_4$ | 0.0601 (<0.001) | 0.0489 (0.006) | 0.0519 (<0.001) |
| $Q_5$ | 0.0600 (<0.001) | 0.0407 (0.041) | 0.0577 (<0.001) |

In the next subsection, for the first period and the second period, we choose the parameter $Q$ corresponding to the maximum $TE$ value in Table 6. In the analysis of the first and third periods, the parameter $Q$ was set to $Q_4$ and $Q_2$, respectively. In addition, the second period is a special period, in which both $Q_1$ and $Q_2$ correspond to insignificant $TE$ values, suggesting that extreme values in this period weakly affect the level of information flow. For the second period, we set $Q_3$ and $Q_4$ for calculation.

4.2.2. Information Flow Dynamics for 2007–2010 Data

Figure 3 shows the $Z$-score series for the 2007–2010 data. If the threshold is 1.645, the series includes three significant periods: 26 November 2007 (28 June 2007)–25 June 2009, 28 October 2009 (1 June 2009)–8 January 2010, and 24 May 2010 (21 December 2009)–19 October 2010. Here, the date in parentheses is the left endpoint of the calculation window for the first $Z$-score greater than 1.645 within the period. In particular, the maximum value (3.9880) is in the first period and the calculation window is 3 November 2008–7 April 2009.

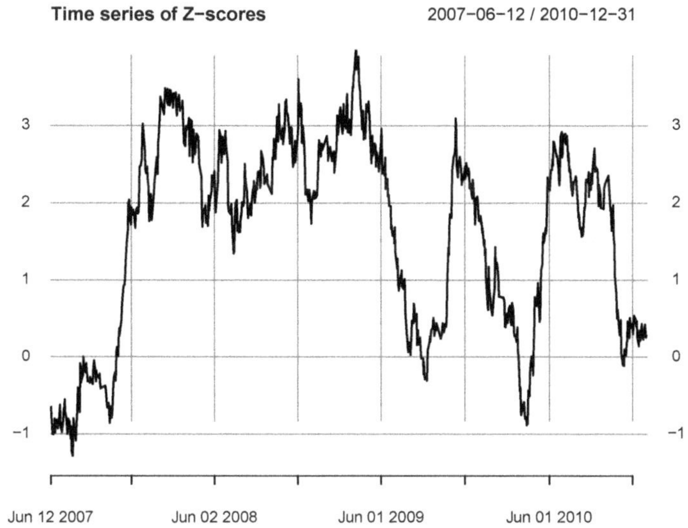


**Figure 3.** The figure shows the $Z$-score time series between $R^1_{us}$ and $R^1_{cn}$, where $L_w = 100$.

The calculation results clearly show that the 2007–2009 crisis in the US market significantly affected the Chinese market. In addition, we also observe that the $us \rightarrow cn$ information flow also exists in the post-crisis period. These results imply not one peak with strong information flow, but multiple peaks with different intensities.

#### 4.2.3. Information Flow Dynamics for 2014–2017 Data

Figure 4 shows the information flow dynamics for the 2014–2017 data. Figure 4a includes a major significant period: 6 January 2016 (6 August 2015)–10 May 2016. The results show that the information flow in the $us \rightarrow cn$ direction is enhanced after the Chinese market bubble burst. Figure 4b shows the calculation results corresponding to $Q_4$. We found that it also includes a major significant period: 30 December 2016 (1 August 2016)–1 March 2017. The period included an important event that affected US markets: the 2016 presidential election. A previous study showed that the event significantly affected correlation dynamics in the US stock market [35]. The analysis here suggests that there was a significant information flow from the US stock market to the Chinese stock market around this event. The two results in Figure 4 show that from 2 January 2014 to 29 December 2017, the US market significantly influenced the Chinese market in two periods, and both periods included major events.

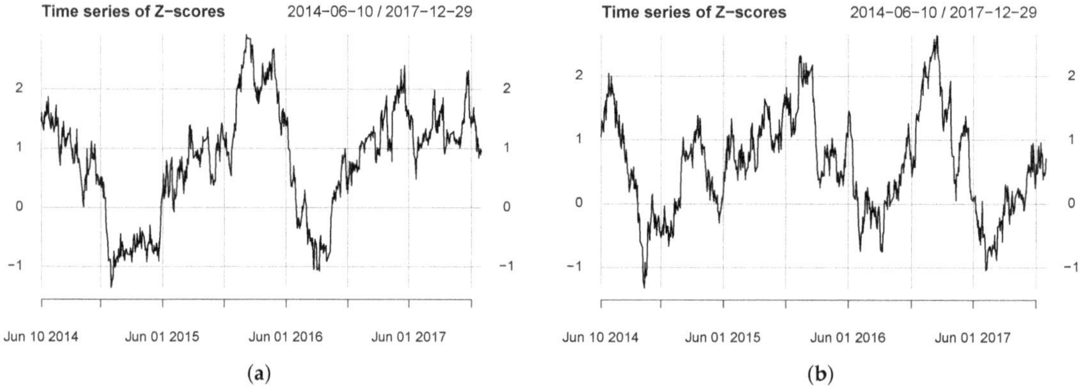


**Figure 4.** The figure shows the $Z$-score time series between $R^2_{us}$ and $R^2_{cn}$, where (**a**,**b**) correspond to parameters $Q_3$ and $Q_4$, respectively.

#### 4.2.4. Information Flow Dynamics for 2018–2021 Data

Figure 5 shows the calculation results for the 2018–2021 data. The period includes several sub-periods greater than the critical $Z$-score, suggesting that from 2 January 2018 to 31 December 2021, the flow of information between the two markets was active and strong. These sub-periods are as follows: 7 June 2018 (3 January 2018)–31 July 2018 ($T_1$), 22 February 2019 (12 September 2018)–15 May 2019 ($T_2$), 19 June 2019 (11 January 2019)–30 July 2019 ($T_3$), 27 August 2019 (29 March 2019)–5 February 2020 ($T_4$), 27 February 2020 (19 September 2019)–29 July 2020 ($T_5$), 3 March 2021 (24 September 2020)–2 August 2021 ($T_6$). We label the different subperiods with $T_i$. It can be found that $T_2$, $T_3$, $T_4$ and $T_5$ can be combined into an important period. There were several major events in this period: Sino-US trade friction, the oil crisis, the COVID-19 pandemic, the US stock market crisis in March 2020. In addition, periods $T_1$ and $T_6$ still belong to the period of Sino-US trade friction and the period of the COVID-19 pandemic, respectively. In particular, the calculation window corresponding to the maximum value (3.4339) is 21 October 2019–23 March 2020. Several major events are included in this period, such as the oil crisis, the COVID-19 pandemic, and the March 2020 market crisis. To sum up, it is reasonable to speculate that the impact of multiple major events led to the active information flow during the period.

**Figure 5.** The figure shows the $Z$-score series for $R^3_{us}$ and $R^3_{cn}$, which includes two main peaks ($Q = Q_2$, $L_w = 100$).

### 4.3. Information Flow between Industry Indices in the Chinese Market and the S&P500 Index

#### 4.3.1. Global Information Flow Analysis between Industry and Market Index

Since the composite index includes stocks belonging to various industries, in this section, we analyze the relationship between the S&P500 index and the industry index in the Chinese market. Here, we consider the case where a larger weight is given to the tail of the distribution, i.e., $Q = Q_1$. Table 7 lists the $TE$ values and $p$-values, where the symbol "ind" represents the industry index, and the code is listed in the second column.

The fourth and sixth columns show that there are several industry indices that are closely related to the S&P500 index. For example, there is a significant bidirectional flow of information between the petrochemical industry and the S&P500 index. In addition, some unidirectional effects can also be found, such as in real estate. It should be noted that not all industry indices are affected by the S&P500 index, such as the public utilities index and the architectural ornament index. To sum up, Table 7 shows that the information flow from S&P500 to Chinese market industry indices is heterogeneous.

**Table 7.** $TE$ values of industry indices and S&P500 index.

| Industry | Code | $TE_{ind \to us}$ | $p_{ind \to us}$ | $TE_{us \to ind}$ | $p_{us \to ind}$ |
|---|---|---|---|---|---|
| AFAF | 801010.SL | 0.0341 | 0.254 | 0.0444 | 0.011 |
| Basic chemicals | 801030.SL | 0.0448 | 0.012 | 0.0394 | 0.061 |
| Steel industry | 801040.SL | 0.0350 | 0.093 | 0.0686 | 0.000 |
| Nonferrous metals | 801050.SL | 0.0353 | 0.125 | 0.0475 | 0.005 |
| Electronics | 801080.SL | 0.0325 | 0.334 | 0.0413 | 0.034 |
| Household Electric Appliance | 801110.SL | 0.0280 | 0.551 | 0.0622 | 0.000 |
| Food and beverage | 801120.SL | 0.0293 | 0.491 | 0.0443 | 0.012 |
| Textile, clothing. | 801130.SL | 0.0415 | 0.055 | 0.0483 | 0.004 |
| Light manufacturing | 801140.SL | 0.0378 | 0.102 | 0.0504 | 0.000 |
| Bio-pharmaceutical industry | 801150.SL | 0.0294 | 0.499 | 0.0418 | 0.036 |
| Public utilities | 801160.SL | 0.0431 | 0.022 | 0.0359 | 0.121 |
| Transportation industry | 801170.SL | 0.0338 | 0.257 | 0.0519 | 0.001 |
| Real estate | 801180.SL | 0.0304 | 0.427 | 0.0469 | 0.008 |
| Retail business | 801200.SL | 0.0388 | 0.079 | 0.0358 | 0.154 |
| Social services | 801210.SL | 0.0407 | 0.040 | 0.0428 | 0.028 |
| Comprehensive company | 801230.SL | 0.0296 | 0.482 | 0.0476 | 0.007 |
| Building material | 801710.SL | 0.0347 | 0.213 | 0.0468 | 0.009 |

**Table 7.** *Cont.*

| Industry | Code | $TE_{ind\to us}$ | $p_{ind\to us}$ | $TE_{us\to ind}$ | $p_{us\to ind}$ |
|---|---|---|---|---|---|
| Architectural ornament | 801720.SL | 0.0393 | 0.072 | 0.0302 | 0.379 |
| Power equipment | 801730.SL | 0.0279 | 0.559 | 0.0347 | 0.160 |
| National defense industry | 801740.SL | 0.0342 | 0.239 | 0.0314 | 0.351 |
| Computer industry | 801750.SL | 0.0253 | 0.719 | 0.0383 | 0.096 |
| Media industry | 801760.SL | 0.0309 | 0.334 | 0.0484 | 0.006 |
| Communication industry | 801770.SL | 0.0423 | 0.047 | 0.0414 | 0.031 |
| Banking | 801780.SL | 0.0457 | 0.021 | 0.0333 | 0.220 |
| Non-bank Financial Institutions | 801790.SL | 0.0251 | 0.775 | 0.0372 | 0.087 |
| Automobile industry | 801880.SL | 0.0270 | 0.591 | 0.0508 | 0.004 |
| Mechanical equipment | 801890.SL | 0.0443 | 0.021 | 0.0447 | 0.017 |
| Coal industry | 801950.SL | 0.0418 | 0.031 | 0.0424 | 0.016 |
| Petrochemical industry | 801960.SL | 0.0510 | 0.001 | 0.0469 | 0.004 |
| Environmental protection | 801970.SL | 0.0282 | 0.595 | 0.0475 | 0.006 |
| Beauty care | 801980.SL | 0.0337 | 0.258 | 0.0436 | 0.017 |

AFAF : agriculture, forestry, animal husbandry, and fishery.

#### 4.3.2. The Dynamics of Information Flow between Industry Index and S&P500 Index

In the previous section, we found diversity in the relationship between the S&P500 index and industry indices, with some industry indices having a stronger flow of information between them and the S&P500 index. A related issue is whether the dynamics of information flow for different industry indices have similar patterns. Here, we select three industry indices and compare information flow dynamics. The three indices are 801040.SL (Steel industry), 801180.SL (Real estate) and 801960.SL (Petrochemical industry), and all have significant $us \to ind$ information flow.

We analyze dynamics using weekly data and always set $Q = Q_1$ and $L_w = 96$ weeks. Figure 6 shows the dynamics of information flow for $us \to 801040.SL$. The Z-score series includes 794 values, and 396 values are greater than the critical threshold, implying that there is a continuous flow of information $us \to 801040.SL$ during the period. In addition, it includes several local peaks, and the corresponding periods of the two main peaks are 18 April 2008 (2 June 2006)–15 April 2011 and 24 October 2014 (21 December 2012)–30 June 2017, respectively.

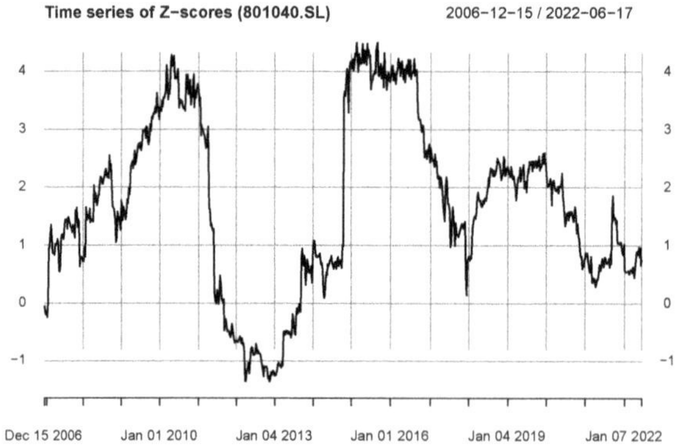


**Figure 6.** The figure shows the Z-score time series for the industry index 801040.SL (Steel industry) and the S&P500 index, including two clear local peaks.

Figure 7 presents the Z-score series for the S&P500 Index → Real estate Index. For this Z-score series, there are only 124 values greater than the critical threshold. It shows a different pattern from Figure 6, such as observations in the period 4 September 2020 (26 October 2018)–16 July 2021 contribute significantly to the information flow. In addition, another major peak corresponds to the period 25 December 2009 (15 February 2008)–24 June 2011.

**Figure 7.** The figure shows the Z-score time series for the industry index 801180.SL (Real estate) and the S&P500 index, with only a few Z-score values above the critical threshold.

The information flow analysis result of $us \rightarrow 801960.SL$ is shown in Figure 8. We find that it included a period with most values greater than the critical threshold: 10 October 2014–9 June 2017. This period is close to that of the second peak in Figure 6. However, for the series shown in Figure 8, it includes some values less than the critical threshold.

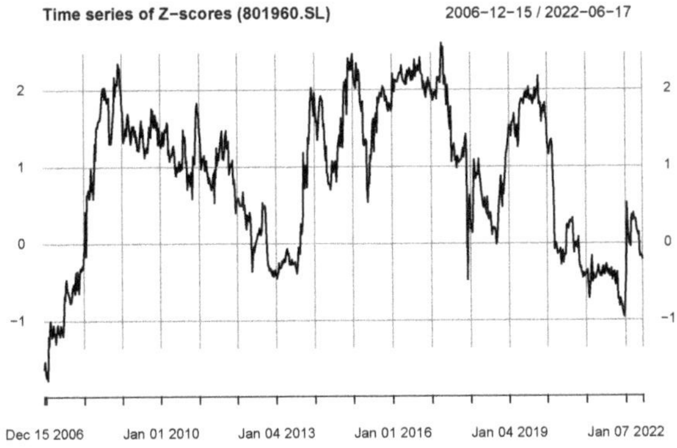


**Figure 8.** The figure shows the Z-score time series of the industry index 801960.SL (Petrochemical industry) and S&P500 index, and most of the Z-score values greater than the critical threshold are in the period 2014–2017.

The results in Figures 6–8 show that the information flow dynamics of $us \rightarrow ind$ exhibit multiple patterns, thus indicating a high complexity of information flow between markets.

## 5. Discussion

This study examines periods of strong information flow between two important markets, where these periods are accompanied by major events. This implies that long-term major events may affect market relations more deeply. This provides a forward-looking analysis for analyzing the relationship between the two markets. In particular, we can analyze the dynamics of information flow in detail to identify changes in the relationship between markets when markets are impacted by major events. A related topic that can be studied is whether non-linear information flow helps predict future returns in the market.

Several previous studies have revealed that major events affect the relationship between markets [4–14]. Unlike previous studies, we analyze long-term data and focus on the dynamics of information flow. In particular, $LRP$-based analysis characterizes the strength of the local information flow, so that the time-varying characteristics of the information flow in a period can be observed in detail.

## 6. Conclusions

We discussed the long-term relationship between the Chinese market and the US market and found that there is an asymmetric information flow, where the US market is the source of information flow. Calculations show that the information flow between the Chinese market and the US market is not stable, but there are some significant periods. The $LRP$-based analysis identified three periods of significant information flow. All three periods include some major events affecting the market, such as the 2008 crisis, the oil crisis, and the COVID-19 pandemic, thus suggesting that the dynamics of information flow may be impacted by major events. In addition, we conducted a detailed dynamic analysis with daily data. Calculations show that each period includes nontrivial dynamics, some of which appear as local peaks, suggesting dramatic changes in information flow.

We also analyzed the information flow between the S&P500 index and some industry indices in the Chinese A-share market. Calculations show that the S&P500 index does not always significantly affect each industry index. Some industries are significantly affected, such as the steel industry and real estate. However, analysis based on weekly data shows that the dynamics of information flow between the industry indices and the S&P500 index exhibits multiple patterns. In addition, there are also some significant information flows from the Chinese market to the US market, thus demonstrating the heterogeneity of information flow.

This study demonstrates the high complexity of nonlinear causal relationships between two important markets and facilitates quantitative studies of the impact of major events. In particular, empirical analysis suggests that the heterogeneity of information flow in industry indices needs to be taken into account when discussing risk spillovers between markets. In this article, we only consider the information flow between the two markets. A related and important topic is the study of how major events affect information flow networks that include multiple markets. Since $LRP$-based methods involve two variables, for the information flow network, we need to construct a method that includes multiple variables. An alternative approach is to characterize changes in network structure by shuffling observations. In addition, the relationship between industry indices within a market can be analyzed in detail through $LRP$-based methods and high-frequency data. For example, for a trading day, we can analyze how important announcements impact the information flow network constructed by multiple industry indices.

**Author Contributions:** Methodology, C.-X.N.; formal analysis, C.-X.N. and J.X.; investigation, C.-X.N. and J.X.; writing—original draft preparation, C.-X.N.; writing—review and editing, C.-X.N.; visualization, C.-X.N. and J.X.; funding acquisition, C.-X.N. All authors have read and agreed to the published version of the manuscript.

**Funding:** The work is supported by the characteristic & preponderant discipline of key construction universities in Zhejiang province (Zhejiang Gongshang University—Statistics).

**Institutional Review Board Statement:** Not applicable.

**Informed Consent Statement:** Not applicable.

**Data Availability Statement:** Not applicable.

**Conflicts of Interest:** The authors declare no conflict of interest.

## References


1. Goh, J.C.; Jiang, F.; Tu, J.; Wang, Y. Can US economic variables predict the Chinese stock market? *Pac.-Basin Financ. J.* **2013**, *22*, 69–87. [CrossRef]
2. Chen, J.; Jiang, F.; Li, H.; Xu, W. Chinese stock market volatility and the role of US economic variables. *Pac.-Basin Financ. J.* **2016**, *39*, 70–83. [CrossRef]
3. Hu, Z.; Kutan, A.M.; Sun, P.W. Is US economic policy uncertainty priced in China's A—shares market? Evidence from market, industry, and individual stocks. *Int. Rev. Financ. Anal.* **2018**, *57*, 207–220. [CrossRef]
4. Kenourgios, D.; Samitas, A.; Paltalidis, N. Financial crises and stock market contagion in a multivariate time-varying asymmetric framework. *J. Int. Financ. Mark. Inst. Money* **2011**, *21*, 92–106. [CrossRef]
5. Cheung, W.; Fung, S.; Tsai, S.C. Global capital market interdependence and spillover effect of credit risk: Evidence from the 2007–2009 global financial crisis. *Appl. Financ. Econ.* **2010**, *20*, 85–103. [CrossRef]
6. Hwang, E.; Min, H.G.; Kim, B.H.; Kim, H. Determinants of stock market comovements among US and emerging economies during the US financial crisis. *Econ. Model.* **2013**, *35*, 338–348. [CrossRef]
7. Samarakoon, L.P. Stock market interdependence, contagion, and the US financial crisis: The case of emerging and frontier markets. *J. Int. Financ. Mark. Inst. Money* **2011**, *21*, 724–742. [CrossRef]
8. Wang, G.J.; Xie, C.; Lin, M.; Stanley, H.E. Stock market contagion during the global financial crisis: A multiscale approach. *Financ. Res. Lett.* **2017**, *22*, 163–168. [CrossRef]
9. Dungey, M.; Gajurel, D. Equity market contagion during the global financial crisis: Evidence from the world's eight largest economies. *Econ. Syst.* **2014**, *38*, 161–177. [CrossRef]
10. Jin, X.; An, X. Global financial crisis and emerging stock market contagion: A volatility impulse response function approach. *Res. Int. Bus. Financ.* **2016**, *36*, 179–195. [CrossRef]
11. Ahmad, W.; Sehgal, S.; Bhanumurthy, N. Eurozone crisis and BRIICKS stock markets: Contagion or market interdependence? *Econ. Model.* **2013**, *33*, 209–225.
12. Sansa, N.A. The Impact of the COVID-19 on the Financial Markets: Evidence from China and USA. *Electron. Res. J. Soc. Sci. Humanit.* **2020**, *2*. [CrossRef]
13. Nguyen, D.T.; Phan, D.H.B.; Ming, T.C. An assessment of how COVID-19 changed the global equity market. *Econ. Anal. Policy* **2021**, *69*, 480–491.
14. Zhang, D.; Hu, M.; Ji, Q. Financial markets under the global pandemic of COVID-19. *Financ. Res. Lett.* **2020**, *36*, 101528. [CrossRef] [PubMed]
15. Schreiber, T. Measuring information transfer. *Phys. Rev. Lett.* **2000**, *85*, 461.
16. Nie, C.X. Dynamics of the price–Volume information flow based on surrogate time series. *Chaos Interdiscip. J. Nonlinear Sci.* **2021**, *31*, 013106. [CrossRef] [PubMed]
17. Theiler, J.; Eubank, S.; Longtin, A.; Galdrikian, B.; Farmer, J.D. Testing for nonlinearity in time series: The method of surrogate data. *Phys. D Nonlinear Phenom.* **1992**, *58*, 77–94.
18. Small, M. *Applied Nonlinear Time Series Analysis: Applications in Physics, Physiology and Finance*; World Scientific: Singapore, 2005; pp. 115–129.
19. Lancaster, G.; Iatsenko, D.; Pidde, A.; Ticcinelli, V.; Stefanovska, A. Surrogate data for hypothesis testing of physical systems. *Phys. Rep.* **2018**, *748*, 1–60.
20. Marschinski, R.; Kantz, H. Analysing the information flow between financial time series. *Eur. Phys. J. B-Condens. Matter Complex Syst.* **2002**, *30*, 275–281.
21. Dimpfl, T.; Peter, F.J. Group transfer entropy with an application to cryptocurrencies. *Phys. A Stat. Mech. Its Appl.* **2019**, *516*, 543–551. [CrossRef]
22. Jizba, P.; Kleinert, H.; Shefaat, M. Rényi's information transfer between financial time series. *Phys. A Stat. Mech. Its Appl.* **2012**, *391*, 2971–2989. [CrossRef]
23. Papana, A.; Kyrtsou, C.; Kugiumtzis, D.; Diks, C. Detecting Causality in Non-stationary Time Series Using Partial Symbolic Transfer Entropy: Evidence in Financial Data. *Comput. Econ.* **2016**, *47*, 341–365. [CrossRef]
24. Kwon, O.; Yang, J.S. Information flow between composite stock index and individual stocks. *Phys. A Stat. Mech. Its Appl.* **2008**, *387*, 2851–2856. [CrossRef]
25. Dimpfl, T.; Peter, F.J. The impact of the financial crisis on transatlantic information flows: An intraday analysis. *J. Int. Financ. Mark. Inst. Money* **2014**, *31*, 1–13. [CrossRef]
26. Dimpfl, T.; Peter, F.J. Using transfer entropy to measure information flows between financial markets. *Stud. Nonlinear Dyn. Econom.* **2013**, *17*, 85–102. [CrossRef]
27. Daugherty, M.S.; Jithendranathan, T. A study of linkages between frontier markets and the U.S. equity markets using multivariate GARCH and transfer entropy. *J. Multinatl. Financ. Manag.* **2015**, *32*, 95–115. [CrossRef]
28. Kwon, O.; Yang, J.S. Information flow between stock indices. *EPL Europhys. Lett.* **2008**, *82*, 68003. [CrossRef]
29. Ong, M.A. An information theoretic analysis of stock returns, volatility and trading volumes. *Appl. Econ.* **2015**, *47*, 3891–3906. [CrossRef]
30. Kyrtsou, C.; Kugiumtzis, D.; Papana, A. Further insights on the relationship between SP500, VIX and volume: A new asymmetric causality test. *Eur. J. Financ.* **2019**, *25*, 1402–1419. [CrossRef]

31. Zheng, J.; Nie, C.X. The Dynamics of Price Volume Information Transfer in the Cryptocurrency Markets. *Adv. Complex Syst.* **2020**, *23*, 2050014. [CrossRef]
32. Sensoy, A.; Sobaci, C.; Sensoy, S.; Alali, F. Effective transfer entropy approach to information flow between exchange rates and stock markets. *Chaos Solitons Fractals* **2014**, *68*, 180–185. [CrossRef]
33. Kim, Y.; Kim, J.; Yook, S.H. Information transfer network of global market indices. *Phys. A Stat. Mech. Its Appl.* **2015**, *430*, 39–45. [CrossRef]
34. Behrendt, S.; Dimpfl, T.; Peter, F.J.; Zimmermann, D.J. RTransferEntropy—Quantifying information flow between different time series using effective transfer entropy. *SoftwareX* **2019**, *10*, 100265. [CrossRef]
35. Nie, C.X. A network-based method for detecting critical events of correlation dynamics in financial markets. *EPL Europhys. Lett.* **2020**, *131*, 50001. [CrossRef]

entropy

MDPI

Article

# Integral Non-Singular Terminal Sliding Mode Consensus Control for Multi-Agent Systems with Disturbance and Actuator Faults Based on Finite-Time Observer

**Pu Yang *, Yu Ding, Ziwei Shen and Kejia Feng**

Department of Automation, Nanjing University of Aeronautics and Astronautics, Nanjing 211106, China; jasondy@nuaa.edu.cn (Y.D.); shenziwei@nuaa.edu.cn (Z.S.); fdaxia@nuaa.edu.cn (K.F.)
* Correspondence: ppyang@nuaa.edu.cn

**Abstract:** This paper studies the consensus fault-tolerant control problem of a class of second-order leader–follower multi-agent systems with unknown disturbance and actuator faults, and proposes an integral non-singular terminal sliding mode control algorithm based on a finite-time observer. First, a finite-time disturbance observer was designed based on a combination of high-order sliding mode and dual layers adaptive rules to realize fast estimation and compensation of disturbance and faults. Then, a sliding surface with additional integral links was designed based on the conventional sliding surface, and an integral non-singular terminal sliding mode controller is proposed to realize the robust consensus in finite time and accurately diminish the chattering phenomena. Finally, a numerical example and simulation verify the effectiveness.

**Keywords:** integral non-singular terminal sliding mode control (INTSMC); multi-agent systems (MASs); finite-time observer; fault-tolerant control

## 1. Introduction

With the rapid development of society, communication, artificial intelligence, and so on, multi-agent systems (MASs) have become a hot research issue in the field of control engineering. Due to the collaboration and cooperation of individuals between MASs, these systems can complete complex tasks that are difficult for a single individual. At present, MASs are widely used in mobile sensor networks [1], mobile robots [2], unmanned aerial vehicle formation [3], satellite cluster attitudes [4], and other engineering fields [5–7].

In recent years, the problem of consensus control has always been an important research direction in the field of MASs. The goal of consensus control is to ensure that each agent subsystem can use its own local information to complete the task and make the states of all subsystems eventually converge to the same value. Much research has been conducted on consensus control, and this research has produced effective results for different objects in different fields. In [8], an adaptive control protocol is proposed to adjust the virtual impedance to compensate for the mismatched line impedance to achieve accurate reactive power distribution. In [9], a control protocol of general linear MASs is designed by using scroll optimization control. To reduce unnecessary calculation pressure, a reduced-order controller is used to achieve the consensus control of output feedback MASs [10]. In [11], the time delay problem of a class of high-order MASs is considered, and a distributed state feedback control strategy is designed to ensure the consensus of the system.

Some adaptive robust fault-tolerant control methods are proposed to solve the consensus problem of nonlinear leader–follower MASs [12–14]. In [15], a static event-triggered protocol is designed to solve the leader–follower consensus problem of first-order MASs with disturbance. In [16], a finite-time fault-tolerant super-twisting algorithm is proposed to solve the effects of actuator faults and unknown disturbance, avoiding the chattering problem. In [17], a novel dynamic sliding mode control protocol is proposed to achieve the finite-time consensus of nonlinear heterogeneous multi-agent systems, which ensures their robustness. In [18], the Gaussian basis function is introduced to deal with the non-strict feedback term, which realizes the leader–follower consensus of multi-agent systems under the unknown switching mechanism.

In fact, MASs are susceptible to disturbance and faults from dynamic models and exosystems during operation. Taking the wheeled robot system as an example, in the actual operation process, the robot will inevitably be affected by external disturbances due to the different road conditions and the surrounding environment and by actuator faults due to incorrect installation or wheel wear. The unknown disturbance and faults will reduce the control effect of the system, bring significant uncertainty, and eventually lead to a decline in the overall performance of the MASs and even the failure of the task. Given that sliding mode control can overcome the uncertainty of the system as well as its strong robustness to interference and unmodeled dynamics, it is widely used in the consensus control of MASs.

Zhao D et al. proposed an adaptive sliding mode control method for a class of second-order systems with a leader to achieve the tracking consensus of the system [19]. Jiang Y L et al. used the integral sliding mode method to achieve a consensus for a class of multi-agent systems with state delays [20]. Dong et al. used the linear sliding mode method to solve the problem of time-varying topology in a second-order nonlinear MASs tracking system [21]. Zhao L et al. studied a class of second-order nonlinear MASs with external disturbance based on the terminal sliding mode method to enable the control system to achieve tracking consistency in a limited time [22]. Sanjoy M et al. studied the consensus of a class of high-order nonlinear MASs based on the integral sliding mode method [23].

Zheng et al. proposed an event-triggered sliding mode method to solve the consensus of a class of leader–follower MASs [24]. Zhao et al. studied the robust consensus of high-order MASs based on distributed protocols, whose considered system models are general, and the designed protocols have strong robustness [25]. However, from the presented results, the chattering phenomenon of this algorithm is not sufficiently weakened. In [26], a distributed fixed-time control algorithm is designed based on backstepping for a class of high-order multi-agent systems, successfully applied to the control of a class of wheeled robot systems. However, the backstepping method needs to continuously differentiate some nonlinear functions, which leads to a high degree of complexity and nonlinearity of the obtained control law, especially when the system order is high.

To solve these problems, the improvements outlined in this paper mainly focus on two aspects. On the one hand, the terminal sliding mode control is selected instead of the backstepping control in the controller design method to overcome the expansion problem of the differential term of the controller. On the other hand, it is necessary to select an appropriate switching function to effectively solve the chattering problem. In addition, the faults and disturbance of the second-order MASs are unpredictable during the operation process, so an observer is the most popular choice to estimate and make up for it based on active control. Zaidi et al. designed a distributed observer to estimate, which takes the external disturbance into account to optimize [27]. To avoid dependence on speed information, Hua et al. proposed two distributed finite-time algorithms that only include a relative position information measurement [28].

Although the idea of an observer is relatively developed, the results of applying it to multi-agent fault diagnosis are relatively scarce, and most of the research stops at fault estimation. In [29], a new controller based on an integral sliding mode and super-twisting sliding mode is designed to solve the fault-tolerant tracking problem of a fault multi-agent system. In [30], based on the idea of the global sliding mode, the disturbance term containing a fault was extracted from the system model, and a distributed observer that can accurately estimate the fault information was designed. Most of the research considers the information of global variables in the design of the observer, and the observer requirements are relatively high, especially for large and complex systems, which should be improved.

The shortcomings of previous research can be summarized in two aspects. On the one hand, some methods combine sliding mode control with other control methods such as adaptive control, which creates a complicated controller with many parameters to be optimized. On the other hand, some methods are sensitive to disturbances and uncertainties, adding unnecessary restrictions or too many parameters with the design. In order to reduce the complexity of the control algorithm and the need for uncertain parameters, this paper, inspired by the existing research, studies the finite-time consensus problem for second-order leader–follower MASs with unknown disturbance and actuator faults by using the integral non-singular terminal sliding mode control algorithm based on the finite-time observer. The main contributions are as follows:

1. A novel adaptive finite-time observer is designed based on a combination of high-order sliding mode and dual-layer adaptive rules, which realizes the centralized estimation and compensation of unknown disturbance and actuator fault in finite time. Additionally, there is no need to obtain the upper bound of the disturbance in advance;
2. On the basis of the conventional sliding mode surface, a new integral part is added to the sliding mode surface, which improves the robustness of the system and sufficiently diminishes the chattering phenomena. The controller proposed in this paper solves the singularity problem and realizes consensus in finite time for disturbed second-order leader–follower MASs;
3. The model studied in this paper comprehensively considers the influence of nonlinear terms, unknown external disturbance, and actuator faults, which improves the practicability of the control algorithm.

The remaining part of this paper is organized as follows: In Section 2, the graph theory and problem formulation are given, and some lemmas, assumptions, definitions, and notations that will be used later are listed. In Section 3, a finite-time observer based on a high-order sliding mode and dual-layer adaptive rule is proposed, and an integral non-singular terminal sliding mode controller is designed, which is analyzed by Lyapunov stability theory. In Section 4, a numerical example and simulation verify the effectiveness of the proposed method in this paper compared with the existing method. Finally, a brief conclusion is given in Section 5.

## 2. Preliminaries and Problem Formulation

### 2.1. Graph Theory

Consider a MAS containing $n$ agents whose topology structure of communication is denoted by the notation $G = (V, E, A)$, where $V = \{v_1, v_2, \cdots, v_n\}$ denotes the set of nodes, $E \subseteq V \times V$ denotes the set of directed edges, and $A = [a_{ij}]$ is defined as the adjacency matrix. An ordered pair of nodes $(v_j, v_i)$ is used to represent that a directed edge exists from $v_j$ to $v_i$, and $v_i$ can obtain information from $v_j$. If $(v_j, v_i) \in E$, then $a_{ij} > 0$, otherwise $a_{ij} = 0$. The main diagonal elements of the adjacency matrix are equal to 0, that is $a_{ii} = 0$. The degree of node $i$ is defined as $d_i = \sum_{j=1}^{n} a_{ij}$, and $D = diag\{d_1, d_2, \cdots, d_n\}$ denotes the degree matrix. The Laplacian matrix of the graph $G$ is defined as $L = D - A = [l_{ij}]$, where $l_{ij}$ can be depicted as $l_{ij} = -a_{ij}(j \neq i)$ and $l_{ii} = d_i$. For a leader–follower multi-agent system with one leader and $n$ followers, the leader is denoted by node 0, and followers are denoted by node $1, 2, \cdots, n$. The topology structure of communication is denoted by $\overline{G} = (\overline{V}, \overline{E}, \overline{A})$, where $\overline{V} = V \cup \{v_0\}$, $\overline{E} \subseteq \overline{V} \times \overline{V}$, and $\overline{A}$ is the adjacency matrix of $\overline{G}$. Define a matrix $B = diag\{b_1, b_2, \cdots, b_n\}$ to describe whether the leader can directly send information to the follower $i$, and if it can, then $b_i > 0$, otherwise $b_i = 0$. In this paper, the leader can only send information to certain followers, but cannot receive information from any follower.

## 2.2. Problem Formulation

Consider a second-order leader–follower multi-agent system with one leader and $n$ followers. The dynamics of the leader subsystem are described as

$$\begin{cases} \dot{x}_0(t) = v_0(t) \\ \dot{v}_0(t) = u_0(t) \end{cases} \tag{1}$$

where $x_0(t) \in R$, $v_0(t) \in R$ represent position and velocity state of the leader agent, respectively. $u_0(t) \in R$ represents the control input.

The dynamics of the follower $i$ $(i = 1, 2, \cdots, n)$ is described as

$$\begin{cases} \dot{x}_i(t) = v_i(t) \\ \dot{v}_i(t) = f_i(x_i(t), v_i(t), t) + u_i^a(t) + d_i(t) \end{cases} \tag{2}$$

where $x_i(t) \in R$ and $v_i(t) \in R$ represent the position and velocity state of the follower agent $i$. $f_i(x_i(t), v_i(t), t) \in R$ is the inherent nonlinear dynamic function. $d_i(t) \in R$ is the external disturbance. $u_i^a(t) \in R$ denotes the actual control input of the follower agent $i$, in which the specific model can be expressed as

$$u_i^a(t) = (1 - c_i)u_i(t) \tag{3}$$

where $u_i(t) \in R$ denotes the ideal control input, $0 \leq c_i < 1$ denotes the failure factor of the actuator of the follower agent $i$, and $c_i = 0$ represents that the control input of the actuator is normal, namely $u_i^a(t) = u_i(t)$. Thus, the system (2) can be rewritten as

$$\begin{cases} \dot{x}_i(t) = v_i(t) \\ \dot{v}_i(t) = f_i(x_i(t), v_i(t), t) + u_i(t) + \omega_i(t) \end{cases} \tag{4}$$

where $\omega_i(t) = d_i(t) - c_i u_i(t)$ denotes the so-called lumped faults, which include external disturbance and actuator faults.

**Remark 1.** *Since the major purpose of the algorithm designed in this paper is to achieve the consensus of MASs, inspired by [31], the actuator failure faults and external disturbances existing in the follower agents are unified into the lumped faults, which are estimated as a whole by the observer. Therefore, during the design process of the control algorithm, there is no need to know the upper limit of the disturbance and the magnitude of the fault, which improves the practicability of the fault-tolerant strategy.*

**Assumption 1** ([32]). *There is a directed spanning tree with the leader as the root node in the graph* $\overline{G}$.

**Assumption 2** ([33]). *For the nonlinear dynamic function* $f_i(x_i(t), v_i(t), t)$, *there is a positive real constant* $\overline{f}$ *that satisfies* $|f_i(x_i(t), v_i(t), t)| \leq \overline{f}_i$

**Assumption 3** ([34]). *For the first derivative of the lumped faults* $\dot{\omega}_i(t)$, *there is a positive real constant* $\overline{\omega}$ *which satisfies* $|\dot{\omega}_i(t)| \leq \overline{\omega}$

**Definition 1** ([35]). *The consensus of the leader–follower MAS is to design a control law* $u_i(t)$ *for each follower, so that the states of the follower tend to those of the leader. The relationship can be described as*

$$\begin{cases} \lim\limits_{t\to\infty} x_i(t) - x_0(t) = 0, \ \forall i = 1, 2, \cdots, n \\ \lim\limits_{t\to\infty} v_i(t) - v_0(t) = 0, \ \forall i = 1, 2, \cdots, n \end{cases} \tag{5}$$

**Definition 2** ([36]). *The origin points of the system (1) and (2) are considered to be globally finite-time stable if they are globally asymptotically stable with a bounded time function* $T(x_0)$, *i.e.,* $T_{\max} > 0$ *such that* $T(x_0)$ *satisfies the term* $T(x_0) < T_{\max}$.

### 2.3. Some Lemmas and Notations

Some notations are given as follows. The given values $x = [x_1, x_2, \cdots, x_n]^T \in R^n$ and $\gamma \in R$, denote $\text{sgn}(x) = [\text{sgn}(x_1), \text{sgn}(x_2), \cdots, \text{sgn}(x_n)]^T$, and $sig^\gamma(x) = [sig^\gamma(x_1), sig^\gamma(x_2), \cdots, sig^\gamma(x_n)]^T$, where $\text{sgn}(\cdot)$ is the standard signum function and $sig^\gamma(x_i) = |x_i|^\gamma \text{sgn}(x_i)$ $i = 1, 2, \cdots n$. $|\cdot|$ is the absolute value. The value $1_n$ is the unit column vector in $R^n$.

**Lemma 1** ([33]). *If there is a directed spanning tree in the graph $\overline{G}$, and the leader is the root node, then the matrix $L + B$ is invertible.*

**Lemma 2** ([37]). *Consider a differential equation*

$$\dot{x} = -\frac{2\alpha_0}{1 + e^{-\eta_0(|x| - \varepsilon_0)}} - \frac{2\beta_0}{1 + e^{\mu_0(|x| - \varepsilon_0)}} sig^{\gamma_0}(x) \tag{6}$$

*where $\alpha_0$, $\beta_0$, $\eta_0$, $\mu_0$ are positive constants, $0 < \gamma_0 < 1$, and $\varepsilon_0 = (\beta_0/\alpha_0)^{1/(1-\gamma_0)}$.*

*Accordingly, the dynamic (6) is declared as finite-time stabilization with respect to the initial term $x(0)$ and the settling time $T_0$ is given by*

$$T_0 < \frac{\ln(|x(0)|) - \ln(\varepsilon_0)}{\alpha_0} + \frac{1}{\beta_0(1 - \gamma_0)} |\varepsilon_0|^{1-\gamma_0} \tag{7}$$

**Lemma 3** ([28,38]). *Consider a nonlinear system $\dot{x} = f(x)$, where$f(0) = 0$. If there is a positive definite continuous function $V(x)$, which makes an open area at the origin $x = 0$ satisfy*

$$\dot{V}(x) + mV^\delta(x) \leq 0 \tag{8}$$

*where $m > 0$ and $0 < \delta < 1$, then the function $V(x)$ will converge to the origin within a certain finite time, and the upper bound of the finite convergence time $T$ depends on the initial state of the system $x(0)$, namely*

$$T(x(0)) \leq \frac{V(x(0))^{(1-\delta)}}{m(1 - \delta)} \tag{9}$$

**Lemma 4** ([39]). *For $x_i \in R (i = 1, 2, \cdots, n)$, $0 < \eta < 2$, then $\sum\limits_{i=1}^{n} |x_i|^\eta \geq \left(\sum\limits_{i=1}^{n} x_i^2\right)^{\frac{\eta}{2}}$.*

## 3. Main Results

### 3.1. Design of Consensus

According to the neighbor information obtained by the follower $i$, the consensus error of position $e_{xi}(t)$ and consensus error of velocity $e_{vi}(t)$ are defined as

$$\begin{cases} e_{xi}(t) = \sum\limits_{j=1}^{n} a_{ij}(x_i(t) - x_j(t)) + b_i(x_i(t) - x_0(t)), \\ e_{vi}(t) = \sum\limits_{j=1}^{n} a_{ij}(v_i(t) - v_j(t)) + b_i(v_i(t) - v_0(t)), \end{cases} i \in \{1, 2, \cdots, n\} \tag{10}$$

Define

$$\overline{x}_i(t) = x_i(t) - x_0(t)$$
$$\overline{v}_i(t) = v_i(t) - v_0(t)$$
$$\overline{x} = [\overline{x}_1(t), \overline{x}_2(t), \cdots, \overline{x}_n(t)]^T$$
$$\overline{v} = [\overline{v}_1(t), \overline{v}_2(t), \cdots, \overline{v}_n(t)]^T$$

$$e_x = \left[e_{x1}, e_{x2}, \cdots, e_{xn}\right]^T$$
$$e_v = \left[e_{v1}, e_{v2}, \cdots, e_{vn}\right]^T$$

Then the global synchronization error can be defined as

$$\begin{cases} e_x = (L+B)\overline{x} \\ e_v = (L+B)\overline{v} \end{cases} \tag{11}$$

Taking the first derivative of $e_{xi}$ and $e_{vi}$ with respect to time, we can obtain

$$\begin{cases} \dot{e}_{xi} & = e_{vi} \\ \dot{e}_{vi} & = \left( \sum\limits_{j=1;j\neq i}^{n} a_{ij} + b_i \right) u_i - \sum\limits_{j=1;j\neq i}^{n} a_{ij} u_j - b_i u_0 \\ & + \left( \sum\limits_{j=1;j\neq i}^{n} a_{ij} + b_i \right) (f_i + \omega_i) - \sum\limits_{j=1;j\neq i}^{n} a_{ij} (f_j + \omega_j) \end{cases} \tag{12}$$

Define

$$F = \left[f_1(x_1(t), v_1(t), t), \cdots, f_n(x_n(t), v_n(t), t)\right]^T$$
$$u = \left[u_1(t), u_2(t), \cdots, u_n(t)\right]^T$$
$$\omega = \left[\omega_1(t), \omega_2(t), \cdots, \omega_n(t)\right]^T$$
$$\dot{e}_x = \left[\dot{e}_{x1}, \dot{e}_{x2}, \cdots, \dot{e}_{xn}\right]^T$$
$$\dot{e}_v = \left[\dot{e}_{v1}, \dot{e}_{v2}, \cdots, \dot{e}_{vn}\right]^T$$

Taking the first derivative of Equation (11) with respect to time, the global form of (12) can be obtained as

$$\begin{cases} \dot{e}_x = e_v \\ \dot{e}_v = (L+B)(F + u - 1_n u_0 + \omega) \end{cases} \tag{13}$$

If the Equation (13) is asymptotically stable according to Definition 1, then the consensus problem of MASs described as (1) and (2) can be solved.

*3.2. Design of the Finite-Time Observer*

Given that the information of velocity and faults is difficult to obtain during the operation of MASs, a novel finite-time observer was designed based on high-order sliding mode observer and dual-layer adaptive rule.

$$\begin{cases} \dot{\hat{v}}_i = f_i + u_i + \hat{\omega}_i - \frac{2\alpha_1}{1+e^{-\eta_1(|\widetilde{v}_i|-\varepsilon_1)}} \widetilde{v}_i - \frac{2\beta_1}{1+e^{\mu_1(|\widetilde{v}_i|-\varepsilon_1)}} sig^{\gamma_1}(\widetilde{v}_i) \\ \dot{\hat{\omega}}_i = -\frac{2\alpha_2}{1+e^{-\eta_2(|\widetilde{\omega}_i|-\varepsilon_2)}} \widetilde{\omega}_i - \frac{2\beta_2}{1+e^{\mu_2(\widetilde{\omega}_i-\varepsilon_2)}} sig^{\gamma_2}(\widetilde{\omega}_i) - \Xi \mathrm{sgn}(\widetilde{\omega}_i) \end{cases} \tag{14}$$

where $\hat{v}_i$ and $\hat{\omega}_i$ are the approximated values of $v_i$ and $\omega_i$. Denote $\widetilde{v} = \hat{v} - v$ and $\widetilde{\omega} = \hat{\omega} - \omega$. $\alpha_k, \beta_k, \eta_k, \mu_k$ are positive constants, $0 < \gamma_k < 1$, and $\varepsilon_k = (\beta_k/\alpha_k)^{1/(1-\gamma_k)}$, where $k = 1, 2$. Additionally, $\Xi$ is an adaptive gain value.

**Theorem 1.** *For the MASs described as (1) and (2), if the observer is designed as (14) and the adaptive gain value satisfies* $\Xi > \left|\dot{\omega}_i\right|$, *then the approximation error will converge to zero in finite time.*

**Proof.** Taking the first derivative of $\widetilde{v}_i$ and $\widetilde{\omega}_i$ based on (14), the following can be obtained

$$\begin{cases} \dot{\widetilde{v}}_i = \widetilde{\omega}_i - \frac{2\alpha_1}{1+e^{-\eta_1(|\widetilde{v}_i|-\varepsilon_1)}} \widetilde{v}_i - \frac{2\beta_1}{1+e^{\mu_1(|\widetilde{v}_i|-\varepsilon_1)}} sig^{\gamma_1}(\widetilde{v}_i) \\ \dot{\widetilde{\omega}}_i = -\dot{\omega}_i - \Xi \mathrm{sgn}(\widetilde{\omega}_i) - \frac{2\alpha_2}{1+e^{-\eta_2(|\widetilde{\omega}_i|-\varepsilon_2)}} \widetilde{\omega}_i - \frac{2\beta_2}{1+e^{\mu_2(\widetilde{\omega}_i-\varepsilon_2)}} sig^{\gamma_2}(\widetilde{\omega}_i) \end{cases} \tag{15}$$

A Lyapunov function $V_{1i} = \frac{1}{2}\widetilde{\omega}^2$ can be defined, and the first derivative of $V_{1i}$ with respect to time can be obtained as follows:

$$\begin{aligned}
\dot{V}_{1i} &= \widetilde{\omega}_i\dot{\widetilde{\omega}}_i \\
&= \widetilde{\omega}_i\left(-\dot{\omega}_i - \Xi\mathrm{sgn}(\widetilde{\omega}_i) - \frac{2\alpha_2}{1+e^{-\eta_2(|\widetilde{\omega}_i|-\varepsilon_2)}}\widetilde{\omega}_i - \frac{2\beta_2}{1+e^{\mu_2(\widetilde{\omega}_i-\varepsilon_2)}}|\widetilde{\omega}_i|^{\gamma_2}\mathrm{sgn}(\widetilde{\omega}_i)\right) \\
&= -\dot{\omega}_i\widetilde{\omega} - \Xi|\widetilde{\omega}_i| - \frac{2\alpha_2}{1+e^{-\eta_2(|\widetilde{\omega}_i|-\varepsilon_2)}}\widetilde{\omega}_i^2 - \frac{2\beta_2}{1+e^{\mu_2(\widetilde{\omega}_i-\varepsilon_2)}}|\widetilde{\omega}_i|^{\gamma_2+1} \\
&\leq -(\Xi - |\dot{\omega}_i|)|\widetilde{\omega}_i| - \frac{2\alpha_2}{1+e^{-\eta_2(|\widetilde{\omega}_i|-\varepsilon_2)}}\widetilde{\omega}_i^2 - \frac{2\beta_2}{1+e^{\mu_2(\widetilde{\omega}_i-\varepsilon_2)}}|\widetilde{\omega}_i|^{\gamma_2+1} \\
&\leq -\frac{2\alpha_2}{1+e^{-\eta_2(|\widetilde{\omega}_i|-\varepsilon_2)}}\widetilde{\omega}_i^2 - \frac{2\beta_2}{1+e^{\mu_2(\widetilde{\omega}_i-\varepsilon_2)}}|\widetilde{\omega}_i|^{\gamma_2+1} \\
&\leq 0
\end{aligned} \tag{16}$$

According to Lemma 2, the proposed observer can estimate the lumped faults in a finite time. □

The convergence time $T_1$ satisfies

$$T_1 < \frac{\ln(|\widetilde{\omega}_i(0)|) - \ln(\varepsilon_2)}{\alpha_2} + \frac{1}{\beta_2(1-\gamma_2)}|\varepsilon_2|^{1-\gamma_2} \tag{17}$$

The design of $\Xi$ depends on the upper bound of $\omega_i$ according to Assumption 3, while in the actual application process, it is often accompanied by uncertain parameters affecting the system. To address the need for all uncertain parameters, a dual layers adaptive law is designed to improve the design of $\Xi$ as follows:

$$\begin{cases} \dot{\Xi} = -(\Delta_1 + \Delta_2)\mathrm{sgn}(\Psi) \\ \dot{\Delta}_2 = \begin{cases} \Delta_d|\Psi|, & |\Psi| > \Psi_0 \\ 0, & |\Psi| \leq \Psi_0 \end{cases} \end{cases} \tag{18}$$

where

$$\begin{cases} \Psi = \Xi - \dfrac{|\Lambda|}{\kappa_1} - \kappa_2 \\ \dot{\Lambda} = \zeta fal(-\Xi\mathrm{sgn}(\widetilde{\omega}) - \Lambda, \chi, \varsigma) \\ fal(\Gamma, \chi, \varsigma) = \begin{cases} |\Gamma|^{\chi}\mathrm{sgn}(\Gamma), & |\Gamma| > \varsigma \\ \frac{\Gamma}{\varsigma^{1-\chi}}, & |\Gamma| \leq \varsigma \end{cases} \end{cases} \tag{19}$$

where $\Delta_1$, $\Delta_d$, $\zeta$, $\kappa_1$, $\kappa_2$, $\chi$ and $\varsigma$ are positive constants. Therefore, the value of $\Xi\mathrm{sgn}(\widetilde{\omega})$ can be obtained by the $fal(\cdot)$ function in real time.

### 3.3. Design of Sliding Mode Controller

From the consensus errors given in Section 3.1, the sliding surface can be described as [40]:

$$s_i = e_{xi} + \int_0^t \left(\xi_1\mathrm{sgn}(e_{vi})|e_{vi}|^{\theta_1} + \xi_2\mathrm{sgn}(e_{xi})|e_{xi}|^{\theta_2}\right)dt \tag{20}$$

where $\xi_1$ and $\xi_2$ are positive constants, $1 < \theta_1 < 2$, and $\theta_2 = \theta_1/(2-\theta_1)$. To ensure that the convergence time of $s \to 0$ is limited, and to eliminate the chattering problem caused by sliding mode control, a novel integral non-singular terminal sliding mode control (INTSMC) surface was designed as follows:

$$\begin{aligned}
\sigma_i &= \dot{s}_i + \lambda s_i \\
&= e_{vi} + \xi_1\mathrm{sgn}(e_{vi})|e_{vi}|^{\theta_1} + \xi_2\mathrm{sgn}(e_{xi})|e_{xi}|^{\theta_2} \\
&\quad + \lambda\left(e_{xi} + \int_0^t \left(\xi_1\mathrm{sgn}(e_{vi})|e_{vi}|^{\theta_1} + \xi_2\mathrm{sgn}(e_{xi})|e_{xi}|^{\theta_2}\right)dt\right)
\end{aligned} \tag{21}$$

where $\lambda > 0$ is a tuning constant. The derivative of $\sigma$ can be obtained as

$$\begin{aligned}\dot{\sigma}_i &= \ddot{s}_i + \lambda \dot{s}_i \\ &= \dot{e}_{vi} + \xi_1\theta_1\dot{e}_{vi}|e_{vi}|^{\theta_1-1} + \xi_2\theta_2 e_{vi}|e_{xi}|^{\theta_2-1} \\ &+\lambda\left(e_{vi} + \xi_1 \mathrm{sgn}(e_{vi})|e_{vi}|^{\theta_1} + \xi_2 \mathrm{sgn}(e_{xi})|e_{xi}|^{\theta_2}\right)\end{aligned} \tag{22}$$

Instituting (12) into (22), we can obtain

$$\begin{aligned}\dot{\sigma} &= \left(\begin{array}{c}\left(\sum\limits_{j=1;j\neq i}^{n} a_{ij} + b_i\right)u_i - \sum\limits_{j=1;j\neq i}^{n} a_{ij}u_j - b_i u_0 \\ +\left(\sum\limits_{j=1;j\neq i}^{n} a_{ij} + b_i\right)(f_i + \omega_i) - \sum\limits_{j=1;j\neq i}^{n} a_{ij}(f_j + \omega_j)\end{array}\right)(1 + \xi_1\theta_1|e_{vi}|^{\theta_1-1}) \\ &+\xi_2\theta_2 e_{vi}|e_{xi}|^{\theta_2-1} + \lambda\left(e_{vi} + \xi_1 \mathrm{sgn}(e_{vi})|e_{vi}|^{\theta_1} + \xi_2 \mathrm{sgn}(e_{xi})|e_{xi}|^{\theta_2}\right)\end{aligned} \tag{23}$$

The controller for the follower agent $i$ is designed as follows:

$$u_i = u_{i1} + u_{i2} \tag{24}$$

with

$$u_{i1} = (d_i + b_i)^{-1}\left\{\begin{array}{c}\sum\limits_{j=1;j\neq i}^{n} a_{ij}u_j + b_i u_0 - (2n + b_i)(\overline{f} + \hat{\omega}_i) + (1 + \xi_1\theta_1|e_{vi}|^{\theta_1-1})^{-1} \\ \xi_2\theta_2 e_{vi}|e_{xi}|^{\theta_2-1} + \lambda\left(e_{vi} + \xi_1 \mathrm{sgn}(e_{vi})|e_{vi}|^{\theta_1} + \xi_2 \mathrm{sgn}(e_{xi})|e_{xi}|^{\theta_2}\right)\end{array}\right\} \tag{25}$$

$$u_{i2} = -(d_i + b_i)^{-1}(1 + \xi_1\theta_1|e_{vi}|^{\theta_1-1})^{-1}\rho \mathrm{sgn}(\sigma) \tag{26}$$

where $\rho$ is a positive constant, $\overline{f}$ is the upper bound of $f$ based on Assumption 2. $u_{i1}$ denotes the part of equivalent control and $u_{i2}$ denotes the part of switching control.

**Theorem 2.** *The origin errors of MASs described as (13) can converge to zero in a finite time by the controller designed as (24)–(26), which means the fault-tolerant consensus goals can be achieved.*

**Proof.** A Lyapunov function $V_2 = \frac{1}{2}\sum\limits_{i=1}^{n} \sigma_i^2$ is defined, and the first derivative of $V_2$ with respect to time can be obtained as follows:

$$\begin{aligned}\dot{V}_2 &= \sum_{i=1}^{n} \sigma_i\dot{\sigma}_i \\ &= \sum_{i=1}^{n} \sigma_i\left(\left(\sum_{j=1;j\neq i}^{n} a_{ij} + b_i\right)u_i - \sum_{j=1;j\neq i}^{n} a_{ij}u_j - b_i u_0\right)(1 + \xi_1\theta_1|e_{vi}|^{\theta_1-1}) \\ &+\sum_{i=1}^{n} \sigma_i\left(\left(\sum_{j=1;j\neq i}^{n} a_{ij} + b_i\right)(f_i + \omega_i) - \sum_{j=1;j\neq i}^{n} a_{ij}(f_j + \omega_j)\right)(1 + \xi_1\theta_1|e_{vi}|^{\theta_1-1}) \\ &+\sum_{i=1}^{n} \sigma_i\left(\xi_2\theta_2 e_{vi}|e_{xi}|^{\theta_2-1} + \lambda\left(e_{vi} + \xi_1 \mathrm{sgn}(e_{vi})|e_{vi}|^{\theta_1} + \xi_2 \mathrm{sgn}(e_{xi})|e_{xi}|^{\theta_2}\right)\right)\end{aligned} \tag{27}$$

Introducing (24)–(26) into (27) can transform it into

$$\begin{aligned}
\dot{V}_2 &= \sum_{i=1}^{n} \sigma_i \left( \left( \sum_{j=1;j\neq i}^{n} a_{ij} + b_i \right)(f_i + \omega_i) - \sum_{j=1;j\neq i}^{n} a_{ij}(f_j + \omega_j) \right) \\
&- \sum_{i=1}^{n} \sigma_i((2n + b_i)(\overline{f} + \hat{\omega}_i)) - \sum_{i=1}^{n} \sigma_i \rho \mathrm{sgn}(\sigma)(1 + \xi_1 \theta_1 |e_{vi}|^{\theta_1 - 1}) \\
&\leq \sum_{i=1}^{n} (2n + b_i)(f_i - \overline{f})|\sigma_i| + \sum_{i=1}^{n} (2n + b_i)(\omega_i - \hat{\omega}_i)|\sigma_i| - \sum_{i=1}^{n} \sigma_i \rho \mathrm{sgn}(\sigma) \\
&\leq -\sum_{i=1}^{n} \rho |\sigma_i| \\
&\leq -\rho \left(\sum_{i=1}^{n} \sigma_i^2\right)^{\frac{1}{2}} \\
&= -\sqrt{2}\rho V_2^{\frac{1}{2}} \\
&< 0
\end{aligned} \tag{28}$$

According to Lemma 3, the INTSMC surface designed as (21) can converge to zero in a finite time. □

The convergence time $T_2$ satisfies

$$T_2 \leq \frac{\sqrt{2}}{\rho}(V_2(0))^{\frac{1}{2}} \tag{29}$$

**Remark 2.** *Consider the finite convergence time $T_1$ and $T_2$, both of which have upper bounds related to the initial states. The value of $T_1$ is determined by the initial estimation error of the observer, and the value of $T_2$ is determined by the initial consensus tracking error of the follower agents. In practical applications, the initial value of the latter is larger, and thus the convergence time is longer.*

## 4. Simulations

In this section, we will take a numerical simulation to verify the effectiveness and superiority of the method proposed in this paper compared with the method proposed in [30,40]. Consider a leader–follower MAS with one leader and five followers, whose topology structure graph is shown in Figure 1.

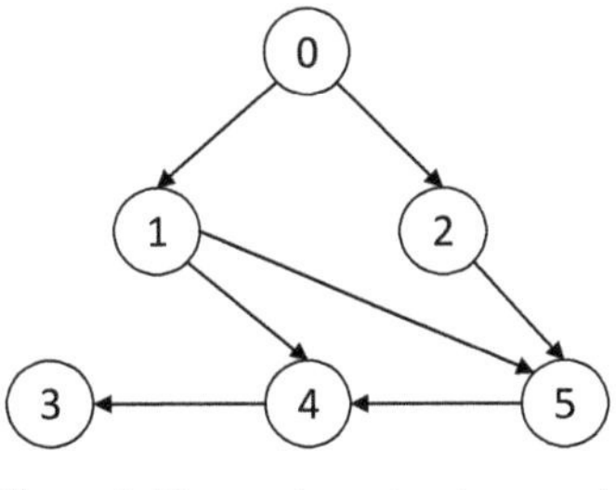


**Figure 1.** The topology structure graph.

The weight of all edges is 1, and the Laplacian matrix $L$ and adjacency matrix $B$ can be calculated according to Figure 1 as

$$L = \begin{bmatrix} 0 & 0 & 0 & 0 & 0 \\ 0 & 0 & 0 & 0 & 0 \\ 0 & 0 & 1 & -1 & 0 \\ -1 & 0 & 0 & 2 & -1 \\ -1 & -1 & 0 & 0 & 2 \end{bmatrix}, B = \begin{bmatrix} 1 & 0 & 0 & 0 & 0 \\ 0 & 1 & 0 & 0 & 0 \\ 0 & 0 & 0 & 0 & 0 \\ 0 & 0 & 0 & 0 & 0 \\ 0 & 0 & 0 & 0 & 0 \end{bmatrix} \tag{30}$$

The dynamic function of the leader agent is described as

$$\begin{cases} \dot{x}_0(t) = \dot{v}_0(t) \\ \dot{v}_0(t) = \sin(t/12) \end{cases} \tag{31}$$

where the values of initial states are set to $x_0(0) = 0$ and $v_0(0) = 1$. The dynamic equation of follower $i$ $(i = 1, 2, \cdots, 5)$ is described as:

$$\begin{cases} x_i(t) = v_i(t) \\ v_i(t) = \cos(x_i(t)) + \cos(v_i(t)) + u_i^a(t) + d_i(t) \end{cases} \tag{32}$$

where $f_i(x_i(t), v_i(t), t) = \cos(x_i(t)) + \cos(v_i(t))$ is the inherent nonlinear dynamics. It is apparent that $f_i(x_i(t), v_i(t), t) \leq 2$, which satisfies Assumption 2. The values of the initial states are set to $x_1(0) = 1$, $x_2(0) = 1.5$, $x_3(0) = -2.2$, $x_4(0) = -1.5$, $x_5(0) = -0.5$, $v_1(0) = 1.2$, $v_2(0) = 0.5$, $v_3(0) = 0$, $v_4(0) = 0.8$, $v_5(0) = 1.5$. Here, we consider that the follower 1 and the follower 4 are with actuator faults and disturbance, while the others are only with disturbance. The failure factors are set to $c_1 = 0.2$ and $c_4 = 0.4$. The disturbance suffered by followers is described as follows

$$d_{1,2} = 0.2 + 0.2\sin(0.5\pi t),\ d_{3,4,5} = 0.3 + 0.1\sin(2\pi t) \tag{33}$$

The tuning parameters are mainly selected based on our design experience and experimental debugging, and a set of parameters that can accurately reflect the tracking and estimation effectiveness of the observer (14) is chosen. The parameter selections of the observers are as followers: $\alpha_1 = \alpha_2 = 6$, $\beta_1 = \beta_2 = 6$, $\eta_1 = \eta_2 = 0.9$, $\mu_1 = \mu_2 = 1.2$, $\gamma_1 = \gamma_2 = 0.7$, $\Delta_1 = 5$, $\Delta_d = 20$, $\Psi_0 = 0.7$, $\zeta = 4$, $\kappa_1 = 0.4$, $\kappa_2 = 3$, $\chi = 0.9$, $\varsigma = 0.7$. The simulations of the lumped faults tracking curve are seen in Figures 2 and 3, and the estimated time of $\omega_1$ and $\omega_4$ is shown in Table 1. To verify the superiority of the observer proposed in this paper, this result is compared with the observer proposed in [30], which can be seen in Figures 2 and 3.

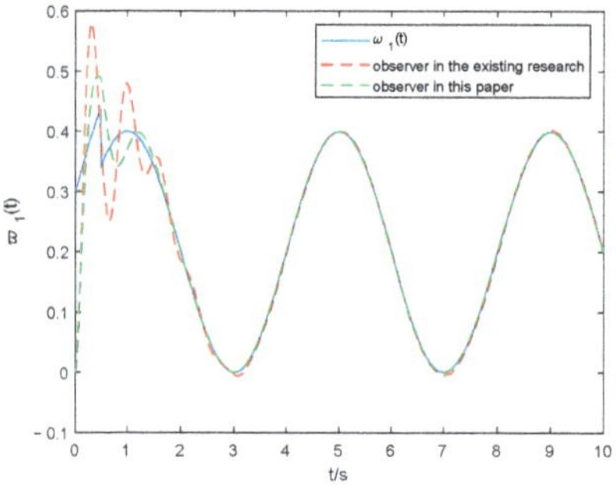


**Figure 2.** Estimation of the lumped faults $\omega_1$ with the observer in [30] and that in this paper.

**Table 1.** Estimated time of the two methods.

| Method | Estimated Time for $\omega_1$ | Estimated Time for $\omega_4$ |
|---|---|---|
| Proposed in this paper | 1.864 s | 0.583 s |
| Proposed in [30] | 2.543 s | 1.064 s |

From Figures 2 and 3, it can be seen that both the observer proposed in this paper and that in [30] have a good estimation. However, it is apparent from Table 1 that the observer proposed in this paper has a better state tracking accuracy and shorter time than that in [30], which verifies that the observer proposed in this paper is able to achieve the consistent tracking of lumped faults in finite time. Therefore, this can help improve performance and reduce dynamic computing load through a timely and accurate estimation.

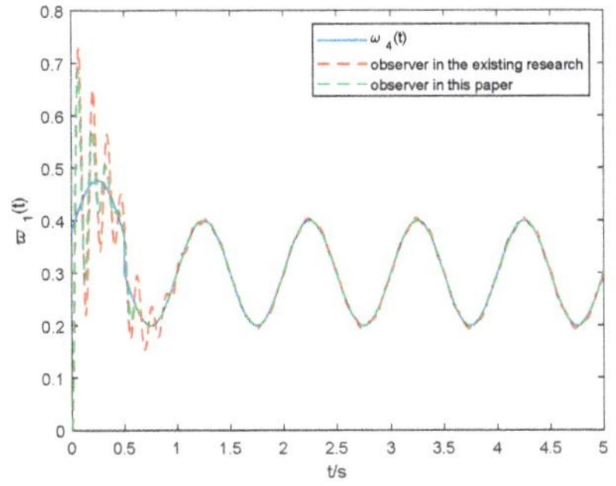


**Figure 3.** Estimation of the lumped faults $\omega_4$ with the observer in [30] and that in this paper.

The choice of parameter settings of the proposed controller is shown below: $\xi_1 = 0.2$, $\xi_2 = 0.2$, $\theta_1 = 1.001$, $\lambda = 10$ and $\rho = 25$. The simulations of the consensus error curve are seen in Figures 4–7. Figures 4 and 5 show the state tracking errors of followers by the controller proposed in this paper. Figures 6 and 7 show the state tracking errors of followers by the protocol in [40]. It can be seen that the tracking error of the position and velocity variables can converge to zero within a finite time by using the controller outlined in this paper. From the comparison of Figures 4 and 5 with Figures 6 and 7, the settling time with the controller in this paper is about 6 s, while that with the protocol in [40] is about 11 s. The overshoot is smaller, and the convergence speed is faster of the curve based on the controller proposed in this paper than that in [40], which proves the superiority of our proposed controller.

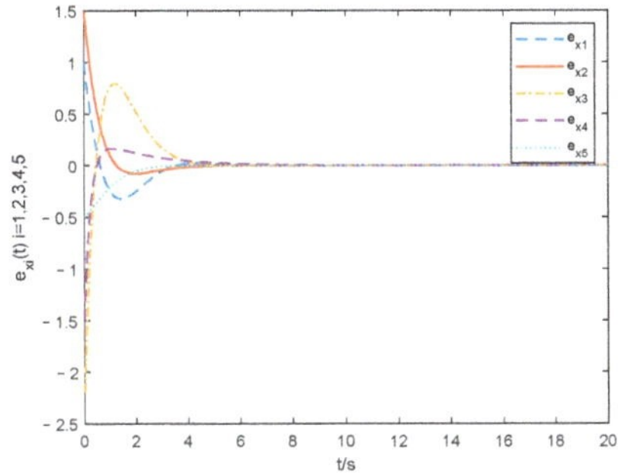


**Figure 4.** Position tracking errors with the method in this paper.

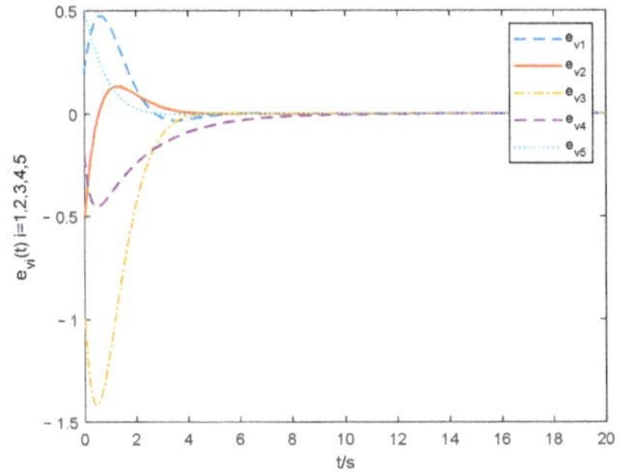


**Figure 5.** Velocity tracking errors with the method in this paper.

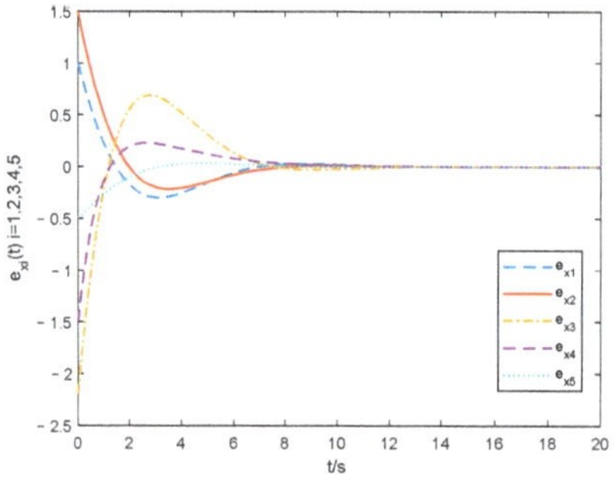


**Figure 6.** Position tracking errors with the method in [40].

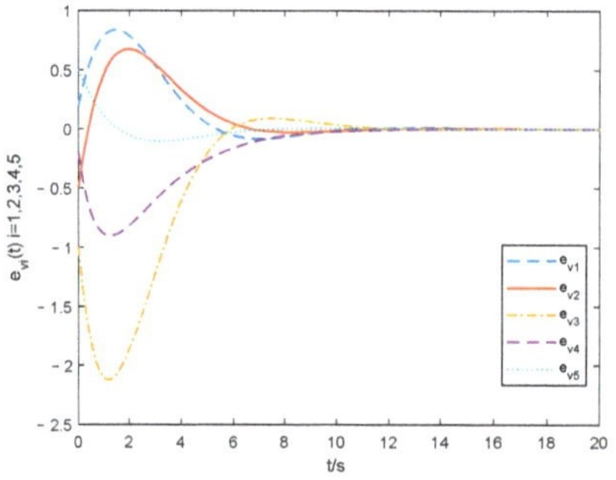


**Figure 7.** Velocity tracking errors with the method in [40].

The curve of the control input $u_i(t)$ by the proposed controller and the protocol in [40] are seen in Figures 8 and 9. The curve of the control input in this paper is smoother and better at eliminating the chattering problem than that in [40].

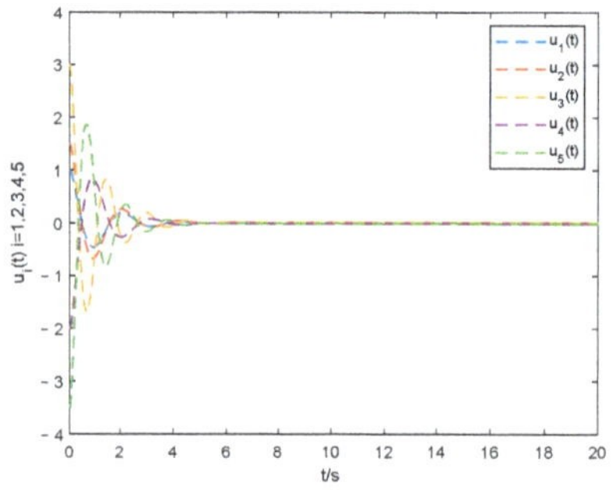


**Figure 8.** Control input with the method in this paper.

From the above simulation comparison results, the proposed controller can effectively solve the problem of chattering, while the controller in [40] cannot. Consequently, the control method proposed in this paper has an improved robustness and can realize rapid and accurate control of consensus.

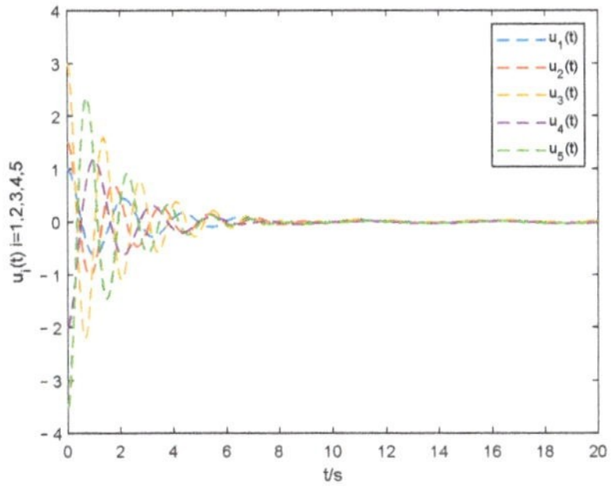


**Figure 9.** Control input with the method in [40].

## 5. Conclusions

To solve the consensus problem of disturbed second-order leader–follower MASs with disturbance and actuator faults, this paper proposes a novel integral non-singular terminal sliding mode control algorithm based on a finite-time observer. The addition of an integral link makes the system states converge faster than the traditional terminal SMC and solves the chattering problem. Additionally, the proposed controller performs well under unknown disturbances and actuator faults. Through the verification of a numerical example and simulation, leader–follower MASs can realize consensus under disturbance and actuator faults in finite time, which proves that the algorithm proposed in this paper can effectively improve the robust consensus of the system.

Owing to the superiority of the proposed algorithm and its ability to handle the disturbances and actuator faults, it can be applied in multi-wheeled robotic systems and multi-UAV systems. Their structure is similar to that of the system model described in this study, but minor changes are required in practical applications. In future research, the consensus control of MASs with unmatched disturbance or sensor faults will be considered.

**Author Contributions:** Conceptualization, P.Y.; methodology, P.Y. and Y.D.; validation, Y.D.; writing—original draft preparation Y.D.; investigation, Z.S. and K.F.; checks and corrections, Z.S. and K.F. All authors have read and agreed to the published version of the manuscript.

**Funding:** This work was funded by Key Laboratories for National Defense Science and Technology (6142605200402), National Key Laboratory of Science and Technology on Helicopter Transmission (No. HTL-O-21G11), the Aeronautical Science Foundation of China (20200007018001), the Aero Engine Corporation of China Industry-University-Research Cooperation Project (HFZL2020CXY011) and the Research Fund of State Key Laboratory of Mechanics and Control of Mechanical Structures (Nanjing University of Aeronautics and Astronautics) (MCMS-I-0121G03).

**Institutional Review Board Statement:** Not applicable.

**Informed Consent Statement:** Not applicable.

**Data Availability Statement:** Not applicable.

**Conflicts of Interest:** The authors declare no conflict of interest.

## References

1. Ye, D.; Zhang, M.; Yang, Y. A multi-agent framework for packet routing in wireless sensor networks. *Sensors* **2015**, *15*, 10026–10047. [CrossRef]
2. Alonso-Mora, J.; Baker, S.; Rus, D. Multi-robot formation control and object transport in dynamic environments via constrained optimization. *Int. J. Robot. Res.* **2017**, *36*, 1000–1021. [CrossRef]
3. Chen, Y.; Yu, R.; Zhang, Y.; Liu, C. Circular formation flight control for unmanned aerial vehicles with directed network and external disturbance. *IEEE/CAA J. Autom. Sin.* **2020**, *7*, 505–516. [CrossRef]
4. Zhang, X. Adaptive Control and Reconfiguration of Mobile Wireless Sensor Networks for Dynamic Multi-Target Tracking. *IEEE Trans. Autom. Control* **2011**, *56*, 2429–2444. [CrossRef]

5. Su, H.; Qiu, Y.; Wang, L. Semi-global output consensus of discrete time multi-agent systems with input saturation and external disturbances. *ISA Trans.* **2017**, *67*, 131–139. [CrossRef]
6. Xu, J. Adaptive fault tolerant tracking control for a class of stochastic nonlinear systems with output constraint and actuator faults. *Syst. Control Lett.* **2017**, *107*, 100–109.
7. Tan, Y.; Fei, S.; Liu, J.; Zhang, D. Asynchronous adaptive event-triggered tracking control for multi-agent systems with stochastic actuator faults. *Appl. Math. Comput.* **2019**, *355*, 482–496. [CrossRef]
8. Zhang, H.; Kim, S.; Sun, Q.; Zhou, J. Distributed Adaptive Virtual Impedance Control for Accurate Reactive Power Sharing Based on Consensus Control in Microgrids. *IEEE Trans. Smart Grid* **2017**, *8*, 1749–1761. [CrossRef]
9. Li, H.; Yan, W. Receding horizon control based consensus scheme in general linear multi-agent systems. *Automatica* **2015**, *56*, 12–18. [CrossRef]
10. Singh, V.; Natarajan, V. Reduced-order controllers for multi-agent consensus using output regulation viewpoint. *Automatica* **2020**, *119*, 109062. [CrossRef]
11. Abdessameud, A.; Tayebi, A. Distributed consensus algorithms for a class of high-order multi-agent systems on directed graphs. *IEEE Trans. Autom. Control* **2018**, *63*, 3464–3470. [CrossRef]
12. Jin, X.; Wang, S.; Qin, J.; Zheng, W.; Kang, Y. Adaptive fault-tolerant consensus for a class of uncertain nonlinear second-order multi-agent systems with circuit implementation. *IEEE Trans. Circuits Syst. I Regul. Pap.* **2018**, *65*, 2243–2255. [CrossRef]
13. Li, X.; Wang, J. Fault-tolerant tracking control for a class of nonlinear multi-agent systems. *Syst. Control Lett.* **2020**, *135*, 104576. [CrossRef]
14. Sader, M.; Chen, Z.; Liu, Z.; Deng, C. Distributed robust fault-tolerant consensus control for a class of nonlinear multi-agent systems with intermittent communications. *Appl. Math. Comput.* **2021**, *403*, 126166. [CrossRef]
15. Li, X.; Yu, Z.; Jiang, H. Event-triggered fixed -time integral sliding mode control for nonlinear multi-agent systems with disturbances. *Entropy* **2021**, *23*, 1412. [CrossRef] [PubMed]
16. Liu, Z.; Jahanshahi, H.; Volos, C.; Bekiros, S.; He, S.; Alassafi, M.O.; Ahmad, A.M. Distributed consensus tracking control of chaotic multi-agent supply chain network: A new fault-tolerant, finite-time, and chatting-free approach. *Entropy* **2021**, *24*, 33. [CrossRef]
17. Derakhshannia, M.; Moosapour, S. Disturbance observer-based sliding mode control for consensus tracking of chaotic nonlinear multi-agent systems. *Math. Comput. Simul.* **2022**, *194*, 610–628. [CrossRef]
18. Li, Z.; Zhao, J. Adaptive consensus of non-strict feedback switched multi-agent systems with input saturations. *IEEE/CAA J. Autom. Sin.* **2022**, *8*, 1752–1761. [CrossRef]
19. Zhao, D.; Zou, T.; Li, S.; Zhu, Q. Adaptive backstepping sliding mode control for leader-follower multi-agent systems. *Control Theory Appl. IET* **2012**, *6*, 1109–1117. [CrossRef]
20. Jiang, Y.; Liu, J.; Wang, S. Robust integral sliding-mode consensus tracking for multi-agent systems with time-varying delay. *Asian J. Control* **2016**, *18*, 224–235. [CrossRef]
21. Dong, L.; Chai, S.; Zhang, B.; Nguang, S. Sliding mode control for multi-agent systems under a time-varying topology. *Int. J. Syst. Sci.* **2016**, *9*, 2193–2200. [CrossRef]
22. Zhao, L.; Yu, J.; Lin, C.; Yu, H. Distributed adaptive fixed-time consensus tracking for second-order multi-agent systems using modified terminal sliding mode. *Appl. Math. Comput.* **2017**, *312*, 23–35. [CrossRef]
23. Sanjoy, M.; Jawhar, G.; Maarouf, S. Homogeneous finite-time consensus control for higher-order multi-agent systems by full order sliding mode. *J. Syst. Sci. Complex.* **2018**, *31*, 1186–1205.
24. Zheng, B.; Guo, L.; Li, K. Event-triggered Sliding Mode Fault-tolerant Consensus for a Class of Leader-follower Multi-agent Systems. *Int. J. Control Autom. Syst.* **2021**, *19*, 2664–2673. [CrossRef]
25. Zhao, N.; Zhu, J. Sliding Mode Control for Robust Consensus of General Linear Uncertain Multi-agent Systems. *Int. J. Control Autom. Syst.* **2020**, *18*, 2170–2175. [CrossRef]
26. Chen, Y.; Zhu, B. Distributed fixed-time control of high-order multi-agent systems with non-holonomic constraints. *J. Frankl. Inst.* **2021**, *358*, 2948–2963. [CrossRef]
27. Zaidi, A.; Kazim, M.; Weng, R.; Wang, D.; Zhang, X. Distributed Observer-Based Leader Following Consensus Tracking Protocol for a Swarm of Drones. *J. Intell. Robot. Syst.* **2021**, *102*, 64. [CrossRef]
28. Hua, C.; Sun, X.; You, X.; Guan, X. Finite-time consensus control for second-order multi-agent systems without velocity measurements. *Int. J. Syst. Sci.* **2017**, *48*, 337–346. [CrossRef]
29. Ren, Y.; Zhou, W.; Li, Z.; Liu, L.; Sun, Y. Prescribed-time leader-following consensus for stochastic second-order multi-agent systems subject to actuator failures via sliding mode control strategy. *Neurocomputing* **2021**, *425*, 82–95. [CrossRef]
30. Taoufik, A.; Defoort, M.; Djemai, M.; Busawon, K.; Diego, J. Distributed global fault detection scheme in multi-agent systems with chained-form dynamics. *Int. J. Robust Nonlinear Control* **2021**, *31*, 3859–3877. [CrossRef]
31. Tang, P.; Lin, D.; Zheng, D.; Fan, S.; Ye, J. Observer based finite-time fault tolerant quadrotor attitude control with actuator faults. *Aerosp. Sci. Technol.* **2020**, *104*, 105968. [CrossRef]
32. Hu, X.; Yang, P.; Ma, B.; Zhang, Z.; Wang, Z. Consensus sliding-mode fault-tolerant control for second-order multi-agent systems. *J. Adv. Comput. Intell. Intell. Inform.* **2021**, *25*, 947–981. [CrossRef]
33. Qin, J.; Zhang, G.; Zheng, W.; Kang, Y. Adaptive sliding mode consensus tracking for second-order nonlinear multiagent systems with actuator faults. *IEEE Trans. Cybern.* **2019**, *49*, 1605–1615. [CrossRef] [PubMed]

34. Zhao, Z.; He, X.; Ahn, C. Boundary disturbance observer-based control of a vibrating single-link flexible manipulator. *IEEE Trans. Syst. Man Cybern. Syst.* **2021**, *51*, 2382–2390. [CrossRef]
35. Tian, X.; Liu, H.; Liu, H. Robust finite-time consensus control for multi-agent systems with disturbances and unknown velocities. *ISA Trans.* **2018**, *80*, 73–80. [CrossRef]
36. Vo, A.; Truong, T.; Kang, H. A novel tracking control algorithm with finite-time disturbance observer for a class of second-order nonlinear systems and its applications. *IEEE Access* **2021**, *9*, 31373–31389. [CrossRef]
37. Pan, H.; Zhang, G.; Ouyang, H.; Mei, L. A novel global fast terminal sliding mode control scheme for second-order systems. *IEEE Access* **2020**, *8*, 22758–22769. [CrossRef]
38. Bhat, S.; Bernstein, D. Finite-time stability of continuous autonomous systems. *SIAM J. Control Optim.* **2000**, *38*, 751–766. [CrossRef]
39. Zhang, K.; Liu, G.; Jiang, B. Robust unknown input observer-based fault estimation of leader-follower linear multi-agent systems. *Circuits Syst. Signal Process.* **2017**, *36*, 525–542. [CrossRef]
40. Asl, R.; Hagh, Y.; Palm, R.; Handroos, H. Integral Non-Singular Terminal Sliding Mode Controller for nth-Order Nonlinear Systems. *IEEE Access* **2019**, *7*, 102792–102802.

MDPI AG
Grosspeteranlage 5
4052 Basel
Switzerland
Tel.: +41 61 683 77 34

MDPI Books Editorial Office
E-mail: books@mdpi.com
www.mdpi.com/books

www.ingramcontent.com/pod-product-compliance
Lightning Source LLC
LaVergne TN
LVHW070504061126
851858LV00011B/287